Table 2: Fixed-End Moments

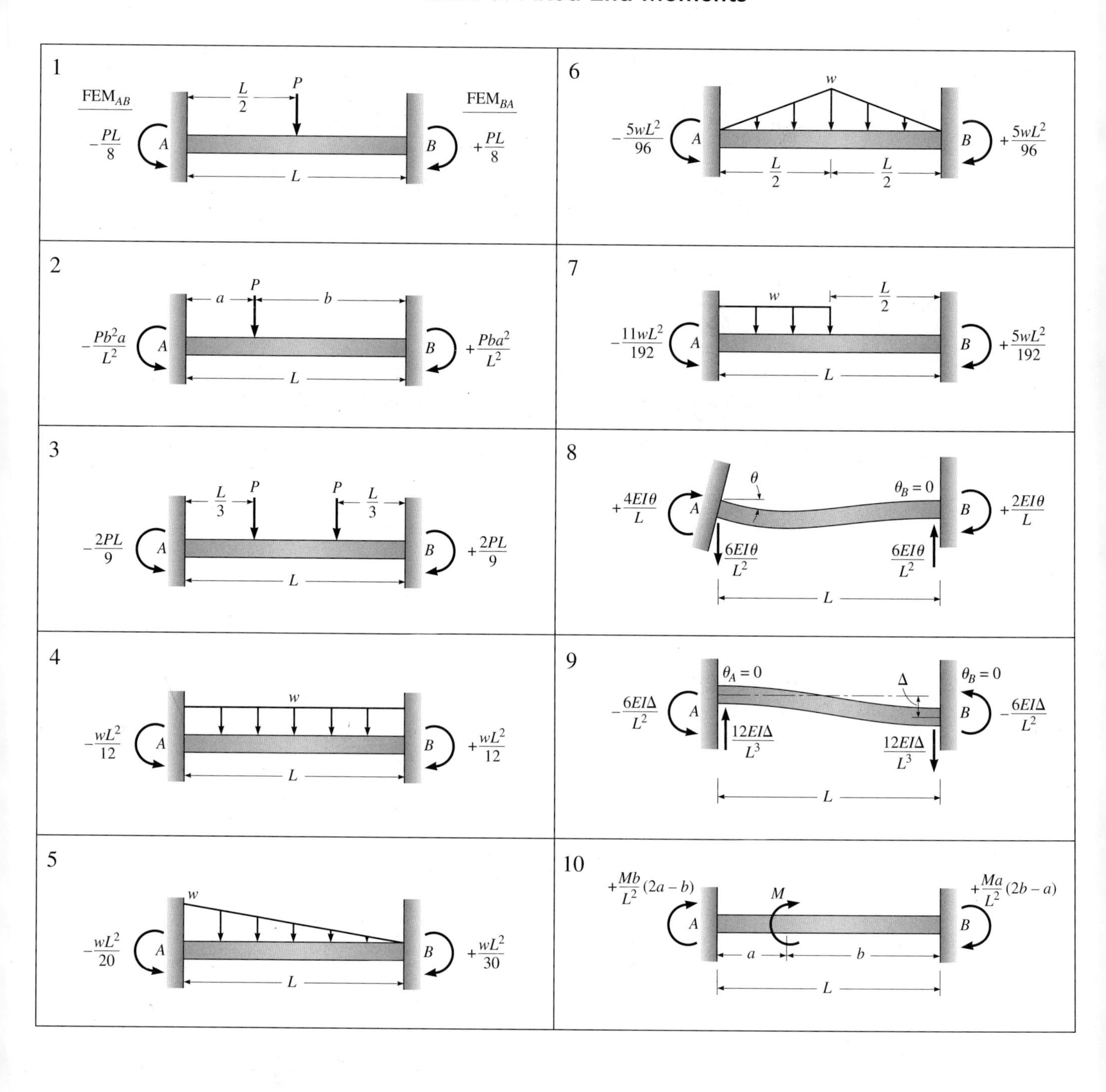

Fundamentals of Structural Analysis

Fundamentals of Structural Analysis

Fourth Edition

Kenneth M. Leet
Professor Emeritus, Northeastern University

Chia-Ming Uang
Professor, University of California, San Diego

Anne M. Gilbert
Adjunct Assistant Professor, Yale University

The McGraw·Hill Companies

FUNDAMENTALS OF STRUCTURAL ANALYSIS, FOURTH EDITION
International Edition 2011

10 09 08 07 06 05
20 15 14 13
CTP SLP

When ordering this title, use ISBN: 978-007-128938-2 or MHID: 007-128938-0

Printed in Singapore

www.mhhe.com

For Judith H. Leet

ABOUT THE AUTHORS

Kenneth Leet is a Professor Emeritus of structural engineering at Northeastern University. He received his Ph.D. in structural engineering from the Massachusetts Institute of Technology. As a professor of civil engineering at Northeastern University, he taught graduate and undergraduate courses in reinforced concrete design, structural analysis, foundations, plates and shells, and capstone courses on comprehensive engineering projects for over thirty years. Professor Leet was given an Excellence in Teaching award at Northeastern University in 1992. He was also a faculty member for ten years at Drexel University in Philadelphia.

In addition to being the author of the first edition of this book on structural analysis, originally published by Macmillan in 1988, he is the author of *Fundamentals of Reinforced Concrete,* published by McGraw-Hill in 1982 and now in its third edition.

Before teaching, he was employed by the Corps of Army Engineers as a construction management engineer, by Catalytic Construction Company as a field engineer, and by several structural engineering firms as a structural designer. He has also served as a structural consultant to a number of government agencies and private firms, including the U.S. Department of Transportation, Procter & Gamble, Teledyne Engineering Services, and the City of Philadelphia and Boston Bridge Departments.

As a member of the American Arbitration Association, the American Concrete Institute, the ASCE, and the Boston Society of Civil Engineers, Professor Leet actively participated in professional societies for many years.

Chia-Ming Uang is a Professor of structural engineering at the University of California, San Diego (UCSD). He received a B.S. degree in civil engineering from National Taiwan University and M.S. and Ph.D. degrees in civil engineering from the University of California, Berkeley. His research areas include seismic analysis and design of steel, composite, and timber structures.

Professor Uang also coauthored the text *Ductile Design of Steel Structures* for McGraw-Hill. He received the UCSD Academic Senate Distinguished Teaching Award in 2004. He is also the recipient of the ASCE Raymond C. Reese Research Prize in 2001, the ASCE Moisseiff Award in 2004, and the AISC Special Achievement Award in 2007.

Anne M. Gilbert, PE, SECB, is an Adjunct Assistant Professor in the School of Architecture at Yale University. She is also a senior project engineer at Spiegel Zamecnik & Shah, Inc., and is a registered Structural Engineer in Connecticut and Washington, D.C. She received a B.A. in architecture at the University of North Carolina, a B.S. in civil engineering from Northeastern University, and a M.S. in civil engineering from the University of Connecticut. Gilbert specializes in structural design of hospitals, laboratories, university and residential buildings, as well as seismic evaluation and renovation of structures in high seismic areas. Her work includes preparation of construction documents and construction administration. Gilbert's architectural design experience includes the design of commercial and residential buildings, and rehabilitation of urban brownstones.

TABLE OF CONTENTS

PREFACE

This text introduces engineering and architectural students to the basic techniques required for analyzing the majority of structures and the elements of which most structures are composed, including beams, frames, arches, trusses, and cables. Although the authors assume that readers have completed basic courses in statics and strength of materials, we briefly review the basic techniques from these courses the first time we mention them. To clarify the discussion, we use many carefully chosen examples to illustrate the various analytic techniques introduced, and whenever possible, we select examples confronting engineers in real-life professional practice.

Features of This Text

1. **Historical Notes.** New to this edition are historical notes that have been added to various chapters providing points in the history of accomplishments and developments in structural analysis methods.
2. **Expanded treatment of design loads.** Chapter 2 is devoted to a comprehensive discussion of loads that include dead and live loads, snow, earthquake, and wind loads Based on the ANSI/ASCE 7 Standard. The presentation aims to provide students with a basic understanding of how design loads are determined for practical design of multistory buildings, bridges, and other structures.
3. **New homework problems.** A substantial number of the problems are new or revised for this edition (in both metric and U.S. Customary System units), and many are typical of analysis problems encountered in practice. The many choices enable the instructor to select problems suited for a particular class or for a particular emphasis.
4. **Computer problems and applications.** Computer problems, some new to this edition, provide readers with a deeper understanding of the structural behavior of trusses, frames, arches, and other structural systems. These carefully tailored problems illustrate significant aspects of structural behavior that, in the past, experienced designers needed many years of practice to understand and to analyze correctly. The computer problems are identified with a computer screen icon and begin in Chapter 4 of the text. The computer problems can be solved using the Educational Version of the commercial software RISA-2D that is available to

users at the textbook website. However, any software that produces deflected shapes as well as shear, moment, and axial load diagrams can be used to solve the problems. An overview on the use of the RISA-2D software and an author-written tutorial are also available at the textbook website.

5. **Improved layout of example problems.** The content of the examples has been clarified by showing them on one page or two facing pages—surrounded by boxes—so students can see the complete problem without turning the pages.
6. **Expanded discussion of the general stiffness method.** Chapter 16, on the general stiffness method, provides a clear transition from classical methods of analysis to those using matrix formulations for computer analysis, as discussed in Chapters 17 and 18.
7. **Realistic, fully drawn illustrations.** The illustrations in the text provide a realistic picture of actual structural elements and a clear understanding of how the designer models joints and boundary conditions. Photographs complement the text to illustrate examples of building and bridge failures.
8. **Problem solutions have been carefully checked for accuracy.** The authors have carried out multiple checks on the problem solutions but would appreciate hearing from users about any ambiguities or errors. Corrections can be sent to Professor Chia-Ming Uang (cmu@ucsd.edu).
9. **Textbook Website.** A text specific website is available to users at www.mhhe.com/leet. The site offers an array of tools, including lecture slides, an image bank of the text's art, helpful Web links, and the RISA-2D educational software.
10. **Hands-on Mechanics.** Hands-on Mechanics is a website designed for instructors who are interested in incorporating three-dimensional, hands-on teaching aids into their lectures. Developed through a partnership between the McGraw-Hill Engineering Team and the Department of Civil and Mechanical Engineering at the United States Military Academy at West Point, this website not only provides detailed instructions for how to build 3-D teaching tools using materials found in any lab or local hardware store, but also provides a community where educators can share ideas, trade best practices, and submit their own original demonstrations for posting on the site. Visit www.handsonmechanics.com for more information.

Contents and Sequence of Chapters

We present the topics in this book in a carefully planned sequence to facilitate the student's study of analysis. In addition, we tailor the explanations to the level of students at an early stage in their engineering education. These explanations are based on the authors' many years of experience teaching analysis.

Chapter 1 provides a historical overview of structural engineering (from earliest post and lintel structures to today's high-rises and cable bridges) and a brief explanation of the interrelationship between analysis and design. We also describe the essential characteristics of basic structures, detailing both their advantages and their disadvantages.

Chapter 2 on loads is described above in *Features of This Text.*

Chapters 3, 4, and 5 cover the basic techniques required to determine bar forces in trusses, and shear and moment in beams and frames. The methods developed in these chapters are used to solve almost every problem in the remainder of the text.

Chapters 6 and 7 interrelate the behavior of arches and cables and cover their special characteristics (of acting largely in direct stress and using materials efficiently).

Chapter 8 covers methods for positioning live load on determinate structures to maximize the internal force at a specific section of a beam, frame, or bars of a truss.

Chapters 9 and 10 provide methods used to compute the deflections of structures to verify that a structure is not excessively flexible and to analyze indeterminate structures by the method of consistent deformations.

Chapters 11, 12, and 13 introduce several classical methods for analyzing indeterminate structures. Although most complex indeterminate structures are now analyzed by computer, certain traditional methods (e.g., moment distribution) are useful to estimate the forces in highly indeterminate beams and frames to establish initial properties of members for the computer analysis.

Chapter 14 extends the influence line method introduced in Chapter 8 to the analysis of indeterminate structures. Engineers use the techniques in both chapters to design bridges or other structures subject to moving loads or to live loads whose position on the structure can change.

Chapter 15 gives approximate methods of analysis, used to estimate the value of forces at selected points in highly indeterminate structures. With approximate methods, designers can verify the accuracy of computer studies or check the results of more traditional, lengthy hand analyses described in earlier chapters.

Chapters 16, 17, and 18 introduce matrix methods of analysis. Chapter 16 extends the general stiffness method to a variety of simple structures. The matrix formulation of the stiffness method is applied to the analysis of trusses (Chapter 17) and to the analysis of beams and frames (Chapter 18).

ACKNOWLEDGMENTS

As the senior author, I would like to acknowledge the many hours of editing and support provided by my wife Judith Leet, over a forty-year period; her help is deeply appreciated.

I would also like to thank Richard Scranton, Saul Namyet, Robert Taylor, and Marilyn Scheffler for their help with the first edition, and Dennis Bernal who wrote Chapter 18, all then of Northeastern University.

For their assistance with the first McGraw-Hill edition, we thank Amy Hill, Gloria Schiesl, Eric Munson and Patti Scott of McGraw-Hill and Jeff Lachina of Lachina Publishing Services.

For their assistance with the second and third editions, we thank Amanda Green, Suzanne Jeans, Jane Mohr and Gloria Schiesl of McGraw-Hill; Rose Kernan of RPK Editorial Services Inc.; and Patti Scott, who edited the second edition.

For their assistance with this fourth edition, we thank Debra Hash, Peter Massar, Lorraine Buczek, Joyce Watters and Robin Reed of McGraw-Hill, and Rose Kernan of RPK Editorial Services Inc.

We also wish to thank Bruce R. Bates of RISA Technologies for providing an educational version of the RISA-2D computer program with its many options for presenting results. Mr. Dong-Won Kim assisted in preparing the answers for the fourth edition.

We would like to thank the following reviewers for their much appreciated comments and advice:

William Cofer, *Washington State University*
Ross B. Corotis, *University of Colorado- Boulder*
Gianluca Cusatis, *Rensselaer Polytechnic Institute*
Robert K. Dowell, *San Diego State University*
Fouad Fanous, *Iowa State University*
Terje Haukaas, *University of British Columbia*
Yue Li, *Michigan Technological University*
Thomas Miller, *Oregon State University*

Kenneth Leet
Emeritus Professor
Northeastern University

Chia-Ming Uang
Professor
University of California,
San Diego

Anne M. Gilbert
Adjunct Assistant Professor
Yale University

GUIDED TOUR

APPROACH

Leet, Uang, and Gilbert combine a mix of classical methods and computer analysis.

A substantial number of the homework problems in this text are new or revised.

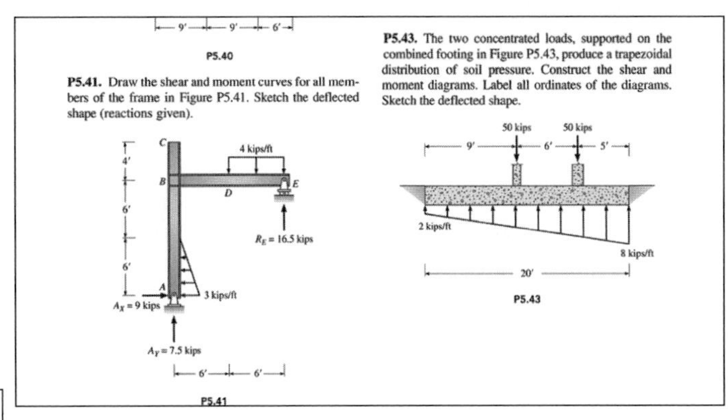

P5.40

P5.41. Draw the shear and moment curves for all members of the frame in Figure P5.41. Sketch the deflected shape (reactions given).

P5.41

P5.43. The two concentrated loads, supported on the combined footing in Figure P5.43, produce a trapezoidal distribution of soil pressure. Construct the shear and moment diagrams. Label all ordinates of the diagrams. Sketch the deflected shape.

P5.43

P9.44. Because of poor foundation conditions, a 30-in-deep steel beam with a cantilever is used to support an exterior building column that carries a dead load of 600 kips and a live load of 150 kips (Figure P9.44). What is the magnitude of the initial camber that should be induced at point *C*, the tip of the cantilever, to eliminate the deflection produced by the total load? Neglect the beam's weight. Given: I = 46,656 in^4 and E_S = 30,000 ksi. See case 5 in Table 9.1 for the deflection equation. The clip angle connection at *A* may be treated as a pin and the cap plate support at *B* as a roller.

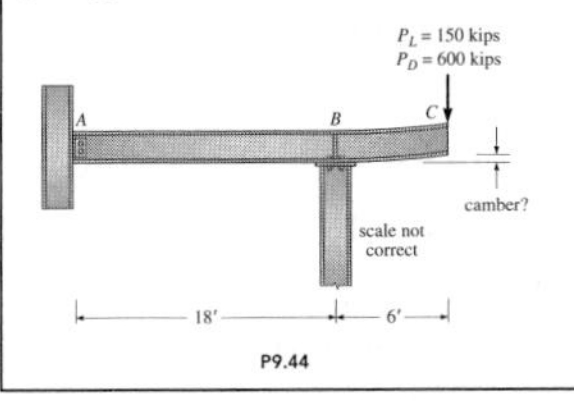

P9.44

P9.46. *Computer study of the behavior of multistory building frames.* The object of this study is to examine the behavior of building frames fabricated with two common types of connections. When open interior spaces and future flexibility of use are prime considerations, building frames can be constructed with *rigid connections* usually fabricated by welding. Rigid joints (see Figure P9.46*b*) are expensive to fabricate and now cost in the range of $700 to $850 depending on the size of members. Since the ability of a welded frame to resist lateral loads depends on the bending stiffness of the beams and columns, heavy members may be required when lateral loads are large or when lateral deflections must be limited. Alternately, frames can be constructed less expensively by connecting the webs of beams to columns by angles or plates, called *shear connections*, which currently cost about $80 each (Figure P9.46*c*). If *shear connections* are used, diagonal bracing, which forms a deep vertical truss with the attached columns and floor beams, is typically required to provide lateral stability (unless floors can be connected to stiff shear walls constructed of reinforced masonry or concrete).

Properties of Members

Computer icons throughout the text point to homework problems that can be analyzed using a computer.

RISA-2D

EDUCATIONAL SOFTWARE

Students may download a free academic version of RISA-2D and an author-written tutorial with the purchase of a new text.

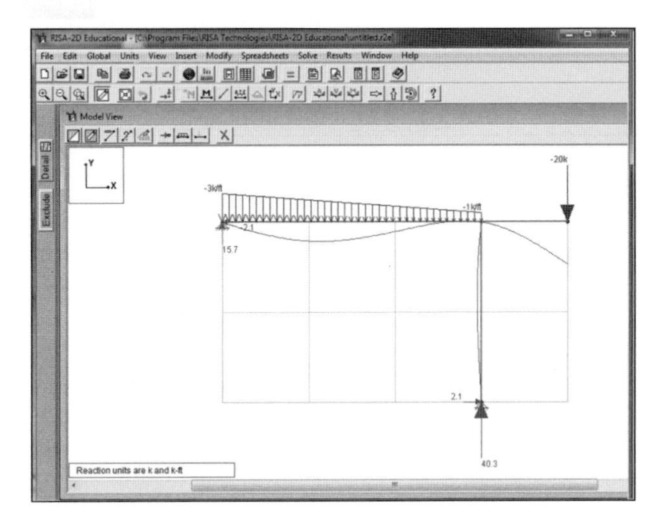

HIGHLY REALISTIC ART PROGRAM

The illustrations in the text provide a realistic picture of actual structural elements.

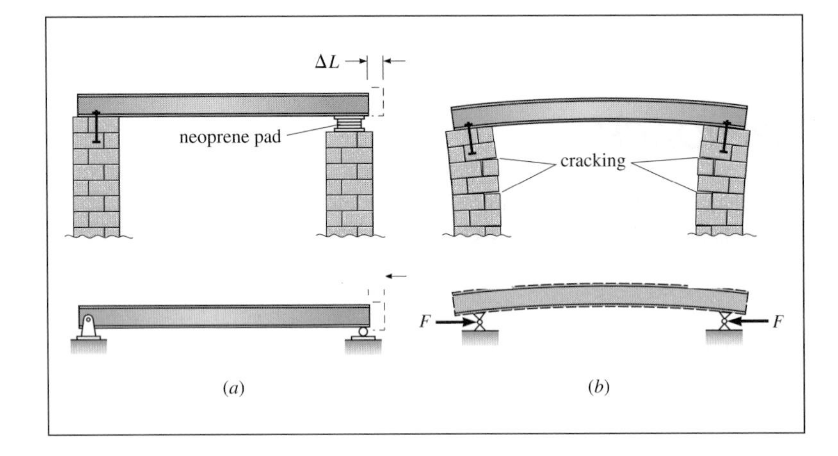

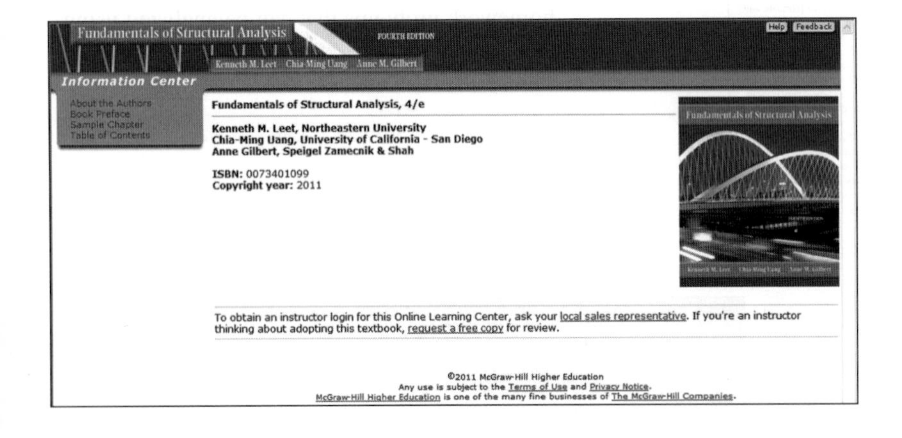

SUPPLEMENTS

Textbook Website. A text specific website is available to users at http://www.mhhe.com/leet. The site offers an array of tools, including lecture slides, an image bank of the text's art, helpful Web links, and the RISA-2D educational software.

Fundamentals of Structural Analysis

Fundamentals of Structural Analysis

The Brooklyn Bridge

Opened in 1883 at a cost of $9 million, this bridge was heralded as the "Eighth Wonder of the World." The center span, which rises 135 ft above the surface of the East River, spans nearly 1600 ft between towers. Designed in part by engineering judgment and in part by calculations, the bridge is able to support more than three times the original design load. The large masonry towers are supported on pneumatic caissons 102 by 168 ft in plan. In 1872 Colonel Washington A. Roebling, the director of the project, was paralyzed by caissons disease while supervising the construction of one of the submerged piers. Crippled for life, he directed the balance of the project from bed with the assistance of his wife and engineering staff.

CHAPTER 1

Introduction

1.1 Overview of the Text

As an engineer or architect involved with the design of buildings, bridges, and other structures, you will be required to make many technical decisions about structural systems. These decisions include (1) selecting an efficient, economical, and attractive structural form; (2) evaluating its safety, that is, its strength and stiffness; and (3) planning its erection under temporary construction loads.

To design a structure, you will learn to carry out a *structural analysis* that establishes the internal forces and deflections at all points produced by the design loads. Designers determine the internal forces in key members in order to size both members and the connections between members. And designers evaluate deflections to ensure a serviceable structure—one that does not deflect or vibrate excessively under load so that its function is impaired.

Analyzing Basic Structural Elements

During previous courses in statics and strength of materials, you developed some background in structural analysis when you computed the bar forces in trusses and constructed shear and moment curves for beams. You will now broaden your background in structural analysis by applying, in a systematic way, a variety of techniques for determining the forces in and the deflections of a number of basic structural elements: beams, trusses, frames, arches, and cables. These elements represent the basic components used to form more complex structural systems.

Moreover, as you work analysis problems and examine the distribution of forces in various types of structures, you will understand more about how structures are stressed and deformed by load. And you will gradually develop a clear sense of which structural configuration is optimal for a particular design situation.

Further, as you develop an almost intuitive sense of how a structure behaves, you will learn to estimate with a few simple computations the approximate values of forces at the most critical sections of the structure. This ability will serve you well, enabling you (1) to verify the accuracy of the results of a computer analysis of large, complex structures and (2) to estimate the preliminary design forces needed to size individual components of multimember structures during the early design phase when the tentative configuration and proportions of the structure are being established.

Analyzing Two-Dimensional Structures

As you may have observed while watching the erection of a multistory building frame, when the structure is fully exposed to view, its structure is a complex three-dimensional system composed of beams, columns, slabs, walls, and diagonal bracing. Although load applied at a particular point in a three-dimensional structure will stress all adjacent members, most of the load is typically transmitted through certain key members directly to other supporting members or into the foundation.

Once the behavior and function of the various components of most three-dimensional structures are understood, the designer can typically simplify the analysis of the actual structure by subdividing it into smaller two-dimensional subsystems that act as beams, trusses, or frames. This procedure also significantly reduces the complexity of the analysis because two-dimensional structures are much easier and faster to analyze than three-dimensional structures. Since with few exceptions (e.g., geodesic domes constructed of light tubular bars) designers typically analyze a series of simple two-dimensional structures—even when they are designing the most complex three-dimensional structures—we will devote a large portion of this book to the analysis of two-dimensional or *planar* structures, those that carry forces lying in the plane of the structure.

Once you clearly understand the basic topics covered in this text, you will have acquired the fundamental techniques required to analyze most buildings, bridges, and structural systems typically encountered in professional practice. Of course, before you can design and analyze with confidence, you will require some months of actual design experience in an engineering office to gain further understanding of the total design process from a practitioner's perspective.

For those of you who plan to specialize in structures, mastery of the topics in this book will provide you with the basic structural principles required in more advanced analysis courses—those covering, for example, matrix methods or plates and shells. Further, because design and analysis are closely interrelated, you will use again many of the analytical procedures in this text for more specialized courses in steel, reinforced concrete, and bridge design.

1.2 The Design Process: Relationship of Analysis to Design

The design of any structure—whether it is the framework for a space vehicle, a high-rise building, a suspension bridge, an offshore oil drilling platform, a tunnel, or whatever—is typically carried out in alternating steps of *design* and *analysis*. Each step supplies new information that permits the designer to proceed to the next phase. The process continues until the analysis indicates that no changes in member sizes are required. The specific steps of the procedure are described below.

Conceptual Design

A project begins with a specific need of a client. For example, a developer may authorize an engineering or architectural firm to prepare plans for a sports complex to house a regulation football field, as well as seating 60,000 people, parking for 4000 cars, and space for essential facilities. In another case, a city may retain an engineer to design a bridge to span a 2000-ft-wide river and to carry a certain hourly volume of traffic.

The designer begins by considering all possible layouts and structural systems that might satisfy the requirements of the project. Often architects and engineers consult as a team at this stage to establish layouts that lend themselves to efficient structural systems in addition to meeting the architectural (functional and aesthetic) requirements of the project. The designer next prepares sketches of an architectural nature showing the main structural elements of each design, although details of the structural system at this point are often sketchy.

Preliminary Design

In the preliminary design phase, the engineer selects from the conceptual design several of the structural systems that appear most promising, and sizes their main components. This preliminary proportioning of structural members requires an understanding of structural behavior and a knowledge of the loading conditions (dead, live, wind, and so forth) that will most likely affect the design. At this point, the experienced designer may make a few rough computations to estimate the proportions of each structure at its critical sections.

Analysis of Preliminary Designs

At this next stage, the precise loads the structure will carry are not known because the exact size of members and the architectural details of the design are not finalized. Using estimated values of load, the engineer carries out an analysis of the several structural systems under consideration to determine the forces at critical sections and the deflections at any point that influences the serviceability of the structure.

The true weight of the members cannot be calculated until the structure is sized exactly, and certain architectural details will be influenced, in turn, by the structure. For example, the size and weight of mechanical equipment cannot be determined until the volume of the building is established, which in turn depends on the structural system. The designer, however, knows from past experience with similar structures how to estimate values for load that are fairly close approximations of final values.

Redesign of the Structures

Using the results of the analysis of preliminary designs, the designer recomputes the proportions of the main elements of all structures. Although each analysis was based on estimated values of load, the forces established at this stage are probably indicative of what a particular structure must carry, so that proportions are unlikely to change significantly even after the details of the final design are established.

Evaluation of Preliminary Designs

The various preliminary designs are next compared with regard to cost, availability of materials, appearance, maintenance, time for construction, and other pertinent considerations. The structure best satisfying the client's established criteria is selected for further refinement in the final design phase.

Final Design and Analysis Phases

In the final phase, the engineer makes any minor adjustments to the selected structure that will improve its economy or appearance. Now the designer carefully estimates dead loads and considers specific positions of the live load that will maximize stresses at specific sections. As part of the final analysis, the strength and stiffness of the structure are evaluated for all significant loads and combinations of load, dead and live, including wind, snow, earthquake, temperature change, and settlements. If the results of the final design confirm that the proportions of the structure are adequate to carry the design forces, the design is complete. On the other hand, if the final design reveals certain deficiencies (e.g., certain members are overstressed, the structure is unable to resist lateral wind loads efficiently, members are excessively flexible, or costs are over budget), the designer will either have to modify the configuration of the structure or consider an alternate structural system.

Steel, reinforced concrete, wood, and metals, such as aluminum, are all analyzed in the same manner. The different properties of materials are taken into account during the design process. When members are sized, designers refer to design codes, which take into account each material's special properties.

This text is concerned primarily with the *analysis* of structures as detailed above. Design is covered in separate courses in most engineering programs; however, since the two topics are so closely interrelated, we will necessarily touch upon some design issues.

1.3 Strength and Serviceability

The designer must proportion structures so that they will neither fail nor deform excessively under any possible loading conditions. Members are always designed with a capacity for load significantly greater than that required to support anticipated *service loads* (the real loads or the loads specified by design code). This additional capacity also provides a factor of safety against accidental overload. Moreover, by limiting the level of stress, the designer indirectly provides some control over the deformations of the structure. The maximum allowable stress permitted in a member is determined either by the tensile or compressive strength of the material or, in the case of slender compression members, by the stress at which a member (or a component of a member) buckles.

Although structures must be designed with an adequate factor of safety to reduce the probability of failure to an acceptable level, the engineer must also ensure that the structure has sufficient stiffness to function usefully under all loading conditions. For example, floor beams must not sag excessively or vibrate under live load. Excessively large deflections of beams may produce cracking of masonry walls and plaster ceilings, or may damage equipment that becomes misaligned. High-rise buildings must not sway excessively under wind loads (or the building may cause motion sickness in the inhabitants of upper floors). Excessive movements of a building not only are disturbing to the occupants, who become concerned about the safety of the structure, but also may lead to cracking of exterior curtain walls and windows. Photo 1.1 shows a modern office building whose facade was constructed of large floor-to-ceiling glass panels. Shortly after the high-rise building was completed, larger than anticipated wind loads caused many glass panels to crack and fall out. The falling glass constituted an obvious danger to pedestrians in the street below. After a thorough investigation and further testing, all the original glass panels were removed. To correct the design deficiencies, the structure of the building was stiffened, and the facade was reconstructed with thicker, tempered glass panels. The dark areas in Photo 1.1 show the temporary plywood panels used to enclose the building during the period in which the original glass panels were removed and replaced by the more durable, tempered glass.

Photo 1.1: Wind damage. Shortly after thermopane windows were installed in this high-rise office building, they began failing and falling out, scattering broken glass on passers-by beneath.

Before the building could be occupied, the structural frame had to be stiffened and all the original glass panels had to be replaced by thicker, tempered glass—costly procedures that delayed the opening of the building for several years.

1.4 Historical Development of Structural Systems

To give you some historical perspective on structural engineering, we will briefly trace the evolution of structural systems from those trial-and-error designs used by the ancient Egyptians and Greeks to the highly sophisticated configurations used today. The evolution of structural forms is closely related to the materials available, the state of construction technology, the designer's knowledge of structural behavior (and much later, analysis), and the skill of the construction worker.

For their great engineering feats, the early Egyptian builders used stone quarried from sites along the Nile to construct temples and pyramids. Since the tensile strength of stone, a brittle material, is low and highly variable (because of a multitude of internal cracks and voids), beam spans in temples had to be short (see Figure 1.1) to prevent bending failures. Since this *post-and-lintel* system—massive stone beams balanced on relatively thick stone columns—has only a limited capacity for horizontal or eccentric vertical

Figure 1.1: Early post-and-lintel construction as seen in an Egyptian Temple.

Figure 1.2: Front of Parthenon, columns taper and were fluted for decoration.

loads, buildings had to be relatively low. For stability, columns had to be thick—a slender column will topple more easily than a thick column.

The Greeks, greatly interested in refining the aesthetic appearance of the stone column, used the same type of post-and-lintel construction in the Parthenon (about 400 B.C.), a temple considered one of the most elegant examples of stone construction of all time (Figure 1.2). Even up to the early twentieth century, long after post-and-lintel construction was superseded by steel and reinforced concrete frames, architects continued to impose the facade of the classic Greek temple on the entrance of public buildings. The classic tradition of the ancient Greeks was influential for centuries after their civilization declined.

Gifted as builders, Roman engineers made extensive use of the arch, often employing it in multiple tiers in coliseums, aqueducts, and bridges (Photo 1.2). The curved shape of the arch allows a departure from rectangular lines and permits much longer clear spans than are possible with masonry post-and-lintel construction. The stability of the masonry arch requires (1) that its entire cross section be stressed in compression under all combinations of load and (2) that abutments or end walls have sufficient strength to absorb the large diagonal thrust at the base of the arch. The Romans also, largely by trial and error, developed a method of enclosing an interior space by a masonry dome, as seen in the Pantheon still standing in Rome.

During the Gothic period of great cathedral buildings (Chartres, Notre Dame), the arch was refined by trimming away excess material, and its shape became far more elongated. The vaulted roof, a three-dimensional form of the arch, also appeared in the construction of cathedral roofs. Arch-like masonry elements, termed *flying buttresses*, were used together with

Photo 1.2: Romans pioneered in the use of arches for bridges, buildings, and aqueducts. Pont-du-Gard. Roman aqueduct built in 19 B.C. to carry water across the Gardon Valley to Nimes. Spans of the first- and second-level arches are 53 to 80 ft. (Near Remoulins, France.)

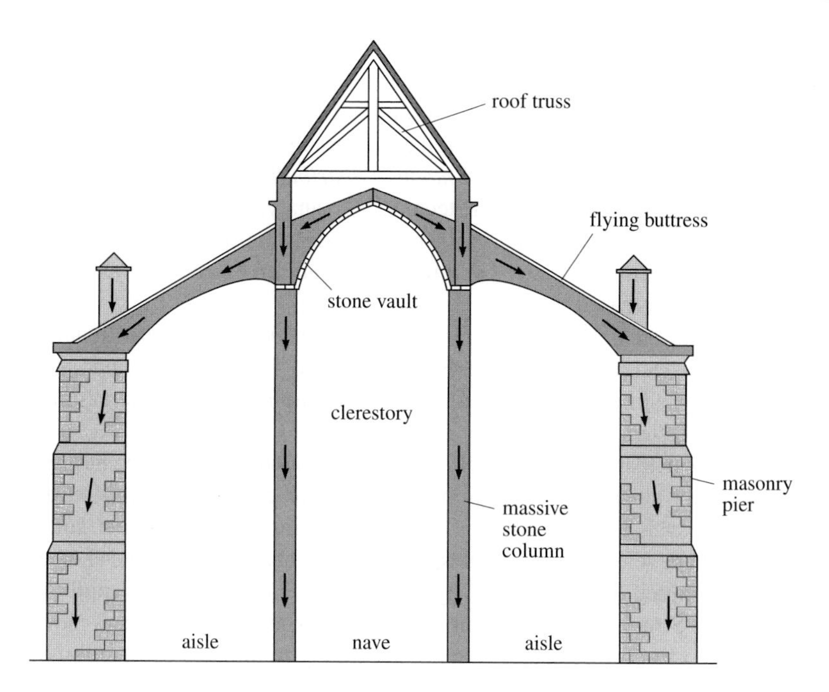

Figure 1.3: Simplified cross section showing the main structural elements of Gothic construction. Exterior masonry arches, called *flying buttresses,* used to stabilize the arched stone vault over the nave. The outward thrust of the arched vault is transmitted through the flying buttresses to deep masonry piers on the exterior of the building. Typically the piers broaden toward the base of the building. For the structure to be stable, the masonry must be stressed in compression throughout. Arrows show the flow of forces.

piers (thick masonry columns) or walls to carry the thrust of vaulted roofs to the ground (Figure 1.3). Engineering in this period was highly empirical based on what master masons learned and passed on to their apprentices; these skills were passed down through the generations.

Although magnificent cathedrals and palaces were constructed for many centuries in Europe by master builders, no significant change occurred in construction technology until cast iron was produced in commercial quantities in the mid-eighteenth century. The introduction of cast iron made it possible for engineers to design buildings with shallow but strong beams, and columns with compact cross sections, permitting the design of lighter structures with longer open spans and larger window areas. The massive bearing walls required for masonry construction were no longer needed. Later, steels with high tensile and compressive strengths permitted the construction of taller structures and eventually led to the skyscraper of today.

In the late nineteenth century, the French engineer Eiffel constructed many long-span steel bridges in addition to his innovative Eiffel Tower, the internationally known landmark in Paris (Photo 1.3). With the development of high-strength steel cables, engineers were able to construct long-span suspension bridges. The Verrazano Bridge at the entrance of New York harbor—one of the longest bridges in the world—spans 4260 ft between towers.

The addition of steel reinforcement to concrete enabled engineers to convert unreinforced concrete (a brittle, stonelike material) into tough, ductile structural members. Reinforced concrete, which takes the shape of the temporary forms into which it is poured, allows a large variety of forms to be constructed. Since reinforced concrete structures are *monolithic*, meaning they act as one continuous unit, they are highly indeterminate.

Photo 1.3: The Eiffel Tower, constructed of wrought iron in 1889, dominates the skyline of Paris in this early photograph. The tower, the forerunner of the modern steel frame building, rises to a height of 984 ft (300 m) from a 330-ft (100.6-m) square base. The broad base and the tapering shaft provide an efficient structural form to resist the large overturning forces of the wind. At the top of the tower where the wind forces are the greatest, the width of the building is smallest.

Until improved methods of indeterminate analysis enabled designers to predict the internal forces in reinforced concrete construction, design remained semi-empirical; that is, simplified computations were based on observed behavior and testing as well as on principles of mechanics. With the introduction in the early 1920s of *moment distribution* by Hardy Cross, engineers acquired a relatively simple technique to analyze continuous structures. As designers became familiar with moment distribution, they were able to analyze indeterminate frames, and the use of reinforced concrete as a building material increased rapidly.

The introduction of welding in the late nineteenth century facilitated the joining of steel members—welding eliminated heavy plates and angles required by earlier riveting methods—and simplified the construction of rigid-jointed steel frames.

In recent years, the computer and research in materials science have produced major changes in the engineer's ability to construct special-purpose structures, such as space vehicles. The introduction of the computer and the subsequent development of stiffness matrices for beams, plates, and shell elements permitted designers to analyze many complex structures rapidly and accurately. Structures that even in the 1950s took teams of engineers months to analyze can now be analyzed more accurately in minutes by one designer using a computer.

1.5 Basic Structural Elements

All structural systems are composed of a number of basic structural elements—beams, columns, hangers, trusses, and so forth. In this section we describe the main characteristics of these basic elements so that you will understand how to use them most effectively.

Hangers, Suspension Cables—Axially Loaded Members in Tension

Since all cross sections of axially loaded members are uniformly stressed, the material is used at optimum efficiency. The capacity of tension members is a direct function of the tensile strength of the material. When members are constructed of high-strength materials, such as alloyed steels, even members with small cross sections have the capacity to support large loads (see Figure 1.4).

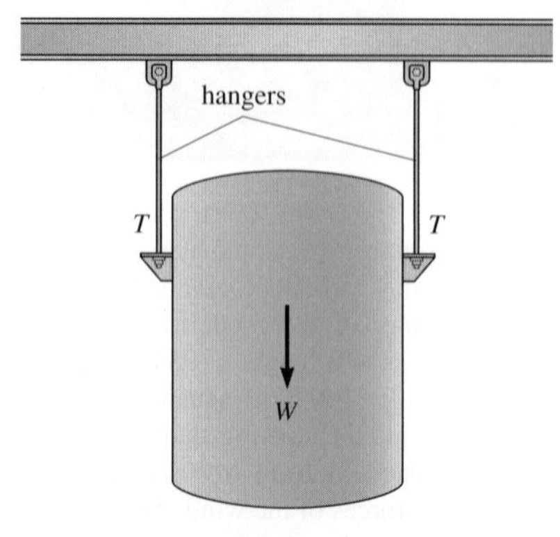

Figure 1.4: Chemical storage tank supported by tension hangers carrying force T.

As a negative feature, members with small cross sections are very flexible and tend to vibrate easily under moving loads. To reduce this tendency to vibrate, most building codes specify that certain types of tension members have a minimum amount of flexural stiffness by placing an upper limit on their *slenderness ratio* l/r, where l is the length of member and r is the radius of gyration. By definition $r = \sqrt{I/A}$ where I equals the moment of inertia and A equals the area of the cross section. If the direction of load suddenly reverses (a condition produced by wind or earthquake), a slender tension member will buckle before it can provide any resistance to the load.

Columns—Axially Loaded Members in Compression

Columns also carry load in direct stress very efficiently. The capacity of a compression member is a function of its slenderness ratio l/r. If l/r is large, the member is slender and will fail by buckling when stresses are low—often with little warning. If l/r is small, the member is stocky. Since stocky members fail by overstress—by crushing or yielding—their capacity for axial load is high. The capacity of a slender column also depends on the restraint supplied at its ends. For example, a slender cantilever column—fixed at one end and free at the other—will support a load that is one-fourth as large as that of an identical column with two pinned ends (Figure 1.5*b*, *c*).

In fact, columns supporting pure axial load occur only in idealized situations. In actual practice, the initial slight crookedness of columns or an eccentricity of the applied load creates bending moments that must be taken into account by the designer. Also in reinforced concrete or welded building frames where beams and columns are connected by rigid joints, columns carry both axial load and bending moment. These members are called *beam-columns* (see Figure 1.5*d*).

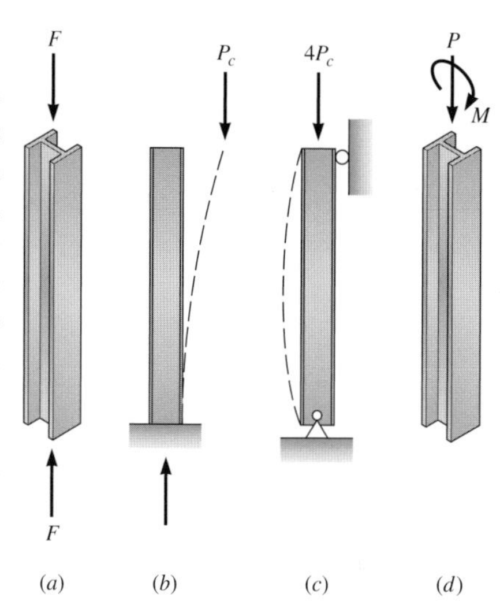

Figure 1.5: (*a*) Axially loaded column; (*b*) cantilever column with buckling load P_c; (*c*) pin-supported column with buckling load $4P_c$; (*d*) beam-column.

Beams—Shear and Bending Moment Created by Loads

Beams are slender members that are loaded perpendicular to their longitudinal axis (see Figure 1.6*a*). As load is applied, a beam bends and deflects into a shallow curve. At a typical section of a beam, internal forces of shear V and moment M develop (Figure 1.6*b*). Except in short, heavily loaded beams, the shear stresses τ produced by V are relatively small, but the longitudinal bending stresses produced by M are large. If the beam behaves elastically, the bending stresses on a cross section (compression on the top and tension on the bottom) vary linearly from a horizontal axis passing through the centroid of the cross section. The bending stresses are directly proportional to the moment, and vary in magnitude along the axis of the beam.

Shallow beams are relatively inefficient in transmitting load because the arm between the forces C and T that make up the internal couple is small. To increase the size of the arm, material is often removed from the center of the cross section and concentrated at the top and bottom surfaces, producing an I-shaped section (Figure 1.6*c* and *d*).

Planar Trusses—All Members Axially Loaded

A truss is a structural element composed of slender bars whose ends are assumed to be connected by frictionless pin joints. If pin-jointed trusses are loaded at the joints only, direct stress develops in all bars. Thus the material is used at optimum efficiency. Typically, truss bars are assembled in a triangular pattern—the simplest stable geometric configuration (Figure 1.7*a*). In the nineteenth century, trusses were often named after the designers who established a particular configuration of bars (see Figure 1.7*b*).

The behavior of a truss is similar to that of a beam in which the solid web (which transmits the shear) is replaced by a series of vertical and diagonal bars. By eliminating the solid web, the designer can reduce the deadweight of the truss significantly. Since trusses are much lighter than beams of the same

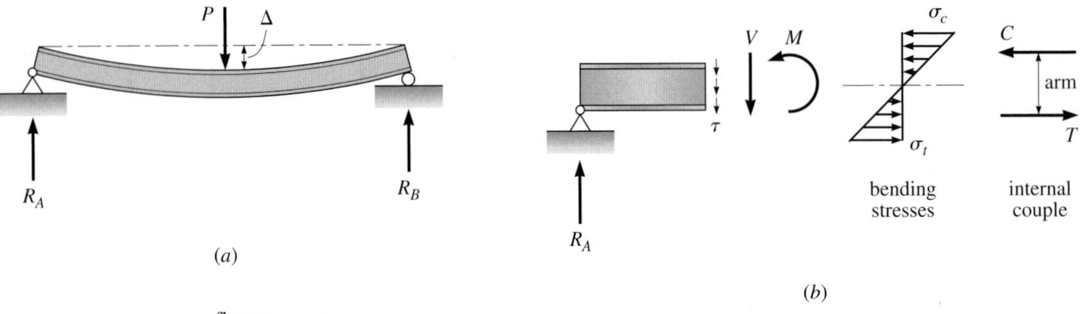

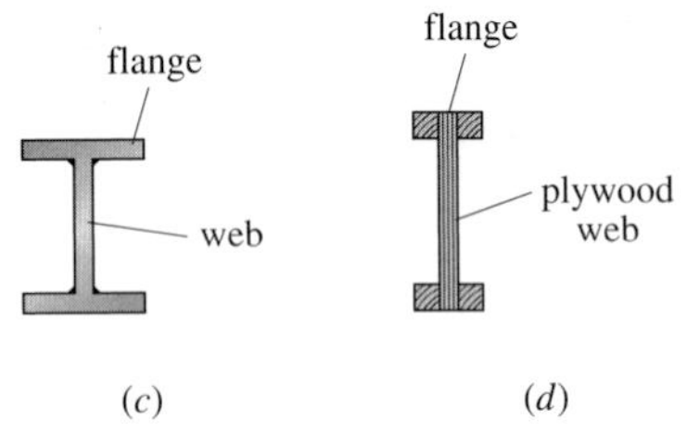

Figure 1.6: (*a*) Beam deflects into a shallow curve; (*b*) internal forces (shear *V* and moment *M*); (*c*) I-shaped steel section; (*d*) glue-laminated wood I-beam.

capacity, trusses are easier to erect. Although most truss joints are formed by welding or bolting the ends of the bars to a connection (or gusset) plate (Figure 1.8*a*), an analysis of the truss based on the assumption of pinned joints produces an acceptable result.

Although trusses are very stiff in their own plane, they are very flexible when loaded perpendicular to their plane. For this reason, the compression

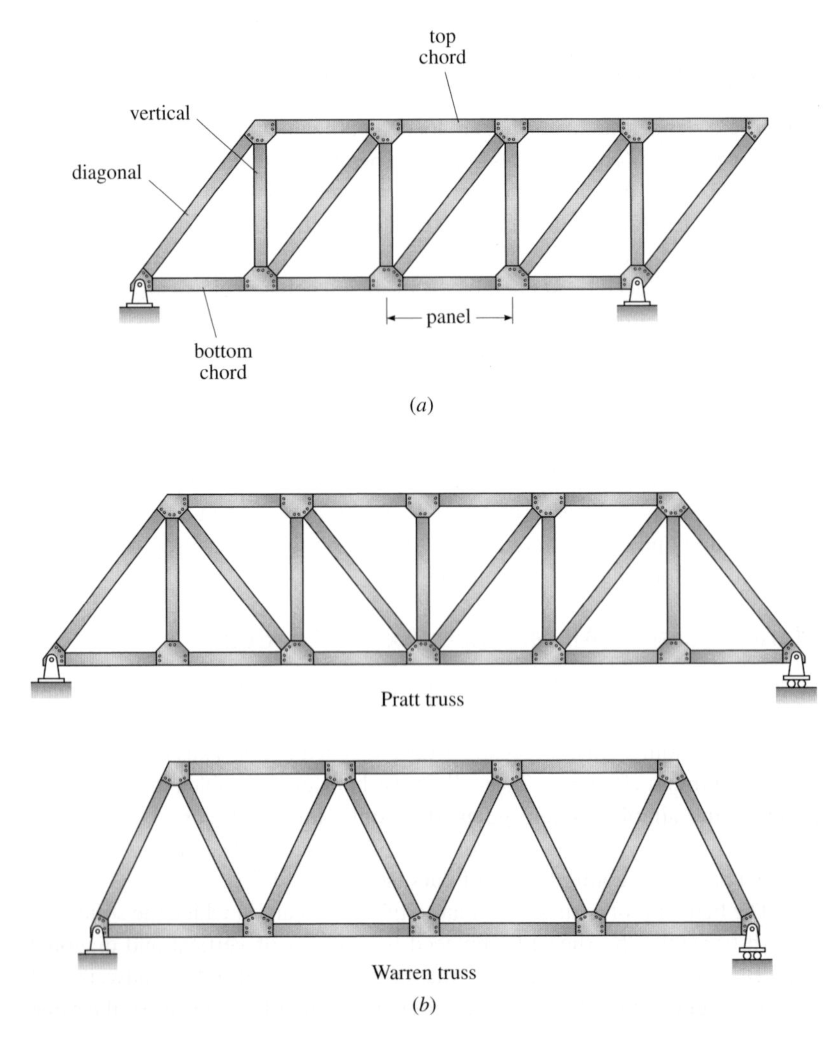

Figure 1.7: (*a*) Assembly of triangular elements to form a truss; (*b*) two common types of trusses named after the original designer.

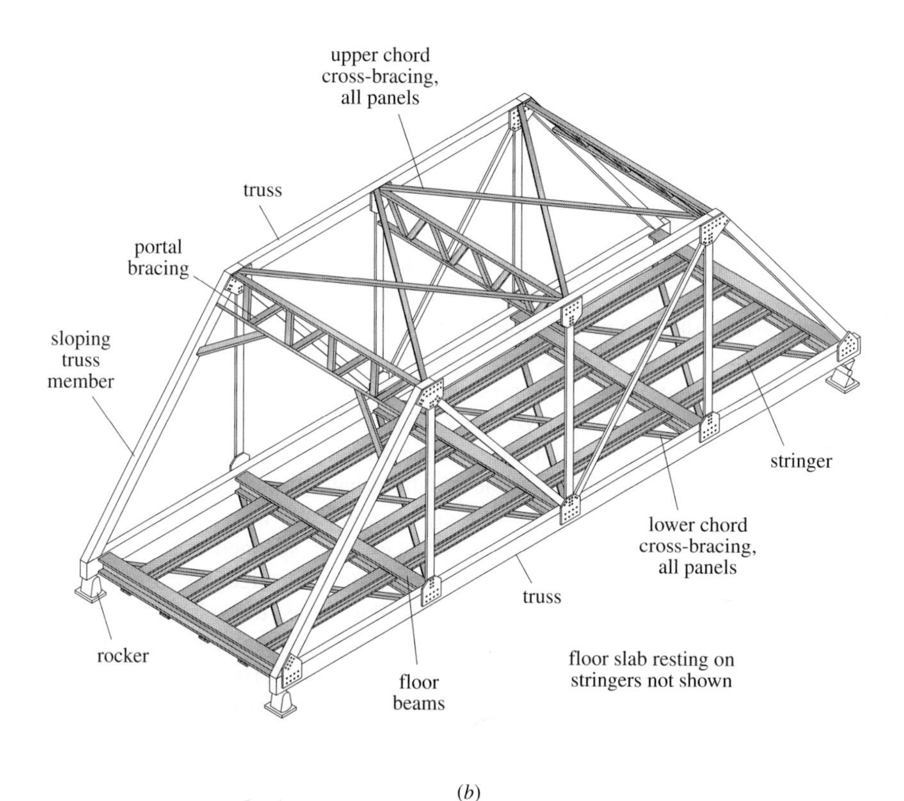

Figure 1.8: (*a*) Bolted joint detail; (*b*) truss bridge showing cross-bracing needed to stabilize the two main trusses.

chords of trusses must be stabilized and aligned by cross-bracing (Figure 1.8*b*). For example, in buildings, the roof or floor systems attached to the joints of the upper chord serve as lateral supports to prevent lateral buckling of this member.

Arches—Curved Members Stressed Heavily in Direct Compression

Arches typically are stressed in compression under their dead load. Because of their efficient use of material, arches have been constructed with spans of more than 2000 ft. To be in pure compression, an efficient state of stress, the arch must be shaped so that the resultant of the internal forces on each section passes through the centroid. For a given span and rise, only one shape of arch exists in

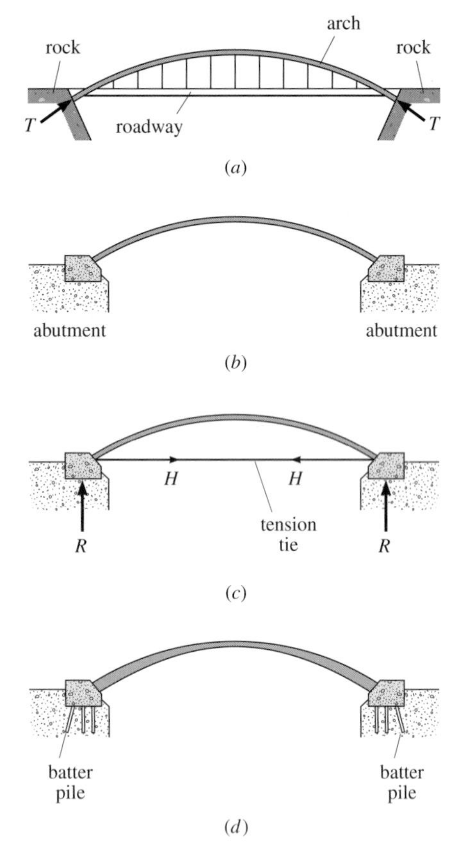

Figure 1.9: (*a*) Fixed-end arch carries roadway over a canyon where rock walls provide a natural support for arch thrust *T;* (*b*) large abutments provided to carry arch thrust; (*c*) tension tie added at base to carry horizontal thrust, foundations designed only for vertical reaction *R;* (*d*) foundation placed on piles, batter piles used to transfer horizontal component of thrust into ground.

which direct stress will occur for a particular force system. For other loading conditions, bending moments develop that can produce large deflections in slender arches. The selection of the appropriate arch shape by the early builders in the Roman and Gothic periods represented a rather sophisticated understanding of structural behavior. (Since historical records report many failures of masonry arches, obviously not all builders understood arch action.)

Because the base of the arch intersects the end supports (called *abutments*) at an acute angle, the internal force at that point exerts a horizontal as well as a vertical thrust on the abutments. When spans are large, when loads are heavy, and when the slope of the arch is shallow, the horizontal component of the thrust is large. Unless natural rock walls exist to absorb the horizontal thrust (Figure 1.9*a*), massive abutments must be constructed (Figure 1.9*b*), the ends of the arch must be tied together by a tension member (Figure 1.9*c*), or the abutment must be supported on piles (Figure 1.9*d*).

Cables—Flexible Members Stressed in Tension by Transverse Loads

Cables are relatively slender, flexible members composed of a group of high-strength steel wires twisted together mechanically. By drawing alloyed steel bars through dies—a process that aligns the molecules of the metal—manufacturers are able to produce wire with a tensile strength reaching as high as 270,000 psi. Since cables have no bending stiffness, they can only carry direct tensile stress (they would obviously buckle under the smallest compressive force). Because of their high tensile strength and efficient manner of transmitting load (by direct stress), cable structures have the strength to support the large loads of long-span structures more economically than most other structural elements. For example, when distances to be spanned exceed 2000 ft, designers usually select suspension or cable-stayed bridges (see Photo 1.4). Cables can be used to construct roofs as well as guyed towers.

Under its own deadweight (a uniform load acting along the arc of the cable), the cable takes the shape of a catenary (Figure 1.10*a*). If the cable carries a load distributed uniformly over the horizontal projection of its span, it will assume the shape of a *parabola* (Figure 1.10*b*). When the *sag* (the vertical distance between the cable chord and the cable at midspan) is small (Figure 1.10*a*), the cable shape produced by its dead load may be closely approximated by a parabola.

Because of a lack of bending stiffness, cables undergo large changes in shape when concentrated loads are applied. The lack of bending stiffness also

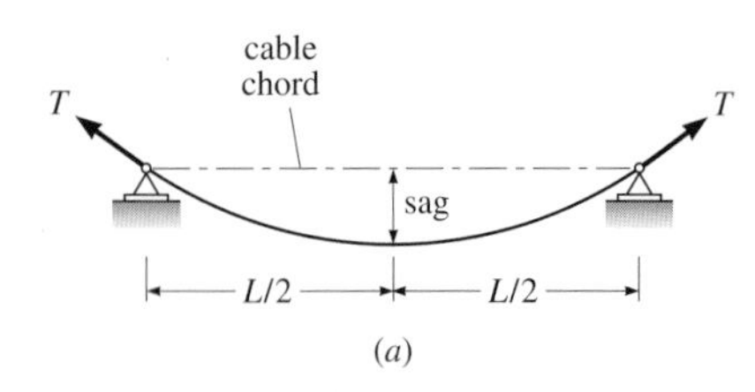

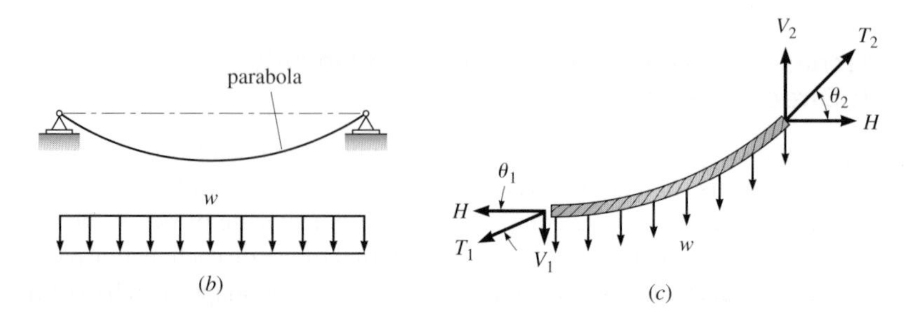

Figure 1.10: (*a*) Cable in the shape of a catenary under dead load; (*b*) parabolic cable produced by a uniform load; (*c*) free-body diagram of a section of cable carrying a uniform vertical load; equilibrium in horizontal direction shows that the horizontal component of cable tension *H* is constant.

(*a*)

(*b*)

Photo 1.4: (*a*) Golden Gate Bridge (San Francisco Bay Area). Opened in 1937, the main span of 4200 ft was the longest single span at that time and retained this distinction for 29 years. Principal designer was Joseph Strauss who had previously collaborated with Ammann on the George Washington Bridge in New York City. (*b*) Rhine River Bridge at Flehe, near Dusseldorf, Germany. Single-tower design. The single line of cables is connected to the center of the deck, and there are three traffic lanes on each side. This arrangement depends on the torsional stiffness of the deck structure for overall stability.

makes it very easy for small disturbing forces (e.g., wind) to induce oscillations (flutter) into cable-supported roofs and bridges. To utilize cables effectively as structural members, engineers have devised a variety of techniques to minimize deformations and vibrations produced by live loads. Techniques to stiffen cables include (1) pretensioning, (2) using tie-down cables, and (3) adding extra dead load (see Figure 1.11).

As part of the cable system, supports must be designed to absorb the cable end reactions. Where solid rock is available, cables can be anchored economically by grouting the anchorage into rock (see Figure 1.12). If rock is not available, heavy foundations must be constructed to anchor the cables.

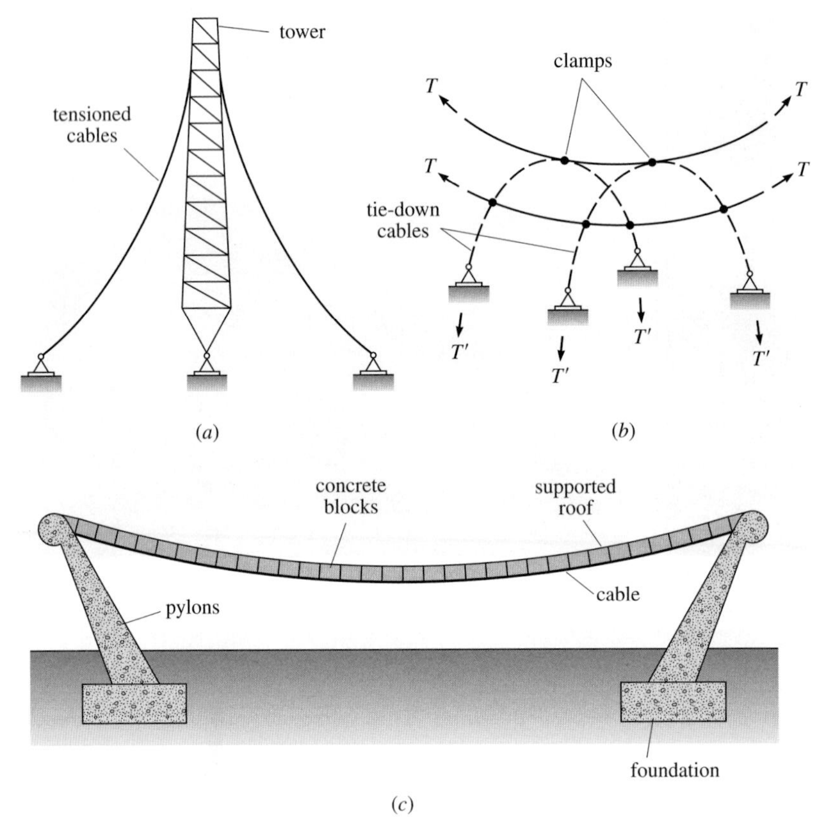

Figure 1.11: Techniques to stiffen cables: (*a*) guyed tower with pretensioned cables stressed to approximately 50 percent of their ultimate tensile strength; (*b*) three-dimensional net of cables; tie-down cables stabilize the upward-sloping cables; (*c*) cable roof paved with concrete blocks to hold down cable to eliminate vibrations. Cables supported by massive pylons (columns) at each end.

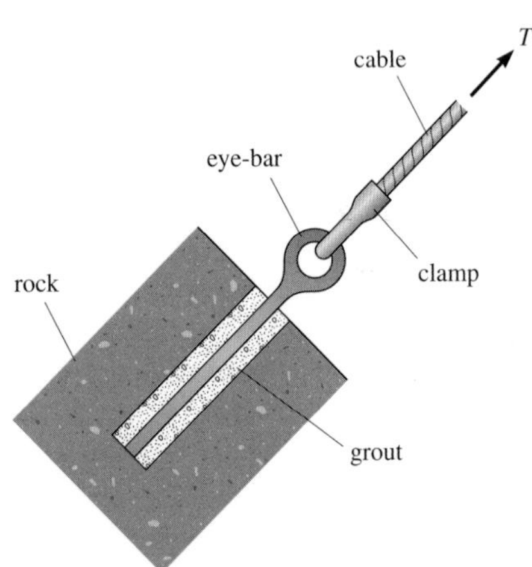

Figure 1.12: Detail of a cable anchorage into rock.

In the case of suspension bridges, large towers are required to support the cable, much as a clothes pole props up a clothesline.

Rigid Frames—Stressed by Axial Load and Moment

Examples of rigid frames (structures with rigid joints) are shown in Figure 1.13*a* and *b*. Members of a rigid frame, which typically carry axial load and moment, are called *beam-columns*. For a joint to be rigid, the angle between the members framing into a joint must not change when the members are loaded. Rigid joints in reinforced concrete structures are simple to construct because of the monolithic nature of poured concrete. However, rigid joints fabricated from steel beams with flanges (Figure 1.6*c*) often require stiffening plates to transfer the large forces in the flanges between members framing into the joint (see Figure 1.13*c*). Although joints can be formed by riveting or bolting, welding greatly simplifies the fabrication of rigid joints in steel frames.

Plates or Slabs—Load Carried by Bending

Plates are planar elements whose depth (or thickness) is small compared to their length and width. They are typically used as floors in buildings and

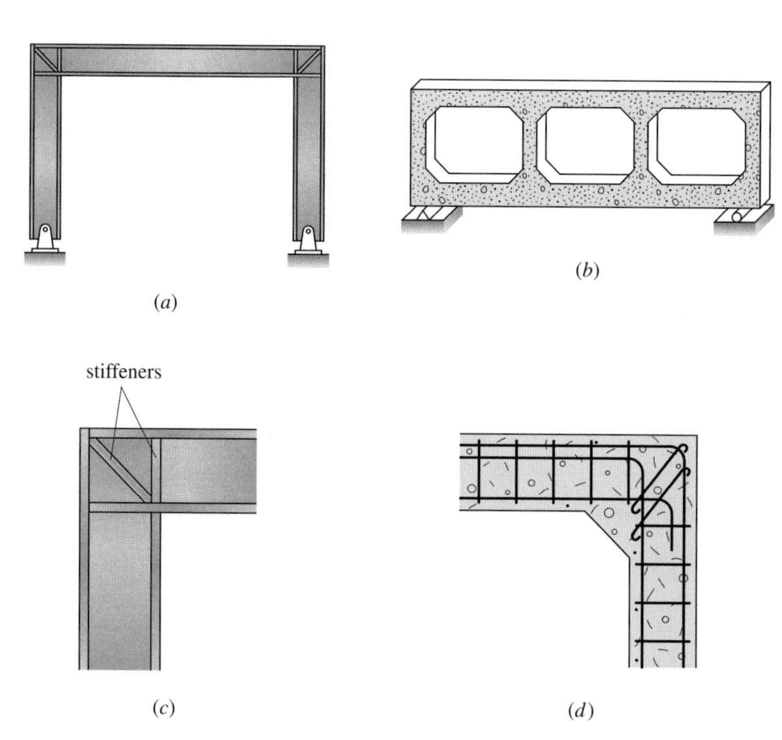

Figure 1.13: Rigid-jointed structures: (*a*) one-story rigid frame; (*b*) Vierendeel truss, loads transmitted both by direct stress and bending; (*c*) details of a welded joint at the corner of a steel rigid frame; (*d*) reinforcing detail for corner of concrete frame in (*b*).

bridges or as walls for storage tanks. The behavior of a plate depends on the position of supports along the boundaries. If rectangular plates are supported on opposite edges, they bend in single curvature (see Figure 1.14*a*). If supports are continuous around the boundaries, double curvature bending occurs.

Since slabs are flexible owing to their small depth, the distance they can span without sagging excessively is relatively small. (For example, reinforced concrete slabs can span approximately 12 to 16 ft.) If spans are large, slabs are typically supported on beams or stiffened by adding ribs (Figure 1.14*b*).

If the connection between a slab and the supporting beam is properly designed, the two elements act together (a condition called *composite action*) to form a T-beam (Figure 1.14*c*). When the slab acts as the flange of a rectangular beam, the stiffness of the beam will increase by a factor of approximately 2.

By corrugating plates, the designer can create a series of deep beams (called *folded plates*) that can span long distances. At Logan Airport in Boston, a prestressed concrete folded plate of the type shown in Figure 1.14*d* spans 270 ft to act as the roof of a hanger.

Thin Shells (Curved Surface Elements)—Stresses Acting Primarily in Plane of Element

Thin shells are three-dimensional curved surfaces. Although their thickness is often small (several inches is common in the case of a reinforced concrete

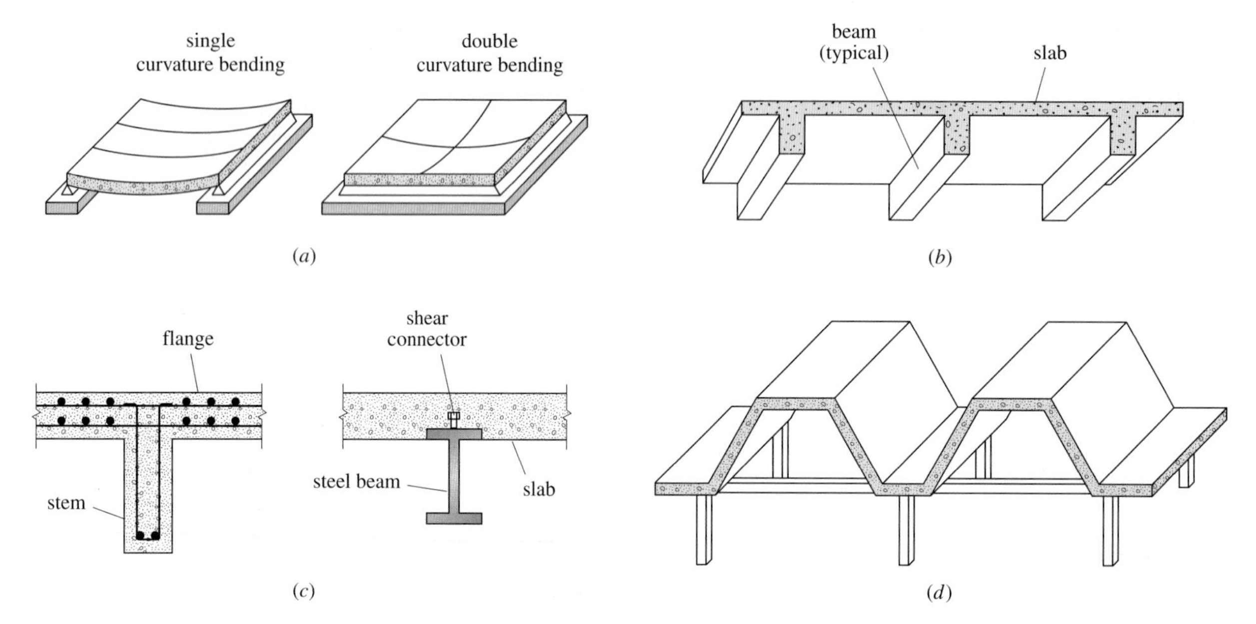

Figure 1.14: (*a*) Influence of boundaries on curvature. (*b*) Beam and slab system. (*c*) Slab and beams act as a unit. On left, concrete slab cast with stem to form a T-beam; right, shear connector joins concrete slab to steel beam, producing a composite beam. (*d*) A folded plate roof.

shell), they can span large distances because of the inherent strength and stiffness of the curved shape. Spherical domes, which are commonly used to cover sports arenas and storage tanks, are one of the most common types of shells built.

Under uniformly distributed loads, shells develop in-plane stresses (called *membrane stresses*) that efficiently support the external load (Figure 1.15). In addition to the membrane stresses, which are typically small in magnitude, shear stresses perpendicular to the plane of the shell, bending moments, and torsional moments also develop. If the shell has boundaries that can equilibrate the membrane stresses at all points (see Figure 1.16*a* and *b*), the majority of the load will be carried by the membrane stresses. But if the shell boundaries cannot supply reactions for the membrane stresses (Figure 1.16*d*), the region of the shell near the boundaries will deform. Since these deformations create shear normal to the surface of the shell as well as moments, the shell must be thickened or an edge member supplied. In most shells, boundary shear and moments drop rapidly with distance from the edge.

The ability of thin shells to span large unobstructed areas has always excited great interest among engineers and architects. However, the great expense of forming the shell, the acoustical problems, the difficulty of producing a watertight roof, and problems of buckling at low stresses have restricted their use. In addition, thin shells are not able to carry heavy concentrated loads without the addition of ribs or other types of stiffeners.

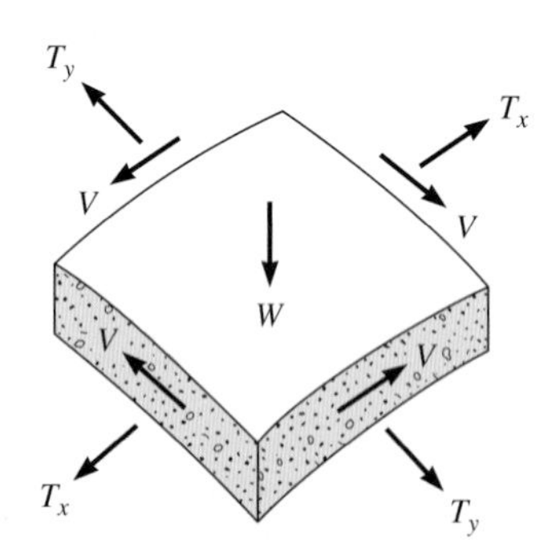

Figure 1.15: Membrane stresses acting on a small shell element.

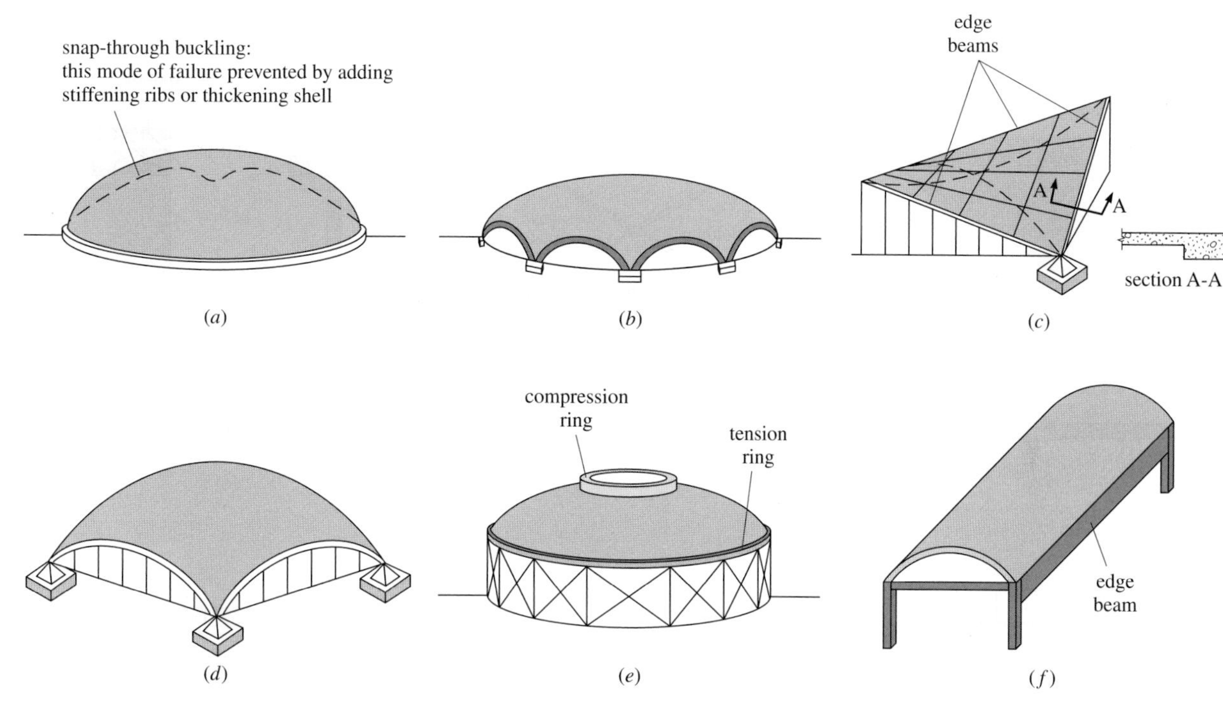

Figure 1.16: Commonly constructed types of shells. (*a*) Spherical dome supported continuously. Boundary condition for membrane action is supplied. (*b*) Modified dome with closely spaced supports. Due to openings, the membrane condition is disturbed somewhat at the boundaries. Shell must be thickened or edge beams supplied at openings. (*c*) Hyperbolic paraboloid. Straight-line generators form this shell. Edge members are needed to supply the reaction for the membrane stresses. (*d*) Dome with widely spaced supports. Membrane forces cannot develop at the boundaries. Edge beams and thickening of shell are required around the perimeter. (*e*) Dome with a compression ring at the top and a tension ring at the bottom. These rings supply reactions for membrane stresses. Columns must carry only vertical load. (*f*) Cylindrical shell.

1.6 Assembling Basic Elements to Form a Stable Structural System

One-Story Building

To illustrate how the designer combines the basic structural elements (described in Section 1.5) into a stable structural system, we will discuss in detail the behavior of a simple structure, considering the one-story, boxlike structure in Figure 1.17*a*. This building, representing a small storage facility, consists of structural steel frames covered with light-gage corregated metal panels. (For simplicity, we neglect windows, doors, and other architectural details.)

In Figure 1.17*b*, we show one of the steel frames located just inside the end wall (labeled *ABCD* in Figure 1.17*a*) of the building. Here the metal roof deck is supported on beam *CD* that spans between two pipe columns located at the

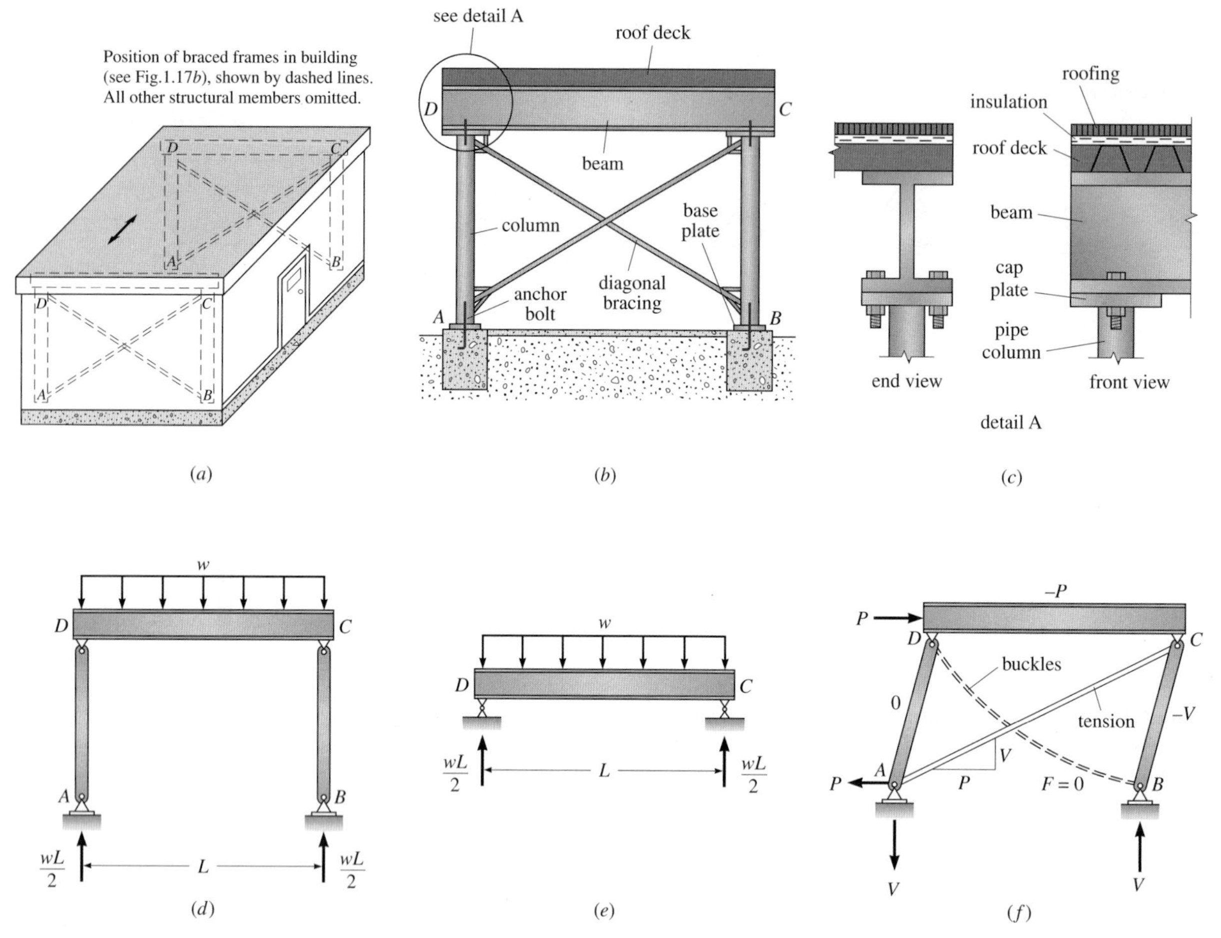

Figure 1.17: (*a*) Three-dimensional view of building (arrow indicates direction in which roof deck spans); (*b*) details of cross-braced frame with bolted joints; (*c*) details of beam-to-column connections; (*d*) idealized model of structural system transmitting gravity loads from roof; (*e*) model of beam *CD*; (*f*) idealized model of truss system for transmitting lateral load acting to the right. Diagonal member *DB* buckles and is ineffective.

corners of the building. As shown in Figure 1.17*c*, the ends of the beam are connected to the tops of the columns by bolts that pass through the bottom flange of the beam and a cap plate welded to the top of the column. Since this type of connection cannot transmit moment effectively between the end of the beam and the top of the column, the designer assumes that this type of a connection acts as a small-diameter hinge.

Because these bolted joints are not rigid, additional light members (often circular bars or steel angle members) are run diagonally between adjacent columns in the plane of the frame, serving to stabilize the structure further. Without this diagonal bracing (Figure 1.17b), the resistance of the frame to lateral loads would be small, and the structure would lack stiffness. Designers insert similar cross-bracing in the other three walls—and sometimes in the plane of the roof.

The frame is connected to the foundation by bolts that pass through a light steel baseplate, welded to the bottom of the column. The bottom ends of the bolts, called *anchor bolts*, are embedded in concrete piers located directly under the column. Typically, designers assume that a simple bolted connection of this type acts as a *pin support*; that is, the connection prevents the base of the column from displacing vertically and horizontally, but it does not have sufficient stiffness to prevent rotation (engineering students often wrongly assume that a flat baseplate bolted to a concrete pier produces a fixed-end condition, but they are not taking into account the large loss of rotational restraint induced by even small flexural deformations of the plate).

Although the bolted connection does have the capacity to apply a small but uncertain amount of rotational restraint to the base of the column, the designer usually treats it conservatively as a *frictionless pin*. However, it is usually unnecessary to achieve a more rigid connection because to do so is expensive, and the additional rigidity can be supplied more simply and economically by increasing the moment of inertia of the columns. If designers wish to produce a fixed support at the base of a column to increase its stiffness, they must use a heavy, stiffened baseplate and the foundation must be massive.

Design of Frame for Gravity Load. To analyze this small frame for gravity load, the designer assumes the weight of the roof and any vertical live load (e.g., snow or ice) are carried by the roof deck (acting as a series of small parallel beams) to the frame shown in Figure 1.17*d*. This frame is idealized by the designer as a beam connected by a pinned joint to the columns. *The designer neglects the diagonal bracing as a secondary member—assumed to be inactive when vertical load acts*. Since no moments are assumed to develop at the ends of the beam, the designer analyzes the beam as a simply supported member with a uniform load (see Figure 1.17*e*). Because the reactions of the beam are applied directly over the centerlines of the columns, the designer assumes that the column carries only direct stress and behaves as an axially loaded compression member.

Design for Lateral Load. The designer next checks for lateral loads. If a lateral load P (produced by wind, for example) is applied to the top of the roof (see Figure 1.17*f*), the designer can assume that one of the diagonals acting together with the roof beam and columns forms a truss. If the diagonals are light flexible members, only the diagonal running from A to C, which stretches and develops tensile stresses as the beam displaces to the right, is assumed to be effective. The opposite diagonal BD is assumed to buckle because it is slender and placed in compression by the lateral movement of the beam. If the wind reverses direction, the other diagonal BD would become effective, and diagonal AC would buckle.

As we have illustrated in this simple problem, under certain types of loads, certain members come into play to transmit the loads into the supports. As long as the designer understands how to pick a logical path for these loads, the analysis can be greatly simplified by eliminating members that are not effective.

1.7 Analyzing by Computer

Until the late 1950s, the analysis of certain types of indeterminate structures was a long, tedious procedure. The analysis of a structure with many joints and members (a space truss, for example) might require many months of computations by a team of experienced structural engineers. Moreover, since a number of simplifying assumptions about structural behavior were often required, the accuracy of the final results was uncertain. Today computer programs are available that can analyze most structures rapidly and accurately. Some exceptions still exist. If the structure is an unusual shape and complex—a thick-walled nuclear containment vessel or the hull of a submarine—the computer analysis can still be complicated and time-consuming.

Most computer programs for analyzing structures are written to produce a *first-order analysis*; that is, they assume (1) linear-elastic behavior, (2) that member forces are unaffected by the deformations (change in geometry) of the structure, and (3) that no reduction in flexural stiffness is produced in columns by compression forces.

The classical methods of analysis covered in this book produce a first-order analysis, suitable for the majority of structures, such as trusses, continuous beams, and frames, encountered in engineering practice. When a first-order analysis is used, structural design codes provide empirical procedures needed to adjust forces that may be underestimated.

While more complicated to use, second-order programs that do account for inelastic behavior, changes in geometry, and other effects influencing the magnitude of forces in members are more precise and produce a more accurate analysis. For example, long slender arches under moving loads can undergo changes in geometry that increase bending moments significantly. For structures of this type, a second-order analysis is essential.

Engineers typically use computer programs prepared by teams of structural specialists who are also skilled programmers and mathematicians. Of course, if the designer does not establish a stable structure, or if a critical loading condition is overlooked, the information supplied by the analysis is obviously not adequate to produce a safe, serviceable structure.

In 1977, the failure of the large three-dimensional space truss (see pages 72 and 714) supporting the 300-ft by 360-ft roof of the Hartford Civic Center Arena is an example of a structural design in which the designers relied on an incomplete computer analysis and failed to produce a safe structure. Among the factors contributing to this disaster were inaccurate data (the designer underestimated the deadweight of the roof by 1.5 million lb), and the inability of the computer program to predict the buckling load of the compression members in the truss. In other words, the presumption existed in the program that the structure was stable—an assumption in the majority of early computer programs used for analyzing structures. Shortly after a winter storm deposited a heavy load of rain-soaked snow and ice on the roof, the buckling of certain slender compression members in the roof truss precipitated a sudden collapse of the entire roof. Fortunately, the failure occurred several hours after a crowd of 5000 sports fans attending a basketball game had left the building. Had the failure taken place several hours sooner (when the building was occupied), hundreds of people would have been killed. Although no loss of life occurred, the facility was unusable for a considerable period, and large sums of money were required to clear the wreckage, to redesign the building, and to reconstruct the arena.

Although the computer has reduced the hours of computations required to analyze structures, the designer must still have a basic insight into all potential failure modes in order to assess the reliability of the solutions generated by the computer. Preparation of a mathematical model that adequately represents the structure remains one of the most important aspects of structural engineering.

1.8 Preparation of Computations

Preparation of a set of clear, complete computations for each analysis is an important responsibility of each engineer. A well-organized set of computations not only will reduce the possibility of computational error, but also will provide essential information if the strength of an existing structure must be investigated at some future time. For example, the owner of a building may wish to determine if one or more additional floors can be added to an existing structure without overstressing the structural frame and foundations. If the original computations are complete and the engineer can determine the design loads, the allowable stresses, and the assumptions upon which the original analysis and design were based, evaluation of the modified structure's strength is facilitated.

Occasionally, a structure fails (in the worst case, lives are lost) or proves unsatisfactory in service (e.g., floors sag or vibrate; walls crack). In these situations, the original computations will be examined closely by all parties to establish the liability of the designer. A sloppy or incomplete set of computations can damage an engineer's reputation.

Since the computations required to solve the homework problems in this book are similar to those made by practicing engineers in design offices, students should consider each assignment as an opportunity to improve the skills required to produce computations of professional quality. With this objective in mind, the following suggestions are offered:

1. State the objective of the analysis in a short sentence.
2. Prepare a clear sketch of the structure, showing all loads and dimensions. Use a sharp pencil and a straightedge to draw lines. Figures and numbers that are neat and clear have a more professional appearance.
3. *Include all steps of your computations.* Computations cannot easily be checked by another engineer unless all steps are shown. Provide a word or two stating what is being done, as needed for clarification.
4. *Check the results* of your computations by making a static check (i.e., writing additional equilibrium equations).
5. If the structure is complex, check the computations by making an approximate analysis (see Chapter 14).
6. Verify that the direction of the deflections is consistent with the direction of the applied forces. If a structure is analyzed by a computer, the displacements of joints (part of the output data) can be plotted to scale to produce a clear picture of the deformed structure.

Summary

- To begin our study of structural analysis, we reviewed the relationship between planning, design, and analysis. In this interrelated process, the structural engineer first establishes one or more initial configurations of possible structural forms, estimates deadweights, selects critical design loads, and analyzes the structure. Once the structure is analyzed, major members are resized. If the results of the design confirm that the initial assumptions were correct, the design is complete. If there are large differences between the initial and final proportions, the design is modified, and the analysis and sizing repeated. This process continues until final results confirm that the proportions of the structure require no modifications.
- Also we reviewed the characteristics of common structural elements that comprise typical buildings and bridges. These include beams, trusses, arches, frames with rigid joints, cables, and shells.
- Although most structures are three-dimensional, the designer who develops an understanding of structural behavior can often divide the structure into a series of simpler planar structures for analysis. The designer is able to select a simplified and idealized model that accurately represents the essentials of the real structure. For example, although the exterior masonry or windows and wall panels of a building, connected to the structural frame, increase the stiffness of the structure, this interaction is typically neglected.
- Since most structures are analyzed by computer, structural engineers must develop an understanding of structural behavior so they can, with a few simple computations, verify that the results of the computer analysis are reasonable. Structural failures not only involve high costs, but also may result in injury to the public or loss of life.

Hanshin Expressway Damaged by the 1995 Kobe Earthquake in Japan

The 1995 Kobe Earthquake (Magnitude 6.9) caused a section of the elevated Hanshin Expressway to topple. The proximity of the epicenter in the highly populated urban cities caused significant casualties and economic losses. Observations of earthquake damage in countries like the United States and Japan show a high correlation between damage and construction era. Unless they are retrofitted, bridges built before 1970 are more vulnerable to earthquake damage.

CHAPTER 2

Design Loads

Chapter Objectives

- Learn the importance of codes for the determination of the governing design loads as they relate to life safety and apply to a building's structural framing systems.
- Understand that code prescribed loads generate minimum design forces, which are either applied statically or dynamically in the analysis of the building's structural systems.
- Become familiar with dead and live loads, calculate floor material's self weight, select live loads based on a building's occupancy use, and learn tributary area method for calculating forces on beams, girders or columns.
- Develop an understanding of lateral loads, calculate wind loads by the simplified method and earthquake loads by the equivalent lateral force procedure, and determine a structure's horizontal base shear and overturning forces due to wind or earthquake loads.

2.1 Building and Design Code

A code is a set of technical specifications and standards that control major details of analysis, design, and construction of buildings, equipment, and bridges. The purpose of codes is to produce safe, economical structures so that the public will be protected from poor or inadequate design and construction.

Two types of codes exist. One type, called a *structural code*, is written by engineers and other specialists who are concerned with the design of a particular class of structure (e.g., buildings, highway bridges, or nuclear power plants) or who are interested in the proper use of a specific material (steel, reinforced concrete, aluminum, or wood). Typically, structural codes specify design loads, allowable stresses for various types of members, design assumptions, and requirements for materials. Examples of structural codes frequently used by structural engineers include the following:

1. *Standard Specifications for Highway Bridges* by the American Association of State Highway and Transportation Officials (AASHTO) covers the design and analysis of highway bridges.

Building codes exist for the purpose of protecting public health, safety, and welfare in the construction and occupancy of buildings and structures, and do so by sets of rules specifying the minimum loads and requirements. One of earliest preserved form of building code is the Code of Hammurabi of Babylon from the Middle Bronze Age. Most of the ancient codes reflect the eye-for-an-eye principle, and were regarded as a set of laws that not even a King could alter. One example from the Code of Hammurabi is "If a builder builds a house for someone, and does not construct it properly, and the house which he built falls in and kills its owner, then that builder shall be put to death."

2. *Manual for Railway Engineering* by the American Railway Engineering and Maintenance of Way Association (AREMA) covers the design and analysis of railroad bridges.
3. *Building Code Requirements for Reinforced Concrete* (ACI 318) by the American Concrete Institute (ACI) covers the analysis and design of concrete structures.
4. *Manual of Steel Construction* by the American Institute of Steel Construction (AISC) covers the analysis and design of steel structures.
5. *National Design Specifications for Wood Construction* by the American Forest & Paper Association (AFPA) covers the analysis and design of wood structures.

The second type of code, called a *building code*, is established to cover construction in a given region (often a city or a state). A building code contains provisions pertaining to architectural, structural, mechanical, and electrical requirements. The objective of a building code is also to protect the public by accounting for the influence of local conditions on construction. Those provisions of particular concern to the structural designer cover such topics as soil conditions (bearing pressures), live loads, wind pressures, snow and ice loads, and earthquake forces. Today many building codes adopt the provisions of the *Standard Minimum Design Loads for Buildings and Other Structures* published by the American Society of Civil Engineers (ASCE) or the more recent *International Building Code* by the International Code Council.

As new systems evolve, as new materials become available, or as repeated failures of accepted systems occur, the contents of codes are reviewed and updated. In recent years the large volume of research on structural behavior and materials has resulted in frequent changes to both types of codes. For example, the AISC issues a revised edition of the code, *Manual of Steel Construction*, every five years.

Most codes make provision for the designer to depart from prescribed standards if the designer can show by tests or analytical studies that such changes produce a safe design.

2.2 Loads

Structures must be proportioned so that they will not fail or deform excessively under load. Therefore, an engineer must take great care to anticipate the probable loads a structure must carry. Although the design loads specified by the codes are generally satisfactory for most buildings, the designer must also decide if these loads apply to the specific structure under consideration. For example, if the shape of a building is unusual (and induces increased wind speeds), wind forces may deviate significantly from the minimum prescribed by a building code. In such cases, the designer should conduct wind tunnel tests on models to establish the appropriate design forces. The designer should also try to foresee if the function of a structure (and consequently the loads it must carry) will change in the future. For example, if the possibility exists that heavier equipment may be introduced into an area that is originally designed for a smaller load, the designer may decide to increase the design loads specified by the code. Designers typically differentiate between two types of load: live load and dead load.

2.3 Dead Loads

The load associated with the weight of the structure and its permanent components (floors, ceilings, ducts, and so forth) is called the *dead load*. Since the dead load must be used in the computations to size members but is not known precisely until the members are sized, its magnitude must be estimated initially. After members are sized and architectural details finalized, the dead load can be computed more accurately. If the computed value of dead load is approximately equal to (or slightly less than) the initial estimate of its value, the analysis is finished. But if a large difference exists between the estimated and computed values of dead load, the designer should revise the computations, using the improved value of deadweight.

In most buildings the space directly under each floor is occupied by a variety of utility lines and supports for fixtures including air ducts, water and sewage pipes, electrical conduit, and lighting fixtures. Rather than attempt to account for the actual weight and position of each item, designers add an additional 10 to 15 lb/ft^2 (0.479 to 0.718 kN/m^2) to the weight of the floor system to ensure that the strength of the floor, columns, and other structural members will be adequate.

Distribution of Dead Load to Framed Floor Systems

Many floor systems consist of a reinforced concrete slab supported on a rectangular grid of beams. The supporting beams reduce the span of the slab and permit the designer to reduce the depth and weight of the floor system. The distribution of load to a floor beam depends on the geometric configuration of the beams forming the grid. To develop an insight into how load from a particular region of a slab is transferred to supporting beams, we will examine the three cases shown in Figure 2.1. In the first case, the edge beams support a uniformly loaded *square slab* (see Figure 2.1*a*). From symmetry we can infer that each of the four beams along the outside edges of the slab carries the same triangular load. In fact, if a slab with the same area of uniformly distributed reinforcement in the x and y directions were loaded to failure by a uniform load, large cracks would open along the main diagonals, confirming that each beam supports the load on a triangular area. The area of slab that is supported by a particular beam is termed the beam's *tributary area*.

In the second case, we consider a rectangular slab supported on opposite sides by two parallel beams (Figure 2.1*b*). In this case, if we imagine a uniformly loaded 1-ft-wide strip of slab that acts as a beam spanning a distance L_s

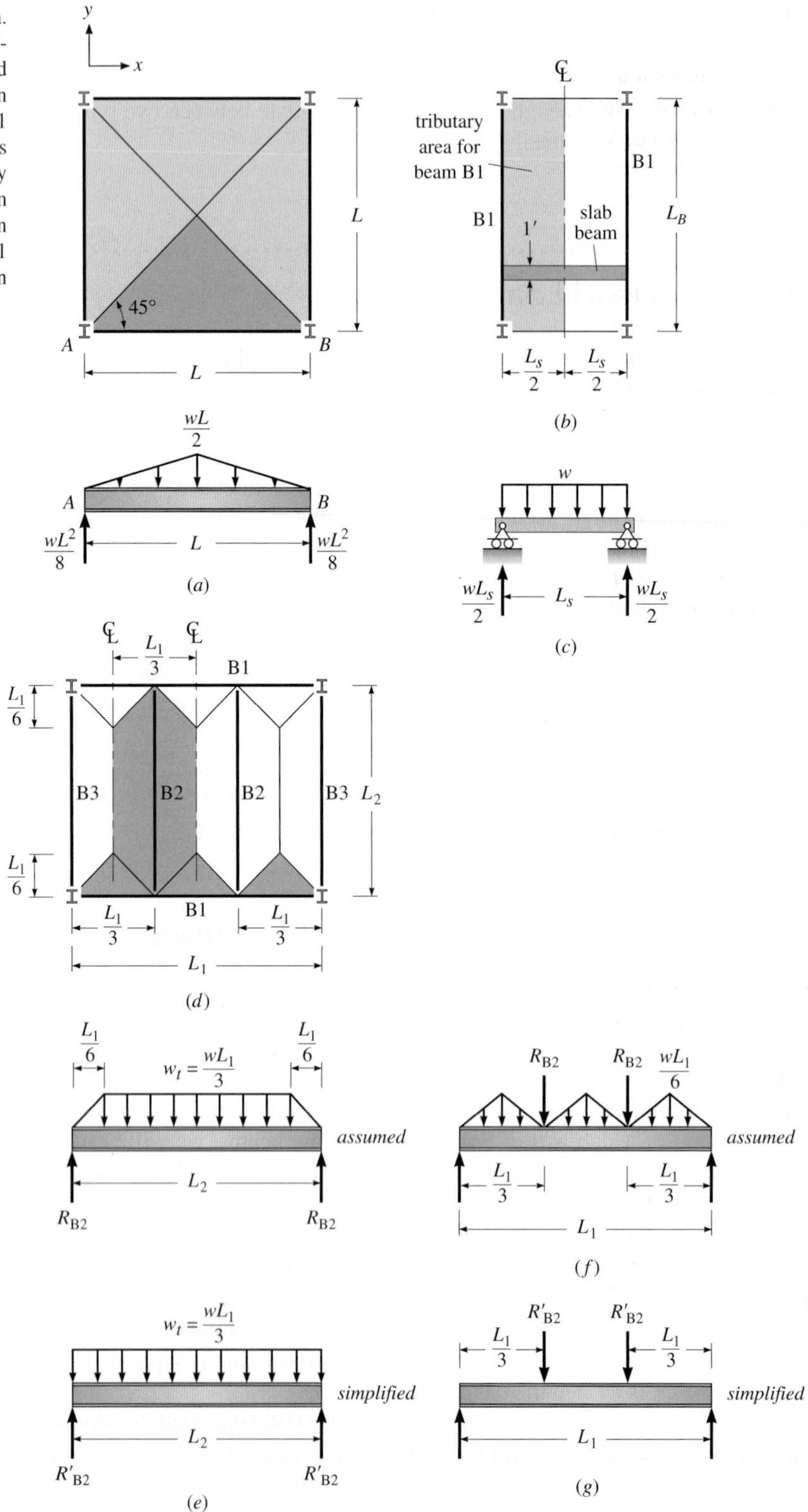

Figure 2.1: Concept of tributary area. (*a*) square slab, all edge beams support a triangular area; (*b*) two edge beams divide load equally; (*c*) load on a 1-ft width of slab in Figure (*b*); (*d*) tributary areas for beams B1 and B2 shown shaded, all diagonal lines slope at 45°; (*e*) top figure shows most likely load on beam B2 in Figure (*d*); bottom figure shows simplified load distribution on beam B2; (*f*) most likely load on beam B1 in Figure (*d*); (*g*) simplified load distribution to beam B1.

between two edge beams B1 and B2 (Figure 2.1*b*), we can see that the load on the slab divides equally between the supporting edge beams; that is, each foot of beam carries a uniformly distributed load of $wL_s/2$ (Figure 2.1*c*), and the tributary area for each beam is a rectangular area that extends out from the beam a distance $L_s/2$ to the centerline of the slab.

For the third case, shown in Figure 2.1*d*, a slab, carrying a uniformly distributed load w, is supported on a rectangular grid of beams. The tributary area for both an interior and an exterior beam is shown shaded in Figure 2.1*d*. Each interior beam B2 (see Figure 2.1*d*) carries a trapezoidal load. The edge beam B1, which is loaded at the third points by the reactions from the two interior beams, also carries smaller amounts of load from three triangular areas of slab (Figure 2.1*f*). If the ratio of the long to short side of a panel is approximately 2 or more, the actual load distributions on beam B2 can be simplified by assuming conservatively that the total load per foot, $w_t = wL_1/3$, is uniformly distributed over the entire length (see Figure 2.1*e*), producing the reaction R'_{B2}. In the case of beam B1, we can simplify the analysis by assuming the reaction R'_{B2} from the uniformly loaded B2 beams is applied as a concentrated load at the third points (see Figure 2.1*g*).

Table 2.1*a* on page 34 lists the unit weights of a number of commonly used construction materials, and Table 2.1*b* contains the weights of building components that are frequently specified in building construction. We will make use of these tables in examples and problems.

Examples 2.1 and 2.2 introduce computations for dead load.

EXAMPLE 2.1

A three-ply asphalt felt and gravel roof over 2-in-thick insulation board is supported by 18-in-deep precast reinforced concrete beams with 3-ft- wide flanges (see Figure 2.2). If the insulation weighs 3 lb/ft² and the asphalt roofing weighs $5\frac{1}{2}$ lb/ft², determine the total dead load, per foot of length, each beam must support.

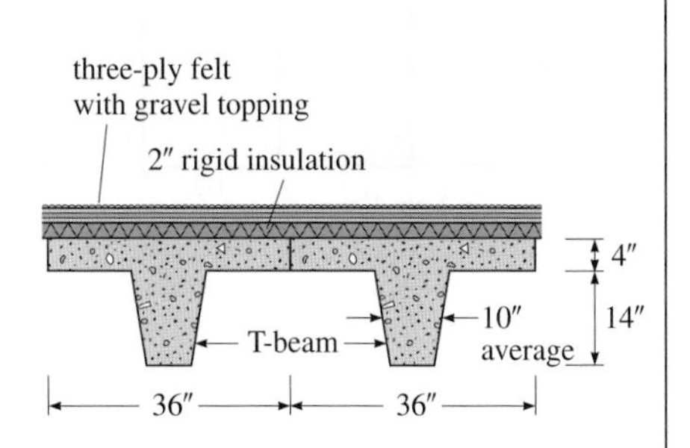

Figure 2.2: Cross section of reinforced concrete beams.

Solution

Weight of beam is as follows:

Flange $\quad \frac{4}{12}\text{ ft} \times \frac{36}{12}\text{ ft} \times 1\text{ ft} \times 150\text{ lb/ft}^3 = 150\text{ lb/ft}$

Stem $\quad \frac{10}{12}\text{ ft} \times \frac{14}{12}\text{ ft} \times 1\text{ ft} \times 150\text{ lb/ft}^3 = 145\text{ lb/ft}$

Insulation $\quad 3\text{ lb/ft}^2 \times 3\text{ ft} \times 1\text{ ft} = 9\text{ lb/ft}$

Roofing $\quad 5\frac{1}{2}\text{ lb/ft}^2 \times 3\text{ ft} \times 1\text{ ft} = 16.5\text{ lb/ft}$

Total = 320.5 lb/ft,
round to 0.321 kip/ft

EXAMPLE 2.2

The steel framing plan of a small building is shown in Figure 2.3*a*. The floor consists of a 5-in-thick reinforced concrete slab supported on steel beams (see section 1-1 in Figure 2.3*b*). Beams are connected to each other and to the corner columns by clip angles; see Figure. 2.3*c*. The clip angles are assumed to provide the equivalent of a pin support for the beams; that is, they can transmit vertical load but no moment. An acoustical board ceiling, which weighs 1.5 lb/ft^2, is suspended from the concrete slab by closely spaced supports, and it can be treated as an additional uniform load on the slab. To account for the weight of ducts, piping, conduit, and so forth, located between the slab and ceiling (and supported by hangers from the slab), an additional dead load allowance of 20 lb/ft^2 is assumed. The designer initially estimates the weight of beams B1 at 30 lb/ft and the 24-ft girders B2 on column lines 1 and 2 at 50 lb/ft. Establish the magnitude of the dead load distribution on beam B1 and girder B2.

Solution

We will assume that all load between panel centerlines on either side of beam B1 (the tributary area) is supported by beam B1 (see the shaded area in Figure 2.3*a*). In other words, as previously discussed, to compute the dead load applied by the

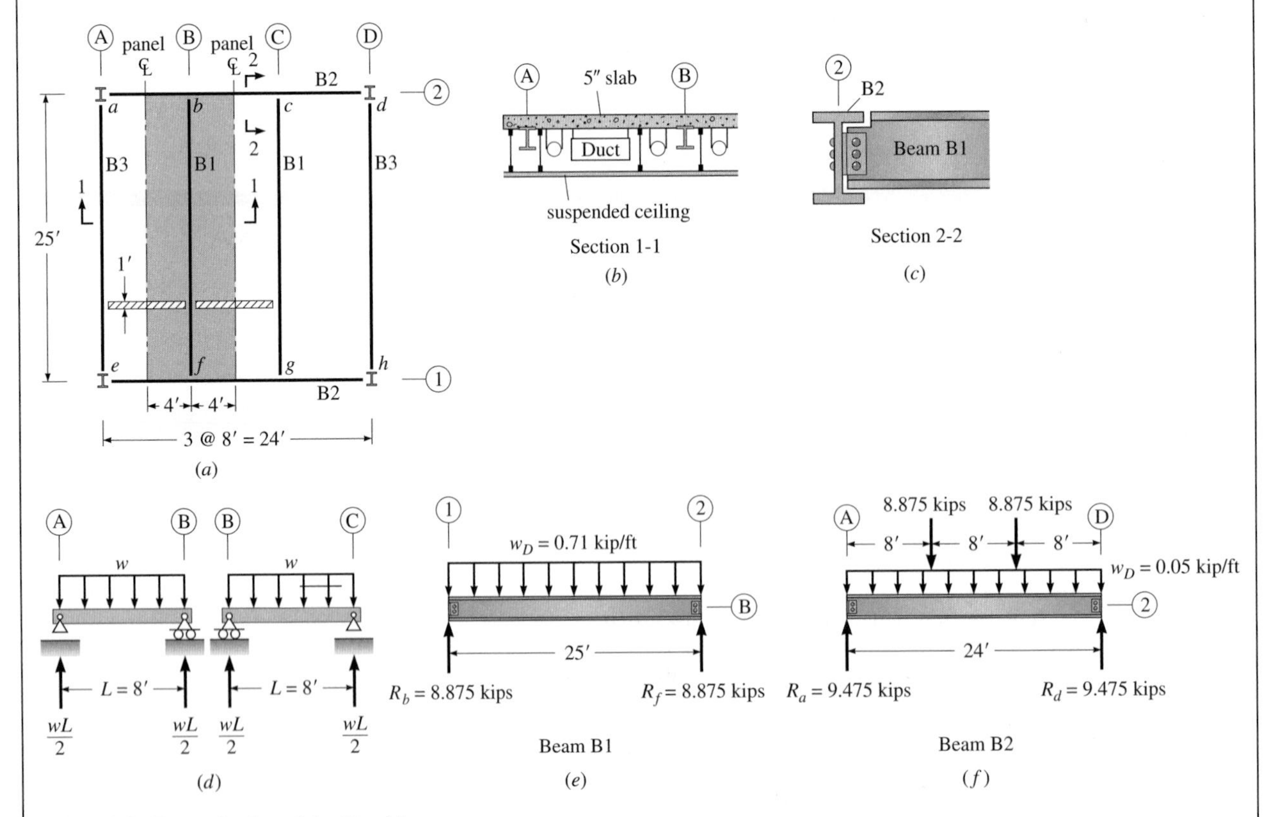

Figure 2.3: Determination of dead load for beam and girder.

slab to the beam, we treat the slab as a series of closely spaced, 1-ft-wide, simply supported beams, spanning between the steel beams on column lines A and B, and between B and C (see the cross-hatched area in Figure 2.3*a*). One-half of the load, $wL/2$, will go to each supporting beam (Figure 2.3*d*), and the total slab reaction applied per foot of steel beam equals $wL = 8w$ (see Figure 2.3*e*).

Total dead load applied per foot to beam B1:

Weight of slab $\quad 1\text{ ft} \times 1\text{ ft} \times \frac{5}{12}\text{ ft} \times 8\text{ ft} \times 150\text{ lb/ft}^3 = 500\text{ lb/ft}$

Weight of ceiling $\quad 1.5\text{ lb/ft}^2 \times 8\text{ ft} = 12\text{ lb/ft}$

Weight of ducts, etc. $\quad 20\text{ lb/ft}^2 \times 8\text{ ft} = 160\text{ lb/ft}$

Estimated weight of beam $\quad = 30\text{ lb/ft}$

Total = 702 lb/ft, round to 0.71 kip/ft

Sketches of each beam with its applied loads are shown in Figure 2.3*e* and *f*. The reactions (8.875 kips) from the B1 beams are applied as concentrated loads to the third points of girder B2 on column line 2 (Figure 2.3*f*). The uniform load of 0.05 kip/ft is the estimated weight of girder B2.

Tributary Areas of Columns

To determine the dead load transmitted into a column from a floor slab, the designer can either (1) determine the reactions of the beams framing into the column or (2) multiply the tributary area of the floor surrounding the column by the magnitude of the dead load per unit area acting on the floor. The *tributary area* of a column is defined as *the area surrounding the column that is bounded by the panel centerlines*. Use of tributary areas is the more common procedure of the two methods for computing column loads. In Figure 2.4 the tributary areas are shaded for corner column A1, interior column B2, and exterior column C1. Exterior columns located on the perimeter of a building also support the exterior walls as well as floor loads.

As you can see by comparing tributary areas for the floor system in Figure 2.4, when column spacing is approximately the same length in both directions, interior columns support approximately 4 times more floor dead load than corner columns. When we use the tributary areas to establish column loads, we do not consider the position of floor beams, but we do include an allowance for their weight.

Use of tributary areas is the more common procedure of the two methods for computing columns loads because designers also need the tributary areas to compute live loads given that design codes specify that the percentage of *live load* transmitted to a column is an inverse function of the tributary areas; that is, as the tributary areas increase, the live load reduction increases. For columns supporting large areas this reduction can reach a maximum of 40 to 50 percent. We will cover the ASCE standard for live load reduction in Section 2.4.

TABLE 2.1

Typical Design Dead Loads

(*a*) Material Weights

Substance	Weight, lb/ft^3 (kN/m^3)
Steel	490 (77.0)
Aluminum	165 (25.9)
Reinforced concrete:	
Normal weight	150 (23.6)
Light weight	90–120 (14.1–18.9)
Brick	120 (18.9)
Wood	
Southern pine	37 (5.8)
Douglas fir	34 (5.3)
Plywood	36 (5.7)

(*b*) Building Component Weights

Component	Weight, lb/ft^2 (kN/m^2)
Ceilings	
Gypsum plaster on suspended metal lath	10 (0.48)
Acoustical fiber tile on metal lath and channel ceiling	5 (0.24)
Floors	
Reinforced concrete slab per inch of thickness	
Normal weight	$12\frac{1}{2}$ (0.60)
Lightweight	7.5–10 (0.36–0.48)
Roofs	
Three-ply felt tar and gravel	$5\frac{1}{2}$ (0.26)
2-in insulation	3 (0.14)
Walls and partitions	
Gypsum board (1-in thick)	4 (0.19)
Brick (per inch of thickness)	10 (0.48)
Hollow concrete masonry unit (12 in thick)	
Heavy aggregate	85 (4.06)
Light aggregate	55 (2.63)
Hollow clay tile (6-in thick)	30 (1.44)
2 × 4 studs at 16 in on center, $\frac{1}{2}$-in gypsum wall on both sides	8 (0.38)

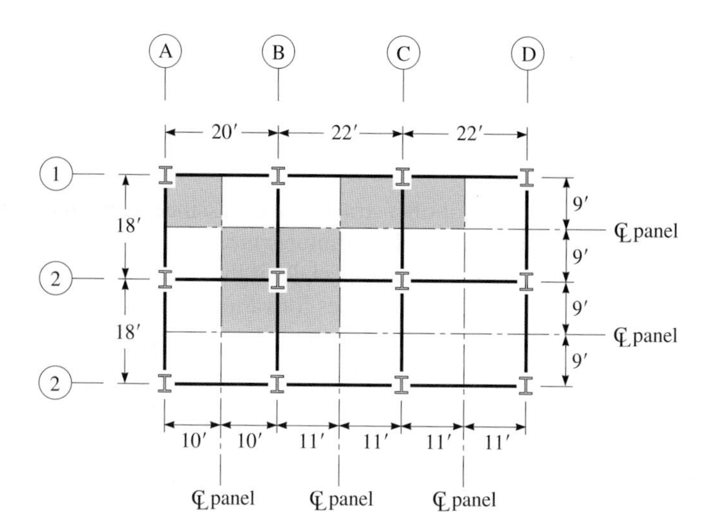

Figure 2.4: Tributary area of columns A1, B2, and C1 shown shaded.

EXAMPLE 2.3

Using the tributary area method, compute the floor dead loads supported by columns A1 and B2 in Figure 2.4. The floor system consists of a 6-in-thick reinforced concrete slab weighing 75 lb/ft^2. Allow 15 lb/ft^2 for the weight of floor beams, utilities, and a ceiling suspended from the floor. The precast exterior wall supported by the perimeter beams weighs 600 lb/ft.

Solution

Total floor dead load is

$$D = 75 + 15 = 90 \text{ lb/ft}^2 = 0.09 \text{ kip/ft}^2$$

Dead load to column A1 is as follows:

Tributary area $A_t = 9 \times 10 = 90 \text{ ft}^2$

Floor dead load $A_t D = 90 \times 0.09 \text{ kip/ft}^2 = 8.1 \text{ kips}$

Weight of exterior wall =
weight/ft (length) = (0.6 kip/ft)(10 + 9) = 11.4 kips

Total = 19.5 kips

Dead load to column B2 is as follows:

Tributary area = $18 \times 21 = 378 \text{ ft}^2$

Total dead load = $378 \text{ ft}^2 \times 0.09 \text{ kip/ft}^2 = 34.02 \text{ kips}$

2.4 Live Loads

Buildings

Loads that can be moved on or off a structure are classified as *live loads*. Live loads include the weight of people, furniture, machinery, and other equipment. Live loads can vary over time especially if the function of the building changes. The live loads specified by codes for various types of buildings represent a conservative estimate of the maximum load likely to be produced by the intended use and occupancy of the building. In each region of the country, building codes typically specify the design live load. Currently, many state and city building codes base the magnitude of live loads and design procedures on the ASCE standard, which has evolved over time by relating the magnitude of the design load to the successful performance of actual buildings. When sizing members, designers must also consider short-term construction live loads, particularly if these loads are large. In the past a number of building failures have occurred during construction when large piles of heavy construction material were concentrated in a small area of a floor or roof of a partially erected building, when the capacity of members, not fully bolted or braced, is below their potential load capacity.

The ASCE standard typically specifies a minimum value of uniformly distributed live load for various types of buildings (a portion of the ASCE minimum live load table is show in Table 2.2). If certain structures, such as parking garages, are also subjected to large concentrated loads, the standard may require

TABLE 2.2

Typical Design Floor Live Loads, L_o

Occupancy Use	Live Load, lb/ft^2 (kN/m^2)
Assembly areas and theaters	
Fixed seats (fastened to floor)	60 (2.87)
Lobbies	100 (4.79)
Stage floors	150 (7.18)
Libraries	
Reading rooms	60 (2.87)
Stack rooms	150 (7.18)
Office buildings	
Lobbies	100 (4.79)
Offices	50 (2.40)
Residential (one- and two-family)	
Habitable attics and sleeping areas	30 (1.44)
Uninhabitable attics with storage	20 (0.96)
All other areas (except balconies)	40 (1.92)
Schools	
Classrooms	40 (1.92)
Corridors above the first floor	80 (3.83)
First-floor corridors	100 (4.79)

that forces in members be investigated for both uniform and concentrated loads, and that the design be based on the loading condition that creates the greatest stresses. For example, the ASCE standard specifies that, in the case of parking garages for passengers vehicles, members be designed to carry either the forces produced by a uniformly distributed live load of 40 lb/ft^2 or a concentrated load of 3,000 lb acting over an area of 4.5 in. by 4.5 in.—whichever is larger.

The ASCE standard specifies wall partitions to be live loads. Normally designers try to position beams directly under heavy masonry walls to carry their weight directly into supports. If an owner requires flexibility to move walls or partitions periodically in order to reconfigure office or laboratory space, the designer can add an appropriate allowance to the floor live load. If partitions are light, (such as stud walls with $\frac{1}{2}$ inch gypsum board each side), a minimum additional uniform floor live load of 15 lb/ft^2 (0.479 kN/m^2) is required. Similarly, in a factory or a laboratory that houses heavy test equipment, the allowance may be 3 or 4 times larger.

The ASCE standard specifies the minimum design live load on roofs as a uniformly distributed 20 psf on ordinary flat, pitched, and curved roofs. However, roof design live loads must also include mechanical equipment, architectural features, as well as potential live loads that can occur during construction, maintenance, and the life of the structure.

Live Load Reduction

Recognizing that a member supporting a large tributary area is less likely to be loaded at all points by the maximum value of live load than a member supporting a smaller floor area, building codes permit live load reductions for members that have a large tributary area. For this situation, the ASCE standard permits a reduction of the design floor live loads L_o, as listed in Table 2.2, by the following equation when the *influence area* $K_{LL}A_T$ is larger than 400 ft^2 (37.2 m^2). However, the reduced live load must not be less than 50 percent of L_o for members supporting one floor or a section of a single floor, nor less than 40 percent of L_o for members supporting two or more floors:

$$L = L_o\left(0.25 + \frac{15}{\sqrt{K_{LL}A_T}}\right) \quad \text{U.S. customary units} \qquad (2.1a)$$

$$L = L_o\left(0.25 + \frac{4.57}{\sqrt{K_{LL}A_T}}\right) \quad \text{SI units} \qquad (2.1b)$$

where L_o = design live load listed in Table 2.2
L = reduced value of live load
A_T = tributary area, ft^2 (m^2)
K_{LL} = live load element factor, equal to 4 for interior columns and exterior columns without cantilever slabs and 2 for interior beams and edge beams without cantilever slabs.

The minimum uniformly distributed roof live loads are permitted to be reduced by ASCE standard as follows:

$$L_r = L_o R_1 R_2 \qquad (2.1c)$$

where L_o = design roof live load

L_r = reduced roof live load, with minimum of 12 psf $\leq L_r \leq$ 20 psf (0.58 m^2 $\leq L_r \leq$ 0.96 m^2 in SI units) for ordinary flat, pitched and curved roofs.

R_1 = 1 for $A_T \leq$ 200 ft^2 (18.58 m^2); and R_1 = 0.6 for $A_T \geq$ 600 ft^2 (55.74 m^2); $R_1 = 1.2 - 0.001A_T$ ($R_1 = 1.2 - 0.011A_T$ in SI units) for 200 ft^2 $< A_T <$ 600 ft^2 (18.58 m^2 $< A_T <$ 55.74m^2);

R_2 = 1.0 for flat roofs $F \leq 4$; $R_2 = 1.2 - 0.05F$ for $4 < F < 12$; and $R_2 = 0.6$ for $F \geq 12$; where F = number of inches of rise per foot of roof slope for pitched roofs. (In SI: $F = 0.12 \times$ slope, with slope expressed in percentage).

For a column or beam supporting more than one floor, the term A_T represents the sum of the tributary areas from all floors.

Note that the ASCE standard limits the amount of live load reduction for special occupancies. Reduction in live load is not permitted for public assembly areas or when the live load is high (>100 psf).

EXAMPLE 2.4

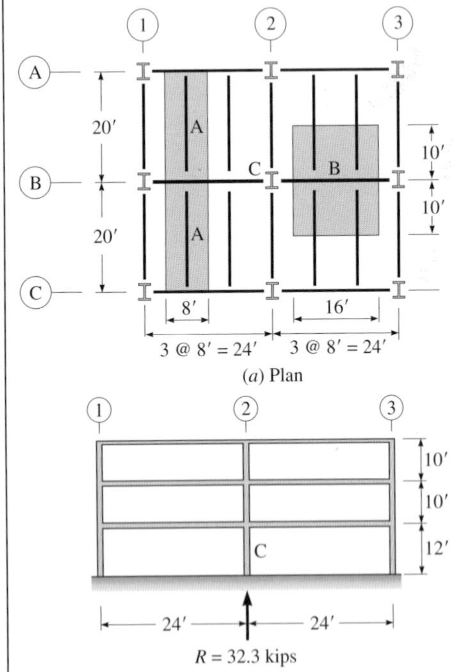

Figure 2.5: Live load reduction (*continues*).

For the three-story building shown in Figure 2.5*a* and *b*, calculate the design live load supported by (1) floor beam A, (2) girder B, and (3) the interior column C located at grid 2-B in the first story. Assume a 50 lb/ft^2 design live load, L_o, on all floors including the roof.

Solution

(1) Floor beam A

Span = 20 ft tributary area $A_T = 8(20) = 160$ ft^2 $K_{LL} = 2$

Determine if live loads can be reduced:

$$K_{LL}A_T = 2A_T = 2(160) = 320 \text{ ft}^2 < 400 \text{ ft}^2$$

therefore, no live load reduction is permitted.

Compute the uniform live load per foot to beam:

$$w = 50(8) = 400 \text{ lb/ft} = 0.4 \text{ kip/ft}$$

See Figure 2.5*d* for loads and reactions.

(2) Girder B

Girder B is loaded at each third point by the reactions of two floor beams. Its tributary area extends outward 10 ft from its longitudinal axis to the midpoint of the panels on each side of the girder (see shaded area in Figure 2.5*a*); therefore $A_T = 20(16) = 320$ ft^2.

$$K_{LL}A_T = 2(320) = 640 \text{ ft}^2$$

Since $K_{LL}A_T = 640\ \text{ft}^2 > 400\ \text{ft}^2$, a live load reduction is permitted. Use Equation 2.1*a*.

$$L = L_o\left(0.25 + \frac{15}{\sqrt{K_{LL}A_T}}\right) = 50\left(0.25 + \frac{15}{\sqrt{640}}\right) = 50(0.843) = 42.1\ \text{lb/ft}^2$$

Since $42.1\ \text{lb/ft}^2 > 0.5(50) = 25\ \text{lb/ft}^2$ (the lower limit), still use $w = 42.1\ \text{lb/ft}^2$.

$$\text{Load at third point} = 2\left[\frac{42.1}{1000}(8)(10)\right] = 6.736\ \text{kips}$$

The resulting design loads are shown in Figure 2.5*e*.

(3) Column C in the first story

The shaded area in Figure. 2.5*c* shows the tributary area of the interior column *for each floor*. Compute the tributary area for roof:

$$A_T = 20(24) = 480\ \text{ft}^2$$

The reduction for roof live load using Equation 2.1*c* is

$$R_1 = 1.2 - 0.001A_T = 0.72$$

$$R_2 = 1.0$$

and the reduced roof live load is

$$L_{\text{roof}} = L_o R_1 R_2 = 50(0.72)(1.0) = 36.0\ \text{psf}$$

Compute the tributary area for the remaining two floors:

$$2A_T = 2(480) = 960\ \text{ft}^2$$

and $$K_{LL}A_T = 4(960) = 3840\ \text{ft}^2 > 400\ \text{ft}^2$$

Therefore, reduce live load for two floors using Equation 2.1*a* (but not less than $0.4L_o$) is

$$L_{\text{floor}} = L_o\left(0.25 + \frac{15}{\sqrt{K_{LL}A_T}}\right) = 50\ \text{lb/ft}^2\left(0.25 + \frac{15}{\sqrt{3840}}\right)$$

$$= 24.6\ \text{lb/ft}^2$$

Since $24.6\ \text{lb/ft}^2 > 0.4 \times 50\ \text{lb/ft}^2 = 20\ \text{lb/ft}^2$ (the lower limit), use $L = 24.6\ \text{lb/ft}^2$.

Load to column $= A_T(L_{\text{roof}}) + 2A_T(L_{\text{floor}}) = 480(36.0) + 960(24.6) = 40{,}896\ \text{lb} = 40.9\ \text{kips}$.

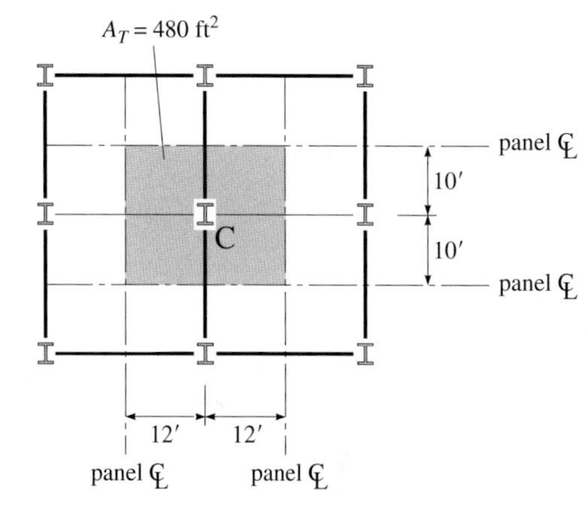

(*c*) Tributary area to column C shown shaded

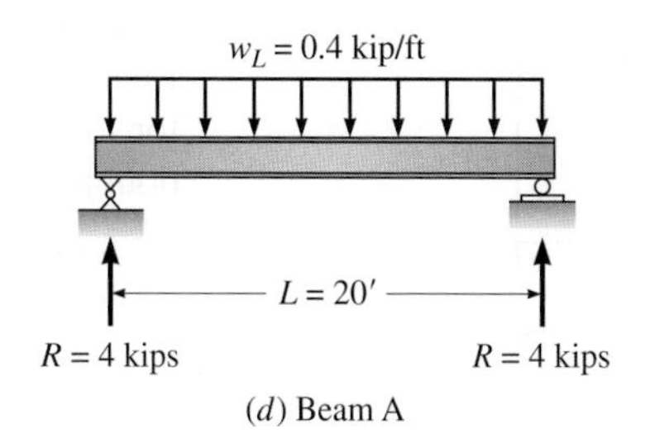

(*d*) Beam A

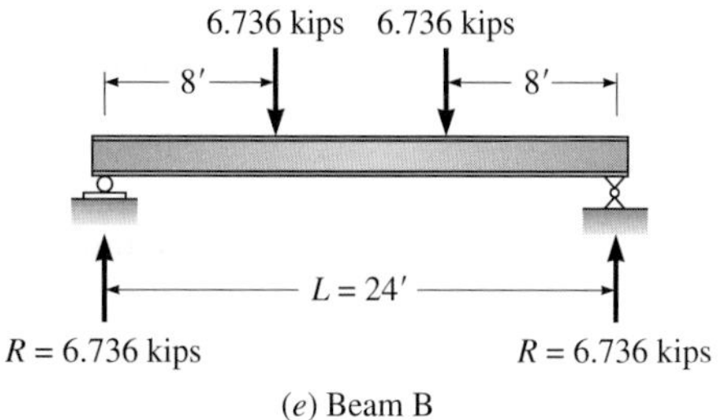

(*e*) Beam B

Figure 2.5: *Continued*

TABLE 2.3
Live Load Impact Factor

Loading Case	Impact Factor I, Percent
Supports of elevators and elevator machinery	100
Supports of light machinery, shaft or motor-driven	20
Supports of reciprocating machinery or power-driven units	50
Hangers supporting floors and balconies	33
Cab-operated traveling crane support girders and their connections	25

Impact

Normally the values of live loads specified by building codes are treated as static loads because the majority of loads (desks, bookcases, filing cabinets, and so forth) are stationary. If loads are applied rapidly, they create additional impact forces. When a moving body (e.g., an elevator coming to a sudden stop) loads a structure, the structure deforms and absorbs the kinetic energy of the moving object. As an alternative to a dynamic analysis, moving loads are often treated as static forces and increased empirically by an impact factor. The magnitude of the impact factor I for a number of common structural supports is listed in Table 2.3.

EXAMPLE 2.5

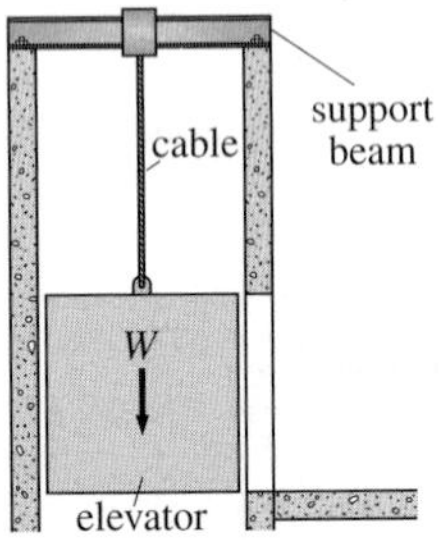

Figure 2.6: Beam supporting an elevator.

Determine the magnitude of the concentrated force for which the beam in Figure 2.6 supporting an elevator must be designed. The elevator, which weighs 3000 lb, can carry a maximum of six people with an average weight of 160 lb.

Solution

Read in Table 2.3 that an impact factor I of 100 percent applies to all elevator loads. Therefore, the weight of the elevator and its passengers must be doubled.

$$\text{Total load} = D + L = 3000 + 6 \times 160 = 3960 \text{ lb}$$
$$\text{Design load} = (D + L)2 = 3960 \times 2 = 7920 \text{ lb}$$

Bridges

Standards for highway bridge design are governed by AASHTO specifications, which require that the engineer consider either a single HS20 truck or the uniformly distributed and concentrated loads shown in Figure 2.7. Typically the HS20 truck governs the design of shorter bridges whose spans do not exceed approximately 145 ft. For longer spans the distributed loading usually controls.

Since moving traffic, particularly when roadway surfaces are uneven, bounces up and down, producing impact forces, truck loads must be increased by an impact factor I given by

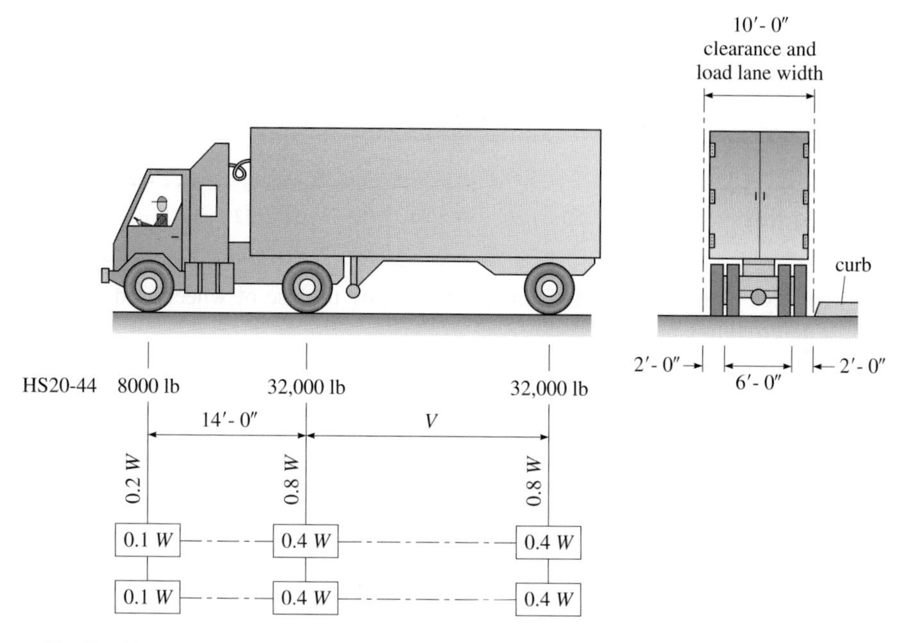

W = Combined weight on the first two axles, which is the same as for the corresponding H truck
V = Variable spacing – 14 ft to 30 ft inclusive. Spacing to be used is that which produces maximum stresses.

(*a*)

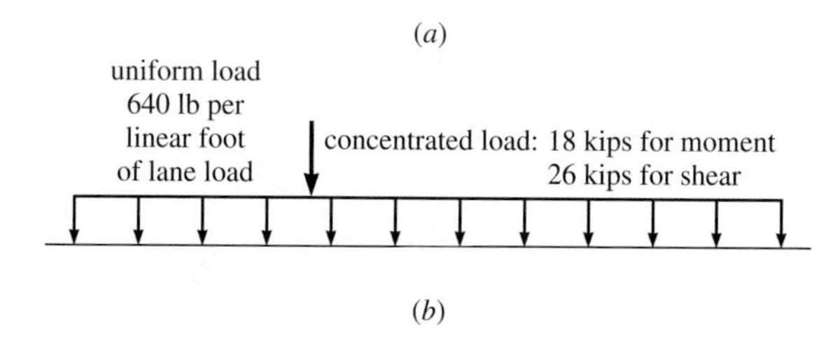

(*b*)

Figure 2.7: AASHTO HS20-44 design live loads.

$$I = \frac{50}{L + 125} \quad \text{U.S. customary units} \tag{2.2a}$$

$$I = \frac{15.2}{L + 38.1} \quad \text{SI units} \tag{2.2b}$$

but the impact factor need not be greater than 30 percent, and L = the length in feet (meters) of the portion of the span that is loaded to produce maximum stress in the member.

The position of the span length L in the denominator of Equation 2.2 indicates that the additional forces created by impact are an inverse function of span length. In other words, since long spans are more massive and have a longer natural period than short spans, dynamic loads produce much larger forces in a short-span bridge than in a long-span bridge.

Railroad bridge design uses the Cooper E80 loading (Figure 2.8) contained in the AREMA *Manual for Railway Engineering*. This loading consists of two locomotives followed by a uniform load representing the weight of freight cars. The AREMA manual also provides an equation for impact. Since the AASHTO and Cooper loadings require the use of influence lines to establish the position of wheels to maximize forces at various positions in a

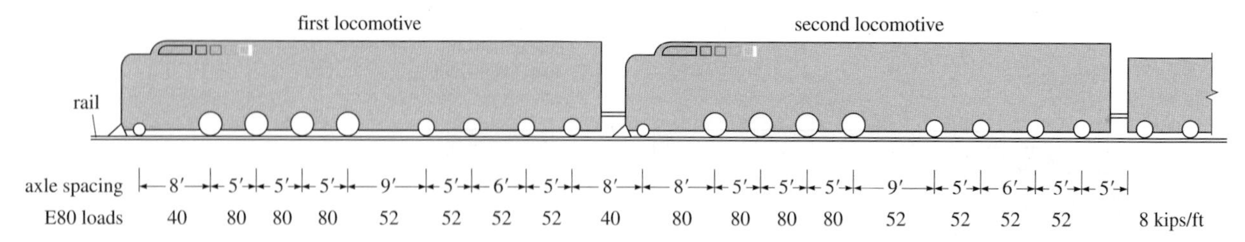

Figure 2.8: AREMA E80 railroad loadings.

bridge member, design examples illustrating the use of wheel loads will be deferred to Chapter 9.

2.5 Snow Loads

Snow load on roofs needs to be considered in cold regions. The design snow load on a flat roof (with slope $\leq 5°$, 1 in/ft = 4.76°) is given by the ASCE standard as follows:

$$p_f = 0.7 C_e C_t I p_g \tag{2.3}$$

where p_g = ground snow load
C_e = exposure factor (0.7 in windy area and 1.3 in sheltered areas with little wind)
C_t = thermal factor (1.2 in unheated buildings and 1.0 in heated buildings, except greenhouses)
I = *occupancy importance factor* is based on an occupancy category that represents how essential a given structure is to the community in the event of a failure; and has different values for the load conditions: snow, ice, wind, and earthquake. There are 4 categories of building occupancies. Category I represents occupancies with low hazard to human life, such as agricultural or minor storage facilities. Category II represents occupancies not included in categories I, III and IV, such as typical office and residential buildings. Category III represents building types that would be a substantial hazard to human life, such as buildings with public assembly spaces for 300 or more people, elementary schools, power stations, and telecommunications facilities. Category IV represents essential facilities, such as hospitals, emergency, fire, and police facilities. For calculating snow loads, I is 0.8, 1.0, 1.1, and 1.2 for the respective categories I, II, III, and IV.

Ground snow loads, p_g, are given by ASCE standard or governed by local building codes for site-specific locations, (for instance, ground snow load is 40 lb/ft^2 in Boston, 25 lb/ft^2 in Chicago).

The design snow load on a sloped roof, p_s, defined as roof slopes greater than 5°, is given by ASCE standard as follows:

$$p_s = C_s p_f \tag{2.4}$$

where C_s = roof slope factor, which reduces from 1.0 as the roof slope increases.

P_f = flat roof snow load (from Equation 2.3)

The roof slope factor, C_s, depends on the slope of the roof, the thermal factor, C_t, and whether the roof surface consists of an unobstructed slippery surface or non-slippery with or without obstructions. For example, asphalt shingles are considered to have a non-slippery surface, and therefore will hold the snow on the roof. Additionally the shape of the roof contributes to the design snow load.

Other snow load conditions that need to be considered in design of roofs include snow drift, partial loading, unbalanced snow load, sliding snow, rain-on-snow surcharge, and ponding instability that can occur due to rain-on-snow or snow meltwater.

2.6 Wind Loads

Introduction

As we have all observed from the damage produced by a hurricane or tornado, high winds exert large forces. These forces can tear off tree limbs, carry away roofs, and break windows. Since the speed and direction of wind are continually changing, the exact pressure or suction applied by winds to structures is difficult to determine. Nevertheless, by recognizing that wind is like a fluid, it is possible to understand many aspects of its behavior and to arrive at reasonable design loads.

The magnitude of wind pressures on a structure depends on the wind velocity, the shape and stiffness of the structure, the roughness and profile of the surrounding ground, and the influence of adjacent structures. When wind strikes an object in its path, the kinetic energy of the moving air particles is converted to a pressure q_s, which is given by

$$q_s = \frac{mV^2}{2} \tag{2.5}$$

where m represents the mass density of the air and V equals the wind velocity. Thus the pressure of the wind varies with the density of the air—a function of temperature—and with the square of the wind velocity.

The friction between the ground surface and the wind strongly influences the wind velocity. For example, winds blowing over large, open, paved areas (e.g., runways of an airport) or water surfaces are not slowed as much as winds blowing over rougher, forest-covered areas where the friction is greater. Also near the ground surface, the friction between the air and ground reduces the velocity, whereas at higher elevations above the ground, friction has little influence and wind velocities are much higher. Figure 2.9*a* shows the approximate variation of wind velocity with height above the ground surface. This information is supplied by *anemometers*—devices that measure wind speeds.

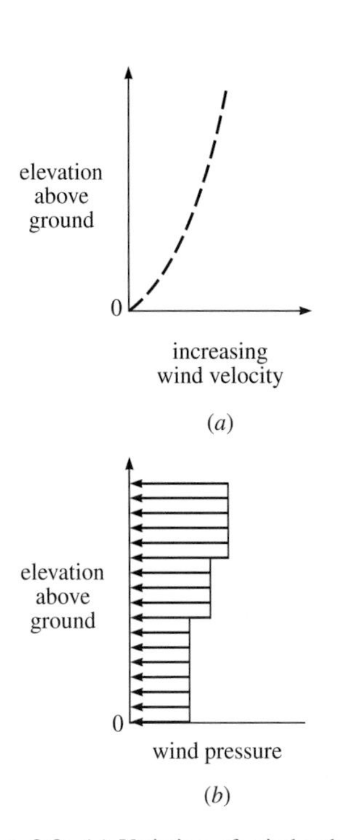

Figure 2.9: (*a*) Variation of wind velocity with distance above ground surface; (*b*) variation of wind pressure specified by typical building codes for windward side of building.

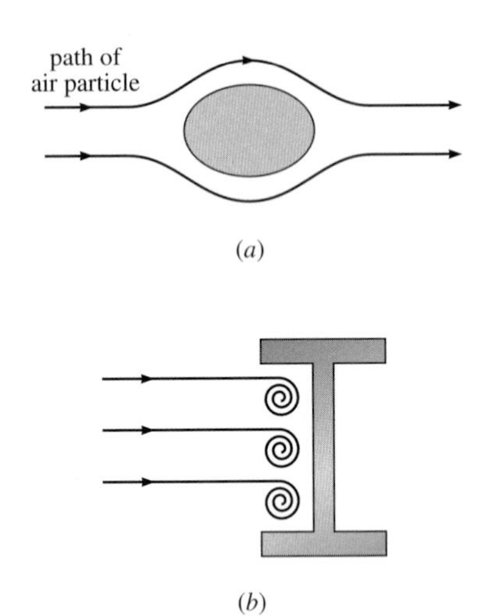

Figure 2.10: Influence of shape on drag factor: (*a*) curved profile permits air to pass around body easily (drag factor is small); (*b*) wind trapped by flanges increases pressure on web of girder (drag factor is large).

Wind pressure also depends on the shape of the surface that the wind strikes. Pressures are smallest when the body has a streamlined cross section and greatest for blunt or concave cross sections that do not allow the wind to pass smoothly around (see Figure 2.10). The influence of shape on wind pressure is accounted for by *drag factors* that are tabulated in certain building codes.

As an alternative to computing wind pressures from wind velocities, some building codes specify an equivalent horizontal wind pressure. This pressure increases with height above the ground surface (Figure 2.9*b*). The force exerted by the wind is assumed to be equal to the product of the wind pressure and the surface area of a building or other structure.

When wind passes over a sloping roof (see Figure 2.11*a*), it must increase its velocity to maintain continuity of the flowing air. As the wind velocity increases, the pressure on the roof reduces (Bernoulli's principle). This reduction in pressure exerts an uplift—much like the wind flowing over the wing of an airplane—that can carry away a poorly anchored roof. A similar negative pressure occurs on both sides of a building parallel to the wind direction and to a smaller extent on the leeward side (see sides *AB* and side *BB* in Figure 2.11*b*) as the wind speeds up to pass around the building.

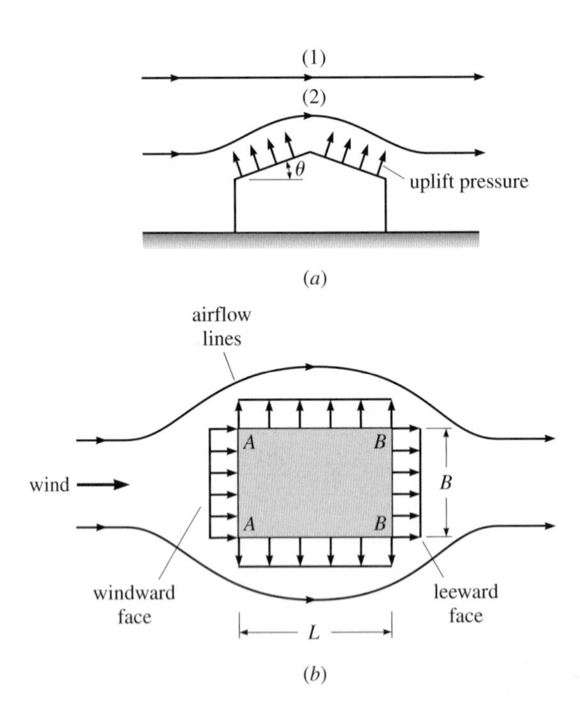

Figure 2.11: (*a*) Uplift pressure on a sloping roof; the wind speed along path 2 is greater than that along path 1 because of the greater length of path. Increased velocity reduces pressure on top of roof, creating a pressure differential between inside and outside of building. The uplift is a function of the roof angle θ; (*b*) Increased velocity creates negative pressure (suction) on sides and leeward face; direct pressure on windward face *AA*.

Photo 2.1: Failure of the Tacoma Narrows Bridge showing the first section of the roadway as it crashes into Puget Sound. The breakup of the narrow, flexible bridge was produced by large oscillation induced by the wind.

Vortex Shedding. As wind moving at constant velocity passes over objects in its path, the air particles are retarded by surface friction. Under certain conditions (critical velocity of wind and shape of surface) small masses of restrained air periodically break off and flow away (see Figure 2.12). This process is called *vortex shedding*. As the air mass moves away, its velocity causes a change in pressure on the discharge surface. If the period (time interval) of the vortices leaving the surface is close to that of the natural period of the structure, oscillations in the structure will be induced by the pressure variations. With time these oscillations will increase and shake a structure vigorously. The Tacoma Narrows Bridge failure shown in Photo 2.1 is a dramatic example of the damage that vortex shedding can wreak. Tall chimneys and suspended pipelines are other structures that are susceptible to wind-induced vibrations. To prevent damage to vibration-sensitive structures by vortex shedding, *spoilers* (see Figure 2.13), which cause the vortices to leave in a random pattern, or *dampers*, which absorb energy, may be attached to the discharge surface. As an alternative solution, the natural period of the structure may be modified so that it is out of the range that is sensitive to vortex shedding. The natural period is usually modified by increasing the stiffness of the structural system.

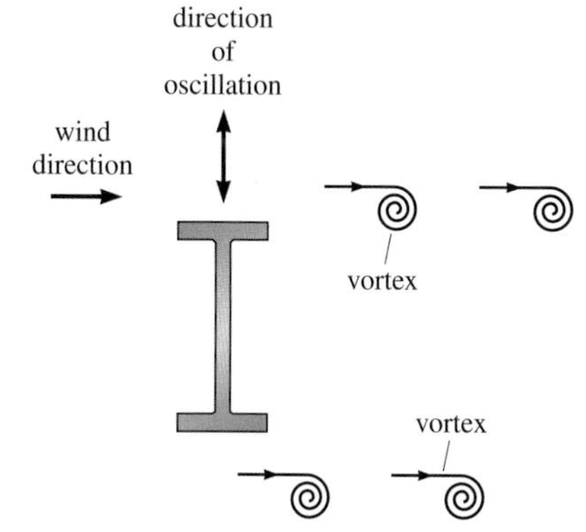

Figure 2.12: Vortices discharging from a steel girder. As vortex speeds off, a reduction in pressure occurs, causing girder to move vertically.

For several decades after the Tacoma Narrows Bridge failure, designers added stiffening trusses to the sides of suspension bridge roadways to minimize bending of the decks (Photo 2.2). Currently designers use stiff aerodynamically shaped box sections that resist wind-induced deflections effectively.

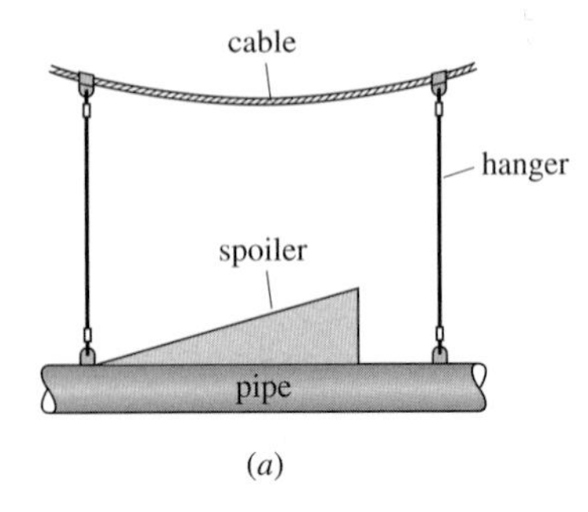

(*a*)

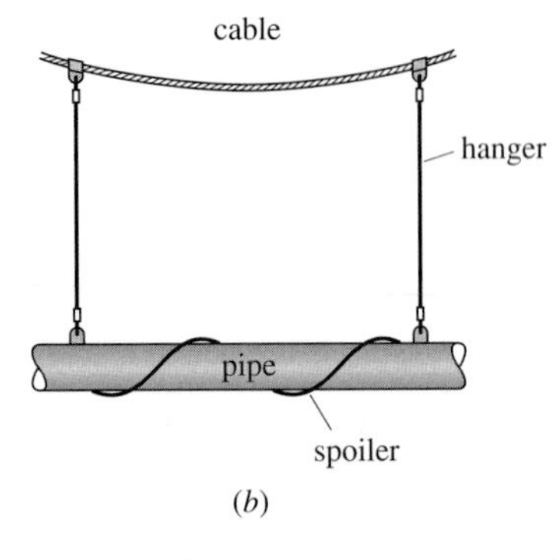

(*b*)

Figure 2.13: Spoilers welded to a suspender pipe to change the period of vortices: (*a*) triangular plate used as a spoiler; (*b*) spiral rod welded to pipe used as spoiler.

Structural Bracing Systems for Wind and Earthquake Forces

The floors of buildings are typically supported on columns. Under dead and live loads that act vertically downward (also called *gravity load*), columns are loaded primarily by axial compression forces. Because columns carry axial load efficiently in direct stress, they have relatively small cross sections—a desirable condition since owners want to maximize usable floor space.

When lateral loads, such as wind or the inertia forces generated by an earthquake, act on a building, lateral displacements occur. These displacements are zero at the base of the building and increase with height. Since slender columns have relatively small cross sections, their bending stiffness is small. As a result, in a building with columns as the only supporting elements, large lateral displacements can occur. These lateral displacements can crack partition walls, damage utility lines, and produce motion sickness in occupants (particularly in the upper floors of multistory buildings where they have the greatest effect).

To limit lateral displacements, structural designers often insert, at appropriate locations within the building, structural walls of reinforced masonry or reinforced concrete. These *shear walls* act in-plane as deep cantilever beam-columns with large bending stiffnesses several orders of magnitude greater than those of all the columns combined. Because of their large stiffness, shear walls often are assumed to carry all transverse loads from wind or earthquake into the foundation. Since the lateral loads act normal to the longitudinal axis of the wall, just as the shear acts in a beam, they are called shear walls (Figure 2.14*a*). In fact, these walls must also be reinforced for bending along

Photo 2.2: The main span of the new San Francisco-Oakland Bay Bridge is a single-tower and self-anchored suspension bridge. The suspension cable loops around the near bent and anchors to the bridge deck at the far bent.

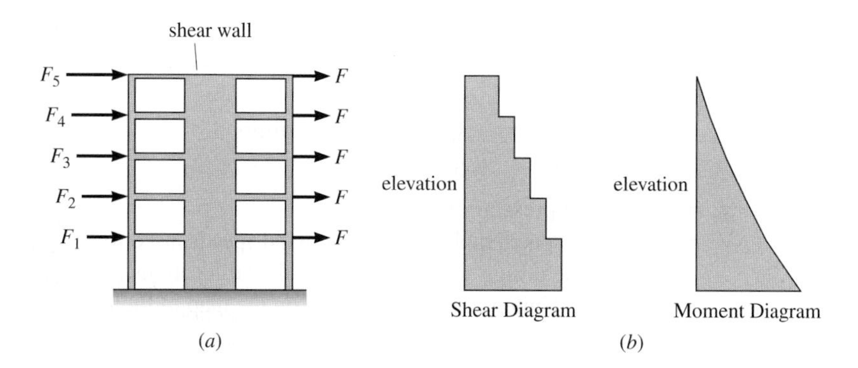

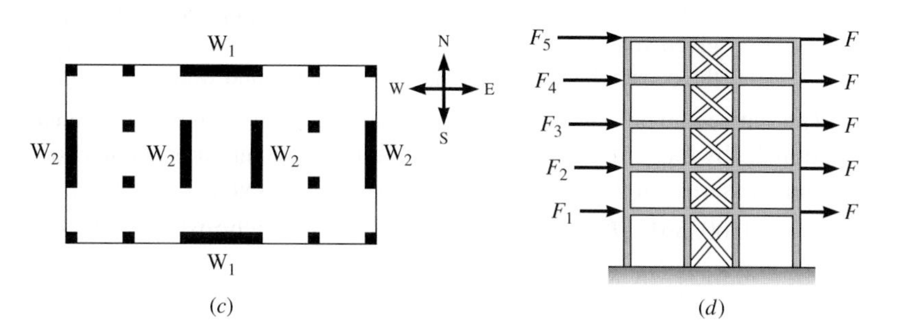

Figure 2.14: Structural systems to resist lateral loads from wind or earthquake. (*a*) Reinforced concrete shear wall carries all lateral wind loads; (*b*) shear and moment diagrams for shear wall produced by the sum of wind loads on the windward and leeward sides of the building in (*a*); (*c*) plan of building showing position of shear walls and columns; (*d*) cross-bracing between steel columns; forms a truss to carry lateral wind loads into the foundations.

both vertical edges since they can bend in either direction. Figure 2.14*b* shows the shear and moment diagrams for a typical shear wall.

Loads are transmitted to the walls by continuous floor slabs that act as rigid plates, termed *diaphragm action* (Figure 2.14*a*). In the case of wind, the floor slabs receive the load from air pressure acting on the exterior walls. In the case of earthquake, the combined mass of the floors and attached construction determines the magnitude of the inertia forces transmitted to the shear walls as the building flexes from the ground motion.

Shear walls may be located in the interior of buildings or in the exterior walls (Figure 2.14*c*). Since only the in-plane flexural stiffness of the wall is significant, walls are required in both directions. In Figure 2.14*c* two shear walls, labeled W_1, are used to resist wind loads acting in the east-west direction on the shorter side of the building; four shear walls, labeled W_2, are used to resist wind load, in the north-south direction, acting on the longer side of the building.

In buildings constructed of structural steel, as an alternative to constructing shear walls, the designer can add X-shaped or V-shaped cross-bracing between columns to form deep wind trusses, which are very stiff in the plane of the truss (Figure 2.14*d* and Photo 2.3).

Equations to Predict Design Wind Pressures

Our primary objective in establishing the wind pressures on a building is to determine the forces that must be used to size the structural members that make up the wind bracing system. In this section we will discuss procedures for establishing wind pressures using a simplified format based on the provisions of the most recent edition of the ASCE standard.

Photo 2.3: The cross-bracing, together with the attached columns and horizontal floor beams in the plane of the bracing, forms a deep continuous, vertical truss that extends the full height of the building (from foundation to roof) and produces a stiff, lightweight structural element for transmitting lateral wind and earthquake forces into the foundation.

If the mass density of air at 59°F (15°C) and at sea level pressure of 29.92 in. of mercury (101.3 kPa) is substituted into Equation 2.3*a*, the equation for the static wind pressure q_s becomes

$$q_s = 0.00256V^2 \qquad \text{U.S. customary units} \qquad (2.6a)$$

$$q_s = 0.613V^2 \qquad \text{SI units} \qquad (2.6b)$$

where q_s = static wind pressure, lb/ft² (N/m²)
V = basic wind speed, mph (m/s). Basic wind speeds, used to establish the design wind force for particular locations in the continental United States, are plotted on the map in Figure 2.15. These wind velocities are measured by anemometers located 33 ft (10 m) above grade in open terrain and represent wind speeds that have only a 2 percent probability of being exceeded in any given year. Notice that the greatest wind velocities occur along the coast where the friction between wind and water is minimal.

The static wind pressure q_s given by Equation 2.6*a* or *b* is next modified in Equation 2.7 by four empirical factors to establish the magnitude of the velocity wind pressure q_z at various elevations above ground level.

$$q_z = 0.00256V^2IK_zK_{zt}K_d \qquad \text{U.S. customary units} \qquad (2.7a)$$

$$q_z = 0.613V^2IK_zK_{zt}K_d \qquad \text{SI units} \qquad (2.7b)$$

Or using Equation 2.6*a*, we can replace the first two terms of Equation 2.7 by q_s to give

$$q_z = q_sIK_zK_{zt}K_d \qquad (2.8)$$

where q_z = velocity wind pressure at height z above ground level
I = *Occupancy importance factor* is described in Section 2.5. For calculating wind loads in non-hurricane regions (V = 85 to 100 mph and in Alaska), I is 0.87, 1.0, 1.15, and 1.15 for the respective categories I, II, III, and IV. For hurricane regions with $V > 100$ mph, I in category I reduces to 0.77.
K_z = *velocity pressure exposure coefficient*, which accounts for both the influence of height above grade and exposure conditions. Three exposure categories (B through D) considered are as follows:

B: Urban and suburban, or wooded areas with low structures.
C: Open terrain with scattered obstructions generally less than 30 ft (9.1 m) high.
D: Flat, unobstructed areas exposed to wind flowing over open water for a distance of at least 5000 ft (1.524 km) or 20 times the building height, whichever is greater.

Values of K_z are tabulated in Table 2.4 and shown graphically in Figure 2.16.
K_{zt} = *topographic factor*, which equals 1 if building is located on level ground; for buildings located on elevated sites (top of hills), K_{zt} increases to account for greater wind speed.
K_d = *wind directionality factor*, which accounts for the reduced probability of maximum winds coming from any given direction and for the reduced probability of the maximum pressure developing for any given wind direction (see Table 2.5).

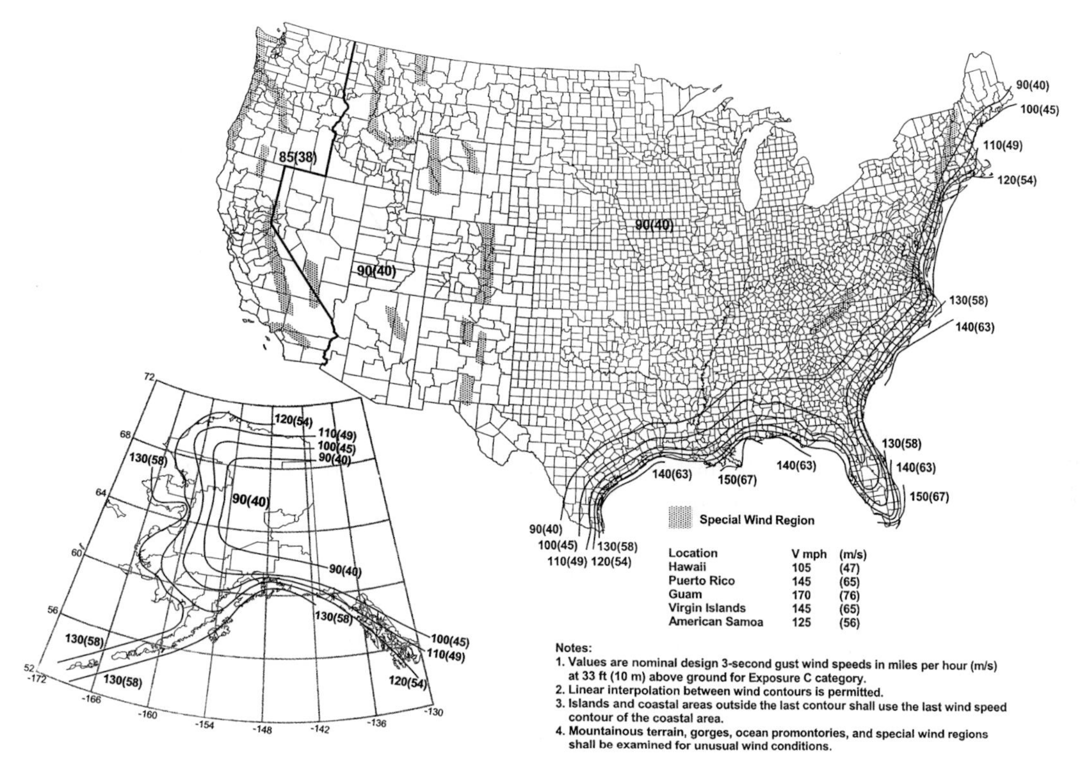

Figure 2.15: ASCE basic wind speed contour map. Larg<None>est wind velocities occur along the eastern and southeastern coasts of the United States.

TABLE 2.4
Velocity Pressure Exposure Coefficient K_z

Height z above Ground Level		Exposure		
ft	(m)	B	C	D
0–15	(0–4.6)	0.57 (0.70)*	0.85	1.03
20	(6.1)	0.62 (0.70)	0.90	1.08
25	(7.6)	0.66 (0.70)	0.94	1.12
30	(9.1)	0.70	0.98	1.16
40	(12.2)	0.76	1.04	1.22
50	(15.2)	0.81	1.09	1.27
60	(18)	0.85	1.13	1.31
70	(21.3)	0.89	1.17	1.34
80	(24.4)	0.93	1.21	1.38
90	(27.4)	0.96	1.24	1.40
100	(30.5)	0.99	1.26	1.43
120	(36.6)	1.04	1.31	1.48
140	(42.7)	1.09	1.36	1.52
160	(48.8)	1.13	1.39	1.55
180	(54.9)	1.17	1.43	1.58

*For low-rise buildings with mean roof height not exceeding 60 ft (18 m) and least horizontal dimension.

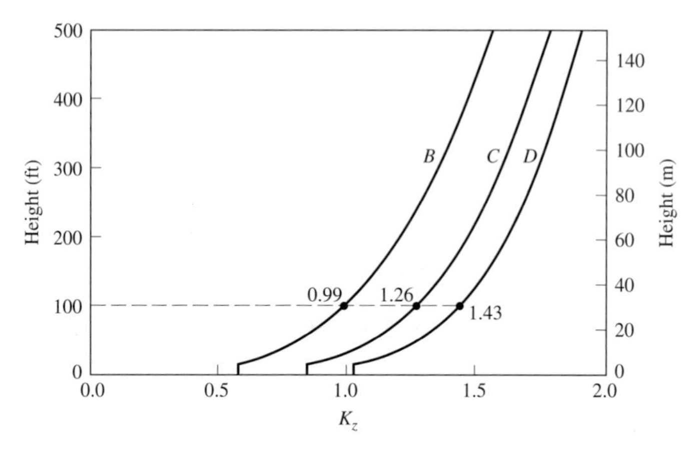

Figure 2.16: Variations of K_z.

The final step for establishing the *design wind pressure p* is to modify q_z, given by Equation 2.7*a* or *b*, by two additional factors, G and C_p:

$$p = q_z G C_p \tag{2.9}$$

where p = design wind pressure on a particular face of the building

G = *gust factor*, which equals 0.85 for rigid structures; that is, the natural period is less than 1 second. For flexible structures with a natural period greater than 1 second, a series of equations for G are available in the ASCE standard.

C_p = *external pressure coefficient*, which establishes how a fraction of the wind pressure (given by Equation 2.7*a* or *b*) is to be distributed to each of the four sides of the building (see Table 2.6). For the wind applied normal to the wall on the windward side of the building $C_p = 0.8$. On the leeward side, $C_p = -0.2$ to -0.5. The minus sign indicates a pressure acting outward from the face of the building. The magnitude of C_p is a function of the ratio of length L in the windward direction to length B in the direction normal to the wind. The main wind bracing system must be sized for the sum of the wind forces on the windward and leeward sides of the building. Finally, on the sides of the building perpendicular to the direction of the wind, where negative pressure also occurs, $C_p = -0.7$.

TABLE 2.5

Wind Directionality Factor K_d

Structural Type	K_d
Buildings	
Main wind force-resisting system	0.85
Components and cladding	0.85
Chimneys, tanks, and similar structures	
Square	0.90
Round or hexagonal	0.95
Trussed towers	
Triangular, square, rectangular	0.85
All other cross sections	0.95

TABLE 2.6

External Pressure Coefficient C_p

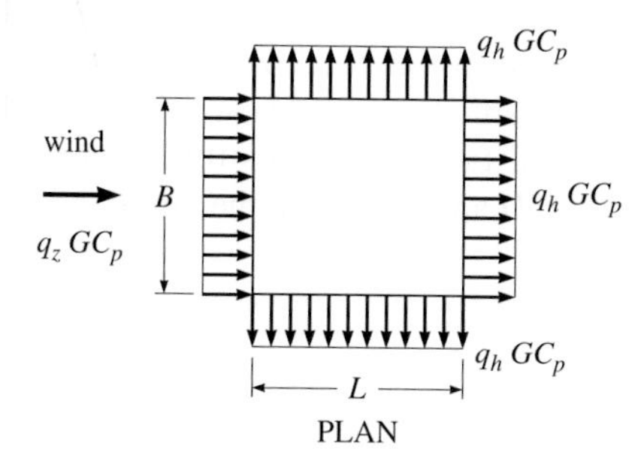

Wall Pressure Coefficients C_p

Surface	L/B	C_p	Use with
Windward wall	All values	0.8	q_z
Leeward wall	0–1	−0.5	
	2	−0.3	q_h
	≥4	−0.2	
Side wall	All values	−0.7	q_h

Notes:

1. Plus and minus signs signify pressures acting toward and away from the surfaces, respectively.
2. Notations: B is the horizontal dimension of the building, in feet (meters) measured normal to wind direction, and L is the horizontal dimension of the building in feet (meters), measured parallel to wind direction.

The wind pressure increases with height only on the windward side of a building where wind pressure acts inward on the walls. On the other three sides the magnitude of the negative wind pressure, acting outward, is constant with height, and the value of K_z is based on the mean roof height h. A typical distribution of wind pressure on a multistory building is shown in Figure 2.17. Example 2.6 illustrates the procedure to evaluate the wind pressure on the four sides of a building 100 ft high.

Since wind can act in any direction, designers must also consider additional possibilities of wind loading a building at various angles. For high-rise buildings in a city—particularly those with an unusual shape—wind tunnel studies using small-scale models are often employed to determine maximum wind pressures. For these studies, adjacent high-rise buildings, which influence the direction of airflow, must be included. Models are typically constructed on a small platform that can be inserted into a wind tunnel and rotated to determine the orientation of the wind that produces the largest values of positive and negative pressure.

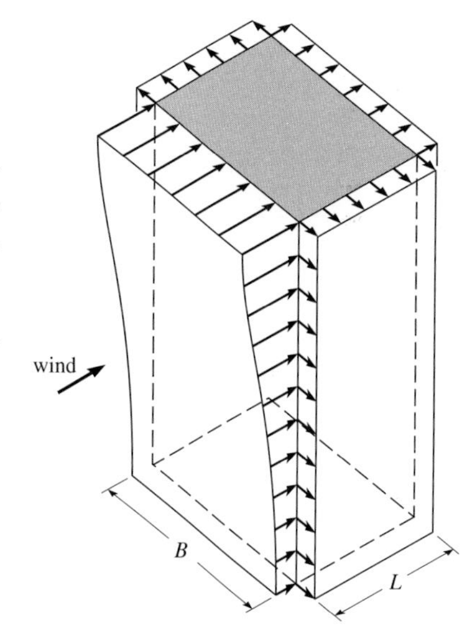

Figure 2.17: Typical wind load distribution on a multistory building.

EXAMPLE 2.6

Determine the wind pressure distribution on the four sides of an eight-story hotel located on level ground; the basic wind speed is 130 mph. Consider the case of a strong wind acting directly on face AB of the building in Figure 2.18a. Assume the building is classified as stiff because its natural period is less than 1 s; therefore, the gust factor G equals 0.85. The importance factor I equals 1.15 and exposure D applies. Since the building is located on level ground, $K_{zt} = 1$.

Solution

STEP 1 Compute the static wind pressure using Equation 2.6a:

$$q_s = 0.00256V^2 = 0.00256(130)^2 = 43.26 \text{ lb/ft}^2$$

STEP 2 Compute the magnitude of wind pressure on the windward side at the top of the building, 100 ft above grade, using Equation 2.7a.

$$I = 1.15$$

$$K_z = 1.43 \qquad \text{(Figure 2.16 or Table 2.4)}$$

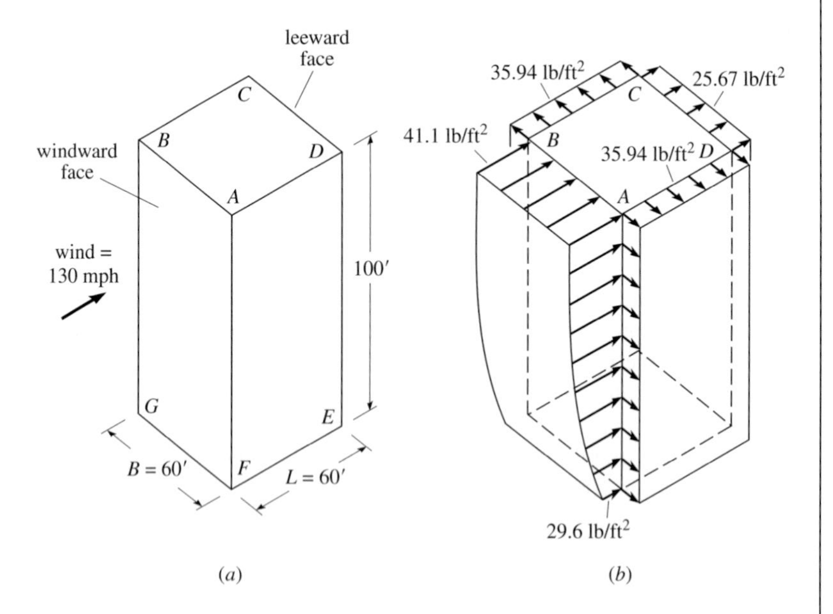

Figure 2.18: Variation of wind pressure on sides of buildings.

$$K_{zt} = 1 \qquad \text{(level ground)}$$

$$K_d = 0.85 \qquad \text{(Table 2.5)}$$

Substituting the above values into Equation 2.8 to determine the design wind pressure at 100 ft above grade gives

$$q_z = q_s I K_z K_{zt} K_d$$

$$= 43.26(1.15)(1.43)(1)(0.85) = 60.4 \text{ lb/ft}^2$$

Note: To compute wind pressures at other elevations on the windward side, the only factor that changes in the above equation is K_z, tabulated in Table 2.4. For example, at an elevation of 50 ft, $K_z = 1.27$ and $q_z = 53.64$ lb/ft^2.

STEP 3 Determine the design wind pressure on the *windward* face *AB*, using Equation 2.9.

Gust factor $G = 0.85$, read $C_p = 0.8$ (from Table 2.6). Substituting into Equation 2.9 produces

$$p = q_z G C_p = 60.4(0.85)(0.8) = 41.1 \text{ lb/ft}^2$$

STEP 4 Determine the wind pressure on the *leeward* side:

$$C_p = -0.5 \qquad \text{(Table 2.6)} \qquad \text{and} \qquad G = 0.85$$

$$p = q_z G C_p = 60.4(0.85)(-0.5) = -25.67 \text{ lb/ft}^2$$

STEP 5 Compute the wind pressure on the two *sides perpendicular* to the wind:

$$C_p = -0.7 \qquad G = 0.85$$

$$p = q_z G C_p = 60.4(0.85)(-0.7) = -35.94 \text{ lb/ft}^2$$

The distribution of wind pressures is shown in Figure 2.18*b*.

Simplified Procedure: Wind Loads for Low-Rise Buildings

In addition to the procedure just discussed for computing wind loads, the ASCE standard provides a simplified procedure to establish wind pressures on enclosed or partially enclosed low-rise buildings of regular shape whose mean roof height h does not exceed neither 60 ft (18.2 m) nor the building's least horizontal dimension and to which the following conditions apply.

1. Floor and roof slabs (diaphragms) must be designed to act as rigid plates and connect to the main wind force-resisting system, which may include shear walls, moment frames, or braced frames.
2. The building has an approximately symmetric cross section, and the roof slope θ does not exceed 45°.
3. The building is classified as rigid; that is, its natural frequency is greater than 1 Hz. (Most low-rise buildings with wind force-resisting systems, such as shear walls, moment frames, or braced frames, fall in this category.)
4. The building is not torsionally sensitive.

For such regular *rectangular* structures, the procedure to establish the design pressures follows:

1. Determine the wind velocity at the building site, using Figure 2.15.
2. Establish the design wind pressure p_s acting on the walls and roof

$$p_s = \lambda K_{zt} I p_{s30} \tag{2.10}$$

where p_{s30} is the *simplified design wind pressure* for exposure B (see Table 2.7), with $h = 30$ ft and the *importance factor* I taken as 1.0. If the importance factor I differs from 1, substitute its value into Equation 2.10. For exposure C or D and for h other than 30 ft, the ASCE standard supplies an *adjustment factor* λ, tabulated in Table 2.8.

The distribution of p_s on the walls and roof for wind load in both the transverse and longitudinal directions is shown in Figure 2.19. Each line in

TABLE 2.7

Simplified Design Wind Pressure p_{S30} (lb/ft^2) (Exposure B at h = 30 ft with I = 1.0 and K_{zt} = 1.0)

Basic Wind Speed (mph)	Roof Angle (degrees)	Zones							
		Horizontal Pressures				Vertical Pressures			
		A	B	C	D	E	F	G	H
90	0 to 5°	12.8	−6.7	8.5	−4.0	−15.4	−8.8	−10.7	−6.8
	10°	14.5	−6.0	9.6	−3.5	−15.4	−9.4	−10.7	−7.2
	15°	16.1	−5.4	10.7	−3.0	−15.4	−10.1	−10.7	−7.7
	20°	17.8	−4.7	11.9	−2.6	−15.4	−10.7	−10.7	−8.1
	25°	16.1	2.6	11.7	2.7	−7.2	−9.8	−5.2	−7.8
		—	—	—	—	−2.7	−5.3	−0.7	−3.4
	30° to 45°	14.4	9.9	11.5	7.9	1.1	−8.8	0.4	−7.5
		14.4	9.9	11.5	7.9	5.6	−4.3	4.8	−3.1

Table 2.7 lists the values of the uniform wind pressure for eight areas of a building's walls and roof.

- Plus and minus signs signify pressures acting toward and away from projected surfaces.
- Pressures for additional wind speeds are given in the ASCE Standard.

These areas, shown in Figure 2.19, are labeled with *circled letters* (A to H). Table 2.7 contains values of p_{s30} for buildings subjected to 90-mph winds; the complete Standard provides data for winds varying from 85 to 170 mph.

The value of a, which defines the extent of *regions of greatest wind pressure* (see areas A, B, E, and F on the walls and roof in Figure 2.19), is evaluated as 10 percent of the smaller horizontal dimension of the building or $0.4h$, whichever is smaller (h is the mean height), but not less than either 4 percent of the least horizontal dimension or 3 ft (0.9 m). Notice that the wind pressures are largest near the corners of walls and the edges of roofs. The ASCE standard also specifies a minimum wind load with 10 psf of p_s acting on zones A, B, C, and D while other zones are not loaded.

Example 2.7 illustrates the use of the simplified procedure to establish the design wind pressures for the wind analysis of a 45-ft-high rectangular building.

TABLE 2.8

Adjustment Factor λ for Building Height and Exposure

Mean roof height h (ft)	Exposure B	Exposure C	Exposure D
15	1.00	1.21	1.47
20	1.00	1.29	1.55
25	1.00	1.35	1.61
30	1.00	1.40	1.66
35	1.05	1.45	1.70
40	1.09	1.49	1.74
45	1.12	1.53	1.78
50	1.16	1.56	1.81
55	1.19	1.59	1.84
60	1.22	1.62	1.87

From ASCE Standard.

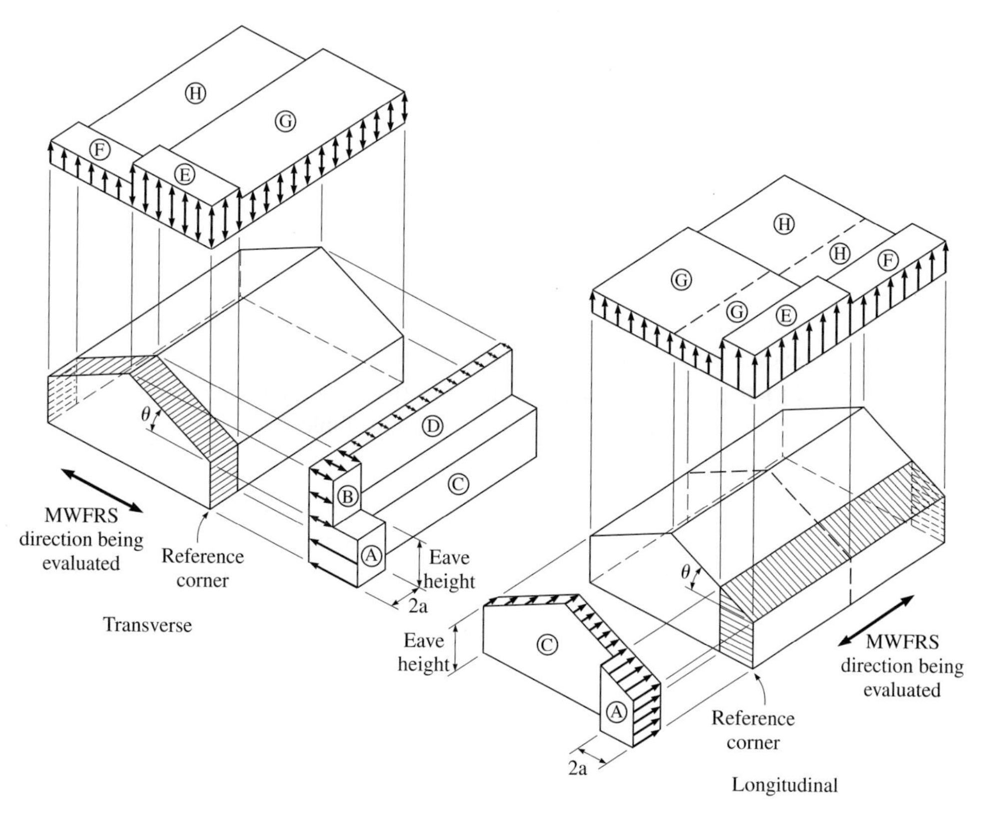

Figure 2.19: Distribution of design wind pressures for the simplified method. See Table 2.7 for the magnitude of the pressures in areas A through H. $h = 60$ ft. (From ASCE standard.)

EXAMPLE 2.7

Figure 2.15 indicates the velocity of the wind acting on the 45-ft-high, three-story building in Figure 2.20*a* is 90 mph. If exposure condition C applies, determine the wind force transmitted to the building's foundations by each of the two large reinforced concrete shear walls that make up the main wind-resisting system. The walls located at the midpoint of each side of the building have identical proportions. The importance factor I equals 1 and $K_{zt} = 1.0$.

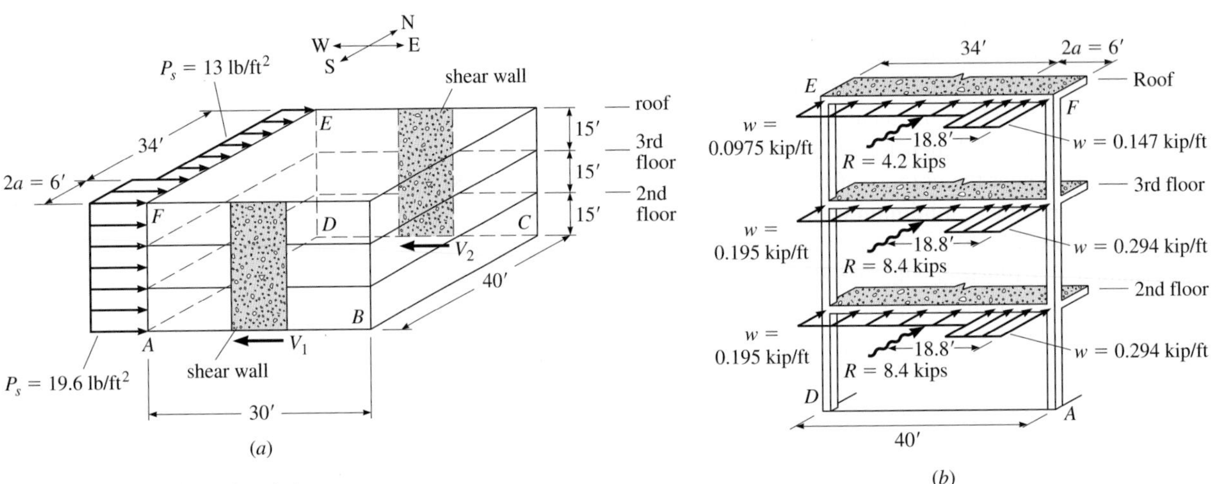

Figure 2.20: Horizontal wind pressure analysis by the simplified method. (*a*) Wind pressure distribution and details of the loaded structure; (*b*) wind forces applied by the exterior walls to the edge of the roof and floor slabs; (*c*) plan view of the resultant wind force and the reactions of the shear walls; (*d*) free body of the shear wall located in plane *ABDF* showing the wind forces applied by the floor slabs and the reactions on base (*continues*).

Solution

Compute the wind load transmitted from the wall on the windward side to the roof and each floor slab. Assume each 1-ft-wide vertical strip of wall acts as a simply supported beam spanning 10 ft between floor slabs; therefore, one-half of the wind load on the wall between floors is carried to the slabs above and below by the fictitious beam (see Figure 2.20*b*).

STEP 1 Since the roof is flat, $\theta = 0$. For the *simplified design wind pressures* p_{s30}, read the top line in Table 2.7.

Region A: $p_{s30} = 12.8 \text{ lb/ft}^2$

Region C: $p_{s30} = 8.5 \text{ lb/ft}^2$

Note: There is no need to compute the values of p_s for zones B and D because the building does not have a sloped roof.

STEP 2 Adjust p_{s30} for exposure C and a mean height of $h = 45$ ft. Read in Table 2.8 that adjustment factor $\lambda = 1.53$. Compute the wind pressure $p_s = \lambda K_{zt} I p_{s30}$.

Region A: $p_s = 1.53(1)(1)(12.8) = 19.584$ round to 19.6 lb/ft²

Region C: $p_s = 1.53(1)(1)(8.5) = 13.005$ round to 13 lb/ft²

STEP 3 Compute the resultant wind forces transmitted from the exterior walls to the edge of the roof and floor slabs.

Load per foot, w, to roof slab (see Figure 2.20b)

$$\text{Region A: } w = \frac{15}{2} \times \frac{19.6}{1000} = 0.147 \text{ kip/ft}$$

$$\text{Region C: } w = \frac{15}{2} \times \frac{13}{1000} = 0.0975 \text{ kip/ft}$$

Load per foot, w, to second- and third-floor slabs

$$\text{Region A: } w = 15 \times \frac{19.6}{1000} = 0.294 \text{ kip/ft}$$

$$\text{Region C: } w = 15 \times \frac{13}{1000} = 0.195 \text{ kip/ft}$$

STEP 4 Compute the resultants of the distributed wind loads.

Roof slab:

$$R_1 = 0.147 \times 6 + 0.0975 \times 34 = 4.197 \quad \text{round to 4.2 kips}$$

Second and third floors:

$$R_2 = 0.294 \times 6 + 0.195 \times 34 = 8.394 \quad \text{round to 8.4 kips}$$

Total horizontal wind force $= 4.2 + 8.4 + 8.4 = 21$ kips

STEP 5 Locate the position of the resultant. Sum moments about a vertical axis through points A and F (see Figure 2.20c).

At the level of the first floor slab

$$R\bar{x} = \Sigma F \cdot x$$
$$4.197\bar{x} = 0.882(3) + 3.315(6 + 34/2)$$
$$\bar{x} = 18.797 \text{ ft} \quad \text{round to 18.8 ft}$$

Since the variation of the pressure distribution is identical at all floor levels on the back of the wall, the resultant of all forces acting on the ends of the roof and floor slabs is located a distance of 18.8 ft from the edge of the building (Figure 2.20b).

STEP 6 Compute the shear force at the base of the shear walls. Sum the moments of all forces about a vertical axis passing through point A at the corner of the building (see Figure 2.20c).

$$\Sigma M_A = 21 \times 18.8 - V_2(40) \quad \text{and} \quad V_2 = 9.87 \text{ kips} \qquad \textbf{Ans.}$$

Compute V_1: $\quad V_2 + V_1 = 21$ kips

$$V_1 = 21 - 9.87 = 11.13 \text{ kips} \qquad \textbf{Ans.}$$

Note: A complete analysis for wind requires that the designer consider the vertical pressures in zones E to H acting on the roof. These pressures are carried by a separate structural system, composed of the roof slabs and beams, to the columns as well as to the shear walls. In the case of a flat roof, the wind flowing over the roof produces upward pressures (uplift) that reduce the axial compression in the columns.

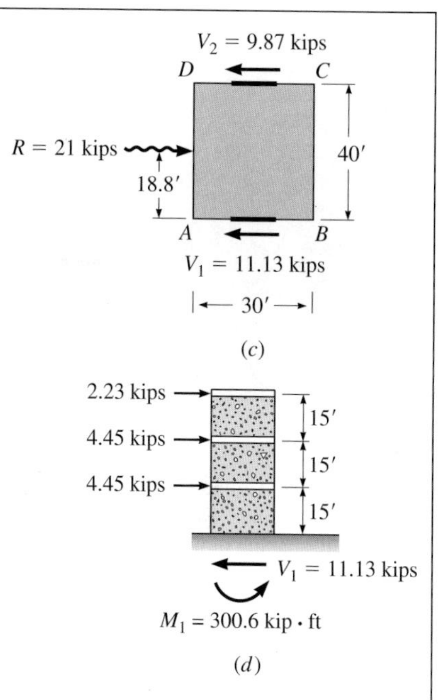

Figure 2.20: *Continued*

(a) (b)

Photo 2.4: (*a*) Collapse of apartment buildings: The 1999 Chi-Chi earthquake (magnitude 7.7) in Taiwan caused the upper floors of the apartment buildings to topple over as a unit due to the formation of a *soft-story* at the ground level. (*b*) Collapse of the Struve Sough Bridge: The severe shaking of the soil by the 1989 Loma Prieta Earthquake in California produced *differential settlements* of the foundations supporting rows of columns that carried the roadway slab. This uneven settlement caused the columns that underwent the largest settlements to transfer the weight of the bridge deck to adjacent columns whose settlement was smaller. The additional load, which had to be transferred into the column by shear stresses in the slab around the column's perimeter, produced the *punching shear* failures shown.

2.7 Earthquake Loads

Earthquakes occur in many regions of the world. In certain locations where the intensity of the ground shaking is small, the designer does not have to consider seismic effects. In other locations—particularly in regions near an active geological fault (a fracture line in the rock structure), such as the San Andreas fault that runs along the western coast of California—large ground motions frequently occur that can damage or destroy buildings and bridges in large areas of cities (see Photo 2.4*a* and *b*). For example, San Francisco was devastated by an earthquake in 1906, before building and bridge codes contained seismic provisions.

The ground motions created by major earthquake forces cause buildings to sway back and forth. Assuming the building is fixed at its base, the displacement of floors will vary from zero at the base to a maximum at the roof (see Figure 2.21*a*). As the floors move laterally, the lateral bracing system is stressed as it acts to resist the lateral displacement of the floors. The forces associated with this motion, *inertia forces*, are a function of both the weight of the floors and attached equipment and partitions as well as the stiffness of the structure. The sum of the lateral inertia forces acting on all floors and transmitted to the foundations is termed the *base shear* and is denoted by V (see Figure 2.21*b*). In most buildings in which the weight of floors is similar in magnitude, the distribution of the inertia forces is similar to that created by wind, as discussed in Section 2.6.

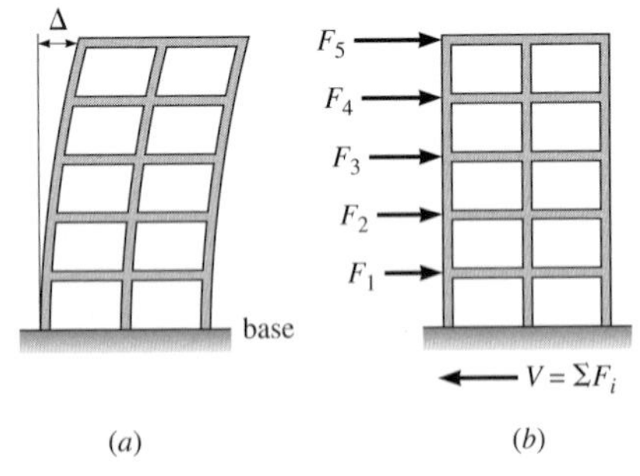

Figure 2.21: (*a*) Displacement of floors as building sways; (*b*) inertia forces produced by motion of floors.

Although there are several analytical procedures to determine the magnitude of the base shear for which buildings must be designed, we will only consider the *equivalent lateral force procedure*, described in the ASCE standard. Using this procedure, we compute the magnitude of the base shear as

$$V = \frac{S_{D1}W}{T(R/I)} \tag{2.10a}$$

but not to exceed

$$V_{max} = \frac{S_{DS}W}{R/I} \tag{2.10b}$$

and not less than

$$V_{min} = 0.044 S_{DS} I W \tag{2.10c}$$

where W = total dead load of building and its permanent equipment and partitions

T = fundamental natural period of building, which can be computed by the following empirical equation

$$T = C_t h_n^x \tag{2.11}$$

h_n = the building height in feet (meters, above the base), $C_t = 0.028$ (or 0.068 in SI units), and $x = 0.8$ for steel rigid frames (moment frames), $C_t = 0.016$ (0.044 SI) and $x = 0.9$ for reinforced concrete rigid frames, and $C_t = 0.02$ (0.055 SI) and $x = 0.75$ for most other systems (for example, systems with braced frames or structural walls). The natural period of a building (the time required for a building to go through one complete cycle of motion) is a function of the lateral stiffness and the mass of the structure. Since the base shear V is inversely proportional to the magnitude of the natural period, it reduces as the lateral stiffness of the structural bracing system increases. Of course, if the stiffness of the lateral bracing system is too small, lateral displacements may become excessive, producing damage to windows, exterior walls, and other nonstructural elements.

S_{D1} = a factor computed using seismic maps that shows intensity of design earthquake for structures with $T = 1$ s. Table 2.9 gives the values for several locations.

S_{DS} = a factor computed using seismic maps that shows intensity of design earthquake at particular locations for structures with $T = 0.2$ s. See Table 2.9 for values at several locations.

R = *response modification factor*, which represents the ability of a structural system to resist seismic forces. This factor, which varies from 8 to 1.25, is tabulated in Table 2.10 for several common structural systems. The highest values are assigned to ductile systems; the lowest values, to brittle systems. Since R occurs in the denominator of Equations 2.10*a* and *b*, a structural system with a large value of R will permit a large reduction in the seismic force the structural system must be designed to support.

TABLE 2.9

Representative Values of S_{DS} and S_{D1} at Selected Cities

City	S_{DS}, g	S_{D1}, g
Los Angeles, California	1.3	0.5
Salt Lake City, Utah	1.2	0.5
Memphis, Tennessee	0.83	0.27
New York, New York	0.27	0.06

Note: Values of S_{DS} and S_{D1} are based on the assumption that foundations are supported on rock of moderate strength. These values increase for weaker soils with lower bearing capacity.

TABLE 2.10

Values of *R* for Several Common Lateral Bracing Structural Systems

Description of Structural System	*R*
Ductile steel or concrete frame with rigid joints	8
Ordinary reinforced concrete shear walls	4
Ordinary reinforced masonry shear wall	2

I = *occupancy importance factor* is described in Section 2.5. For calculating earthquake loads, I is 1.0, 1.0, 1.25 and 1.5 for the respective categories I, II, III, and IV.

Note: The upper limit given by Equation 2.10*b* is required because Equation 2.10*a* produces values of base shear that are too conservative for very stiff structures that have short natural periods. The ASCE standard also sets a lower limit (Equation 2.10*c*) to ensure that the building is designed for a minimum seismic force.

Distribution of Seismic Base Shear *V* to Each Floor Level

The distribution of the *seismic base shear* V to each floor is computed using Equation 2.12.

$$F_x = \frac{w_x h_x^k}{\sum_{i=1}^{n} w_i h_i^k} V \tag{2.12}$$

where
F_x = the lateral seismic force at level x
w_i and w_x = deadweight of floor at levels i and x
h_i and h_x = height from base to floors at levels i and x
k = 1 for $T \le 0.5$ s, 2 for $T \ge 2.5$ s. For structures with a period between 0.5 and 2.5 s, k is determined by linear interpolation between T equal to 1 and 2 as

$$k = 1 + \frac{T - 0.5}{2} \tag{2.13}$$

See Figure 2.22 for graphical representation of Equation 2.13.

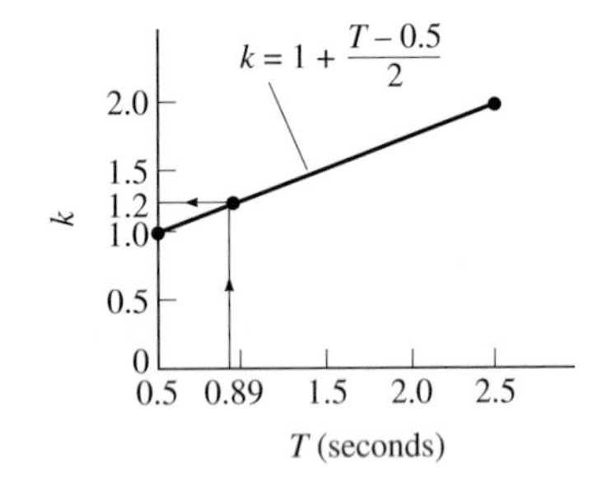

Figure 2.22: Interpolate for k value.

EXAMPLE 2.8

Determine the design seismic forces acting at each floor of the six-story office building in Figure 2.23. The structure of the building consists of steel moment frames (all joints are rigid) that have an R value of 8. The 75-ft-tall building is located in a high seismic region with $S_{D1} = 0.4g$ and $S_{D1} = 1.0g$ for a building supported on rock, where g is the gravitational acceleration. The deadweight of each floor is 700 kips.

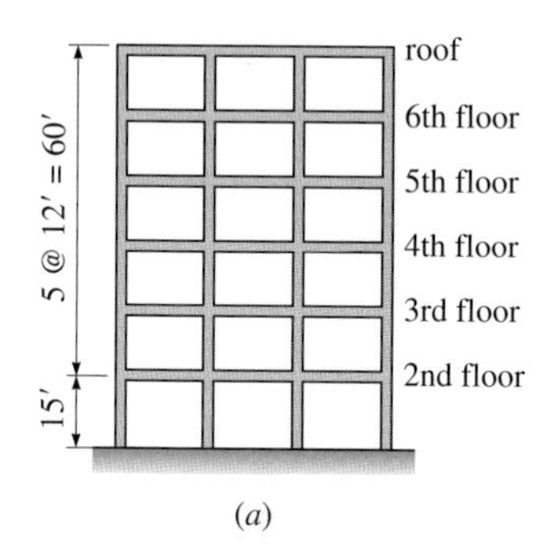

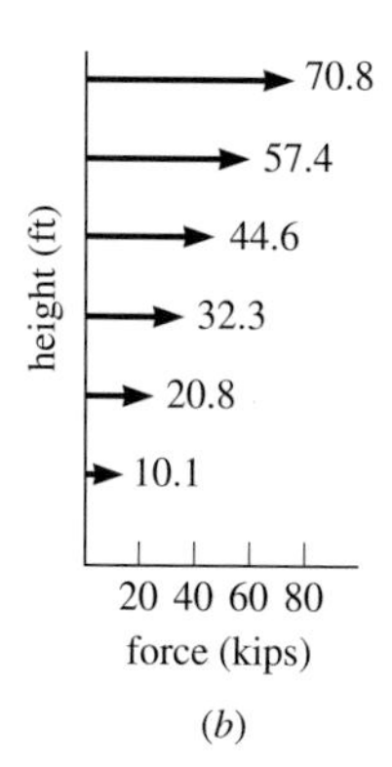

Figure 2.23: (*a*) Six-story building; (*b*) lateral load profile.

Solution

Compute the fundamental period, using Equation 2.11:

$$T = C_t h_n^x = 0.028(75)^{0.8} = 0.89 \text{ s}$$

Assuming that the floor deadweight contains an allowance for the weight of columns, beams, partitions, ceiling, etc., the total weight W of the building is $W = 700(6) = 4200$ kips.

The occupancy importance factor I is 1 for office buildings. Compute the base shear V using Equations 2.10a and c:

$$V = \frac{S_{D1}}{T(R/I)} W = \frac{0.4}{0.89(8/1)}(4200) = 236 \text{ kips} \qquad (2.10a)$$

but not more than

$$V_{\max} = \frac{S_{DS}}{R/I} W = \frac{1.0}{8/1}(4200) = 525 \text{ kips} \qquad (2.10b)$$

and not less than

$$V_{\min} = 0.044 S_{DS} I W = 0.044 \times 1.0 \times 1 \times 4200 = 184.8 \text{ kips} \qquad (2.10c)$$

Therefore, use $V = 236$ kips.

Computations of the lateral seismic force at each floor level are summarized in Table 2.11. To illustrate these computations, we compute the load at the third floor. Since $T = 0.89$ s lies between 0.5 and 2.5 s, we must interpolate using Equation 2.13 to compute the k value (see Figure 2.22):

$$k = 1 + \frac{T - 0.5}{2} = 1 + \frac{0.89 - 0.5}{2} = 1.2$$

$$F_{\text{3rd floor}} = \frac{w_3 h_3^k}{\sum_{i=1}^{n} w_i h_i^k} V = \frac{36{,}537}{415{,}262}(236) = 20.8 \text{ kips}$$

[*continued on next page*]

Example 2.8 continues . . .

TABLE 2.11

Computation of Seismic Lateral Forces

Floor	Weight w_i (kips)	Floor Height h_i ft	$w_i h_i^k$	$\frac{w_x h_x^k}{\sum_{i=1}^{6} w_i h_i^k}$	F_x (kips)
Roof	700	75	124,501	0.300	70.8
6th	700	63	100,997	0.243	57.4
5th	700	51	78,376	0.189	44.6
4th	700	39	56,804	0.137	32.3
3rd	700	27	36,537	0.088	20.8
2nd	700	15	18,047	0.043	10.1

$W = \sum_{i=1}^{6} w_i = 4200$ $\qquad \sum_{i=1}^{6} w_i h_i^k = 415{,}262$ $\qquad V = \sum_{i=1}^{6} F_i = 236$

2.8 Other Loads

There are a number of other loads specified in ASCE that need to be considered when appropriate. These include loads on structures built in flood zone areas, structures built below grade, structures that can accumulate rain or ice, structures subject to impact due to explosions or vehicular impact, structures that support reciprocating mechanical systems, structures exposed to variable or extreme thermal or moist conditions, among others.

Flood loads occur in flood hazard areas, such as along seacoasts or rivers. The flood zone areas are defined by the appropriate authority having jurisdiction. Floods produce hydrostatic, hydrodynamic or wave loads on structures and can cause damage or failure due to scour and erosion of a structure.

Similarly, soils develop hydrostatic pressure below grade. These can cause lateral earth pressures on walls or uplift pressures on floor slabs and foundations.

Flats roofs need to be properly drained to avoid the ponding of rain water. The ASCE standard requires that each portion of the roof be designed to support the weight of all rainwater that could accumulate on it if the primary drainage system for that portion were blocked. If not properly considered in design, rain loads may produce excessive deflections of roof beams, producing an instability problem (called *ponding*), causing the roof to collapse.

These types of loads can have an adverse affect on structural behavior, strength, stability and must be appropriately combined with all other possible loads to determine the worst case design forces acting on a given structure or structural member.

2.9 Load Combinations

The forces (e.g., axial force, moment, shear) produced by various combinations of loads discussed need to be combined in a proper manner and increased by a factor of safety (load factor) to produce the desired level of safety. The combined load effect, sometimes called the *required factored strength*, represents the minimum strength for which members need to be designed. Considering the load effect produced by the dead load D, live load L, roof live load L_r, wind load W, earthquake load E, and snow load S, the ASCE standard requires that the following load combinations be considered:

$$1.4D \tag{2.15}$$

$$1.2D + 1.6L + 0.5(L_r \text{ or } S) \tag{2.16}$$

$$1.2D + 1.6(L_r \text{ or } S) + (L \text{ or } 0.8W) \tag{2.17}$$

$$1.2D + 1.6W + L + 0.5(L_r \text{ or } S) \tag{2.18}$$

$$1.2D + 1.0E + L + 0.2S \tag{2.19}$$

The load combination that produces the *largest* value of force represents the load for which the member must be designed.

EXAMPLE 2.9

A column in a building is subject to gravity load only. Using the tributary area concept, the axial loads produced by the dead load, live load, and roof live load are

$$P_D = 90 \text{ kips}$$
$$P_L = 120 \text{ kips}$$
$$P_{L_r} = 20 \text{ kips}$$

What is the required axial strength of the column?

Solution

$$1.4P_D = 1.4(90) = 126 \text{ kips} \tag{2.15}$$

$$1.2P_D + 1.6P_L + 0.5P_{L_r} = 1.2(90) + 1.6(120) + 0.5(20) = 310 \text{ kips} \tag{2.16}$$

$$1.2P_D + 1.6P_{L_r} + 0.5P_L = 1.2(90) + 1.6(20) + 0.5(120) = 200 \text{ kips} \tag{2.17}$$

Therefore, the required axial load is 310 kips. In this case, the load combination in Equation 2.16 governs. However, if the dead load is significantly larger than the live loads, Equation 2.15 may govern.

EXAMPLE 2.10

To determine the required flexural strength at one end of a beam in a concrete frame, the moments produced by dead, live, and wind load are:

$$M_D = -100 \text{ kip·ft}$$

$$M_L = -50 \text{ kip·ft}$$

$$M_w = \pm 200 \text{ kip·ft}$$

where the minus sign indicates that the beam end is subject to counterclockwise moment while the plus sign indicates clockwise moment. Both the plus and minus signs are assigned to M_w because the wind load can act on the building in either direction. Compute the required flexural strength for both positive and negative bending.

Solution

Negative bending:

$$1.4M_D = 1.4(-100) = -140 \text{ kip·ft} \quad (2.15)$$

$$1.2M_D + 1.6M_L = 1.2(-100) + 1.6(-50) = -200 \text{ kip·ft} \quad (2.16)$$

$$\begin{aligned} 1.2M_D + 1.6M_w + M_L &= 1.2(-100) + 1.6(-200) + (-50) \\ &= -490 \text{ kip·ft} \quad \text{(governs)} \end{aligned} \quad (2.18)$$

Positive bending: Load combinations from Equations 2.15 and 2.16 need not be considered because both produce negative moments.

$$\begin{aligned} 1.2M_D + 1.6M_w + M_L &= 1.2(-100) + 1.6(+200) + (-50) \\ &= +150 \text{ kip·ft} \end{aligned} \quad (2.18)$$

Therefore, the beam needs to be designed for a positive moment of 150 kip·ft and a negative moment of 490 kip·ft.

Summary

- Loads that engineers must consider in the design of buildings and bridges include dead loads, live loads, and environmental forces—wind, earthquake, snow, and rain. Other types of structures such as dams, water tanks, and foundations must resist fluid and soil pressures, and for these cases specialists are often consulted to evaluate these forces.
- The loads that govern the design of structures are specified by national and local building codes. Structural codes also specify additional loading provisions that apply specifically to construction materials such as steel, reinforced concrete, aluminum, and wood.
- Since it is unlikely that maximum values of live load, snow, wind, earthquake, and so forth will act simultaneously, codes permit a reduction in the values of loads when various load combinations are considered.

Dead load, however, is not reduced unless it provides an adverse effect, such as when determining uplift force on a footing.

- To account for dynamic effects from moving vehicles, elevators, supports for reciprocating machinery, and so forth, *impact factors* that increase the live load are specified in building codes.
- In zones where wind or earthquake forces are small, low-rise buildings are initially proportioned for live and dead load, and then checked for wind or earthquake, or both, depending on the region; the design can be easily modified as needed.

  On the other hand, for high-rise buildings located in regions where large earthquakes or high winds are common, designers must give high priority in the preliminary design phase to select structural systems (for example, shear walls or braced frames) that resist lateral loads efficiently.
- Wind velocities increase with height above the ground. Values of positive wind pressures are given by the velocity pressure exposure coefficient K_z tabulated in Table 2.4.
- Negative pressures of uniform intensity develop on three sides of rectangular buildings that are evaluated by multiplying the magnitude of the positive windward pressure at the top of the building by the coefficients in Table 2.6.
- The wind bracing system in each direction must be designed to carry the sum of the wind forces on the windward and leeward sides of the building.
- For tall buildings or for buildings with an unusual profile, wind tunnel studies using instrumented small-scale models often establish the magnitude and distribution of wind pressures. The model must also include adjacent buildings, which influence the magnitude and the direction of the air pressure on the building being studied.
- The ground motions produced by earthquakes cause buildings, bridges, and other structures to sway. In buildings this motion creates lateral inertia forces that are assumed to be concentrated at each floor. The inertia forces are greatest at the top of buildings where the displacements are greatest.
- The magnitude of the inertia forces depends on the size of the earthquake, the weight of the building, the natural period of the building, the stiffness and ductility of the structural frame, and the soil type.
- Buildings with a ductile frame (that can undergo large deformations without collapsing) may be designed for much smaller seismic forces than structures that depend on a brittle structural system (for example, unreinforced masonry).

PROBLEMS

P2.1. Determine the deadweight of a 1-ft-long segment of the prestressed, reinforced concrete tee-beam whose cross section is shown in Figure P2.1. Beam is constructed with lightweight concrete which weighs 120 lbs/ft^3.

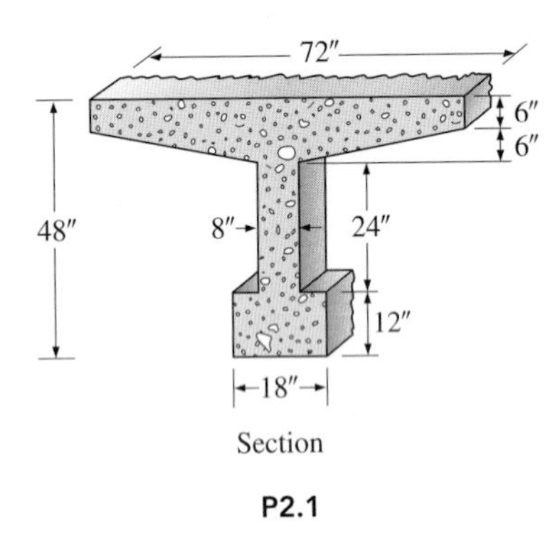

P2.1

P2.2. Determine the deadweight of a 1-ft-long segment of a typical 20-in-wide unit of a roof supported on a nominal 2 in × 16 in southern pine beam (the actual dimensions are $\frac{1}{2}$ in smaller). The $\frac{3}{4}$-in plywood weighs 3 lb/ft^2.

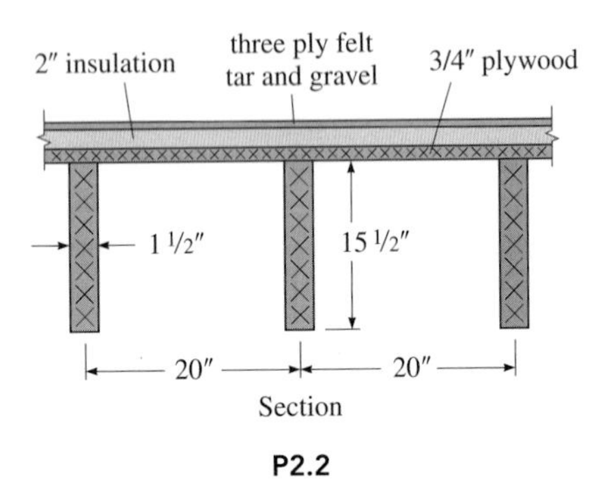

P2.2

P2.3. A wide flange steel beam shown in Figure P2.3 supports a permanent concrete masonry wall, floor slab, architectural finishes, mechanical and electrical systems. Determine the uniform dead load in kips per linear foot acting on the beam.

The wall is 9.5 ft high, non-load bearing and laterally braced at the top to upper floor framing framing (not shown). The wall consists of 8 inch lightweight reinforced concrete masonry units with an average weight of 90 psf. The composite concrete floor slab construction spans over simply supported steel beams, with a tributary width of 10 ft, and weighs 50 psf.

The estimated uniform dead load for structural steel framing, fireproofing, architectural features, floor finish and ceiling tiles equals 24 psf, and for mechanical ducting, piping and electrical systems equals 6 psf.

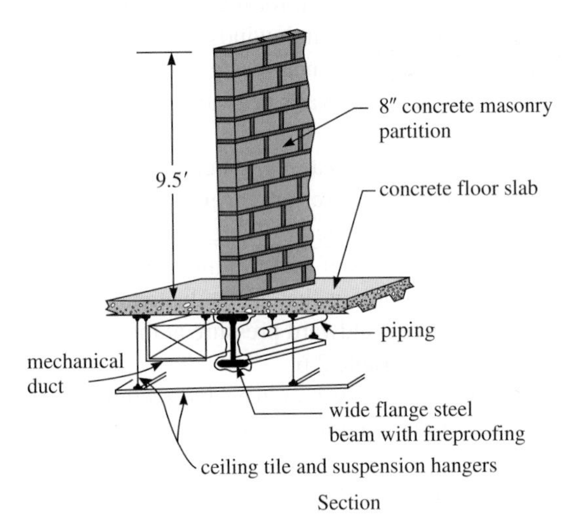

P2.3

P2.4. Consider the floor plan shown in Figure P2.4. Compute the tributary areas for (*a*) floor beam B1, (*b*) girder G1, (*c*) girder G2, (*d*) corner column C1, and (*e*) interior column C2.

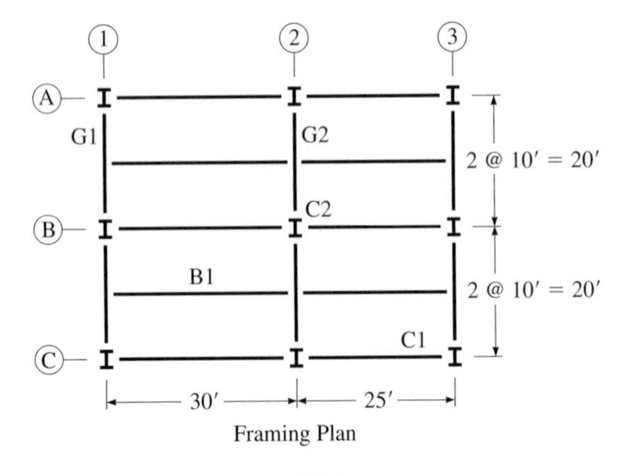

P2.4

P2.5. Refer to Figure P2.4 for the floor plan. Calculate the tributary areas for (*a*) floor beam B1, (*b*) girder G1, (*c*) girder G2, (*d*) corner column C1, and (*e*) interior column C2.

P2.6. The uniformly distributed live load on the floor plan in Figure P2.4 is 60 lb/ft^2. Establish the loading for members (*a*) floor beam B1, (*b*) girder G1, and (*c*) girder G2. Consider the live load reduction if permitted by the ASCE standard.

P2.7. The building section associated with the floor plan in Figure P2.4 is shown in Figure P2.7. Assume a live load of 60 lb/ft^2 on all three floors. Calculate the axial forces produced by the live load in column C2 in the third and first stories. Consider any live load reduction if permitted by the ASCE standard.

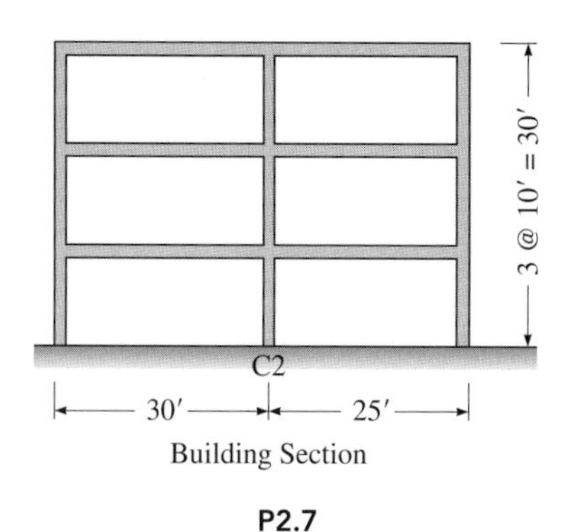

P2.7

P2.8. A five-story building is shown in Figure P2.8. Following the ASCE standard, the wind pressure along the height on the windward side has been established as shown in Figure P2.8(*c*). (*a*) Considering the windward pressure in the east-west direction, use the tributary area concept to compute the resultant wind force at each floor level. (*b*) Compute the horizontal base shear and the overturning moment of the building.

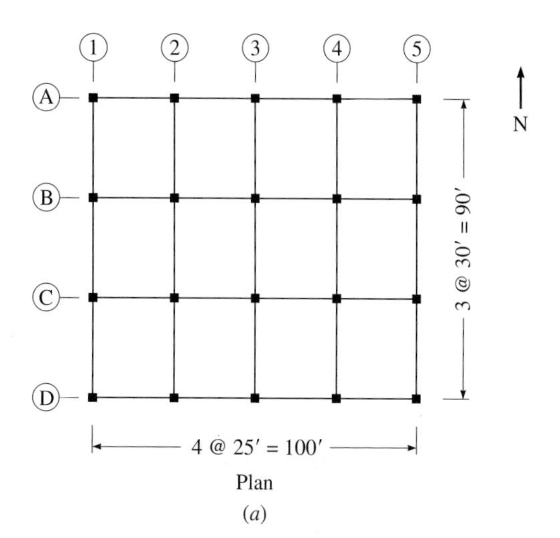

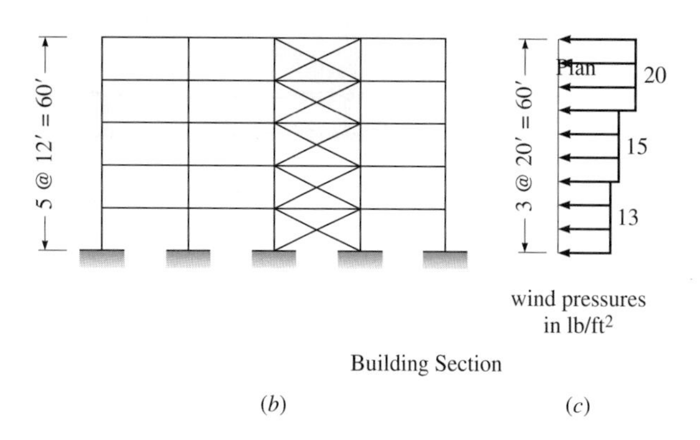

P2.8

P2.9. A mechanical support framing system is shown in Figure P2.9. The framing consists of steel floor grating over steel beams and entirely supported by four tension hangers that are connected to floor framing above it. It supports light machinery with an operating weight of 4,000 lbs, centrally located. (*a*) Determine the impact factor *I* from the *Live Load Impact Factor*, Table 2.3. (*b*) Calculate the total live load acting on one hanger due to the machinery and uniform live load of 40 psf around the machine. (*c*) Calculate the total dead load acting on one hanger. The floor framing dead load is 25 psf. Ignore the weight of the hangers. Lateral bracing is located on all 4 edges of the mechanical floor framing for stability and transfer of lateral loads.

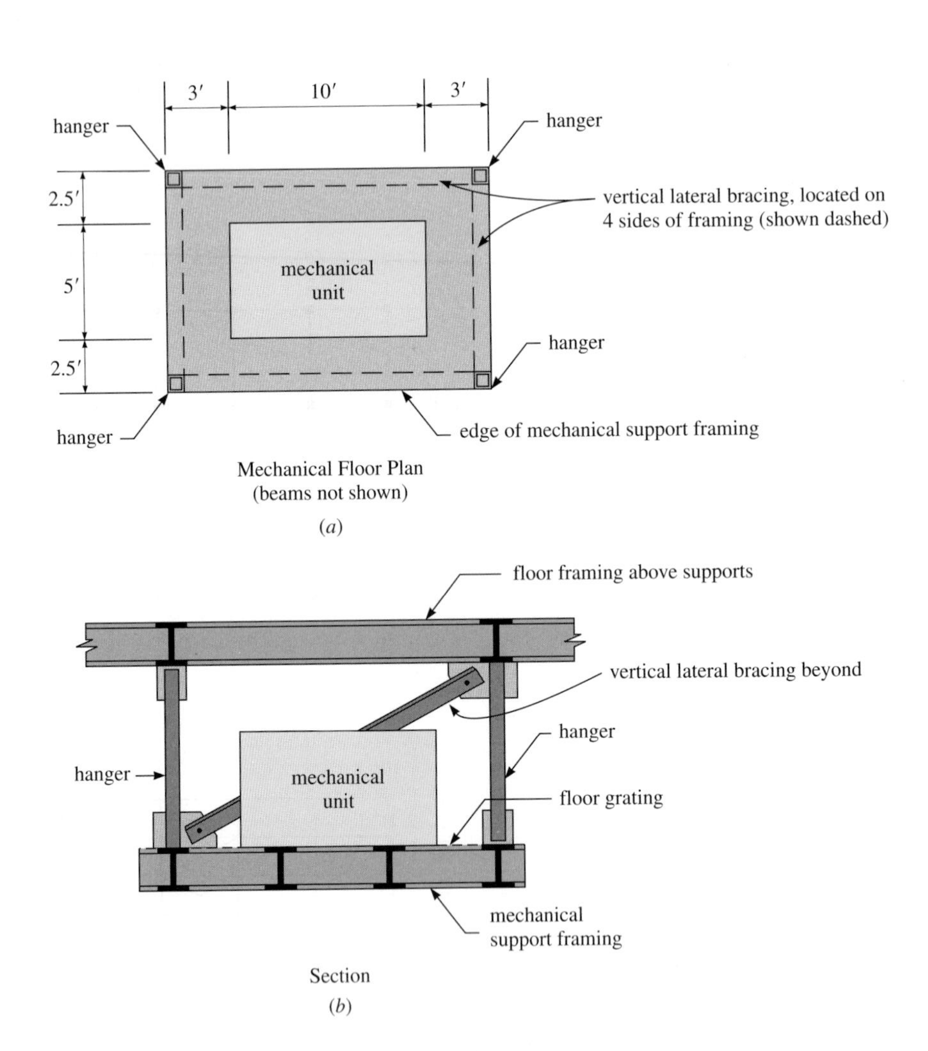

P2.9

P2.10. The dimensions of a 9-m-high warehouse are shown in Figure P2.10. The windward and leeward wind pressure profiles in the long direction of the warehouse are also shown. Establish the wind forces based on the following information: basic wind speed = 40 m/s, wind exposure category = C, $K_d = 0.85$, $K_{zt} = 1.0$, $G = 0.85$, and $C_p = 0.8$ for windward wall and -0.2 for leeward wall. Use the K_z values listed in Table 2.4. What is the total wind force acting in the long direction of the warehouse?

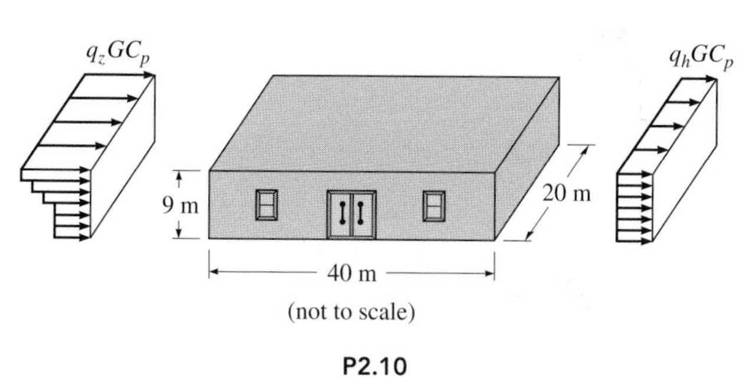

P2.10

P2.11. The dimensions of an enclosed gabled building are shown in Figure P2.11*a*. The external pressures for the wind load perpendicular to the ridge of the building are shown in Figure P2.11*b*. Note that the wind pressure can act toward or away from the windward roof surface. For the particular building dimensions given, the C_p value for the roof based on the ASCE standard can be determined from Table P2.11, where plus and minus signs signify pressures acting toward and away from the surfaces, respectively. Where two values of C_p are listed, this indicates that the windward roof slope is subjected to either positive or negative pressures, and the roof structure should be designed for both loading conditions. The ASCE standard permits linear interpolation for the value of the inclined angle of roof θ. But interpolation should only be carried out between values of the same sign. Establish the wind pressures on the building when positive pressure acts on the windward roof. Use the following data: basic wind speed = 100 mi/h, wind exposure category = B, $K_d = 0.85$, $K_{zt} = 1.0$, $G = 0.85$, and $C_p = 0.8$ for windward wall and –0.2 for leeward wall.

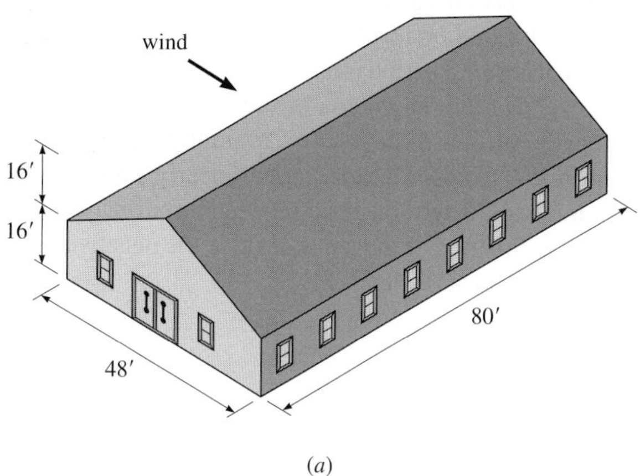

(*a*)

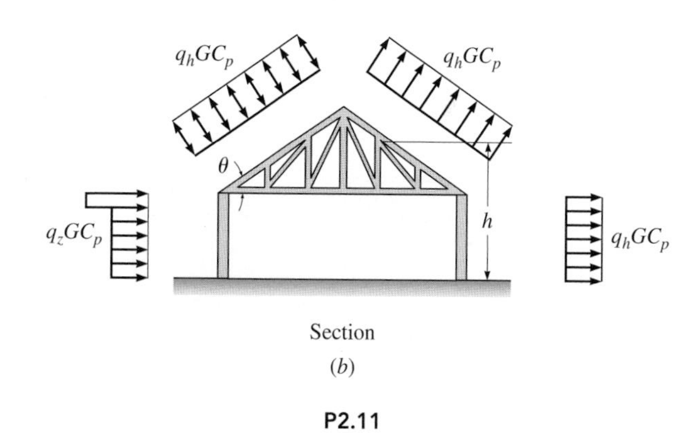

(*b*)

P2.11

TABLE P2.11
Roof Pressure Coefficient C_p

	Windward								Leeward		
Angle θ	10	15	20	25	30	35	45	≥60	10	15	≥20
C_p	−0.9	−0.7	−0.4	−0.3	−0.2	−0.2	0.0	0.01θ*	−0.5	−0.5	−0.6
			0.0	0.2	0.2	0.3	0.4				

*θ defined in Figure P2.11

P2.12. Establish the wind pressures on the building in Problem P2.11 when the windward roof is subjected to an uplift wind force.

P2.13. (*a*) Determine the wind pressure distribution on the four sides of the 10-story hospital shown in Figure P2.13. The building is located near the Georgia coast where the wind velocity contour map in Figure 2.15 of the text specifies a design wind speed of 140 mph. The building, located on level flat ground, is classified as *stiff* because its natural period is less than 1 s. On the windward side, evaluate the magnitude of the wind pressure every 35 ft in the vertical direction. (*b*) Assuming the wind pressure on the windward side varies linearly between the 35-ft intervals, determine the total wind force on the building in the direction of the wind. Include the negative pressure on the leeward side.

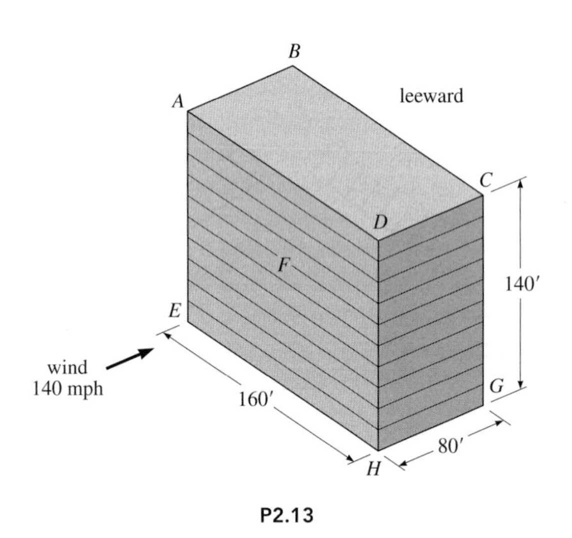

P2.13

P2.14. Consider the five-story building shown in Figure P2.8. The average weights of the floor and roof are 90 lb/ft^2 and 70 lb/ft^2, respectively. The values of S_{DS} and S_{D1} are equal to $0.9g$ and $0.4g$, respectively. Since *steel* moment frames are used in the north-south direction to resist the seismic forces, the value of R equals 8. Compute the seismic base shear V. Then distribute the base shear along the height of the building.

P2.15. When a moment frame does not exceed 12 stories in height and the story height is at least 10 ft, the ASCE standard provides a simpler expression to compute the approximate fundamental period:

$$T = 0.1N$$

where N = number of stories. Recompute T with the above expression and compare it with that obtained from Problem P2.14. Which method produces a larger seismic base shear?

P2.16. (*a*) A two-story hospital facility shown in Figure P2.16 is being designed in New York with a basic wind speed of 90 mi/h and wind exposure D. The importance factor I is 1.15 and K_z = 1.0. Use the simplified procedure to determine the design wind load, base shear, and building overturning moment. (*b*) Use the equivalent lateral force procedure to determine the seismic base shear and overturning moment. The facility, with an average weight of 90 lb/ft^2 for both the floor and roof, is to be designed for the following seismic factors: $S_{DS} = 0.27g$ and $S_{D1} = 0.06g$; reinforced concrete frames with an R value of 8 are to be used. The importance factor I is 1.5. (*c*) Do wind forces or seismic forces govern the strength design of the building?

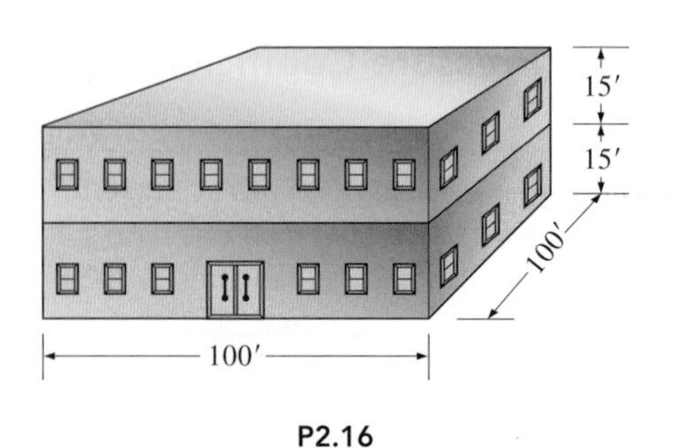

P2.16

P2.17. In the gabled roof structure shown in Figure P2.11, determine the sloped roof snow load P_s. The building is heated and is located in a windy area in Boston. Its roof consists of asphalt shingles. The building is used for a manufacturing facility, placing it in a type II occupancy category. Determine the roof slope factor, C_s using the ASCE graph shown in Figure P2.17. If roof trusses are spaced at 16 ft on center, what is the uniform snow load along a truss?

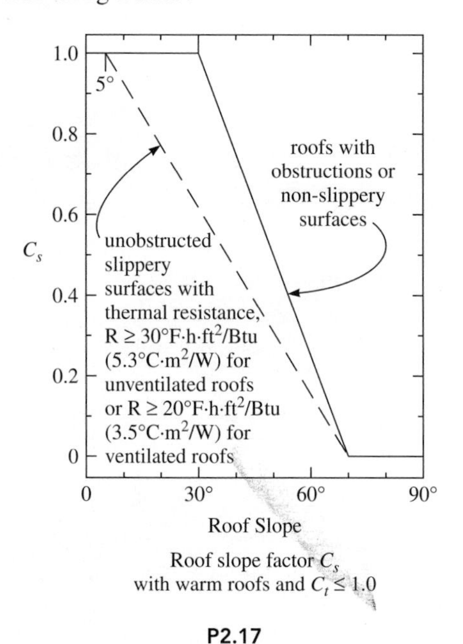

Roof slope factor C_s with warm roofs and $C_t \leq 1.0$

P2.17

P2.18. A beam that is part of a rigid frame has end moments and mid-span moments for dead, live and earthquake loads shown below. Determine the governing load combination for both negative and positive moments at the ends and mid-span of the beam. Earthquake load can act in either direction, generating both negative and positive moments in the beam.

End Moments (ft-kip)		**Mid-Span Moments (ft-kip)**
Dead Load	−180	+90
Live Load	−150	+150
Earthquake	±80	0

Space Roof Truss of Hartford Civic Center Arena in Connecticut

This immense structure, which covered a rectangular area 300 by 360 ft, was supported on four corner columns. To speed construction, the truss was assembled on the ground before being lifted into place. In the photo, the space truss has been raised a short distance to permit workers to install ducts, conduit, and other fixtures from the ground. Only a few hours after five thousand basketball fans had left, the roof collapsed under the weight of a heavy, wet snow load in 1978, shown in the opening photo in Chapter 18.

C H A P T E R 3

Statics of Structures—Reactions

Chapter Objectives

- Review statics, prepare idealized structures and identify appropriate free-body diagrams. Utilize the principle of transmissibility, equations of static equilibrium and equations of condition in the analysis of structures.
- Study support conditions and their restraints, which include the prevention or allowance of translational and rotational movements.
- Calculate reactions for beams, bent frames, multi-story frames and trusses.
- Classify determinate and indeterminate structures, and determine the degree of indeterminacy for the latter. Understand and compare determinate and indeterminate structures in terms of safety through redundancy and proper locations of support conditions.
- Determine if a structure is stable or unstable. Understand instability caused by concurrent and parallel force sytems.

3.1 Introduction

With few exceptions, structures must be stable under all conditions of load; that is, they must be able to support applied loads (their own weight, anticipated live loads, wind, and so forth) without changing shape, undergoing large displacements, or collapsing. Since structures that are stable do not move perceptibly when loaded, their analysis—the determination of both internal and external forces (reactions)—is based in large part on the principles and techniques contained in the branch of engineering mecha-nics called *statics*. The subject of statics, which you have studied previously, covers force systems acting on rigid bodies at rest (the most common case) or moving at constant velocity; that is, in either case the acceleration of the body is zero.

Although the structures we will study in this book are not absolutely rigid because they undergo small elastic deformations when loaded, in most situations the deflections are so small that we can (1) treat the structure or its components as rigid bodies and (2) base the analysis on the initial dimensions of the structure.

We begin this chapter with a brief review of statics. In this review we consider the characteristics of forces, discuss the equations of static equilibrium for two-dimensional (planar) structures, and use the equations of static equilibrium to determine the reactions and internal forces in a variety of simple determinate structures such as beams, trusses, and simple frames.

We conclude this chapter with a discussion of *determinacy* and *stability*. By determinacy, we mean procedures to establish if the equations of statics alone are sufficient to permit a complete analysis of a structure. If the structure cannot be analyzed by the equations of statics, the structure is termed *indeterminate*. To analyze an indeterminate structure, we must supply additional equations by considering the geometry of the deflected shape. Indeterminate structures will be discussed in later chapters.

By *stability*, we mean the geometric arrangement of members and supports required to produce a stable structure, that is, a structure that can resist load from any direction without undergoing either a radical change in shape or large rigid-body displacements. In this chapter, we consider the stability and determinacy of structures that can be treated as either a single rigid body or as several interconnected rigid bodies. The principles that we establish for these simple structures will be extended to more complex structures in later chapters.

3.2 Forces

To solve typical structural problems, we use equations involving forces or their components. Forces may consist of either a *linear force* that tends to produce translation or a *couple* that tends to produce rotation of the body on which it acts. Since a force has magnitude and direction, it can be represented by a vector. For example, Figure 3.1a shows a force **F** with a magnitude F lying in the xy plane and passing through point A.

A couple consists of a pair of equal and oppositely directed forces lying in the same plane (see Figure 3.1b). The moment **M** associated with the couple equals the product of the force **F** and the perpendicular distance (or arm) d between forces. Since a moment is a vector, it has magnitude as well as

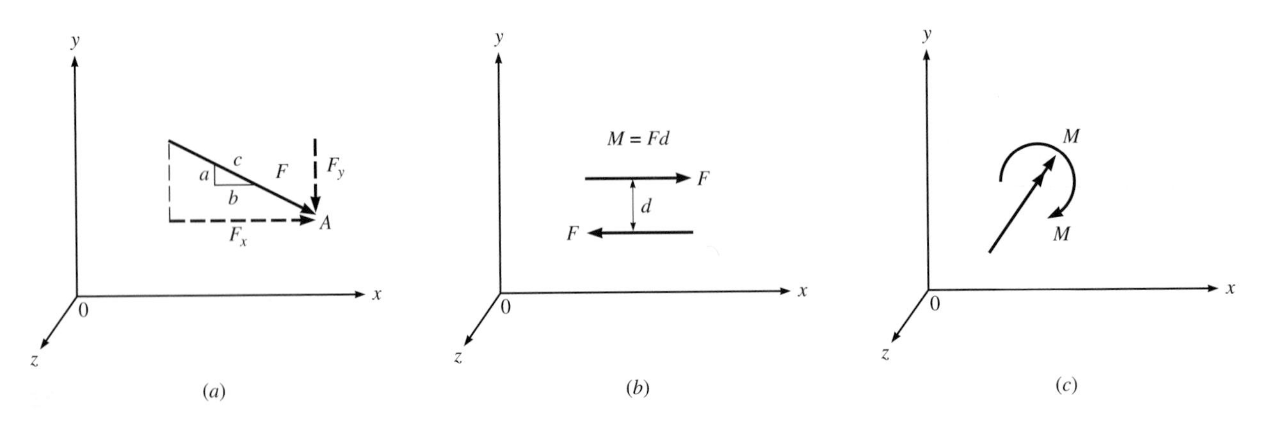

Figure 3.1: Force and moment vectors; (a) linear force vector resolved into x and y components; (b) couple of magnitude Fd; (c) alternative representation of moment M, by a vector using the right-hand rule.

direction. Although we often represent a moment by a curved arrow to show that it acts in the clockwise or counterclockwise direction (see Figure 3.1*c*), we can also represent a moment by a vector—usually a double-headed arrow—using the *right-hand rule*. In the right-hand rule we curl the fingers of the right hand in the direction of the moment, and the direction in which the thumb points indicates the direction of the vector.

We must frequently carry out computations that require either resolving a force into its components or combining several forces to produce a single resultant force. To facilitate these calculations, it is convenient to select arbitrarily horizontal and vertical axes—an x-y coordinate system—as the basic reference directions.

A force can be resolved into components by using the geometric relationship—similar triangles—that exists between the vector components and the slope of the vector. For example, to express the vertical component F_y of the vector F in Figure 3.1*a* in terms of the slope of the vector, we write, using similar triangles,

$$\frac{F_y}{a} = \frac{F}{c}$$

and

$$F_y = \frac{a}{c}F$$

Similarly, if we set up a proportion between the horizontal component F_x and F and the sides of the slope triangle noted on the vector, we can write

$$F_x = \frac{b}{c}F$$

If a force is to be resolved into components that are not parallel to an x-y coordinate system, the *law of sines* provides a simple relationship between length of sides and interior angles opposite the respective sides. For the triangle shown in Figure 3.2, we can state the law of sines as

$$\frac{a}{\sin A} = \frac{b}{\sin B} = \frac{c}{\sin C}$$

where A is the angle opposite side a, B is the angle opposite side b, and C is the angle opposite side c.

Example 3.1 illustrates the use of the law of sines to compute the orthogonal components of a vertical force in arbitrary directions.

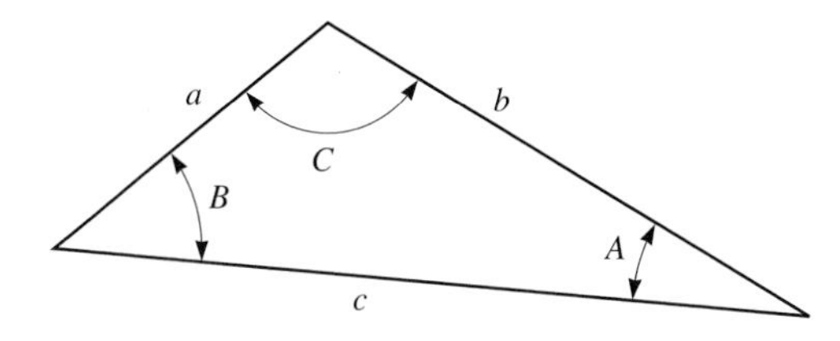

Figure 3.2: Diagram to illustrate law of sines.

EXAMPLE 3.1

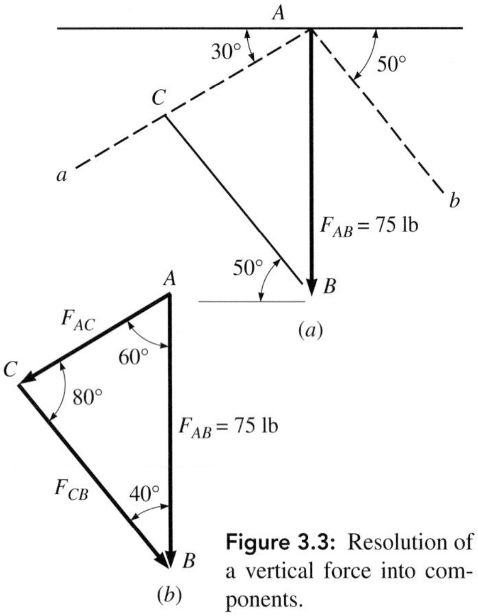

Using the law of sines, resolve the 75-lb vertical force $\mathbf{F}_{AB}$ in Figure 3.3*a* into components directed along lines *a* and *b*.

Solution

Through point *B* draw a line parallel to line *b*, forming triangle *ABC*. The interior angles of the triangle are easily computed from the information given. Vectors *AC* and *CB* (Figure 3.3*b*) represent the required components of force $\mathbf{F}_{AB}$. From the law of sines we can write

$$\frac{\sin 80°}{75} = \frac{\sin 40°}{F_{AC}} = \frac{\sin 60°}{F_{CB}}$$

where sin 80° = 0.985, sin 60° = 0.866, and sin 40° = 0.643. Solving for F_{AC} and F_{CB} yields

$$F_{AC} = \frac{\sin 40°}{\sin 80°}(75) = 48.96 \text{ lb} \qquad \textbf{Ans.}$$

$$F_{CB} = \frac{\sin 60°}{\sin 80°}(75) = 65.94 \text{ lb} \qquad \textbf{Ans.}$$

Figure 3.3: Resolution of a vertical force into components.

Resultant of a Planar Force System

In certain structural problems we will need to determine the magnitude and location of the *resultant* of a force system. Since the resultant is a single force that produces the same external effect on a body as the original force system, the resultant *R* must satisfy the following three conditions:

1. The horizontal component of the resultant R_x must equal the algebraic sum of the horizontal components of all forces:

$$R_x = \Sigma F_x \tag{3.1a}$$

2. The vertical component of the resultant R_y must equal the algebraic sum of the vertical components of all forces:

$$R_y = \Sigma F_y \tag{3.1b}$$

3. The moment M_o produced by the resultant about a reference axis through point *o* must equal the moment about point *o* produced by all forces and couples that make up the original force system.

$$M_o = Rd = \Sigma F_i d_i + \Sigma M_i \tag{3.1c}$$

where R = resultant force $= \sqrt{R_x^2 + R_y^2}$

d = perpendicular distance from line of action of resultant to axis about which moments are computed (3.1*d*)

$\Sigma F_i d_i$ = moment of all forces about reference axis

ΣM_i = moment of all couples about reference axis

EXAMPLE 3.2

Computation of a Resultant

Determine the magnitude and location of the resultant R of the three wheel loads shown in Figure 3.4.

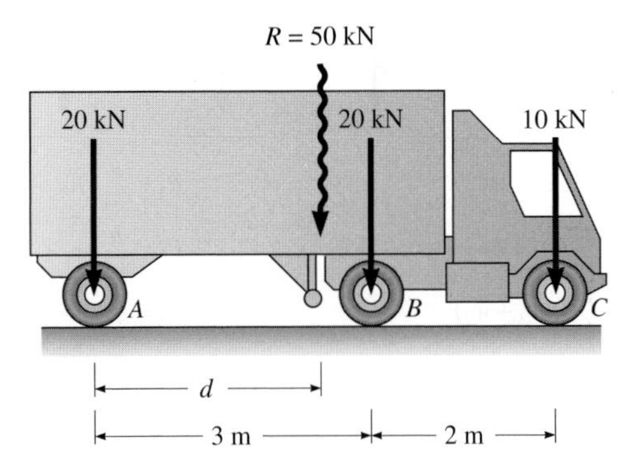

Figure 3.4

Solution

Since none of the forces act in the horizontal direction or have components in the horizontal direction,

$$R_x = 0$$

Using Equation 3.1*b* gives

$$R = R_y = \Sigma F_y = 20 + 20 + 10 = 50 \text{ kN} \qquad \textbf{Ans.}$$

Locate the position of the resultant, using Equation 3.1*c*; that is, equate the moment produced by the original force system to the moment produced by the resultant R. Select a reference axis through point A (choice of A arbitrary).

$$Rd = \Sigma F_i d_i$$

$$50d = 20(0) + 20(3) + 10(5)$$

$$d = 2.2 \text{ m} \qquad \textbf{Ans.}$$

Resultant of a Distributed Load

In addition to concentrated loads and couples, many structures carry distributed loads. The external effect of a distributed load (the computation of reactions it produces, for example) is most easily handled by replacing the distributed loads by an equivalent resultant force. As you have learned previously in statics and mechanics of materials courses, the magnitude of the resultant of a distributed load equals the area under the load curve and acts at its centroid (see Table A.1 for values of area and location of the centroid for several common geometric shapes). Example 3.3 illustrates the use of integration to compute the magnitude and location of the resultant of a distributed load with a parabolic variation.

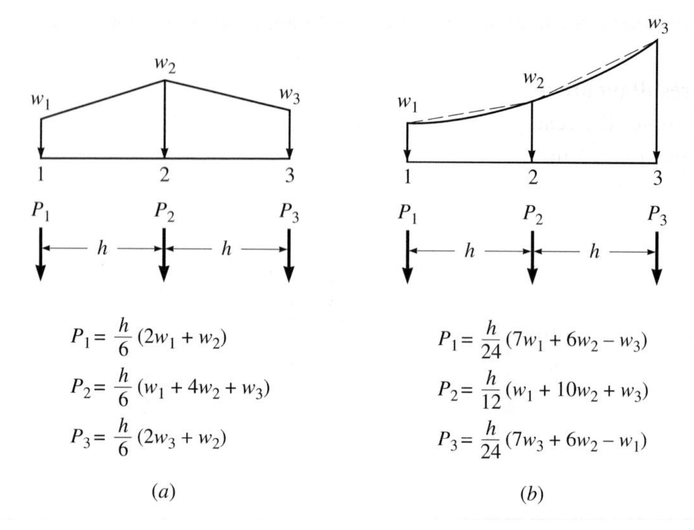

Figure 3.5: (*a*) Expressions to convert a trapezoidal variation of load to a set of statically equivalent, equally spaced, concentrated loads; (*b*) equations to convert a parabolic variation of load to a statically equivalent set of concentrated loads. Equations are valid for concave downward parabolas also, and will give a close approximation for higher-order curves.

If the shape of a distributed load is complex, the designer can often simplify the computation of the magnitude and position of the resultant by subdividing the area into several smaller geometric areas whose properties are known. In most cases distributed loads are uniform or vary linearly. For the latter case, you can divide the area into triangular and rectangular areas (see Example 3.7).

As an alternative procedure the designer may replace a distributed load that varies in a complex manner by a *statically equivalent* set of concentrated loads using the equations in Figure 3.5. To use these equations, we divide the distributed loads into an arbitrary number of segments of length h. The ends of the segments are termed the *nodes*. Figure 3.5 shows two typical segments. The nodes are labeled 1, 2, and 3. The number of segments into which the load is divided depends on the length and shape of the distributed load and the quantity we will compute. If the distributed load varies *linearly* between nodes, the equivalent concentrated force at each node is given by the equations in Figure 3.5*a*. The equations for forces labeled P_1 and P_3 apply at an exterior node—a segment is located on only one side of the node, and P_2 applies to an interior node—segments are located on both sides of a node.

For a distributed load with a *parabolic* variation (either concave up or concave down), the equations in Figure 3.5*b* should be used. These equations will also give good results (within 1 or 2 percent of the exact values) for distributed loads whose shape is represented by a higher-order curve. If the length of the segments is not too large, the simpler equations in Figure 3.5*a* can also be applied to a distributed load whose ordinates lie on a curve such as shown in Figure 3.5*b*. When they are applied in this fashion, we are in effect replacing the actual loading curve by a series of trapezoidal elements, as shown by the dashed line in Figure 3.5*b*. As we reduce the distance h between nodes (or equivalently increase the number of segments), the trapezoidal approximation approaches the actual curve. Example 3.4 illustrates the use of the equations of Figure 3.5.

Although the resultant of a distributed load produces the same external effect on a body as the original loading, the internal stresses produced by the resultant are not the same as those produced by the distributed load. For example, the resultant force can be used to compute the reactions of a beam, but the computations for internal forces—for example, shear and moment—must be based on the actual loading.

EXAMPLE 3.3

Compute the magnitude and location of the resultant of the parabolic loading shown in Figure 3.6. The slope of the parabola is zero at the origin.

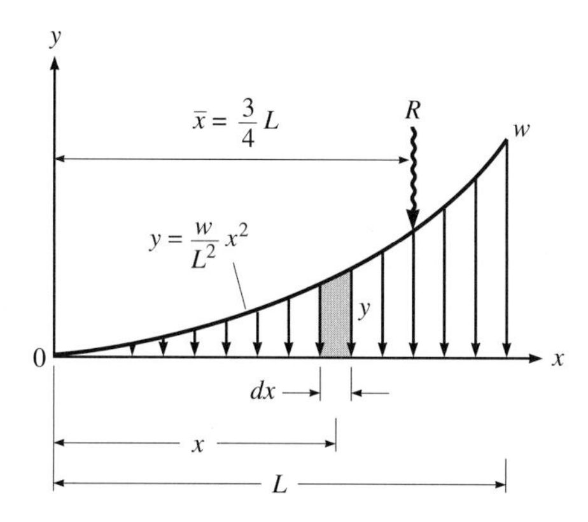

Figure 3.6

Solution

Compute R by integrating the area under the parabola $y = (w/L^2)x^2$.

$$R = \int_0^L y\,dx = \int_0^L \frac{wx^2}{L^2}\,dx = \left[\frac{wx^3}{3L^2}\right]_0^L = \frac{wL}{3} \qquad \textbf{Ans.}$$

Locate the position of the centroid. Using Equation 3.1c and summing moments about the origin o gives

$$R\bar{x} = \int_0^L y\,dx(x) = \int_0^L \frac{w}{L^2}x^3\,dx = \left[\frac{wx^4}{4L^2}\right]_0^L = \frac{wL^2}{4}$$

Substituting $R = wL/3$ and solving the equation above for $\bar{x}$ yield

$$\bar{x} = \frac{3}{4}L \qquad \textbf{Ans.}$$

EXAMPLE 3.4

The beam in Figure 3.7*a* supports a distributed load whose ordinates lie on a parabolic curve. Replace the distributed load by a statically equivalent set of concentrated loads.

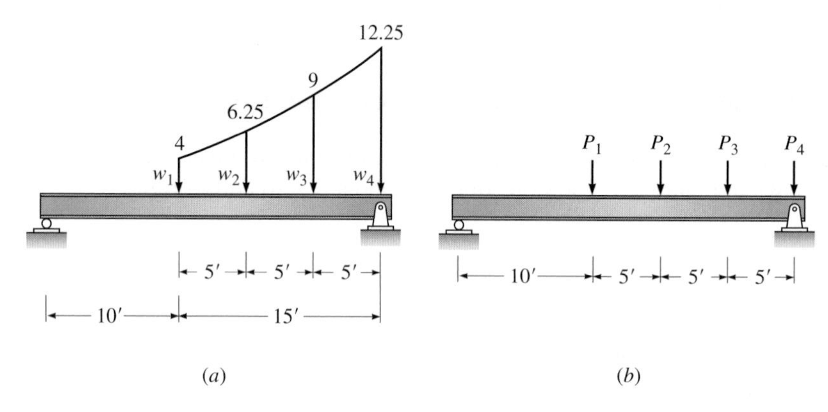

Figure 3.7: (*a*) Beam with a distributed load (units of load in kips per foot); (*b*) beam with equivalent concentrated loads.

Solution

Divide the load into three segments where $h = 5$ ft. Evaluate the equivalent loads, using the equations in Figure 3.5*b*.

$$P_1 = \frac{h}{24}(7w_1 + 6w_2 - w_3) = \frac{5}{24}[7(4) + 6(6.25) - 9] = 11.77 \text{ kips}$$

$$P_2 = \frac{h}{12}(w_1 + 10w_2 + w_3) = \frac{5}{12}[4 + 10(6.25) + 9] = 31.46 \text{ kips}$$

$$P_3 = \frac{h}{12}(w_2 + 10w_3 + w_4) = \frac{5}{12}[6.25 + 10(9) + 12.25] = 45.21 \text{ kips}$$

$$P_4 = \frac{h}{24}(7w_4 + 6w_3 - w_2) = \frac{5}{24}[7(12.25) + 6(9) - 6.25] = 27.81 \text{ kips}$$

Also compute the approximate values of loads P_1 and P_2, using the equations in Figure 3.5*a* for a trapezoidal distribution of load.

$$P_1 = \frac{h}{6}(2w_1 + w_2) = \frac{5}{6}[2(4) + 6.25] = 11.88 \text{ kips}$$

$$P_2 = \frac{h}{6}(w_1 + 4w_2 + w_3) = \frac{5}{6}[4 + 4(6.25) + 9] = 31.67 \text{ kips}$$

The analysis above indicates that for this case the approximate values of P_1 and P_2 deviate less than 1 percent from the exact values.

Principle of Transmissibility

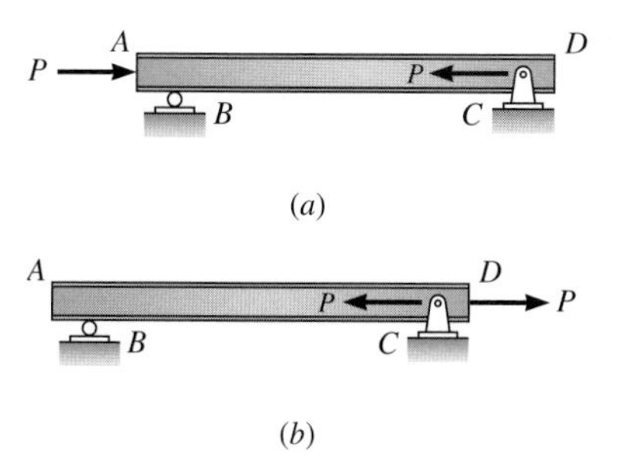

Figure 3.8: Principle of transmissibility.

The principle of transmissibility states that a force may be moved along its line of action without changing the external effect that it produces on a body. For example, in Figure 3.8*a* we can see from a consideration of equilibrium in the x direction that the horizontal force P applied to the beam at point A creates a horizontal reaction at support C equal to P. If the force at point A is moved along its line of action to point D at the right end of the beam (see Figure 3.8*b*), the same horizontal reaction P develops at C. Although the effect of moving the force along its line of action produces no change in the reactions, we can see that the internal force in the member is affected by the position of the load. For example, in Figure 3.8*a* compression stresses develop between points A and C. On the other hand, if the load acts at D, the stress between points A and C is zero and tensile stresses are created between C and D (see Figure 3.8*b*).

The ability of the engineer to move vectors along their line of action is used frequently in structural analysis to simplify computations, to solve problems involving vectors graphically, and to develop a better understanding of behavior. For example, in Figure 3.9 the forces acting on a retaining wall consist of the weight W of the wall and the thrust of the soil pressure T on the back of the wall. These force vectors can be added on the figure by sliding T and W along their lines of actions until they intersect at point A. At that point the vectors can be combined to produce the resultant force R acting on the wall. The magnitude and direction of R are evaluated graphically in Figure 3.9*b*. Now—in accordance with the principle of transmissibility—the resultant can be moved along its line of action until it intersects the base at point x. If the resultant intersects the base within the middle third, it can be shown that compressive stresses exist over the entire base—a desirable state of stress because soil cannot transmit tension. On the other hand, if the resultant falls outside the middle third of the base, compression will exist under only a portion of the base, and the stability of the wall—the possibility the wall will overturn or overstress the soil—must be investigated.

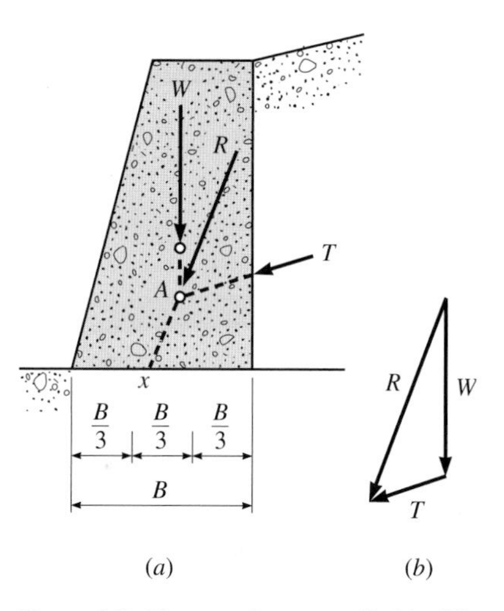

Figure 3.9: Forces acting on a wall: (*a*) addition of weight W and soil pressure (thrust) T; (*b*) vector addition of W and T to produce R.

3.3 Supports

To ensure that a structure or a structural element remains in its required position under all loading conditions, it is attached to a foundation or connected to other structural members by supports. In certain cases of light construction, supports are provided by nailing or bolting members to supporting walls, beams, or columns. Such supports are simple to construct, and little attention is given to design details. In other cases where large, heavily loaded structures must be supported, large complex mechanical devices that allow certain displacements to occur while preventing others must be designed to transmit large loads.

Although the devices used as supports can vary widely in shape and form, we can classify most supports in one of four major categories based on the

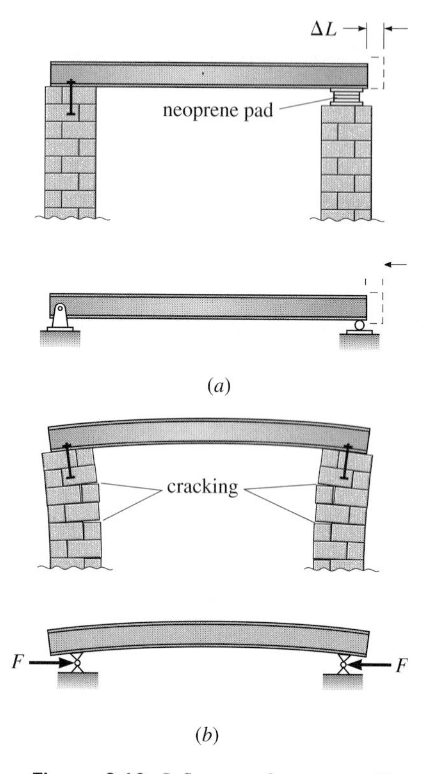

Figure 3.10: Influence of supports: Idealized representation shown below actual construction condition: (*a*) right end is free to expand laterally, no stresses created by temperature change; (*b*) both ends are restrained, compressive and bending stresses develop in beam. Walls crack.

restraints or *reactions* the supports exert on the structure. The most common supports, whose characteristics are summarized in Table 3.1, include the pin, the roller, the fixed support, and the link.

The pin support shown in Table 3.1, case (*a*), represents a device that connects a member to a fixed point by a frictionless pin. Although this support prevents displacement in any direction, it allows the end of the member to rotate freely. Fixed supports [see Table 3.1 case (*f*)], although not common, occasionally exist when the end of a member is deeply embedded in a massive block of concrete or grouted into solid rock (Figure 3.11 on page 84).

The system of supports a designer selects will influence the forces that develop in a structure and also the forces transmitted to the supporting elements. For example, in Figure 3.10*a* the left end of a beam is connected to a wall by a bolt that prevents relative displacement between the beam and the wall while the right end is supported on a neoprene pad that allows the end of the beam to move laterally without developing any significant restraining force. If the temperature of the beam increases, the beam will expand. Since no longitudinal restraint develops at the right end to resist the expansion, no stresses are created in either the beam or the walls. On the other hand, if both ends of the same beam are bolted to masonry walls (see Figure 3.10*b*), an expansion of the beam produced by an increase in temperature will push the walls outward and possibly crack them. If the walls are stiff, they will exert a restraining force on the beam that will create compressive stresses (and possibly bending stresses if the supports are eccentric to the centroid of the member) in the beam. Although these effects typically have little effect on structures when spans are short or temperature changes moderate, they can produce undesirable effects (buckle or overstress members) when spans are long or temperature changes large.

Photo 3.1: Pin support for the 2.1-mile long steel box-girder San Diego-Coronado Bridge.

Photo 3.2: Roller support for the San Diego-Coronado Bridge.

TABLE 3.1 Characteristics of Supports

Type	Sketch	Symbol	Movements Allowed or Prevented	Reaction Forces	Unknowns Created
(*a*) Pin	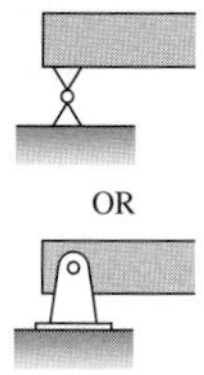 OR	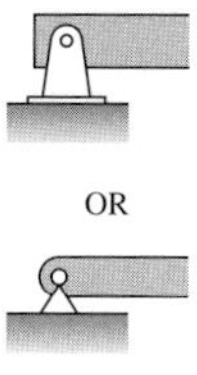 OR	*Prevented:* horizontal translation, vertical translation *Allowed:* rotation	A single linear force of unknown direction or equivalently A horizontal force and a vertical force which are the components of the single force of unknown direction	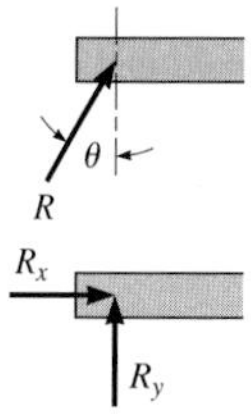 θ, R; R_x, R_y
(*b*) Hinge			*Prevented:* relative displacement of member ends *Allowed:* both rotation and horizontal and vertical displacement	Equal and oppositely directed horizontal and vertical forces	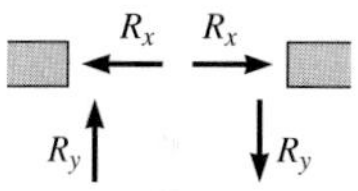 R_x, R_x, R_y, R_y
(*c*) Roller		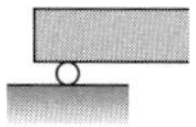 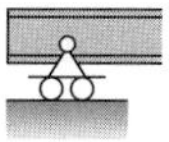	*Prevented:* vertical translation *Allowed:* horizontal translation, rotation	A single linear force (either upward or downward*)	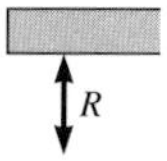 R
(*d*) Rocker		OR			
(*e*) Elastomeric pad	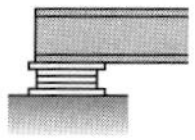				
(*f*) Fixed end	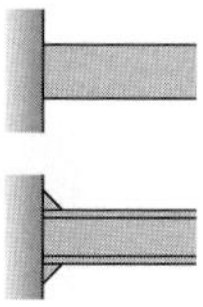	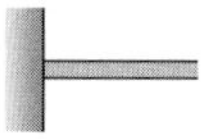	*Prevented:* horizontal translation, vertical translation, rotation *Allowed:* none	Horizontal and vertical components of a linear resultant; moment	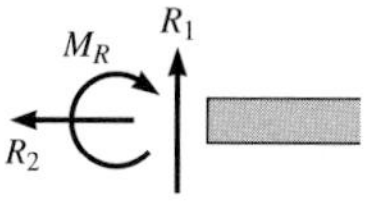 M_R, R_1, R_2
(*g*) Link	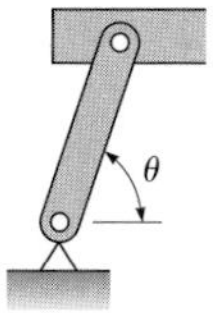 θ	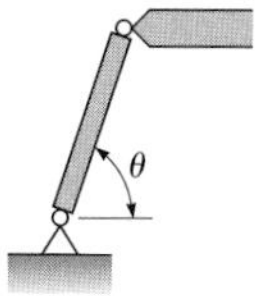 θ	*Prevented:* translation in the direction of link *Allowed:* translation perpendicular to link, rotation	A single linear force in the direction of the link	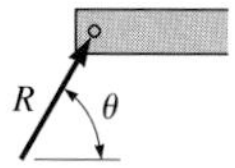 R, θ
(*h*) Guide	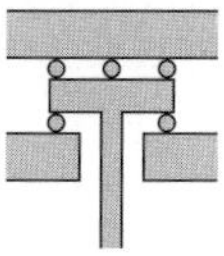	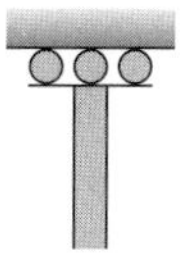	*Prevented:* vertical translation, rotation *Allowed:* horizontal translation	A single vertical linear force; moment	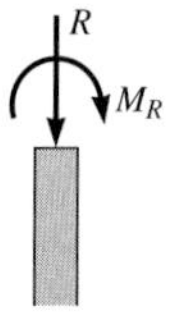 R, M_R

*Although the symbol for a roller support, for the sake of simplicity, shows no restraint against upward movement, it is intended that a roller can provide a downward reaction force if necessary.

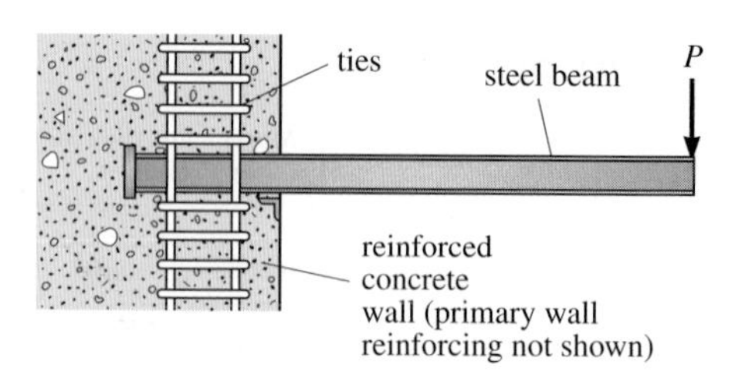

Figure 3.11: Fixed-end beam produced by embedding its left end in a reinforced concrete wall.

To produce a fixed-end condition for a steel beam or column is expensive and rarely done. For a steel beam a fixed-end condition can be created by embedding one end of the beam in a massive block of reinforced concrete (see Figure 3.11).

To produce a fixed-end condition at the base of a steel column, the designer must specify a thick steel baseplate, reinforced by vertical steel stiffener plates connected to the column and the baseplate (see Figure 3.12). In addition, the baseplate must be anchored to the support by heavily tensioned anchor bolts.

On the other hand, when structural members are constructed of reinforced concrete, a fixed end or a pin end can be produced more easily. In the case of a beam, a fixed end is produced by extending reinforcing bars a specified distance into a supporting element (see Figure 3.13*a*).

For a reinforced concrete column, the designer can create a hinge at its base by (1) notching the bottom of the column just above the supporting wall or footing and (2) crossing the reinforcing bars as shown in Figure 3.13*b*. If the axial force in the column is large, to ensure that the concrete in the region of the notch does not fail by crushing, additional vertical reinforcing bars must be added at the centerline of the column to transfer the axial force.

P
M
steel
column
stiffener plates
each side
anchor
bolt
base plate
foundation
elevation

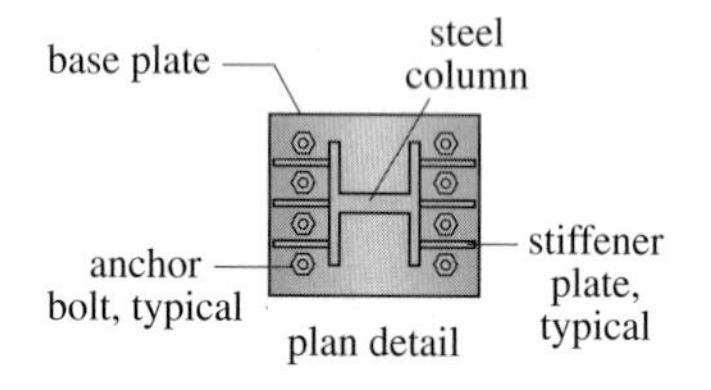

Figure 3.12: A steel column supported on a stiffened baseplate, which is bolted to a concrete foundation, producing a fixed-end condition at its base.

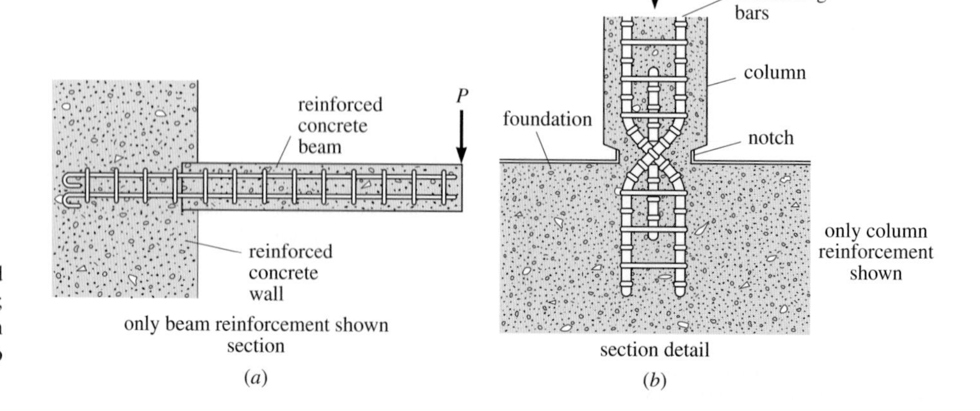

Figure 3.13: (*a*) A reinforced concrete beam with a fixed end; (*b*) a reinforced concrete column whose lower end is detailed to act as a pin.

3.4 Idealizing Structures

Before a structure can be analyzed, the designer must develop a simplified physical model of the structure and its supports as well as the applied loads. This model is typically represented by a simple line drawing. To illustrate this procedure, we will consider the structural steel rigid frame in Figure 3.14*a*. For purposes of analysis, the designer would probably represent the rigid frame by the simplified sketch in Figure 3.14*b*. In this sketch the columns and girders are represented by the centerlines of the actual members. Although the maximum load applied to the girder of the frame may be created by a deep uneven pile of heavy, wet snow, the designer, following code specifications, will design the frame for an equivalent uniform load w. As long as the equivalent load produces, in the members, forces of the same magnitude as the real load, the designer will be able to size the members with the strength required to support the real load.

In the actual structure, plates welded to the base of the columns, are bolted to foundation walls to support the frame. Sometimes a tension rod is also run between the bases of the columns to carry the lateral thrust that is produced by the vertical load on the girder. By using the tension rod to carry the horizontal forces tending to move the bases of the columns, supported on foundation walls, outward, the designers can size the walls and foundations for vertical load only, a condition that reduces the cost of the walls significantly. Although some rotational restraint obviously develops at the base of the columns, designers typically neglect it and assume that the actual supports can be represented by frictionless pins. This assumption is made for the following reasons:

1. The designer has no simple procedure to evaluate rotational restraint.
2. The rotational restraint is modest because of the flexural deformation of the plate, the elongation of the bolts, and small lateral movements of the wall.
3. Finally, the assumption of a pin support at the base is conservative (restraints of any type stiffen the structure).

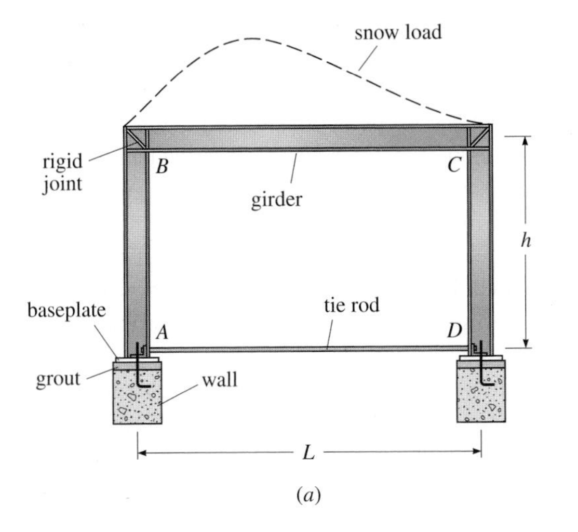

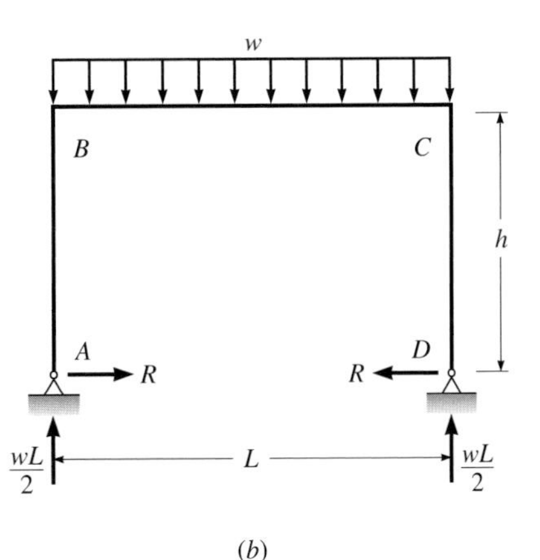

Figure 3.14: (*a*) Welded rigid frame with snow load; (*b*) idealized frame on which analysis is based.

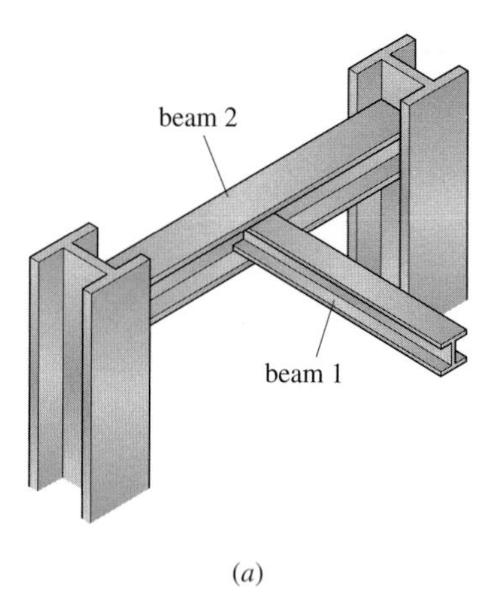

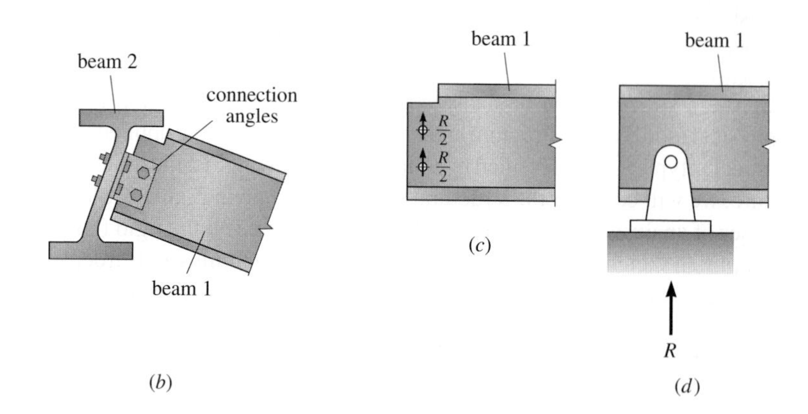

Figure 3.15: Bolted web connection idealized as a pin support: (*a*) Perspective of joint. (*b*) Details of connection shown to an exaggerated scale: slope of beam 1 bends the flexible web of beam 2. The flexible joint is assumed to supply no rotational restraint. (*c*) Since the connection supplies only vertical restraint (its capacity for lateral restraint is not mobilized), we are free to model the joint as a pin or roller support as shown in (*d*).

As an example, we will consider the behavior of the standard web connection between the two steel beams in Figure 3.15*a*. As shown in Figure 3.15*b*, the upper flange of beam 1 is cut back so that the top flanges are at the same elevation. The connection between the two beams is made by means of a pair of angles that are bolted (or welded) to the webs of both beams. The forces applied to the members by the bolts are shown in Figure 3.15*c*. Since the web of beam 2 is relatively flexible, the connection is typically designed to transfer only vertical load between the two members. Although the connection has a limited capacity for horizontal load, this capacity is not utilized because beam 1 carries primarily gravity load and little or no axial load. Designers typically model this type of connection as a pin or roller (Figure 3.15d).

3.5 Free-Body Diagrams

As a first step in the analysis of a structure, the designer will typically draw a simplified sketch of the structure or the portion of the structure under consideration. This sketch, which shows the required dimensions together with all the external and internal forces acting on the structure, is called a *free-body diagram* (*FBD*). For example, Figure 3.16*a* shows a free-body diagram of a three-hinged arch that carries two concentrated loads. Since the reactions at supports *A* and *C* are unknown, their directions must be assumed.

The designer could also represent the arch by the sketch in Figure 3.16*b*. Although the supports are not shown (as they are in Figure 3.16*a*) and the arch is represented by a single line, the free-body diagram contains all the information required to analyze the arch. However, since the pin supports at *A* and *C* are not shown, it is not obvious to someone unfamiliar with the problem (and seeing the sketch for the first time) that points *A* and *B* are not free to displace because of the pins at those locations. In each case, designers must use their judgment to decide what details are required for clarity. If the internal forces at the center hinge at *B* are to be computed, either of the free bodies shown in Figure 3.16*c* could be used.

When the direction of a force acting on a free body is unknown, the designer is free to assume its direction. If the direction of the force is assumed correctly, the analysis, using the equations of equilibrium, will produce a positive value of the force. On the other hand, if the analysis produces a negative value of an unknown force, the initial direction was assumed incorrectly, and the designer must reverse the direction of the force (see Example 3.5).

Free-body diagrams can also be used to determine the internal forces in structures. At the section to be studied, we imagine the structure is cut apart by passing an imaginary plane through the element. If the plane is oriented perpendicular to the longitudinal axis of the member and if the internal force on the cross section is resolved into components parallel and perpendicular to the cut, in the most general case the forces acting on the cut surface will consist of an axial force **F**, a shear **V**, and a moment **M** (in this book we will not consider members that carry torsion). Once **F**, **V**, and **M** are evaluated, we can use standard equations (developed in a basic *strength of materials* course) to compute the axial, shear, and bending stresses on the cross section.

For example, if we wished to determine the internal forces at section 1-1 in the left arch segment (see Figure 3.16*c*), we would use the free bodies shown in Figure 3.16*d*. Following Newton's third law, "for each action there

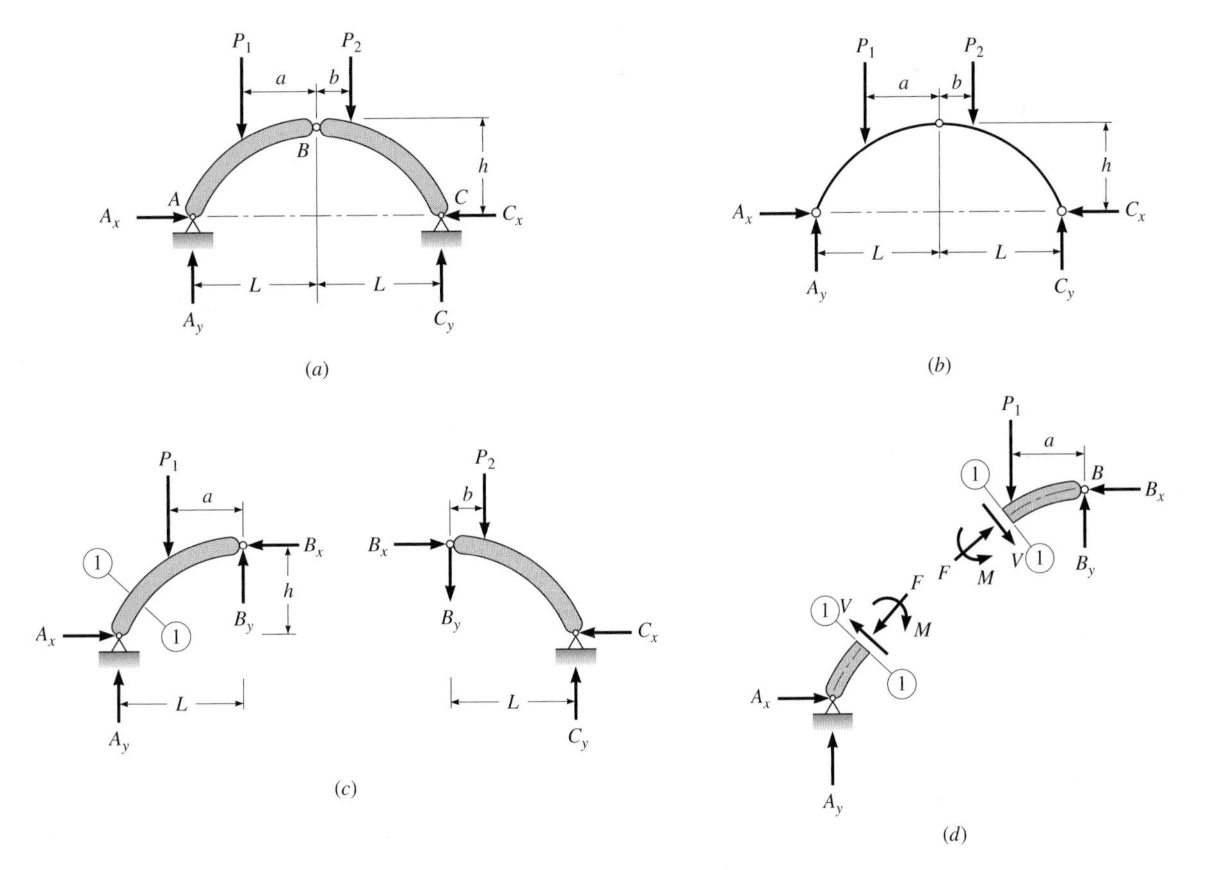

Figure 3.16: Free-body diagrams: (*a*) free-body diagram of three-hinged arch; (*b*) simplified free body of arch in (*a*); (*c*) free-body diagrams of arch segments; (*d*) free-body diagrams to analyze internal forces at section 1-1.

exists an equal and opposite reaction," we recognize that the internal forces on each side of the cut are equal in magnitude and oppositely directed. Assuming that the reactions at the base of the arch and the hinge forces at B have been computed, the shear, moment, and axial forces can be determined by applying the three equations of statics to either of the free bodies in Figure 3.16*d*.

3.6 Equations of Static Equilibrium

Although, the concept of static equilibrium was understood well over 2,000 years ago, as evidenced in Archimedes' (287–212 BC) experiments with equilibrium of the lever, it was during the height of scientific revolution that Sir Isaac Newton (1642–1727), developed the three physical laws of motion in his publication "Philosophiae Naturalis Principia Mathematica" (1687), which formed the foundation for classical mechanics and paved the way for modern structural analysis.
Attributable to Isaac Newton, Simon Stevin (1548–1620) and Pierre Varignon (1654–1722) are the synthesis of statics, equations of static equilibrium, force vector analysis, graphic statics, and parallelogram law for the addition of force vectors.

As you learned in dynamics, a system of *planar forces* acting on a rigid structure (see Figure 3.17) can always be reduced to two resultant forces:

1. A linear force R passing through the center of gravity of the structure where R equals the vector sum of the linear forces.
2. A moment M about the center of gravity. The moment M is evaluated by summing the moments of all forces and couples acting on the structure with respect to an axis through the center of gravity and perpendicular to the plane of the structure.

The linear acceleration a of the center of gravity and the angular accelerations α of the body about the center of gravity are related to the resultant forces R and M by Newton's second law, which can be stated as follows.

$$R = ma \tag{3.2a}$$
$$M = I\alpha \tag{3.2b}$$

where m is the mass of the body and I is the mass moment of inertia of the body with respect to its center of gravity.

If the body is at rest—termed a state of *static equilibrium*—both the linear acceleration a and the angular acceleration α equal zero. For this condition, Equations 3.2*a* and 3.2*b* become

$$R = 0 \tag{3.3a}$$
$$M = 0 \tag{3.3b}$$

If R is replaced by its components R_x and R_y, which can be expressed in terms of the components of the actual force system by Equations 3.1*a* and 3.1*b*, we can write the equations of static equilibrium for a planar force system as

$$\Sigma F_x = 0 \tag{3.4a}$$
$$\Sigma F_y = 0 \tag{3.4b}$$
$$\Sigma M_z = 0 \tag{3.4c}$$

Equations 3.4*a* and 3.4*b* establish that the structure is not moving in either the x or y direction, while Equation 3.4*c* ensures that the structure is not

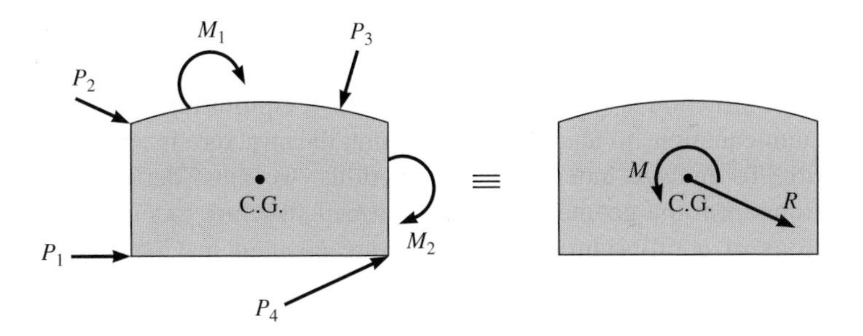

Figure 3.17: Equivalent planar force systems acting on a rigid body.

rotating. Although Equation 3.4*c* was based on a summation of moments about the center of gravity of the structure because we were considering the angular acceleration of the body, this restriction can be removed for structures in static equilibrium. Obviously, if a structure is at rest, the resultant force is zero. Since the actual force system can be replaced by its resultant, it follows that summing moments about any axis parallel to the z-reference axis and normal to the plane of the structure must equal zero because the resultant is zero.

As you may remember from your course in statics, either or both of Equations 3.4*a* and 3.4*b* can also be replaced by moment equations. Several equally valid sets of equilibrium equations are

$$\Sigma F_x = 0 \quad (3.5a)$$

$$\Sigma M_A = 0 \quad (3.5b)$$

$$\Sigma M_z = 0 \quad (3.5c)$$

or

$$\Sigma M_A = 0 \quad (3.6a)$$

$$\Sigma M_B = 0 \quad (3.6b)$$

$$\Sigma M_z = 0 \quad (3.6c)$$

where points A, B, and z do not lie on the same straight line.

Since the deformations that occur in real structures are generally very small, we typically write the equations of equilibrium in terms of the initial dimensions of the structure. In the analysis of flexible columns, long-span arches, or other flexible structures subject to buckling, the deformations of the structural elements or the structure under certain loading conditions may be large enough to increase the internal forces by a significant amount. In these situations, the equilibrium equations must be written in terms of the geometry of the deformed structure if the analysis is to give accurate results. Structures experiencing large deflections of this type are not covered in this text.

If the forces acting on a structure—including both the reactions and the internal forces—can be computed using any of the foregoing sets of equations of static equilibrium, the structure is said to be *statically determinate* or, more simply, *determinate*. Examples 3.5 to 3.7 illustrate the use of the equations of static equilibrium to compute the reactions of a determinate structure that can be treated as a single rigid body.

If the structure is stable but the equations of equilibrium do not provide sufficient equations to analyze the structure, the structure is termed *indeterminate*. To analyze indeterminate structures, we must derive additional equations from the geometry of the deformed structure to supplement the equations of equilibrium. These topics are covered in Chapters 11, 12, and 13.

EXAMPLE 3.5

Compute the reactions for the beam in Figure 3.18*a*.

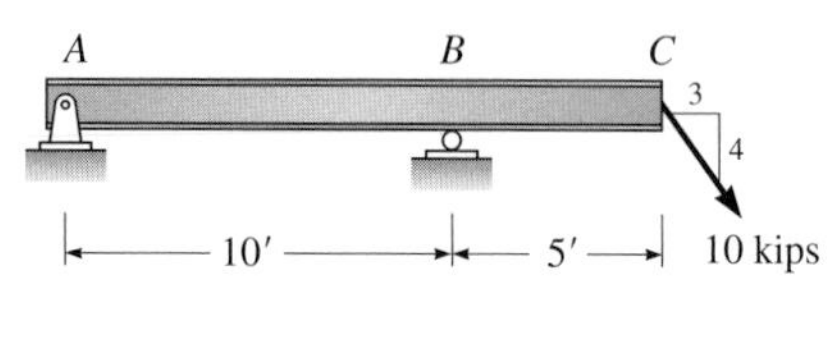

(*a*)

A_x 6 kips 10′ 5′ A_y B_y 8 kips

(*b*)

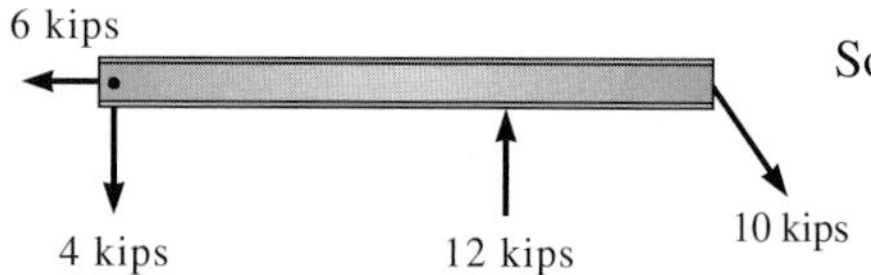

(*c*)

Figure 3.18

Solution

Resolve the force at C into components and assume directions for the reactions at A and B (see Figure 3.18*b*). Ignore the depth of the beam.

Method 1. Solve for reactions using Equations 3.4*a* to 3.4*c*. Assume a positive direction for forces as indicated by arrows:

$$\xrightarrow{+} \quad \Sigma F_x = 0 \qquad -A_x + 6 = 0 \qquad (1)$$

$$\overset{+}{\uparrow} \quad \Sigma F_y = 0 \qquad A_y + B_y - 8 = 0 \qquad (2)$$

$$\circlearrowright^{+} \quad \Sigma M_A = 0 \qquad -10B_y + 8(15) = 0 \qquad (3)$$

Solving Equations 1, 2, and 3 gives

$$A_x = 6 \text{ kips} \qquad B_y = 12 \text{ kips} \qquad A_y = -4 \text{ kips} \qquad \textbf{Ans.}$$

where a plus sign indicates that the assumed direction is correct and a minus sign establishes that the assumed direction is incorrect and the reaction must be reversed. See Figure 3.18*c* for final results.

Method 2. Recompute reactions, using equilibrium equations that contain only one unknown. One possibility is

$$\circlearrowright^{+} \quad \Sigma M_A = 0 \qquad -B_y(10) + 8(15) = 0$$

$$\circlearrowright^{+} \quad \Sigma M_B = 0 \qquad A_y(10) + 8(5) = 0$$

$$\xrightarrow{+} \quad \Sigma F_x = 0 \qquad -A_x + 6 = 0$$

Solving again gives $A_x = 6$ kips, $B_y = 12$ kips, $A_y = -4$ kips.

EXAMPLE 3.6

Compute the reactions for the truss in Figure 3.19.

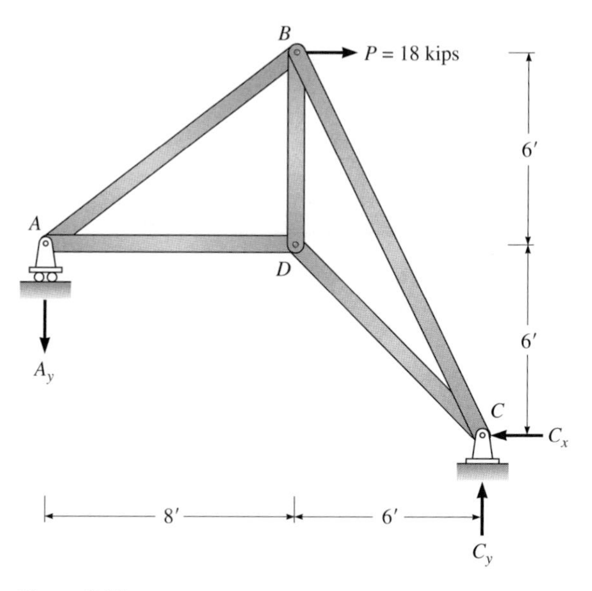

Figure 3.19

Solution

Treat the truss as a rigid body. Assume directions for reactions (see Figure 3.19). Use equations of static equilibrium.

$$\circlearrowright^{+} \quad \Sigma M_C = 0 \qquad 18(12) - A_y(14) = 0 \qquad (1)$$

$$\rightarrow + \quad \Sigma F_x = 0 \qquad 18 - C_x = 0 \qquad (2)$$

$$\overset{+}{\uparrow} \quad \Sigma F_y = 0 \qquad -A_y + C_y = 0 \qquad (3)$$

Solving Equations 1, 2, and 3 gives

$$C_x = 18 \text{ kips} \qquad A_y = 15.43 \text{ kips} \qquad C_y = 15.43 \text{ kips} \qquad \textbf{Ans.}$$

NOTE. The reactions were computed using the initial dimensions of the unloaded structure. Since displacements in well-designed structures are small, no significant change in the magnitude of the reactions would result if we had used the dimensions of the deformed structure.

For example, suppose support A moves 0.5 inches to the right and joint B moves upward 0.25 in. when the 18-kip load is applied; the moment arms for A_y and the 18-kip load in Equation 1 would equal 13.96 ft and 12.02 ft, respectively. Substituting these dimensions into Equation 1, we would compute $A_y = 15.47$ kips. As you can see, the value of A_y does not change enough (0.3 percent in this problem) to justify using the dimensions of the deformed structure, which are time-consuming to compute.

EXAMPLE 3.7

The frame in Figure 3.20 carries a distributed load that varies from 4 to 10 kN/m. Compute the reactions.

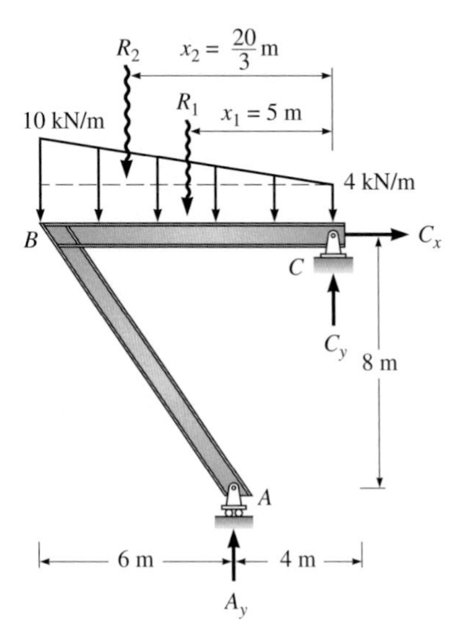

Figure 3.20

Solution

Divide the distributed load into a triangular and a rectangular distributed load (see the dashed line). Replace the distributed loads by their resultant.

$$R_1 = 10(4) = 40 \text{ kN}$$

$$R_2 = \frac{1}{2}(10)(6) = 30 \text{ kN}$$

Compute A_y.

$$\circlearrowright^{+} \quad \Sigma M_C = 0$$

$$A_y(4) - R_1(5) - R_2\left(\frac{20}{3}\right) = 0$$

$$A_y = 100 \text{ kN} \qquad \textbf{Ans.}$$

Compute C_y.

$$\overset{+}{\uparrow} \quad \Sigma F_y = 0$$

$$100 - R_1 - R_2 + C_y = 0$$

$$C_y = -30 \text{ kN} \downarrow$$

(minus sign indicates initial direction incorrectly assumed)

Compute C_x.

$$\rightarrow + \quad \Sigma F_x = 0$$

$$C_x = 0 \qquad \textbf{Ans.}$$

EXAMPLE 3.8

Compute the reactions for the beam in Figure 3.21*a*, treating member *AB* as a link.

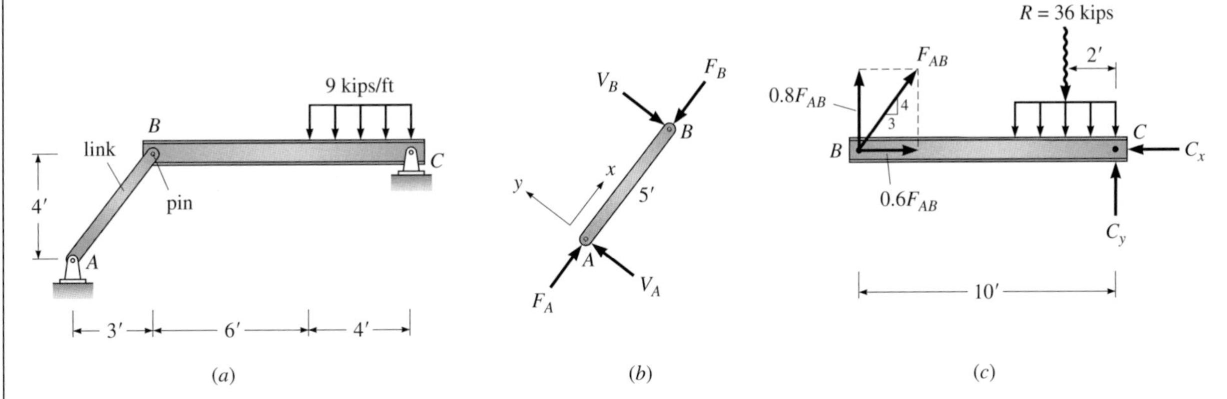

Figure 3.21: (*a*) Beam *BC* supported by link *AB*; (*b*) free body of link *AB*; (*c*) free body of beam *BC*.

Solution

First compute the forces in the link. Since link *AB* is pinned at *A* and *B*, no moments exist at these points. Assume initially that both shear *V* and axial force *F* are transmitted through the pins (see Figure 3.21*b*). Using a coordinate system with an *x* axis along the longitudinal axis of the member, we write the following equilibrium equations:

$$\xrightarrow{+} \quad \Sigma F_x = 0 \qquad 0 = F_A - F_B \qquad (1)$$

$$\overset{+}{\uparrow} \quad \Sigma F_y = 0 \qquad 0 = V_A - V_B \qquad (2)$$

$$\circlearrowright^{+} \quad M_A = 0 \qquad 0 = V_B(5) \qquad (3)$$

Solving the equations above gives

$$F_A = F_B \text{ (call } F_{AB}) \quad \text{and} \quad V_A = V_B = 0$$

These computations show that a member pinned at both ends and not loaded between its ends carries only axial load, that is, is a *two-force* member.

Now compute F_{AB}. Consider beam *BC* as a free body (see Figure 3.21*c*). Resolve F_{AB} into components at *B* and sum moments about *C*.

$$\circlearrowright^{+} \quad \Sigma M_c = 0 \qquad 0 = 0.8F_{AB}(10) - 36(2)$$

$$\xrightarrow{+} \quad \Sigma F_x = 0 \qquad 0 = 0.6F_{AB} - C_x$$

$$\overset{+}{\uparrow} \quad \Sigma F_y = 0 \qquad 0 = 0.8F_{AB} - 36 + C_y$$

Solving gives $F_{AB} = 9$ kips, $C_x = 5.4$ kips, and $C_y = 28.8$ kips.

3.7 Equations of Condition

The reactions of many structures can be determined by treating the structure as a single rigid body. Other stable determinate structures, which consist of several rigid elements connected by a hinge or which contain other devices or construction conditions that release certain internal restraints, require that the structure be divided into several rigid bodies in order to evaluate the reactions.

Consider, for example, the three-hinged arch shown in Figure 3.16*a*. If we write the equations of equilibrium for the entire structure, we will find that only three equations are available to solve for the four unknown reaction components A_x, A_y, C_x, and C_y. To obtain a solution, we must establish an additional equation of equilibrium without introducing any new variables. We can write a fourth independent equilibrium equation by considering the equilibrium of either arch segment between the hinge at B and an end support (see Figure 3.16*c*). Since the hinge at B can transfer a force with horizontal and vertical components, but has no capacity to transfer moment (that is, $M_B = 0$), we can sum moments about the hinge at B to produce an additional equation in terms of the support reactions and applied loads. This additional equation is called an *equation of condition* or an *equation of construction*.

If the arch were continuous (no hinge existed at B), an internal moment could develop at B and we could not write an additional equation without introducing an additional unknown—M_B, the moment at B.

As an alternative approach, we could determine both the reactions at the supports and the forces at the center hinge by writing and solving three equations of equilibrium for each segment of the arch in Figure 3.16*c*. Considering both free bodies, we have six equilibrium equations available to solve for six unknown forces (A_x, A_y, B_x, B_y, C_x, and C_y). Examples 3.9 and 3.10 illustrate the procedure to analyze structures with devices (a hinge in one case and a roller in the other) that release internal restraints.

EXAMPLE 3.9

Compute the reactions for the beam in Figure 3.22*a*. A load of 12 kips is applied directly to the hinge at *C*.

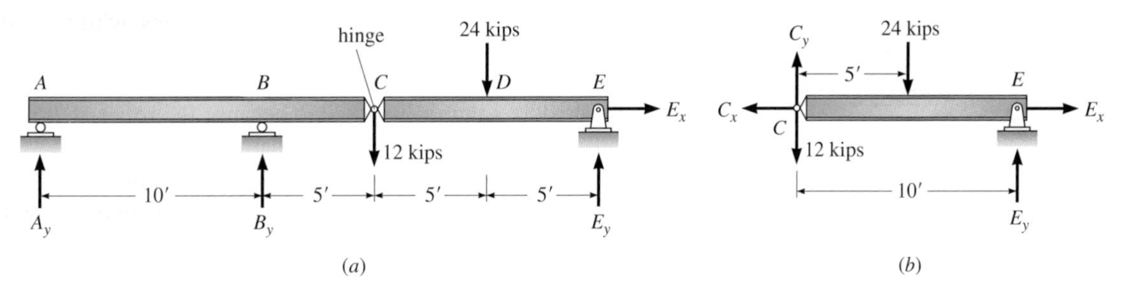

Figure 3.22

Solution

The supports provide four reactions. Since three equations of equilibrium are available for the entire structure in Figure 3.22*a* and the hinge at *C* provides one condition equation, the structure is determinate. Compute E_y by summing moments about *C* (see Figure 3.22*b*).

$$\circlearrowright^{+} \quad \Sigma M_c = 0$$

$$0 = 24(5) - E_y(10) \quad \text{and} \quad E_y = 12 \text{ kips} \qquad \textbf{Ans.}$$

Complete the analysis, using the free body in Figure 3.22*a*.

$$\rightarrow + \quad \Sigma F_x = 0 \qquad 0 + E_x = 0$$

$$E_x = 0 \qquad \textbf{Ans.}$$

$$\circlearrowright^{+} \quad \Sigma M_A = 0 \qquad 0 = -B_y(10) + 12(15) + 24(20) - 12(25)$$

$$B_y = 36 \text{ kips} \qquad \textbf{Ans.}$$

$$\overset{+}{\uparrow} \quad \Sigma F_y = 0 \qquad 0 = A_y + B_y - 12 - 24 + E_y$$

Substituting $B_y = 36$ kips and $E_y = 12$ kips, we compute $A_y = -12$ kips (down).

EXAMPLE 3.10

Compute the reactions for the beams in Figure 3.23*a*.

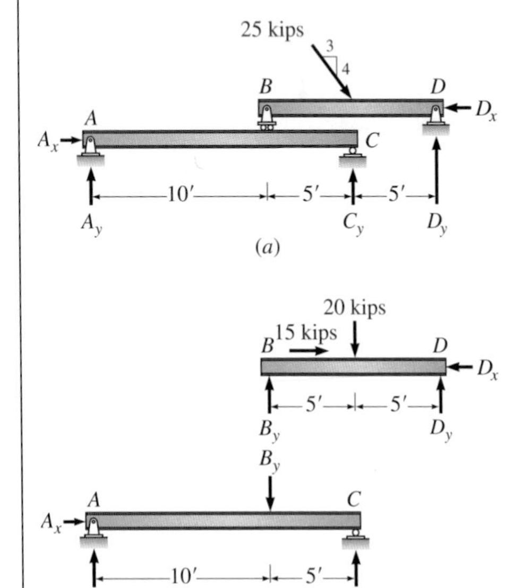

Figure 3.23

Solution

If we treat the entire structure in Figure 3.23*a* as a single rigid body, the external supports supply five reactions: A_x, A_y, C_y, D_x, and D_y. Since only three equations of equilibrium are available, the reactions cannot be established. A solution is possible because the roller at *B* supplies two additional pieces of information (that is, $M_B = 0$ and $B_x = 0$). By separating the structure into two free bodies (see Figure 3.23*b*), we can write a total of six equilibrium equations (three for each free body) to determine the six unknown forces exerted by the external reactions and the roller at *B*.

Applying the equations of equilibrium to member *BD* in Figure 3.23*b*, we have

$$\rightarrow+ \quad \Sigma F_x = 0 \qquad 0 = 15 - D_x \tag{1}$$

$$\circlearrowright^+ \quad \Sigma M_D = 0 \qquad 0 = B_y(10) - 20(5) \tag{2}$$

$$\overset{+}{\uparrow} \quad \Sigma F_y = 0 \qquad 0 = B_y - 20 + D_y \tag{3}$$

Solving Equations 1, 2, and 3, we compute $D_x = 15$ kips, $B_y = 10$ kips, and $D_y = 10$ kips.

With B_y evaluated, we can determine the balance of the reactions by applying the equations of equilibrium to member *AC* in Figure 3.23*b*.

$$\rightarrow+ \quad \Sigma F_x = 0 \qquad 0 = A_x \tag{4}$$

$$\circlearrowright^+ \quad \Sigma M_A = 0 \qquad 0 = 10(10) - 15C_y \tag{5}$$

$$\overset{+}{\uparrow} \quad \Sigma F_y = 0 \qquad 0 = A_y - 10 + C_y \tag{6}$$

Solving Equations 4, 5, and 6, we find $A_x = 0$, $C_y = 20/3$ kips, and $A_y = 10/3$ kips.

Since the roller at *B* cannot transfer a horizontal force between beams, we recognize that the 15-kip horizontal component of the load applied to *BD* must be equilibrated by the reaction D_x. Since no horizontal forces act on member *AC*, $A_x = 0$.

Static check: To verify the accuracy of the computations, we apply $\Sigma F_y = 0$ to the entire structure in Figure 3.23*a*.

$$A_y + C_y + D_y - 0.8(25) = 0$$

$$\frac{10}{3} + \frac{20}{3} + 10 - 20 = 0$$

$$0 = 0 \qquad \text{OK}$$

3.8 Influence of Reactions on Stability and Determinacy of Structures

To produce a stable structure, the designer must supply a set of supports that prevents the structure or any of its components from moving as a rigid body. The number and types of supports required to stabilize a structure depend on the geometric arrangement of members, on any construction conditions built into the structure (hinges, for example), and on the position of supports. The equations of equilibrium in Section 3.6 provide the theory required to understand the influence of reactions on (1) stability, and (2) determinacy (the ability to compute reactions using the equations of statics). We begin this discussion by considering structures composed of a *single rigid body*, and then we extend the results to structures composed of several interconnected bodies.

For a set of supports to prevent motion of a structure under all possible loading conditions, the applied loads and the reactions supplied by the supports must satisfy the three equations of static equilibrium

$$\Sigma F_x = 0 \tag{3.4a}$$

$$\Sigma F_y = 0 \tag{3.4b}$$

$$\Sigma M_z = 0 \tag{3.4c}$$

To develop criteria for establishing the stability and the determinacy of a structure, we will divide this discussion into three cases that are a function of the number of reactions.

Case 1. Supports Supply Less Than Three Restraints: $R < 3$ (R = number of restraints or reactions)

Since three equations of equilibrium must be satisfied for a rigid body to be in equilibrium, the designer must apply at least three reactions to produce a stable structure. If the supports supply less than three reactions, then one or more of the equations of equilibrium cannot be satisfied, and the structure is not in equilibrium. A structure not in equilibrium is *unstable*. For example, let us use the equations of equilibrium to determine the reactions of the beam in Figure 3.24a. The beam, supported on two rollers, carries a vertical load P at midspan and a horizontal force Q.

$$\overset{+}{\uparrow} \quad \Sigma F_y = 0 \qquad 0 = R_1 + R_2 - P \tag{1}$$

$$\circlearrowright^{+} \quad \Sigma M_A = 0 \qquad 0 = \frac{PL}{2} - R_2 L \tag{2}$$

$$\rightarrow + \quad \Sigma F_x = 0 \qquad 0 = Q \quad \text{inconsistent; unstable} \tag{3}$$

Equations 1 and 2 can be satisfied if $R_1 = R_2 = P/2$; however, Equation 3 is not satisfied because Q is a real force and is not equal to zero. Since equilibrium is not satisfied, the beam is unstable and will move to the right under the unbalanced force. Mathematicians would say the set of equations above is *inconsistent* or *incompatible*.

As a second example, we will apply the equations of equilibrium to the beam supported by a pin at point A in Figure 3.23*b*.

$$\xrightarrow{+} \quad \Sigma F_x = 0 \qquad 0 = R_1 - 3 \tag{4}$$

$$\overset{+}{\uparrow} \quad \Sigma F_y = 0 \qquad 0 = R_2 - 4 \tag{5}$$

$$\circlearrowright^{+} \quad \Sigma M_A = 0 \qquad 0 = 4(10) - 3(1) = 37 \tag{6}$$

Examination of Equations 4 through 6 shows that Equations 4 and 5 can be satisfied if $R_1 = 3$ kips and $R_2 = 4$ kips; however, Equation 6 is not satisfied since the right side equals 37 kip·ft and the left side equals zero. Because the equation of moment equilibrium is not satisfied, the structure is unstable; that is, the beam will rotate about the pin at A.

As a final example, we apply the equations of equilibrium to the column in Figure 3.24*c*.

$$\xrightarrow{+} \quad \Sigma F_x = 0 \qquad 0 = R_x \tag{7}$$

$$\overset{+}{\uparrow} \quad \Sigma F_y = 0 \qquad 0 = R_y - P \tag{8}$$

$$\circlearrowright^{+} \quad \Sigma M_A = 0 \qquad 0 = 0 \tag{9}$$

Examination of the equilibrium equations shows that if $R_x = 0$ and $R_y = P$, all equations are satisfied and the structure is in equilibrium. (Equation 9 is automatically satisfied because all forces pass through the moment center.) Even though the equations of equilibrium are satisfied when the column carries a vertical force, we intuitively recognize that the structure is unstable. Although

Figure 3.24: (*a*) Unstable, horizontal restraint missing; (*b*) unstable, free to rotate about A; (*c*) unstable, free to rotate about A; (*d*) and (*e*) unbalanced moments produce failure; (*f*) and (*g*) stable structures.

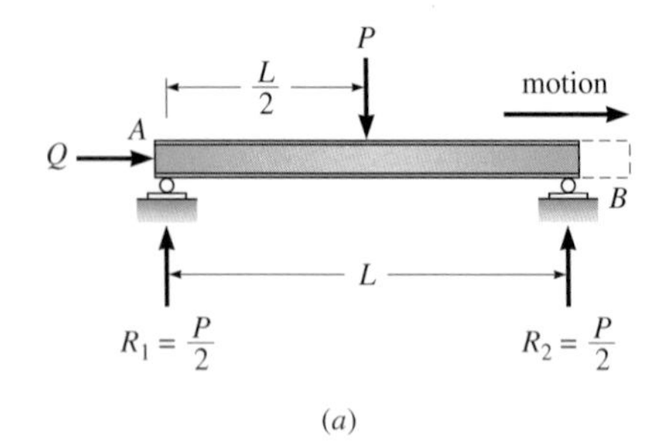

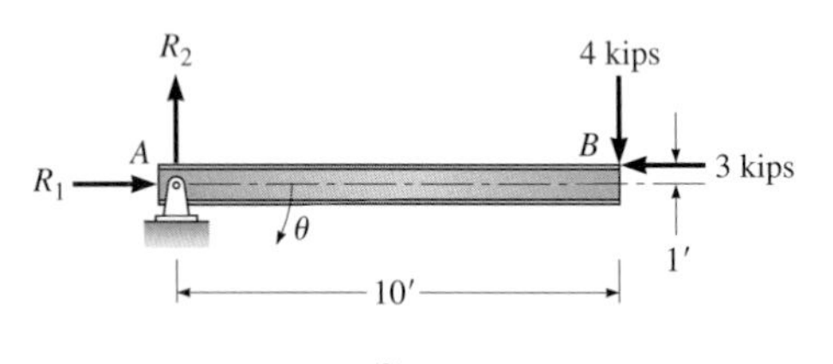

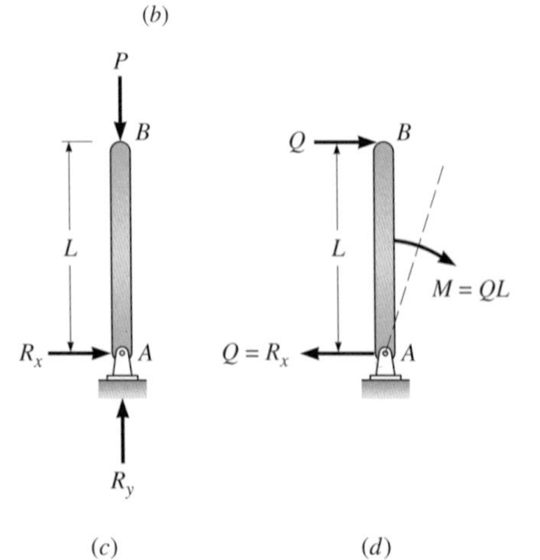

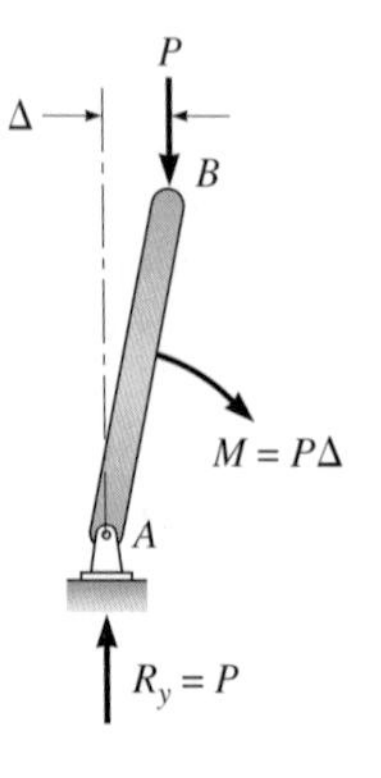

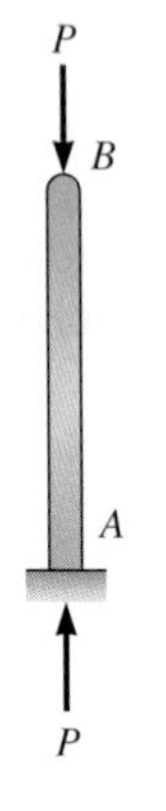

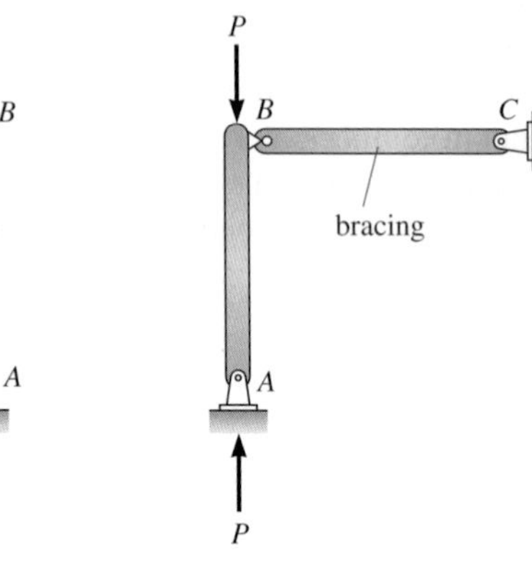

the pin support at A prevents the base of the column from displacing in any direction, it does not supply any rotational restraint to the column. Therefore, either the application of a small lateral force Q (see Figure 3.24*d*) or a small deviation of the top joint from the vertical axis passing through the pin at A while the vertical load P acts (see Figure 3.24*e*) will produce an overturning moment that will cause the column to collapse by rotating about the hinge at A. From this example we see that to be classified as stable, a structure must have the capacity to resist load from any direction.

To supply restraint against rotation, thereby stabilizing the column, the designer could do either of the following.

1. Replace the pin at A by a fixed support that can supply a restraining moment to the base of the column (see Figure 3.24*f*).
2. As shown in Figure 3.24*g*, connect the top of the column to a stable support at C with a horizontal member BC (a member such as BC, whose primary function is to align the column vertically and not to carry load, is termed *bracing*, or a *secondary member*).

In summary, we conclude that a structure is unstable if the supports supply less than three reactions.

Case 2. Supports Supply Three Reactions: $R = 3$

If supports supply three reactions, it will usually be possible to satisfy the three equations of equilibrium (the number of unknowns equals the number of equations). Obviously, if the three equations of static equilibrium are satisfied, the structure is in equilibrium (i.e., is *stable*). Further, if the equations of equilibrium are satisfied, the values of the three reactions are uniquely determined, and we say that the structure is *externally determinate*. Finally, since three equations of equilibrium must be satisfied, it follows that a minimum of three restraints are required to produce a stable structure under any loading condition.

If a system of supports supplies three reactions that are configured in such a way that the equations of equilibrium cannot be satisfied, the structure is called *geometrically unstable*. For example, in Figure 3.25*a*, member ABC, which carries a vertical load P and a horizontal force Q, is supported by a link and two rollers that apply three restraints to member ABC. Since all restraints act vertically, they offer no resistance to displacement in the horizontal direction (i.e., the reactions form a parallel force system). Writing the equation of equilibrium for beam ABC in the x direction, we find

$$\xrightarrow{+} \quad \Sigma F_x = 0$$

$$Q = 0 \qquad \text{(not consistent)}$$

Since Q is a real force and is not equal to zero, the equilibrium equation is not satisfied. Therefore, the structure is unstable. Under the action of force Q, the structure will move to the right until the link develops a horizontal component (because of a change in geometry) to equilibrate Q (see Figure 3.25*b*). Thus for it to be classified as a stable structure, we require that the applied loads be equilibrated by the original direction of the reactions in the unloaded structure. A structure that must undergo a change in geometry before its reactions are mobilized to balance applied loads is classified as unstable.

As a second example of an unstable structure restrained by three reactions, we consider in Figure 3.25*c* a beam supported by a pin at *A* and a roller at *B* whose reaction is directed horizontally. Although equilibrium in the *x* and *y* directions can be satisfied by the horizontal and vertical restraints supplied by the supports, the restraints are not positioned to prevent rotation of the structure about point *A*. Writing the equilibrium equation for moment about point *A* gives

$$\circlearrowright^{+} \quad \Sigma M_A = 0 \tag{3.4c}$$

$$Pa = 0 \qquad \text{(not consistent)}$$

Because neither P nor a is zero, the product Pa cannot equal zero. Thus an equation of equilibrium is not satisfied—a sign that the structure is unstable. Since the lines of action of all reactions pass through the pin at *A* (i.e., the reactions are equivalent to a *concurrent* force system), they are not able to prevent rotation initially.

In summary, we conclude that for a single rigid body a minimum of three restraints is necessary to produce a stable structure (one that is in equilibrium)—subject to the restriction that the restraints not be equivalent to either a parallel or a concurrent force system.

We have also demonstrated that the stability of a structure may always be verified by analyzing the structure with the equations of equilibrium for various arbitrary loading conditions. If the analysis produces an inconsistent result, that is, the equations of equilibrium are not satisfied for any portion of the structure, we can conclude the structure is unstable. This procedure is illustrated in Example 3.11.

Case 3. Restraints Greater Than 3: $R > 3$

If a system of supports, which is not equivalent to either a parallel or a concurrent force system, supplies more than three restraints to a *single* rigid structure, the values of the restraints cannot be uniquely determined because the number of unknowns exceeds the three equilibrium equations available for their solution. Since one or more of the reactions cannot be determined, the

Figure 3.25: (*a*) Geometrically unstable, reactions form a parallel force system; (*b*) equilibrium position, horizontal reaction develops as link elongates and changes slope; (*c*) geometrically unstable—reactions form a concurrent force system passing through the pin at *A*; (*d*) indeterminate beam.

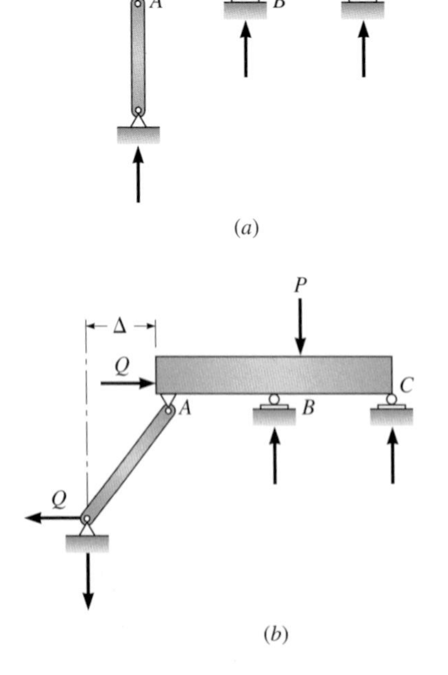

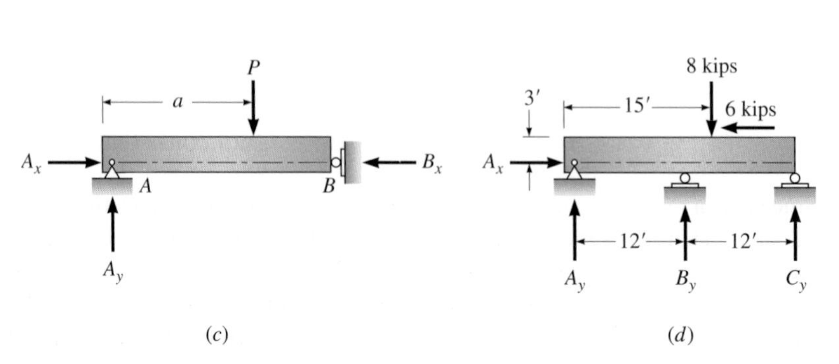

structure is termed *indeterminate*, and the *degree of indeterminacy* equals the number of restraints in excess of 3, that is,

$$\text{Degree of indeterminacy} = R - 3 \tag{3.7}$$

where R equals the number of reactions and 3 represents the number of equations of statics.

As an example, in Figure 3.25*d* a beam is supported by a pin at A and rollers at points B and C. Applying the three equations of equilibrium gives

$$\rightarrow + \quad \Sigma F_x = 0 \qquad A_x - 6 = 0$$

$$+\uparrow \quad \Sigma F_y = 0 \qquad -8 + A_y + B_y + C_y = 0$$

$$\circlearrowright + \quad \Sigma M_A = 0 \qquad -6(3) + 8(15) - 12B_y - 24C_y = 0$$

Since the four unknowns A_x, A_y, B_y, and C_y exist and only three equations are available, a complete solution (A_x can be determined from the first equation) is not possible, and we say that the structure is indeterminate to the first degree.

If the roller support at B were removed, we would have a stable determinate structure since now the number of unknowns would equal the number of equilibrium equations. This observation forms the basis of a common procedure for establishing the degree of indeterminacy. In this method we establish the degree of indeterminacy by removing restraints until a stable determinate structure remains. The number of restraints removed is equal to the degree of indeterminacy. As an example, we will establish the degree of indeterminacy of the beam in Figure 3.26*a* by removing restraints. Although a variety of choices are available, we first remove the rotational restraint (M_A) at support A

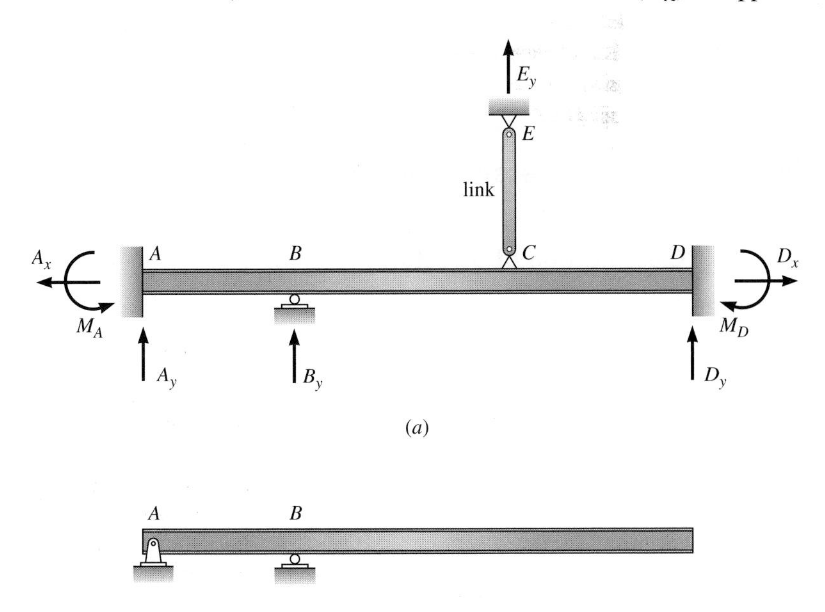

Figure 3.26: (*a*) Indeterminate structure; (*b*) base (or released) structure remaining after redundant supports removed.

but retain the horizontal and vertical restraint. This step is equivalent to replacing the fixed support with a pin. If we now remove the link at C and the fixed support at D, we have removed a total of five restraints, producing the stable, determinate *base* or *released* structure shown in Figure 3.26*b* (the restraints removed are referred to as *redundants*). Thus we conclude that the original structure was indeterminate to the fifth degree.

Determinacy and Stability of Structures Composed of Several Rigid Bodies

If a structure consists of several rigid bodies interconnected by devices (hinges, for example) that release C internal restraints, C additional equations of equilibrium (also called condition equations) can be written to solve for the reactions (see Sec. 3.7). For structures in this category, the criteria developed for establishing the stability and determinacy of a single rigid structure must be modified as follows:

1. If $R < 3 + C$, the structure is unstable.
2. If $R = 3 + C$ and if neither the reactions for the entire structure nor those for a component of the structure are equivalent to a parallel or a concurrent force system, the structure is stable and determinate.
3. If $R > 3 + C$ and the reactions are not equivalent to a parallel or a concurrent force system, the structure is stable and indeterminate; moreover, the degree of indeterminacy for this condition given by Equation 3.7 must be modified by subtracting from the number of reactions the number $(3 + C)$, which represents the number of equilibrium equations available to solve for the reactions; that is,

$$\text{Degree of indeterminacy} = R - (3 + C) \tag{3.8}$$

Table 3.2 summarizes the discussion of the influence of reactions on the stability and determinacy of structures.

TABLE 3.2*a*

Summary of the Criteria for Stability and Determinacy of a Single Rigid Structure

	Classification of Structure		
	Stable		
Condition*	Determinate	Indeterminate	Unstable
$R < 3$	—	—	Yes; three equations of equilibrium cannot be satisfied for all possible conditions of load
$R = 3$	Yes, if reactions are uniquely determined	—	Only if reactions form a parallel or concurrent force system
$R > 3$	—	Yes; degree of indeterminacy $= R - 3$	Only if reactions form a parallel or concurrent force system

*R is the number of reactions.

TABLE 3.2b

Summary of the Criteria for Stability and Determinacy of Several Interconnected Rigid Structures

	Classification of Structure		
	Stable		
Condition*	**Determinate**	**Indeterminate**	**Unstable**
$R < 3 + C$	—	—	Yes; equations of equilibrium cannot be satisfied for all possible loading conditions
$R = 3 + C$	Yes, if reactions can be uniquely determined	—	Only if reactions form a parallel or concurrent force system
$R > 3 + C$	—	Yes, degree of indeterminacy $= R - (3 + C)$	Only if reactions form a parallel or concurrent force system

*Here R is the number of reactions; C is the number of conditions.

EXAMPLE 3.11

Investigate the stability of the structure in Figure 3.27*a*. Hinges at joints B and D.

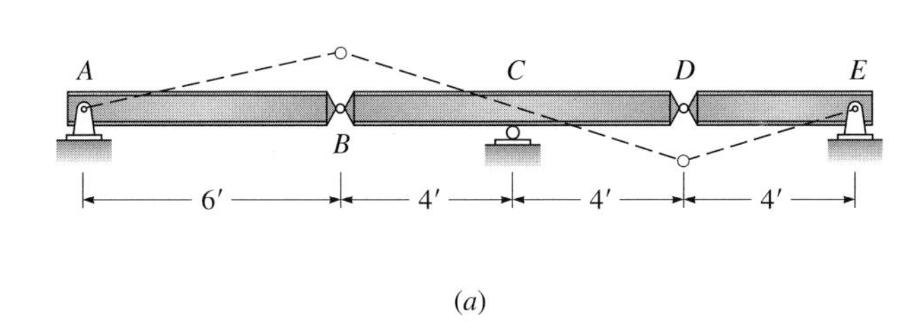

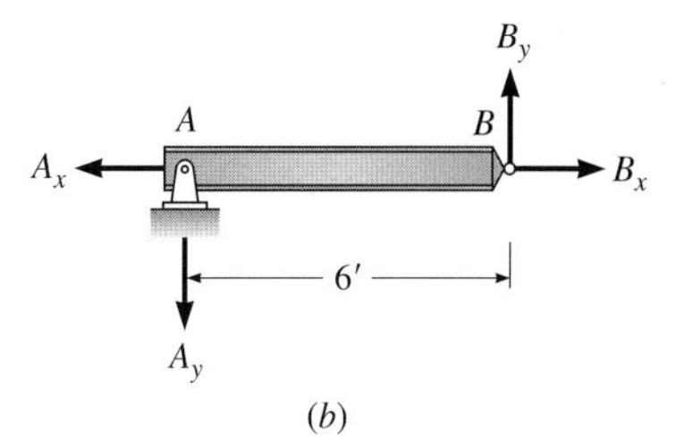

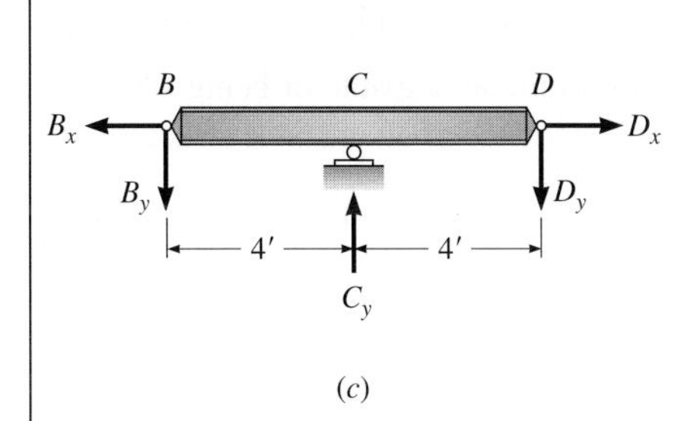

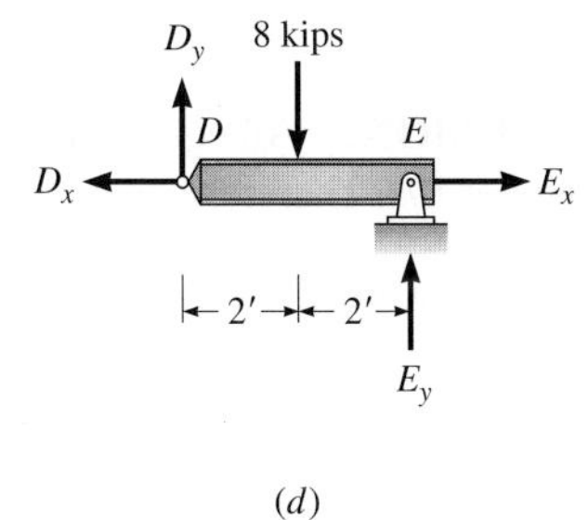

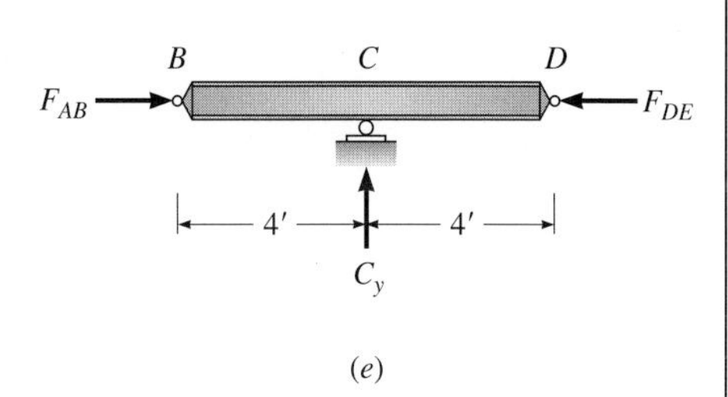

Figure 3.27: (*a*) Details of structure; (*b*) free body of member *AB*; (*c*) free body of member *BD*; (*d*) free body of member *DE*; (*e*) unstable structure (if *AB* and *DE* treated as links, i.e., reactions form a concurrent force system).

Solution

A necessary condition for stability requires

$$R = 3 + C$$

Since R, the number of reactions, equals 5 and C, the number of condition equations, equals 2, the necessary condition is satisfied. However, because

[*continues on next page*]

Example 3.11 continues . . .

the structure has so many hinges and pins, the possibility exists that the structure is geometrically unstable. To investigate this possibility, we will apply an arbitrary load to the structure to verify that the equations of equilibrium can be satisfied for each segment. Imagine that we apply a vertical load of 8 kips to the center of member *DE* (see Figure 3.27*d*).

STEP 1 Check the equilibrium of *DE*.

$$\xrightarrow{+} \quad \Sigma F_x = 0 \qquad E_x - D_x = 0$$

$$E_x = D_x$$

$$\circlearrowright^{+} \quad \Sigma M_D = 0 \qquad 8(2) - 4E_y = 0$$

$$E_y = 4 \text{ kips}$$

$$\overset{+}{\uparrow} \quad \Sigma F_y = 0 \qquad D_y + E_y - 8 = 0$$

$$D_y = 4 \text{ kips}$$

CONCLUSION. Although we were not able to determine either D_x or E_x, the equations of equilibrium are satisfied. Also, because the forces acting on the free body do not comprise either a parallel or a concurrent force system, there is no indication at this stage that the structure is unstable.

STEP 2 Check the equilibrium of member *BD* (see Figure 3.27*c*).

$$\circlearrowright^{+} \quad \Sigma M_c = 0 \qquad 4D_y - 4B_y = 0$$

$$B_y = D_y = 4 \text{ kips} \qquad \textbf{Ans.}$$

$$\xrightarrow{+} \quad \Sigma F_x = 0 \qquad D_x - B_x = 0$$

$$D_x = B_x$$

$$\overset{+}{\uparrow} \quad \Sigma F_y = 0 \qquad -B_y + C_y - D_y = 0$$

$$C_y = 8 \text{ kips} \qquad \textbf{Ans.}$$

CONCLUSION. All equations of equilibrium are capable of being satisfied for member *BD*. Therefore, there is still no evidence of an unstable structure.

STEP 3 Check the equilibrium of member *AB*. (See Figure 3.27*b*.)

$$\circlearrowright^{+} \quad \Sigma M_A = 0 \qquad 0 = -B_y(6) \quad \text{(inconsistent equation)}$$

CONCLUSION. Since previous computations for member *BD* established that $B_y = 4$ kips, the right side of the equilibrium equation equals -24 kip·ft—not zero. Therefore, the equilibrium equation is not satisfied, indicating that the structure is *unstable*. A closer examination of member *BCD* (see Figure 3.27*e*) shows that the structure is unstable because it is possible for the reactions supplied by members *AB* and *DE* and the roller *C* to form a concurrent force system. The dashed line in Figure 3.27*a* shows one possible deflected shape of the structure as an unstable mechanism.

3.9 Classifying Structures

One of the major goals of this chapter is to establish guidelines for constructing a stable structure. In this process we have seen that the designer must consider both the geometry of the structure and the number, position, and type of supports supplied. To conclude this section, we will examine the structures in Figure 3.28 and Figure 3.29 to establish if they are stable or unstable with respect to external reactions. For those structures that are stable, we will also establish if they are determinate or indeterminate. Finally, if a structure is indeterminate, we will establish the degree of indeterminacy. All the structures in this section will be treated as a single rigid body that may or may not contain devices that release internal restraints. The effect of internal hinges or rollers will be taken into account by considering the number of associated condition equations.

In the majority of cases, to establish if a structure is determinate or indeterminate, we simply compare the number of external reactions to the equilibrium equations available for the solution—that is, three equations of statics plus any condition equations. Next, we check for stability by verifying that the reactions are not equivalent to a *parallel* or a *concurrent* force system. If any doubt still exists, as a final test, we apply a load to the structure and carry out an analysis using the equations of static equilibrium. If a solution is possible—indicating that the equations of equilibrium are satisfied—the structure is stable. Alternatively, if an inconsistency develops, we recognize that the structure is unstable.

In Figure 3.28*a* the beam is restrained by four reactions—three at the fixed support and one at the roller. Since only three equations of equilibrium are available, the structure is indeterminate to the first degree. The structure is

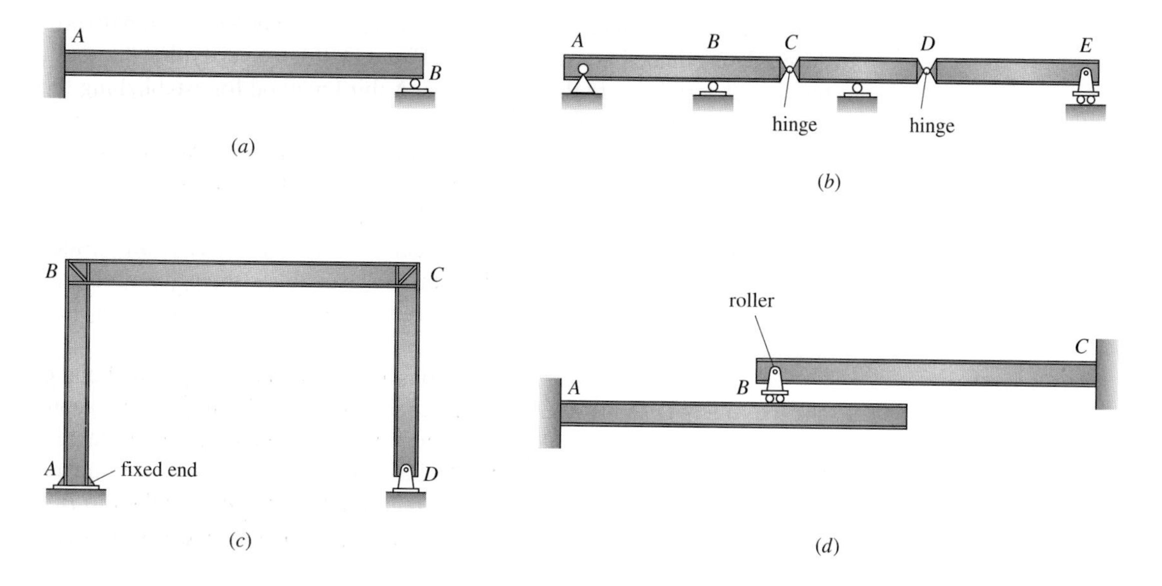

Figure 3.28: Examples of stable and unstable structures: (*a*) indeterminate to first degree; (*b*) stable and determinate; (*c*) indeterminate second degree; (*d*) indeterminate to first degree.

obviously stable since the reactions are not equivalent to either a parallel or a concurrent force system.

The structure in Figure 3.28*b* is stable and determinate because the number of reactions equals the number of equilibrium equations. Five reactions are supplied—two from the pin at *A* and one from each the three rollers. To solve for the reactions, three equations of equilibrium are available for the entire structure, and the hinges at *C* and *D* supply two condition equations. We can also deduce that the structure is stable by observing that member *ABC*—supported by a pin at *A* and a roller at *B*—is stable. Therefore, the hinge at *C*, which is attached to member *ABC*, is a stable point in space and, like a pin support, can apply both a horizontal and vertical restraint to member *CD*. The fact that the hinge at *C* may undergo a small displacement due to the elastic deformations of the structure does not affect its ability to restrain member *CD*. Since a third restraint is supplied to *CD* by the roller at midspan, we conclude that it is a stable element; that is, it is supported by three restraints that are equivalent to neither a parallel nor a concurrent force system. Recognizing that the hinge at *D* is attached to a stable structure, we can see that member *DE* is also supported in a stable manner, that is, two restraints from the hinge and one from the roller at *E*.

Figure 3.28*c* shows a rigid frame restrained by a fixed support at *A* and a pin at *D*. Since three equations of equilibrium are available but five restraints are applied by the supports, the structure is indeterminate to the second degree.

The structure in Figure 3.28*d* consists of two cantilever beams joined by a roller at *B*. If the system is treated as a single rigid body, the fixed supports at *A* and *C* supply a total of six restraints. Since the roller provides two equations of condition (the moment at *B* is zero and no horizontal force can be transmitted through joint *B*) and three equations of statics are available, the structure is indeterminate to the first degree. As a second approach, we could establish the degree of indeterminacy by removing the roller at *B*, which supplies a single vertical reaction, to produce two stable determinate cantilever beams. Since it was necessary to remove only one restraint to produce a determinate base structure (see Figure 3.26), we verify that the structure is indeterminate to the first degree. A third method for establishing the degree of indeterminacy would be to separate the structure into two free-body diagrams and to count the unknown reactions applied by the supports and the internal roller. Each free body would be acted upon by three reactions from the fixed supports at *A* or *C* as well as a single vertical reaction from the roller at *B*—a total of seven reactions for the two free bodies. Since a total of six equations of equilibrium are available—three for each free body—we again conclude that the structure is indeterminate to the first degree.

In Figure 3.29*a* six *external* reactions are supplied by the pins at *A* and *C* and the rollers at *D* and *E*. Since three equations of equilibrium and two condition equations are available, the structure is indeterminate to the first degree. Beam *BC*, supported by a pin at *C* and a roller at *B*, is a stable determinate component of the structure; therefore, regardless of the load applied to *BC*, the vertical reaction at the roller at *B* can always be computed. The structure

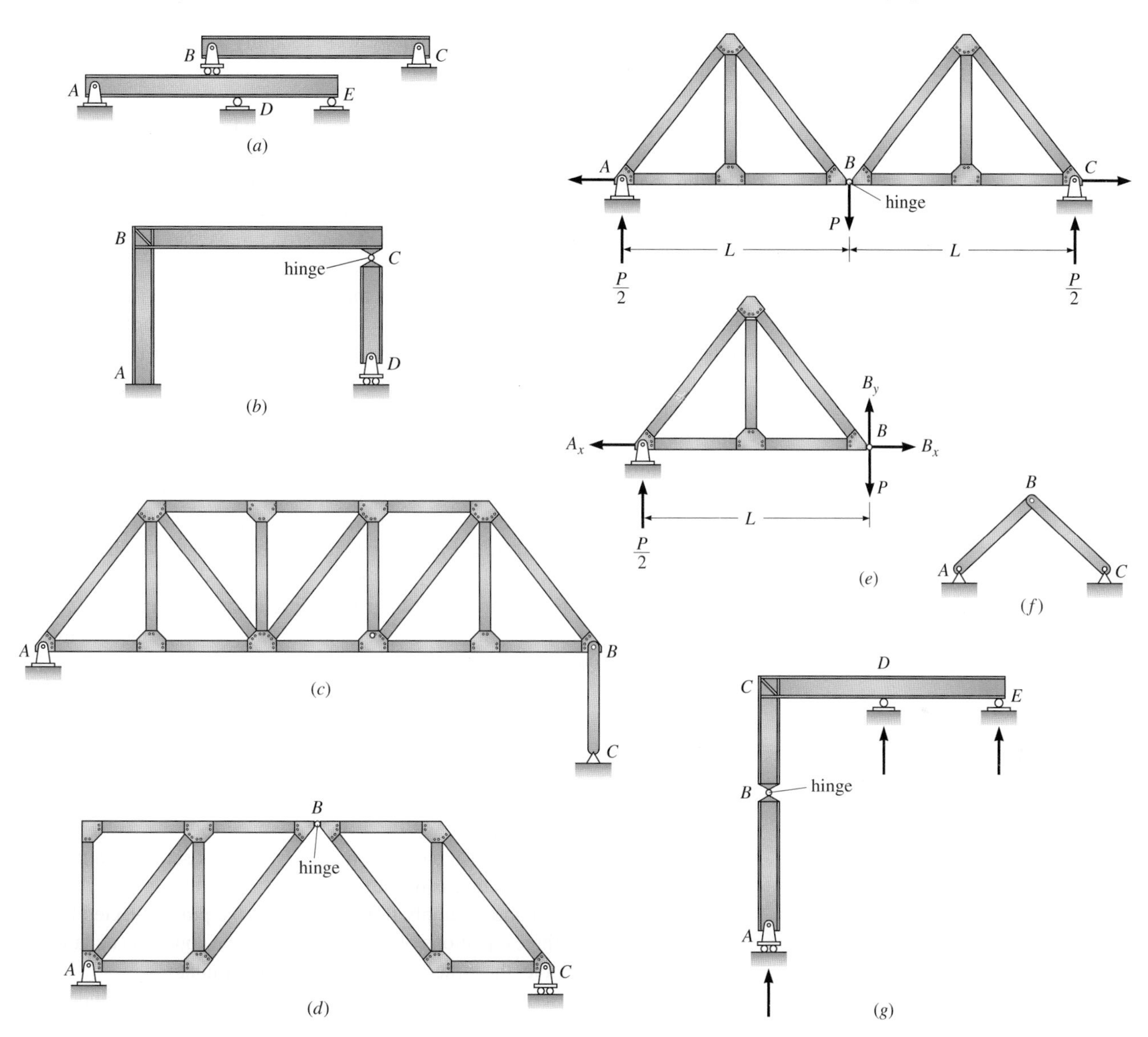

Figure 3.29: (*a*) Indeterminate first degree; (*b*) unstable—reactions applied to *CD* form a concurrent force system; (*c*) stable and determinate; (*d*) unstable $R < 3 + C$; (*e*) unstable, reactions applied to each truss form a concurrent force system; (*f*) stable and indeterminate; (*g*) unstable, reactions on *BCDE* equivalent to a parallel force system.

is indeterminate because member *ADE* is restrained by four reactions—two from the pin at *A* and one each from the rollers at *D* and *E*.

The frame in Figure 3.29*b* is restrained by four reactions—three from the fixed support *A* and one from the roller at *D*. Since three equilibrium equations and one condition equation ($M_c = 0$ from the hinge at *C*) are available, it appears that the structure may be stable and determinate. However, while member *ABC* is definitely stable because it consists of a single L-shaped member connected to a fixed support at *A*, member *CD* is not supported in a stable manner because the vertical reaction from the roller at *D* passes through the hinge at *C*. Thus the reactions applied to member *CD* make up a concurrent force system, indicating that the member is unstable. For example, if we were to apply a horizontal force to member *CD* and then sum moments about the hinge at *C*, an inconsistent equilibrium equation would result.

In Figure 3.29*c* a truss, which may be considered a rigid body, is supported by a pin at *A* and a link *BC*. Since the reactions apply three restraints that are equivalent to neither a parallel nor a concurrent force system, the structure is externally stable and determinate. (As we will show in Chap. 4 when we examine trusses in greater detail, the structure is also internally determinate.)

In Figure 3.29*d* we consider a truss that is composed of two rigid bodies joined by a hinge at *B*. Considering the structure as a unit, we note that the supports at *A* and *C* supply three restraints. However, since four equilibrium equations must be satisfied (three for the structure plus a condition equation at *B*), we conclude that the structure is unstable; that is, there are more equations of equilibrium than reactions.

Treating the truss in Figure 3.29*e* as a single rigid body containing a hinge at *B*, we find that the pins at *A* and *C* supply four reactions. Since three equations of equilibrium are available for the entire structure and one condition equation is supplied by the hinge at *B*, the structure appears to be stable and determinate. However, if a vertical load *P* were applied to the hinge at *B*, symmetry requires that vertical reactions of *P*/2 develop at both supports *A* and *C*. If we now take out the truss between *A* and *B* as a free body and sum moments about the hinge at *B*, we find

$$\circlearrowright^{+} \quad \Sigma M_B = 0$$

$$\frac{P}{2}L = 0 \qquad \text{(inconsistent)}$$

Thus we find that the equilibrium equation $\Sigma M_B = 0$ is not satisfied, and we now conclude that the structure is unstable.

Since the pins at *A* and *C* supply four reactions to the pin-connected bars in Figure 3.29*f*, and three equations of equilibrium and one condition equation (at joint *B*) are available, the structure is stable and determinate.

In Figure 3.29*g* a rigid frame is supported by a link (member *AB*) and two rollers. Since all reactions applied to member *BCDE* act in the vertical direction (they constitute a parallel force system), member *BCDE* has no capacity to resist horizontal load, and we conclude that the structure is unstable.

3.10 Comparison between Determinate and Indeterminate Structures

Since determinate and indeterminate structures are used extensively, it is important that designers be aware of the difference in their behavior in order to anticipate problems that might arise during construction or later when the structure is in service.

If a determinate structure loses a support, immediate failure occurs because the structure is no longer stable. An example of the collapse of a bridge composed of simply supported beams during the 1964 Nigata earthquake is shown in Photo 3.3. As the earthquake caused the structure to sway, in each span the ends of the beams that were supported on rollers slipped off the piers and fell into the water. *Had the ends of girders been continuous or connected, the bridge in all probability would have survived with minimum damage.* In response to the collapse of similar, simply supported highway bridges in California during earthquakes, design codes have been modified to ensure that bridge girders are connected at supports.

On the other hand, in an indeterminate structure alternative paths exist for load to be transmitted to supports. Loss of one or more supports in an indeterminate structure can still leave a stable structure as long as the remaining supports supply three or more restraints properly arranged. Although loss of a support in an indeterminate structure can produce in certain members a significant increase in stress that can lead to large deflections or even to a partial failure locally, a carefully detailed structure, which behaves in a ductile manner, may have sufficient strength to resist complete collapse. Even though a damaged, deformed structure may no longer be functional, its occupants will probably escape injury.

During World War II, when cities were bombed or shelled, a number of buildings with highly indeterminate frames remained standing even though primary structural members—beams and columns—were heavily damaged or destroyed. For example, if support C in Figure 3.30a is lost, the stable, determinate

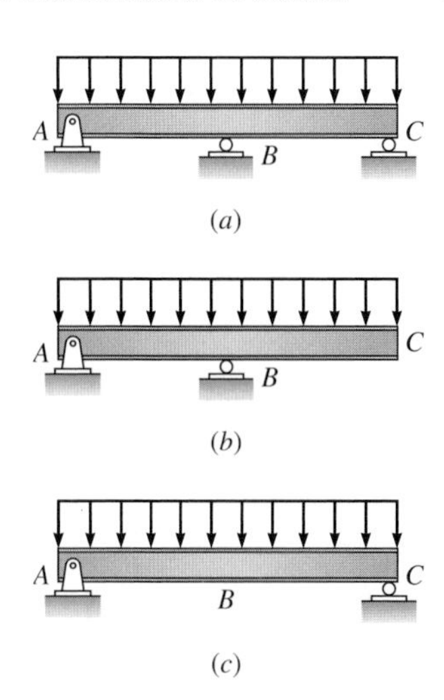

Figure 3.30: Alternative modes of transmitting load to supports.

Photo 3.3: An example of the collapse of a bridge composed of simply supported beams during the 1964 Nigata earthquake is shown here.

cantilever beam shown in Figure 3.30*b* remains. Alternatively, loss of support *B* leaves the stable simple beam shown in Figure 3.30*c*.

Indeterminate structures are also stiffer than determinate structures of the same span because of the additional support supplied by the extra restraints. For example, if we compare the magnitude of the deflections of two beams with identical properties in Figure 3.31, we will find that the midspan deflection of the simply supported determinate beam is 5 times larger than that of the indeterminate fixed-end beam. Although the vertical reactions at the supports are the same for both beams, in the fixed-end beam, negative moments at the end supports resist the vertical displacements produced by the applied load.

Since indeterminate structures are more heavily restrained than determinate structures, support settlements, creep, temperature change, and fabrication errors may increase the difficulty of erection during construction or may produce undesirable stresses during the service life of the structure. For example, if girder *AB* in Figure 3.32*a* is fabricated too long or increases in length due to a rise in temperature, the bottom end of the structure will extend beyond the support at *C*. In order to erect the frame the field crew, using jacks or other loading devices, must deform the structure until it can be connected to its supports (see Figure 3.32*b*). As a result of the erection procedure, the members will be stressed and reactions will develop even when no loads are applied to the structure.

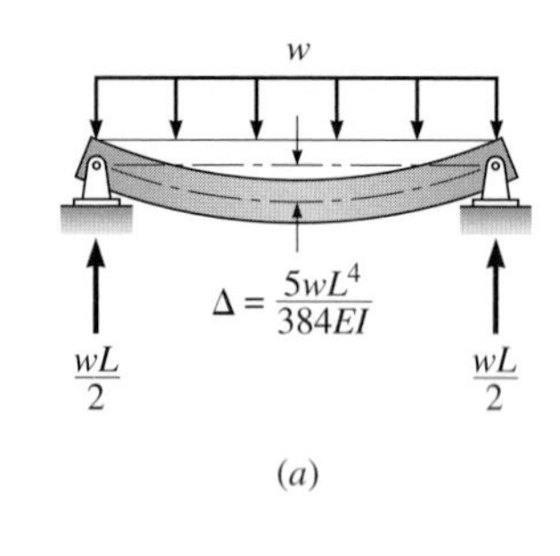

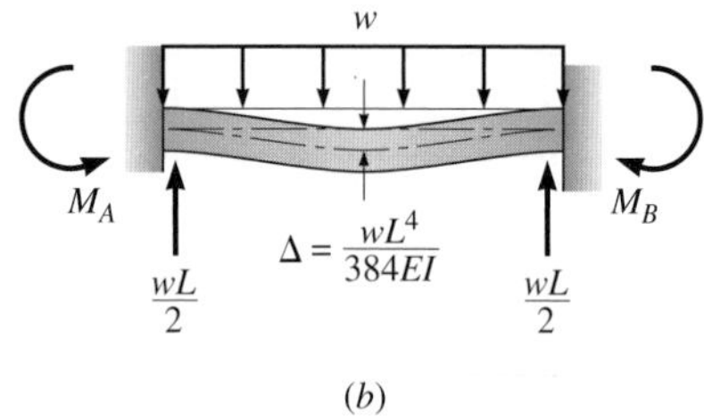

Figure 3.31: Comparison of flexibility between a determinate and indeterminate structure. Deflection of determinate beam in (*a*) is 5 times greater than indeterminate beam in (*b*).

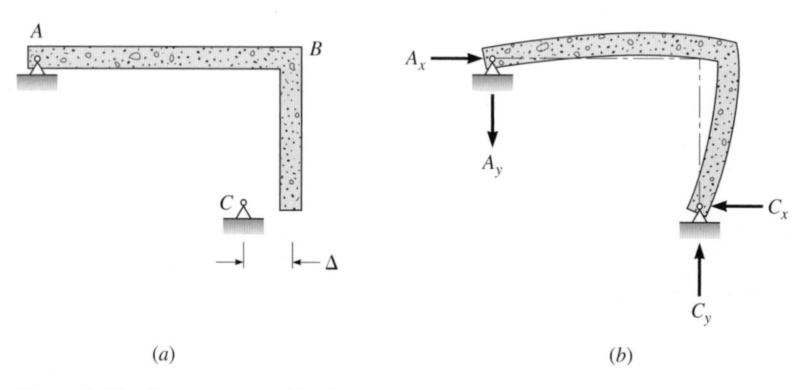

Figure 3.32: Consequences of fabrication error: (*a*) column extends beyond support because girder is too long; (*b*) reactions produced by forcing the bottom of the column into the supports.

Figure 3.33 shows the forces that develop in a continuous beam when the center support settles. Since no load acts on the beam—neglecting the beam's own weight—a set of self-balancing reactions is created. If this were a reinforced concrete beam, the moment created by the support settlement when added to those produced by the service loads could produce a radical change in the design moments at critical sections. Depending on how the beam is reinforced, the changes in moment could overstress the beam or produce extensive cracking at certain sections along the axis of the beam.

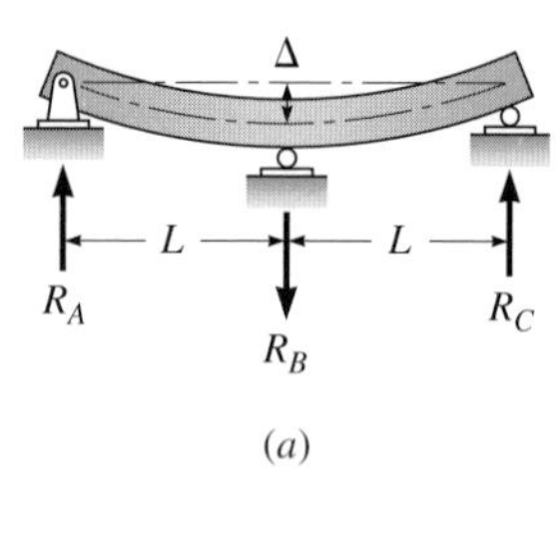

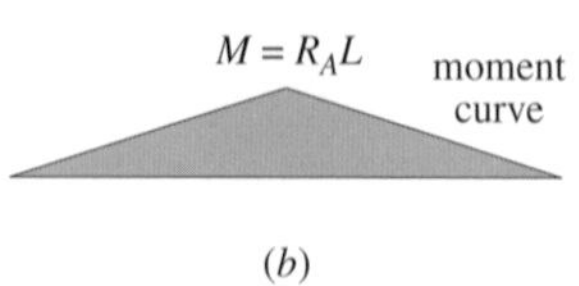

Figure 3.33: (*a*) Support *B* settles, creating reactions; (*b*) moment curve produced by support settlement.

Summary

- Since most loaded structures are at rest and restrained against displacements by their supports, their behavior is governed by the laws of statics, which for planar structures can be stated as follows:

$$\Sigma F_x = 0$$

$$\Sigma F_y = 0$$

$$\Sigma M_o = 0$$

- Planar structures whose reactions and internal forces can be determined by applying these three equations of statics are called *determinate structures*. Highly restrained structures that cannot be analyzed by the three equations of statics are termed *indeterminate structures*. These structures require additional equations based on the geometry of the deflected shape. If the equations of statics cannot be satisfied for a structure or any part of a structure, the structure is considered unstable.
- Designers use a variety of symbols to represent actual supports as summarized in Table 3.1. These symbols represent the primary action of a particular support; but to simplify analysis, neglect small secondary effects. For example, a pin support is assumed to apply restraint against displacement in any direction but to provide no rotational restraint when, in fact, it may supply a small degree of rotational restraint because of friction in the joint.
- Because indeterminate structures have more supports or members than the minimum required to produce a stable determinate structure, they are therefore generally stiffer than determinate structures and less likely to collapse if a single support or member fails.
- Analysis by computer is equally simple for both determinate and indeterminate structures. However, if a computer analysis produces illogical results, designers should consider the strong possibility they are analyzing an unstable structure.

PROBLEMS

P3.1 to **P3.6.** Determine the reactions of each structure in Figures P3.1 to P3.6.

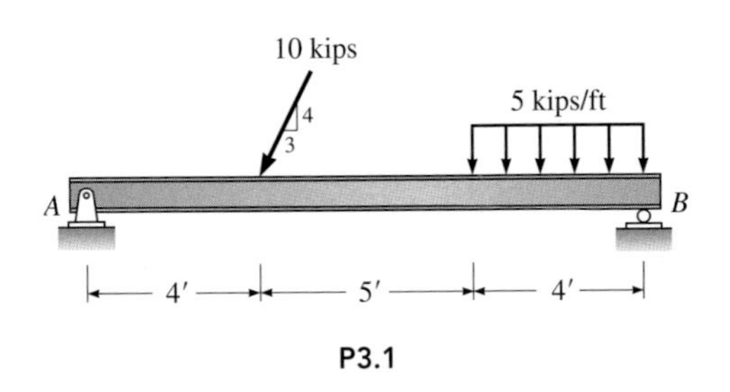

P3.1

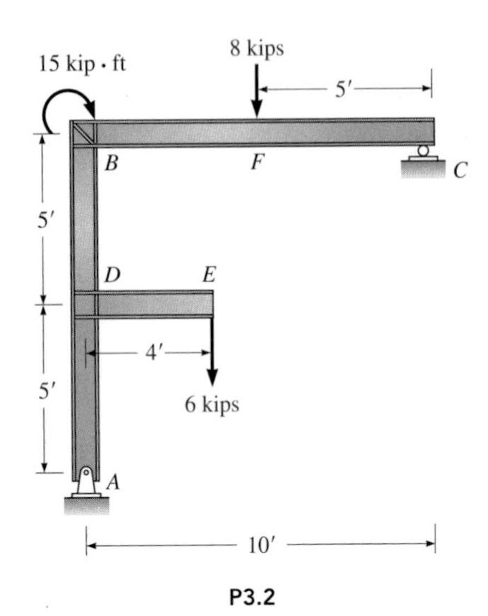

P3.2

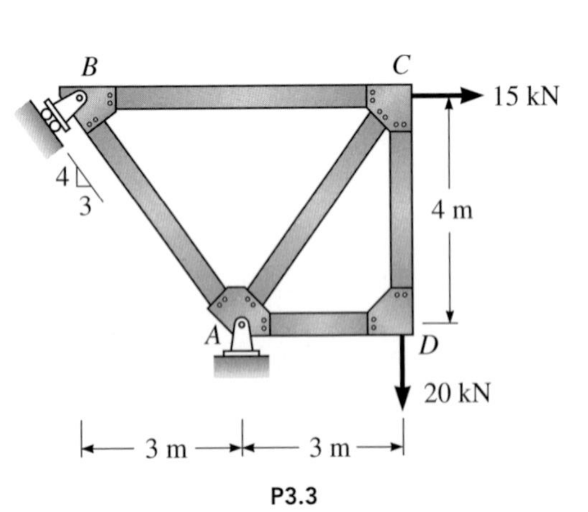

P3.3

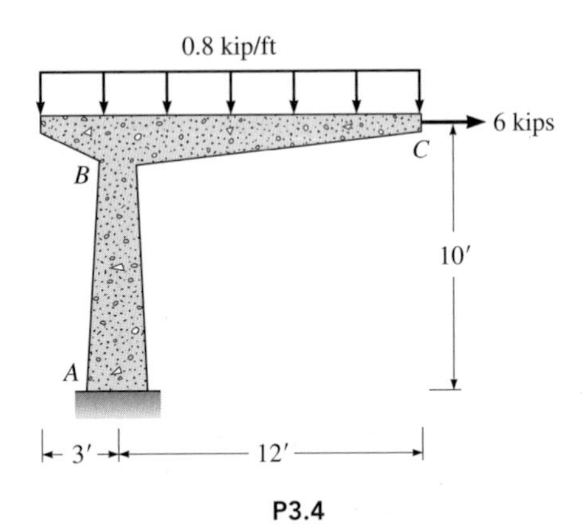

P3.4

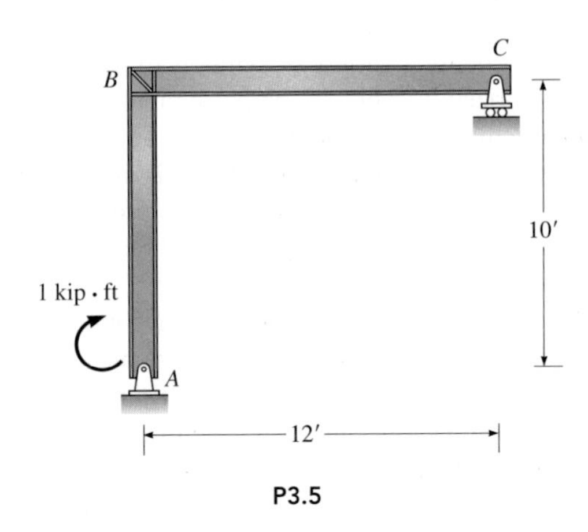

P3.5

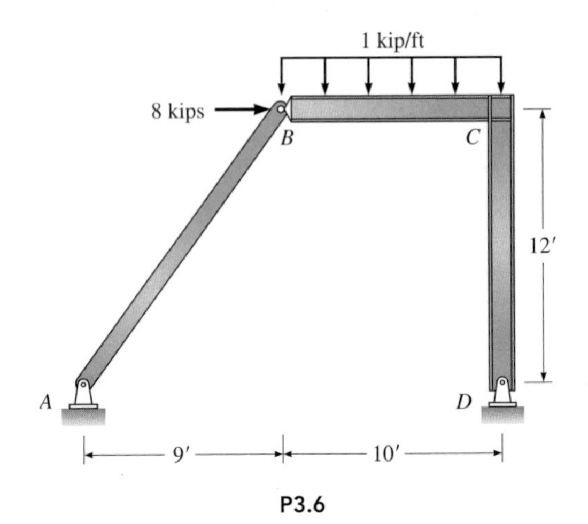

P3.6

P3.7. The support at A prevents rotation and horizontal displacement, but permits vertical displacement. The shear plate at B is assumed to act as a hinge. Determine the moment at A and the reactions at C and D.

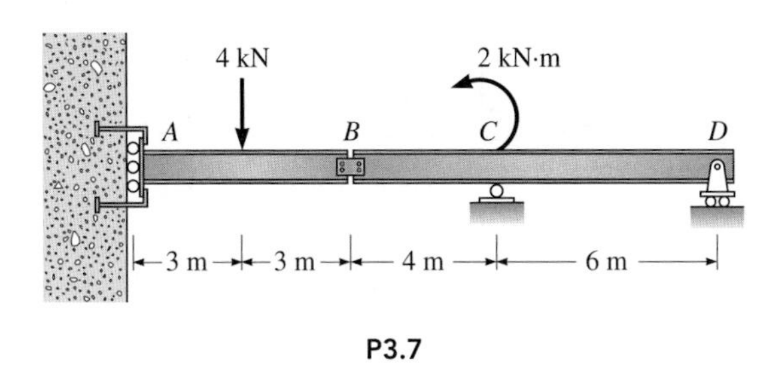

P3.7

P3.8. Determine the reactions at all supports and the force transmitted through the hinge at B.

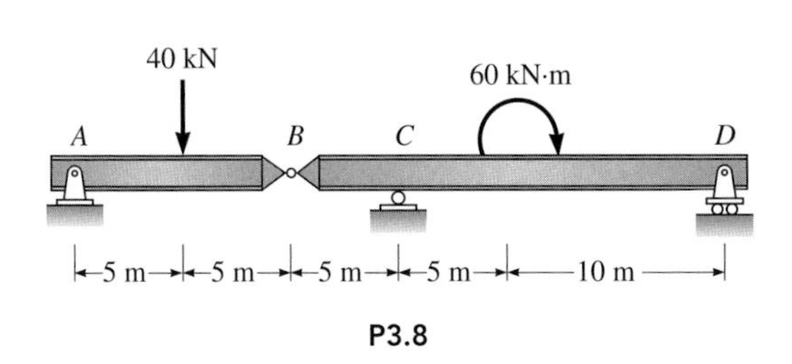

P3.8

P3.9 to **P3.11.** Determine the reactions for each structure. All dimensions are measured from the centerlines of members.

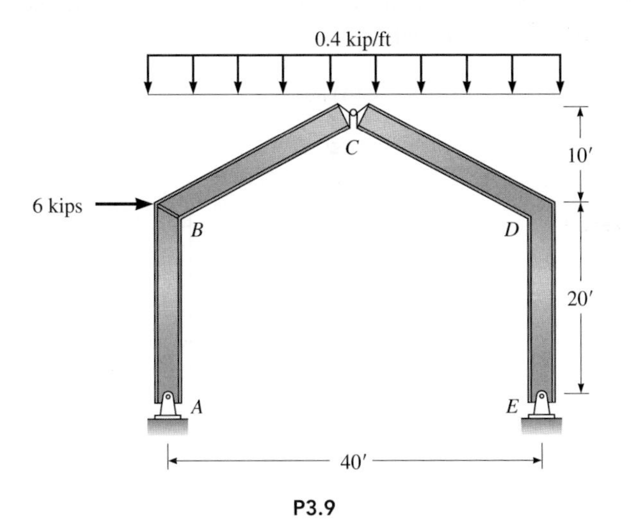

P3.9

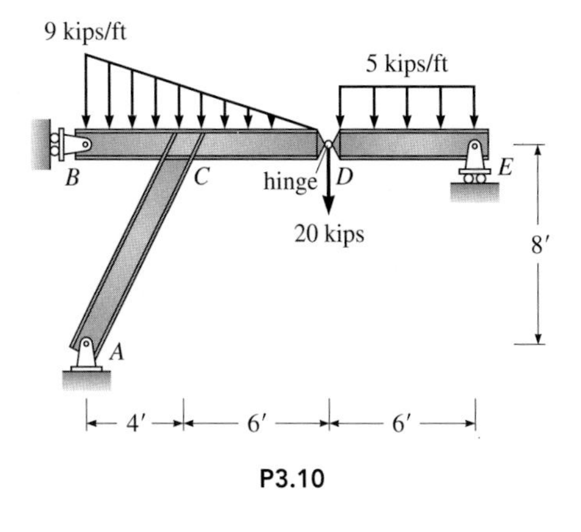

P3.10

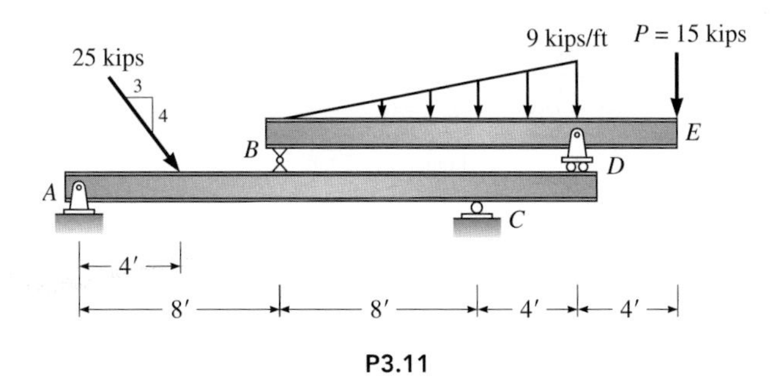

P3.11

P3.12. Determine all reactions. The pin joint at C can be treated as a hinge.

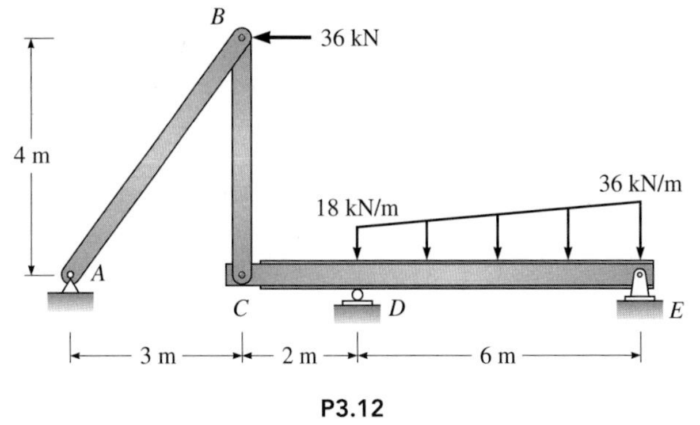

P3.12

P3.13. Determine all reactions. The pin joint at D acts as a hinge.

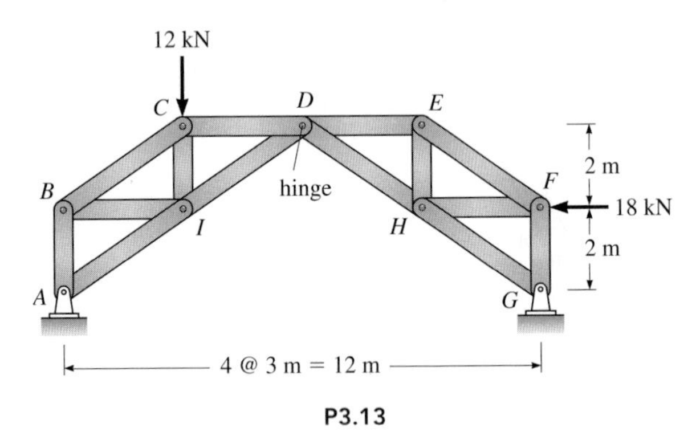

P3.13

P3.14. Determine the reactions at all supports and the force transmitted through the hinge at C.

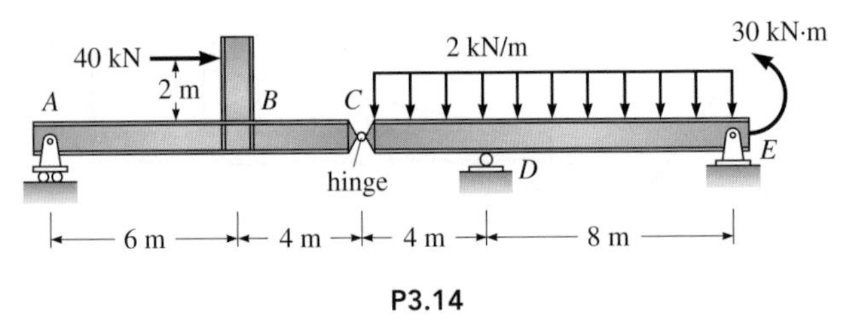

P3.14

P3.15. Determine the reactions at supports A, C, and E.

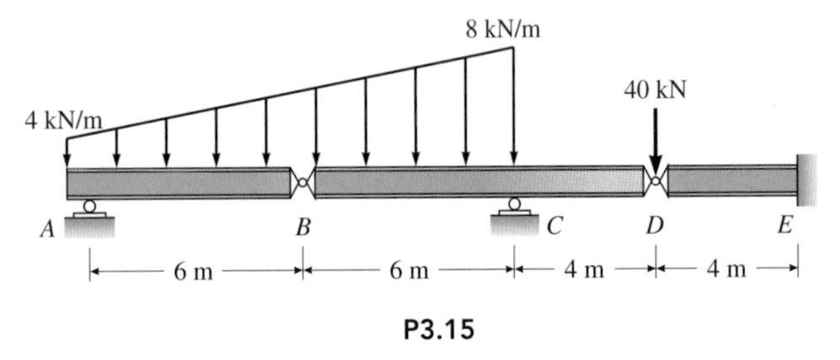

P3.15

P3.16. Determine all reactions. Joint C can be assumed to act as a hinge.

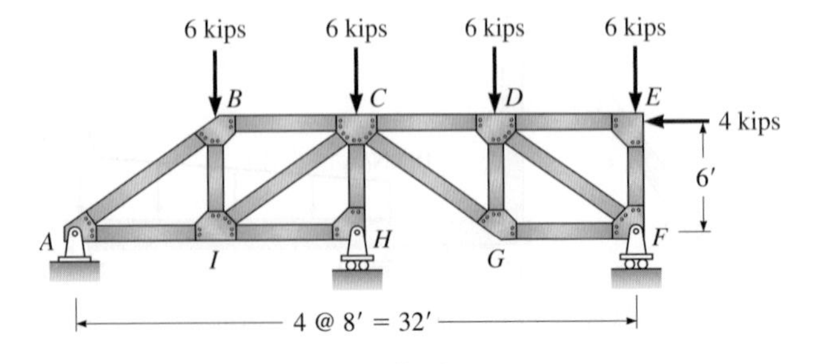

P3.16

P3.17. Determine all reactions. The uniform load on all girders extends to the centerlines of the columns.

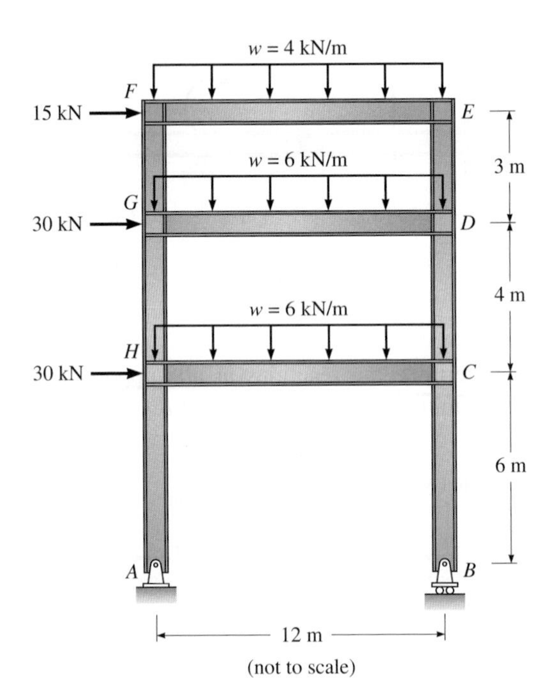

P3.17

P3.18. The bent frame $BCDE$ in figure P3.18 is laterally braced by member AC, which acts like a link. Determine reactions at A, B and E.

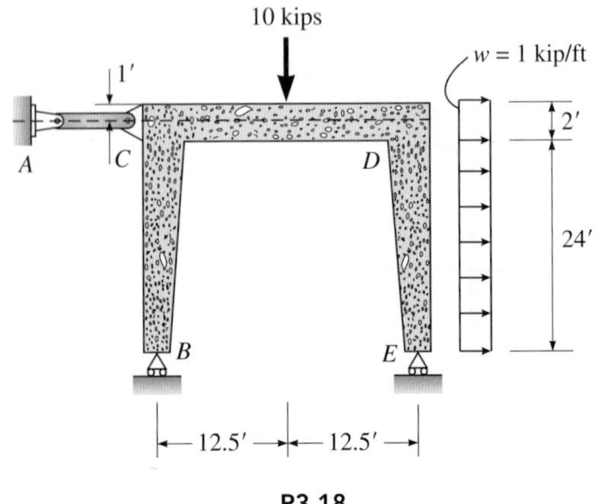

P3.18

P3.19 to **P3.21.** Determine all reactions.

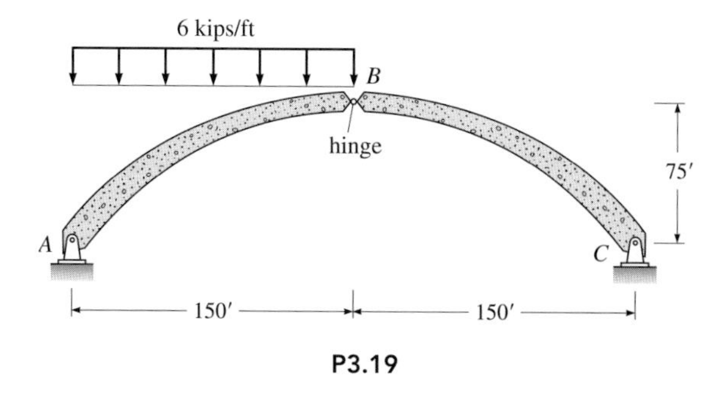

P3.19

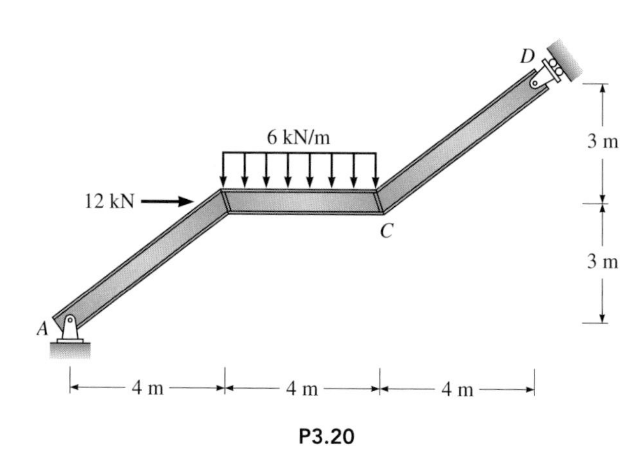

P3.20

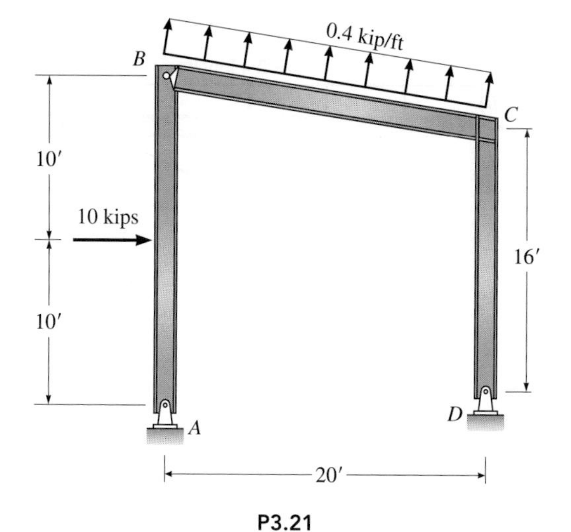

P3.21

P3.22. Determine all reactions. The pin joint at E acts as a hinge.

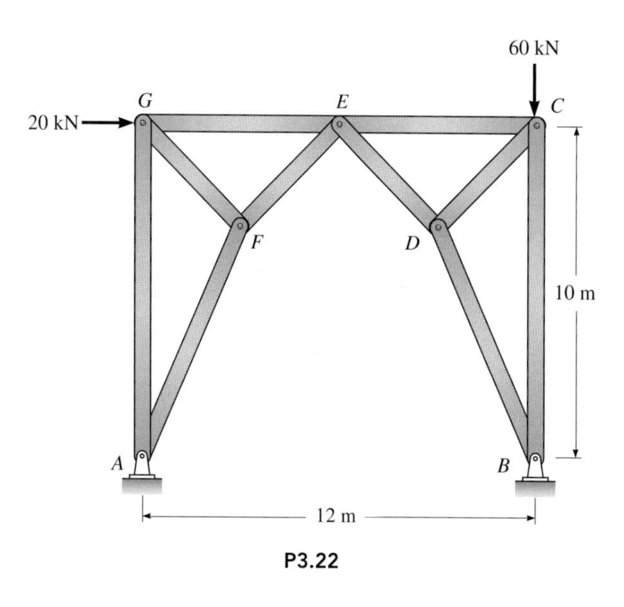

P3.22

P3.23. The roof truss is bolted to a reinforced masonry pier at A and connected to an elastomeric pad at C. The pad, which can apply vertical restraint in either direction but no horizontal restraint, can be treated as a roller. The support at A can be treated as a pin. Compute the reactions at supports A and C produced by the wind load.

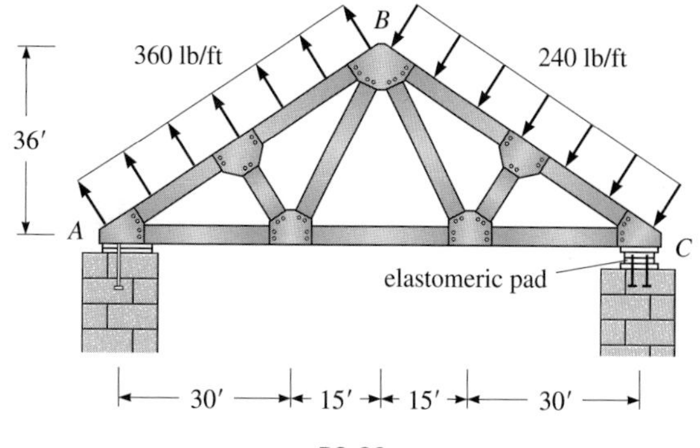

P3.23

P3.24. The clip angle connecting the beam's web at A to the column may be assumed equivalent to a pin support. Assume member BD acts as an axially loaded pin-end compression strut. Compute the reactions at points A and D.

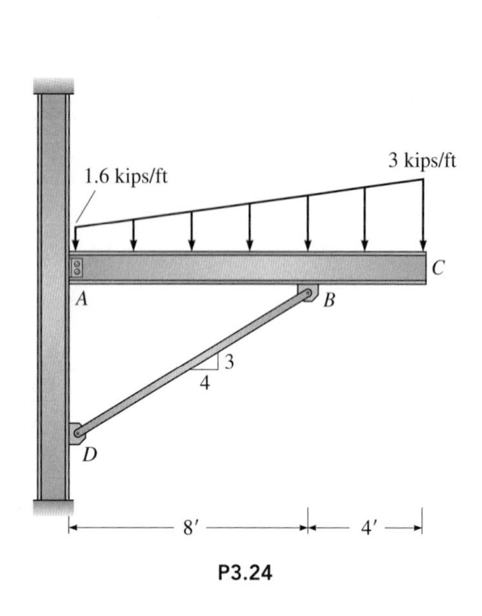

P3.24

P3.25. Compute all reactions.

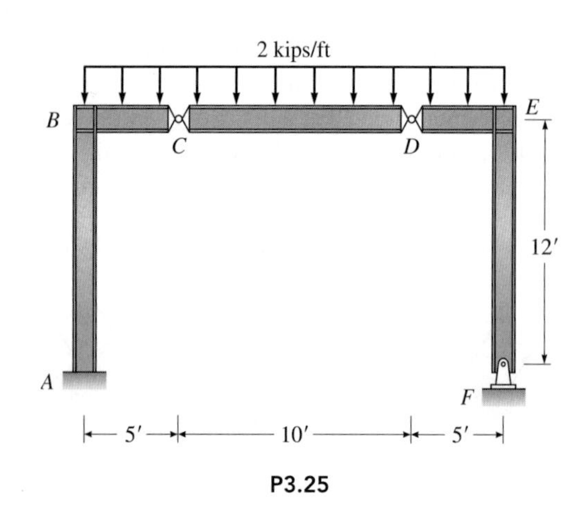

P3.25

P3.26. Compute the reactions at supports A and G, and the force applied by the hinge to member AD.

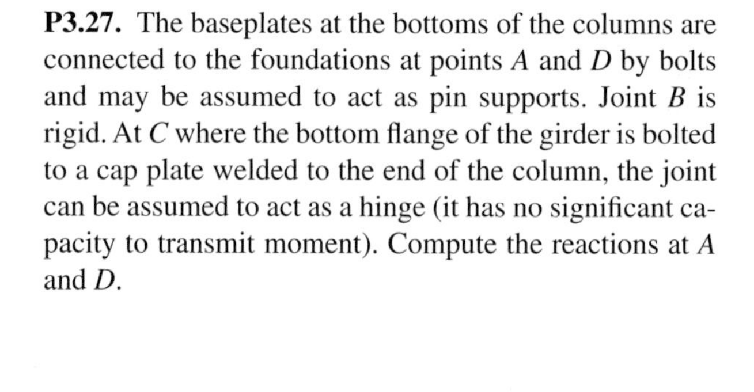

P3.26

P3.27. The baseplates at the bottoms of the columns are connected to the foundations at points A and D by bolts and may be assumed to act as pin supports. Joint B is rigid. At C where the bottom flange of the girder is bolted to a cap plate welded to the end of the column, the joint can be assumed to act as a hinge (it has no significant capacity to transmit moment). Compute the reactions at A and D.

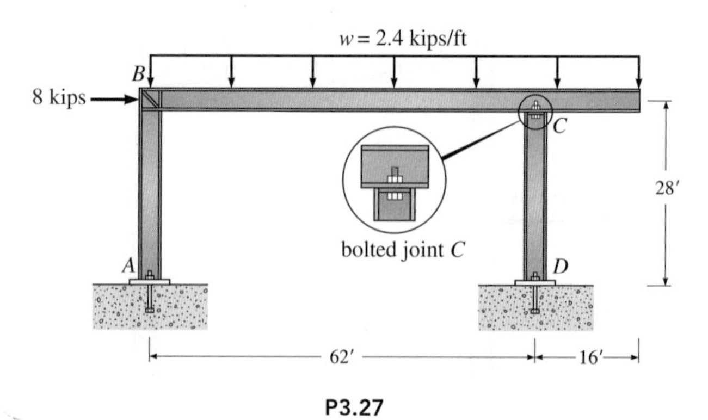

P3.27

P3.28. Draw free-body diagrams of column AB and beam BC and joint B by passing cutting planes through the rigid frame an infinitesimal distance above support A and to the right and immediately below joint B. Evaluate the internal forces on each free body.

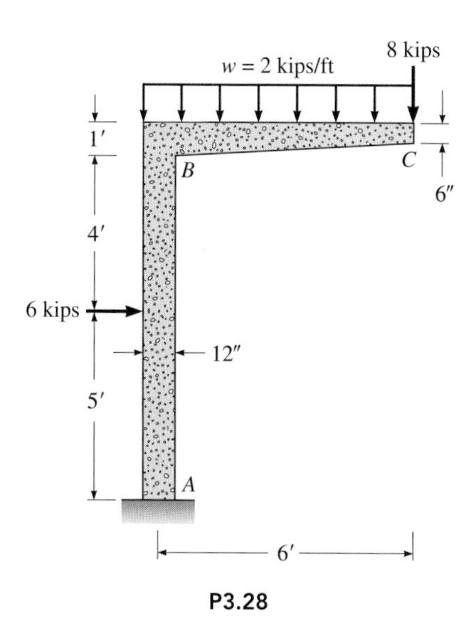

P3.28

P3.29. The frame is composed of members connected by frictionless pins. Draw free-body diagrams of each member and determine the forces applied by the pins to the members.

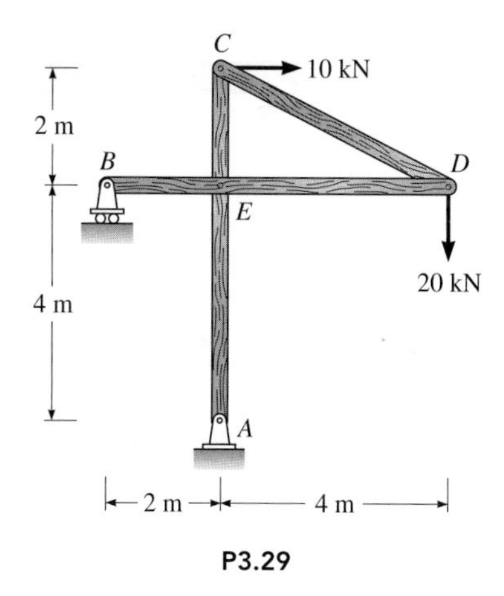

P3.29

P3.30. The truss in Figure P3.30 is composed of pin-jointed members that carry only axial load. Determine the forces in members, a, b, and c by passing vertical section 1-1 through the center of the truss.

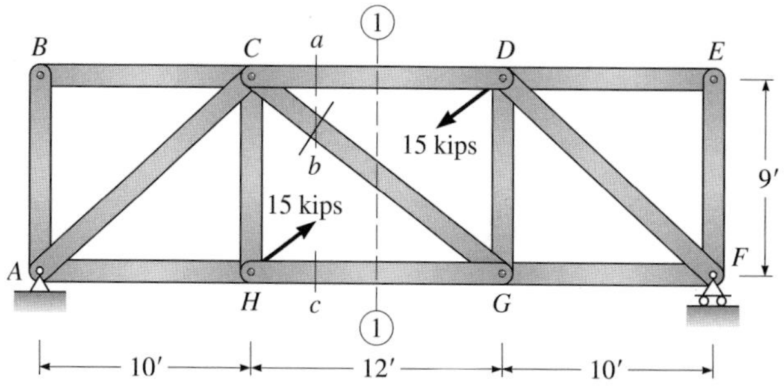

P3.30

P3.31. (a) in Figure P3.31 trusses 1 and 2 are stable elements that can be treated as rigid bodies. Compute all reactions. (b) Draw free-body diagrams of each truss and evaluate the forces applied to the trusses at joints C, B, and D.

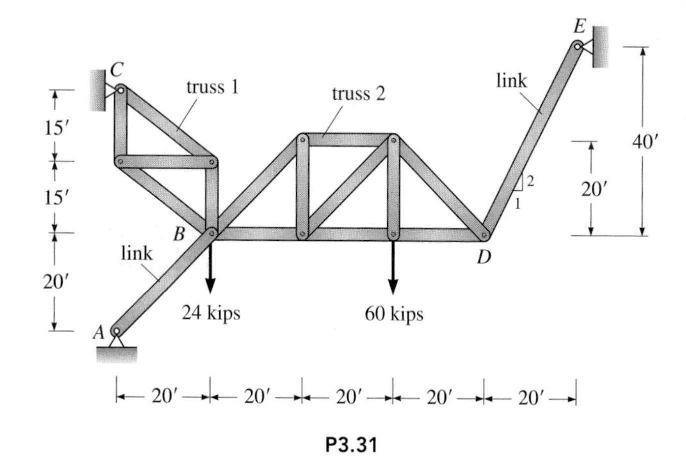

P3.31

P3.32 and **P3.33.** Classify the structures in Figures P3.32 and P3.33. Indicate if stable or unstable. If unstable, indicate the reason. If the structure is stable, indicate if determinate or indeterminate. If indeterminate, specify the degree.

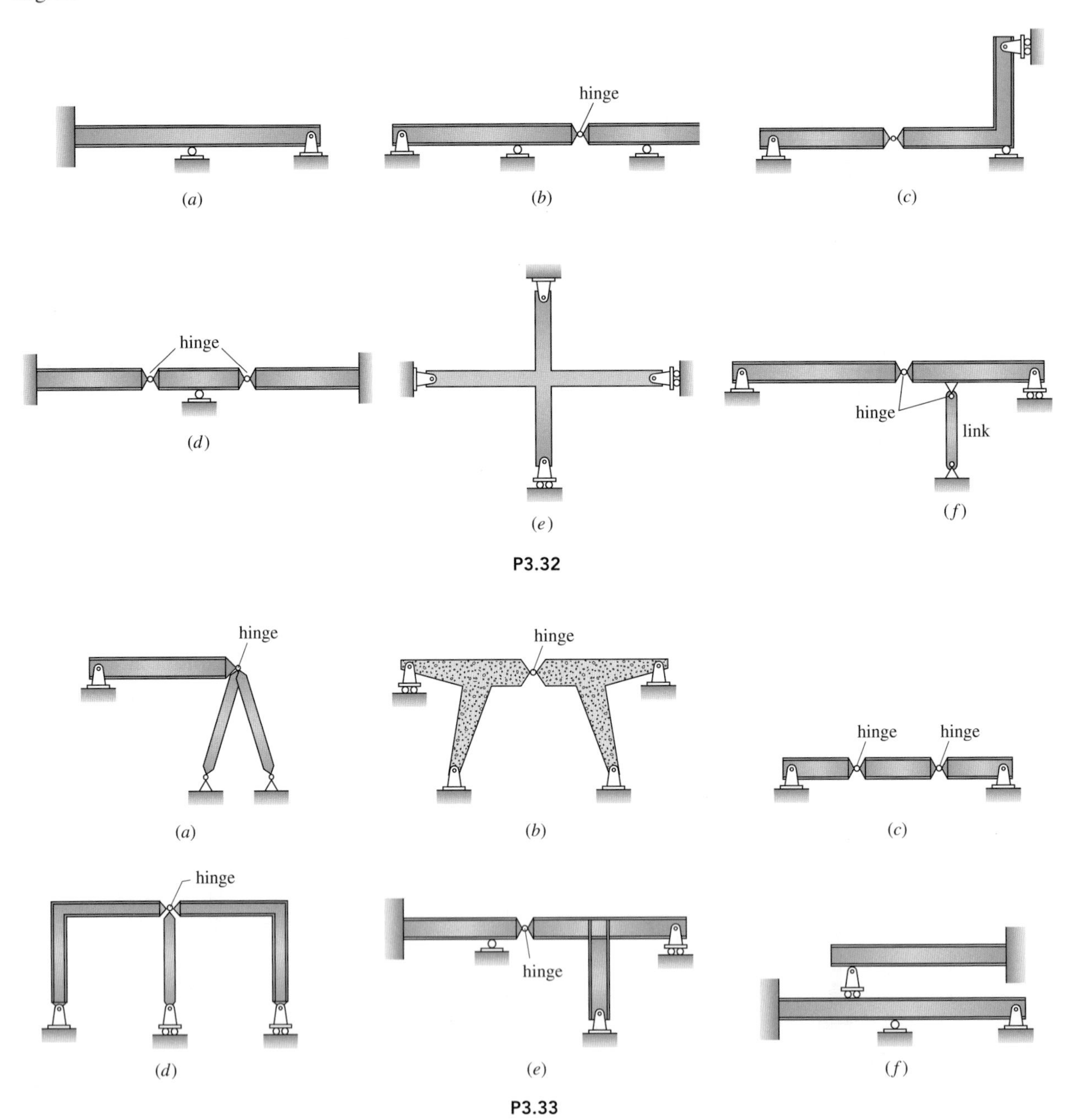

P3.32

P3.33

P3.34. *Practical application*: A one-lane bridge consists of a 10-in-thick, 16-ft-wide reinforced concrete slab supported on two steel girders spaced 10 ft apart. The girders are 62 ft long and weigh 400 lb/ft. The bridge is to be designed for a uniform live load of 700 lb/ft acting over the entire length of the bridge. Determine the maximum reaction applied to an end support due to dead, live, and impact loads. The live load may be assumed to act along the centerline of the deck slab and divide equally between the two girders. Each concrete curb weighs 240 lb/ft and each rail 120 lb/ft. Stone concrete has a unit weight of 150 lb/ft^3. Assume an impact factor of 0.29.

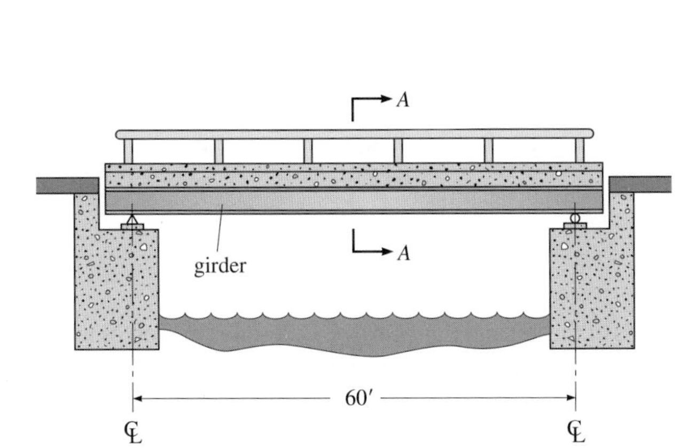

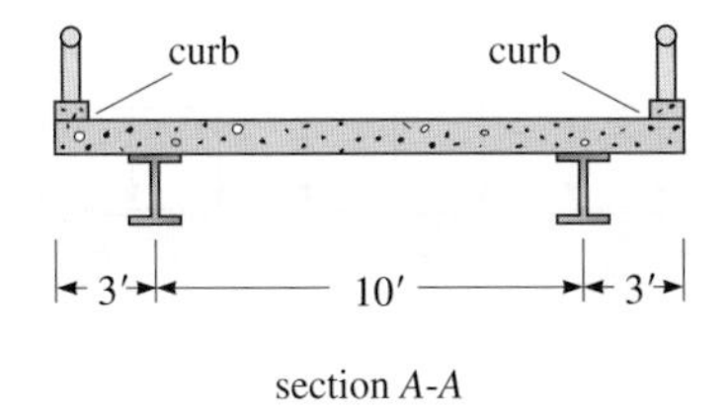

section *A*-*A*

P3.34

P3.35. A timber member supported by three steel links to a concrete frame has to carry the loads shown in Figure P3.35. (*a*) Calculate the reactions at support *A*. (*b*) Determine the axial forces in all links. Indicate if each link is in compression or tension.

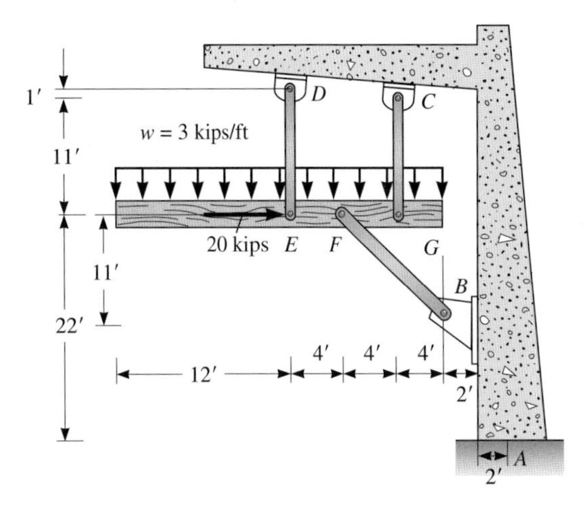

P3.35

P3.36. The three bay, one-story frame consists of beams pin connected to columns and column bases pinned to the foundation in Figure P3.36. The diagonal brace member *CH* is pinned at each end. Determine the reactions at *A*, *B*, *C* and *D*, and calculate the force in the brace.

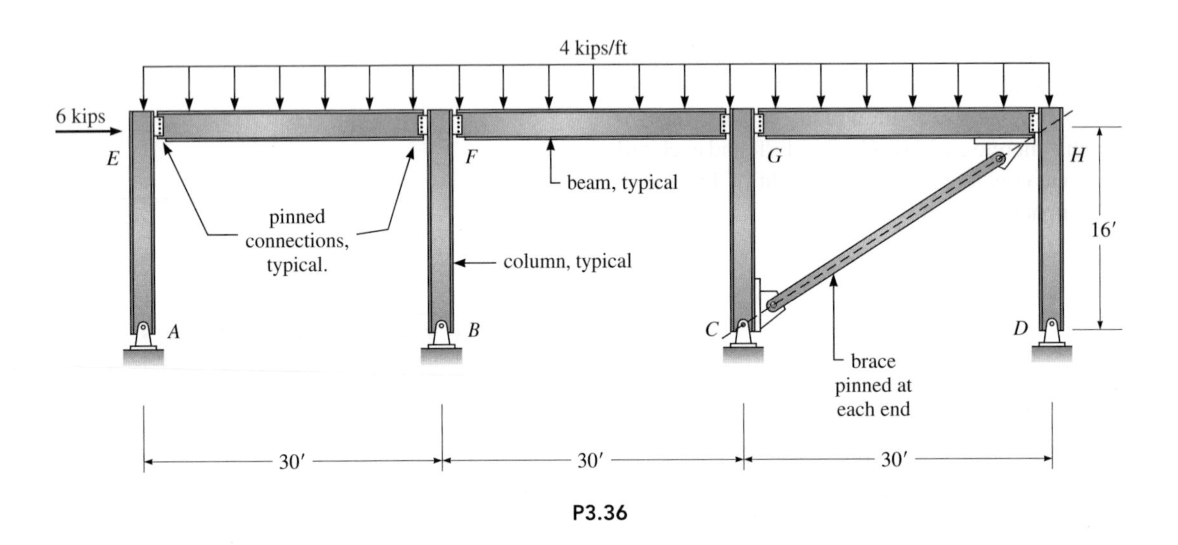

P3.36

P3.37. The multi-span girder in Figure P3.37 has two shear plate connections that act as hinges at *C* and *D*. The mid-span girder *CD* is simply supported on the cantilevered ends of the left and right girders. Determine the forces in the hinges and the reactions at supports *A*, *B*, *E* and *F*.

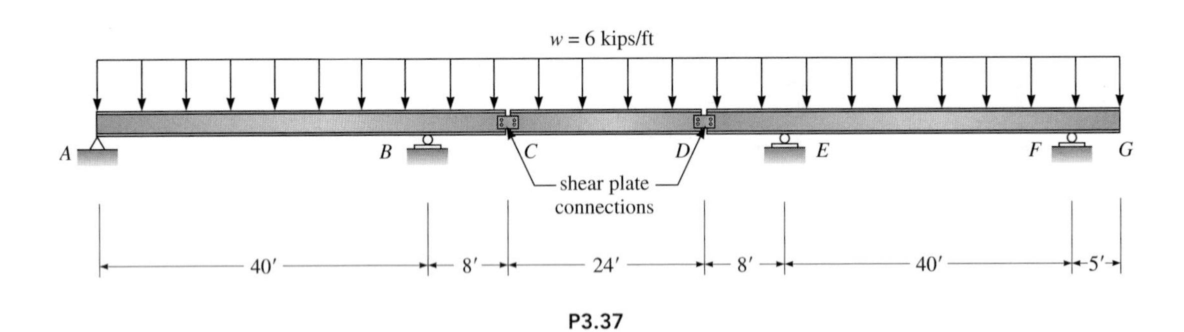

P3.37

Outerbridge Crossing between Staten Island and New Jersey

Constructed in 1928, the Outerbridge Crossing is a cantilever truss bridge consisting of a 750-ft main span, two 375-ft anchor arms, and a 300-ft through truss span at either end. The 143-ft clearance at the suspended midspan permits large ships to pass under the bridge. It is the outermost crossing in the district of The Port Authority of New York and New Jersey. Replaced by newer, stronger materials and structural systems, truss bridges have diminished in popularity in recent years.

CHAPTER

4 Trusses

Chapter Objectives

- Study the characteristics and behavior of trusses. Since truss members carry only axial loads, the configuration of the bars is key to a truss' efficiency and use.
- Analyze determinate trusses by method of joints and method of sections to determine bar forces. Also learn to visually identify bars with zero force.
- Classify determinate and indeterminate truss structures, and determine the degree of indeterminacy.
- Determine if a truss structure is stable or unstable.

4.1 Introduction

A truss is a structural element composed of a stable arrangement of slender interconnected bars (see Figure 4.1*a*). The pattern of bars, which often subdivides the truss into triangular areas, is selected to produce an efficient, lightweight, load-bearing member. Although joints, typically formed by welding or bolting truss bars to gusset plates, are rigid (see Figure 4.1*b*), the designer normally assumes that members are connected at joints by frictionless pins, as shown in Figure 4.1*c*. (Example 4.9 on page 149 clarifies the effect of this assumption.) Since no moment can be transferred through a frictionless pin joint, truss members are assumed to carry only axial force—either tension or compression. Because truss members act in direct stress, they carry load efficiently and often have relatively small cross sections.

As shown in Figure 4.1*a*, the upper and lower members, which are either horizontal or sloping, are called the top and bottom chords. The chords are connected by vertical and diagonal members.

The structural action of many trusses is similar to that of a beam. As a matter of fact, a truss can often be viewed as a beam in which excess material has been removed to reduce weight. The chords of a truss correspond to the flanges of a beam. The forces that develop in these members make up the internal couple that carries the moment produced by the applied loads. The

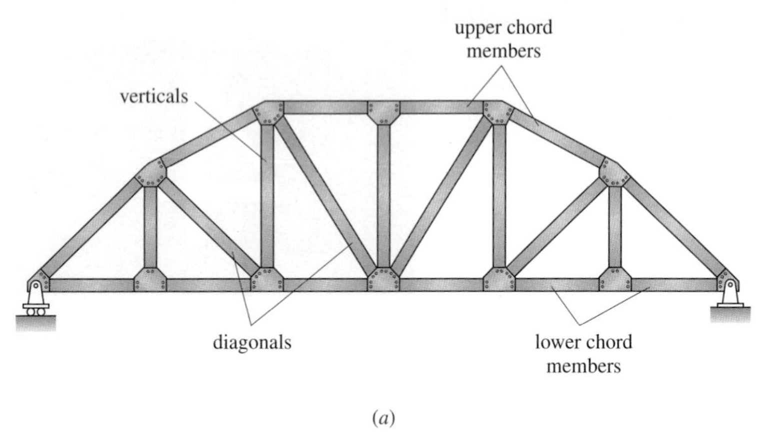

Figure 4.1: (*a*) Details of a truss; (*b*) welded joint; (*c*) idealized joint, members connected by a frictionless pin.

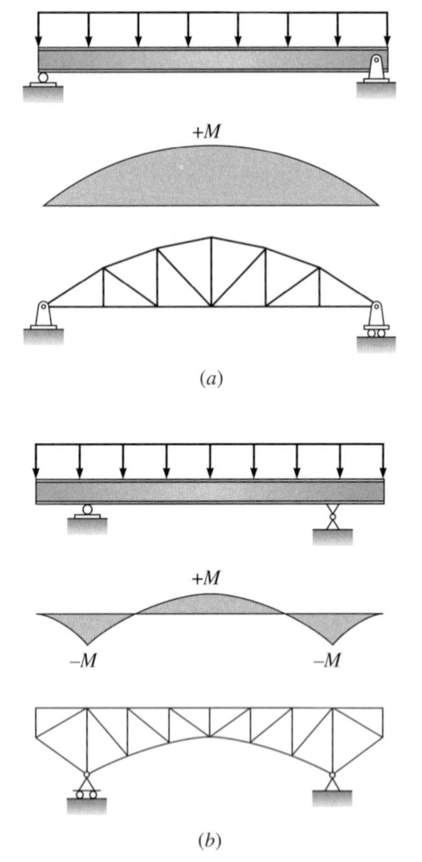

Figure 4.2: (*a*) and (*b*) depth of truss varied to conform to ordinates of moment curve.

primary function of the vertical and diagonal members is to transfer vertical force (shear) to the supports at the ends of the truss. Generally, on a per pound basis it costs more to fabricate a truss than to roll a steel beam; however, the truss will require less material because the material is used more efficiently. In a long-span structure, say 200 ft or more, the weight of the structure can represent the major portion (on the order of 75 to 85 percent) of the design load to be carried by the structure. By using a truss instead of a beam, the engineer can often design a lighter, stiffer structure at a reduced cost.

Even when spans are short, shallow trusses called bar joists are often used as substitutes for beams when loads are relatively light. For short spans these members are often easier to erect than beams of comparable capacity because of their lighter weight. Moreover, the openings between the web members provide large areas of unobstructed space between the floor above and the ceiling below the joist through which the mechanical engineer can run heating and air-conditioning ducts, water and waste pipes, electrical conduit, and other essential utilities.

In addition to varying the area of truss members, the designer can vary the truss depth to reduce its weight. In regions where the bending moment is large—at the center of a simply supported structure or at the supports in a continuous structure—the truss can be deepened (see Figure 4.2).

The diagonals of a truss typically slope upward at an angle that ranges from 45 to 60°. In a long-span truss the distance between panel points should not exceed 15 to 20 ft (5 to 7 m) to limit the unsupported length of the compression chords, which must be designed as columns. As the slenderness of a compression chord increases, it becomes more susceptible to buckling. The slenderness of tension members must be limited also to reduce vibrations produced by wind and live load.

If a truss carries equal or nearly equal loads at all panel points, the direction in which the diagonals slope will determine if they carry tension or compression forces. Figure 4.3, for example, shows the difference in forces set up in the diagonals of two trusses that are identical in all respects (same span, same loads, and so forth) except for the direction in which the diagonals slope (T represents tension and C indicates compression).

Although trusses are very stiff in their own plane, they are very flexible out of plane and must be braced or stiffened for stability. Since trusses are often used in pairs or spaced side by side, it is usually possible to connect several trusses together to form a rigid-box type of structure. For example, Figure 4.4 shows a bridge constructed from two trusses. In the horizontal planes of the top

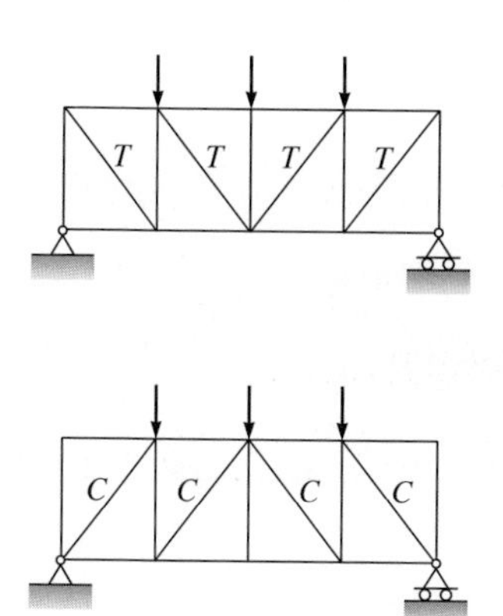

Figure 4.3: T represents tension and C compression.

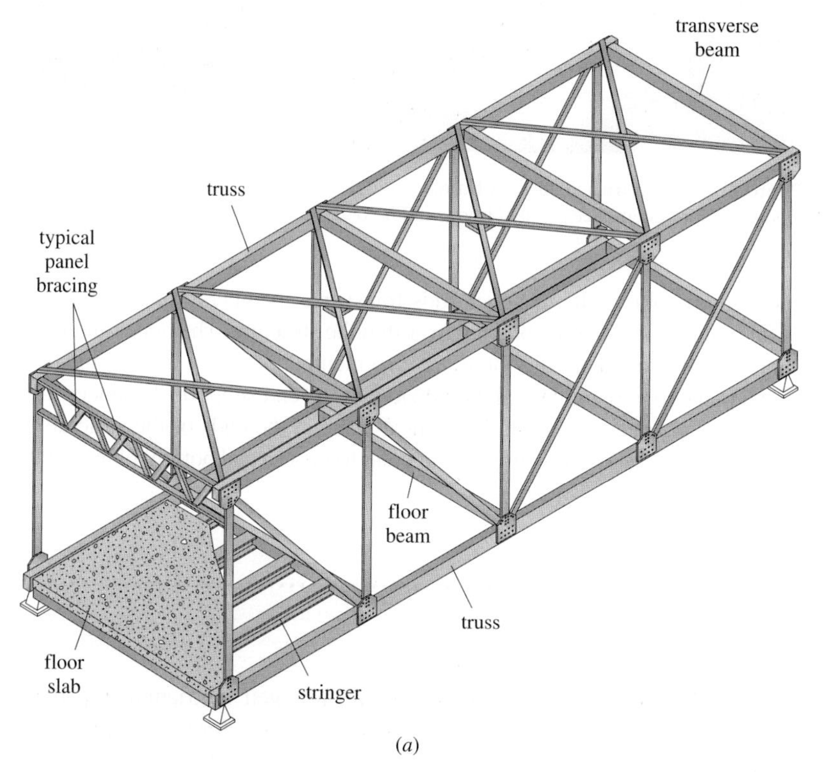

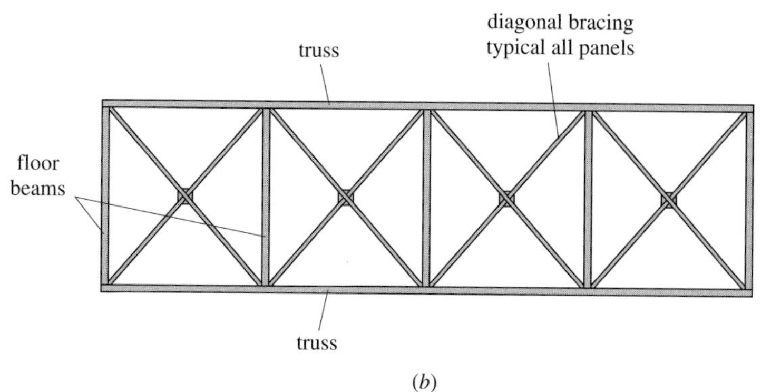

Figure 4.4: Truss with floor beams and secondary bracing: (*a*) perspective showing truss interconnected by transverse beams and diagonal bracing; diagonal bracing in bottom plane, omitted for clarity, is shown in (*b*); (*b*) bottom view showing floor beams and diagonal bracing. Lighter beams and bracing are also required in the top plane to stiffen trusses laterally.

Photo 4.1: Massive roof trusses with bolted joints and gusset plates.

Photo 4.2: Reconstructed Tacoma Narrows bridge showing trusses used to stiffen the roadway floor system. See original bridge in Photo 2.1.

and bottom chords, the designer adds transverse members, running between panel points, and diagonal bracing to stiffen the structure. The upper and lower chord bracing together with the transverse members forms a truss in the horizontal plane to transmit lateral wind load into the end supports. Engineers also add diagonal knee bracing in the vertical plane at the ends of the structure to ensure that the trusses remain perpendicular to the top and bottom planes of the structure.

4.2 Types of Trusses

The members of most modern trusses are arranged in triangular patterns because even when the joints are pinned, the triangular form is geometrically stable and will not collapse under load (see Figure 4.5*a*). On the other hand, a pin-connected rectangular element, which acts like an unstable linkage (see Figure 4.5*b*), will collapse under the smallest lateral load.

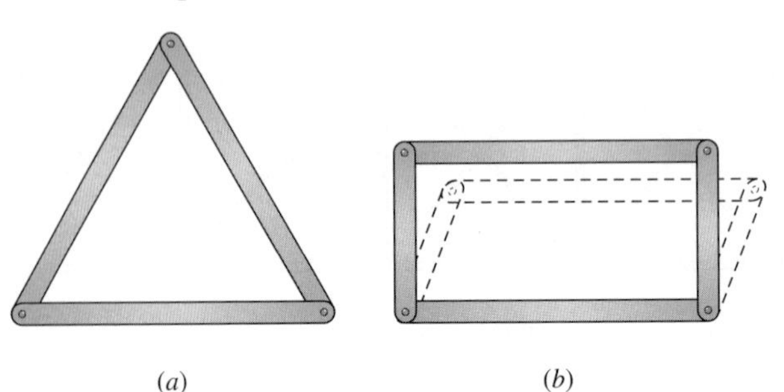

Figure 4.5: Pin-jointed frames: (*a*) stable; (*b*) unstable.

One method to establish a stable truss is to construct a basic triangular unit (see the shaded triangular element ABC in Figure 4.6) and then establish additional joints by extending bars from the joints of the first triangular element. For example, we can form joint D by extending bars from joints B and C. Similarly, we can imagine that joint E is formed by extending bars from joints C and D. Trusses formed in this manner are called *simple trusses*.

If two or more simple trusses are connected by a pin or a pin and a tie, the resulting truss is termed a *compound truss* (see Figure 4.7). Finally, if a truss—usually one with an unusual shape—is neither a simple nor a compound truss, it is termed a *complex truss* (see Figure 4.8). In current practice, where computers are used to analyze, these classifications are not of great significance.

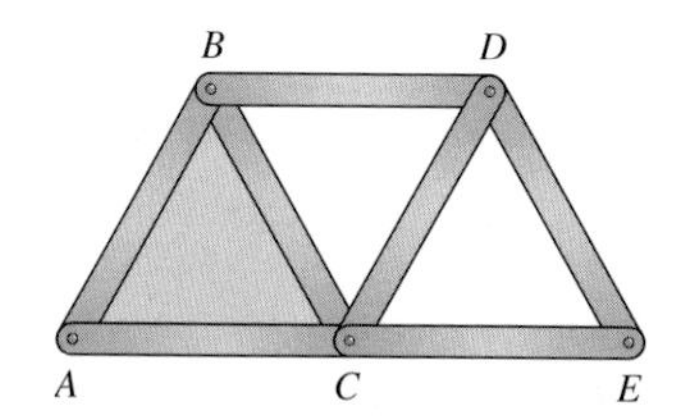

Figure 4.6: Simple truss.

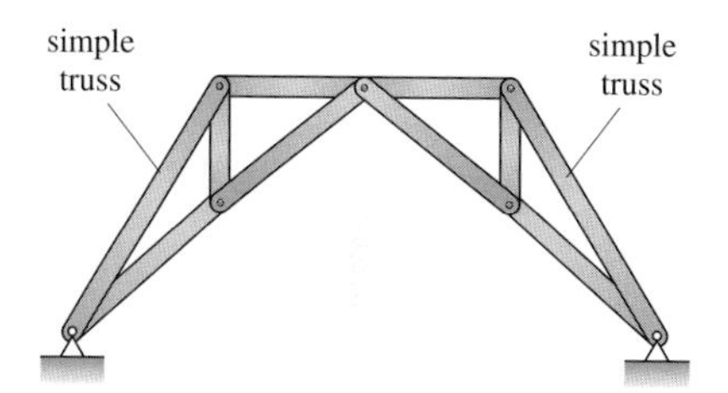

Figure 4.7: Compound truss is made up of simple trusses.

4.3 Analysis of Trusses

A truss is completely analyzed when the magnitude and sense (tension or compression) of all bar forces and reactions are determined. To compute the reactions of a determinate truss, we treat the entire structure as a rigid body and, as discussed in Section 3.6, apply the equations of static equilibrium together with any condition equations that may exist. The analysis used to evaluate the bar forces is based on the following three assumptions:

1. *Bars are straight and carry only axial load* (i.e., bar forces are directed along the longitudinal axis of truss members). This assumption also implies that we have neglected the deadweight of the bar. If the weight of the bar is significant, we can approximate its effect by applying one-half of the bar weight as a concentrated load to the joints at each end of the bar.
2. *Members are connected to joints by frictionless pins*. That is, no moments can be transferred between the end of a bar and the joint to which it connects. (If joints are rigid and members stiff, the structure should be analyzed as a rigid frame.)
3. *Loads are applied only at joints*.

As a sign convention (after the sense of a bar force is established) we label a *tensile force positive* and a *compression force negative*. Alternatively, we can

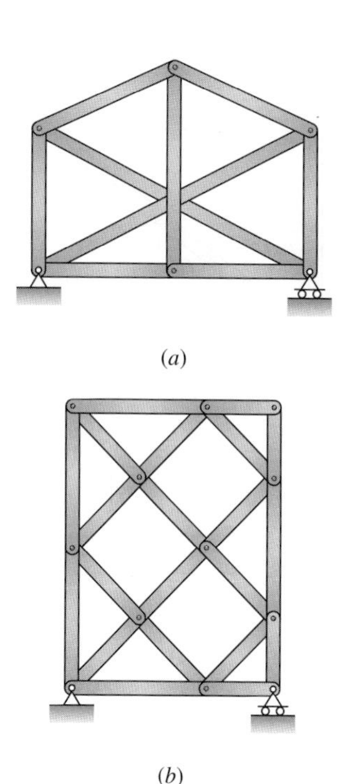

Figure 4.8: Complex trusses.

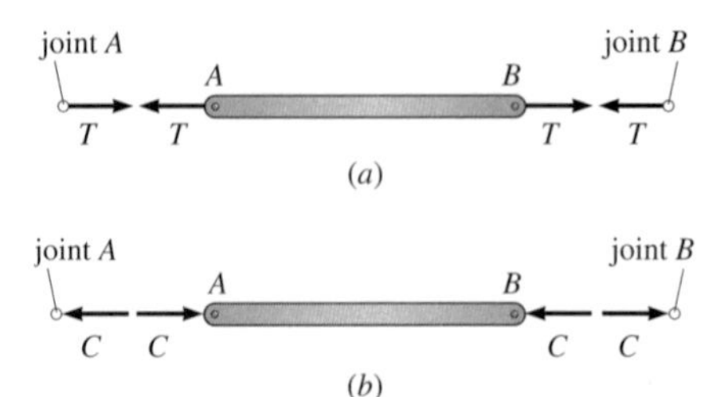

Figure 4.9: Free-body diagrams of axially loaded bars and adjacent joints: (*a*) bar *AB* in tension; (*b*) bar *AB* in compression.

denote the sense of a force by adding after its numerical value a *T* to indicate a tension force or a *C* to indicate a compression force.

If a bar is in tension, the axial forces at the ends of the bar act outward (see Figure 4.9*a*) and tend to elongate the bar. The equal and opposite forces on the ends of the bar represent the action of the joints on the bar. Since the bar applies equal and opposite forces to the joints, a tension bar will apply a force that acts outward from the center of the joint.

If a bar is in compression, the axial forces at the ends of the bar act inward and compress the bar (see Figure 4.9*b*). Correspondingly, a bar in compression pushes against a joint (i.e., applies a force directed inward toward the center of the joint).

Bar forces may be analyzed by considering the equilibrium of a joint—the *method of joints*—or by considering the equilibrium of a section of a truss—the *method of sections*. In the latter method, the section is produced by passing an imaginary cutting plane through the truss. The method of joints is discussed in Section 4.4; the method of sections is treated in Section 4.6.

4.4 Method of Joints

To determine bar forces by the method of joints, we analyze free-body diagrams of joints. The free-body diagram is established by imagining that we cut the bars by an imaginary section just before the joint. For example, in Figure 4.10*a* to determine the bar forces in members *AB* and *BC*, we use the free body of joint *B* shown in Figure 4.10*b*. Since the bars carry axial force, the line of action of each bar force is directed along the longitudinal axis of the bar.

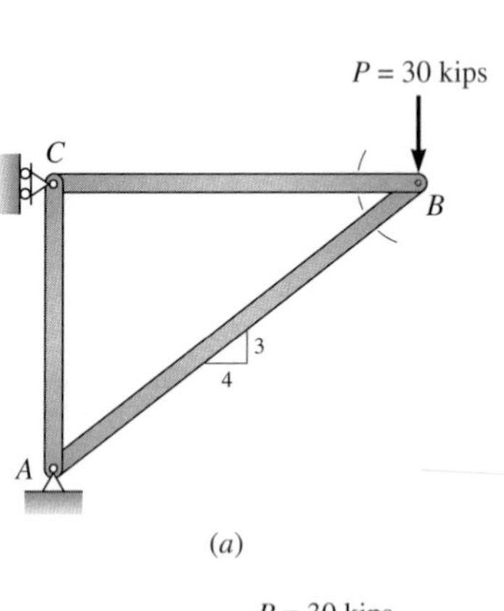

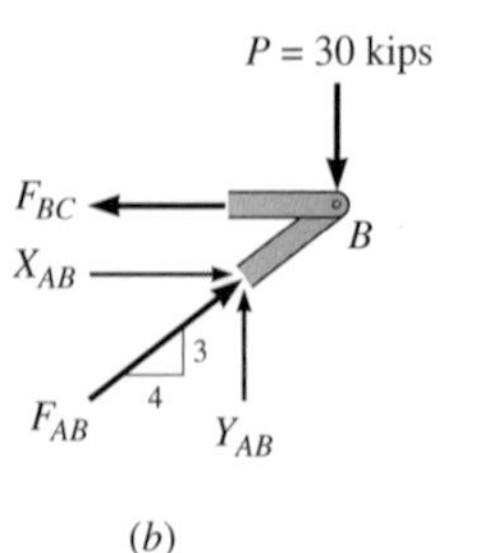

Figure 4.10: (*a*) Truss (dashed lines show location of circular cutting plane used to isolate joint *B*); (*b*) free body of joint *B*.

Because all forces acting at a joint pass through the pin, they constitute a concurrent force system. For this type of force system, only two equations of statics (that is, $\Sigma F_x = 0$ and $\Sigma F_y = 0$) are available to evaluate unknown bar forces. Since only two equations of equilibrium are available, we can only analyze joints that contain a maximum of two unknown bar forces.

The analyst can follow several procedures in the method of joints. For the student who has not analyzed many trusses, it may be best initially to write the equilibrium equations in terms of the components of the bar forces. On the other hand, as one gains experience and becomes familiar with the method, it is possible, without formally writing out the equilibrium equations, to determine bar forces at a joint that contains only one sloping bar by observing the magnitude and direction of the components of the bar forces required to produce equilibrium in a particular direction. The latter method permits a more rapid analysis of a truss. We discuss both procedures in this section.

To determine bar forces by writing out the equilibrium equations, we must assume a direction for each *unknown* bar force (*known* bar forces must be shown in their correct sense). The analyst is free to assume either tension or compression for any unknown bar force (many engineers like to assume that all bars are in tension, that is, they show all unknown bar forces acting outward from the center of the joint). Next, the forces are resolved into their

X and Y (rectangular) components. As shown in Figure 4.10*b*, the force or the components of a force in a particular bar are subscripted with the letters used to label the joints at each end of the bar. To complete the solution, we write and solve the two equations of equilibrium.

If only one unknown force acts in a particular direction, the computations are most expeditiously carried out by summing forces in that direction. After a component is computed, the other component can be established by setting up a proportion between the components of the force and the slope of the bar (the slope of the bar and the bar force are obviously identical).

If the solution of an equilibrium equation produces a positive value of force, the direction initially assumed for the force was correct. On the other hand, if the value of force is negative, its magnitude is correct, but the direction initially assumed is incorrect, and the direction of the force must be reversed on the sketch of the free-body diagram. After the bar forces are established at a joint, the engineer proceeds to adjacent joints and repeats the preceding computation until all bar forces are evaluated. This procedure is illustrated in Example 4.1.

Determination of Bar Forces by Inspection

Trusses can often be analyzed rapidly by inspection of the bar forces and loads acting on a joint that contains one sloping bar in which the force is unknown. In many cases the direction of certain bar forces will be obvious after the resultant of the known force or forces is established. For example, since the applied load of 30 kips at joint B in Figure 4.10*b* is directed downward, the y-component, Y_{AB} of the force in member AB—the only bar with a vertical component—must be equal to 30 kips and directed upward to satisfy equilibrium in the vertical direction. If Y_{AB} is directed upward, force F_{AB} must act upward and to the right, and its horizontal component X_{AB} must be directed to the right. Since X_{AB} is directed to the right, equilibrium in the horizontal direction requires that F_{BC} act to the left. The value of X_{AB} is easily computed from similar triangles because the slopes of the bars and the bar forces are identical (see Section 3.2).

$$\frac{X_{AB}}{4} = \frac{Y_{AB}}{3}$$

and

$$X_{AB} = \frac{4}{3}Y_{AB} = \frac{4}{3}(30)$$

$$X_{AB} = 40 \text{ kips} \qquad \textbf{Ans.}$$

To determine the force F_{BC}, we mentally sum forces in the x direction.

$$\rightarrow + \quad \Sigma F_x = 0$$

$$0 = -F_{BC} + 40$$

$$F_{BC} = 40 \text{ kips} \qquad \textbf{Ans.}$$

EXAMPLE 4.1

Analyze the truss in Figure 4.11a by the method of joints. Reactions are given.

Solution

The slopes of the various members are computed and shown on the sketch. For example, the top chord ABC, which rises 12 ft in 16 ft, is on a slope of 3 : 4.

To begin the analysis, we must start at a joint with a maximum of two bars. Either joint A or C is acceptable. Since the computations are simplest at a joint with one sloping member, we start at A. On a free body of joint A (see Figure 4.11b), we arbitrarily assume that bar forces F_{AB} and F_{AD} are tensile forces and show them acting outward on the joint. We next replace F_{AB} by its rectangular components X_{AB} and Y_{AB}. Writing the equilibrium equation in the y-direction, we compute Y_{AB}.

$$\overset{+}{\uparrow} \quad \Sigma F_y = 0$$

$$0 = -24 + Y_{AB} \qquad \text{and} \qquad Y_{AB} = 24 \text{ kips} \qquad \textbf{Ans.}$$

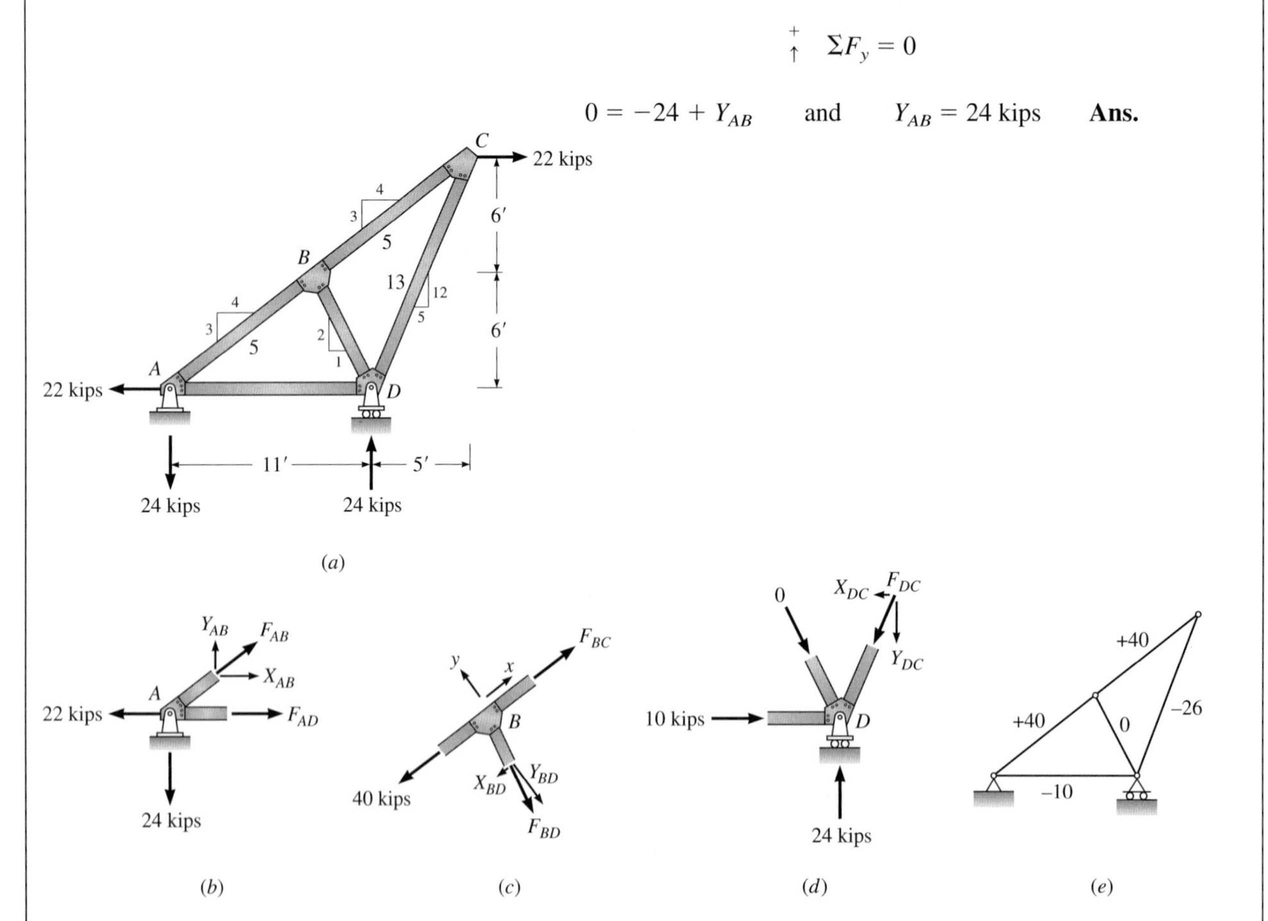

Figure 4.11: (a) Truss; (b) joint A; (c) joint B; (d) joint D; (e) summary of bar forces (units in kips).

Since Y_{AB} is positive, it is a tensile force, and the assumed direction on the sketch is correct. Compute X_{AB} and F_{AB} by proportion, considering the slope of the bar.

$$\frac{Y_{AB}}{3} = \frac{X_{AB}}{4} = \frac{F_{AB}}{5}$$

and

$$X_{AB} = \frac{4}{3}Y_{AB} = \frac{4}{3}(24) = 32 \text{ kips}$$

$$F_{AB} = \frac{5}{3}Y_{AB} = \frac{5}{3}(24) = 40 \text{ kips} \qquad \textbf{Ans.}$$

Compute F_{AD}.

$$\rightarrow + \quad \Sigma F_x = 0$$

$$0 = -22 + X_{AB} + F_{AD}$$

$$F_{AD} = -32 + 22 = -10 \text{ kips} \qquad \textbf{Ans.}$$

Since the minus sign indicates that the direction of force F_{AD} was assumed incorrectly, the force in member *AD* is compression, not tension.

We next isolate joint *B* and show all forces acting on the joint (see Figure 4.11*c*). Since we determined $F_{AB} = 40$ kips tension from the analysis of joint *A*, it is shown on the sketch acting outward from joint *B*. Superimposing an *x-y* coordinate system on the joint and resolving F_{BD} into rectangular components, we evaluate Y_{BD} by summing forces in the *y* direction.

$$\overset{+}{\uparrow} \quad \Sigma F_y = 0$$

$$Y_{BD} = 0$$

Since $Y_{BD} = 0$, it follows that $F_{BD} = 0$. From the discussion to be presented in Section 4.5 on zero bars, this result could have been anticipated.

Compute F_{BC}.

$$\rightarrow + \quad \Sigma F_x = 0$$

$$0 = F_{BC} - 40$$

$$F_{BC} = 40 \text{ kips tension} \qquad \textbf{Ans.}$$

Analyze joint *D* with $F_{BD} = 0$ and F_{DC} shown as a compressive force (see Figure 4.11*d*).

$$\rightarrow + \quad \Sigma F_x = 0 \quad 0 = 10 - X_{DC} \quad \text{and} \quad X_{DC} = 10 \text{ kips}$$

$$\overset{+}{\uparrow} \quad \Sigma F_y = 0 \quad 0 = 24 - Y_{DC} \quad \text{and} \quad Y_{DC} = 24 \text{ kips}$$

As a check of the results, we observe that the components of F_{DC} are proportional to the slope of the bar. Since all bar forces are known at this point, we can also verify that joint *C* is in equilibrium, as an alternative check. The results of the analysis are summarized in Figure 4.11*e* on a sketch of the truss. A tension force is indicated with a plus sign, a compressive force with a minus sign.

4.5 Zero Bars

Trusses, such as those used in highway bridges, typically support moving loads. As the load moves from one point to another, forces in truss members vary. For one or more positions of the load, certain bars may remain unstressed. The unstressed bars are termed *zero bars*. The designer can often speed the analysis of a truss by identifying bars in which the forces are zero. In this section we discuss two cases in which bar forces are zero.

Case 1. If No External Load Is Applied to a Joint That Consists of Two Bars, the Force in Both Bars Must Be Zero

To demonstrate the validity of this statement, we will first assume that forces F_1 and F_2 exist in both bars of the two-bar joint in Figure 4.12*a*, and then we demonstrate that the joint cannot be in equilibrium unless both forces equal zero. We begin by superimposing on the joint a rectangular coordinate system with an x axis oriented in the direction of force F_1, and we resolve force F_2 into components X_2 and Y_2 that are parallel to the x and y axes of the coordinate system, respectively. If we sum forces in the y direction, it is evident that the joint cannot be in equilibrium unless Y_2 equals zero because no other force is available to balance Y_2. If Y_2 equals zero, then F_2 is zero, and equilibrium requires that F_1 also equal zero.

A second case in which a bar force must equal zero occurs when a joint is composed of three bars—two of which are collinear.

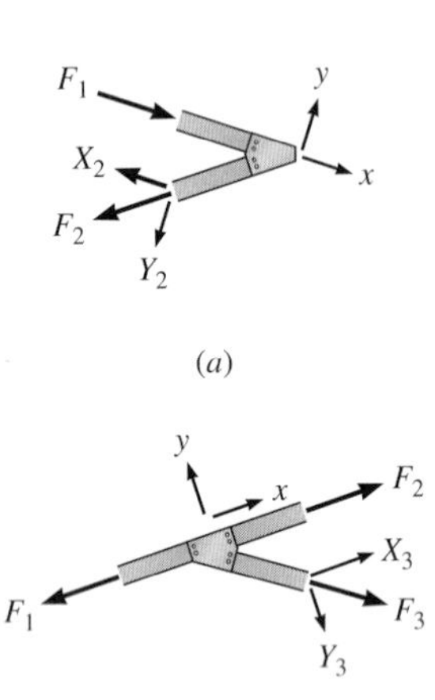

Figure 4.12: Conditions that produce zero forces in bars: (*a*) two bars and no external loads, F_1 and F_2 equal zero; (*b*) two collinear bars and no external loads, force in third bar (F_3) is zero.

Case 2. If No External Load Acts at a Joint Composed of Three Bars—Two of Which Are Collinear—the Force in the Bar That Is Not Collinear Is Zero

To demonstrate this conclusion, we again superimpose a rectangular coordinate system on the joint with the x axis oriented along the axis of the two collinear bars. If we sum forces in the y direction, the equilibrium equation can be satisfied only if F_3 equals zero because there is no other force to balance its y-component Y_3 (see Figure 4.12*b*).

Although a bar may have zero force under a certain loading condition, under other loadings the bar may carry stress. Thus the fact that the force in a bar is zero does not indicate that the bar is not essential and may be eliminated.

EXAMPLE 4.2

Based on the earlier discussion in Section 4.5, label all the bars in the truss of Figure 4.13 that are unstressed when the 60-kip load acts.

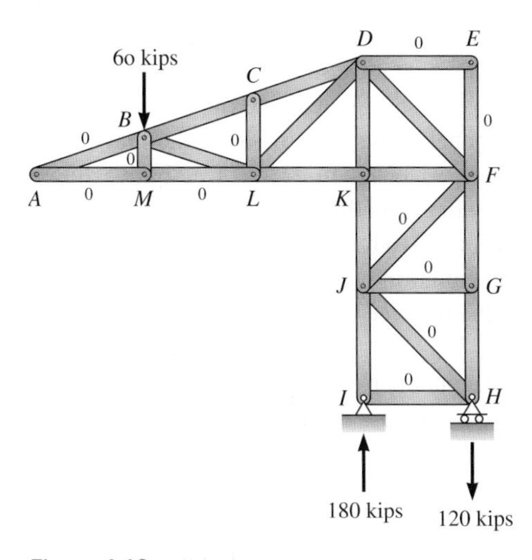

Figure 4.13

Solution

Although the two cases discussed in this section apply to many of the bars, we will examine only joints *A*, *E*, *I*, and *H*. The verification of the remaining zero bars is left to the student. Since joints *A* and *E* are composed of only two bars and no external load acts on the joints, the forces in the bars are zero (see Case 1).

Because no horizontal loads act on the truss, the horizontal reaction at *I* is zero. At joint *I* the force in bar *IJ* and the 180-kip reaction are collinear; therefore, the force in bar *IH* must equal zero because no other horizontal force acts at the joint. A similar condition exists at joint *H*. Since the force in bar *IH* is zero, the horizontal component of bar *HJ* must be zero. If a component of a force is zero, the force must also be zero.

4.6 Method of Sections

To analyze a stable truss by the method of sections, we imagine that the truss is divided into two free bodies by passing an imaginary cutting plane through the structure. The cutting plane must, of course, pass through the bar whose force is to be determined. At each point where a bar is cut, the internal force in the bar is applied to the face of the cut as an external load. Although there is no restriction on the number of bars that can be cut, we often use sections that cut three bars since three equations of static equilibrium are available to analyze a free body. For example, if we wish to determine the bar forces in the chords and diagonal of an interior panel of the truss in Figure 4.14*a*, we can pass a vertical section through the truss, producing the free-body diagram shown in Figure 4.14*b*. As we saw in the method of joints, the engineer is free to assume the direction of the bar force. If a force is assumed in the correct direction, solution of the equilibrium equation will produce a positive value of force. Alternatively, a negative value of force indicates that the direction of the force was assumed incorrectly.

If the force in a diagonal bar of a truss with parallel chords is to be computed, we cut a free body by passing a vertical section through the diagonal bar to be analyzed. An equilibrium equation based on summing forces in the y-direction will permit us to determine the vertical component of force in the diagonal bar.

If three bars are cut, the force in a particular bar can be determined by extending the forces in the other two bars along their line of action until they intersect. By summing moments about the axis through the point of intersection, we can write an equation involving the third force or one of its components. Example 4.3 illustrates the analysis of typical bars in a truss with parallel chords. Example 4.4 on page 136, which covers the analysis of a determinate truss with four restraints, illustrates a general approach to the analysis of a complicated truss using both the method of sections and the method of joints.

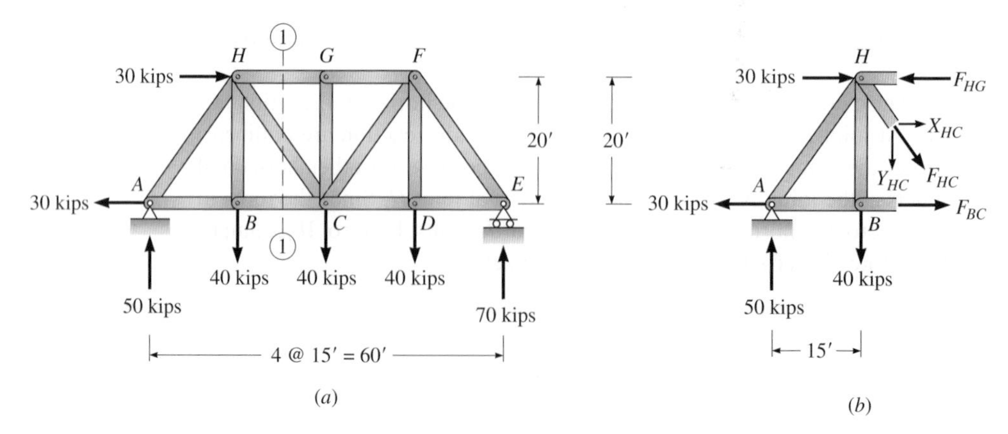

Figure 4.14

EXAMPLE 4.3

Using the method of sections, compute the forces or components of force in bars *HC*, *HG*, and *BC* of the truss in Figure 4.14*a*.

Solution

Pass section 1-1 through the truss cutting the free body shown in Figure 4.14*b*. The direction of the axial force in each member is arbitrarily assumed. To simplify the computations, force F_{HC} is resolved into vertical and horizontal components.

Compute Y_{HC} (see Figure 4.14*b*).

$$\overset{+}{\uparrow} \quad \Sigma F_y = 0$$

$$0 = 50 - 40 - Y_{HC}$$

$$Y_{HC} = 10 \text{ kips tension} \qquad \textbf{Ans.}$$

From the slope relationship,

$$\frac{X_{HC}}{3} = \frac{Y_{HC}}{4}$$

$$X_{HC} = \frac{3}{4}Y_{HC} = 7.5 \text{ kips} \qquad \textbf{Ans.}$$

Compute F_{BC}. Sum moments about an axis through *H* at the intersection of forces F_{HG} and F_{HC}.

$$\circlearrowright^{+} \quad \Sigma M_H = 0$$

$$0 = 30(20) + 50(15) - F_{BC}(20)$$

$$F_{BC} = 67.5 \text{ kips tension} \qquad \textbf{Ans.}$$

Compute F_{HG}.

$$\rightarrow + \quad \Sigma F_x = 0$$

$$0 = 30 - F_{HG} + X_{HC} + F_{BC} - 30$$

$$F_{HG} = 75 \text{ kips compression} \qquad \textbf{Ans.}$$

Since the solution of the equilibrium equations above produced positive values of force, the directions of the forces shown in Figure 4.14*b* are correct.

EXAMPLE 4.4

Analyze the determinate truss in Figure 4.15a to determine all bar forces and reactions.

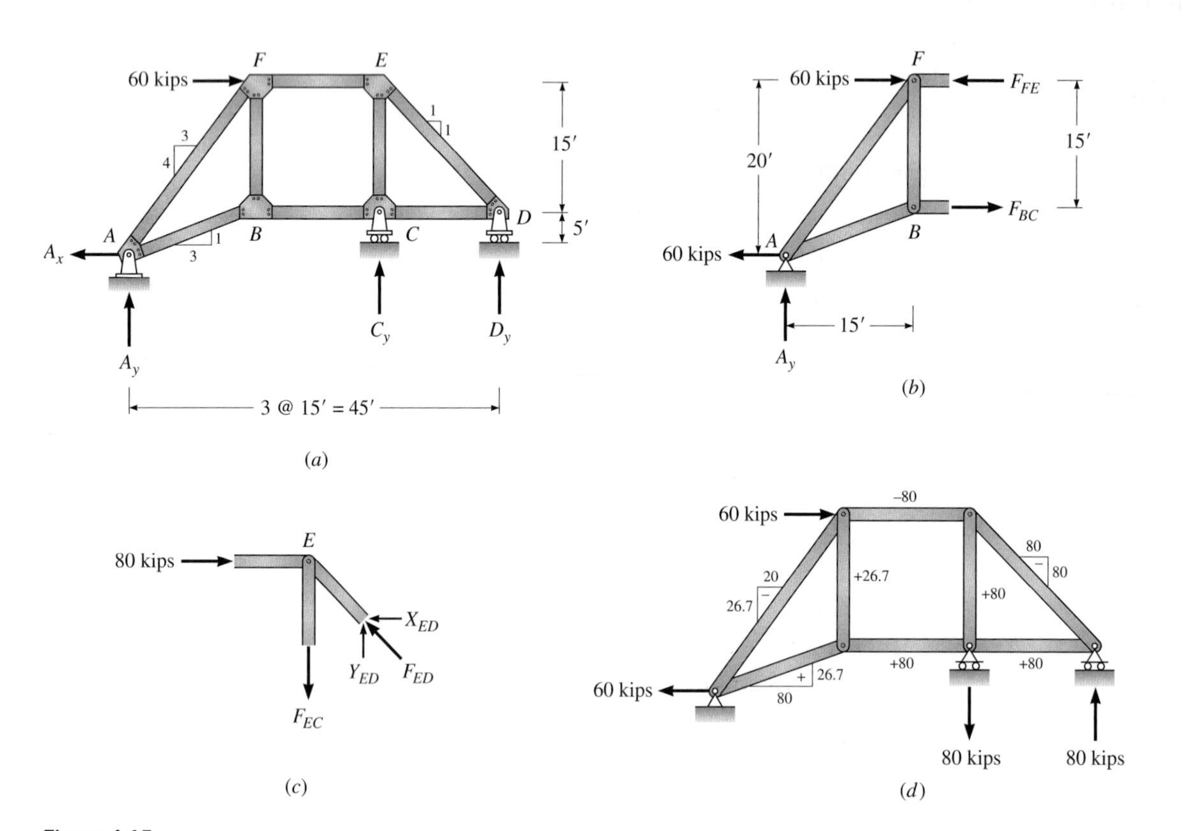

Figure 4.15

Solution

Since the supports at A, C, and D supply four restraints to the truss in Figure 4.15a, and only three equations of equilibrium are available, we cannot determine the value of all the reactions by applying the three equations of static equilibrium to a free body of the entire structure. However, recognizing that only one horizontal restraint exists at support A, we can determine its value by summing forces in the x-direction.

$$\rightarrow + \quad \Sigma F_x = 0$$

$$-A_x + 60 = 0$$

$$A_x = 60 \text{ kips} \qquad \textbf{Ans.}$$

Since the remaining reactions cannot be determined by the equations of statics, we must consider using the method either of joints or of sections. At this stage the method of joints cannot be applied because three or more unknown forces act at each joint. Therefore, we will pass a vertical section through the center panel of the truss to produce the free body shown in Figure 4.15*b*. We must use the free body to the left of the section because the free body to the right of the section cannot be analyzed since the reactions at *C* and *D* and the bar forces in members *BC* and *FE* are unknown.

Compute A_y (see Figure 4.15*b*).

$$\overset{+}{\uparrow} \quad \Sigma F_y = 0$$

$$A_y = 0 \qquad \textbf{Ans.}$$

Compute F_{BC}. Sum moments about an axis through joint *F*.

$$\circlearrowright^{+} \quad \Sigma M_F = 0$$

$$60(20) - F_{BC}(15) = 0$$

$$F_{BC} = 80 \text{ kips (tension)} \qquad \textbf{Ans.}$$

Compute F_{FE}.

$$\rightarrow + \quad \Sigma F_x = 0$$

$$+60 - 60 + F_{BC} - F_{FE} = 0$$

$$F_{FE} = F_{BC} = 80 \text{ kips (compression)} \qquad \textbf{Ans.}$$

Now that several internal bar forces are known, we can complete the analysis using the method of joints. Isolate joint *E* (Figure 4.15*c*).

$$\rightarrow + \quad \Sigma F_x = 0$$

$$80 - X_{ED} = 0$$

$$X_{ED} = 80 \text{ kips (compression)} \qquad \textbf{Ans.}$$

Since the slope of bar *ED* is 1:1, $Y_{ED} = X_{ED} = 80$ kips.

$$\overset{+}{\uparrow} \quad \Sigma F_y = 0$$

$$F_{EC} - Y_{ED} = 0$$

$$F_{EC} = 80 \text{ kips (tension)} \qquad \textbf{Ans.}$$

The balance of the bar forces and the reactions at *C* and *D* can be determined by the method of joints. Final results are shown on a sketch of the truss in Figure 4.15*d*.

EXAMPLE 4.5

Determine the forces in bars HG and HC of the truss in Figure 4.16a by the method of sections.

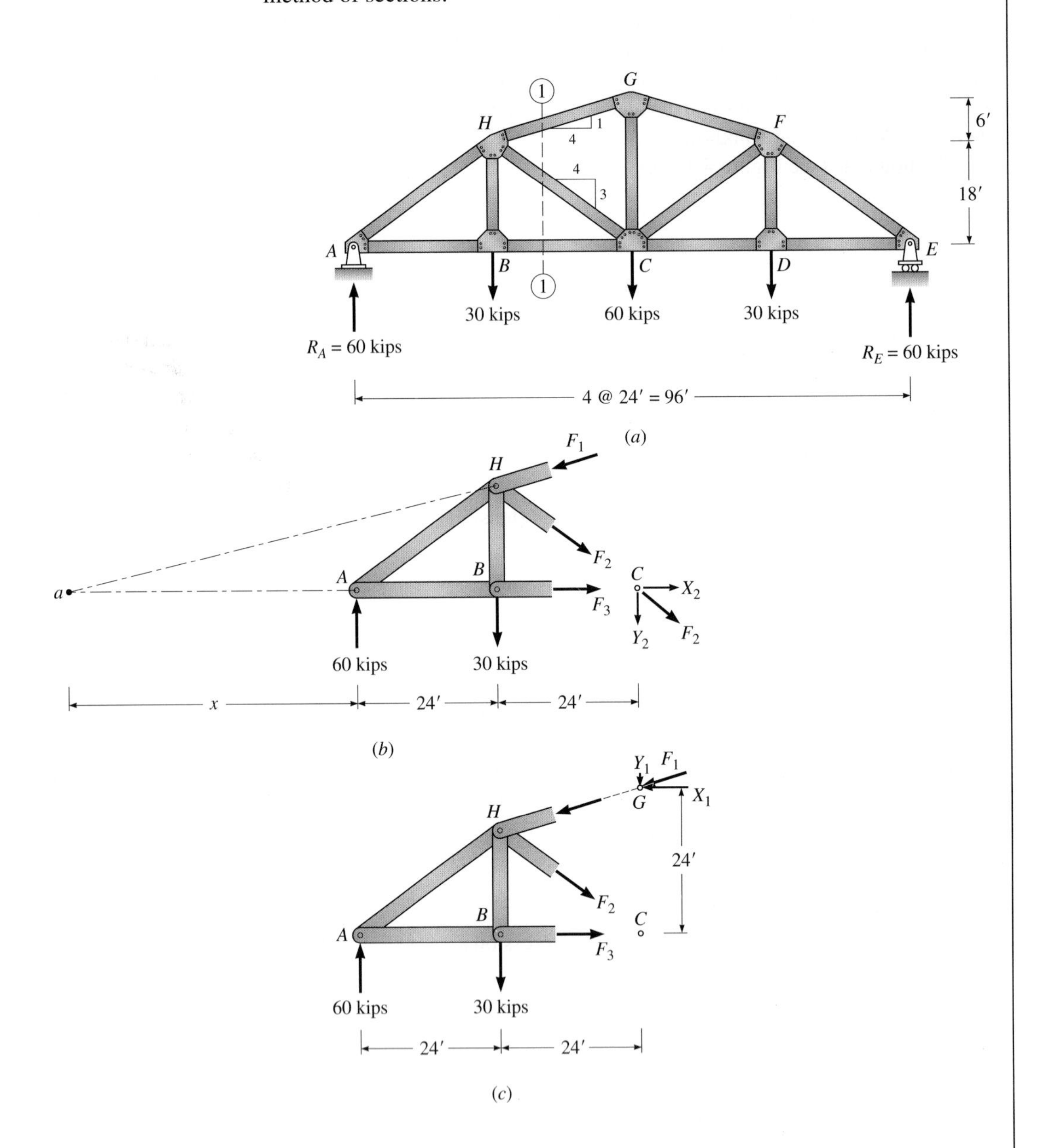

Figure 4.16: (a) Details of truss; (b) free body to compute force in bar HC; (c) free body to compute force in bar HG.

Solution

First compute the force in bar HC. Pass vertical section 1-1 through the truss, and consider the free body to the left of the section (see Figure 4.16*b*). The bar forces are applied as external loads to the ends of the bars at the cut. Since three equations of statics are available, all bar forces can be determined by the equations of statics. Let F_2 represent the force in bar HC. To simplify the computations, we select a moment center (point a that lies at the intersection of the lines of action of forces F_1 and F_3). Force F_2 is next extended along its line of action to point C and replaced by its rectangular components X_2 and Y_2. The distance x between a and the left support is established by proportion using similar triangles, that is, aHB and the slope (1: 4) of force F_1.

$$\frac{1}{18} = \frac{4}{x + 24}$$

$$x = 48 \text{ ft}$$

Sum moments of the forces about point a and solve for Y_2.

$$\circlearrowright^{+} \quad \Sigma M_a = 0$$

$$0 = -60(48) + 30(72) + Y_2(96)$$

$$Y_2 = 7.5 \text{ kips tension} \qquad \textbf{Ans.}$$

Based on the slope of bar HC, establish X_2 by proportion.

$$\frac{Y_2}{3} = \frac{X_2}{4}$$

$$X_2 = \frac{4}{3}Y_2 = 10 \text{ kips} \qquad \textbf{Ans.}$$

Now compute the force F_1 in bar HG. Select a moment center at the intersection of the lines of action of forces F_2 and F_3, that is, at point C (see Figure 4.16*c*). Extend force F_1 to point G and break into rectangular components. Sum moments about point C.

$$\circlearrowright^{+} \quad \Sigma M_c = 0$$

$$0 = 60(48) - 30(24) - X_1(24)$$

$$X_1 = 90 \text{ kips compression} \qquad \textbf{Ans.}$$

Establish Y_1 by proportion.

$$\frac{X_1}{4} = \frac{Y_1}{1}$$

$$Y_1 = \frac{X_1}{4} = 22.5 \text{ kips} \qquad \textbf{Ans.}$$

EXAMPLE 4.6

Using the method of sections, compute the forces in bars BC and JC of the K truss in Figure 4.17a.

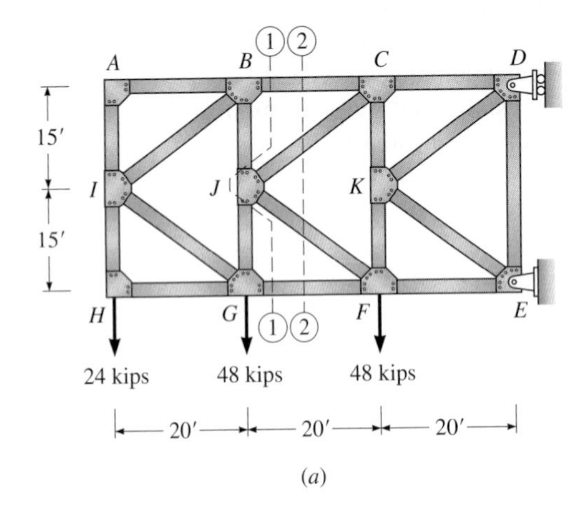

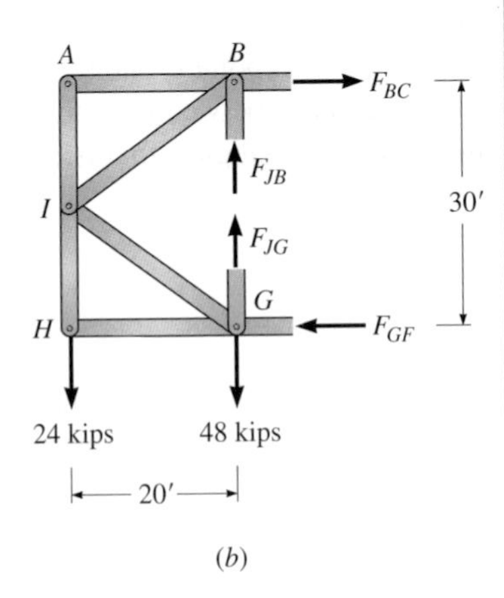

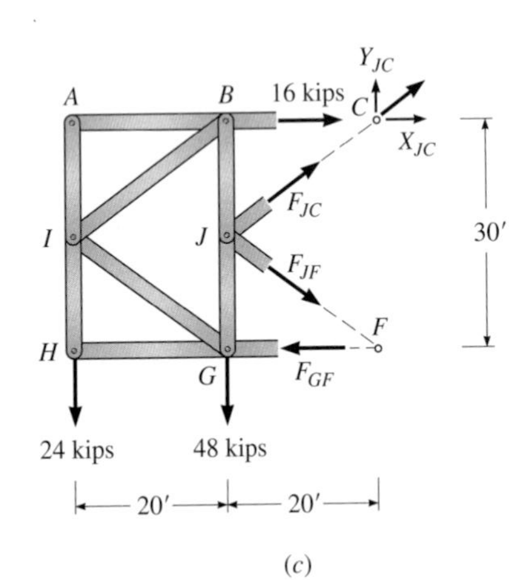

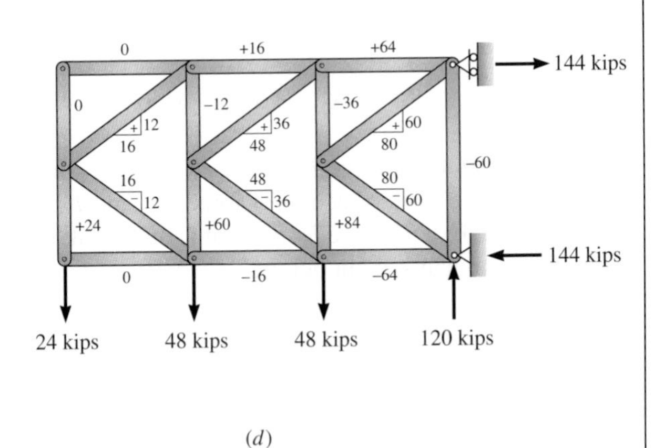

Figure 4.17: (a) K truss; (b) free body to the left of section 1-1 used to evaluate F_{BC}; (c) free body used to compute F_{JC}; (d) bar forces.

Solution

Since any *vertical* section passing through the panel of a K truss cuts four bars, it is not possible to compute bar forces by the method of sections because the number of unknowns exceeds the number of equations of statics. Since no moment center exists through which three of the bar forces pass, not even a partial solution is possible using a standard vertical section. As we illustrate in this example, it is possible to analyze a K truss by using two sections in sequence, the first of which is a special section curving around an interior joint.

To compute the force in bar BC, we pass Section 1–1 through the truss in Figure 4.17*a*. The free body to the left of the section is shown in Figure 4.17*b*. Summing moments about the bottom joint G gives

$$\circlearrowleft^{+} \quad \Sigma M_G = 0$$

$$30F_{BC} - 24(20) = 0$$

$$F_{BC} = 16 \text{ kips tension} \qquad \textbf{Ans.}$$

To compute F_{JC}, we pass section 2-2 through the panel and consider again the free body to the left (see Figure 4.17*c*). Since the force in bar BC has been evaluated, the three unknown bar forces can be determined by the equations of statics. Use a moment center at F. Extend the force in bar JC to point C and break into rectangular components.

$$\circlearrowleft^{+} \quad \Sigma M_F = 0$$

$$0 = 16(30) + X_{JC}(30) - 20(48) - 40(24)$$

$$X_{JC} = 48 \text{ kips}$$

$$F_{JC} = \frac{5}{4} X_{JC} = 60 \text{ kips tension} \qquad \textbf{Ans.}$$

NOTE. The K truss can also be analyzed by the method of joints by starting from an outside joint such as A or H. The results of this analysis are shown in Figure 4.17*d*. The K bracing is typically used in deep trusses to reduce the length of the diagonal members. As you can see from the results in Figure 4.17*d*, the shear in a panel divides equally between the top and bottom diagonals. One diagonal carries compression, and the other carries tension.

4.7 Determinacy and Stability

Thus far the trusses we have analyzed in this chapter have all been stable determinate structures; that is, we knew in advance that we could carry out a complete analysis using the equations of statics alone. Since indeterminate trusses are also used in practice, an engineer must be able to recognize a structure of this type because indeterminate trusses require a special type of analysis. As we will discuss in Chapter 11, compatibility equations must be used to supplement equilibrium equations.

If you are investigating a truss designed by another engineer, you will have to establish if the structure is determinate or indeterminate before you begin the analysis. Further, if you are responsible for establishing the configuration of a truss for a special situation, you must obviously be able to select an arrangement of bars that is stable. The purpose of this section is to extend to trusses the introductory discussion of stability and determinacy in Sections 3.8 and 3.9—topics you may wish to review before proceeding to the next paragraph.

If a loaded truss is in equilibrium, all members and joints of the truss must also be in equilibrium. If load is applied only at the joints and if all truss members are assumed to carry only axial load (an assumption that implies the dead load of members may be neglected or applied at the joints as an equivalent concentrated load), the forces acting on a free-body diagram of a joint will constitute a concurrent force system. To be in equilibrium, a concurrent force system must satisfy the following two equilibrium equations:

$$\Sigma F_x = 0$$
$$\Sigma F_y = 0$$

Since we can write two equilibrium equations for each joint in a truss, the total number of equilibrium equations available to solve for the unknown bar forces b and reactions r equals $2n$ (where n represents the total number of joints). Therefore, it must follow that if a truss is *stable* and *determinate*, the relationship between bars, reactions, and joints must satisfy the following criteria:

$$r + b = 2n \tag{4.1}$$

In addition, as we discussed in Section 3.8, *the restraints exerted by the reactions must not constitute either a parallel or a concurrent force system.*

Although three equations of statics are available to compute the reactions of a determinate truss, these equations are not independent and they cannot be added to the $2n$ joint equations. Obviously, if all joints of a truss are in equilibrium, the entire structure must also be equilibrium; that is, the resultant of the external forces acting on the truss equals zero. If the resultant is zero, the equations of static equilibrium are automatically satisfied when applied to the entire structure and thus do not supply additional independent equilibrium equations.

If

$$r + b > 2n$$

then the number of unknown forces exceed the available equations of statics and the truss is indeterminate. The degree of indeterminacy D equals

$$D = r + b - 2n \tag{4.2}$$

Finally, if

$$r + b < 2n$$

there are insufficient bar forces and reactions to satisfy the equations of equilibrium, and the structure is unstable.

Moreover, as we discussed in Section 3.8, you will always find that the analysis of an unstable structure leads to an inconsistent equilibrium equation. Therefore, if you are uncertain about the stability of a structure, analyze the structure for any arbitrary loading. If a solution that satisfies statics results, the structure is stable.

To illustrate the criteria for stability and determinacy for trusses introduced in this section, we will classify the trusses in Figure 4.18 as stable or unstable. For those structures that are stable, we will establish whether they are determinate or indeterminate. Finally, if a structure is indeterminate, we will also establish the degree of indeterminacy.

A B

(*a*)

A B

(*b*)

A B C D

(*c*)

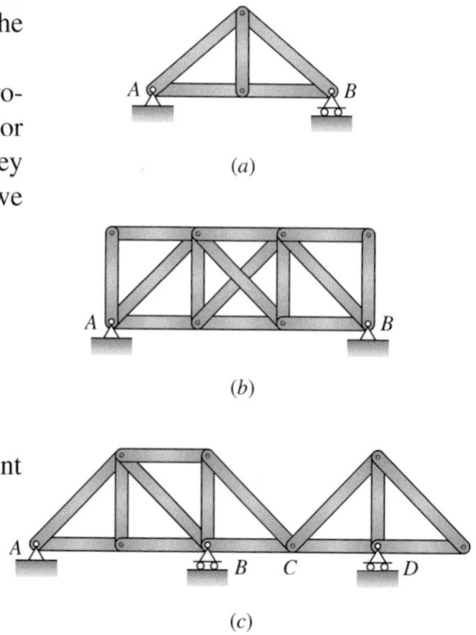

Figure 4.18: Classifying trusses: (*a*) stable determinate; (*b*) indeterminate second degree; (*c*) determinate (*continues*).

Figure 4.18*a*

$$b + r = 5 + 3 = 8 \qquad 2n = 2(4) = 8$$

Since $b + r = 2n$ and the reactions are not equivalent to either a concurrent or a parallel force system, the truss is stable and determinate.

Figure 4.18*b*

$$b + r = 14 + 4 = 18 \qquad 2n = 2(8) = 16$$

Since $b + r$ exceeds $2n$ ($18 > 16$), the structure is indeterminate to the second degree. The structure is one degree *externally* indeterminate because the supports supply four restraints, and *internally* indeterminate to the first degree because an extra diagonal is supplied in the middle panel to transmit shear.

Figure 4.18*c*

$$b + r = 14 + 4 = 18 \qquad 2n = 2(9) = 18$$

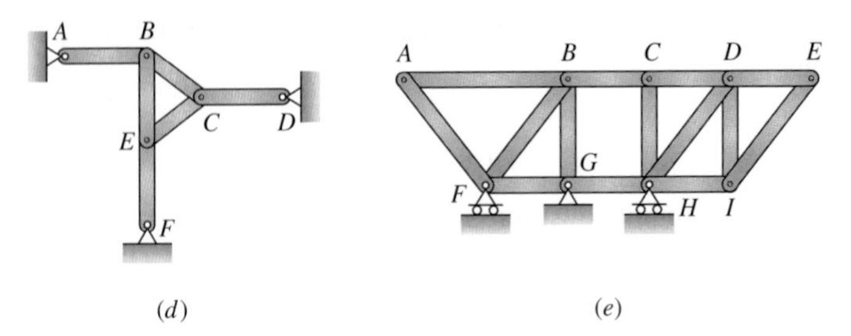

Figure 4.18: *Continued*: (*d*) determinate; (*e*) determinate.

Because $b + r = 2n = 18$, and the supports are not equivalent to either a parallel or a concurrent force system, the structure appears stable. We can confirm this conclusion by observing that truss *ABC* is obviously a stable component of the structure because it is a simple truss (composed of triangles) that is supported by three restraints—two supplied by the pin at *A* and one supplied by the roller at *B*. Since the hinge at *C* is attached to the stable truss on the left, it, too, is a stable point in space. Like a pin support, it can supply both horizontal and vertical restraint to the truss on the right. Thus we can reason that truss *CD* must also be stable since it, too, is a simple truss supported by three restraints, that is, two supplied by the hinge at *C* and one by the roller at *D*.

Figure 4.18*d* Two approaches are possible to classify the structure in Figure 4.18*d*. In the first approach, we can treat triangular element *BCE* as a three-bar truss ($b = 3$) supported by three links—*AB*, *EF*, and *CD* ($r = 3$). Since the truss has three joints (*B*, *C*, and *E*), $n = 3$. And $b + r = 6$ equals $2n = 2(3) = 6$, and the structure is determinate and stable.

Alternatively, we can treat the entire structure as a six-bar truss ($b = 6$), with six joints ($n = 6$), supported by three pins ($r = 6$), $b + r = 12$ equals $2n = 2(6) = 12$. Again we conclude that the structure is stable and determinate.

Figure 4.18*e*

$$b + r = 14 + 4 = 18 \qquad 2n = 2(9) = 18$$

Since $b + r = 2n$, it appears the structure is stable and determinate; however, since a rectangular panel exists between joints *B*, *C*, *G*, and *H*, we will verify that the structure is stable by analyzing the truss for an arbitrary load of 4 kips applied vertically at joint *F* (see Example 4.7). Since analysis by the method of joints produces unique values of bar force in all members, we conclude that the structure is both stable and determinate.

Figure 4.18*f*

$$b + r = 8 + 4 = 12 \qquad 2n = 2(6) = 12$$

Although the bar count above satisfies the necessary condition for a stable determinate structure, the structure appears to be unstable because the center panel, lacking a diagonal bar, cannot transmit vertical force. To confirm this conclusion, we will analyze the truss, using the equations of statics. (The analysis is carried out in Example 4.8.) Since the analysis leads to an inconsistent equilibrium equation, we conclude that the structure is unstable.

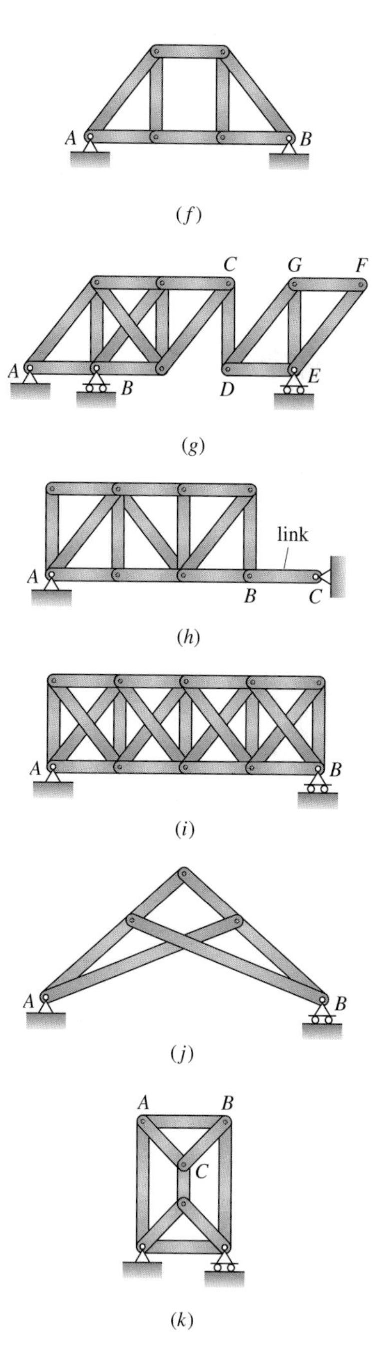

Figure 4.18*g*

$$b = 16 \qquad r = 4 \qquad n = 10$$

Although $b + r = 2n$, the small truss on the right ($DEFG$) is unstable because its supports—the link CD and the roller at E—constitute a parallel force system.

Figure 4.18*h* Truss is geometrically unstable because the reactions constitute a concurrent force system; that is, the reaction supplied by the link BC passes through the pin at A.

Figure 4.18*i*

$$b = 21 \qquad r = 3 \qquad n = 10$$

And $b + r = 24$, $2n = 20$; therefore, truss is indeterminate to the fourth degree. Although the reactions can be computed for any loading, the indeterminacy is due to the inclusion of double diagonals in all interior panels.

Figure 4.18*j*

$$b = 6 \qquad r = 3 \qquad n = 5$$

And $b + r = 9$, $2n = 10$; the structure is unstable because there are fewer restraints than required by the equations of statics. To produce a stable structure, the reaction at B should be changed from a roller to a pin.

Figure 4.18*k* Now $b = 9$, $r = 3$, and $n = 6$; also $b + r = 12$, $2n = 12$. However, the structure is unstable because the small triangular truss ABC at the top is supported by three parallel links, which provide no lateral restraint.

Figure 4.18: Classifying trusses: (*f*) unstable; (*g*) unstable; (*h*) unstable; (*i*) indeterminate fourth degree; (*j*) unstable; (*k*) unstable.

EXAMPLE 4.7

Verify that the truss in Figure 4.19 is stable and determinate by demonstrating that it can be completely analyzed by the equations of statics for a force of 4 kips at joint F.

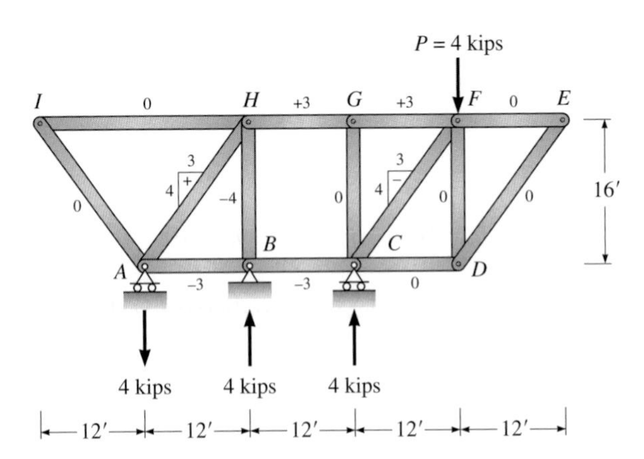

Figure 4.19: Analysis by *method of joints* to verify that truss is stable.

Solution

Since the structure has four reactions, we cannot start the analysis by computing reactions, but instead must analyze it by the method of joints. We first determine the zero bars.

Since joints E and I are connected to only two bars and no external load acts on the joints, the forces in these bars are zero (see Case 1 of Section 4.5). With the remaining two bars connecting to joint D, applying the same argument would indicate that these two members are also zero bars. Applying Case 2 of Section 4.5 to joint G would indicate that bar CG is a zero bar.

Next we analyze in sequence joints F, C, G, H, A, and B. Since all bar forces and reactions can be determined by the equations of statics (results are shown on Figure 4.19), we conclude that the truss is stable and determinate.

Prove that the truss in Figure 4.20a is unstable by demonstrating that its analysis for a load of arbitrary magnitude leads to an inconsistent equation of equilibrium.

EXAMPLE 4.8

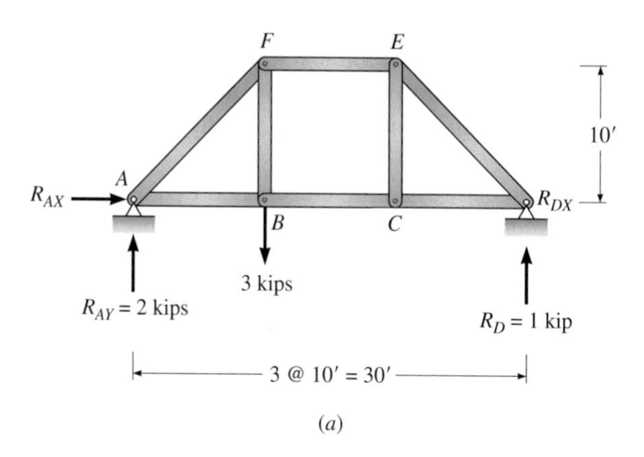

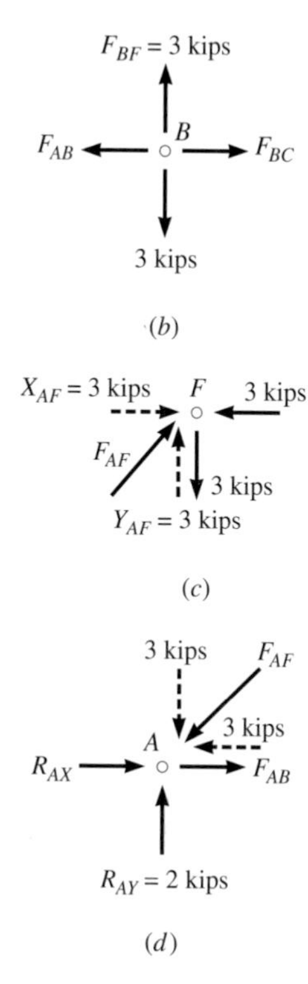

Figure 4.20: Check of truss stability: (*a*) details of truss; (*b*) free body of joint *B*; (*c*) free body of joint *F*; (*d*) free body of support *A*.

Solution

Apply a load at joint *B*, say 3 kips, and compute the reactions, considering the entire structure as a free body.

$$\circlearrowright^{+} \quad \Sigma M_A = 0$$

$$3(10) - 30R_D = 0 \qquad R_D = 1 \text{ kip}$$

$$\overset{+}{\uparrow} \quad \Sigma F_y = 0$$

$$R_{AY} - 3 + R_D = 0 \qquad R_{AY} = 2 \text{ kips}$$

Equilibrium of joint *B* (see Figure 4.20*b*) requires that $F_{BF} = 3$ kips tension. Equilibrium in the *x* direction is possible if $F_{AB} = F_{BC}$.

We next consider joint *F* (see Figure 4.20*c*). To be in equilibrium in the *y*-direction, the vertical component of F_{AF} must equal 3 kips and be directed upward, indicating that bar *AF* is in compression. Since the slope of bar *AF* is 1:1, its horizontal component also equals 3 kips. Equilibrium of joint *F* in the *x* direction requires that the force in bar *FE* equal 3 kips and act to the left.

We now examine support *A* (Figure 4.20*d*). The reaction R_A and the components of force in bar *AF*, determined previously, are applied to the joint. Writing the equation of equilibrium in the *y*-direction, we find

$$\overset{+}{\uparrow} \quad \Sigma F_y = 0$$

$$2 - 3 \neq 0 \qquad \text{(inconsistent)}$$

Since the equilibrium equation is not satisfied, the structure is not stable.

4.8 Computer Analysis of Trusses

The preceding sections of this chapter have covered the analysis of trusses based on the assumptions that (1) members are connected at joints by frictionless pins and (2) loads are applied at joints only. When design loads are conservatively chosen, and deflections are not excessive, over the years these simplifying assumptions have generally produced satisfactory designs.

Since joints in most trusses are constructed by connecting members to gusset plates by welds, rivet, or high-strength bolts, joints are usually *rigid*. To analyze a truss with rigid joints (a highly indeterminate structure) would be a lengthy computation by the classical methods of analysis. That is why, in the past, truss analysis has been simplified by allowing designers to assume pinned joints. Now that computer programs are available, we can analyze both determinate and indeterminate trusses as a rigid-jointed structure to provide a more precise analysis, and the limitation that loads must be applied at joints is no longer a restriction.

Because computer programs require values of cross-sectional properties of members—area and moment of inertia—members must be initially sized. Procedures to estimate the approximate size of members are discussed in Chapter 15 of the text. In the case of a truss with rigid joints, the assumption of pin joints will permit you to compute axial forces that can be used to select the initial cross-sectional areas of members.

To carry out the computer analyses, we will use the RISA-2D computer program that is located on the website of this textbook; that is, http://www.mhhe.com/leet. Although a tutorial is provided on the website to explain, step by step, how to use the RISA-2D program, a brief overview of the procedure is given below.

1. Number all joints and members.
2. After the RISA-2D program is opened, click **Global** at the top of the screen. Insert a descriptive title, your name, and the number of sections.
3. Click **Units**. Use either Standard Metric or Standard Imperial for U.S. Customary System units.
4. Click **Modify**. Set the scale of the grid so the figure of the structure lies within the grid.
5. Fill in tables in **Data Entry Box**. These include Joint Coordinates, Boundary Conditions, Member Properties, Joint Loads, etc. Click **View** to label members and joints. The figure on the screen permits you to check visually that all required information has been supplied correctly.
6. Click **Solve** to initiate the analysis.
7. Click **Results** to produce tables listing bar forces, joint defections, and support reactions. The program will also plot a deflected shape.

EXAMPLE 4.9

Using the RISA-2D computer program, analyze the determinate truss in Figure 4.21, and compare the magnitude of the bar forces and joint displacements, assuming (1) joints are *rigid* and (2) joints are *pinned*. Joints are denoted by numbers in a circle; members, by numbers in a rectangular box. A preliminary analysis of the truss was used to establish initial values of each member's cross-sectional properties (see Table 4.1). For the case of pinned joints, the member data are similar, but the word *pinned* appears in the columns titled **End Releases**.

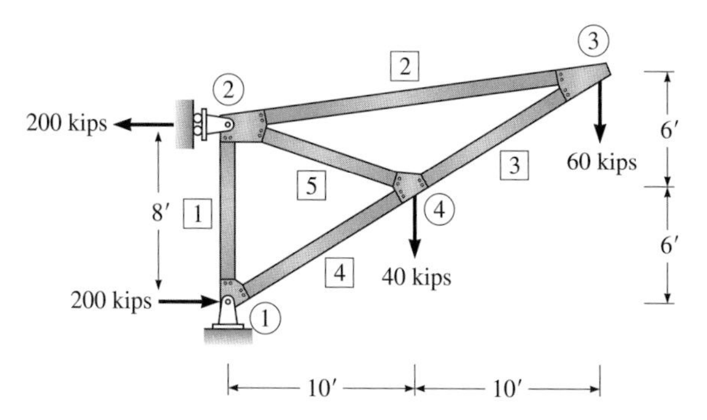

Figure 4.21: Cantilever truss.

TABLE 4.1
Member Data for Case of Rigid Joints

Member Label	I Joint	J Joint	Area (in^2)	Moment of Inertia (in^4)	Elastic Modulus (ksi)	End Releases I-End	End Releases J-End	Length (ft)
1	1	2	5.72	14.7	29,000			8
2	2	3	11.5	77	29,000			20.396
3	3	4	11.5	77	29,000			11.662
4	4	1	15.4	75.6	29,000			11.662
5	2	4	5.72	14.7	29,000			10.198

TABLE 4.2
Comparision of Joint Displacements

Rigid Joints			Pinned Joints		
Joint Label	X Translation (in)	Y Translation (in)	Joint Label	X Translation (in)	Y Translation (in)
1	0	0	1	0	0
2	0	0.011	2	0	0.012
3	0.257	−0.71	3	0.266	−0.738
4	0.007	−0.153	4	0	−0.15

[*continues on next page*]

Example 4.9 continues . . .

TABLE 4.3

Comparison of Member Forces

Rigid Joints					Pin Joints		
Member Label	Section*	Axial (kips)	Shear (kips)	Moment (kip · ft)	Member Label	Section*	Axial (kips)
1	1	−19.256	−0.36	0.918	1	1	−20
	2	−19.256	−0.36	−1.965		2	−20
2	1	−150.325	0.024	−2.81	2	1	−152.971
	2	−150.325	0.024	−2.314		2	−152.971
3	1	172.429	0.867	−2.314	3	1	174.929
	2	172.429	0.867	7.797		2	174.929
4	1	232.546	−0.452	6.193	4	1	233.238
	2	232.546	−0.452	0.918		2	233.238
5	1	−53.216	−0.24	0.845	5	1	−50.99
	2	−53.216	−0.24	−1.604		2	−50.99

*Sections 1 and 2 refer to member ends.

To facilitate the connection of the members to the gusset plates, the truss members are often fabricated from pairs of double angles oriented back to back. The cross-sectional properties of these structural shapes, tabulated in the *AISC Manual of Steel Construction*, are used in this example.

CONCLUSIONS: The results of the computer analysis shown in Tables 4.2 and 4.3 indicate that the magnitude of the axial forces in the truss bars, as well as the joint displacements, are approximately the same for both pinned and rigid joints. The axial forces are slightly smaller in most bars when *rigid* joints are assumed because a portion of the load is transmitted by shear and bending.

Since members in direct stress carry axial load efficiently, cross-sectional areas tend to be small when sized for axial load alone. However, the flexural stiffness of small compact cross sections is also small. Therefore, when joints are *rigid*, bending stress in truss members may be *significant even when the magnitude of the moments is relatively small.* If we check stresses in member M3, which is constructed from two $8 \times 4 \times {}^{1}/_{2}$ in angles, at the section where the moment is 7.797 kip·ft, the axial stress is $P/A = 14.99$ kips/in^2 and the bending stress $Mc/I = 6.24$ kips/in^2. In this case, we conclude that bending stresses are significant in several truss members when the analysis is carried out assuming joints are *rigid*, and the designer must verify that the combined stress of 21.23 kips/in^2 does not exceed the allowable value specified by the AISC design specifications.

Summary

- Trusses are composed of slender bars that are assumed to carry only axial force. Joints in large trusses are formed by welding or bolting members to gusset plates. If members are relatively small and lightly stressed, joints are often formed by welding the ends of vertical and diagonal members to the top and bottom chords.
- Although trusses are stiff in their own plane, they have little lateral stiffness; therefore, they must be braced against lateral displacement at all panel points.
- To be *stable* and *determinate*, the following relationship must exist among the number of bars b, reactions r, and joints n:

  $$b + r = 2n$$

  In addition, the restraints exerted by the reactions must *not* constitute either a parallel or a concurrent force system.

  If $b + r < 2n$, the truss is unstable. If $b + r > 2n$, the truss is indeterminate.
- Determinate trusses can be analyzed either by the method of joints or by the method of sections. The method of sections is used when the force in one or two bars is required. The method of joints is used when all bar forces are required.
- If the analysis of a truss results in an inconsistent value of forces, that is, one or more joints are not in equilibrium, then the truss is unstable.

PROBLEMS

P4.1. Classify the trusses in Figure P4.1 as stable or unstable. If stable, indicate if determinate or indeterminate. If indeterminate, indicate the degree of indeterminacy.

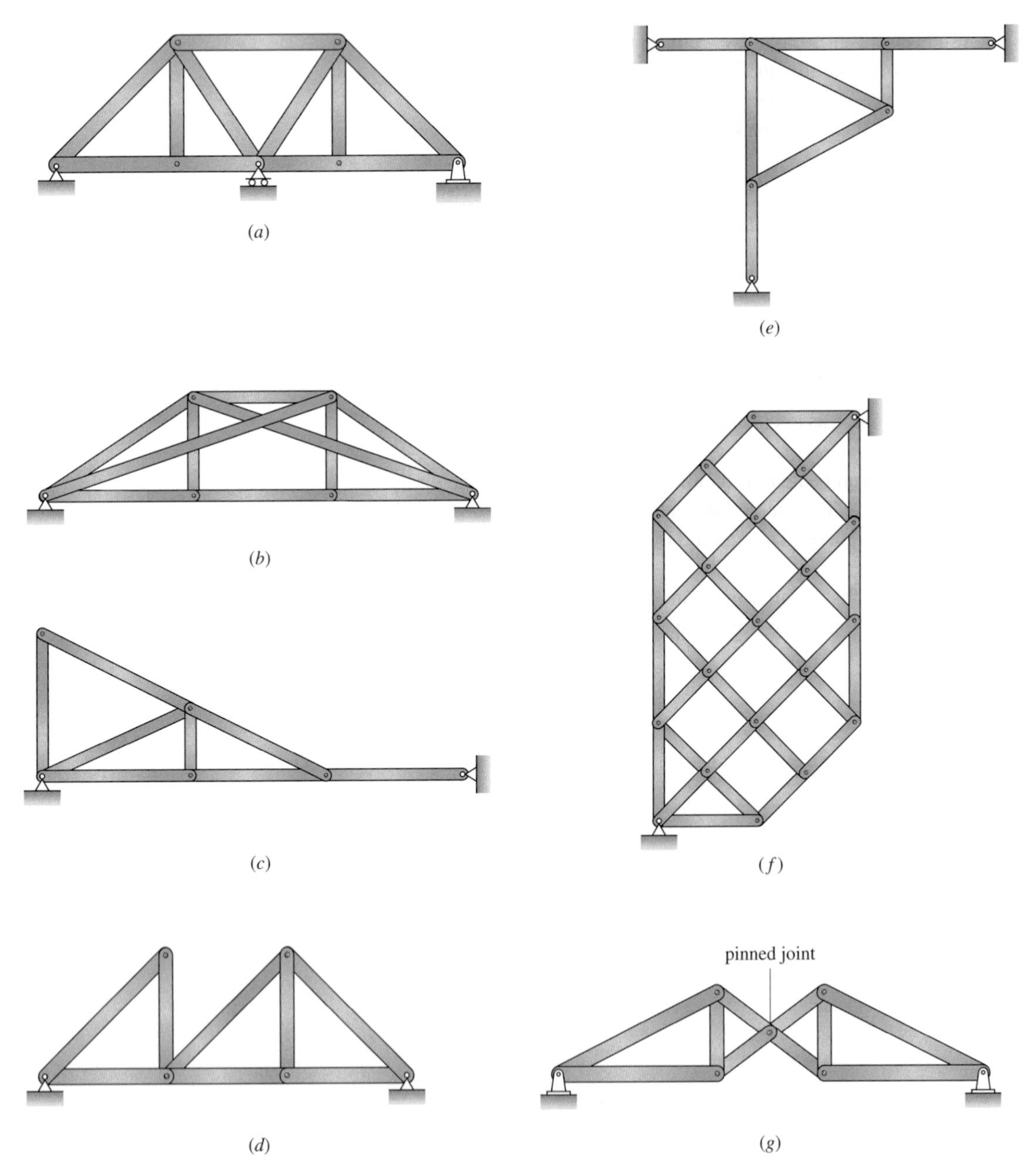

P4.1

P4.2. Classify the trusses in Figure P4.2 as stable or unstable. If stable, indicate if determinate or indeterminate. If indeterminate, indicate the degree.

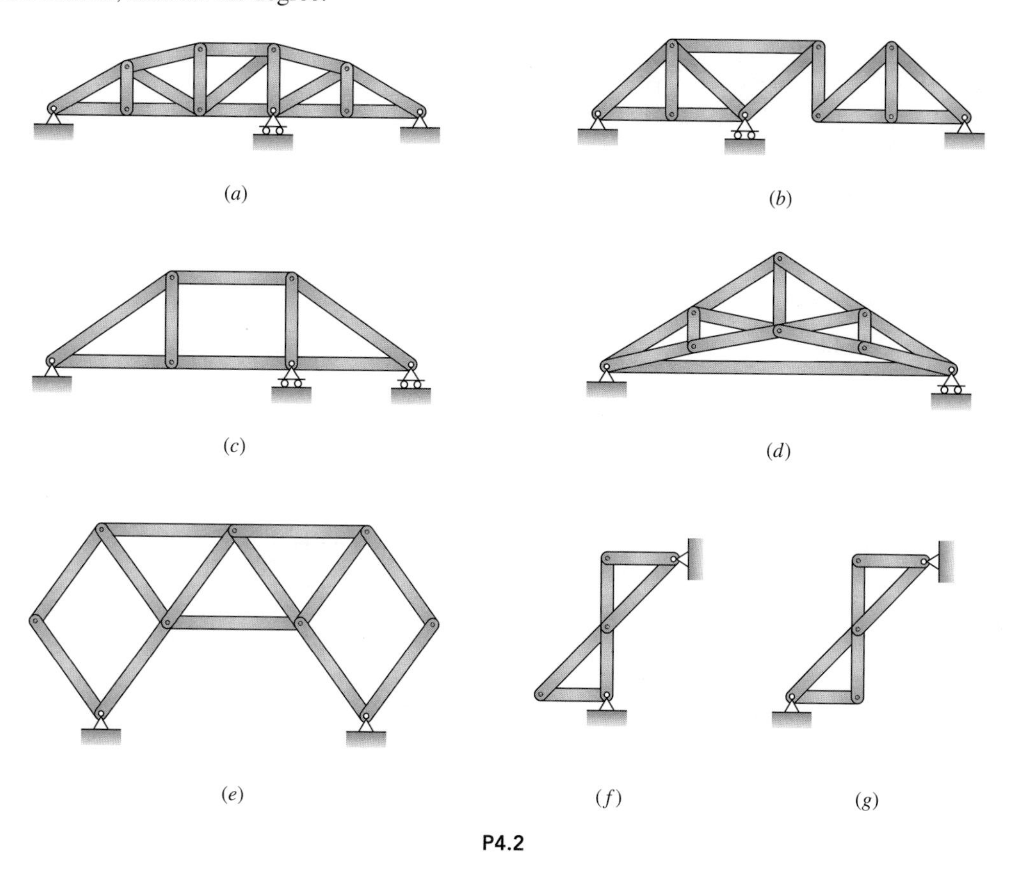

P4.2

P4.3 and **P4.4.** Determine the forces in all bars of the trusses. Indicate tension or compression.

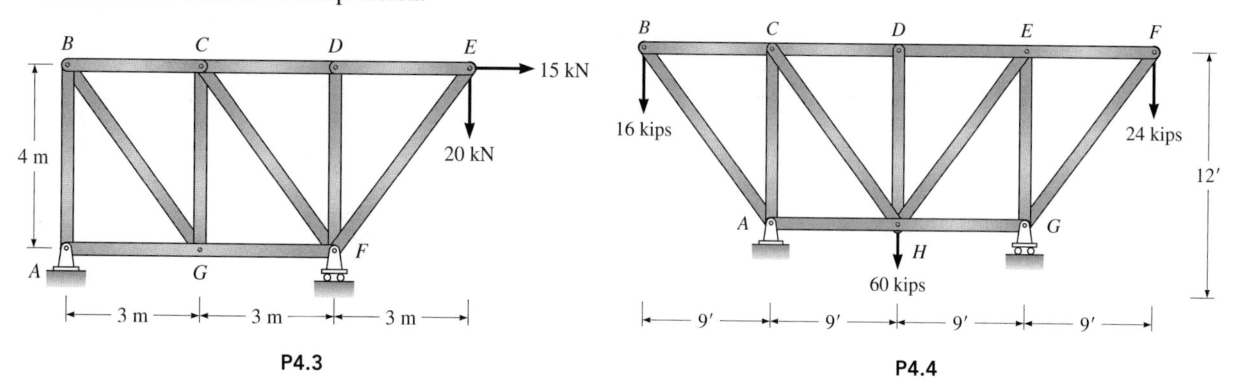

P4.5 to **P4.10.** Determine the forces in all bars of the trusses. Indicate tension or compression.

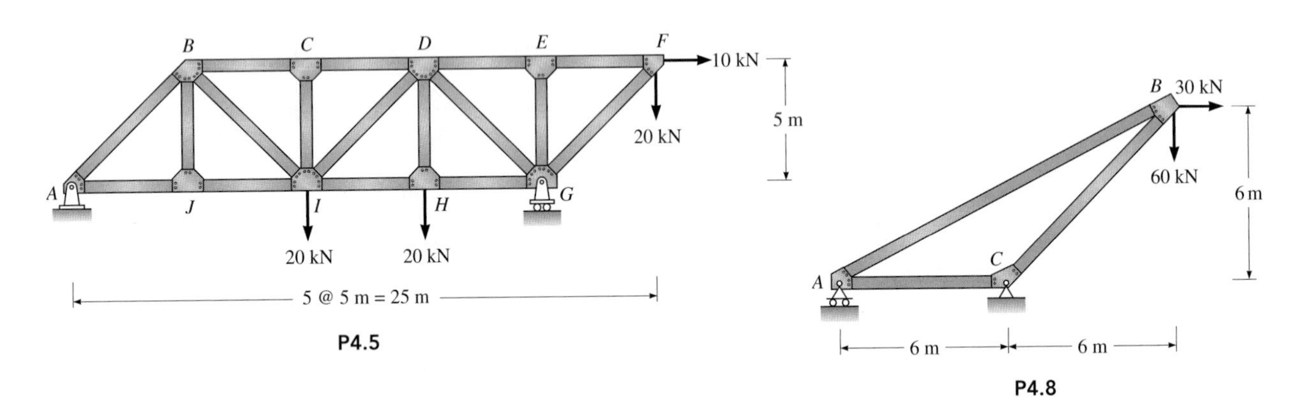

P4.5

P4.8

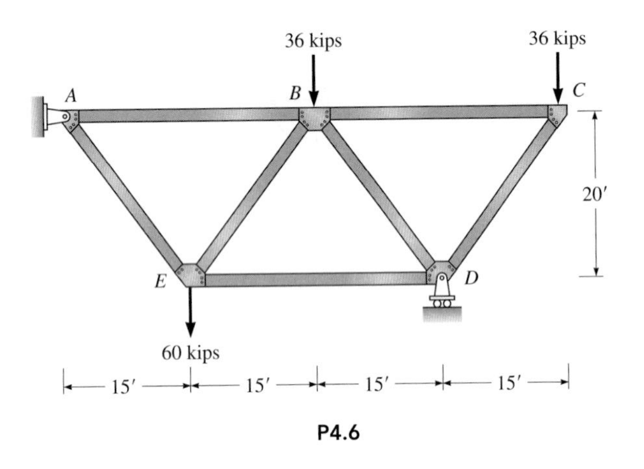

P4.6

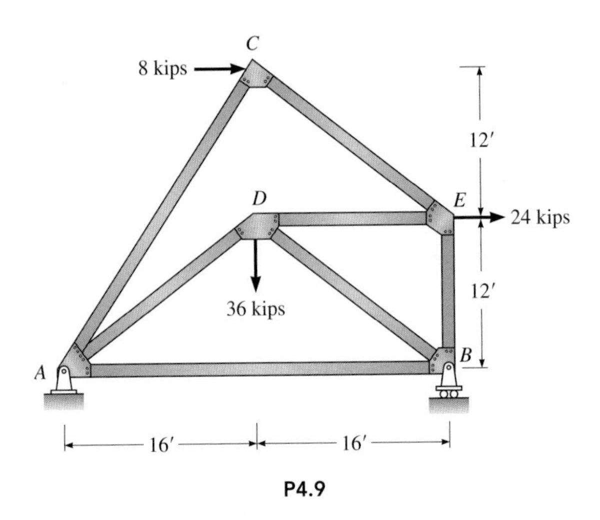

P4.9

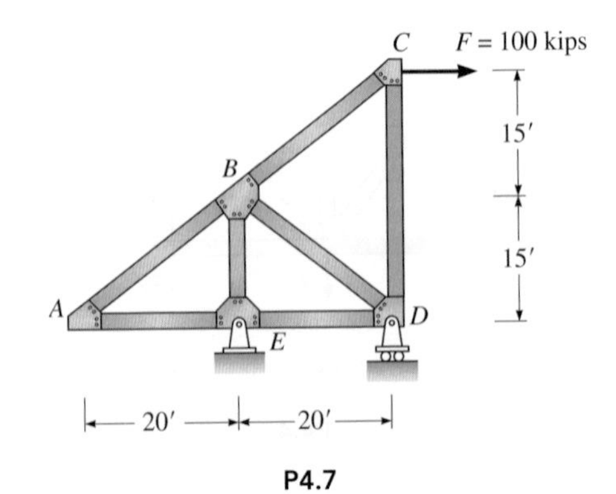

P4.7

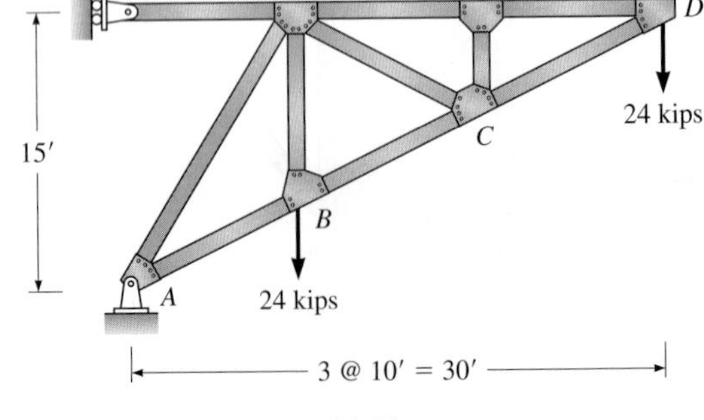

P4.10

P4.11 to **P4.15.** Determine the forces in all bars of the trusses. Indicate if tension or compression.

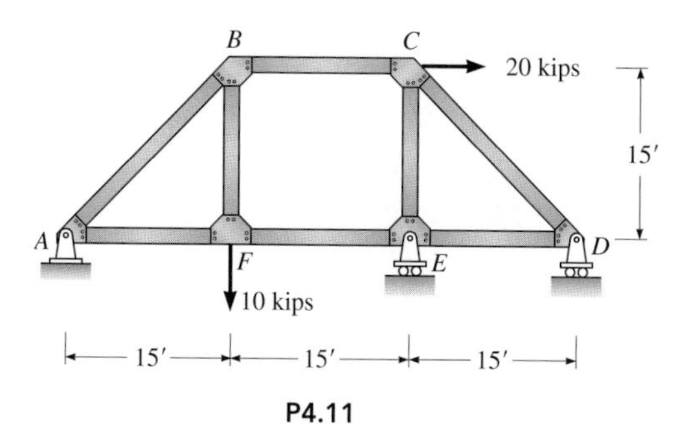

P4.11

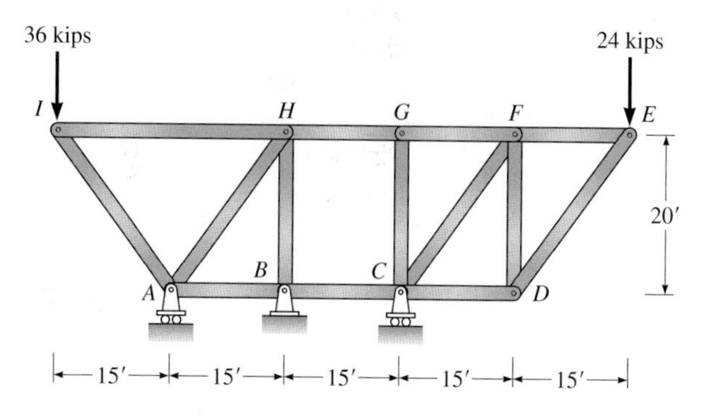

P4.12

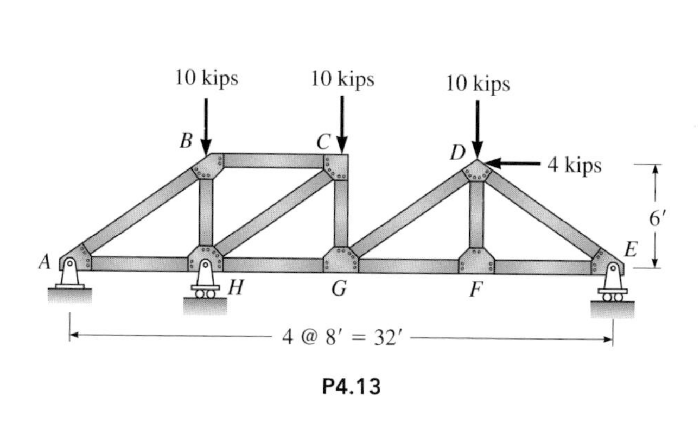

P4.13

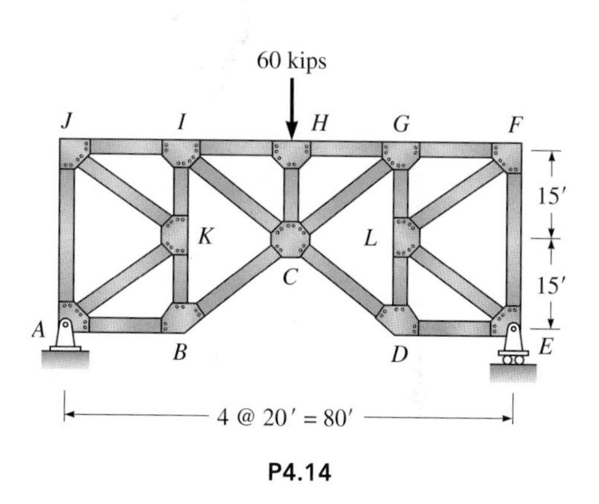

P4.14

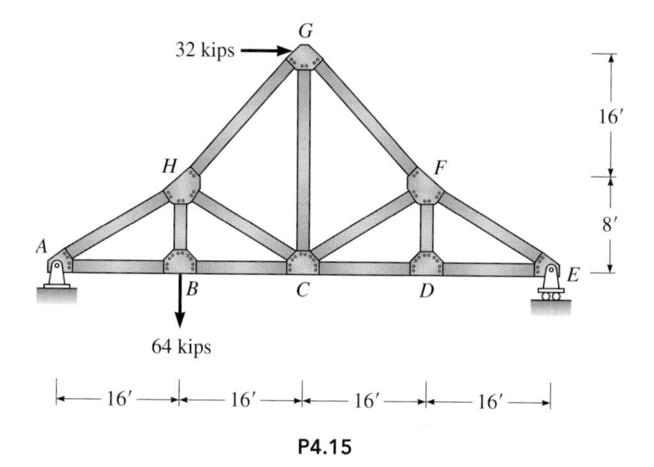

P4.15

P4.16. Determine the forces in all bars of the truss. *Hint*: If you have trouble computing bar forces, review *K* truss analysis in Example 4.6.

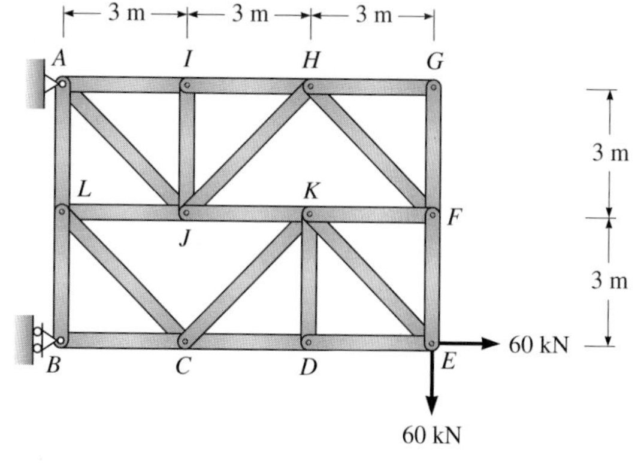

P4.16

P4.17 to **P4.21.** Determine the forces in all bars of the trusses. Indicate if tension or compression.

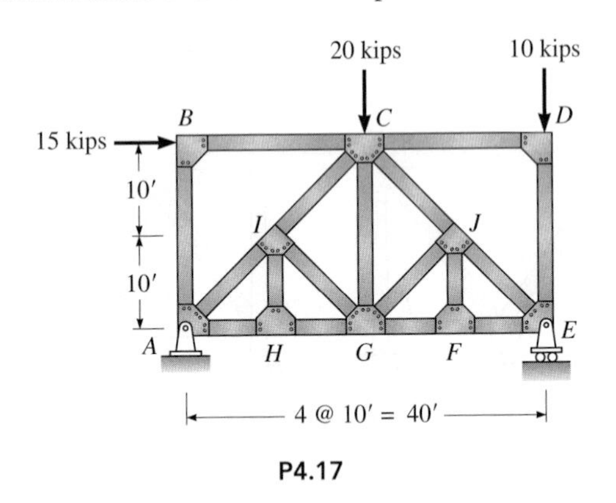

P4.17

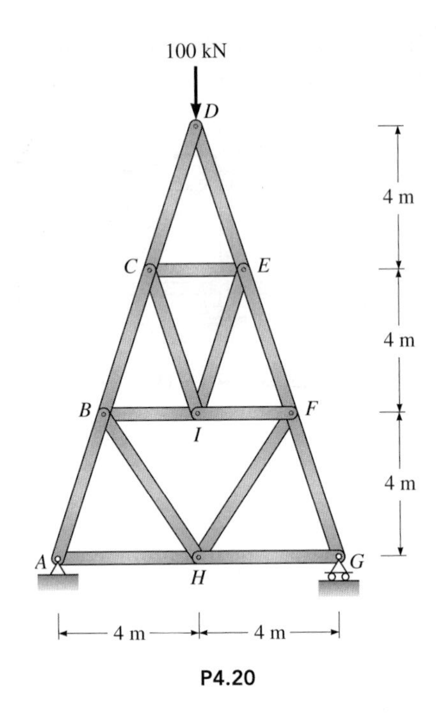

P4.20

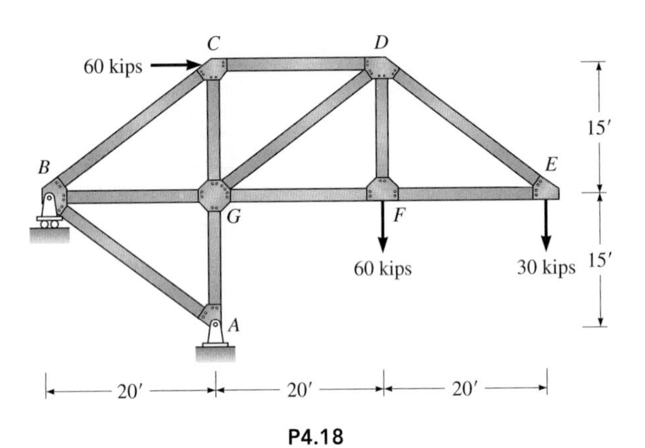

P4.18

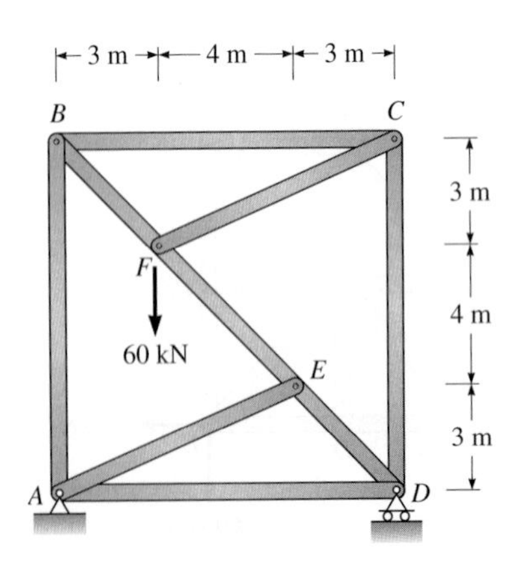

P4.19

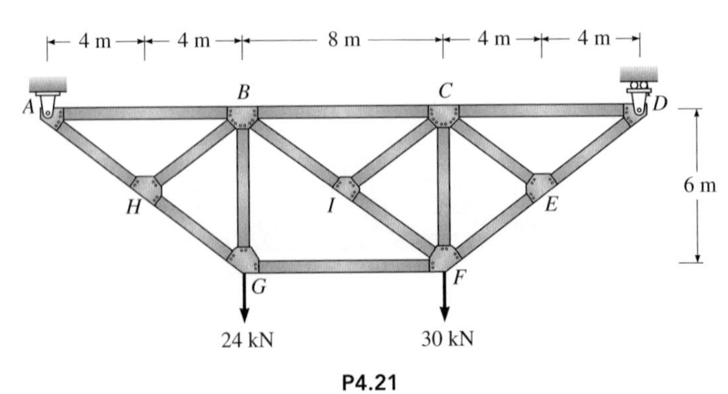

P4.21

P4.22 to **P4.26.** Determine the forces in all truss bars. Indicate tension or compression.

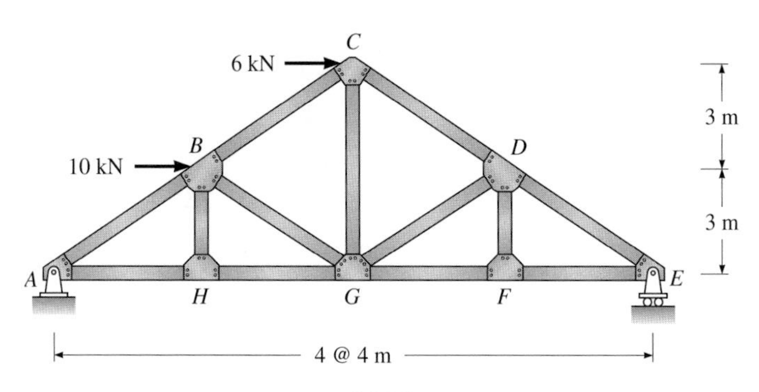

P4.22

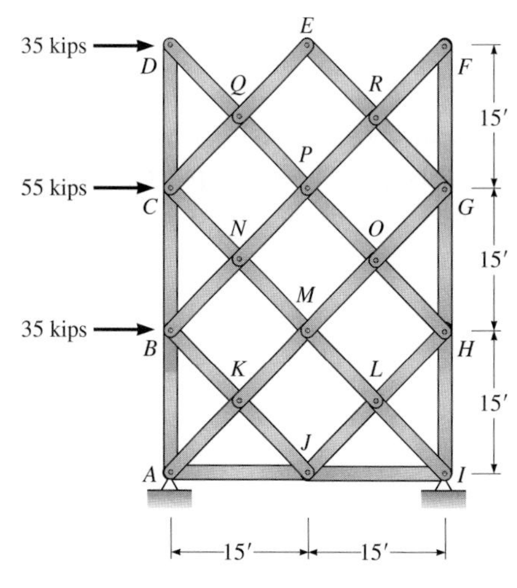

P4.25

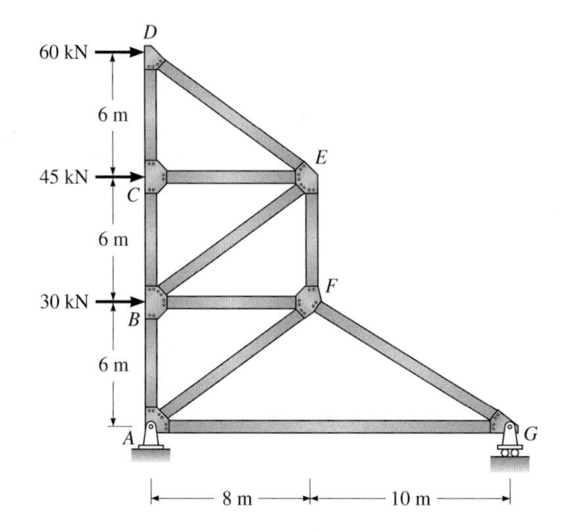

P4.23

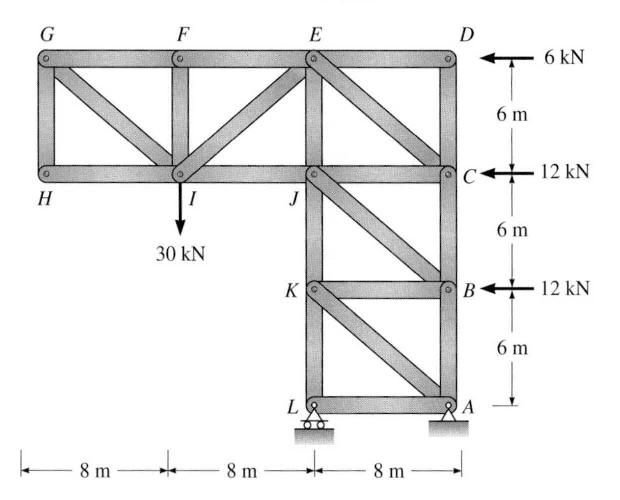

P4.24

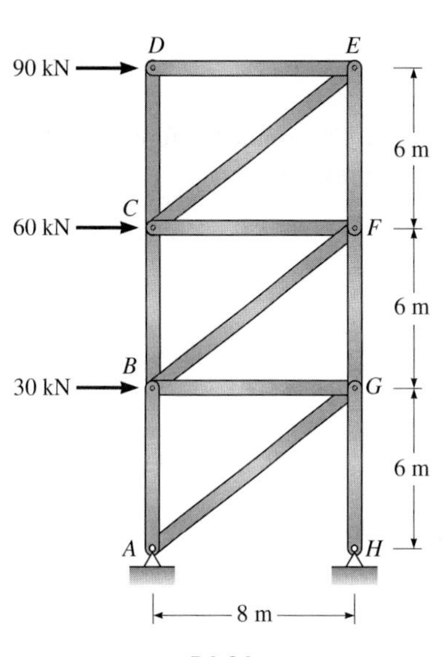

P4.26

P4.27. Determine the forces in all bars of the truss in Figure P4.27. If your solution is statically inconsistent, what conclusions can you draw about the truss? How might you modify the truss to improve its behavior? Also, analyze the truss with your computer program. Explain your results.

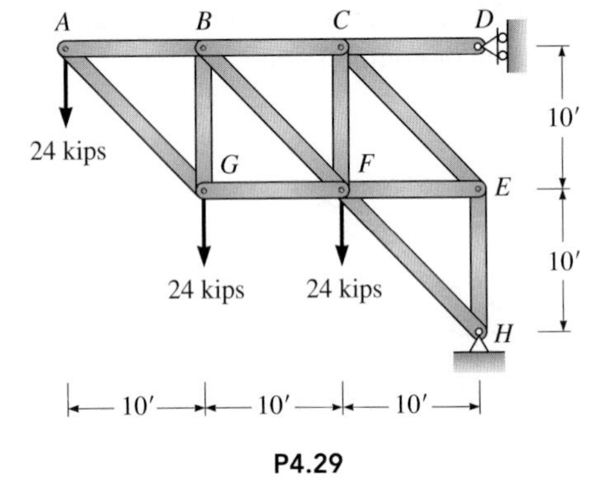

P4.29

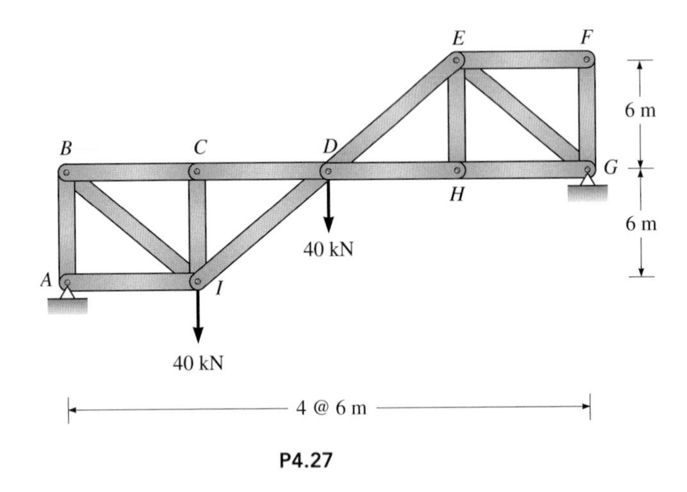

P4.27

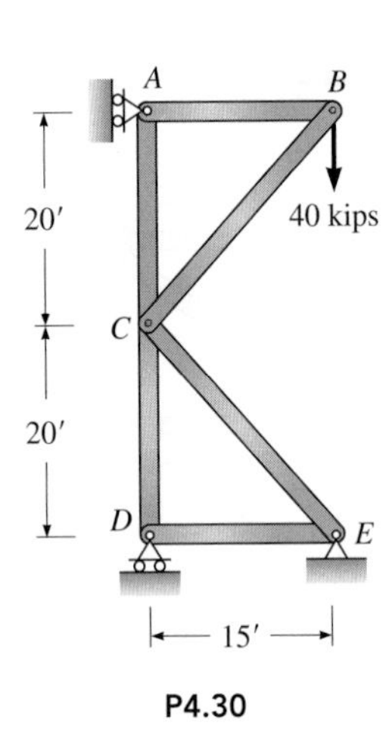

P4.30

P4.28 to **P4.31.** Determine the forces in all bars. Indicate tension or compression.

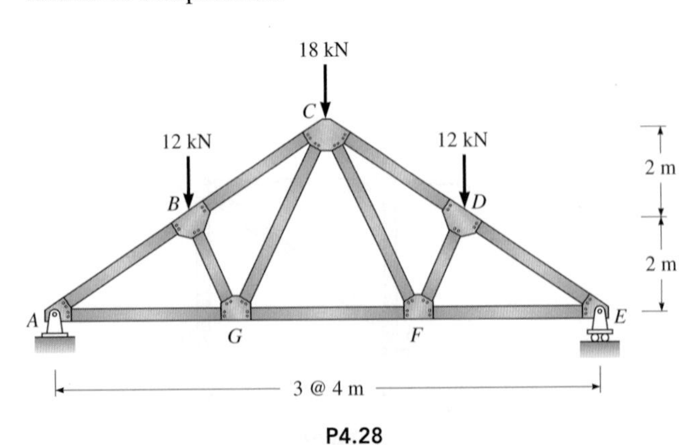

P4.28

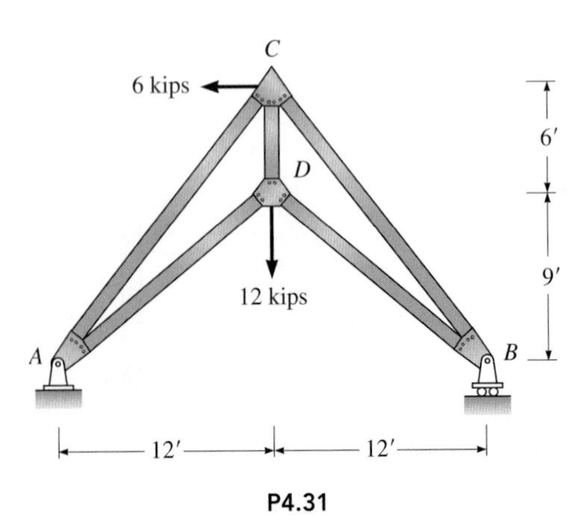

P4.31

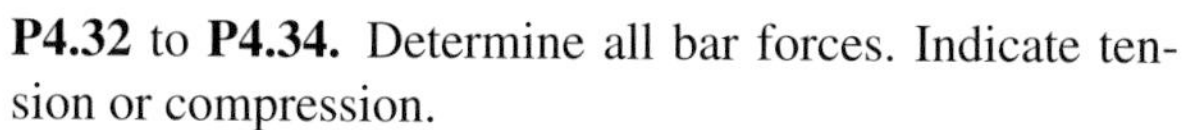

P4.32 to **P4.34.** Determine all bar forces. Indicate tension or compression.

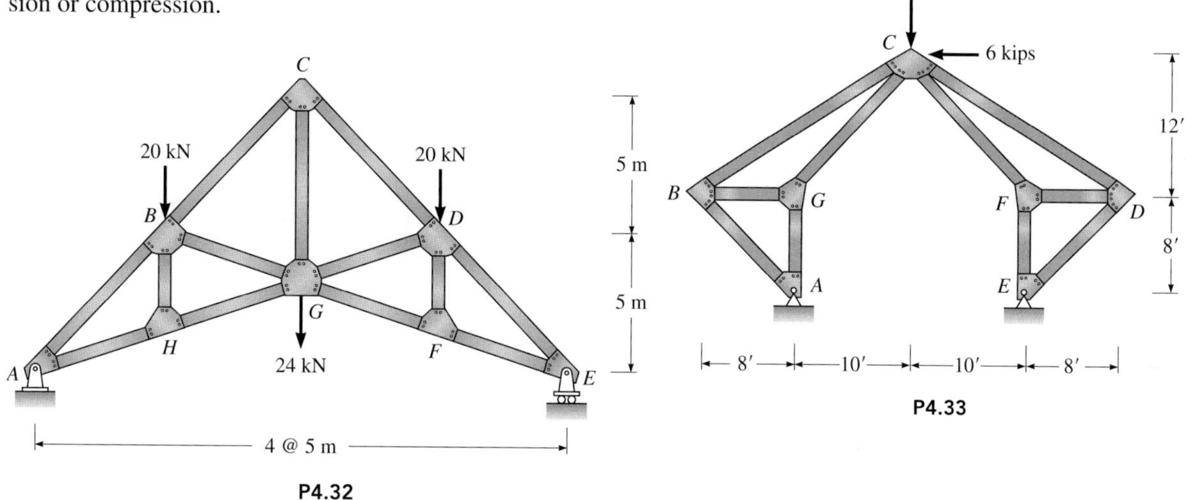

P4.32

P4.33

20 kN 40 kN 40 kN 40 kN 40 kN 40 kN 20 kN

B C D E F G H

M N O P

A L K J I

3 m

4 m

6 @ 4 m

P4.34

P4.35 to **P4.36.** Using the method of sections, determine the forces in the bars listed below each figure.

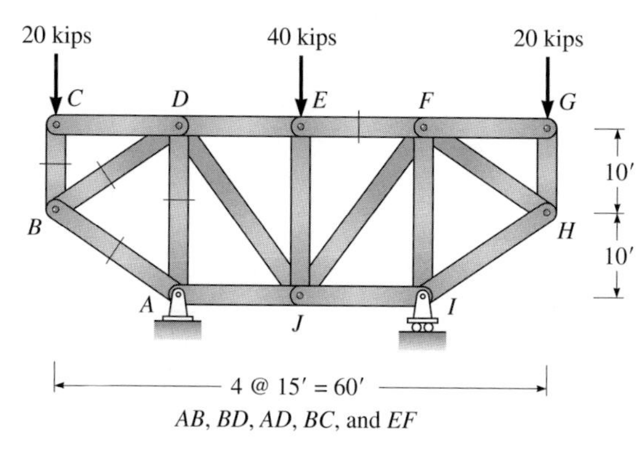

AB, *BD*, *AD*, *BC*, and *EF*

P4.35

J

K I

L H

A B C D E F G

3′

6′

12′

30 kips 90 kips 30 kips

6 @ 15′ = 90′

BL, *KJ*, *JD*, and *LC*

P4.36

P4.37 and **P4.39.** Using the method of sections, determine the forces in the bars listed below each figure.

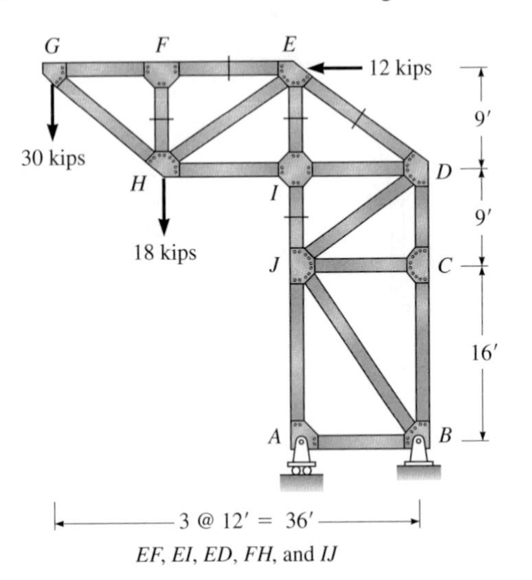

EF, EI, ED, FH, and *IJ*

P4.37

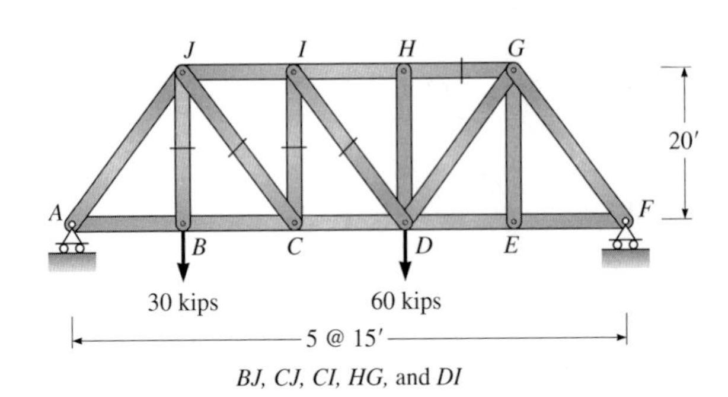

BJ, CJ, CI, HG, and *DI*

P4.38

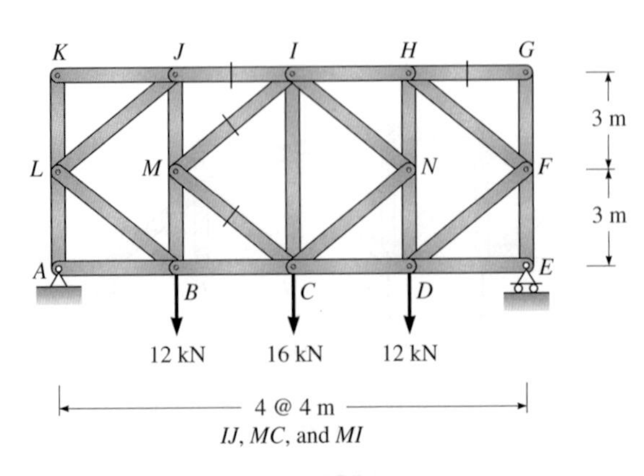

IJ, MC, and *MI*

P4.39

P4.40 to **P4.42.** Determine the forces in all bars of the trusses in Figures P4.40 to P4.42. Indicate if bar forces are tension or compression. *Hint*: Start with the method of sections.

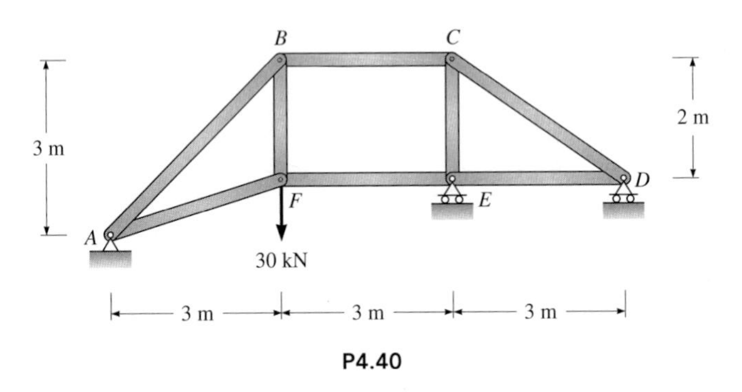

P4.40

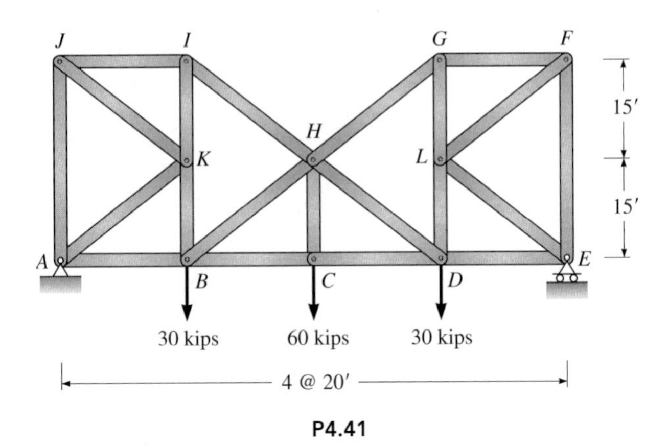

P4.41

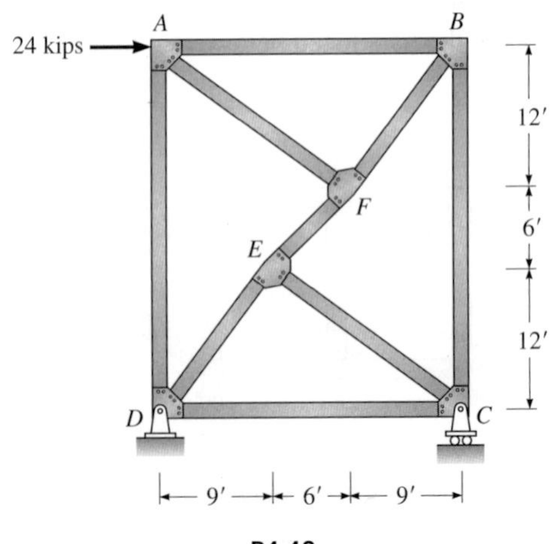

P4.42

P4.43 to **P4.47.** Determine the forces or components of force in all bars of the trusses in Figures P4.43 to P4.47. Indicate tension or compression.

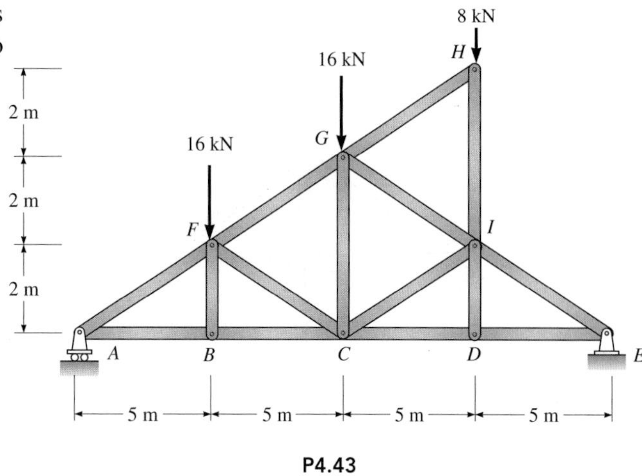

P4.43

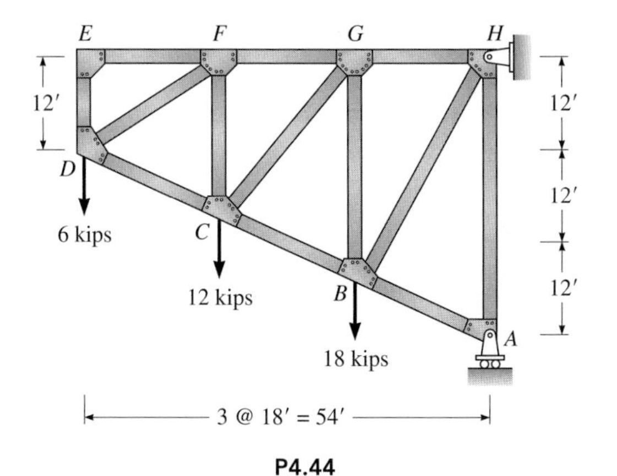

P4.44

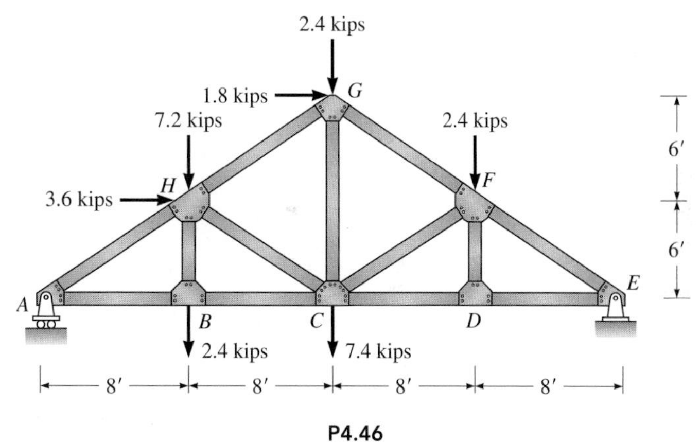

P4.46

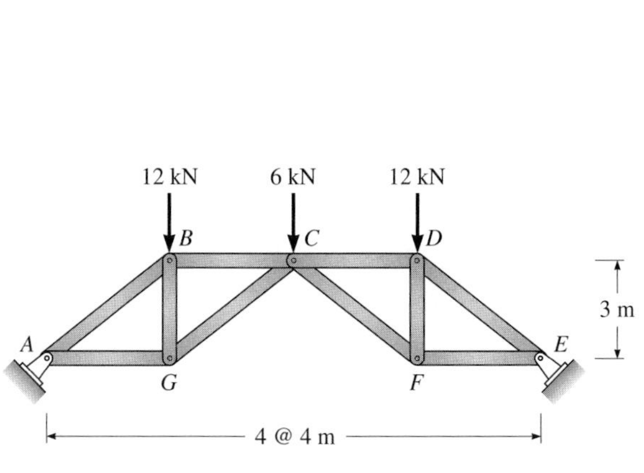

P4.45

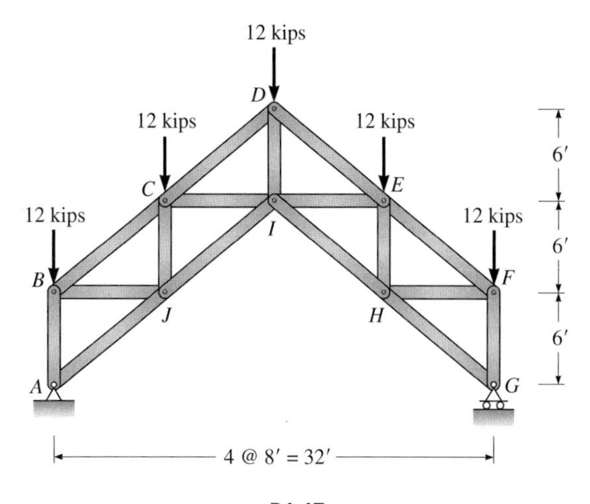

P4.47

P4.48 to **P4.51.** Determine the forces or components of force in all bars of the trusses in Figures P4.48 to P4.51. Indicate tension or compression.

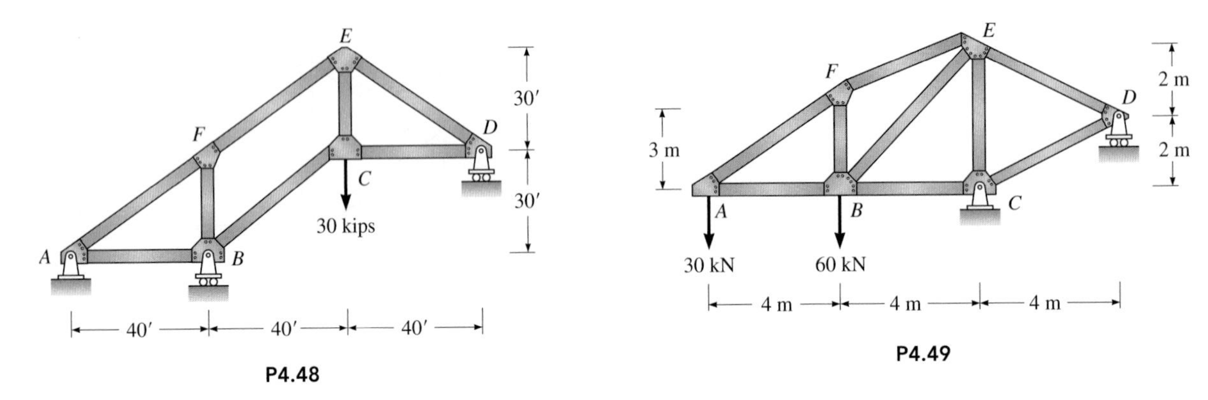

P4.48

P4.49

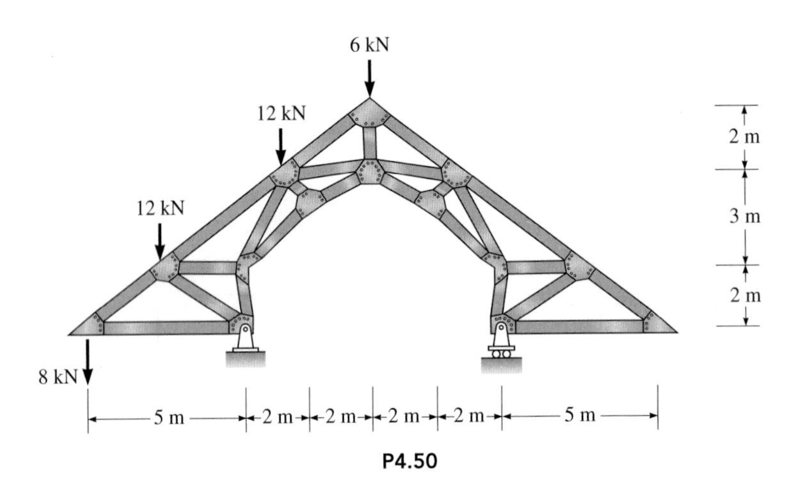

P4.50

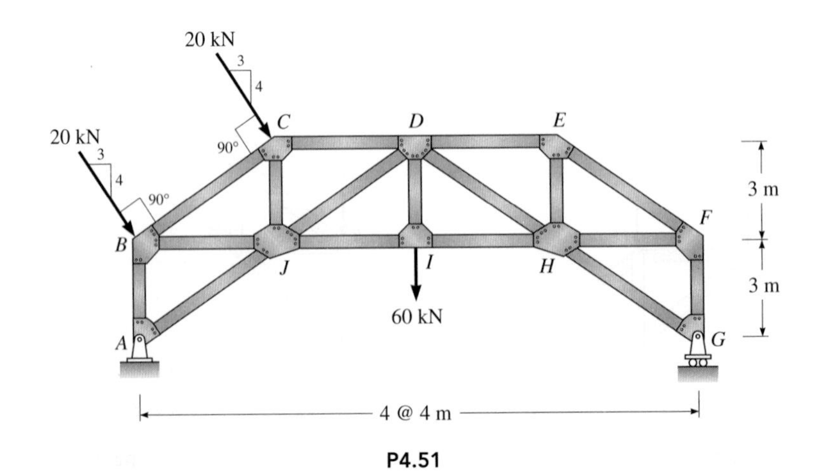

P4.51

P4.52. A two-lane highway bridge, supported on two deck trusses that span 64 ft, consists of an 8-in reinforced concrete slab supported on four steel stringers. The slab is protected by a 2-in wearing surface of asphalt. The 16-ft-long stringers frame into the floor beams, which in turn transfer the live and dead loads to the panel points of each truss. The truss, bolted to the left abutment at point A, may be treated as pin supported. The right end of the truss rests on an elastomeric pad at G. The elastomeric pad, which permits only horizontal displacement of the joint, can be treated as a roller. The loads shown represent the total dead and live loads. The 18-kip load is an additional live load that represents a heavy wheel load. Determine the force in the lower chord between panel points I and J, the force in member JB, and the reaction applied to the abutment at support A.

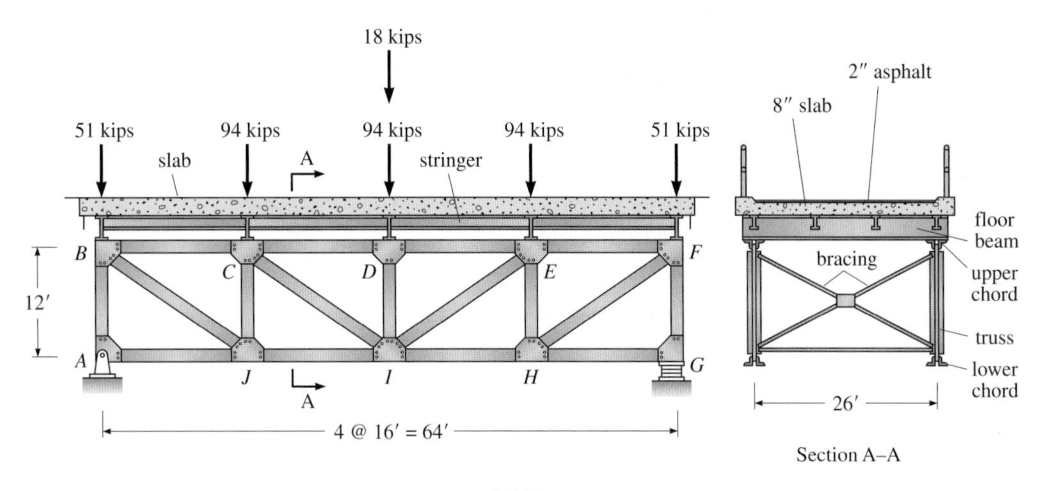

P4.52

P4.53. *Computer analysis of a truss.* The purpose of this study is to show that the *magnitude of the joint displacements* as well as the magnitude of the forces in members may control the proportions of structural members. For example, building codes typically specify maximum permitted displacements to ensure that excessive cracking of attached construction, such as exterior walls and windows, does not occur (see Photo 1.1 in Section 1.3).

A preliminary design of the truss in Figure P4.53 produces the following bar areas: member 1, 2.5 in^2; member 2, 3 in^2; and member 3, 2 in^2. Also $E = 29{,}000$ kips/in^2.

Case 1: Determine all bar forces, joint reactions, and joint displacements, assuming pin joints. Use the computer program to plot the deflected shape.

Case 2: If the maximum horizontal displacement of joint 2 is not to exceed 0.25 in, determine the minimum required area of the truss bars. For this case assume that all truss members have the *same* cross-sectional area. Round the area to the nearest whole number.

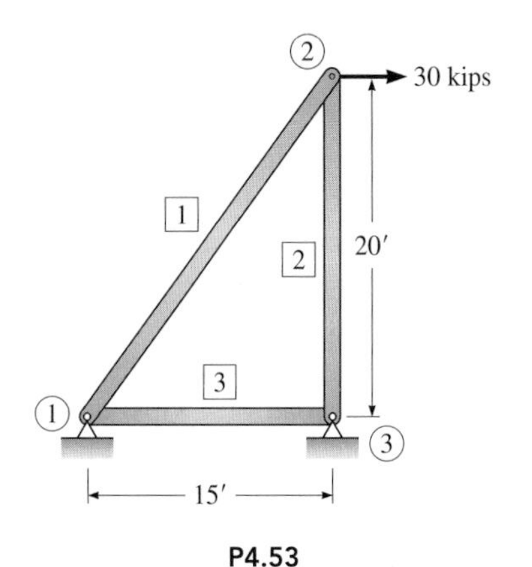

P4.53

P4.54. *Computer study.* The objective is to compare the behavior of a determinate and an indeterminate structure.

The forces in members of *determinate* trusses are not affected by member stiffness. Therefore, there was no need to specify the cross-sectional properties of the bars of the determinate trusses we analyzed by hand computations earlier in this chapter. In a *determinate* structure, for a given set of loads, only one load path is available to transmit the loads into the supports, whereas in an *indeterminate structure*, multiple load paths exist (see Section 3.10). In the case of trusses, the axial stiffness of members (a function of a member's cross-sectional area) that make up each load path will influence the magnitude of the force in each member of the load path. We examine this aspect of behavior by varying the properties of certain members of the indeterminate truss shown in Figure P4.54. Use $E = 29{,}000$ kips/in^2.

Case 1: Determine the reactions and the forces in members 4 and 5 if the area of all bars is 10 in^2.

Case 2: Repeat the analysis in **Case 1**, this time increasing the area of member 4 to 20 in^2. The area of all other bars remains 10 in^2.

Case 3: Repeat the analysis in **Case 1**, increasing the area of member 5 to 20 in^2. The area of all other bars remains 10 in^2.

What conclusions do you reach from the above study?

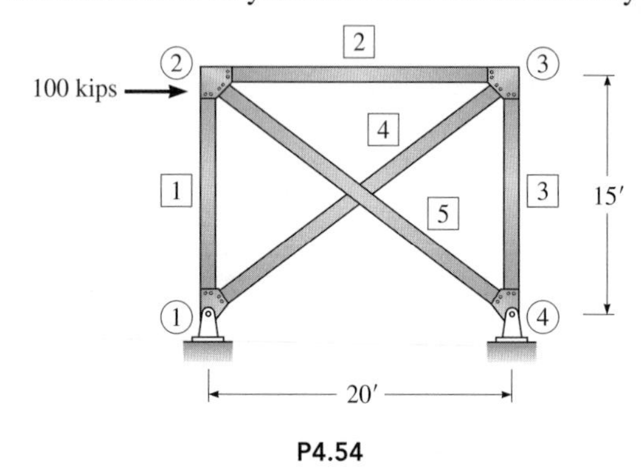

P4.54

Practical Application

P4.55. *Computer analysis of a truss with rigid joints.* The truss in Figure P4.55 is constructed of square steel tubes *welded* to form a structure with rigid joints. The top chord members 1, 2, 3, and 4 are $4 \times 4 \times {}^1/_4$ inch square tubes with $A = 3.59$ in^2 and $I = 8.22$ in^4. All other members are $3 \times 3 \times {}^1/_4$ inch square tubes with $A = 2.59$ in^2 and $I = 3.16$ in^4. Use $E = 29{,}000$ kips/in^2.

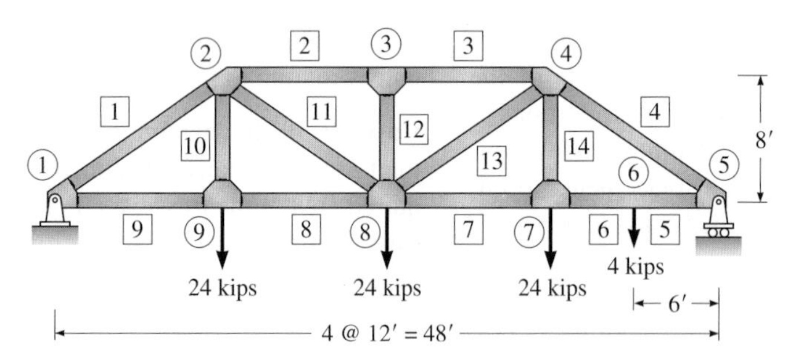

P4.55

(*a*) Considering all joints as rigid, compute the axial forces and moments in all bars and the deflection at midspan when the three 24-kip design loads act at joints 7, 8, and 9. (Ignore the 4-kip load.)

(*b*) If a hoist is also attached to the lower chord at the midpoint of the end panel on the right (labeled joint 6*) to raise a concentrated load of 4 kips, determine the forces and moments in the lower chord (members 5 and 6). If the maximum stress is not to exceed 25 kips/in^2, can the lower chord support the 4-kip load safely in addition to the three 24-kip loads? Compute the maximum stress, using the equation

$$\sigma = \frac{F}{A} + \frac{Mc}{I}$$

where $c = 1.5$ in (one-half the depth of the lower chord).

**Note*: If you wish to compute the forces or deflection at a particular point of a member, designate the point as a joint.

Practical Application

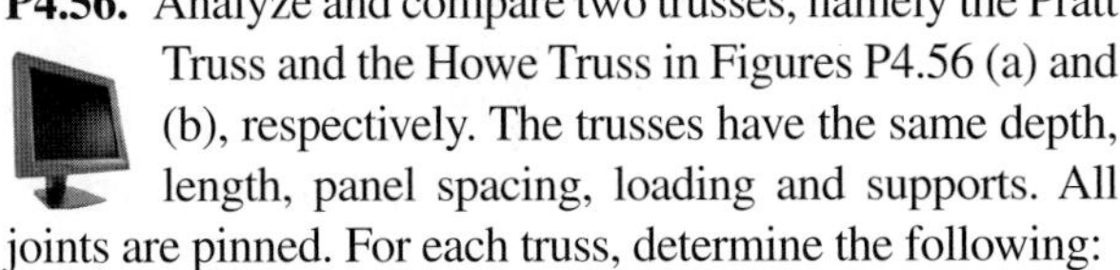

P4.56. Analyze and compare two trusses, namely the Pratt Truss and the Howe Truss in Figures P4.56 (a) and (b), respectively. The trusses have the same depth, length, panel spacing, loading and supports. All joints are pinned. For each truss, determine the following:

a) All bar forces, indicate tension or compression.

b) The required cross sectional areas for each bar, given an allowable tensile stress of 45 ksi, and an allowable compressive stress of 24 ksi. Note that allowable compressive stress is lower due to buckling.

c) Tabulate your results showing bar forces, cross sectional areas, and lengths.

d) Calculate the total weight of each truss and determine which truss has a more efficient configuration. Explain your results.

e) What other conclusions can you draw from the study?

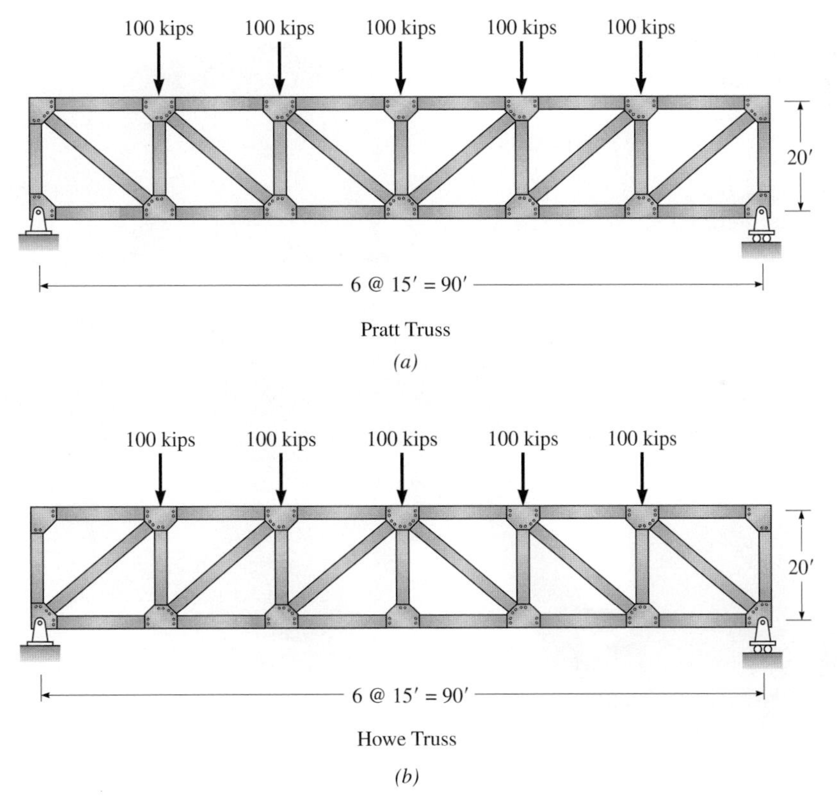

Pratt Truss

(*a*)

Howe Truss

(*b*)

P4.56

Brazos River Bridge in Brazos, Texas

The Brazos River Bridge collapsed in 1956 during erection of the 973-ft, continuous steel plate girders that support the roadway. The failure was initiated by overstress of the connections between the web and flange during erection. Structures are particularly vulnerable to failure during erection because stiffening elements—for example, floor slabs and bracing—may not be in place. In addition, the structure's strength may be reduced when certain connections are partially bolted or not fully welded to permit precise alignment of members during construction.

C H A P T E R 5

Beams and Frames

Chapter Objectives

- Learn the structural characteristics of beams and frames of various configurations and support conditions.
- Review the beam theory and the relationship between load, shear, and moment using first-order analysis, as previously learned in statics and mechanics of materials.
- Solve reactions and write equations for shear and moment, construct shear and moment curves, and sketch the deflected shapes of loaded beams and frames.
- Classify determinate and indeterminate beam or frame structures, determine the degree of indeterminacy for the latter, and determine if a beam or frame is stable or unstable.

5.1 Introduction

Beams

Beams are one of the most common elements found in structures. When a beam is loaded perpendicular to its longitudinal axis, internal forces—shear and moment—develop to transmit the applied loads into the supports. If the ends of a beam are restrained longitudinally by its supports, or if a beam is a component of a continuous frame, axial force may also develop. If the axial force is small—the typical situation for most beams— it can be neglected when the member is designed. In the case of reinforced concrete beams, small values of axial compression actually produce a modest increase (on the order of 5 to 10 percent) in the flexural strength of the member.

To design a beam, the engineer must construct the shear and moment curves to determine the location and magnitude of the maximum values of these forces. Except for short, heavily loaded beams whose dimensions are controlled by shear requirements, the proportions of the cross section are determined by the magnitude of the maximum moment in the span. After the

Earliest demonstration of beam theory were ascribed to Leonardo da Vinci (1452–1519) and later by Galileo Galilei (1564–1642) during the Renaissance Era. Beam theory (e.g., Eq. 5.1) was not formalized with accuracy until Jacob Bernoulli's (1654–1705) work and later by Daniel Bernoulli (1700–1782) and Leonard Euler's (1707–1783) development of the beam theory equation as well as by Robert Hooke's Law of Elasticity (1635–1705). However, it required a synergy of theories to develop practical applications in engineering design, for which is credited to Claude Louis Navier (1785–1836). In his publication Résumé des Leçons in 1826, Navier combined elastic theory and beam theory and moved engineering design methods from empirical to practical methods for linearly elastic members.

section is sized at the point of maximum moment, the design is completed by verifying that the shear stresses at the point of maximum shear—usually adjacent to a support—are equal to or less than the allowable shear strength of the material. Finally, the deflections produced by service loads must be checked to ensure that the member has adequate stiffness. Limits on deflection are set by structural codes.

If behavior is elastic (as, for example, when members are made of steel or aluminum), and if allowable stress design is used, the required cross section can be established using the basic beam equation.

$$\sigma = \frac{Mc}{I} \tag{5.1}$$

where σ = flexural stress produced by service load moment M
c = distance from neutral axis to the outside fiber where the flexural stress σ is to be evaluated
I = moment of inertia of the cross section with respect to the centroidal axis of the section

To select a cross section, σ in Equation 5.1 is set equal to the allowable flexural stress σ_{allow}, and the equation is solved for I/c, which is termed the *section modulus* and denoted by S_x.

$$S_x = \frac{I}{c} = \frac{M}{\sigma_{\text{allow}}} \tag{5.2}$$

S_x, a measure of a cross section's flexural capacity, is tabulated in design handbooks for standard shapes of beams produced by various manufacturers.

After a cross section is sized for moment, the designer checks shear stress at the section where the shear force V is maximum. For beams that behave elastically, shear stresses are computed by the equation

$$\tau = \frac{VQ}{Ib} \tag{5.3}$$

where τ = shear stress produced by shear force V
V = maximum shear (from shear curve)
Q = static moment of that part of area that lies above or below point where shear stress is to be computed; for a rectangular or an I-shaped beam, maximum shear stress occurs at middepth
I = moment of inertia of cross-sectional area about the centroid of section
b = thickness of cross section at elevation where τ is computed

When a beam has a rectangular cross section, the maximum shear stress occurs at middepth. For this case Equation 5.3 reduces to

$$\tau_{\max} = \frac{3V}{2A} \tag{5.4}$$

where A equals the area of the cross section.

If *strength design* (which has largely replaced working stress design) is used, members are sized for *factored loads*. Factored loads are produced by multiplying service loads by *load factors*—numbers that are typically greater than 1. Using factored loads, the designer carries out an elastic analysis—the subject of this text. The forces produced by factored loads represent the *required strength*. The member is sized so that its *design strength* is equal to the required strength. The design strength, evaluated by considering the state of stress associated with a particular mode of failure, is a function of the properties of the cross section, the stress condition at failure (for example, steel yields or concrete crushes), and a *reduction factor*—a number less than 1.

The final step in the design of a beam is to verify that it does not deflect excessively (i.e., that deflections are within the limits specified by the applicable design code). Beams that are excessively flexible undergo large deflections that can damage attached nonstructural construction: plaster ceilings, masonry walls, and rigid piping, for example, may crack.

Since most beams that span short distances, say up to 30 or 40 ft, are manufactured with a constant cross section, to minimize cost, they have excess flexural capacity at all sections except the one at which maximum moment occurs. If spans are long, in the range 150 to 200 ft or more, and if loads are large, then deep heavy girders are required to support the design loads. For this situation, in which the weight of the girder may represent as much as 75 to 80 percent of the total load, some economy may be achieved by shaping the beam to conform to the ordinates of the moment curve. For these largest girders, the moment capacity of the cross section can be adjusted either by varying the depth of the beam or by changing the thickness of the flange (see Figure 5.1). In addition, reducing the weight of the girders may result in smaller piers and foundations.

Beams are typically classified by the manner in which they are supported. A beam supported by a pin at one end and a roller at the other end is called a *simply supported* beam (see Figure 5.2*a*). If the end of a simply supported beam extends over a support, it is referred to as a beam with an *overhang* (see Figure 5.2*b*). A *cantilever* beam is fixed at one end against translation and rotation (Figure 5.2*c*). Beams that are supported by several intermediate supports are called *continuous* beams (Figure 5.2*d*). If both ends of a beam are fixed by the supports, the beam is termed *fixed ended* (see Figure 5.2*e*). Fixed-end beams are not commonly constructed in practice, but the values of end moments in them produced by various types of load are used extensively as the starting point in several methods of analysis for indeterminate structures (see Figure 2.5). In this chapter we discuss only determinate beams that can be analyzed by the three equations of statics. Beams of this type are common in wood and bolted or riveted steel construction. On the other hand, continuous beams (analyzed in Chaps. 11 to 13) are commonly found in structures with rigid joints—welded steel or reinforced concrete frames, for example.

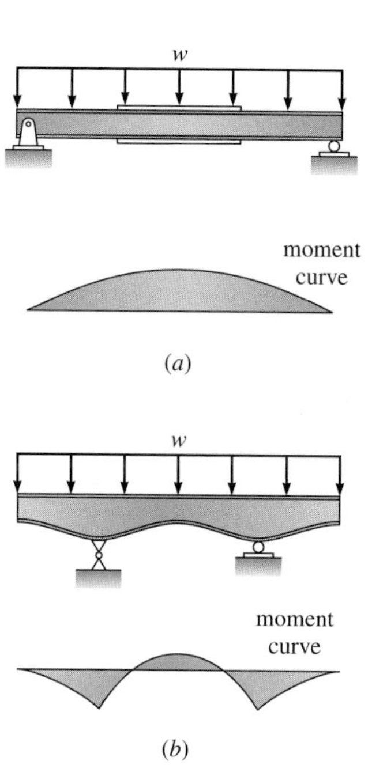

Figure 5.1: (*a*) Flange thickness varied to increase flexural capacity; (*b*) depth varied to modify flexural capacity.

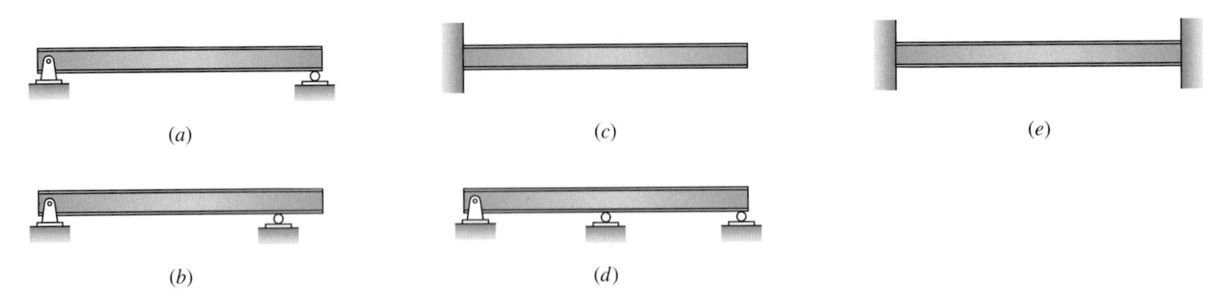

Figure 5.2: Common beam types: (*a*) simply supported; (*b*) beam with overhang; (*c*) cantilever; (*d*) two-span continuous; (*e*) fixed ended.

Frames

Frames, as discussed in Chapter 1, are structural elements composed of beams and columns connected by rigid joints. The angle between the beam and column is usually 90°. As shown in Figure 5.3*a* and *b*, frames may consist of a single column and girder or, as in the case of a multistory building, of many columns and beams.

Frames may be divided into two categories: braced and unbraced. A *braced frame* is one in which the joints at each level are free to rotate but are prevented from moving laterally by attachment to a rigid element that can supply lateral restraint to the frame. For example, in a multistory building, structural frames are often attached to shear walls (stiff structural walls often constructed of reinforced concrete or reinforced masonry; see Figure 5.3*c*). In simple one-bay frames, light diagonal cross-bracing connected to the base of columns can be used to resist lateral displacement of top joints (see Figure 5.3*d*).

An *unbraced frame* (see Figure 5.3*e*) is one in which lateral resistance to displacement is supplied by the flexural stiffness of the beams and columns. In unbraced frames, joints are free to displace laterally as well as to rotate. Since unbraced frames tend to be relatively flexible compared to braced frames, under lateral load they may undergo large transverse deflections that

Photo. 5.1: Harvard Bridge. This bridge is composed of variable depth girders with overhangs at each end.

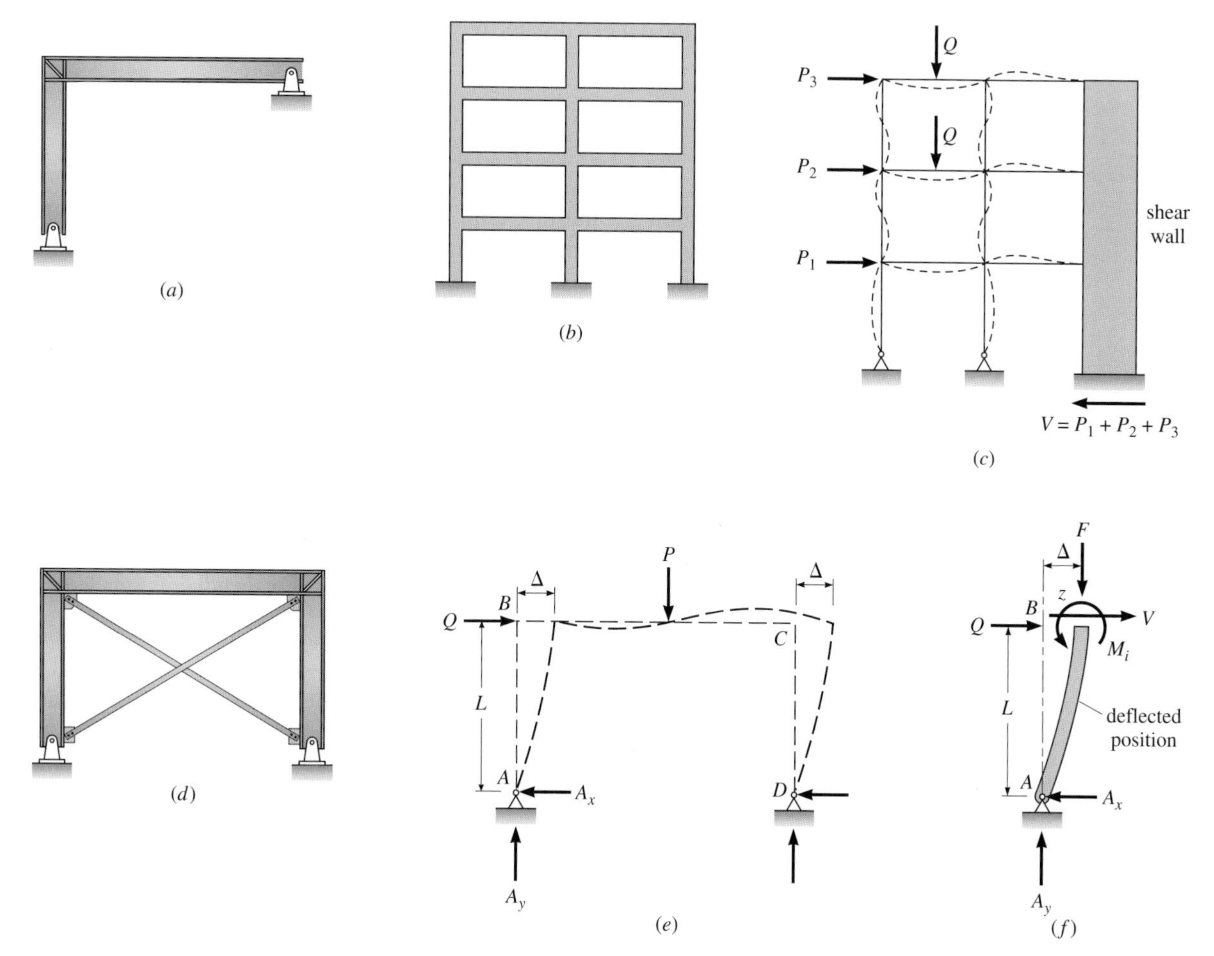

Figure 5.3: (*a*) Simple frame; (*b*) multistory continuous building frame; (*c*) frame braced by a shear wall; (*d*) frame braced by diagonal bracing; (*e*) sidesway of an unbraced frame; (*f*) free body of column in deflected position.

damage attached nonstructural elements, for example, walls, windows, and so forth.

Although both beams and columns of rigid frames carry axial force, shear, and moment, the axial force in beams is usually so small that it can be neglected and the beam sized for moment only. On the other hand, in columns, the axial force—particularly in the lower interior columns of multistory frames—is often large, and the moments are small. For columns of this type, proportions are determined primarily by the axial capacity of members.

If frames are flexible, additional bending moment is created by the lateral displacement of the member. For example, the tops of the columns in the unbraced frame in Figure 5.3*e* displace a distance Δ to the right. To evaluate the forces in the column, we consider a free body of column *AB* in its deflected position (see Figure 5.3*f*). The free body is cut by passing an imaginary plane through the column just below joint *B*. The cutting plane is perpendicular to the longitudinal axis of the column. We can express the internal moment M_i acting on the cut in terms of the reactions at the base of the column and the geometry of the deflected shape by summing moments about a z axis through the centerline of the column.

$$M_i = \Sigma M_z$$

$$M_i = A_x(L) + A_y(\Delta) \tag{5.5}$$

In Equation 5.5 the first term represents the moment produced by the applied loads, neglecting the lateral deflection of the column's axis. This moment is called the *primary moment* and associated with a *first-order* analysis (described in Section 1.7). The second term, $A_y(\Delta)$, which represents the additional moment produced by the eccentricity of the axial load, is termed the *secondary moment* or the *P-delta moment*. The secondary moment will be small and can be neglected without significant error under the following two conditions:

1. The axial forces are small (say, less than 10 percent of the axial capacity of the cross section).
2. The flexural stiffness of the column is large, so that the lateral displacement of the column's longitudinal axis produced by bending is small.

In this book we will only make a *first-order analysis*; that is, we do not consider the computation of the secondary moment—a subject usually covered in advanced courses in structural mechanics. Since we neglect secondary moments, the analysis of frames is similar to that of beams; that is, the analysis is complete when we establish the shear and moment curves (also the axial force) based on the initial geometry of the unloaded frame.

5.2 Scope of Chapter

We begin the study of beams and frames by discussing a number of basic operations that will be used frequently in deflection computations and in the analysis of indeterminate structures. These operations include

1. Writing expressions for shear and moment at a section in terms of the applied loads.
2. Constructing shear and moment curves.
3. Sketching the deflected shapes of loaded beams and frames.

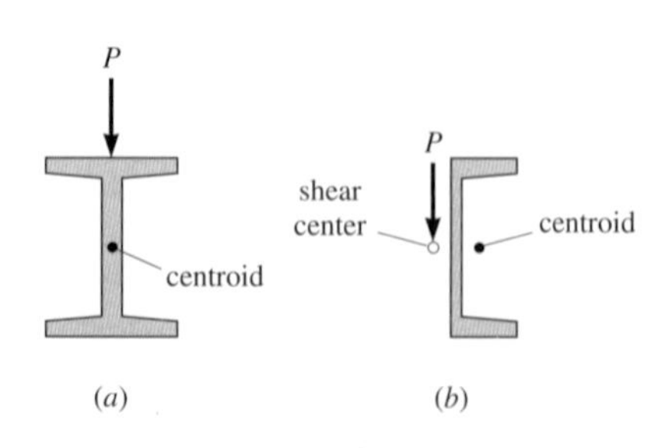

Figure 5.4: (*a*) Beam loaded through centroid of symmetric section; (*b*) unsymmetric section loaded through shear center.

Since many of these procedures were introduced previously in *statics* and *strength of materials courses*, much of this chapter for most students is a review of basic topics.

In the examples in this chapter, we assume that all beams and frames are two-dimensional structures supporting in-plane loads that produce shear, moment, and possibly axial forces, but no torsion. For this condition—one of the most common in actual practice—to exist, the *in-plane* loads must pass through the centroid of a symmetric section or through the shear center of an unsymmetric section (see Figure 5.4).

5.3 Equations for Shear and Moment

We begin the study of beams by writing equations that express the shear V and the moment M at sections along the longitudinal axis of a beam or frame in terms of the applied loads and the distance from a reference origin. Although equations for shear have limited use, those for moment are required in deflection computations for beams and frames by both the double-integration method (see Chapter 9) and *work-energy* methods (see Chapter 10).

As you may remember from the study of beams in mechanics of materials or statics courses, *shear* and *moment* are the internal forces in a beam or frame produced by the applied transverse loads. The shear acts perpendicular to the longitudinal axis, and the moment represents the internal couple produced by the bending stresses. These forces are evaluated at a particular point along the beam's axis by cutting the beam with an imaginary section perpendicular to the longitudinal axis (see Figure 5.5*b*) and then writing equilibrium equations for the free body to either the left or the right of the cut. Since the shear force produces equilibrium in the direction normal to the longitudinal axis of the member, it is evaluated by summing forces perpendicular to the longitudinal axis; that is, for a horizontal beam, we sum forces in the vertical direction. In this book, shear in a horizontal member will be considered positive if it acts downward on the face of the free body to the left of the section (see Figure 5.5*c*). Alternately, we can define shear as positive if it tends to produce clockwise rotation of the free body on which it acts. Shear acting downward on the face of the free body to the left of the section indicates the *resultant of the external forces* acting on the same free body is up. Since the shear acting on the section to the left represents the force applied by the free body to the right of the section, an equal but oppositely directed value of shear force must act upward on the face of the free body to the right of the section.

The internal moment M at a section is evaluated by summing moments of the external forces acting on the free body to either side of the section about an axis (perpendicular to the plane of the member) that passes through the centroid of the cross section. Moment will be considered positive if it produces compression stresses in the top fibers of the cross section and tension in the bottom fibers (see Figure 5.5*d*). Negative moment, on the other hand, bends a member concave down (see Figure 5.5*e*).

If a flexural member is vertical, the engineer is free to define the positive and negative sense of both the shear and moment. For the case of a single vertical member, one possible approach for establishing the positive direction for shear and moment is to rotate the computation sheet containing the sketch 90° clockwise so that the member is horizontal, and then apply the conventions shown in Figure 5.5.

For single-bay frames many analysts define moment as positive when it produces compression stresses on the outside surface of the member, where *inside* is defined as the region within the frame (see Figure 5.6). The positive direction for shear is then arbitrarily defined, as shown by the arrows on Figure 5.6.

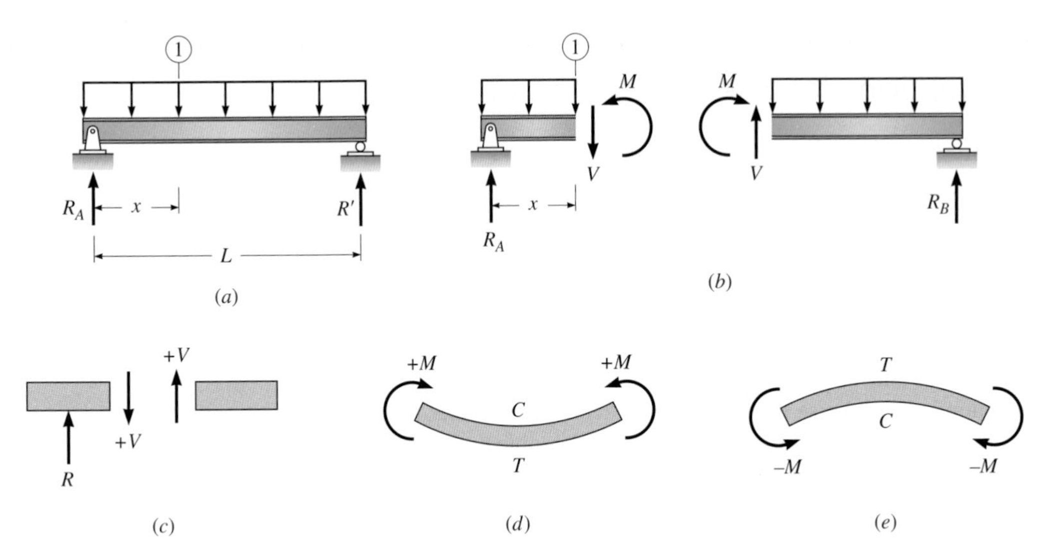

Figure 5.5: Sign conventions for shear and moment: (*a*) beam cut by section 1; (*b*) shear V and moment M occur as pairs of internal forces; (*c*) positive shear: resultant R of external forces on free body to left of section acts up; (*d*) positive moment; (*e*) negative moment.

Axial force on a cross section is evaluated by summing all forces perpendicular to the cross section. Forces acting outward from the cross section are tension forces T; those directed toward the cross section are compression forces C (see Figure 5.6).

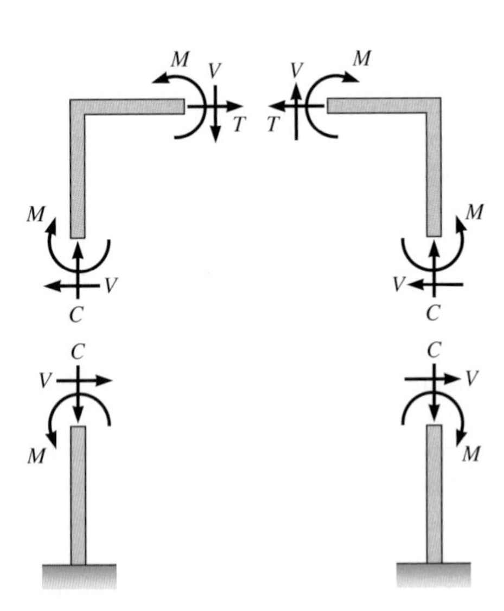

Figure 5.6: Internal forces acting on sections of the frame.

EXAMPLE 5.1

Write the equations for the variation of shear V and moment M along the axis of the cantilever beam in Figure 5.7. Using the equation, compute the moment at Section 1–1, 4 ft to the right of point B.

Solution

Determine the equation for shear V between points A and B (see Figure 5.7b); show V and M in the positive sense. Set origin at A ($0 \leq x_1 \leq 6$).

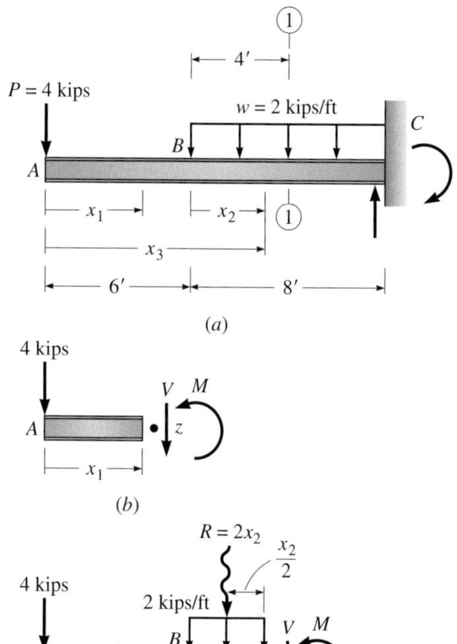

$$\overset{+}{\uparrow} \quad \Sigma F_y = 0$$

$$0 = -4 - V$$

$$V = -4 \text{ kips}$$

Determine the equation of moment M between points A and B. Set the origin at A. Sum the moments about the section.

$$\circlearrowright^{+} \quad \Sigma M_z = 0$$

$$0 = -4x_1 - M$$

$$M = -4x_1 \text{ kip} \cdot \text{ft}$$

The minus sign indicates V and M act opposite in sense to the directions shown in Figure 5.7b.

Determine the equation for shear V between points B and C (see Figure 5.7c). Set the origin at B, $0 \leq x_2 \leq 8$.

$$\overset{+}{\uparrow} \quad \Sigma F_y = 0$$

$$0 = -4 - 2x_2 - V$$

$$V = -4 - 2x_2$$

The moment M between B and C is

$$\circlearrowright^{+} \quad \Sigma M_z = 0$$

$$0 = -4(6 + x_2) - 2x_2\left(\frac{x_2}{2}\right) - M$$

$$M = -24 - 4x_2 - x_2^2$$

For M at section 1-1, 4 ft to right of B, set $x_2 = 4$ ft.

$$M = -24 - 16 - 16 = -56 \text{ kip} \cdot \text{ft}$$

Alternatively, compute M between points B and C, using an origin at A, and measure distance with x_3 (see Figure 5.7d), where $6 \leq x_3 \leq 14$.

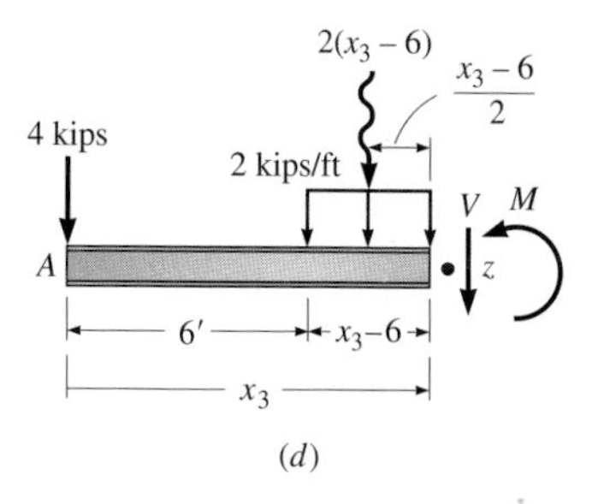

Figure 5.7

$$\circlearrowright^{+} \quad \Sigma M_z = 0$$

$$0 = -4x_3 - 2(x_3 - 6)\left(\frac{x_3 - 6}{2}\right) - M$$

$$M = -x_3^2 + 8x_3 - 36$$

Recompute the moment at section 1-1; set $x_3 = 10$ ft.

$$M = -10^2 + 8(10) - 36 = -56 \text{ kip} \cdot \text{ft}$$

EXAMPLE 5.2

For the beam in Figure 5.8 write the expressions for moment between points B and C, using an origin located at (a) support A, (b) support D, and (c) point B. Using each of the expressions above, evaluate the moment at Section 1–1. Shear force on sections is omitted for clarity.

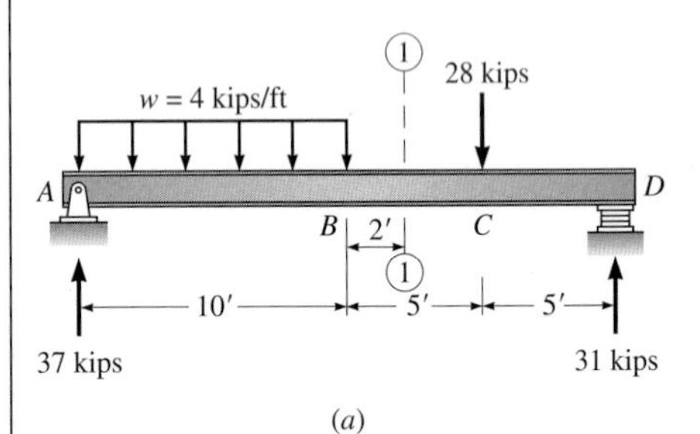

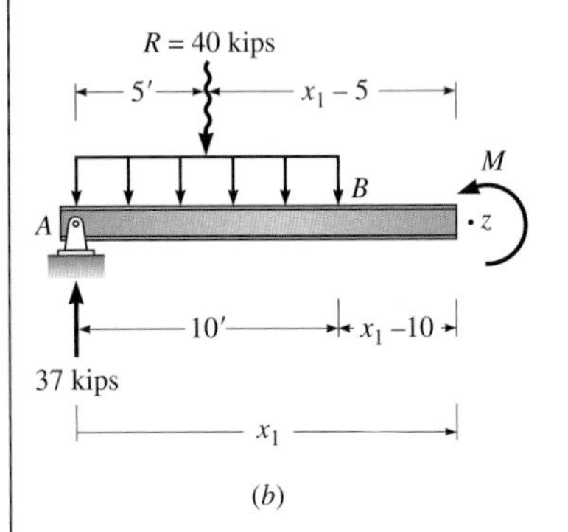

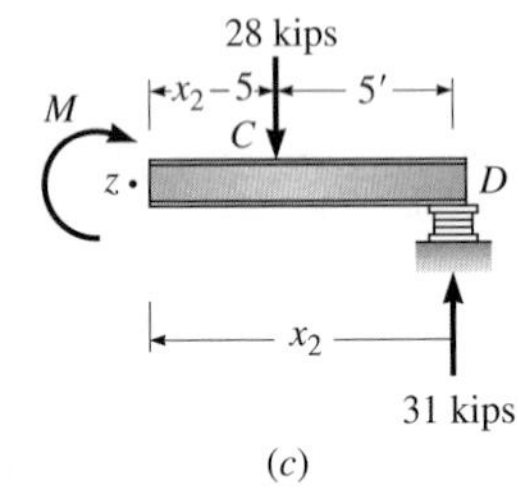

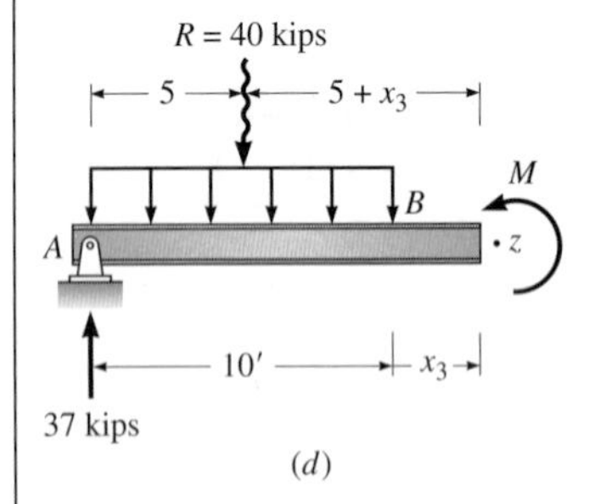

Figure 5.8

Solution

(a) See Figure 5.8b; summing moments about the cut gives

$$\circlearrowleft^{+} \quad \Sigma M_z = 0$$

$$0 = 37x_1 - 40(x_1 - 5) - M$$

$$M = 200 - 3x_1$$

At Section 1–1, $x_1 = 12$ ft; therefore,

$$M = 200 - 3(12) = 164 \text{ kip}\cdot\text{ft}$$

(b) See Figure 5.8c; summing moments about the cut yields

$$\circlearrowleft^{+} \quad \Sigma M_z = 0$$

$$0 = M + 28(x_2 - 5) - 31x_2$$

$$M = 3x_2 + 140$$

At Section 1–1, $x_2 = 8$ ft; therefore,

$$M = 3(8) + 140 = 164 \text{ kip}\cdot\text{ft}$$

(c) See Figure 5.8d; summing moments about the cut, we have

$$\circlearrowleft^{+} \quad \Sigma M_z = 0$$

$$37(10 + x_3) - 40(5 + x_3) - M = 0$$

$$M = 170 - 3x_3$$

At Section 1–1, $x_3 = 2$ ft; therefore,

$$M = 170 - 3(2) = 164 \text{ kip}\cdot\text{ft}$$

NOTE. As this example demonstrates, the moment at a section is single-valued and based on equilibrium requirements. The value of the moment does not depend on the location of the origin of the coordinate system.

EXAMPLE 5.3

Write the equations for shear and moment as a function of distance x along the axis of the beam in Figure 5.9. Select the origin at support A. Plot the individual terms in the equation for moment as a function of the distance x.

Solution

Pass an imaginary section through the beam a distance x to the right of support A to produce the free body shown in Figure 5.9b (the shear V and the moment M are shown in the positive sense). To solve for V, sum forces in the y direction.

$$\overset{+}{\uparrow} \quad \Sigma F_y = 0$$

$$\frac{wL}{2} - wx - V = 0$$

$$V = \frac{wL}{2} - wx \tag{1}$$

To solve for M, sum moments at the cut about a z axis passing through the centroid.

$$\circlearrowright^{+} \quad \Sigma M_z = 0$$

$$0 = \frac{wL}{2}(x) - wx\left(\frac{x}{2}\right) - M$$

$$M = \frac{wL}{2}(x) - \frac{wx^2}{2} \tag{2}$$

where in both equations $0 \leq x \leq L$.

A plot of the two terms in Equation 2 is shown in Figure 5.9c. The first term in Equation 2 (the moment produced by the vertical reaction R_A at support A) is a linear function of x and plots as a straight line sloping upward to the right. The second term, which represents the moment due to the uniformly distributed load, is a function of x^2 and plots as a parabola sloping downward. When a moment curve is plotted in this manner, we say that it is plotted by *cantilever parts*. In Figure 5.9d, the two curves are combined to give a parabolic curve whose ordinate at midspan equals the familiar $wL^2/8$.

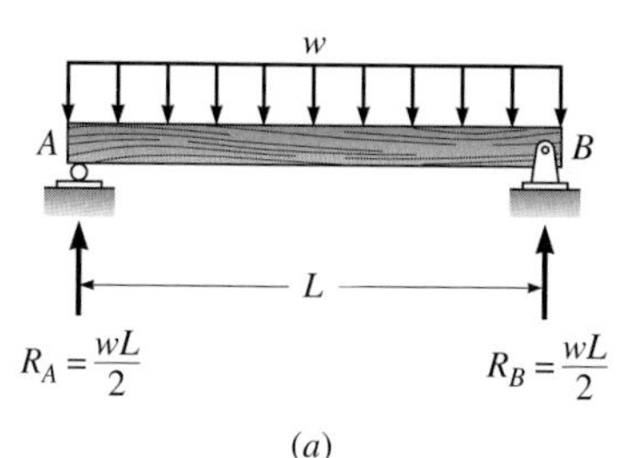

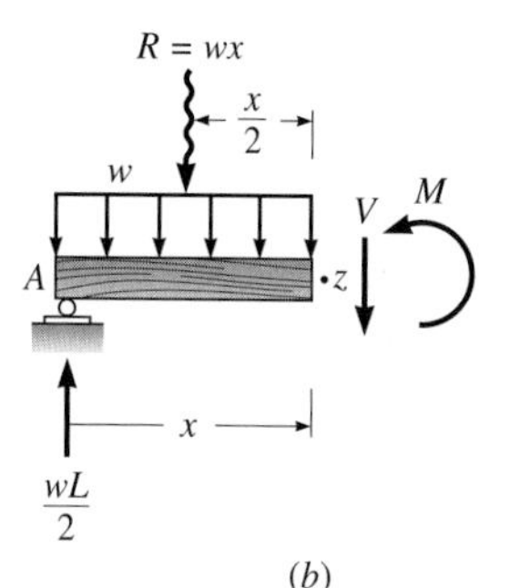

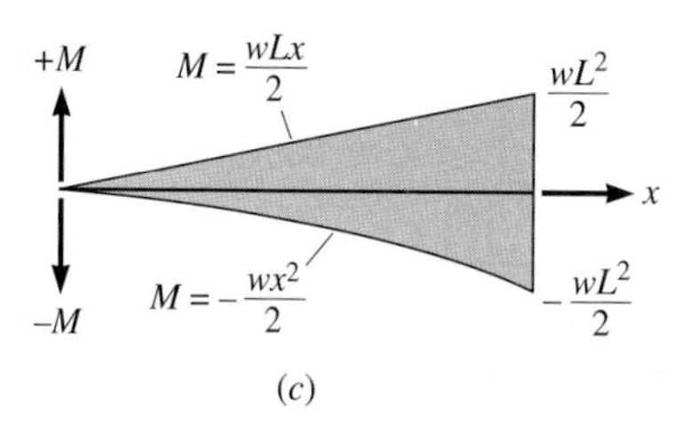

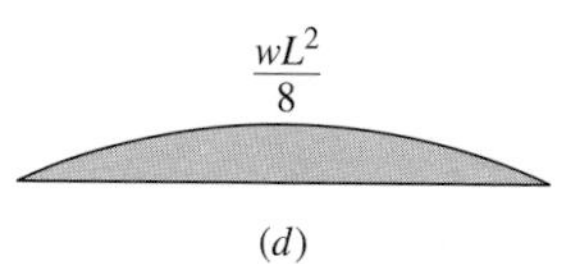

Figure 5.9: (a) Uniformly loaded beam; (b) free body of beam segment; (c) moment curve plotted by "parts;" (d) combined moment diagram, a symmetric parabola.

EXAMPLE 5.4

(*a*) Write the equations for shear and moment on a vertical section between supports *B* and *C* for the beam in Figure 5.10*a*.

(*b*) Using the equation for shear in part (a), determine the point where the shear is zero (the point of maximum moment).

(*c*) Plot the variation of the shear and moment between *B* and *C*.

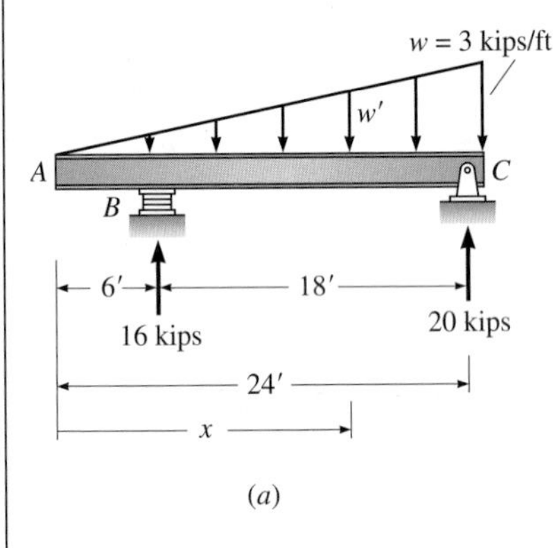

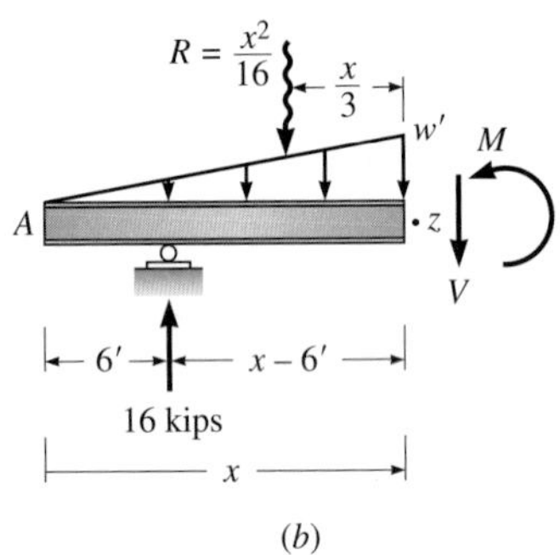

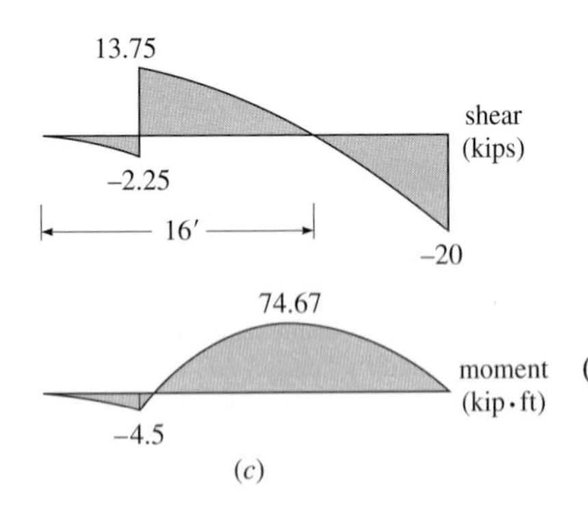

Figure 5.10

Solution

(*a*) Cut the free body shown in Figure 5.10*b* by passing a section through the beam a distance *x* from point *A* at the left end. Using similar triangles, express *w*′, the ordinate of the triangular load at the cut (consider the triangular load on the free body and on the beam), in terms of *x* and the ordinate of the load curve at support *C*.

$$\frac{w'}{x} = \frac{3}{24} \qquad \text{therefore} \qquad w' = \frac{x}{8}$$

Compute the resultant of the triangular load on the free body in Figure 5.10*b*.

$$R = \frac{1}{2}xw' = \frac{1}{2}(x)\left(\frac{x}{8}\right) = \frac{x^2}{16}$$

Compute *V* by summing forces in the vertical direction.

$$\overset{+}{\uparrow} \quad \Sigma F_y = 0$$

$$0 = 16 - \frac{x^2}{16} - V$$

$$V = 16 - \frac{x^2}{16} \tag{1}$$

Compute *M* by summing moments about the cut.

$$\circlearrowright^{+} \quad \Sigma M_z = 0$$

$$0 = 16(x - 6) - \frac{x^2}{16}\left(\frac{x}{3}\right) - M$$

$$M = -96 + 16x - \frac{x^3}{48} \tag{2}$$

(*b*) Set $V = 0$ and solve Equation 1 for *x*.

$$0 = 16 - \frac{x^2}{16} \qquad \text{and} \qquad x = 16 \text{ ft}$$

(*c*) See Figure 5.10*c* for a plot of *V* and *M*.

EXAMPLE 5.5

Write the equations for moment in members *AC* and *CD* of the frame in Figure 5.11. Draw a free body of joint *C*, showing all forces.

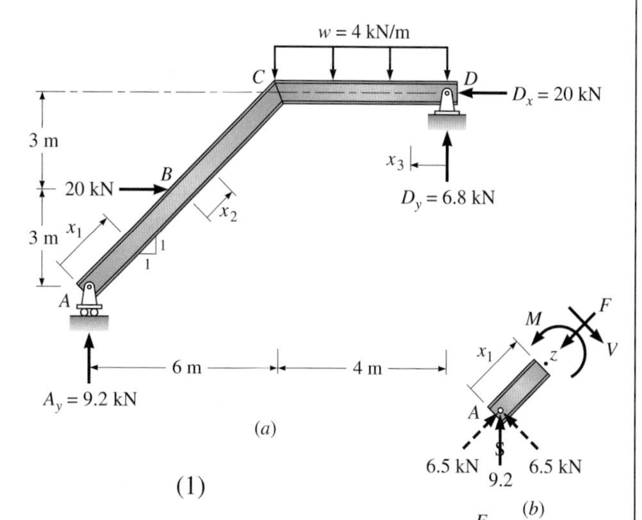

Solution

Two equations are needed to express the moment in member *AC*. To compute the moment between *A* and *B*, use the free body in Figure 5.11*b*. Take the origin for x_1 at support *A*. Break the vertical reaction into components parallel and perpendicular to the longitudinal axis of the sloping member. Sum moments about the cut.

$$\circlearrowright^{+} \quad \Sigma M_z = 0$$

$$0 = 6.5x_1 - M$$

$$M = 6.5x_1 \qquad (1)$$

where $0 \leq x_1 \leq 3\sqrt{2}$.

Compute the moment between *B* and *C*, using the free body in Figure 5.11*c*. Select an origin at *B*. Break 20 kN force into components. Sum moments about the cut.

$$\circlearrowright^{+} \quad \Sigma M_z = 0$$

$$0 = 6.5(3\sqrt{2} + x_2) - 14.14x_2 - M$$

$$M = 19.5\sqrt{2} - 7.64x_2 \qquad (2)$$

where $0 \leq x_2 \leq 3\sqrt{2}$.

Compute the moment between *D* and *C*, using the free body in Figure 5.11*d*. Select an origin at *D*.

$$^{+}\circlearrowleft \quad \Sigma M_z = 0$$

$$0 = 6.8x_3 - 4x_3\left(\frac{x_3}{2}\right) - M$$

$$M = 6.8x_3 - 2x_3^2 \qquad (3)$$

The free body of joint *C* is shown in Figure 5.11*e*. The moment at the joint can be evaluated with Equation 3 by setting $x_3 = 4$ m.

$$M = 6.8(4) - 2(4)^2 = -4.8 \text{ kN}\cdot\text{m}$$

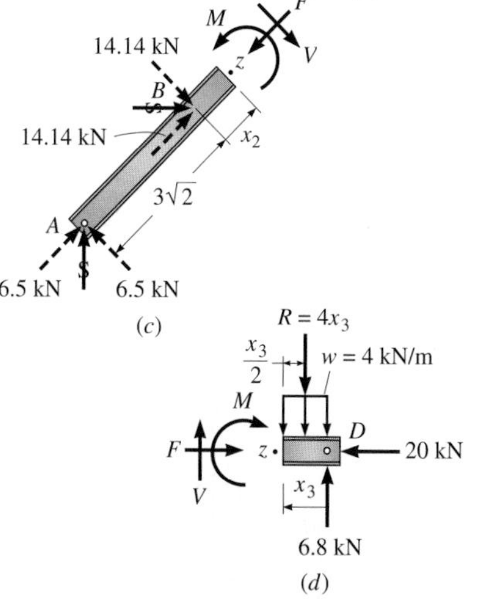

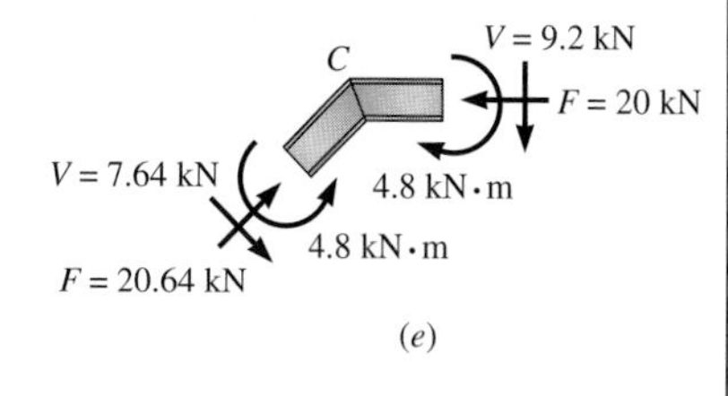

Figure 5.11

5.4 Shear and Moment Curves

To design a beam, we must establish the magnitude of the shear and moment (and axial load if it is significant) at all sections along the axis of the member. If the cross section of a beam is constant along its length, it is designed for the maximum values of moment and shear within the span. If the cross section varies, the designer must investigate additional sections to verify that the member's capacity is adequate to carry the shear and moment.

To provide this information graphically, we construct shear and moment curves. These curves, which preferably should be drawn to scale, consist of values of shear and moment plotted as ordinates against distance along the axis of the beam. Although we can construct shear and moment curves by cutting free bodies at intervals along the axis of a beam and writing equations of equilibrium to establish the values of shear and moment at particular sections, it is much simpler to construct these curves from the basic relationships that exist between load, shear, and moment.

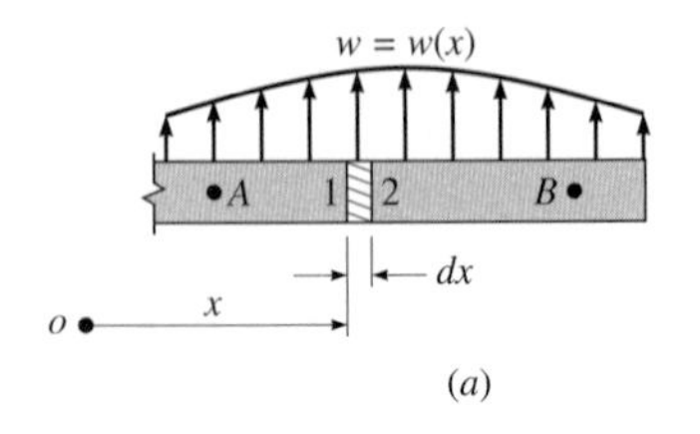

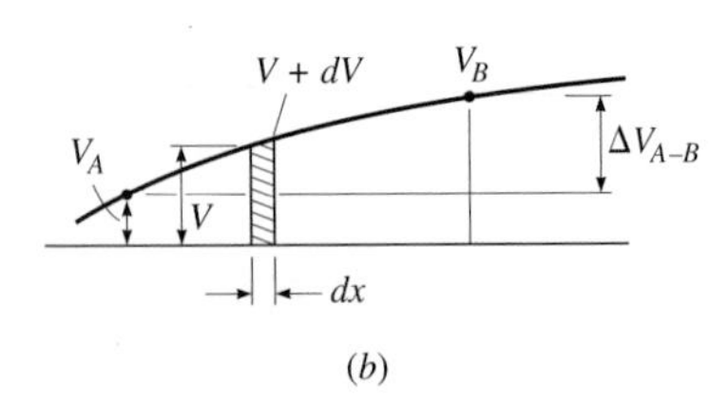

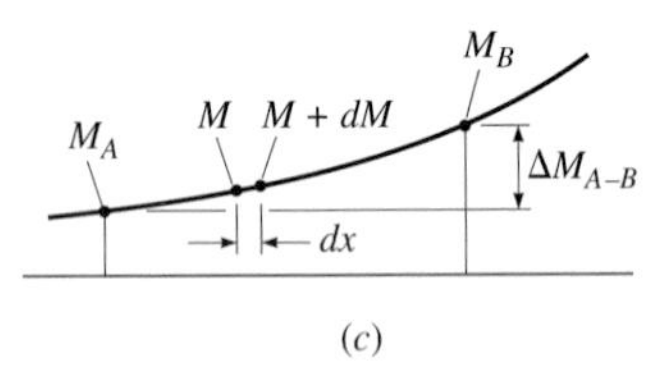

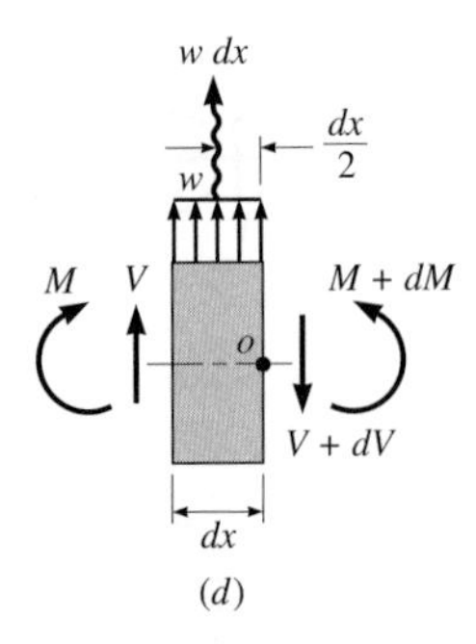

Figure 5.12: (*a*) Segment of beam with a distributed load; (*b*) shear curve; (*c*) moment curve; (*d*) infinitesimal element located between points 1 and 2.

Relationship Between Load, Shear, and Moment

To establish the relationship between load, shear, and moment, we will consider the beam segment shown in Figure 5.12*a*. The segment is loaded by a distributed load $w = w(x)$ whose ordinates vary with distance x from an origin o located to the left of the segment. The load will be considered positive when it acts upward, as shown in Figure 5.12*a*.

To derive the relationship between load, shear, and moment, we will consider the equilibrium of the beam element shown in Figure 5.12*d*. The element, cut by passing imaginary vertical planes through the segment at points 1 and 2 in Figure 5.12*a*, is located a distance x from the origin. Since dx is infinitesimally small, the slight variation in the distributed load acting over the length of the element may be neglected. Therefore, we can assume that the distributed load is constant over the length of the element. Based on this assumption, the resultant of the distributed load is located at the midpoint of the element.

The curves representing the variation of the shear and the moment along the axis of the member are shown in Figure 5.12*b* and *c*. We will denote the shear and moment on the left face of the element in Figure 5.12*d* by V and M respectively. To denote that a small change in shear and moment occurs over the length dx of the element, we add the differential quantities dV and dM to the shear V and the moment M to establish the values of shear and moment on the right face. All forces shown on the element act in the positive sense as defined in Figure 5.5*c* and *d*.

Considering equilibrium of forces acting in the y direction on the element, we can write

$$\overset{+}{\uparrow} \quad \Sigma F_y = 0$$

$$0 = V + w\,dx - (V + dV)$$

Simplifying and solving for dV gives

$$dV = w\,dx \tag{5.6}$$

To establish the difference in shear ΔV_{A-B} between points A and B along the axis of the beam in Figure 5.12a, we must integrate Equation 5.6.

$$\Delta V_{A-B} = V_B - V_A = \int_A^B dV = \int_A^B w\,dx \tag{5.7}$$

The integral on the left side of Equation 5.7 represents the change in shear ΔV_{A-B} between points A and B. In the integral on the right, the quantity $w\,dx$ can be interpreted as an infinitesimal area under the load curve. The integral or sum of these infinitesimal areas represents the area under the load curve between points A and B. Therefore, we can state Equation 5.7 as

$$\Delta V_{A-B} = \text{area under load curve between } A \text{ and } B \tag{5.7a}$$

where an upward load produces a positive change in shear and a downward load a negative change, moving from left to right.

Dividing both sides of Equation 5.6 by dx produces

$$\frac{dV}{dx} = w \tag{5.8}$$

Equation 5.8 states that *the slope of the shear curve at a particular point along the axis of a member equals the ordinate of the load curve at that point.*

If the load acts upward, the slope is positive (upward to the right). If the load acts downward, the slope is negative (downward to the right). In a region of the beam in which no load acts, $w = 0$. For this condition Equation 5.8 indicates the slope of the shear curve is zero—indicating that the shear remains constant.

To establish the relationship between shear and moment, we sum moments of the forces acting on the element about an axis normal to the plane of the beam and passing through point o (see Figure 5.12d). Point o is located at the level of the centroid of the cross section

$$\circlearrowright^{+} \quad \Sigma M_o = 0$$

$$M + V\,dx - (M + dM) + w\,dx\,\frac{dx}{2} = 0$$

Since the last term $w\,(dx)^2/2$ contains the product of a differential quantity squared, it is many orders of magnitude smaller that the terms containing a single differential. Therefore, we drop the term. Simplifying the equation yields

$$dM = V\,dx \tag{5.9}$$

To establish the change in moment ΔM_{A-B} between points A and B, we will integrate both sides of Equation 5.9.

$$\Delta M_{A-B} = M_B - M_A = \int_A^B dM = \int_A^B V\,dx \tag{5.10}$$

The center term in Equation 5.10 represents the difference in moment ΔM_{A-B} between point A and B. Since the term $V\,dx$ can be interpreted as an infinitesimal area under the shear curve between points 1 and 2 (see Figure 5.12*b*), the integral on the right, the sum of all the infinitesimal areas between points A and B, represents the total area under the shear curve between points A and B. Based on the observations above, we can state Equation 5.10 as

$$\Delta M_{A-B} = \text{area under shear curve between } A \text{ and } B \tag{5.10a}$$

where a positive area under the shear curve produces a positive change in moment and a negative area under the shear curve produces a negative change; ΔM_{A-B} is shown graphically in Figure 5.12*c*.

Dividing both sides of Equation 5.9 by dx gives

$$\frac{dM}{dx} = V \tag{5.11}$$

Equation 5.11 states that *the slope of the moment curve at any point along the axis of a member is the shear at that point.*

If the ordinates of the shear curve are positive, the slope of the moment curve is positive (directed upward to the right). Similarly, if the ordinates of the shear curve are negative, the slope of the moment curve is negative (directed downward to the right).

At a section where $V = 0$, Equation 5.11 indicates that the slope of the moment curve is zero—a condition that establishes the location of a maximum value of moment. If the shear is zero at several sections in a span, the designer must compute the moment at each section and compare results to establish the absolute maximum value of moment in the span.

Equations 5.6 to 5.11 do not account for the effect of a concentrated load or moment. A concentrated force produces a sharp change in the ordinate of a shear curve. If we consider equilibrium in the vertical direction of the element in Figure 5.13*a*, the change in shear between the two faces of the element equals the magnitude of the concentrated force. Similarly, the change in moment at a point equals the magnitude of the concentrated moment M_1 at the point (see Figure 5.13*b*). In Figure 5.13 all forces are shown acting in the positive sense. Examples 5.6 to 5.8 illustrate the use of Equations 5.6 to 5.11 to construct shear and moment curves.

To construct the shear and moment curves for a beam supporting distributed and concentrated loads, we first compute the shear and moment at the left end of the member. We then proceed to the right, locating the next point on the shear curve by adding algebraically, to the shear at the left, the force represented by (1) the area under the load curve between the two points or (2) a concentrated load. To establish a third point, load is added to or subtracted from the value of shear at the second point. The process of locating additional points is continued until the shear curve is completed. Typically, we evaluate the ordinates of the shear curve at each point where a concentrated load acts or where a distributed load begins or ends.

In a similar manner, points on the moment curve are established by adding algebraically to the moment, at a particular point, the increment of moment represented by the area under the shear curve between a second point.

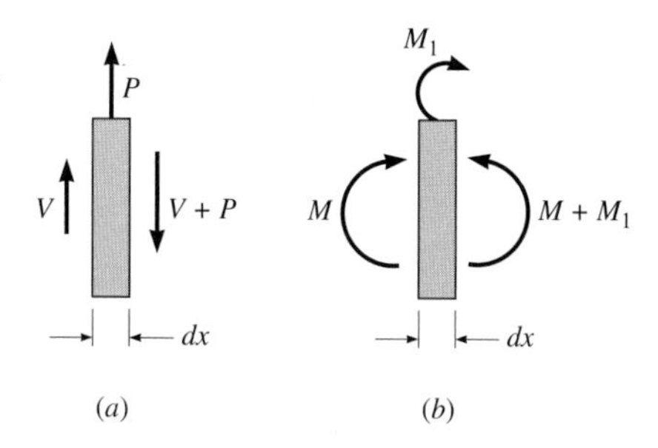

Figure 5.13: (*a*) Effect of a concentrated load on the change in shear; (*b*) change in internal moment produced by the applied moment M_1.

Sketching Deflected Shapes of Beams

After the shear and moment curves are constructed, the designer may wish to draw a sketch of the beam's deflected shape. Although we will discuss this topic in great detail in Section 5.6, the procedure is introduced briefly at this point. The deflected shape of a beam must be consistent with (1) the restraints imposed by the supports, and (2) the curvature produced by the moment. *Positive moment bends the beam concave upward*, and *negative moment bends the beam concave downward.*

The restraints imposed by various types of supports are summarized in Table 3.1. For example, at a fixed support, the beam's longitudinal axis is restrained against rotation and deflection. At a pin support, the beam is free to rotate but not to deflect. Sketches of deflected shapes to an *exaggerated* vertical scale are included in Examples 5.6 to 5.8.

EXAMPLE 5.6

Draw the shear and moment curves for the simply supported beam in Figure 5.14.

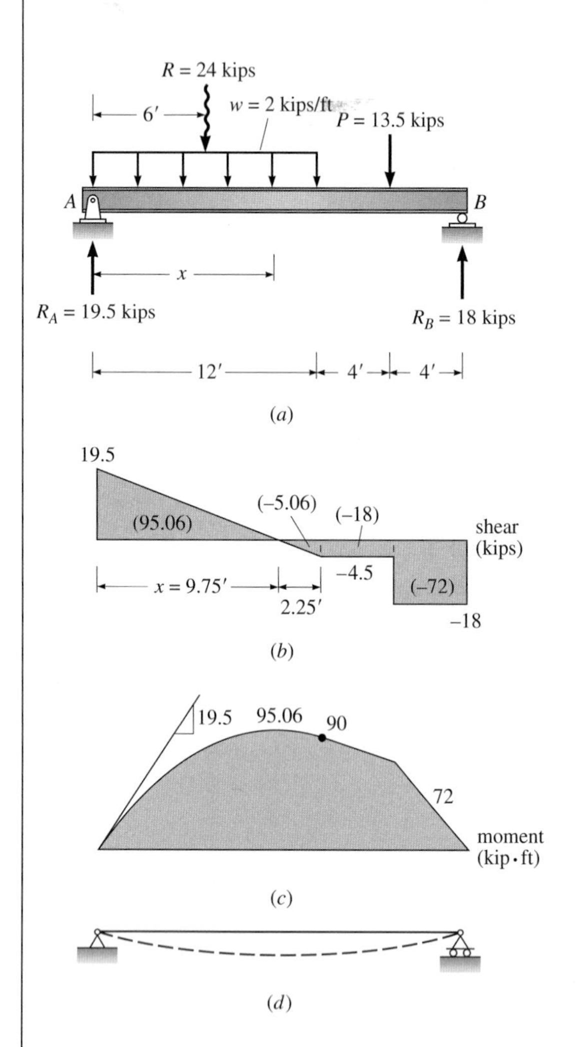

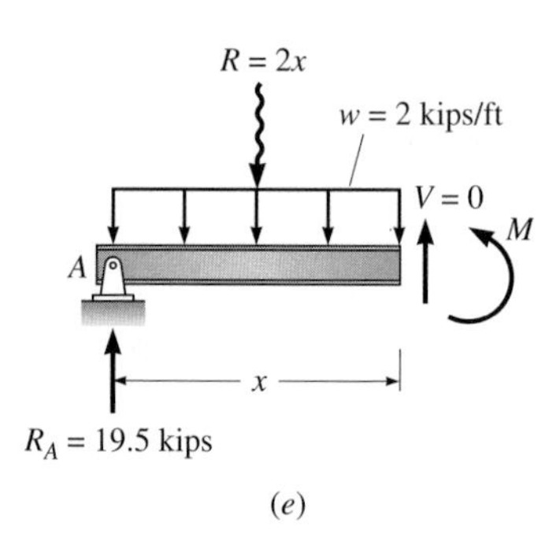

Figure 5.14: (*a*) Beam details; (*b*) shear curve (numbers in brackets represent areas under shear curve); (*c*) moment curve; (*d*) deflected shape; (*e*) free body used to establish location of point of zero shear and maximum moment.

Solution

Compute reactions (use the resultant of the distributed load).

$$\circlearrowright^{+} \quad \Sigma M_A = 0$$

$$24(6) + 13.5(16) - 20R_B = 0$$

$$R_B = 18 \text{ kips}$$

$$\overset{+}{\uparrow} \quad \Sigma F_y = 0$$

$$R_A + R_B - 24 - 13.5 = 0$$

$$R_A = 19.5 \text{ kips}$$

Shear Curve. The shear just to the right of support A equals the reaction of 19.5 kips. Since the reaction acts upward, the shear is positive. To the right of the support the uniformly distributed load acting downward reduces the shear linearly. At the end of the distributed load—12 ft to the right of the support—the shear equals

$$V_{@12} = 19.5 - (2)(12) = -4.5 \text{ kips}$$

At the 13.5-kip concentrated load, the shear drops to -18 kips. The shear diagram is shown in Figure 5.14*b*. The maximum value of moment occurs where the shear equals zero. To compute the location of the point of zero shear, denoted by the distance x from the left support, we consider the forces acting on the free body in Figure 5.14*e*.

$$\overset{+}{\uparrow} \quad \Sigma F_y = 0$$

$$0 = R_A - wx \qquad \text{where } w = 2 \text{ kips/ft}$$

$$0 = 19.5 - 2x \qquad \text{and} \qquad x = 9.75 \text{ ft}$$

Moment Curve. Points along the moment curve are evaluated by adding to the moment, at the left end, the change in moment between selected points. The change in moment between any two points is equal to the area under the shear curve between the two points. For this purpose, the shear curve is divided into two triangular and two rectangular areas. The values of the respective areas (in units of kip·ft) are given by the numbers in parentheses in Figure 5.14*b*. Because the ends of the beam are supported on a roller and a pin, supports that offer no rotational restraint, the moments at the ends are zero. Since the moment starts at zero at the left and ends at zero on the right, the algebraic sum of the areas under the shear curve between ends must equal zero. Because of rounding errors, you will find the ordinates of the moment curve do not always satisfy the boundary conditions exactly.

At the left end of the beam, the slope of the moment curve is equal to 19.5 kips—the ordinate of the shear curve. The slope is positive because the shear is positive. As the distance to the right of support A increases, the ordinates of the shear curve reduce, and correspondingly the slope of the moment curve reduces. The maximum moment of 95.06 kip·ft occurs at the point of zero shear. To the right of the point of zero shear, the shear is negative, and the slope of the moment curve is downward to the right. The moment curve is plotted in Figure 5.14*c*. Since the moment is positive over the entire length, the member is bent concave upward, as shown by the dashed line in Figure 5.14*d*.

EXAMPLE 5.7

Draw the shear and moment curves for the uniformly loaded beam in Figure 5.15a. Sketch the deflected shape.

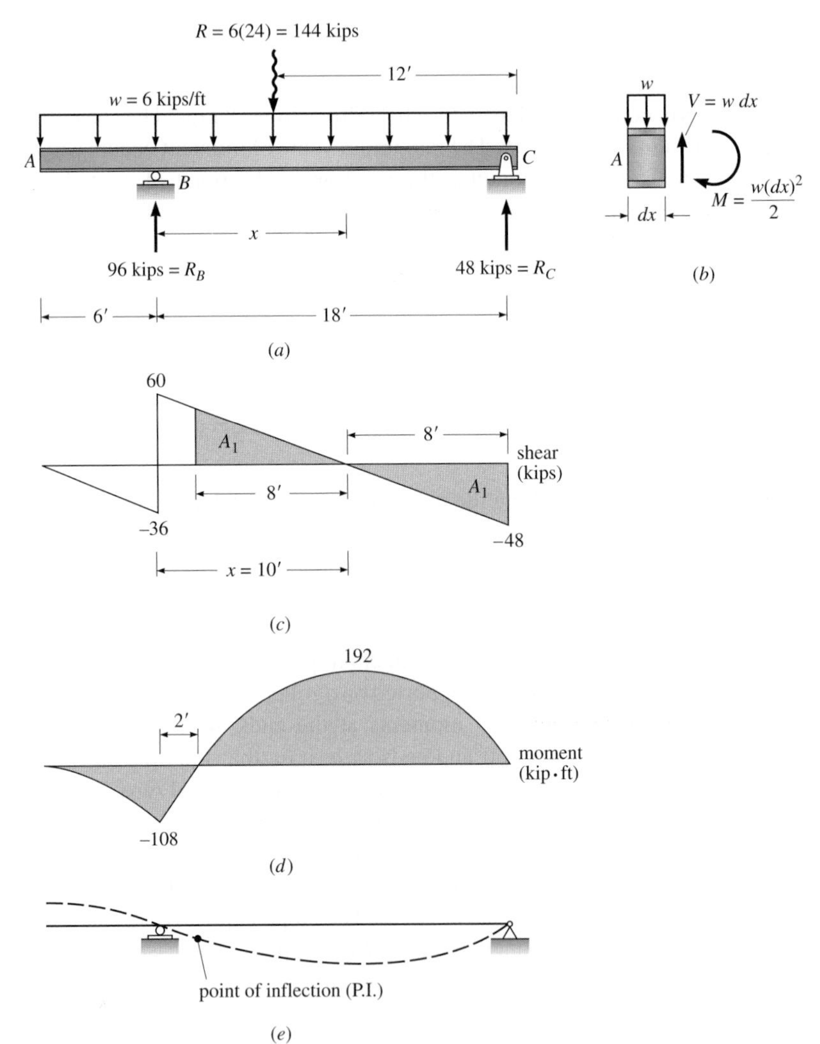

Figure 5.15: (a) Beam with uniform load; (b) infinitesimal element used to establish that V and M equal zero at the left end of the beam; (c) shear curve (units in kips); (d) moment curve (units in kip·ft); (e) approximate deflected shape (vertical deflections shown to exaggerated scale by dashed line).

Solution

Compute R_B by summing moments of forces about support C. The distributed load is represented by its resultant of 144 kips.

$$\circlearrowleft^{+} \quad \Sigma M_c = 0$$

$$18R_B - 144(12) = 0 \qquad R_B = 96 \text{ kips}$$

Compute R_C.

$$\overset{+}{\uparrow} \quad \Sigma F_y = 0$$

$$96 - 144 + R_C \qquad R_C = 48 \text{ kips}$$

Verify equilibrium; check $\circlearrowright^{+} \quad \Sigma M_B = 0$.

$$144(6) - 48(18) = 0 \qquad \text{OK}$$

We begin by establishing the values of shear and moment at the left end of the beam. For this purpose we consider the forces on an infinitesimal element cut from the left end (at point A) by a vertical section (see Figure 5.15*b*). Expressing the shear and moment in terms of the uniform load w and the length dx, we observe that as dx approaches zero, both the shear and the moment reduce to zero.

Shear Curve. Because the magnitude of the load is constant over the entire length of the beam and directed downward, Equation 5.8 establishes that the shear curve will be a straight line with a constant slope of -6 kips/ft at all points (see Figure 5.15*c*). Starting from $V = 0$ at point A, we compute the shear just to the left of support B by evaluating the area under the load curve between points A and B (Equation 5.7*a*).

$$V_B = V_A + \Delta V_{A-B} = 0 + (-6 \text{ kips/ft})(6 \text{ ft}) = -36 \text{ kips}$$

Between the left and right sides of the support at B, the reaction, acting upward, produces a positive 96-kip change in shear; therefore, to the right of support B the ordinate of the shear curve rises to $+60$ kips. Between points B and C, the change in shear (given by the area under the load curve) equals $(-6 \text{ kips/ft})(18 \text{ ft}) = -108$ kips. Thus the shear drops linearly from 60 kips at B to -48 kips at C.

To establish the distance x to the right of point B, where the shear equals zero, we equate the area wx under the load curve in Figure 5.15*a* to the 60 kip shear at B.

$$60 - wx = 0$$

$$60 - 6x = 0 \qquad x = 10 \text{ ft}$$

Moment Curve. To sketch the moment curve, we will locate the points of maximum moment, using Equation 5.10*a*; that is, the area under the shear diagram between two points equals the change in moment between the points. Thus we must evaluate in sequence the alternate positive and negative areas (triangles in this example) under the shear curve. We then use Equation 5.11 to establish the correct slope of the curve between points of maximum moment.

$$M_B = M_A + \Delta M_{A-B} = 0 + \frac{1}{2}(6)(-36) = -108 \text{ kip}\cdot\text{ft}$$

[*continues on next page*]

Example 5.7 continues . . .

Compute the value of the maximum positive moment between B and C. The maximum moment occurs 10 ft to the right of support B where $V = 0$.

$$M_{max} = M_B + \text{area under } V\text{-curve between } x = 0 \text{ and } x = 10$$

$$= -108 + \frac{1}{2}(60)(10) = +192 \text{ kip}\cdot\text{ft}$$

Since the slope of the moment curve is equal to the ordinate of the shear curve, the slope of the moment curve is zero at point A. To the right of point A, the slope of the moment curve becomes progressively steeper because the ordinates of the shear curve increase. Since the shear is negative between points A and B, the slope is negative (i.e., downward to the right). Thus to be consistent with the ordinates of the shear curve, the moment curve must be concave downward between points A and B.

Since the shear is positive to the right of support B, the slope of the moment curve reverses direction and becomes positive (slopes upward to the right). Between support B and the point of maximum positive moment, the slope of the moment curve reduces progressively from 60 kips to zero, and the moment curve is concave down. To the right of the point of maximum moment, the shear is negative, and the slope of the moment curve again changes direction and becomes progressively steeper in the negative sense toward support C.

Point of Inflection. A point of inflection occurs at a point of zero moment. Here the curvature changes from concave up to concave down. To locate a point of inflection, we use the areas under the shear curve. Since the triangular area A_1 of the shear diagram between support C and the point of maximum positive moment produces a change in moment of 192 kip·ft, an equal area under the shear curve (see Figure 5.15*c*), extending 8 ft to the left of the point of maximum moment, will drop the moment to zero. Thus the point of inflection is located 16 ft to the left of support C or equivalently 2 ft to the right of support B.

Sketching the Deflected Shape. The approximate deflected shape of the beam is shown in Figure 5.15*e*. At the left end where the moment is negative, the beam is bent concave downward. On the right side, where the moment is positive, the beam is bent concave upward. Although we can easily establish the curvature at all sections along the axis of the beam, the deflected position of certain points must be assumed. For example, at point A the left end of the loaded beam is arbitrarily assumed to deflect upward above the initial undeflected position represented by the straight line. On the other hand, it is also possible that point A is located below the undeflected position of the beam's axis if the cantilever is flexible. The actual elevation of point A must be established by computation.

EXAMPLE 5.8

Draw the shear and moment curves for the inclined beam in Figure 5.16*a*.

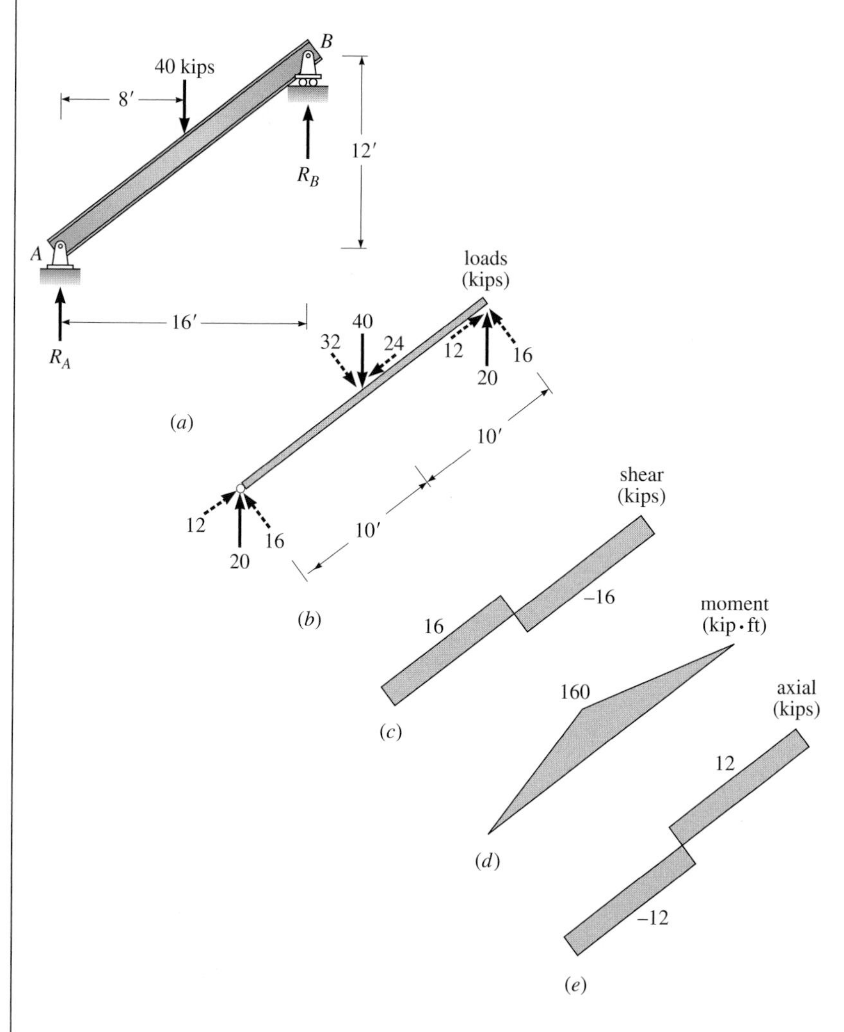

Figure 5.16: (*a*) Sloping beam; (*b*) forces and reactions broken into components parallel and perpendicular to the longitudinal axis; (*c*) shear curve; (*d*) moment curve; (*e*) variation of axial load—tension is positive and compression is negative.

Solution

We begin the analysis by computing the reactions in the usual manner with the equations of statics. Since shear and moment are produced only by loads acting perpendicular to the member's longitudinal axis, all forces are broken into components parallel and perpendicular to the longitudinal axis (Figure 5.16*b*). The longitudinal components produce axial compression in the lower half of the member and tension in the upper half (see Figure 5.16*e*). The transverse components produce the shear and moment curves shown in Figure 5.16*c* and *d*.

EXAMPLE 5.9

Draw the shear and moment curves for the beam in Figure 5.17*a*. Sketch the deflected shape.

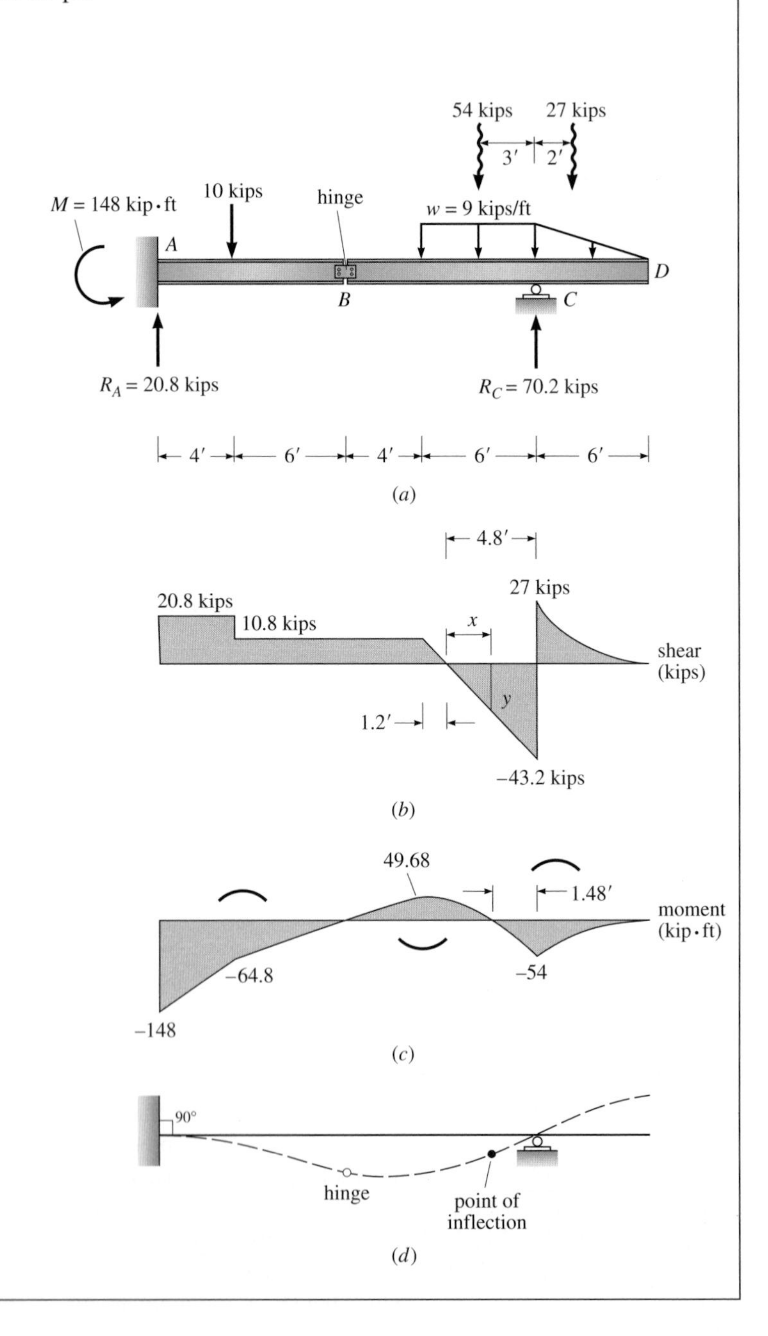

Figure 5.17: (*a*) Beam (reactions given); (*b*) shear curve (kips); (*c*) moment curve (kip·ft); (*d*) deflected shape.

Solution

We begin the analysis by computing the reaction at support C, using a free body of member BCD. Summing moments of the applied forces (resultants of the distributed load are shown by wavy arrows) about the hinge at B, we compute

$$\circlearrowleft^{+} \quad \Sigma M_B = 0$$

$$0 = 54(7) + 27(12) - R_C(10)$$

$$R_C = 70.2 \text{ kips}$$

After R_C is computed, the balance of the reactions are computed using the entire structure as a free body. Even though a hinge is present, the structure is stable because of the restraints supplied by the supports. The shear and moment curves are plotted in Figure 5.17*b* and *c*. As a check of the accuracy of the computations, we observe the moment at the hinge is zero. The curvature (concave up or concave down) associated with positive and negative moments is indicated by the short curved lines above or below the moment curve.

To locate the point of inflection (zero moment) to the left of support C, we equate the triangular area under the shear curve between the points of maximum and zero moment to the change in moment of 49.68 kip·ft. The base of the triangle is denoted by x and the altitude by y in Figure 5.17*b*. Using similar triangles, we express y in terms of x.

$$\frac{x}{y} = \frac{4.8}{43.2}$$

$$y = \frac{43.2x}{4.8}$$

$$\text{Area under shear curve} = \Delta M = 49.68 \text{ kip} \cdot \text{ft}$$

$$\left(\frac{1}{2}x\right)\left(\frac{43.2x}{4.8}\right) = 49.68 \text{ kip} \cdot \text{ft}$$

$$x = 3.32 \text{ ft}$$

The distance of the point of inflection from support C is

$$4.8 - 3.32 = 1.48 \text{ ft}$$

The sketch of the deflected shape is shown in Figure 5.17*d*. Since the fixed support at A prevents rotation, the longitudinal axis of the beam is horizontal at support A (i.e., makes an angle of 90° with the vertical face of the support). Because the moment is negative between A and B, the beam bends concave downward and the hinge displaces downward. Since the moment changes from positive to negative just to the left of support C, the curvature of member BCD reverses. Although the general shape of member BCD is consistent with the moment curve, the exact position of the end of the member at point D must be established by computation.

EXAMPLE 5.10

Draw the shear and moment curves for beam *ABC* in Figure 5.18*a*. Also sketch the deflected shape. Rigid joints connect the vertical members to the beam. Elastomeric pad at *C* equivalent to a roller.

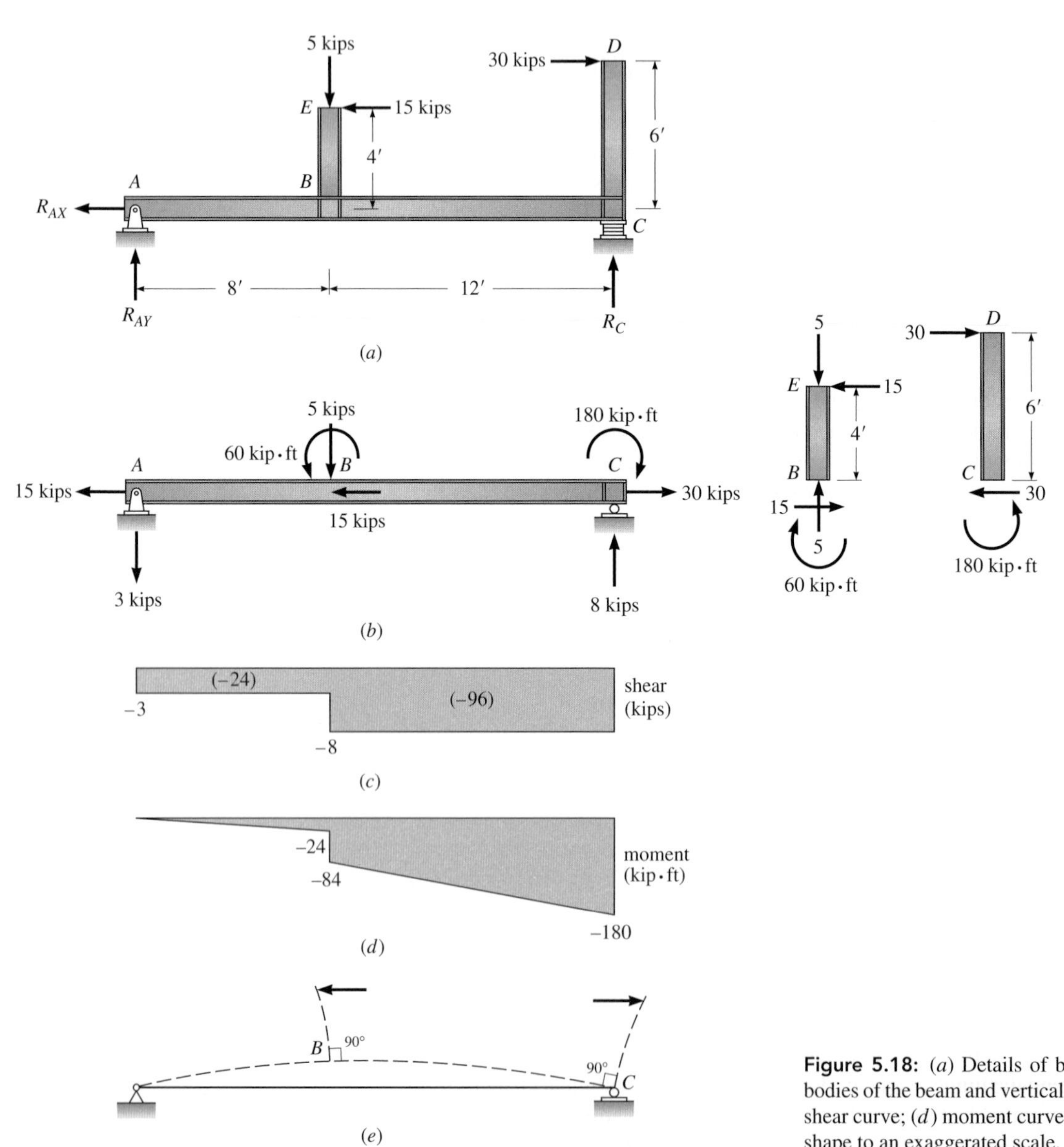

Figure 5.18: (*a*) Details of beam; (*b*) free bodies of the beam and vertical members; (*c*) shear curve; (*d*) moment curve; (*e*) deflected shape to an exaggerated scale.

Solution

Compute the reaction at C; sum moments about A of all forces acting on Figure 5.18a.

$$\circlearrowright^{+} \quad \Sigma M_A = 0$$

$$0 = 5(8) - 15(4) + 30(6) - 20R_C$$

$$R_C = 8 \text{ kips}$$

$$\overset{+}{\uparrow} \quad \Sigma F_y = 0 = 8 - 5 + R_{AY}$$

$$R_{AY} = -3 \text{ kips}$$

$$\rightarrow + \quad \Sigma F_x = 0$$

$$30 - 15 - R_{AX} = 0$$

$$R_{AX} = 15 \text{ kips}$$

Figure 5.18b shows free-body diagrams of the beam and the vertical members. The forces on the bottom of the vertical members represent forces applied by the beam. The verticals, in turn, exert equal and oppositely directed forces on the beam. The shear and moment curves are constructed next. Because the shear at a section is equal to the sum of the vertical forces to either side of the section, the concentrated moment and longitudinal forces do not contribute to the shear.

Since a pin support is located at the left end, the end moment starts at zero. Between points A and B the change in moment, given by the area under the shear curve, equals −24 kip·ft. At B the counterclockwise concentrated moment of 60 kip·ft causes the moment curve to drop sharply to −84 kip·ft. The action of a concentrated moment that produces a positive change in moment in the section just to the right of the concentrated moment is illustrated in Figure 5.13b. Because the moment at B is opposite in sense to the moment in Figure 5.13b, it produces a negative change. Between B and C the change in moment is again equal to the area under the shear curve. The end moment in the beam at C must balance the 180 kip·ft applied by member CD.

Since the moment is negative over the entire length of the beam, the entire beam bends concave downward, as shown in Figure 5.18e. The axis of the beam remains a smooth curve throughout.

EXAMPLE 5.11

Draw the shear and moment curves and sketch the deflected shape of the continuous beam in Figure 5.19a. The support reactions are given.

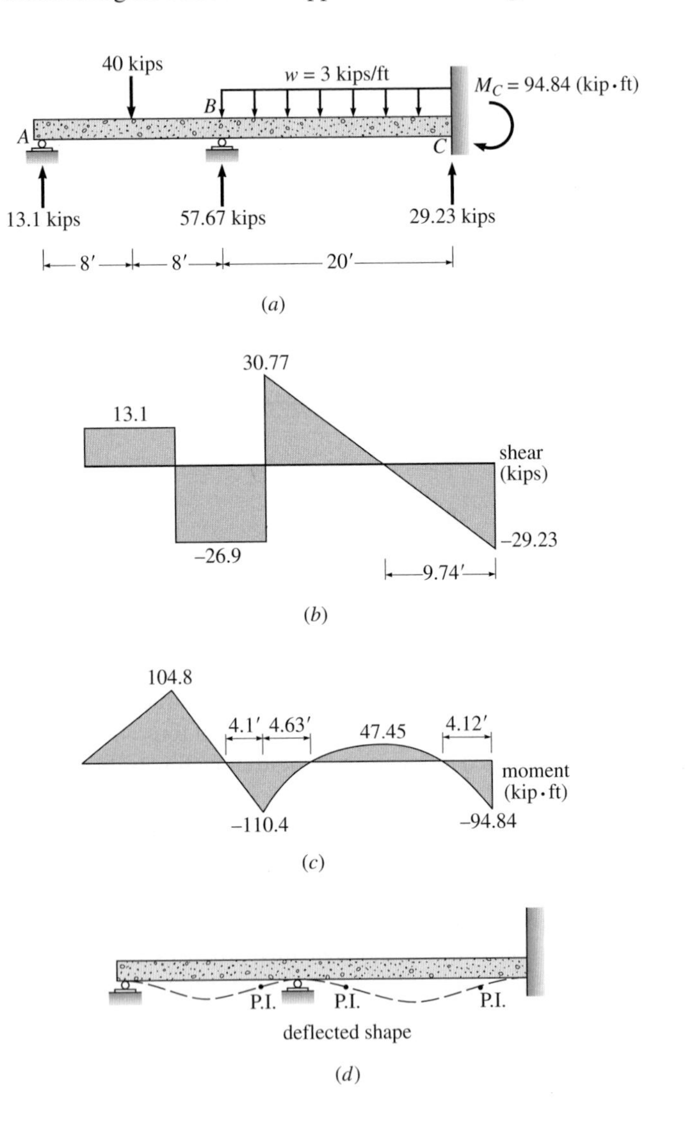

Figure 5.19: (a) Beam (reactions given); (b) shear curve (kips); (c) moment curve (kip·ft); (d) deflected shape.

Solution

Because the beam is indeterminate to the second degree, the reactions must be determined by one of the methods of indeterminate analysis covered in Chapters 11 through 13. Once the reactions are established, the procedure to draw the shear and moment curves is identical to that used in Examples 5.6 to 5.10. Figure 5.19d shows the deflected shape of the structure. Points of inflection are indicated by small black dots.

EXAMPLE 5.12

Analyze the girder AB supporting a floor system in Figure 5.20. Stringers, FE and EDC, the small longitudinal beams that support the floor, are supported by girder AB. Draw the shear and moment curves for the girder.

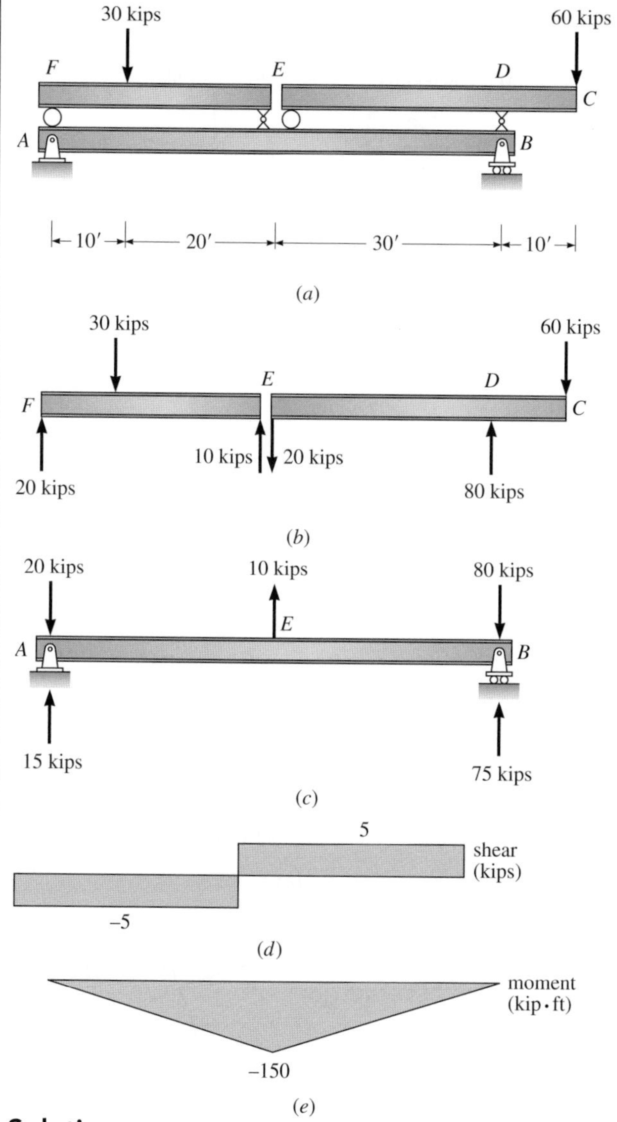

Figure 5.20: (*a*) Details of girder-stringer system; (*b*) free bodies of stringers; (*c*) free body of girder; (*d*) shear curve of girder; (*e*) moment curve of girder.

Solution

Since the stringers FE and EDC are statically determinate, their reactions can be determined by statics using the free bodies shown in Figure 5.20*b*. After the reactions of the stringers are computed, they are applied in the opposite direction to the free body of the girder in Figure 5.20*c*. At point E we can combine the reactions and apply a net load of 10 kips upward to the girder. After the reactions of the girder are computed, the shear and moment curves are drawn (see Figure 5.20*d* and *e*).

EXAMPLE 5.13

Draw the shear and moment curves for each member of the frame in Figure 5.21*a*. Also sketch the deflected shape and show the forces acting on a free body of joint *C*. Treat the connection at *B* as a hinge.

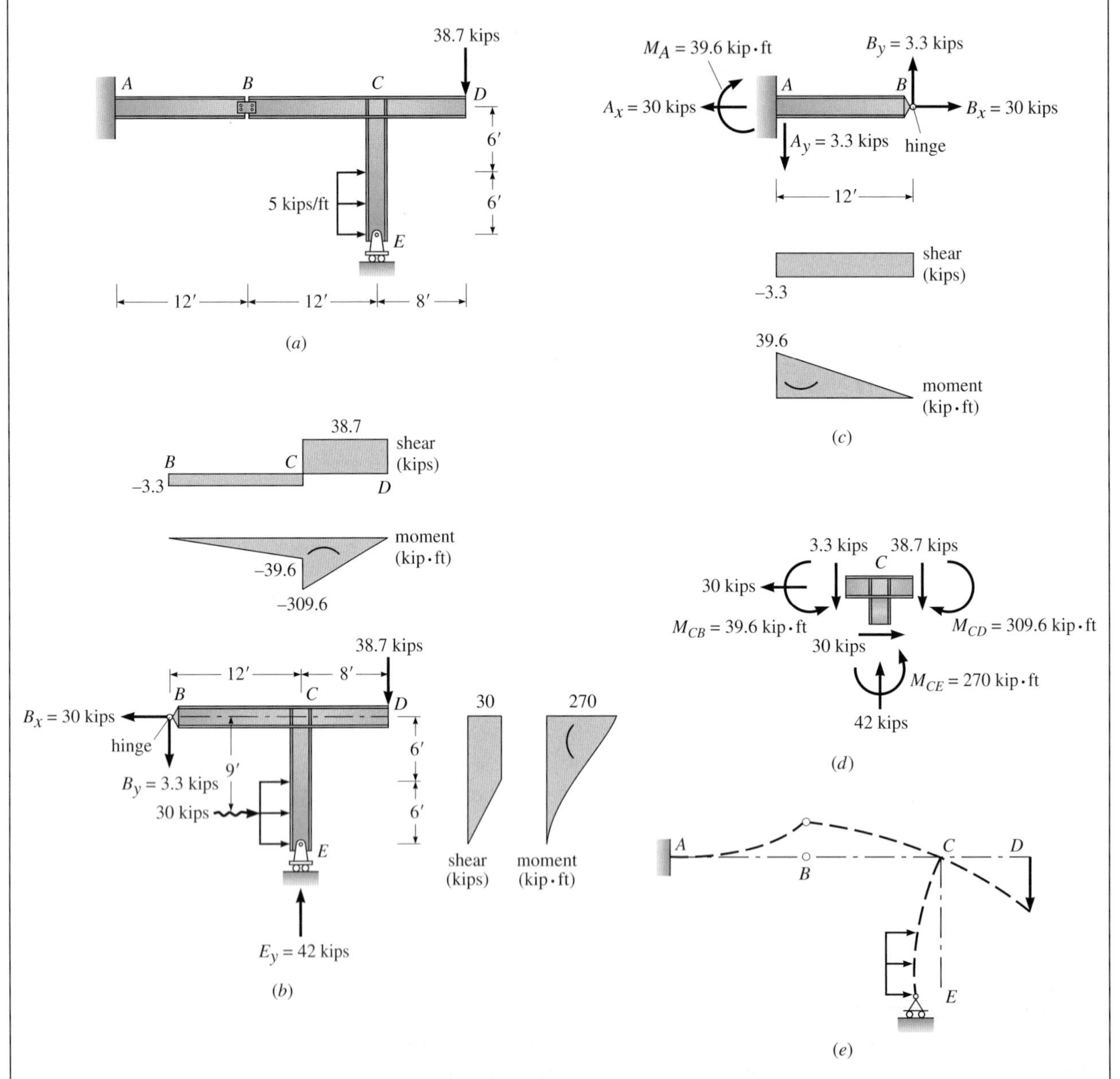

Figure 5.21: (*a*) Determinate frame; (*b*) shear and moment curves for frame *BCDE*; (*c*) shear and moment curves for cantilever *AB*; (*d*) free body of joint *C*; (*e*) deflected shape of frame.

Solution

We begin the analysis of the frame by analyzing free bodies of the structure on either side of the hinge at B to compute the reactions. To compute the vertical reaction at the roller (point E), we sum moments about B of the forces acting on the free body in Figure 5.21b.

$$\circlearrowleft^{+} \quad \Sigma M_B = 0$$

$$0 = 38.7(20) - 30(9) - E_y(12)$$

$$E_y = 42 \text{ kips}$$

The components of the hinge forces at B can now be determined by summing forces in the x and y directions.

$$\rightarrow + \quad \Sigma F_x = 0$$

$$30 - B_x = 0 \qquad B_x = 30 \text{ kips}$$

$$\overset{+}{\uparrow} \quad \Sigma F_y = 0$$

$$-B_y + 42 - 38.7 = 0 \qquad B_y = 3.3 \text{ kips}$$

After the hinge forces at B are established, the cantilever in Figure 5.21c can be analyzed by the equations of statics. The results are shown on the sketch. With the forces known at the ends of all members, we draw the shear and moment curves for each member. These results are plotted next to each member. The curvature associated with each moment curve is shown by a curved line on the moment diagram.

The free body of joint C is shown in Figure 5.21d. As you can verify by using the equations of statics (that is, $\Sigma F_y = 0$, $\Sigma F_x = 0$, $\Sigma M = 0$), the joint is in equilibrium.

A sketch of the deflected shape is shown in Figure 5.21e. Since A is a fixed support, the longitudinal axis of the cantilever beam is horizontal at that point. If we recognize that neither axial forces nor the curvature produced by moment produces any significant change in the length of members, then joint C is restrained against horizontal and vertical displacement by members CE and ABC, which connect to supports that prevent displacement along the axes of these members. Joint C is free to rotate. As you can see, the concentrated load at D tends to rotate joint C clockwise. On the other hand, the distributed load of 30 kips on member CE tries to rotate the joint counterclockwise. Since member BCD is bent concave downward over its entire length, the clockwise rotation dominates.

Although the curvature of member CE is consistent with that indicated by the moment diagram, the final deflected position of the roller at E in the horizontal direction is uncertain. Although we show that the roller has displaced to the left of its initial position, it is possible that it could also be located to the right of its undeflected position if the column is flexible. Techniques to compute displacements will be introduced in Chapters 9 and 10.

5.5 Principle of Superposition

Many of the analytical techniques that we develop in this book are based on the *principle of superposition*. This principle states:

> **If a structure behaves in a linearly elastic manner, the force or displacement at a particular point produced by a set of loads acting simultaneously can be evaluated by adding (superimposing) the forces or displacements at the particular point produced by each load of the set acting individually. In other words, the response of a linear, elastic structure is the same if all loads are applied simultaneously or if the effects of the individual loads are combined.**

The principle of superposition may be illustrated by considering the forces and deflections produced in the cantilever beam shown in Figure 5.22. Figure 5.22*a* shows the reactions and the deflected shape produced by forces P_1 and P_2. Figures 5.22*b* and *c* show the reactions and the deflected shapes produced by the loads acting separately on the beam. The principle of superposition states that the *algebraic sum* of the reactions, or internal forces, or displacements at any particular point in Figures 5.22*b* and *c* will be equal to the reaction, or internal force, or displacement at the corresponding point in Figure 5.22*a*. In other words, the following expressions are valid:

$$R_A = R_{A1} + R_{A2}$$

$$M_A = M_{A1} + M_{A2}$$

$$\Delta_C = \Delta_{C1} + \Delta_{C2}$$

The principle of superposition does not apply to beam-columns or to structures that undergo large changes in geometry when loaded. For example, Figure 5.23*a* shows a cantilever column loaded by an axial force P.

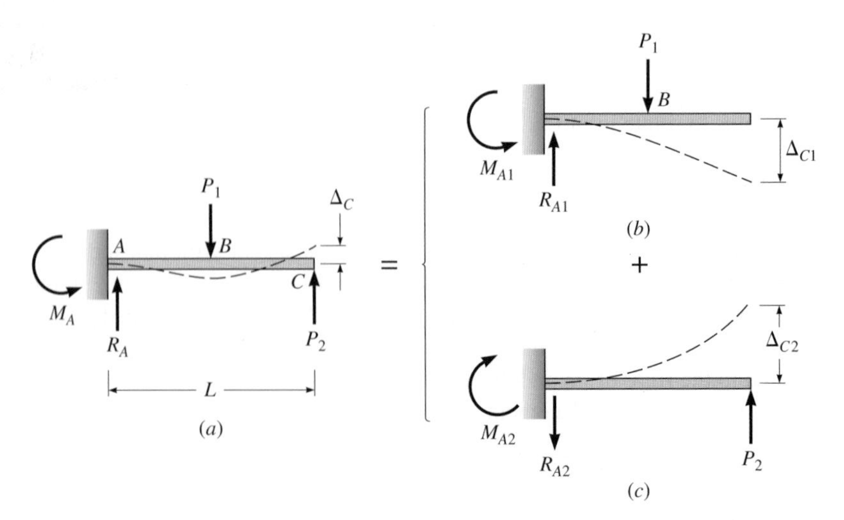

Figure 5.22

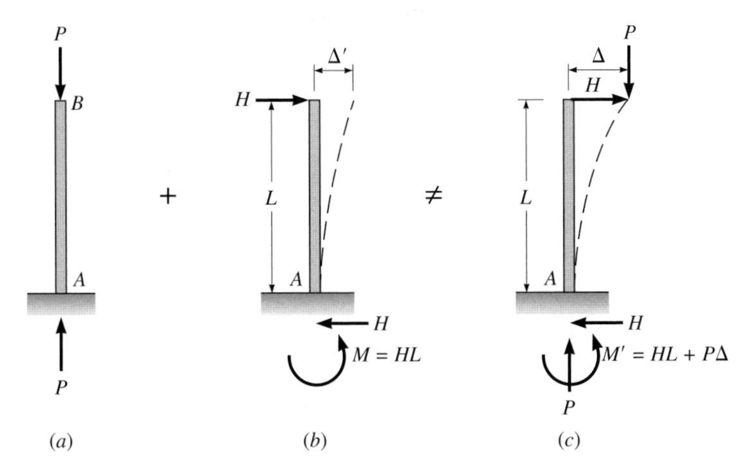

Figure 5.23: Superposition not applicable. (*a*) Axial force produces direct stress; (*b*) lateral force produces moment; (*c*) axial force produces $P\Delta$ moment.

The effect of the axial load P is to produce only direct stress in the column; P produces no moment. Figure 5.23*b* shows a horizontal force H applied to the top of the same column. This load produces both shear and moment.

In Figure 5.23*c*, the loads in Figure 5.23*a* and *b* are applied simultaneously to the column. If we sum moments about A to evaluate the moment at the base of the column in its deflected position (the top has deflected horizontally a distance Δ), the moment at the base can be expressed as

$$M' = HL + P\Delta$$

The first term represents the *primary moment* produced by the transverse load H. The second term, called the $P\Delta$ *moment*, represents moment produced by the eccentricity of the axial load P. The total moment at the base obviously exceeds the moment produced by summing cases a and b. Since the lateral displacement of the top of the column produced by the lateral load creates additional moment at all sections along the length of the column, the flexural deformations of the column in Figure 5.23*c* are greater than those in Figure 5.23*b*. Because the presence of axial load increases the deflection of the column, we see that the axial load has the effect of reducing the flexural stiffness of the column. If the flexural stiffness of the column is large and Δ is small or if P is small, the $P\Delta$ moment will be small and in most practical cases may be neglected.

Figure 5.24 shows a second case in which superposition is *invalid*. In Figure 5.24*a* a flexible cable supports two loads of magnitude P at the third points of the span. These loads deflect the cable into a symmetric shape. The sag of the cable at B is denoted by h. If the loads are applied separately, they produce the deflected shapes shown in Figure 5.24*b* and *c*. Although the sum of the vertical components of the reactions at the supports in *b* and *c* equals those in *a*, computations clearly indicate that the sum of the horizontal components H_1 and H_2 does not equal H. It is also evident that the sum of the vertical deflections at B, h_1, and h_2 is much greater than the value of h in case a.

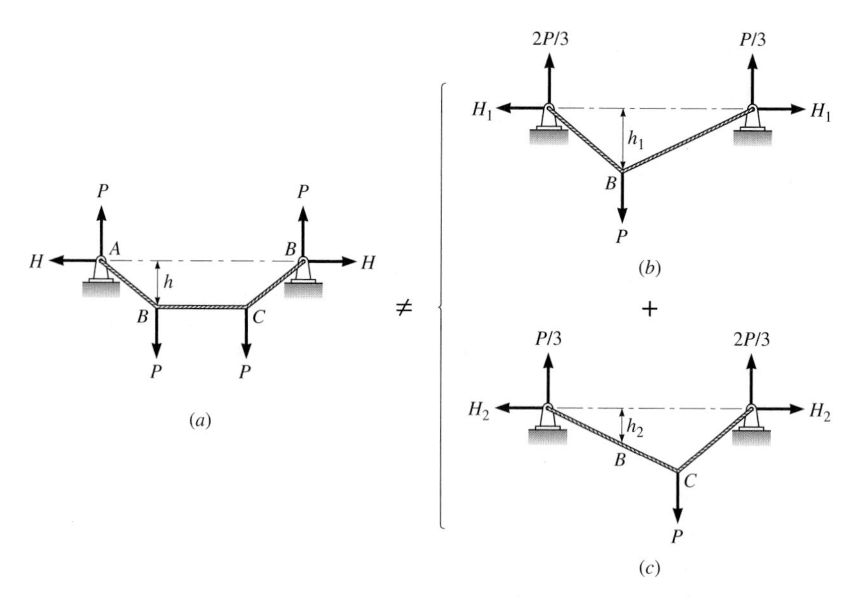

Figure 5.24: Superposition not applicable: (*a*) cable with two equal loads at the third points of the span; (*b*) cable with single load at *B*; (*c*) cable with single load at *C*.

The principle of superposition provides the basis for the analysis of indeterminate structures by the flexibility method discussed in Chapter 11 as well as matrix methods in Chapters 16, 17, and 18. Superposition is also used frequently to simplify computations involving the moment curves of beams that carry several loads. For example, in the moment-area method (a procedure to compute the slope or deflection at a point along the axis of a beam) we must evaluate the product of an area and the distance between the area's centroid and a reference axis. If several loads are supported by the beam, the shape of the moment diagram may be complicated. If no simple equations are available to evaluate either the area under the moment diagram or the position of the area's centroid, the required computation can be carried out only by integrating a complicated function. To avoid this time-consuming operation, we can analyze the beam separately for the action of each load. In this way we produce several moment curves with simple geometric shapes whose area and centroids can be evaluated and located by standard equations (see back inside cover). Example 5.14 illustrates the use of superposition to establish the reactions and moment curve of a beam loaded with both a uniform load and end moments.

EXAMPLE 5.14

(a) Evaluate the reactions and construct the moment diagram for the beam in Figure 5.25a by superposition of the reactions and moment curves associated with the individual loads in parts (b), (c), and (d).
(b) Calculate the moment of the area under the moment diagram between the left support and the center of the beam with respect to an axis through support A.

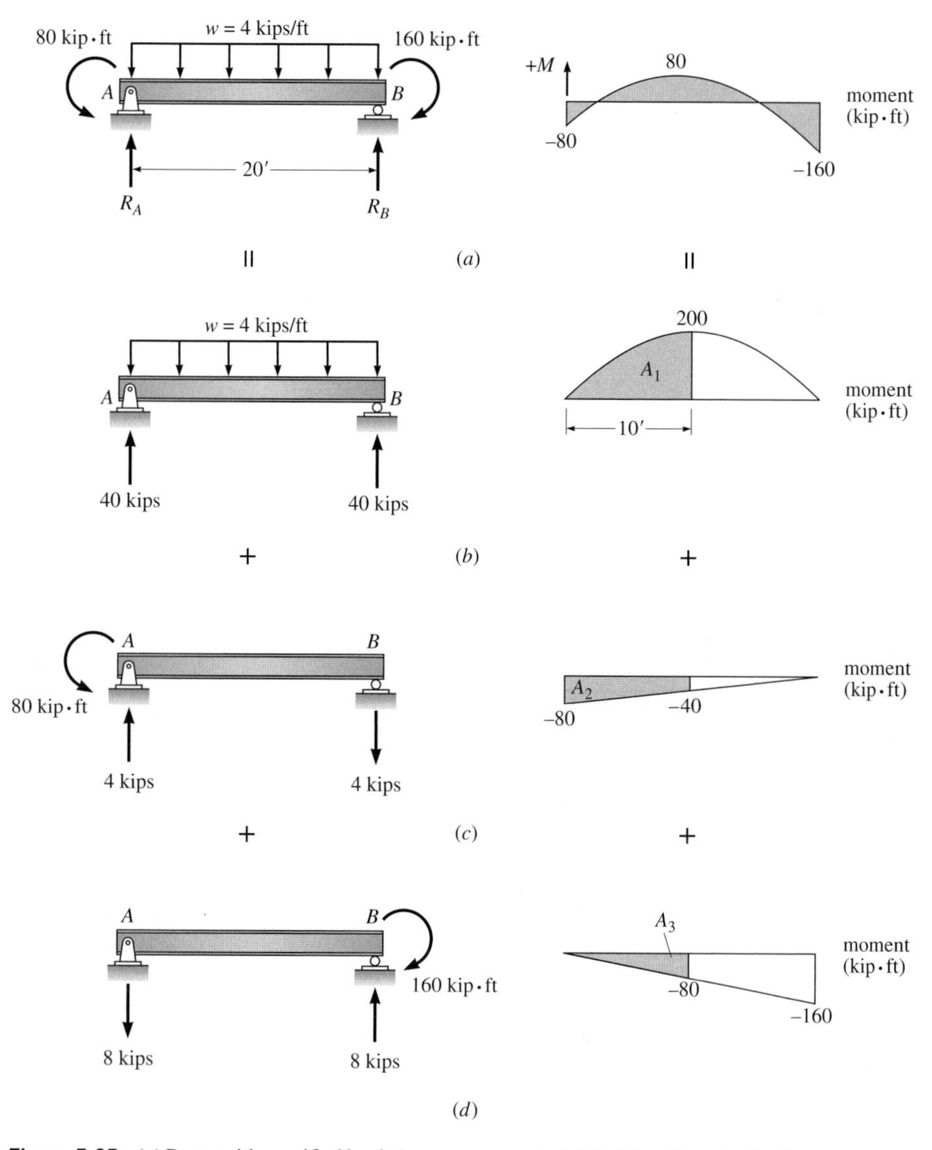

Figure 5.25: (a) Beam with specified loads (moment curve to right); (b) uniform load only applied; (c) reactions and moment curve associated with 80 kip · ft moment; (d) reactions and moment curve produced by end moment of 160 kip·ft at B.

[*continues on next page*]

Example 5.14 continues . . .

Solution

(*a*) To solve by superposition, also called *moment curves by parts*, we analyze the beam separately for the individual loads. (The reactions and moment diagrams are shown in Figure 5.25*b*, *c*, and *d*.) The reactions and the ordinates of the moment diagram produced by all loads acting simultaneously (Figure 5.25*a*) are then established by summing algebraically the contribution of the individual cases.

$$R_A = 40 + 4 + (-8) = 36 \text{ kips}$$

$$R_B = 40 + (-4) + 8 = 44 \text{ kips}$$

$$M_A = 0 + (-80) + 0 = -80 \text{ kip} \cdot \text{ft}$$

$$M_{\text{center}} = 200 + (-40) + (-80) = 80 \text{ kip} \cdot \text{ft}$$

(*b*) $$\text{Moment of area} = \sum_{n=1}^{3} A_n \cdot \bar{x}$$

(see Table 3 inside the back cover)

$$= \frac{2}{3}(10)(200)\left(\frac{5}{8} \times 10\right) + (-40 \times 10)(5)$$

$$+ \frac{1}{2}(-40)(10)\left(\frac{10}{3}\right) + \frac{1}{2}(10)(-80)\left[\frac{2}{3}(10)\right]$$

$$= 3000 \text{ kip} \cdot \text{ft}^3$$

5.6 Sketching the Deflected Shape of a Beam or Frame

To ensure that structures are serviceable—that is, their function is not impaired because of excessive flexibility that permits large deflections or vibrations under service loads—designers must be able to compute deflections at all critical points in a structure and compare them to allowable values specified by building codes. As a first step in this procedure, the designer must be able to draw an accurate sketch of the deflected shape of the beam or frame. Deflections in well-designed beams and frames are usually small compared to the dimensions of the structure. For example, many building codes limit the maximum deflection of a simply supported beam under live load to 1/360 of the span length. Therefore, if a simple beam spans 20 ft (240 in), the maximum deflection at midspan due to the live load must not exceed $\frac{2}{3}$ in.

If we represent a beam spanning 20 ft by a line 2 in long, we are reducing the dimension along the beam's axis by a factor of 120 (or we can say that we are using a scale factor of $\frac{1}{120}$ with respect to the distance along the beam's axis). If we were to use the same scale to show the deflection at midspan, the $\frac{2}{3}$ in displacement would have to be plotted as 0.0055 in. A distance of this dimension, which is about the size of a period, would not be perceptible to the naked eye. To produce a clear picture of the deflected shape, we must exaggerate the deflections by using a vertical scale 50 to 100 times greater than the scale applied to the longitudinal dimensions of the member. Since we use different horizontal and vertical scales to sketch the deflected shapes of beams and frames, the designer must be aware of the distortions that must be introduced into the sketch to ensure that the deflected shape is an accurate representation of the loaded structure.

An accurate sketch must satisfy the following rules:

1. The curvature must be consistent with the moment curve.
2. The deflected shape must satisfy the constraints of the boundaries.
3. The original angle (usually 90°) at a rigid joint must be preserved.
4. The length of the deformed member is the same as the original length of the unloaded member.
5. The horizontal projection of a beam or the vertical projection of a column is equal to the original length of the member.
6. Axial deformations, trivial compared to flexural deformations, are neglected.

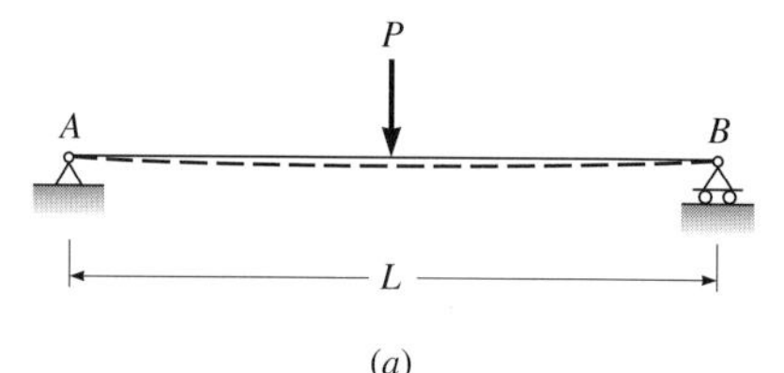

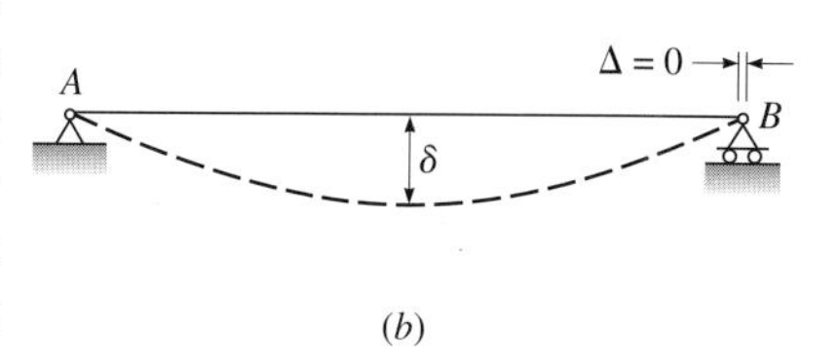

Figure 5.26

For example, in Figure 5.26*a* the deflected shape of a simply supported beam with the service load in place is shown by the dashed line. Since the deflection is almost imperceptible to the naked eye, a sketch of this type would not be useful to a designer who was interested in computing slopes or deflections at a particular point along the axis of the beam. Instead, to show the deflected shape clearly, we will draw the *distorted* sketch shown in Figure 5.26*b*. In Figure 5.26*b* the scale used to draw the deflection δ at midspan is about 75 times greater than the scale used in the longitudinal direction to show the length of the member. When we show the length of the bent member to a distorted scale, the distance along the deflected axis of the member appears much greater than the length of the chord connecting the ends of the member. If a designer were inexperienced, he or she might assume that the roller at the right end of the beam moves to the left a distance Δ. Since the midspan deflection is very small (see Figure 5.26*a*), rule 4 applies. Recognizing that there is no significant difference in length between the loaded and unloaded members, we conclude that the horizontal displacement of the roller at B equals zero, and we show the member spanning to the original position of support B.

As a second example, we draw the deflected shape of the vertical cantilever beam in Figure 5.27*a*. The moment curve produced by the horizontal load at joint B is shown in Figure 5.27*b*. The short curved line within the moment curve indicates the sense of the member's curvature. In Figure 5.27*c* the deflected shape of the cantilever is drawn to an exaggerated scale in the horizontal direction. Since the base of the column is attached to a fixed support, the elastic curve must rise initially from the support at an angle of 90°. Because the vertical projection of the column is assumed equal to the initial length (rule 5), the vertical deflection of the top

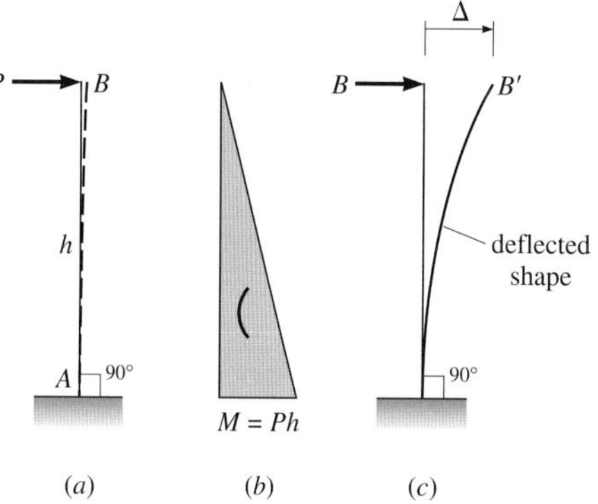

Figure 5.27: (*a*) Deflected shape shown by dashed line to actual scale; (*b*) moment curve for cantilever in (*a*); (*c*) horizontal deflections exaggerated for clarity.

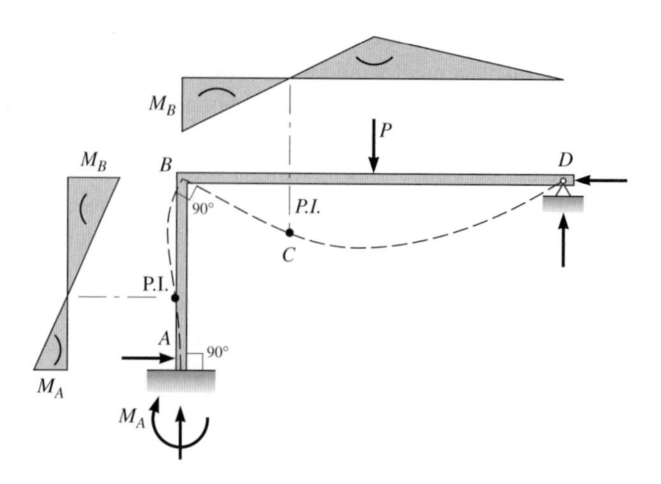

Figure 5.28: Deflected shape of a braced frame. Moment diagrams shown above and to the left of frame.

of the cantilever is assumed to be zero; that is, B moves horizontally to B'. To be consistent with the curvature produced by the moment, the top of the cantilever must displace laterally to the right.

In Figure 5.28 we show with dashed lines the deflected shape produced by a single concentrated load applied at midspan to girder BD of a *braced* frame. In a braced frame all joints are restrained against lateral displacement by supports or by members connected to immovable supports. For example, joint B does not move laterally because it is connected by girder BD to a pin at joint D. We can assume that the length of BD does not change because (1) axial deformations are trivial, and (2) no change in length is produced by bending. To plot the deflected shape, we show the column leaving the fixed support at A in the vertical direction. The curvature produced by the moment indicates the lower section of the column develops compressive stresses on the outside face and tension on the inside face. At the point where the moment reduces to zero—the point of inflection (P.I.)—the curvature reverses and the column curves back toward joint B. The applied load bends the girder downward, causing joint B to rotate in the clockwise direction and joint D in the counterclockwise direction. Since joint B is rigid, the angle between the column and the girder remains 90°.

In Figure 5.29*a* we show an L-shaped cantilever with a horizontal load applied to the top of the column at B. The moment produced by the horizontal force at joint B (see Figure 5.29*b*) bends the column to the right. Since no moments develop in beam BC, it remains straight. Figure 5.29*c* shows the deflected shape to an exaggerated scale. We start the sketch from the fixed support at A because both the slope (90°) and the deflection (zero) are known at that point. Because the angular rotation of joint B is small, the horizontal projection of beam BC can be assumed equal to the original length L of the member. Notice that both joints B and C displace the same horizontal distance Δ to the right. As was the case with the top of the column in Figure 5.27, joint B is assumed to move horizontally only. On the other hand, joint C, in addition to moving the same distance Δ to the right as joint B, moves downward a distance $\Delta_v = \theta L$ due to the rotation of member BC through an angle θ. As shown in Figure 5.29*d*, the clockwise rotation of joint B (which is rigid) can be measured from either the x or the y axis.

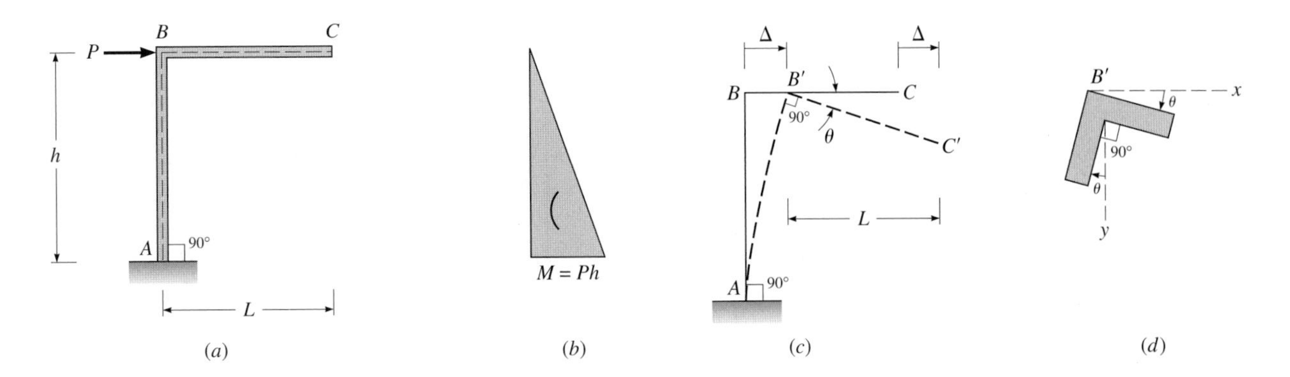

Figure 5.29: (*a*) Deflected shape shown to scale by dashed line; (*b*) moment diagram; (*c*) deflected shape drawn to an exaggerated scale; (*d*) rotation of joint *B*.

The lateral load at joint *B* of the frame in Figure 5.30*a* produces moment that creates compression on the outside faces of both column *AB* and girder *BC*. To begin the sketch of the deflected shape, we start at the pin at *A*—the only point on the deflected frame whose final position is known. We will arbitrarily assume that the bottom of column *AB* rises vertically from the pin support at *A*. Since the moment curve indicates that the column bends to the left, joint *B* will move horizontally to *B′* (Figure 5.30*b*). Because joint *B* is rigid, we draw the *B* end of member *BC* perpendicular to the top of the column. Since member *BC* curves concave upward, joint *C* will move to point *C′*. Although the frame has the *correct deformed shape in every respect*, the position of joint *C* violates the boundary conditions imposed by the roller at *C*. Since *C* is constrained to move horizontally only, it cannot displace vertically to *C″*.

We can establish the correct position of the frame by imagining that the entire structure is rotated clockwise as a rigid body about the pin at *A* until joint *C* drops to the level of the plane (at *C″*) on which the roller moves. The path followed by *C* during the rotation about *A* is indicated by the arrow between *C′* and *C″*. As the rigid body rotation occurs, joint *B* moves horizontally to the right to point *B″*.

As shown in Figure 5.30*c*, an *incorrect* sketch, the *B* end of member *AB* *cannot* enter joint *B* with a slope that is upward and to the left because the 90° angle could not be preserved at joint *B* if the upward curvature of the girder is also maintained. Since joint *B* is free to move laterally as the column bends, the frame is termed an *unbraced* frame.

Figure 5.30: (*a*) Moment curves for frame *ABC*; (*b*) deformed frame in final position; (*c*) incorrect deflected shape: 90° angle at *B* not preserved.

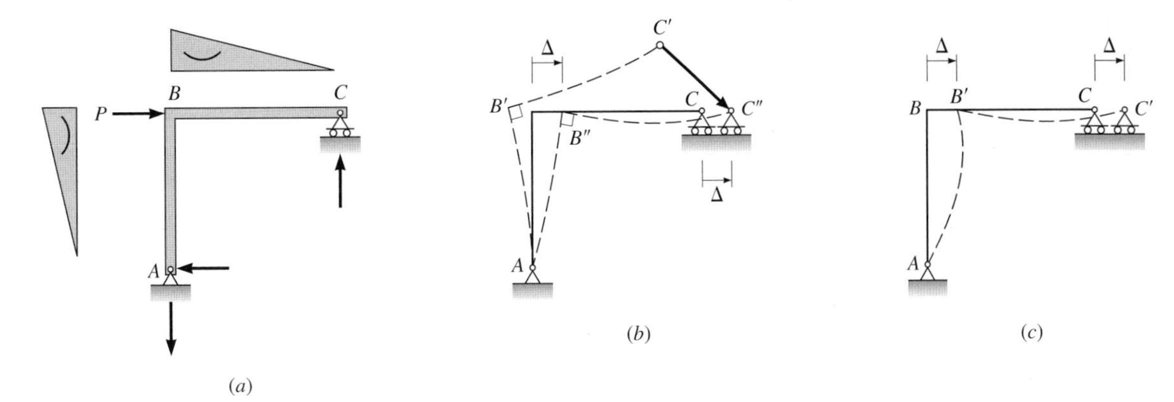

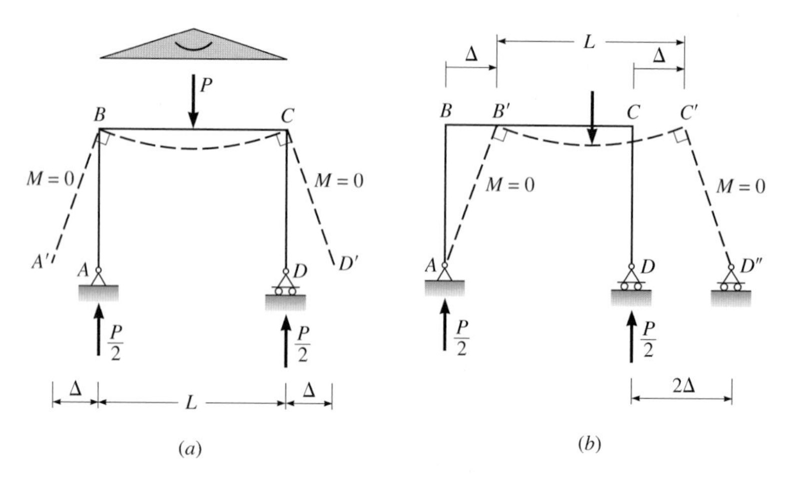

Figure 5.31: (*a*) Deformations produced by load shown by dashed line; (*b*) position required by constraints of supports.

In Figure 5.31*a* a symmetrically loaded unbraced frame carries a concentrated load at the midspan of girder *BC*. Based on the initial dimensions, we find that the reactions at the pin at *A* and the roller at *D* are both equal to *P*/2. Since no horizontal reactions develop at the supports, the moment in both columns is zero (they carry only axial load), and the columns remain straight. Girder *BC*, which acts as a simply supported beam, bends concave upward. If we sketch the deflected shape of the girder assuming that it does not displace laterally, the deflected shape shown by the dashed lines results. Since the right angles must be preserved at joints *B* and *C*, the bottom ends of the columns will displace outward horizontally at *A'* and *D'*. Although the deflected shape is correct, joint *A* cannot move because it is connected to the pin at *A*. The correct position of the frame is established by shifting the entire deformed frame as a rigid body to the right an amount Δ (see Figure 5.31*b*). As shown in this figure, joints *B* and *C* move horizontally only, and the length of the loaded girder is the same as its initial undeformed length of *L*.

Figure 5.32 shows a frame with a hinge at *C*. Since the curvature of member *AB* and the final position of joints *A* and *B* are known, we begin the

Photo 5.2: Two-legs of a reinforced concrete rigid frame. Frame supports a cable-stayed bridge.

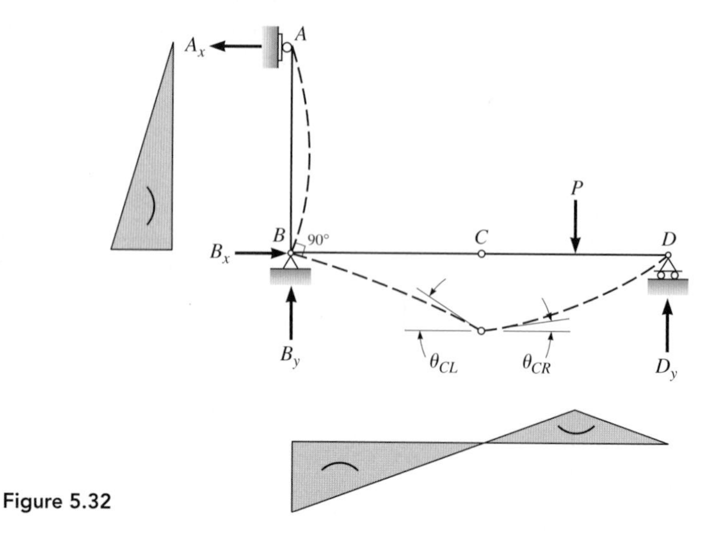

Figure 5.32

sketch by drawing the deflected shape of member AB. Since joint B is rigid, the 90° angle is preserved at B, and member BC must slope downward to the right. Since the hinge at C provides no rotational restraint, the members must frame into each side of the hinge with different slopes because of the difference in curvature indicated by the moment curves.

5.7 Degree of Indeterminacy

In our previous discussion of stability and indeterminacy in Chapter 3, we considered a group of structures that could be treated as a single rigid body or as several rigid bodies with internal releases provided by hinges or rollers. We now want to extend our discussion to include indeterminate frames—structures composed of members that carry shear, axial load, and moment at a given section. The basic approaches we discussed in Chapter 3 still apply. We begin our discussion by considering the rectangular frame in Figure 5.33*a*. This rigid jointed structure, fabricated from a single member, is supported by a pin support at A and a roller at B. At point D a small gap exists between the ends of the members which cantilever out from joints C and E. Since the supports supply three restraints that are neither a parallel nor a concurrent force system, we conclude that the structure is stable and determinate; that is, three equations of statics are available to compute the three support reactions. After the reactions are evaluated, internal forces—shear, axial, and moment—at any section can be evaluated by passing a cutting plane through the section and applying the equations of equilibrium to the free-body diagram on either side of the cut.

Photo 5.3: Legs of a rigid frame fabricated from steel plates.

If the two ends of the cantilever were now connected by inserting a hinge at D (see Figure 5.33*b*), the structure would no longer be statically determinate. Although the equations of statics permit us to compute the reactions for any loading, the internal forces within the structure cannot be determined because it is not possible to isolate a section of the structure as a free body that has only three unknown forces. For example, if we attempt to compute the internal forces at section 1–1 at the center of member AC in Figure 5.33*b* by considering the equilibrium of the free body that extends from section 1–1 to the hinge at D (see Figure 5.33*c*), five internal forces—three at section 1–1 and two at the hinge—must be evaluated. Since only three equations of statics are available for their solution, we conclude that the structure is indeterminate to the second degree. We can reach this same conclusion by recognizing that if we remove the hinge at D, the structure reduces to the determinate frame in Figure 5.33*a*. In other words, when we connect the two ends of the structure together with a hinge, both horizontal restraint and vertical restraint are added at D. These restraints, which provide alternative load paths, make the structure indeterminate. For example, if a horizontal force is applied at C to determinate frame in Figure 5.33*a*, the entire load must be transmitted through member CA to the pin at A and the roller at B. On the other hand, if the same force is applied to the frame in Figure 5.33*b*, a certain percentage of the force is transferred through the hinge to the right side of the structure to member DE and then through member EB to the pin at B.

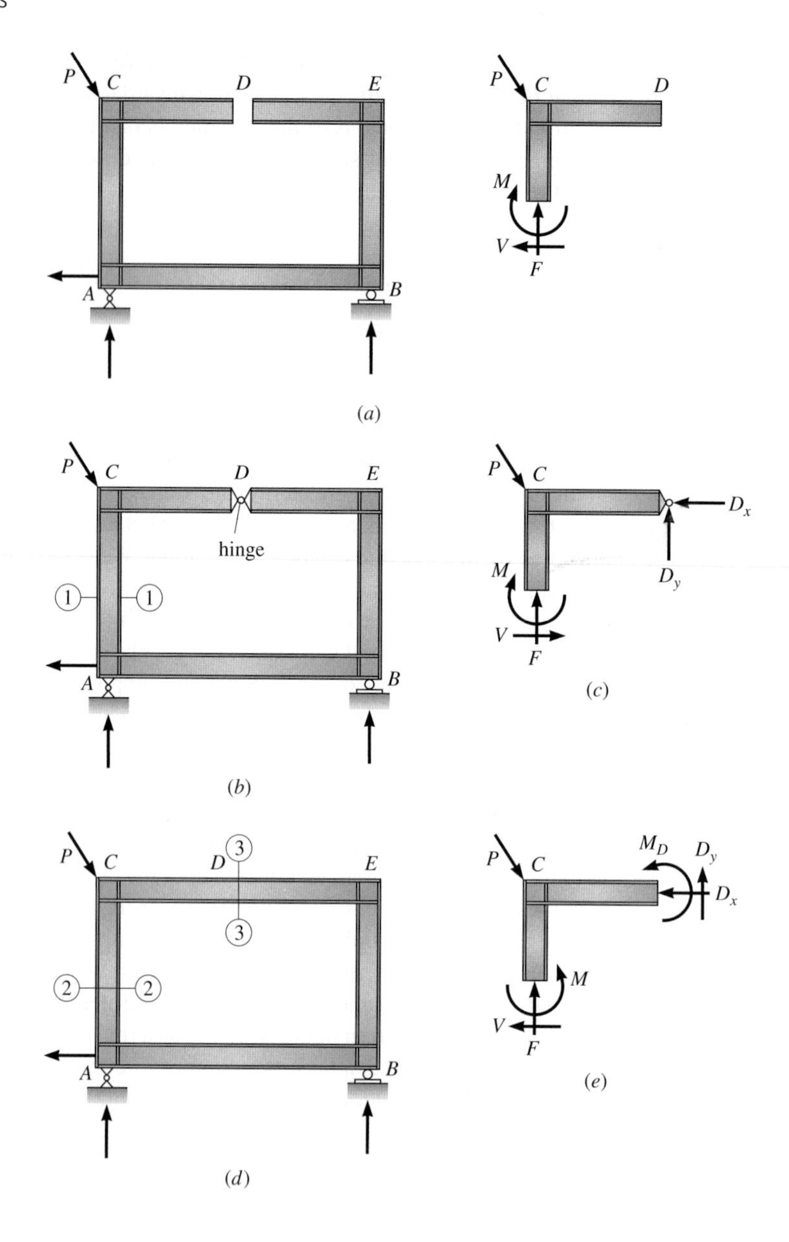

Figure 5.33: (*a*) Stable, *externally* determinate frame; (*b*) *internally* indeterminate frame to second degree; (*c*) free body of upper left corner of hinged frame; (*d*) closed ring *internally* indeterminate to the third degree; (*e*) free body of upper left corner of closed ring (see *d*).

If the two ends of the frame at D are welded to form a solid continuous member (see Figure 5.33*d*), that section will have the capacity to transmit moment as well as shear and axial load. The addition of flexural restraint at D raises the degree of indeterminacy of the frame to three. As shown in Figure 5.33*e*, a typical free body of any portion of the structure can develop six unknown internal forces. With only three equations of equilibrium, the structure is indeterminate internally to the third degree. In summary, a closed ring is statically indeterminate internally to the third degree. To establish the degree of indeterminacy of a structure composed of a number of closed rings (e.g., a welded steel building frame) we can remove restraints—either internal or external—until a stable *base* structure remains. *The number of*

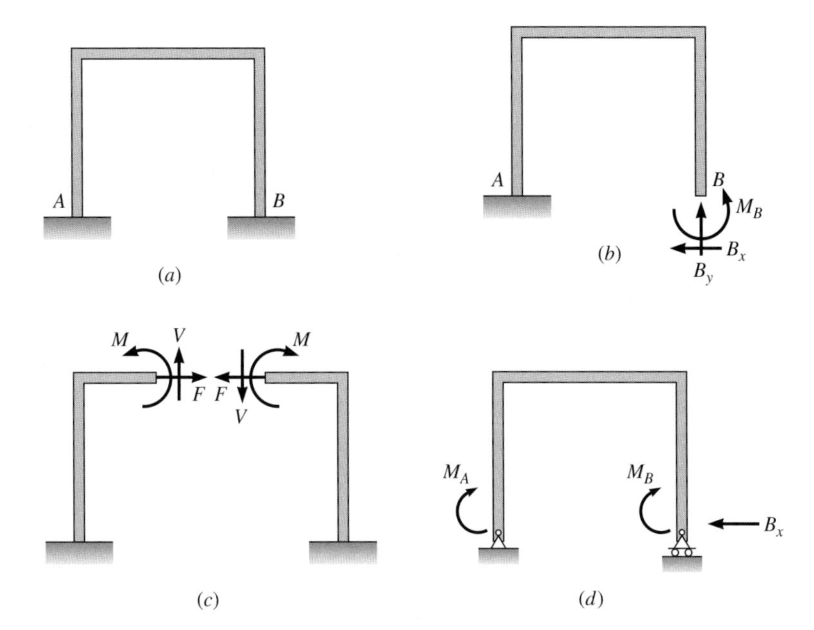

Figure 5.34: Establishing the degree of indeterminacy by removing supports until a stable determinate structure remains. (*a*) A fixed-end frame; (*b*) the fixed support at *B* removed; (*c*) the girder cut; (*d*) roller and pin used to eliminate moment and horizontal restraint at *B* and the moment at *A*.

restraints removed equals the degree of indeterminacy. This procedure was introduced in Section 3.7; see Case 3.

To illustrate this procedure for establishing the degree of indeterminacy of a rigid frame by removing restraints, we will consider the frame in Figure 5.34*a*. When evaluating the degree of indeterminacy of a structure the designer always has a variety of choices with regard to which restraints are to be removed. For example, in Figure 5.34*b* we can imagine the frame is cut just above the fixed support at *B*. Since this action removes three restraints B_x, B_y, and M_B, but leaves a stable U-shaped structure connected to the fixed support at *A*, we conclude that the original structure is indeterminate to the third degree. As an alternative procedure, we can eliminate three restraints (*M*, *V*, and *F*) by cutting the girder at midspan and leaving two stable determinate L-shaped cantilevers (see Figure 5.34*c*). As a final example (see Figure 5.34*d*), a stable determinate base structure can be established by removing the moment restraint at *A* (physically equivalent to replacing the fixed support by a pin support) and by removing moment and horizontal restraint at *B* (the fixed support is replaced by a roller).

As a second example, we will establish the degree of indeterminacy of the frame in Figure 5.35*a* by removing both internal and external restraints. As one of many possible procedures (see Figure 5.35*b*), we can eliminate two restraints by removing the pin at *C* completely. A third external restraint (resistance to horizontal displacement) can be removed by replacing the pin at *B* with a roller. At this stage we have removed sufficient restraints to produce a structure that is *externally* determinate. If we now cut girders *EF* and *ED*, removing six additional restraints, a stable determinate structure remains. Since a total of nine restraints was removed, the structure is indeterminate to the ninth degree. Figure 5.36 shows several additional

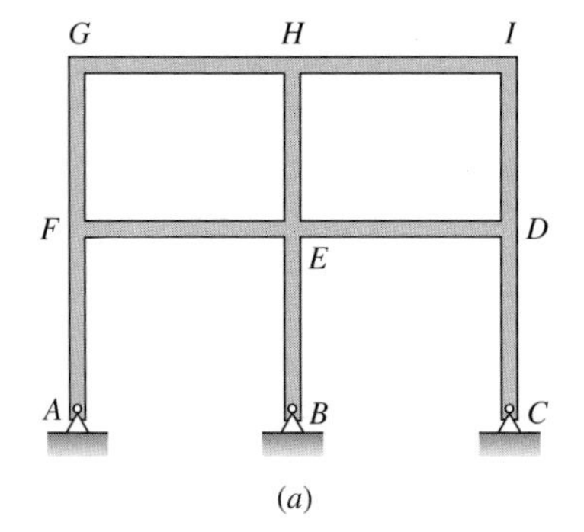

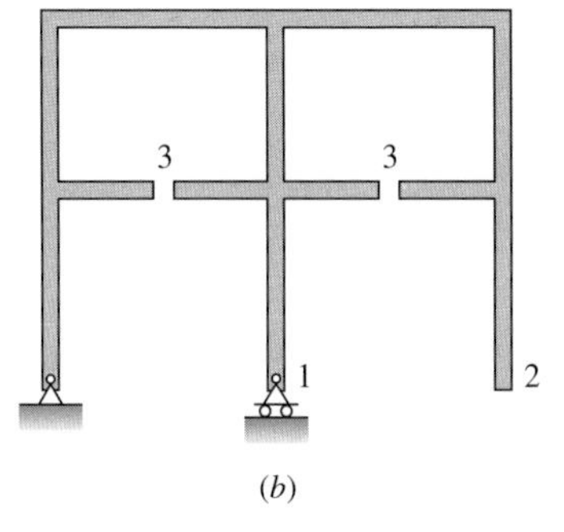

Figure 5.35: (*a*) Frame to be evaluated; (*b*) removing restraints (numbers on figure refer to number of restraints removed at that point to produce the base structure).

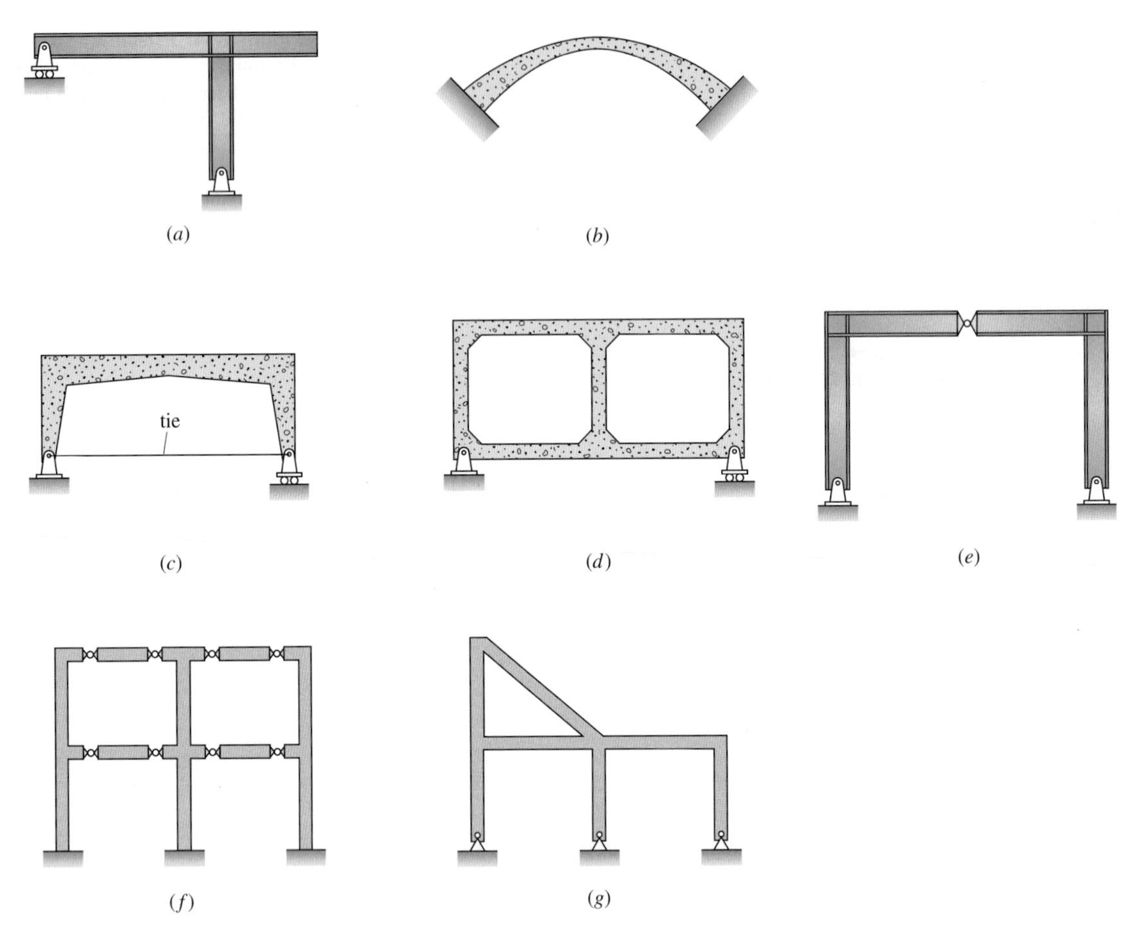

Figure 5.36: Classifying rigid frames: (*a*) stable and determinate, 3 reactions, 3 equations of statics; (*b*) hingeless arch, indeterminate to third degree, 6 reactions, and 3 equations of statics; (*c*) indeterminate first degree, 3 reactions and 1 unknown force in tie, 3 equations of statics; (*d*) indeterminate sixth degree (internally); (*e*) stable determinate structure, 4 reactions, 3 equations of statics and 1 condition equation at hinge; (*f*) indeterminate fourth degree; (*g*) indeterminate sixth degree.

structures whose degree of indeterminacy has been evaluated by the same method. Students should verify the results to check their understanding of this procedure.

For the frame in Figure 5.36*f*, one method of establishing the degree of indeterminacy is to consider the structure in Figure 5.35*a* with the three pins at *A*, *B*, and *C* replaced by fixed supports. This modification would produce a structure similar to the one shown in Figure 5.36*f* except without internal hinges. This modification would increase the previously established ninth degree of indeterminacy to 12 degrees. Now, the addition of eight hinges to produce the structure in Figure 5.36*f* would remove eight internal moment restraints, producing a stable structure that was indeterminate to the fourth degree.

Summary

- In our discussion of beams and frames, we considered members loaded primarily by forces (or components of forces) acting perpendicular to a member's longitudinal axis. These forces bend the member and produce internal forces of shear and moment on sections normal to the longitudinal axis.
- We compute the magnitude of the moment on a section by summing moments of all external forces on a free body to either side of the section. Moments of forces are computed about a horizontal axis passing through the centroid of the cross section. The summation must include any reactions acting on the free body. For horizontal members we assume moments are positive when they produce curvature that is concave up and negative when curvature is concave down.
- Shear is the resultant force acting parallel to the surface of a section through the beam. We compute its magnitude by summing forces or components of forces that are parallel to the section, on either side of the cross section.
- We established procedures to write equations for shear and moment at all sections along a member's axis. These equations will be required in Chapter 10 to compute deflections of beams and frames by the method of virtual work.
- We also established four relationships among load, shear, and moment that facilitate the construction of shear and moment diagrams:

  1. The change in shear ΔV between two points equals the area under the load curve between the two points.
  2. The slope of the shear curve at a given point equals the ordinate of the load curve at that point.
  3. The change in moment ΔM between two points equals the area under the shear curve between the two points.
  4. The slope of the moment curve at a given point equals the ordinate of the shear curve at that point.

- We also established that points of inflection (where curvature changes from positive to negative) in a beam's deflected shape occur where values of moment equal zero.
- We also learned to use moment diagrams to supply information required to draw accurate sketches of the deflected shapes of beams and frames. The ability of the designer to construct accurate deflected shapes is required in the moment-area method covered in Chapter 9. The moment-area method is used to compute slopes and deflections at a selected point along the axis of a beam or frame.
- Finally we established a procedure for determining if a beam or frame is statically determinate or indeterminate, and if indeterminate, then the degree of indeterminacy.

PROBLEMS

P5.1. Write the equations for shear and moment between points B and C as a function of distance x along the longitudinal axis of the beam in Figure P5.1 for (a) origin of x at point A, and (b) origin of x at D.

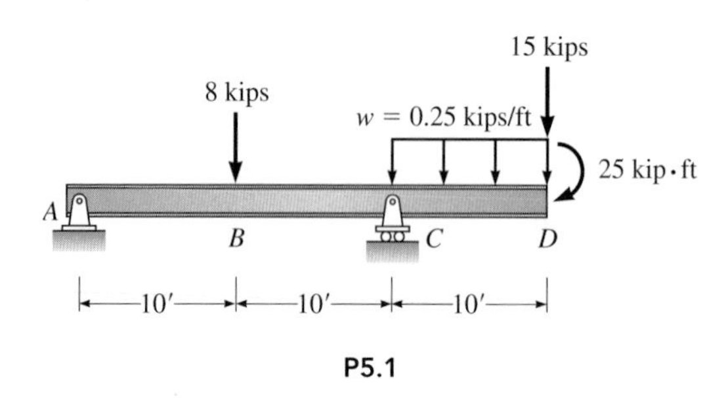

P5.1

P5.2. Write the equations for shear and moment between points D and E. Select the origin at D.

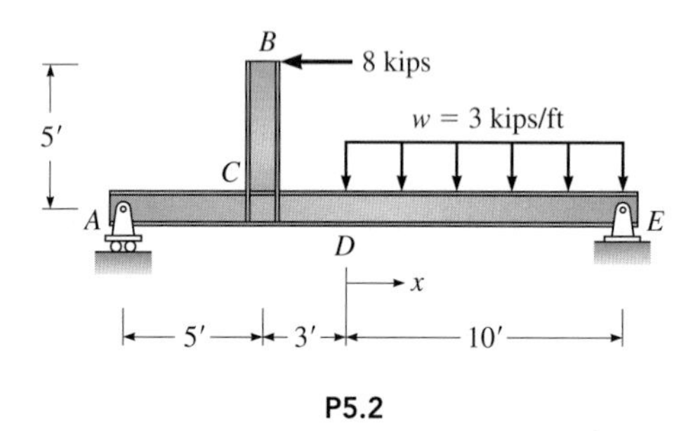

P5.2

P5.3. Write the equations for shear and moment between points A and B. Select the origin at A. Plot the graph of each force under a sketch of the beam. The rocker at A is equivalent to a roller.

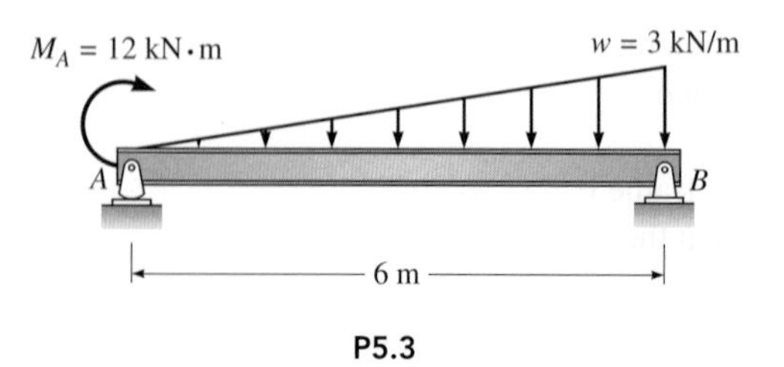

P5.3

P5.4. Write the equations for shear V and moment M between points B and C. Take the origin at point A. Evaluate V and M at point C using the equations.

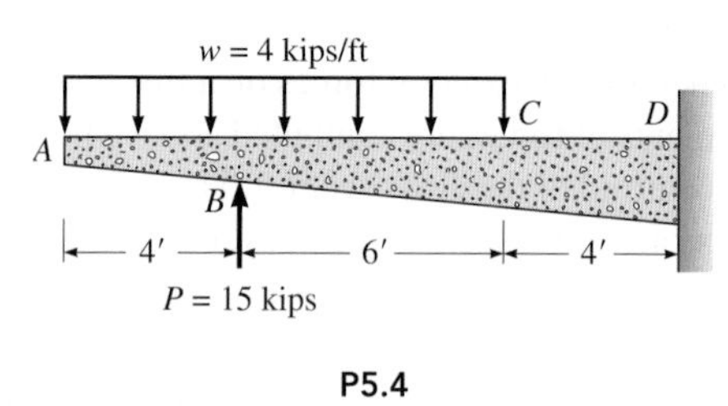

P5.4

P5.5. Write the equations for moment between points B and C as a function of distance x along the longitudinal axis of the beam in Figure P5.5 for (a) origin of x at A and (b) origin of x at B.

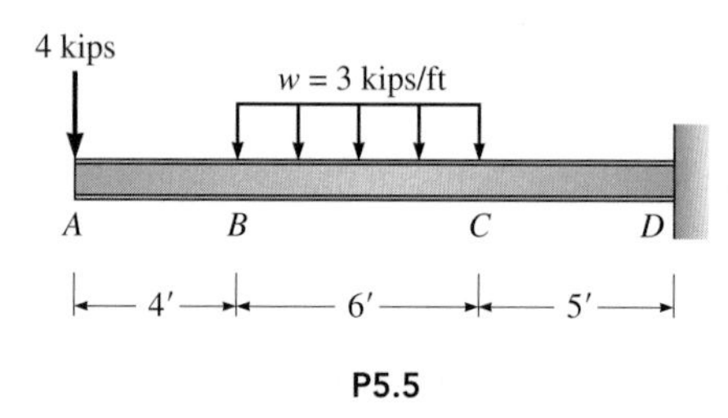

P5.5

P5.6. Write the equations required to express the moment along the entire length of beam in Figure P5.6. Use an origin at point A, and then repeat computations using an origin at point D. Verify that both procedures give the same value of moment at point C.

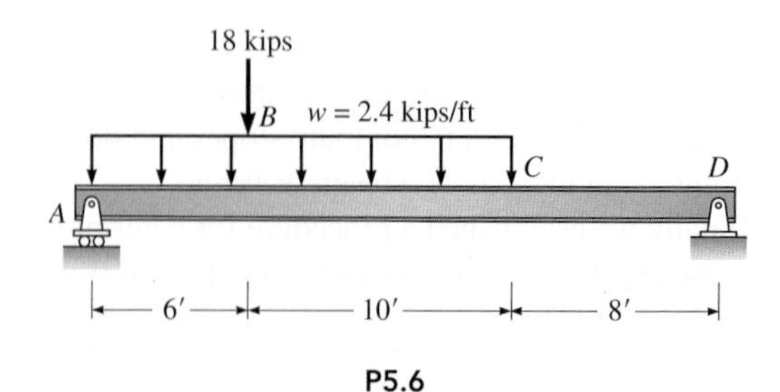

P5.6

P5.7. Write the equations for shear and moment using the origins shown in the figure. Evaluate the shear and moment at C, using the equations based on the origin at point D.

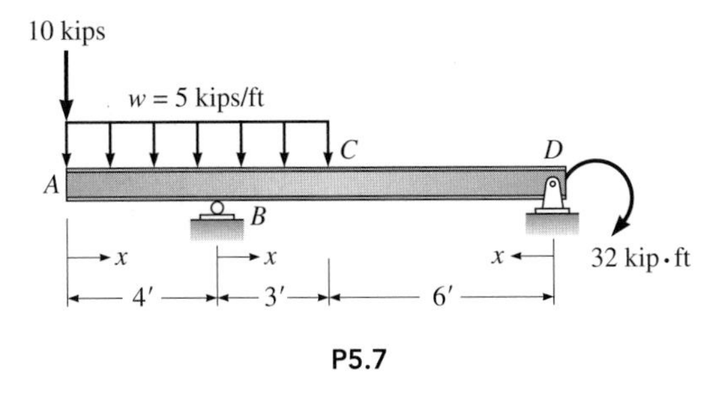

P5.7

P5.8. Write the equations for shear V and moment M in terms of distance x along the length of the beam in Figure P5.8. Take the origin at point A.

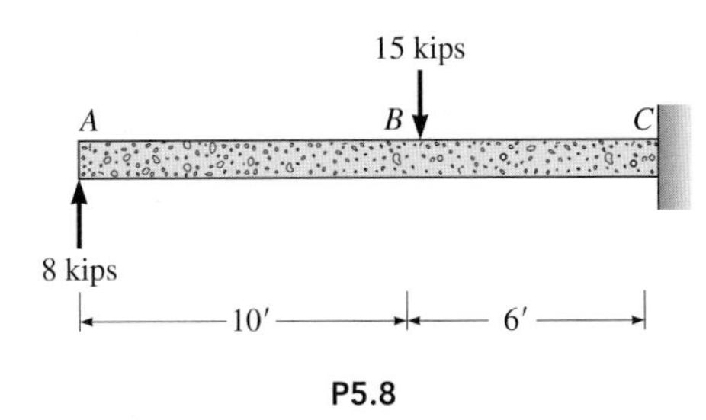

P5.8

P5.9. Write the equation for moment between points B and C for the rigid frame in Figure P5.9.

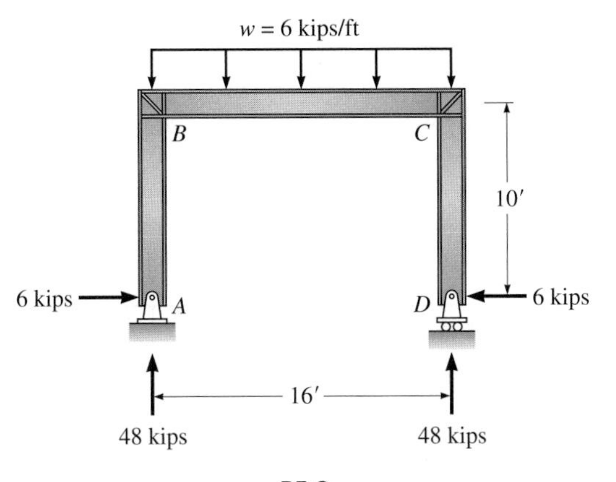

P5.9

P5.10. Write the equations for moment as a function of distance along the longitudinal axes for members AB and BC of the frame in Figure P5.10. Origins for each member are shown.

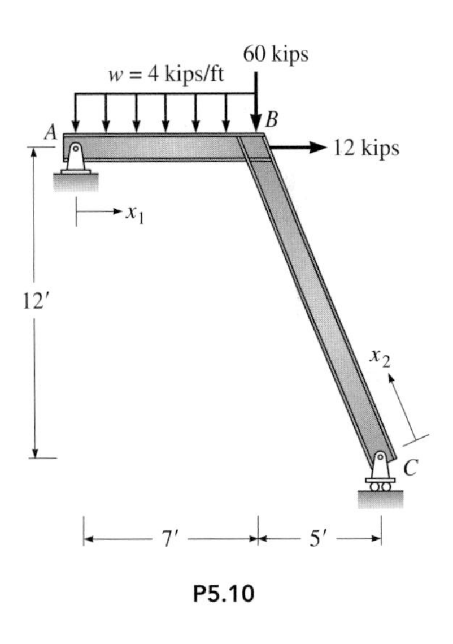

P5.10

P5.11. Write the equations for shear and moment between points B and C for the rigid frame in Figure P5.11. Select the origin at point C.

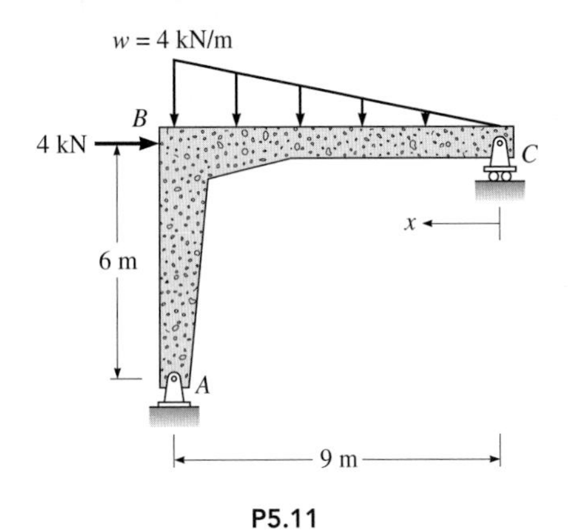

P5.11

P5.12. Consider the beam shown in Figure P5.12.

(a) Write the equations for shear and moment using an origin at end A.

(b) Using the equations, evaluate the moment at Section 1.

(c) Locate the point of zero shear between B and C.

(d) Evaluate the maximum moment between points B and C.

(e) Write the equations for shear and moment using an origin at C.

(f) Evaluate the moment at Section 1.

(g) Locate the section of maximum moment and evaluate M_{max}.

(h) Write the equations for shear and moment between B and C using an origin at B.

(i) Evaluate the moment at Section 1.

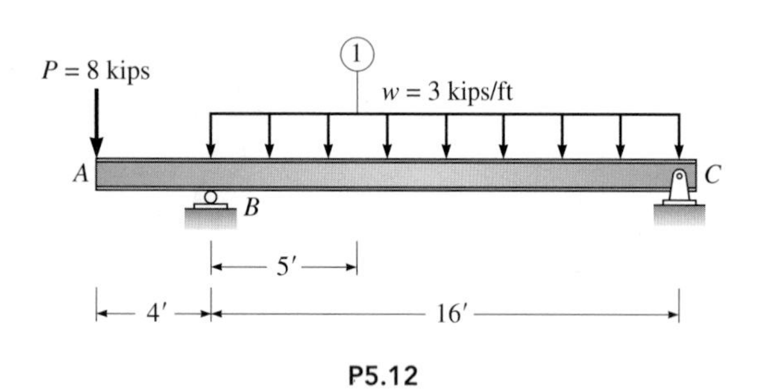

P5.12

P5.13 to P5.19. For each beam, draw the shear and moment diagrams, label the maximum values of shear and moment, locate points of inflection, and sketch the deflected shape.

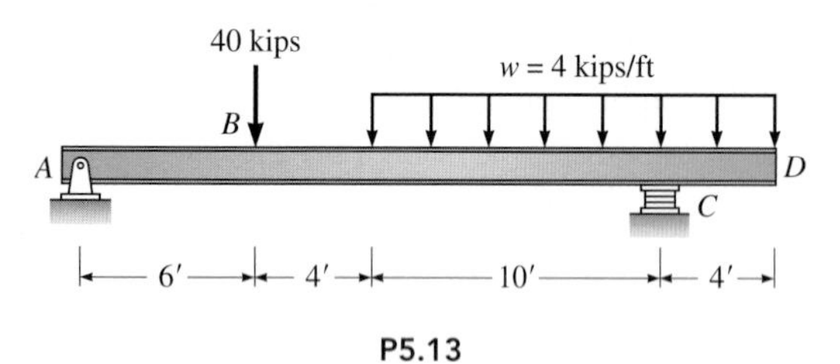

P5.13

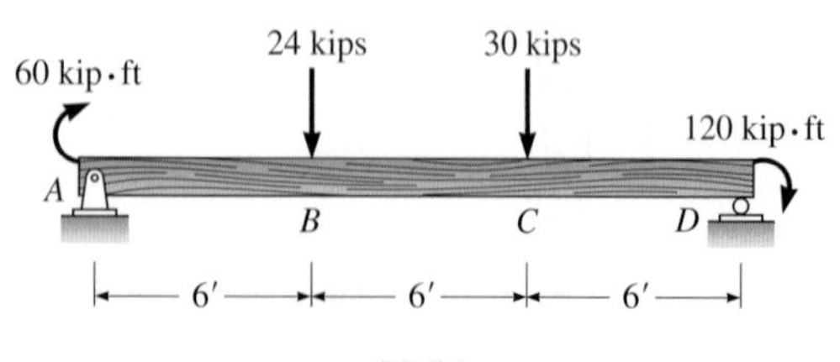

P5.14

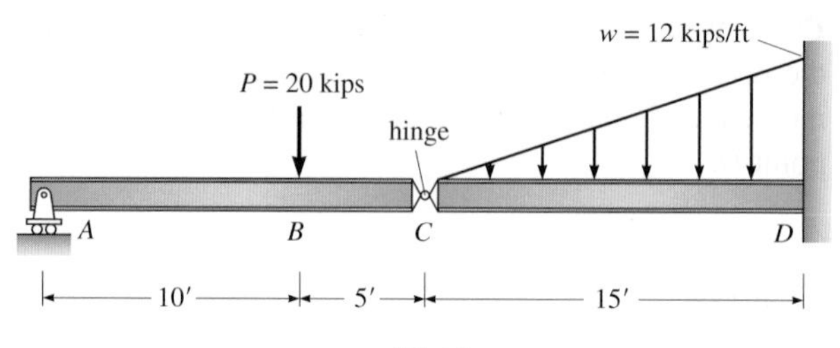

P5.15

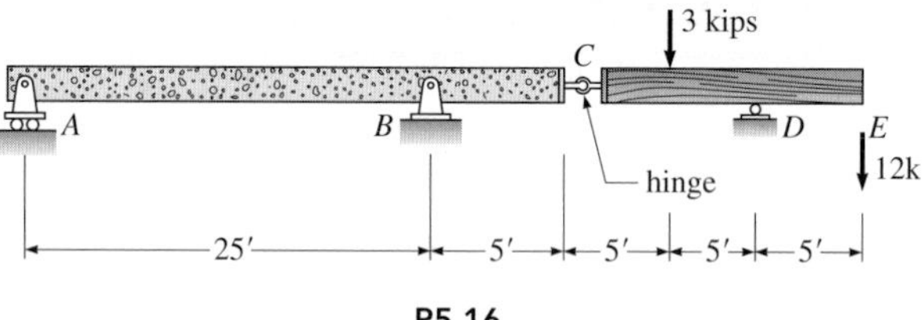

P5.16

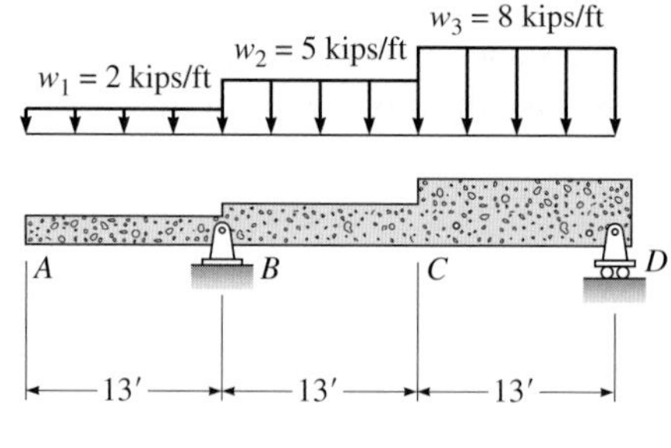

P5.17

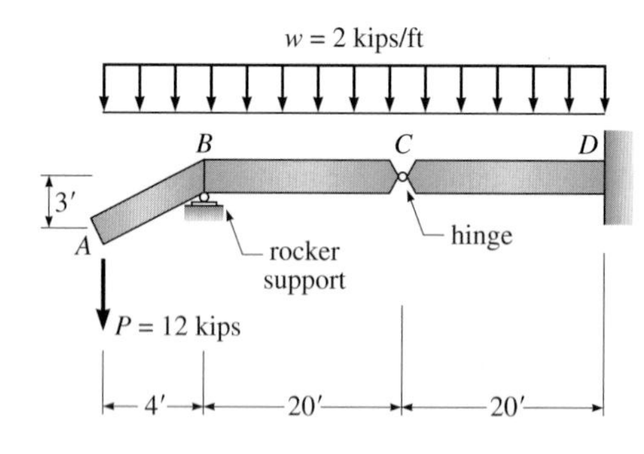

P5.18

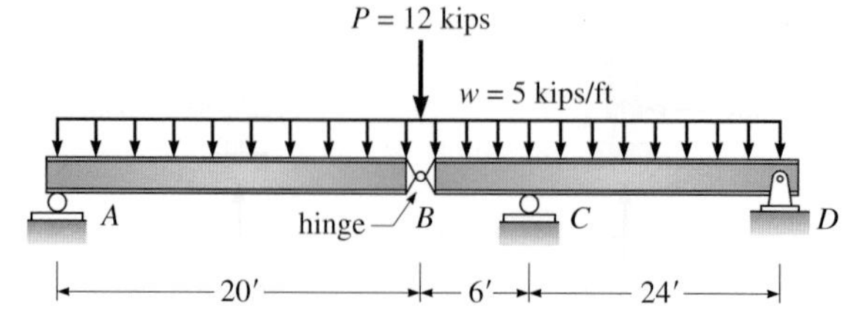

P5.19

P5.20. Draw the shear and moment diagrams for girder $BCDE$ and sketch its deflected shape. The support at E may be treated as a roller, and connections at joints A, C, D, and F may be treated as frictionless pins.

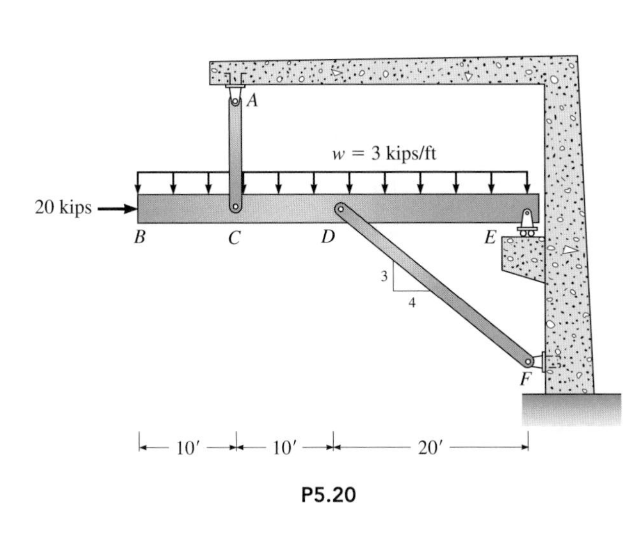

P5.20

P5.21. Draw the shear and moment curves for each member of the frame in Figure P5.21. Sketch the deflected shape.

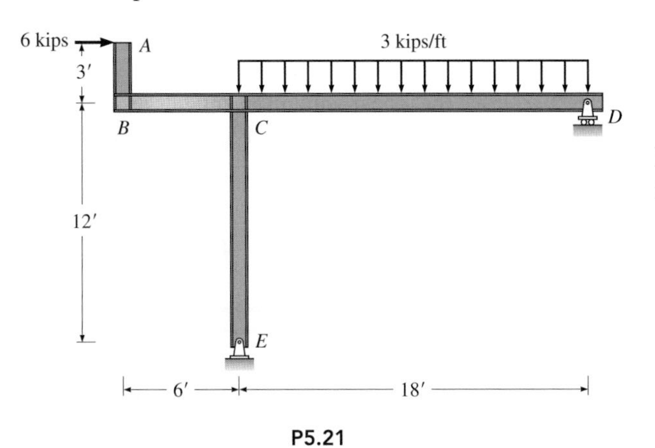

P5.21

P5.22. Draw the shear and moment curves for each member of the frame in Figure P5.22. Sketch the deflected shape.

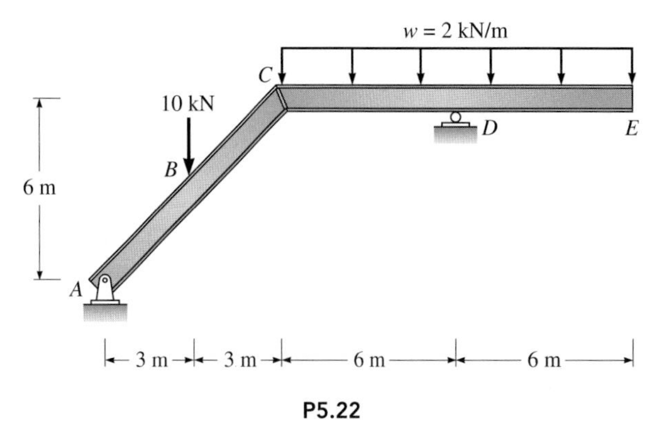

P5.22

P5.23. Draw the shear and moment curves for each member of the frame in Figure P5.23. Sketch the deflected shape hinges at B and C.

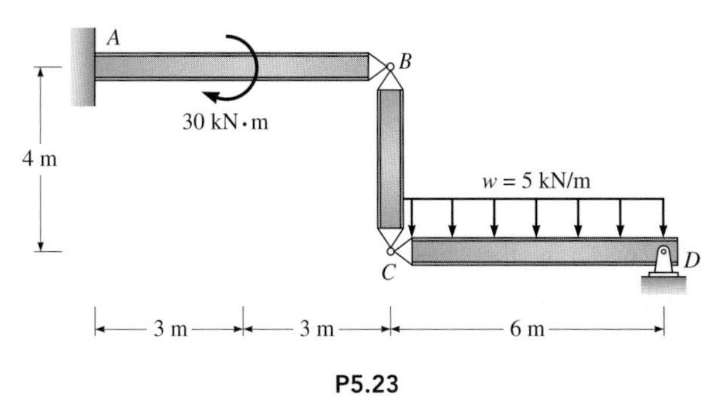

P5.23

P5.24. Draw the shear and moment curves for the beam in Figure P5.24. Sketch the deflected shape.

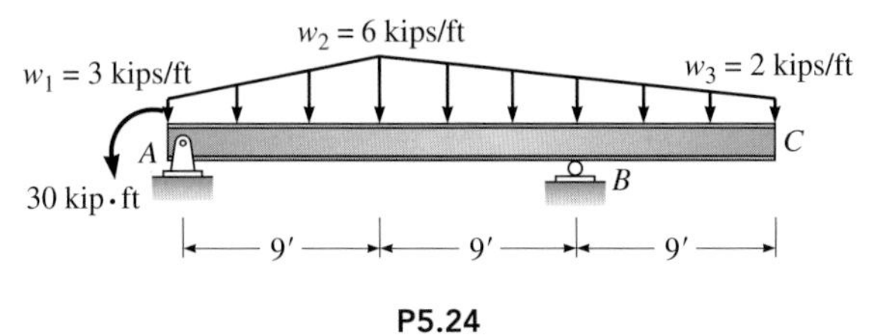

P5.24

P5.25. Draw the shear and moment curves for each member of the frame in Figure P5.25. Sketch the deflected shape.

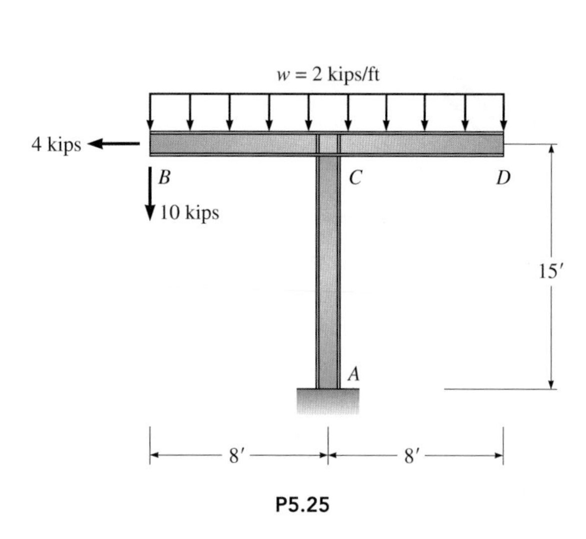

P5.25

P5.26. Draw the shear and moment curves for each member of the frame in Figure P5.26. Sketch the deflected shape.

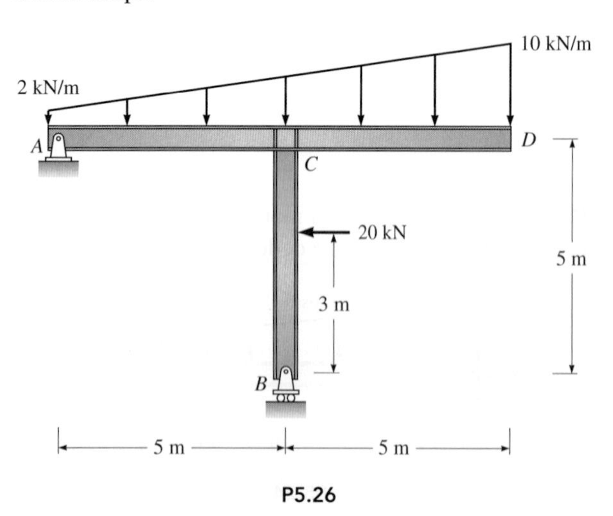

P5.26

P5.27. Draw the shear and moment curves for each member of the frame in Figure P5.27. Sketch the deflected shape.

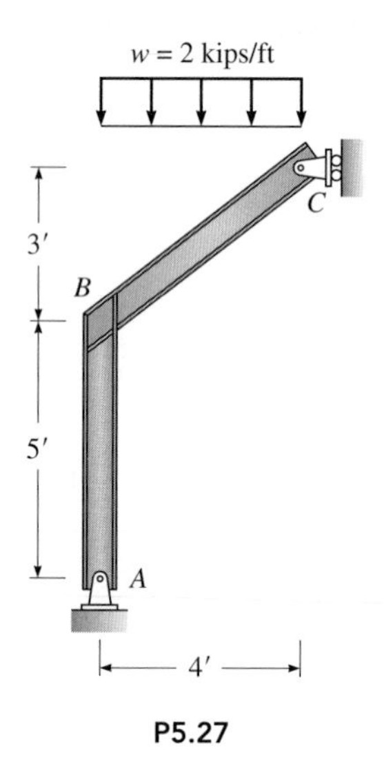

P5.27

P5.28. Draw the shear and moment curves for each member of the beam in Figure P5.28. Sketch the deflected shape. The shear connection at *B* acts as a hinge.

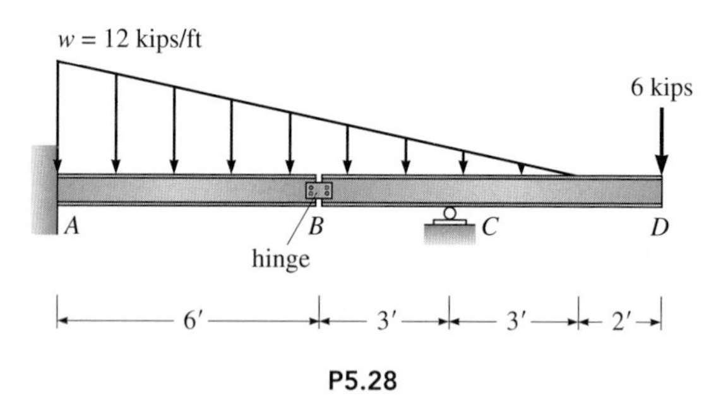

P5.28

P5.29. Draw the shear and moment curves for each member of the beam in Figure P5.29. Sketch the deflected shape.

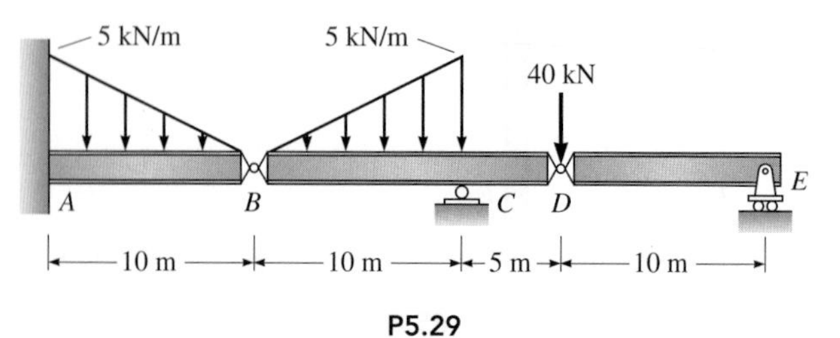

P5.29

P5.30. Draw the shear and moment curves for the beam in Figure P5.30. Sketch the deflected shape.

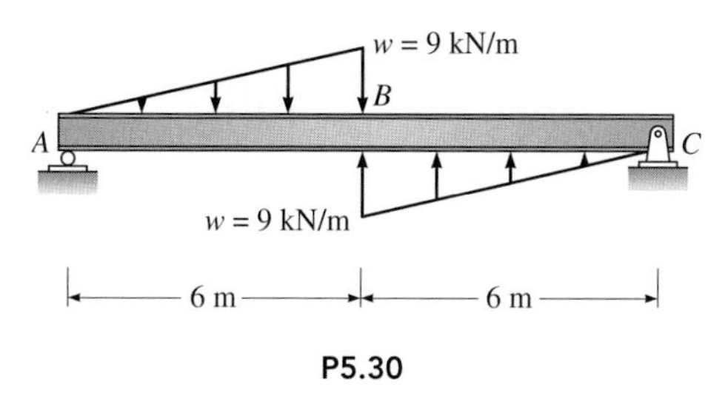

P5.30

P5.31. Draw the shear and moment curves for the indeterminate beam in Figure P5.31. Reactions at support *A* are given. Sketch the deflected shape.

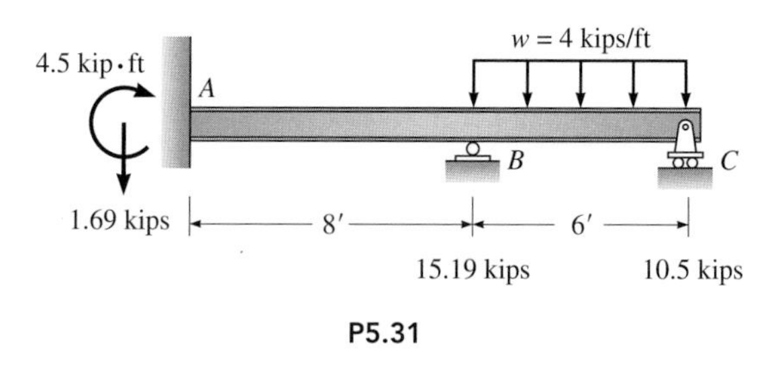

P5.31

P5.32. Draw the shear and moment curves for the beam in Figure P5.32. Sketch the deflected shape.

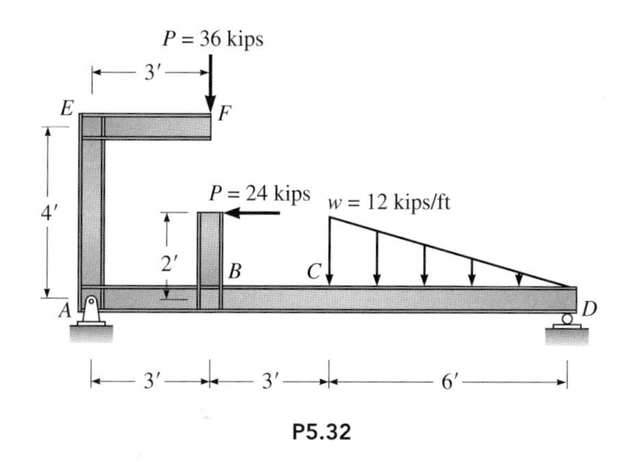

P5.32

P5.33. Draw the shear and moment curves for the beam in Figure P5.33. Reaction at support *B* is given. Locate all points of zero shear and moment. Sketch the deflected shape.

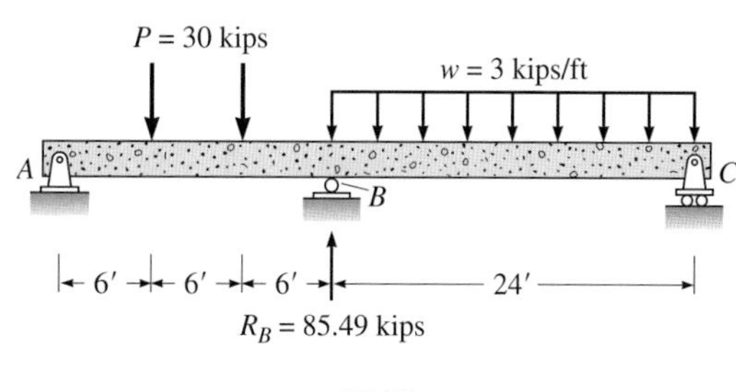

P5.33

P5.34 and **P5.35.** Draw the shear and moment diagrams for each indeterminate beam. Reactions are given. Label maximum values of shear and moment. Locate all inflection points, and sketch the deflected shape.

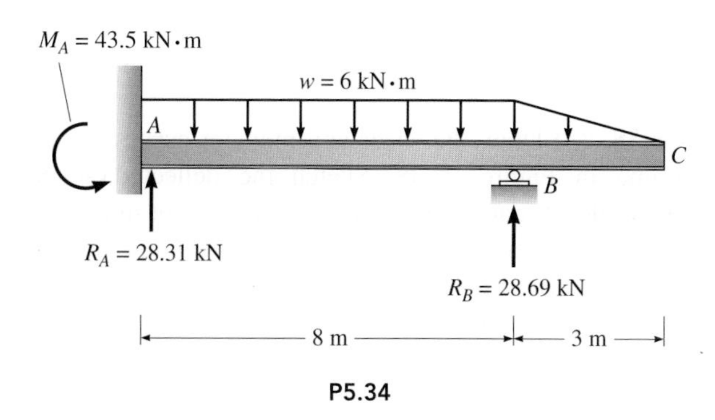

P5.34

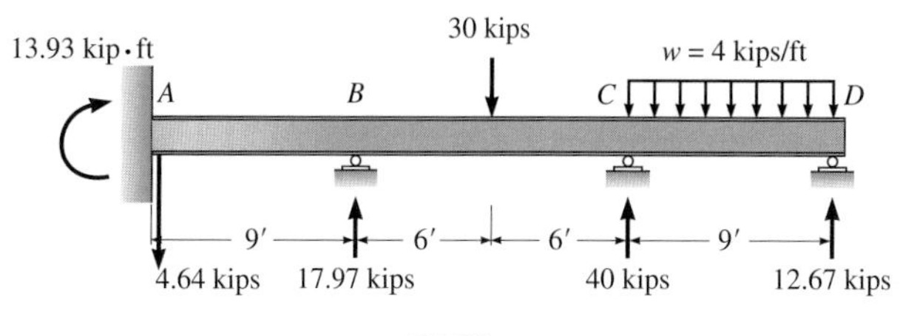

P5.35

P5.36 and **P5.37.** Draw the shear and moment diagrams and sketch the deflected shape.

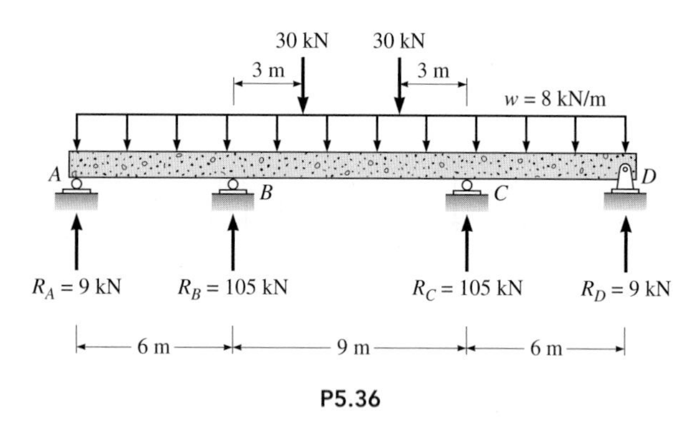

P5.36

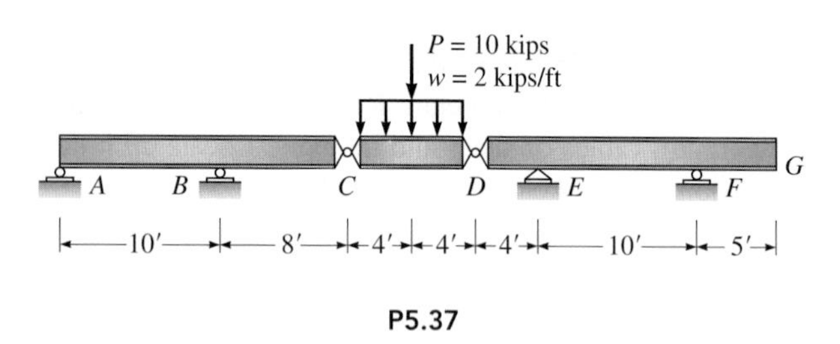

P5.37

P5.38. (*a*) Draw the shear and moment curves for the frame in Figure P5.38. Sketch the deflected shape. (*b*) Write the equations for shear and moment in column *AB*. Take the origin at *A*. (*c*) Write the shear and moment equations for girder *BC*. Take the origin at joint *B*.

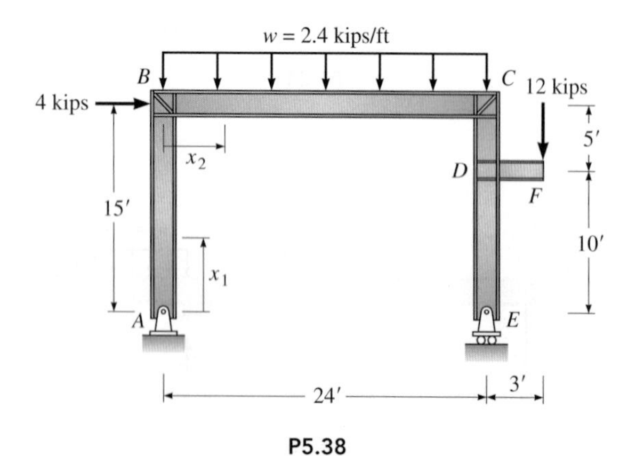

P5.38

P5.39. Draw the shear and moment curves for each member of the rigid frame in Figure P5.39. Sketch the deflected shape.

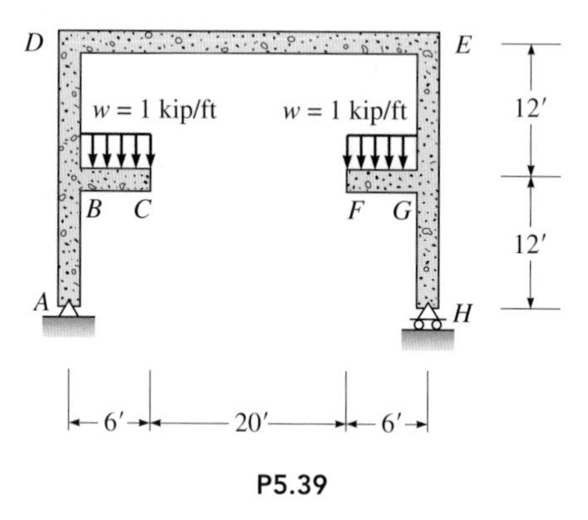

P5.39

P5.40. Draw the shear and moment curves for each member of the frame in Figure P5.40. Sketch the deflected shape. Joints *B* and *D* are rigid.

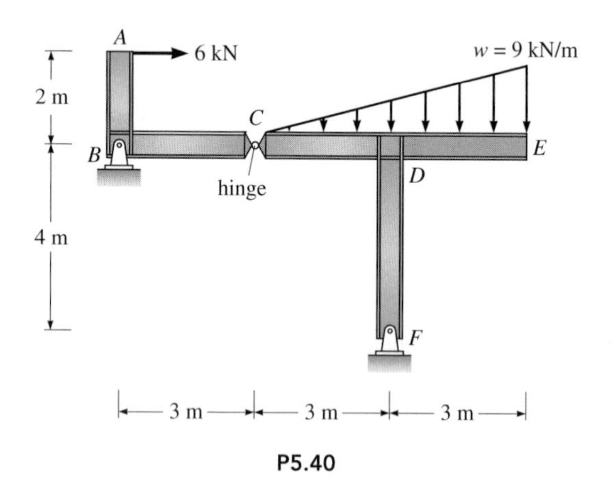

P5.40

P5.41. Draw the moment diagrams for each member of the frame in Figure P5.41. Sketch the deflected shape of the frame. Joints *B* and *C* are rigid.

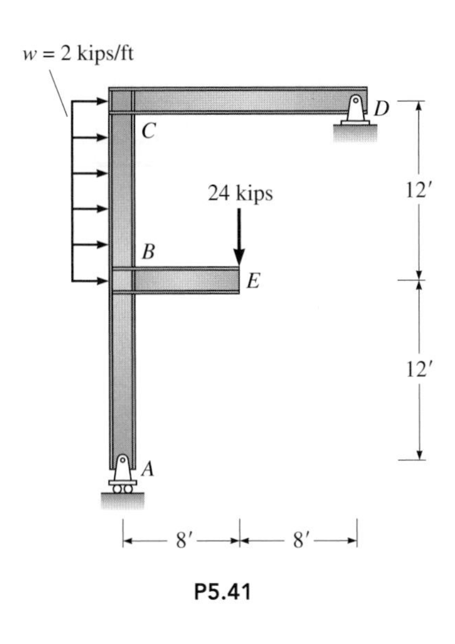

P5.41

P5.42. Draw the shear and moment curves for each member of the frame in Figure P5.42. Sketch the deflected shape. Treat the shear plate connection at *C* as a hinge.

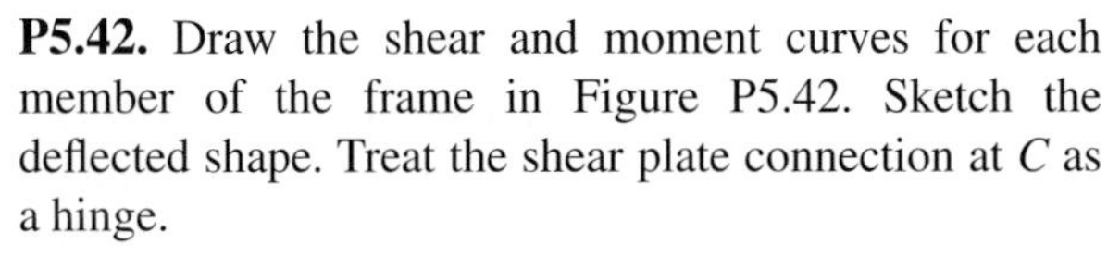

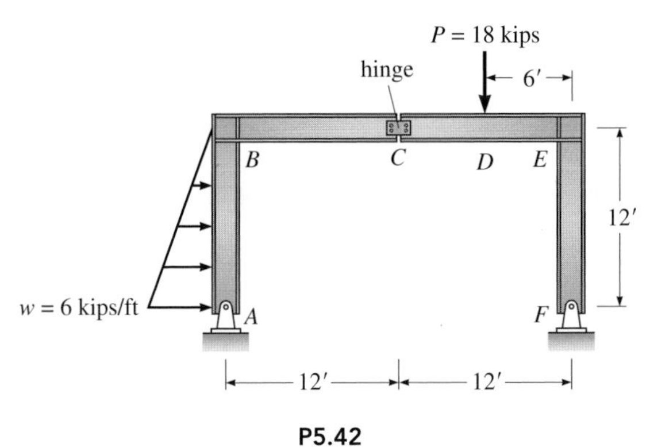

P5.42

P5.43. Draw the shear and moment curves for each member of the frame and draw the deflected shape. Joints *B* and *C* are rigid.

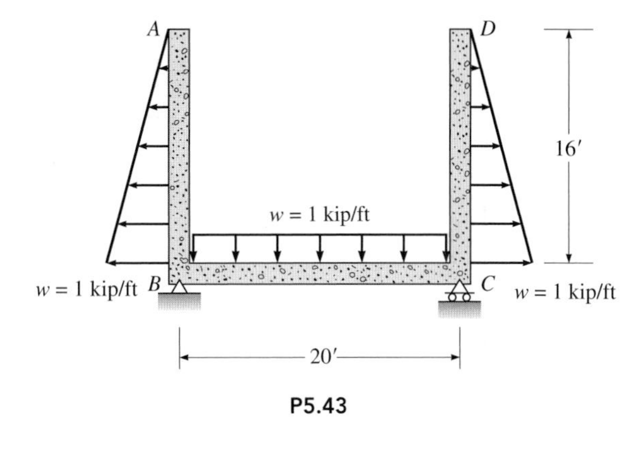

P5.43

P5.44. Draw the shear and moment curves for the column in Figure P5.44. Sketch the deflected shape. The load *P* is equal to 55 kips, and the load is eccentric from the column centerline with an eccentricity of 10 inches.

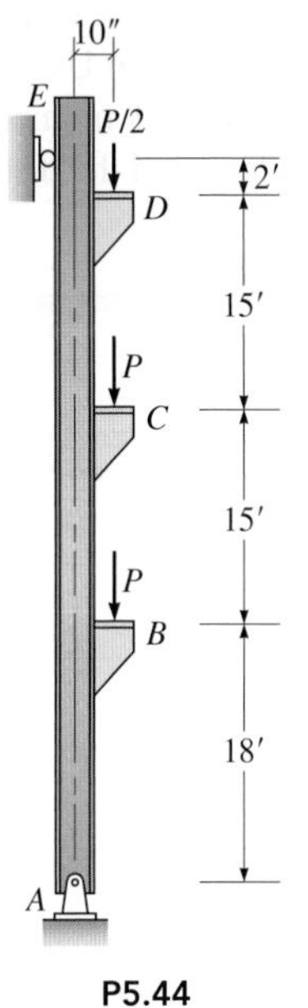

P5.44

P5.45. For the frame in Figure P5.45, draw the shear and moment curves for all members. Next sketch the deflected shape of the frame. Show all forces acting on a free-body diagram of joint C.

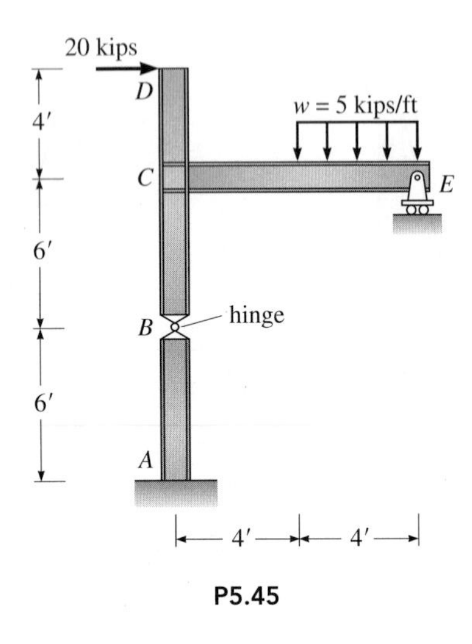

P5.45

P5.46. (a) Sketch the deflected shape of the frame in Figure P5.46. Reactions and moment curves are given. Curvature is also indicated. Joints B and D are rigid. The hinge is located at point C. (b) Using an origin at A, write the equations for shear and moment in member AB in terms of the applied load and the distance x.

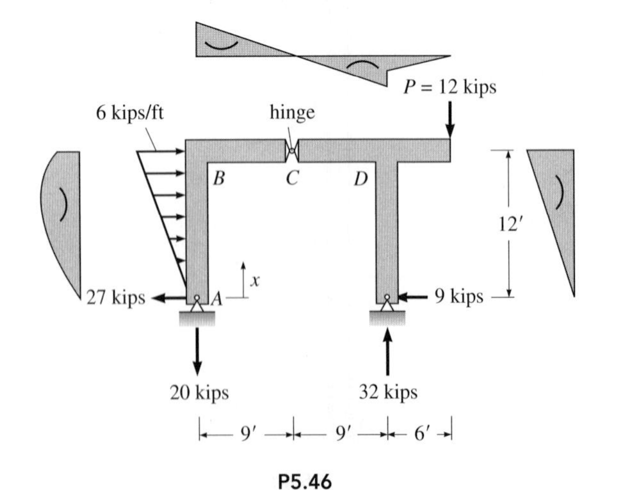

P5.46

P5.47. Draw the shear and moment curves for all members of the frame in Figure P5.47. Sketch the deflected shape.

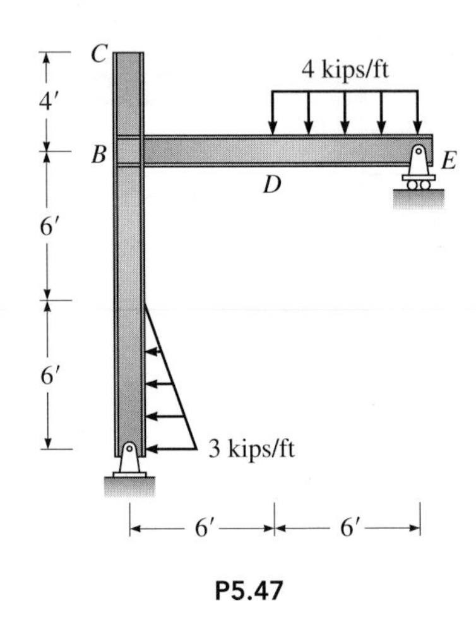

P5.47

P5.48. The hollow structural section beam $ABCD$ in Figure P5.48 is supported by a roller at point D and two links BE and CE. Compute all reaction, draw the shear and moment curves for the beam, and sketch the deflected shape of the structure.

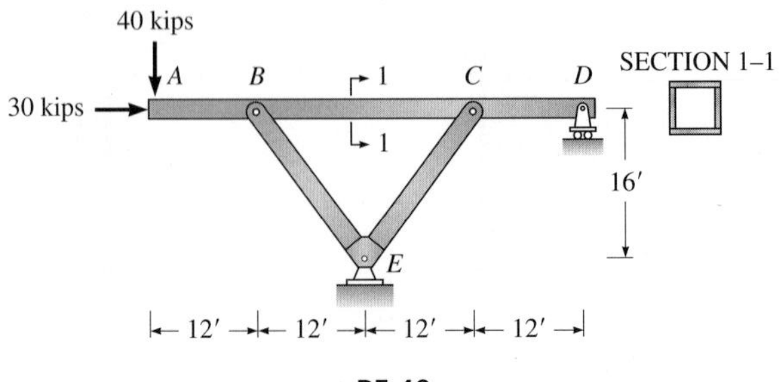

P5.48

Practical Application

P5.49. The combined footing shown in Figure P5.49 is designed as a narrow reinforced concrete beam. The footing has been proportioned so that the resultant of the column loads passes through the centroid of the footing, producing a uniformly distributed soil pressure on the base of the footing. Draw the shear and moment diagrams for the footing in the longitudinal direction. The width of the footing is controlled by the allowable\soil pressure and does not affect the analysis.

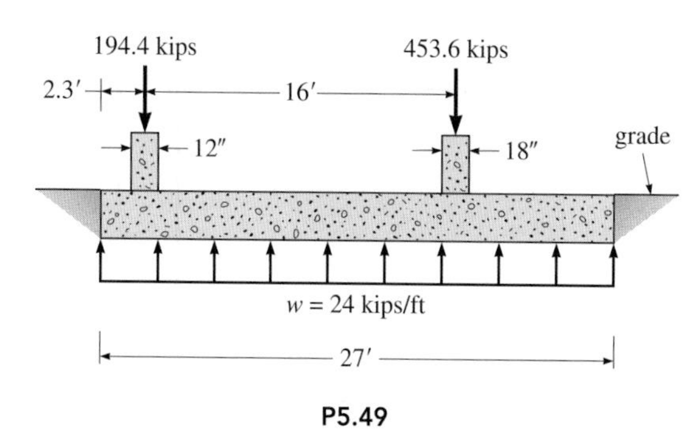

P5.49

P5.50. The two concentrated loads, supported on the combined footing in Figure P5.50, produce a trapezoidal distribution of soil pressure. Construct the shear and moment diagrams. Label all ordinates of the diagrams. Sketch the deflected shape.

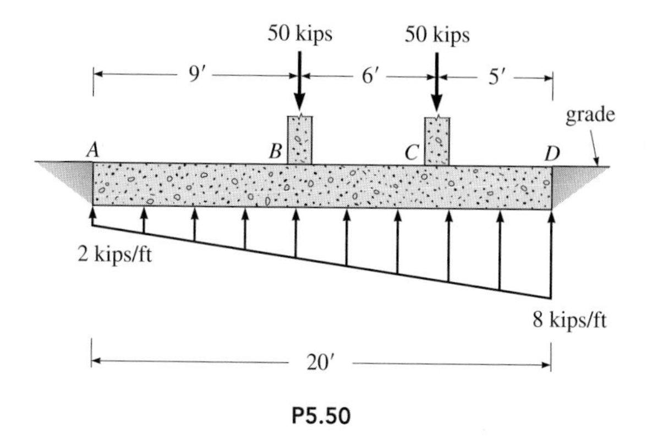

P5.50

P5.51 and **P5.52.** Classify the structures in Figures P5.51 and P5.52. Indicate whether stable or unstable. If stable, indicate whether determinate or indeterminate. If indeterminate, give the degree.

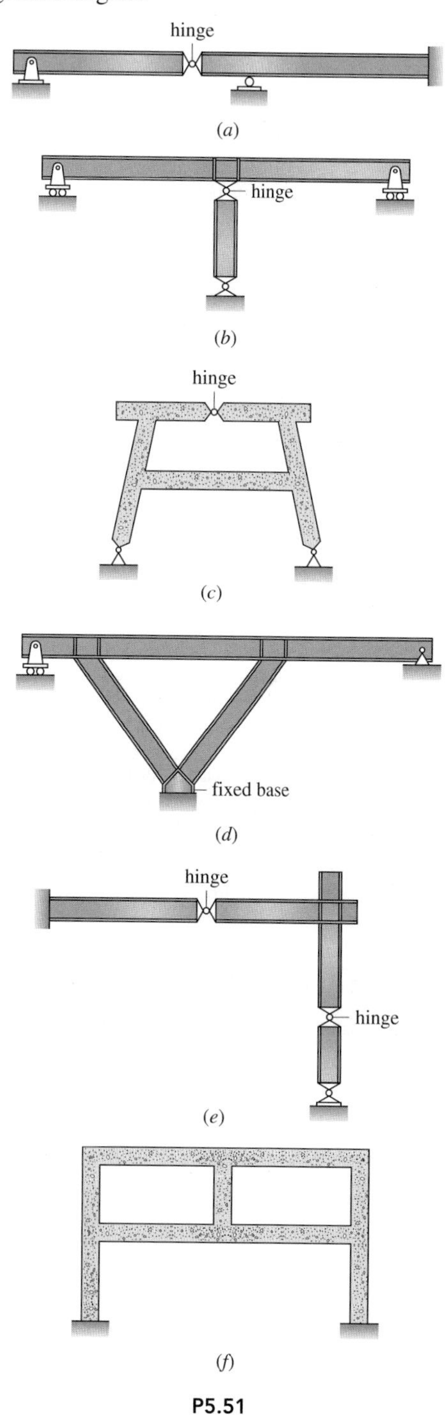

P5.51

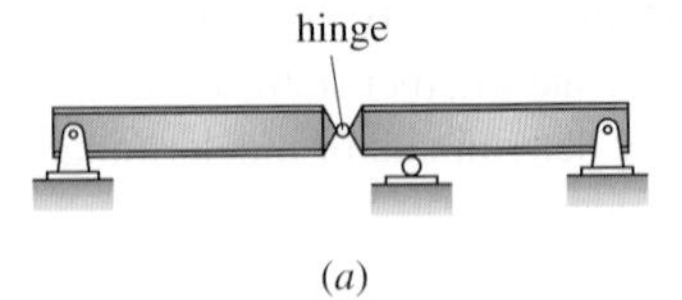

(a)

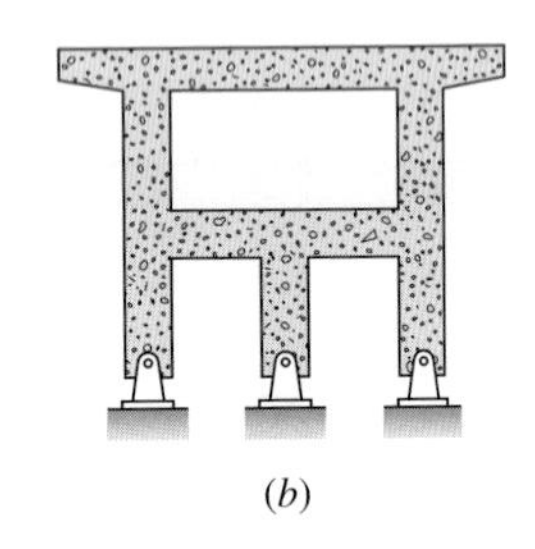

(b)

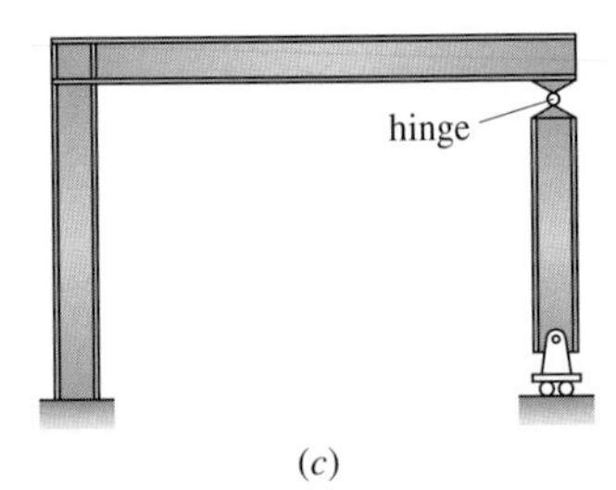

(c)

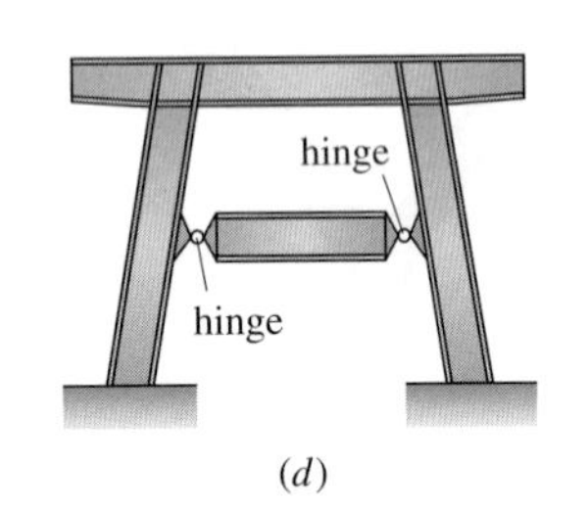

(d)

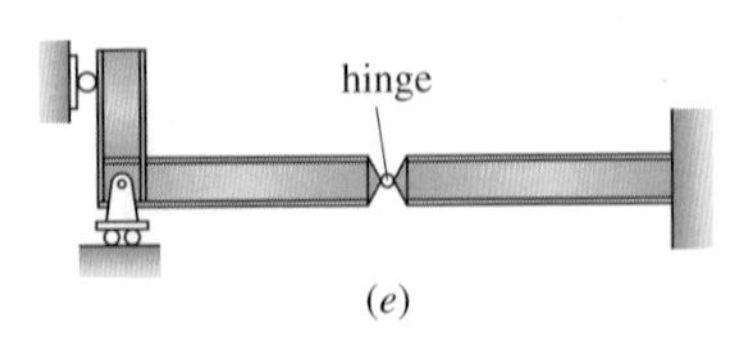

(e)

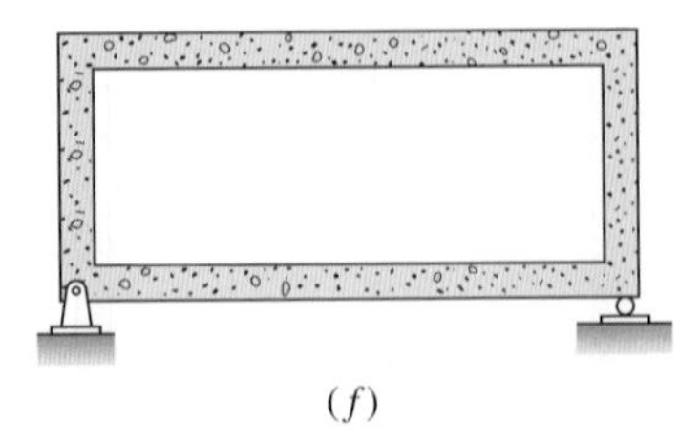

(f)

P5.52

Practical Application

P5.53. The corner panel of a typical floor of a warehouse is shown in Figure P5.53. It consists of a 10-in-thick reinforced concrete slab supported on steel beams. The slab weighs 125 lb/ft^2. The weight of light fixtures and utilities suspended from the bottom of the slab is estimated to be 5 lb/ft^2. The exterior beams B_1 and B_2 support a 14-ft-high masonry wall constructed of lightweight, hollow concrete block that weighs 38 lb/ft^2. We assume that the tributary area for each beam is shown by the dashed lines in Figure P5.46, and the weight of the beams and their fireproofing is estimated to be 80 lb/ft. Draw the shear and moment diagrams produced by the total dead load for beams B_1 and B_2.

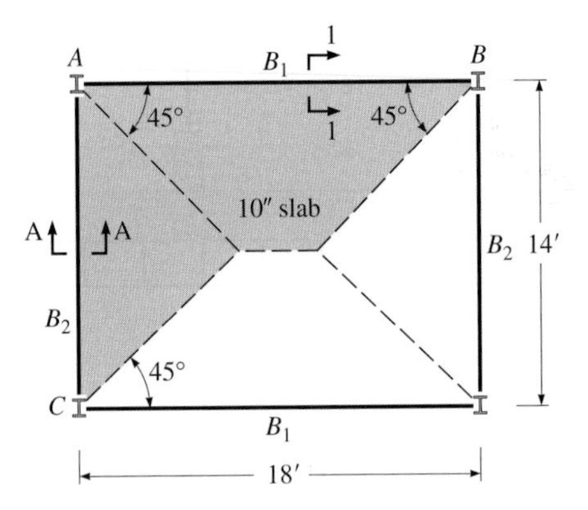

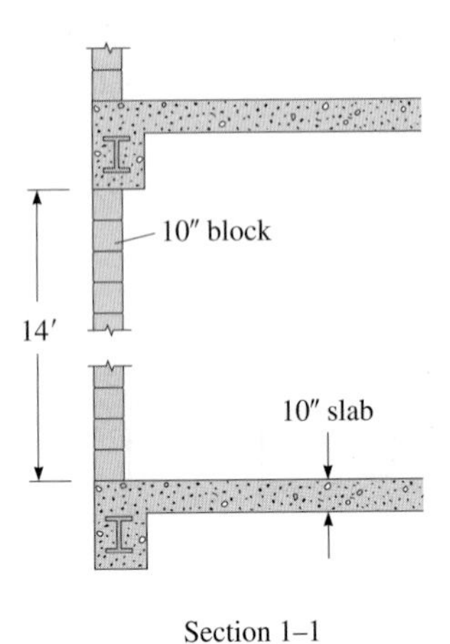

Section 1–1

P5.53

P5.54. *Computer analysis of a continuous beam.* The continuous beam in Figure P5.54 is constructed from a W18×106 wide flange steel section with $A = 31.1 \text{ in}^2$ and $I = 1910 \text{ in}^4$. Determine the reactions, plot the shear and moment diagrams and the deflected shape. Evaluate the deflections. Neglect weight of beam. $E = 29{,}000$ ksi.

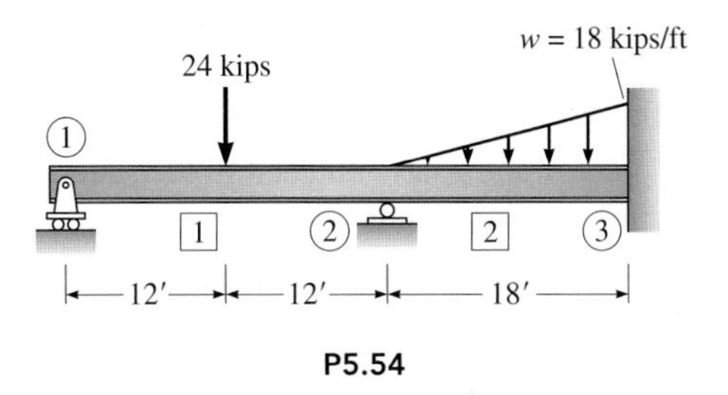

P5.54

P5.55. *Computer analysis.* The columns and girder of the indeterminate rigid frame in Figure P5.55*a* are fabricated from a W18×130 wide flange steel section: $A = 38.2 \text{ in}^2$ and $I = 2460 \text{ in}^4$. The frame is to be designed for a uniform load of 4 kips/ft and a lateral wind load of 6 kips; use $E = 29{,}000 \text{ kips/in}^2$. The weight of the girder is included in the 4 kips/ft uniform load.

(*a*) Compute the reactions, plot the deflected shape, and draw the shear and moment curves for the columns and girder, using the computer program.

(*b*) To avoid ponding* of rainwater on the roof, the girder is to be fabricated with a camber equal to the deflection at midspan of the roof girder produced by the uniform load. Determine the camber (see Figure P5.55*b*).

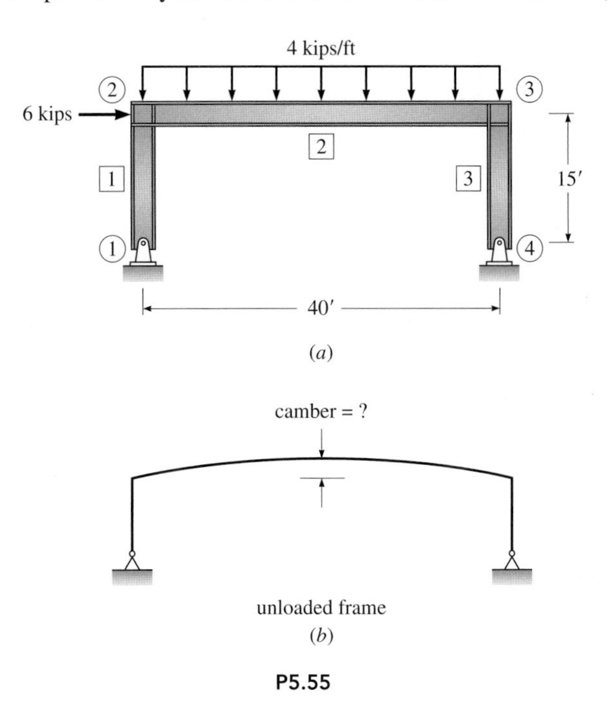

P5.55

*Ponding refers to the pool of water that can collect on a roof when the roof drains are not adequate to carry away rain water or become clogged. This condition has resulted in the collapse of flat roofs. To avoid ponding, beams may be cambered upward so rain water cannot accumulate at the center regions of the roof. See Figure P5.55*b*.

P5.56. Computer investigation of wind load on a building frame.

Case 1: The columns and girders of the rigid building frame in Figure P5.56 have been designed initially for vertical load as specified by the building code. Floor beams are connected to columns by rigid joints. As part of the design, the building frame must be checked for lateral deflection under the 0.8 kip/ft wind load to ensure that lateral displacement will not damage the exterior walls attached to the structural frame. If the code requires that the maximum lateral deflection at the top of the roof not exceed 0.48 in. to prevent damage to the exterior walls, is the building frame sufficiently stiff to satisfy this requirement?

Case 2: If the bases of the columns at point A and F are attached to the foundations by fixed supports instead of pin supports, how much is the lateral deflection at joint D reduced?

Case 3: If a pin-connected diagonal bar bracing with a 2 in. × 2 in. square cross section running from support A to joint E is added, determine the lateral deflection at joint D. Assume pin supports at joints A and F.

For the columns, $I = 640$ in^4 and $A = 17.9$ in^2; for the girders, $I = 800$ in^4 and $A = 11.8$ in^2; for the diagonal brace, $A = 4$ in^2. Use $E = 29{,}000$ ksi.

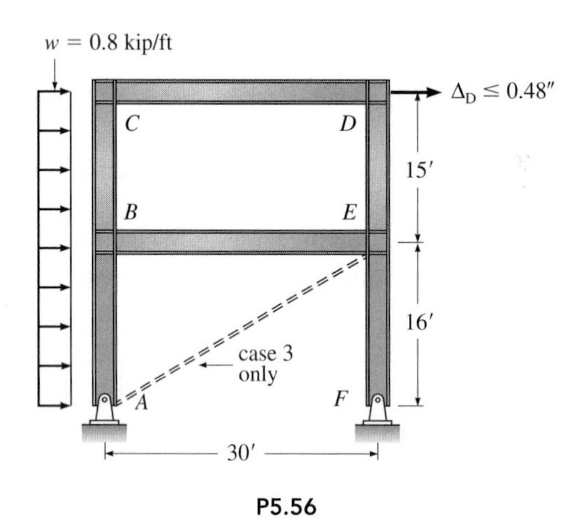

P5.56

Cooper River Bridge in Charleston, South Carolina

The new Cooper River Bridge is a cable-stayed bridge with a main span of 1,546 ft. Completed in 2006, the bridge provides a 1,000-ft navigational channel with a minimum vertical clearance of 186 ft above the river. Each 574-ft tall, diamond-shaped concrete tower is protected from ship collision by a rock island surrounding the base of the tower.

C H A P T E R 6

Cables

Chapter Objectives

- Study the characteristics and behavior of cable structures. Since cables are flexible, they carry tension forces only and compression and bending stresses equal zero along the cable.
- Analyze determinate cable structures and calculate support reactions by two methods, namely by equations of static equilibrium and by the general cable theorem, as well as determine the cable forces at specific points along its length.
- Generate an efficient funicular shape of an arch in direct compressive stress by utilizing an inverted cable under its self-weight.

6.1 Introduction

As we discussed in Section 1.5, cables constructed of high-strength steel wires are completely flexible and have a tensile strength four or five times greater than that of structural steel. Because of their great strength-to-weight ratio, designers use cables to construct long-span structures, including suspension bridges and roofs over large arenas and convention halls. To use cable construction effectively, the designer must deal with two problems:

1. Preventing large displacements and oscillations from developing in cables that carry live loads whose magnitude or direction changes with time.
2. Providing an efficient means of anchoring the large tensile force carried by cables.

To take advantage of the cable's high strength while minimizing its negative features, designers must use greater inventiveness and imagination than are required in conventional beam and column structures. For example, Figure 6.1 shows a schematic drawing of a roof composed of cables connected to a center tension ring and an outer compression ring. The small center ring, loaded symmetrically by the cable reactions, is stressed primarily in direct tension while the outer ring carries mostly axial compression. By creating a self-balancing system composed of members in direct stress, the

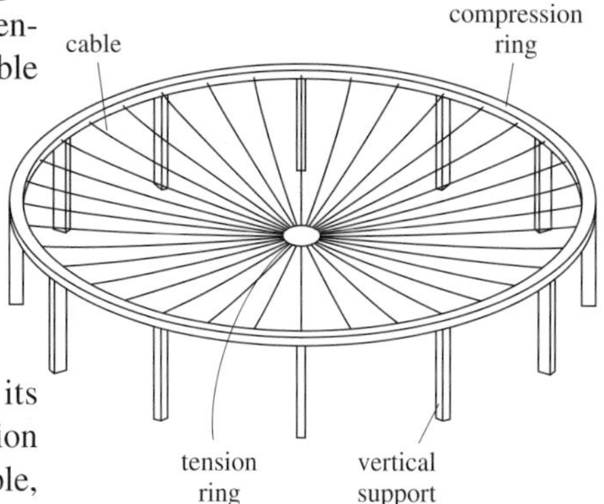

Figure 6.1: Cable-supported roof composed of three elements: cables, a center tension ring, and an outer compression ring.

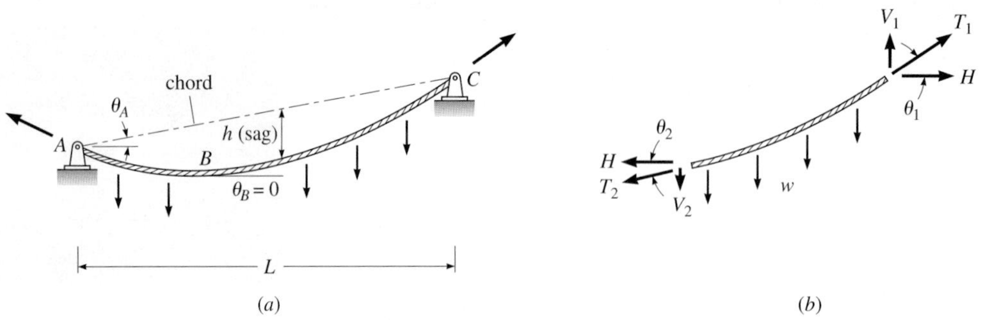

Figure 6.2: Vertically loaded cables: (*a*) cable with an inclined chord—the vertical distance between the chord and the cable, h, is called the sag; (*b*) free body of a cable segment carrying vertical loads; although the resultant cable force T varies with the slope of the cable, $\Sigma F_x = 0$ requires that H, the horizontal component of T, is constant from section to section.

Photo 6.1: Terminal building at Dulles airport. Roof supported on a net of steel cables spanning between massive, sloping, reinforced concrete pylons.

designer creates an efficient structural form for gravity loads that requires only vertical supports around its perimeter. A number of sports arenas, including Madison Square Garden in New York City, are roofed with a cable system of this type.

In a typical cable analysis the designer establishes the position of the end supports, the magnitude of the applied loads, and the elevation of one other point on the cable axis (often the sag at midspan; see Figure 6.2*a*). Based on these parameters, the designer applies cable theory to compute the end reactions, the force in the cable at all other points, and the position of other points along the cable axis.

6.2 Characteristics of Cables

Cables, which are made of a group of high-strength wires twisted together to form a strand, have an ultimate tensile strength of approximately 270 kips/in^2 (1862 MPa). The twisting operation imparts a spiral pattern to the individual wires.

Photo 6.2: Cable-stayed bridge over Tampa bay.

While the drawing of wires through dies during the manufacturing process raises the yield point of the steel, it also reduces its ductility. Wires can undergo an ultimate elongation of 7 or 8 percent compared to 30 to 40 percent for structural steel with a moderate yield point, say, 36 kips/in^2 (248 MPa). Steel cables have a modulus of elasticity of approximately 26,000 kips/in^2 (179 GPa) compared to a modulus of 29,000 kips/in^2 (200 GPa) for structural steel bars. The lower modulus of the cable is due to the uncoiling of the wire's spiral structure under load.

Since a cable carries only direct stress, the resultant axial force T on all sections must act tangentially to the longitudinal axis of the cable (see Figure 6.2*b*). Because a cable lacks flexural rigidity, designers must use great care when designing cable structures to ensure that live loads do not induce either large deflections or vibrations. In early prototypes, many cable-supported bridges and roofs developed large wind-induced displacements (flutter) that resulted in failure of the structure. The complete destruction of the Tacoma Narrows Bridge on November 7, 1940 by wind-induced oscillations is one of the most spectacular examples of a structural failure of a large cable-supported structure. The bridge, which spanned 5939 ft (1810 m) over Puget Sound near the City of Tacoma, Washington, developed vibrations that reached a maximum amplitude in the vertical direction of 28 ft (8.53 m) before the floor system broke up and dropped into the water below (see Photo 2.1).

6.3 Variation of Cable Force

If a cable supports vertical load only, the horizontal component H of the cable tension T is constant at all sections along the axis of the cable. This conclusion can be demonstrated by applying the equilibrium equation $\Sigma F_x = 0$ to a segment of cable (see Figure 6.2*b*). If the cable tension is expressed in terms of the horizontal component H and the cable slope θ,

$$T = \frac{H}{\cos\theta} \tag{6.1}$$

At a point where the cable is horizontal (e.g., see point B in Figure 6.2*a*), θ equals zero. Since $\cos 0 = 1$, Equation 6.1 shows that $T = H$. The maximum value of T typically occurs at the support where the cable slope is largest.

6.4 Analysis of a Cable Supporting Gravity (Vertical) Loads

When a set of concentrated loads is applied to a cable of negligible weight, the cable deflects into a series of linear segments (Figure 6.3*a*). The resulting shape is called the *funicular polygon*. Figure 6.3*b* shows the forces acting at point B on a cable segment of infinitesimal length. Since the segment is in equilibrium, the vector diagram consisting of the cable forces and the applied load forms a closed force polygon (see Figure 6.3*c*).

A cable supporting vertical load (e.g., see Figure 6.3*a*) is a *determinate* member. Four equilibrium equations are available to compute the four reaction components supplied by the supports. These equations include the three equations of static equilibrium applied to the free body of the cable and a condition equation, $\Sigma M_z = 0$. Since the moment at all sections of the cable is zero, the condition equation can be written at any section as long as the cable sag (the vertical distance between the cable chord and the cable) is known. Typically, the designer sets the maximum sag to ensure both a required clearance and an economical design.

To illustrate the computations of the support reactions and the forces at various points along the cable axis, we will analyze the cable in Figure 6.4*a*. The cable sag at the location of the 12-kip load is set at 6 ft. In this analysis we will assume that the weight of the cable is trivial (compared to the load) and neglect it.

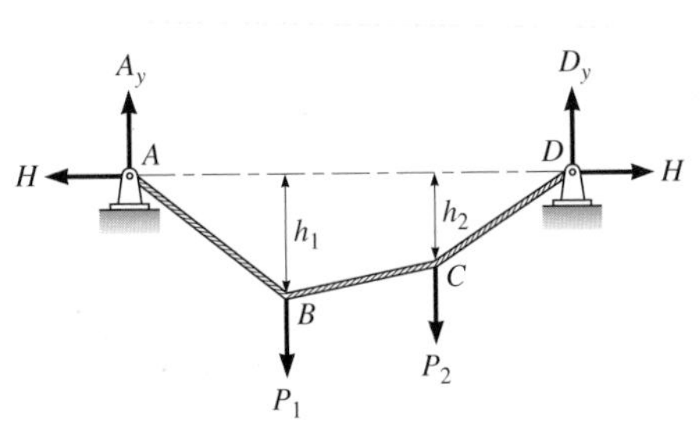

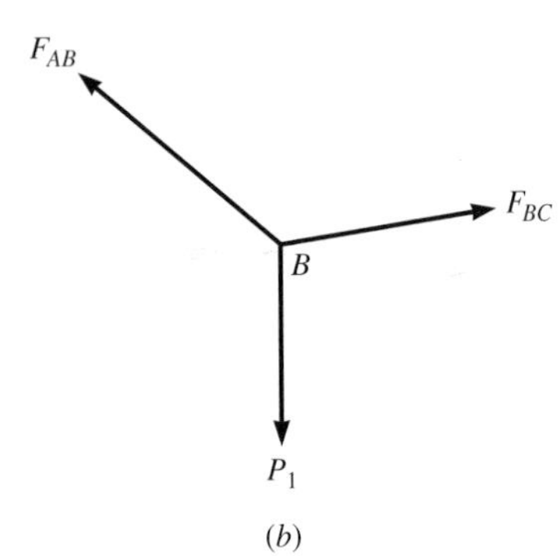

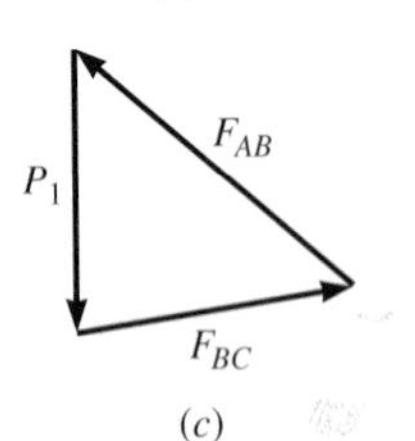

Figure 6.3: Vector diagrams: (*a*) cable with two vertical loads; (*b*) forces acting on an infinitesimal segment of cable at B; (*c*) force polygon for vectors in (*b*).

STEP 1 Compute D_y by summing moments about support A.

$$\circlearrowright^{+} \quad \Sigma M_A = 0$$

$$(12 \text{ kips})(30) + (6 \text{ kips})(70) - D_y(100) = 0$$

$$D_y = 7.8 \text{ kips} \tag{6.2}$$

STEP 2 Compute A_y.

$$\overset{+}{\uparrow} \quad \Sigma F_y = 0$$

$$0 = A_y - 12 - 6 + 7.8$$

$$A_y = 10.2 \text{ kips} \tag{6.3}$$

STEP 3 Compute H; sum moments about B (Figure 6.4b).

$$\circlearrowleft^{+} \quad \Sigma M_B = 0$$

$$0 = A_y(30) - Hh_B$$

$$h_B H = (10.2)(30) \tag{6.4}$$

Setting $h_B = 6$ ft yields

$$H = 51 \text{ kips}$$

After H is computed, we can establish the cable sag at C by considering a free body of the cable just to the right of C (Figure 6.4c).

STEP 4

$$\circlearrowleft^{+} \quad \Sigma M_C = 0$$

$$-D_y(30) + Hh_c = 0$$

$$h_c = \frac{30D_y}{H} = \frac{30(7.8)}{51} = 4.6 \text{ ft} \tag{6.5}$$

To compute the force in the three cable segments, we establish θ_A, θ_B, and θ_C and then use Equation 6.1.

Compute T_{AB}.

$$\tan\theta_A = \frac{6}{30} \quad \text{and} \quad \theta_A = 11.31°$$

$$T_{AB} = \frac{H}{\cos\theta_A} = \frac{51}{0.981} = 51.98 \text{ kips}$$

Compute T_{BC}.

$$\tan\theta_B = \frac{6 - 4.6}{40} = 0.035 \quad \text{and} \quad \theta_B = 2°$$

$$T_{BC} = \frac{H}{\cos\theta_B} = \frac{51}{0.999} = 51.03 \text{ kips}$$

Compute T_{CD}.

$$\tan\theta_C = \frac{4.6}{30} = 0.153 \quad \text{and} \quad \theta_C = 8.7°$$

$$T_{CD} = \frac{H}{\cos\theta_C} = \frac{51}{0.988} = 51.62 \text{ kips}$$

Since the slopes of all cable segments in Figure 6.4a are relatively small, the computations above show that the difference in magnitude between the horizontal component of cable tension H and the total cable force T is small.

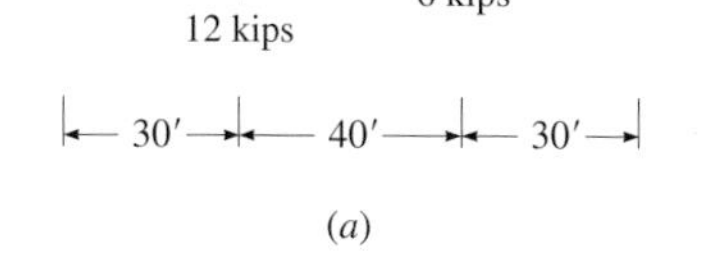

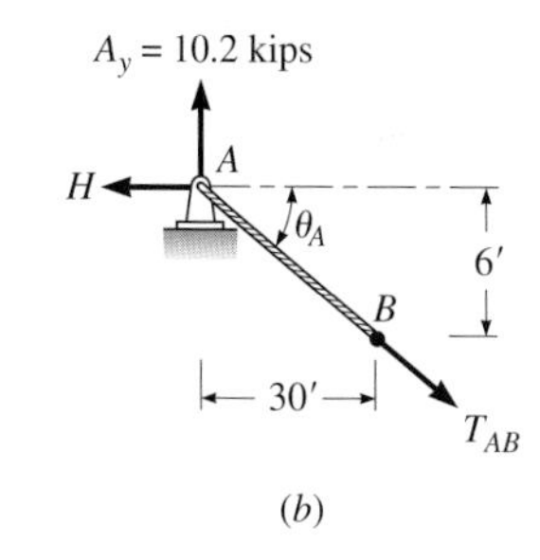

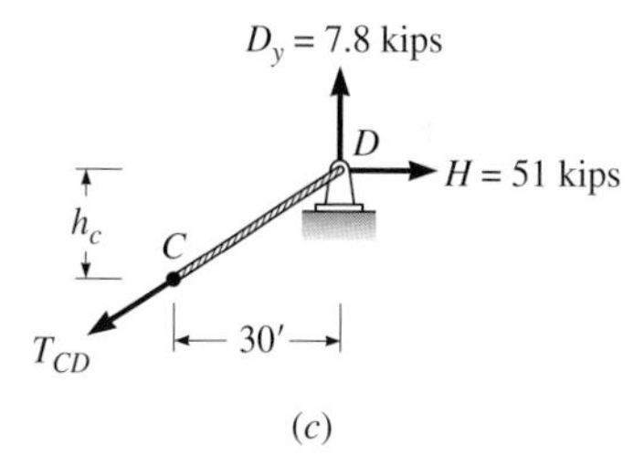

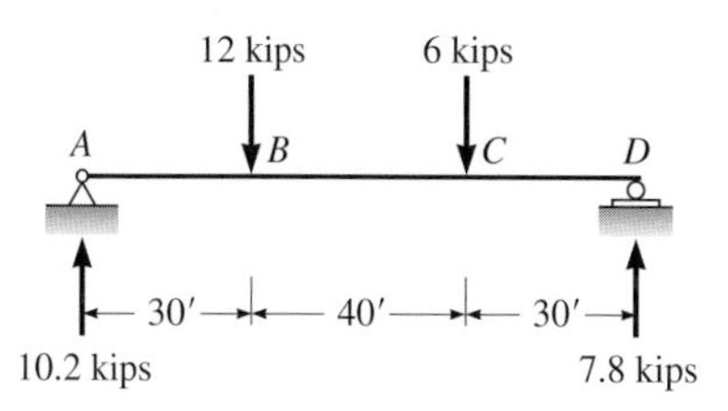

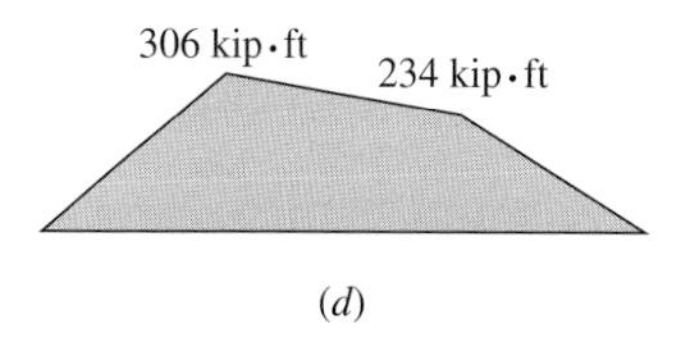

Figure 6.4: (a) Cable loaded with vertical forces, cable sag at B set at 6 ft; (b) free body of cable to left of B; (c) free body of cable to right of C; (d) a simply supported beam with same loads and span as cable (moment diagram below).

6.5 General Cable Theorem

As we carried out the computations for the analysis of the cable in Figure 6.4*a*, you may have observed that certain of the computations are similar to those you would make in analyzing a simply supported beam with a span equal to that of the cable and carrying the same loads applied to the cable. For example, in Figure 6.4*c* we apply the cable loads to a beam whose span equals that of the cable. If we sum moments about support A to compute the vertical reaction D_y at the right support, the moment equation is identical to Equation 6.2 previously written to compute the vertical reaction at the right support of the cable. In addition, you will notice that the shape of the cable and the moment curve for the beam in Figure 6.4 are identical. A comparison of the computations between those for a cable and those for a simply supported beam that supports the cable loads leads to the following statement of the *general cable theorem*:

> **At any point on a cable supporting vertical loads, the product of the cable sag h and the horizontal component H of the cable tension equals the bending moment at the same point in a simply supported beam that carries the same loads in the same position as those on the cable. The span of the beam is equal to that of the cable.**

The relationship above can be stated by the following equation:

$$Hh_z = M_z \tag{6.6}$$

where H = horizontal component of cable tension
h_z = cable sag at point z where M_z is evaluated
M_z = moment at point z in a simply supported beam carrying the loads applied to the cable

Since H is constant at all sections, Equation 6.6 shows that the cable sag h is proportional to the ordinates of the moment curve.

To verify the general cable theorem given by Equation 6.6, we will show that at an arbitrary point z on the cable axis *the product of the horizontal component H of cable thrust and the cable sag h_z equals the moment at the same point in a simply supported beam carrying the cable loads* (see Figure 6.5). We will also assume that the end supports of the cable are located at different elevations. The vertical distance between the two supports can be expressed in terms of α, the slope of the cable chord, and the cable span L as

$$y = L \tan \alpha \tag{6.7}$$

Directly below the cable we show a simply supported beam to which we apply the cable loads. The distance between loads is the same in both members. In both the cable and the beam, the arbitrary section at which we will evaluate the terms in Equation 6.6 is located a distance x to the right of the left support.

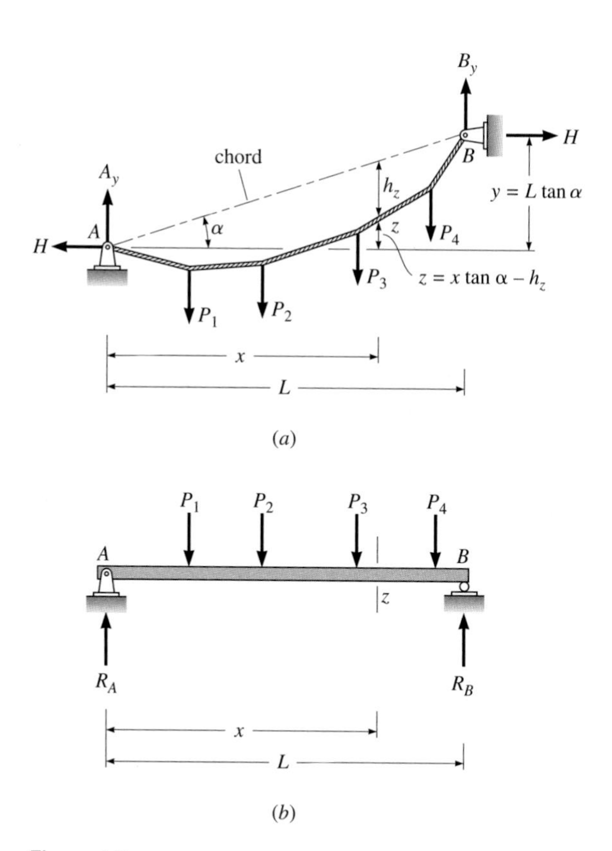

Figure 6.5

We begin by expressing the vertical reaction of the cable at support A in terms of the vertical loads and H (Figure 6.5*a*).

$$\circlearrowright^{+} \quad \Sigma M_B = 0$$

$$0 = A_y L - \Sigma m_B + H(L \tan \alpha) \tag{6.8}$$

where Σm_B represents the moment about support B of the vertical loads (P_1 through P_4) applied to the cable.

In Equation 6.8 the forces A_y and H are the unknowns. Considering a free body to the left of point z, we sum moments about point z to produce a second equation in terms of the unknown reactions A_y and H.

$$\circlearrowright^{+} \quad \Sigma M_z = 0$$

$$0 = A_y x + H(x \tan \alpha - h_z) - \Sigma m_z \tag{6.9}$$

where Σm_z represents the moment about z of the loads on a free body of the cable to the left of point z. Solving Equation 6.8 for A_y gives

$$A_y = \frac{\Sigma m_B - H(L \tan \alpha)}{L} \tag{6.10}$$

Substituting A_y from Equation 6.10 into Equation 6.9 and simplifying, we find

$$Hh_z = \frac{x}{L}\Sigma m_B - \Sigma m_z \tag{6.11}$$

We next evaluate M_z, the bending moment in the beam at point z (see Figure 6.5*b*):

$$M_z = R_A x - \Sigma m_z \tag{6.12}$$

To evaluate R_A in Equation 6.12, we sum moments of the forces about the roller at B. Since the loads on the beam and the cable are identical, as are the spans of the two structures, the moment of the applied loads (P_1 through P_4) about B also equals Σm_B.

$$\circlearrowright^{+} \quad \Sigma M_B = 0$$

$$0 = R_A L - \Sigma m_B$$

$$R_A = \frac{\Sigma m_B}{L} \tag{6.13}$$

Substituting R_A from Equation 6.13 into Equation 6.12 gives

$$M_z = x\frac{\Sigma m_B}{L} - \Sigma m_z \tag{6.14}$$

Since the right sides of Equations 6.11 and 6.14 are identical, we can equate the left sides, giving, $Hh_z = M_z$, and Equation 6.6 is verified.

6.6 Establishing the Funicular Shape of an Arch

The material required to construct an arch is minimized when all sections along the axis of the arch are in direct stress. For a particular set of loads the arch profile in direct stress is called the *funicular* arch. By imagining that the loads carried by the arch are applied to a cable, the designer can automatically generate a funicular shape for the loads. If the cable shape is turned upside down, the designer produces a funicular arch. Since dead loads are usually much greater than the live loads, a designer might use them to establish the funicular shape (see Figure 6.6).

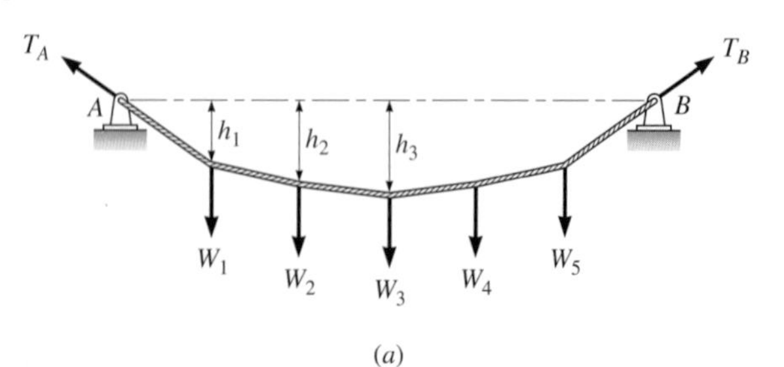

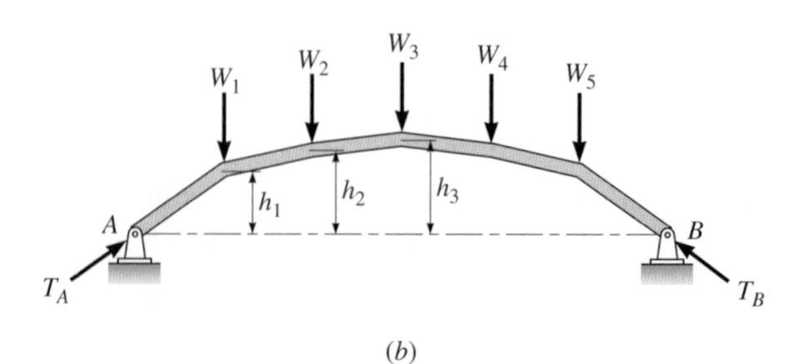

Figure 6.6: Establishing the shape of the funicular arch: (*a*) loads supported by arch applied to a cable whose sag h_3 at midspan equals the midspan height of the arch; (*b*) arch (produced by inverting the cable profile) in direct stress.

EXAMPLE 6.1

Determine the reactions at the supports produced by the 120-kip load at midspan (Figure 6.7) (*a*) using the equations of static equilibrium and (*b*) using the general cable theorem. Neglect the weight of the cable.

Solution

(*a*) Since supports are not on the same level, we must write two equilibrium equations to solve for the unknown reactions at support *C*. First consider Figure 6.7*a*.

$$\circlearrowright^{+} \quad \Sigma M_A = 0$$

$$0 = 120(50) + 5H - 100C_y \qquad (1)$$

Next consider Figure 6.7*b*.

$$\circlearrowright^{+} \quad \Sigma M_B = 0$$

$$0 = 10.5H - 50C_y$$

$$H = \frac{50}{10.5} C_y \qquad (2)$$

Substitute *H* from Equation 2 into Equation 1.

$$0 = 6000 + 5\left(\frac{50}{10.5} C_y\right) - 100C_y$$

$$C_y = 78.757 \text{ kips} \qquad \textbf{Ans.}$$

Substituting C_y into Equation 2 yields

$$H = \frac{50}{10.5}(78.757) = 375 \text{ kips} \qquad \textbf{Ans.}$$

(*b*) Using the general cable theorem, apply Equation 6.6 at midspan where the cable sag $h_z = 8$ ft and $M_z = 3000$ kip · ft (see Figure 6.7*c*).

$$Hh_z = M_z$$

$$H(8) = 3000$$

$$H = 375 \text{ kips} \qquad \textbf{Ans.}$$

After *H* is evaluated, sum moments about *A* in Figure 6.7*a* to compute $C_y = 78.757$ kips.

NOTE. Although the vertical reactions at the supports for the cable in Figure 6.7*a* and the beam in Figure 6.7*c* are not the same, the final results are identical.

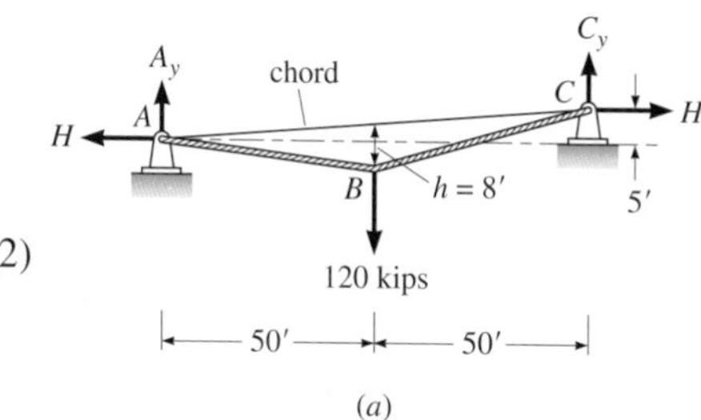

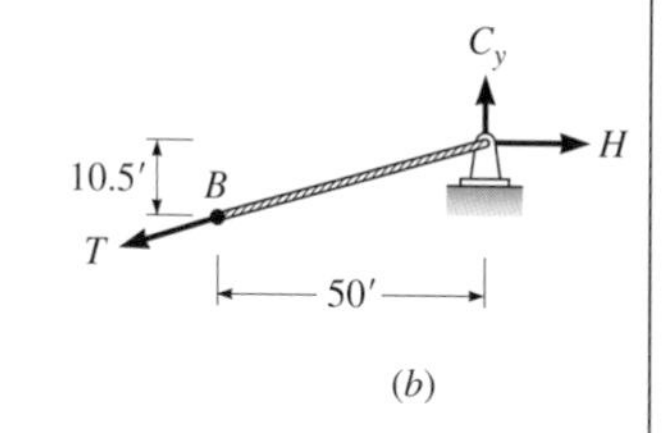

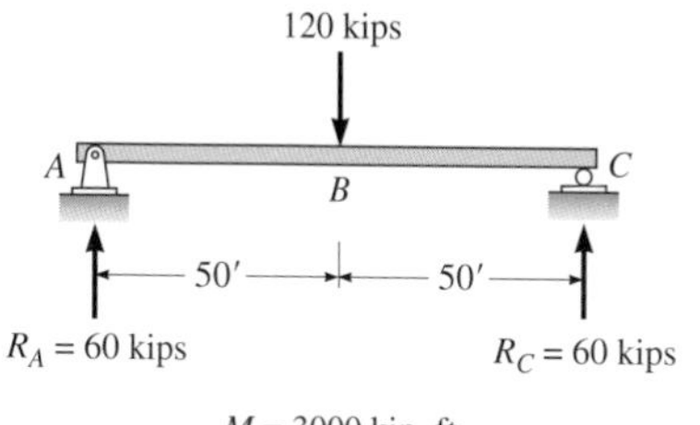

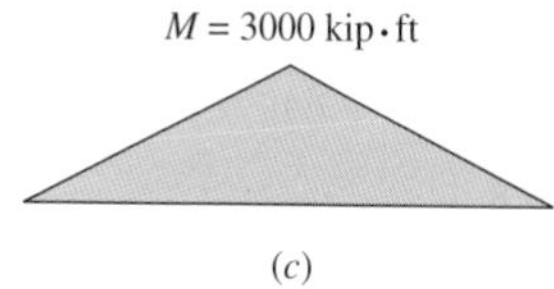

Figure 6.7: (*a*) Cable with a vertical load at midspan; (*b*) free body to the right of *B*; (*c*) simply supported beam with same length as cable. Beam supports cable load.

EXAMPLE 6.2

A cable-supported roof carries a uniform load $w = 0.6$ kip/ft (see Figure 6.8*a*). If the cable sag at midspan is set at 10 ft, what is the maximum tension in the cable (*a*) between points *B* and *D* and (*b*) between points *A* and *B*?

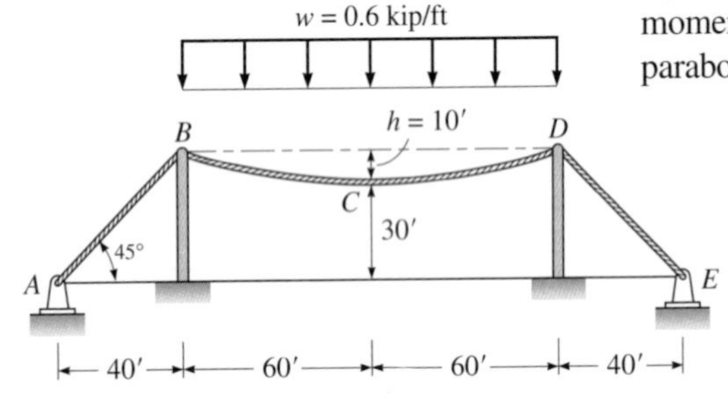

(*a*)

$y = \frac{4hx^2}{L^2}$

(*b*)

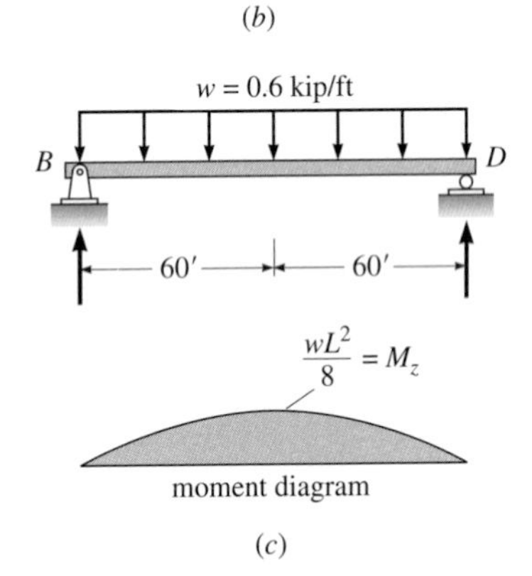

(*c*)

Figure 6.8

Solution

(*a*) Apply Equation 6.6 at midspan to analyze the cable between points *B* and *D*. Apply the uniform load to a simply supported beam and compute the moment M_z at midspan (see Figure 6.8*c*). Since the moment curve is a parabola, the cable is also a parabola between points *B* and *D*.

$$Hh = M_z = \frac{wL^2}{8}$$

$$H(10) = \frac{0.6(120)^2}{8}$$

$$H = 108 \text{ kips}$$

The maximum cable tension in span *BD* occurs at the supports where the slope is maximum. To establish the slope at the supports, we differentiate the equation of the cable $y = 4hx^2/L^2$ (see Figure 6.8*b*).

$$\tan\theta = \frac{dy}{dx} = \frac{8hx}{L^2}$$

at $x = 60$ ft, $\tan\theta = 8(10)(60)/(120)^2 = \frac{1}{3}$, and $\theta = 18.43°$:

$$\cos\theta = 0.949$$

Substituting into

$$T = \frac{H}{\cos\theta} \tag{6.1}$$

$$T = \frac{108}{0.949} = 113.8 \text{ kips}$$ **Ans.**

(*b*) If we neglect the weight of the cable between points *A* and *B*, the cable can be treated as a straight member. Since the cable slope θ is 45°, the cable tension equals

$$T = \frac{H}{\cos\theta} = \frac{108}{0.707} = 152.76 \text{ kips}$$ **Ans.**

Summary

- Cables, composed of multiple strands of cold-drawn, high-strength steel wires twisted together, have tensile strengths varying from 250 to 270 ksi. Cables are used to construct long-span structures such as suspension and cable-stayed bridges, as well as roofs over large arenas (sports stadiums and exhibition halls) that require column-free space.
- Since cables are flexible, they can undergo large changes in geometry under moving loads; therefore, designers must provide stabilizing elements to prevent excessive deformations. Also the supports at the ends of cables must be capable of anchoring large forces. If bedrock is not present for anchoring the ends of suspension bridge cables, massive abutments of reinforced concrete may be required.
- Because cables (due to their flexibility) have no bending stiffness, the moment is zero at all sections along the cable.
- The general cable theorem establishes a simple equation to relate the horizontal thrust H and the cable sag h to the moment that develops in a fictitious, simply supported beam with the same span as the cable

$$Hh_z = M_z$$

where H = horizontal component of cable tension
h_z = sag at point z where M_z is evaluated. The sag is the vertical distance from the cable chord to the cable.
M_z = moment at point z in a simply supported beam with the same span as the cable and carrying the same loads as the cable

- When cables are used in suspension bridges, floor systems must be very stiff to distribute the concentrated wheel loads of trucks to multiple suspenders, thereby minimizing deflections of the roadway.
- Since a cable is in direct stress under a given loading (usually the dead load), the cable shape can be used to generate the funicular shape of an arch by turning it upside down.

PROBLEMS

P6.1. Determine the reactions at the supports, the magnitude of the cable sag at joints B and E, the magnitude of the tension force in each segment of the cable, and the total length of the cable in Figure P6.1.

P6.2. The cable in Figure P6.2 supports four simply supported girders uniformly loaded with 4 kips/ft. (*a*) Determine the minimum required area of the main cable $ABCDE$ if the allowable stress is 60 kips/in^2. (*b*) Determine the cable sag at point B.

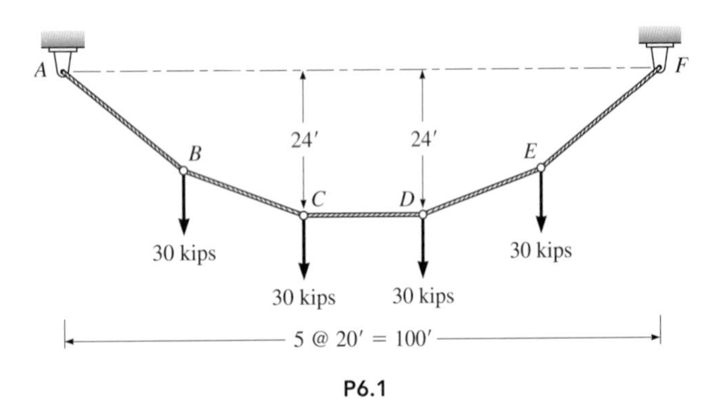

P6.1

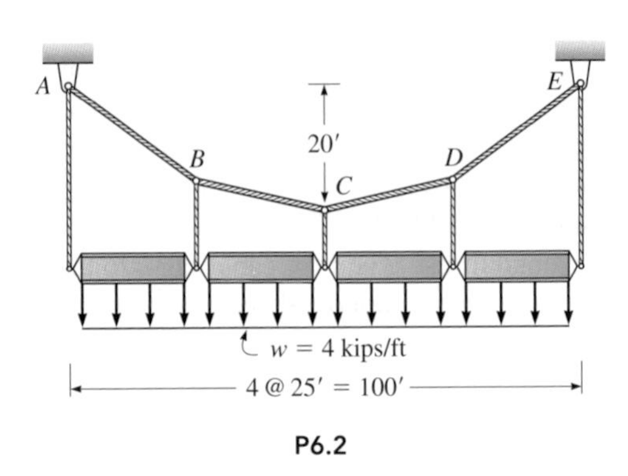

P6.2

P6.3. The cable in Figure P6.3 supports girder DE uniformly loaded with 4 kips/ft. The supporting hangers are closely spaced, generating a smooth curved cable. Determine the support reactions at A and C. If the maximum tensile force in the cable cannot exceed 600 kips, determine the sag h_B at midspan.

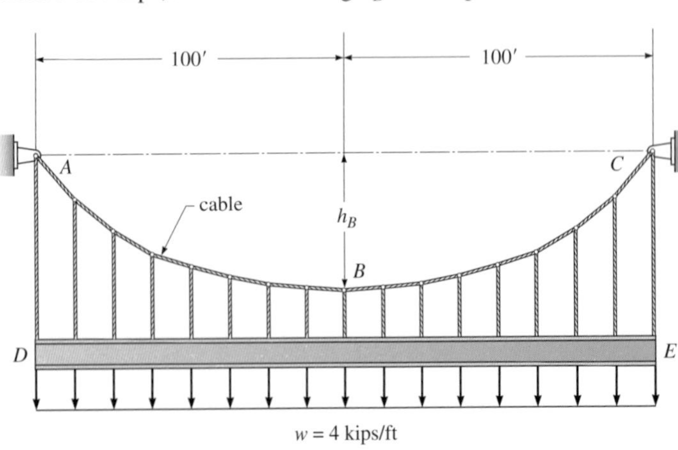

P6.3

P6.4. (*a*) Determine the reactions at supports A and E and the maximum tension in the cable in Figure P6.4. (*b*) Establish the cable sag at points C and D.

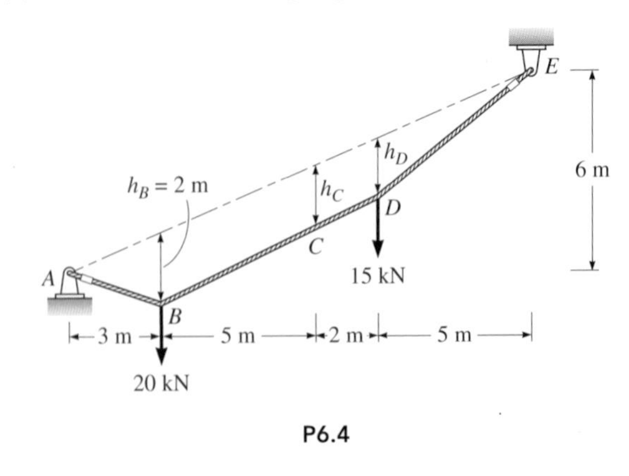

P6.4

P6.5. Compute the support reactions and the maximum tension in the main cable in Figure P6.5. The hangers can be assumed to provide a simple support for the suspended beams.

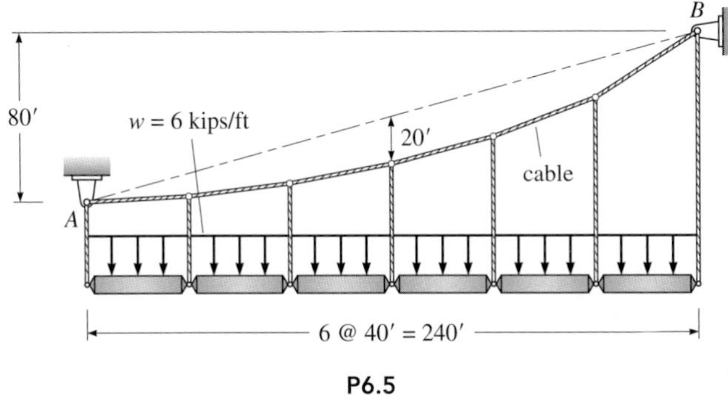

P6.5

P6.6. What value of θ is associated with the minimum volume of cable material required to support the 100-kip load in Figure P6.6? The allowable stress in the cable is 150 kips/in^2.

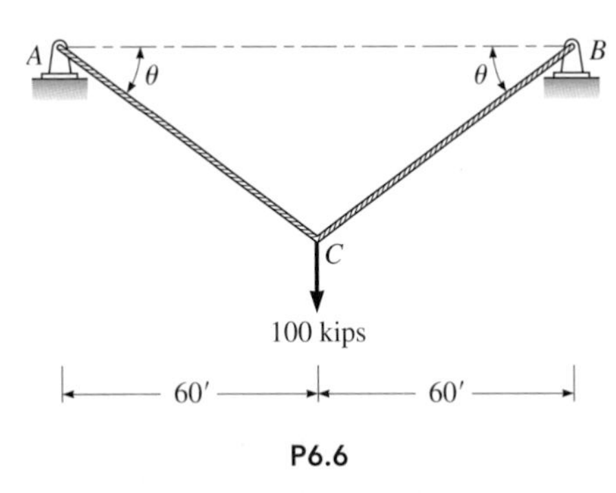

P6.6

P6.7. The cables in Figure P6.7 have been dimensioned so that a 3-kip tension force develops in each vertical strand when the main cables are tensioned. What value of jacking force T must be applied at supports B and C to tension the system?

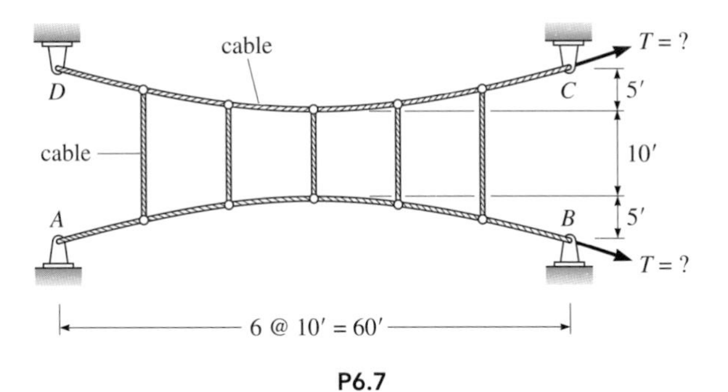

P6.7

P6.8. Compute the support reactions and the maximum tension in the cable in Figure P6.8.

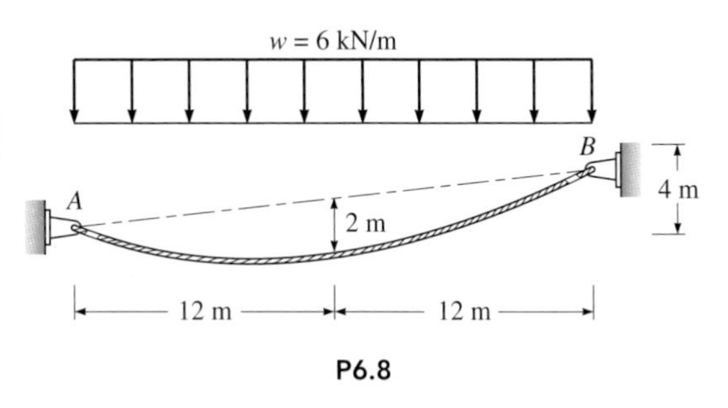

P6.8

P6.9. A uniformly distributed load on beam ABC in Figure P6.9 causes it to sag. To counteract this sag, a cable and post are added beneath the beam. The cable is tensioned until the force in the post causes a moment equal in magnitude, but opposite in direction, to the moment in the beam. Determine the forces in the cable and the post, and determine the reactions at A and C.

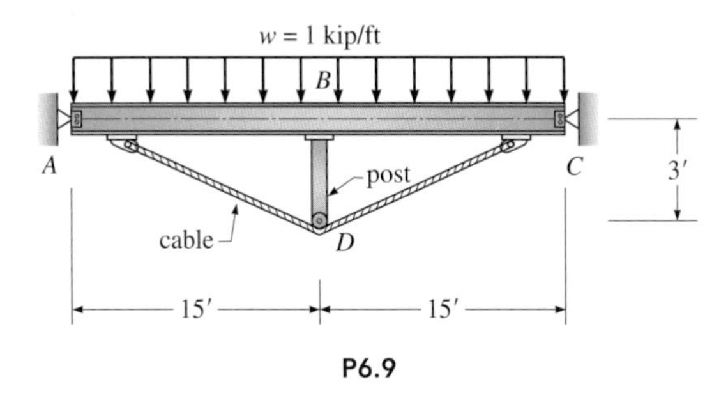

P6.9

P6.10. A point load P of 60 kips exists on beam AE in Figure P6.10. A cable and two posts are added beneath the beam to counteract this load. Determine the required force in the cable such that it generates a moment in the beam that is the equal in magnitude but opposite in direction to the moment due to load P. Determine the axial forces in posts, the axial force in the beam AE and reactions at A and E.

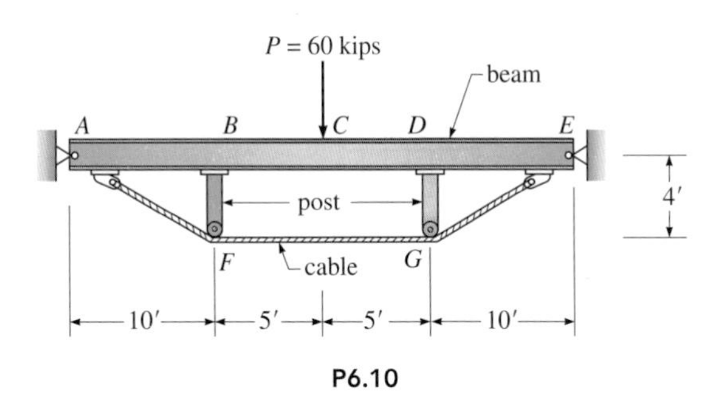

P6.10

P6.11. Compute the support reactions and the maximum tension in the cable in Figure P6.11.

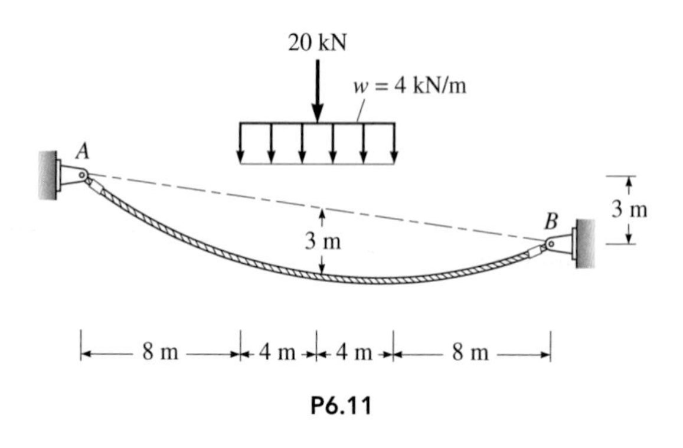

P6.11

P6.12. A cable *ABCD* is pulled at end *E* by a force *P* (Figure P6.12). The cable is supported at point *D* by a rigid member *DF*. Compute the force *P* that produces a sag of 2 m at points *B* and *C*. The horizontal reaction at support *F* is zero. Compute the vertical reaction at *F*.

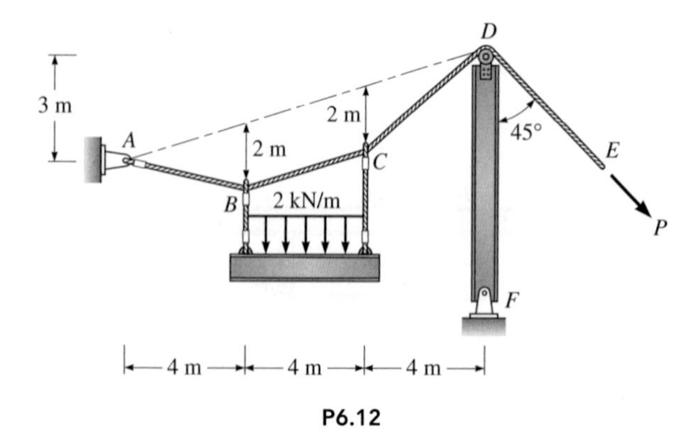

P6.12

P6.13. Compute the support reactions and the maximum tension in the cable in Figure P6.13. The sag at midspan is 12 ft. Each hanger can be assumed to provide a simple support for the suspended beam. Determine the sag at points *B* and *D*.

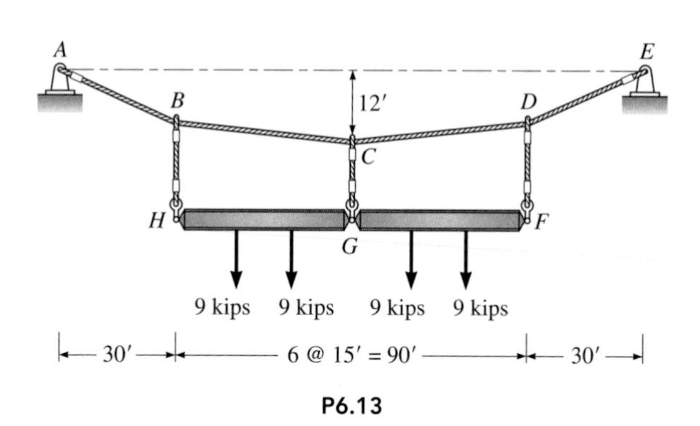

P6.13

P6.14. Determine the location of the 40-kN load such that sags at points *B* and *C* in Figure P6.14 are 3 m and 2 m, respectively. Determine the maximum tension in the cable and the reactions at supports *A* and *D*.

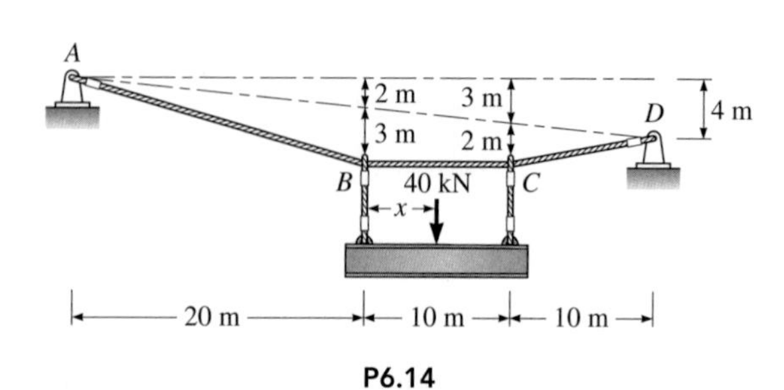

P6.14

Practical Application

P6.15. The cable-supported roof for a summer theater, shown in Figure P6.15, is composed of 24 equally spaced cables that span from a tension ring at the center to a compression ring on the perimeter. The tension ring lies 12 ft below the compression ring. The roof weighs 25 lb/ft^2 based on the horizontal projection of the roof area. If the sag at midspan of each cable is 4 ft, determine the tensile force each cable applies to the compression ring. What is the required area of each cable if the allowable stress is 110 kips/in^2? Determine the weight of the tension ring required to balance the vertical components of the cable forces.

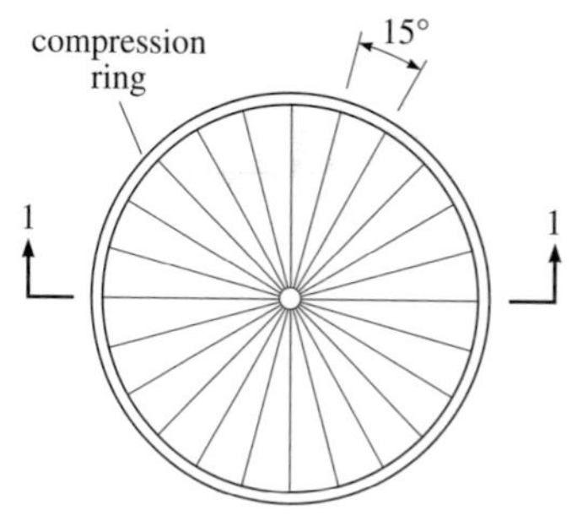

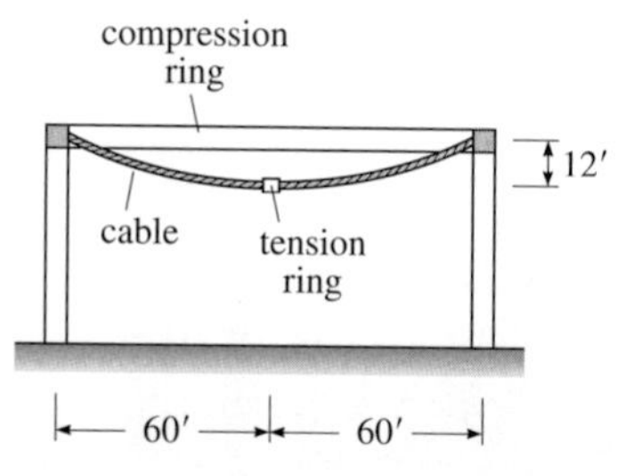

P6.15

P6.16. *Computer study of a cable-stayed bridge.* The deck and tower making up the two-span, cable-stayed bridge in Figure P6.16 are constructed of reinforced concrete. The cross-section of the bridge is constant with an area of 15 ft^2 and a moment of inertia of 19 ft^4. The dead weight of the girders is 4 kips/ft. In addition the girders are to be designed to support a live load of 0.6 kips/ft that is to be positioned to maximize the design forces in individual members. The vertical cable tower, located at the center support, has a cross-sectional area of 24 ft^2 and a moment of inertia of 128 ft^4. Four cables, each with an area of 13 in^2 and an effective modulus of elasticity of 26,000 kips/in^2, are used to support the deck at the third points of each 120 ft span. The modulus of elasticity of the concrete is 5,000 kips/in^2. The cable reaction may be assumed to be applied to the underside of the roadway. Members have been detailed such that the support at D acts as a simple support for both the tower and the roadway girders.

(*a*) Analyze the structure for full live and dead loads on both spans, that is, establish the shear, moment, and axial load diagrams for the girders, the forces in the cables, and the maximum deflection of the girders.

(*b*) With the dead load on both spans and the live load on the left span $ABCD$, determine the shear, moment, and axial load diagrams for both spans, the axial force in the cables, and the shear, moment, and axial load in the vertical cable tower. Also determine the lateral deflection of the cable tower.

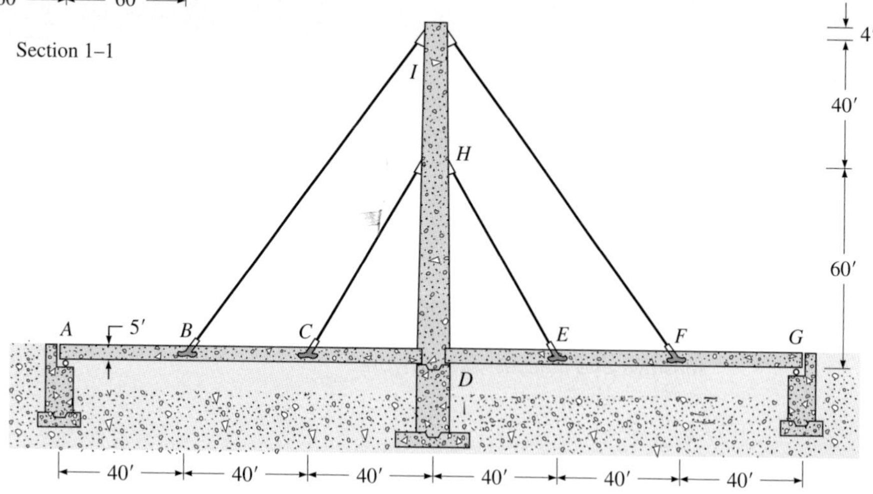

P6.16

Bixby Creek Bridge in Big Sur, California

The wide use of elegant open-spandrel, reinforced concrete arch bridges in the first half of the 20th century has resulted in many notable historic bridges in the United States, including the Bixby Creek Bridge. It is 714 ft long with a 320-ft main span and is over 280 ft high. The seismic retrofit, completed in 1996, ensures not only the bridge will remain stable during intense earthquake shaking but also addresses important aesthetic and environmental concerns during the retrofit construction over the ecologically sensitive and inaccessible canyon.

CHAPTER

7 Arches

Chapter Objectives

- Study the characteristics, types, and behavior of arch structures.
- Analyze determinate three-hinged arches and trussed arches.
- Establish a funicular arch such that the forces are in direct compression along the arch, producing an efficient minimum weight arch utilizing the *general cable theorem*.

7.1 Introduction

As we discussed in Section 1.5, the arch uses material efficiently because applied loads create mostly axial compression on all cross sections. In this chapter we show that for a particular set of loads, the designer can establish one shape of arch—the *funicular shape*—in which all sections are in direct compression (moments are zero).

Typically, dead load constitutes the major load supported by the arch. If a funicular shape is based on the dead load distribution, moments will be created on cross sections by live loads whose distribution differs from that of the dead load. But normally in most arches, the bending stresses produced by live load moments are so small compared to the axial stresses that net compression stresses exist on all sections. Because arches use material efficiently, designers often use them as the main structural elements in long-span bridges (say, 400 to 1800 ft) or buildings that require large column-free areas, for example, airplane hangers, field houses, or convention halls.

In this chapter we consider the behavior and analysis of three-hinged arches. As part of this study, we derive the equation for the shape of a funicular arch that supports a uniformly distributed load, and we apply the *general cable theory* (Section 6.5) to produce the funicular arch for an arbitrary set of concentrated loads. Finally, we apply the concept of *structural optimization* to establish the minimum weight of a simple three-hinged arch carrying a concentrated load.

7.2 Types of Arches

The Romans mastered the construction of arch structures thousands of years ago by empirical methods of proportioning, as exampled in Photo 1.2. However, the theory and analysis of masonry arches was formalized much later. In particular, Philippe de LaHire (1640–1718) applied statics to geometrical solution of funicular polygons (1695), and found that semicircular arches are unstable and rely on grout bond or friction between masonry or stone wedges to prevent sliding. Further important development was made by Charles Coulomb (1736–1806), in which he established design equations for determining the limiting values of arch thrust in order to achieve stability.

Arches are often classified by the number of hinges they contain or by the manner in which their bases are constructed. Figure 7.1 shows the three main types: three-hinged, two-hinged, and fixed-ended. The three-hinged arch is statically determinate; the other two types are indeterminate. The three-hinged arch is the easiest to analyze and construct. Since it is determinate, temperature changes, support settlements, and fabrication errors do not create stresses. On the other hand, because it contains three hinges, it is more flexible than the other arch types.

Fixed-ended arches are often constructed of masonry or concrete when the base of an arch bears on rock, massive blocks of masonry, or heavy reinforced concrete foundations. Indeterminate arches can be analyzed by the flexibility method covered in Chapter 11 or more simply and rapidly by any general-purpose computer program. To determine the forces and displacements at arbitrary points along the axis of the arch using a computer, the designer treats the points as joints that are free to displace.

In long-span bridges, two main arch ribs are used to support the roadway beams. The roadway beams can be supported either by tension hangers from the arch (Figure 1.9*a*) or by columns that bear on the arch (Photo 7.1). Since the arch rib is mostly in compression, the designer must also consider the possibility of its buckling—particularly if it is slender (Figure 7.2*a*). If the arch is constructed of steel members, a built-up rib or a box section may be used to increase the bending stiffness of the cross section and to reduce the likelihood of buckling. In many arches, the floor system or wind bracing is used to stiffen the arch against lateral buckling. In the case of the trussed arch shown in Figure 7.2*b*, the vertical and diagonal members brace the arch rib against buckling in the vertical plane.

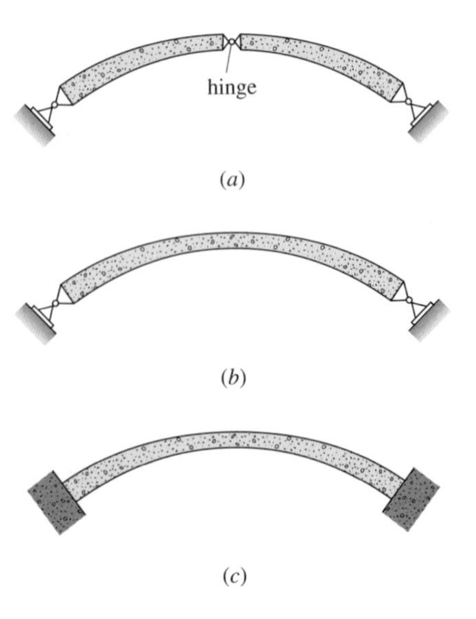

Figure 7.1 Types of arches: (*a*) three-hinged arch, stable and determinate; (*b*) two-hinged arch, indeterminate to the first degree; (*c*) fixed-end arch, indeterminate to the third degree.

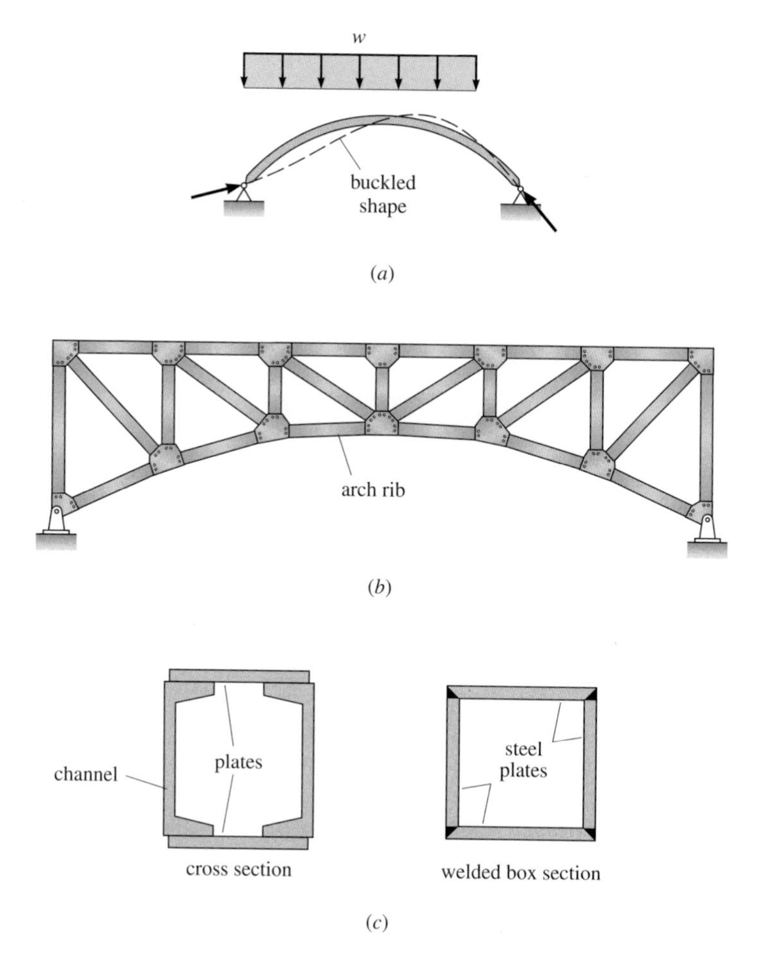

Figure 7.2: (*a*) Buckling of an unsupported arch; (*b*) trussed arch, the vertical and diagonal members brace the arch rib against buckling in the vertical plane; (*c*) two types of built-up steel cross sections used to construct an arch rib.

Since many people find the arch form aesthetically pleasing, designers often use low arches to span small rivers or roads in parks and other public places. At sites where rock sidewalls exist, designers often construct short-span highway bridges using *barrel arches* (see Figure 7.3). Constructed of accurately fitted masonry blocks or reinforced concrete, the barrel arch consists of a wide, shallow arch that supports a heavy, compacted fill on which the engineer places the roadway slab. The large weight of the fill induces sufficient compression in the barrel arch to neutralize any tensile bending stresses created by even the heaviest vehicles. Although the loads supported by the barrel arch may be large, direct stresses in the arch itself are typically low—on the order of 300 to 500 psi because the cross-sectional area of the arch is large. A study by the senior author of a number of masonry barrel-arch bridges built in Philadelphia in the mid-nineteenth century showed that they have the capacity to support vehicles three to five times heavier than the standard AASHTO truck (see Figure 2.7), which highway bridges are currently designed to support. Moreover, while many steel and reinforced concrete bridges built in the past 100 years are no longer serviceable because of corrosion produced by salts used to melt snow, many masonry arches, constructed of good-quality stone, show no deterioration.

Photo 7.1: Railroad bridge (1909) over the Landwasser Gorge, near Wiesen, Switzerland. Masonry construction. The main arch is parabolic, has a span of 55 m and a rise of 33 m. The bridge is narrow as the railway is single-track. The arch ribs are a mere 4.8 m at the crown, tapering to 6 m at the supports.

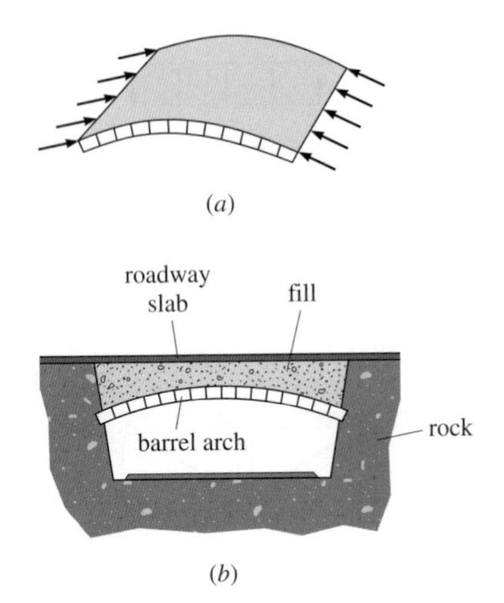

Figure 7.3: (*a*) Barrel arch resembles a curved slab; (*b*) barrel arch used to support a compacted fill and roadway slab.

7.3 Three-Hinged Arches

To demonstrate certain of the characteristics of arches, we will consider how the bar forces vary as the slope θ of the bars changes in the pin-jointed arch in Figure 7.4*a*. Since the members carry axial load only, this configuration represents the funicular shape for an arch supporting a single concentrated load at midspan.

Because of symmetry, the vertical components of the reactions at supports A and C are identical in magnitude and equal to $P/2$. Denoting the slope of bars AB and CB by angle θ, we can express the bar forces F_{AB} and F_{CB} in terms of P and the slope angle θ (see Figure 7.4*b*) as

$$\sin\theta = \frac{P/2}{F_{AB}} = \frac{P/2}{F_{CB}}$$

$$F_{AB} = F_{CB} = \frac{P/2}{\sin\theta} \tag{7.1}$$

Equation 7.1 shows that as θ increases from 0 to 90°, the force in each bar decreases from infinity to $P/2$. We can also observe that as the slope angle θ increases, the length of the bars—and consequently the material required—also

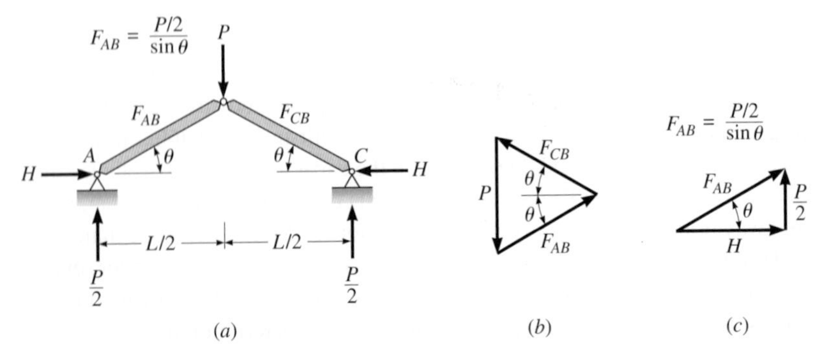

Figure 7.4: (*a*) Three-hinged arch with a concentrated load; (*b*) vector diagram of forces acting on the hinge at B, forces F_{CB} and F_{AB} are equal because of symmetry; (*c*) components of force in bar AB.

increases. To establish the slope that produces the most economical structure for a given span L, we will express the volume V of bar material required to support the load P in terms of the geometry of the structure and the compressive strength of the material

$$V = 2AL_B \tag{7.2}$$

where A is the area of one bar and L_B is the length of a bar.

To express the required area of the bars in terms of load P, we divide the bar forces given by Equation 7.1 by the allowable compressive stress σ_{allow}:

$$A = \frac{P/2}{(\sin\theta)\sigma_{\text{allow}}} \tag{7.3}$$

We will also express the bar length L_B in terms of θ and the span length L as

$$L_B = \frac{L/2}{\cos\theta} \tag{7.4}$$

Substituting A and L_B given by Equations 7.3 and 7.4 into Equation 7.2, simplifying, and using the trigonometric identity $\sin 2\theta = 2\sin\theta\cos\theta$, we calculate

$$V = \frac{PL}{2\sigma_{\text{allow}}\sin 2\theta} \tag{7.5}$$

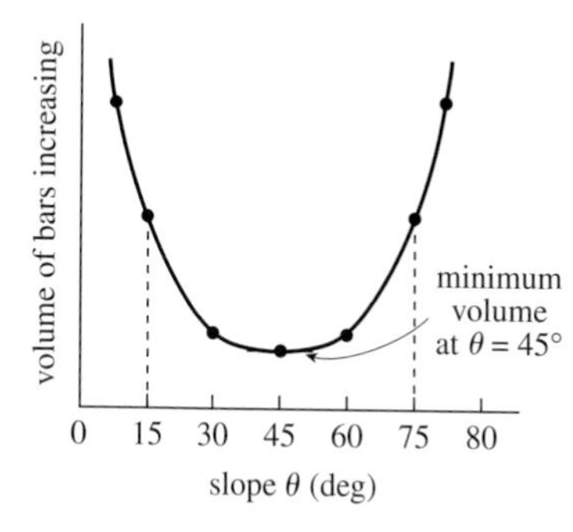

Figure 7.5: Variation of volume of material with slope of bars in Figure 7.4*a*.

If V in Equation 7.5 is plotted as a function of θ (see Figure 7.5), we observe that the minimum volume of material is associated with an angle of $\theta = 45°$. Figure 7.5 also shows that very shallow arches ($\theta \leq 15°$) and very deep arches ($\theta \geq 75°$) require a large volume of material; on the other hand, the flat curvature in Figure 7.5 when θ varies between 30 and 60° indicates that the volume of the bars is not sensitive to the slope between these limits. Therefore, the designer can vary the shape of the structure within this range without significantly affecting either its weight or its cost.

In the case of a curved arch carrying a distributed load, the engineer will also find that the volume of material required in the structure, within a certain range, is not sensitive to the depth of the arch. Of course, the cost of a very shallow or very deep arch will be greater than that of an arch of moderate depth. Finally, in establishing the shape of an arch, the designer will also consider the profile of the site, the location of solid bearing material for the foundations, and the architectural and functional requirements of the project.

7.4 Funicular Shape for an Arch That Supports a Uniformly Distributed Load

Many arches carry dead loads that have a uniform or nearly uniform distribution over the span of the structure. For example, the weight per unit length of the floor system of a bridge will typically be constant. To establish for a uniformly loaded arch the funicular shape—the form required if only direct stress is to develop at all points along the axis of an arch—we will consider the symmetric three-hinged arch in Figure 7.6*a*. The height (or rise) of the arch is denoted by h. Because of symmetry, the vertical reactions at supports A and C are equal to $wL/2$ (one-half the total load supported by the structure).

The horizontal thrust H at the base of the arch can be expressed in terms of the applied load w and the geometry of the arch by considering the free body to the right of the center hinge in Figure 7.6*b*. Summing moments about the center hinge at B, we find

$$\circlearrowright^{+} \quad \Sigma M_B = 0$$

$$0 = \left(\frac{wL}{2}\right)\frac{L}{4} - \left(\frac{wL}{2}\right)\frac{L}{2} + Hh$$

$$H = \frac{wL^2}{8h} \tag{7.6}$$

To establish the equation of the axis of the arch, we superimpose a rectangular coordinate system, with an origin o located at B, on the arch. The positive sense of the vertical y axis is directed downward. We next express the moment M at an arbitrary section (point D on the arch's axis) by considering the free body of the arch between D and the pin at C.

$$\circlearrowright^{+} \quad \Sigma M_D = 0$$

$$0 = \left(\frac{L}{2} - x\right)^2 \frac{w}{2} - \frac{wL}{2}\left(\frac{L}{2} - x\right) + H(h - y) + M$$

Solving for M gives

$$M = \frac{wL^2 y}{8h} - \frac{wx^2}{2} \tag{7.7}$$

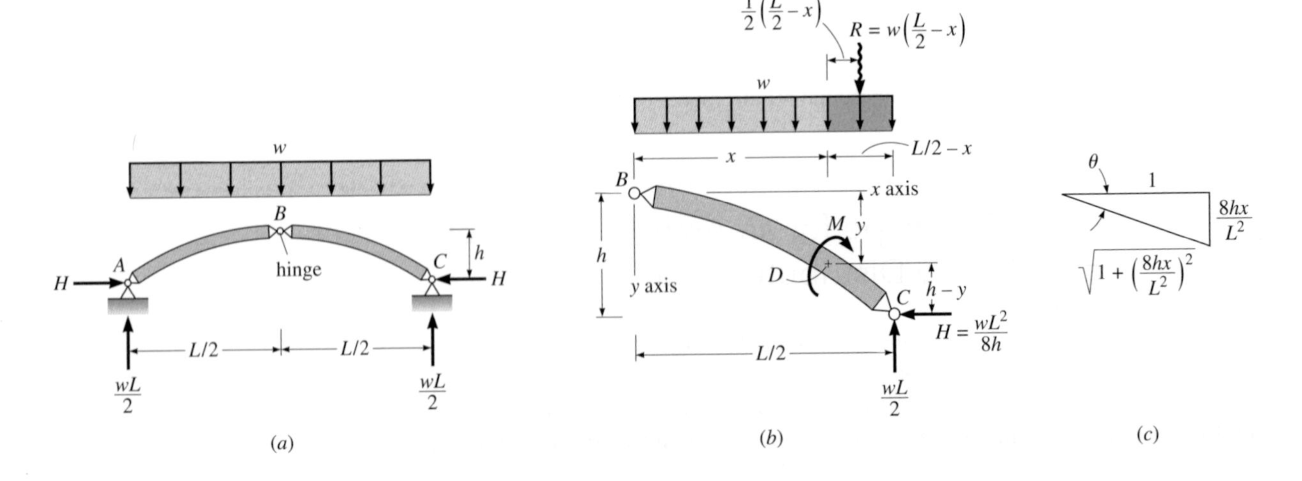

Figure 7.6: Establishing the funicular shape for a uniformly loaded arch.

If the arch axis follows the funicular shape, $M = 0$ at all sections. Substituting this value for M into Equation 7.7 and solving for y establishes the following mathematical relationship between y and x:

$$y = \frac{4h}{L^2}x^2 \tag{7.8}$$

Equation 7.8, of course, represents the equation of a parabola. Even if the parabolic arch in Figure 7.6 were a fixed-ended arch, a uniformly distributed load—assuming no significant change in geometry from axial shortening—would still produce direct stress at all sections because the arch conforms to the funicular shape for a uniform load.

From a consideration of equilibrium in the horizontal direction, we can see that the horizontal thrust at any section of an arch equals H, the horizontal reaction at the support. In the case of a uniformly loaded parabolic arch, the total axial thrust T at any section, a distance x from the origin at B (see Figure 7.6*b*), can be expressed in terms of H and the slope at the given section as

$$T = \frac{H}{\cos\theta} \tag{7.9}$$

To evaluate $\cos\theta$, we first differentiate Equation 7.8 with respect to x to give

$$\tan\theta = \frac{dy}{dx} = \frac{8hx}{L^2} \tag{7.10}$$

The tangent of θ can be shown graphically by the triangle in Figure 7.6*c*. From this triangle we can compute the hypotenuse r using $r^2 = x^2 + y^2$:

$$r = \sqrt{1 + \left(\frac{8hx}{L^2}\right)^2} \tag{7.11}$$

From the relationship between the sides of the triangle in Figure 7.6*c* and the cosine function, we can write

$$\cos\theta = \frac{1}{\sqrt{1 + \left(\frac{8hx}{L^2}\right)^2}} \tag{7.12}$$

Substituting Equation 7.12 into Equation 7.9 gives

$$T = H\sqrt{1 + \left(\frac{8hx}{L^2}\right)^2} \tag{7.13}$$

Equation 7.13 shows that the largest value of thrust occurs at the supports where x has its maximum value of $L/2$. If w or the span of the arch is large, the designer may wish to vary (taper) the cross section in direct proportion to the value of T so that the stress on the cross section is constant.

Example 7.1 illustrates the analysis of a three-hinged trussed arch for both a set of loads that corresponds to the funicular shape of the arch as well as for a single concentrated load. Example 7.2 illustrates the use of cable theory to establish a funicular shape for the set of vertical loads in Example 7.1.

Robert Hooke (1635–1703) established that the shape of a hanging chain, when inverted, would generate an efficient funicular shape of an arch in direct compressive stress. Antonio Gaudi (1852–1926) utilized Hooke's theory and built funicular arches by scaling physical models that consisted of a network of strings with hanging weights. Gaudi used this method in the construction of Colonia Güell in Santa Coloma, Spain. Another iconic example of a funicular arch is St Louis Gateway Arch in St Louis, Missouri, completed in 1967.

EXAMPLE 7.1

The geometry of the bottom chord of the arch is the funicular shape for the loads shown. Analyze the three-hinged trussed arch in Figure 7.7*a* for the dead loads applied at the top chord. Member *KJ*, which is detailed so that it cannot transmit axial force, acts as a simple beam instead of a member of the truss. Assume joint *D* acts as a hinge.

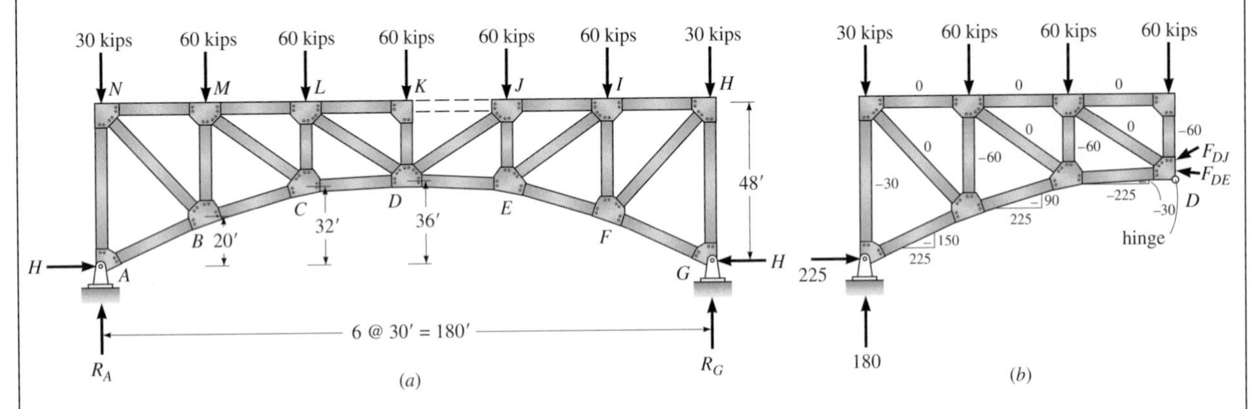

Figure 7.7

Solution

Because the arch and its loads are symmetric, the vertical reactions at *A* and *G* are equal to 180 kips (one-half the applied load). Compute the horizontal reaction at support *G*.

Consider the free body of the arch to the left of the hinge at *D* (Figure 7.7*b*), and sum moments about *D*.

$$\circlearrowright^{+} \quad \Sigma M_D = 0$$

$$0 = -60(30) - 60(60) - 30(90) + 180(90) - 36H$$

$$H = 225 \text{ kips}$$

We now analyze the truss by the method of joints starting at support *A*. Results of the analysis are shown on a sketch of the truss in Figure 7.7*b*.

Figure 7.8

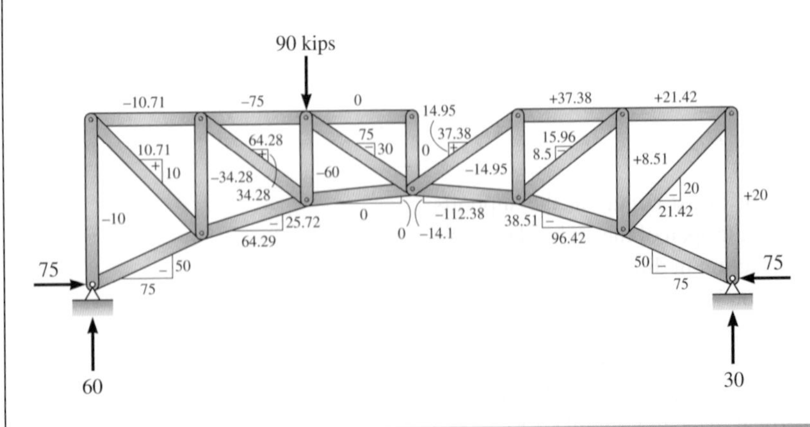

NOTE. Since the arch rib is the funicular shape for the loads applied at the top chord, the only members that carry load—other than the rib—are the vertical columns, which transmit the load down to the arch. The diagonals and top chords will be stressed when a loading pattern that does not conform to the funicular shape acts. Figure 7.8 shows the forces produced in the same truss by a single concentrated load at joint *L*.

EXAMPLE 7.2

Establish the shape of the funicular arch for the set of loads acting on the trussed arch in Figure 7.7. The rise of the arch at midspan is set at 36 ft.

Solution

We imagine that the set of loads is applied to a cable that spans the same distance as the arch (see Figure 7.9*a*). The sag of the cable is set at 36 ft—the height of the arch at midspan. Since the 30-kip loads at each end of the span act directly at the supports, they do not affect the force or the shape of the cable and may be neglected. Applying the general cable theory, we imagine that the loads supported by the cable are applied to an imaginary simply supported beam with a span equal to that of the cable (Figure 7.9*b*). We next construct the shear and moment curves. According to the general cable theorem at every point,

$$M = Hy \tag{6.6}$$

where M = moment at an arbitrary point in the beam
H = horizontal component of support reaction
y = cable sag at an arbitrary point

Since $y = 36$ ft at midspan and $M = 8100$ kip·ft, we can apply Equation 6.6 at that point to establish H.

$$H = \frac{M}{y} = \frac{8100}{36} = 225 \text{ kips}$$

With H established we next apply Equation 6.6 at 30 and 60 ft from the supports. Compute y_1 at 30 ft:

$$y_1 = \frac{M}{H} = \frac{4500}{225} = 20 \text{ ft}$$

Compute y_2 at 60 ft:

$$y_2 = \frac{M}{H} = \frac{7200}{225} = 32 \text{ ft}$$

A cable profile is always a funicular structure because a cable can only carry direct stress. If the cable profile is turned upside down, a funicular arch is produced. When the vertical loads acting on the cable are applied to the arch, they produce compression forces at all sections equal in magnitude to the tension forces in the cable at the corresponding sections.

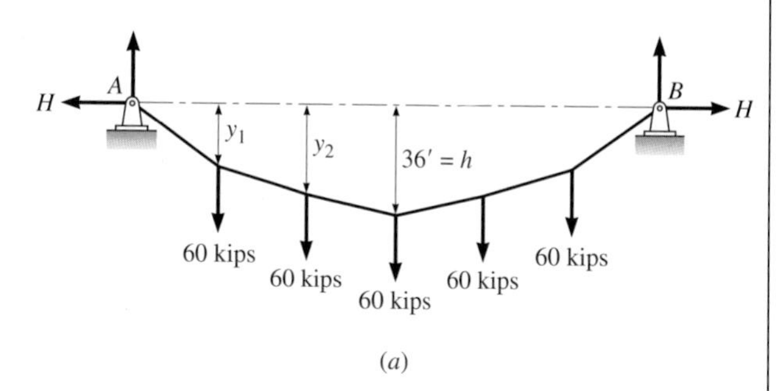

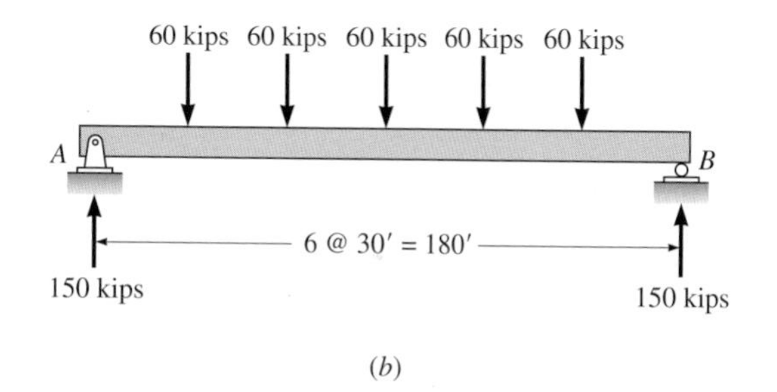

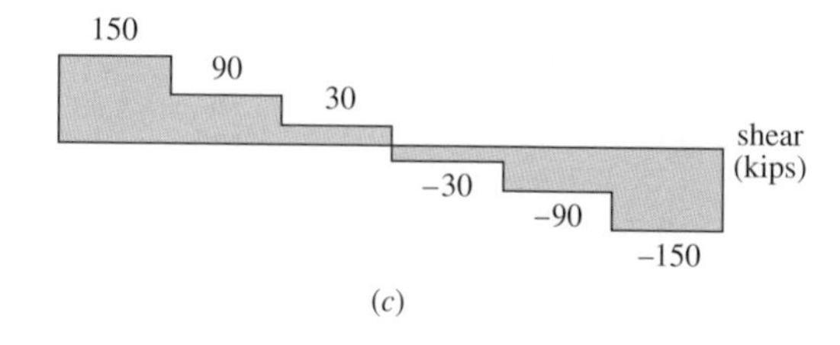

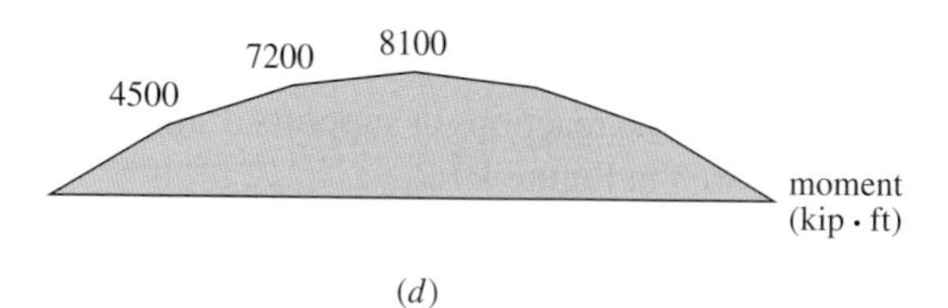

Figure 7.9: Use of cable theory to establish the funicular shape of an arch.

Summary

- Although short masonry arches are often used in scenic locations because of their attractive form, they also produce economical designs for long-span structures that (1) support large, uniformly distributed dead load and (2) provide a large unobstructed space under the arch (suitable for convention halls or sports arenas or in the case of a bridge providing clearance for tall ships).
- Arches can be shaped (termed a *funicular* arch) so that dead load produces only direct stress—a condition that leads to a minimum weight structure.
- For a given set of loads, the *funicular shape* of arch can be established using cable theory.

PROBLEMS

P7.1. For the parabolic arch in Figure P7.1, plot the variation of the thrust T at support A for values of $h = 12, 24, 36, 48$, and 60 ft.

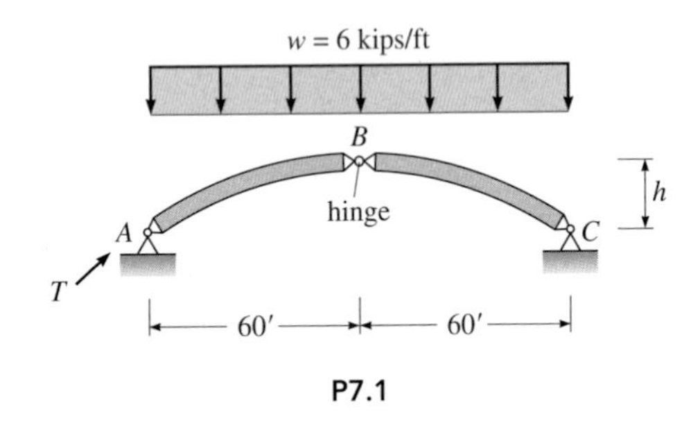

P7.1

P7.2. Determine the reactions at supports A and E of the three-hinged arch in Figure P7.2.

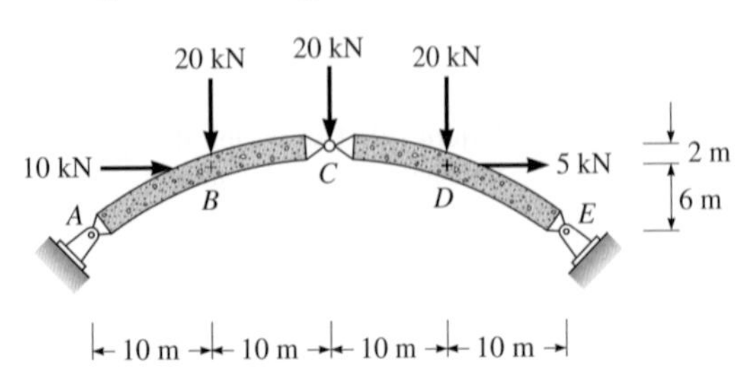

P7.2

P7.3. Determine the reactions at supports A and C of the three-hinged circular arch.

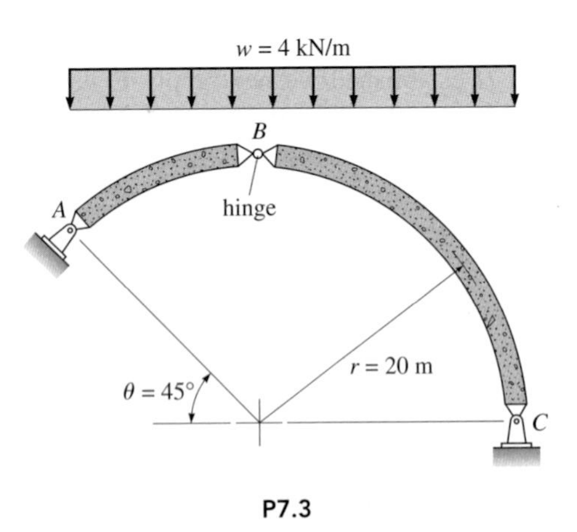

P7.3

P7.4. For the arch shown in Figure P7.4, the thrust force exceeded the abutment's lateral support capacity, which is represented by a roller at C. Load P was removed temporarily and tension rod AC was added. If the maximum compression in members AB and BC will be 750 kips, what size diameter tension rod is required? (Ignore dead load of the arch.) The allowable tensile capacity of the rod is 32 ksi. Determine the reactions.

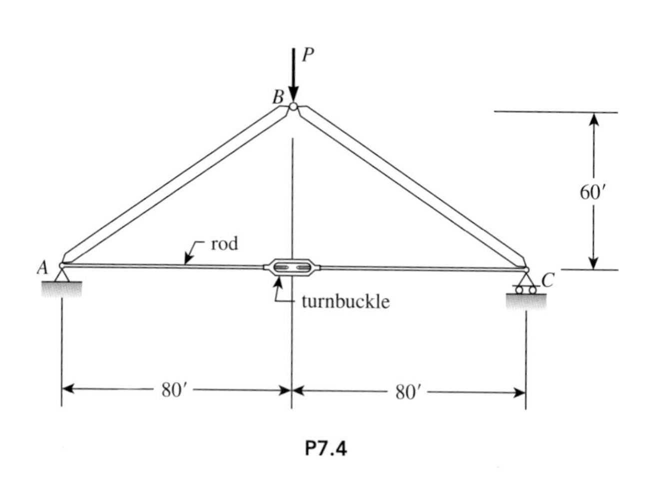

P7.4

P7.5. The arch shown in Figure P7.5 has a pin support at A and a roller at C. A tension rod connects A and C. Determine the reactions at A and C and the tension in rod AC.

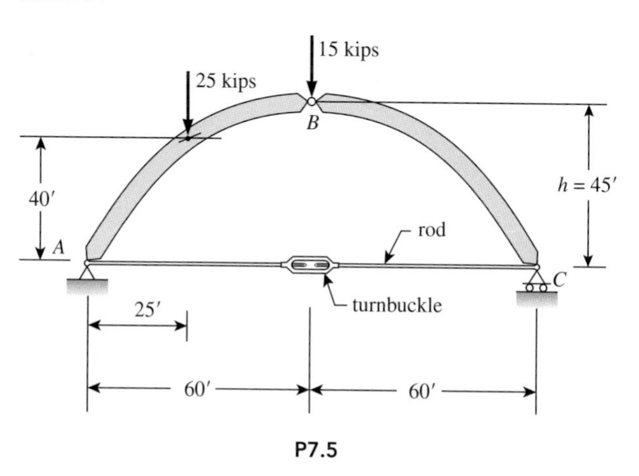

P7.5

P7.6. For the three-hinged arch shown in Figure P7.6, compute the reactions at A and C. Determine the axial force, shear and moment at D.

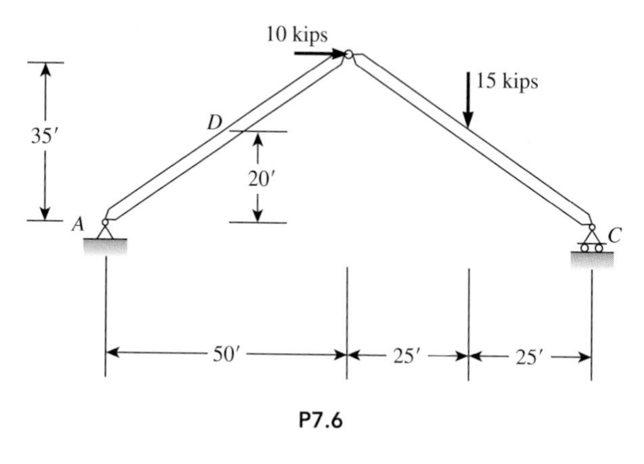

P7.6

P7.7. Compute the reactions at supports A and E of the three-hinged parabolic arch in Figure P7.7. Next compute the shear, axial load, and moment at points B and D, located at the quarter points.

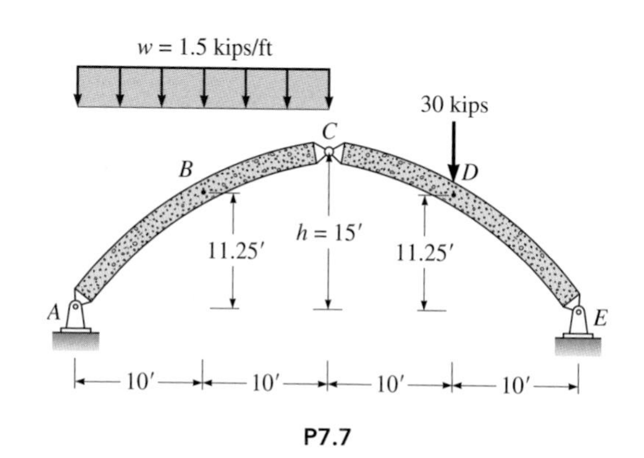

P7.7

P7.8. The three-hinged parabolic arch in Figure P7.8 supports 60-kip loads at the quarter points. Determine the shear, axial load, and moment on sections an infinitesimal distance to the left and right of the loads. The equation for the arch axis is $y = 4hx^2/L^2$.

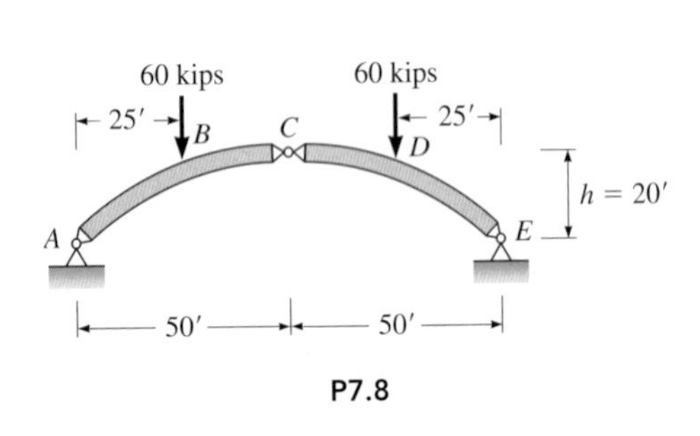

P7.8

P7.9. Compute the support reactions for the arch in Figure P7.9. (*Hint*: You will need two moment equations: Consider the entire free body for one, and a free body of the portion of truss to either the left or right of the hinge at *B*.)

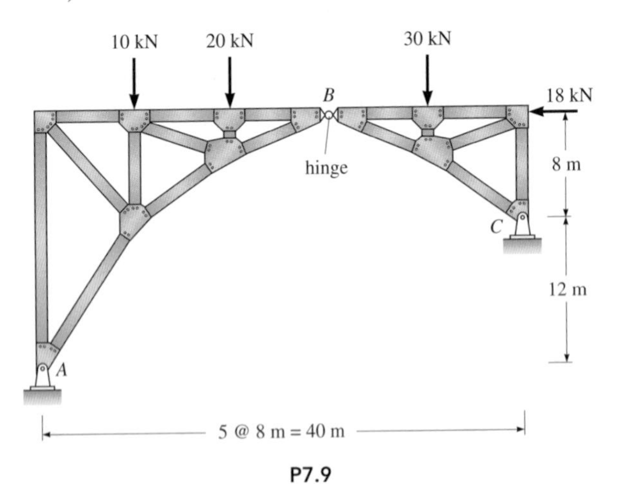

P7.9

P7.10. (*a*) In Figure P7.10 compute the horizontal reaction A_x at support *A* for a 10-kip load at joint *B*. (*b*) Repeat the computation if the 10-kip load is also located at joints *C* and *D*, respectively.

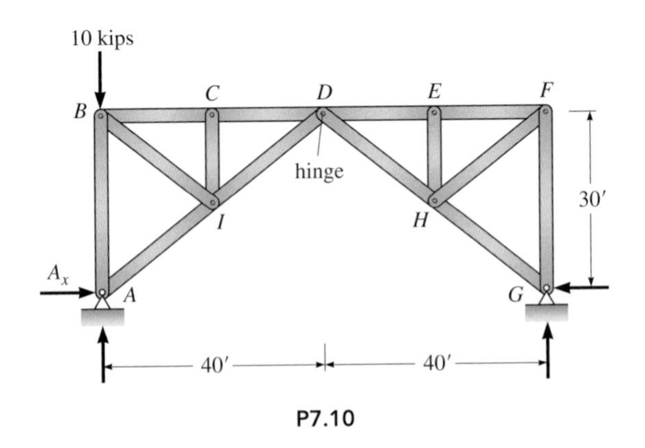

P7.10

P7.11. (*a*) Determine the reactions and all bar forces of the three-hinged, trussed arch in Figure P7.11 for the following cases.
Case A: Only the 90-kN force at joint *D* acts.
Case B: Both the 90-kN and 60-kN forces at joints *D* and *M* act. (*b*) Determine the maximum axial force in the arch in *Case B*.

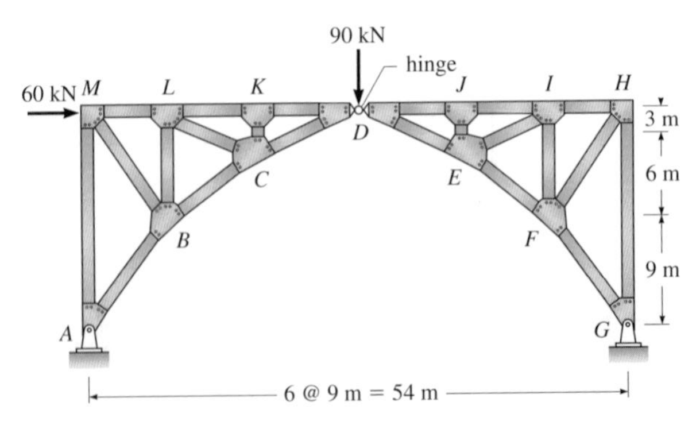

P7.11

P7.12. Establish the funicular arch for the system of loads in Figure P7.12.

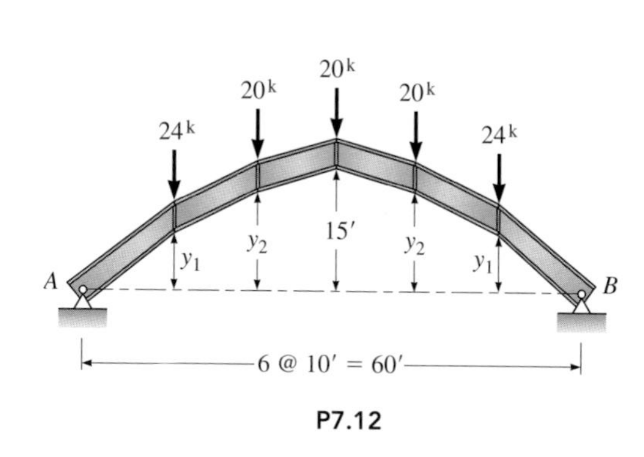

P7.12

P7.13. Determine the load P such that all the members in the three-hinged arch in Figure P7.13 are in pure compression. What is the value of y_1?

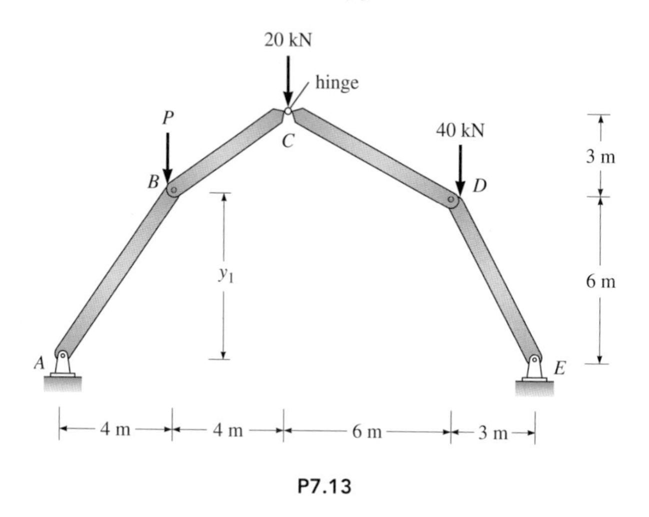

P7.13

P7.14. If the arch rib $ABCDE$ in Figure P7.14 is to be funicular for the dead loads shown at the top joints, establish the elevation of the lower chord joints at B and D.

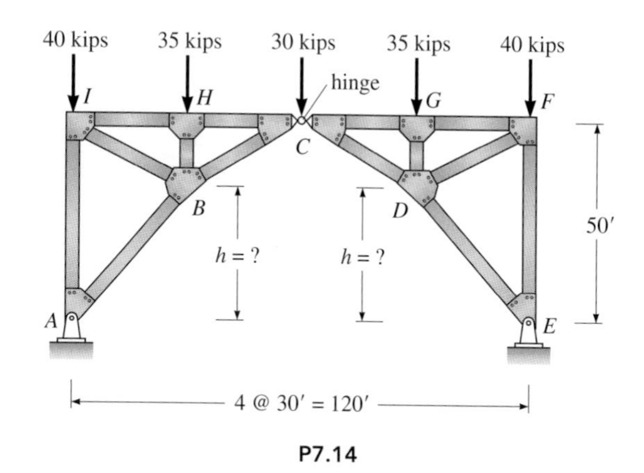

P7.14

P7.15. For the arch rib to be funicular for the dead loads shown, establish the elevation of the lower chord joints B, C, and E.

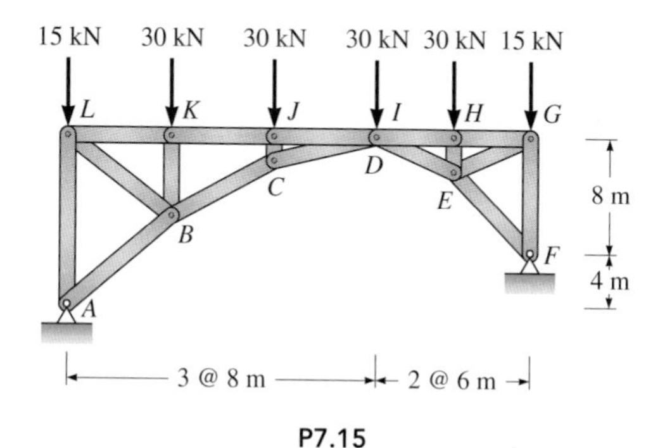

P7.15

P7.16. *Computer study of a two-hinged arch.* The objective is to establish the difference in response of a *parabolic* arch to (1) uniformly distributed loads and (2) a single concentrated load.

(*a*) The arch in Figure P7.16 supports a roadway consisting of simply supported beams connected to the arch by high-strength cables with area $A = 2 \text{ in}^2$ and $E = 26{,}000$ ksi. (Each cable transmits a dead load from the beams of 36 kips to the arch.) Determine the reactions, the axial force, shear, and moment at each joint of the arch, and the joint displacements. Plot the deflected shape. Represent the arch by a series of straight segments between joints. The arch has a constant cross section with $A = 24 \text{ in}^2$, $I = 2654 \text{ in}^4$, and $E = 29{,}000$ ksi.

(*b*) Repeat the analysis of the arch if a single 48-kips vertical load acts downward at joint 18. Again, determine all the forces acting at each joint of the arch, the joint displacements, etc., and compare results with those in (*a*). Briefly describe the difference in behavior.

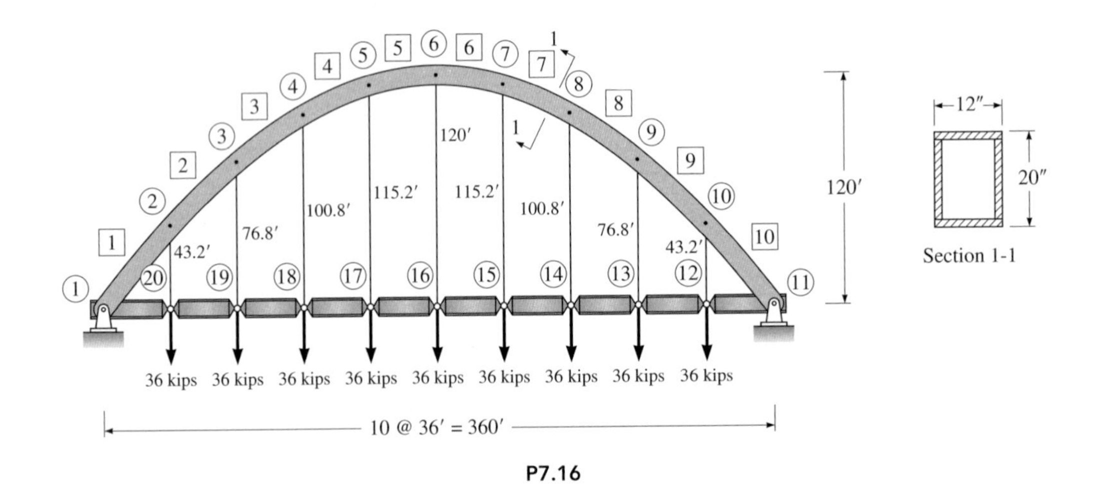

P7.16

P7.17. *Computer study of arch with a continuous floor girder.* Repeat part (b) in problem P7.16 if a continuous girder with $A = 102.5 \text{ in}^2$ and $I = 40{,}087 \text{ in}^4$, as shown in Figure P7.17, is provided to support the floor system. For both the girder and the arch, determine all forces acting on the arch joints as well as the joint displacements. Discuss the results of your study of P7.16 and P7.17 with particular emphasis on the magnitude of the forces and displacements produced by the 48-kip load.

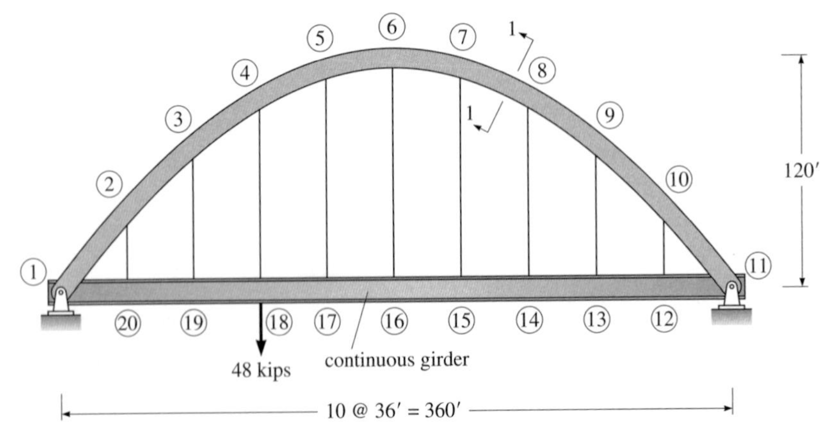

P7.17

P7.18. To reduce the vertical displacement of the roadway floor system of the arch (shown in P7.16, part *b*) produced by the 48-kip load at joint 18, diagonal cables of 2-in. in diameter are added as shown in Figure P7.18. For this configuration, determine the vertical displacement of all the floor system joints. Compare the results of this analysis with part *b* of P7.16 by plotting to scale the vertical deflections of all joints along the roadway from joints 1 to 11. Properties of the diagonal cables are the same as those of the vertical cables.

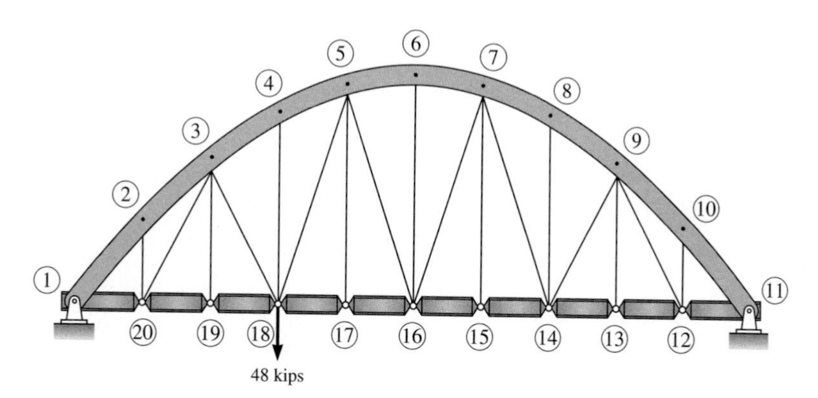

P7.18

Rion-Antirion Bridge in Greece

The 7,388-ft long Rion-Antirion Bridge spanning the Corinthian Gulf is the world's longest multi-span cable-stayed bridge, and was completed in 2004. The adverse conditions the designer had to consider included a water depth of 213 ft, poor soil conditions, strong seismicity, and the potential collision of a tanker to the structure. The fully suspended continuous deck is designed to move as a pendulum during an earthquake; dampers are used to reduce the sway of the deck caused by strong wind. Each of the four Pylons rests on a 297-ft diameter reinforced concrete caisson, which sits on a gravel layer on the sea floor such that the structure would remain flexible and free to slide on the gravel layer during a seismic event.

CHAPTER 8

Live Load Forces: Influence Lines for Determinate Structures

Chapter Objectives

- Understand the concept of influence lines for analyzing structures with moving loads.
- Use the basic concept together with statics to construct influence lines.
- Learn the Müller-Breslau principle to graphically construct influence lines of determinate structures.
- Learn to position moving loads to maximize the influence of moving loads (e.g., internal forces and reactions) based on the influence lines.

8.1 Introduction

Thus far we have analyzed structures for a variety of loads without considering how the position of a concentrated load or the distribution of a uniform load was established. Further, we have not distinguished between dead load, which is fixed in position, and live load, which can change position. In this chapter our objective is to establish how to position live load (for example, a truck or a train) to maximize the value of a certain type of force (*shear* or *moment* in a beam or *axial* force in a truss) at a designated section of a structure.

8.2 Influence Lines

As a moving load passes over a structure, the internal forces at each point in the structure vary. We intuitively recognize that a concentrated load applied to a beam at midspan produces much greater bending stresses and deflection than the same load applied near a support. For example, suppose that you had to cross a small stream filled with alligators by walking over an old, flexible, partially cracked plank. You would be more concerned about the plank's capacity to support your weight as you approached midspan than you would be when you were standing on the end of the plank at the support (see Figure 8.1).

If a structure is to be safely designed, we must proportion its members and joints so that the maximum force at each section produced by live and dead load is less than or equal to the available capacity of the section. To establish maximum design forces at critical sections produced by moving loads, we frequently construct *influence lines*.

An influence line is a diagram whose ordinates, which are plotted as a function of distance along the span, give the value of an internal force, a reaction, or a displacement at a particular point in a structure as a unit load of 1 kip or 1 kN moves across the structure.

Once the influence line is constructed, we can use it (1) to determine where to place live load on a structure to maximize the force (shear, moment, etc.) for which the influence line is drawn, and (2) to evaluate the magnitude of the force (represented by the influence line) produced by the live load. Although an influence line represents the action of a single moving load, it can also be used to establish the force at a point produced by several concentrated loads or by a uniformly distributed load.

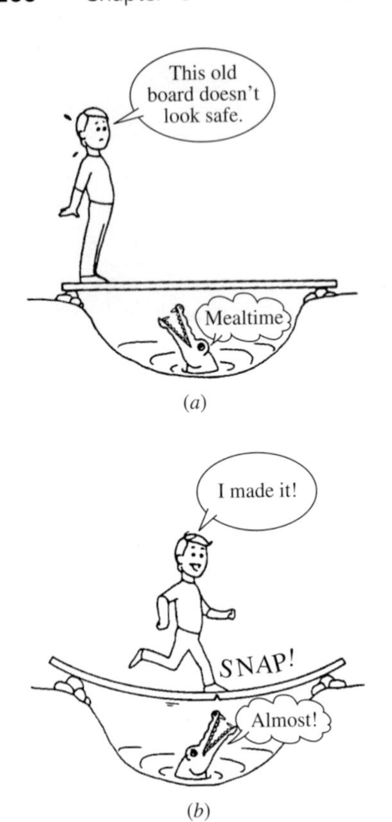

Figure 8.1: Variation of bending with position of load; (*a*) no bending at midspan, load at support; (*b*) maximum bending and deflection, load at midspan. Board fails.

8.3 Construction of an Influence Line

To introduce the procedure for constructing influence lines, we will discuss in detail the steps required to draw the influence line for the reaction R_A at support A of the simply supported beam in Figure 8.2*a*.

As noted previously, we can establish the ordinates of the influence lines for the reaction at A by computing the value of R_A for successive positions of a unit load as it moves across the span. We begin by placing the unit load at support A. By summing moments about support B (Figure 8.2*b*), we compute $R_A = 1$ kip. We then arbitrarily move the unit load to a second position located a distance $L/4$ to the right of support A. Again, summing moments about B, we compute $R_A = \frac{3}{4}$ kip (Figure 8.2*c*). Next, we move the load to midspan and compute $R_A = \frac{1}{2}$ kip (Figure 8.2*d*). For the final computation, we position the 1-kip load directly over support B, and we compute $R_A = 0$ (Figure 8.2*e*). To construct the influence line, we now plot the numerical values of R_A directly below each position of the unit load associated with the corresponding value of R_A. The resulting influence line diagram is shown in Figure 8.2*f*. The influence line shows that the reaction at A varies linearly from 1 kip when the load is at A to a value of 0 when the load is at B. Since the reaction at A is in kips, the ordinates of the influence line have units of kips per 1 kip of load.

As you become familiar with the construction of influence lines, you will only have to place the unit load at two or three positions along the axis of the beam to establish the correct shape of the influence line. Several points to remember about Figure 8.2*f* are summarized here:

1. All ordinates of the influence line represent values of R_A.
2. Each value of R_A is plotted directly below the position of the unit load that produced it.
3. The maximum value of R_A occurs when the unit load acts at A.
4. Since all ordinates of the influence line are positive, a load acting vertically downward anywhere on the span produces a reaction at A directed upward. (A negative ordinate would indicate the reaction at A is directed downward.)
5. The influence line is a straight line. As you will see, influence lines for determinate structures are either straight lines or composed of linear segments.

By plotting values of the reaction of B for various positions of the unit load, we generate the influence line for R_B shown in Figure 8.2*g*. Since the sum of the reactions at A and B must always equal 1 (the value of the applied load) for all positions of the unit load, the sum of the ordinates of the two influence lines at any section must also equal 1 kip.

In Example 8.1 we construct influence lines for the reactions of a beam with an overhang. Example 8.2 illustrates the construction of influence lines for shear and moment in a beam. If the influence lines for the reactions are drawn first, they will facilitate the construction of influence lines for other forces in the same structure.

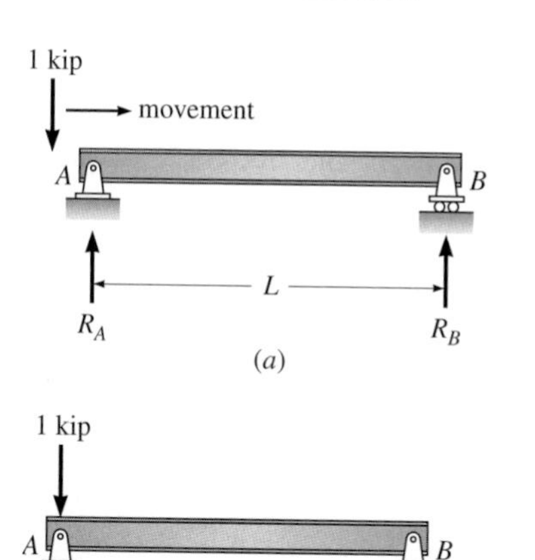

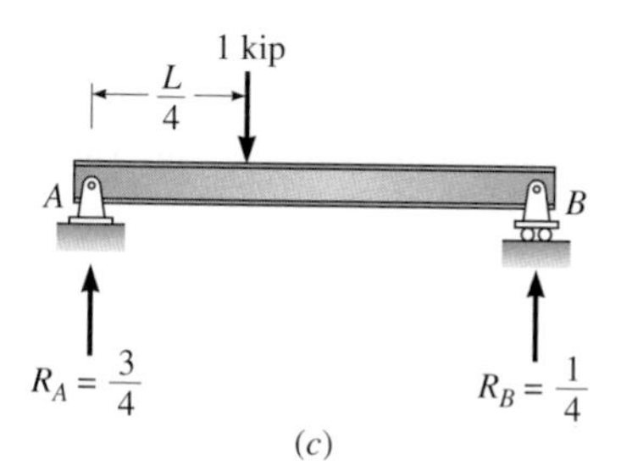

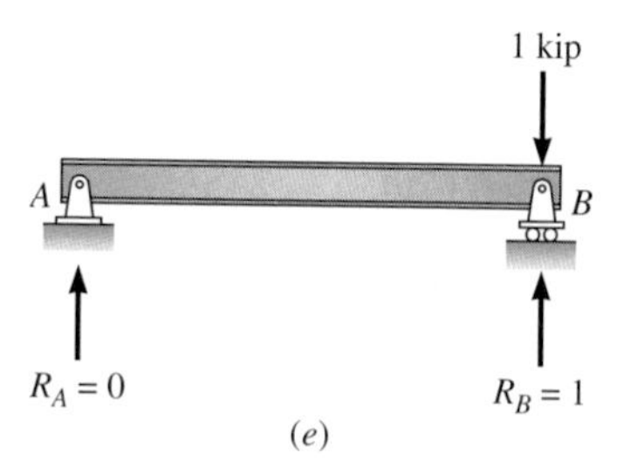

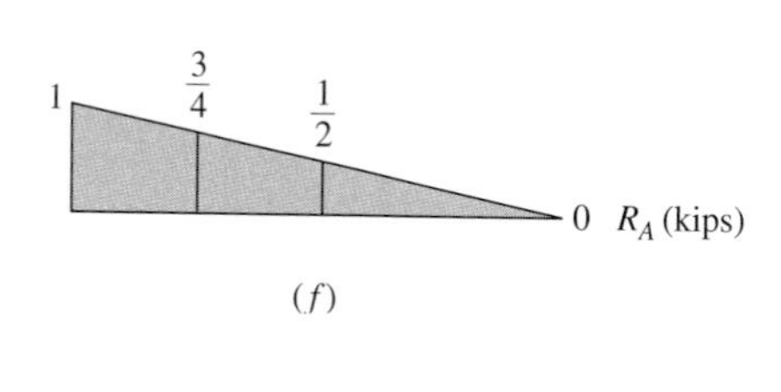

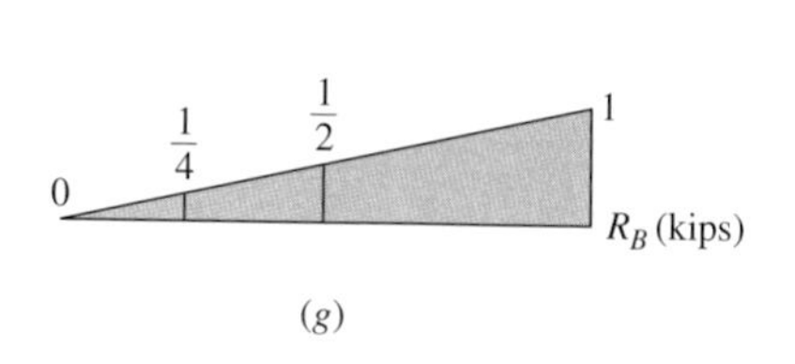

Figure 8.2: Influence lines for reactions at A and B; (*a*) beam; (*b*), (*c*), (*d*), and (*e*) show successive positions of unit load; (*f*) influence line for R_A; (*g*) influence line for R_B.

EXAMPLE 8.1

Construct the influence lines for the reactions at A and C for the beam in Figure 8.3a.

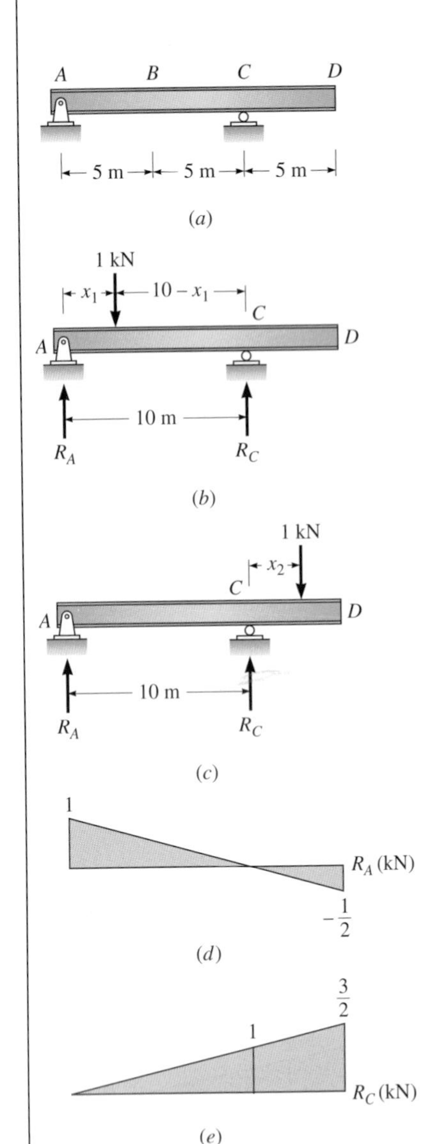

Figure 8.3: Influence lines for reactions at supports A and C; (a) beam; (b) load between A and C; (c) unit load between C and D; (d) influence line for R_A; (e) influence line for R_C.

Solution

To establish a general expression for values of R_A for any position of the unit load between supports A and C, we place the unit load a distance x_1 to the right of support A (see Figure 8.3b) and sum moments about support C.

$$\circlearrowright^{+} \quad \Sigma M_C = 0$$

$$10R_A - (1\text{ kN})(10 - x_1) = 0$$

$$R_A = 1 - \frac{x_1}{10} \tag{1}$$

where $0 \leqslant x_1 \leqslant 10$.

Evaluate R_A for $x_1 = 0, 5$, and 10 m.

x_1	R_A
0	1
5	$\frac{1}{2}$
10	0

A general expression for R_A, when the unit load is located between C and D, can be written by summing moments about C for the free-body diagram shown in Figure 8.3c.

$$\circlearrowright^{+} \quad \Sigma M_C = 0$$

$$10R_A + (1\text{ kN})(x_2) = 0$$

$$R_A = -\frac{x_2}{10} \tag{2}$$

where $0 \leqslant x_2 \leqslant 5$.

The minus sign in Equation 2 indicates that R_A acts downward when the unit load is between points C and D. For $x_2 = 0, R_A = 0$; for $x_2 = 5, R_A = -\frac{1}{2}$. Using the foregoing values of R_A from Equations 1 and 2, we draw the influence line shown in Figure 8.3d.

To draw the influence line for R_C (see Figure 8.3e), either we can compute the values of the reaction at C as the unit load moves across the span, or we can subtract the ordinates of the influence line in Figure 8.3d from 1, because the sum of the reactions for each position of the unit load must equal 1—the value of the applied load.

EXAMPLE 8.2

Draw the influence lines for shear and moment at section B of the beam in Figure 8.4a.

Solution

The influence lines for shear and moment at section B are drawn in Figure 8.4c and d. The ordinates of these influence lines were evaluated for the five positions of the unit load indicated by the circled numbers along the span of the beam in Figure 8.4a. To evaluate the shear and moment at B produced by the unit load, we will pass an imaginary cut through the beam at B and consider the equilibrium of the free body to the left of the section. (The positive directions for shear and moment are defined in Figure 8.4b.)

To establish the ordinates of the influence lines for V_B and M_B at the left end (support A), we place the unit load directly over the support at A and compute the shear and moment at section B. Since the entire unit load is carried by the reaction at support A, the beam is unstressed; thus the shear and moment at section B are zero. We next position the unit load at point 2, an infinitesimal distance to the left of section B, and evaluate the shear V_B and moment M_B at the section (see Figure 8.4e). Summing moments about an axis through section B to evaluate the moment, we see that the unit load, which passes through the moment center, does not contribute to M_B. On the other hand, when we sum forces in the vertical direction to evaluate the shear V_B, the unit load appears in the summation.

We next move the unit load to position 3, an infinitesimal distance to the right of section B. Although the reaction at A remains the same, the unit load is no longer on the free body to the left of the section (see Figure 8.4f). Therefore, the shear reverses direction and undergoes a 1-kip change in magnitude (from $-\frac{1}{4}$ to $+\frac{3}{4}$ kip). The 1-kip jump that occurs between sides of a cut is a characteristic of influence lines for shear. On the other hand, the moment does not change as the unit load moves an infinitesimal distance from one side of the section to the other.

As the unit load moves from B to D, the ordinates of the influence lines reduce linearly to zero at support D because both the shear and the moment at B are a direct function of the reaction at A, which in turn varies linearly with the position of the load between B and D.

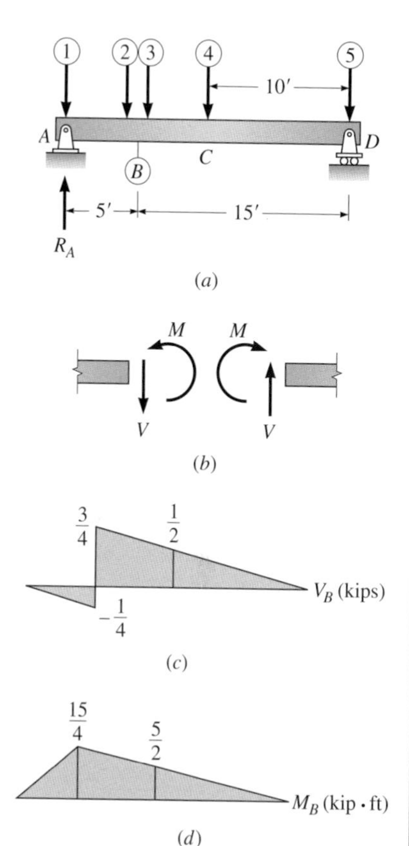

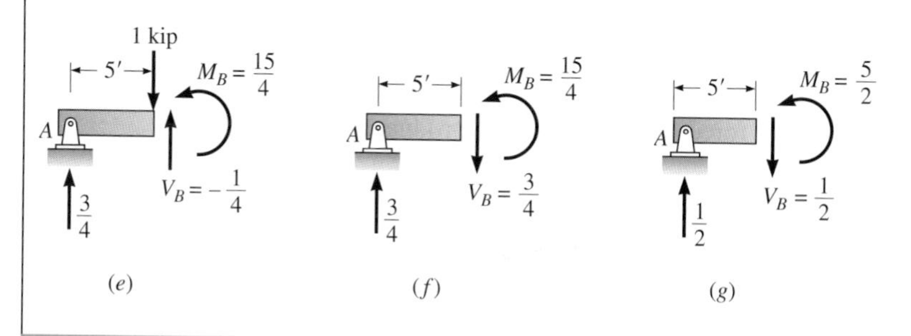

Figure 8.4: Influence lines for shear and moment at section B; (a) position of unit load; (b) positive sense of shear and moment defined; (c) influence line for shear at B; (d) influence line for moment at B; (e) free body for unit load to left of section B; (f) free body for unit load to right of section B; (g) free body for unit load at midspan.

EXAMPLE 8.3

For the frame in Figure 8.5, construct the influence lines for the horizontal and vertical components of the reactions A_x and A_y at support A and for the vertical component of force F_{By} applied by member BD to joint B. The bolted connection of member BD to the girder may be treated as a pin connection, making BD a two-force member (or a link).

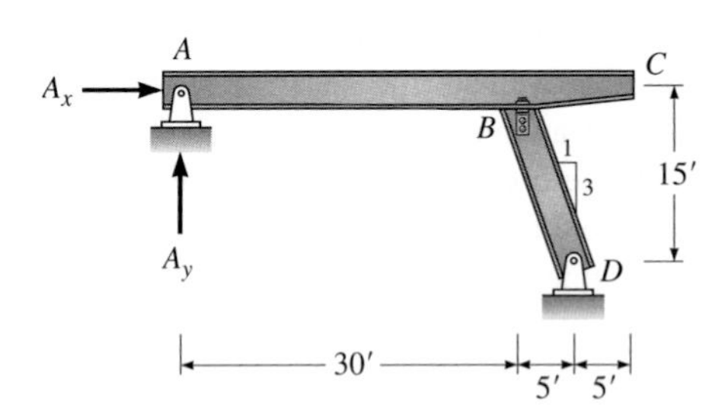

Figure 8.5

Solution

To establish the ordinates of the influence lines, we position a unit load a distance x_1 from support A on a free body of member ABC (Figure 8.6a). Next we apply the three equations of equilibrium to express the reactions at points A and B in terms of the unit load and the distance x_1.

Since the force F_B in member BD acts along the axis of the member, the horizontal and vertical components of F_B are proportional to the slope of the member; therefore,

$$\frac{F_{Bx}}{1} = \frac{F_{By}}{3}$$

and

$$F_{Bx} = \frac{F_{By}}{3} \tag{1}$$

Summing forces acting on member ABC (Figure 8.6a) in the y direction gives

$$\overset{+}{\uparrow} \quad \Sigma F_y = 0$$

$$0 = A_y + F_{By} - 1 \text{ kip}$$

$$A_y = 1 \text{ kip} - F_{By} \tag{2}$$

Next, a sum of forces in the x direction produces

$$\rightarrow + \quad \Sigma F_x = 0$$

$$A_x - F_{Bx} = 0$$

$$A_x = F_{Bx} \tag{3}$$

Substituting Equation 1 into Equation 3, we can express A_x in terms of F_{By} as

$$A_x = \frac{F_{By}}{3} \tag{4}$$

To express F_{By} in terms of x_1, we sum moments of forces on member ABC about the pin at support A:

$$\circlearrowright^{+} \quad \Sigma M_A = 0$$

$$(1 \text{ kip})x_1 - F_{By}(30) = 0$$

$$F_{By} = \frac{x_1}{30} \tag{5}$$

Substituting F_{By} given by Equation 5 into Equations 2 and 4 permits us to express A_y and A_x in terms of the distance x_1:

$$A_y = 1 \text{ kip} - \frac{x_1}{30} \tag{6}$$

$$A_x = \frac{x_1}{90} \tag{7}$$

To construct the influence lines for the reactions shown in Figure 8.6*b*, *c*, and *d*, we evaluate F_{By}, A_y, and A_x, given by Equations 5, 6, and 7, for values of $x_1 = 0, 30$, and 40 ft.

x_1	F_{By}	A_y	A_x
0	0	1	0
30	1	0	$\frac{1}{3}$
40	$\frac{4}{3}$	$-\frac{1}{3}$	$\frac{4}{9}$

As we can observe from examining the shape of the influence lines in Examples 8.1 through 8.3, influence lines for determinate structures consist of a series of straight lines; therefore, we can define most influence lines by connecting the ordinates at a few critical points along the axis of a beam where the slope of the influence line changes or is discontinuous. These points are located at supports, hinges, ends of cantilevers, and, in the case of shear forces, on each side of the section on which they act. To illustrate this procedure, we will construct the influence lines for the reactions at the supports of the beam in Example 8.4.

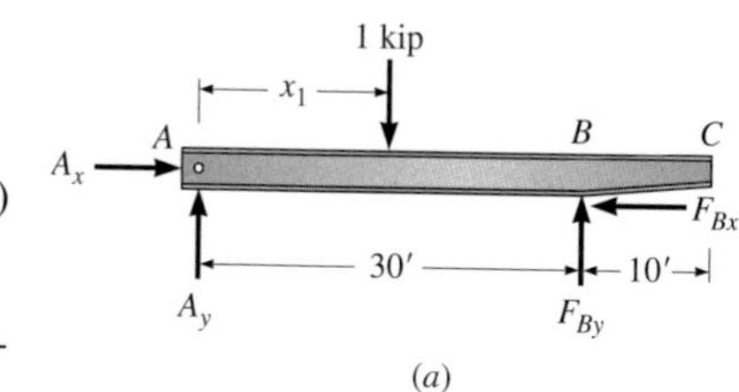

(*a*)

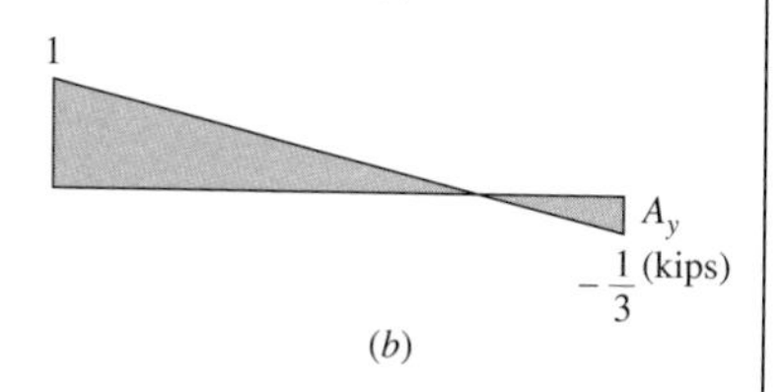

(*b*)

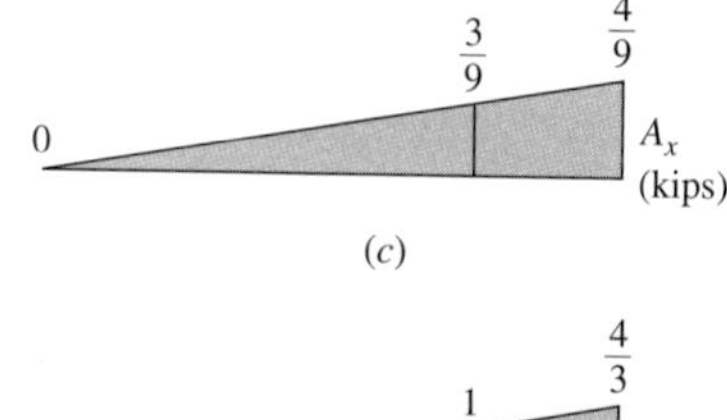

(*c*)

(*d*)

Figure 8.6: Influence lines.

EXAMPLE 8.4

Draw the influence lines for reactions R_A and M_A at the fixed support at A and for reaction R_C at the roller support at C (see Figure 8.7*a*). The arrows shown in Figure 8.7*a* indicate the positive sense for each reaction.

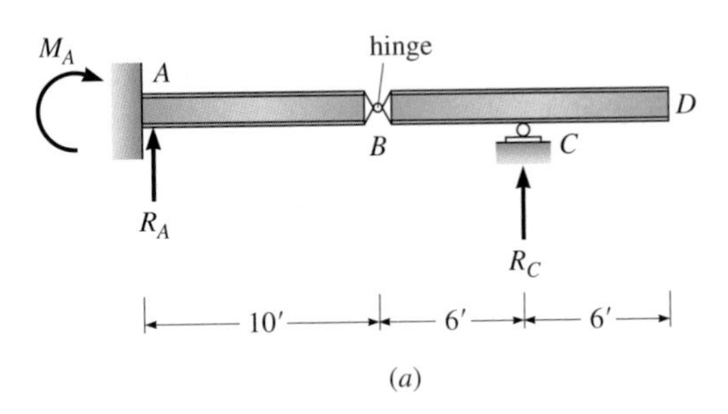

(*a*)

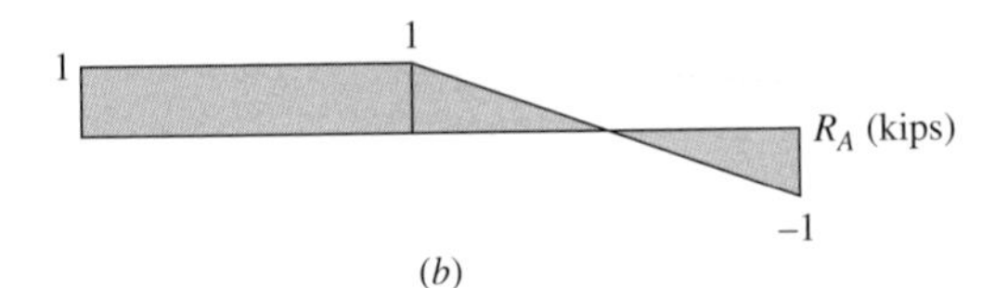

(*b*)

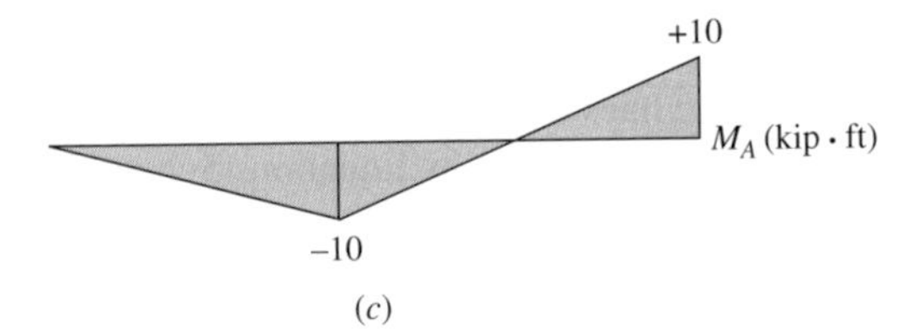

(*c*)

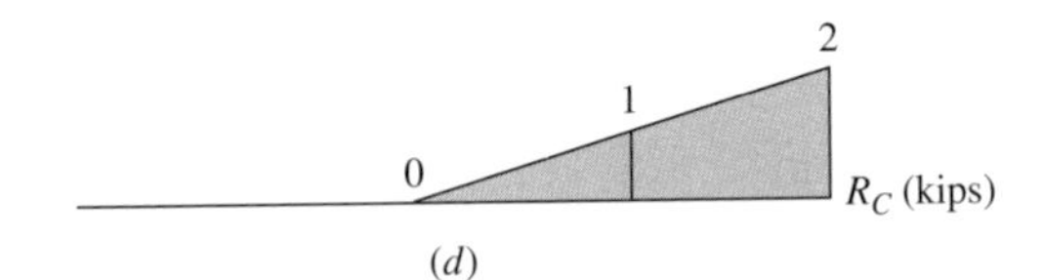

(*d*)

Figure 8.7

Solution

In Figure 8.8*a*, *b*, *d*, and *e*, we position the unit load at four points to supply the forces required to draw the influence lines for the support reactions. In Figure 8.8*a*, we place the unit load at the face of the fixed support at point A. In this position the entire load flows directly into the support, producing the reaction R_A. Since no load is transmitted through the rest of the structure, and all other reactions are equal to zero, the structure is unstressed.

We next move the unit load to the hinge at point B (Figure 8.8b). If we consider a free body of beam BCD to the right of the hinge (Figure 8.8c) and sum moments about the hinge at B, the reaction R_C must be equal to zero because no external loads act on beam BD. If we sum forces in the vertical direction, it follows that the force R_B applied by the hinge also equals zero. Therefore, we conclude that the entire load is supported by cantilever AB and produces the reactions at A shown in Figure 8.8b.

We next position the unit load directly over support C (Figure 8.8d). In this position the entire force is transmitted through the beam into the support at C, and the balance of the beam is unstressed. In the final position, we move the unit load to the end of the cantilever at point D (Figure 8.8e). Summing moments about the hinge at B gives

$$\circlearrowright^{+} \quad \Sigma M_B = 0$$

$$0 = 1\text{ kip}(12\text{ ft}) - R_C(6\text{ ft})$$

$$R_C = 2\text{ kips}$$

Summing forces on member BCD in the vertical direction, we establish that the pin at B applies a force of 1 kip downward on member BCD. In turn, an equal and opposite force of 1 kip must act upward at the B end of member AB, producing the reactions shown at support A.

We now have all the information required to plot the influence lines shown in Figure 8.7b, c, and d. Figure 8.8a supplies the values of the influence line ordinates at support A for the three influence lines; i.e., in Figure 8.7b, $R_A = 1$ kip, in Figure 8.7c, $M_A = 0$, and in Figure 8.7d, $R_C = 0$.

Figure 8.8b supplies the values of the three influence line ordinates at point B, that is, $R_A = 1$ kip, $M_A = -10$ kip·ft (counterclockwise), and $R_C = 0$. Figure 8.8d supplies the ordinates of the influence lines at support C, and Figure 8.8e gives the value of the influence line ordinates at point D, the cantilever tip. Drawing straight lines between the four points completes the construction of the influence lines for the three reactions.

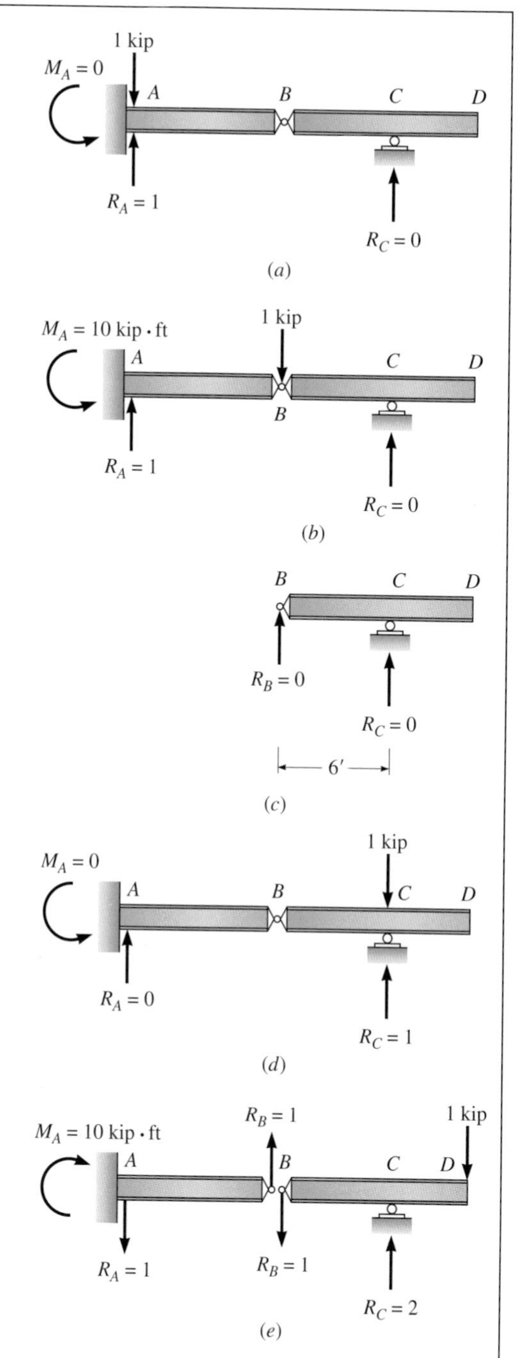

Figure 8.8

8.4 The Müller–Breslau Principle

The Müller–Breslau principle provides a simple procedure for establishing the shape of influence lines for the reactions or the internal forces (shear and moment) in beams. The qualitative influence lines, which can be quickly sketched, can be used in the following three ways:

1. To verify that the shape of an influence line, produced by moving a unit load across a structure, is correct.
2. To establish where to position live load on a structure to maximize a particular function without evaluating the ordinates of the influence line. Once the critical position of the load is established, it is simpler to analyze certain types of structures directly for the specified live load than to draw the influence line.
3. To determine the location of the maximum and minimum ordinates of an influence line so that only a few positions of the unit load must be considered when the influence line ordinates are computed.

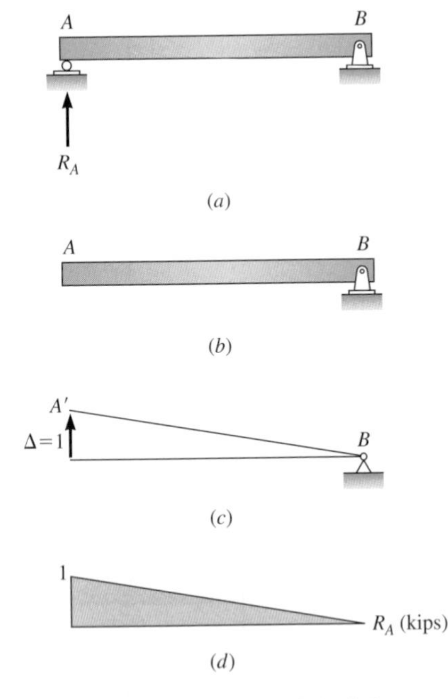

Figure 8.9: Construction of the influence line for R_A by the Müller–Breslau principle. (*a*) Simply supported beam. (*b*) The *released* structure. (*c*) Displacement introduced that corresponds to reaction at A. The deflected shape is the influence line to some unknown scale. (*d*) The influence line for R_A.

Although the Müller–Breslau method applies to both determinate and indeterminate beams, we limit the discussion in this chapter to determinate members. Influence lines for indeterminate beams are covered in Chapter 14. Since the derivation of the method requires an understanding of work-energy, covered in Chapter 10, the proof is deferred to Chapter 14.

The *Müller–Breslau principle* states:

The influence line for any reaction or internal force (shear, moment) corresponds to the deflected shape of the structure produced by removing the capacity of the structure to carry that force and then introducing into the modified (or released) structure a unit deformation that corresponds to the restraint removed.

The unit deformation refers to a unit displacement for reaction, a relative unit displacement for shear, and a relative unit rotation for moment. To introduce the method, we will draw the influence line for the reaction at A of the simply supported beam in Figure 8.9*a*. We begin by removing the vertical restraint supplied by the reaction at A, producing the *released* structure shown in Figure 8.9*b*. We next displace the left end of the beam vertically upward, in the direction of R_A, a unit displacement (see Figure 8.9*c*). Since the beam must rotate about the pin at B, its deflected shape, which is the influence line, is a triangle that varies from 0 at B to 1.0 at A'. This result confirms the shape of the influence line for the reaction at A that we constructed in Section 8.2 (see Figure 8.2*f*).

As a second example, we will draw the influence line for the reaction at B for the beam in Figure 8.10*a*. Figure 8.10*b* shows the released structure produced by removing the support at B. We now introduce a unit vertical displacement Δ that corresponds to the reaction at B producing the deflected shape, which is the influence line (see Figure 8.10*c*). From similar triangles, we compute the value of the ordinate of the influence line at point *C* as $\frac{3}{2}$.

To construct an influence line for shear at a section of a beam by the Müller–Breslau method, we must remove the capacity of the cross section to transmit shear but not axial force or moment. We will imagine that the device constructed of plates and rollers in Figure 8.11*a* permits this modification when introduced into a beam.

To illustrate the Müller–Breslau method, we will construct the influence line for shear at point *C* of the beam in Figure 8.11*b*. In Figure 8.11*c* we insert the plate and roller device at section *C* to release the shear capacity of the cross section. We then offset the beam segments to the left and right of section *C* by Δ_1 and Δ_2 such that a unit relative displacement ($\Delta_1 + \Delta_2 = 1$) is introduced (see Figure 8.11*c*). Since the sliding device inserted at *C* still maintains moment capacity, no relative rotation is allowed. That is, segments *AC* and *CD* should remain parallel, and the rotation (θ) of these two segments is identical. From geometry in Figure 8.11*d*,

$$\Delta_1 = 5\theta, \qquad \Delta_2 = 15\theta$$

and

$$\Delta_1 + \Delta_2 = 5\theta + 15\theta = 20\theta = 1$$

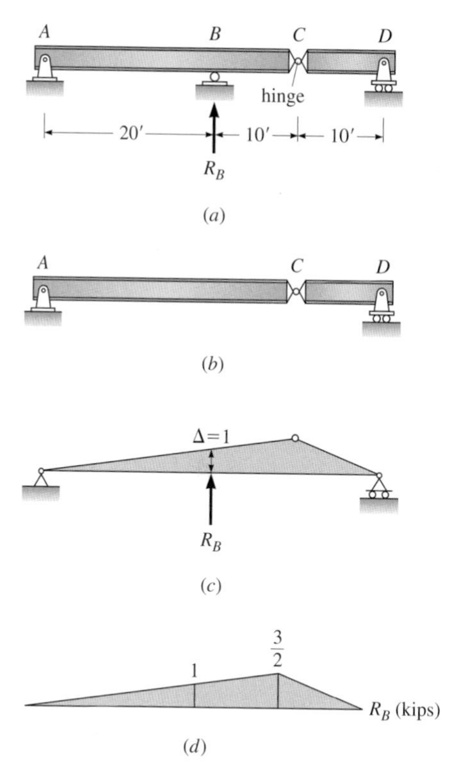

Figure 8.10: Influence line for the reaction at *B*: (*a*) cantilever beam with hinge at *C*; (*b*) reaction removed, producing the released structure; (*c*) displacement of released structure by reaction at *B* establishes the shape of the influence line; (*d*) influence line for reaction at *B*.

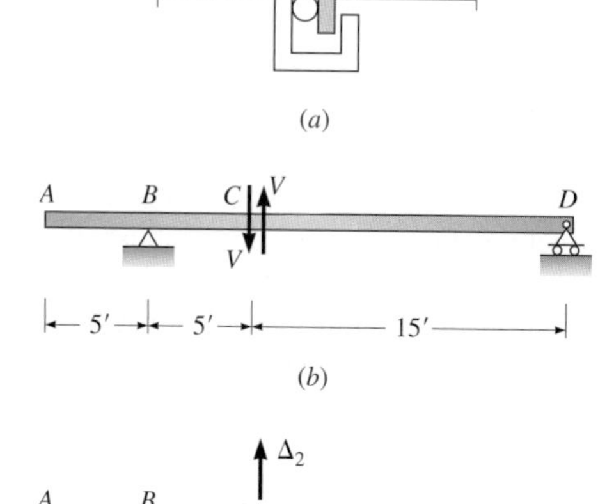

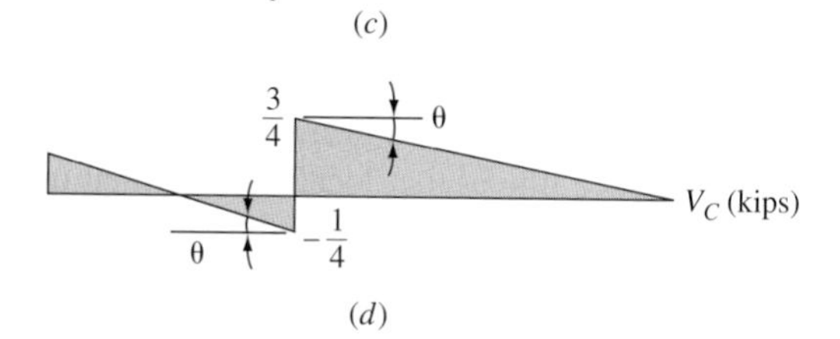

Figure 8.11: Influence line for shear using Müller–Breslau method; (*a*) device to release shear capacity of cross section; (*b*) beam details; (*c*) shear capacity released at section *C*; (*d*) influence line for shear at section *C*.

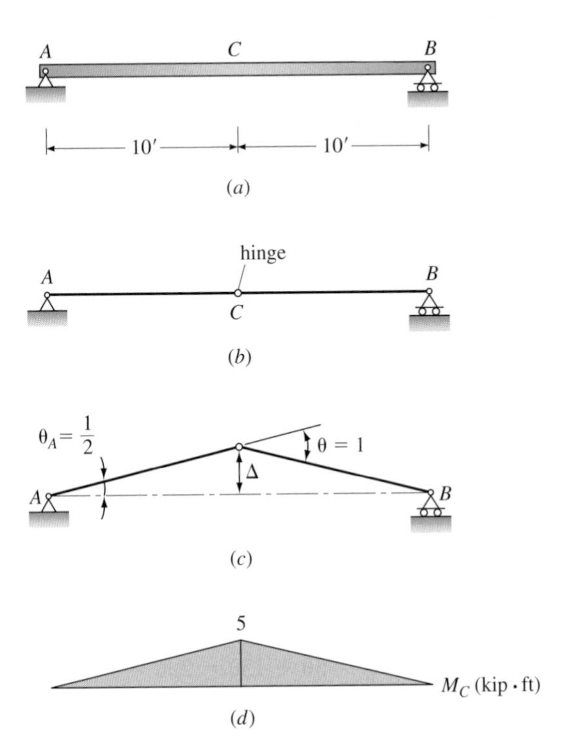

Figure 8.12: Influence line for moment: (*a*) details of beam; (*b*) released structure—hinge inserted at midspan; (*c*) displacement of released structure by moment; (*d*) influence line for moment at midspan.

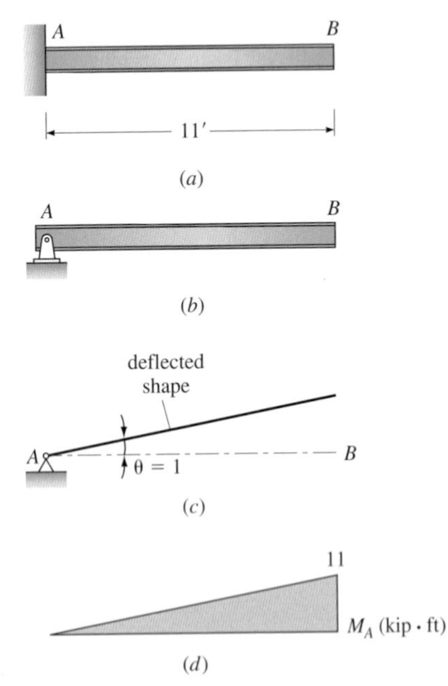

Figure 8.13: Influence line for moment at support *A*: (*a*) details of structure; (*b*) released structure; (*c*) deformation produced by moment at support *A*; (*d*) influence line for moment at *A*.

It follows that $\theta = \frac{1}{20}$, and $\Delta_1 = \frac{1}{4}$ (but with a minus sign), $\Delta_2 = \frac{3}{4}$.

To draw an influence line for moment at an arbitrary section of a beam using the Müller–Breslau method, we introduce a hinge at the section to produce the released structure. For example, to establish the shape of the influence line for moment at midspan of the simply supported beam in Figure 8.12*a*, we introduce a hinge at midspan as shown in Figure 8.12*b*. We then move the hinge at C up by an amount Δ such that a unit relative rotation (or a "kink") of $\theta = 1$ between segments AC and CB is achieved. From geometry in Figure 8.12*c*, $\theta_A = \frac{1}{2}$, and Δ is computed as $\frac{1}{2}(10) = 5$, which is the ordinate of the influence line at C. The final influence line is shown in Figure 8.12*d*.

In Figure 8.13 we use the Müller–Breslau method to construct the influence line for the moment M at the fixed support of a cantilever beam. The released structure is established by introducing a pin at the left support. Introducing a unit relative rotation between the fixed support and the released beam produces a deflected shape with a beam tip deflection of 11, which is the ordinate of the influence line at that location. The final influence line is shown in Figure 8.13*d*. The theoretical basis of the Müller–Breslau principle, which requires the Maxwell-Betti law in Section 10.9, will be described in Section 14.3.

8.5 Use of Influence Lines

As noted previously, we construct influence lines to establish the maximum value of reactions or internal forces produced by live load. In this section we describe how to use an influence line to compute the maximum value of a function when the live load, which can act anywhere on the structure, is either a *single concentrated load* or a *uniformly distributed load of variable length*.

Since the ordinate of an influence line represents the value of a certain function produced by a unit load, the value produced by a concentrated load can be established by multiplying the influence line ordinate by the magnitude of the concentrated load. This computation simply recognizes that the forces created in an elastic structure are directly proportional to the magnitude of the applied load.

If the influence line is positive in certain regions and negative in others, the function represented by the influence line reverses direction for certain positions of the live load. To design members in which the direction of the force has a significant influence on behavior, we must establish the value of the largest force in each direction by multiplying both the maximum positive and the maximum negative ordinates of the influence line by the magnitude of the concentrated load. For example, if a support reaction reverses direction, the support must be detailed to transmit the largest values of tension (uplift) as well as the largest value of compression into the foundation.

In the design of buildings and bridges, live load is frequently represented by a uniformly distributed load. For example, a building code may require that floors of parking garages be designed for a uniformly distributed live load of a certain magnitude instead of a specified set of wheel loads.

To establish the maximum value of a function produced by a uniform load w of variable length, we must distribute the load over the member in the region or regions in which the ordinates of the influence line are either positive or negative. We will demonstrate next that the value of the function produced by a *distributed load* w acting over a certain region of an influence line is equal to the area under the influence line in that region multiplied by the magnitude w of the distributed load.

To establish the value of a function F produced by a uniform load w acting over a section of beam of length a between points A and B (see Figure 8.14), we will replace the distributed load by a series of infinitesimal forces dP, and then sum the increments of the function (dF) produced by the infinitesimal forces. As shown in Figure 8.14, the force dP produced by the uniform load w acting on an infinitesimal beam segment of length dx equals the product of the distributed load and the length of the segment, that is,

$$dP = w\,dx \tag{8.1}$$

During the Industrial Revolution, railroads expanded as did the need for longer span bridges. Structural engineers responded to the concern for simulating moving train loads in terms of safety and economy in bridge design. The concept of influence lines originated through the analysis of elastic arch in 1868 by Emil Winkler (1835–1888) and Otto Mohr (1835–1918), and later in Johann Weyrauch's (1845–1917) publication in 1878, where he coined the method as 'influence lines'. In 1883, Henrich Müller-Breslau (1851–1925) first published his graphical method for determining maximum influence (i.e., internal forces) for continuous beams.

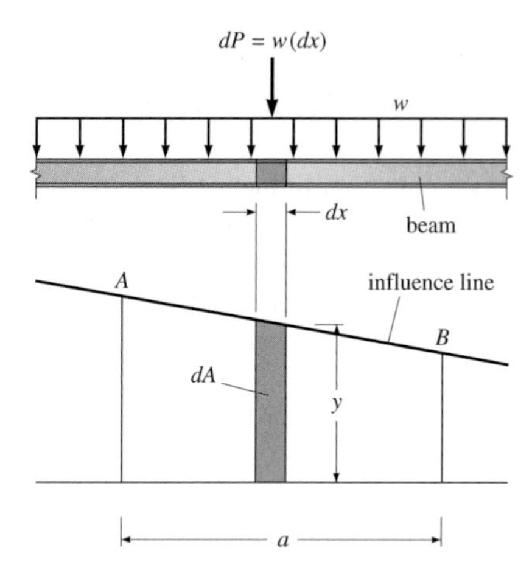

Figure 8.14

To establish the increment of the function dF produced by the force dP, we multiply dP by the ordinate y of the influence line at the same point, to give

$$dF = (dP)y \tag{8.2}$$

Substituting dP given in Equation 8.1 into Equation 8.2 gives

$$dF = w\ dx\ y \tag{8.3}$$

To evaluate the magnitude of the function F between any two points A and B, we integrate both sides of Equation 8.3 between those limits to give

$$F = \int_A^B dF = \int_A^B w\ dx\ y \tag{8.4}$$

Since the value of w is a constant, we can factor it out of the integral, producing

$$F = w \int_A^B y\ dx \tag{8.5}$$

Recognizing that $y\ dx$ represents an infinitesimal area dA under the influence line, we can interpret the integral on the right side of Equation 8.5 as the area under the influence line between points A and B. Thus,

$$F = w(\text{area}_{AB}) \tag{8.6}$$

where area_{AB} is the area under the influence line between A and B.

In Example 8.5 we apply the principles established in this section to evaluate the maximum values of positive and negative moment at midspan of a beam that supports both a distributed load of variable length and a concentrated force.

EXAMPLE 8.5

The beam in Figure 8.15*a* is to be designed to support its deadweight of 0.45 kip/ft and a live load that consists of a 30-kip concentrated load and a variable length, uniformly distributed load of 0.8 kip/ft. The live loads can act anywhere on the span. The influence line for moment at point C is given in Figure 8.15*b*. Compute (*a*) the maximum positive and negative values of live load moment at section C and (*b*) the moment at C produced by the beam's weight.

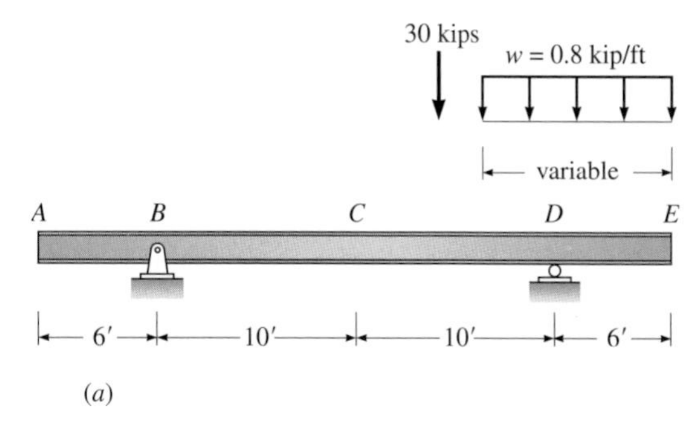

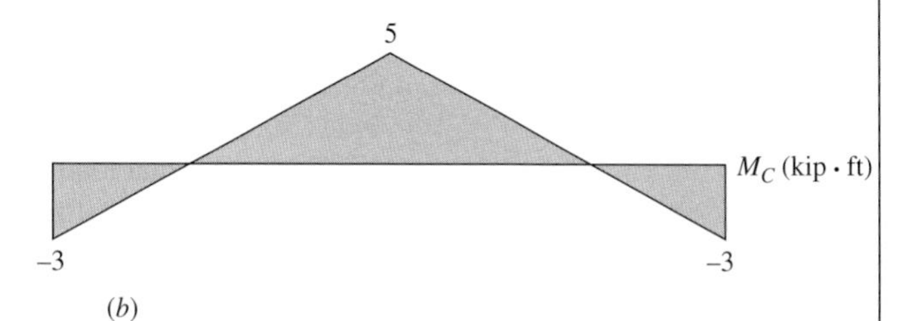

Solution

(*a*) To compute the maximum positive live load moment, we load the region of the beam where the ordinates of the influence line are positive (see Figure 8.15*c*). The concentrated load is positioned at the maximum positive ordinate of the influence line:

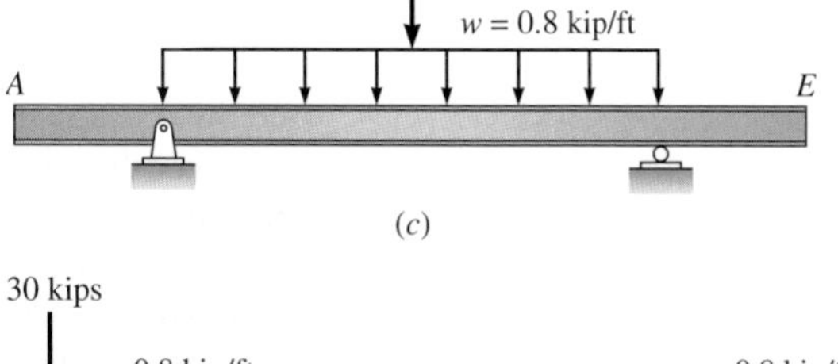

$$\text{Max.} + M_C = 30(5) + 0.8\left[\tfrac{1}{2}(20)5\right] = 190 \text{ kip} \cdot \text{ft}$$

(*b*) For maximum negative live load moment at C, we position the loads as shown in Figure 8.15(*d*). Because of symmetry, the same result occurs if the 30-kip load is positioned at E.

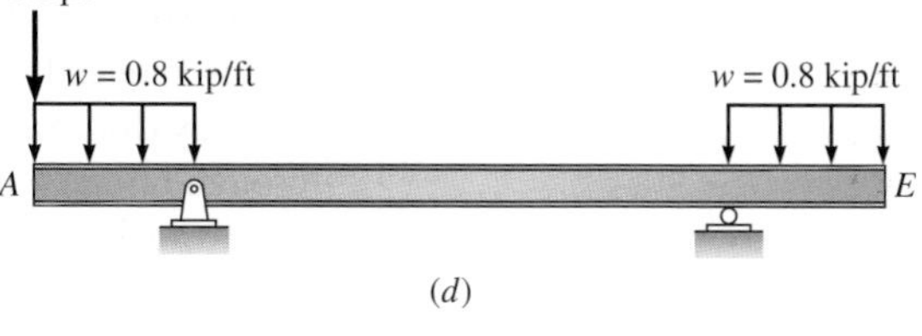

Figure 8.15: (*a*) Dimensions of beam with design live loads indicated at the left end; (*b*) influence line for moment at C; (*c*) position of live load to maximize positive moment at C; (*d*) position of live load to maximize negative moment at C. Alternately, the 30-kip load could be positioned at E.

$$\text{Max.} - M_C = (30 \text{ kips})(-3) + 0.8\left[\tfrac{1}{2}(6)(-3)\right](2) = -104.4 \text{ kip} \cdot \text{ft}$$

(*c*) For the moment at C due to dead load, multiply the area under the entire influence line by the magnitude of the dead load.

$$M_C = 0.45\left[\tfrac{1}{2}(6)(-3)\right](2) + 0.45\left[\tfrac{1}{2}(20)5\right]$$

$$= -8.1 + 22.5 = +14.4 \text{ kip} \cdot \text{ft}$$

8.6 Influence Lines for Girders Supporting Floor Systems

Figure 8.16*a* shows a schematic drawing of a structural framing system commonly used to support a bridge deck. The system is composed of three types of beams: stringers, floor beams, and girders. To show the main flexural members clearly, we simplify the sketch by omitting the deck, cross-bracing, and connection details between members.

In this system a relatively flexible slab is supported on a series of small longitudinal beams—the stringers—that span between transverse floor beams. Stringers are typically spaced about 8 to 10 ft apart. The thickness of the slab depends on the spacing between stringers. If the span of the slab is reduced by spacing the stringers close together, the designer can reduce the depth of the slab. As the spacing between stringers increases, increasing the span of the slab, the slab depth must be increased to carry larger design moments and to limit deflections.

The load from the stringers is transferred to the floor beams, which in turn transmit that load together with their own weight to the girders. In the case of a steel bridge, if the connections of both the stringers to the floor beams and the floor beams to the girders are made with standard steel clip angles, we assume that the connections can transfer only vertical load (no moment) and treat them (the connections) as simple supports. Except for the weight of the girder, all loads are transferred into the girders by the floor beams. The points at which the floor beams connect to the girders are termed *panel points*.

In a deck-type bridge, the roadway is positioned at the top of the girders (see the cross section in Figure 8.16*b*). In this configuration it is possible to cantilever the slab beyond the girders to increase the width of the roadway. Often the cantilevers support pedestrian walkways. If the floor beams are positioned near the bottom flange of the girders (see Figure 8.16*c*)—a *half-through* bridge—the distance from the bottom of the bridge to the top of vehicles is reduced. If a bridge must run under a second bridge and over a highway (for example, at an intersection where three highways cross), a half-through bridge will reduce the required headroom.

To analyze the girder, it is modeled as shown in Figure 8.16*d*. In this figure the stringers are shown as simply supported beams. For clarity we often omit the rollers and pins under the stringers and just show them resting on the floor beams. Recognizing that the girder in Figure 8.16*d* actually represents both the girders in Figure 8.16*a*, we must make an additional computation to establish the proportion of the vehicle's wheel loads that is distributed to each girder. For example, if a single vehicle is centered between girders in the middle of the roadway, both girders will carry one-half the vehicle weight. On the other hand, if the resultant of the wheel loads is located at the quarter point of a floor beam, three-fourths of the load will go to the near girder and one-fourth to the far girder (see Figure 8.16*e*). Establishing the portion of the vehicle loads that go to each girder is a separate computation that we make after the influence lines are drawn.

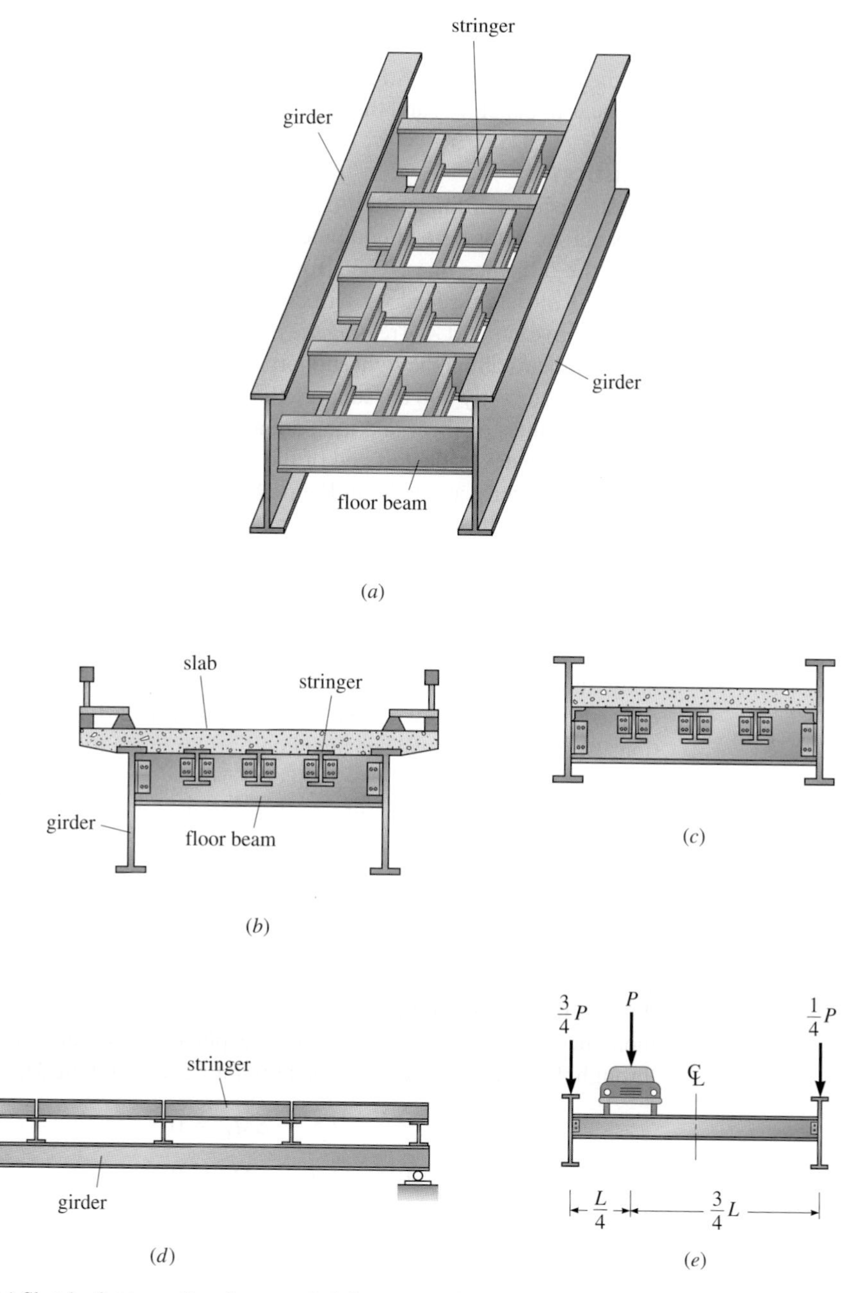

Figure 8.16: (*a*) Sketch of stringer, floor beam, and girder system; (*b*) deck bridge; (*c*) half-through bridge; (*d*) schematic representation of (*a*); (*e*) one lane loaded.

EXAMPLE 8.6

For the girder in Figure 8.17*a*, draw the influence lines for the reaction at *A*, the shear in panel *BC*, and the moment at *C*.

Solution

To establish the ordinates of the influence lines, we will move a unit load of 1 kN across stringers and compute the forces and reactions required to construct the influence lines. The arrows above the stringers denote the various positions of the unit load we will consider. We start with the unit load positioned above support *A*. Treating the entire structure as a rigid body, and summing moments about the right support, we compute $R_A = 1$ kN. Since the unit load passes directly into the support, the balance of the structure is unstressed. Thus the values of shear and moment at all points within the girder are zero, and the ordinates at the left end of the influence lines for shear V_{BC} and moment M_C are zero, as shown in Figure 8.17*c* and *d*.

To compute the ordinates of the influence lines at *B*, we next move the unit load to panel point *B*, and we compute $R_A = \frac{4}{5}$ kN (Figure 8.17*e*). Since the unit load is directly at the floor beam, 1 kN is transmitted into the girder at panel point *B* and the reactions at all floor beams are zero. To compute the shear in panel *BC*, we pass section 1 through the girder, producing the free body shown in Figure 8.17*e*. Following the convention for positive shear defined in Section 5.3, we show V_{BC} acting downward on the face of the section. To compute V_{BC}, we consider equilibrium of the forces in the *y* direction

$$\overset{+}{\uparrow} \quad \Sigma F_y = 0 = \tfrac{4}{5} - 1 - V_{BC}$$

$$V_{BC} = -\tfrac{1}{5} \text{ kN}$$

where the minus sign indicates that the shear is opposite in sense to that shown on the free body (Figure 8.17*e*).

To compute the moment at *C* with the unit load at *B*, we pass Section 2 through the girder, producing the free body shown in Figure 8.17*f*. Summing moments about an axis, normal to the plane of the member and passing through the centroid of the section at point *C*, we compute M_C.

$$\circlearrowright^{+} \quad \Sigma M_C = 0$$

$$\tfrac{4}{5}(12) - 1(6) - M_C = 0$$

$$M_C = \tfrac{18}{5} \text{ kN} \cdot \text{m}$$

We now shift the unit load to panel point *C* and compute $R_A = \frac{3}{5}$ kN. To compute V_{BC}, we consider equilibrium of the free body to the left of Section 1 (Figure 8.17*g*). Since the unit load is at *C*, no forces are applied to the girder by the floor beams at *A* and *B*, and the reaction at *A* is the only external force applied to the free body. Summing forces in the *y* direction gives us

$$\overset{+}{\uparrow} \quad \Sigma F_y = 0 = \tfrac{3}{5} - V_{BC} \qquad \text{and} \qquad V_{BC} = \tfrac{3}{5} \text{ kN}$$

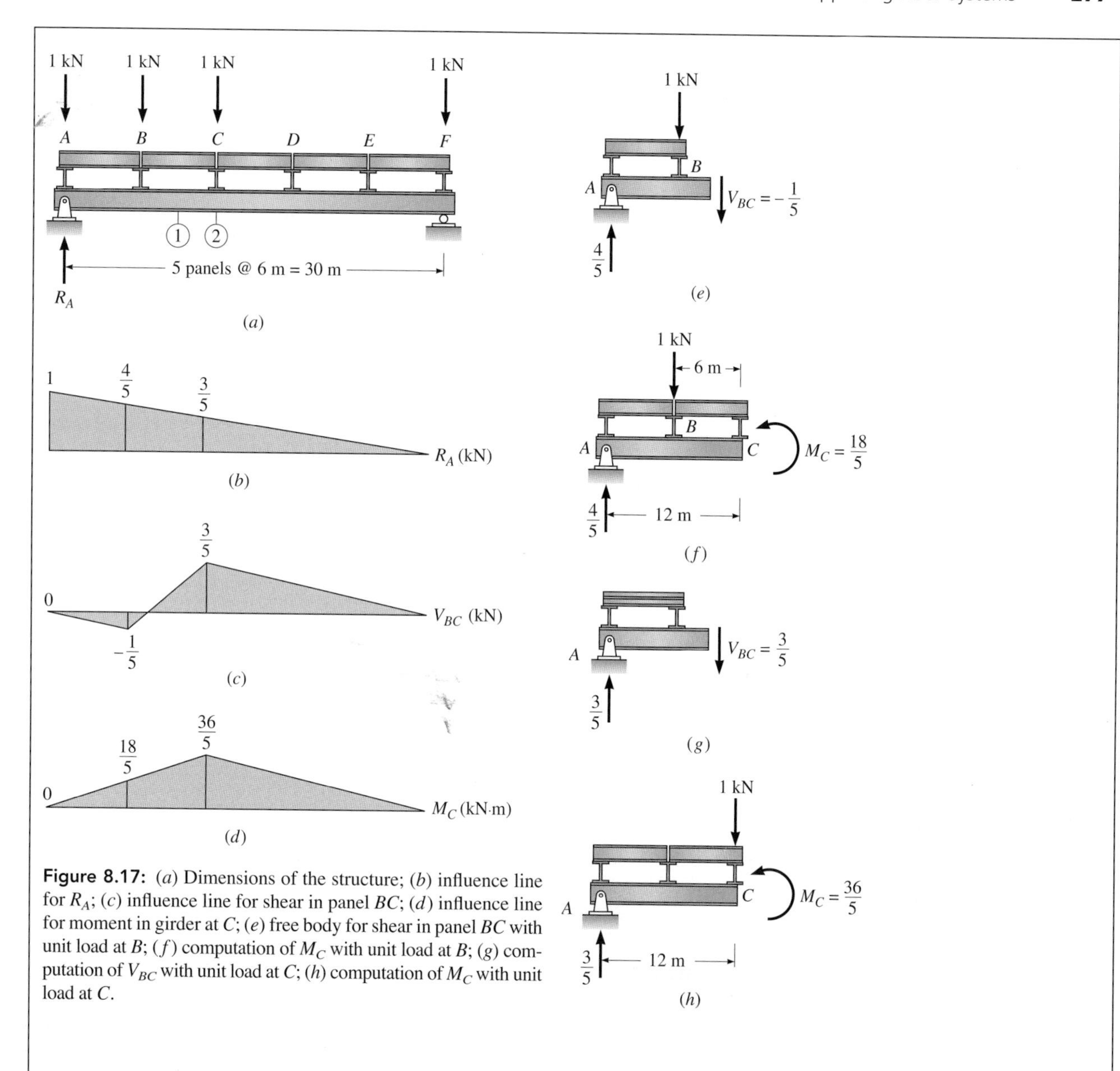

Figure 8.17: (*a*) Dimensions of the structure; (*b*) influence line for R_A; (*c*) influence line for shear in panel *BC*; (*d*) influence line for moment in girder at *C*; (*e*) free body for shear in panel *BC* with unit load at *B*; (*f*) computation of M_C with unit load at *B*; (*g*) computation of V_{BC} with unit load at *C*; (*h*) computation of M_C with unit load at *C*.

Using the free body in Figure 8.17*h*, we sum moments about *C* to compute $M_C = \frac{36}{5}$ kN·m.

When the unit load is positioned to the right of panel point *C*, the reactions of the floor beams on the free-body diagrams to the left of Sections 1 and 2 are zero (the reaction at *A* is the only external force). Since the reaction at *A* varies linearly as the load moves from point *C* to point *F*, V_{BC} and M_C—both linear functions of the reaction at *A*—also vary linearly, reducing to zero at the right end of the girder.

EXAMPLE 8.7

Construct the influence line for the bending moment M_C at point C in the girder shown in Figure 8.18*a*. The influence line for the support reaction R_G is given in Figure 8.18*b*.

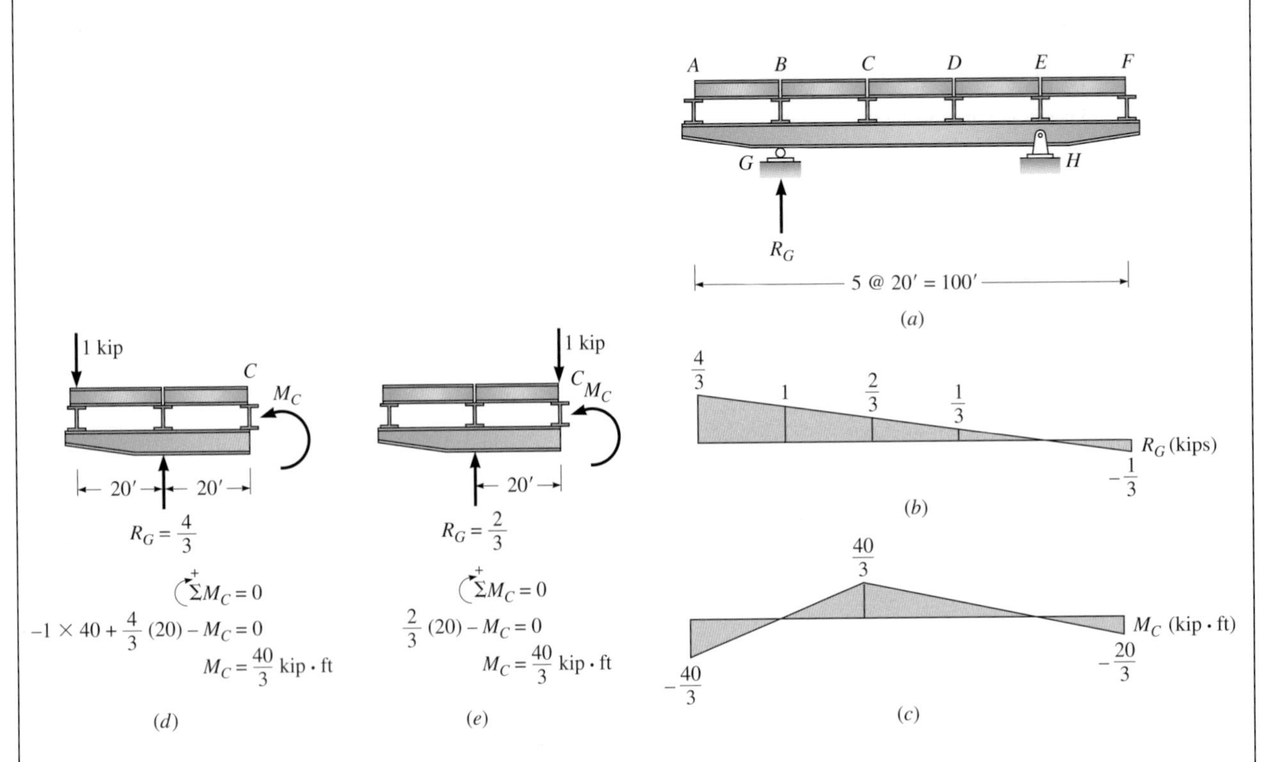

Figure 8.18: Influence lines for cantilever bridge girder. (*a*) details of floor system; (*b*) influence line for R_G; (*c*) influence line for M_C.

Solution

To establish the influence line showing the variation of M_C, we position the unit load at each panel point (the location of the floor beams). The moment in the girder is computed using a free body cut by passing a vertical plane through the floor system at point C. The value of the girder reaction R_G at the left support is read from the influence line for R_G shown in Figure 8.18*b*.

We can establish two points on the influence line without computation by observing that when the unit load is positioned over the girder supports at points B and E, the entire load passes directly into the supports, no stresses develop in the girder, and accordingly the moment on a section through point C is zero. The free bodies and the computation of M_C for the unit load at points A and C are shown in Figure 8.18*d* and *e*. The complete influence line for M_C is shown in Figure 8.18*c*. Again, we observe that the influence lines for a determinate structure are composed of straight lines.

EXAMPLE 8.8

Draw the influence line for the bending moment on a vertical section through point B on the girder (Figure 8.19a). At points A and F the connection of the stringers to the floor beam is equivalent to a pin. At points B and E, the connections of the stringers to the floor beam are equivalent to a roller. The influence line for the reaction at A is given in Figure 8.19b.

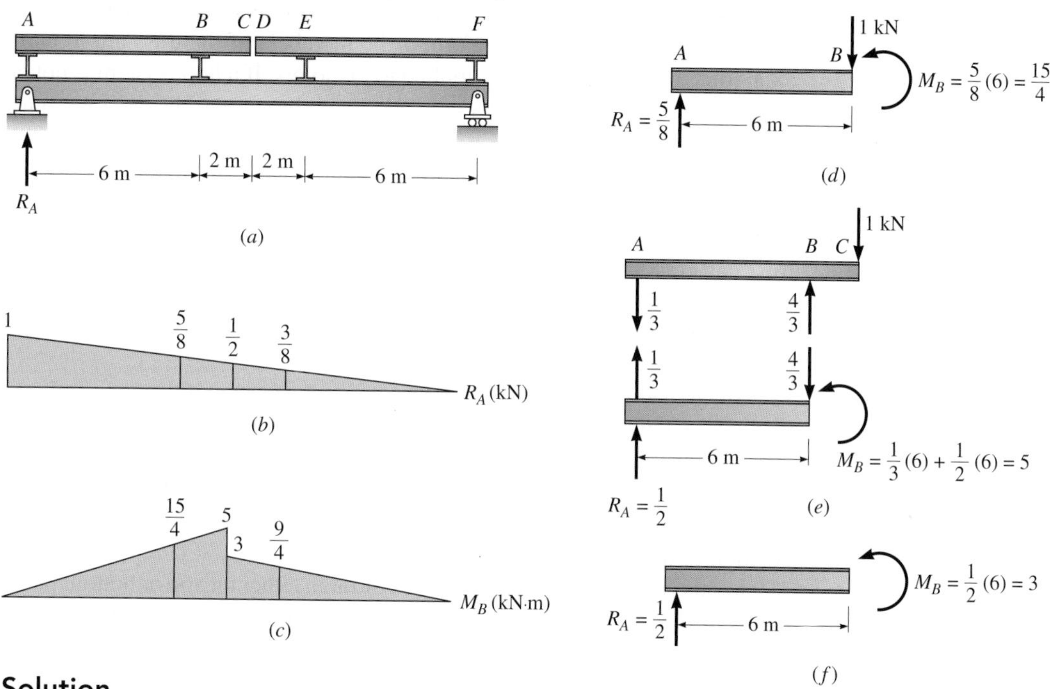

Figure 8.19: Influence lines for bridge girder loaded by stringers with cantilevers.

Solution

When the unit load is positioned at point A, the entire load passes directly through the floor beam into the pin support at point A. Since no stresses develop in sections of the girder away from the support, the bending moment on the section at point B is zero.

We next move the unit load to point B, producing a reaction R_A of $\frac{5}{8}$ kN (Figure 8.19b). Summing moments of the applied loads, about the section at point B, we compute $M_B = \frac{15}{4}$ kN·m (Figure 8.19d).

Next, the unit load is moved to point C, the tip of the cantilever, producing the stringer reactions shown in Figure 8.19e. The forces on the girder are equal in magnitude to the reactions on the stringer but directed in the opposite direction. Again summing moments about the vertical section at point B, we compute $M_B = 5$ kN·m. When the unit load is moved an infinitesimal distance across the gap to point D at the tip of the cantilever on the right, stringer ABC is no longer loaded; however, the reaction at A, the only force acting on the free body of the girder to the left of section B, remains equal to $\frac{1}{2}$ kN. We now sum moments about B and find that M_B has reduced to 3 kN·m (Figure 8.19f). As the unit load moves from point D to point F, computations show that the moment at Section B reduces linearly to zero.

8.7 Influence Lines for Trusses

Since truss members are typically designed for axial force, their cross sections are relatively small because of the efficient use of material in direct stress. Because a truss member with a small cross section bends easily, transverse loads applied directly to the member between its joints would produce excessive flexural deflections. Therefore, if the members of the truss are to carry axial force only, loads must be applied to the joints. If a floor system is not an integral part of the structural system supported by a truss, the designer must add a set of secondary beams to carry load into the joints (see Figure 8.20). These members, together with light diagonal bracing in the top and bottom planes, form a rigid horizontal truss that stabilizes the main vertical truss and prevents its compression chord from buckling laterally. Although an isolated truss has great stiffness in its own plane, it has no significant lateral stiffness. Without the lateral bracing system, the compression chord of the truss would buckle at a low level of stress, limiting the capacity of the truss for vertical load.

Since load is transmitted to a truss through a system of beams similar to those shown in Figure 8.16*a* for girders supporting a floor system, the procedure to construct influence lines for the bars of a truss is similar to that for a girder with a floor system; that is, the unit load is positioned at successive panel points, and the corresponding bar forces are plotted directly below the position of the load.

Loads can be transmitted to trusses through either the top or bottom panel points. If load is applied to the joints of the top chord, the truss is known as a *deck* truss. Alternatively, if load is applied to the bottom chord panel points, the truss is termed a *through* truss.

Construction of Influence Lines for a Truss

To illustrate the procedure for constructing influence lines for a truss, we will compute the ordinates of the influence lines for the reaction at A and for bars BK, CK, and CD of the truss in Figure 8.21*a*. In this example we will assume that load is transmitted to the truss through the lower chord panel points.

We begin by constructing the influence line for the reaction at A. Since the truss is a rigid body, we compute the ordinate of the influence line at any panel point by placing the unit load at that point and summing moments about an axis through the right support. The computations show that the influence line for the reaction at A is a straight line whose ordinates vary from 1 at the left support to zero at the right support (see Figure 8.21*b*). This example shows that the influence lines for the support reactions of simply supported beams and trusses are identical.

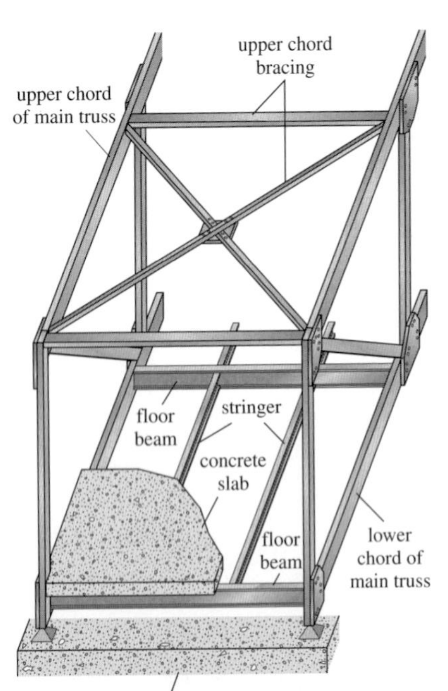

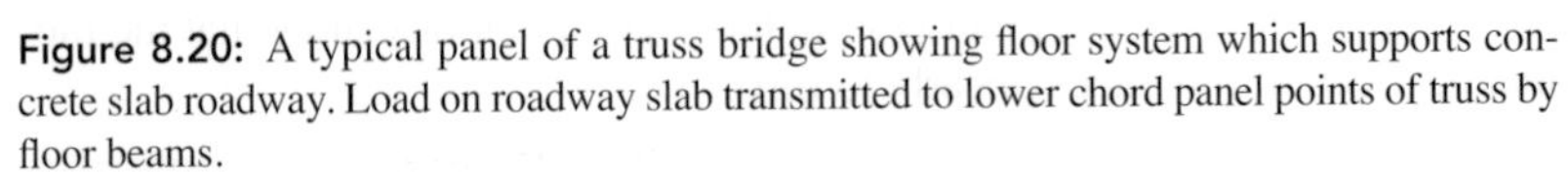

Figure 8.20: A typical panel of a truss bridge showing floor system which supports concrete slab roadway. Load on roadway slab transmitted to lower chord panel points of truss by floor beams.

To construct the influence line for the force in bar BK, we apply the unit load to a panel point and then determine the force in bar BK by analyzing a free body of the truss cut by a vertical section passing through the second panel of the truss (see Section 1 in Figure 8.21a). Figure 8.22a shows the free body of the truss to the left of Section 1 when the unit load is at the first panel

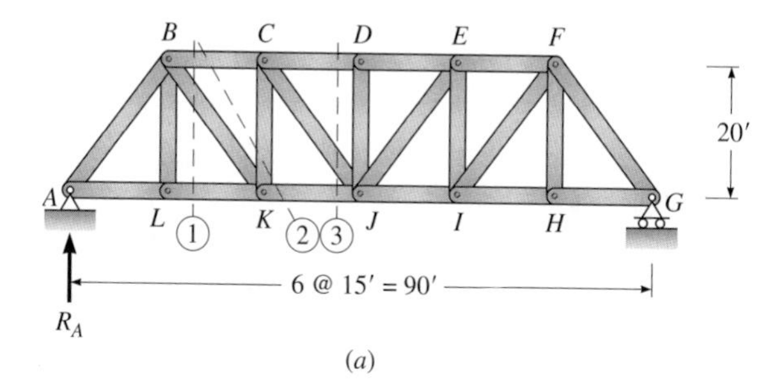

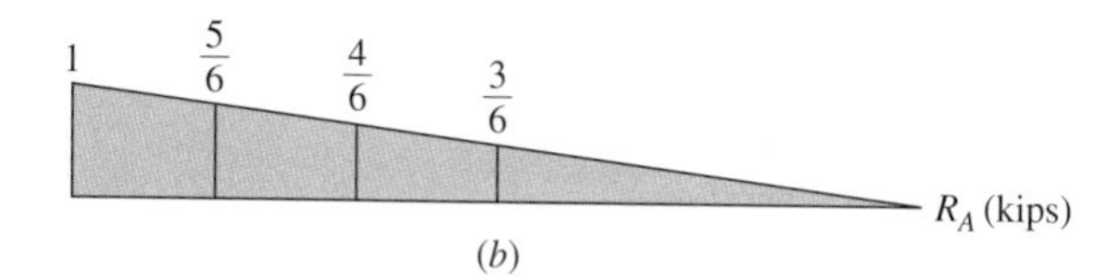

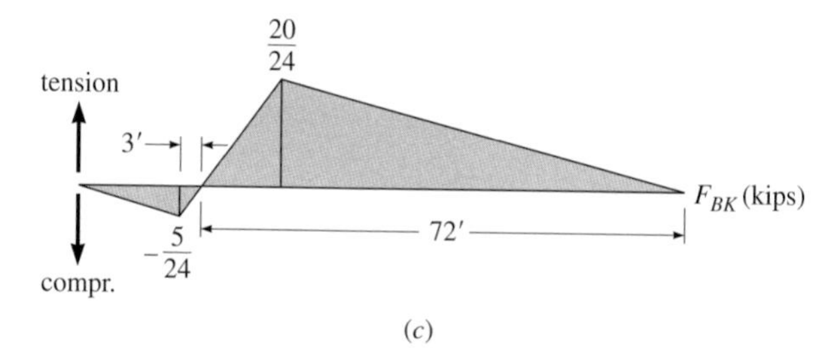

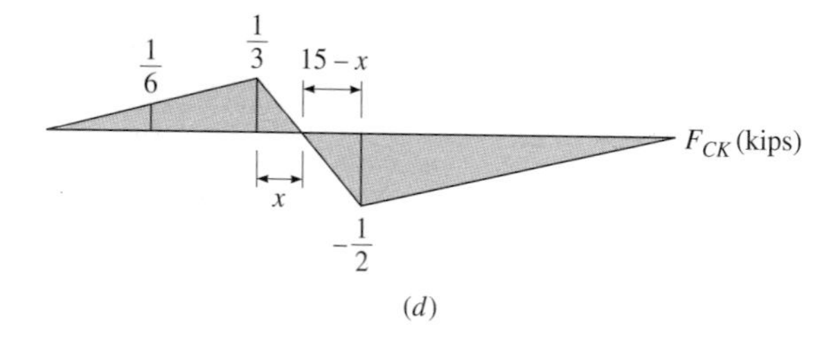

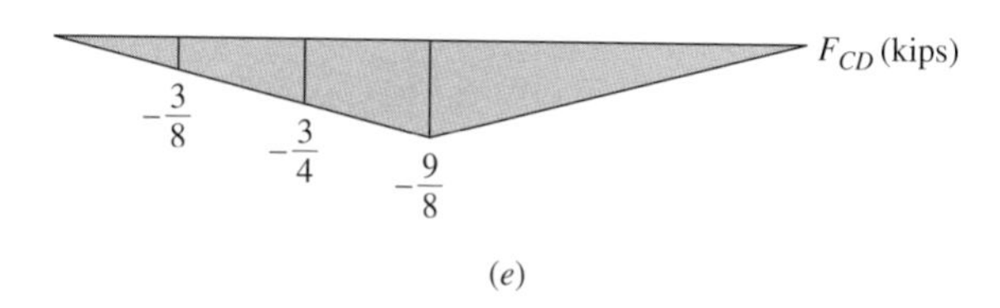

Figure 8.21: Influence lines for truss: (a) details of truss; (b) influence line for reaction at A; (c) influence line for bar BK; (d) influence line for bar CK; (e) influence line for bar CD.

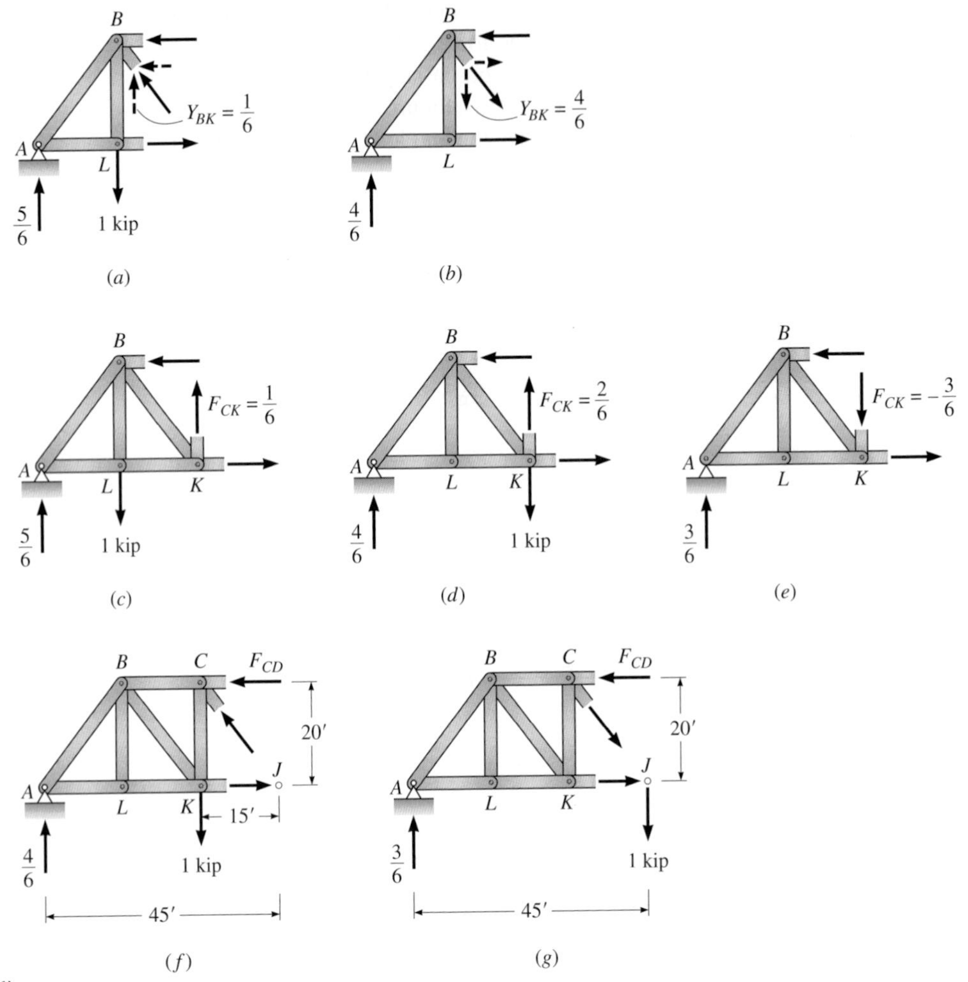

Figure 8.22: Free-body diagrams to construct influence lines.

point. By summing forces in the y direction, we compute the vertical component Y_{BK} of the force in bar BK.

$$\overset{+}{\uparrow} \quad \Sigma F_y = 0$$

$$\tfrac{5}{6} - 1 + Y_{BK} = 0$$

$$Y_{BK} = \tfrac{1}{6} \text{ kip (compression)}$$

Since the sides of the slope triangle of the bar are in a ratio of 3:4:5, we compute F_{BK} by simple proportion.

$$\frac{F_{BK}}{5} = \frac{Y_{BK}}{4}$$

$$F_{BK} = \frac{5}{4} Y_{BK} = \frac{5}{24} \text{ kip}$$

Because F_{BK} is a compression force, we plot it as a *negative* influence line ordinate (see Figure 8.21*c*).

Figure 8.22*b* shows the free body to the left of Section 1 when the unit load acts at joint *K*. Since the unit load is no longer on the free body, the vertical component of force in bar *BK* must equal $\frac{4}{6}$ kip and act downward to balance the reaction at support *A*. Multiplying Y_{BK} by $\frac{5}{4}$, we compute a tensile force F_{BK} equal to $\frac{20}{24}$ kip. Since the reaction of *A* reduces linearly to zero as the unit load moves to the right support, the influence line for the force in bar *BK* must also reduce linearly to zero at the right support.

To evaluate the ordinates of the influence line for the force in bar *CK*, we will analyze the free body of the truss to the left of Section 2, shown in Figure 8.21*a*. Figure 8.22*c*, *d*, and *e* shows free bodies of this section for three successive positions of the unit load. The force in the bar *CK*, which changes from tension to compression as the unit load moves from panel point *K* to *J*, is evaluated by summing forces in the *y* direction. The resulting influence line for bar *CK* is shown in Figure 8.21*d*. To the right of point *K* the distance at which the influence line passes through zero is determined by similar triangles:

$$\frac{\frac{1}{3}}{x} = \frac{\frac{1}{2}}{15 - x}$$

$$x = 6 \text{ ft}$$

The influence line for the force in bar *CD* is computed by analyzing a free body of the truss cut by a vertical section through the third panel (see Section 3 in Figure 8.21*a*). Figure 8.22*f* shows a free body of the truss to the left of Section 3 when the unit load is at panel point *K*. The force in *CD* is evaluated by summing moments about the intersection of the other two bar forces at *J*.

$$\circlearrowright^{+} \quad \Sigma M_J = 0$$

$$\tfrac{4}{6}(45) - 1(15) - F_{CD}(20) = 0$$

$$F_{CD} = \tfrac{3}{4} \text{ kip (compression)}$$

Figure 8.22*g* shows the free body of the truss to the left of Section 3 when the unit load is at joint *J*. Again we evaluate F_{CD} by summing moments about *J*.

$$\circlearrowright^{+} \quad \Sigma M_J = 0$$

$$0 = \tfrac{3}{6}(45) - F_{CD}(20)$$

$$F_{CD} = \tfrac{9}{8} \text{ kips (compression)}$$

The influence line for bar *CD* is shown in Figure 8.21*e*.

Influence Lines for a Trussed Arch

As another example, we will construct the influence lines for the reactions at *A* and for the forces in bars *AI*, *BI*, and *CD* of the three-hinged trussed arch in Figure 8.23*a*. The arch is constructed by joining two truss segments with a pin

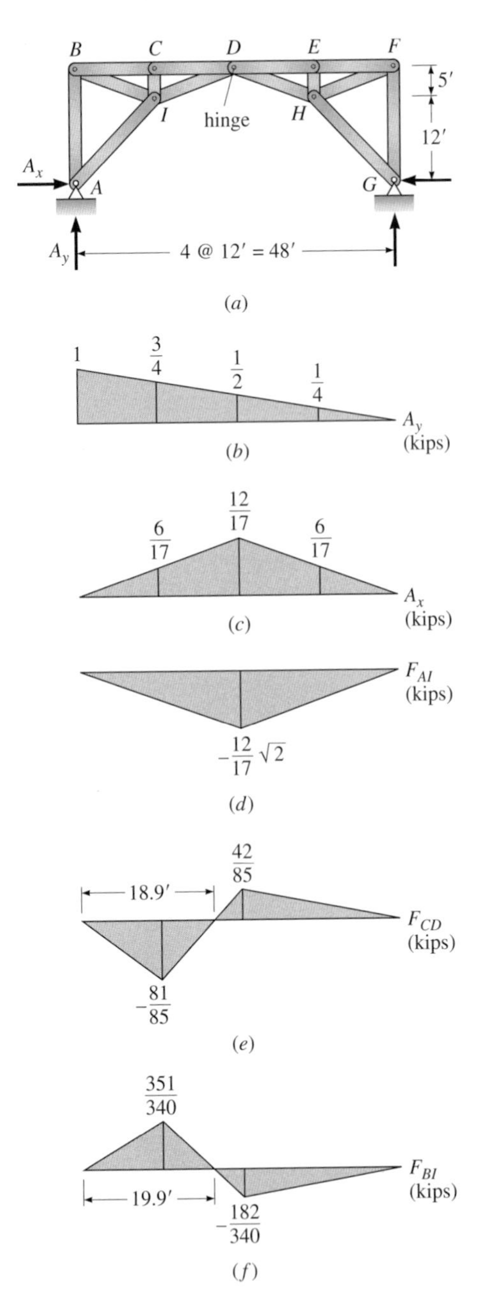

Figure 8.23: Influence lines for a trussed arch: (*a*) truss details; (*b*) reaction A_y; (*c*) reaction A_x; (*d*) force in bar *AI*; (*e*) force in bar *CD*; (*f*) force in bar *BI*.

at midspan. We assume that loads are transmitted through the upper chord panel points.

To begin the analysis, we construct the influence line for A_y, the vertical reaction at *A*, by summing moments of forces about an axis through the pin support at *G*. Since the horizontal reactions at both supports pass through *G*, the computations for the ordinates of the influence line are identical to those of a simply supported beam. The influence line for A_y is shown in Figure 8.23*b*.

Now that A_y is established for all positions of the unit load, we next compute the influence line for A_x, the horizontal reaction at *A*. In this computation we will analyze a free body of the truss to the left of the center hinge at point *D*. For example, Figure 8.24*a* shows the free body used to compute A_x when the unit load is positioned at the second panel point. By summing moments about the hinge at *D*, we write an equation in which A_x is the only unknown.

$$\circlearrowleft^{+} \quad M_D = 0$$

$$0 = \tfrac{3}{4}(24) - A_x(17) - 1(12)$$

$$A_x = \tfrac{6}{17} \text{ kip}$$

The complete influence line for A_x is shown in Figure 8.23*c*.

To evaluate the axial force in bar *AI*, we isolate the support at *A* (see Figure 8.24*b*). Since the horizontal component of the force in bar *AI* must equal A_x, the ordinates of the influence line for *AI* will be proportional to those of A_x. Because bar *AI* is on a slope of 45°, $F_{AI} = \sqrt{2}X_{AI} = \sqrt{2}A_x$. The influence line for F_{AI} is shown in Figure 8.23*d*.

Figure 8.24*c* shows the free body used to determine the influence line for the force in bar *CD*. This free body is cut from the truss by a vertical section through the center of the second panel. Using the values of A_x and A_y from the influence lines in Figure 8.23*b* and *c*, we can solve for the force in bar *CD* by summing moments about a reference axis through joint *I*. Plotting the ordinates of F_{CD} for various positions of the unit load, we draw the influence line shown in Figure 8.23*e*.

To determine the force in bar *BI*, we consider a free body of the truss to the left of a vertical section passing through the first panel (see Figure 8.24*d*). By summing moments of the forces about an axis at point *X* (the intersection of the lines of action of the forces in bars *AI* and *BC*), we can write a moment equation in terms of the force F_{BI}. We can further simplify the computation by extending force F_{BI} along its line of action to joint *B* and resolving the force into rectangular components. Since X_{BI} passes through the moment center at point *X*, only the *y-component* of F_{BI} appears in the moment equation. From the slope relationship, we can express F_{BI} as

$$F_{BI} = \tfrac{13}{5} Y_{BI}$$

The influence line for F_{BI} is plotted in Figure 8.23*f*.

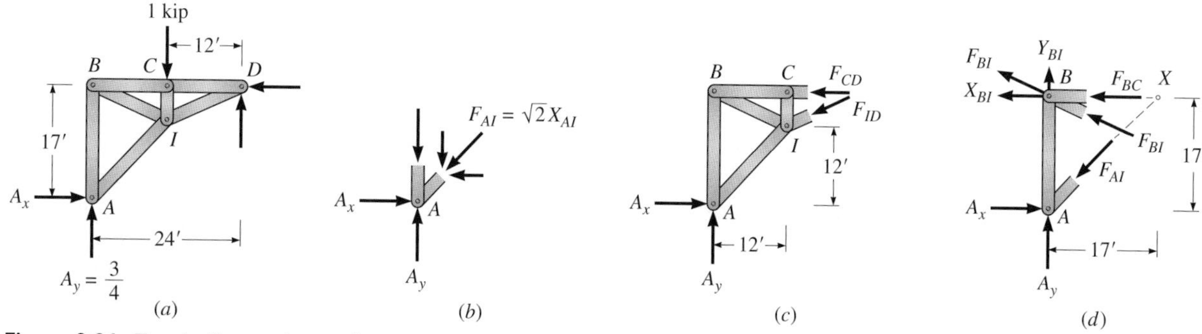

Figure 8.24: Free bodies used to analyze the three-hinged arch in Figure 8.23*a*.

8.8 Live Loads for Highway and Railroad Bridges

In Section 8.5 we established how to use an influence line to evaluate the force at a section produced by either a uniformly distributed or a concentrated live load. We now will extend the discussion to include establishing the maximum force at a section produced by a set of moving loads such as those applied by the wheels of a truck or train. In this section we describe briefly the characteristics of the live loads (the standard trucks and trains) for which highway and railroad bridges are designed. In Section 8.9 we describe the increase–decrease method for positioning the wheel loads.

Highway Bridges

The live loads for which highway bridges in the United States must be designed are specified by the American Association of State Highway and Transportation Officials (AASHTO). At present major highway bridges must be designed to carry in *each lane* either the standard 72-kip six-wheel HS 20-44 truck shown in Figure 8.25*a* or a lane loading consisting of the uniformly distributed and concentrated loads shown in Figure 8.25*b*. The forces produced by a standard truck usually control the design of members whose spans are less than 145 ft. When spans exceed 145 ft, the forces created by a lane loading generally exceed those produced by a standard truck. If a bridge is to be constructed over a secondary road and only light vehicles are expected to traverse the bridge, the standard truck and lanes loads can be reduced by either 25 or 50 percent, depending on the anticipated weight of vehicles. These reduced vehicle loads are termed HS 15 and HS 10 loadings, respectively.

Although not used extensively by engineers, the AASHTO code also specifies a lighter (40 kips) *four-wheeled* HS 20 truck for secondary-road bridges that do not carry heavy trucks. Since a bridge will often have a life of 50 to 100 years or even more, and since it is difficult to predict the types of vehicles that will use a particular bridge in the future, use of live load based on a heavier truck may be prudent. Moreover, because a heavier truck also results in thicker members, the useful life of bridges that are subject to corrosion from salting or acid rain will be longer than those designed for lighter trucks.

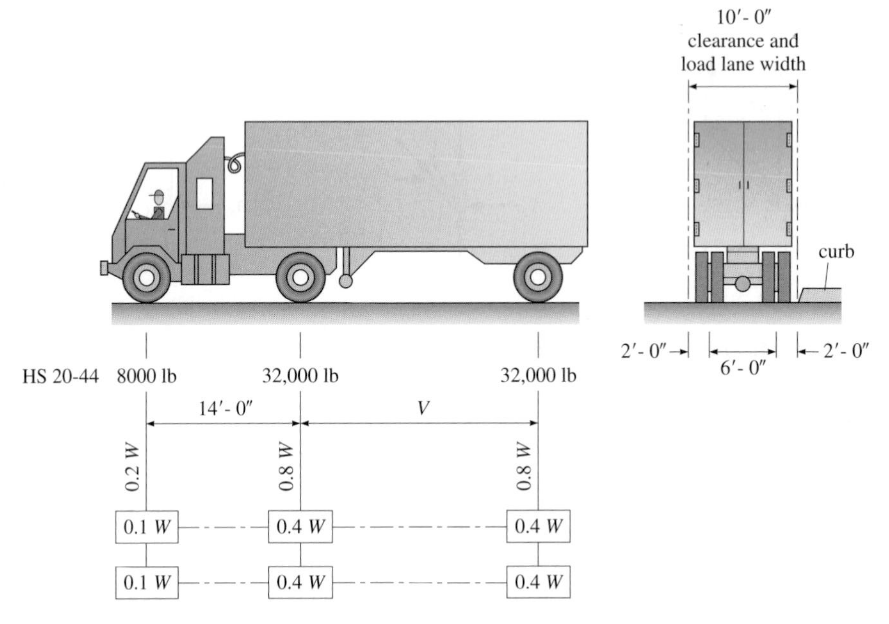

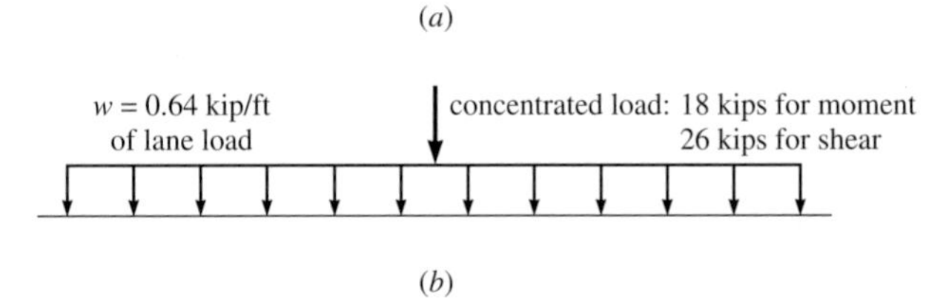

Figure 8.25: Lane loads used to design highway bridges; (*a*) standard 72-kip, HS 20-44 truck; or (*b*) uniform load plus concentrated load which is positioned to maximize force in structure.

Although the distance between the front and middle wheels of the standard HS truck (see Figure 8.25*a*) is fixed at 14 ft, the designer is free to set a value of *V* between 14 and 30 ft for the spacing between the middle and rear wheels. The wheel spacing the designer selects should maximize the value of the design force being computed. In all designs, the engineer should consider the possibility of the truck moving in either direction across the span.

Although it might seem logical to consider two or more trucks acting on the span of bridges spanning 100 ft or more, the AASHTO specifications require only that the designer consider a single truck or, alternatively, the lane loading. Although highway bridges fail occasionally because of deterioration, faulty construction, material defects, and so forth, no recorded cases exist of bridge failures from overstress when the members have been sized for either an HS 15 or an HS 20 truck.

Railroad Bridges

The design loads for railroad bridges are contained in the specifications of the American Railway Engineering and Maintenance of Way Association (AREMA). The AREMA specifications require that bridges be designed for a train composed of two engines followed by a line of railroad cars. As shown in Figure 8.26, the wheels of the engines are represented by concentrated loads and the railroad cars by a uniformly distributed load. The live load representing the weight of trains is

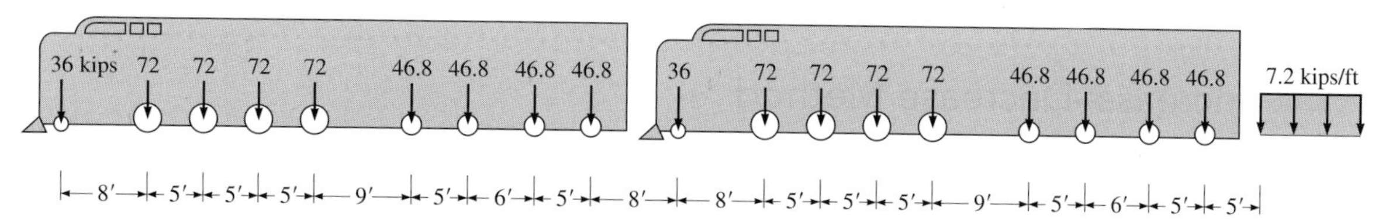

Figure 8.26: Cooper E-72 train for design of railroad bridges (wheel loads in kips).

specified in terms of a Cooper E loading. Most bridges today are designed for the Cooper E-72 loading shown in Figure 8.26. The number 72 in the Cooper designation represents the axle load in units of kips applied by the main drive wheels of the locomotive. Other Cooper loadings are also used. These loadings are proportional to those of the Cooper E-72. For example, to establish a Cooper E-80 loading, all forces in Figure 8.26 should be multiplied by the ratio 80/72.

Impact

If you have traveled by truck or car, you probably recognize that moving vehicles bounce up and down as they move over a roadway—springs are supplied to dampen these oscillations. The vertical motion of a vehicle is a function of the roughness of the roadway surface. Bumps, an uneven surface, expansion joints, potholes, spalls, and so forth all contribute to vertical sinusoidal motion of the vehicle. The downward vertical movement of the vehicle's mass increases the force applied to the bridge through the wheels. Since the dynamic force, a function of the natural periods of both the bridge and the vehicle, is difficult to predict, we account for it by increasing the value of the live load stresses by an impact factor I. For highway bridges the AASHTO specifications require that for a particular member

$$I = \frac{50}{L + 125} \quad \text{but not more than } 0.3 \tag{8.7}$$

where L is the length in feet of the section of span that must be loaded to produce the maximum stress in a particular member.

For example, to compute the impact factor for the tension force in member BK of the truss in Figure 8.21a, we use the influence line in Figure 8.21c to establish $L = 72$ ft (the length of the region in which the ordinates of the influence line are positive). Substituting this length into the equation for I, we compute

$$I = \frac{50}{72 + 125} = 0.254$$

Therefore, the force in bar BK produced by the live load must be multiplied by 1.254 to establish the total force due to live load and impact.

If we were computing the maximum live load compression force in bar BK, the impact factor would change. As indicated by the influence line in Figure 8.21c, compression is created in the bar when load acts on the truss over a distance of 18 ft to the right of support A. Substituting $L = 18$ ft into the impact equation, we compute

$$I = \frac{50}{18 + 125} = 0.35 \qquad (0.3 \text{ control})$$

Since 0.35 exceeds 0.3, we use the upper limit of 0.3.

The dead load stresses are not increased by the impact factor. Other bridge codes have similar equations for impact.

8.9 Increase–Decrease Method

In Section 8.5 we discussed how to use an influence line to evaluate the maximum value of a function when the live load is represented by either a single concentrated load or a uniformly distributed load. We now want to extend the discussion to include maximizing a function when the live load consists of a set of concentrated loads *whose relative position is fixed*. Such a set of loads might represent the forces exerted by the wheels of a truck or a train.

In the increase–decrease method, we position the set of loads on the structure so that the leading load is located at the maximum ordinate of the influence line. For example, in Figure 8.27 we show a beam that is to be designed to carry a live load applied by five wheels. To begin the analysis, we imagine that the loads have been moved onto the structure so that force F_1 is directly below the maximum ordinate y of the influence line. In this case the last load F_5 is not on the structure. We make no computations at this stage.

We now shift the entire set of loads forward a distance x_1 so that the second wheel is located at the maximum ordinate of the influence line. As a result of the shift, the value of the function (represented by the influence line) changes. The contribution of the first wheel F_1 to the function decreases (i.e., at the new location the ordinate of the influence line y' is smaller than the former ordinate y). On the other hand, the contribution of F_2, F_3, and F_4 increases because they have moved to a position where the ordinates of the influence line are larger. Since wheel F_5 is now on the structure, it too stresses the member. If the net change is a *decrease* in the value of the function, the first position of the loads is more critical than the second position, and we can evaluate the function by multiplying the loads in position 1 (see Figure 8.27*c*) by the corresponding ordinates of the influence line (that is, F_1 is multiplied by y). However, if the shift of loads to position 2 (see Figure 8.27*d*) produces an *increase* in value of the function, the second position is more critical than the first.

To ensure that the second position is the most critical, we will shift all loads forward again a distance x_2 so that force F_3 is at the maximum ordinate (see Figure 8.27*e*). We again compute the change in magnitude of the function produced by the shift. If the function decreases, the previous position is critical. If the function increases, we again shift the loads. This procedure is continued until a shift of the loads results in a decrease in value of the function. Once we secure this result, we establish that the previous position of the loads maximizes the function.

The change in value of the function produced by the movement of a particular wheel equals the difference between the product of the wheel load and the ordinate of the influence line in the two positions. For example, the change in the function Δf due to wheel F_1 as it moves forward a distance x_1 equals

$$\Delta f = F_1 y - F_1 y'$$

$$\Delta f = F_1(y - y') = F_1(\Delta y) \tag{8.8}$$

where the difference in ordinates of the influence line $\Delta y = y - y'$.

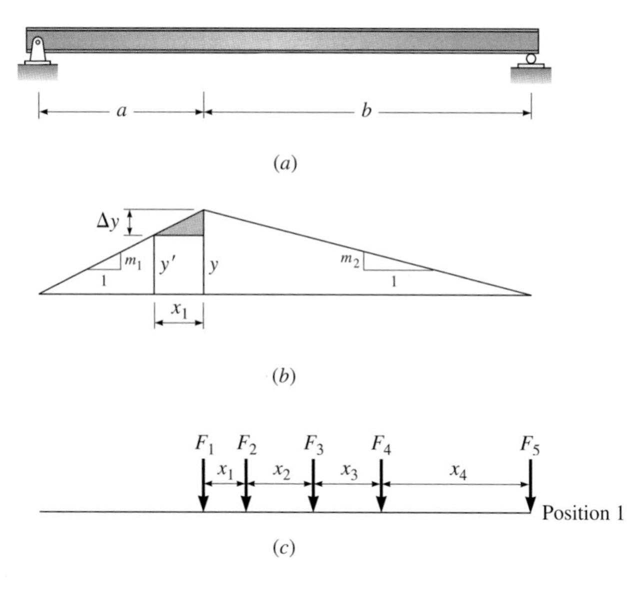

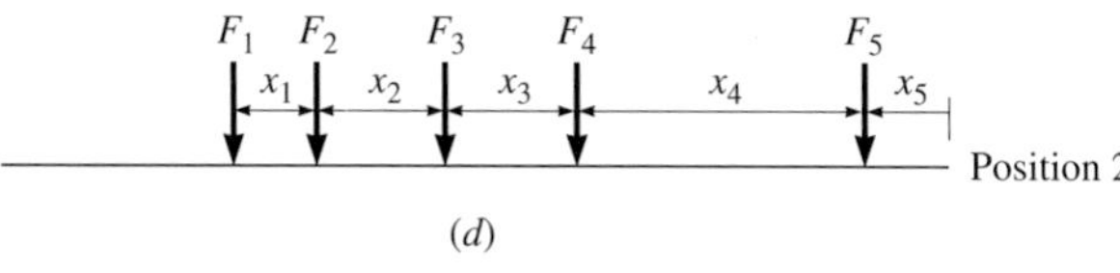

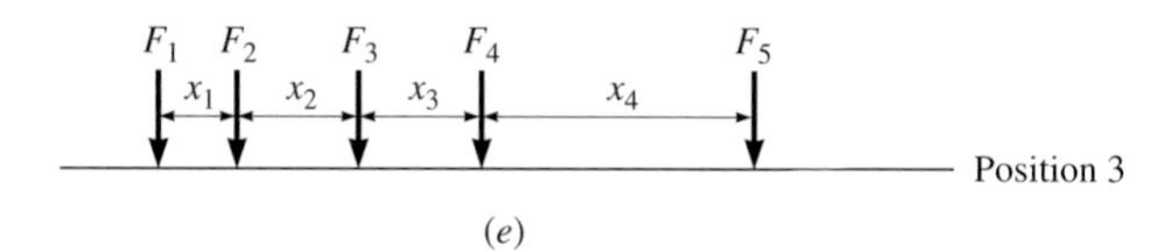

Figure 8.27: Increase–decrease method for establishing the maximum values of a function produced by a set of concentrated live loads. (*a*) Beam. (*b*) Influence line for some function whose maximum ordinate equals y. (*c*) Position 1: the first wheel load F_1 is located at maximum ordinate y. (*d*) In position 2: all wheel loads moved forward a distance x_1, bringing wheel F_2 up to the maximum ordinate. (*e*) Position 3: all wheels moved forward a distance x_2, bringing wheel F_3 up to the maximum ordinate.

If m_1 is the slope of the influence line in the region of the shift, we can express Δy as a function of the slope and the magnitude of the shift by considering the proportions between the slope triangle and the shaded area shown in Figure 8.27*b*:

$$\frac{\Delta y}{x_1} = \frac{m_1}{1}$$

$$\Delta_y = m_1 x_1 \tag{8.9}$$

Substituting Equation 8.9 into Equation 8.8 gives

$$\Delta f = F_1 m_1 x_1 \tag{8.10}$$

where the slope m_1 can be negative or positive and F_1 is the wheel load.

If a load moves on or off the structure, its contribution Δf to the function would be evaluated by substituting the actual distance it moves into Equation 8.10. For example, the contribution of force F_5 (see Figure 8.27*d*) as it moves on to the structure would be equal to

$$\Delta f = F_5 m_2 x_5$$

where x_5 is the distance from the end of the beam to load F_5. The increase–decrease method is illustrated in Example 8.9.

EXAMPLE 8.9

The 80-ft bridge girder in Figure 8.28*b* must be designed to support the wheel loads shown in Figure 8.28*a*. Using the increase–decrease method, determine the maximum value of moment at panel point *B*. The wheels can move in either direction. The influence line for moment at panel point *B* is given in Figure 8.28*b*.

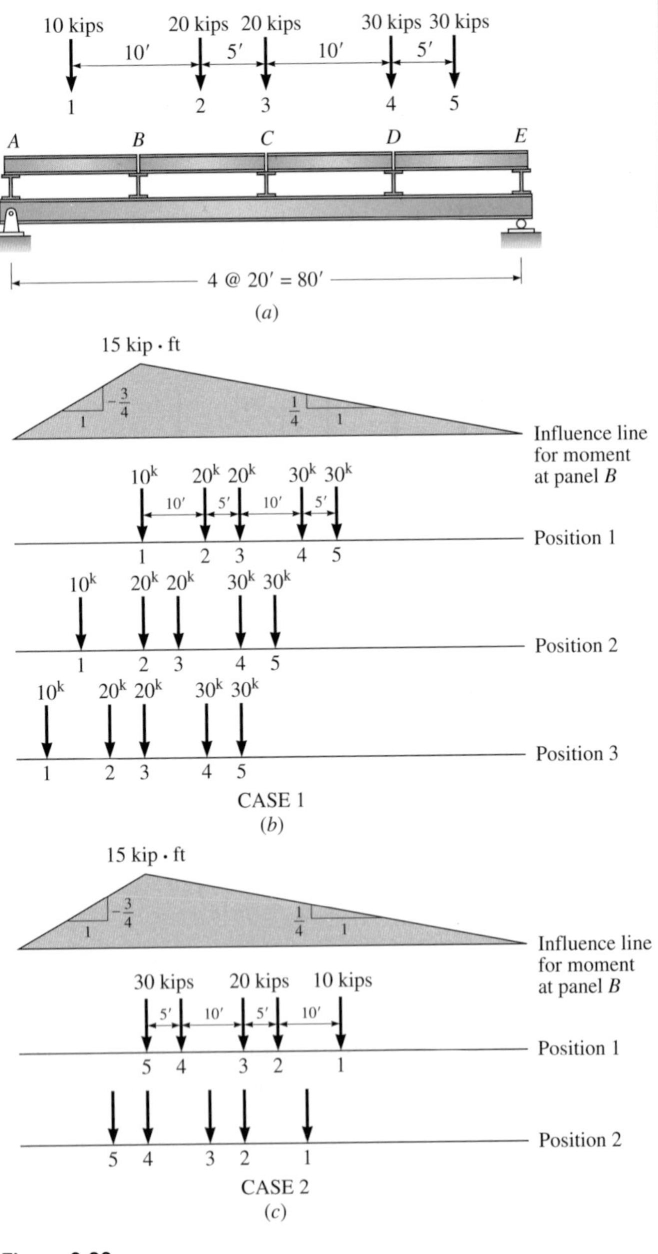

Figure 8.28

Solution

Case 1. *A 10-kip load moves from right to left*. Begin with the 10-kip load at panel B (see position in Figure 8.28b). Calculate the change in moment as all loads shift left 10 ft; that is, load 2 moves up to panel point B (see position 2). Use Equation 8.10.

$$\text{Increase in moment (loads 2, 3, 4, and 5)} = (20 + 20 + 30 + 30)\left(\frac{1}{4}\right)(10) = +250 \text{ kip}\cdot\text{ft}$$

$$\text{Decrease in moment (load 1)} = 10\left(-\frac{3}{4}\right)(10) = -75 \text{ kip}\cdot\text{ft}$$

$$\text{Net change} = +175 \text{ kip}\cdot\text{ft}$$

Therefore, position 2 is more critical than position 1.

Shift the loads again to determine if the moment continues to increase. Calculate the change in moment as the loads move 5 ft to the left to position 3; that is, load 3 moves up to panel point B.

$$\text{Increase in moment (loads 3, 4, and 5)} = (20 + 30 + 30)(5)\left(\frac{1}{4}\right) = +100.0 \text{ kip}\cdot\text{ft}$$

$$\text{Decrease in moment (loads 2 and 3)} = (10 + 20)(5)\left(-\frac{3}{4}\right) = -112.5 \text{ kip}\cdot\text{ft}$$

$$\text{Net change} = -12.5 \text{ kip}\cdot\text{ft}$$

Therefore, position 2 is more critical than position 3.

Evaluate the maximum moment at panel point B. Multiply each load by the corresponding influence line ordinate (number in parentheses).

$$\begin{aligned} M_B &= 10(7.5) + 20(15) + 20(13.75) + 30(11.25) + 30(10) \\ &= 1287.5 \text{ kip}\cdot\text{ft} \end{aligned}$$

Case 2. *The 30-kip load moves from right to left*. Begin with a 30-kip load at panel B (see position 1 in Figure 8.28c). Compute the change in moment as loads move 5 ft left to position 2.

$$\text{Increase in moment (loads 4, 3, 2, and 1)} = (80 \text{ kips})(5)\left(\frac{1}{4}\right) = +100.0 \text{ kip}\cdot\text{ft}$$

$$\text{Decrease in moment (load 5)} = (30 \text{ kips})(5)\left(-\frac{3}{4}\right) = -112.5 \text{ kip}\cdot\text{ft}$$

$$\text{Net change} = -12.5 \text{ kip}\cdot\text{ft}$$

Therefore, position 1 is more critical than position 2.

Compute the moment at panel point 2, using influence line ordinates.

$$\begin{aligned} M_B &= 30(15) + 30(13.75) + 20(11.25) + 20(10) + 10(7.5) \\ &= 1362.5 \text{ kip}\cdot\text{ft} \textit{ controls design} > 1287.5 \text{ kip}\cdot\text{ft} \end{aligned}$$

8.10 Absolute Maximum Live Load Moment

Case 1. Single Concentrated Load

A single concentrated load acting on a beam produces a triangular moment curve whose maximum ordinate occurs directly at the load. As a concentrated load moves across a simply supported beam, the value of the maximum moment directly under the load increases from zero when the load is at either support to $0.25PL$ when the load is at midspan. Figure 8.29*b*, *c*, and *d* shows the moment curves produced by a single concentrated load P for three loading positions, a distance $L/6$, $L/3$, and $L/2$ from the left support, respectively. In Figure 8.29*e*, the dashed line, termed the *moment envelope*, represents the maximum value of live load moment produced by the concentrated load that can develop at each section of the simply supported beam in Figure 8.29*a*. The moment envelope is established by plotting the ordinates of the moment curves in Figure 8.29*b* to *d*. Since a beam must be designed to carry the maximum moment at each section, the flexural capacity of the member must equal or exceed that given by the moment envelope (rather than by the moment curve shown in Figure 8.29*d*). The *absolute maximum live load moment* due to a single load on a simple beam occurs at midspan.

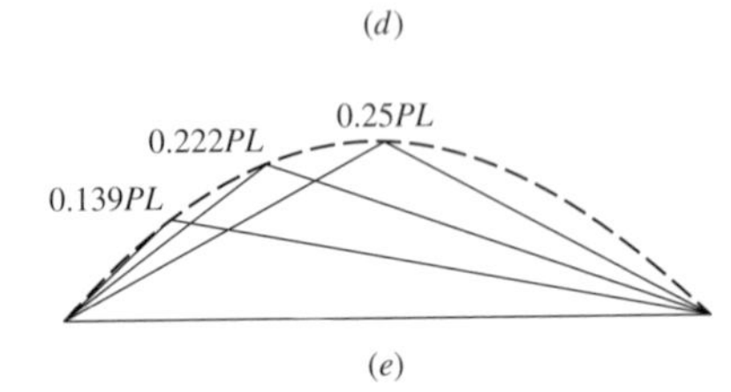

Figure 8.29: Moment envelope for a concentrated load on a simply supported beam: (*a*) four loading positions (*A* through *D*) considered for construction of moment envelope; (*b*) moment curve for load at point *B*; (*c*) moment curve for load at point *C*; (*d*) moment curve for load at point *D* (midspan); (*e*) moment envelope, curve showing maximum value of moment at each section.

Case 2. Series of Wheel Loads

The increase–decrease method provides a procedure to establish the maximum moment produced at an arbitrary section of a beam by a set of moving loads. To use this method, we must first construct the influence line for moment at the section where the moment is to be evaluated. Although we recognize that the maximum moment produced by a set of wheel loads will be larger for sections at or near midspan than for sections located near a support, thus far we have not established how to locate the *one* section in the span at which the wheel loads produce the greatest value of moment. To locate this section for a *simply supported beam* and to establish the value of the *absolute maximum moment* produced by a particular set of wheel loads, we will investigate the moment produced by the wheel loads acting on the beam in Figure 8.30. In this discussion we will assume that the resultant R of the wheel loads is located a distance d to the right of wheel 2. (The procedure to locate the resultant of a set of concentrated loads is covered in Example 3.2.)

Although we cannot specify with absolute certainty the wheel at which the maximum moment occurs, experience indicates that it will probably occur under one of the wheels adjacent to the resultant of the force system. From our experience with the moment produced by a single concentrated load, we recognize that the maximum moment occurs when the wheel loads are located near the center of the beam. We will arbitrarily assume that the maximum moment occurs under wheel 2, which is located a distance x to the left of the beam's centerline. To determine the value of x that maximizes the moment under wheel 2, we will express the moment in the beam under wheel 2 as a function of x.

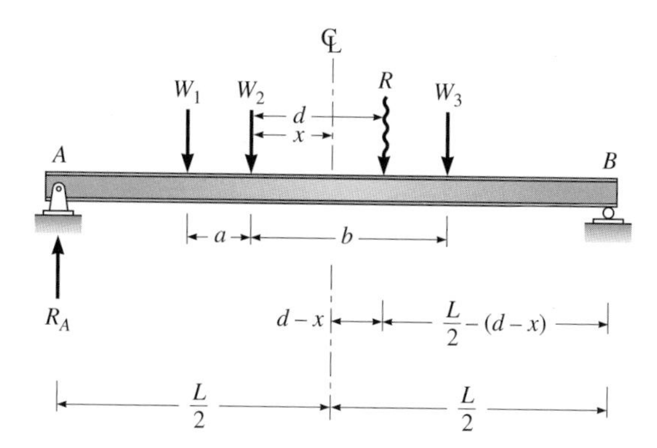

Figure 8.30: Set of wheel loads with a resultant R.

By differentiating the expression for moment with respect to x and setting the derivative equal to zero, we will establish the position of wheel 2 that maximizes the moment. To compute the moment under wheel 2, we use the resultant R of the wheel loads to establish the reaction at support A. Summing moments about support B gives

$$\circlearrowright^{+} \quad \Sigma M_B = 0$$

$$R_A L - R\left[\frac{L}{2} - (d - x)\right] = 0$$

$$R_A = \frac{R}{L}\left(\frac{L}{2} - d + x\right) \qquad (8.11)$$

To compute the moment M in the beam at wheel 2 by summing moments about a section through the beam at that point, we write

$$M = R_A\left(\frac{L}{2} - x\right) - W_1 a \qquad (8.12)$$

where a is the distance between W_1 and W_2. Substituting R_A given by Equation 8.11 into Equation 8.12 and simplifying give

$$M = \frac{RL}{4} - \frac{Rd}{2} + \frac{xRd}{L} - x^2\frac{R}{L} - W_1 a \qquad (8.13)$$

To establish the maximum value M, we differentiate Equation 8.13 with respect to x and set the derivative equal to zero.

$$0 = \frac{dM}{dx} = d\frac{R}{L} - 2x\frac{R}{L}$$

and

$$x = \frac{d}{2} \qquad (8.14)$$

For x to equal $d/2$ requires that we position the loads so that the centerline of the beam splits the distance between the resultant and the wheel under which the maximum moment is assumed to occur. In Example 8.10 we will use the foregoing principle to establish the absolute maximum moment produced in a simply supported beam by a set of wheel loads.

EXAMPLE 8.10

Determine the absolute maximum moment produced in a simply supported beam with a span of 30 ft by the set of loads shown in Figure 8.31*a*.

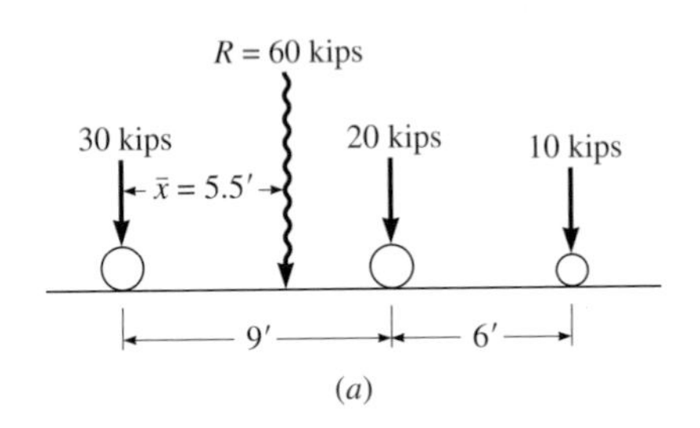

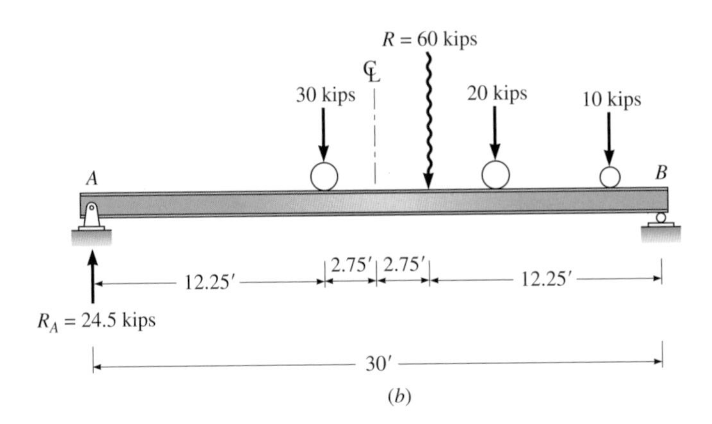

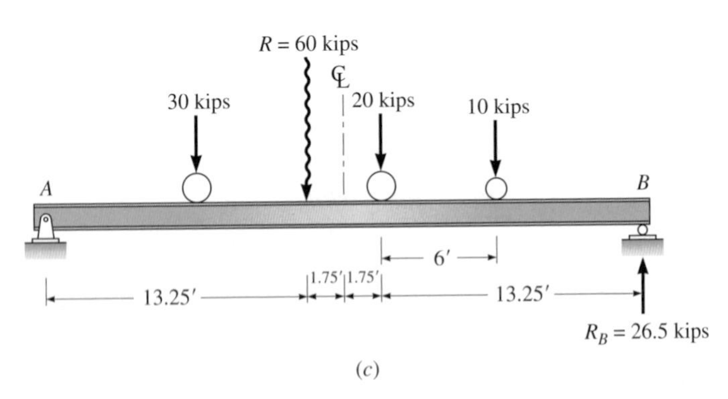

Figure 8.31: (*a*) Wheel loads; (*b*) position of loads to check maximum moment under 30-kip load; (*c*) position of loads to check maximum moment under 20-kip load.

Solution

Compute the magnitude and location of the resultant of the loads shown in Figure 8.31*a*.

$$R = \Sigma F_y = 30 + 20 + 10 = 60 \text{ kips}$$

Locate the position of the resultant by summing moments about the 30-kip load.

$$R \cdot \bar{x} = \Sigma F_n \cdot x_n$$

$$60\bar{x} = 20(9) + 10(15)$$

$$\bar{x} = 5.5 \text{ ft}$$

Assume that the maximum moment occurs under the 30-kip load. Position the loads as shown in Figure 8.31*b*; that is, the beam's centerline divides the distance between the 30-kip load and the resultant. Compute R_A by summing moments about B.

$$\circlearrowleft^{+} \quad \Sigma M_B = 0 = R_A(30) - 60(12.25)$$

$$R_A = 24.5 \text{ kips}$$

$$\text{Moment at 30-kip load} = 24.5(12.25)$$

$$= 300 \text{ kip} \cdot \text{ft}$$

Assume that the maximum moment occurs under the 20-kip load. Position the loads as shown in Figure 8.31*c*; that is, the centerline of the beam is located halfway between the 20-kip load and the resultant.

Compute R_B by summing moments about A.

$$\circlearrowleft^{+} \quad \Sigma M_A = 0 = 60(13.25) - R_B(30)$$

$$R_B = 26.5 \text{ kips}$$

$$\text{Moment at 20-kip load} = 13.25(26.5) - 10(6) = 291.1 \text{ kip} \cdot \text{ft}$$

$$\text{Absolute maximum moment} = 300 \text{ kip} \cdot \text{ft under 30-kip load}$$ **Ans.**

8.11 Maximum Shear

The maximum value of shear in a beam (simply supported or continuous) typically occurs adjacent to a support. In a simply supported beam, the shear at the end of a beam will be equal to the reaction; therefore, to maximize the shear, we position loads to maximize the reaction. The influence line for the reaction (see Figure 8.32*b*) indicates that load should be placed as close to the support as possible and that the entire span should be loaded. If a simple beam carries a set of moving loads, the increase–decrease method of Section 8.9 can be used to establish the position of the loads on the member to maximize the reaction.

To maximize the shear at a particular section *B-B*, the influence line in Figure 8.32*c* indicates that load should be placed (1) only on one side of the section and (2) on the side that is most distant from the support. For example, if the beam in Figure 8.32*a* supports a uniformly distributed live load of variable length, to maximize the shear at section *B*, the live load should be placed between *B* and *C*.

If a simply supported beam carries a uniform live load of variable length, the designer may wish to establish the critical live load shear at sections along the beam's axis by constructing an envelope of maximum shear. An acceptable envelope can be produced by running a straight line between the maximum shear at the support and the maximum shear at midspan (see Figure 8.33). The maximum shear at the support equals $wL/2$ and occurs when the entire span is loaded. The maximum shear at midspan equals $wL/8$ and occurs when load is placed on either half of the span.

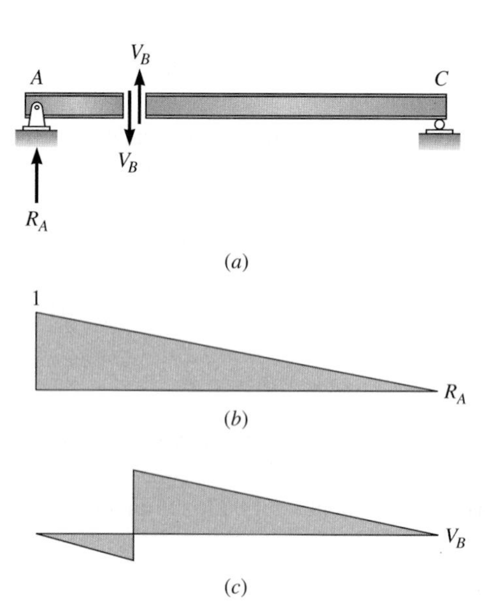

Figure 8.32: Maximum shear in a simply supported beam: (*a*) positive sense of shear at *B*; (*b*) influence line for R_A; (*c*) influence line for shear at section *B*.

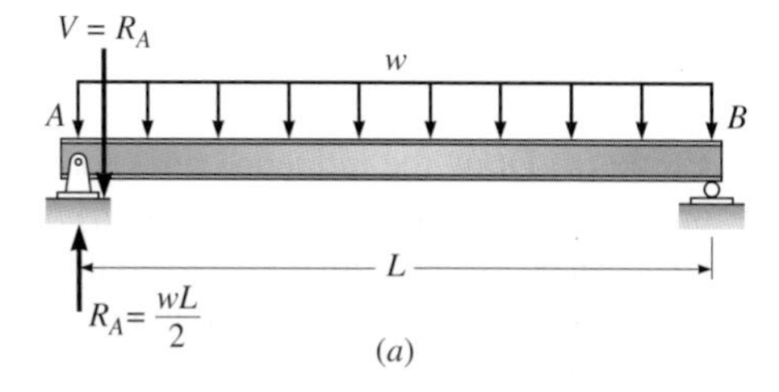

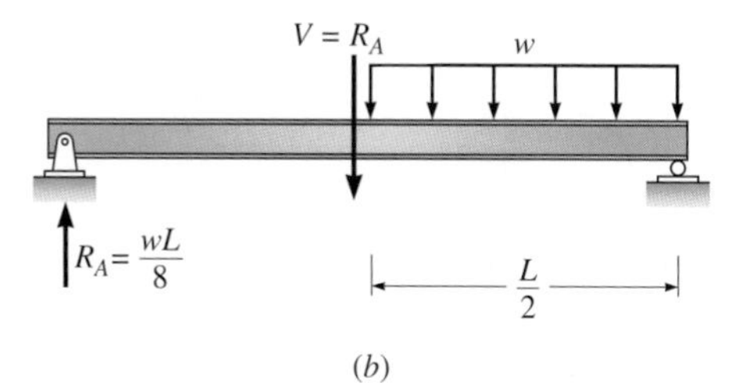

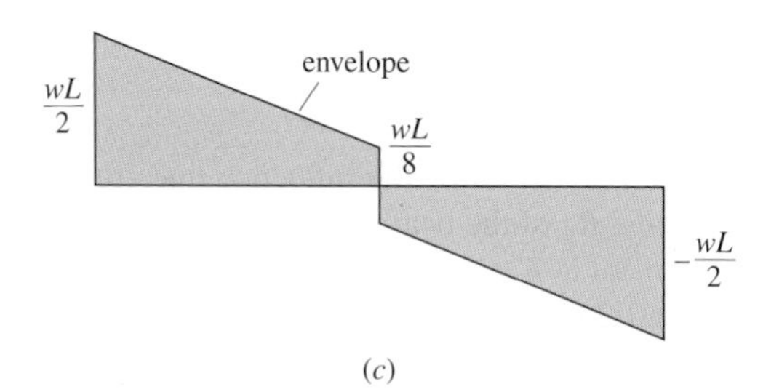

Figure 8.33: Loading conditions to establish the shear envelope for a beam supporting a uniform live load of variable length: (*a*) entire span loaded to maximum shear at support; (*b*) maximum shear at midspan produced by loading on half of span; (*c*) shear envelope.

Summary

- Influence lines are used to establish where to position a moving load or a variable length of uniformly distributed live load on a structure to maximize the value of an internal force at a particular section of a beam, truss, or other type of structure.
- Influence lines are constructed for an internal force or a reaction at a particular point in a structure by evaluating the value of the force at the particular point as a unit load moves over the structure. The value of the internal force for each position of the unit load is plotted directly below the position of the unit load.
- Influence lines consist of a series of straight lines for determinate structures and curved lines for indeterminate structures.
- The Müller–Breslau principle provides a simple procedure for establishing the qualitative shape of an influence line. The principle states: *The influence line for any reaction or internal force (shear, moment) corresponds to the deflected shape of the structure produced by removing the capacity of the structure to carry that force and then introducing into the modified (or released) structure a unit deformation that corresponds to the restraint removed.*

PROBLEMS

P8.1. Draw the influence lines for the reaction at A and for the shear and moment at points B and C. The rocker at D is equivalent to a roller.

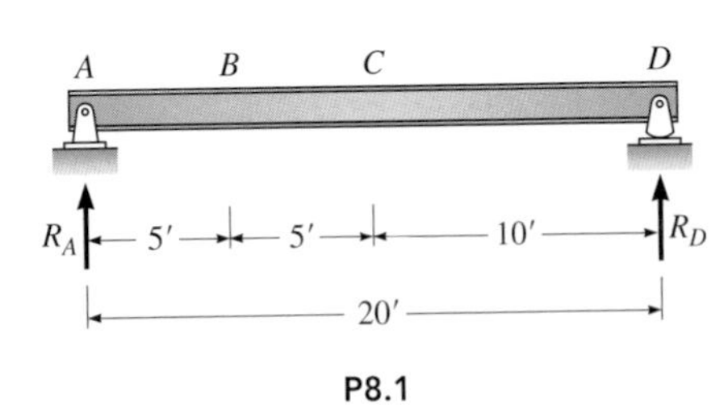

P8.1

P8.2. For the beam shown in Figure P8.2, draw the influence lines for the reactions M_A and R_A and the shear and moment at point B.

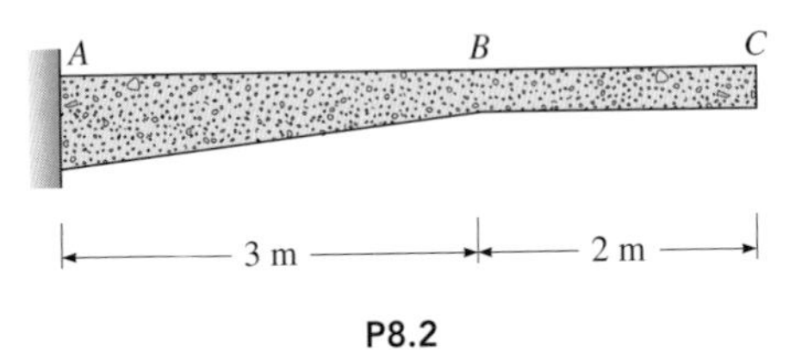

P8.2

P8.3. Draw the influence lines for the reactions at supports A and C, the shear and moment at section B, and the shear just to the left of support C.

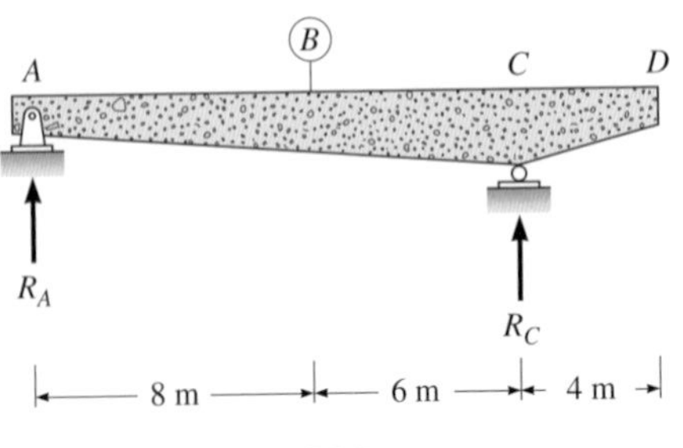

P8.3

P8.4. (*a*) Draw the influence lines for reactions M_A, R_A, and R_c of the beam in Figure P8.4. (*b*) Assuming that the span can be loaded with a 1.2 kips/ft uniform load of variable length, determine the maximum positive and negative values of the reactions.

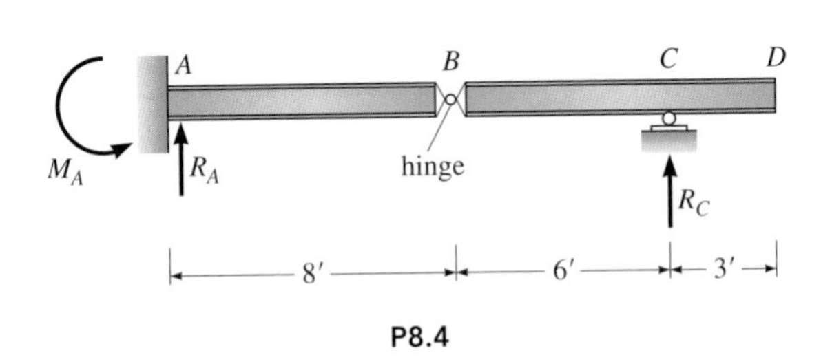

P8.4

P8.5. (*a*) Draw the influence lines for reactions R_B, R_D, and R_F of the beam in Figure P8.5 and the shear and moment at E. (*b*) Assuming that the span can be loaded with a 1.2 kips/ft uniform load of variable length, determine the maximum positive and negative values of the reactions.

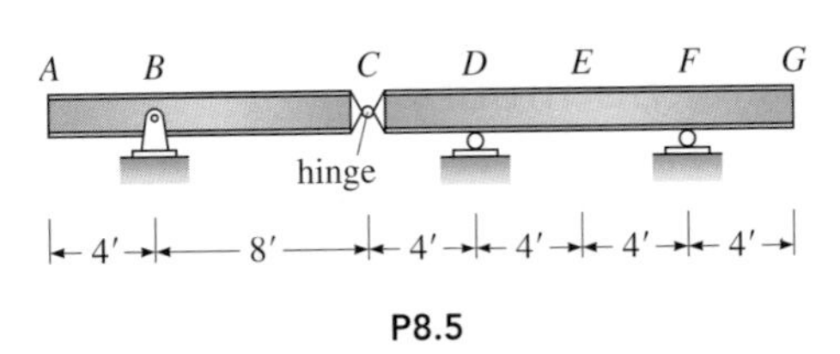

P8.5

P8.6. For the beam in Figure P8.6, draw the influence lines for reactions at B, C, E, and G, and moments at C and E. If a uniform load of 2 kips/ft is applied over the entire length of the beam, compute the reactions at B, C, D and E, moments at C and D.

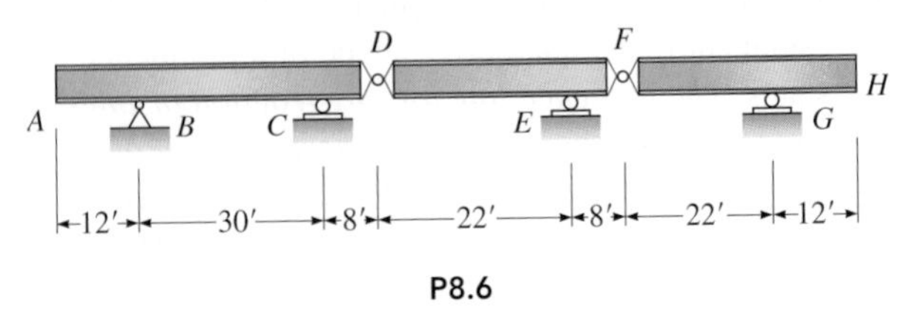

P8.6

P8.7. Load moves along girder $BCDE$. Draw the influence lines for the reactions at supports A and D, the shear and moment at section C, and the moment at D. Point C is located directly above support A.

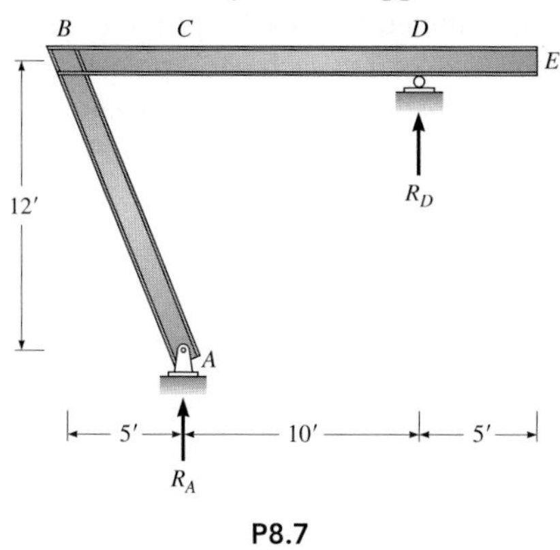

P8.7

P8.8. Hoist load moves along beam AB shown in Figure P8.8. Draw the influence lines for the vertical reaction at C, and moments at B and C.

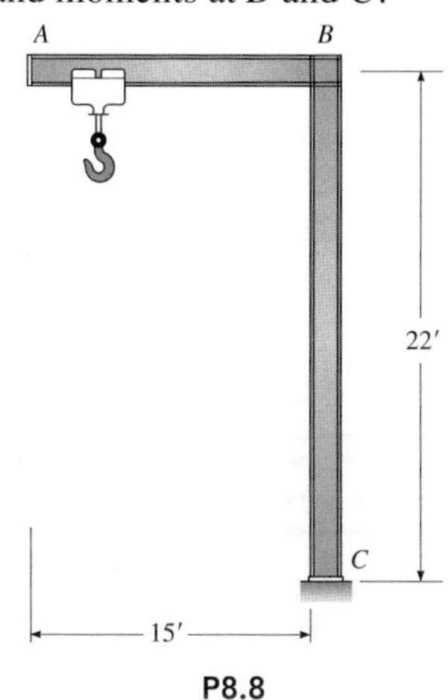

P8.8

P8.9. Beam AD is connected to a cable at C. Draw the influence lines for the force in cable CE, the vertical reaction at support A, and the moment at B.

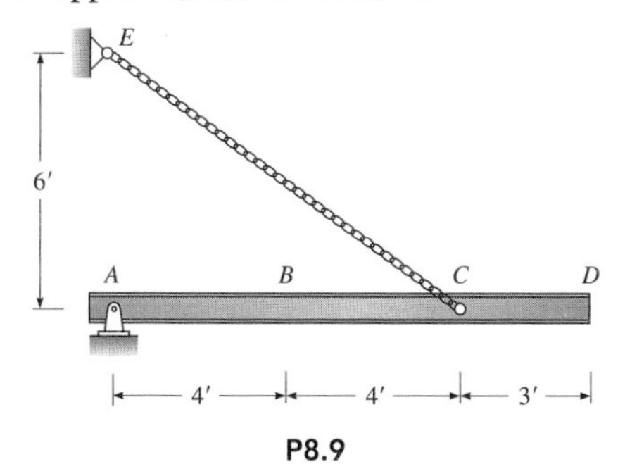

P8.9

P8.10 to **P8.13.** Using the Müller–Breslau principle, draw the influence lines for the reactions and internal forces noted below each structure.

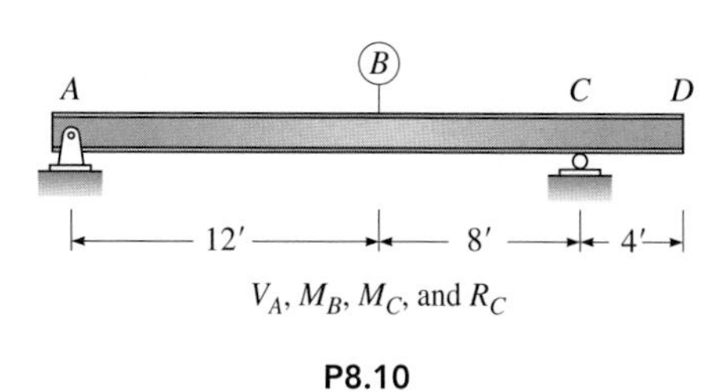

V_A, M_B, M_C, and R_C

P8.10

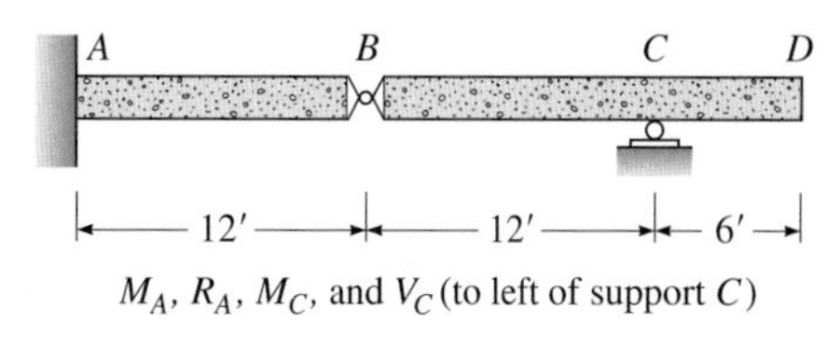

M_A, R_A, M_C, and V_C (to left of support C)

P8.11

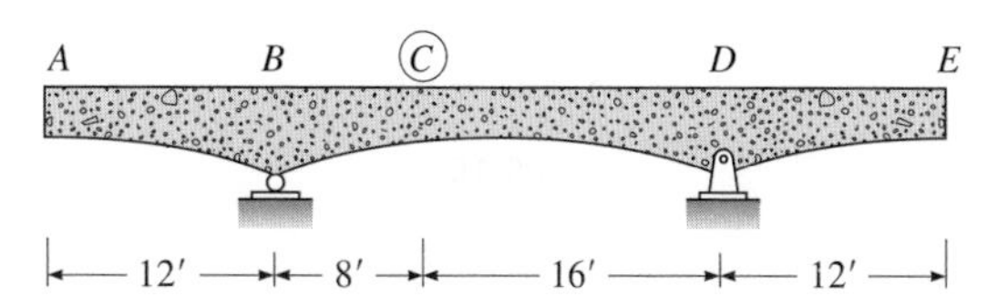

R_B, V_B (to left of support B), V_B (to right of support B), M_C, and V_C

P8.12

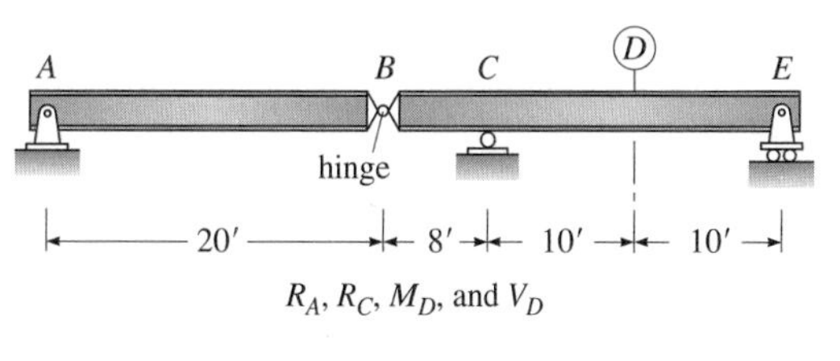

R_A, R_C, M_D, and V_D

P8.13

P8.14. For the beam shown in Figure P8.14, draw the influence lines for the reactions at A, B, and F, the end moment at F, shears to the left and right of support B, and shear at E.

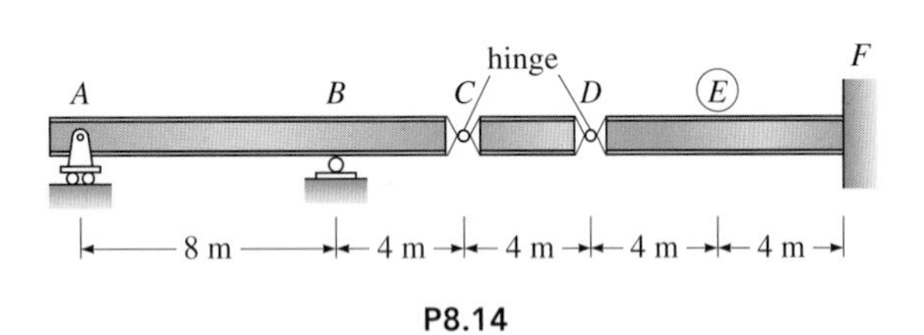

P8.14

P8.15. Draw the influence lines for the shear between points A and B and for moment at point E in the girder GH shown in Figure P8.15.

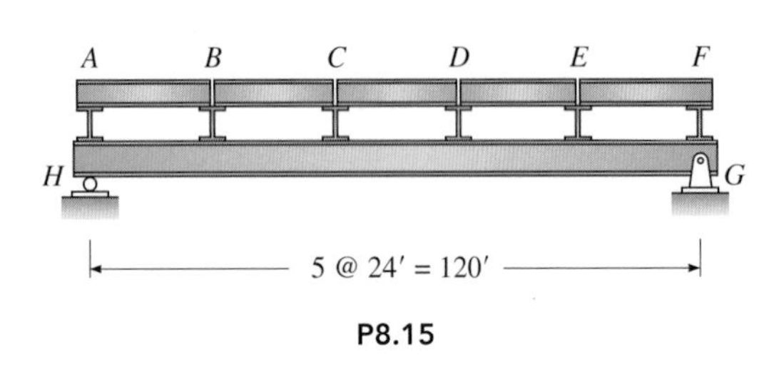

P8.15

P8.16. For the floor system shown in Figure P8.15, draw the influence lines for shear between points B and C and for the moment at points C and E in the girder.

P8.17. For the girder in Figure P8.17, draw the influence lines for the reaction at A, the moment at point C, and the shear between points B and C in girder AE.

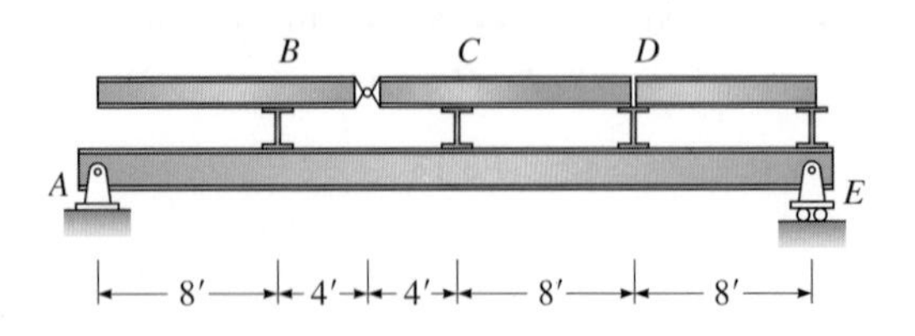

P8.17

P8.18. (*a*) Draw the influence lines for the reactions at B and E, the shear between CD, the moment at B and D for the girder HG in Figure P8.18. (*b*) If the dead load of the floor system (stringers and slab) is approximated by a uniformly distributed load of 3 kip/ft, the reaction of the floor beam's dead load to each panel point equals 1.5 kips, and the deadweight of the girder is 2.4 kips/ft, determine the moment in the girder at D and the shear just to the right of C.

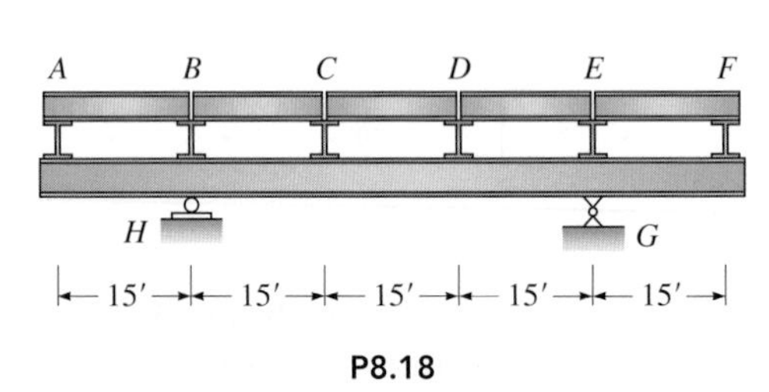

P8.18

P8.19. For the girder in Figure P8.19, draw the influence lines for the reaction at I, the shear to the right of support I, the moment at C, and the shear between CE.

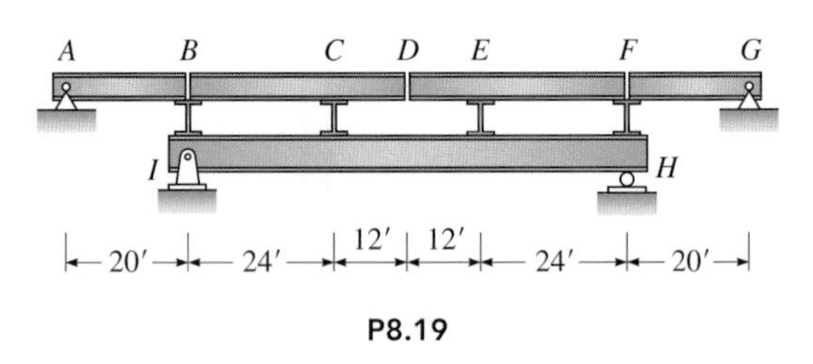

P8.19

P8.20. (*a*) For the girder HIJ shown in Figure P8.20, draw the influence line for moment at C. (*b*) Draw the influence line for the reactions at support H and K.

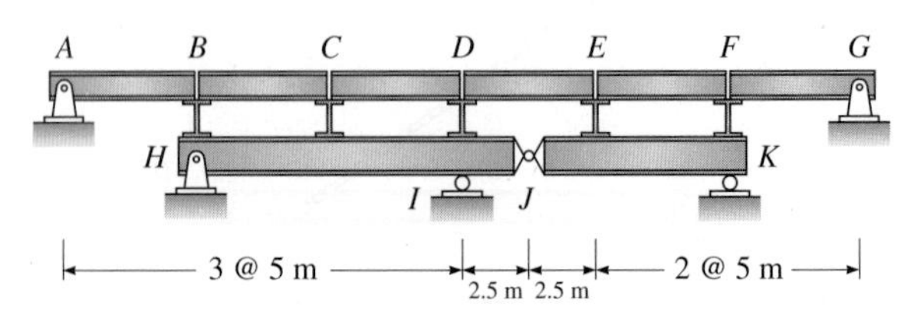

P8.20

P8.21. The load can only be applied between points B and D of the girder shown in Figure P8.21. Draw the influence lines for the reaction at A, the moment at D, and the shear to the right of support A.

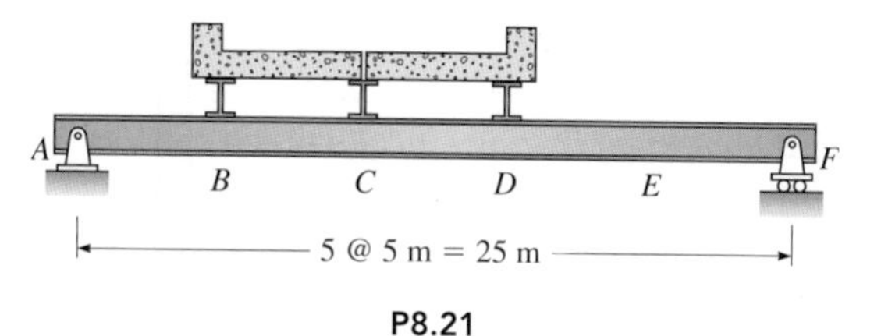

P8.21

P8.22. (*a*) The three-hinged arch shown in Figure P8.22 has a parabolic profile. Draw the influence lines for both the horizontal and vertical reactions at A and the moment at D. (*b*) Compute the horizontal and vertical reactions at support A if the arch is loaded by a uniform load of 10 kN/m. (*c*) Compute the maximum moment at point D.

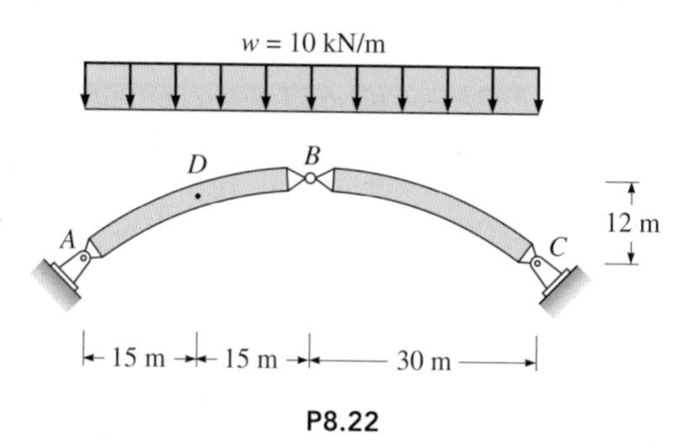

P8.22

P8.23. For the semi-circular, three-hinged arch ABC, shown in Figure P8.23, construct the influence lines for reactions at A and C, and shear, axial load and moment at F. A uniform live load of 2 kips/ft is applied along the top deck girder DE, assume that the load acts uniformly on the arch through the vertical struts. Compute the reactions at A and C, and shear, axial load and moment at F using the influence lines.

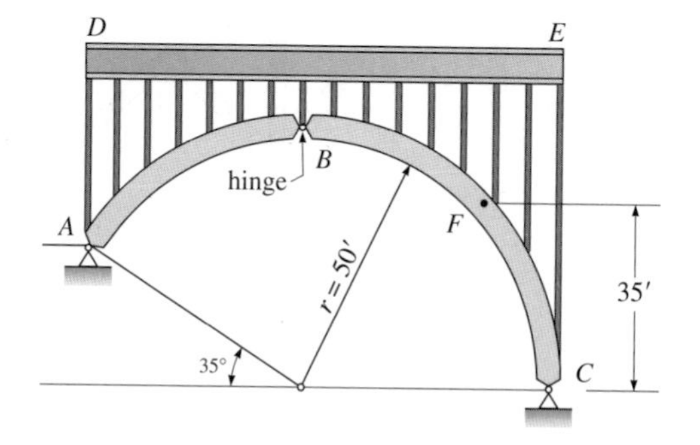

P8.23

P8.24. Load moves along the three-hinged, parabolic arch ABC, shown in Figure P8.24. Construct the influence lines for the reactions at C, and shear, axial load and moment at point D. The equation for the parabolic arch is $y = 4hx^2/L^2$. If a point load $P = 3$ kips is applied at B, compute shear, axial load and moment at D.

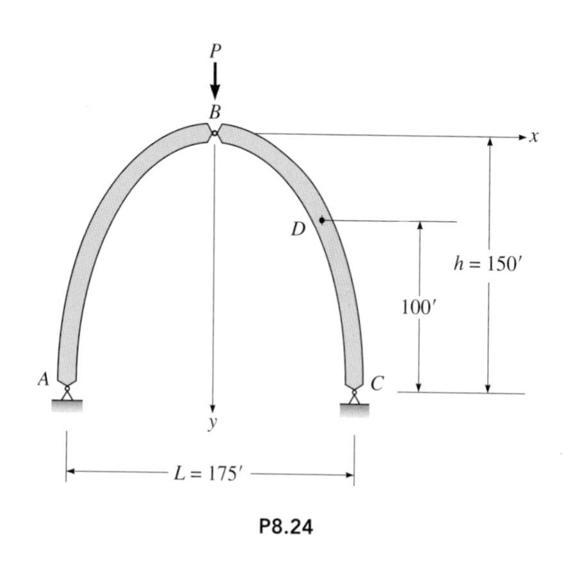

P8.24

P8.25. Draw the influence lines for the reactions at A and F and for the shear and moment at Section 1. Using the influence lines, determine the reactions at supports A and F if the dead load of the floor system can be approximated by a uniform load of 10 kN/m. See Figure P8.25.

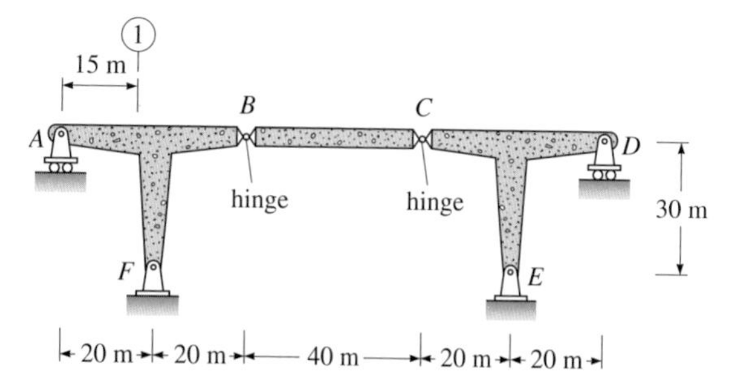

P8.25

P8.26. The horizontal load P can act at any location along the length of member AC shown in Figure P8.26. Draw the influence lines for the moment and shear at Section 1, and the moment at Section 2.

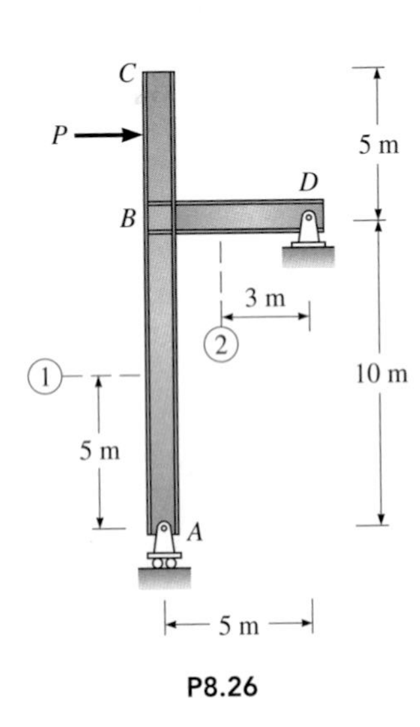

P8.26

P8.27. Load moves along the inverted kingpost truss. Construct the influence lines for the force in the cable, axial force in post BD, and moment at B in beam ABC. If the uniform live load of 2 kips/ft is applied along the beam ABC, compute the forces in the cable and post, and moment at B.

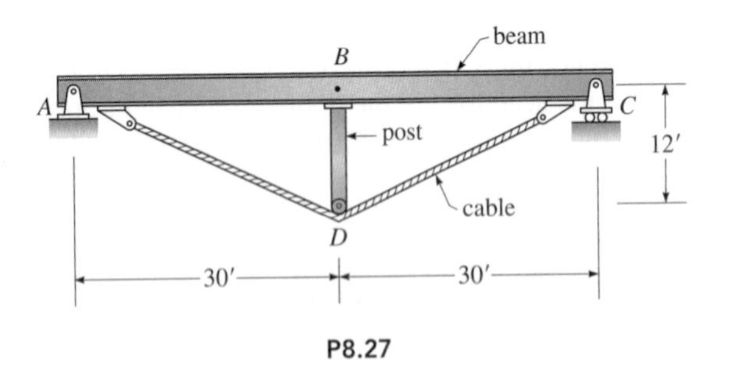

P8.27

P8.28. Load moves along girder BC. Draw the influence lines for the reactions at A and the bending moment on Section 1 located 1 ft from the centerline of column AB.

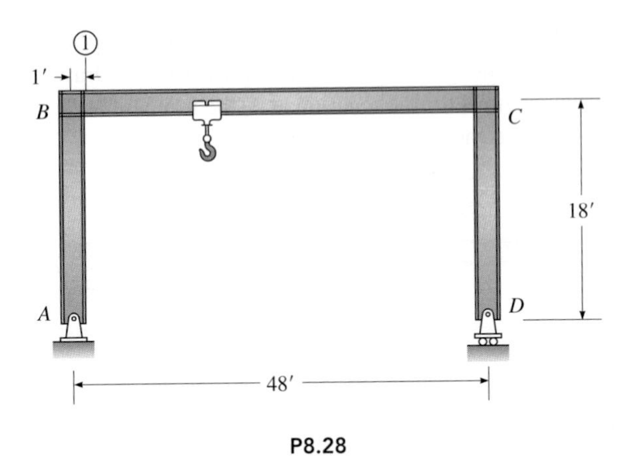

P8.28

P8.29. Draw the influence lines for the reactions A_x and A_y at the left pin support and the bending moment on Section 1 located 1 ft from the centerline of column AB.

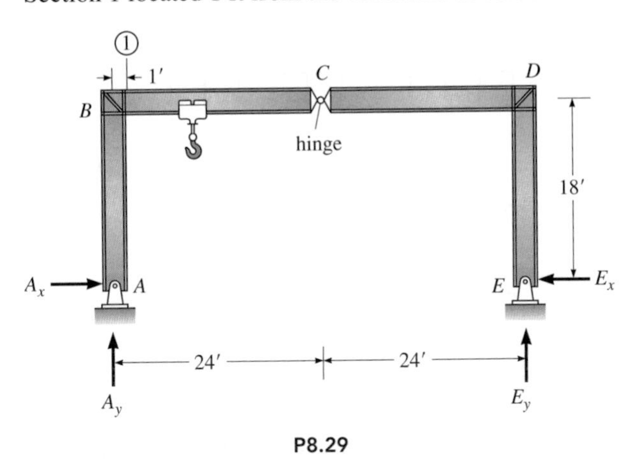

P8.29

P8.30. Draw the influence lines for the bar forces in members AB, BK, BC, and LK if the live load is applied to the truss in Figure P8.30 through the lower chord.

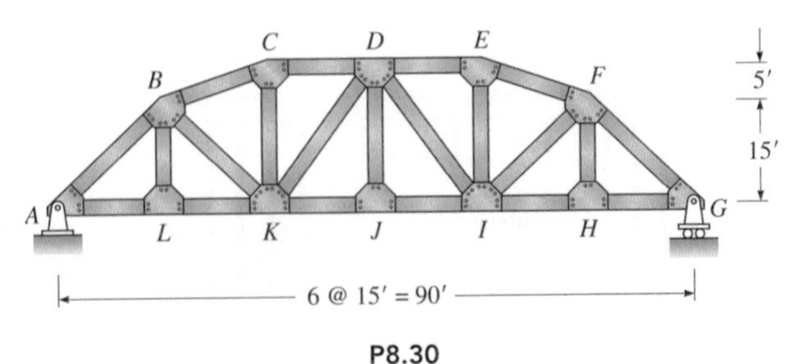

P8.30

P8.31. Draw the influence lines for the bar forces in members DE, DI, EI, and IJ if the live load in Figure P8.30 is applied through the lower chord panel points.

P8.32. (*a*) Draw the influence lines for the bar forces in members *HC*, *HG*, and *CD* of the truss shown in Figure P8.32. The load moves along the bottom chord of the truss. (*b*) Compute the force in member *HC* if panel points *B*, *C*, and *D* are each loaded by a concentrated vertical load of 12 kips.

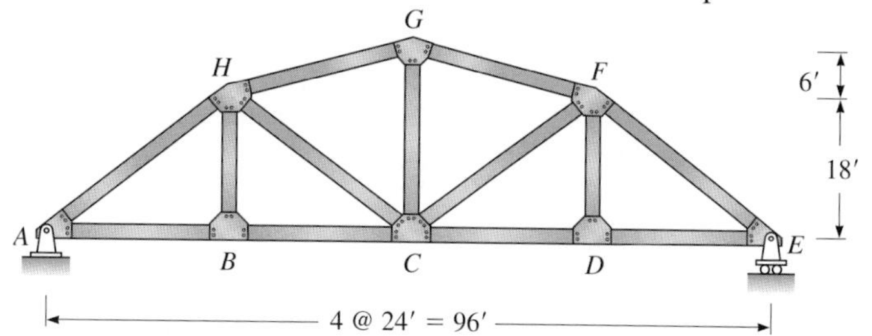

P8.32

P8.33. Draw the influence lines for R_A and the bar forces in members *AD*, *EF*, *EM*, and *NM*. Loads are transmitted into the truss through the lower chord panel points. Vertical members *EN* and *GL* are 18 ft long, *FM* is 16 ft.

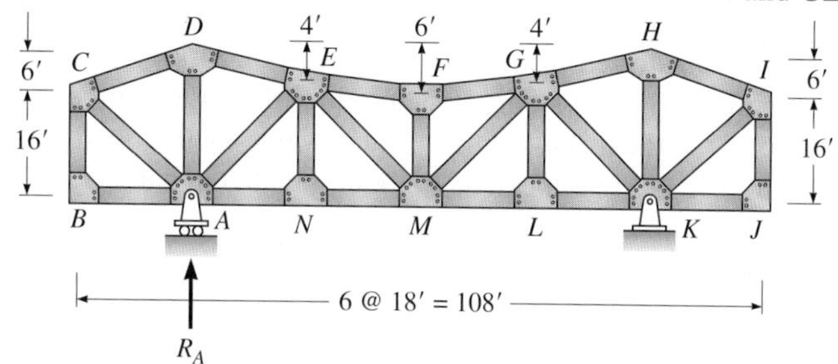

P8.33

P8.34. Draw the influence lines for bar forces in members *CD*, *EL*, and *ML* of the truss shown in Figure P8.34. The load moves along *BH* of the truss.

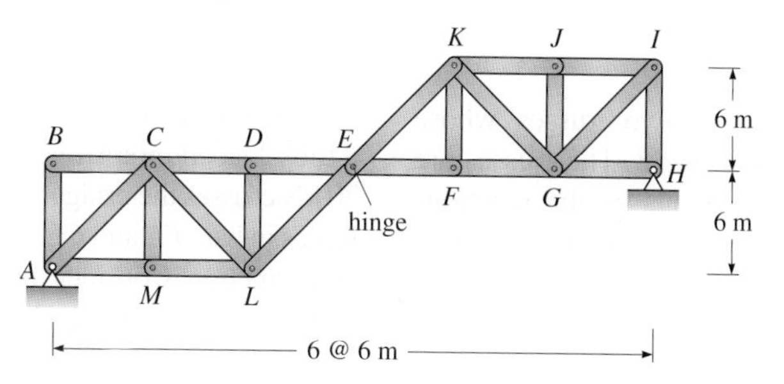

P8.34

P8.35. Draw the influence lines for bar forces in members *ML*, *BL*, *CD*, *EJ*, *DJ*, and *FH* of the cantilever truss in Figure P8.35 if the live load is applied through the lower chord panel points.

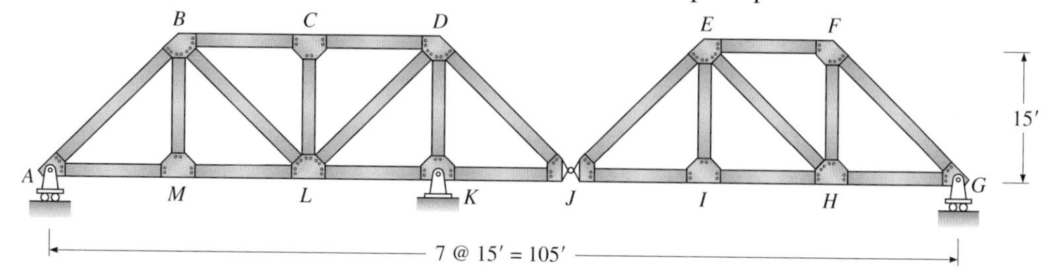

P8.35

P8.36. Draw the influence lines for the vertical and horizontal reactions, A_X and A_Y, at support A and the bar forces in members AD, CD, and BC. If the truss is loaded by a uniform dead load of 4 kips/ft over the entire length of the top chord, determine the magnitude of the bar forces in members AD and CD.

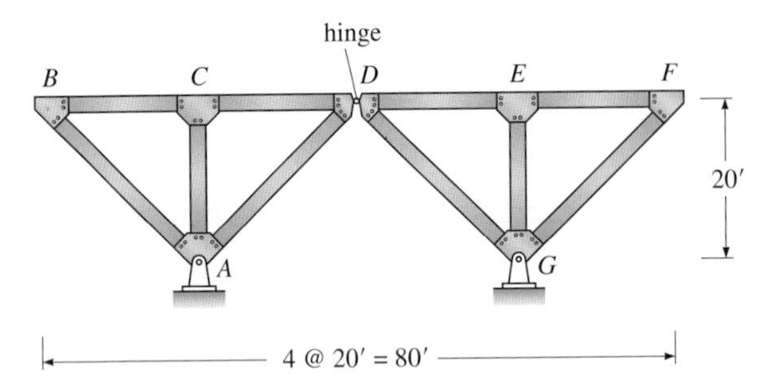

P8.36

P8.37. Draw the influence lines for the forces in members BC, AC, CD, and CG. Load is transferred from the roadway to the upper panel points by a system of stringers and floor beams (not shown). If the truss is to be designed for a uniform live load of 0.32 kip/ft that can be placed anywhere on the span in addition to a concentrated live load of 24 kips that can be positioned where it will produce the largest force in bar CG, determine the maximum value of live load force (tension, compression, or both) created in bar CG.

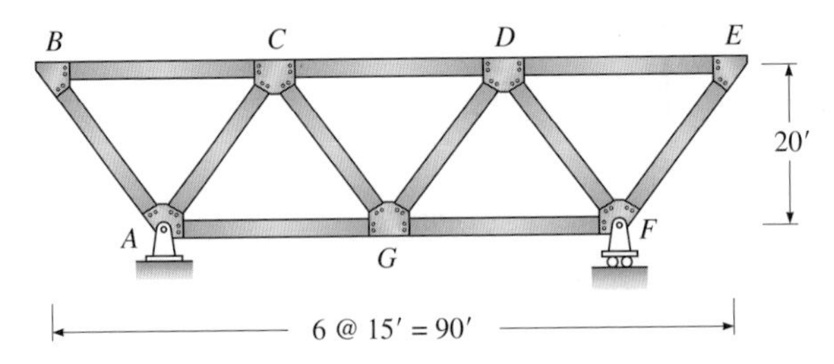

P8.37

P8.38. A bridge is composed of two trusses whose configuration is shown in Figure P8.38. The trusses are loaded at their top chord panel points by the reactions from a stringer and floor beam system that supports a roadway slab. Draw the influence lines for forces in bars FE and CE. Assume that vehicles move along the center of the roadway so one-half the load is carried by each truss. If a fully loaded motorized ore carrier with a total weight of 70 kN crosses the bridge, determine the maximum live load forces in bars FE and CE. Assume the truck can move in either direction. Consider the possibility of both tension and compression force in each bar.

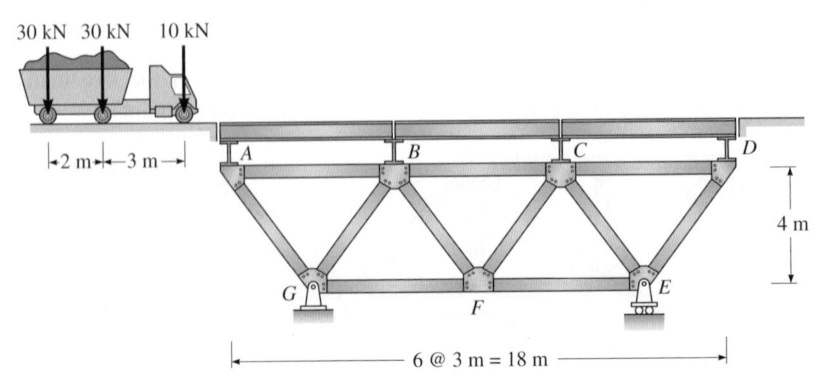

P8.38

P8.39. Draw the influence lines for forces in bars *AL* and *KJ* in Figure P8.39. Using the influence lines, determine the maximum live load force (consider both tension and compression) produced by the 54-kip truck as it transverses the bridge, which consists of two trusses. Assume the truck moves along the center of the roadway so that one-half of the truck load is carried by each truss. Assume the truck can travel in either direction.

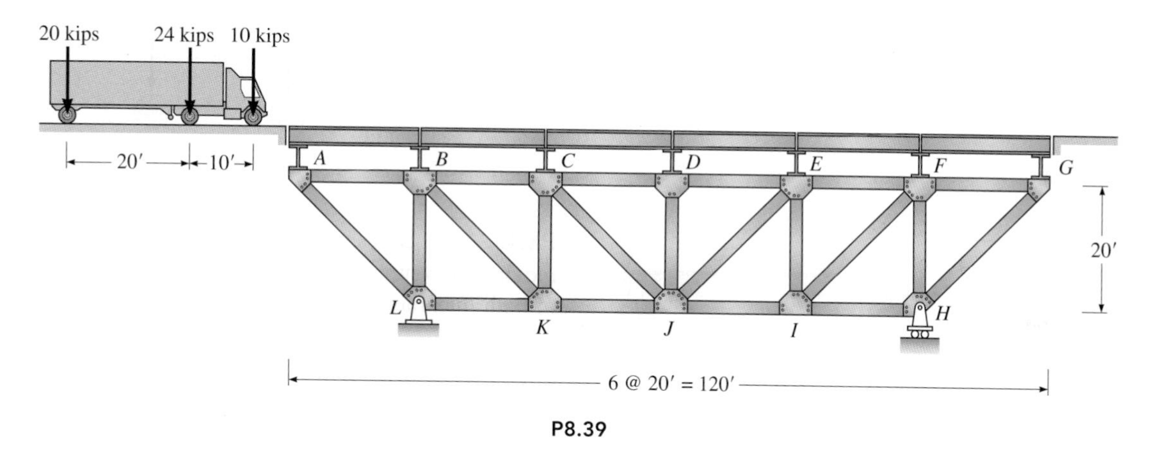

P8.39

P8.40. (*a*) Load is applied to the three-hinged trussed arch in Figure P8.40 through the upper chord panel points by a floor beam and stringer floor system. Draw the influence lines for the horizontal and vertical reactions at support *A* and the forces or components of force in members *BC*, *CM*, and *ML*. (*b*) Assuming that the dead load of the arch and floor system can be represented by a uniform load of 4.8 kip/ft, determine the forces in bars *CM* and *ML* produced by the dead load. (*c*) If the live load is represented by a uniformly distributed load of 0.8 kip/ft of variable length and a concentrated load of 20 kips, determine the maximum force in bar *CM* produced by the live load. Consider both tension and compression. Joint *E* acts as a hinge.

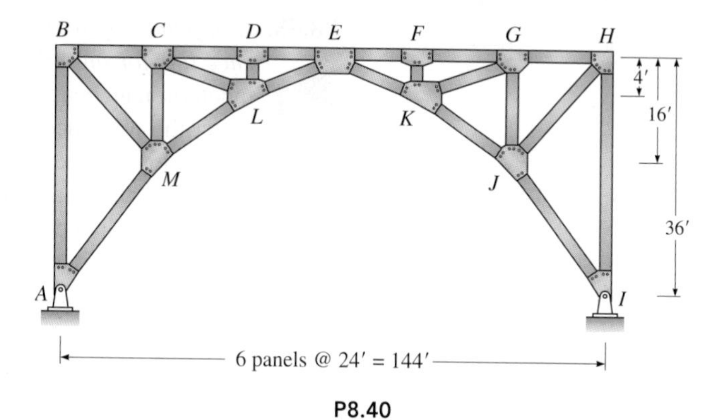

P8.40

P8.41. Compute the absolute maximum shear and moment produced in a simply supported beam by two concentrated live loads of 20 kips spaced 10 ft apart. The beam span is 30 ft.

P8.42. Draw the envelopes for maximum shear and moment in a 24-ft-long simply supported beam produced by a live load that consists of both a uniformly distributed load of 0.4 kip/ft of variable length and a concentrated load of 10 kips (Figure P8.42). The 10-kip load can act at any point. Compute values of the envelope at the supports, quarter points, and midspan.

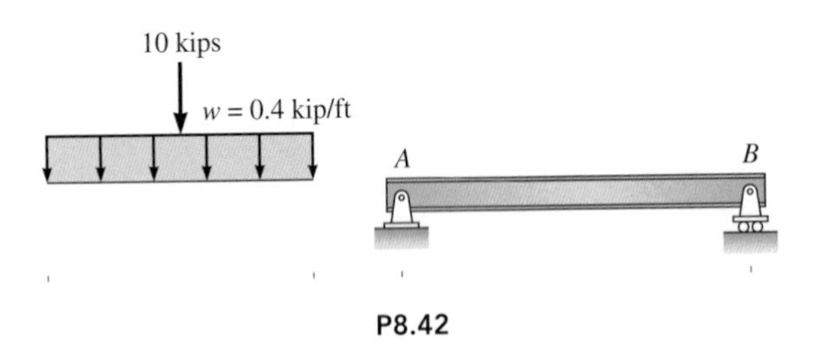

P8.42

P8.43. Determine (*a*) the absolute maximum values of shear and moment in the beam produced by the wheel loads and (*b*) the maximum value of moment when the middle wheel is positioned at the center of the beam in Figure P8.43.

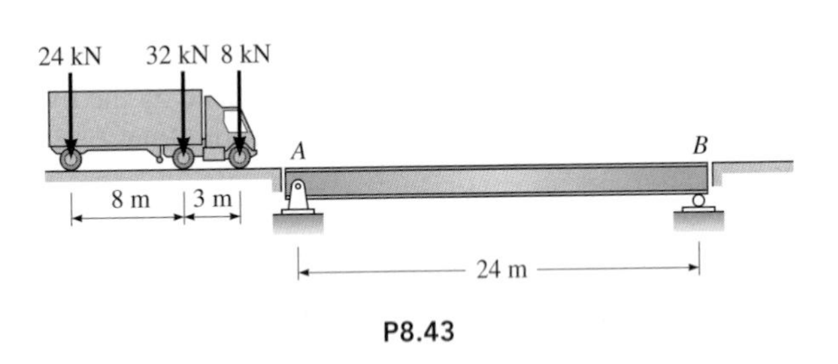

P8.43

P8.44. Determine (*a*) the absolute maximum value of live load moment and shear produced in the 50-ft girder and (*b*) the maximum value of moment at midspan (Figure P8.44). *Hint*: For part (*b*) use the influence line for moment.

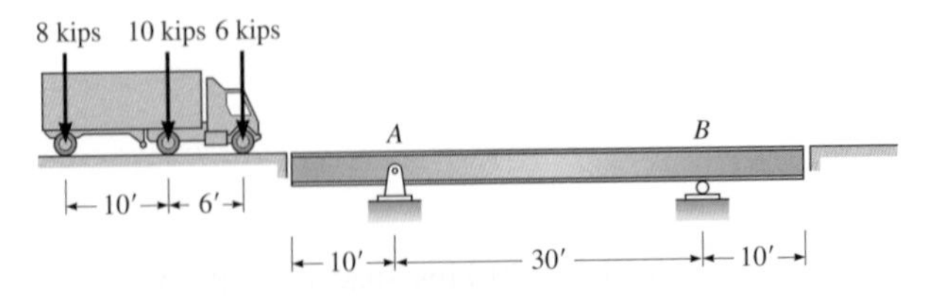

P8.44

P8.45. Determine the absolute maximum value of live load shear and moment produced in a simply supported beam spanning 40 ft by the wheel loads shown in Figure P8.45.

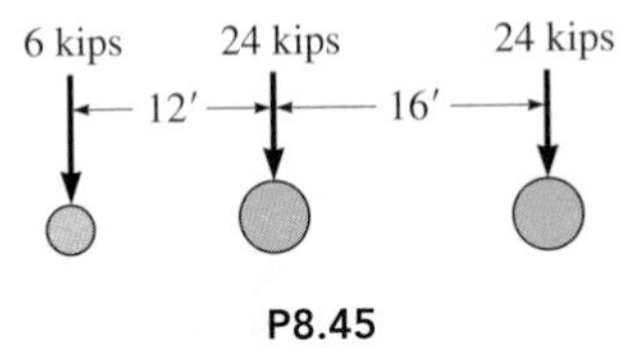

P8.45

P8.46. The beam shown in Figure P8.46 is subjected to a moving concentrated load of 80 kN. Construct the envelope of both maximum positive and negative moments for the beam.

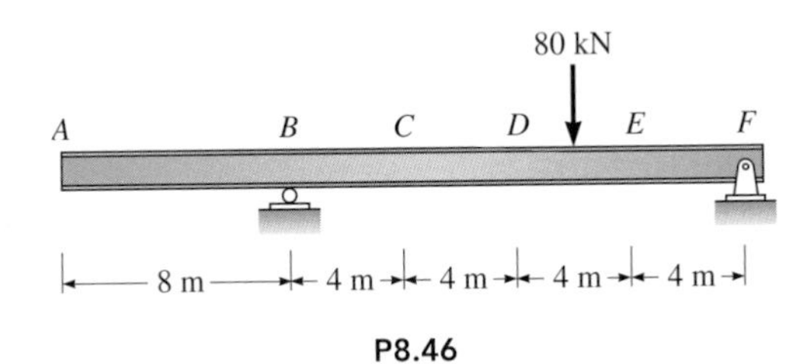

P8.46

P8.47. Consider the beam shown in Figure P8.46 without the 80 kN load. Construct the envelope of maximum positive shear assuming the beam supports a 6 kN/m uniformly distributed load of variable length.

P8.48. *Computer application. Construction of an influence line for an indeterminate beam.* (*a*) For the indeterminate beam shown in Figure P8.48, construct the influence lines for M_A, R_A, and R_B by applying a unit load to the beam at 4-ft intervals to compute the corresponding magnitudes of the reactions.

(*b*) Using the influence line in part (*a*), determine the maximum value of the reaction R_B produced by two concentrated 20-kip wheel loads spaced 8 ft apart.

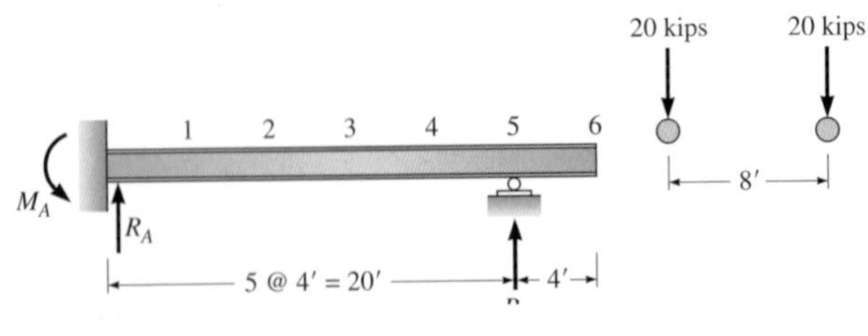

P8.48

P8.49. A simply supported crane runway girder has to support a moving load shown in Figure P8.49. The moving load shown has to be increased by an impact factor listed in Table 2.3. (*a*) Position the moving load to compute the maximum moment. Also compute the maximum deflection produced by the load. (*b*) Re-position the moving load symmetrically on the span and compute the maximum moment and the maximum deflection. Which case produces a larger deflection?

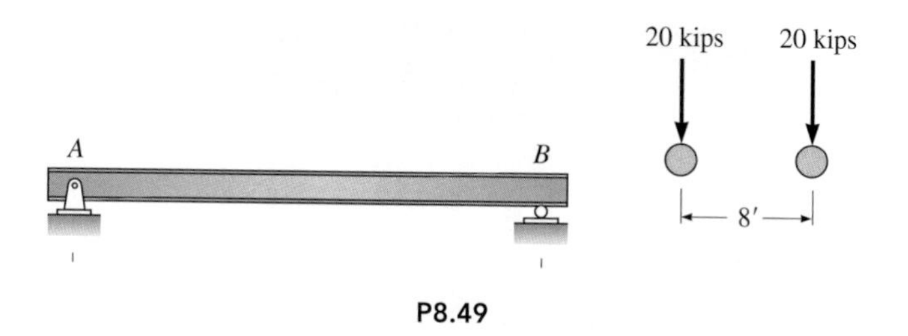

P8.49

Taipei 101 in Taiwan

Taipei 101 in Taiwan was the tallest building before Burj Khalif in Dubai exceeded its height by more than 1,000 ft in 2010. Taipei 101 is designed to resist severe earthquakes and typhoons. The building's fundamental period is about 7 seconds. A building of this height tends to vibrate during windy days, causing discomfort to the occupants. To reduce wind-induced vibration, a pendulum-type, 18-ft diameter steel passive tuned mass damper weighing about 1,450 kips is suspended from Floor 92 to Floor 87. The vibration energy of the mass is then dissipated by eight diagonally placed hydraulic viscous damping devices.

CHAPTER 9

Deflections of Beams and Frames

Chapter Objectives

- Introduce several methods to compute deflections and slopes of elastic beams and frames.
- Learn the double integration method based on the basic differential equation of the elastic curve, which relates the curvature to M/EI along the member's longitudinal axis.
- Learn the moment-area method based on the M/EI diagram between two points along the member's axis. It is a geometric method requiring the deflected shape to be properly drawn.
- Learn the elastic load (i.e., M/EI) and the more powerful conjugate beam methods to compute slope and deflection at any point along the member axis.

9.1 Introduction

When a structure is loaded, its stressed elements deform. In a truss, bars in tension elongate and bars in compression shorten. Beams bend and cables stretch. As these deformations occur, the structure changes shape and points on the structure displace. Although these deflections are normally small, as part of the total design, the engineer must verify that these deflections are within the limits specified by the governing design code to ensure that the structure is serviceable. For example, large deflections of beams can lead to cracking of nonstructural elements such as plaster ceilings, tile walls, or brittle pipes. The lateral displacement of buildings produced by wind forces must be limited to prevent cracking of walls and windows. Since the magnitude of deflections is also a measure of a member's stiffness, limiting deflections also ensures that excessive vibrations of building floors and bridge decks are not created by moving loads.

Deflection computations are also an integral part of a number of analytical procedures for analyzing indeterminate structures, computing buckling loads, and determining the natural periods of vibrating members.

In this chapter we consider several methods of computing deflections and slopes at points along the axis of beams and frames. These methods are based on the differential equation of the elastic curve of a beam. This equation relates curvature at a point along the beam's longitudinal axis to the bending moment at that point and the properties of the cross section and the material.

9.2 Double Integration Method

The double integration method is a procedure to establish the equations for slope and deflection at points along the longitudinal axis (elastic curve) of a loaded beam. The equations are derived by integrating the differential equation of the elastic curve twice, hence the name *double integration*. The method assumes that all deformations are produced by moment. Shear deformations, which are typically less than 1 percent of the flexural deformations in beams of normal proportions, are not usually included. But if beams are deep, have thin webs, or are constructed of a material with a low modulus of rigidity (plywood, for example), the magnitude of the shear deformations can be significant and should be investigated.

To understand the principles on which the double integration method is based, we first review the geometry of curves. Next, we derive the differential equation of the elastic curve—the equation that relates the curvature at a point on the elastic curve to the moment and the flexural stiffness of the cross section. In the final step we integrate the differential equation of the elastic curve twice and then evaluate the constants of integration by considering the boundary conditions imposed by the supports. The first integration produces the equation for slope; the second integration establishes the equation for deflection. Although the method is not used extensively in practice since evaluating the constants of integration is time-consuming for many types of beams, we begin our study of deflections with this method because several other important procedures for computing deflections in beams and frames are based on the differential equation of the elastic curve.

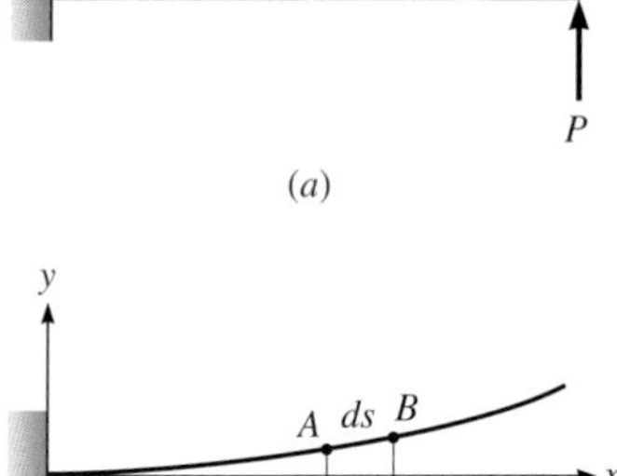

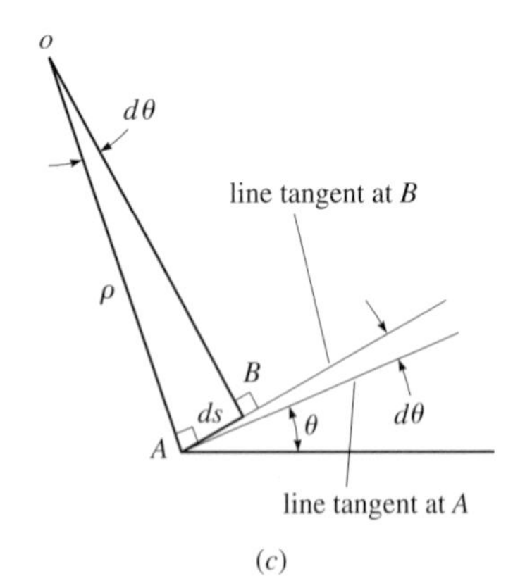

Figure 9.1

Geometry of Shallow Curves

To establish the geometric relationships required to derive the differential equation of the elastic curve, we will consider the deformations of the cantilever beam in Figure 9.1*a*. The deflected shape is represented in Figure 9.1*b* by the displaced position of the longitudinal axis (also called the *elastic curve*). As reference axes, we establish an x-y coordinate system whose origin is located at the fixed end. For clarity, vertical distances in this figure are greatly exaggerated. Slopes, for example, are typically very small—on the order of a few tenths of a degree. If we were to show the deflected shape to scale, it would appear as a straight line.

To establish the geometry of a curved element, we will consider an infinitesimal element of length ds located a distance x from the fixed end. As shown in Figure 9.1*c*, we denote the radius of the curved segment by ρ. At points A and B we draw tangent lines to the curve. The infinitesimal angle between these tangents is denoted by $d\theta$. Since the tangents to the curve are perpendicular to the radii at points A and B, it follows that the angle between the radii is also $d\theta$. The slope of the curve at point A equals

$$\frac{dy}{dx} = \tan\theta$$

If the angles are small ($\tan\theta \approx \theta$ radians), the slope can be written

$$\frac{dy}{dx} = \theta \tag{9.1}$$

From the geometry of the triangular segment *ABo* in Figure 9.1*c*, we can write

$$\rho \, d\theta = ds \tag{9.2}$$

Dividing each side of the equation above by ds and rearranging terms give

$$\psi = \frac{d\theta}{ds} = \frac{1}{\rho} \tag{9.3}$$

where $d\theta/ds$, representing the change in slope per unit length of distance along the curve, is called the *curvature* and denoted by the symbol ψ. Since slopes are small in actual beams, $ds \approx dx$, and we can express the curvature in Equation 9.3 as

$$\psi = \frac{d\theta}{dx} = \frac{1}{\rho} \tag{9.4}$$

Differentiating both sides of Equation 9.1 with respect to x, we can express the curvature $d\theta/dx$ in Equation 9.4 in terms of rectangular coordinates as

$$\frac{d\theta}{dx} = \frac{d^2y}{dx^2} \tag{9.5}$$

Differential Equation of the Elastic Curve

To express the curvature of a beam at a particular point in terms of the moment acting at that point and the properties of the cross section, we will consider the flexural deformations of the small beam segment of length dx, shown with darker shading in Figure 9.2*a*. The two vertical lines representing the sides of the element are perpendicular to the longitudinal axis of the unloaded beam. As load is applied, moment is created, and the beam bends (see Figure 9.2*b*); the element deforms into a trapezoid as the sides of the segment, which remain straight, rotate about a horizontal axis (the neutral axis) passing through the centroid of the section (Figure 9.2*c*).

In Figure 9.2*d* the deformed element is superimposed on the original unstressed element of length dx. The left sides are aligned so that the deformations are shown on the right. As shown in this figure, the longitudinal fibers of the segment located above the neutral axis shorten because they are stressed in compression. Below the neutral axis the longitudinal fibers, stressed in tension, lengthen. Since the change in length of the longitudinal fibers (flexural deformations) is zero at the neutral axis (N.A.), the strains and stresses at that level equal zero. The variation of longitudinal strain with depth is shown in Figure 9.2*e*. Since the strain is equal to the longitudinal deformations divided by the original length dx, it also varies linearly with distance from the neutral axis.

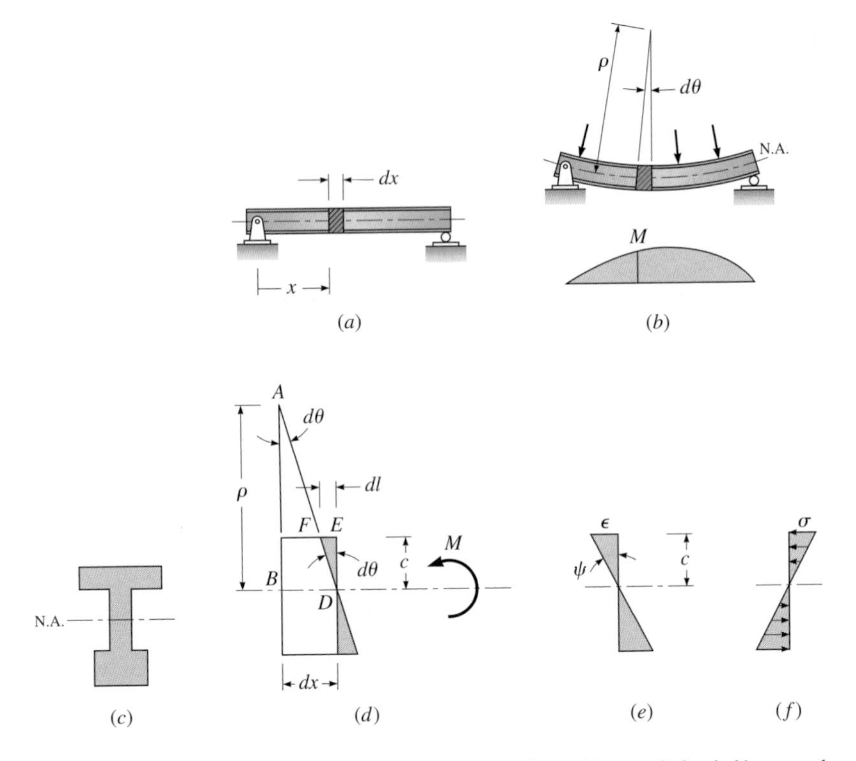

Figure 9.2: Flexural deformations of segment dx: (*a*) unloaded beam; (*b*) loaded beam and moment curve; (*c*) cross section of beam; (*d*) flexural deformations of the small beam segment; (*e*) longitudinal strain; (*f*) flexural stresses.

Considering triangle DFE in Figure 9.2*d*, we can express the change in length of the top fiber dl in terms of $d\theta$ and the distance c from the neutral axis to the top fiber as

$$dl = d\theta \ c \tag{9.6}$$

By definition, the strain ϵ at the top surface can be expressed as

$$\epsilon = \frac{dl}{dx} \tag{9.7}$$

Using Equation 9.6 to eliminate dl in Equation 9.7 gives

$$\epsilon = \frac{d\theta}{dx} c \tag{9.8}$$

Using Equation 9.5 to express the curvature $d\theta/dx$ in rectangular coordinates, we can write Equation 9.8 as

$$\frac{d^2y}{dx^2} = \frac{\epsilon}{c} \tag{9.9}$$

If behavior is elastic, the flexural stress, σ, can be related to the strain ϵ at the top fiber by *Hooke's law*, which states that

$$\sigma = E\epsilon$$

where $E =$ the modulus of elasticity

Solving for ϵ gives

$$\epsilon = \frac{\sigma}{E} \tag{9.10}$$

Using Equation 9.10 to eliminate ϵ in Equation 9.9 produces

$$\frac{d^2y}{dx^2} = \frac{\sigma}{Ec} \tag{9.11}$$

For elastic behavior the relationship between the flexural stress at the top fiber and the moment acting on the cross section is given by

$$\sigma = \frac{Mc}{I} \tag{5.1}$$

Substituting the value of σ given by Equation 5.1 into Equation 9.11 produces the basic differential equation of the elastic curve

$$\frac{d^2y}{dx^2} = \frac{M}{EI} \tag{9.12}$$

In Examples 9.1 and 9.2 we use Equation 9.12 to establish the equations for both the slope and the deflection of the elastic curve of a beam. This operation is carried out by expressing the bending moment in terms of the applied load and distance x along the beam's axis, substituting the equation for moment in Equation 9.12, and integrating twice. The method is simplest to apply when the loading and support conditions permit the moment to be expressed by a single equation that is valid over the entire length of the member—the case for Examples 9.1 and 9.2. For beams of constant cross section, E and I are constant along the length of the member. If E or I varies, it must also be expressed as a function of x in order to carry out the integration of Equation 9.12. If the loads or the cross section varies in a complex manner along the axis of the member, the equations for moment or for I may be difficult to integrate. For this situation approximate procedures can be used to facilitate the solution (see, for example, the finite summation in Example 10.16).

EXAMPLE 9.1

Using the double integration method, establish the equations for slope and deflection for the uniformly loaded beam in Figure 9.3. Evaluate the deflection at midspan and the slope at support A. EI is constant.

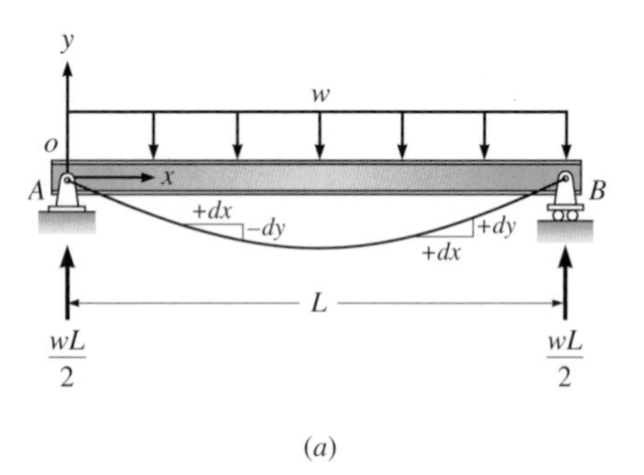

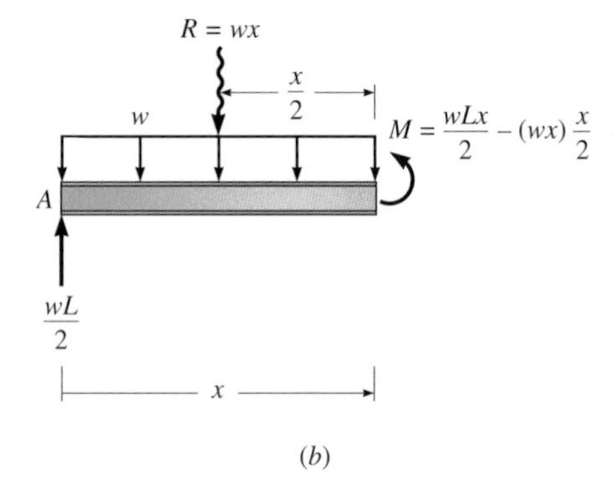

Figure 9.3: (*a*) beam with deflected shape; (*b*) free-body diagram.

Solution

Establish a rectangular coordinate system with the origin at support A. Since the slope increases as x increases (the slope is negative at A, zero at midspan, and positive at B), the curvature is positive. If we consider a free body of the beam cut by a vertical section located a distance x from the origin at A (see Figure 9.3*b*), we can write the internal moment at the section as

$$M = \frac{wLx}{2} - \frac{wx^2}{2}$$

Substituting M into Equation 9.12 gives

$$EI\frac{d^2y}{dx^2} = \frac{wLx}{2} - \frac{wx^2}{2} \tag{1}$$

Integrating twice with respect to x yields

$$EI\frac{dy}{dx} = \frac{wLx^2}{4} - \frac{wx^3}{6} + C_1 \tag{2}$$

$$EIy = \frac{wLx^3}{12} - \frac{wx^4}{24} + C_1x + C_2 \tag{3}$$

To evaluate the constants of integration C_1 and C_2, we use the boundary conditions at supports A and B. At A, $x = 0$ and $y = 0$. Substituting these values into Equation 3, we find that $C_2 = 0$. At B, $x = L$ and $y = 0$. Substituting these values into Equation 3 and solving for C_1 gives

$$0 = \frac{wL^4}{12} - \frac{wL^4}{24} + C_1L$$

$$C_1 = -\frac{wL^3}{24}$$

Substituting C_1 and C_2 into Equations 2 and 3 and dividing both sides by EI yields

$$\theta = \frac{dy}{dx} = \frac{wLx^2}{4EI} - \frac{wx^3}{6EI} - \frac{wL^3}{24EI} \tag{4}$$

$$y = \frac{wLx^3}{12EI} - \frac{wx^4}{24EI} - \frac{wL^3x}{24EI} \tag{5}$$

Compute the deflection at midspan by substituting $x = L/2$ into Equation 5.

$$y = \frac{5wL^4}{384EI} \qquad \textbf{Ans.}$$

Compute the slope at A by substituting $x = 0$ into Equation 4.

$$\theta_A = \frac{dy}{dx} = -\frac{wL^3}{24EI} \qquad \textbf{Ans.}$$

EXAMPLE 9.2

For the cantilever beam in Figure 9.4a, establish the equations for slope and deflection by the double integration method. Also determine the magnitude of the slope θ_B and deflection Δ_B at the tip of the cantilever. EI is constant.

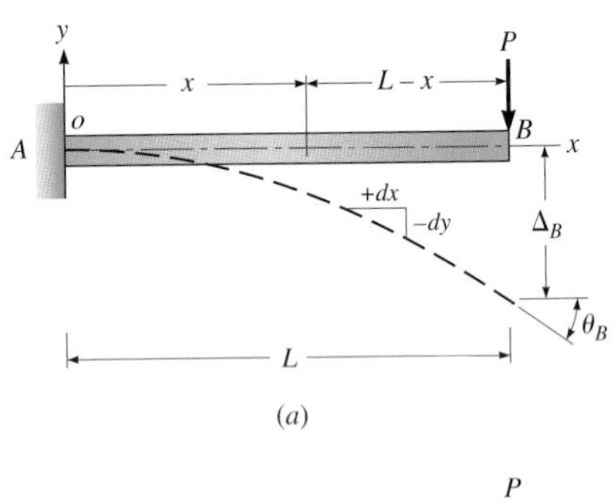

Figure 9.4: (a) beam with deflected shape; (b) free-body diagram.

Solution

Establish a rectangular coordinate system with the origin at the fixed support A. Positive directions for the axes are up (y-axis) and to the right (x-axis). Since the slope is negative and becomes steeper in the positive x direction, the curvature is *negative*. Passing a section through the beam a distance x from the origin and considering a free body to the right of the cut (see Figure 9.4b), we can express the bending moment at the cut as

$$M = P(L - x)$$

Substituting M into Equation 9.12 and adding a minus sign because the curvature is negative lead to

$$\frac{d^2y}{dx^2} = \frac{M}{EI} = \frac{-P(L - x)}{EI}$$

Integrating twice to establish the equations for slope and deflection yields

$$\frac{dy}{dx} = \frac{-PLx}{EI} + \frac{Px^2}{2EI} + C_1 \tag{1}$$

$$y = \frac{-PLx^2}{2EI} + \frac{Px^3}{6EI} + C_1x + C_2 \tag{2}$$

To evaluate the constants of integration C_1 and C_2 in Equations 1 and 2, we use the boundary conditions imposed by the fixed support at A:

1. When $x = 0$, $y = 0$; then from Equation 2, $C_2 = 0$.
2. When $x = 0$, $dy/dx = 0$; then from Equation 1, $C_1 = 0$.

The final equations are

$$\theta = \frac{dy}{dx} = \frac{-PLx}{EI} + \frac{Px^2}{2EI} \tag{3}$$

$$y = \frac{-PLx^2}{2EI} + \frac{Px^3}{6EI} \tag{4}$$

To establish θ_B and Δ_B, we substitute $x = L$ in Equations 3 and 4 to compute

$$\theta_B = \frac{-PL^2}{2EI} \qquad \textbf{Ans.}$$

$$\Delta_B = \frac{-PL^3}{3EI} \qquad \textbf{Ans.}$$

9.3 Moment-Area Method

As we observed in the double integration method, based on Equation 9.12, the slope and deflection of points along the elastic curve of a beam or a frame are functions of the bending moment M, moment of inertia I, and modulus of elasticity E. In the moment-area method we will establish a procedure that utilizes the area of the moment diagrams [actually, the M/EI diagrams] to evaluate the slope or deflection at selected points along the axis of a beam or frame.

This method, which requires an accurate sketch of the deflected shape, employs two theorems. One theorem is used to calculate a *change in slope* between two points on the elastic curve. The other theorem is used to compute the vertical distance (called a *tangential deviation*) between a point on the elastic curve and a line tangent to the elastic curve at a second point. These quantities are illustrated in Figure 9.5. At points A and B, tangent lines, which make a slope of θ_A and θ_B with the horizontal axis, are drawn to the elastic curve. For the coordinate system shown, the slope at A is negative and the slope at B is positive. The change in slope between points A and B is denoted by $\Delta\theta_{AB}$. The tangential deviation at point B—the vertical distance between point B on the elastic curve and point C on the line drawn tangent to the elastic curve at A—is denoted as t_{BA}. We will use two subscripts to label all tangential deviations. The first subscript indicates the location of the tangential deviation; the second subscript specifies the point at which the tangent line is drawn. As you can see in Figure 9.5, t_{BA} is not the deflection of point B (v_B is the deflection). With some guidance you will quickly learn to use tangential deviations and changes in slope to compute values of slope and deflection at any desired point on the elastic curve. In the next section we develop the two moment-area theorems and illustrate their application to a variety of beams and frames.

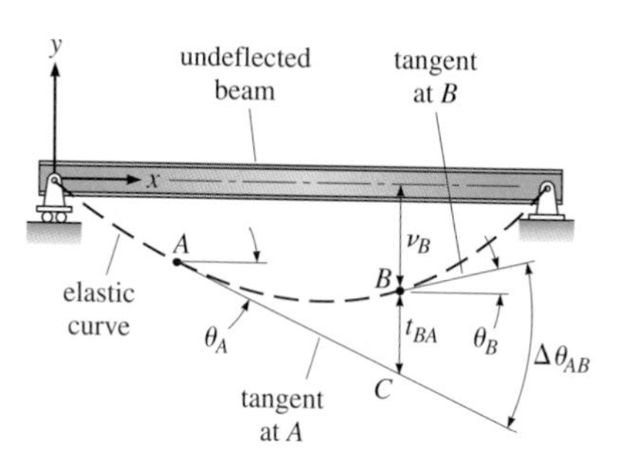

Figure 9.5: Change in slope and tangential deviation between points A and B.

Derivation of the Moment-Area Theorems

Figure 9.6*b* shows a portion of the elastic curve of a loaded beam. At points A and B tangent lines are drawn to the curve. The total angle between the two tangents is denoted by $\Delta\theta_{AB}$. To express $\Delta\theta_{AB}$ in terms of the properties of the cross section and the moment produced by the applied loads, we will consider the increment of angle change $d\theta$ that occurs over the length ds of the infinitesimal segment located a distance x to the left of point B. Previously, we established that the curvature at a point on the elastic curve can be expressed as

$$\frac{d\theta}{dx} = \frac{M}{EI} \tag{9.12}$$

where E is the modulus of elasticity and I is the moment of inertia. Multiplying both sides of Equation 9.12 by dx gives

$$d\theta = \frac{M}{EI}\,dx \tag{9.13}$$

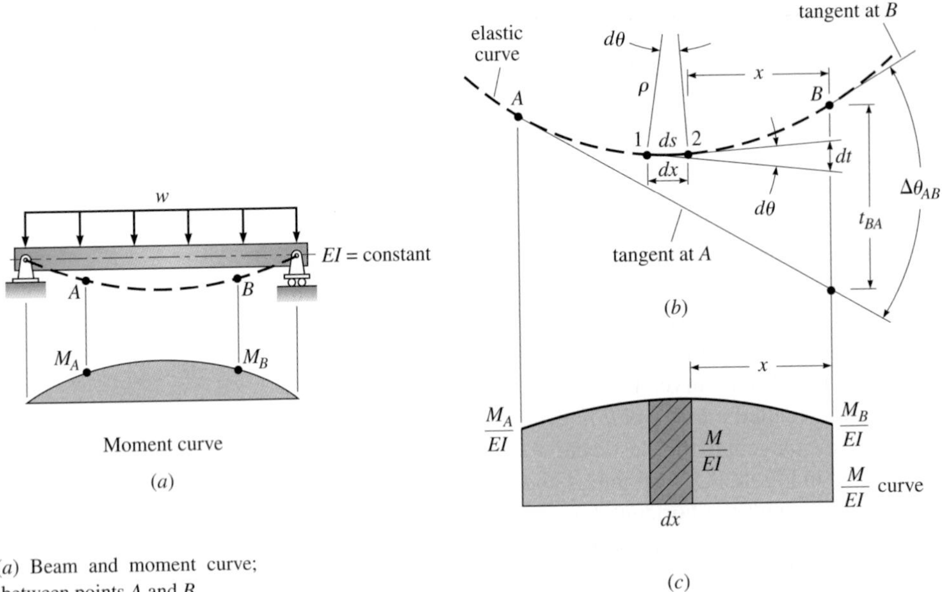

Figure 9.6: (*a*) Beam and moment curve; (*b*) *M/EI* curve between points *A* and *B*.

To establish the total angle change $\Delta\theta_{AB}$, we must sum up the $d\theta$ increments for all segments of length *ds* between points *A* and *B* by integration.

$$\Delta\theta_{AB} = \int_A^B d\theta = \int_A^B \frac{M\,dx}{EI} \tag{9.14}$$

We can evaluate the quantity $M\,dx/EI$ in the integral of Equation 9.14 graphically by dividing the ordinates of the moment curve by *EI* to produce an M/EI curve (see Figure 9.6*c*). If *EI* is constant along the beam's axis the (most common case), the M/EI curve has the same shape as the moment diagram. Recognizing that the quantity $M\,dx/EI$ represents an infinitesimal area of height M/EI and length *dx* (see the crosshatched area in Figure 9.6*c*), we can interpret the integral in Equation 9.14 as representing the area under the M/EI diagram between points *A* and *B*. This relationship constitutes the first moment-area principle, which can be stated as

The change in slope between any two points on a smooth continuous elastic curve is equal to the area under the M/EI curve between these points.

You will notice that the first moment-area theorem applies only to the case where the elastic curve is continuous and smooth between two points. If a hinge occurs between two points, the area under the M/EI diagram will not account for the difference in slope that can exist on either side of the hinge. Therefore, we must determine the slopes at a hinge by working with the elastic curve on either side.

To establish the second moment-area theorem, which enables us to evaluate a tangential deviation, we must sum the infinitesimal increments of length *dt* that make up the total tangential deviation t_{BA} (see Figure 9.6*b*). The magnitude of a typical increment *dt*, when contributing to the tangential deviation t_{BA}

by the curvature of a typical segment ds between points 1 and 2 on the elastic curve, can be expressed in terms of the angle between the lines tangent to the ends of the segment and the distance x between the segment and point B as

$$dt = d\theta \, x \tag{9.15}$$

Expressing $d\theta$ in Equation 9.15 by Equation 9.13, we can write

$$dt = \frac{M\,dx}{EI}x \tag{9.16}$$

To evaluate t_{BA}, we must sum all increments of dt by integrating the contribution of all the infinitesimal segments between points A and B:

$$t_{BA} = \int_A^B dt = \int_A^B \frac{Mx}{EI}\,dx \tag{9.17}$$

Remembering that the quantity $M\,dx/EI$ represents an infinitesimal area under the M/EI diagram and that x is the distance from that area to point B, we can interpret the integral in Equation 9.17 as the moment about point B of the area under the M/EI diagram between points A and B. This result constitutes the second moment-area theorem, which can be stated as follows:

> **The tangential deviation at a point B on a smooth continuous elastic curve from the tangent line drawn to the elastic curve at a second point A is equal to the moment about B of the area under the M/EI curve between the two points.**

Although it is possible to evaluate the integral in Equation 9.17 by expressing the moment M as a function of x and integrating, it is faster and simpler to carry out the computation graphically. In this procedure we divide the area of the M/EI diagram into simple geometric shapes—rectangles, triangles, parabolas, and so forth. Then the moment of each area is evaluated by multiplying each area by the distance from its centroid to the point at which the tangential deviation is to be computed. For this computation, we can use Table 3 inside the back cover, which tabulates properties of areas you will frequently encounter.

Application of the Moment-Area Theorems

The first step in computing the slope or deflection of a point on the elastic curve of a member is to draw an accurate sketch of the deflected shape. As discussed in Section 5.6, the curvature of the elastic curve must be consistent with the moment curve, and the ends of members must satisfy the constraints imposed by the supports. Once you have constructed a sketch of the deflected shape, the next step is to find a point on the elastic curve where the slope of a tangent to the curve is known. After this reference tangent is established, the slope or deflection at any other point on the continuous elastic curve can easily be established by using the moment-area theorems.

The strategy for computing slopes and deflections by the moment-area method will depend on how a structure is supported and loaded. Most continuous members will fall into one of the following three categories:

1. Cantilevers
2. Structures with a vertical axis of symmetry that are loaded symmetrically
3. Structures that contain a member whose ends do not displace in the direction normal to the original position of the member's longitudinal axis

If a member is not continuous because of an internal hinge, the deflection at the hinge must be computed initially to establish the position of the endpoints of the member. This procedure is illustrated in Example 9.10 on page 331. In the next sections we discuss the procedure for computing slopes and deflections for members in each of the foregoing categories.

Case 1. In a cantilever, a tangent line of known slope can be drawn to the elastic curve at the fixed support. For example, in Figure 9.7*a* the line tangent to the elastic curve at the fixed support is horizontal (i.e., the slope of the elastic curve at *A* is zero because the fixed support prevents the end of the member from rotating). The slope at a second point *B* on the elastic curve can then be computed by adding algebraically, to the slope at *A*, the change in slope $\Delta\theta_{AB}$ between the two points. This relationship can be stated as

$$\theta_B = \theta_A + \Delta\theta_{AB} \tag{9.18}$$

where θ_A is the slope at the fixed end (that is, $\theta_A = 0$) and $\Delta\theta_{AB}$ is equal to the area under the M/EI diagram between points *A* and *B*.

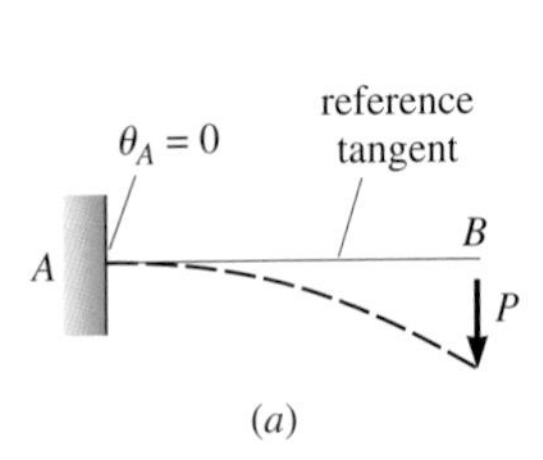

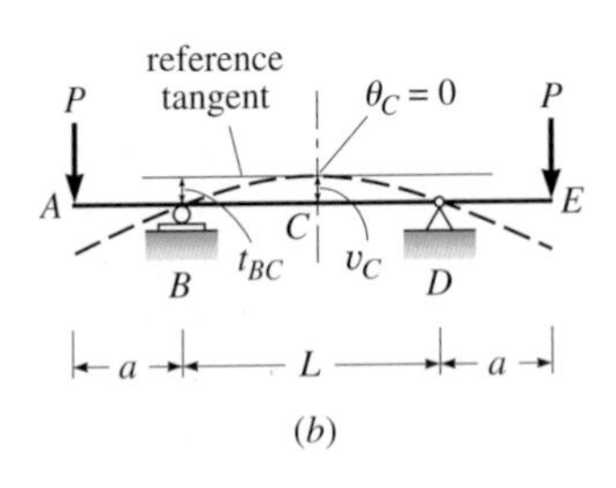

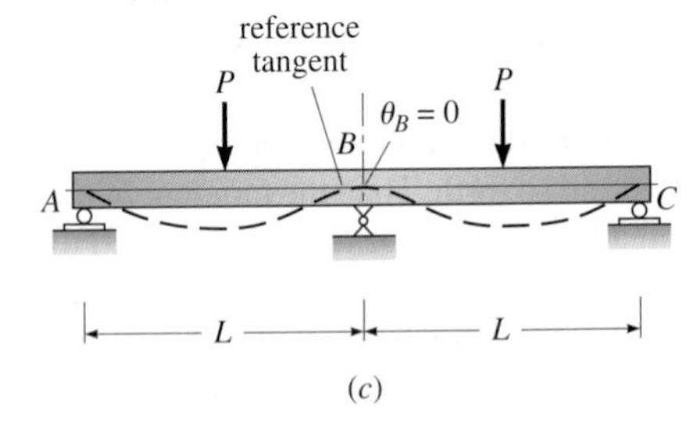

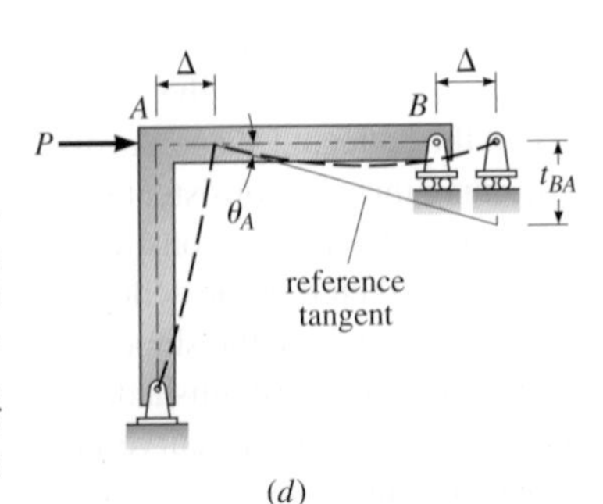

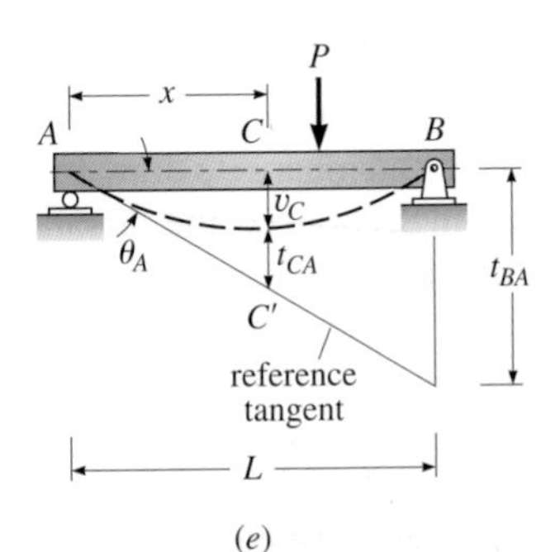

Figure 9.7: Position of tangent line: (*a*) cantilever, point of tangency at fixed support; Position of tangent line: (*a*) cantilever, point of tangency at fixed support; (*b*) and (*c*) symmetric members with symmetric loading, point of tangency at intersection of axis of symmetry and elastic curve; and (*d*) and (*e*) point of tangency at left end of member *AB*.

Since the reference tangent is horizontal, tangential deviations—the vertical distance between the tangent line and the elastic curve—are, in fact, displacements. Examples 9.3 to 9.5 (pages 323 and 325) cover the computation of slopes and deflections in cantilevers. Example 9.4 on page 324 illustrates how to modify an M/EI curve for a member whose moment of inertia varies. In Example 9.5 the moment curves produced by both a uniform and a concentrated load are plotted separately in order to produce moment curves with a known geometry. (See Table 3 inside the back cover for the properties of these areas.)

Case 2. Figures 9.7*b* and *c* show examples of symmetric structures loaded symmetrically with respect to the vertical axis of symmetry at the center of the structure. Because of symmetry the slope of the elastic curve is zero at the point where the axis of symmetry intersects the elastic curve. At this point the tangent to the elastic curve is horizontal. For the beams in Figure 9.7*b* and *c* we conclude, based on the first moment-area principle, that the slope at any point on the elastic curve equals the area under the M/EI curve between that point and the axis of symmetry.

The computation of deflections for points along the axis of the beam in Figure 9.7*c*, which has an *even* number of spans, is similar to that of the cantilever in Figure 9.7*a*. At the point of tangency (point B), both the deflection and slope of the elastic curve equal zero. Since the tangent to the elastic curve is horizontal, deflections at any other point are equal to tangential deviations from the tangent line drawn to the elastic curve at support B.

When a symmetric structure consists of an *odd* number of spans (one, three, and so on), the foregoing procedure must be modified slightly. For example, in Figure 9.7*b* we observe that the tangent to the elastic curve is horizontal at the axis of symmetry. Computation of slopes will again be referenced from the point of tangency at C. However, the centerline of the beam has displaced upward a distance v_C; therefore, tangential deviations from the reference tangents are usually not deflections. We can compute v_C by noting that the vertical distance between the tangent line and the elastic curve at either support B or C is a tangential deviation that equals v_C. For example, in Figure 9.7*b* v_C equals t_{BC}. After v_C is computed, the deflection of any other point that lies above the original position of the unloaded member equals v_C minus the tangential deviation of the point from the reference tangent. If a point lies below the undeflected position of the beam (for example, the tips of the cantilever at A or E), the deflection is equal to the tangential deviation of the point minus v_C. Examples 9.6 and 9.7 illustrate the computation of deflections in a symmetric structure.

Case 3. The structure is not symmetric but contains a member whose ends do not displace in a direction normal to the member's longitudinal axis. Examples of this case are shown in Figure 9.7*d* and *e*. Since the frame in Figure 9.7*d* is not symmetric and the beam in Figure 9.7*e* is not symmetrically loaded, the point at which a tangent to the elastic curve is horizontal is not initially known. Therefore, we must use a sloping tangent line as a reference for computing both slopes and deflections at points along the elastic curve. For this case we establish the slope of the elastic curve at either end of the member. At one end of the member, we draw a tangent to the curve and compute the tangential deviation at the opposite end. For example, in either Figure 9.7*d* or *e*, because deflections are small the slope of the tangent to the elastic curve at *A* can be written

$$\tan \theta_A = \frac{t_{BA}}{L} \tag{9.19}$$

Since $\tan \theta_A \approx \theta_A$ in radians, we can write Equation 9.19 as

$$\theta_A = \frac{t_{BA}}{L}$$

At a second point *C*, the slope would equal

$$\theta_C = \theta_A + \Delta\theta_{AC}$$

where $\Delta\theta_{AC}$ equals the area under the M/EI curve between points *A* and *C*.

To compute the displacements of a point *C* located a distance *x* to the right of support *A* (see Figure 9.7*e*), we first compute the vertical distance CC' between the initial position of the longitudinal axis and the reference tangent. Since θ_A is small, we can write

$$CC' = \theta_A(x)$$

The difference between CC' and the tangential deviation t_{CA} equals the deflection v_C:

$$v_C = CC' - t_{CA}$$

Examples 9.8 to 9.12 illustrate the procedure to compute slopes and deflections in members with inclined reference tangents.

If the M/EI curve between two points on the elastic curve contains both positive and negative areas, the net angle change in slope between those points equals the algebraic sum of the areas. If an accurate sketch of the deflected shape is drawn, the direction of both the angle changes and the deflections are generally apparent, and the student does not have to be concerned with establishing a formal sign convention to establish if a slope or deflection increases or decreases. Where the moment is positive (see Figure 9.8*a*), the member bends concave upward, and a tangent drawn to either end of the elastic curve will lie below the curve. In other words, we can interpret a positive value of tangential deviation as an indication that we move upward from the tangent line to the elastic curve. Conversely, if the tangential deviation is associated with a negative area under the M/EI curve, the tangent line lies above the elastic curve (see Figure 9.8*b*), and we move downward vertically from the tangent line to reach the elastic curve.

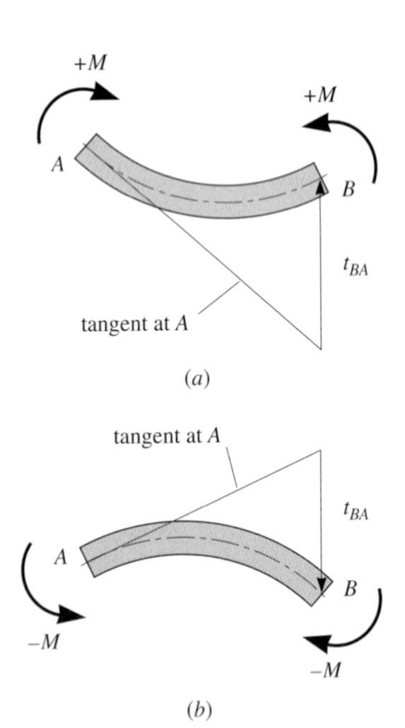

Figure 9.8: Position of reference tangent: (*a*) positive moment; (*b*) negative moment.

EXAMPLE 9.3

Compute the slope θ_B and the deflection v_B at the tip of the cantilever beam in Figure 9.9*a*. *EI* is constant.

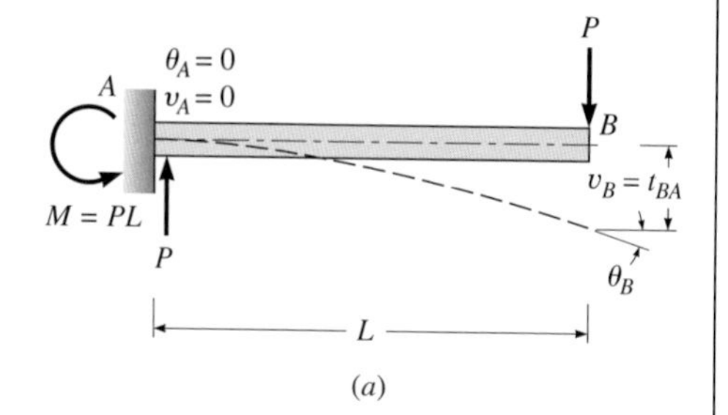

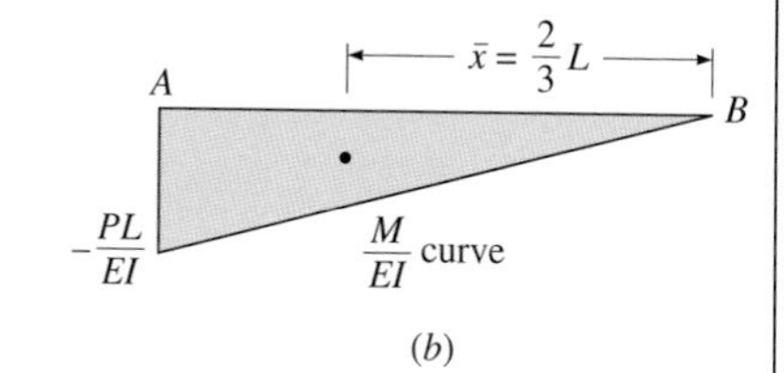

Figure 9.9: (*a*) beam; (*b*) *M/EI* curve.

Solution

Draw the moment curve and divide all ordinates by *EI* (Figure 9.9*b*).

Compute θ_B by adding to the slope at *A* the change in slope $\Delta\theta_{AB}$ between points *A* and *B*. Since the fixed support prevents rotation, $\theta_A = 0$.

$$\theta_B = \theta_A + \Delta\theta_{AB} = \Delta\theta_{AB} \tag{1}$$

By the first moment-area theorem, $\Delta\theta_{AB}$ equals the area under the triangular M/EI curve between points *A* and *B*.

$$\Delta\theta_{AB} = \frac{1}{2}(L)\left(\frac{-PL}{EI}\right) = \frac{-PL^2}{2EI} \tag{2}$$

Substituting Equation 2 into Equation 1 gives

$$\theta_B = -\frac{PL^2}{2EI} \qquad \textbf{Ans.}$$

Since the tangent line at *B* slopes downward to the right, its slope is negative. In this case the negative ordinate of the M/EI curve gave the correct sign. In most problems the direction of the slope is evident from the sketch of the deflected shape.

Compute the deflection v_B at the tip of the cantilever using the second moment-area theorem. The black dot in the M/EI curve denotes the centroid of the area.

$$v_B = t_{BA} = \text{moment of triangular area of } M/EI \text{ diagram about point } B$$

$$v_B = \frac{1}{2}L\left(\frac{-PL}{EI}\right)\frac{2L}{3} = -\frac{PL^3}{3EI} \quad \text{(minus sign indicates that the tangent line lies above elastic curve)} \qquad \textbf{Ans.}$$

EXAMPLE 9.4

Beam with a Variable Moment of Inertia

Compute the deflection of point C at the tip of the cantilever beam in Figure 9.10 if $E = 29{,}000$ kips/in^2, $I_{AB} = 2I$, and $I_{BC} = I$, where $I = 400$ in^4.

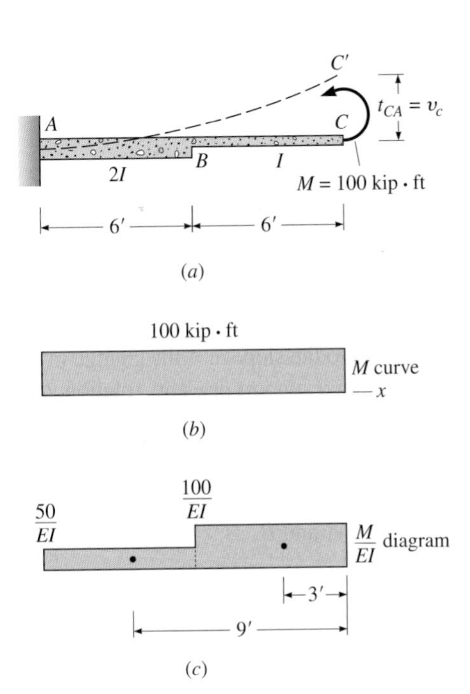

Figure 9.10: (a) Deflected shape; (b) moment curve; (c) M/EI curve divided into two rectangular areas.

Solution

To produce the M/EI curve, the ordinates of the moment curve are divided by the respective moments of inertia. Since I_{AB} is twice as large as I_{BC}, the ordinates of the M/EI curve between A and B will be one-half the size of those between B and C. Since the deflection at C, denoted by v_C, equals t_{CA}, we compute the moment of the area of the M/EI diagram about point C. For this computation, we divide the M/EI diagram into two rectangular areas.

$$v_C = t_{CA} = \frac{100}{2EI}(6)(9) + \frac{100}{EI}(6)(3) = \frac{4500}{EI}$$

$$v_C = \frac{4500(1728)}{29{,}000(400)} = 0.67 \text{ in} \qquad \textbf{Ans.}$$

where 1728 converts cubic feet to cubic inches.

EXAMPLE 9.5

Use of Moment Curve by "Parts"

Compute the slope of the elastic curve at B and C and the deflection at C for the cantilever beam in Figure 9.11*a*; EI is constant.

Solution

To produce simple geometric shapes in which the location of the centroid is known, the moment curves produced by the concentrated load P and the uniform load w are plotted separately and divided by EI in Figure 9.11*b* and *c*. Table 3 on the back inside cover provides equations for evaluating the areas of common geometric shapes and the position of their centroids.

Compute the slope at C where $\Delta\theta_{AC}$ is given by the sum of the areas under the M/EI diagrams in Figure 9.11*b* and *c*; $\theta_A = 0$ (see Figure 9.11*d*).

$$\theta_C = \theta_A + \Delta\theta_{AC}$$

$$= 0 + \frac{1}{2}(6)\left(\frac{-48}{EI}\right) + \frac{1}{3}(12)\left(\frac{-72}{EI}\right)$$

$$\theta_c = -\frac{432}{EI} \text{ radians} \qquad \textbf{Ans.}$$

Compute the slope at B. The area between A and B in Figure 9.11*c* is computed by deducting the parabolic area between B and C in Figure 9.11*c* from the total area between A and C. Since the slope at B is smaller than the slope at C, the area between B and C will be treated as a positive quantity to reduce the negative slope at C.

$$\theta_B = \theta_C + \Delta\theta_{BC}$$

$$= -\frac{432}{EI} + \frac{1}{3}(6)\left(\frac{18}{EI}\right)$$

$$\theta_B = -\frac{396}{EI} \text{ radians} \qquad \textbf{Ans.}$$

Compute Δ_C, the deflection at C. The deflection at C equals the tangential deviation of C from the tangent to the elastic curve at A (see Figure 9.11*d*).

$$\Delta_C = t_{CA} = \text{moments of areas under } M/EI \text{ curves between } A \text{ and } C \text{ in Figure 9.11}b \text{ and } c$$

$$= \frac{1}{2}(6)\left(\frac{-48}{EI}\right)(6 + 4) + \frac{1}{3}(12)\left(\frac{-72}{EI}\right)(9)$$

$$\Delta_c = \frac{-4032}{EI} \qquad \textbf{Ans.}$$

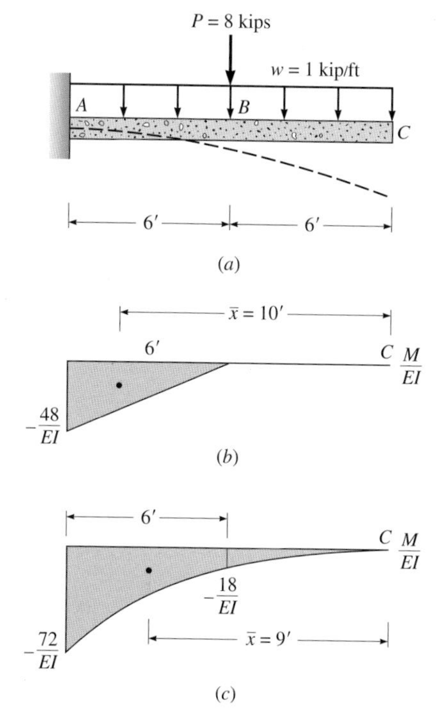

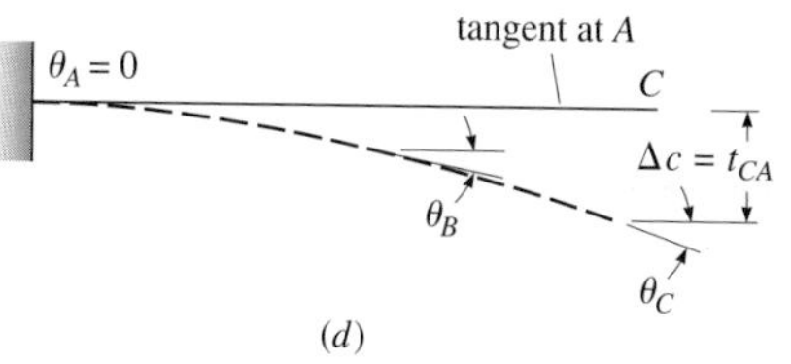

Figure 9.11: Moment curve by "parts": (*a*) beam; (*b*) M/EI curve associated with P; (*c*) M/EI curve associated with uniform load w; (*d*) deflected shape.

EXAMPLE 9.6

Analysis of a Symmetric Beam

For the beam in Figure 9.12*a*, compute the slope at *B* and the deflections at midspan and at point *A*. Also *EI* is constant.

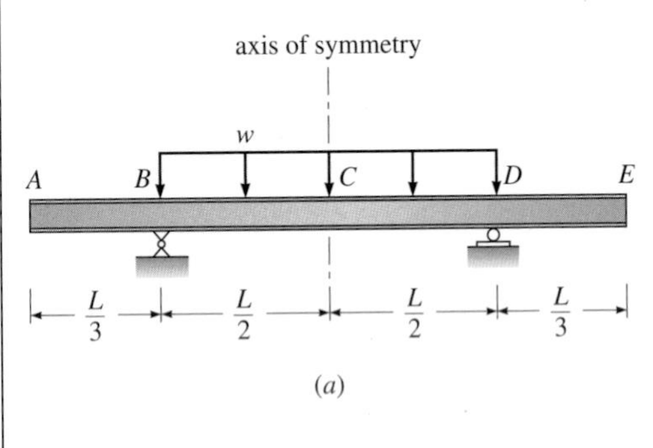

$\bar{x} = \frac{5L}{16}$ $\frac{wL^2}{8EI}$

$\frac{M}{EI}$

(b)

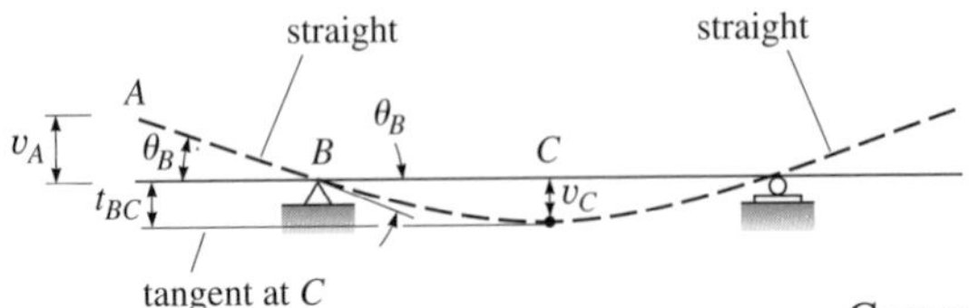

Figure 9.12: (*a*) Symmetric beam; (*b*) M/EI curve; (*c*) geometry of the deflected shape.

Solution

Because both the beam and its loading are symmetric with respect to the vertical axis of symmetry at midspan, the slope of the elastic curve is zero at midspan and the tangent line at that point is horizontal. Since no bending moments develop in the cantilevers (they are unloaded), the elastic curve is a straight line between points *A* and *B* and points *D* and *E*. See Appendix for geometric properties of a parabolic area.

Compute θ_B.

$$\theta_B = \theta_C + \Delta\theta_{CB}$$

$$= 0 + \frac{2}{3}\left(\frac{L}{2}\right)\left(\frac{wL^2}{8EI}\right)$$

$$= \frac{wL^3}{24EI} \qquad \textbf{Ans.}$$

Compute v_C. Since the tangent at *C* is horizontal, v_C equals t_{BC}. Using the second moment-area theorem, we compute the moment of the parabolic area between *B* and *C* about *B*.

$$v_C = t_{BC} = \frac{2}{3}\left(\frac{L}{2}\right)\left(\frac{wL^2}{8EI}\right)\left(\frac{5L}{16}\right) = \frac{5wL^4}{384EI} \qquad \textbf{Ans.}$$

Compute v_A. Since the cantilever *AB* is straight,

$$v_A = \theta_B \frac{L}{3} = \frac{wL^3}{24EI}\,\frac{L}{3} = \frac{wL^4}{72EI} \qquad \textbf{Ans.}$$

where θ_B is evaluated in the first computation.

EXAMPLE 9.7

The beam in Figure 9.13*a* supports a concentrated load P at midspan (point C). Compute the deflections at points B and C. Also compute the slope at A. EI is constant.

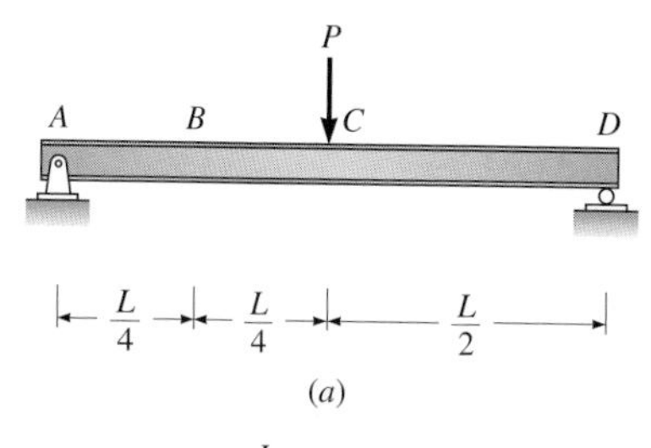

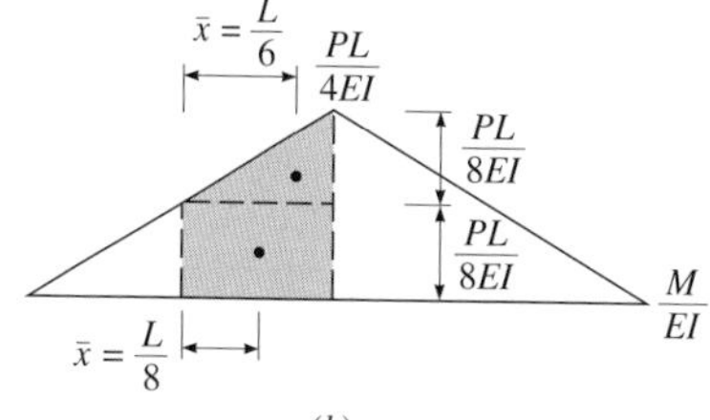

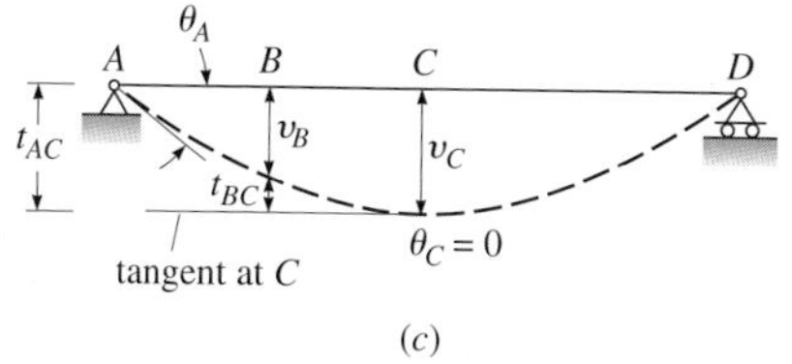

Figure 9.13: (*a*) Beam details; (*b*) M/EI curve; (*c*) deflected shape.

Solution

Compute θ_A. Since the structure is symmetrically loaded, the slope of the line tangent to the elastic curve at midspan is zero; that is, $\theta_C = 0$ (see Figure 9.13*c*).

$$\theta_A = \theta_C + \Delta\theta_{AC}$$

where $\Delta\theta_{AC}$ is equal to the area under M/EI curve between A and C.

$$\theta_A = 0 + \frac{1}{2}\left(\frac{L}{2}\right)\left(\frac{PL}{4EI}\right) = \frac{PL^2}{16EI} \quad \text{radians} \qquad \textbf{Ans.}$$

Compute v_C, the deflection at midspan. Since the tangent at C is horizontal, $v_C = t_{AC}$, where t_{AC} equals the moment about A of the triangular area under the M/EI curve between A and C.

$$v_C = \frac{1}{2}\left(\frac{L}{2}\right)\left(\frac{PL}{4EI}\right)\left(\frac{2}{3}\,\frac{L}{2}\right) = \frac{PL^3}{48EI} \tag{1}$$

Compute v_B, the deflection at the quarter point. As shown in Figure 9.13*c*,

$$v_B + t_{BC} = v_C = \frac{PL^3}{48EI} \tag{2}$$

where t_{BC} is the moment about B of the area under the M/EI curve between B and C. For convenience, we divide this area into a triangle and a rectangle. See the shaded area in Figure 9.13*b*.

$$t_{BC} = \frac{1}{2}\left(\frac{L}{4}\right)\left(\frac{PL}{8EI}\right)\left(\frac{L}{6}\right) + \frac{L}{4}\left(\frac{PL}{8EI}\right)\left(\frac{L}{8}\right) = \frac{5PL^3}{768EI}$$

Substituting t_{BC} into Equation 2, we compute v_B.

$$v_B = \frac{11PL^3}{768EI} \qquad \textbf{Ans.}$$

EXAMPLE 9.8

For the beam in Figure 9.14*a*, compute the slope of the elastic curve at points *A* and *C*. Also determine the deflection at *A*. Assume rocker at *C* equivalent to a roller.

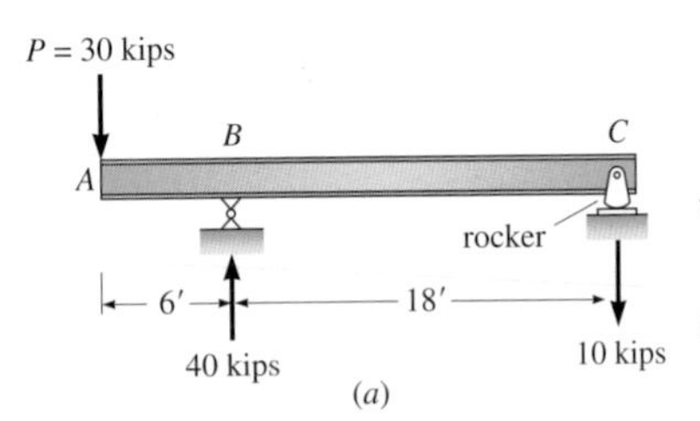

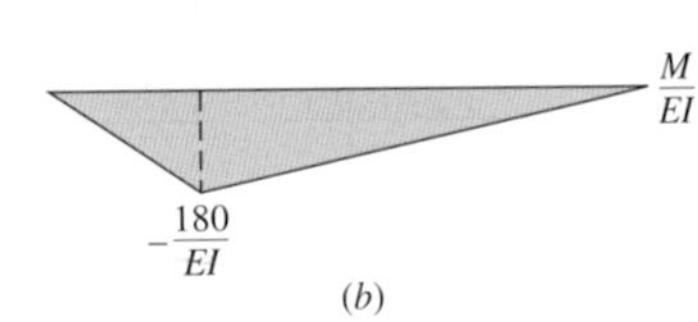

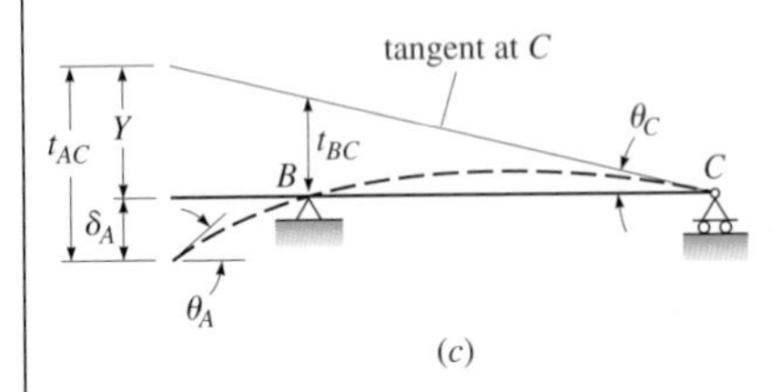

Figure 9.14: (*a*) Beam; (*b*) M/EI curve; (*c*) geometry of deflected shape.

Solution

Since the moment curve is negative at all sections along the axis of the beam, it is bent concave downward (see the dashed line in Figure 9.14*c*). To compute θ_C, we draw a tangent to the elastic curve at point *C* and compute t_{BC}.

$$\theta_C = \frac{t_{BC}}{18} = \frac{9720}{EI}\left(\frac{1}{18}\right) = -\frac{540}{EI} \qquad \textbf{Ans.}$$

$$\text{where} \quad t_{BC} = \text{area}_{BC} \cdot \bar{x} = \frac{1}{2}(18)\left(-\frac{180}{EI}\right)\left(\frac{18}{3}\right) = -\frac{9720}{EI}$$

(Since the tangent line slopes downward to the right, the slope θ_C is negative.)

. Compute θ_A.

$$\theta_A = \theta_C + \Delta\theta_{AC}$$

where $\Delta\theta_{AC}$ is the area under the M/EI curve between *A* and *C*. Since the elastic curve is concave downward between points *A* and *C*, the slope at *A* must be opposite in sense to the slope at *C*; therefore, $\Delta\theta_{AC}$ must be treated as a positive quantity.

$$\theta_A = -\frac{540}{EI} + \frac{1}{2}(24)\left(\frac{180}{EI}\right) = \frac{1620}{EI} \qquad \textbf{Ans.}$$

Compute δ_A.

$$\delta_A = t_{AC} - Y \text{ (see Figure 9.14}c\text{)} = \frac{8640}{EI} \qquad \textbf{Ans.}$$

$$\text{where} \quad t_{AC} = \text{area}_{AC} \cdot \bar{x} = \frac{1}{2}(24)\left(\frac{180}{EI}\right)\left(\frac{6 + 24}{3}\right) = \frac{21{,}600}{EI}$$

[See case (a) in Table 3 on the back inside cover for the equation for $\bar{x}$.]

$$Y = 24\theta_C = 24\left(\frac{540}{EI}\right) = \frac{12{,}960}{EI}$$

EXAMPLE 9.9

Analysis Using a Sloping Reference Tangent

For the steel beam in Figure 9.15*a*, compute the slope at *A* and *C*. Also determine the location and value of the maximum deflection. If the maximum deflection is not to exceed 0.6 in, what is the minimum required value of *I*? *EI* is constant and $E = 29{,}000$ kips/in^2.

Solution

Compute the slope θ_A at support *A* by drawing a line tangent to the elastic curve at that point. This will establish a reference line of known direction (see Figure 9.15*c*).

$$\tan \theta_A = \frac{t_{CA}}{L} \tag{1}$$

Since for small angles $\tan \theta_A \approx \theta_A$ (radians), Equation 1 can be written

$$\theta_A = \frac{t_{CA}}{L} \tag{2}$$

$$t_{CA} = \text{moment of } M/EI \text{ area between } A \text{ and } C \text{ about } C$$

$$= \frac{1}{2}(18)\left(\frac{96}{EI}\right)\left(\frac{18+6}{3}\right) = \frac{6912}{EI}$$

where the expression for the moment arm is given in Table 3 on the back inside cover, case (a). Substituting t_{CA} into Equation 2 gives

$$\theta_A = \frac{-6912/EI}{18} = -\frac{384}{EI} \text{ radians} \qquad \textbf{Ans.}$$

A minus sign is added, because moving in the positive *x* direction, the tangent line, directed downward, has a negative slope.

Compute θ_C.

$$\theta_C = \theta_A + \Delta\theta_{AC}$$

where $\Delta\theta_{AC}$ equals area under M/EI curve between *A* and *C*.

$$\theta_C = -\frac{384}{EI} + \frac{1}{2}(18)\left(\frac{96}{EI}\right) = \frac{480}{EI} \text{ radians} \qquad \textbf{Ans.}$$

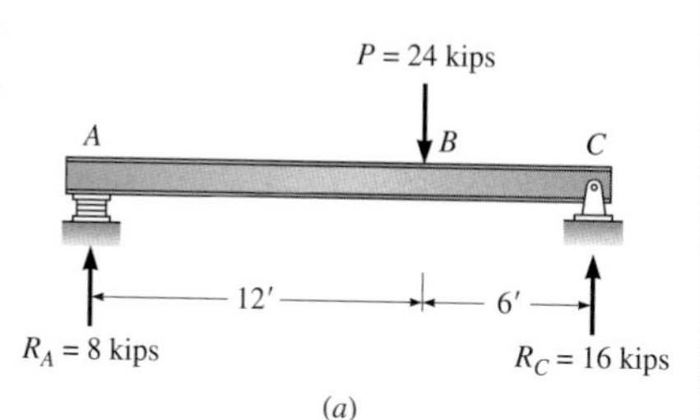

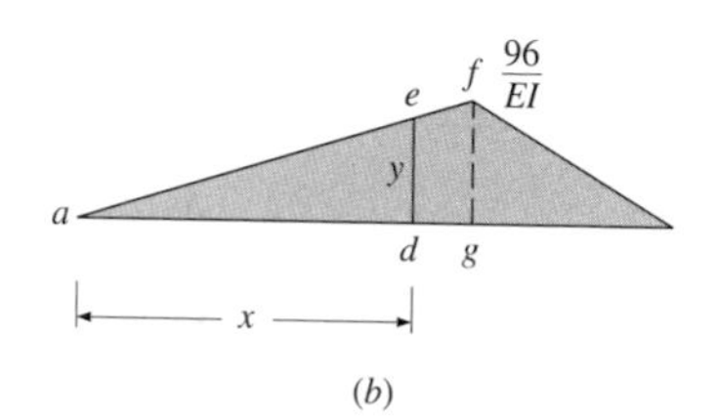

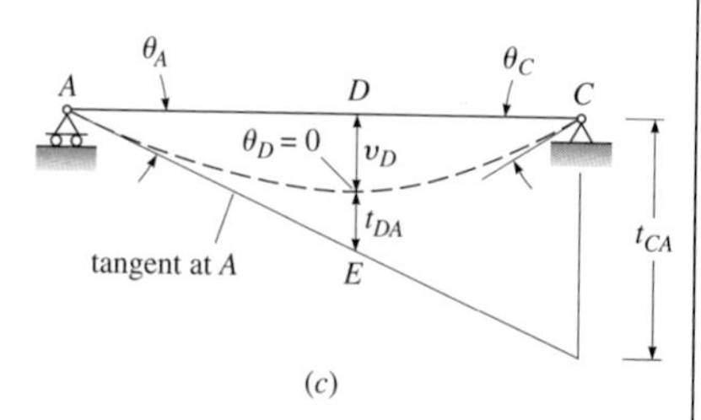

Figure 9.15: (*a*) Beam; (*b*) *M/EI* curve; (*c*) geometry of deflected shape.

[*continues on next page*]

Example 9.9 continues . . .

Compute the maximum deflection. The point of maximum deflection occurs at point D where the slope of the elastic curve equals zero (that is, $\theta_D = 0$). To determine this point, located an unknown distance x from support A, we must determine the area under the M/EI curve between A and D that equals the slope at A. Letting y equal the ordinate of the M/EI curve at D (Figure 9.15b) gives

$$\theta_D = \theta_A + \Delta\theta_{AD}$$

$$0 = -\frac{384}{EI} + \frac{1}{2}xy \tag{3}$$

Expressing y in terms of x by using similar triangles afg and aed (see Figure 9.15b) yields

$$\frac{96/(EI)}{12} = \frac{y}{x}$$

$$y = \frac{8x}{EI} \tag{4}$$

Substituting the foregoing value of y into Equation 3 and solving for x give

$$x = 9.8 \text{ ft}$$

Substituting x into Equation 4 gives

$$y = \frac{78.4}{EI}$$

Compute the maximum deflection v_D at $x = 9.8$ ft

$$v_D = DE - t_{DA} \tag{5}$$

where the terms in Equation 5 are illustrated in Figure 9.15c.

$$DE = \theta_A \cdot x = \frac{384}{EI}(9.8) = \frac{3763.2}{EI}$$

$$t_{DA} = (\text{area}_{AD})\bar{x} = \frac{1}{2}(9.8)\left(\frac{78.4}{EI}\right)\left(\frac{9.8}{3}\right) = \frac{1254.9}{EI}$$

Substituting DE and t_{DA} into Equation 5 gives

$$v_D = \frac{3763.2}{EI} - \frac{1254.9}{EI} = \frac{2508.3}{EI} \tag{6}$$

Compute I_{min} if v_D is not to exceed 0.6 in; in Equation 6 set $v_D = 0.6$ in and solve for I_{min}.

$$v_D = \frac{2508.3(1728)}{29{,}000 I_{min}} = 0.6 \text{ in} \qquad \textbf{Ans.}$$

$$I_{min} = 249.1 \text{ in}^4 \qquad \textbf{Ans.}$$

EXAMPLE 9.10

The beam in Figure 9.16*a* contains a hinge at *B*. Compute the deflection v_B of the hinge, the slope of the elastic curve at support *E*, and the end slopes θ_{BL} and θ_{BR} of the beams on either side of the hinge (see Figure 9.16*d*). Also locate the point of maximum deflection in span *BE*. *EI* is constant. The elastomeric pad at *E* is equivalent to a roller.

Solution

The deflection of the hinge at *B*, denoted by v_B, equals t_{BA}, the tangential deviation of *B* from the tangent to the fixed support at *A*. Deflection t_{BA} equals the moment of the area under the *M*/*EI* curve between *A* and *B* about *B* (see Figure 9.16*b*).

$$v_B = t_{BA} = \text{area} \cdot \bar{x} = \frac{1}{2}\left(-\frac{108}{EI}\right)(9)(6) = -\frac{2916}{EI}$$

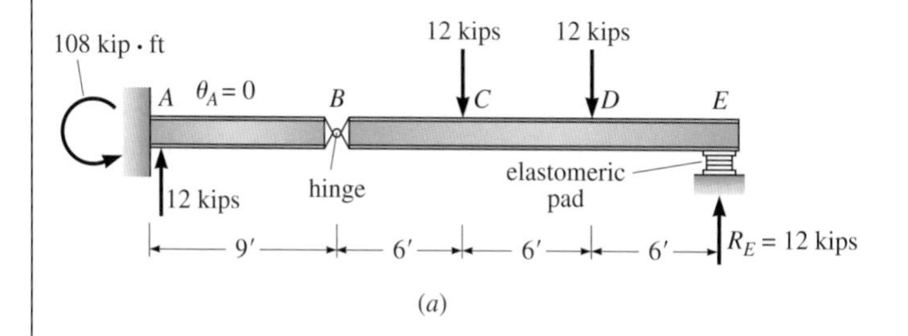

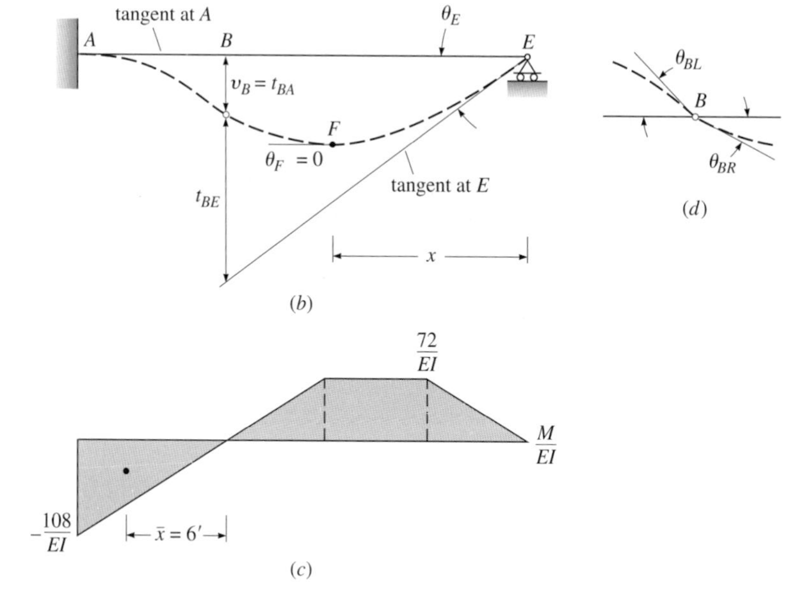

Figure 9.16: (*a*) Beam with hinge at *B*; (*b*) deflected shape; (*c*) *M*/*EI* curve; (*d*) detail showing the difference in slope of the elastic curve on each side of the hinge.

[*continues on next page*]

Example 9.10 continues . . .

Compute θ_{BL}, the slope of the B end of cantilever AB.

$$\theta_{BL} = \theta_A + \Delta\theta_{AB}$$

$$= 0 + \frac{1}{2}(9)\left(\frac{-108}{EI}\right) = \frac{-486}{EI} \quad \text{radians}$$

where $\Delta\theta_{AB}$ is equal to the triangular area under the M/EI curve between A and B and $\theta_A = 0$ because the fixed support at A prevents rotation.

Compute θ_E, the slope of the elastic curve at E (see Figure 9.16*b*).

$$\theta_E = \frac{v_B + t_{BE}}{18} = \left(\frac{2916}{EI} + \frac{7776}{EI}\right)\left(\frac{1}{18}\right) = \frac{594}{EI} \quad \text{radians}$$

where t_{BE} equals the moment of the area under the M/EI curve between B and E about B. This computation is simplified by dividing the trapezoidal area into two triangles and a rectangle (see the dashed lines in Figure 9.16*c*).

$$t_{BE} = \frac{1}{2}(6)\left(\frac{72}{EI}\right)(4) + (6)\left(\frac{72}{EI}\right)(9) + \frac{1}{2}(6)\left(\frac{72}{EI}\right)(14) = \frac{7776}{EI} \quad \text{radians}$$

Locate the point of maximum deflection in span BE. The point of maximum deflection, labeled point F, is located at the point in span BE where the tangent to the elastic curve is zero. Between F and support E, a distance x, the slope goes from 0 to θ_E. Since the change in slope is given by the area under the M/EI curve between these two points, we can write

$$\theta_E = \theta_F + \Delta\theta_{EF} \tag{1}$$

where $\theta_F = 0$ and $\theta_E = 594/EI$ rad. Between points D and E the change in slope produced by the area under the M/EI curve equals $216/EI$. Since this value is less than θ_E, the slope at D has a positive value of

$$\theta_D = \theta_E - \Delta\theta_{ED} = \frac{594}{EI} - \frac{216}{EI} = \frac{378}{EI} \quad \text{radians} \tag{2}$$

Between D and C the area under the M/EI curve equals $432/EI$. Since this value of change in slope exceeds $378/EI$, the point of zero slope must lie between C and D. We can now use Equation 1 to solve for distance x.

$$\frac{594}{EI} = 0 + \frac{1}{2}\left(\frac{72}{EI}\right)(6) + \frac{72}{EI}(x - 6)$$

$$x = 11.25 \text{ ft}$$ **Ans.**

Compute θ_{BR}.

$$\theta_{BR} = \theta_E - \Delta\theta_{BE}$$

$$= \frac{594}{EI} - \left[\frac{72}{EI}(6) + \frac{1}{2}(6)\left(\frac{72}{EI}\right)(2)\right]$$

$$= -\frac{270}{EI} \quad \text{radians}$$ **Ans.**

EXAMPLE 9.11

Determine the deflection of the hinge at C and the rotation of joint B for the frame in Figure 9.17*a*. For all members EI is constant.

Solution

To establish the angular rotation of joint B, we consider the deflected shape of member AB in Figure 9.17*b*. (Because member BCD contains a hinge, its elastic curve is not continuous, and it is not possible initially to compute the slope at any point along its axis.)

$$\theta_B = \frac{t_{AB}}{12} = \frac{\frac{1}{2}12\frac{72}{EI}(8)}{12} = \frac{288}{EI} \qquad \textbf{Ans.}$$

Deflection of hinge:

$$\Delta = 6\theta_B + t_{CB}$$

$$= (6)\left(\frac{288}{EI}\right) + \frac{1}{2}(6)\left(\frac{72}{EI}\right)(4) = \frac{2592}{EI} \qquad \textbf{Ans.}$$

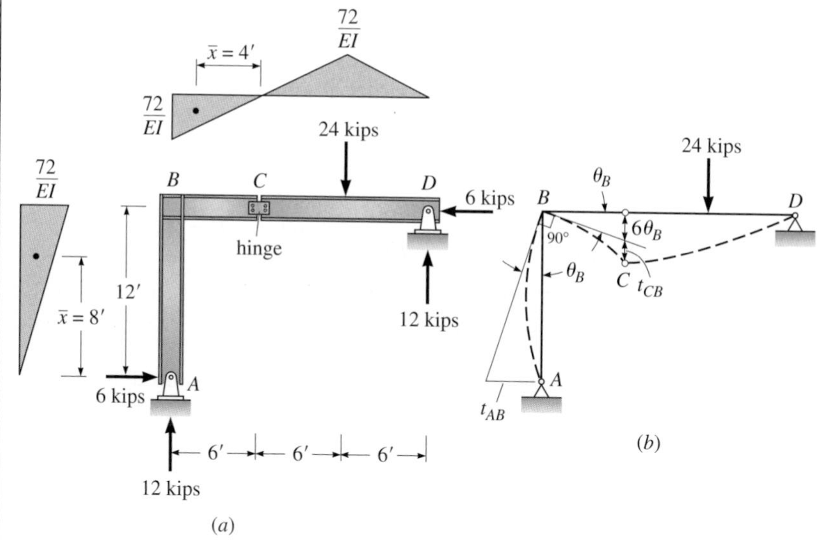

Figure 9.17: (*a*) Frame and M/EI curves; (*b*) deflected shape.

EXAMPLE 9.12

Compute the horizontal deflection of joint B of the frame shown in Figure 9.18*a*. EI is constant all members. Assume elastomeric pad at C acts as a roller.

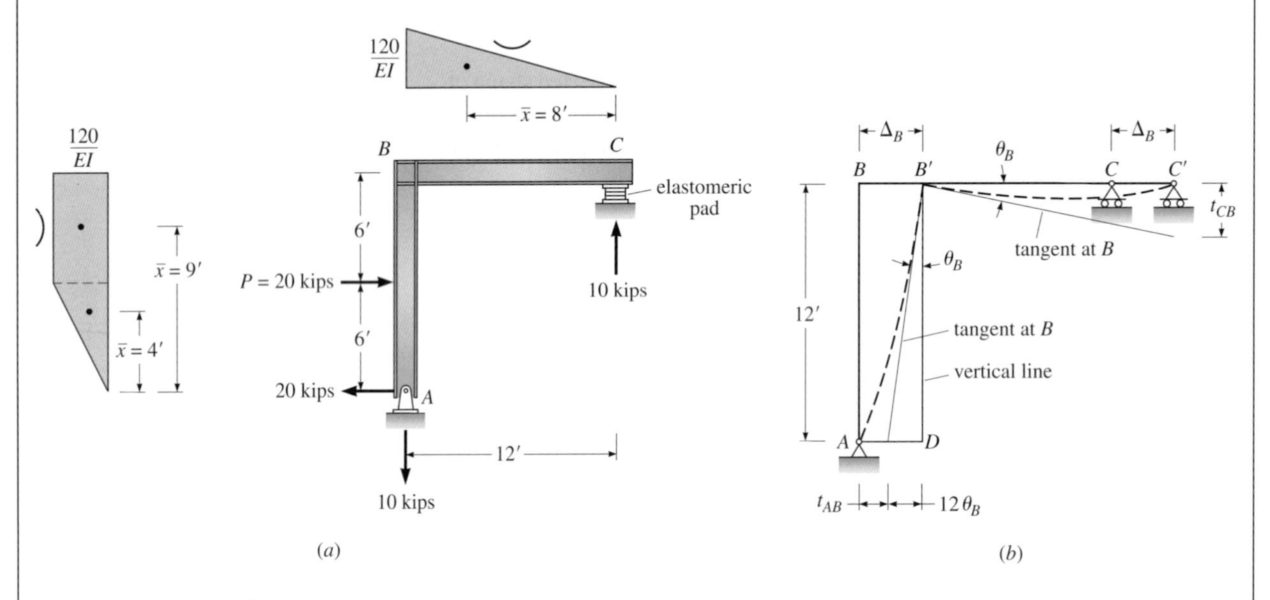

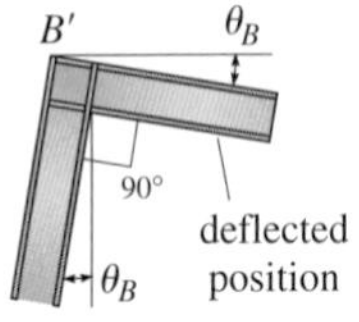

Figure 9.18: (*a*) Frame and M/EI curves; (*b*) deflected shape; (*c*) detail of joint B in deflected position.

Solution

Begin by establishing the slope of the girder at joint B.

$$\theta_B = \frac{t_{CB}}{L} \tag{1}$$

where $t_{CB} = \frac{1}{2}\left(\frac{120}{EI}\right)(12)(8) = \frac{5760}{EI}$ and $L = 12$ ft

Thus $\theta_B = \frac{5760}{EI}\left(\frac{1}{12}\right) = \frac{480}{EI}$ radians

Because joint B is rigid, the top of column AB also rotates through an angle θ_B (see Figure 9.18*c*). Since the deflection Δ_B at joint B is equal to the horizontal distance AD at the base of the column, we can write

$$\begin{aligned}\Delta_B &= AD = t_{AB} + 12\theta_B\\ &= \frac{120}{EI}(6)(9) + \frac{1}{2}\left(\frac{120}{EI}\right)(6)(4) + (12)\left(\frac{480}{EI}\right)\\ &= \frac{13{,}680}{EI} \quad \textbf{Ans.}\end{aligned}$$

where t_{AB} equals the moment of the M/EI diagram between A and B about A, and the M/EI diagram is broken into two areas.

9.4 Elastic Load Method

The elastic load method is a procedure for computing slopes and deflections in simply supported beams. Although the calculations in this method are identical to those of the moment-area method, the procedure appears simpler because we replace computations of tangential deviations and changes in slope with the more familiar procedure of constructing shear and moment curves for a beam. Thus the elastic load method eliminates the need (1) to draw an accurate sketch of the member's deflected shape and (2) to consider which tangential deviations and angle changes to evaluate in order to establish the deflection or the slope at a specific point.

In the elastic load method, we imagine that the M/EI diagram, whose ordinates represent angle change per unit length, is applied to the beam as a load (the *elastic load*). We then compute the shear and moment curves. As we will demonstrate next, the ordinates of the shear and the moment curves at each point equal the slope and deflection, respectively, in the real beam.

To illustrate that the shear and moment at a section produced by an *angle change*, applied to a simply supported beam as a fictitious load, equal the slope and deflection at the same section, we examine the deflected shape of a beam whose longitudinal axis is composed of two straight segments that intersect at a small angle θ. The geometry of the bent member is shown by the solid line in Figure 9.19.

If the beam ABC' is connected to the support at A so that segment AB is horizontal, the right end of the beam at C' will be located a distance Δ_C above support C. In terms of the dimensions of the beam and the angle θ (see triangle $C'BC$), we find

$$\Delta_C = \theta(L - x) \tag{1}$$

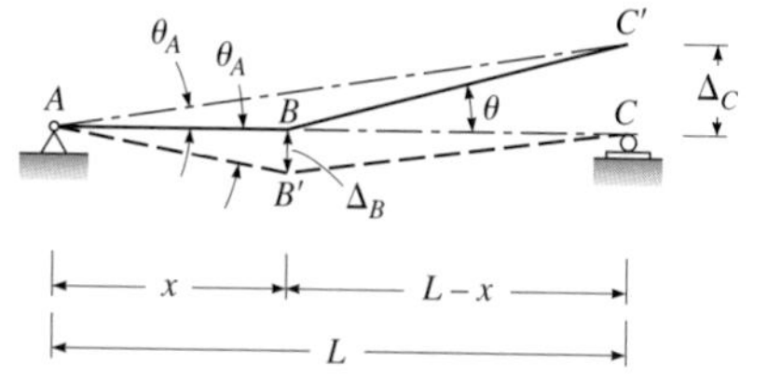

Figure 9.19: Beam with an angle change of θ at point B.

The sloping line AC', which connects the ends of the beam, makes an angle θ_A with a horizontal axis through A. Considering the right triangle ACC', we can express θ_A in terms of Δ_C as

$$\theta_A = \frac{\Delta_C}{L} \tag{2}$$

Substituting Equation 1 into Equation 2 leads to

$$\theta_A = \frac{\theta(L - x)}{L} \tag{3}$$

We now rotate member ABC' clockwise about the pin at A until chord AC' coincides with the horizontal line AC and point C' rests on the roller at C. The final position of the beam is shown by the heavy dashed line $AB'C$. As a result of the rotation, segment AB slopes downward to the right at an angle θ_A.

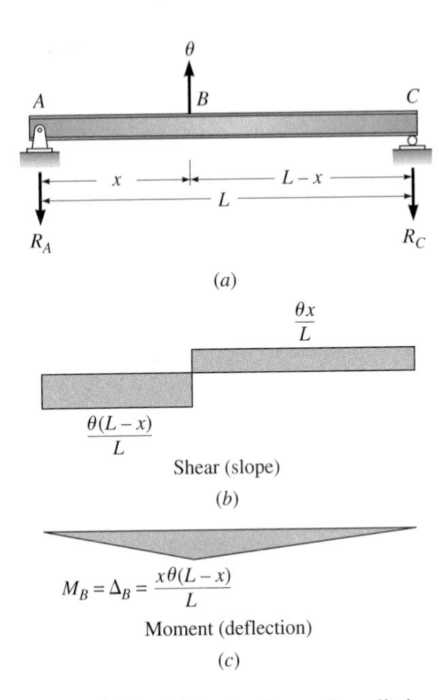

Figure 9.20: (*a*) Angle change θ applied as a load at point B; (*b*) shear produced by load θ equals slope in real beam; (*c*) moment produced by θ equals deflection in real beam (see Figure 9.19).

To express Δ_B, the vertical deflection at B, in terms of the geometry of the deflected member, we consider triangle ABB'. Assuming that angles are small, we can write

$$\Delta_B = \theta_A x \tag{4}$$

Substituting θ_A given by Equation 3 into Equation 4 gives

$$\Delta_B = \frac{\theta(L - x)x}{L} \tag{5}$$

Alternatively, we can compute identical values of θ_A and Δ_B by computing the shear and moment produced by the angle change θ applied as an *elastic* load to the beam at point B (see Figure 9.20*a*). Summing moments about support C to compute R_A produces

$$\circlearrowright^{+} \quad \Sigma M_C = 0$$

$$\theta(L - x) - R_A L = 0$$

$$R_A = \frac{\theta(L - x)}{L} \tag{6}$$

After R_A is computed, we draw the shear and moment curves in the usual manner (see Figure 9.20*b* and *c*). Since the shear just to the right of support A equals R_A, we observe that the shear given by Equation 6 is equal to the slope given by Equation 3. Further, because the shear is constant between the support and point B, the slope of the real structure must also be constant in the same region.

Recognizing that the moment M_B at point B equals the area under the shear curve between A and B, we find

$$\Delta_B = M_B = \frac{\theta(L - x)x}{L} \tag{7}$$

Comparing the value of deflections at B given by Equations 5 and 7, we verify that the moment M_B produced by load θ is equal to the value of Δ_B based on the geometry of the bent beam. We also observe that the maximum deflection occurs at the section where the shear produced by the elastic load is zero.

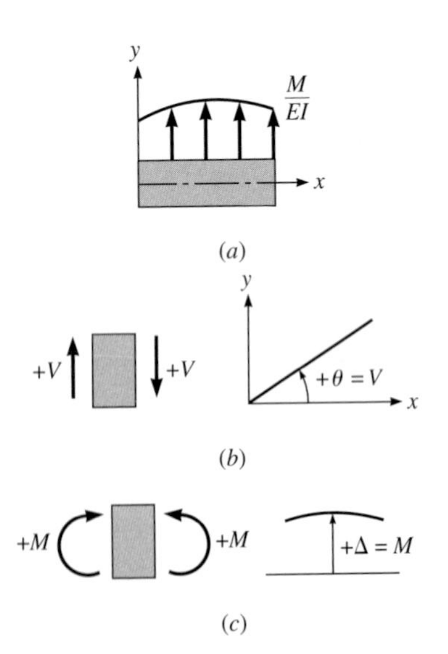

Figure 9.21: (*a*) Positive elastic load; (*b*) positive shear and positive slope; (*c*) positive moment and positive (upward) deflection.

Sign Convention

If we treat positive values of the M/EI diagram applied to the beam as a distributed load acting upward and negative values of M/EI as a downward load, positive shear denotes a positive slope and negative shear a negative slope (see Figure 9.21). Further, negative values of moment indicate a downward deflection and positive values of moment an upward deflection.

Examples 9.13 and 9.14 illustrate the use of the elastic load method to compute deflections of simply supported beams.

EXAMPLE 9.13

Compute the maximum deflection and the slope at each support for the beam in Figure 9.22*a*. Note that *EI* is a constant.

Solution

As shown in Figure 9.22*b*, the M/EI diagram is applied to the beam as an upward load. The resultants of the triangular distributed loads between *AB* and *BC*, which equal $720/EI$ and $360/EI$, respectively, are shown by heavy arrows. That is,

$$\frac{1}{2}(12)\left(\frac{120}{EI}\right) = \frac{720}{EI} \quad \text{and} \quad \frac{1}{2}(6)\left(\frac{120}{EI}\right) = \frac{360}{EI}$$

Using the resultants, we compute the reactions at supports *A* and *C*. The shear and moment curves, drawn in the conventional manner, are plotted in Figure 9.22*c* and *d*. To establish the point of maximum deflection, we locate the point of zero shear by determining the area under the load curve (shown shaded) required to balance the left reaction of $480/EI$.

$$\frac{1}{2}xy = \frac{480}{EI} \tag{1}$$

Using similar triangles (see Figure 9.22*b*) yields

$$\frac{y}{120/(EI)} = \frac{x}{12}$$

and

$$y = \frac{10}{EI}x \tag{2}$$

Substituting Equation 2 into Equation 1 and solving for *x* give

$$x = \sqrt{96} = 9.8 \text{ ft}$$

To evaluate the maximum deflection, we compute the moment at $x = 9.8$ ft by summing moments of the forces acting on the free body to the left of a section through the beam at that point. (See shaded area in Figure 9.22*b*.)

$$\Delta_{\max} = M = -\frac{480}{EI}(9.8) + \frac{1}{2}xy\left(\frac{x}{3}\right)$$

Using Equation 2 to express *y* in terms of *x* and substituting $x = 9.8$ ft, we compute

$$\Delta_{\max} = -\frac{3135.3}{EI} \downarrow \qquad \textbf{Ans.}$$

The values of the end slopes, read directly from the shear curve in Figure 9.22*c*, are

$$\theta_A = -\frac{480}{EI} \qquad \theta_C = \frac{600}{EI} \qquad \textbf{Ans.}$$

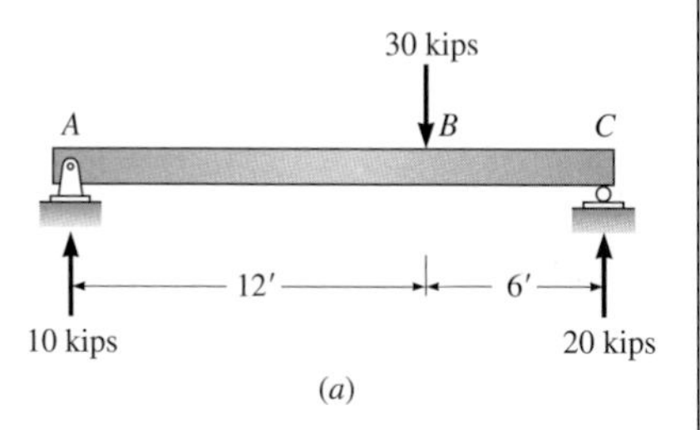

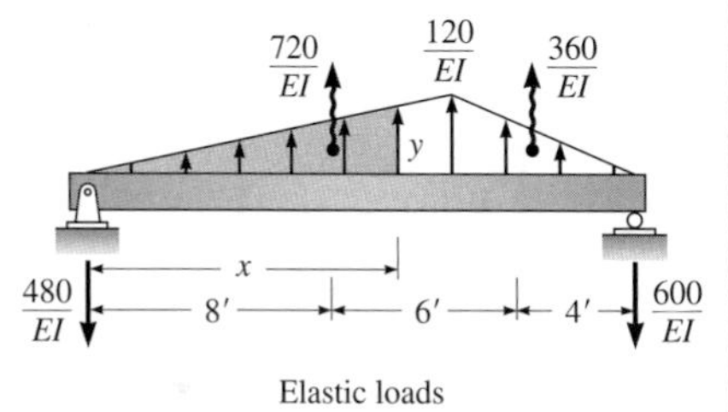

600/EI

−480/EI

x = 9.8′

Shear (slope)

(c)

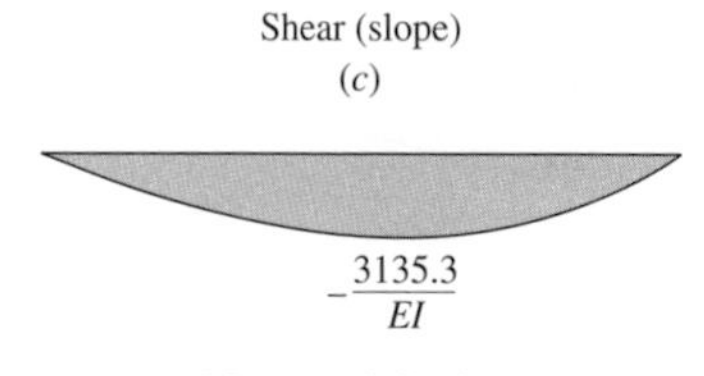

Figure 9.22: (*a*) Beam; (*b*) beam loaded by M/EI diagram; (*c*) variation of slope; (*d*) deflected shape.

EXAMPLE 9.14

Compute the deflection at point B of the beam in Figure 9.23a. Also locate the point of maximum deflection; E is a constant, but I varies as shown on the figure.

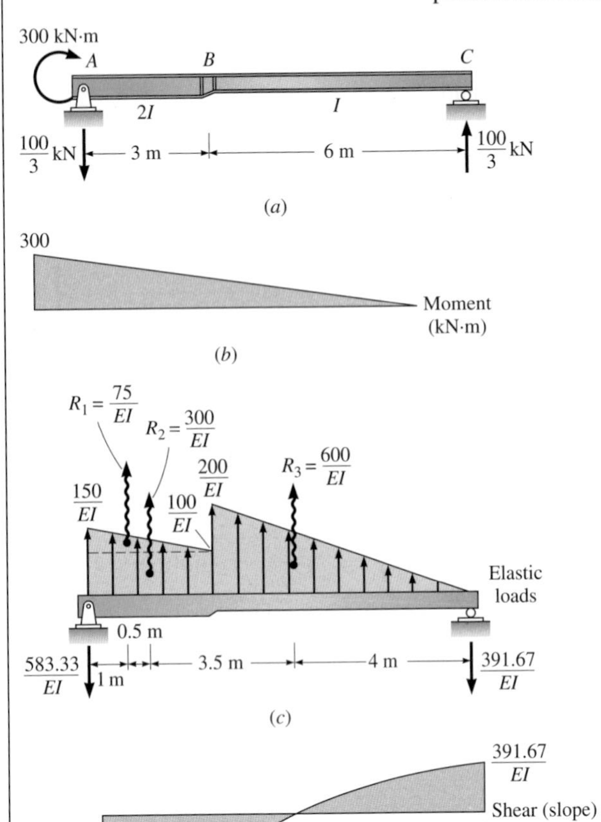

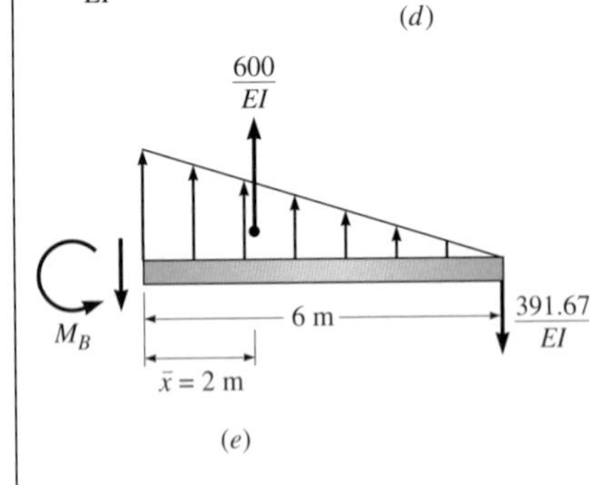

Figure 9.23

Solution

To establish the M/EI curve, we divide the ordinates of the moment curve (see Figure 9.23b) by $2EI$ between A and B and by EI between B and C. The resulting M/EI diagram is applied to the beam as an upward load in Figure 9.23c. The maximum deflection occurs 4.85 m to the left of support C, where the elastic shear equals zero (Figure 9.23d).

To compute the deflection at B, we compute the moment produced at that point by the elastic loads using the free body shown in Figure 9.23e. Summing moments of the applied loads about B, we compute

$$\Delta_B = M_B = \frac{600}{EI}(2) - \frac{391.67}{EI}(6)$$

$$\Delta_B = -\frac{1150}{EI} \downarrow \qquad \textbf{Ans.}$$

9.5 Conjugate Beam Method

In Section 9.4 we used the elastic load method to compute slopes and deflections at points in a simply supported beam. The conjugate beam method, the topic of this section, permits us to extend the elastic load method to beams with other types of supports and boundary conditions by replacing the actual supports with *conjugate supports* to produce a conjugate beam. The effect of these fictitious supports is to impose boundary conditions which ensure that the *shear* and *moment*, produced in a beam loaded by the M/EI diagram, are equal to the *slope* and the *deflection*, respectively, in the real beam.

To explain the method, we consider the relationship between the shear and moment (produced by the elastic loads) and the deflected shape of the cantilever beam shown in Figure 9.24*a*. The M/EI curve associated with the concentrated load P acting on the real structure establishes the curvature at all points along the axis of the beam (see Figure 9.24*b*). For example, at B, where the moment is zero, the curvature is zero. On the other hand, at A the curvature is greatest and equal to $-PL/EI$. Since the curvature is negative at all sections along the axis of the member, the beam is bent concave downward over its entire length, as shown by the curve labeled 1 in Figure 9.24*c*. Although the deflected shape given by curve 1 is consistent with the M/EI diagram, we recognize that it does not represent the correct deflected shape of the cantilever because the slope at the left end is not consistent with the boundary conditions imposed by the fixed support at A; that is, the slope (and the deflection) at A must be zero, as shown by the curve labeled 2.

Therefore, we can reason that if the slope and deflection at A must be zero, the values of *elastic shear* and *elastic moment* at A must also equal zero. Since the only boundary condition that satisfies this requirement is a free end, we must imagine that the support A is removed—if no support exists, no reactions can develop. By establishing the correct slope and deflection at the end of the member, we ensure that the member is oriented correctly.

On the other hand, since both slope and deflection can exist at the free end of the actual cantilever, a support that has a capacity for shear and moment must be provided at B. Therefore, in the conjugate beam we must introduce an *imaginary fixed support* at B. Figure 9.24*d* shows the conjugate beam loaded by the M/EI diagram. The reactions at B in the conjugate beam produced by the elastic load [M/EI diagram] give the slope and deflection in the real beam.

Figure 9.25 shows the conjugate supports that correspond to a variety of standard supports. Two supports that we have not discussed previously—the interior roller and the hinge—are shown in Figure 9.25*d* and *e*. Since an interior roller (Figure 9.25*d*) provides vertical restraint only, the deflection at the roller is zero but the member is free to rotate. Because the member is continuous, the slope is the same on each side of the joint. To satisfy these geometric requirements, the conjugate support must have zero capacity for moment (thus, zero deflection), but must permit equal values of shear to exist on each side of the support—hence the hinge.

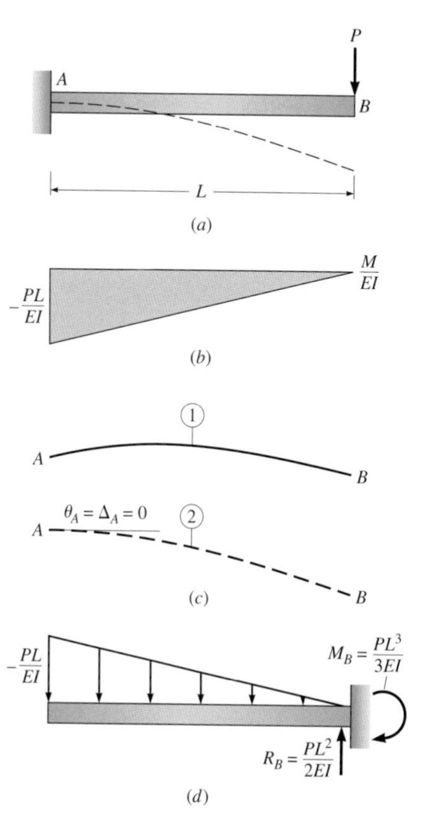

Figure 9.24: (*a*) Deflected shape of a cantilever beam; (*b*) M/EI diagram which establishes variation of curvature; (*c*) Curve 1 shows a deflected shape consistent with M/EI diagram in (*b*) but not with the boundary conditions at A. Curve 2 shows curve 1 rotated clockwise as a rigid body until the slope at A is horizontal; (*d*) Conjugate beam with elastic load.

Conjugate Beam

The conjugate beam method was developed by Harold M. Westergaard (1888–1950) in 1921 to compute deflections of beams. Westergaard made significant contributions in the analysis of reinforced concrete slab, concrete pavement for roads, buckling theory, water pressure on Hoover Dam during earthquakes, elasticity and plasticity theories.

Since a hinge provides no restraint against deflection or rotation in a real structure (see Figure 9.25*e*), the device introduced into the conjugate structure must ensure that moment as well as different values of shear on each side of the joint can develop. These conditions are supplied by using an interior roller in the conjugate structure. Moment can develop because the beam is continuous over the support, and the shear obviously can have different values on each side of the roller.

	Real Support	Conjugate Support
(*a*)	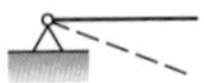 Pin or roller $\Delta = 0$ $\theta \neq 0$	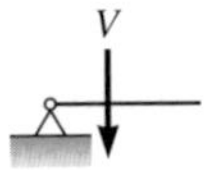 Pin or roller $M = 0$ $V \neq 0$
(*b*)	 Free end $\Delta \neq 0$ $\theta \neq 0$	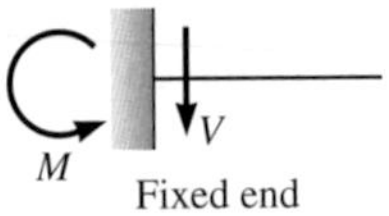 Fixed end $M \neq 0$ $V \neq 0$
(*c*)	 Fixed end $\Delta = 0$ $\theta = 0$	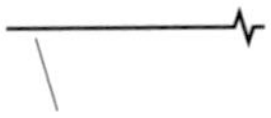 Free end $M = 0$ $V = 0$
(*d*)	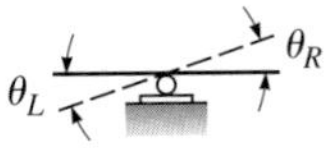 Interior support $\Delta = 0$ $\theta_L = \theta_R \neq 0$	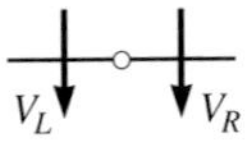 Hinge $M = 0$ $V_L = V_R \neq 0$
(*e*)	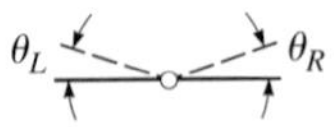 Hinge $\Delta \neq 0$ θ_L and θ_R may have different values	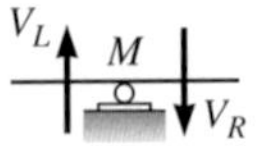 Interior roller $M \neq 0$ V_L and V_R may have different values

Figure 9.25: Conjugate supports.

Figure 9.26 shows the conjugate structures that correspond to eight examples of real structures. If the real structure is indeterminate, the conjugate structure will be unstable (see Figure 9.26*e* to *h*). You do not have to be concerned about this condition because you will find that the M/EI diagram produced by the forces acting on the real structure produces elastic loads that hold the conjugate structure in equilibrium. For example, in Figure 9.27*b* we show the conjugate structure of a fixed-end beam loaded by the M/EI diagram associated with a concentrated load applied at midspan to the real beam. Applying the equations to the entire structure, we can verify that the conjugate structure is in equilibrium with respect to both a summation of forces in the vertical direction and a summation of moments about any point.

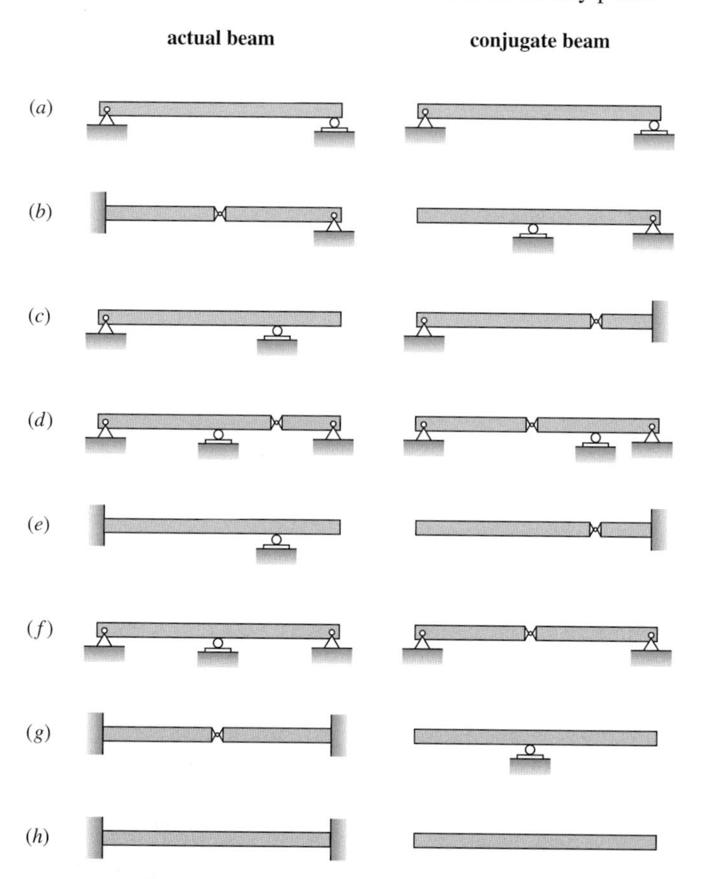

Figure 9.26: Examples of conjugate beams.

In summary, to compute deflections in any type of beam by the conjugate beam method, we proceed as follows.

1. Establish the moment curve for the real structure.
2. Produce the M/EI curve by dividing all ordinates by EI. Variation of E or I may be taken into account in this step.
3. Establish the conjugate beam by replacing actual supports or hinges with the corresponding conjugate supports shown in Figure 9.25.
4. Apply the M/EI diagram to the conjugate structure as the load, and compute the shear and moment at those points where either slope or deflection is required.

Examples 9.15 to 9.17 illustrate the conjugate beam method.

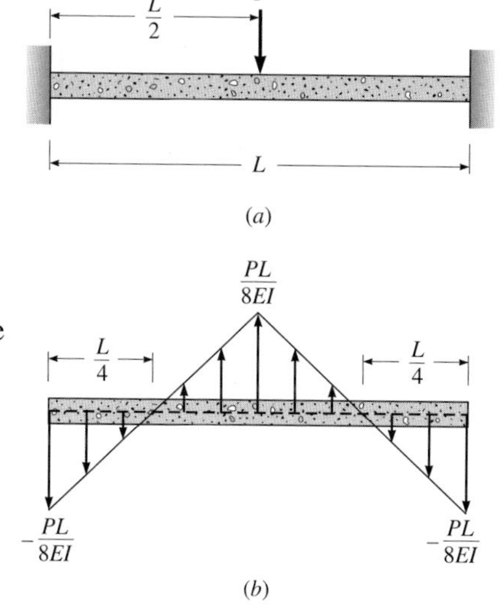

Figure 9.27: (*a*) Fixed-ended beam with concentrated load at midspan; (*b*) conjugate beam loaded with M/EI curve. The conjugate beam, which has no supports, is held in equilibrium by the applied loads.

EXAMPLE 9.15

For the beam in Figure 9.28 use the conjugate beam method to determine the maximum value of deflection between supports A and C and at the tip of the cantilever. EI is constant.

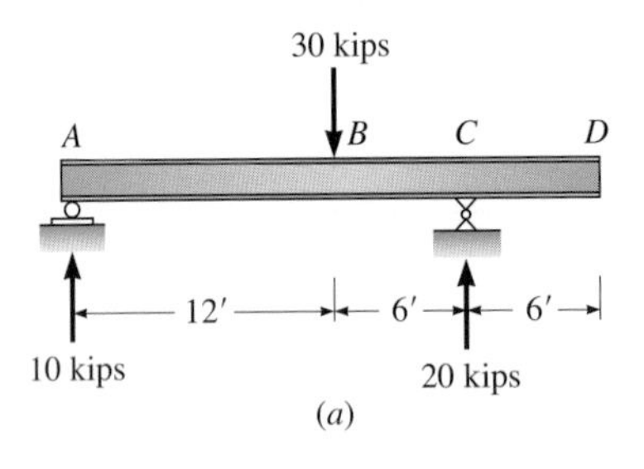

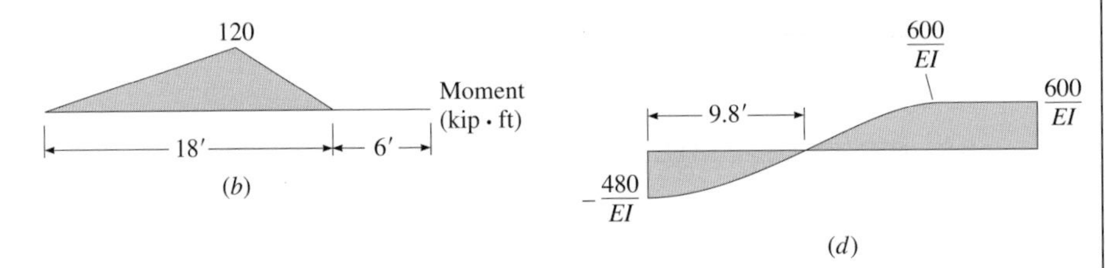

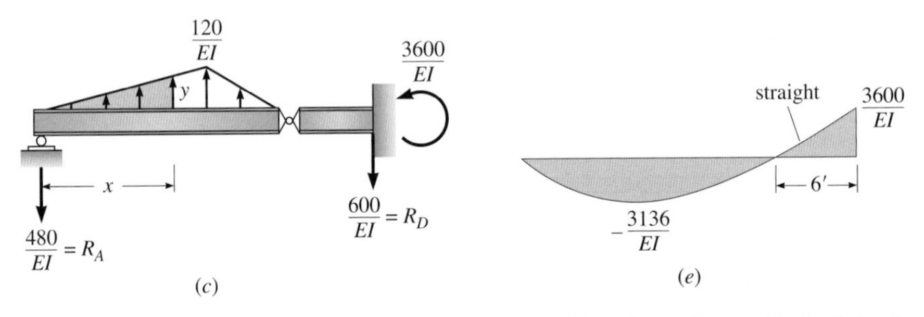

Figure 9.28: (*a*) Beam details; (*b*) moment curve; (*c*) conjugate beam with elastic loads; (*d*) elastic shear (slope); (*e*) elastic moment (deflection).

Solution

The conjugate beam with the M/EI diagram applied as an upward load is shown in Figure 9.28*c*. (See Figure 9.25 for the correspondence between real and conjugate supports.) Compute the reaction at A by summing moments about the hinge.

$$\circlearrowright^{+} \quad \Sigma M_{\text{hinge}} = 0$$

$$-18R_A + \frac{720(10)}{EI} + \frac{360(4)}{EI} = 0$$

$$R_A = \frac{480}{EI}$$

Compute R_D.

$$\overset{+}{\uparrow} \quad \Sigma F_y = 0$$

$$\frac{720}{EI} + \frac{360}{EI} - \frac{480}{EI} - R_D = 0$$

$$R_D = \frac{600}{EI}$$

Draw the shear and moment curves (see Figure 9.28*d* and *e*). Moment at *D* (equals area under shear curve between *C* and *D*) is

$$M_D = \frac{600}{EI}(6) = \frac{3600}{EI}$$

Locate the point of zero shear to the right of support *A* to establish the location of the maximum deflection by determining the area (shown shaded) under the load curve required to balance R_A.

$$\frac{1}{2}xy = \frac{480}{EI} \tag{1}$$

From similar triangles (see Figure 9.28*c*),

$$\frac{y}{\frac{120}{EI}} = \frac{x}{12} \quad \text{and } y = \frac{10}{EI}x \tag{2}$$

Substituting Equation 2 into Equation 1 and solving for *x* give

$$x = \sqrt{96} = 9.8 \text{ ft}$$

Compute the maximum value of negative moment. Since the shear curve to the right of support *A* is parabolic, area $= \frac{2}{3}bh$.

$$\Delta_{\max} = M_{\max} = \frac{2}{3}(9.8)\left(-\frac{480}{EI}\right) = -\frac{3136}{EI} \qquad \textbf{Ans.}$$

Compute the deflection at *D*.

$$\Delta_D = M_D = \frac{3600}{EI} \qquad \textbf{Ans.}$$

EXAMPLE 9.16

Compare the magnitude of the moment required to produced a unit value of rotation ($\theta_A = 1$ rad) at the left end of the beams in Figure 9.29*a* and *c*. Except for the supports at the right end—a pin versus a fixed end—the dimensions and properties of both beams are identical, and *EI* is constant. Analysis indicates that a clockwise moment *M* applied at the left end of the beam in Figure 9.29*c* produces a clockwise moment of $M/2$ at the fixed support.

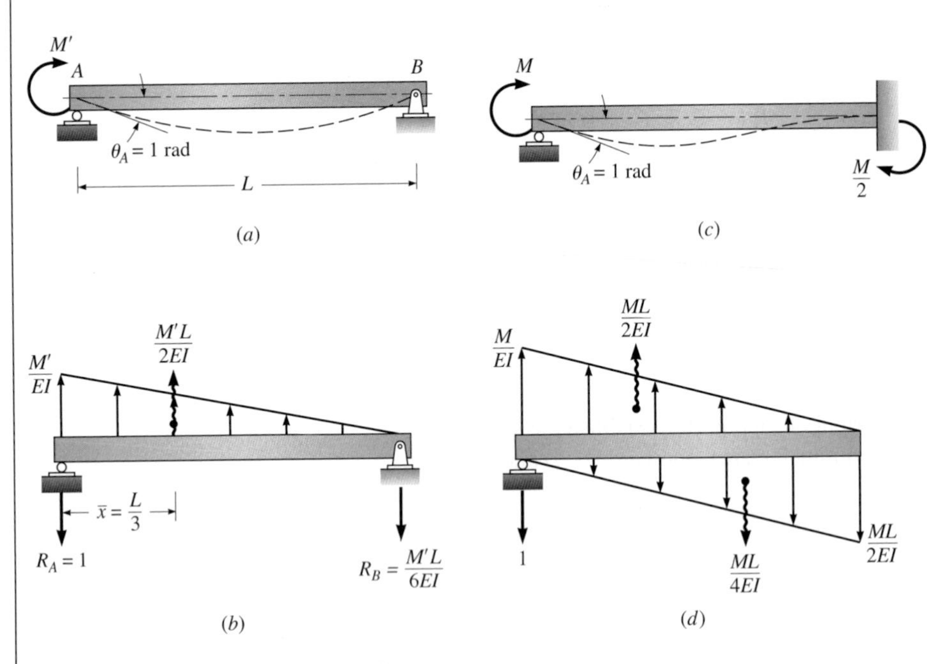

Figure 9.29: Effect of end restraint on flexural stiffness. (*a*) Beam loaded at *A* with far end pinned; (*b*) conjugate structure for beam in (*a*) loaded with M/EI; (*c*) beam loaded at *A* with far end fixed; (*d*) conjugate structure for beam in (*c*) loaded with M/EI.

Solution

The conjugate beam for the pin-ended beam in Figure 9.29*a* is shown in Figure 9.29*b*. Since the applied moment M' produces a clockwise rotation of 1 rad at *A*, the reaction at the left support equals 1. Because the slope at *A* is negative, the reaction acts downward.

To compute the reaction at *B*, we sum moments about support *A*.

$$\circlearrowright^{+} \quad \Sigma M_A = 0$$

$$0 = R_B L - \frac{M'L}{2EI}\left(\frac{L}{3}\right)$$

$$R_B = \frac{M'L}{6EI}$$

Summing forces in the y direction, we express M' in terms of the properties of the member as

$$\overset{+}{\uparrow} \quad \Sigma F_y = 0$$

$$0 = -1 + \frac{M'L}{2EI} - \frac{M'L}{6EI}$$

$$M' = \frac{3EI}{L} \qquad \textbf{Ans.} \qquad (1)$$

The conjugate beam for the fixed-end beam in Figure 9.29*c* is shown in Figure 9.29*d*. The M/EI diagram for each end moment is drawn separately. To express M in terms of the properties of the beam, we sum forces in the y direction.

$$\overset{+}{\uparrow} \quad \Sigma F_y = 0$$

$$0 = -1 + \frac{ML}{2EI} - \frac{1}{2}\frac{ML}{2EI}$$

$$M = \frac{4EI}{L} \qquad \textbf{Ans.} \qquad (2)$$

NOTE. The absolute flexural stiffness of a beam can be defined as the value of end moment required to rotate the end of a beam—supported on a roller at one end and fixed at the other end (see Figure 9.29*c*)—through an angle of 1 radian. Although the choice of boundary conditions is somewhat arbitrary, this particular set of boundary conditions is convenient because it is similar to the end conditions of beams that are analyzed by *moment distribution*—a technique for analyzing indeterminate beams and frames covered in Chapter 13. The stiffer the beam, the larger the moment required to produce a unit rotation.

If a pin support is substituted for a fixed support as shown in Figure 9.29*a*, the flexural stiffness of the beam reduces because the roller does not apply a restraining moment to the end of the member. As this example shows by comparing the values of moment required to produce a unit rotation (see Equations 1 and 2), the flexural stiffness of a pin-ended beam is three-fourths that of a fixed-end beam.

$$\frac{M'}{M} = \frac{3EI/L}{4EI/L}$$

$$M' = \frac{3}{4}M$$

EXAMPLE 9.17

Determine the maximum deflection of the beam in Figure 9.30. EI is a constant.

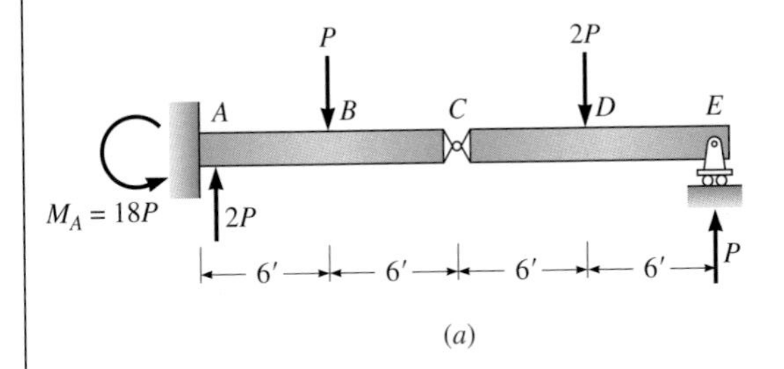

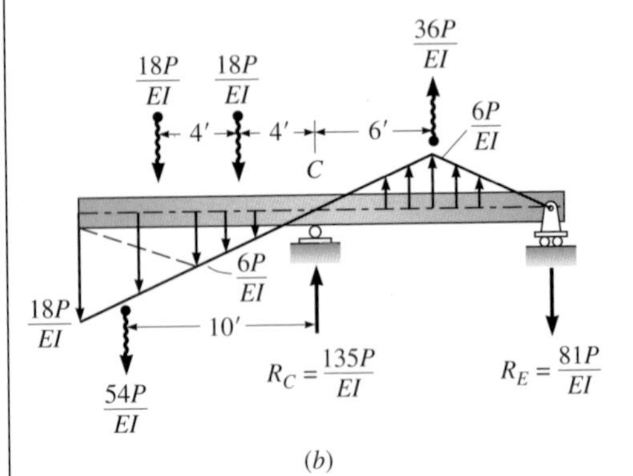

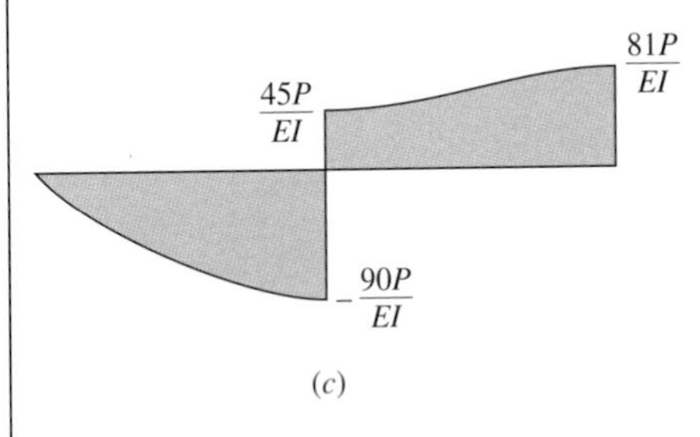

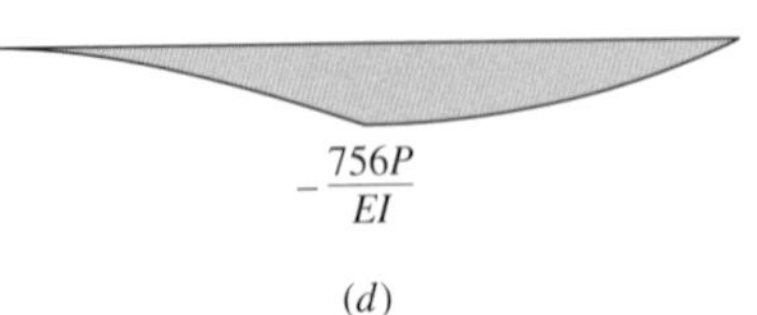

Figure 9.30: (*a*) Beam; (*b*) conjugate beam with elastic loads; (*c*) elastic shear (slope); (*d*) elastic moment (deflection).

Solution

The ordinates of the moment diagram produced by the concentrated loads acting on the real structure in Figure 9.30*a* are divided by EI and applied as a distributed load to the conjugate beam in Figure 9.30*b*. We next divide the distributed load into triangular areas and compute the resultant (shown by heavy arrows) of each area.

Compute R_E.

$$\overset{+}{\circlearrowleft} \quad \Sigma M_C = 0$$

$$\frac{36P}{EI}(6) + \frac{18P}{EI}(4) + \frac{18P}{EI}(8) + \frac{54P}{EI}(10) - 12R_E = 0$$

$$R_E = \frac{81P}{EI}$$

Compute R_C.

$$\overset{+}{\uparrow} \quad \Sigma F_y = 0$$

$$-\frac{54P}{EI} - \frac{18P}{EI} - \frac{18P}{EI} - \frac{81P}{EI} + \frac{36P}{EI} + R_C = 0$$

$$R_C = \frac{135P}{EI}$$

To establish the variation of slope and deflection along the axis of the beam, we construct the shear and moment diagrams for the conjugate beam (see Figure 9.30*c* and *d*). The maximum deflection, which occurs at point C (the location of the real hinge), equals $756P/EI$. This value is established by evaluating the moment produced by the forces acting on the conjugate beam to the left of a section through C (see Figure 9.30*b*).

9.6 Design Aids for Beams

To be designed properly, beams must have adequate stiffness as well as strength. Under service loads, deflections must be limited so that attached non-structural elements—partitions, pipes, plaster ceilings, and windows— will not be damaged or rendered inoperative by large deflections. Obviously floor beams that sag excessively or vibrate as live loads are applied are not satisfactory. To limit deflections under live load, most building codes specify a maximum value of live load deflection as a fraction of the span length—a limit between 1/360 to 1/240 of the span length is common.

If steel beams sag excessively under dead load, they may be cambered. That is, they are fabricated with initial curvature by either rolling or by heat treatment so that the center of the beam is raised an amount equal to or slightly less than the dead load deflection (Figure 9.31). Example 10.12 illustrates a simple procedure to relate curvature to camber. To camber reinforced concrete beams, the center of the forms may be raised an amount to cancel out the dead load deflections.

In practice, designers usually make use of tables in handbooks and design manuals to evaluate deflections of beams for a variety of loading and support conditions. The *Manual of Steel Construction* published by the American Institute of Steel Construction (AISC) is an excellent source of information.

Table 9.1 gives values of maximum deflections as well as moment diagrams for a number of support and loading conditions of beams. We will make use of these equations in Example 9.18.

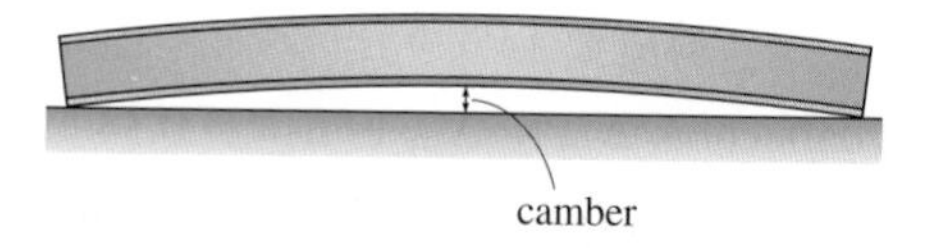

Figure 9.31: Beam fabricated with camber.

EXAMPLE 9.18

A simply supported steel beam spanning 30 ft carries a uniform dead load of 0.4 kip/ft that includes the weight of the beam and a portion of the floor and ceiling supported directly on the beam (Figure 9.32). The beam is also loaded at its third points by two equal concentrated loads that consist of 14.4 kips of dead load and 8.2 kips of live load. To support these loads, the designer selects a 16-in-deep steel wide-flange beam with a modulus of elasticity $E = 29{,}000$ ksi and a moment of inertia $I = 758$ in^4.

(*a*) Specify the required camber of the beam to compensate for the total dead load deflection and 10 percent of the live load deflection.

(*b*) Verify that under *live load* only, the beam does not deflect more than 1/360 of its span length. (This provision ensures the beam will not be excessively flexible and vibrate when the live load acts.)

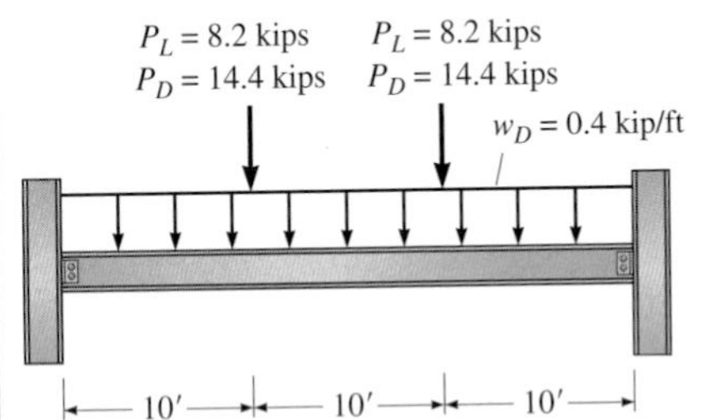

Figure 9.32: Beam, connected to columns by clip angles attached to web, is analyzed as a simply supported determinate beam.

Solution

We first compute the required camber for dead load, using equations for deflection given by cases 1 and 3 in Table 9.1.

(*a*) Dead load deflection produced by uniform load is

$$\Delta_{D1} = \frac{5wL^4}{384EI} = \frac{5(0.4)(30)^4(1728)}{384(29{,}000)(758)} = 0.33 \text{ in}$$

Dead load deflection produced by concentrated loads is

$$\Delta_{D2} = \frac{Pa(3L^2 - 4a^2)}{24EI} = \frac{14.4(10)[3(30)^2 - 4(10)^2](1728)}{24(29{,}000)(758)}$$

$$\Delta_{D2} = 1.08 \text{ in}$$

Total dead load deflection, $\Delta_{DT} = \Delta_{D1} + \Delta_{D2} = 0.33 + 1.08 = 1.41$ in

$$\text{Live load deflection, } \Delta_L = \frac{Pa(3L^2 - 4a^2)}{24EI} = \frac{8.2(10)[3(30)^2 - 4(10)^2](1728)}{24(29{,}000)(758)}$$

$$\Delta_L = 0.62 \text{ in}$$

$$\text{Required camber} = \Delta_{DT} + 0.1\Delta_L = 1.41 + 0.1(0.62) = 1.47 \text{ in}$$

Since real connections are not theoretical pins and have some fixity to them, some designers use 80% of the theoretical beam deflections when specifying camber:

Camber = 0.8(1.47) = 1.18 in, round to 1¼ in **Ans.**

(*b*) Allowable live load deflection is

$$\frac{L}{360} = \frac{30 \times 12}{360} = 1 \text{ in} > 0.62 \text{ in}$$ **Ans.**

Therefore, it is OK.

TABLE 9.1
Moment Diagrams and Equations for Maximum Deflection

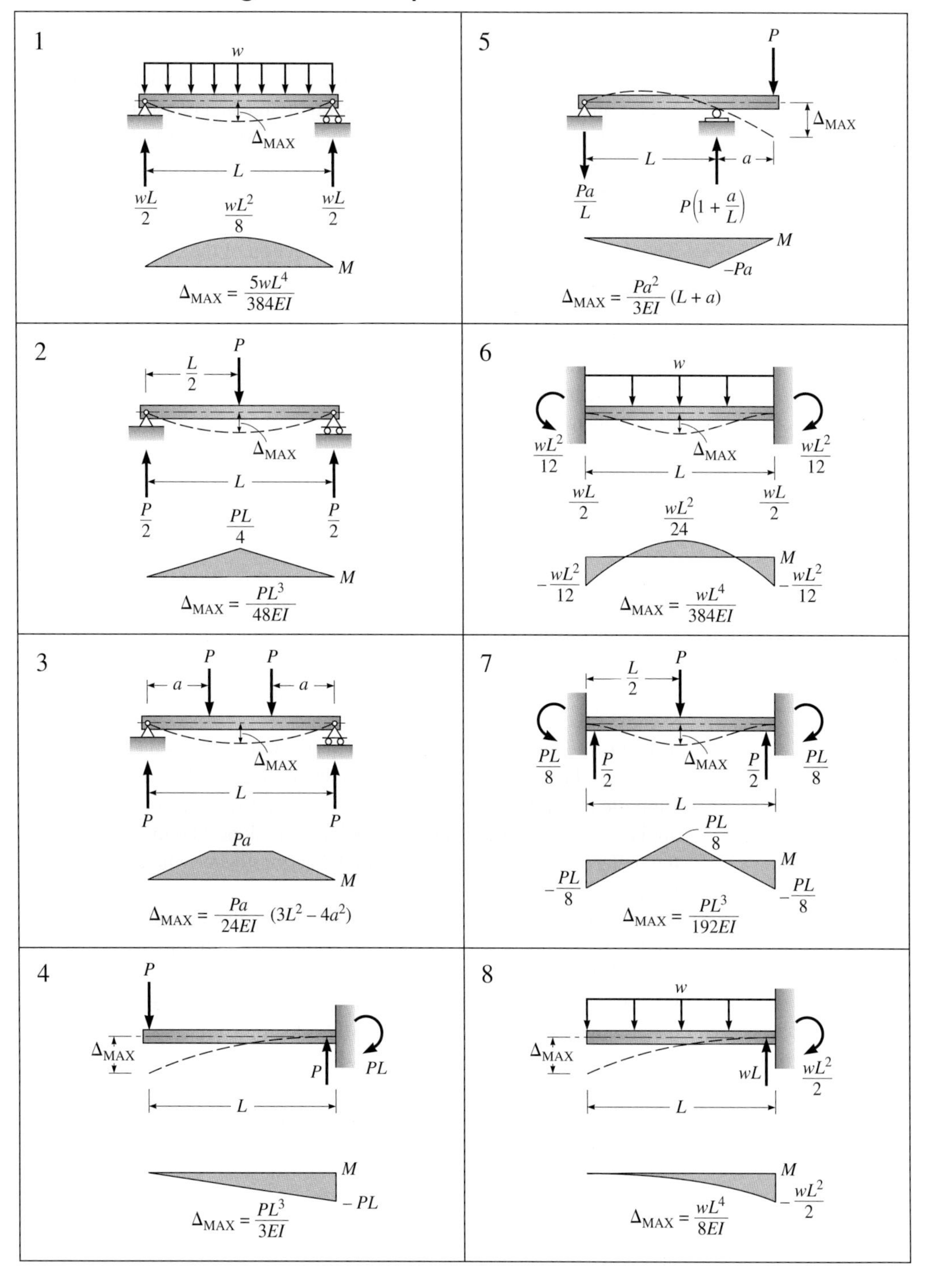

Summary

- The maximum deflections of beams and frames must be checked to ensure that structures are not excessively flexible. Large deflections of beams and frames can produce cracking of attached nonstructural elements (masonry and tile walls, windows, and so forth) as well as excessive vibrations of floor and bridge decks under moving loads. Deflection is also needed to solve indeterminate structures by the flexibility method in Chapter 11.
- The deflection of a beam or frame is a function of the bending moment M and the member's flexural stiffness, which is related to a member's moment of inertia I and modulus of elasticity E. Deflections due to shear are typically neglected unless members are very deep, shear stresses are high, and the shear modulus G is low.
- To establish equations for the slope and deflection of the elastic curve (the deflected shape of the beam's centerline), we begin the study of deflections by integrating the differential equation of the elastic curve

$$\frac{d^2y}{dx^2} = \frac{M}{EI}$$

  This method becomes cumbersome when loads vary in a complex manner.
- Next we consider the *moment-area method*, which utilizes the M/EI diagram to compute slopes and deflections at selected points along the beam's axis. This method, described in Section 9.3, requires an accurate sketch of the deflected shape.
- The *elastic load method* (a variation of the moment-area method), which can be used to compute slopes and deflections in simply supported beams, is reviewed. In this method, the M/EI diagram is applied as a load. The shear at any point is the slope, and the moment is the deflections. Points of maximum deflections occur where the shear is zero.
- The *conjugate beam method*, a variation of the elastic load method, applies to members with a variety of boundary conditions. This method requires that actual supports be replaced by fictitious supports to impose boundary conditions that ensure that the values of shear and moment in the conjugate beam, loaded by the M/EI diagram, are equal at each point to the slope and deflection, respectively, of the real beam.
- Once equations for evaluating maximum deflections are established for a particular beam and loading, tables available in structural engineering reference books (see Table 9.1) supply all the important data required to analyze and design beams.

PROBLEMS

Solve Problems P9.1 to P9.6 by the double integration method. EI is constant for all beams.

P9.1. Derive the equations for slope and deflection for the cantilever beam in Figure P9.1. Compute the slope and deflection at B. Express answer in terms of EI.

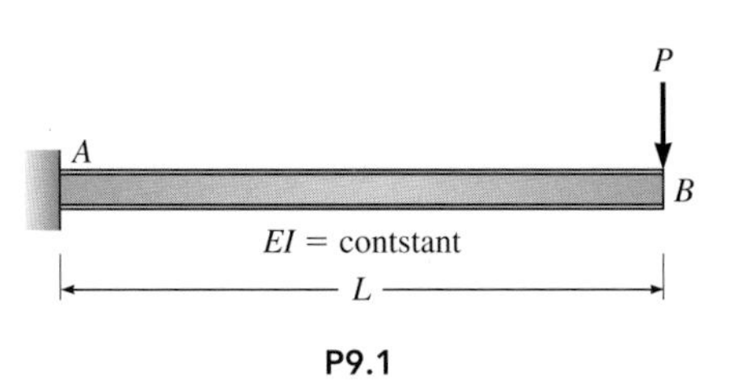

P9.1

P9.2. Derive the equations for slope and deflection for the beam in Figure P9.2. Compare the deflection at B with the deflection at midspan.

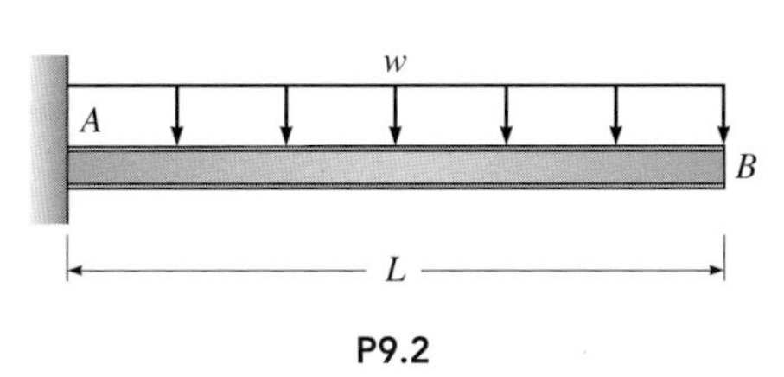

P9.2

P9.3. Derive the equations for slope and deflection for the beam in Figure P9.3. Compute the maximum deflection. *Hint*: Maximum deflection occurs at point of zero slope.

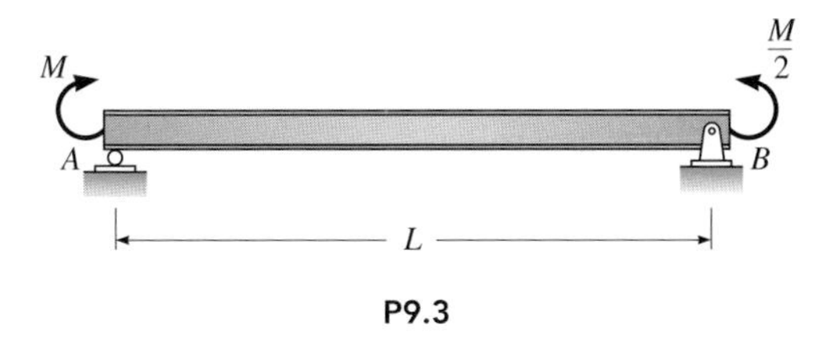

P9.3

P9.4. Derive the equations for slope and deflection for the beam in Figure P9.4. Locate the point of maximum deflection and compute its magnitude.

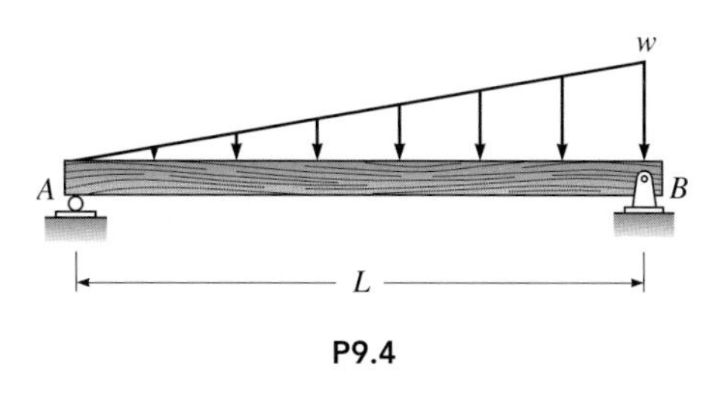

P9.4

P9.5. Establish the equations for slope and deflection for the beam in Figure P9.5. Evaluate the magnitude of the slope at each support. Express answer in terms of EI.

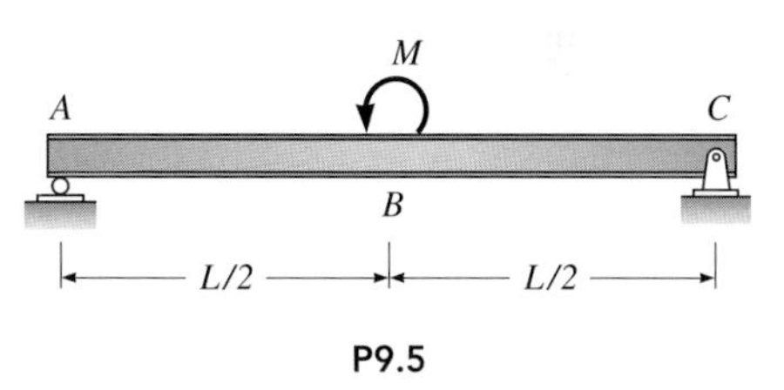

P9.5

P9.6. Derive the equations for slope and deflection for the beam in Figure P9.6. Determine the slope at each support and the value of the deflection at midspan. *Hint*: Take advantage of symmetry; slope is zero at midspan.

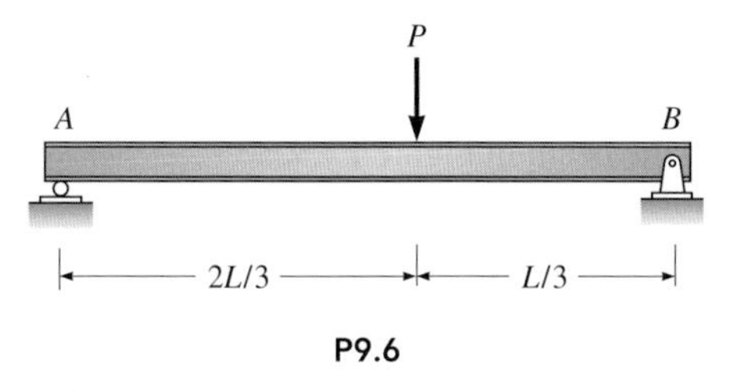

P9.6

Solve Problems P9.7 to P9.12 by the moment-area method. Unless noted otherwise, EI is a constant for all members. Answers may be expressed in terms of EI unless otherwise noted.

P9.7. Compute the slope and deflection at points B and C in Figure P9.7.

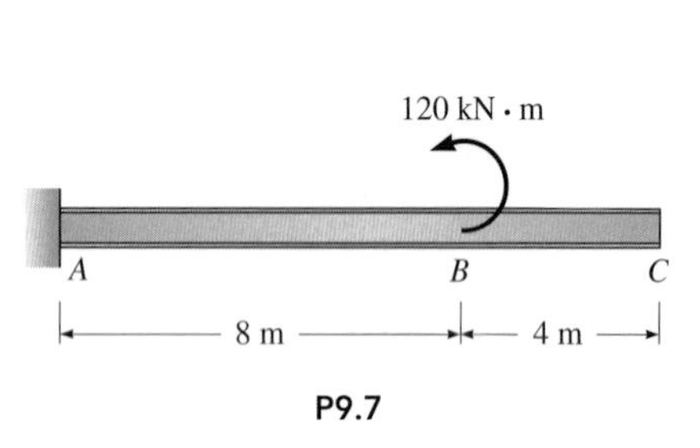

P9.7

P9.8. (*a*) Compute the slopes at A and C and the deflection at D in Figure P9.8. (*b*) Locate and compute the magnitude of the maximum deflection.

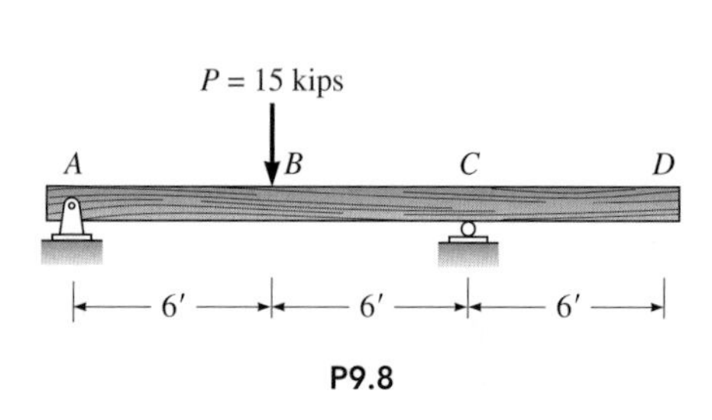

P9.8

P9.9. Compute the slopes at A and C and the deflection at B for the beam in Figure P9.9.

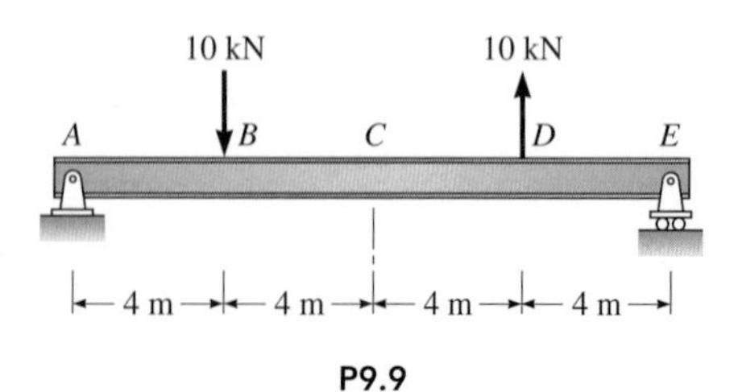

P9.9

P9.10. (*a*) Compute the slope at A and the deflection at midspan in Figure P9.10. (*b*) If the deflection at midspan is not to exceed 1.2 in, what is the minimum required value of I? $E = 29{,}000$ kips/in^2.

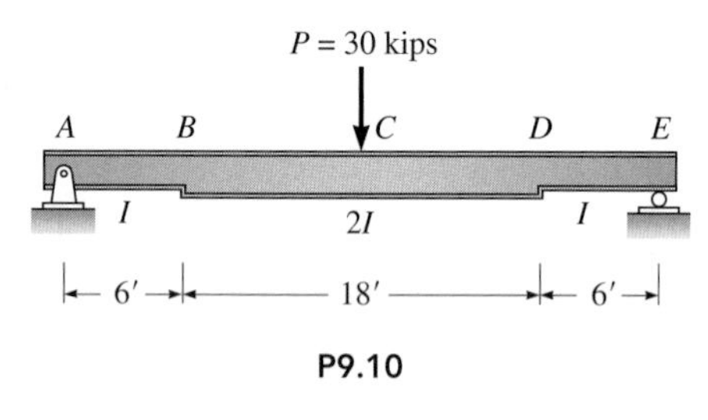

P9.10

P9.11. (*a*) Find the slope and deflection at A in Figure P9.11. (*b*) Determine the location and the magnitude of the maximum deflection in span BC.

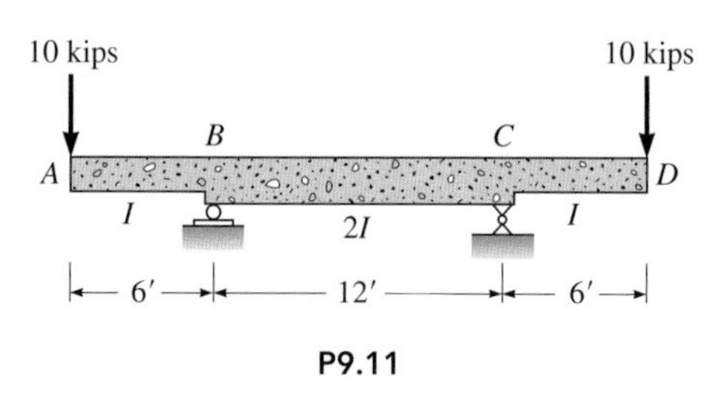

P9.11

P9.12. Compute the slopes of the beam in Figure P9.12 on each side of the hinge at B, the deflection of the hinge, and the maximum deflection in span BC. The elastomeric support at C acts as a roller.

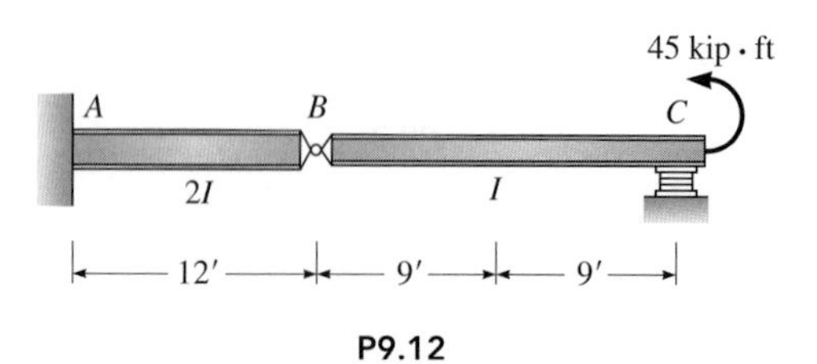

P9.12

Solve Problems P9.13 to P9.18 by the moment-area method. EI is constant.

P9.13. Compute the slope at support A and the deflection at point B. Treat the rocker at D as a roller. Express the answer in terms of EI.

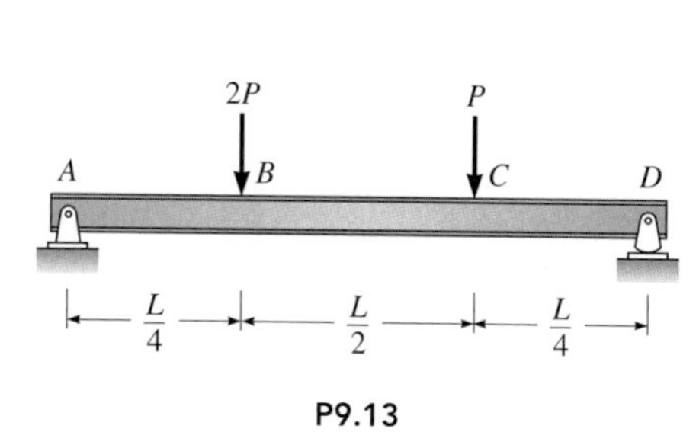

P9.13

P9.14. Determine the slopes at A and B and the deflection at C in Figure P9.14. Express answers in terms of M, E, I, and L.

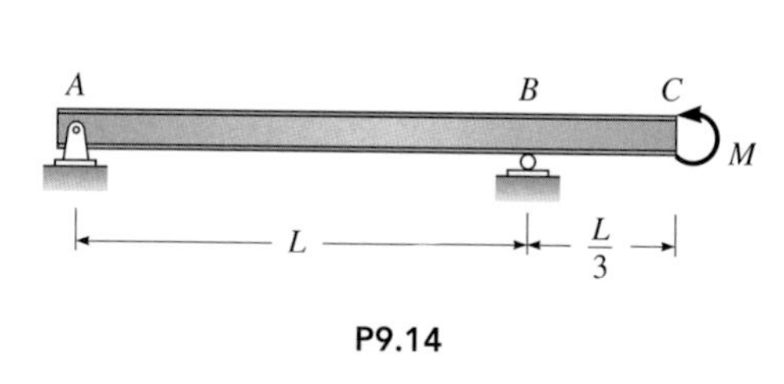

P9.14

P9.15. Determine the slope and deflection of point C in Figure P9.15. *Hint*: Draw moment curves by parts.

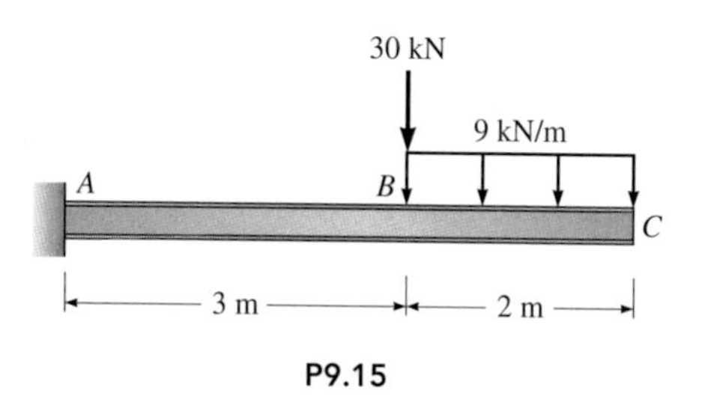

P9.15

P9.16. The roof beam of a building is subjected to the loading shown in Figure P9.16. If a ⅜-in. deflection is permitted at the cantilever end before the ceiling and roofing materials would be damaged, calculate the required moment of inertia for the beam. Use $E = 29{,}000$ ksi.

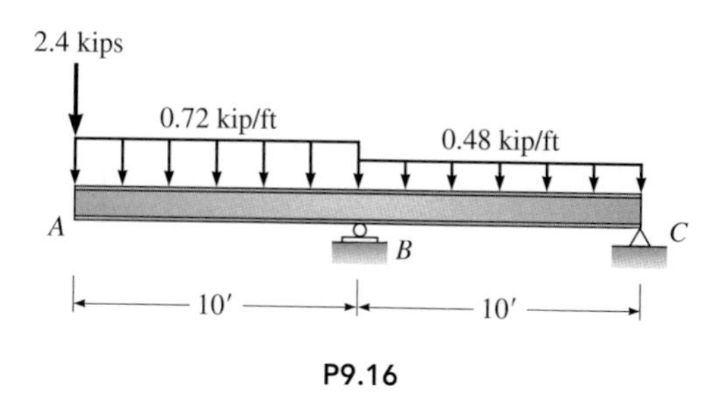

P9.16

P9.17. Compute the slope and deflection under the 32-kip load at B and D. Reactions are given. $I = 510$ in^4 and $E = 29{,}000$ kips/in^2. Sketch the deflected shape.

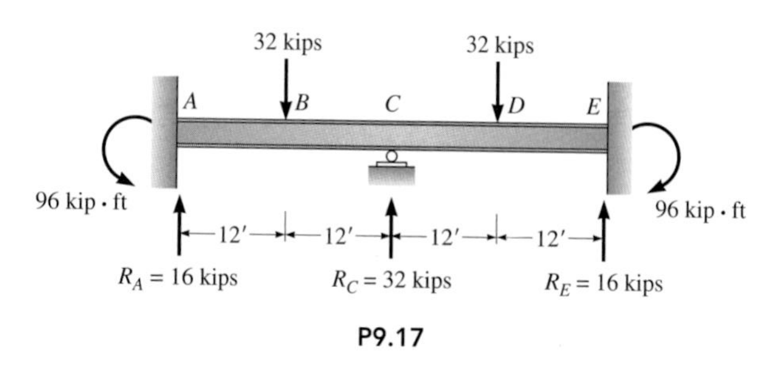

P9.17

P9.18. The vertical reactions at supports A and D of the indeterminate beam in Figure P9.18 are given. Compute the slope at B and the deflection at C. EI is constant.

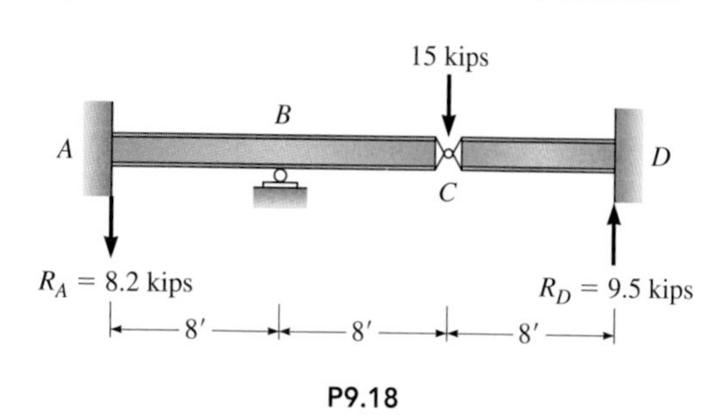

P9.18

Solve Problems P9.19 to P9.23 by the moment-area method. *EI* is constant unless otherwise noted.

P9.19. What value of force *P* is required at *C* in Figure P9.19 if the vertical deflection at *C* is to be zero?

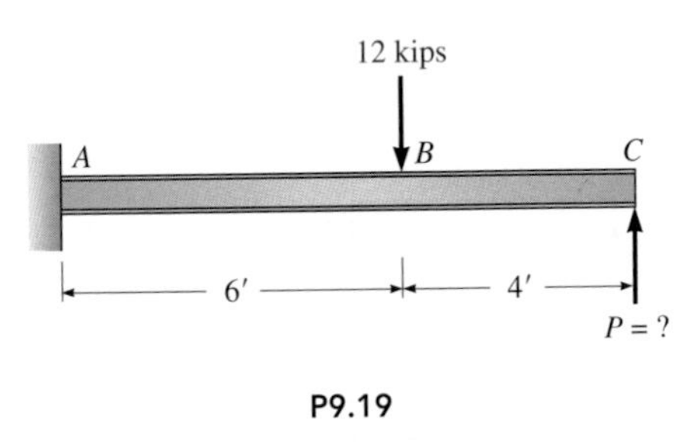

P9.19

P9.20. If the vertical deflection of the beam at midspan (i.e., point *C*) is to be zero, determine the magnitude of force *F*. *EI* is constant. Express *F* in terms of *P* and *EI*.

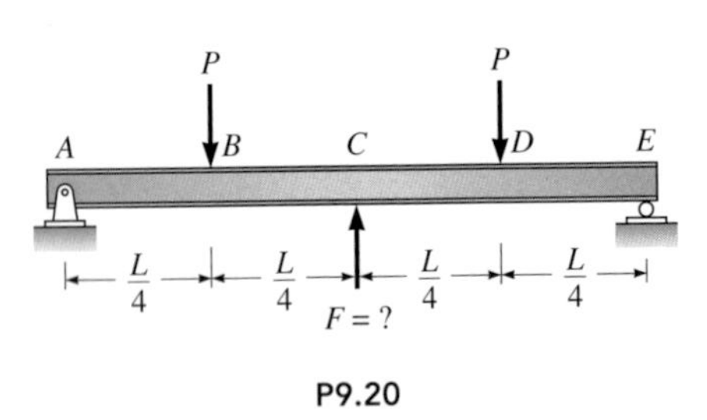

P9.20

P9.21. Compute the horizontal deflection at *D* and vertical deflection at *B* in Figure P9.21. The elastomeric pad at *C* acts as a roller.

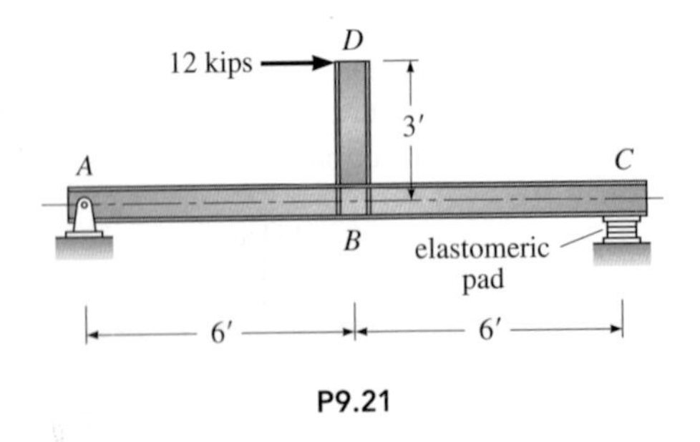

P9.21

P9.22. Compute the horizontal and vertical deflections at *C* of the frame in Figure P9.22. *EI* is constant.

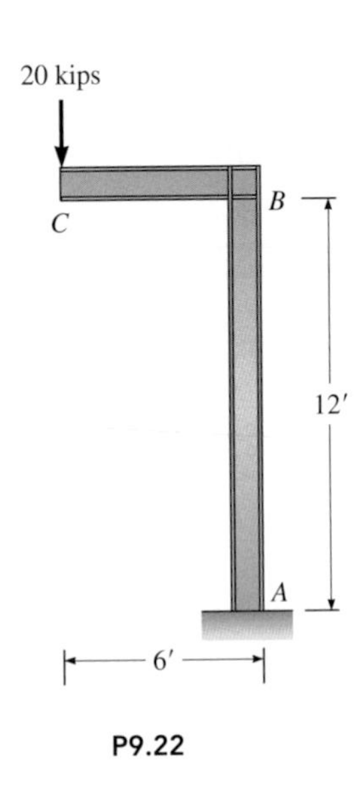

P9.22

P9.23. Compute the slope and the vertical deflection at point *C* and the horizontal displacement at point *D*. $I_{AC} = 800$ in^4, $I_{CD} = 120$ in^4, and $E = 29{,}000$ kips/in^2.

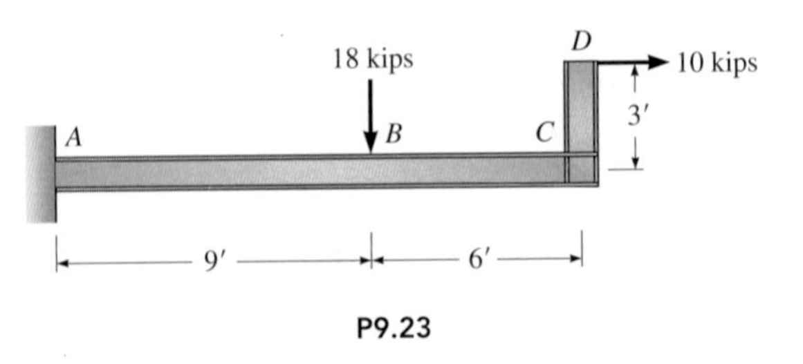

P9.23

Solve Problems P9.24 to P9.27 by the moment-area method. EI is constant.

P9.24. The moment of inertia of the girder in Figure P9.24 is twice that of the column. If the vertical deflection at D is not to exceed 1 in and if the horizontal deflection at C is not to exceed 0.5 in, what is the minimum required value of the moment of inertia? $E = 29{,}000$ kips/in^2. The elastomeric pad at B is equivalent to a roller.

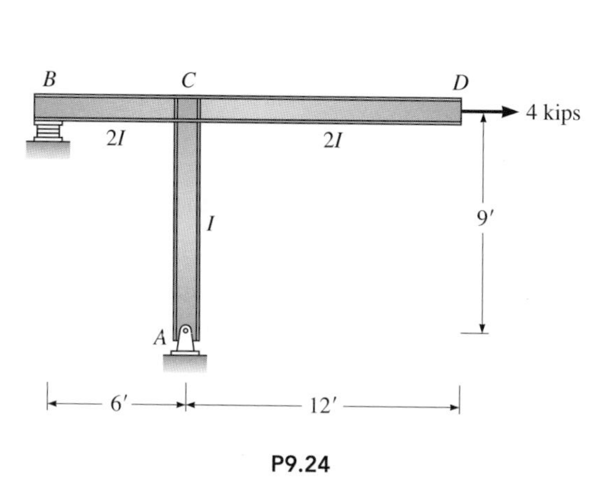

P9.24

P9.25. Compute the vertical displacement of the hinge at C in Figure P9.25. EI is constant.

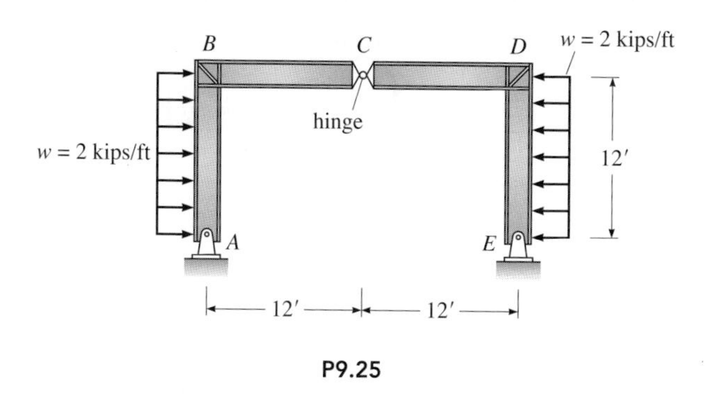

P9.25

P9.26. The loading acting on a column that supports a stair and exterior veneer is shown in Figure P9.26. Determine the required moment of inertia for the column such that the maximum lateral deflection does not exceed ¼ in, a criterion set by the veneer manufacturer. Use $E = 29{,}000$ kips/in^2.

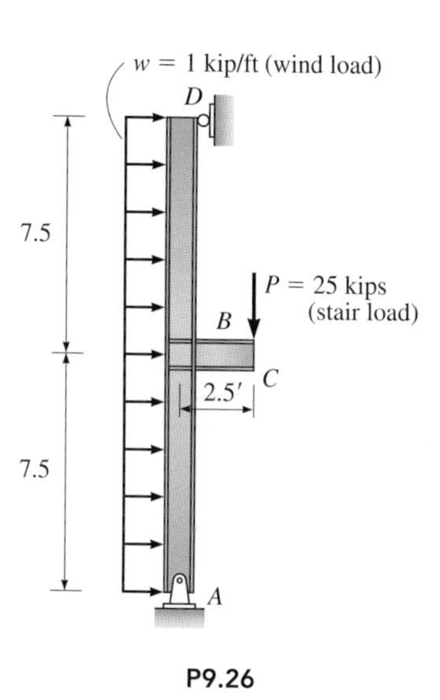

P9.26

P9.27. Compute the slope at A and the horizontal and vertical components of deflection at point D in Figure P9.27.

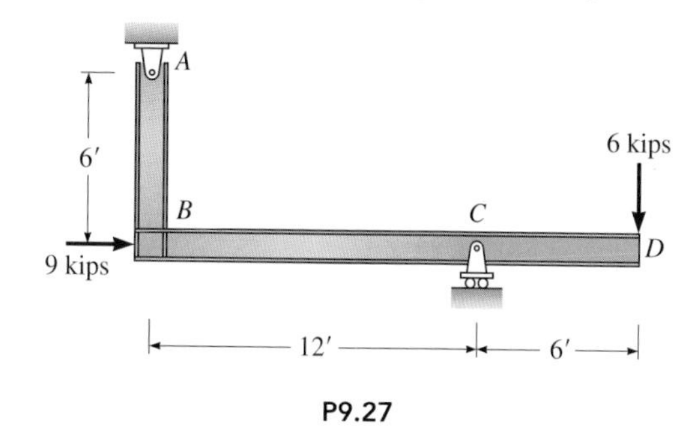

P9.27

Solve Problems P9.28 to P9.30 by the moment-area method. EI is constant.

P9.28. Compute the horizontal displacement of joint B in Figure P9.28. The moment diagram produced by the 12-kip load is given. The base of the columns at points A and E may be treated as fixed supports. *Hint*: Begin by sketching the deflected shape, using the moment diagrams to establish the curvature of members. Moments in units of kip · ft.

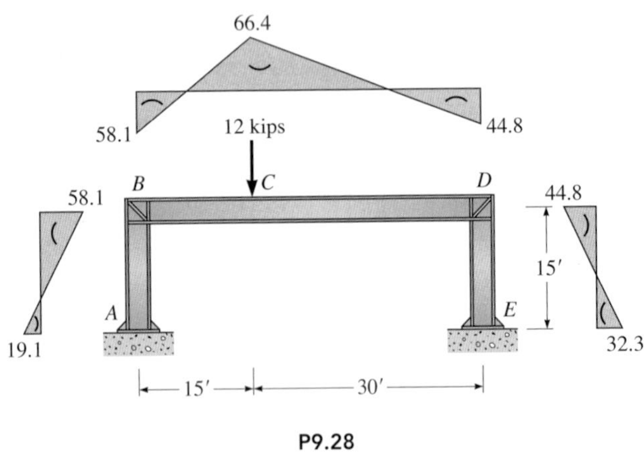

P9.28

P9.29. Compute the rotation at B and the vertical deflection at D. Given: $E = 200$ GPa, $I_{AC} = 400 \times 10^6$ mm^4, and $I_{BD} = 800 \times 10^6$ mm^4.

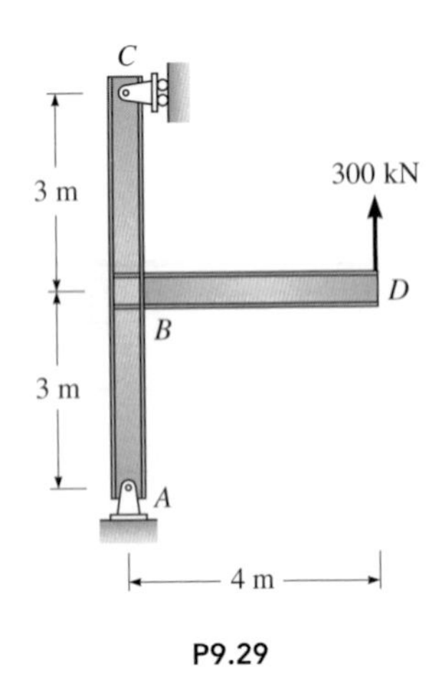

P9.29

P9.30. The frame shown in Figure P9.30 is loaded by a horizontal load at B. Compute the horizontal displacements at B and D. For all members $E = 200$ GPa and $I = 500 \times 10^6$ mm^4.

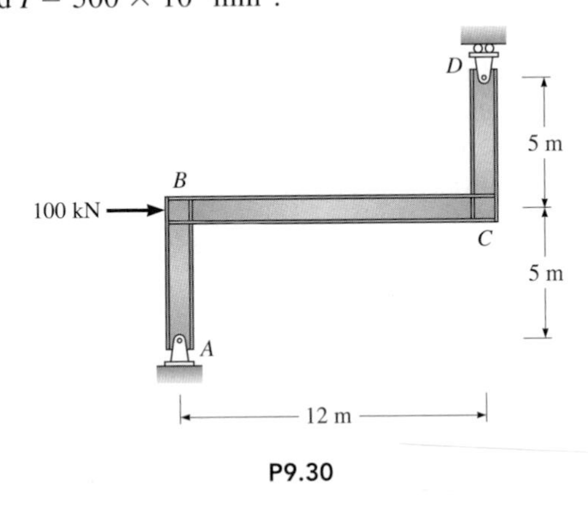

P9.30

Solve Problems P9.31 to P9.37 by the conjugate beam method.

P9.31. Compute the slope and deflection at point B of the cantilever beam in Figure P9.31. EI is constant.

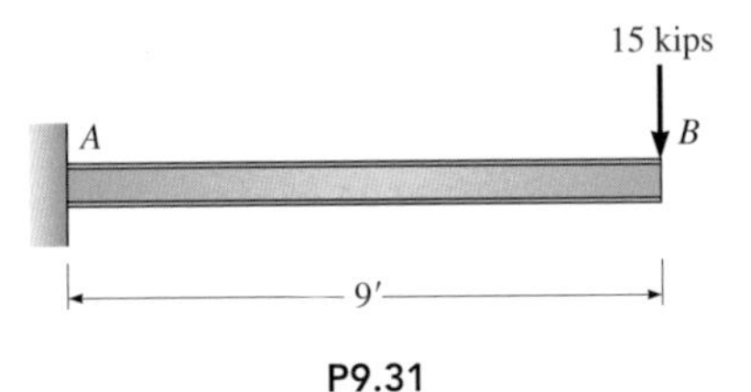

P9.31

P9.32. The moment diagram of a fix-ended beam with an external moment of 200 kip-ft applied at midspan is shown in Figure P9.32. Determine the maximum vertical deflection and maximum slope and their locations.

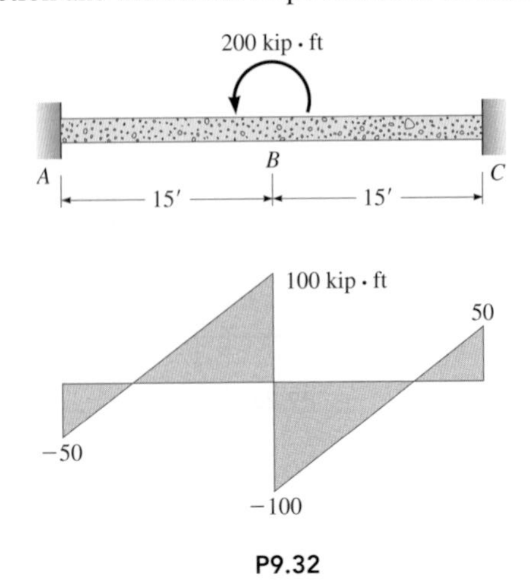

P9.32

P9.33. Compute the slope and deflection at point C and the maximum deflection between A and B for the beam in Figure P9.33. The reactions are given, and EI is constant. The elastomeric pad at B is equivalent to a roller.

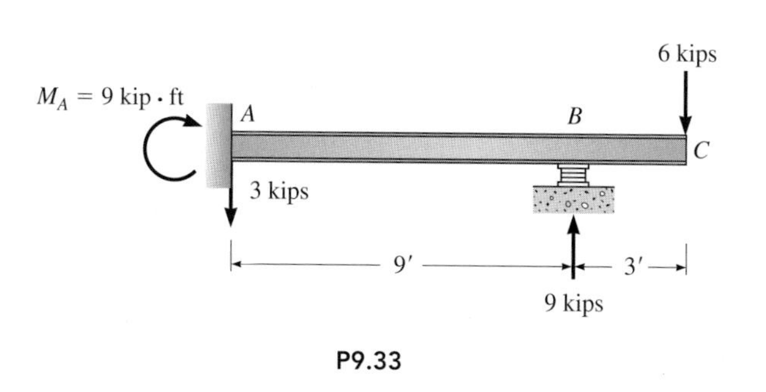

P9.33

P9.34. Compute the slope at A and the deflection at C of the beam in Figure P9.34. E is constant.

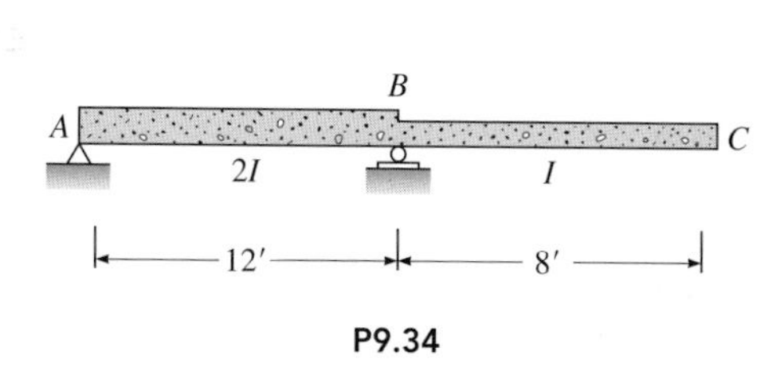

P9.34

P9.35. Determine the flexural stiffness of the beam in Figure P9.35 (see Example 9.16 for criteria) for (*a*) moment applied at A, and (*b*) moment applied at C. E is constant.

A B C
I 2I
$\frac{L}{2}$ $\frac{L}{2}$

P9.35

P9.36. Compute the deflection and the slopes on both sides of the hinge at B in Figure P9.36. EI is constant.

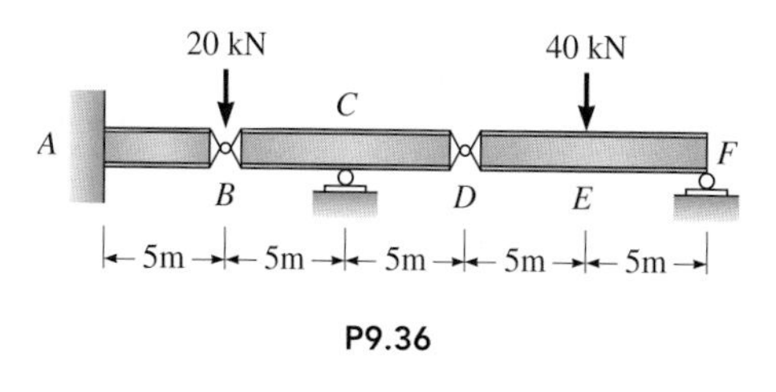

P9.36

P9.37. Compute the maximum deflection in span BD of the beam in Figure P9.37 and the slope on each side of the hinge.

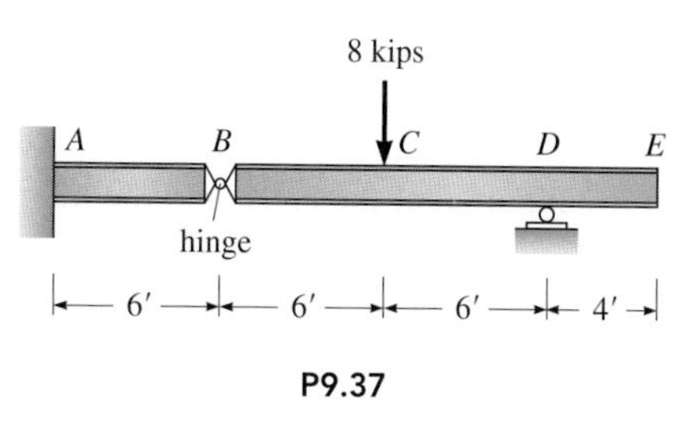

P9.37

P9.38. Solve Problem P9.11 by the conjugate beam method.

P9.39. Solve Problem P9.12 by the conjugate beam method.

P9.40. Solve Problem P9.17 by the conjugate beam method.

P9.41. Solve Problem P9.18 by the conjugate beam method.

P9.42. For the beam shown in Figure P9.42, use the conjugate beam method to compute the vertical deflection and the rotation to the left and the right of the hinge at *C*. Given: $E = 200$ GPa, $I_{AC} = 100 \times 10^6$ mm^4, and $I_{CF} = 50 \times 10^6$ mm^4.

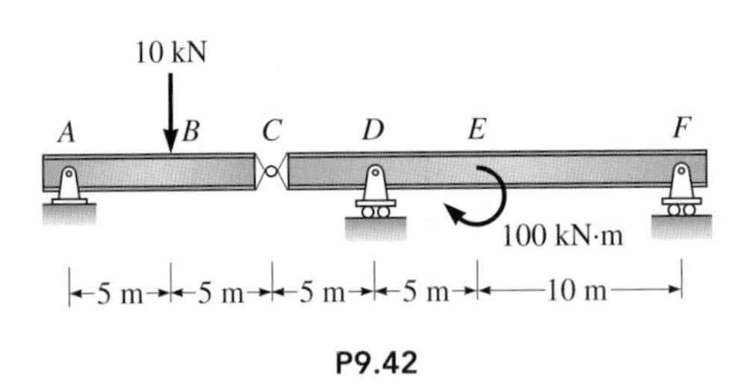

P9.42

Practical Applications of Deflection Computations

P9.43. The reinforced concrete girder shown in Figure P9.43*a* is prestressed by a steel cable that induces a compression force of 450 kips with an eccentricity of 7 in. The external effect of the prestressing is to apply an axial force of 450 kips and equal end moment $M_P = 262.5$ kip·ft at the ends of the girder (Figure P9.43*b*). The axial force causes the beam to shorten but produces no bending deflections. The end moments M_P bend the beam upward (Figure P9.43*c*) so that the entire weight of the beam is supported at the ends, and the member acts as a simply supported beam. As the beam bends upward, the weight of the beam acts as a uniform load to produce downward deflection. Determine the initial camber of the beam at midspan immediately after the cable is tensioned. *Note*: Over time the initial deflection will increase due to creep by a factor of approximately 100 to 200 percent. The deflection at midspan due to the two end moments equals $ML^2/(8EI)$. Given: $I = 46{,}656$ in^4, $A = 432$ in^2, beam weight $w_G = 0.45$ kip/ft, and $E = 5000$ kips/in^2.

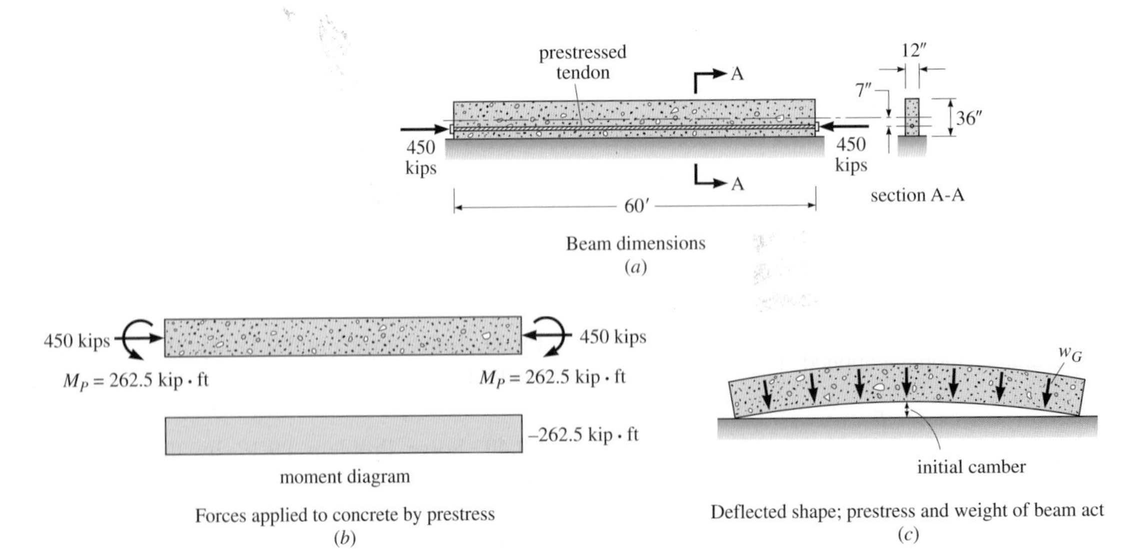

P9.43

P9.44. Because of poor foundation conditions, a 30-in-deep steel beam with a cantilever is used to support an exterior building column that carries a dead load of 600 kips and a live load of 150 kips (Figure P9.44). What is the magnitude of the initial camber that should be induced at point C, the tip of the cantilever, to eliminate the deflection produced by the total load? Neglect the beam's weight. Given: $I = 46{,}656$ in^4 and $E_S = 30{,}000$ ksi. See case 5 in Table 9.1 for the deflection equation. The clip angle connection at A may be treated as a pin and the cap plate support at B as a roller.

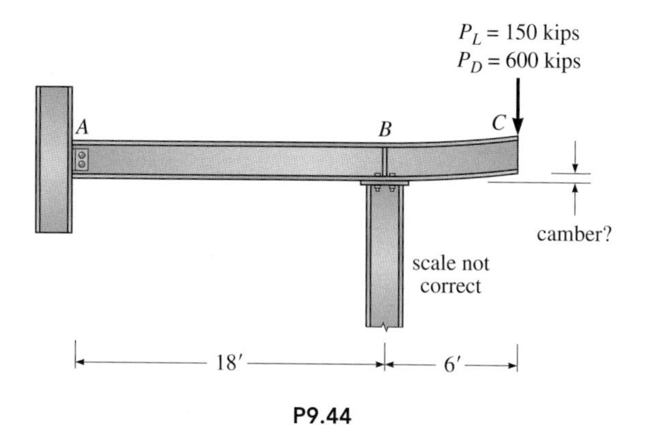

P9.44

P9.45. The rigid jointed steel frame with a fixed base at support A has to carry both the dead and live loads shown in Figure P9.45. Both the column and the girder are constructed from the same size members. What is the minimum required moment of inertia of the frame members if the vertical deflection at D produced by these loads cannot exceed 0.5 in.? Use $E = 29{,}000$ ksi.

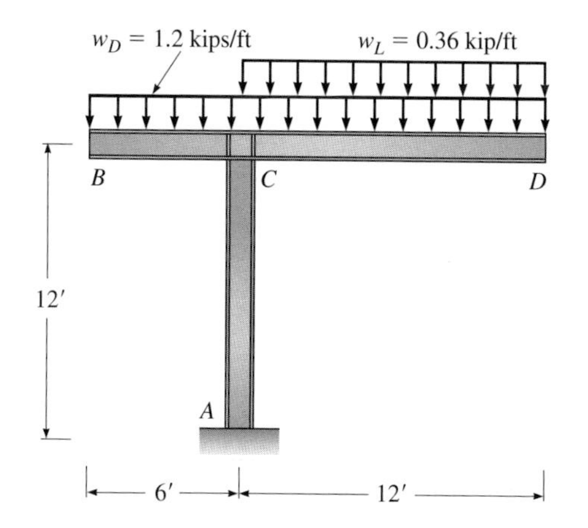

P9.45

P9.46. *Computer study of the behavior of multistory building frames*. The object of this study is to examine the behavior of building frames fabricated with two common types of connections. When open interior spaces and future flexibility of use are prime considerations, building frames can be constructed with *rigid connections* usually fabricated by welding. Rigid joints (see Figure P9.46*b*) are expensive to fabricate and now cost in the range of \$700 to \$850 depending on the size of members. Since the ability of a welded frame to resist lateral loads depends on the bending stiffness of the beams and columns, heavy members may be required when lateral loads are large or when lateral deflections must be limited. Alternately, frames can be constructed less expensively by connecting the webs of beams to columns by angles or plates, called *shear connections*, which currently cost about \$80 each (Figure P9.46*c*). If *shear connections* are used, diagonal bracing, which forms a deep vertical truss with the attached columns and floor beams, is typically required to provide lateral stability (unless floors can be connected to stiff shear walls constructed of reinforced masonry or concrete).

Properties of Members

In this study all members are constructed of steel with $E = 29{,}000$ kips/in^2.

All beams: $I = 300$ in^4 and $A = 10$ in^2
All columns: $I = 170$ in^4 and $A = 12$ in^2

Diagonal bracing using 2.5 in square hollow structural tubes (Case 3 only—see dashed lines in Figure P9.46*a*), $A = 3.11$ in^2, $I = 3.58$ in^4

Using the RISA-2D computer program, analyze the structural frames for gravity and wind loads in the following three cases.

Case 1 Unbraced Frame with Rigid Connections

(*a*) Analyze the frame for the loads shown in Figure P9.46*a*. Determine the forces and displacements at 7 sections along the axis of each member. Use the computer program to plot shear and moment diagrams.

(*b*) Determine if the relative lateral displacement between adjacent floors exceeds $^3/_8$ in—a limit specified to prevent cracking of the exterior façade.

(*c*) Using the computer program, plot the deflected shape of the frame.

(*d*) Note the difference between the magnitudes of the vertical and lateral displacements of joints 4 and 9. What are your conclusions?

Case 2 Unbraced Frame with Shear Connections

(*a*) Repeat steps *a*, *b*, and *c* in Case 1, assuming that the shear connections act as hinges, that is, can transmit shear and axial load, but no moments.

(*b*) What do you conclude about the unbraced frame's resistance to lateral displacements?

Case 3 Braced Frame with Shear Connections

As in case 2, all beams are connected to columns with shear connectors, but diagonal bracing is added to form a vertical truss with floor beams and columns (see dashed lines in Figure P9.46*a*).

(*a*) Repeat steps *a*, *b*, and *c* in Case 1.

(*b*) Compute the lateral deflections of the frame if the area and moment of inertia of the diagonal members are *doubled*. Compare results to the original lighter bracing in (*a*) to establish the effectiveness of heavier bracing.

(*c*) Make up a table comparing lateral displacements of joints 4 and 9 for the three cases. Discuss briefly the results of this study.

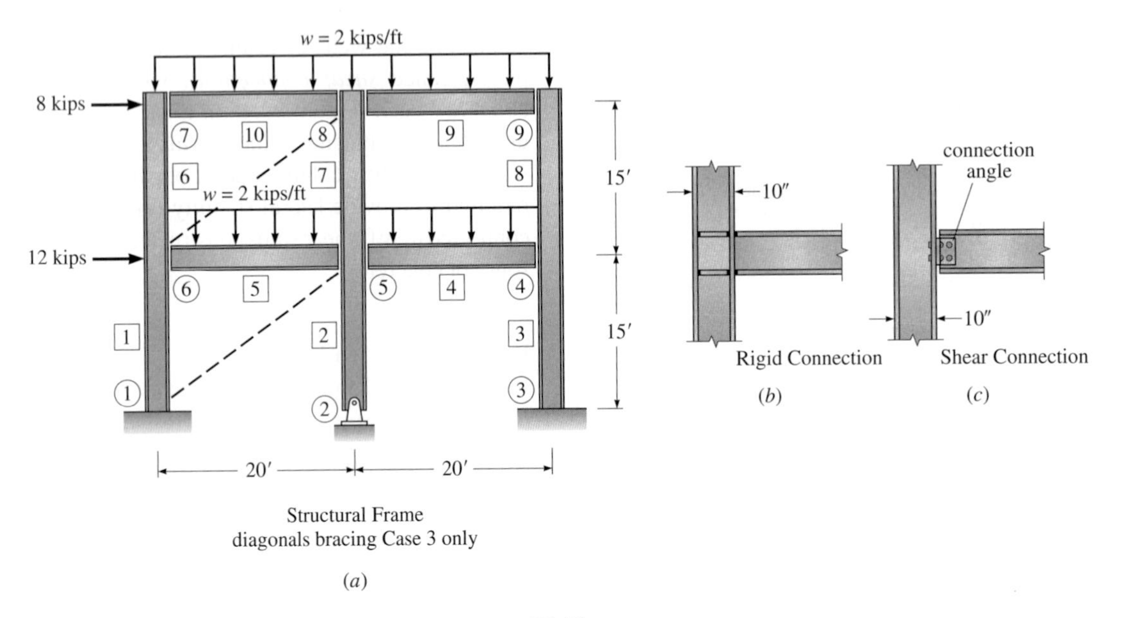

P9.46

NASA Rocket Engine Test Stand in Mississippi

A 235-ft high, A-3 Test Stand at Stennis Space Center in Gulfport, Mississippi, is to test NASA's rocket engine J-2X of the Ares I Spacecraft, which will propel manned missions to the moon and Mars. Structural design challenges included designing for the 1,000-kip thrust force of the engine, lateral forces from 150 mph winds and provide structural stiffness to maintain a maximum lateral deflection of $^1/_4$ inch under a 300-kip lateral force.

CHAPTER 10

Work-Energy Methods for Computing Deflections

Chapter Objectives

- Understand the concept of energy (external work and internal strain energy).
- Use the principle of the conservation of energy to derive the method of real work. Study the limitation of this method for deflection calculation.
- Use the dummy load system (or *Q-system*) and the actual load system (or *P-system*) to derive the virtual work method for calculating deflections. Then apply this very powerful method to calculate deflections of trusses, beams, and frames. The method can also include the effect of temperature change, support settlements, and fabrication errors.
- Study the Bernoulli's principle of virtual displacements.
- Derive the Maxwell-Betti law of reciprocal deflections.

10.1 Introduction

When a structure is loaded, its stressed elements deform. As these deformations occur, the structure changes shape and points on the structure displace. In a well-designed structure, these displacements are small. For example, Figure 10.1*a* shows an unloaded cantilever beam that has been divided arbitrarily into four rectangular elements. When a vertical load is applied at point B, moment develops along the length of the member. This moment creates longitudinal tensile and compressive bending stresses that deform the rectangular elements into trapezoids and cause point B at the tip of the cantilever to displace vertically downward to B'. This displacement, Δ_B, is shown to an exaggerated scale in Figure 10.1*b*.

Similarly, in the example of the truss shown in Figure 10.1*c*, the applied load P produces axial forces F_1, F_2, and F_3 in the members. These forces cause the

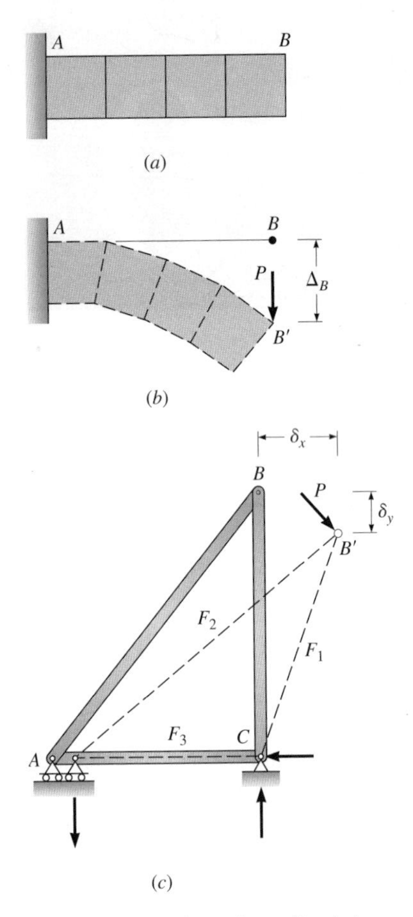

Figure 10.1: Deformations of loaded structures: (*a*) beam before load is applied; (*b*) bending deformations produced by a load at *B*; (*c*) deformations of a truss after load is applied.

members to deform axially as shown by the dashed lines. As a result of these deformations, joint *B* of the truss displaces diagonally to *B'*.

Work-energy methods provide the basis for several procedures used to calculate displacements. Work-energy lends itself to the computation of deflections because the unknown displacements can be incorporated directly into the expression for *work—the product of a force and a displacement*. In the typical deflection computation, the magnitude and direction of the design forces are specified, and the proportions of the members are known. Therefore, once the member forces are computed, the energy stored in each element of the structure can be evaluated and equated to the work done by the external forces applied to the structure. Since the *principle of the conservation of energy* states that the work done by a system of forces applied to a structure equals the strain energy stored in the structure, loads are assumed to be applied slowly *so that neither kinetic nor heat energy is produced*.

We will begin our study of work-energy by reviewing the work done by a force or moment moving through a small displacement. Then we will derive the equations for the energy stored in both an axially loaded bar and a beam. Finally, we will illustrate the work-energy method—also called the *method of real work*—by computing a component of the deflection of a joint of a simple truss. Since the method of real work has serious limitations (i.e., deflections can be computed only at a point where a force acts and only a single concentrated load can be applied to the structure), the major emphasis in this chapter will be placed on the method of *virtual work*.

Virtual work, one of the most useful, versatile methods of computing deflections, is applicable to many types of structural members from simple beams and trusses to complex plates and shells. Although virtual work can be applied to structures that behave either *elastically* or *inelastically*, the method does require that changes in geometry be small (the method could not be applied to a cable that undergoes a large change in geometry by application of a concentrated load). As an additional advantage, virtual work permits the designer to include in deflection computations the influence of support settlements, temperature changes, creep, and fabrication errors.

10.2 Work

Work is defined as the product of a *force* times a *displacement* in the direction of the force. In deflection computations we will be concerned with the work done by both forces and moments. If a force *F* remains constant in magnitude as it moves from point *A* to *B* (see Figure 10.2*a*), the work *W* may be expressed as

$$W = F\delta \tag{10.1}$$

where δ is the component of displacement in the direction of the force. Work is positive when the force and displacement are in the same direction and negative when the force acts opposite in direction to the displacement.

When a force moves perpendicular to its line of action, as shown in Figure 10.2*b*, the work is zero. If the magnitude and direction of a force remain constant as the force moves through a displacement δ that is *not* collinear with the line of action of the force, the total work can be evaluated by summing the work done by each component of the force moving through the corresponding collinear displacement components δ_x and δ_y. For example, in Figure 10.2*c* we can express the work W done by the force F as it moves from point A to B as

$$W = F_x\delta_x + F_y\delta_y$$

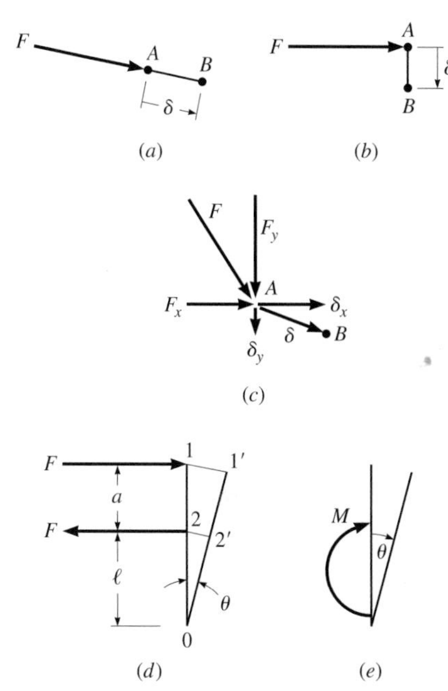

Figure 10.2: Work done by forces and moments: (*a*) force with a collinear displacement; (*b*) force with a displacement perpendicular to line of action of force; (*c*) a noncollinear displacement; (*d*) a couple moving through an angular displacement θ; (*e*) alternative representation of a couple.

Similarly, if a moment remains constant as it is given an angular displacement θ (see Figure 10.2*d* and *e*), the work done equals the product of the moment and the angular displacement θ:

$$W = M\theta \tag{10.2}$$

The expression for work done by a couple can be derived by summing the work done by each force F of the couple in Figure 10.2*d* as it moves on a circular arc during the angular displacement θ. This work equals

$$W = -F\ell\theta + F(\ell + a)\theta$$

Simplifying gives

$$W = Fa\theta$$

Since $Fa = M$,

$$W = M\theta$$

If a force varies in magnitude during a displacement and if the functional relationship between the force F and the collinear displacement δ is known, the work can be evaluated by integration. In this procedure, shown graphically in Figure 10.3*a*, the displacement is divided into a series of small increments of length $d\delta$. The increment of work dW associated with each infinitesimal displacement $d\delta$ equals $F\,d\delta$. The total work is then evaluated by summing all increments:

$$W = \int_0^{\delta} F\,d\delta \tag{10.3}$$

Similarly, for a variable moment that moves through a series of infinitesimal angular displacements $d\theta$, the total work is given as

$$W = \int_0^{\theta} M\,d\theta \tag{10.4}$$

When force is plotted against displacement (see Figure 10.3*a*), the term within the integrals of Equation 10.3 or 10.4 may be interpreted as an infinitesimal area under the curve. The total work done—the sum of all the

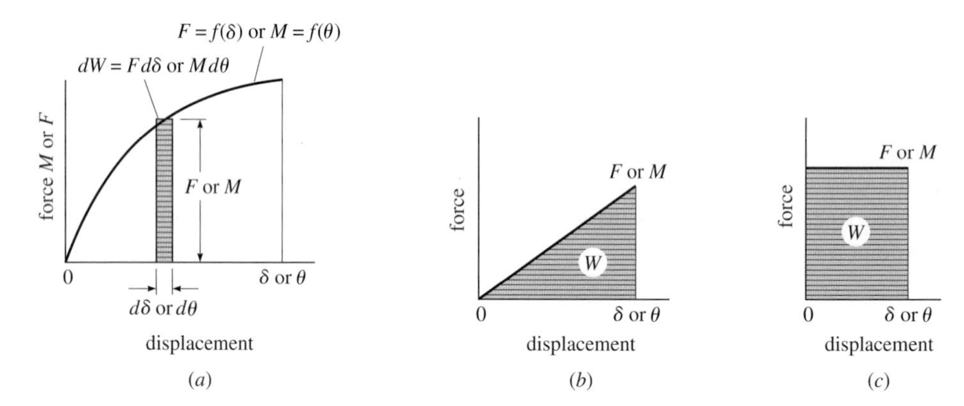

Figure 10.3: Force versus displacement curves: (*a*) increment of work dW produced by a variable force shown crosshatched; (*b*) work (shown by crosshatched area) done by a force or moment that varies linearly from zero to F or M, (*c*) work done by a force or moment that remains constant during a displacement.

infinitesimal areas—equals the total area under the curve. If a force or moment varies linearly with displacement, as it increases from zero to its final value of F or M, respectively, the work can be represented by the triangular area under the linear load-deflection curve (see Figure 10.3*b*). For this condition the work can be expressed as

For force:
$$W = \frac{F}{2}\delta \tag{10.5}$$

For moment:
$$W = \frac{M}{2}\theta \tag{10.6}$$

where F and M are the maximum values of force or moment and δ and θ are the total linear or rotational displacement.

When a *linear* relationship exists between force and displacement and when the force increases from zero to its final value, expressions for work will always contain a *one-half* term, as shown by Equations 10.5 and 10.6. On the other hand, if the magnitude of a force or moment is constant during a displacement (Equations 10.1 and 10.2), the work plots as a rectangular area (see Figure 10.3*c*) and the one-half term is absent.

10.3 Strain Energy

Truss Bars

When a bar is loaded axially, it will deform and store strain energy U. For example, in the bar shown in Figure 10.4*a*, the externally applied load P induces an axial force F of equal magnitude (that is, $F = P$). If the bar behaves

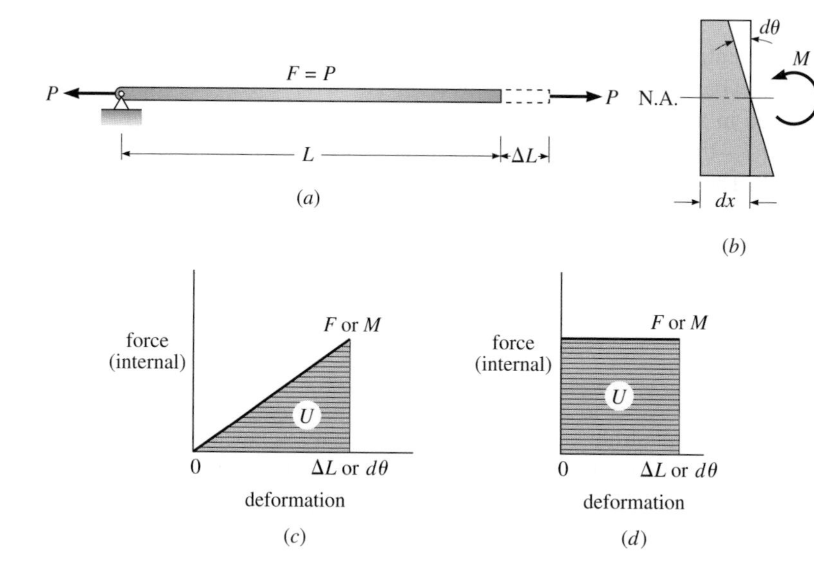

Figure 10.4: Strain energy stored in a bar or beam element. (*a*) Deformation of an axially loaded bar; (*b*) rotational deformation of infinitesimal beam element by moment M; (*c*) plot of load versus deformation for element in which load increases linearly from zero to a final value; (*d*) load deformation curve for member that deforms under a constant load.

elastically (Hooke's law applies), the magnitude of the strain energy U stored in a bar by a force that increases linearly from zero to a final value F as the bar undergoes a change in length ΔL equals

$$U = \frac{F}{2}\,\Delta L \tag{10.7}$$

where

$$\Delta L = \frac{FL}{AE} \tag{10.8}$$

where L = length of bar
A = cross-sectional area of bar
E = modulus of elasticity
F = final value of axial force

Substituting Equation 10.8 into Equation 10.7, we can express U in terms of the bar force F and the properties of the member as

$$U = \frac{F}{2}\,\frac{FL}{AE} = \frac{F^2 L}{2AE} \tag{10.9}$$

If the magnitude of the axial force remains constant as a bar undergoes a change in length ΔL from some outside effect (for example, a temperature change), the strain energy stored in the member equals

$$U = F\,\Delta L \tag{10.10}$$

Notice that when a force remains constant as the axial deformation of a bar occurs, the *one-half* factor does not appear in the expression for U (compare Equations 10.7 and 10.10).

Energy stored in a body, like work done by a force (see Figure 10.3) can be represented graphically. If the variation of bar force is plotted against the change in bar length ΔL, the area under the curve represents the strain energy U stored in the member. Figure 10.4c is the graphic representation of Equation 10.7—the case in which a bar force increases linearly from zero to a final value F. The graphic representation of Equation 10.10—the case in which the bar force remains constant as the bar changes length—is shown in Figure 10.4d. Similar force versus deformation curves can be plotted for beam elements such as the one shown in Figure 10.4b. In the case of the beam element, we plot moment M versus rotation $d\theta$.

Beams

The increment of strain energy dU stored in a beam segment of infinitesimal length dx (see Figure 10.4b) by a moment M that *increases linearly from zero to a final value of* M as the sides of the segment rotate through an angle $d\theta$ equals

$$dU = \frac{M}{2}\, d\theta \tag{10.11}$$

As we have shown previously, $d\theta$ may be expressed as

$$d\theta = \frac{M\, dx}{EI} \tag{9.13}$$

where E equals the modulus of elasticity and I equals the moment of inertia of the cross section with respect to the neutral axis.

Substituting Equation 9.13 into 10.11 gives the increment of strain energy stored in a beam segment of length dx as

$$dU = \frac{M}{2}\frac{M\, dx}{EI} = \frac{M^2\, dx}{2EI} \tag{10.12}$$

To evaluate the total strain energy U stored in a beam of constant EI, the strain energy must be summed for all infinitesimal segments by integrating both sides of Equation 10.12.

$$U = \int_0^L \frac{M^2\, dx}{2EI} \tag{10.13}$$

To integrate the right side of Equation 10.13, M must be expressed in terms of the applied loads and the distance x along the span (see Section 5.3). At each section where the load changes, a new expression for moment is required. If I varies along the axis of the member, it must also be expressed as a function of x.

If the moment M remains *constant* as a segment of beam undergoes a rotation $d\theta$ from *another effect*, the increment of strain energy stored in the element equals

$$dU = M\, d\theta \tag{10.14}$$

When $d\theta$ in Equation 10.14 is produced by a moment of magnitude M_P, we can, using Equation 9.13 to eliminate $d\theta$, and express dU as

$$dU = \frac{MM_P\,dx}{EI} \tag{10.14a}$$

10.4 Deflections by the Work-Energy Method (Real Work)

To establish an equation for computing the deflection of a point on a structure by the work-energy method, we can write according to the principle of conservation of energy that

$$W = U \tag{10.15}$$

where W is the work done by the external force applied to the structure and U is the strain energy stored in the stressed members of the structure.

Equation 10.15 assumes that all work done by an external force is converted to strain energy. To satisfy this requirement, a load theoretically must be applied slowly so that neither kinetic nor heat energy is produced. In the design of buildings and bridges for normal design loads, we will always assume that this condition is satisfied so that Equation 10.15 is valid. Because a single equation permits the solution of only one unknown variable, Equation 10.15—the basis of the method of real *work—can only be applied to structures that are loaded by a single force*.

Work-Energy Applied to a Truss

To establish an equation that can be used to compute the deflection of a point on a truss due to a load P that increases linearly from zero to a final value P, we substitute Equations 10.5 and 10.9 into Equation 10.15 to give

$$\frac{P}{2}\delta = \sum \frac{F^2L}{2AE} \tag{10.16}$$

where P and δ are collinear and the summation sign Σ indicates that the energy in all bars must be summed. The use of Equation 10.16 to compute the horizontal displacement of joint B of the truss in Figure 10.5 is illustrated in Example 10.1.

As shown in Figure 10.5, joint B displaces both horizontally and vertically. Since the applied load of 30 kips is horizontal, we are able to compute the horizontal component of displacement. However, we are not able to compute the vertical component of the displacement of joint B by the method of real work because the applied force does not act in the vertical direction. The *method of virtual work*, which we discuss next, permits us to compute a single displacement component in any direction of any joint for any type of loading and thereby overcomes the major limitations of the method of real work.

EXAMPLE 10.1

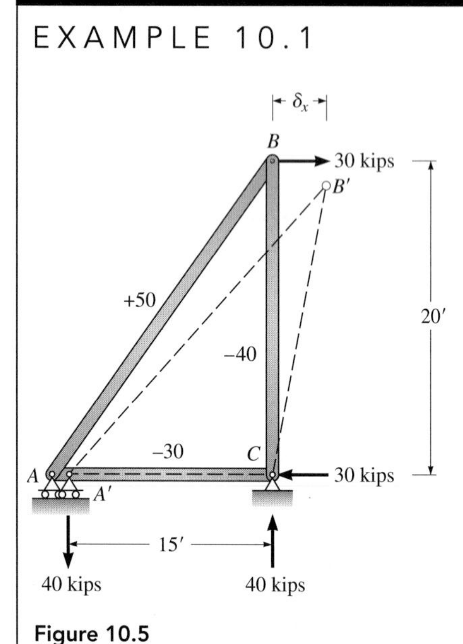

Figure 10.5

Using the method of real work, determine the horizontal deflection δ_x of joint B of the truss shown in Figure 10.5. For all bars, $A = 2.4$ in^2 and $E = 30{,}000$ kips/in^2. The deflected shape is shown by the dashed lines.

Solution

Since the applied force of $P = 30$ kips acts in the direction of the required displacement, the method of real work is valid and Equation 10.16 applies.

$$\frac{P}{2}\delta_x = \sum \frac{F^2L}{2AE} \tag{10.16}$$

Values of bar force F are shown on the truss in Figure 10.5.

$$\frac{30}{2}\delta_x = \frac{(50)^2(25)(12)}{2(2.4)(30{,}000)} + \frac{(-40)^2(20)(12)}{2(2.4)(30{,}000)} + \frac{(-30)^2(15)(12)}{2(2.4)(30{,}000)}$$

$\delta_x = 0.6$ in **Ans.**

10.5 Virtual Work: Trusses

Virtual Work Method

Virtual work is a procedure for computing a single component of deflection at any point on a structure. The method is applicable to many types of structures, from simple beams to complex plates and shells. Moreover, the method permits the designer to include in deflection computations the influence of support settlements, temperature change, and fabrication errors.

To compute a component of deflection by the method of virtual work, the designer applies a force to the structure at the point and in the direction of the desired displacement. This force is often called a *dummy load* because like a ventriloquist's dummy (or puppet), the displacement it will undergo is produced by other effects. These other effects include the real loads, temperature change, support settlements, and so forth. The dummy load and the reactions and internal forces it creates are termed a *Q-system*. Forces, work, displacements, or energy associated with the Q system will be subscripted with a Q. Although the analyst is free to assign any arbitrary value to a dummy load, typically we use a 1-kip or a 1-kN force to compute a linear displacement and a 1 kip · ft or a 1 kN · m moment to determine a rotation or slope.

With the dummy load in place, the *actual loads*—called the *P-system*, are applied to the structure. Forces, deformations, work, and energy associated with the *P*-system will be subscripted with a *P*. As the structure deforms under the actual loads, *external virtual work* W_Q is done by the dummy load (or loads) as it moves through the real displacement of the structure. In accordance with the principle of conservation of energy, an equivalent quantity of *virtual strain energy* U_Q is stored in the structure; that is,

$$W_Q = U_Q \tag{10.17}$$

The virtual strain energy stored in the structure equals the product of the internal forces produced by the dummy load and the distortions (changes in length of axially loaded bars, for example) of the elements of the structure produced by the real loads (i.e., the P-system).

Analysis of Trusses by Virtual Work

To clarify the variables that appear in the expressions for work and energy in Equation 10.17, we will apply the method of virtual work to the one-bar truss in Figure 10.6*a* to determine the horizontal displacement δ_P of the roller at B. The bar, which carries axial load only, has a cross-sectional area A and modulus of elasticity E. Figure 10.6*a* shows the bar force F_P, the elongation of the bar ΔL_P, and the horizontal displacement δ_P of joint B produced by the P system (the actual load). Since the bar is in tension, it elongates an amount ΔL_P, where

$$\Delta L_P = \frac{F_P L}{AE} \tag{10.8}$$

Assuming that the horizontal load at joint B is applied slowly (so that all work is converted to strain energy) and increases from zero to a final value P, we can use Equation 10.5 to express the real work W_P done by force P as

$$W_P = \tfrac{1}{2} P \delta_P \tag{10.18}$$

Although a vertical reaction P_v develops at B, it does no work as the roller displaces because it acts normal to the displacement of joint B. A plot of the deflection of joint B versus the applied load P is shown in Figure 10.6*b*. As we established in Section 10.2, the triangular area W_P under the load-deflection curve represents the real work done on the structure by load P.

As a result of the real work done by P, strain energy U_P of equal magnitude is stored in bar AB. Using Equation 10.7, we can express this strain energy as

$$U_P = \tfrac{1}{2} F_P \, \Delta L_P \tag{10.19}$$

A plot of the strain energy stored in the bar as a function of the bar force F_P and the elongation ΔL_P of the bar is shown in Figure 10.6*c*. In accordance with the conservation of energy, W_P equals U_P, so the shaded areas W_P and U_P under the sloping lines in Figure 10.6*b* and *c* must be equal.

We next consider the work done on the strain energy stored in the bar by applying in sequence the dummy load Q followed by the real load P. Figure 10.6*d* shows the bar force F_Q, the bar deformation ΔL_Q, and the

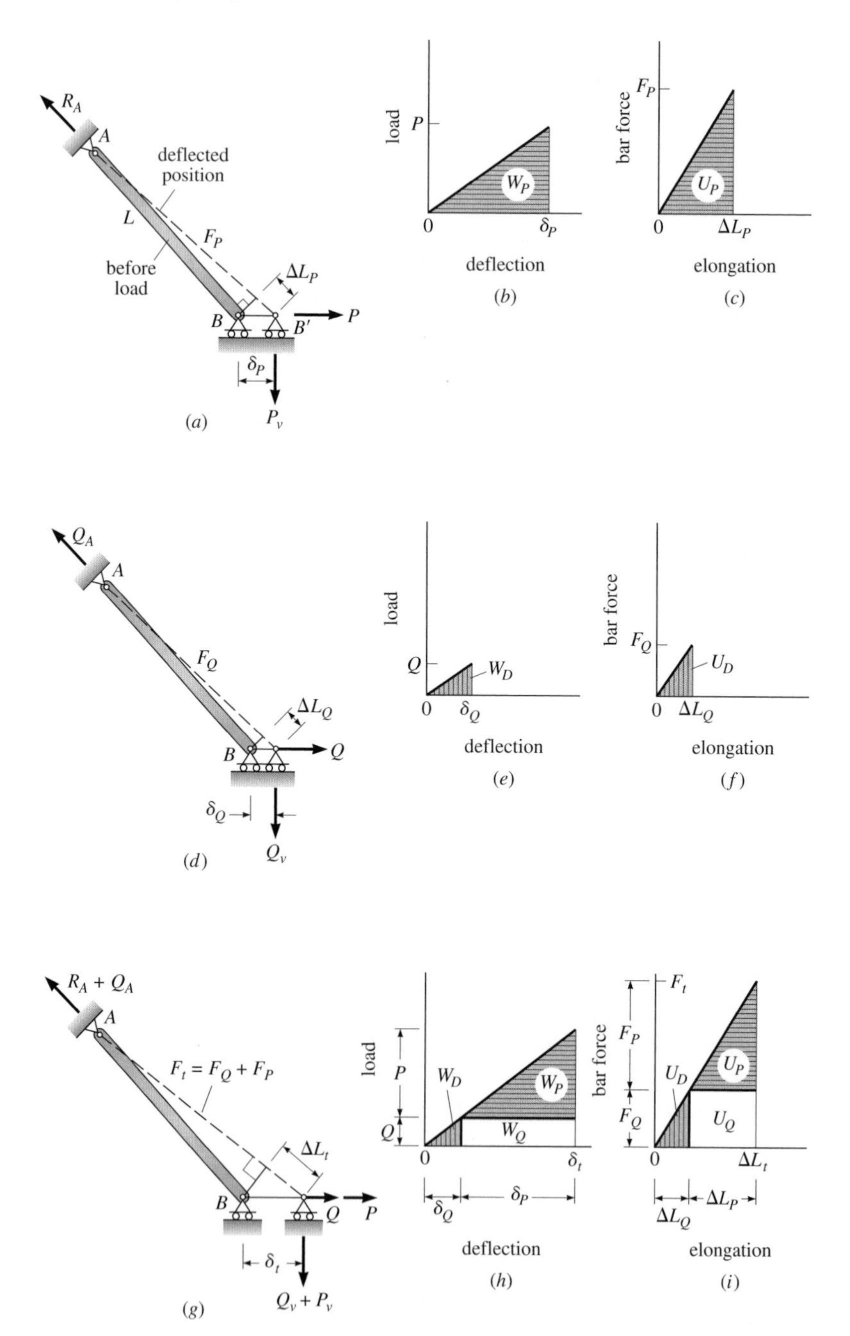

Figure 10.6: Graphical representation of work and energy in the method of virtual work. (*a*) *P*-system: forces and deformations produced by real load *P*; (*b*) graphical representation of real work W_P done by force *P* as roller in (*a*) moves from *B* to *B*′; (*c*) graphical representation of real strain energy U_P stored in bar *AB* as it elongates an amount ΔL_P ($U_P = W_P$); (*d*) forces and displacements produced by dummy load *Q*; (*e*) graphical representation of real work W_D done by dummy load *Q*; (*f*) graphical representation of real strain energy U_D stored in bar *AB* by dummy load; (*g*) forces and deformations produced by forces *Q* and *P* acting together; (*h*) graphical representation of total work W_t done by *Q* and *P*; (*i*) graphical representation of total strain energy U_t stored in bar by *Q* and *P*.

horizontal displacement δ_Q of joint B produced by the dummy load Q. Assuming that the dummy load is applied slowly and increases from zero to its final value Q, we can express the real work W_D done by the dummy load as

$$W_D = \tfrac{1}{2}Q\delta_Q \tag{10.20a}$$

The load-deflection curve associated with the dummy load is shown in Figure 10.6*e*. The triangular area under the sloping line represents the real work W_D done by the dummy load Q. The corresponding strain energy U_D stored in the bar as it elongates is equal to

$$U_D = \tfrac{1}{2}F_Q\,\Delta L_Q \tag{10.20b}$$

Figure 10.6*f* shows the strain energy stored in the structure due to the elongation of bar AB by the dummy load. In accordance with the principle of conservation of energy, W_D must equal U_D. Therefore, the crosshatched triangular areas in Figure 10.6*e* and *f* are equal.

With the dummy load in place we now imagine that the real load P is applied (see Figure 10.6*g*). Because we assume that behavior is elastic, the principle of superposition requires that the final deformations, bar forces, reactions, and so forth (but not the work or the strain energy, as we will shortly establish) equal the sum of those produced by Q and P acting separately (see Figure 10.6*a* and *d*). Figure 10.6*h* shows the total work W_t done by forces Q and P as point B displaces horizontally an amount $\delta_t = \delta_Q + \delta_P$. Figure 10.6*i* shows the total strain energy U_t stored in the structure by the action of forces Q and P.

To clarify the physical significance of virtual work and virtual strain energy, we subdivide the areas in Figure 10.6*h* and *i* that represent the total work and total strain energy into the following three areas:

1. Triangular areas W_D and U_D (shown in vertical crosshatching)
2. Triangular areas W_P and U_P (shown in horizontal crosshatching)
3. Two rectangular areas labeled W_Q and U_Q

Since $W_D = U_D$, $W_P = U_P$, and $W_t = U_t$ by the principle of conservation of energy, it follows that the two rectangular areas W_Q and U_Q, which represent the external virtual work and the virtual strain energy, respectively, must be equal, and we can write

$$W_Q = U_Q \tag{10.17}$$

As shown in Figure 10.6*h*, we can express W_Q as

$$W_Q = Q\delta_p \tag{10.21a}$$

where Q equals the magnitude of the dummy load and δ_P the displacement or component of displacement in the direction of Q produced by the P-system. As indicated in Figure 10.6*i*, we can express U_Q as

$$U_Q = F_Q\,\Delta L_P \tag{10.21b}$$

where F_Q is the bar force produced by the dummy load Q and ΔL_P is the change in length of the bar produced by the P-system.

Substituting Equations 10.21*a* and 10.21*b* into Equation 10.17, we can write the virtual work equation for the one-bar truss as

$$Q \cdot \delta_P = F_Q \, \Delta L_P \tag{10.22}$$

By adding summation signs to each side of Equation 10.22, we produce Equation 10.23, the general virtual work equation for the analysis of any type of truss.

$$\Sigma Q \delta_P = \Sigma F_Q \, \Delta L_P \tag{10.23}$$

The summation sign on the left side of Equation 10.23 indicates that in certain cases (see Example 10.7 on page 384, for example) more than one external Q force contributes to the virtual work. The summation sign on the right side of Equation 10.23 is added because most trusses contain more than one bar.

Equation 10.23 shows that both the internal and external forces are supplied by the Q system and that the displacements and deformations of the structure are supplied by the P system. The term *virtual* signifies that the displacements of the dummy load are produced by an outside effect (i.e., the P system).

When the bar deformations are produced by load, we can use Equation 10.8 to express the bar deformations ΔL_P in terms of the bar force F_P and the properties of the members. For this case we can write Equation 10.23 as

$$\Sigma Q \delta_P = \Sigma F_Q \frac{F_P L}{AE} \tag{10.24}$$

We will illustrate the use of Equation 10.24 by computing the deflection of joint B in the simple two-bar truss shown in Example 10.2. Since the direction of the resultant displacement at B is unknown, we do not know how to orient the dummy load to compute it. Therefore, we will carry out the analysis in two separate computations. First, we compute the component of displacement in the x direction, using a horizontal dummy load (see Figure 10.7*b*). Then we compute the y component of displacement, using a vertical dummy load (see Figure 10.7*c*). If we wish to establish the magnitude and direction of the actual displacement, the components can be combined by vector addition.

EXAMPLE 10.2

Under the action of the 30-kip load, joint B of the truss in Figure 10.7a displaces to B' (the deflected shape is shown by the dashed lines). Using virtual work, compute the components of displacement of joint B. For all bars, $A = 2\text{ in}^2$ and $E = 30{,}000\text{ kips/in}^2$.

Solution

To compute the horizontal displacement δ_x of joint B, we apply a dummy load of 1 kip horizontally at B. Figure 10.7b shows the reactions and bar forces F_Q produced by the dummy load. With the dummy load in place, we apply the real load of 30 kips to joint B (indicated by the dashed arrow). The 30-kip load produces bar forces F_P, which deform the truss. Although both the dummy and the real loading now act dependently on the structure, for clarity we show the forces and deformations produced by the real load, $P = 30$ kips, separately on the sketch in Figure 10.7a. With the bar forces established, we use Equation 10.24 to compute δ_x:

$$\Sigma Q\delta_P = \Sigma F_Q \frac{F_P L}{AE} \qquad (10.24)$$

$$(1\text{ kip})(\delta_x) = \frac{5}{3}\frac{50(20 \times 12)}{2(30{,}000)} + \left(-\frac{4}{3}\right)\frac{(-40)(16 \times 12)}{2(30{,}000)}$$

$$\delta_x = 0.5\text{ in} \rightarrow \qquad \textbf{Ans.}$$

To compute the vertical displacement δ_y of joint B, we apply a dummy load of 1 kip vertically at joint B (see Figure 10.7c) and then apply the real load. Since the value of F_Q in bar AB is zero (see Figure 10.7c), no energy is stored in that bar and we only have to evaluate the strain energy stored in bar BC. Using Equation 10.24, we compute

$$\Sigma Q\delta_P = \Sigma F_Q \frac{F_P L}{AE} \qquad (10.24)$$

$$(1\text{ kip})(\delta_y) = \frac{(-1)(-40)(16 \times 12)}{2(30{,}000)} = 0.128\text{ in} \downarrow \qquad \textbf{Ans.}$$

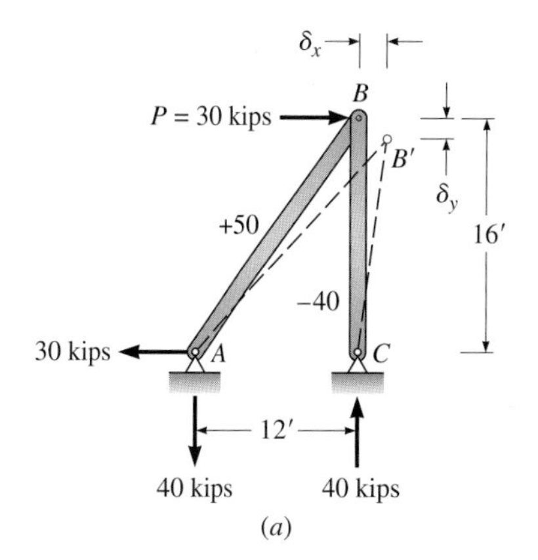

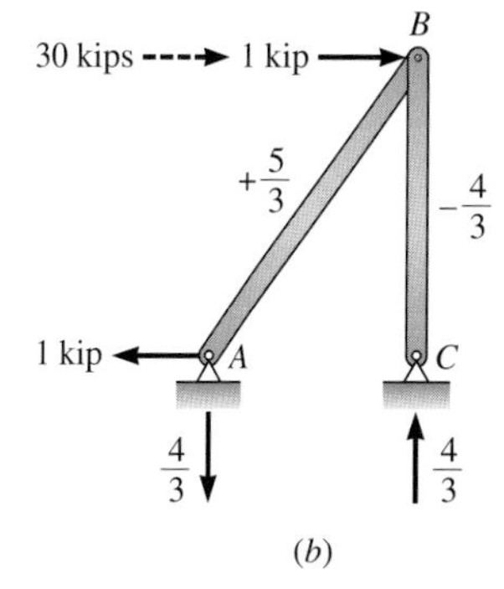

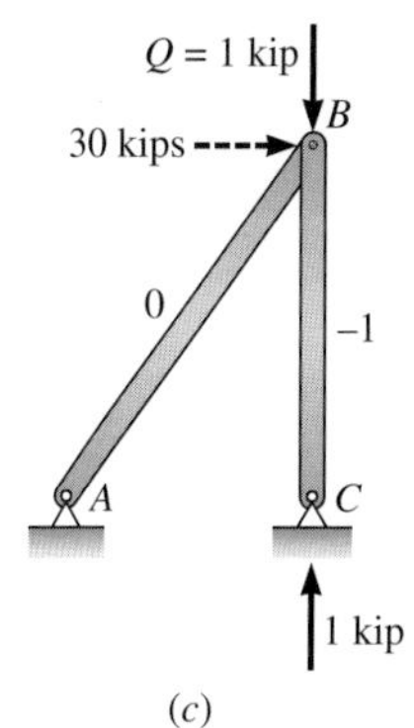

Figure 10.7: (a) Real loads (P-system producing bar forces F_P). (b) Dummy load (Q-system producing F_Q forces) used to compute the horizontal displacement of B. The dashed arrow indicates the actual load that creates the forces F_P shown in (a). (c) Dummy load (Q-system) used to compute the vertical displacement of B.

As you can see, if a bar is unstressed in either the P system or the Q system, its contribution to the virtual strain energy stored in a truss is zero.

NOTE. The use of a 1-kip dummy load in Figure 10.7*b* and *c* was arbitrary, and the same results could have been achieved by applying a dummy force of any value. For example, if the dummy load in Figure 10.7*b* were doubled to 2 kips, the bar forces F_Q would be twice as large as those shown on the figure. When the forces produced by the 2-kip dummy are substituted into Equation 10.24, the external work—a direct function of Q—and the internal strain energy—a direct function of F_Q—will both double. As a result, the computation produces the same value of deflection as that produced by the 1-kip dummy.

Positive values of δ_x and δ_y indicate that both displacements are in the same direction as the dummy loads. If the solution of the virtual work equation produces a negative value of displacement, the direction of the displacement is opposite in sense to the direction of the dummy load. Therefore, it is not necessary to guess the actual direction of the displacement being computed. *The direction of the dummy force may be selected arbitrarily, and the sign of the answer will automatically indicate the correct direction of the displacement.* A positive sign signifies the displacement is in the direction of the dummy force; a negative sign indicates the displacement is opposite in sense to the direction of the dummy load.

To evaluate the expression for virtual strain energy $(F_Q F_P L)/(AE)$ on the right side of Equation 10.24 (particularly when a truss is composed of many bars), many engineers use a table to organize the computations (see Table 10.1 in Example 10.3). Terms in columns 6 of the Table 10.1 equal the product of F_Q, F_P, and L divided by A. If this product is divided by E, the strain energy stored in the bar is established.

The total virtual strain energy stored in the truss equals the sum of the terms in column 6 divided by E. The value of the sum is written at the bottom of column 6. If E is a constant for all bars, it can be omitted from the summation and then introduced in the final step of the deflection computation. *If the value of either F_Q or F_P for any bar is zero, the strain energy in that bar is zero, and the bar can be omitted from the summation.*

If several displacement components are required, more columns for F_Q produced by other dummy loads are added to the table. Extra columns for F_P are also required when deflections are computed for several loadings.

EXAMPLE 10.3

Compute the horizontal displacement δ_x of joint B of the truss shown in Figure 10.8a. Given: $E = 30{,}000$ kips/in^2, area of bars AD and $BC = 5$ in^2; area of all other bars $= 4$ in^2.

Solution

The F_P bar forces produced by the P system are shown in Figure 10.8a, and the F_Q bar forces and reactions produced by a dummy load of 1 kip directed horizontally at joint B are shown in Figure 10.8b. Table 10.1 lists the terms required to evaluate the strain energy U_Q given by the right side of Equation 10.24. Since E is constant, it is factored out of the summation and not included in the table.

Substituting $\Sigma F_Q F_P L/A = 1025$ into Equation 10.24 and multiplying the right side by 12 to convert feet to inches give

$$\Sigma Q\delta_P = \Sigma F_Q \frac{F_P L}{AE} = \frac{1}{E}\Sigma F_Q \frac{F_P L}{A} \qquad (10.24)$$

$$1\ \text{kip}(\delta_x) = \frac{1}{30{,}000}(1025)(12)$$

$$\delta_x = 0.41\ \text{in} \rightarrow \qquad \textbf{Ans.}$$

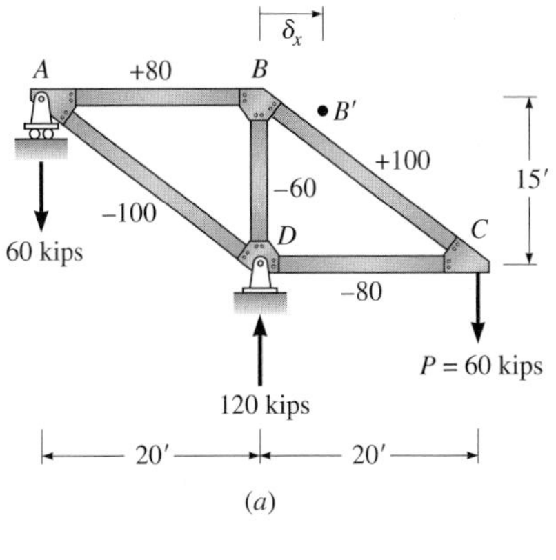

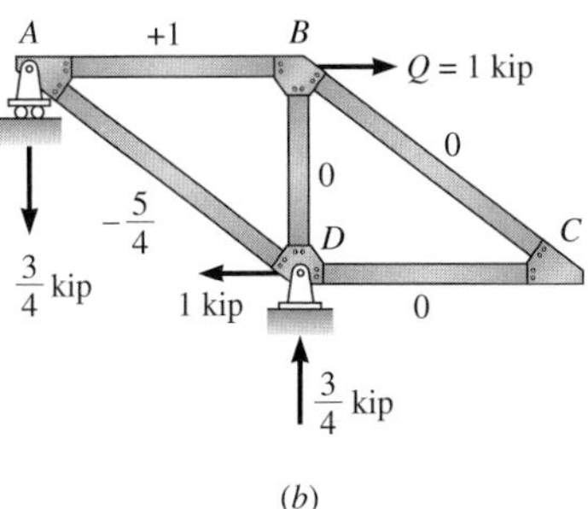

Figure 10.8: (a) P system actually loads; (b) Q system.

TABLE 10.1

Member (1)	F_Q kips (2)	F_P kips (3)	L ft (4)	A in^2 (5)	F_QF_PL/A kips^2·ft/in^2 (6)
AB	+1	+80	20	4	+400
BC	0	+100	25	5	0
CD	0	−80	20	4	0
AD	$-\frac{5}{4}$	−100	25	5	+625
BD	0	−60	15	4	0
					$\Sigma F_Q F_P L/A = 1025$

Truss Deflections Produced by Temperature and Fabrication Error

As the temperature of a member varies, its length changes. An increase in temperature causes a member to expand; a decrease in temperature produces a contraction. In either case the change in length ΔL_{temp} can be expressed as

$$\Delta L_{temp} = \alpha \, \Delta T \, L \tag{10.25}$$

where α = coefficient of thermal expansion, in/in per degree
ΔT = change in temperature
L = length of bar

To compute a component of joint deflection due to a change in temperature of a truss, first we apply a dummy load. Then we assume that the change in length of the bars produced by the temperature change occurs. As the bars change in length and the truss distorts, external virtual work is done as the dummy load displaces. Internally, the change in length of the truss bars results in a change in strain energy U_Q equal to the product of the bar forces F_Q (produced by the dummy load) and the deformation ΔL_{temp} of the bars. The virtual work equation for computing a joint displacement can be established by substituting ΔL_{temp} for ΔL_P in Equation 10.23.

A change in bar length ΔL_{fabr} due to a fabrication error is handled in exactly the same manner as a temperature change. Example 10.4 illustrates the computation of a component of truss displacement for both a temperature change and a fabrication error.

If the bars of a truss change in length simultaneously due to load, temperature change, and a fabrication error, then ΔL_P in Equation 10.23 is equal to the sum of the various effects; that is,

$$\Delta L_P = \frac{F_P L}{AE} + \alpha \, \Delta T \, L + \Delta L_{fabr} \tag{10.26}$$

When ΔL_P given by Equation 10.26 is substituted into Equation 10.23, the general form of the virtual work equation for trusses becomes

$$\Sigma Q \delta_P = \Sigma F_Q \left(\frac{F_P L}{AE} + \alpha \, \Delta T \, L + \Delta L_{fabr} \right) \tag{10.27}$$

EXAMPLE 10.4

For the truss shown in Figure 10.9*a*, determine the horizontal displacement δ_x of joint *B* for a 60°F increase in temperature and the following fabrication errors: (1) bar *BC* fabricated 0.8 in too short and (2) bar *AB* fabricated 0.2 in too long. Given: $\alpha = 6.5 \times 10^{-6}$ in/in per °F.

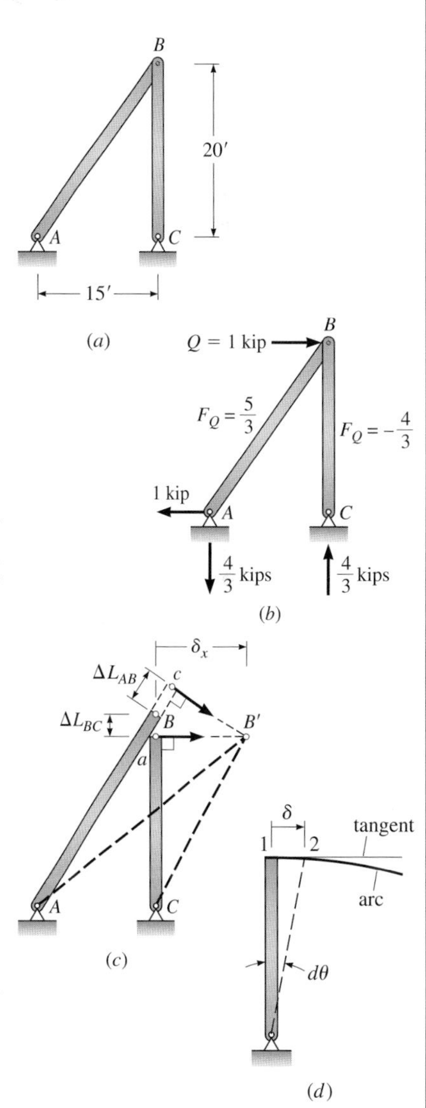

Figure 10.9: (*a*) Truss; (*b*) *Q* system; (*c*) displacement of joint *B* produced by changes in length of bars; (*d*) for small displacements, the free end initially moves perpendicular to the bar's axis.

Solution

Because the structure is determinate, no bar forces are created by either a temperature change or a fabrication error. If the lengths of the bars change, they can still be connected to the supports and joined together at *B* by a pin. For the conditions specified in this example, bar *AB* will elongate and bar *BC* will shorten. If we imagine that the bars in their deformed state are connected to the pin supports at *A* and *C* (see Figure 10.9*c*), bar *AB* will extend beyond point *B* a distance ΔL_{AB} to point *c* and the top of bar *BC* will be located a distance ΔL_{BC} below joint *B* at point *a*. If the bars are rotated about the pins, the upper ends of each bar will move on the arcs of circles that intersect at *B′*. The deflected position of the truss is shown by the dashed lines. Since the initial displacement of each bar is directed tangent to the circle, we can assume for small displacements that the bars initially move in the direction of the tangent lines (i.e., perpendicular to the radii). For example, as shown in Figure 10.9*d* in the region between points 1 and 2, the tangent line and the arc coincide closely.

Changes in length of bars due to temperature increase:

$$\Delta L_{\text{temp}} = \alpha(\Delta T)L \tag{10.25}$$

Bar *AB*: $$\Delta L_{\text{temp}} = 6.5 \times 10^{-6}(60)25 \times 12 = 0.117 \text{ in}$$

Bar *BC*: $$\Delta L_{\text{temp}} = 6.5 \times 10^{-6}(60)20 \times 12 = 0.094 \text{ in}$$

To determine δ_x, we first apply a dummy load of 1 kip at *B* (Figure 10.9*b*) and then allow the specified bar deformations to take place. Using Equation 10.27, we compute

$$\Sigma Q\delta_P = \Sigma F_Q\,\Delta L_P = \Sigma F_Q\,(\Delta L_{\text{temp}} + \Delta L_{\text{fabr}})$$

$$(1 \text{ kip})(\delta_x) = \tfrac{5}{3}(0.117 + 0.2) + (-\tfrac{4}{3})(0.094 - 0.8)$$

$$\delta_x = 1.47 \text{ in} \rightarrow \qquad \textbf{Ans.}$$

Computation of Displacements Produced by Support Settlements

Structures founded on compressible soils (soft clays or loose sand, for example) often undergo significant settlements. These settlements can produce rotation of members and displacement of joints. If a structure is determinate, no internal stresses are created by a support movement because the structure is free to adjust to the new position of the supports. On the other hand, differential support settlements can induce large internal forces in indeterminate structures. The magnitude of these forces is a function of the member's stiffness.

Virtual work provides a simple method for evaluating both the displacements and rotations produced by support movements. To compute a displacement due to a support movement, a dummy load is applied at the point and in the direction of the desired displacement. The dummy load together with its reactions constitute the Q system. As the structure is subjected to the specified support movements, external work is done by both the dummy load and those of its reactions that displace. Since a support movement produces no internal distortion of members or structural elements if the structure is determinate, the virtual strain energy is zero.

Example 10.5 illustrates the use of virtual work to compute joint displacements and rotations produced by the settlements of the supports of a simple truss. The same procedure is applicable to determinate beams and frames.

Inelastic Behavior

The expression for strain energy given by the right side of Equation 10.24 is based on the assumption that all truss bars behave elastically; that is, the level of stress does not exceed the proportional limit σ_{PL} of the material. To extend virtual work to trusses that contain bars stressed beyond the proportional limit into the inelastic region, we must have the stress-strain curve of the material. To establish the axial deformation of a bar, we compute the stress in the bar, use the stress to establish the strain, and then evaluate the change in length ΔL_P using the basic relationship

$$\Delta L_P = \epsilon L \tag{10.28}$$

Example 10.8 on page 386 illustrates the procedure to calculate the deflection of a joint in a truss that contains a bar stressed into the inelastic region.

EXAMPLE 10.5

If support A of the truss in Figure 10.10a settles 0.6 in and moves to the left 0.2 in, determine (a) the horizontal displacement δ_x of joint B and (b) the rotation θ of bar BC.

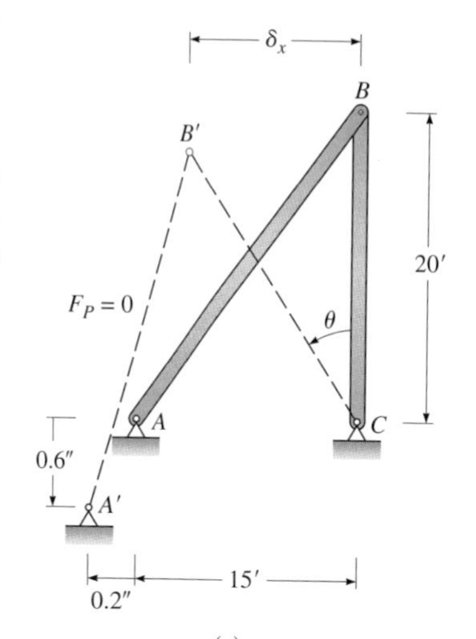

Solution

(a) To compute δ_x, apply a 1-kip dummy load horizontally at B (see Figure 10.10b) and compute all reactions. Assume that the support movements occur, evaluate the external virtual work, and equate to zero. Since no F_P bar forces are produced by the support movement, $F_P = 0$ in Equation 10.24, yielding

$$\Sigma Q\delta_P = 0$$

$$(1 \text{ kip})(\delta_x) + 1(0.2 \text{ in}) + \tfrac{4}{3}(0.6 \text{ in}) = 0$$

$$\delta_x = -1 \text{ in} \qquad \textbf{Ans.}$$

The minus sign indicates δ_x is directed to the left.

(b) To compute the rotation θ of member BC, we apply a dummy load of 1 kip·ft to bar BC anywhere between its ends and compute the support reactions (see Figure 10.10c). As the support movements shown in Figure 10.10a occur, virtual work is done by both the dummy load and the reactions at those supports that displace in the direction of the reactions. In accordance with Equation 10.2, the virtual work produced by a unit moment M_Q used as a dummy load equals $M_Q\theta$. With this term added to W_Q and with $U_Q = 0$, the expression for virtual work equals

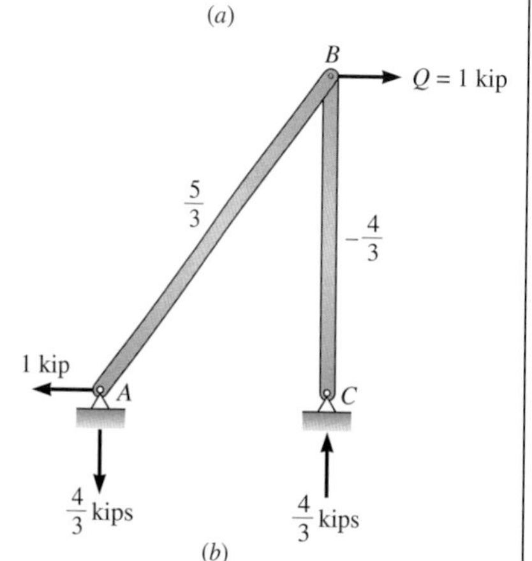

$$W_Q = \Sigma(Q\delta_P + M_Q\theta_P) = 0$$

Expressing all terms in units of kips·in (multiply M_Q by 12) gives

$$1(12)(\theta_P) - \tfrac{1}{15}(0.6) - \tfrac{1}{20}(0.2) = 0$$

$$\theta_P = 0.00417 \text{ rad} \qquad \textbf{Ans.}$$

To verify the computation of θ for bar BC, we can also divide δ_x by 20 ft:

$$\theta_P = \frac{\delta_x}{L} = \frac{1 \text{ in}}{[20(12)] \text{ in}} = 0.00417 \text{ rad}$$

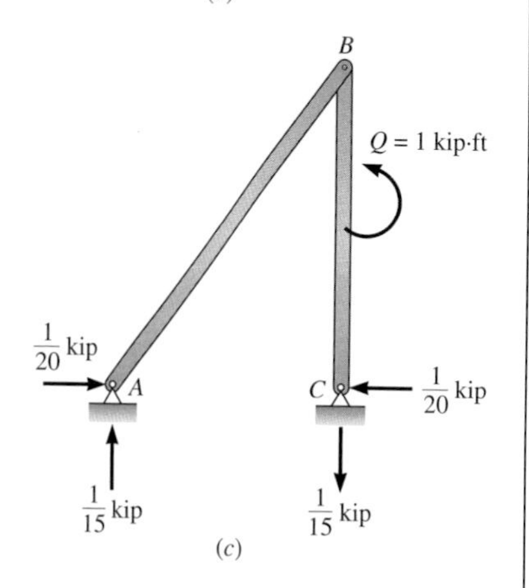

Figure 10.10: (a) Deflected shape (see dashed line) produced by the movement of support A (no F_P forces created); (b) Q system to compute the horizontal displacement of joint B; (c) Q system to compute the rotation of bar BC.

EXAMPLE 10.6

Determine the horizontal displacement δ_{CX} of joint C of the truss in Figure 10.11a. In addition to the 48-kip load applied at joint B, bars AB and BC are subjected to a temperature change ΔT of $+100°F$ [$\alpha = 6.5 \times 10^{-6}$ in/in/°F)], bars AB and CD are each constructed $\frac{3}{4}$ in too long, and support A isconstructed $\frac{3}{5}$ in below point A. For all bars $A = 2$ in^2 and $E = 30{,}000$ kips/in^2. How much should bars CD and DE each be lengthened or shortened if the net horizontal displacement at joint C is to be zero after the various actions listed above occur?

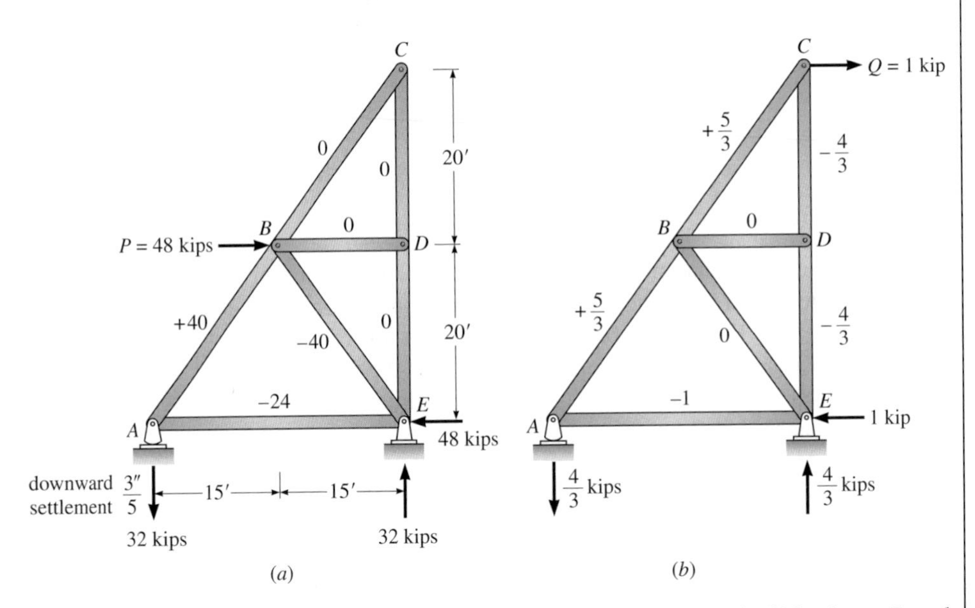

Figure 10.11: (a) Truss with F_P forces shown on bars (P system); (b) bar forces F_Q and reactions produced by dummy load of 1 kip at joint C (Q system).

Solution

Apply a dummy load of 1 kip horizontally at C, as shown in Figure 10.11b, and compute the bar forces F_Q and the reactions. With the dummy load in place, the 48-kip load is applied at B and the support settlement at A, and the changes in bar lengths due to the various effects are assumed to occur. The support settlement produces external virtual work; the load, temperature change, and fabrication errors create virtual strain energy as bars stressed by F_Q forces deform. The virtual strain energy will be zero in any bar in which F_Q is zero or in which the change in length is zero. Therefore, we only have to evaluate the virtual strain energy in bars AB, AE, CD, and BC using Equation 10.27.

$$\Sigma Q\delta_p = \Sigma F_Q\left(\frac{F_P L}{AE} + \alpha\, \Delta T\, L + \Delta L_{\text{fabr}}\right) \qquad (10.27)$$

$$(1\text{ kip})(\delta_{CX}) + \frac{4}{3}\text{kips}\left(\frac{3}{5}\right) = \frac{5}{3}\text{kips}\underbrace{\left[\frac{40(25\times 12)}{2(30{,}000)} + 6.5\times 10^{-6}(100)(25\times 12) + \frac{3}{4}\right]}_{\text{Bar } AB}$$

$$-\ (1\text{ kip})\underbrace{\left[\frac{(-24)(30\times 12)}{2(30{,}000)}\right]}_{\text{Bar } AE} + \underbrace{\left(-\frac{4}{3}\text{kips}\right)\left(\frac{3}{4}\right)}_{\text{Bar } CD}$$

$$+\ \frac{5}{3}\text{kips}\underbrace{\left[6.5\times 10^{-6}(100)(25\times 12)\right]}_{\text{Bar } BC}$$

$$\delta_{CX} = 0.577\text{ in to the right} \qquad \textbf{Ans.}$$

Compute the change in length of bars DE and CD to produce zero horizontal displacement at joint C.

$$\Sigma Q\delta_P = \Sigma F_Q\, \Delta L_P \qquad (10.23)$$

$$1\text{ kip}(-0.577\text{ in}) = -\tfrac{4}{3}(\Delta L_P)2$$

$$\Delta L_P = 0.22\text{ in} \qquad \textbf{Ans.}$$

Since ΔL is positive, bars should be lengthened.

EXAMPLE 10.7

(*a*) Determine the relative movement between joints *B* and *E*, along the diagonal line between them, produced by the 60-kip load at joint *F* (see Figure 10.12*a*). Area of bars *AF*, *FE*, and *ED* = 1.5 in^2, area of all other bars = 2 in^2, and E = 30,000 kips/in^2.

(*b*) Determine the vertical deflection of joint *F* produced by the 60-kip load.

(*c*) If the initial elevation of joint *F* in the unstressed truss is to be 1.2 in above a horizontal line connecting supports *A* and *D*, determine the amount each bar of the bottom chord should be shortened.

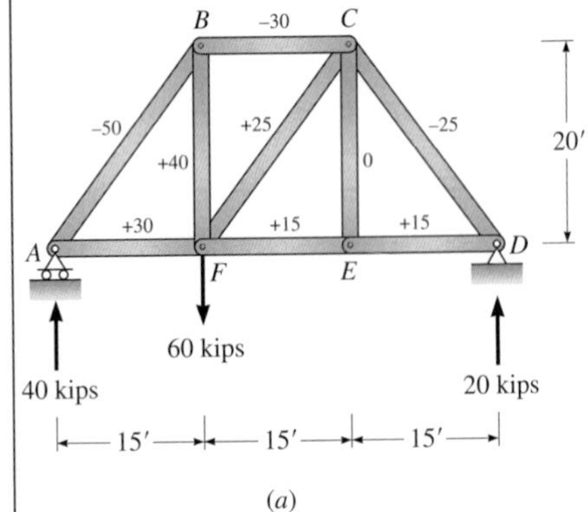

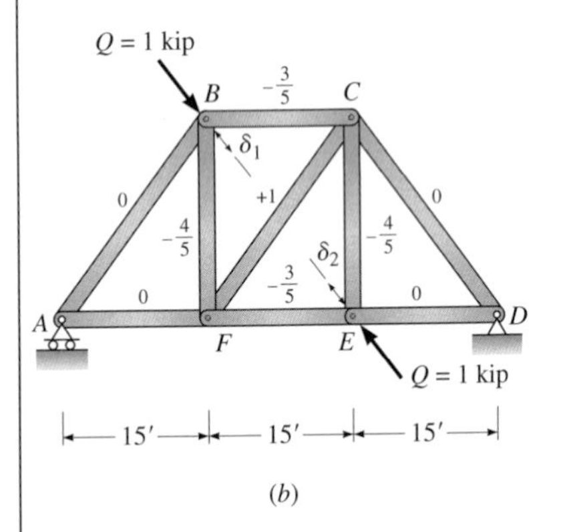

Figure 10.12: (*a*) *P* system with bar forces F_P; (*b*) *Q* system with F_Q forces shown on bars.

Solution

(*a*) To determine the relative displacement between joints *B* and *E*, we use a dummy load consisting of two 1-kip collinear forces at joints *B* and *E*, as shown in Figure 10.12*b*. Since *E* is a constant for all bars, it can be factored out of the summation on the right side of Equation 10.24, producing

$$\Sigma Q\delta_P = \Sigma F_Q \frac{F_P L}{AE} = \frac{1}{E}\Sigma F_Q \frac{F_P L}{A} \tag{10.24}$$

where the quantity $\Sigma F_Q(F_P L/A)$ is evaluated in column 6 of Table 10.2. Substituting into Equation 10.24 and expressing units in kips and inches yield

$$1 \text{ kip}(\delta_1) + 1 \text{ kip}(\delta_2) = \frac{1}{30{,}000}(37.5)(12)$$

Factoring out 1 kip on the left side of the equation and letting $\delta_1 + \delta_2 = \delta_{\text{Rel}}$ give

$$\delta_{\text{Rel}} = \delta_1 + \delta_2 = 0.015 \text{ in} \quad \textbf{Ans.}$$

Since the sign of the relative displacement is positive, joints *B* and *E* move toward each other. In this example we are not able to establish the absolute values of δ_1 and δ_2 because we cannot solve for two unknowns with one equation. To compute δ_1, for example, we must apply a single diagonal dummy load to joint *B* and apply the virtual work equation.

(*b*) To determine the vertical deflection of joint *F* produced by the 60-kip load in Figure 10.12*a*, we must apply a dummy load at joint *F* in the vertical direction. Although we typically use a 1-kip dummy load (as previously discussed in Example 10.2), the magnitude of the dummy load is arbitrary. Therefore, the actual 60-kip load can also serve as the dummy load, and the truss analysis for the *P* system shown in Figure 10.12*a* also

TABLE 10.2

Member (1)	F_Q kips (2)	F_P kips (3)	L ft (4)	A in² (5)	$F_Q F_P \frac{L}{A}$ (kips²·ft)/in² (6)	$F_P^2 \frac{L}{A}$ (kips²·ft/in²) (7)
AB	0	−50	25	2	0	31,250
BC	$-\frac{3}{5}$	−30	15	2	+135	6,750
CD	0	−25	25	2	0	7,812.5
DE	0	+15	15	1.5	0	2,250
EF	$-\frac{3}{5}$	+15	15	1.5	−90	2,250
FA	0	+30	15	1.5	0	9,000
BF	$-\frac{4}{5}$	+40	20	2	−320	16,000
FC	+1	+25	25	2	+312.5	7,812.5
CE	$-\frac{4}{5}$	0	20	2	0	0
					$\Sigma F_Q F_P \frac{L}{A} = +37.5$	$\Sigma F_P^2 \frac{L}{A} = 83,125$

supplies the values of F_Q. Using Equation 10.24 with $F_Q = F_P$, we obtain

$$\Sigma Q\delta_P = \Sigma F_Q \frac{F_P L}{AE} = \frac{1}{E}\Sigma F_P^2 \frac{L}{A}$$

where $\Sigma F_P^2\ (L/A)$, evaluated in column 7 of Table 10.2, equals 83,125. Solving for δ_P gives

$$60\ \delta_P = \frac{1}{30{,}000}(83{,}125)(12)$$

$$\delta_P = 0.554 \text{ in} \downarrow \qquad \textbf{Ans.}$$

(*c*) Since the applied load of 60 kips in Figure 10.12*a* acts in the vertical direction, we can use it as the dummy load to evaluate the vertical displacement (camber) of joint *F* due to shortening of the bottom chord bars. Using Equation 10.23, in which ΔL_P represents the amount each of the three lower chord bars is shortened, we find for $\delta_P = -1.2$ in

$$\Sigma Q\delta_P = \Sigma F_Q\ \Delta L$$

$$(60 \text{ kips})(-1.2) = (30 \text{ kips})(\Delta L_P) + (15 \text{ kips})(\Delta L_P) + (15 \text{ kips})(\Delta L_P)$$

$$\Delta L_P = -1.2 \text{ in} \qquad \textbf{Ans.}$$

A negative 1.2 in is used for δ_P on the left-hand side of Equation 10.23 because the displacement of the joint is opposite in sense to the 60-kip load.

EXAMPLE 10.8

Compute the vertical displacement δ_y of joint C for the truss shown in Figure 10.13a. The truss bars are fabricated from an aluminum alloy whose stress-strain curve (see Figure 10.13c) is valid for both uniaxial tension and compression. The proportional limit, which occurs at a stress of 20 kips/in^2, divides elastic from inelastic behavior. Area of bar $AC = 1$ in^2, and area of bar $BC = 0.5$ in^2. In the elastic region $E = 10{,}000$ kips/in^2.

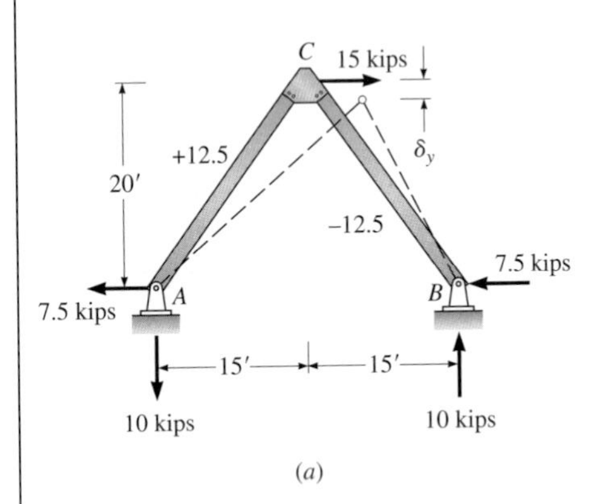

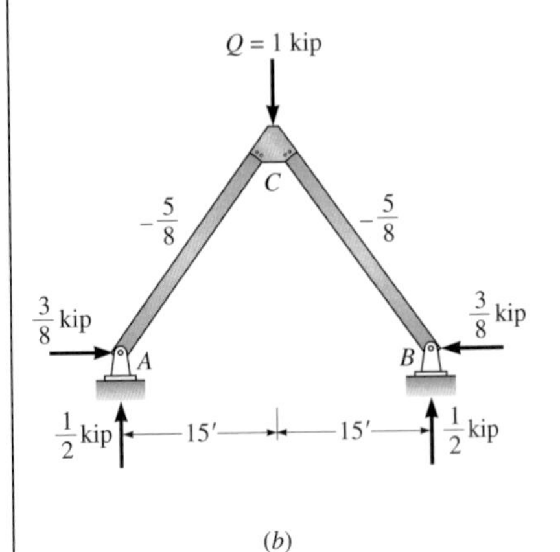

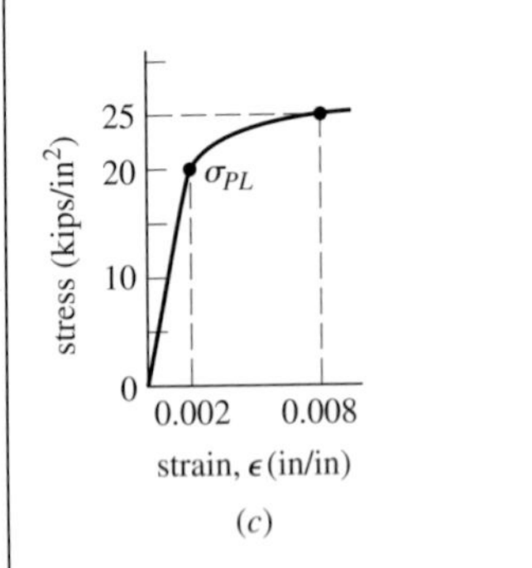

Figure 10.13: (a) P system showing bar forces F_P; (b) Q system showing F_Q bar forces; (c) stress-strain curve (inelastic behavior occurs when stress exceeds 20 kips/in^2).

Solution

The P system with the F_P forces noted on the bars is shown in Figure 10.13a. The Q system with the F_Q forces is shown in Figure 10.13b. To establish if bars behave elastically or are stressed into the inelastic region, we compute the axial stress and compare it to the proportional limit stress.

For bar AC,

$$\sigma_{AC} = \frac{F_P}{A} = \frac{12.5}{1} = 12.5 \text{ kips/in}^2 < \sigma_{PL} \qquad \text{behavior elastic}$$

Using Equation 10.8 gives

$$\Delta L_{AC} = \frac{F_P L}{AE} = \frac{12.5(25 \times 12)}{1(10{,}000)} = 0.375 \text{ in}$$

For bar BC,

$$\sigma_{BC} = \frac{F}{A} = \frac{12.5}{0.5}$$

$$= 25.0 \text{ kips/in}^2 > \sigma_{PL} \qquad \text{bar stressed into inelastic region}$$

To compute ΔL_P, we use Figure 10.13c to establish ϵ. For $\sigma = 25$ ksi, we read $\epsilon = 0.008$ in/in.

$$\Delta L_{BC} = \epsilon L = -0.008(25 \times 12) = -2.4 \text{ in} \qquad \text{(shortens)} \quad \textbf{Ans.}$$

Compute δ_y, using Equation 10.23.

$$(1 \text{ kip})(\delta_y) = \Sigma F_Q \, \Delta L_P$$

$$\delta_y = (-\tfrac{5}{8})(-2.4) + (-\tfrac{5}{8})(0.375)$$

$$= 1.27 \text{ in} \downarrow \qquad \textbf{Ans.}$$

10.6 Virtual Work: Beams and Frames

Both shear and moment contribute to the deformations of beams. However, because the deformations produced by shear forces in beams of normal proportions are small (typically, less than 1 percent of the flexural deformations), we will neglect them in this book (the standard practice of designers) and consider only deformations produced by moment. If a beam is deep (the ratio of span to depth is on the order of 2 or 3), or if a beam web is thin or constructed from a material (wood, for example) with a low shear modulus, shear deformations may be significant and should be investigated.

The procedure to compute a deflection component of a beam by virtual work is similar to that for a truss (except that the expression for strain energy is obviously different). The analyst applies a dummy load Q at the point where the deflection is to be evaluated. Although the dummy load can have any value, typically we use a unit load of 1 kip or 1 kN to compute a linear displacement and a unit moment of 1 kip·ft or 1 kN·m to compute a rotational displacement. For example, to compute the deflection at point C of the beam in Figure 10.14, we apply a 1-kip dummy load Q at C. The dummy load produces a moment M_Q on a typical infinitesimal beam element of length dx, as shown in Figure 10.14*b*. With the dummy load in place, the real loads (the P system) are applied to the beam. The M_P moments produced by the P system bend the beam into its equilibrium position, as shown by the dashed line in Figure 10.14*a*. Figure 10.14*c* shows a short segment of the beam cut from the unstressed member by two vertical planes a distance dx apart. The element is located a distance x from support A. As the forces of the P system increase, the sides of the element rotate through an angle $d\theta$ because of the M_P moments. Neglecting shear deformations, we assume that plane sections before bending remain plane after bending; therefore, longitudinal deformations of the element vary linearly from the neutral axis of the cross section. Using Equation 9.13, we can express $d\theta$ as

$$d\theta = M_P \frac{dx}{EI} \tag{9.13}$$

As the beam deflects, external virtual work W_Q is done by the dummy load Q (and its reactions if supports displace in the direction of the reactions) moving through a distance equal to the actual displacement δ_P in the direction of the dummy load, and we can write

$$W_Q = \Sigma Q \delta_P \tag{10.20}$$

Virtual strain energy dU_Q is stored in each infinitesimal element as the moment M_Q moves through the angle $d\theta$ produced by the P system; thus we can write

$$dU_Q = M_Q \, d\theta \tag{10.14}$$

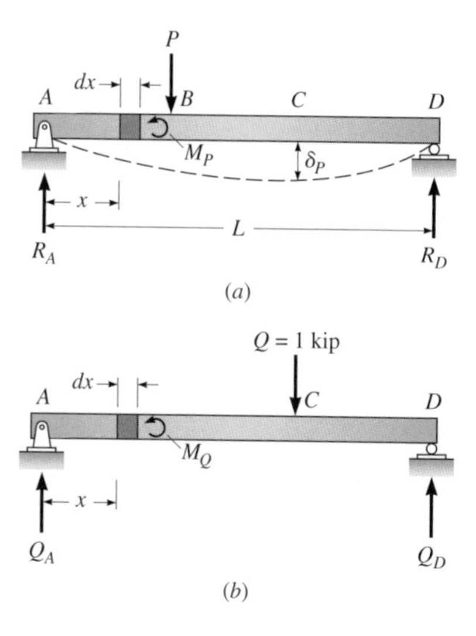

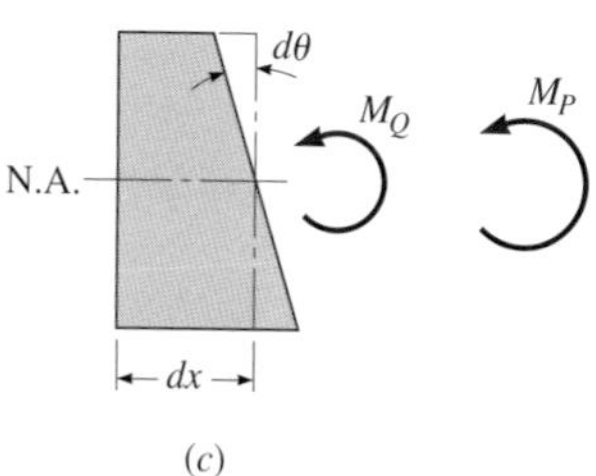

Figure 10.14: (*a*) P system; (*b*) Q system with dummy load at C; (*c*) infinitesimal element; $d\theta$ produced by M_P.

To establish the magnitude of the total virtual strain energy U_Q stored in the beam, we must sum—typically by integration—the energy contained in all the infinitesimal elements of the beam. Integrating both sides of Equation 10.14 over the length L of the beam gives

$$U_Q = \int_{x=0}^{x=L} M_Q \, d\theta \tag{10.29}$$

Since the principle of conservation of energy requires that the external virtual work W_Q equal the virtual strain energy U_Q, we can equate W_Q given by Equation 10.21*a* and U_Q given by Equation 10.29 to produce Equation 10.30, the basic virtual work equation for beams

$$\Sigma Q\delta_P = \int_{x=0}^{x=L} M_Q \, d\theta \tag{10.30}$$

or using Equation 9.13 to express $d\theta$ in terms of the moment M_P and the properties of the cross section, we have

$$\Sigma Q\delta_P = \int_{x=0}^{x=L} M_Q \frac{M_P \, dx}{EI} \tag{10.31}$$

where Q = dummy load and its reactions
δ_P = actual displacement or component of displacement in direction of dummy load produced by real loads (the P system)
M_Q = moment produced by dummy load
M_P = moment produced by real loads
E = modulus of elasticity
I = moment of inertia of beam's cross section with respect to an axis through centroid

If a unit moment $Q_M = 1$ kip·ft is used as a dummy load to establish the change in slope θ_P produced at a point on the axis of a beam by the actual loads, the external virtual work W_Q equals $Q_M\theta_P$ and the virtual work equation is written as

$$\Sigma Q_M\theta_P = \int_{x=0}^{x=L} M_Q \frac{M_P \, dx}{EI} \tag{10.32}$$

To solve Equation 10.31 or 10.32 for the deflection δ_P or the change in slope θ_P, the moments M_Q and M_P must be expressed as a function of x, the distance along the beam's axis, so the right side of the virtual work equation can be integrated. If the cross section of the beam is constant along its length, and if the beam is fabricated from a single material whose properties are uniform, EI is a constant.

Alternate Procedure to Compute U_Q

As an alternate procedure to evaluate the integral on the right-hand side of Equation 10.32 for a variety of M_Q and M_P diagrams of *simple geometric shapes* and for members with a *constant value* of EI, a graphical method entitled "Values of Product Integrals" is provided on the back inside cover of the text. For example, if both M_Q and M_P vary linearly within the span and EI is constant, then the integral can be expressed as follows.

$$\int_{x=0}^{x=L} M_Q M_P \frac{dx}{EI} = \frac{1}{EI}(CM_1M_3L) \tag{10.33}$$

where C = constant listed in product integrals table (Table 4)
M_1 = magnitude of M_Q
M_3 = magnitude of M_P
L = length of member

See Table 4 for other cases of M_Q and M_P distributions. This procedure, together with the classical methods of integration, is illustrated in Examples 10.10 and 10.11.

If the depth of the member varies along the longitudinal axis or if the properties of the material change with distance along the axis, then EI is not a constant and must be expressed as a function of x to permit the integral for virtual strain energy to be evaluated. As an alternative to integration, which may be difficult, the beam may be divided into a number of segments and a finite summation used. This procedure is illustrated in Example 10.16.

In the examples that follow, we will use Equations 10.31, 10.32, and 10.33 to compute the deflections and slopes at various points along the axis of determinate beams and frames. The method can also be used to compute deflections of indeterminate beams after the structure is analyzed and the moment diagrams established.

EXAMPLE 10.9

Using virtual work, compute (*a*) the deflection δ_B and (*b*) the slope θ_B at the tip of the uniformly loaded cantilever beam in Figure 10.15*a*. *EI* is constant.

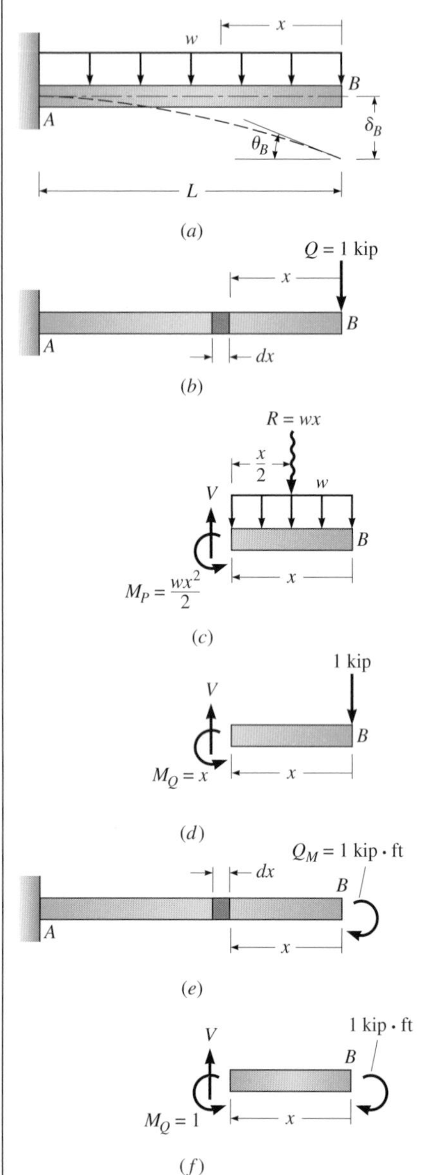

Figure 10.15: (*a*) *P* system; (*b*) *Q* system for computation of δ_B; (*c*) free body to evaluate M_P; (*d*) free body to evaluate M_Q required for computation of δ_B (*e*) *Q* system for computation of θ_B; (*f*) free body to evaluate M_Q for computation of θ_B.

Solution

(*a*) To compute the vertical deflection at *B*, we apply a dummy load of 1 kip vertically at point *B* (see Figure 10.15*b*). The moment M_Q, produced by the dummy load on an element of infinitesimal length *dx* located a distance *x* from point *B*, is evaluated by cutting the free body shown in Figure 10.15*d*. Summing moments about the cut gives

$$M_Q = (1 \text{ kip})(x) = x \text{ kip} \cdot \text{ft} \tag{1}$$

In this computation we arbitrarily assume that the moment is positive when it acts counterclockwise on the end of the section.

With the dummy load on the beam, we imagine that the uniform load *w* (shown in Figure 10.15*a*) is applied to the beam—the uniform load and the dummy load are shown separately for clarity. The dummy load, moving through a displacement δ_B, does virtual work equal to $W_Q = (1 \text{ kip})(\delta_B)$.

We evaluate M_P, the moment produced by the uniform load, with the free body shown in Figure 10.15*c*. Summing moments about the cut, we find

$$M_P = wx\frac{x}{2} = \frac{wx^2}{2} \tag{2}$$

Substituting M_Q and M_P given by Equations 1 and 2 in to Equation 10.31 and integrating, we compute δ_B.

$$W_Q = U_Q; \ \Sigma Q\delta_P = \int_0^L M_Q \frac{M_P\,dx}{EI} = \int_0^L x\frac{wx^2\,dx}{2EI}$$

$$1 \text{ kip}(\delta_B) = \frac{w}{2EI}\left[\frac{x^4}{4}\right]_0^L; \ \delta_B = \frac{wL^4}{8EI} \downarrow \qquad \textbf{Ans.}$$

(*b*) To compute the slope at *B*, we apply a 1 kip·ft dummy load at *B* (see Figure 10.15*e*). Cutting the free body shown in Figure 10.15*f*, we sum moments about the cut to evaluate M_Q as $M_Q = 1$ kip · ft.

Since the initial slope at *B* was zero before load was applied, θ_B, the final slope, will equal the change in slope given by Equation 10.32.

$$\Sigma Q_M\theta_P = \int_0^L M_Q\frac{M_P\,dx}{EI} = \int_0^L \frac{(1)(wx^2)}{2EI}dx$$

$$1 \text{ kip}(\theta_B) = \left[\frac{wx^3}{6EI}\right]_0^L$$

$$\theta_B = \frac{wL^3}{6EI} \curvearrowright \qquad \textbf{Ans.}$$

EXAMPLE 10.10

Compute the vertical displacement and the slope at B produced by the uniformly distributed load w in Figure 10.16*a*. EI is constant. Use the table of product integrals on the back inside cover.

Solution

Evaluate the strain energy for the computation of the vertical deflection at point B in Figure 10.16*a*.

$$1 \text{ kip}(\delta_B) = \frac{1}{EI}(CM_1M_3L) \tag{10.33}$$

$$= \frac{1}{EI}\left[\frac{1}{4}(-L)\left(\frac{-wL^2}{2}\right)(L)\right] = \frac{wL^4}{8EI} \qquad \textbf{Ans.}$$

Evaluate the strain energy for the computation of the slope at point B in Figure 10.16*a*.

$$1 \text{ kip}\cdot\text{ft}(\theta_B) = \frac{1}{EI}(CM_1M_3L) \tag{10.33}$$

$$= \frac{1}{EI}\left[\frac{1}{3}(-1)\left(-\frac{wL^2}{2}\right)(L)\right] = \frac{wL^3}{6EI} \qquad \textbf{Ans.}$$

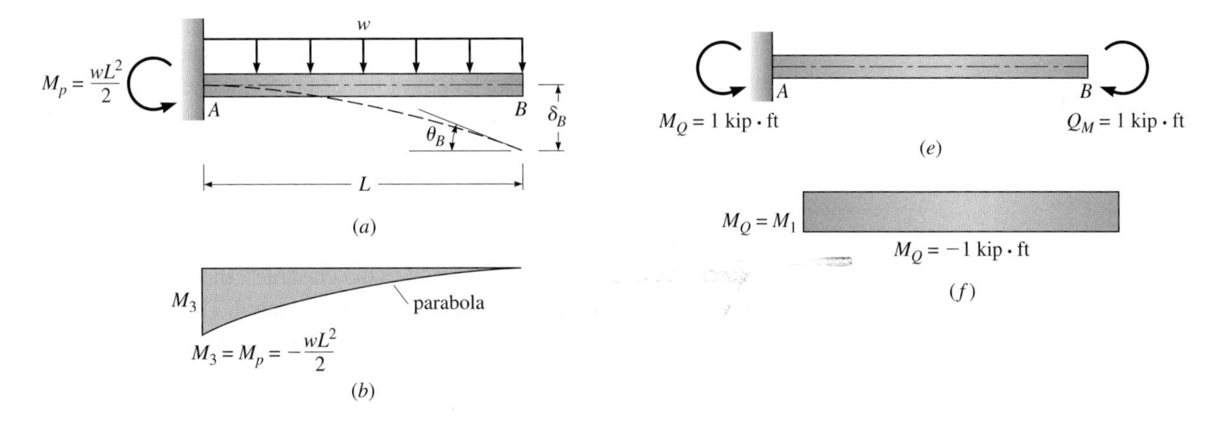

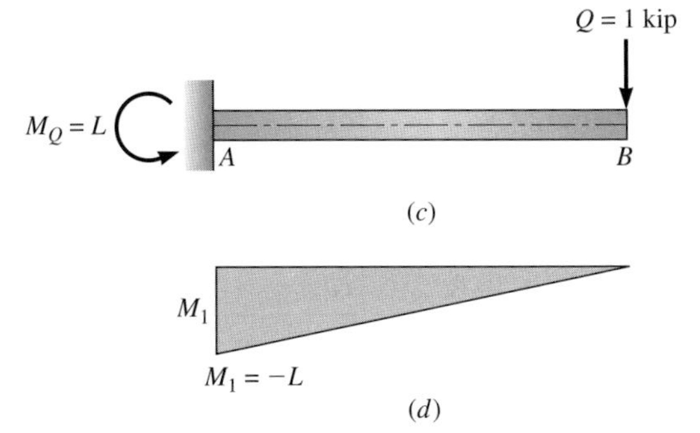

Figure 10.16: Computation of strain energy using Product Integrals Table: (*a*) *P* system; (*b*) moment diagram for the uniformly loaded cantilever beam in (*a*); (*c*) *Q* system for deflection at point *B*; (*d*) moment diagram produced by the *Q* system in (*c*); (*e*) *Q* system for slope at B; (*f*) moment diagram for *Q* system in (*e*).

EXAMPLE 10.11

(*a*) Compute the vertical deflection at midspan δ_C for the beam in Figure 10.17*a*, using virtual work. Given: *EI* is constant, $I = 240$ in^4, $E = 29{,}000$ kips/in^2. (*b*) Recompute δ_c using Equation 10.33 to evaluate U_Q.

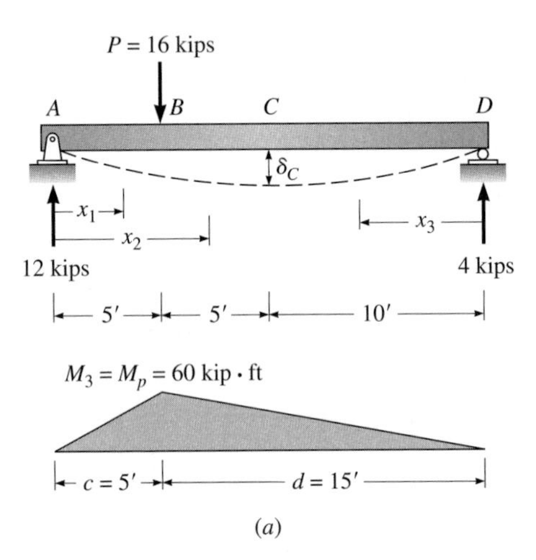

Figure 10.17: (*a*) Real beam (the *P* system).

Solution

(*a*) In this example it is not possible to write a single expression for M_Q and M_P that is valid over the entire length of the beam. Since the loads on the free bodies change with distance along the beam axis, the expression for either M_Q or M_P at a section will change each time the section passes a load in either the real or the dummy system. Therefore, for the beam in Figure 10.17, we must use three integrals to evaluate the total virtual strain energy. For clarity we will denote the region in which a particular free body is valid by adding a subscript to the variable *x* that represents the position of the section where the moment is evaluated. The origins shown in Figure 10.17 are arbitrary. If other positions were selected for the origins, the results would be the same, only the limits of a particular *x* would change. The expressions for M_Q and M_P in each section of the beam are as follows:

Segment	Origin	Range of x	M_Q	M_P
AB	*A*	$0 \le x_1 \le 5$ ft	$\frac{1}{2}x_1$	$12x_1$
BC	*A*	$5 \le x_2 \le 10$ ft	$\frac{1}{2}x_2$	$12x_2 - 16(x_2 - 5)$
DC	*D*	$0 \le x_3 \le 10$ ft	$\frac{1}{2}x_3$	$4x_3$

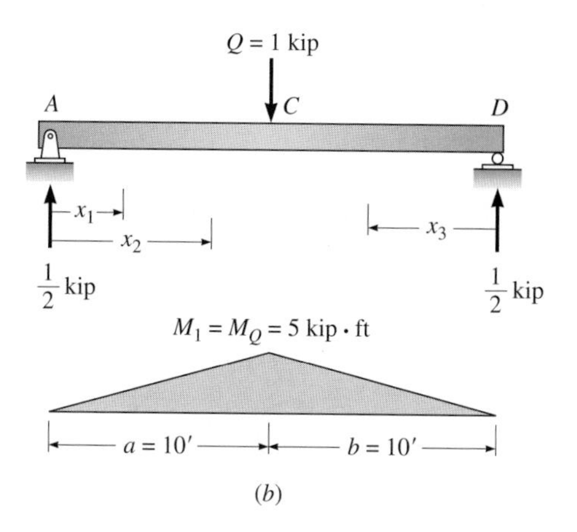

Figure 10.17: (*b*) Dummy load and reactions (the Q system).

In the expressions for M_Q and M_P, positive moment is defined as moment that produces compression on the top fibers of the cross section. Using Equation 10.31, we solve for the deflection.

$$Q\delta_C = \sum_{i=1}^{3} \int M_Q \frac{M_P\, dx}{EI}$$

$$(1\text{ kip})(\delta_C) = \int_0^5 \frac{x_1}{2}(12x_1)\frac{dx}{EI} + \int_5^{10} \frac{x_2}{2}\left[12x_2 - 16(x_2 - 5)\right]\frac{dx}{EI} + \int_0^{10} \frac{x_3}{2}(4x_3)\frac{dx}{EI}$$

$$\delta_C = \frac{250}{EI} + \frac{916.666}{EI} + \frac{666.666}{EI}$$

$$= \frac{1833.33}{EI} = \frac{1833.33(1728)}{240(29{,}000)} = 0.455\text{ in} \qquad \textbf{Ans.}$$

(*b*) Recompute δ_c, using Equation 10.33 (see the product integral in the fifth row and fourth column of the table on the overleaf of the back cover).

$$Q \cdot \delta_c = U_Q = \frac{1}{EI}\left[\frac{1}{3} - \frac{(a-c)^2}{6ad}\right] M_1 M_3 L$$

$$1 \cdot \delta_c = \frac{1}{29{,}000(240)}\left[\frac{1}{3} - \frac{(10-5)^2}{6 \times 10 \times 15}\right] 5 \times 60 \times 20 \times 1728$$

$$\delta_c = 0.455\text{ in} \qquad \textbf{Ans.}$$

EXAMPLE 10.12

Compute the deflection at point C for the beam shown in Figure 10.18a. Given: EI is constant.

Solution

Use Equation 10.31. To evaluate the virtual strain energy U_Q, we must divide the beam into three segments. The following tabulation summarizes the expressions for M_P and M_Q.

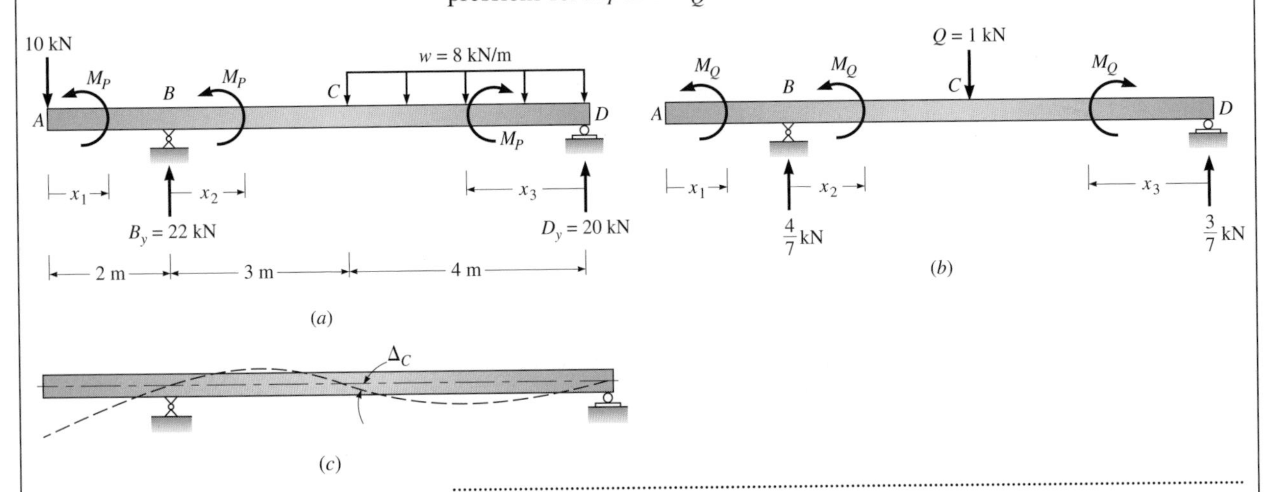

Figure 10.18: (a) P system showing the origins for the coordinate system; (b) Q system; (c) the deflected shape.

Segment	x Origin	Range m	M_P kN·m	M_Q kN·m
AB	A	0–2	$-10x_1$	0
BC	B	0–3	$-10(x_2 + 2) + 22x_2$	$\frac{4}{7}x_2$
DC	D	0–4	$20x_3 - 8x_3(x_3/2)$	$\frac{3}{7}x_3$

Since $M_Q = 0$ in segment AB, the entire integral for this segment will equal zero; therefore, we only have to evaluate the integrals for segments BC and CD:

$$(1\text{ kip})(\Delta_C) = \sum \int M_Q \frac{M_P\,dx}{EI} \tag{10.31}$$

$$\Delta_C = \int_0^2 (0)(-10x_1)\frac{dx}{EI} + \int_0^3 \frac{4}{7}x_2(12x_2 - 20)\frac{dx}{EI} + \int_0^4 \frac{3}{7}x_3(20x_3 - 4x_3^2)\frac{dx}{EI}$$

Integrating and substituting the limits yield

$$\Delta_C = 0 + \frac{10.29}{EI} + \frac{73.14}{EI} = \frac{83.43}{EI} \downarrow \quad \textbf{Ans.}$$

The positive value of Δ_C indicates that the deflection is down (in the direction of the dummy load). A sketch of the beam's deflected shape is shown in Figure 10.18c.

EXAMPLE 10.13

The beam in Figure 10.19 is to be fabricated in the factory with a constant radius of curvature so that a camber of 1.5 in is created at midspan. Using virtual work, determine the required radius of curvature R. Given: EI is constant.

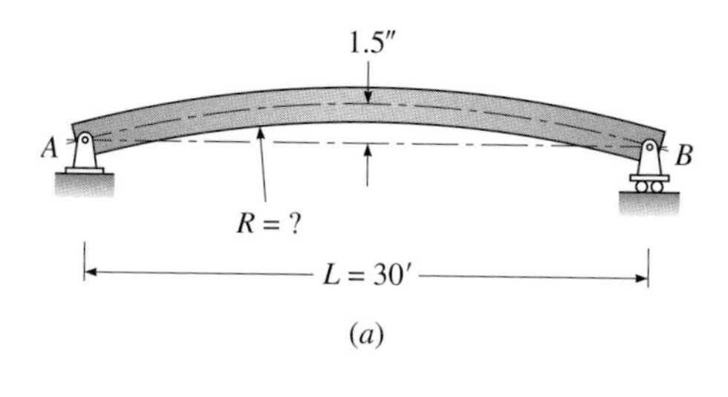

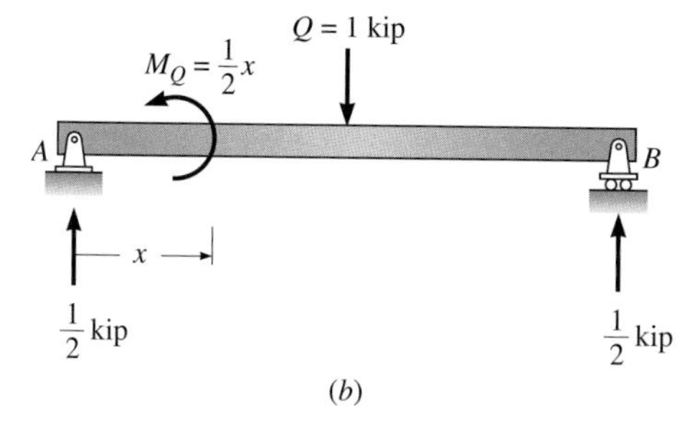

Figure 10.19: (*a*) Beam rolled with a constant radius of curvature R to produce a 1.5-in camber at midspan (P system); (*b*) Q system.

Solution

Use Equation 10.30.

$$\Sigma Q\delta_P = \int M_Q \, d\theta \qquad (10.30)$$

Since $d\theta/dx = 1/R$ and $d\theta = dx/R$ (see Equation 9.4)

$$\delta_P = \frac{1.5 \text{ in}}{12} = 0.125 \text{ ft} \qquad M_Q = \tfrac{1}{2}x \qquad \text{(see Figure 10.19}b\text{)}$$

Substituting $d\theta$, δ_P, and M_Q into Equation 10.30 (because of symmetry we can integrate from 0 to 15 and double the value) gives

$$(1 \text{ kip})(0.125 \text{ ft}) = 2\int_0^{15} \frac{x}{2}\frac{dx}{R}$$

Integrating and substituting limits then give

$$0.125 = \frac{225}{2R}$$

$$R = 900 \text{ ft} \qquad \textbf{Ans.}$$

EXAMPLE 10.14

Considering the strain energy associated with both axial load and moment, compute the horizontal deflection of joint C of the frame in Figure 10.20a. Members are of constant cross section with $I = 600$ in^4, $A = 13$ in^2, and $E = 29{,}000$ kips/in^2.

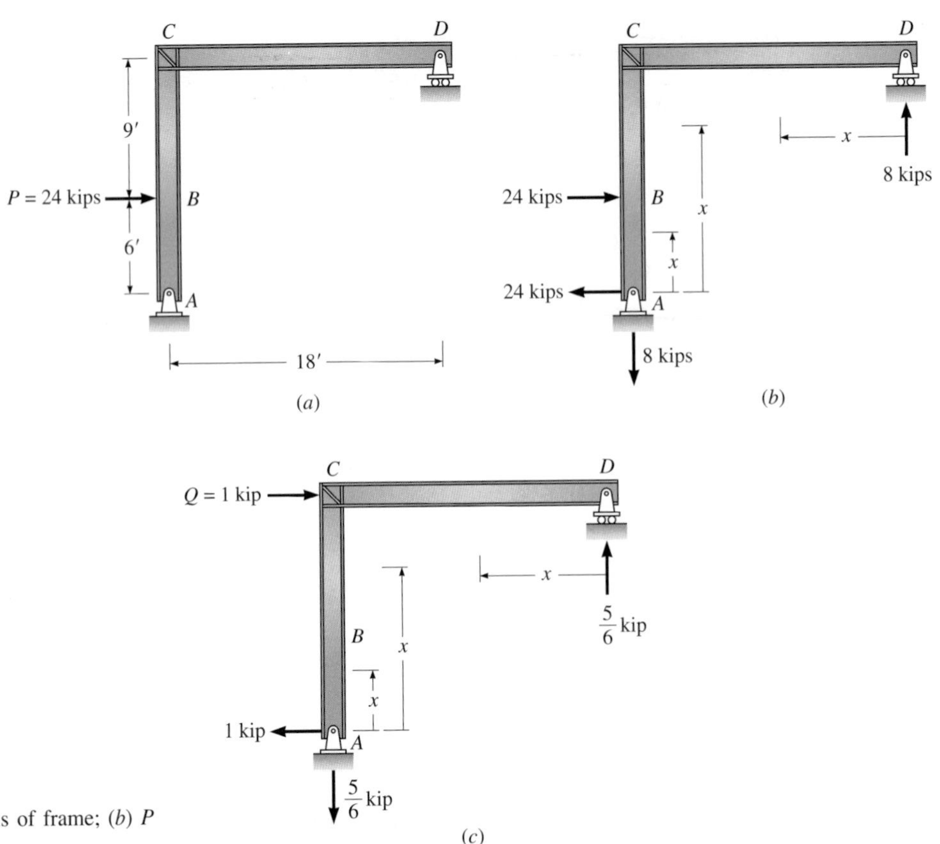

Figure 10.20: (a) Details of frame; (b) P system; (c) Q system.

Solution

Determine the internal forces produced by the P and Q systems (see Figure 10.20b and c).

From A to B, $x = 0$ to $x = 6$ ft:

$$M_P = 24 \cdot x \qquad F_P = +8 \text{ kips (tension)}$$

$$M_Q = 1 \cdot x \qquad F_Q = +\frac{5 \text{ kips}}{6} \text{ (tension)}$$

From B to C, $x = 6$ to $x = 15$ ft:

$$M_P = 24x - 24(x - 6) = 144 \text{ kip} \cdot \text{ft} \qquad F_P = 8 \text{ kips}$$

$$M_Q = 1 \cdot x \qquad F_Q = \frac{5 \text{ kips}}{6}$$

From D to C, $x = 0$ to $x = 18$ ft:

$$M_P = 8x \qquad F_P = 0$$

$$M_Q = \frac{5}{6}x \qquad F_Q = 0$$

Compute the horizontal displacement δ_{CH} using virtual work. Consider both flexural and axial deformations in evaluating U_Q. Only member AC carries axial load:

$$W_Q = U_Q$$

$$\sum Q\delta_{CH} = \sum \int \frac{M_Q M_P \, dx}{EI} + \sum \frac{F_Q F_P L}{AE}$$

$$1 \text{ kip} \cdot \delta_{CH} = \int_0^6 \frac{x(24x)dx}{EI} + \int_6^{15} \frac{x(144)dx}{EI} + \int_0^{18} \frac{(5x/6)(8x)dx}{EI}$$

$$+ \frac{(5/6)(8)(15 \times 12)}{AE}$$

$$= \left[\frac{8x^3}{EI}\right]_0^6 + \left[\frac{72x^2}{EI}\right]_6^{15} + \left[\frac{20x^3 \, dx}{9EI}\right]_0^{18} + \frac{1200}{AE}$$

$$= \frac{28{,}296(1728)}{600(29{,}000)} + \frac{1200}{13(29{,}000)}$$

$$= 2.8 \text{ in} + 0.0032 \text{ in} \qquad \text{round to 2.8 in} \qquad \textbf{Ans.}$$

In the equation above, 2.8 in represents the deflection produced by the flexural deformations, and 0.0032 in is the increment of deflection produced by the axial deformation of the column. In the majority of structures in which deformations are produced by both axial load and flexure, the axial deformations, which are very small compared to the flexural deformations, may be neglected.

EXAMPLE 10.15

Under the 5-kip load the support at A rotates 0.002 rad clockwise and settles 0.26 in (Figure 10.21a). Determine the total vertical deflection at D due to all effects. Consider bending deformations of the member only (i.e., neglect axial deformations). Given: $I = 1200 \text{ in}^4$, $E = 29{,}000 \text{ kips/in}^2$.

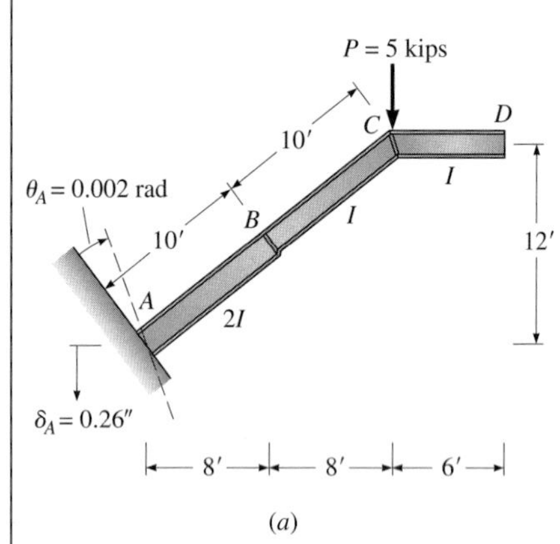

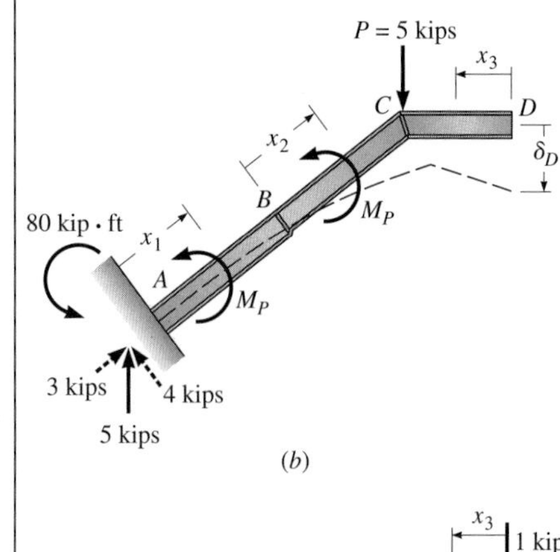

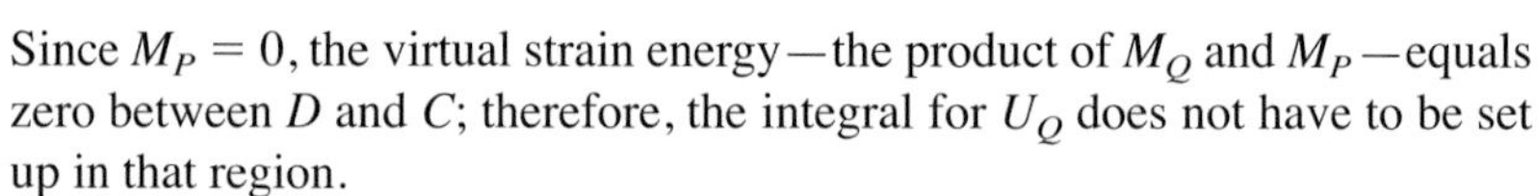

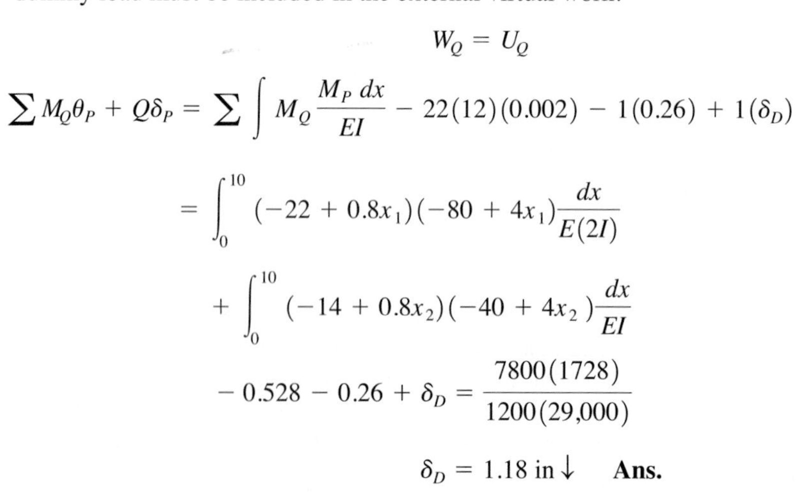

Figure 10.21: (a) A 5-kip load produces settlement and rotation of support A and bending of member ABC; (b) P system [support A also rotates and settles as shown in (a)]; (c) Q system with dummy load of 1 kip downward at D.

Solution

Since the moment of inertia between points A and B is twice as large as that of the balance of the bent member, we must set up separate integrals for the internal virtual strain energy between points AB, BC, and DC. Figure 10.21b and c shows the origins of the x's used to express M_Q and M_P in terms of the applied forces. The expressions for M_Q and M_P to be substituted into Equation 10.31 follow.

Segment	x: Origin	x: Range, ft	M_P, kip·ft	M_Q, kip·ft
AB	A	0–10	$-80 + 4x_1$	$-22 + 0.8x_1$
BC	B	0–10	$-40 + 4x_2$	$-14 + 0.8x_2$
DC	D	0–6	0	$-x_3$

Since $M_P = 0$, the virtual strain energy—the product of M_Q and M_P—equals zero between D and C; therefore, the integral for U_Q does not have to be set up in that region.

Compute δ_D using Equation 10.31. Since support A rotates 0.002 rad and settles 0.26 in, the external virtual work at A done by the reactions of the dummy load must be included in the external virtual work.

$$W_Q = U_Q$$

$$\sum M_Q\theta_P + Q\delta_P = \sum \int M_Q \frac{M_P\,dx}{EI} - 22(12)(0.002) - 1(0.26) + 1(\delta_D)$$

$$= \int_0^{10} (-22 + 0.8x_1)(-80 + 4x_1)\frac{dx}{E(2I)}$$

$$+ \int_0^{10} (-14 + 0.8x_2)(-40 + 4x_2)\frac{dx}{EI}$$

$$-0.528 - 0.26 + \delta_D = \frac{7800(1728)}{1200(29{,}000)}$$

$$\delta_D = 1.18 \text{ in} \downarrow \qquad \textbf{Ans.}$$

10.7 Finite Summation

The structures that we have previously analyzed by virtual work were composed of members of constant cross section (i.e., *prismatic members*) or of members that consisted of several segments of constant cross section. If the depth or width of a member varies with distance along the member's axis, the member is *nonprismatic*. The moment of inertia I of a nonprismatic member will, of course, vary with distance along the member's longitudinal axis. If deflections of beams or frames containing nonprismatic members are to be computed by virtual work using Equation 10.31 or 10.32, the moment of inertia in the strain energy term must be expressed as a function of x in order to carry out the integration. If the functional relationship for the moment of inertia is complex, expressing it as a function of x may be difficult. In this situation, we can simplify the computation of the strain energy by replacing the integration (an infinitesimal summation) by a finite summation.

In a finite summation we divide a member into a series of segments, often of identical length. The properties of each segment are assumed to be constant over the length of a segment, and the moment of inertia or any other property is based on the area of the cross section at the midpoint of the segment. To evaluate the virtual strain energy U_Q contained in the member, we sum the contributions of all segments. We further simplify the summation by assuming that moments M_Q and M_P are constant over the length of the segment and equal to the values at the center of the segment. We can represent the virtual strain energy in a finite summation by the following equation:

$$U_Q = \sum_{1}^{N} M_Q M_P \frac{\Delta x_n}{EI_n} \tag{10.34}$$

where Δx_n = length of segment n
I_n = moment of inertia of a segment based on area of midpoint cross section
M_Q = moment at midpoint of segment produced by dummy load (Q system)
M_P = moment at midpoint of segment produced by real loads (P system)
E = modulus of elasticity
N = number of segments

Although a finite summation produces an approximate value of strain energy, the accuracy of the result is usually good even when a small number of segments (say, five or six) are used. If the cross section of a member changes rapidly in a certain region, smaller length segments should be used to model the variation in moment of inertia. On the other hand, if the variation in cross section is small along the length of a member, the number of segments can be reduced. If all segments are the same length, the computations can be simplified by factoring Δx_n out of the summation.

EXAMPLE 10.16

Using a finite summation, compute the deflection δ_B of the tip of the cantilever beam in Figure 10.22*a*. The 12-in-wide beam has a uniform taper, and $E = 3000$ kips/in^2.

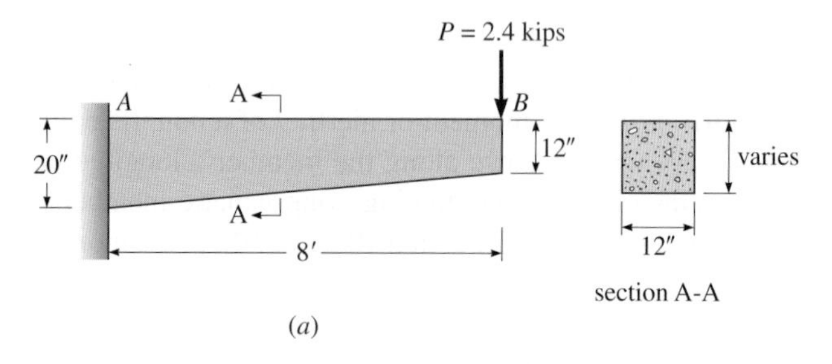

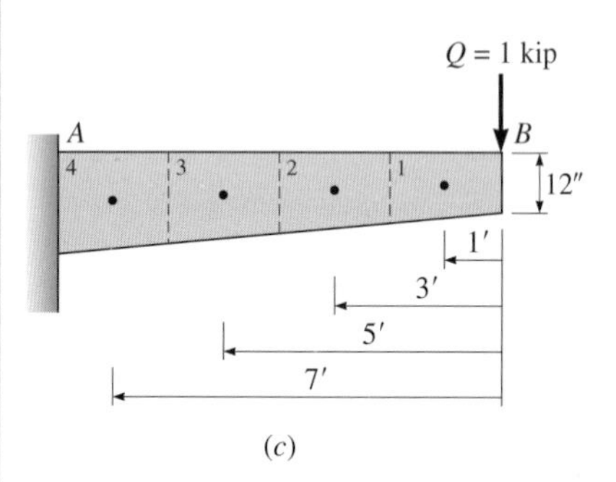

Figure 10.22: (*a*) Details of tapered beam; (*b*) *P* system; (*c*) *Q* system.

Solution

Divide the beam length into four segments of equal length ($\Delta x_n = 2$ ft). Base the moment of inertia of each segment on the depth at the center of each segment (see columns 2 and 3 in Table 10.3). Values of M_Q and M_P are tabulated in columns 4 and 5 of Table 10.3. Using Equation 10.34 to evaluate the right side of Equation 10.31, solve for δ_B.

$$W_Q = U_Q$$

$$(1 \text{ kip})(\delta_B) = \sum_{n=1}^{4} \frac{M_Q M_P \, \Delta x_n}{EI} = \frac{\Delta x_n}{E} \sum \frac{M_Q M_P}{I}$$

Substituting $\Sigma M_Q M_P / I = 5.307$ (from the bottom of column 6 in Table 10.3), $\Delta x_n = 2$ ft, and $E = 3000$ kips/in into Equation 10.34 for U_Q gives

$$\delta_B = \frac{2(12)(5.307)}{3000} = 0.042 \text{ in} \quad \textbf{Ans.}$$

TABLE 10.3

Segment (1)	Depth in (2)	$I = bh^3/12$ in^4 (3)	M_Q kip·ft (4)	M_P kip·ft (5)	$M_Q M_P(144)/I$ kips^2/in^2 (6)
1	13	2197	1	2.4	0.157
2	15	3375	3	7.2	0.922
3	17	4913	5	12	1.759
4	19	6859	7	16.8	2.469

$$\sum \frac{M_Q M_P}{I} = 5.307$$

NOTE. Moments in column 6 are multiplied by 144 to express M_Q and M_P in kip-inches.

10.8 Bernoulli's Principle of Virtual Displacements

Bernoulli's principle of virtual displacements, a basic structural theorem, is a variation of the principle of virtual work. The principle is used in theoretical derivations and can also be used to compute the deflection of points on a determinate structure that undergoes rigid body movement, for example, a support settlement or a fabrication error. Bernoulli's principle, which seems almost self-evident once it is stated, says:

If a rigid body, loaded by a system of forces in equilibrium, is given a small virtual displacement by an outside effect, the virtual work W_Q done by the force system equals zero.

In this statement a *virtual displacement* is a real or hypothetical displacement produced by an action that is separate from the force system acting on the structure. Also, a virtual displacement must be sufficiently small that the geometry and magnitude of the original force system do not change significantly as the structure is displaced from its initial to its final position. Since the body is *rigid*, $U_Q = 0$.

In Bernoulli's principle, virtual work equals the product of each force or moment and the component of the virtual displacement through which it moves. Thus it can be expressed by the equation

$$W_Q = U_Q = 0; \qquad \Sigma Q\delta_P + \Sigma Q_m\theta_P = 0 \tag{10.35}$$

where Q = force that is part of equilibrium force system; δ_P = virtual displacement that is collinear with Q; Q_m = moment that is part of equilibrium force system; θ_P = virtual rotational displacement.

The rationale behind Bernoulli's principle can be explained by considering a rigid body in equilibrium under a coplanar Q force system (the reactions are also considered part of the force system). In the most general case, the force system may consist of both forces and moments. As we discussed in Section 3.6, the external effect of a system of forces acting on a body can always be replaced by a resultant force R through any point and a moment M. If the body is in static equilibrium, the resultant force equals zero, and it follows that

$$R = 0 \qquad M = 0$$

or by expressing R in terms of its rectangular components,

$$R_x = 0 \qquad R_y = 0 \qquad M = 0 \tag{10.36}$$

If we now assume that the rigid body is given a small virtual displacement consisting of a linear displacement ΔL and an angular displacement θ, where ΔL has components Δ_x in the x direction and Δ_y in the y direction, the virtual work W_Q produced by these displacements equals

$$W_Q = R_x\,\Delta_x + R_y\,\Delta_y + M\theta$$

Since Equation 10.36 establishes that R_x, R_y, and M equal zero in the equation above, we verify Bernoulli's principle that

$$W_Q = 0 \tag{10.36a}$$

EXAMPLE 10.17

If support B of the L-shaped beam in Figure 10.23a settles 1.2 in, determine (a) the vertical displacement δ_C of point C, (b) the horizontal displacement δ_D of point D, and (c) the slope θ_A at point A.

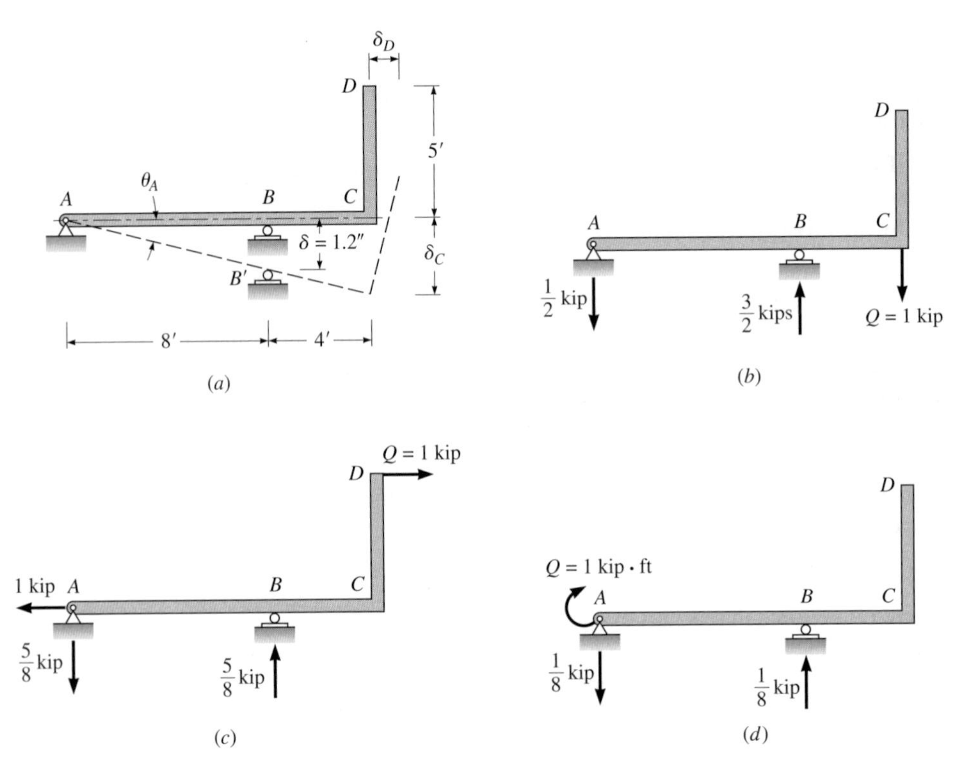

Figure 10.23: (a) Deflected shape produced by the settlement of support B; (b) Q system used to compute the deflection at C; (c) Q system used to compute the horizontal deflection of D; (d) Q system used to compute the slope at A.

Solution

(a) In this example the beam acts as a rigid body because no internal stresses, and consequently no deformations, develop when the beam (a determinate structure) is displaced due to the settlement of support B. To compute the vertical displacement at C, we apply a 1-kip dummy load in the vertical direction at C (see Figure 10.23b). We next compute the reactions at the supports, using the equations of statics. The dummy load and its reactions constitute a force system in equilibrium—a Q system. We now imagine that the loaded beam in Figure 10.23*b* undergoes the support settlement indicated in Figure 10.23*a*. In accordance with Bernoulli's principle, to determine δ_C, we equate to zero the sum of the virtual work done by the Q system forces.

$$W_Q = 0$$

$$1 \text{ kip}(\delta_C) - \left(\frac{3}{2} \text{ kips}\right)(1.2) = 0$$

$$\delta_C = 1.8 \text{ in} \qquad \textbf{Ans.}$$

In the equation above, the virtual work done by the reaction at B is negative because the downward displacement of 1.2 in is opposite in sense to the reaction of $\frac{3}{2}$ kips. Since support A does not move, its reaction produces no virtual work.

(*b*) To compute the horizontal displacement of joint D, we establish a Q system by applying a 1-kip dummy load horizontally at D and computing the support reactions (see Figure 10.23*c*). Then δ_D is computed by subjecting the Q system in Figure 10.23*c* to the virtual displacement shown in Figure 10.23*a*. We then compute the virtual work and set it equal to zero.

$$W_Q = 0$$

$$1 \text{ kip}(\delta_D) - \left(\frac{5}{8} \text{ kip}\right)(1.2) = 0$$

$$\delta_D = 0.75 \text{ in} \qquad \textbf{Ans.}$$

(*c*) We compute θ_A by applying a dummy moment of 1 kip·ft at A (see Figure 10.23*d*). The force system is then given the virtual displacement shown in Figure 10.23*a*, and the virtual work is evaluated. To express θ_A in radians, the 1 kip·ft moment is multiplied by 12 to convert kip-feet to kip-inches.

$$W_Q = 0$$

$$(1 \text{ kip} \cdot \text{ft})(12)\theta_A - \left(\frac{1}{8} \text{ kip}\right)1.2 = 0$$

$$\theta_A = \frac{1}{80} \text{ rad} \qquad \textbf{Ans.}$$

10.9 Maxwell-Betti Law of Reciprocal Deflections

Using the method of real work, we will derive the Maxwell-Betti law of reciprocal deflections, a basic structural theorem. Using this theorem, we will establish in Chapter 11 that the flexibility coefficients in compatibility equations, formulated to solve indeterminate structures of two or more degrees of indeterminacy by the flexibility method, form a symmetric matrix. This observation permits us to reduce the number of deflection computations required in this type of analysis. The Maxwell-Betti law also has applications in the construction of indeterminate influence lines.

The Maxwell-Betti law, which applies to any stable elastic structure (a beam, truss, or frame, for example) on unyielding supports and at constant temperature, states:

> **A linear deflection component at a point *A* in direction 1 produced by the application of a unit load at a second point *B* in direction 2 is equal in magnitude to the linear deflection component at point *B* in direction 2 produced by a unit load applied at *A* in direction 1.**

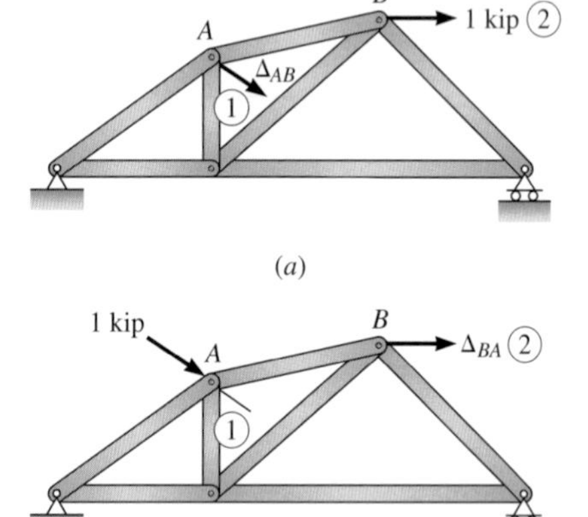

Figure 10.24

Figure 10.24 illustrates the components of truss displacements Δ_{BA} and Δ_{AB} that are equal according to Maxwell's law. Directions 1 and 2 are indicated by circled numbers. Displacements are labeled with two subscripts. The first subscript indicates the location of the displacement. The second subscript indicates the point at which the load producing the displacement acts.

We can establish Maxwell's law by considering the deflections at points *A* and *B* of the beam in Figure 10.25*a* and *b*. In Figure 10.25*a* application of a vertical force F_B at point *B* produces a vertical deflection Δ_{AB} at point *A* and Δ_{BB} at point *B*. Similarly, in Figure 10.25*b* the application of a vertical force F_A at point *A* produces a vertical deflection Δ_{AA} at point *A* and a deflection Δ_{BA} at point *B*. We next evaluate the total work done by the two forces F_A and F_B when they are applied in different order to the simply supported beam. The forces are assumed to increase linearly from zero to their final value. In the first case, we apply F_B first and then F_A. In the second case, we apply F_A first and then F_B. Since the final deflected position of the beam produced by the two loads is the same regardless of the order in which the loads are applied, the total work done by the forces is also the same regardless of the order in which the loads are applied.

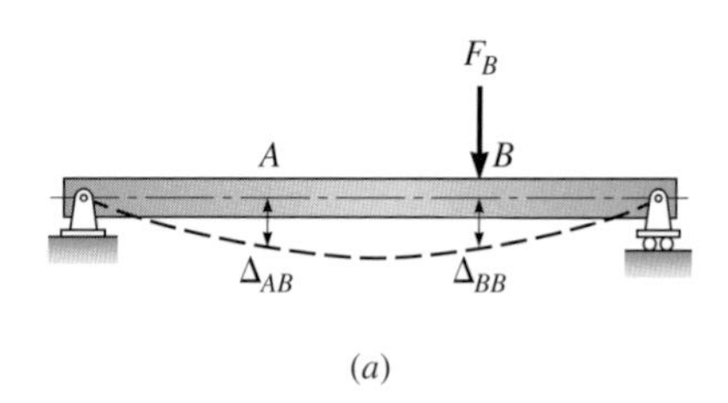

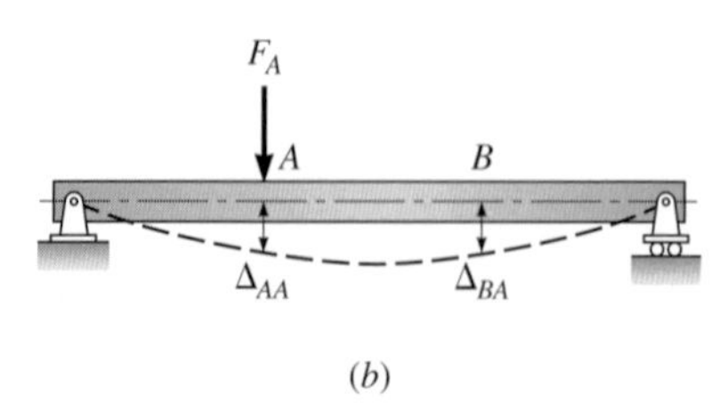

Figure 10.25

Case 1. F_B Applied Followed by F_A

(*a*) Work done when F_B is applied:

$$W_B = \tfrac{1}{2}F_B\,\Delta_{BB}$$

(*b*) Work done when F_A is applied with F_B in place:

$$W_A = \tfrac{1}{2}F_A\,\Delta_{AA} + F_B\,\Delta_{BA}$$

Since the magnitude of F_B does not change as the beam deflects under the action of F_A, the additional work done by F_B (the second term in the equation above) equals the full value of F_B times the deflection Δ_{BA} produced by F_A.

$$\begin{aligned} W_{\text{total}} &= W_B + W_A \\ &= \tfrac{1}{2} F_B\,\Delta_{BB} + \tfrac{1}{2} F_A\,\Delta_{AA} + F_B\,\Delta_{BA} \end{aligned} \tag{10.37}$$

Case 2. F_A Applied Followed by F_B

(*c*) Work done when F_A is applied:

$$W'_A = \tfrac{1}{2} F_A\,\Delta_{AA}$$

(*d*) Work done when F_B is applied with F_A in place:

$$W'_B = \tfrac{1}{2} F_B\,\Delta_{BB} + F_A\,\Delta_{AB}$$

$$\begin{aligned} W'_{\text{total}} &= W'_A + W'_B \\ &= \tfrac{1}{2} F_A\,\Delta_{AA} + \tfrac{1}{2} F_B\,\Delta_{BB} + F_A\,\Delta_{AB} \end{aligned} \tag{10.38}$$

Equating the total work of cases 1 and 2 given by Equations 10.37 and 10.38 and simplifying give

$$\tfrac{1}{2} F_B\,\Delta_{BB} + \tfrac{1}{2} F_A\,\Delta_{AA} + F_B\,\Delta_{BA} = \tfrac{1}{2} F_A\,\Delta_{AA} + \tfrac{1}{2} F_B\,\Delta_{BB} + F_A\,\Delta_{AB}$$

$$F_B\,\Delta_{BA} = F_A\,\Delta_{AB} \tag{10.39}$$

When F_A and $F_B = 1$ kip, Equation 10.39 reduces to the statement of the Maxwell-Betti law:

$$\Delta_{BA} = \Delta_{AB} \tag{10.40}$$

The Maxwell-Betti theorem also holds for rotations as well as rotations and linear displacements. In other words, by equating the total work done by a moment M_A at point A followed by a moment M_B at point B and then reversing the order in which the moments are applied to the same member, we can also state the Maxwell-Betti law as follows:

The rotation at point *A* in direction 1 due to a unit couple at *B* in direction 2 is equal to the rotation at *B* in direction 2 due to a unit couple at *A* in direction 1.

In accordance with the foregoing statement of the Maxwell-Betti law, α_{BA} in Figure 10.26*a* equals α_{AB} in Figure 10.26*b*. Moreover, the couple at *A* and the rotation at *A* produced by the couple at *B* are in the same direction (counterclockwise). Similarly, the moment at *B* and the rotation at *B* produced by the moment at *A* are also in the same direction (clockwise).

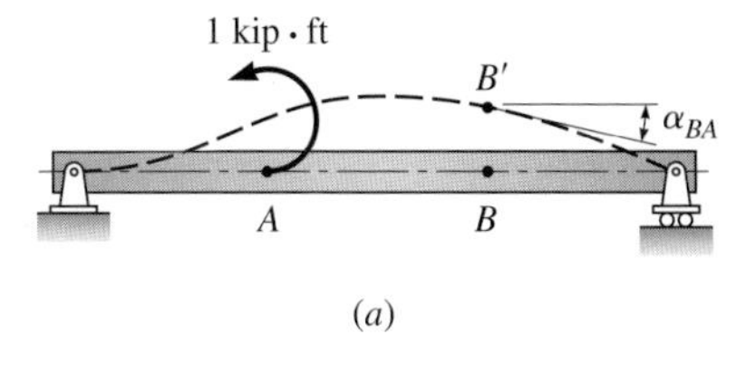

(*a*)

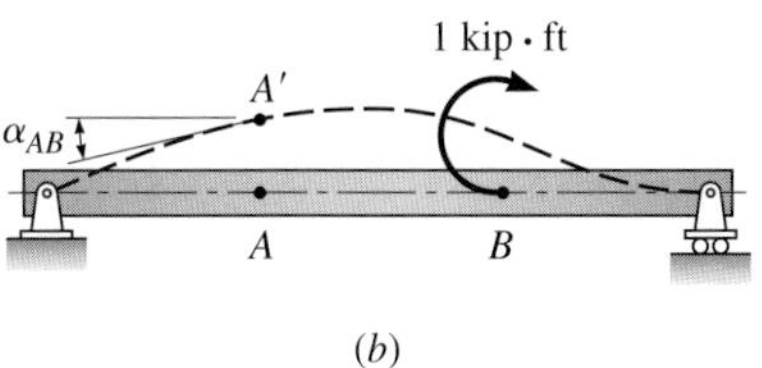

(*b*)

Figure 10.26

As a third variation of the Maxwell-Betti law, we can also state:

Any linear component of deflection at a point *A in direction 1* produced by a unit moment at B *in direction 2* is equal in magnitude to the rotation at *B* (in radians) *in direction 2* due to a unit load at *A in direction 1*.

Figure 10.27 illustrates the foregoing statement of the Maxwell-Betti law; that is, the rotation α_{BA} at point B in Figure 10.27*a* produced by the unit load at A in the vertical direction is equal in magnitude to the vertical deflection Δ_{AB} at A produced by the unit moment at point B in Figure 10.27*b*. Figure 10.27 also shows that Δ_{AB} is the same direction as the load at A, and the rotation α_{BA} and the moment at B are in the same counterclockwise direction.

In its most general form, the Maxwell-Betti law can also be applied to a structure that is supported in two different ways. The previous applications of this law are subsets of the following theorem:

Given a stable linear elastic structure on which arbitrary points have been selected, forces or moments may be acting at some of or all these points in either of two different loading systems. The virtual work done by the forces of the first system acting through the displacements of the second system is equal to the virtual work done by the forces of the second system acting through the corresponding displacements of the first system. If a support displaces in either system, the work associated with the reaction in the other system must be included. Moreover, internal forces at a given section may be included in either system by imagining that the restraint corresponding to the forces is removed from the structure but the internal forces are applied as external loads to each side of the section.

The statement above, illustrated in Example 10.18, may be represented by the following equation:

$$\Sigma F_1\delta_2 = \Sigma F_2\delta_1 \tag{10.41}$$

where F_1 represents a force or moment in system 1 and δ_2 is the displacement in system 2 that corresponds to F_1. Similarly, F_2 represents a force or moment in system 2, and δ_1 is the displacement in system 1 that corresponds to F_2.

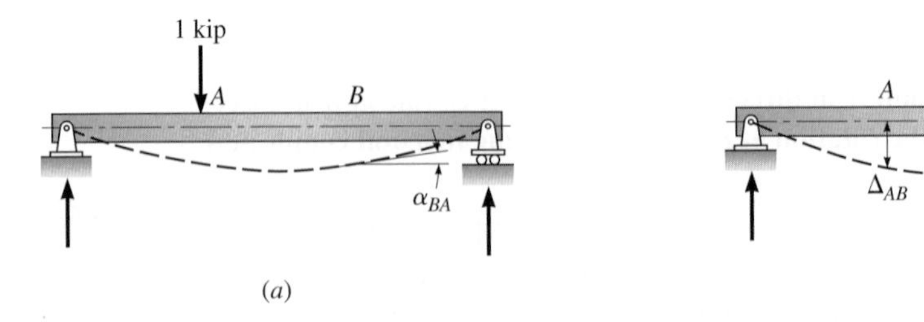

Figure 10.27

EXAMPLE 10.18

Figure 10.28 shows the same beam supported and loaded in two different ways. Demonstrate the validity of Equation 10.41. Required displacements are noted on the figure.

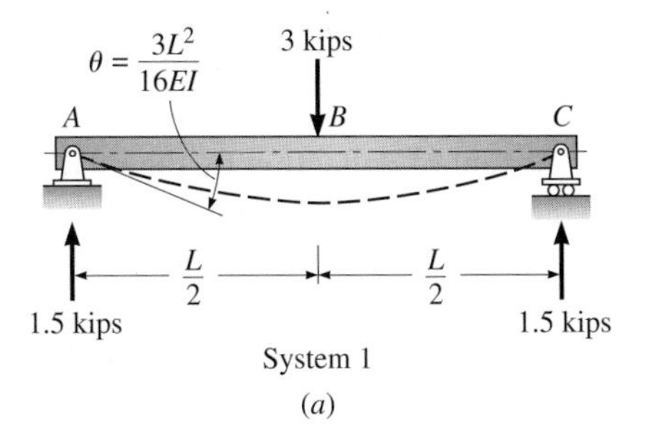

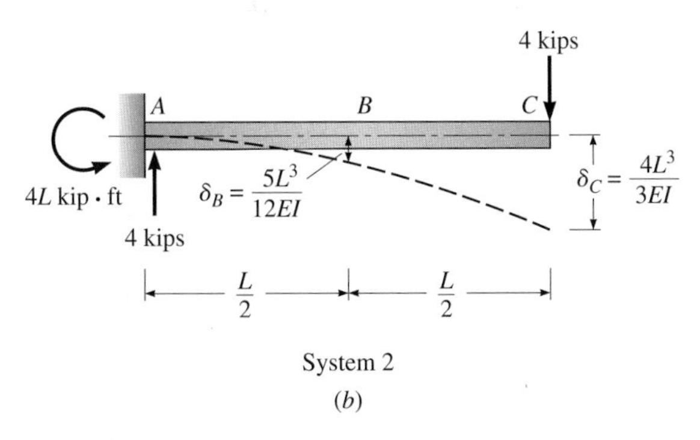

Figure 10.28: Identical beams with two different conditions of support.

Solution

$$\Sigma F_1\delta_2 = \Sigma F_2\delta_1 \qquad (10.41)$$

$$1.5\text{ kips}(0) + (3\text{ kips})\frac{5L^3}{12EI} - (1.5\text{ kips})\frac{4L^3}{3EI} = -(4L\text{ kip}\cdot\text{ft})\frac{3L^2}{16EI} + (4\text{ kips})(0) + (4\text{ kips})(0)$$

$$-\frac{3L^3}{4EI} = -\frac{3L^3}{4EI} \qquad \textbf{Ans.}$$

Summary

- Virtual work is the primary topic of Chapter 10. This method permits the engineer to compute a single component of deflection with each application of the method.
- Based on the *principle of the conservation of energy*, virtual work assumes loads are applied slowly so that neither kinetic nor heat energy is produced.
- To compute a component of deflection by the method of virtual work, we apply a force (also termed the dummy load) to the structure at the point of, as well as in the direction of, the desired displacement. The force and its associated reactions are called the *Q system*. If a slope or angle change is required, the force is a moment. With the dummy load in place the actual loads—called the *P system*—are applied to the structure. As the structure deforms under the actual loads, external virtual work W_Q is done by the dummy loads as they move through the real displacements produced by the P system. Simultaneously an equivalent quantity of virtual strain energy U_Q is stored in the structure. That is,

$$W_Q = U_Q$$

- Although virtual work can be applied to all types of structures including trusses, beams, frames, slabs, and shells, here we limit the application of the method to three of the most common types of planar structures: trusses, beams, and frames. We also neglect the effects of shear since its contribution to the deflections of slender beams and frames is negligible. The effect of shear on deflections is only significant in short, heavily loaded deep beams or beams with a low modulus of rigidity. The method also permits the engineer to include deflections due to temperature change, support settlements, and fabrication errors.
- If a deflection has both vertical and horizontal components, two separate analyses by virtual work are required; the unit load is applied first in the vertical direction and then in the horizontal direction. The actual deflection is the vector sum of the two orthogonal components. In the case of beams or trusses, designers are generally interested only in the maximum vertical deflection under live load, because this component is limited by design codes.
- The use of a unit load to establish a Q system is arbitrary. However, since deflections due to unit loads (called flexibility coefficients) are

utilized in the analysis of indeterminate structures (see Chapter 11), use of unit loads is common practice among structural engineers.

- To determine the virtual strain energy when the depth of a beam varies along its length, changes in cross-sectional properties can be taken into account by dividing the beam into segments and carrying out a finite summation (see Section 10.7).
- In Section 10.9, we introduce the Maxwell-Betti law of reciprocal deflections. This law will be useful when we set up the terms of the symmetric matrices required to solve indeterminant structures by the flexibility method in Chapter 11.

PROBLEMS

P10.1. For the truss in Figure P10.1, compute the horizontal and vertical components of displacement of joint B produced by the 100-kip load. The area of all bars $= 4$ in^2 and $E = 24{,}000$ kips/in^2.

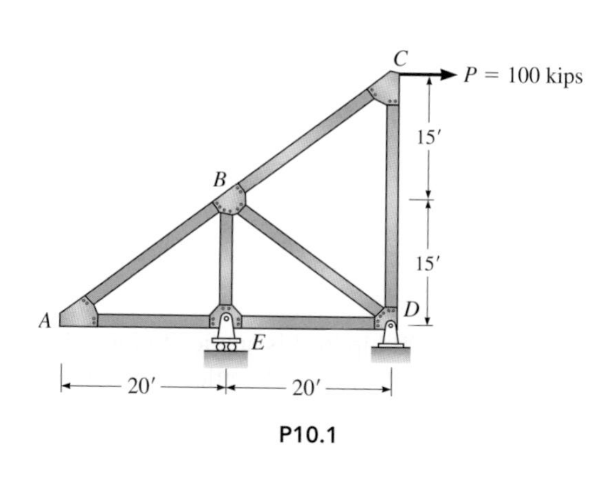

P10.1

P10.2. For the truss in Figure P10.1, compute the vertical displacement of joint A and the horizontal displacement of joint C.

P10.3. For the truss in Figure P10.3, compute the horizontal and vertical components of the displacement of joint C. The area of all bars $= 2500$ mm^2, and $E = 200$ GPa.

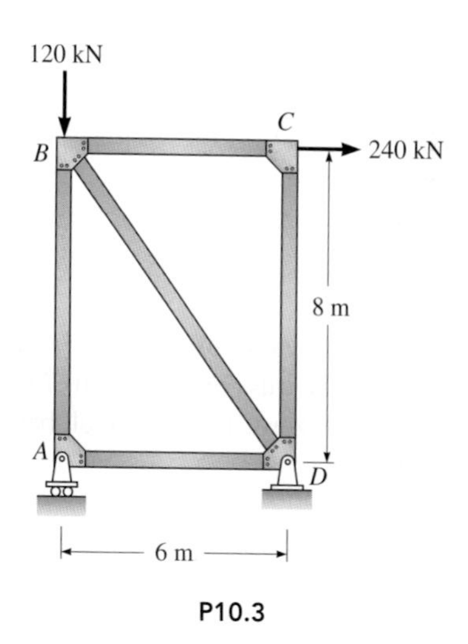

P10.3

P10.4. For the truss in Figure P10.3, compute the vertical displacement of joint B and the horizontal displacement of the roller at joint A.

P10.5. The pin-connected frame in Figure P10.3 is subjected to two vertical loads. Compute the vertical displacement of joint *B*. Will the frame sway horizontally? If yes, compute the horizontal displacement of joint *B*. The area of all bars = 5 in^2 and E = 29,000 kips/in^2.

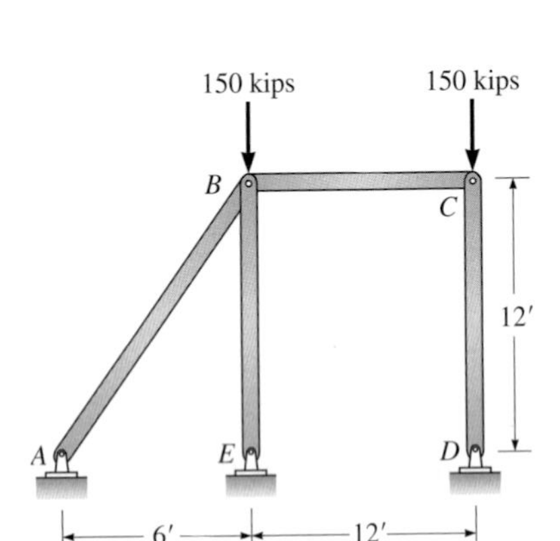

P10.5

P10.6. For the pin-connected frame in Figure P10.5, in addition to the vertical loads a lateral load of 30 kips also acts to the right at joint *B*. Compute the vertical and horizontal displacements at joint *B*.

P10.7. Determine the value of the force *P* that must be applied to joint *C* of the truss in Figure P10.7 if the vertical deflection at *C* is to be zero. The area of all bars = 1.8 in^2; E = 30,000 kips/in^2.

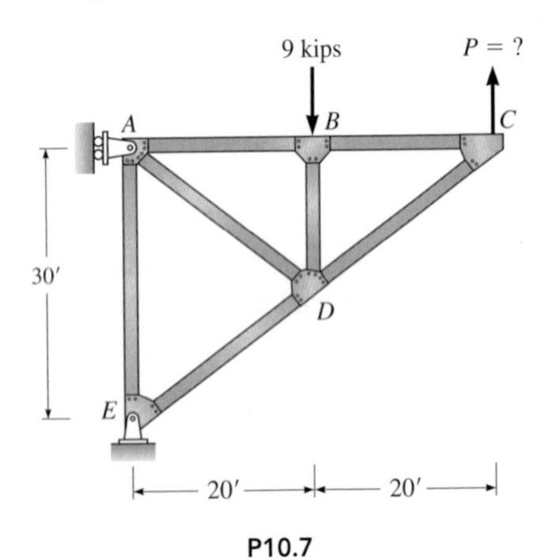

P10.7

P10.8. When the truss in Figure P10.8 is loaded, the support at *E* displaces 0.6 in vertically downward and the support at *A* moves 0.4 in to the right. Compute the horizontal and vertical components of displacement of joint *C*. For all bars the area = 2 in^2 and E = 29,000 kips/in^2.

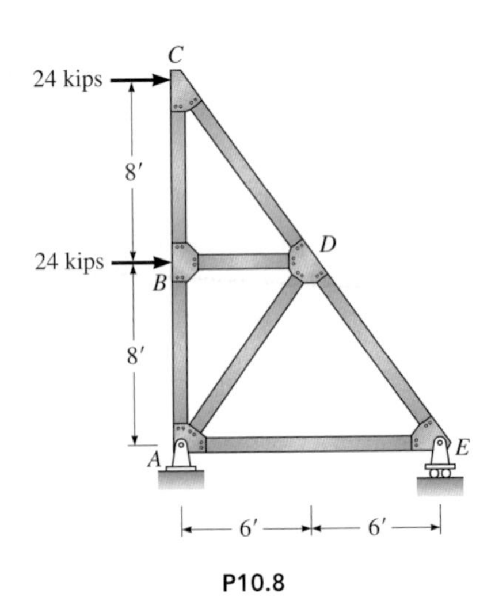

P10.8

P10.9. When the 20-kip load is applied to joint *B* of the truss in Figure P10.9, support *A* settles vertically downward $\frac{3}{4}$ in and displaces $\frac{1}{2}$ in horizontally to the right. Determine the vertical displacement of joint *B* due to all effects. The area of all bars = 2 in^2; E = 30,000 kips/in^2.

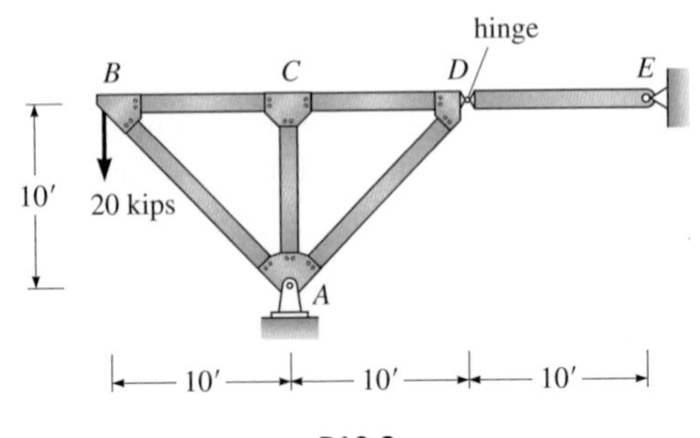

P10.9

P10.10. In Figure P10.10 if support A moves horizontally 2 in to the right and support F settles 1 in vertically, compute the horizontal deflection of the roller at support G.

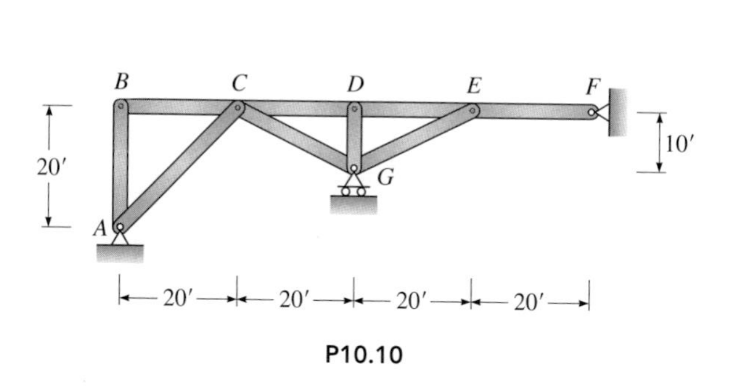

P10.10

P10.11. Determine the horizontal and vertical deflection of joint C of the truss in Figure P10.11. In addition to the load at joint C, the temperature of member BD is subject to a temperature increase of 60°F. For all bars, $E = 29{,}000$ kips/in^2, $A = 4$ in^2, and $\alpha = 6.5 \times 10^{-6}$ (in/in)/°F.

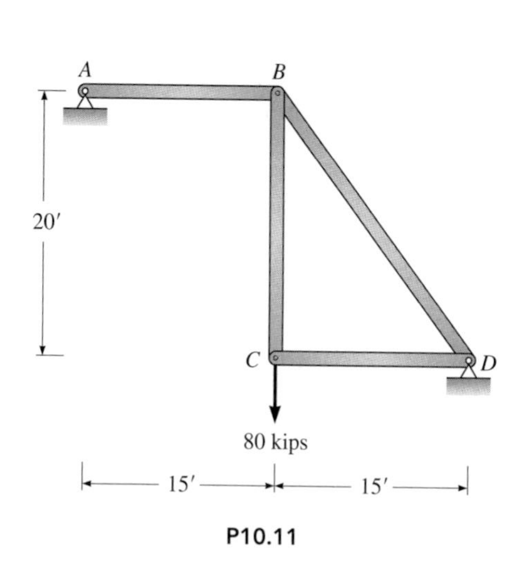

P10.11

P10.12. For the truss in Figure P10.12, compute the vertical displacement at joint G. The area of all bars $= 5$ in^2, and $E = 29{,}000$ kips/in^2.

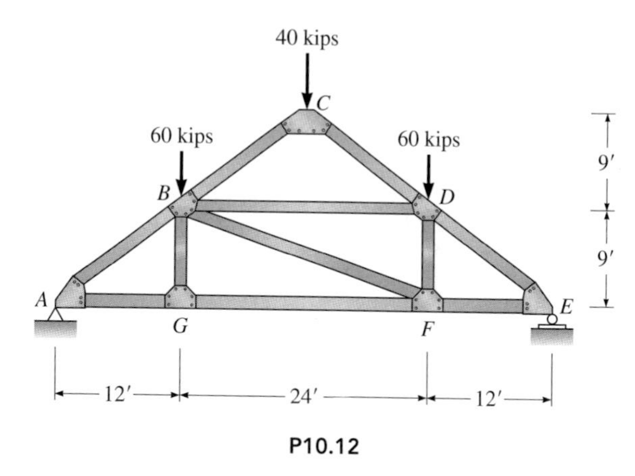

P10.12

P10.13. (*a*) Compute the vertical deflection of joint D produced by the 30-kip load in Figure P10.13. For all bars area $= 2$ in^2 and $E = 9000$ kips/in^2, (*b*) Assume that the truss is not loaded. If bar AE is fabricated $\frac{8}{5}$ in too long, how far to the right must the roller at B be displaced horizontally so that no vertical deflection occurs at joint D?

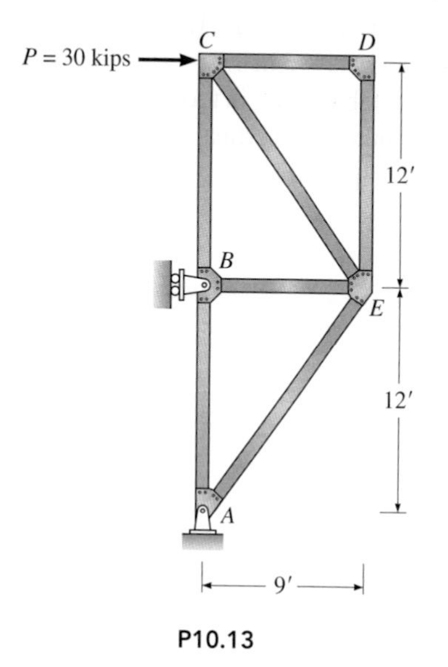

P10.13

P10.14. (*a*) Find the horizontal deflection at joint B produced by the 40-kip load in Figure P10.14. The area of all bars is shown on the sketch of the truss; $E = 30{,}000$ kips/in^2. (*b*) To restore joint B to its initial position in the horizontal direction, how much must bar AB be shortened? (*c*) If the temperature of bars AB and BC increases 80°F, determine the vertical displacement of joint C. $\alpha_t = 6.5 \times 10^{-6}$ (in/in)/°F. The rocker at support A is equivalent to a roller.

P10.16. Compute the vertical displacement of the hinge at C for the funicular loading shown in Figure P10.16. The funicular loading produces direct stress on all sections of the arch. Columns transmit only axial load from the roadway beams to the arch. Also assume that the roadway beams and the columns do not restrain the arch. All reactions are given. For all segments of the arch $A = 70$ in^2, $I = 7{,}800$ in^4, and $E = 30{,}000$ kips/in^2.

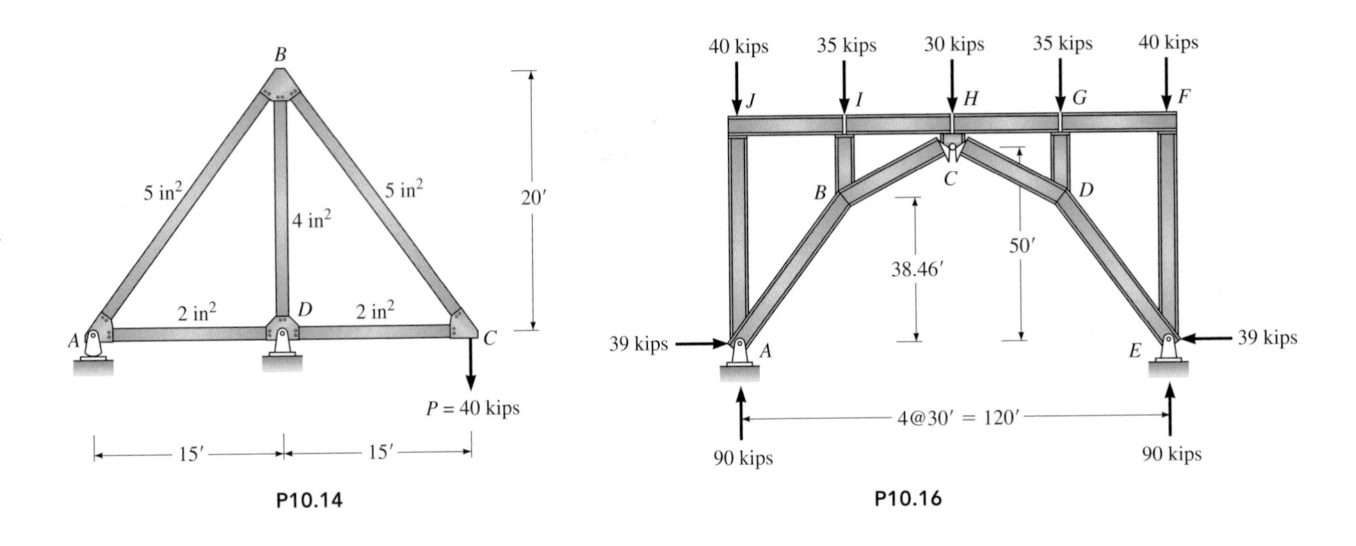

P10.14

P10.16

P10.15. (*a*) In Figure P10.15 compute the vertical and horizontal components of displacement of joint E produced by the loads. The area of bars AB, BD, and $CD = 5$ in^2; the area of all other bars $= 3$ in^2. $E = 30{,}000$ kips/in^2. (*b*) If bars AB and BD are fabricated $\frac{3}{4}$ in too long and support D settles 0.25 in, compute the vertical displacement of joint E. Neglect all the applied loads.

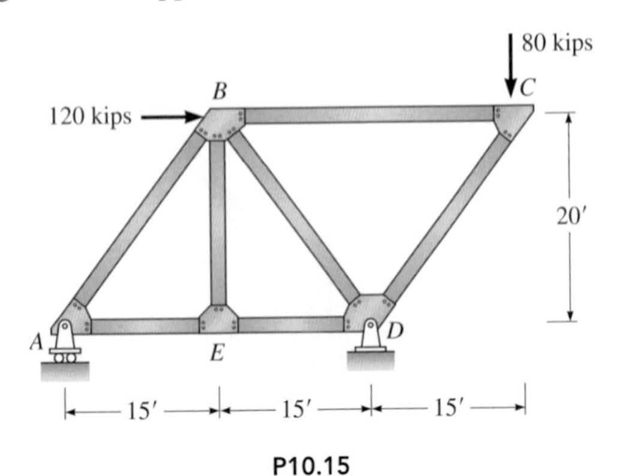

P10.15

P10.17. Determine the horizontal and vertical deflection of the hinge at point C of the arch in Figure P10.16 for a single concentrated load of 60 kips applied at joint B in the vertical direction.

P10.18. Compute the slope at support A and the deflection at B in Figure P10.18. EI is constant. Express your answer in terms of E, I, L, and M.

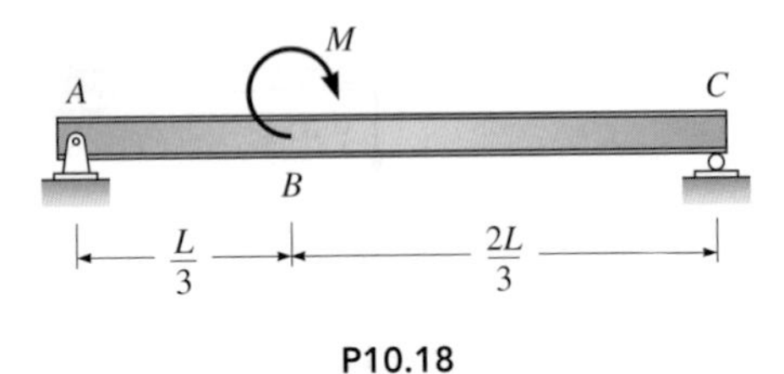

P10.18

P10.19. Compute the deflection at midspan and the slope at A in Figure P10.19. EI is constant. Express the slope in degrees and the deflection in inches. Assume a pin support at A and a roller at D. $E = 29{,}000$ kips/in^2, $I = 2000$ in^4.

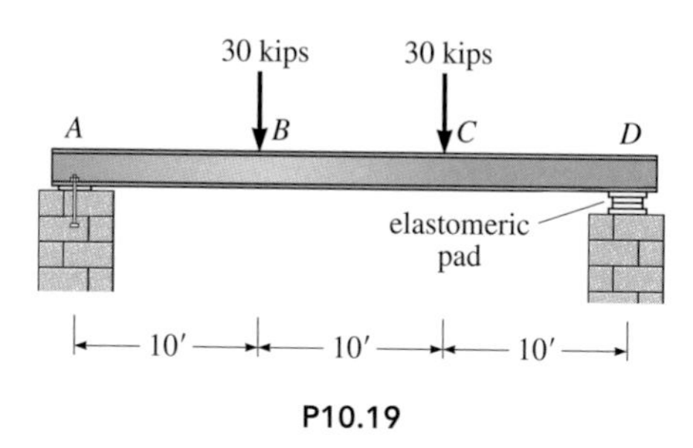

P10.19

P10.20. (*a*) Compute the vertical deflection and slope of the cantilever beam at points B and C in Figure P10.20. Given: EI is constant throughout, $L = 12$ ft, and $E = 4000$ kips/in^2. What is the minimum required value of I if the deflection of point C is not to exceed 0.4 in?

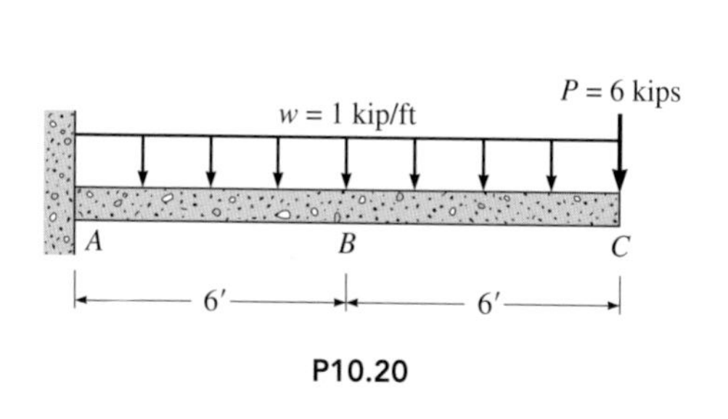

P10.20

P10.21. Compute the deflection at B and the slope at C in Figure P10.21. Given: EI is constant.

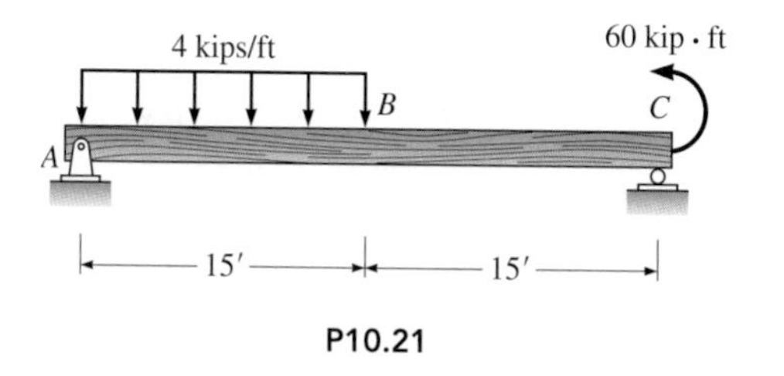

P10.21

P10.22. Determine the value of moment that must be applied to the left end of the beam in Figure P10.22 if the slope at A is to be zero. EI is constant. Assume rocker at support D acts as a roller.

$\theta_A = 0$ A, 24 kN, 24 kN, B, C, D, M_A, 3 m, 3 m, 3 m

P10.22

P10.23. Compute the vertical deflection of point C in Figure P10.23. Given: $I = 1200$ in^4, $E = 29{,}000$ kips/in^2.

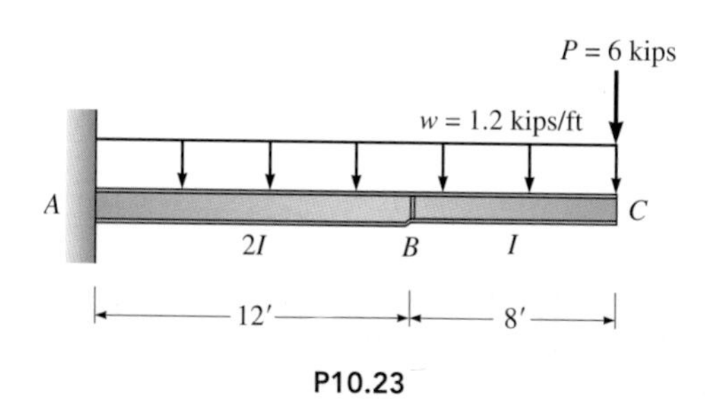

P10.23

P10.24. Compute the deflection at midspan of the beam in Figure P10.24. Given: $I = 46 \times 10^6$ mm^4, $E = 200$ GPa. Treat rocker at E as a roller.

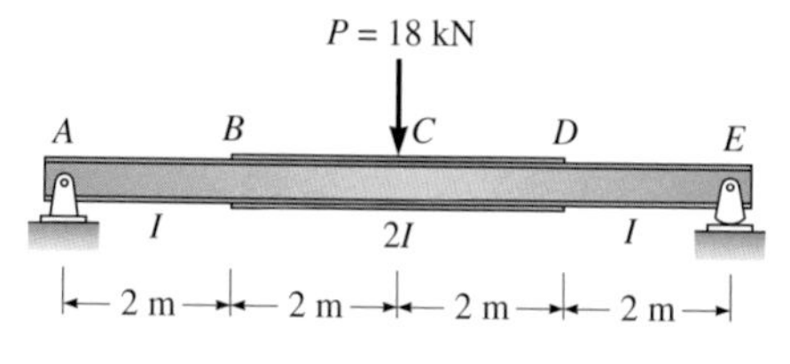

P10.24

P10.25. Under the dead load of the arch in Figure P10.25, the hinge at B is expected to displace 3 in downward. To eliminate the 3-in displacement, the designers will shorten the distance between supports by moving support A to the right. How far should support A be moved?

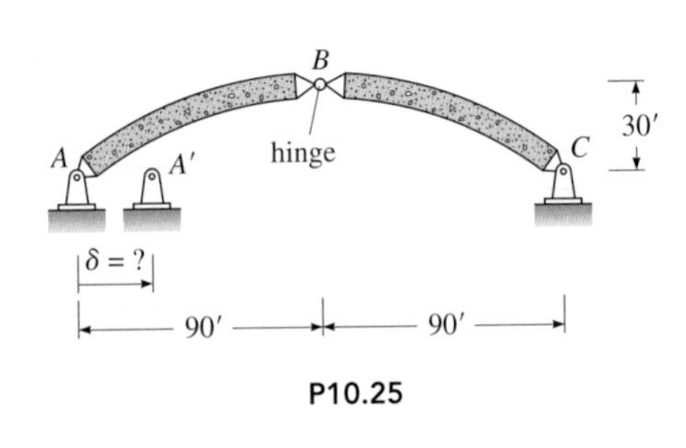

P10.25

P10.26. If supports A and E in Figure P10.26 are constructed 30 ft and 2 in apart instead of 30 ft apart, and if support E is also 0.75 in above its specified elevation, determine the vertical and horizontal components of deflections of the hinge at C and the slope of member AB when the frame is erected.

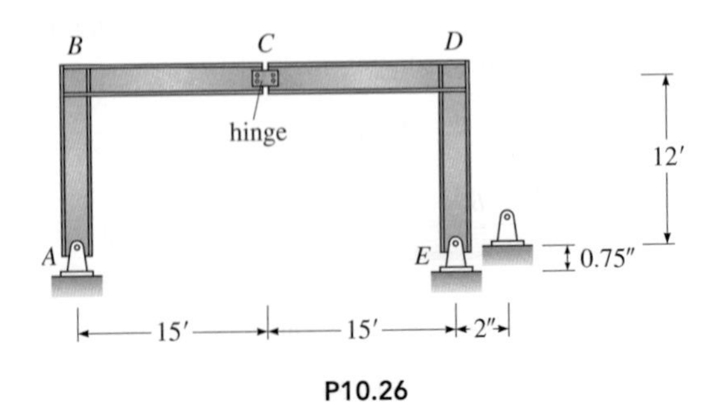

P10.26

P10.27. In Figure P10.27 support D is constructed 1.5 in to the right of its specified location. Using Bernoulli's principle in Section 10.8, compute (*a*) the horizontal and vertical components of the displacement of joint B and (*b*) the change in slope of member BC.

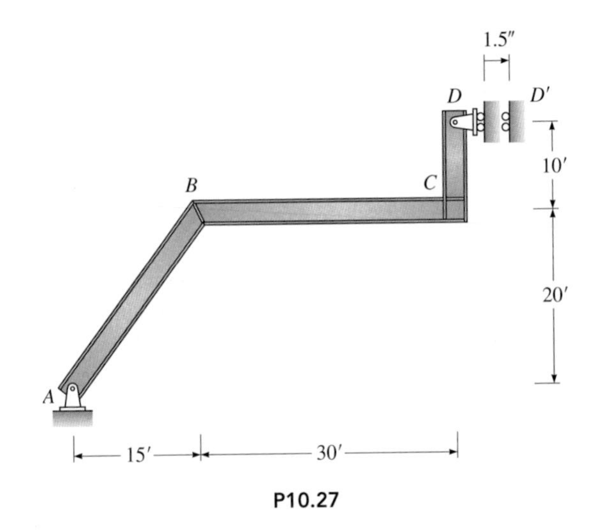

P10.27

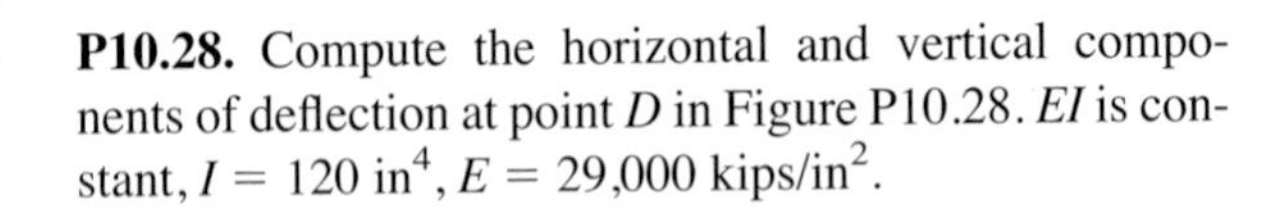

P10.28. Compute the horizontal and vertical components of deflection at point D in Figure P10.28. EI is constant, $I = 120$ in^4, $E = 29{,}000$ kips/in^2.

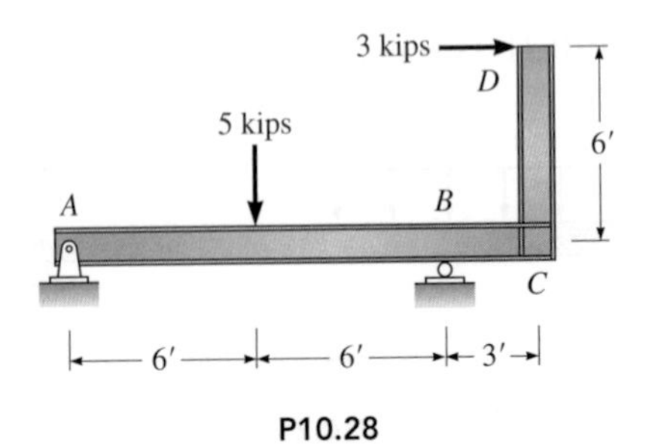

P10.28

P10.29. Compute the horizontal and vertical components of the deflection at C in Figure P10.29. $E = 200$ GPa, $A = 25 \times 10^3$ mm^2, and $I = 240 \times 10^6$ mm^4.

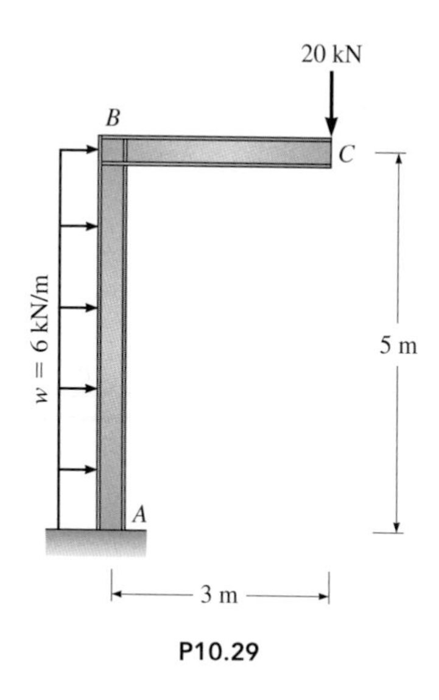

P10.29

P10.31. For the steel rigid frame in Figure P10.31, compute the rotation of joint B and the horizontal displacement of support C. Given: $E = 200$ GPa, $A = 500$ mm^2, $I = 200 \times 10^6$ mm^4.

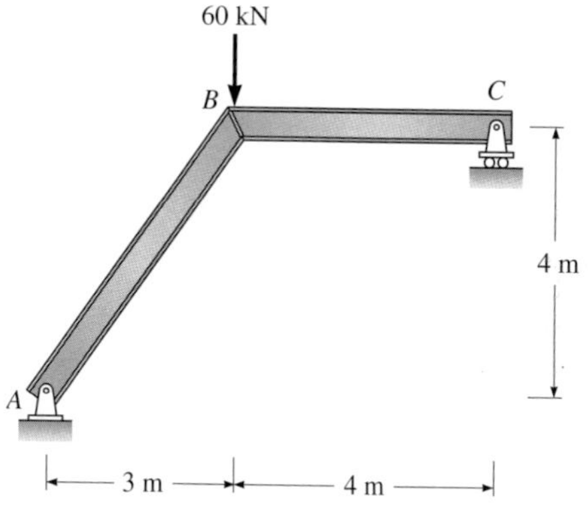

60 kN
B
C
4 m
A
3 m
4 m

P10.31

P10.30. Compute the vertical displacement of joints B and C for the frame shown in Figure P10.30. Given: $I = 360$ in^4, $E = 30{,}000$ kips/in^2. Consider only flexural deformations.

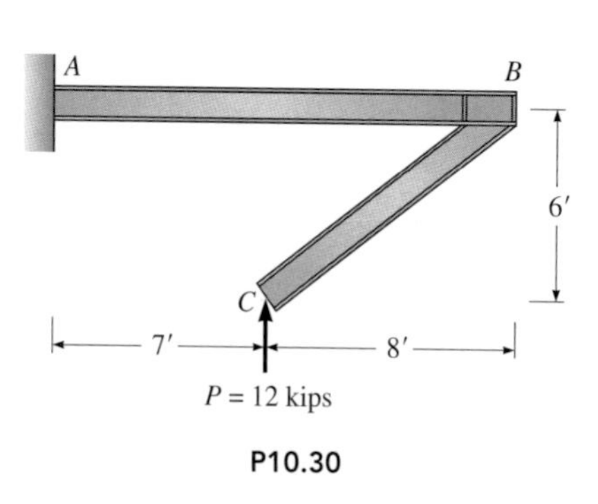

P10.30

P10.32. (a) Compute the slope at A and the horizontal displacement of joint B in Figure P10.32. EI is constant for all members. Consider only bending deformations. Given: $I = 100$ in^4, $E = 29{,}000$ kips/in^2. (b) If the horizontal displacement at joint B is not to exceed $\frac{3}{8}$ in, what is the minimum required value of I?

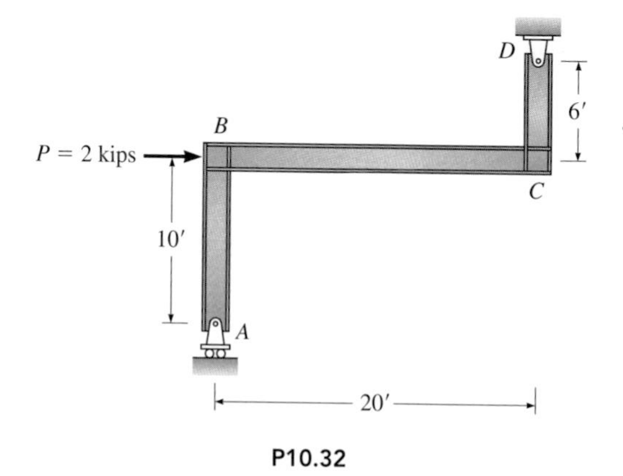

P10.32

P10.33. For the frame in Figure P10.33, compute the horizontal and vertical displacements at joint B. Given: $I = 150$ in^4, $E = 29{,}000$ kips/in^2. Consider only the flexural deformations.

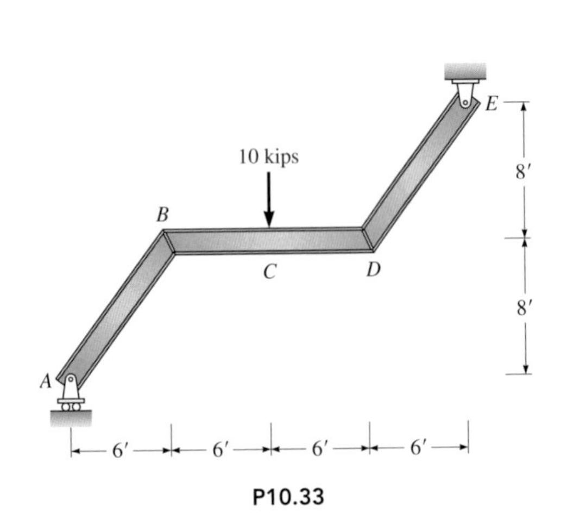

P10.33

P10.34. (*a*) Compute the vertical displacement of the hinge at C in Figure P10.34. EI is constant for all members, $E = 200$ GPa, $I = 1800 \times 10^6$ mm^4. (*b*) The designer would like to offset the vertical displacement of the hinge at C by moving the support A. How far should support A be moved horizontally?

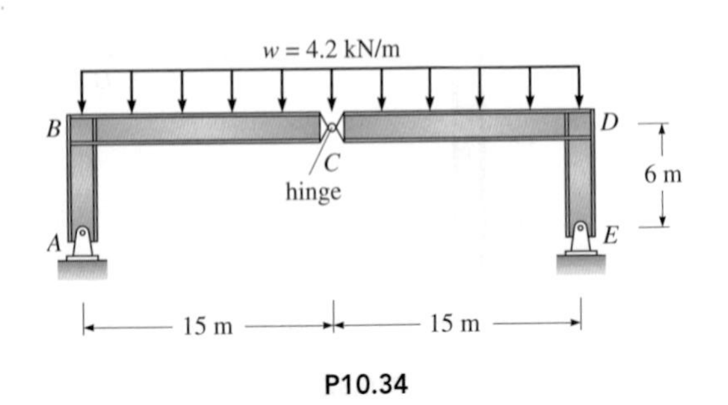

P10.34

P10.35. Compute the vertical displacement of point C for the beam in Figure P10.35. For the beam $A = 5{,}000$ mm^2, $I = 360 \times 10^6$ mm^4, and $E = 200$ GPa. For the cable $A = 6{,}000$ mm^2 and $E = 150$ GPa.

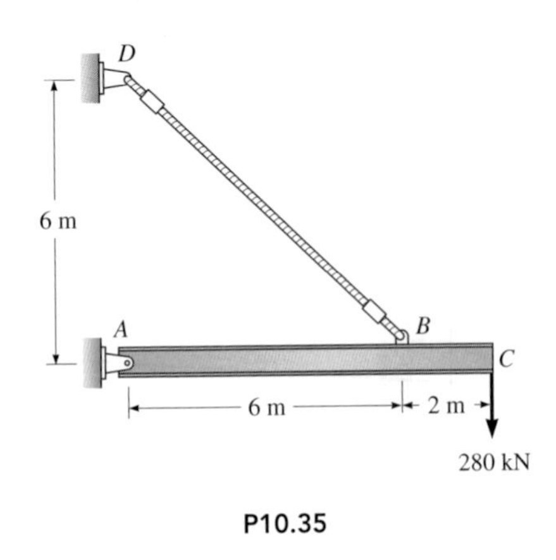

P10.35

P10.36. Compute the vertical deflection of joint C in Figure P10.36. In member ABC consider only the strain energy associated with bending. Given: $I_{AC} = 340$ in^4 and $A_{BD} = 5$ in^2. How much should bar BD be lengthened to eliminate the vertical deflection of point C when the 16-kip load acts?

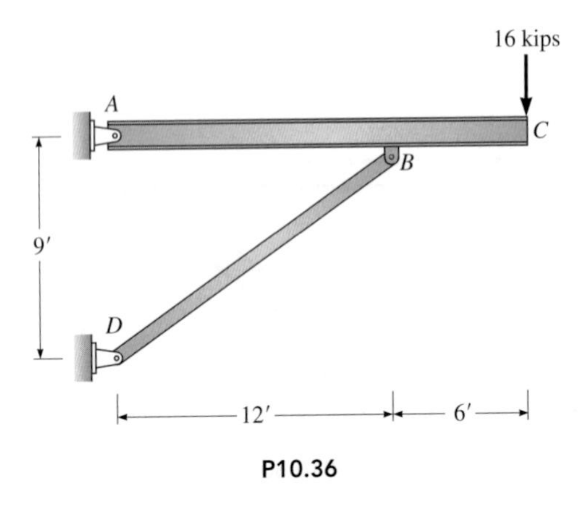

P10.36

P10.37. Compute the vertical deflection at B and the horizontal deflection at C in Figure P10.37. Given: $A_{CD} = 3$ in^2, $I_{AC} = 160$ in^4, $A_{AC} = 4$ in^2, and $E = 29{,}000$ kips/in^2. Consider the strain energy produced by both axial and flexural deformations.

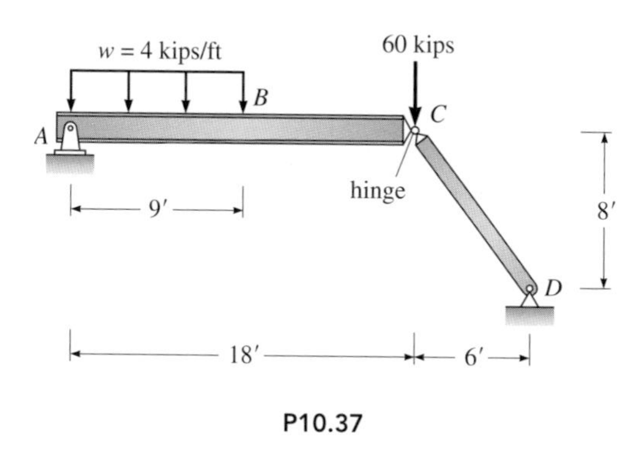

P10.37

P10.38. Compute the vertical and horizontal deflection at B and at the midspan of member CD in Figure P10.38. Consider both axial and bending deformations. Given: $E = 29{,}000$ kips/in^2, $I = 180$ in^4, area of column $= 6$ in^2, area of girder $= 10$ in^2.

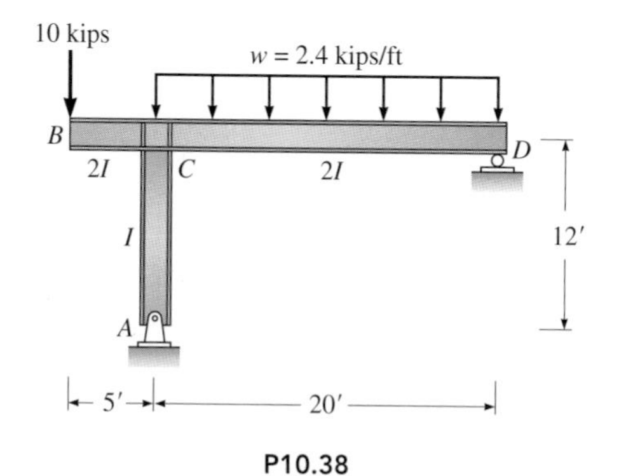

P10.38

P10.39. Beam ABC is supported by a three-bar truss at point C and at A by an elastomeric pad that is equivalent to a roller. (*a*) Compute the vertical deflection of point B in Figure P10.39 due to the applied load. (*b*) Compute the change in length of member DE required to displace point B upward 0.75 in. Is this a shortening or lengthening of the bar? Given: $E = 29{,}000$ kips/in^2, area of all truss bars $= 1$ in^2, area of beam $= 16$ in^2, I of beam $= 1200$ in^4.

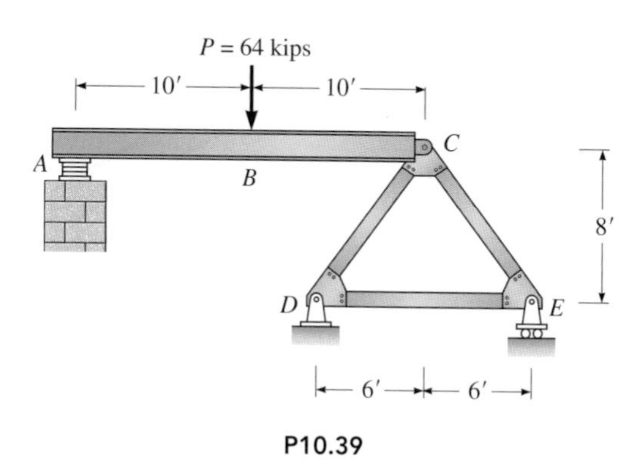

P10.39

P10.40. If the horizontal displacement of joint B of the frame in Figure P10.40 is not to exceed 0.36 in, what is the required I of the members? Bar CD has an area of 4 in^2, and $E = 29{,}000$ kips/in^2. Consider only the bending deformations of members AB and BC and the axial deformation of CD.

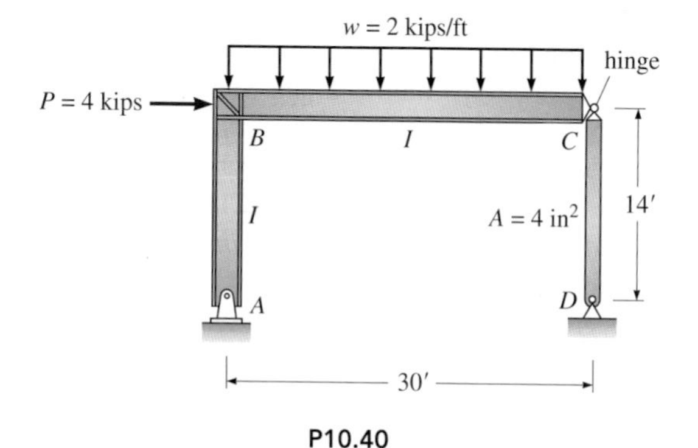

P10.40

P10.41. For the steel frame in Figure P10.41, compute the horizontal displacement of joint B. For member BCD, $A = 6{,}000\ \text{mm}^2$ and $I = 600 \times 10^6\ \text{mm}^4$. For member AB, $A = 3{,}000\ \text{mm}^2$. $E = 200$ GPa for all members.

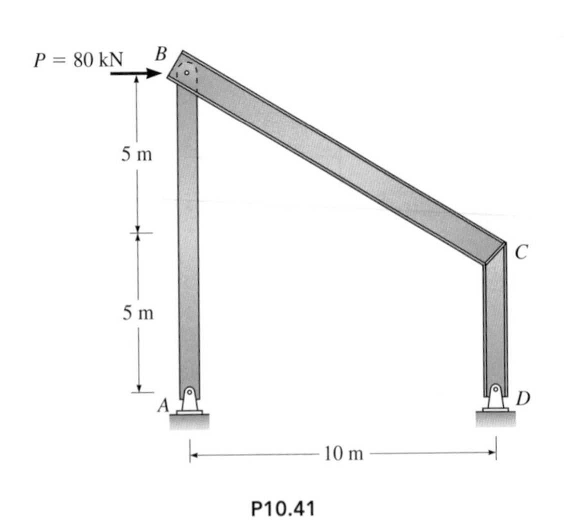

P10.41

P10.42. Using a finite summation, compute the initial deflection at midspan for the beam in Figure P10.42. Given: $E = 3000\ \text{kips/in}^2$. Use 3-ft segments. Assume $I = 0.5I_G$.

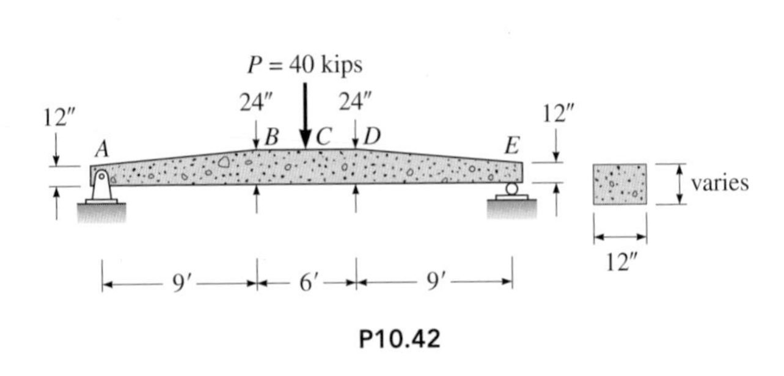

P10.42

Effective Moment of Inertia of a Reinforced Concrete Beam

NOTE: This note applies to Problems P10.42 to P10.44. Because reinforced concrete beams crack due to tensile stresses created by moment and shear, *initial* elastic deflections are based on an *empirical equation for moment of inertia* established from experimental studies of full-size beams (Section 9.5.2.3 of ACI Code). This equation produces an *effective moment of inertia* I_e that varies from about 0.35 to 0.5 of the moment of inertia I_G based on the gross area of the cross section. The additional deflection due to creep and shrinkage that occurs over time, which can exceed the initial deflection, is not considered.

P10.43. Using a finite summation, compute the initial deflection at point C for the tapered beam in Figure P10.43. $E = 24$ GPa. Base your analysis on the properties of $0.5I_G$.

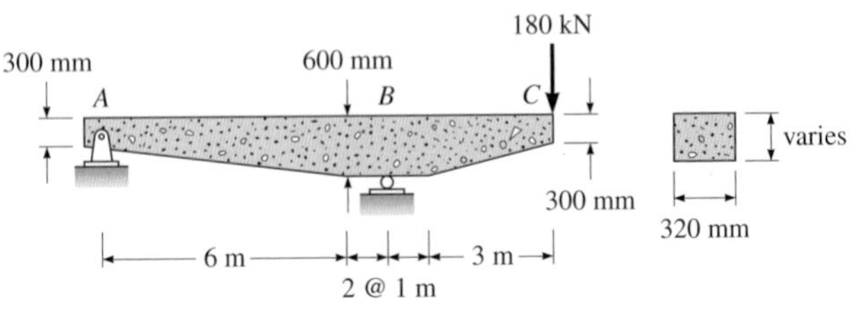

P10.43

P10.44. *Computer study—Influence of supports on frame behavior.* (*a*) Using the RISA-2D computer program, compute the initial elastic deflection at midspan of the girder in Figure P10.44, given that the support at D is a roller. For the computer analysis, replace the tapered members by 3 ft-long segments of constant depth whose properties are based on each segment's midspan dimensions; that is, there will be 9 *members* and 10 *joints*. When you set up the problem, specify in GLOBAL that forces are to be computed at three sections. This will produce values of forces at both ends and at the center of each segment. To account for cracking of the reinforced concrete, assume for girder BCD that $I_e = 0.35I_G$; for column AB assume $I_e = 0.7I_G$ (compression forces in columns reduce cracking). Since deflections of beams and one-story rigid frames are due almost entirely to moment and not significantly affected by the area of the member's cross-section, substitute the gross area in the Member Properties Table.

(*b*) Replace the roller at support D in Figure P10.44 by a pin to prevent horizontal displacement of joint D, and repeat the analysis of the frame. The frame is now an *indeterminate structure*. Compare your results with those in part (*a*), and briefly discuss differences in behavior with respect to the magnitude of deflections and moments.

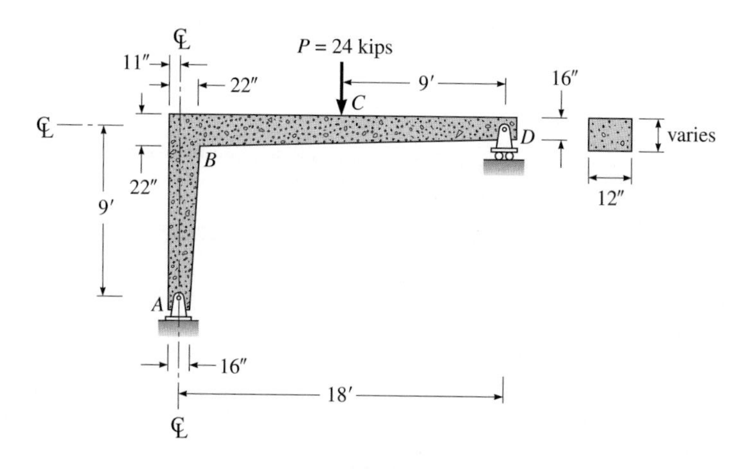

P10.44

East Huntington Bridge over the Ohio River

A 1500-ft-long cable-stayed bridge with a roadway constructed of hybrid concrete and steel girders 5 ft deep. The bridge, opened in 1985, is constructed of high-strength steel and concrete. Note the contrast of the slender lines of the roadway and tower of this modern bridge with those of the Brooklyn Bridge (see the photo at the beginning of Chapter 1).

CHAPTER 11

Analysis of Indeterminate Structures by the Flexibility Method

Chapter Objectives

- Show that the equations of static equilibrium alone are not enough to analyze indeterminate structures; additional equations are needed.
- Learn in the flexibility method to establish additional equations (i.e., compatibility equations) by using extra unknown reactions or internal forces as the redundants.
- Identify redundants and then use any method learned in Chapters 9 or 10 to compute the deflections produced by both external loads and redundants on a released structure to establish the compatibility equations.

11.1 Introduction

The flexibility method, also called the *method of consistent deformations* or the *method of superposition*, is a procedure for analyzing *linear elastic indeterminate* structures. Although the method can be applied to almost any type of structure (beams, trusses, frames, shells, and so forth), the computational effort increases exponentially with the degree of indeterminancy. Therefore, the method is most attractive when applied to structures with a low degree of indeterminancy.

All methods of indeterminate analysis require that the solution satisfy *equilibrium* and *compatibility* requirements. By compatibility we mean that the structure must fit together—no gaps can exist—and the deflected shape must be consistent with the constraints imposed by the supports. In the flexibility method, we will satisfy the equilibrium requirement by using the equations of static equilibrium in each step of the analysis. The compatibility requirement will be satisfied by writing one or more equations (i.e., *compatibility equations*) which state either that no gaps exist internally or that deflections are consistent with the geometry imposed by the supports.

As a key step in the flexibility method, the analysis of an indeterminate structure is replaced by the analysis of a stable determinate structure. This structure—called the *released* or *base structure*—is established from the original indeterminate structure by imagining that certain restraints (supports, for example) are temporarily removed.

11.2 Concept of a Redundant

We have seen in Section 3.7 that a minimum of three restraints, which are not equivalent to either a parallel or a concurrent force system, are required to produce a stable structure, that is, to prevent rigid-body displacement under any condition of load. For example, in Figure 11.1*a* the horizontal and vertical reactions of the pin at A and the vertical reaction of the roller at C prevent both translation and rotation of the beam regardless of the type of force system applied. Since three equations of equilibrium are available to determine the three reactions, the structure is *statically determinate*.

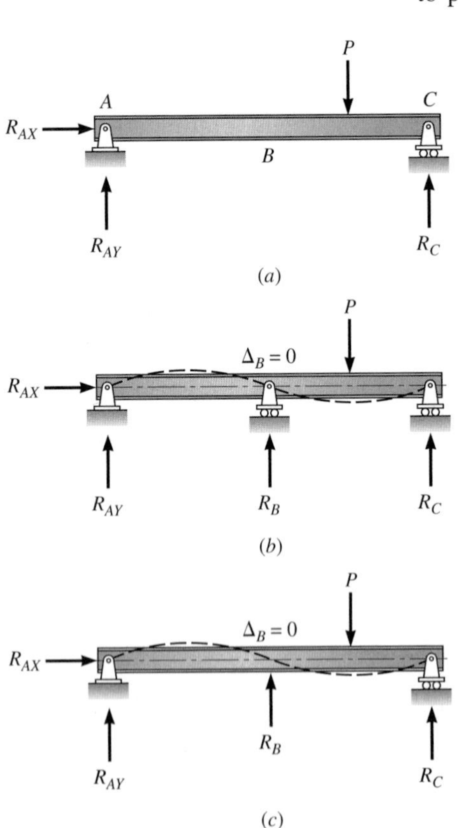

Figure 11.1: (*a*) Determinate beam; (*b*) indeterminate beam with R_B considered the redundant; (*c*) the released structure for the beam in (*b*) with the reaction at B applied as an external force.

If a third support is constructed at B (see Figure 11.1*b*), an additional reaction R_B is available to support the beam. Since the reaction at B is not absolutely essential for the stability of the structure, it is termed a *redundant*. In many structures the designation of a particular reaction as a redundant is arbitrary. For example, the reaction at C in Figure 11.1*b* could just as logically be considered a redundant because the pin at A and the roller at B also provide sufficient restraints to produce a stable determinate structure.

Although the addition of the roller at B produces a structure that is indeterminate to the first degree (four reactions exist but only three equations of statics are available), the roller also imposes the geometric requirement that the vertical displacement at B be zero. This geometric condition permits us to write an additional equation that can be used together with the equations of statics to determine the magnitude of all reactions. In Section 11.3 we outline the main features of the flexibility method and illustrate its use by analyzing a variety of indeterminate structures.

11.3 Fundamentals of the Flexibility Method

In the flexibility method, one imagines that sufficient redundants (supports, for example) are removed from an indeterminate structure to produce a stable, determinate *released* structure. The number of restraints removed equals the degree of indeterminancy. The design loads, which are specified, and the redundants, whose magnitude are unknown at this state, are then applied to the released structure. For example, Figure 11.1*c* shows the determinate released structure for the beam in Figure 11.1*b* when the reaction at B is taken as the redundant. Since the released structure in Figure 11.1*c* is loaded exactly like the original structure, the internal forces and deformations of the released structure are identical to those of the original indeterminate structure.

We next analyze the determinate released structure for the applied loads and redundants. In this step the analysis is divided into separate cases for (1) the applied loads and (2) for each unknown redundant. For each case, deflections are computed at each point where a redundant acts. Since the structure is assumed to behave elastically, these individual analyses can be combined—superimposed—to produce an analysis that includes the effect of all forces and redundants. To solve for the redundants, the deflections are summed at each point where a redundant acts and set equal to the known value of deflection. For example, if a redundant is supplied by a roller, the deflection will be zero in the direction normal to the plane along which the roller moves. This procedure produces a set of *compatibility equations* equal in number to the redundants. Once we determine the values of the redundants, the balance of the structure can be analyzed with the equations of statics. We begin the study of the flexibility method by considering structures that are indeterminate to the first degree. Section 11.7 covers indeterminate structures of higher order.

To illustrate the foregoing procedure, we will consider the analysis of the uniformly loaded beam in Figure 11.2*a*. Since only three equations of statics are available to solve for the four restraints supplied by the fixed support and roller, the structure is indeterminate to the first degree. To determine the reactions, one additional equation is needed to supplement the three equations of statics. To establish this equation, we arbitrarily select as the redundant the reaction R_B exerted by the roller at the right end. In Figure 11.2*b* the free-body diagram of the beam in Figure 11.2*a* is redrawn showing the reaction R_B exerted by the roller at support B but not the roller. By imagining that the roller has been removed, we can treat the indeterminate beam as a simple determinate

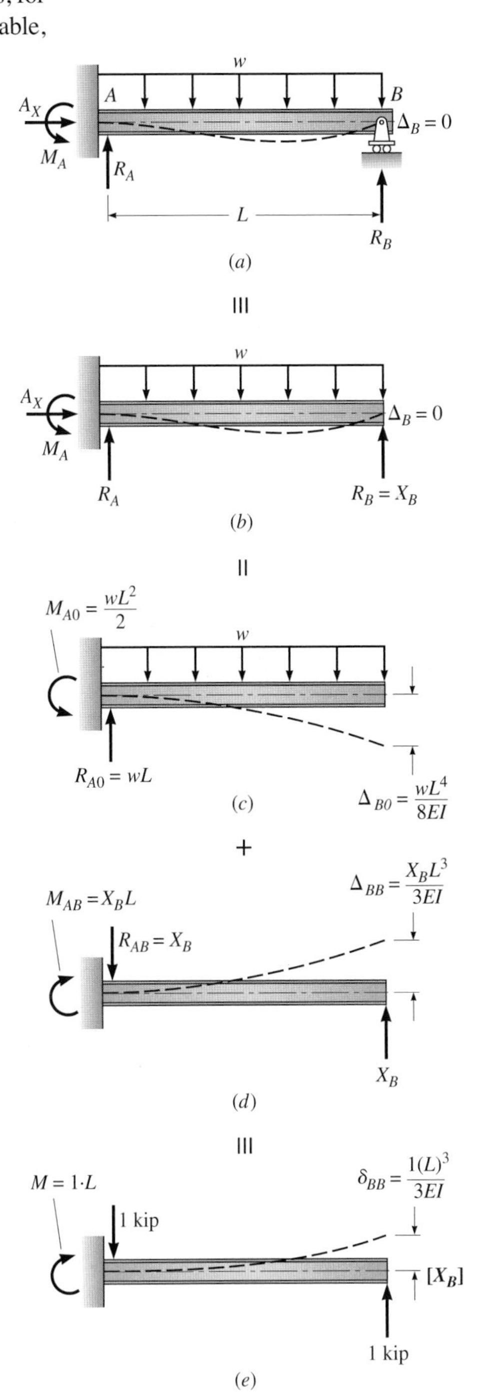

Figure 11.2: Analysis by the flexibility method: (*a*) beam indeterminate to the first degree; (*b*) released structure loaded with load w and redundant R_B; (*c*) forces and displacements produced by load w in the released structure; (*d*) forces and displacements of released structure produced by redundant X_B; (*e*) forces and displacements in released structure produced by a unit value of the redundant.

cantilever beam carrying a uniformly distributed load w and an unknown force R_B at its free end. By adopting this point of view, we have produced a determinate structure that can be analyzed by statics. Since the beams in Figure 11.2*a* and *b* carry exactly the same loads, their shear and moment curves are identical and they both deform in the same manner. In particular, the vertical deflection Δ_B at support B equals zero. To call attention to the fact that the reaction supplied by the roller is the redundant, we now denote R_B by the symbol X_B (see Figure 11.2*b*).

We next divide the analysis of the cantilever beam into the two parts shown in Figure 11.2*c* and *d*. Figure 11.2*c* shows the reactions and the deflections at B, Δ_{B0}, produced by the uniform load whose magnitude is specified. Deflections of the *released* structure produced by the applied loads will be denoted by two subscripts. The first will indicate the location of the deflection; the second subscript will be a zero, to distinguish the released structure from the actual structure. Figure 11.2*d* shows the reactions and the deflection at B, Δ_{BB}, produced by the redundant X_B whose magnitude is unknown. Assuming that the structure behaves elastically, we can add (superimpose) the two cases in Figure 11.2*c* and *d* to give the original case shown in Figure 11.2*b* or *a*. Since the roller in the real structure establishes the geometric requirement that the vertical displacement at B equal zero, the algebraic sum of the vertical displacements at B in Figure 11.2*c* and *d* must equal zero. This condition of geometry or compatibility can be expressed as

$$\Delta_B = 0 \tag{11.1}$$

Superimposing the deflections at point B produced by the applied load in Figure 11.2*c* and the redundant in Figure 11.2*d*, we can write Equation 11.1 as

$$\Delta_{B0} + \Delta_{BB} = 0 \tag{11.2}$$

The deflections Δ_{B0} and Δ_{BB} can be evaluated by the moment-area method or by virtual work, or from tabulated values shown in Figure 11.3*a* and *b*.

As a sign convention, we will assume that displacements are positive when they are in the direction of the redundant. In this procedure you are free to assume the direction in which the redundant acts. If you have chosen the correct direction, the solution will produce a positive value of the redundant. On the other hand, if the solution results in a negative value for the redundant, its magnitude is correct, but its direction is opposite to that initially assumed.

Expressing the deflections in terms of the applied loads and the properties of the members, we can write Equation 11.2 as

$$-\frac{wL^4}{8EI} + \frac{X_B L^3}{3EI} = 0$$

Solving for X_B gives

$$X_B = \frac{3wL}{8} \tag{11.3}$$

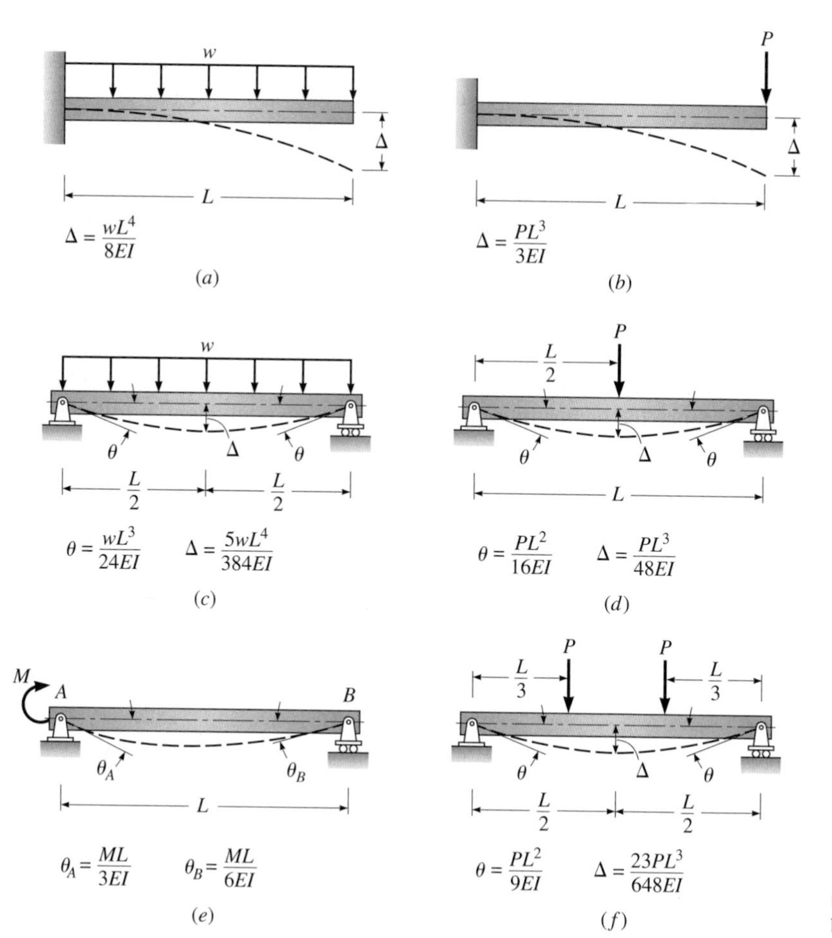

Figure 11.3: Displacements of prismatic beams.

After X_B is computed, it can be applied to the structure in Figure 11.2*a* and the reactions at *A* determined by statics; or as an alternative procedure, the reactions may be computed by summing the corresponding reaction components in Figure 11.2*c* and *d*. For example, the vertical reaction at support *A* equals

$$R_A = wL - X_B = wL - \frac{3wL}{8} = \frac{5wL}{8}$$

Similarly, the moment at *A* equals

$$M_A = \frac{wL^2}{2} - X_B L = \frac{wL^2}{2} - \frac{3wL(L)}{8} = \frac{wL^2}{8}$$

Once the reactions are computed, the shear and the moment curves can be constructed using the sign conventions established in Section 5.3 (see Figure 11.4).

In the preceding analysis, Equation 11.2, the compatibility equation, was expressed in terms of two deflections Δ_{B0} and Δ_{BB}. In setting up the compatibility equations for structures that are indeterminate to more than one degree, it is desirable to display the redundants as unknowns. To write a compatibility equation in this form, we can apply a unit value of the redundant (1 kip in this case) at point B (see Figure 11.2e) and then multiply this case by X_B, the *actual magnitude* of the redundant. To indicate that the unit load (as well as all forces and displacements it produces) is multiplied by the redundant, we show the redundant in brackets next to the unit load on the sketch of the member (Figure 11.2e). The deflection δ_{BB} produced by the unit value of the redundant is called a *flexibility coefficient*. In other words, the units of a flexibility coefficient are in distance per unit load, for example, in/kip or mm/kN. Since the beams in Figure 11.2d and e are equivalent, it follows that

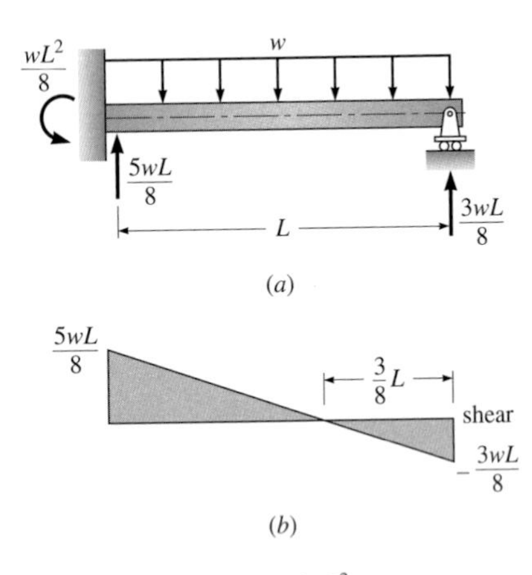

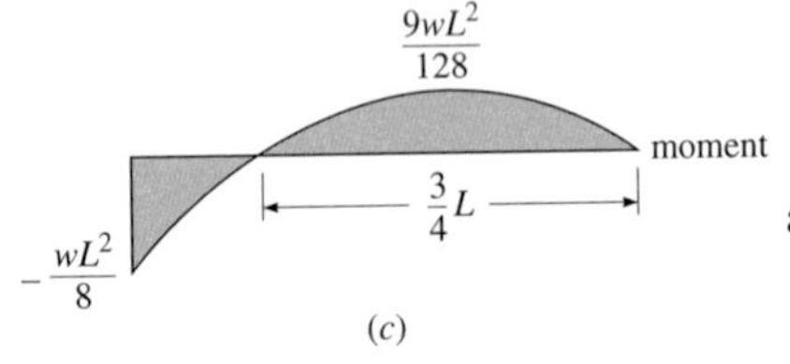

Figure 11.4: Shear and moment curves for beam in Figure 11.2a.

$$\Delta_{BB} = X_B\delta_{BB} \tag{11.4}$$

Substituting Equation 11.4 into Equation 11.2 gives

$$\Delta_{B0} + X_B\delta_{BB} = 0 \tag{11.5}$$

and

$$X_B = -\frac{\Delta_{B0}}{\delta_{BB}} \tag{11.5a}$$

Applying Equation 11.5a to the beam in Figure 11.2, we compute X_B as

$$X_B = -\frac{\Delta_{B0}}{\delta_{BB}} = -\frac{-wL^4/(8EI)}{L^3/(3EI)} = \frac{3wL}{8}$$

After X_B is determined, the reactions or internal forces at any point in the original beam can be determined by combining the corresponding forces in Figure 11.2c with those in Figure 11.2e multiplied by X_B. For example, M_A, the moment at the fixed support, equals

$$M_A = \frac{wL^2}{2} - (1L)X_B = \frac{wL^2}{2} - L\frac{3wL}{8} = \frac{wL^2}{8}$$

11.4 Alternative View of the Flexibility Method (Closing a Gap)

In certain types of problems—particularly those in which we make *internal* releases to establish the released structure—it may be easier for the student to set up the compatibility equation (or equations when several redundants are involved) by considering that the redundant represents the force needed to *close a gap*.

As an example, in Figure 11.5a we again consider a uniformly loaded beam whose right end is supported on an unyielding roller. Since the beam rests on the roller, the gap between the bottom of the beam and the top of the roller is zero. As in the previous case, we select the reaction at B as the redundant and consider the determinate cantilever beam in Figure 11.5*b* as the released structure. Our first step is to apply the uniformly distributed load $w = 2$ kips/ft to the released structure (see Figure 11.5*c*) and compute Δ_{BO}, which represents the 7.96-in gap between the original position of the support and the tip of the cantilever (for clarity, the support is shown displaced horizontally to the right). To indicate that the support has not moved, we show the horizontal distance between the end of the beam and the roller equal to zero inches.

We now apply a 1-kip load upward at B and compute the vertical deflection of the tip $\delta_{BB} = 0.442$ in (see Figure 11.5*d*). Deflection δ_{BB} represents the amount the gap is closed by a unit value of the redundant. Since behavior is elastic, the displacement is directly proportional to the load. If we had applied 10 kips instead of 1 kip, the gap would have closed 4.42 in (that is, 10 times as much). If we consider that the redundant X_B represents the factor with which we must multiply the 1-kip case to close the gap Δ_{BO}, that is,

$$\Delta_B = 0$$

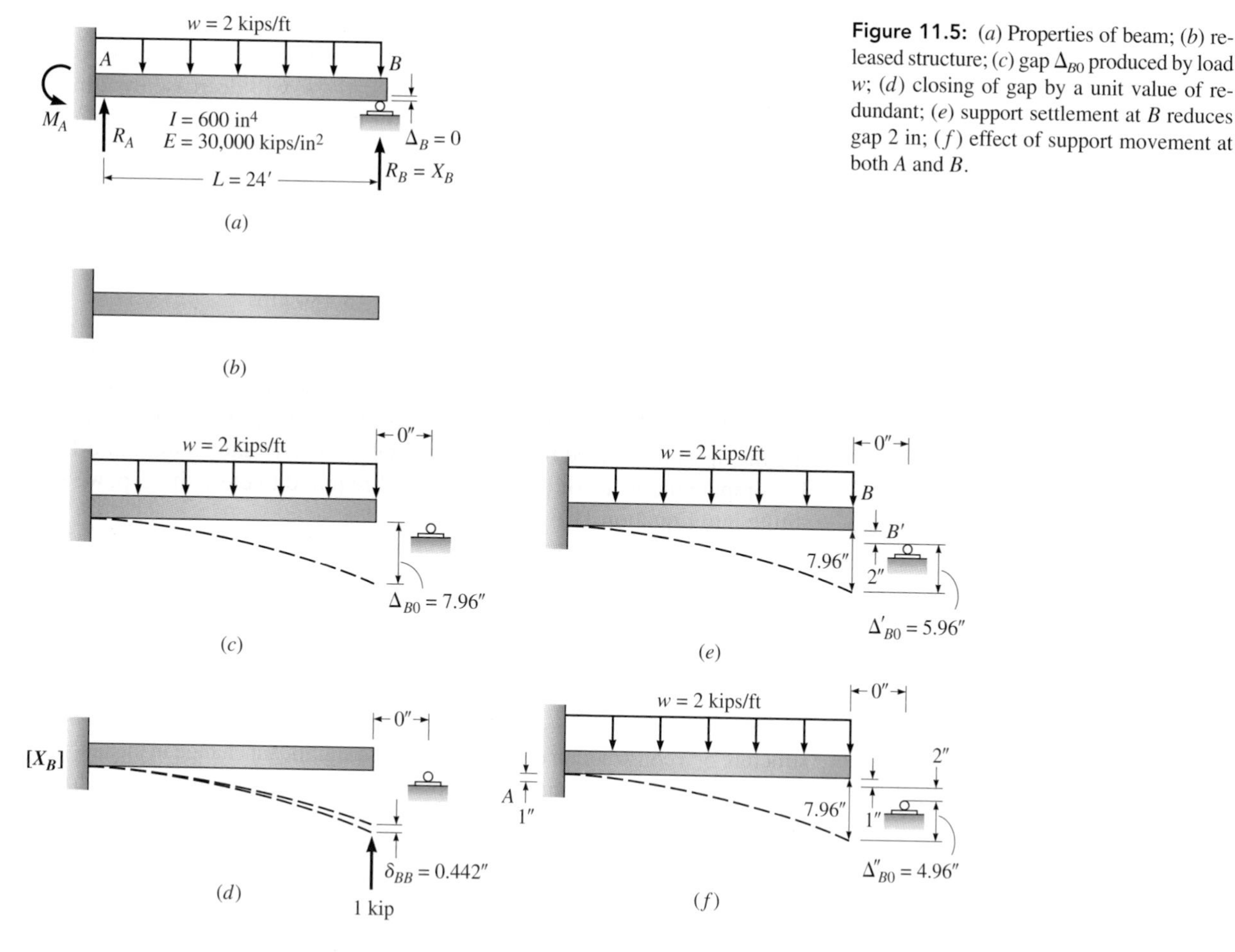

Figure 11.5: (*a*) Properties of beam; (*b*) released structure; (*c*) gap Δ_{BO} produced by load w; (*d*) closing of gap by a unit value of redundant; (*e*) support settlement at B reduces gap 2 in; (*f*) effect of support movement at both A and B.

where Δ_B represents the gap between the bottom of the beam and the roller, we can express this requirement as

$$\Delta_{B0} + \delta_{BB}X_B = 0 \tag{11.6}$$

where Δ_{B0} = gap produced by applied loads or in more general case by load and other effects (support movements, for example)
δ_{BB} = amount the gap is closed by a unit value of redundant
X_B = number by which unit load case must be multiplied to close the gap, or equivalently the value of redundant

As a sign convention, we will assume that any displacement that causes the gap to open is a negative displacement and any displacement that closes the gap is positive. Based on this criterion, δ_{BB} is always positive. Equation 11.6 is, of course, identical to Equation 11.5. Using Figure 11.3 to compute Δ_{B0} and δ_{BB}, we substitute them into Equation 11.6 and solve for X_B, yielding

$$\Delta_{B0} + \delta_{BB}X_B = 0$$
$$-7.96 + 0.442X_B = 0$$
$$X_B = 18.0 \text{ kips}$$

If we are told that support B settles 2 in downward to B' when the load is applied (see Figure 11.5*e*), the size of the gap Δ'_{B0} will decrease by 2 in and equal 5.96 in. To compute the new value for the redundant X'_B now required to close the gap, we again substitute into Equation 11.6 and find

$$\Delta'_{B0} + \delta_{BB}X'_B = 0$$
$$-5.96 + 0.442X'_B = 0$$
$$X'_B = 13.484 \text{ kips}$$

As a final example, if the fixed support at A were accidentally constructed 1 in above its intended position at point A', and if a 2-in settlement also occurred at B when the beam was loaded, the gap Δ''_{B0} between the support and the tip of the loaded beam would equal 4.96 in, as shown in Figure 11.5*f*. To compute the value of the redundant X''_B required to close the gap, we substitute into Equation 11.6 and compute

$$\Delta''_{B0} + \delta_{BB}X''_B = 0$$
$$-4.96 + 0.442X''_B = 0$$
$$X''_B = 11.22 \text{ kips}$$

As you can see from this example, the settlement of a support of an indeterminate structure or a construction error can produce a significant change in the reactions (see Figure 11.6 for a comparison between the shear and moment curves for the case of no settlement versus a 2-in settlement at B). Although an indeterminate beam or structure may often be overstressed locally by moments created by unexpected support settlements, a ductile structure usually possesses a reserve of strength that permits it to deform without collapsing.

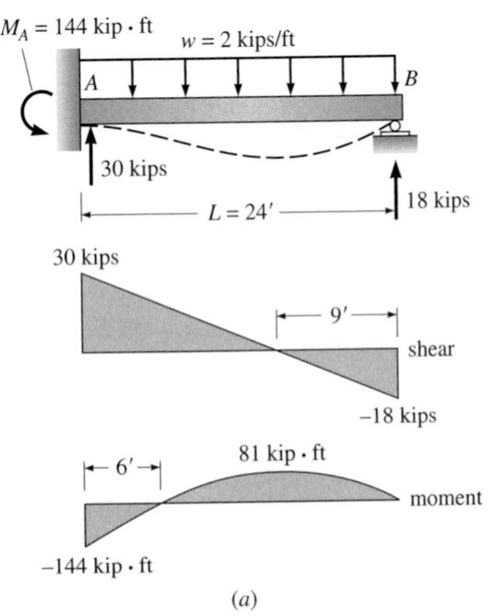

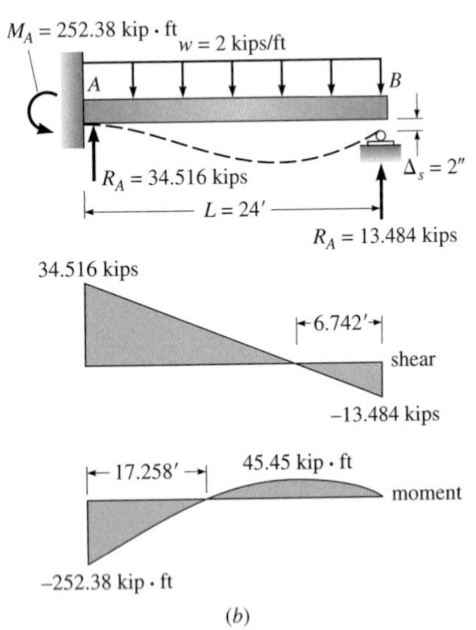

Figure 11.6: Influence of support settlements on shear and moment: (*a*) no settlement; (*b*) support B settles 2 in.

EXAMPLE 11.1

Using the moment M_A at the fixed support as the redundant, analyze the beam in Figure 11.7*a* by the flexibility method.

Solution

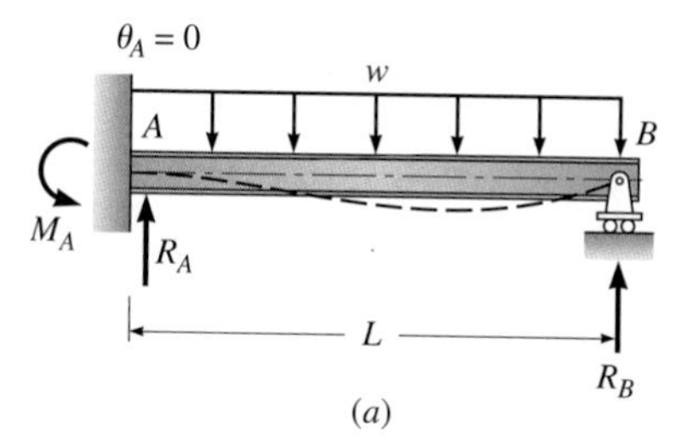

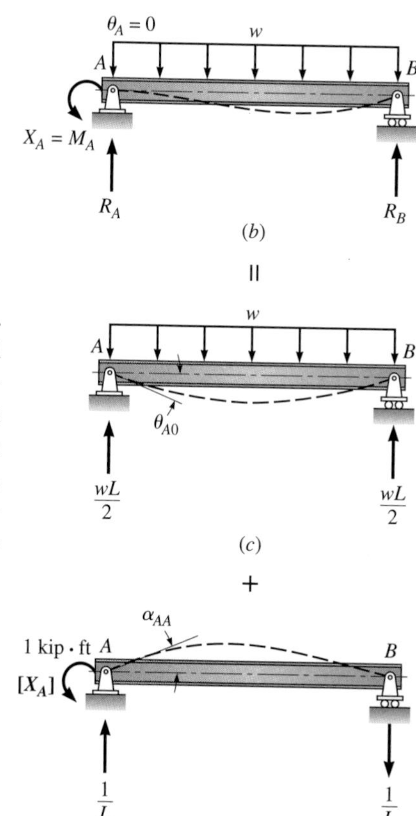

Figure 11.7: Analysis by the flexibility method using M_A as the redundant. (*a*) Beam indeterminate to the first degree; (*b*) released structure with uniform load and redundant M_A applied as external loads; (*c*) released structure with actual load; (*d*) released structure with reactions produced by unit value of redundant.

The fixed support at *A* prevents the left end of the beam from rotating. Removing the rotational restraint while retaining the horizontal and vertical restraints is equivalent to replacing the fixed support by a pin support. The released structure loaded by the redundant and the actual load is shown in Figure 11.7*b*. We now analyze the released structure for the actual load in Figure 11.7*c* and the redundant in Figure 11.7*d*. Since $\theta_A = 0$, the rotation θ_{A0} produced by the uniform load and the rotation $\alpha_{AA}X_A$ produced by the redundant must add to zero. From this geometric requirement we write the compatibility equation as

$$\theta_{A0} + \alpha_{AA}X_A = 0 \tag{1}$$

where θ_{A0} = rotation at *A* produced by uniform load
α_{AA} = rotation at *A* produced by a unit value of redundant (1 kip·ft)
X_A = redundant (moment at *A*)

Substituting into Equation 1 the values of θ_{A0} and α_{AA} given by the equations in Figure 11.3, we find that

$$-\frac{wL^3}{24EI} + \frac{L}{3EI}X_A = 0$$

$$X_A = M_A = \frac{wL^2}{8} \qquad \textbf{Ans.} \tag{2}$$

Since M_A is positive, the assumed direction (counterclockwise) of the redundant was correct. The value of M_A verifies the previous solution shown in Figure 11.4.

EXAMPLE 11.2

Determine the bar forces and reactions in the truss shown in Figure 11.8*a*. Note that *AE* is constant for all bars.

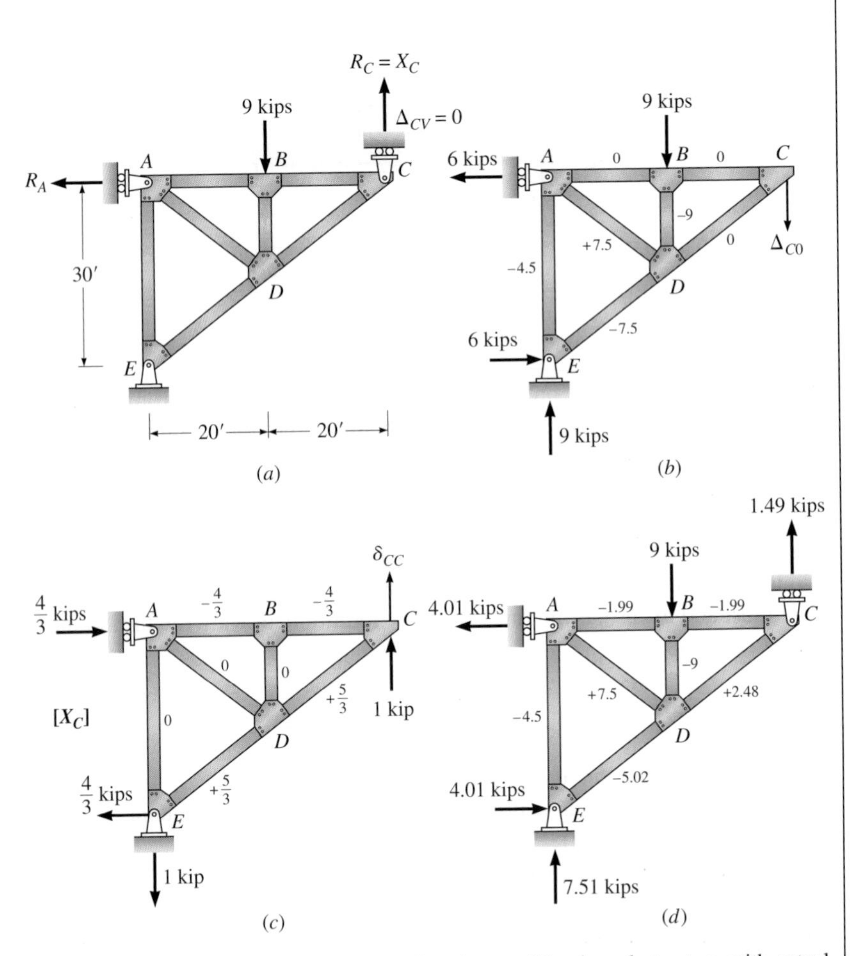

Figure 11.8: (*a*) Truss indeterminate to first degree; (*b*) released structure with actual loads; (*c*) released structure loaded by unit value of redundant; (*d*) final values of bar forces and reactions by superimposing case (*b*) and X_C times case (*c*). All bar forces are in kips.

Solution

Since the truss is externally indeterminate to the first degree (reactions supply four restraints), one compatibility equation is required. Arbitrarily select as the redundant the roller reaction at *C*. We now load the released structure with the actual loading (Figure 11.8*b*) and the redundant (Figure 11.8*c*). Since the roller prevents vertical displacement (that is, $\Delta_{CV} = 0$), superposition of the deflections at C gives the following compatibility equation:

$$\Delta_{C0} + X_C \delta_{CC} = 0 \qquad (1)$$

where Δ_{C0} is the deflection in the released structure produced by the actual load and δ_{CC} is the deflection in the released structure produced by a unit value of the redundant. (Displacements and forces directed upward are positive.)

Evaluate Δ_{C0} and δ_{CC} by virtual work using Equation 10.24. To compute Δ_{C0} (Figure 11.8*b*), use loading in Figure 11.8*c* as the Q system.

$$\Sigma Q \delta_P = \Sigma F_Q \frac{F_P L}{AE}$$

$$1 \text{ kip}(\Delta_{C0}) = \left(\frac{5}{3}\right) \frac{-7.5(25 \times 12)}{AE}$$

$$\Delta_{C0} = -\frac{3750}{AE} \downarrow$$

To compute δ_{CC} produced by the 1-kip load at C (see Figure 11.8*c*), we also use the loading in Figure 11.8*c* as a Q system.

$$1 \text{ kip}(\delta_{CC}) = \sum \frac{F_Q^2 L}{AE}$$

$$\delta_{CC} = \left(-\frac{4}{3}\right)^2 \frac{20 \times 12}{AE}(2) + \left(\frac{5}{3}\right)^2 \frac{25 \times 12}{AE}(2) = \frac{2520}{AE} \uparrow$$

Substituting Δ_{C0} and δ_{CC} into Equation 1 yields

$$-\frac{3750}{AE} + \frac{2520}{AE} X_C = 0$$

$$X_C = 1.49 \qquad \textbf{Ans.}$$

The final reactions and bar forces shown in Figure 11.8*d* are computed by superimposing those in Figure 11.8*b* with 1.49 times those produced by the unit load in Figure 11.8*c*. For example,

$$R_A = 6 - \frac{4}{3}(1.49) = 4.01 \text{ kips} \qquad F_{ED} = -7.5 + \frac{5}{3}(1.49) = -5.02 \text{ kips}$$

EXAMPLE 11.3

Determine the reactions and draw the moment curves for the frame members

Solution

To produce a stable determinate released structure, we arbitrarily select the horizontal reaction R_{CX} as the redundant. Removing the horizontal restraint exerted by the pin at C while retaining its capacity to transmit vertical load is equivalent to introducing a roller. The deformations and reactions in the released structure produced by the applied load are shown in Figure 11.9*b*. The action of the redundant on the released structure is shown in Figure 11.9*c*. Since the horizontal displacement Δ_{CH} in the real structure at joint C is zero, the compatibility equation is

$$\Delta_{C0} + \delta_{CC}X_C = 0 \tag{1}$$

Compute Δ_{C0} using moment-area principles (see the deflected shape in Figure 11.9*b*). From Figure 11.3*d* we can evaluate the slope at the right end of the girder as

$$\theta_{B0} = \frac{PL^2}{16EI} = \frac{10(12)^2}{16EI} = \frac{90}{EI}$$

Since joint B is rigid, the rotation of the top of column BC also equals θ_{B0}. Because the column carries no moment, it remains straight and

$$\Delta_{C0} = 6\theta_{B0} = \frac{540}{EI}$$

Compute δ_{CC} by virtual work (see Figure 11.9*c*). Use the loading in Figure 11.9*c* as both the Q system and the P system (i.e., the P and Q systems are identical). To evaluate M_Q and M_P, we select coordinate systems with origins at A in the girder and C in the column.

$$1\text{ kip}(\delta_{CC}) = \int M_Q M_P \frac{dx}{EI} = \int_0^{12} \frac{x}{2}\left(\frac{x}{2}\right)\frac{dx}{EI} + \int_0^{6} x(x)\frac{dx}{EI} \tag{10.31}$$

Integrating and substituting the limits give

$$\delta_{CC} = \frac{216}{EI}$$

Substituting Δ_{C0} and δ_{CC} into Equation 1 gives

$$-\frac{540}{EI} + \frac{216}{EI}(X_C) = 0$$

$$X_C = 2.5 \qquad \textbf{Ans.}$$

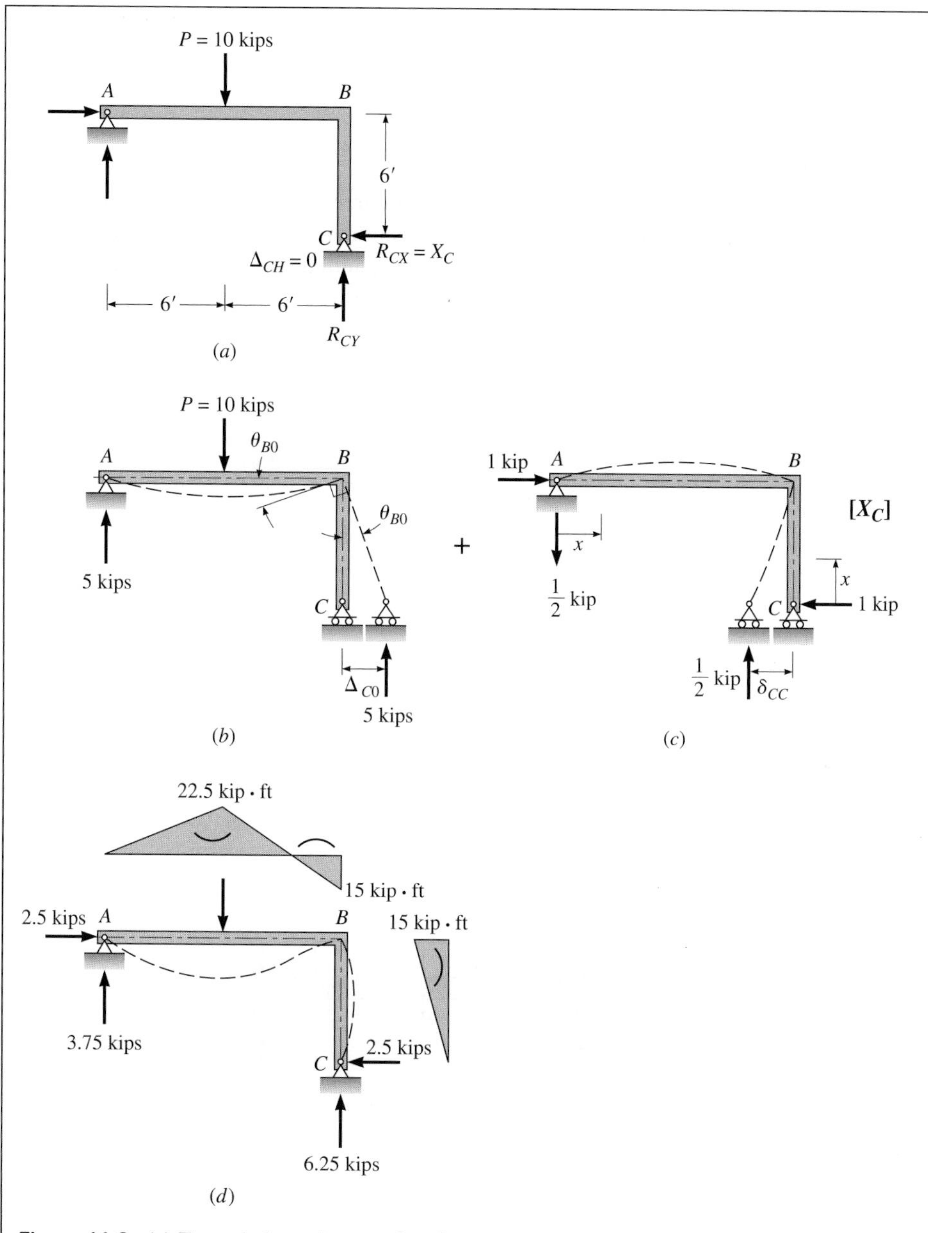

Figure 11.9: (*a*) Frame indeterminate to first degree, R_{CX} selected as redundant; (*b*) design load applied to released structure; (*c*) reactions and deformations in released structure due to unit value of redundant; (*d*) final forces by superposition of values in (*b*) plus (X_C) times values in (*c*). Moment curves (in kip·ft) also shown.

The final reactions (see Figure 11.9*d*) are established by superimposing the forces in Figure 11.9*b* and those in Figure 11.9*c* multiplied by $X_C = 2.5$.

EXAMPLE 11.4

Determine the reactions of the continuous beam in Figure 11.10*a* by the flexibility method. Given: EI is constant.

Solution

The beam is indeterminate to the first degree (i.e., four reactions and three equations of statics). We arbitrarily select the reaction at B as the redundant. The released structure is a simple beam spanning from A to C. The released structure loaded by the specified loads and the redundant X_B is shown in Figure 11.10*b*. Since the roller prevents vertical deflection at B, the geometric equation stating this fact is

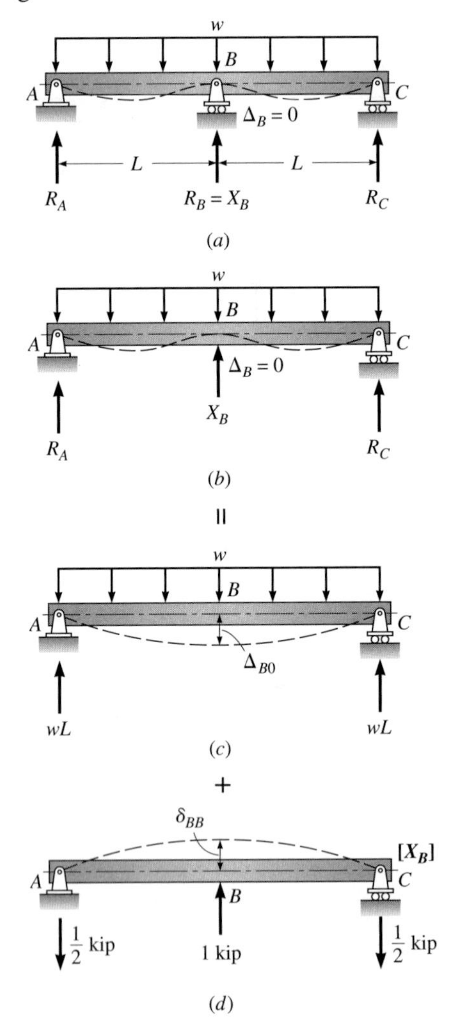

Figure 11.10: Analysis by consistent deformations: (*a*) continuous beam indeterminate to the first degree, and reaction at B taken as redundant; (*b*) released structure loaded by external load and redundant; (*c*) released structure with external load; (*d*) released structure loaded by redundant; (*e*) shear and moment curves.

$$\Delta_B = 0 \tag{1}$$

To determine the redundant, we superimpose the deflections at B produced by (1) the external load (see Figure 11.10c) and (2) a unit value of the redundant multiplied by the magnitude of the redundant X_B (see Figure 11.10d). Expressing Equation 1 in terms of these displacements yields

$$\Delta_{B0} + \delta_{BB}X_B = 0 \tag{2}$$

Using Figure 11.3c and d, we compute the displacements at B.

$$\Delta_{B0} = -\frac{5w(2L)^4}{384EI} \qquad \delta_{BB} = \frac{(1\text{ kip})(2L)^3}{48EI}$$

Substituting Δ_{B0} and δ_{BB} into Equation 2 and solving for X_B give

$$R_B = X_B = 1.25wL \qquad \textbf{Ans.}$$

We compute the balance of the reactions by adding, at the corresponding points, the forces in Figure 11.10c to those in Figure 11.10d multiplied by X_B:

$$R_A = wL - \tfrac{1}{2}(1.25wL) = \tfrac{3}{8}wL \qquad \textbf{Ans.}$$

$$R_C = wL - \tfrac{1}{2}(1.25wL) = \tfrac{3}{8}wL \qquad \textbf{Ans.}$$

The shear and moment curves are plotted in Figure 11.10e.

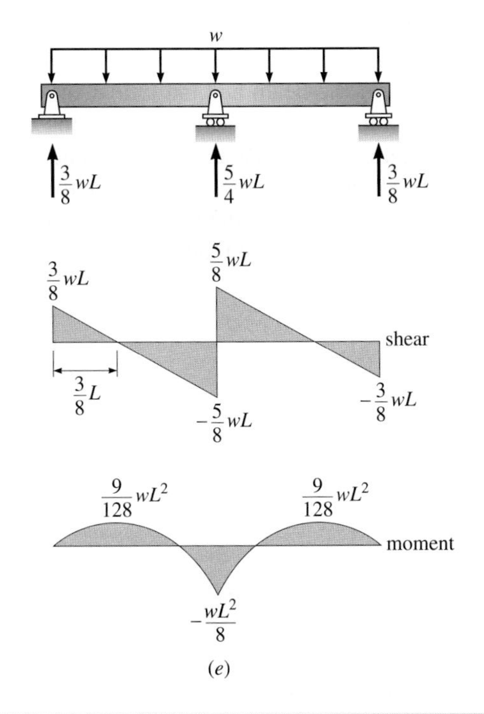

(e)

11.5 Analysis Using Internal Releases

In previous examples of indeterminate structures analyzed by the flexibility method, support reactions were selected as the redundants. If the supports do not settle, the compatibility equations express the geometric condition that the displacement in the direction of the redundant is zero. We will now extend the flexibility method to a group of structures in which the released structure is established by removing an *internal* restraint. For this condition, redundants are taken as *pairs of internal forces*, and the compatibility equation is based on the geometric condition that no *relative displacement* (i.e., no gap) occurs between the ends of the section on which the redundants act.

We begin our study by considering the analysis of a cantilever beam whose free end is supported by an elastic link (see Figure 11.11*a*). Since the fixed end and the link apply a total of four restraints to the beam, but only three equations of equilibrium are available for a planar structure, the structure is indeterminate to the first degree. To analyze this structure, we select as the redundant the tension force T in bar BC. The released structure with both the actual load of 6 kips and the redundant applied as an external load is shown in Figure 11.11*b*. As we have noted previously, you are free to assume the direction in which the redundant acts. If the solution of the compatibility equation produces a positive value of the redundant, the assumed direction is correct. A negative value indicates that the direction of the redundant must be reversed. Since the redundant T is assumed to act up on the beam and down on the link, upward displacements of the beam are positive and downward displacements are negative. For the link a downward displacement at B is positive and an upward displacement negative.

In Figure 11.11*c* the design load is applied to the released structure, producing a gap Δ_{B0} between the end of the beam and the unloaded link. Figure 11.11*d* shows the action of the internal redundant T in closing the gap. The unit values of the redundant elongate the bar an amount δ_1 and displace the tip of the cantilever upward an amount δ_2. To account for the actual value of the redundant, the forces and displacements produced by the unit loads are multiplied by T—the magnitude of the redundant.

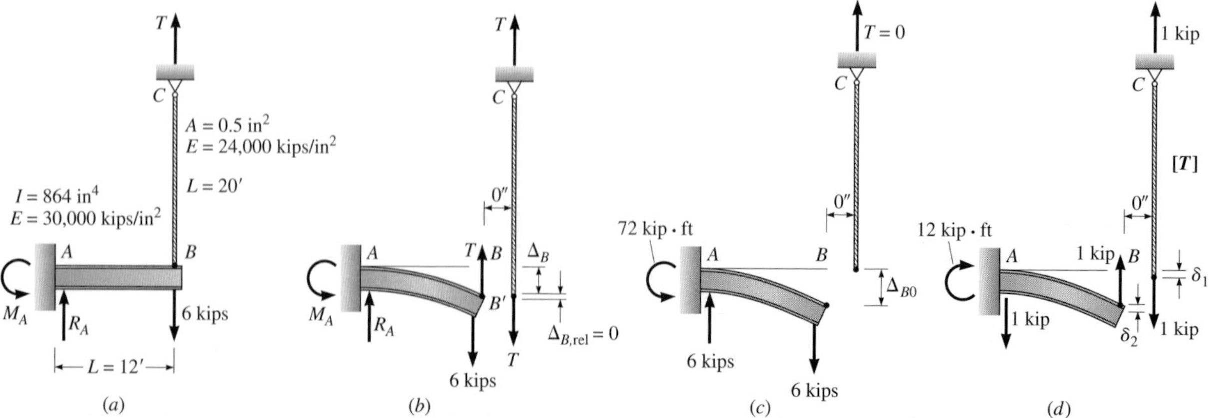

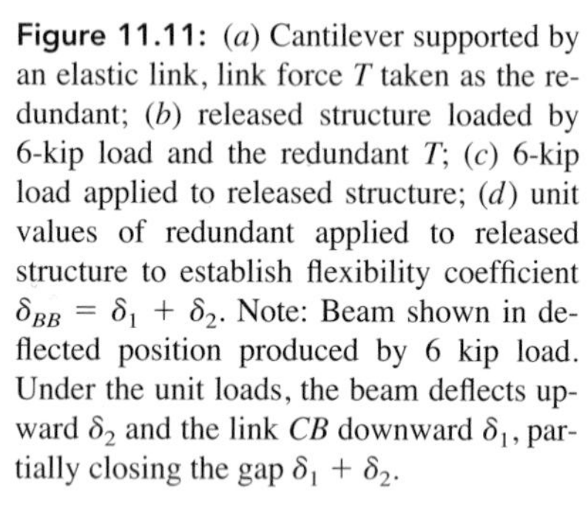
Figure 11.11: (*a*) Cantilever supported by an elastic link, link force T taken as the redundant; (*b*) released structure loaded by 6-kip load and the redundant T; (*c*) 6-kip load applied to released structure; (*d*) unit values of redundant applied to released structure to establish flexibility coefficient $\delta_{BB} = \delta_1 + \delta_2$. Note: Beam shown in deflected position produced by 6 kip load. Under the unit loads, the beam deflects upward δ_2 and the link CB downward δ_1, partially closing the gap $\delta_1 + \delta_2$.

The compatibility equation required to solve for the redundant is based on the observation that the right end of the beam and the link BC both deflect the same amount Δ_B because they are connected by a pin. Alternatively, we can state that the *relative displacement* $\Delta_{B,\text{Rel}}$ between the top of the beam and the link is zero (see Figure 11.11*b*). This latter approach is adopted in this section.

Superimposing the deflections at B in Figure 11.11*c* and *d*, we can write the compatibility equation as

$$\Delta_{B,\text{Rel}} = 0 \tag{11.7}$$

$$\Delta_{B0} + \delta_{BB}(T) = 0$$

where Δ_{B0} is the downward displacement of the beam (i.e., the opening of the gap in the released structure by the 6-kip load) and δ_{BB} is the distance the gap is closed by the unit values of the redundant (that is, $\delta_{BB} = \delta_1 + \delta_2$; see Figure 11.11*d*).

In Figure 11.11*c*, Δ_{B0} may be evaluated from Figure 11.3*b* as

$$\Delta_{B0} = -\frac{PL^3}{3EI} = -\frac{6(12 \times 12)^3}{3(30{,}000)864} = -0.2304 \text{ in}$$

And $\delta_{BB} = \delta_1 + \delta_2$, where $\delta_1 = FL/(AE)$ and δ_2 is given by Figure 11.3*b*.

$$\delta_1 = \frac{FL}{AE} = \frac{1 \text{ kip}(20 \times 12)}{0.5(24{,}000)} = 0.02 \text{ in} \qquad \delta_2 = \frac{PL^3}{3EI} = \frac{1 \text{ kip}(12)^3(1728)}{3 \times 30{,}000 \times 864}$$
$$= 0.0384$$

$$\delta_{BB} = \delta_1 + \delta_2 = 0.02 + 0.0384 = 0.0584 \text{ in}$$

Substituting Δ_{B0} and δ_{BB} into Equation 11.7, we compute the redundant T as

$$-0.2304 + 0.0584\,T = 0$$
$$T = 3.945 \text{ kips}$$

The actual deflection at B (see Figure 11.11*b*) may be computed either by evaluating the change in length of the link

$$\Delta_B = \frac{FL}{AE} = \frac{3.945(20 \times 12)}{0.5(24{,}000)} = 0.0789 \text{ in}$$

or by adding the deflections at the tip of the beam in Figure 11.11*c* and *d*,

$$\Delta_B = \Delta_{B0} - T\delta_2 = 0.2304 - 3.945(0.0384) = 0.0789 \text{ in}$$

After the redundant is established, the reactions and internal forces can be computed by superimposition of forces in Figure 11.11*c* and *d*; for example,

$$R_A = 6 - 1(T) = 6 - 3.945 = 2.055 \text{ kips} \qquad \textbf{Ans.}$$

$$M_A = 72 - 12(T) = 72 - 12(3.945) = 24.66 \text{ kip} \cdot \text{ft}$$

EXAMPLE 11.5

Analyze the continuous beam in Figure 11.12*a* by selecting the internal moment at *B* as the redundant. The beam is indeterminate to the first degree. *EI* is constant.

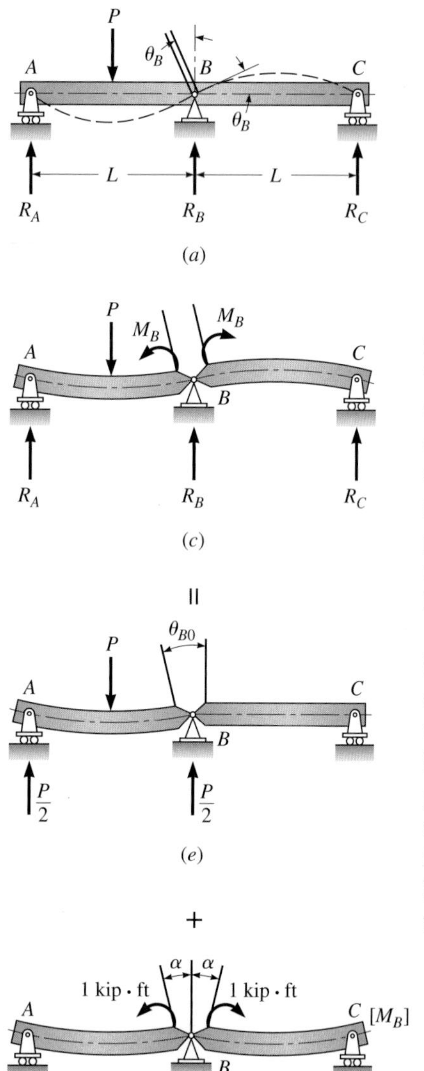

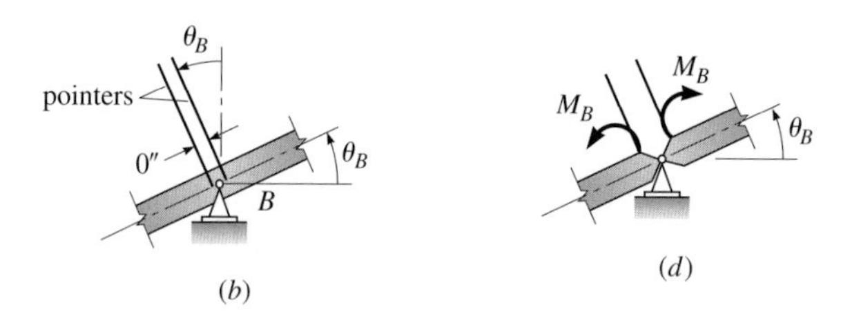

Solution

To clarify the angular deformations involved in the solution, we will imagine that two pointers are welded to the beam on each side of joint *B*. The pointers, which are spaced zero inches apart, are perpendicular to the longitudinal axis of the beam. When the concentrated load is applied to span *AB*, joint *B* rotates counterclockwise, and both the longitudinal axis of the beam and the pointers move through the angle θ_B, as shown in Figure 11.12*a* and *b*. Since the pointers are located at the same point, they remain parallel (i.e., the angle between them is zero).

We now imagine that a hinge, which can transmit axial load and shear but not moment, is introduced into the continuous beam at support *B*, producing a released structure that consists of two simply supported beams (see Figure 11.12*c*). At the same time that the hinge is introduced, we imagine that the actual value of internal moment M_B in the original beam is applied as an external load to the ends of the beam on either side of the hinge at *B* (see Figure 11.12*c* and *d*). Since each member of the released structure is supported and loaded in the same manner as in the original continuous beam, the internal forces in the released structure are identical to those in the original structure.

Figure 11.12: (*a*) Continuous beam indeterminate to the first degree; (*b*) detail of joint *B* showing rotation θ_B of longitudinal axis; (*c*) released structure loaded by actual load *P* and the redundant moment M_B; (*d*) detail of joint *B* in (*c*); (*e*) released structure with actual load; (*f*) released structure loaded by redundant; forces shown are produced by a unit value of the redundant M_B.

To complete the solution, we analyze the released structure separately for (1) the actual loading (see Figure 11.12*e*) and (2) the redundant (see Figure 11.12*f*), and we superimpose the two cases.

The compatibility equation is based on the geometric requirement that no angular gaps exist between the ends of the continuous beam at support B; or equivalently that the angle between the pointers is zero. Thus we can write the compatibility equation as

$$\theta_{B,\text{Rel}} = 0$$

$$\theta_{B0} + 2\alpha M_B = 0 \tag{11.8}$$

Evaluate θ_{B0}, using Figure 11.3*d*:

$$\theta_{B0} = \frac{PL^2}{16EI}$$

Evaluate α, using Figure 11.3*e*:

$$\alpha = \frac{1L}{3EI}$$

Substituting θ_{B0} and α into Equation 11.8 and solving for the redundant give

$$\frac{PL^2}{16EI} + 2\frac{L}{3EI}M_B = 0$$

$$M_B = -\frac{3}{32}(PL) \qquad \textbf{Ans.}$$

Superimposing forces in Figure 11.12*e* and *f*, we compute

$$R_A = \frac{P}{2} + \frac{1}{L}M_B = \frac{P}{2} + \frac{1}{L}\left(-\frac{3}{32}PL\right) = \frac{13}{32}P \uparrow$$

$$R_C = 0 + \frac{1}{L}\left(-\frac{3}{32}PL\right) = -\frac{3}{32}P \downarrow \quad \text{(minus sign indicates that assumed direction up is wrong)}$$

Similarly, θ_B can be evaluated by summing rotations at the right end of AB to give

$$\theta_B = \theta_{B0} + \alpha M_B = \frac{PL^2}{16EI} + \frac{L}{3EI}\left(-\frac{3}{32}PL\right) = \frac{PL^2}{32EI} \circlearrowright$$

or by summing rotations of the left end of BC:

$$\theta_B = 0 + \alpha M_B = \frac{L}{3EI}\left(-\frac{3}{32}PL\right) = -\frac{PL^2}{32EI} \circlearrowright$$

EXAMPLE 11.6

Determine the forces in all members of the truss in Figure 11.13. AE is constant for all bars.

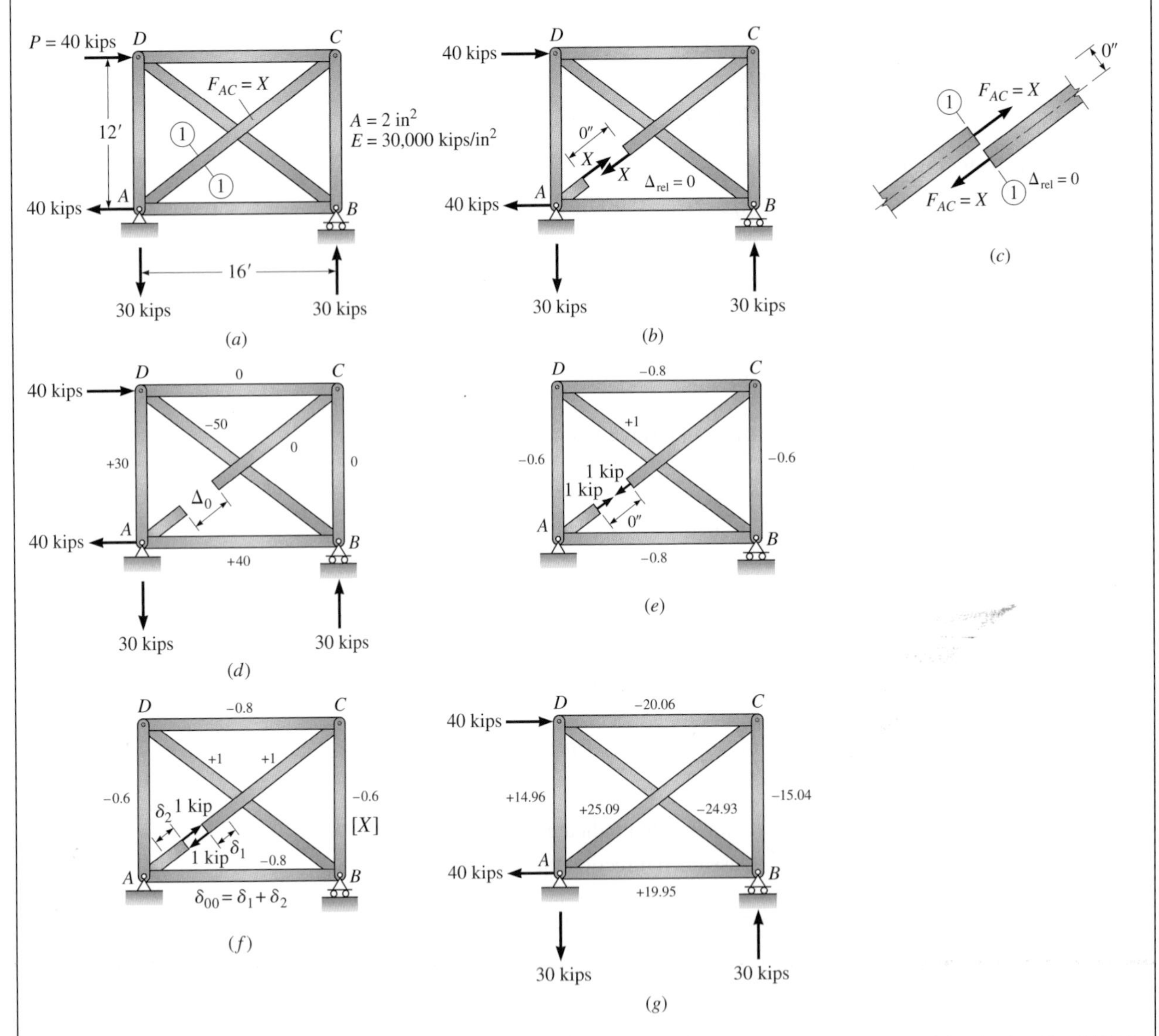

Figure 11.13: (*a*) Details of truss; (*b*) released structure loaded with redundant X and 40-kip load; (*c*) detail showing redundant; (*d*) 40-kip load applied to released structure; (*e*) Q system for Δ_0; (*f*) unit value of redundant applied to released structure; (*g*) final results.

Solution

The truss in Figure 11.13*a* is internally indeterminate to the first degree. The unknown forces—bars and reactions—total nine, but only $2n = 8$ equations are available for their solution. From a physical point of view, an extra diagonal member that is not required for stability has been added to transmit lateral load into support A.

Application of the 40-kip horizontal force at D produces forces in all bars of the truss. We will select the axial force F_{AC} in bar AC as the redundant and represent it by the symbol X. We now imagine that bar AC is cut by passing an imaginary Section 1-1 through the bar. On each side of the cut, the redundant X is applied to the ends of the bar as an external load (see Figure 11.13*b*). A detail at the cut is shown in Figure 11.13*c*. To show the action of the internal forces on each side of the cut, the bars have been offset. The zero dimension between the longitudinal axis of the bars indicates that the bars are actually collinear. To show that no gap exists between the ends of the bars, we have noted on the sketch that the relative displacement between the ends of the bars Δ_{Rel} equals zero.

$$\Delta_{\text{Rel}} = 0 \qquad (11.9)$$

The requirement that no gap exists between the ends of the bars in the actual structure forms the basis of the compatibility equation.

As in previous examples, we next divide the analysis into two parts. In Figure 11.13*d* the released structure is analyzed for the applied load of 40 kips. As the stressed bars of the released structure deform, a gap Δ_0 opens between the ends of the bars at Section 1-1. The Q system required to compute Δ_0 is shown in Figure 11.13*e*. In Figure 11.13*f*, the released structure is analyzed for the action of the redundant. The relative displacement δ_{00} of the ends of the bar produced by the unit value of the redundant equals the sum of the displacements δ_1 and δ_2. To compute δ_{00}, we again use the force system shown in Figure 11.13*e* as the Q system. In this case the Q system and the P system are identical.

Expressing the geometric condition given by Equation 11.9 in terms of the displacements produced by the applied loads and the redundant, we can write

$$\Delta_0 + X\delta_{00} = 0 \qquad (11.10)$$

Substituting numerical values of Δ_0 and δ_{00} into Equation 11.10 and solving for X give

$$-0.346 + 0.0138X = 0$$

$$X = 25.07 \text{ kips}$$

[*continues on next page*]

Example 11.6 continues . . .

The computations for Δ_0 and δ_{00} using virtual work are given below.

Δ_0:

Use the P system in Figure 11.13*d* and Q system in Figure 11.13*e*:

$$W_Q = \sum \frac{F_Q F_P L}{AE}$$

$$1 \text{ kip}(\Delta_0) = \overset{\text{bar } DB}{\frac{1(-50)(20 \times 12)}{AE}} + \overset{\text{bar } AB}{\frac{-0.8(40)(16 \times 12)}{AE}} + \overset{\text{bar } AD}{\frac{-0.6(30)(12 \times 12)}{AE}}$$

$$\Delta_0 = -\frac{20{,}736}{AE} = -\frac{20{,}736}{2(30{,}000)} = -0.346 \text{ in}$$

δ_{00}:

P system in Figure 11.13*f* and Q system in Figure 11.13*e* (*note*: P and Q systems are the same; therefore, $F_Q = F_P$):

$$W_Q = \sum \frac{F_Q^2 L}{AE}$$

$$1 \text{ kip}(\delta_1) + 1 \text{ kip}(\delta_2) = \frac{(-0.6)^2(12 \times 12)}{AE}(2) + \frac{(-0.8)^2(16 \times 12)}{AE}(2) + \frac{1^2(20 \times 12)}{AE}(2)$$

Since $\delta_1 + \delta_2 = \delta_{00}$,

$$\delta_{00} = \frac{829.44}{AE} = \frac{829.44}{2(30{,}000)} = 0.0138 \text{ in}$$

Bar forces are established by superposition of the forces in Figure 11.13*d* and *f*. For example, the forces in bars DC, AB, and DB are

$$F_{DC} = 0 + (-0.8)(25.07) = -20.06 \text{ kips}$$

$$F_{AB} = 40 + (-0.8)(25.07) = 19.95 \text{ kips} \qquad \textbf{Ans.}$$

$$F_{DB} = -50 + 1(25.07) = -24.93 \text{ kips}$$

Final results are summarized in Figure 11.13*g*.

11.6 Support Settlements, Temperature Change, and Fabrication Errors

Support settlements, fabrication errors, temperature changes, creep, shrinkage, and so forth create forces in indeterminate structures. To ensure that such structures are safely designed and do not deflect excessively, the designer should investigate the influence of these effects—particularly when the structure is unconventional or when the designer is unfamiliar with the behavior of a structure.

Since it is standard practice for designers to assume that members will be fabricated to the exact length and that supports will be constructed at the precise location and elevation specified on the construction drawings, few engineers consider the effects of fabrication or construction errors when designing routine structures. If problems do arise during construction, they are typically handled by the field crew. For example, if supports are constructed too low, steel plates—shims—can be inserted under the base plates of columns. If problems arise after construction is complete and the client is inconvenienced or is not able to use the structure, lawsuits often follow.

On the other hand, most building codes require that engineers consider the forces created by differential settlement of structures constructed on compressible soils (soft clays and loose sands), and the AASHTO specifications requires that bridge designers evaluate the forces created by temperature change, shrinkage, and so forth.

The effects of support settlements, fabrication errors, and so forth can easily be included in the flexibility method by modifying certain terms of the compatibility equations. We begin our discussion by considering support settlements. Once you understand how to incorporate these effects into the compatibility equation, other effects can easily be included.

Case 1. A Support Movement Corresponds to a Redundant

If a predetermined support movement occurs that corresponds to a redundant, the compatibility equation (normally set equal to zero for the case of no support settlements) is simply set equal to the value of the support movement. For example, if support B of the cantilever beam in Figure 11.14 settles 1 in when the member is loaded, we write the compatibility equation as

$$\Delta_B = -1 \text{ in}$$

Superimposing displacements at B yields

$$\Delta_{B0} + \delta_{BB} X_B = -1$$

where Δ_{B0}, the deflection at B in the released structure produced by the applied load, and δ_{BB}, the deflection at B in the released structure produced by a unit value of the redundant, are shown in Figure 11.2.

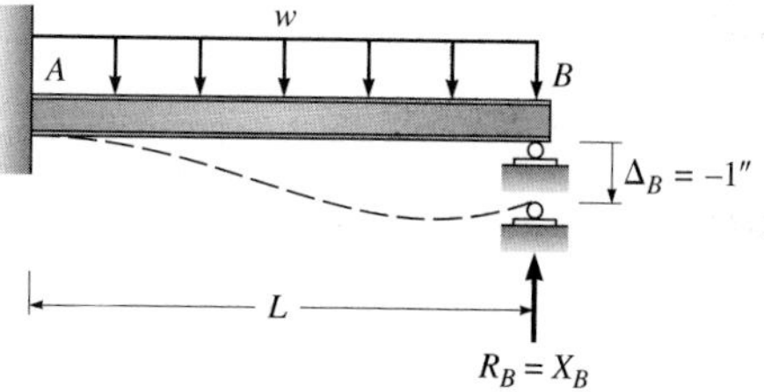

Figure 11.14: Support settlement at location of the redundant.

Following the convention established previously, the support settlement Δ_B is considered negative because it is opposite in sense to the assumed direction of the redundant.

Case 2. The Support Settlement Does Not Correspond to a Redundant

If a support movement occurs that does not correspond to a redundant, its effect can be included as part of the analysis of the *released* structure for the applied loads. In this step you evaluate the displacement that corresponds to the redundant produced by the movement of the other support. When the geometry of the structure is simple, a sketch of the released structure in which the support movements are shown will often suffice to establish the displacement that corresponds to the redundant. If the geometry of the structure is complex, you can use virtual work to compute the displacement. As an example, we will set up the compatibility equation for the cantilever beam in Figure 11.14, assuming that support A settles 0.5 in and rotates clockwise 0.01 rad and support B settles 1 in. Figure 11.15*a* shows the deflection at B, denoted by Δ_{BS}, due to the -0.5 in settlement and the 0.01-rad rotation of support A. Figure 11.15*b* shows the deflection at B due to the applied load. We can then write the compatibility equation required to solve for the redundant X as

$$\Delta_B = -1$$

$$(\Delta_{B0} + \Delta_{BS}) + \delta_{BB}X_B = -1$$

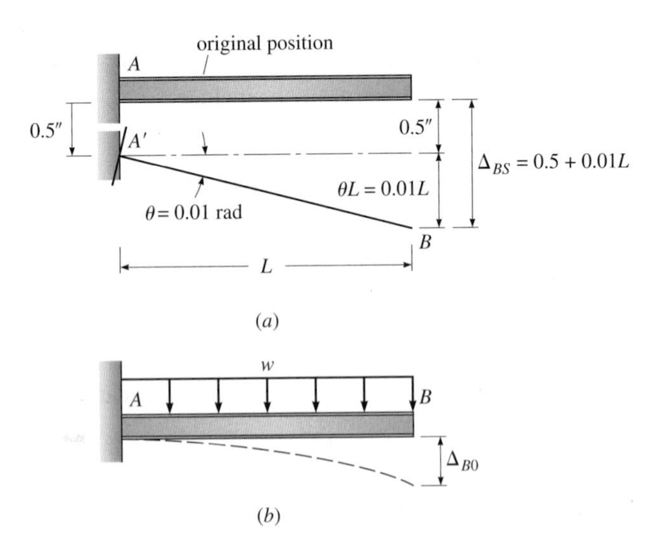

Figure 11.15: (*a*) Deflection at B produced by settlement and rotation at support A; (*b*) deflection at B produced by applied load.

EXAMPLE 11.7

Determine the reactions induced in the continuous beam shown in Figure 11.16*a* if support *B* settles 0.72 in and support *C* settles 0.48 in. Given: *EI* is constant, $E = 29{,}000$ kips/in^2, and $I = 288$ in^4.

Solution

Arbitrarily select the reaction at support *B* as the redundant. Figure 11.16*b* shows the *released* structure with support *C* in its displaced position. Because the released structure is determinate, it is not stressed by the settlement of support *C* and remains straight. Since the displacement of the beam's axis varies linearly from A, $\Delta_{BO} = -0.24$ in. Because support *B* in its final position lies below the axis of the beam in Figure 11.16*b*, it is evident that the reaction at *B* must act downward to pull the beam down to the support. The forces and displacements produced by a unit value of the redundant are shown in Figure 11.16*c*. Using Figure 11.3*d* to evaluate δ_{BB} gives

$$\delta_{BB} = \frac{PL^3}{48EI} = \frac{1(32)^3(1728)}{48(29{,}000)(288)} = 0.141 \text{ in}$$

Since support *B* settles 0.72 in, the compatibility equation is

$$\Delta_B = -0.72 \text{ in} \tag{1}$$

The displacement is negative because the positive direction for displacements is established by the direction assumed for the redundant. Superimposing the displacements at *B* in Figure 11.16*b* and *c*, we write Equation 1 as

$$\Delta_{BO} + \delta_{BB}X = -0.72$$

Substituting the numerical values of Δ_{BO} and δ_{BB}, we compute *X* as

$$-0.24 + 0.141X = -0.72$$

$$X = -3.4 \text{ kips} \downarrow \quad \textbf{Ans.}$$

The final reactions, which can be computed by statics or by superposition of forces in Figure 11.6*b* and *c*, are shown in Figure 11.16*d*.

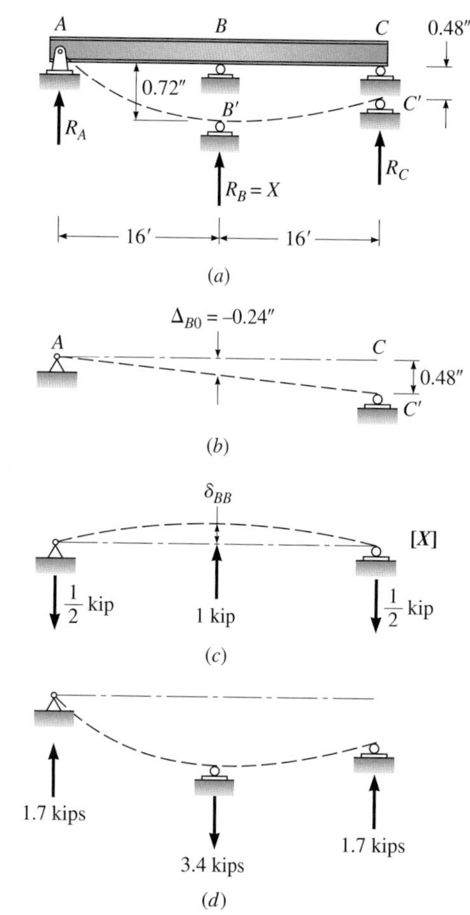

Figure 11.16: (*a*) Continuous beam with specified support settlements; (*b*) released structure with support *C* in displaced position (no reactions or forces in the member develop); (*c*) unit value of redundant applied; (*d*) final reactions computed by superposition of (*b*) and [*X*] times (*c*).

EXAMPLE 11.8

Compute the reaction at support C of the truss in Figure 11.17a if the temperature of bar AB increases 50°F, member ED is fabricated 0.3 in too short, support A is constructed 0.48 in to the right of its intended position, and support C is constructed 0.24 in too high. For all bars $A = 2\ \text{in}^2$, $E = 30{,}000\ \text{kips/in}^2$, and the coefficient of temperature expansion $\alpha = 6 \times 10^{-6}$ (in/in)/°F.

Solution

We arbitrarily select the reaction at support C as the redundant. Figure 11.17b shows the deflected shape of the released structure to an exaggerated scale. The deflected shape results from the 0.48-in displacement of support A to the right, the expansion of bar AB, and the shortening of bar ED. Since the released structure is determinate, no forces are created in the bars or at the reactions due to the displacement of support A or the small changes in length of the bars; however, joint C displaces vertically a distance Δ_{C0}. In Figure 11.17b the

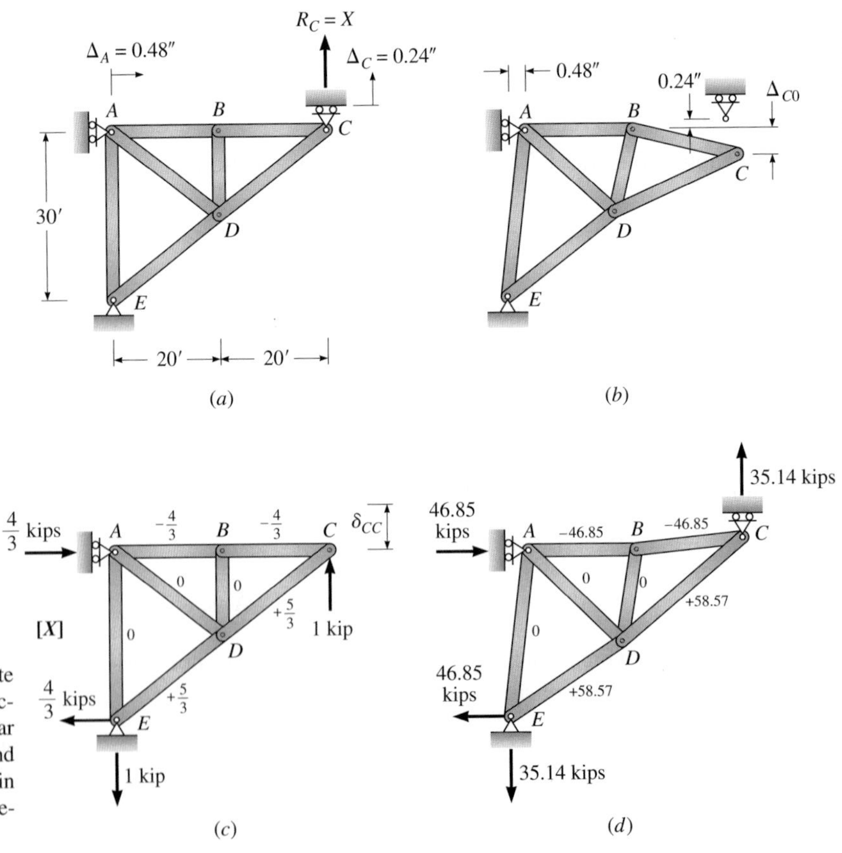

Figure 11.17: (a) Details of indeterminate truss; (b) deflected shape of released structure after displacement of support A and bar deformations due to temperature change and fabrication error; (c) forces and reactions in released structure due to a unit value of redundant; (d) final results of analysis.

support at C is shown in its "as-constructed position." Figure 11.17c shows the forces and deflections produced by a unit value of the redundant.

Since support C is constructed 0.24 in above its intended position, the compatibility equation is

$$\Delta_C = 0.24 \text{ in} \tag{1}$$

Superimposing the deflections at C in Figure 11.17b and c, we can write

$$\Delta_{C0} + \delta_{CC} X = 0.24 \tag{2}$$

To determine X, we compute Δ_{C0} and δ_{CC} by the method of virtual work.

To compute Δ_{C0} (see Figure 11.17b), use the force system in Figure 11.17c as the Q system. Compute ΔL_{temp} of bar AB using Equation 10.25:

$$\Delta L_{\text{temp}} = \alpha(\Delta T)L = (6 \times 10^{-6})50(20 \times 12) = 0.072 \text{ in}$$

$$\Sigma Q \delta_P = \Sigma F_Q \, \Delta L_P \tag{10.23}$$

where ΔL_P is given by Equation 10.26

$$1 \text{ kip}(\Delta_{C0}) + \frac{4}{3} \text{ kips } (0.48) = \frac{5}{3}(-0.3) + \left(-\frac{4}{3}\right)(0.072)$$

$$\Delta_{C0} = -1.236 \text{ in} \downarrow$$

In Example 11.2, δ_{CC} was evaluated as

$$\delta_{CC} = \frac{2520}{AE} = \frac{2520}{2(30{,}000)} = 0.042 \text{ in}$$

Substituting Δ_{C0} and δ_{CC} into Equation 2 and solving for X give

$$-1.236 + 0.042X = 0.24$$

$$X = 35.14 \text{ kips} \qquad \textbf{Ans.}$$

Final bar forces and reactions from all effects, established by superimposing the forces (all equal to zero) in Figure 11.17b and those in Figure 11.17c multiplied by the redundant X, are shown in Figure 11.17d.

11.7 Analysis of Structures with Several Degrees of Indeterminacy

The analysis of a structure that is indeterminate to more than one degree follows the same format as that for a structure with a single degree of indeterminacy. The designer establishes a determinate released structure by selecting certain reactions or internal forces as redundants. The unknown redundants are applied to the released structure as loads together with the actual loads. The structure is then analyzed separately for each redundant as well as for the actual load. Finally, compatibility equations equal in number to the redundants are written in terms of the displacements that correspond to the redundants. The solution of these equations permits us to evaluate the redundants. Once the redundants are known, the balance of the analysis can be completed by using the equations of static equilibrium or by superposition.

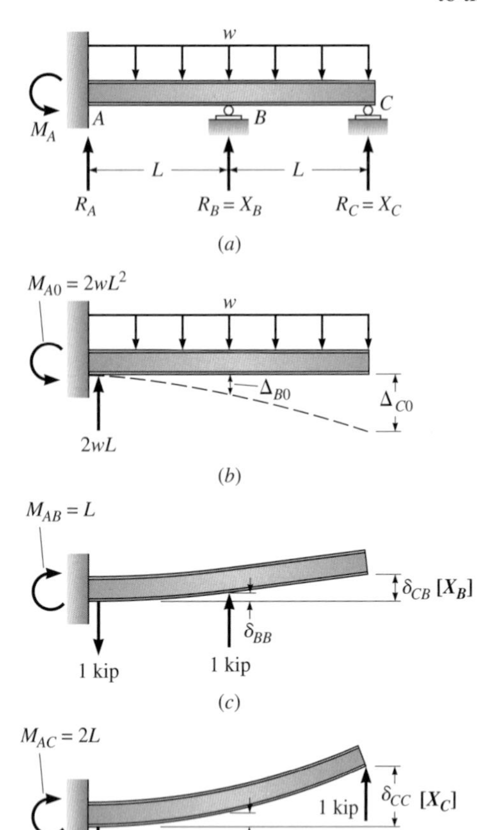

Figure 11.18: (*a*) Beam indeterminate to second degree with R_B and R_C selected as redundants; (*b*) deflections in released structure due to actual load; (*c*) deflection of released structure due to a unit value of the redundant at *B*; (*d*) deflection of released structure due to a unit value of the redundant at *C*.

To illustrate the method, we consider the analysis of the two-span continuous beam in Figure 11.18*a*. Since the reactions exert five restraints on the beam and only three equations of statics are available, the beam is indeterminate to the second degree. To produce a released structure (in this case a determinate cantilever fixed at *A*), we will select the reactions at supports *B* and *C* as the redundants. Since the supports do not move, the vertical deflection at both *B* and *C* must equal zero. Next, we divide the analysis of the beam into three cases, which will be superimposed. First, the released structure is analyzed for the applied loads (see Figure 11.18*b*). Then, separate analyses are carried out for each redundant (see Figure 11.18*c* and *d*). The effect of each redundant is determined by applying a unit value of the redundant to the released structure and then multiplying all forces and deflections it produces by the magnitude of the redundant. To indicate that the unit load is multiplied by the redundant, we show the redundant in brackets next to the sketch of the loaded member.

To evaluate the redundants, we next write compatibility equations at supports *B* and *C*. These equations state that the sum of the deflections at points *B* and *C* from the cases shown in Figure 11.18*b* to *d* must total zero. This requirement leads to the following compatibility equations:

$$\begin{aligned} \Delta_B &= 0 = \Delta_{B0} + X_B\delta_{BB} + X_C\delta_{BC} \\ \Delta_C &= 0 = \Delta_{C0} + X_B\delta_{CB} + X_C\delta_{CC} \end{aligned} \tag{11.11}$$

Once the numerical values of the six deflections are evaluated and substituted into Equations 11.11, the redundants can be determined. A small saving in computational effort can be realized by using the Maxwell-Betti law (see Section 10.9), which requires that $\delta_{CB} = \delta_{BC}$. As you can see, the magnitude of the computations increases rapidly as the degree of indeterminacy increases. For a structure that was indeterminate to the third degree, you would have to write three compatibility equations and evaluate 12 deflections (use of the Maxwell-Betti law would reduce the number of unknown deflections to nine).

EXAMPLE 11.9

Analyze the two-span continuous beam in Figure 11.19*a*, using the moments at supports A and B as the redundants; EI is constant. Loads on the beam act at midspan.

Solution

The released structure—two simply supported beams—is formed by inserting a hinge in the beam at B and replacing the fixed support at A by a pin. Two pointers, perpendicular to the beam's longitudinal axis, are attached to the beam at B. This device is used to clarify the rotation of the ends of the beam connecting to the hinge. The released structure, loaded with the applied loads and redundants, is shown in Figure 11.19*c*. The compatibility equations are based on the following conditions of geometry:

(*a*) The slope is zero at the fixed support at A.

$$\theta_A = 0 \tag{1}$$

(*b*) The slope of the beam is the same on either side of the center support (see Figure 11.19*b*). Equivalently, we can say that the relative rotation between the ends is zero (i.e., the pointers are parallel).

$$\theta_{B,\text{Rel}} = 0 \tag{2}$$

The released structure is analyzed for the applied loads in Figure 11.19*d*, a unit value of the redundant at A in Figure 11.19*e*, and a unit value of the redundant

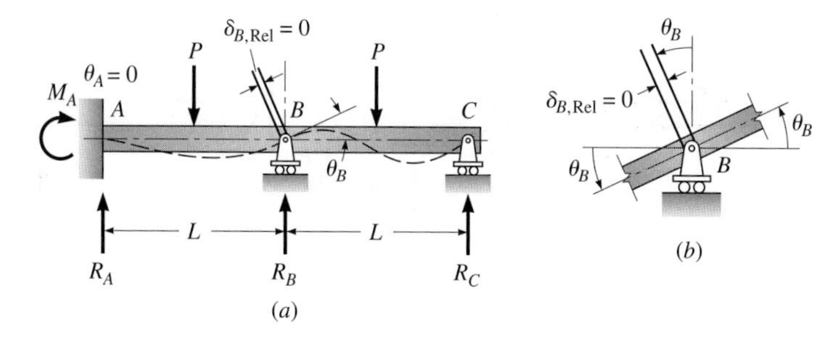

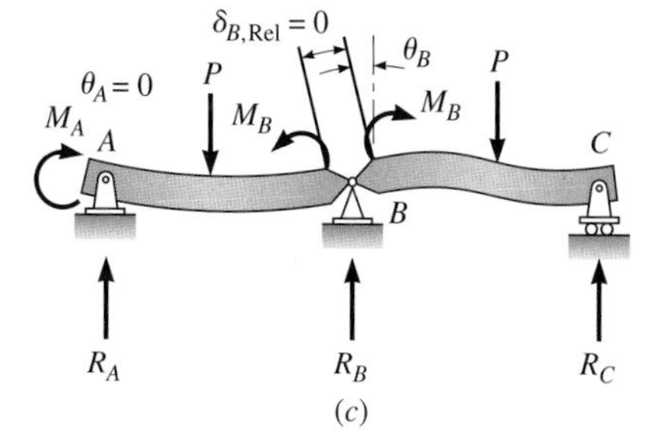

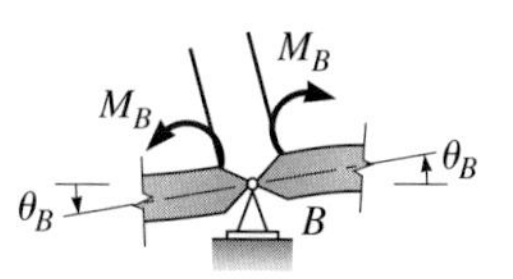

Figure 11.19: (*a*) Beam indeterminate to second degree; (*b*) detail of joint B showing the difference between the rotation of B and the relative rotation of the ends of members; (*c*) released structure with actual loads and redundants applied as external forces.

[*continues on next page*]

Example 11.9 continues . . .

at B in Figure 11.19f. Superimposing the angular deformations in accordance with the compatibility Equations 1 and 2, we can write

$$\theta_A = 0 = \theta_{A0} + \alpha_{AA}M_A + \alpha_{AB}M_B \quad (3)$$

$$\theta_{B,\mathrm{Rel}} = 0 = \theta_{B0} + \alpha_{BA}M_A + \alpha_{BB}M_B \quad (4)$$

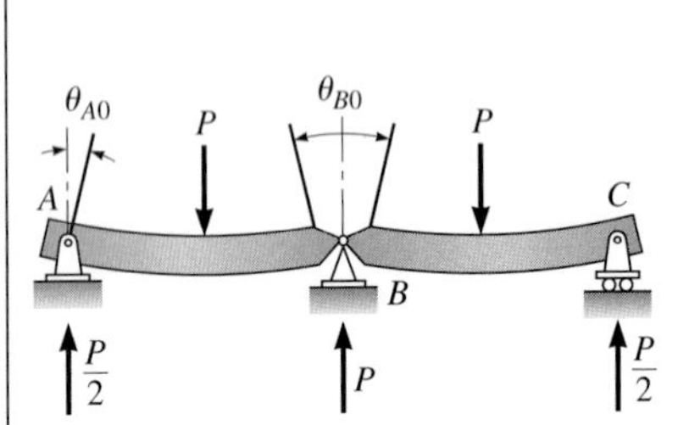

Using Figure 11.3d and e, evaluate the angular deformations.

$$\theta_{A0} = \frac{PL^2}{16EI} \qquad \theta_{B0} = 2\left(\frac{PL^2}{16EI}\right) \qquad \alpha_{AA} = \frac{L}{3EI}$$

$$\alpha_{BA} = \frac{L}{6EI} \qquad \alpha_{AB} = \frac{L}{6EI} \qquad \alpha_{BB} = 2\left(\frac{L}{3EI}\right)$$

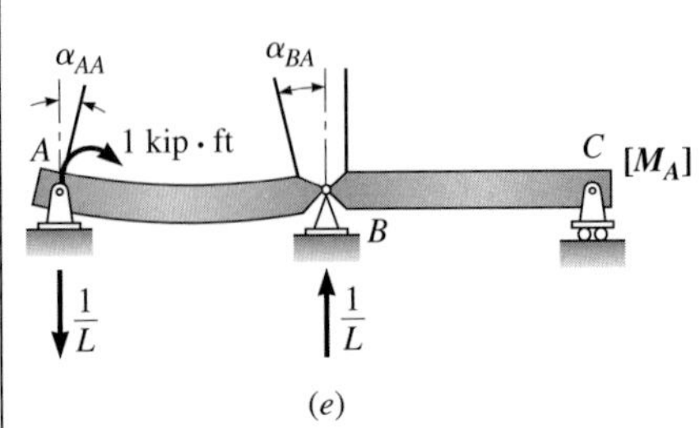

Substituting the angular displacements into Equations 3 and 4 and solving for the redundants give

$$M_A = -\frac{3PL}{28} \qquad M_B = -\frac{9PL}{56} \qquad \textbf{Ans.}$$

The minus signs indicate that the actual directions of the redundants are opposite in sense to those initially assumed in Figure 11.19c. Figure 11.20 shows the free-body diagrams of the beams used to evaluate the end shears and also the final shear and moment curves.

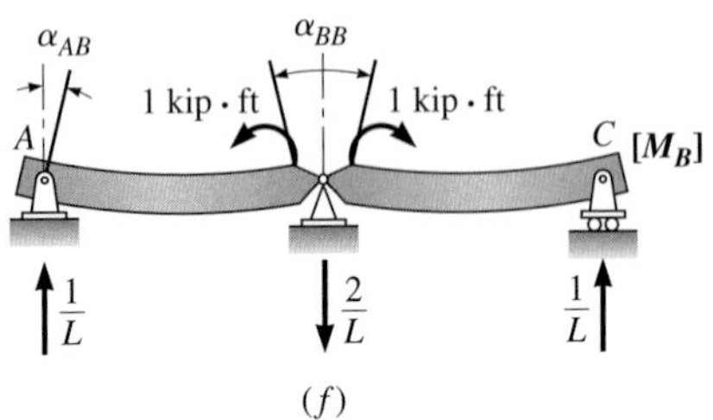

Figure 11.19: (d) Actual loads applied to released structure; (e) unit value of redundant at A applied to released structure; (f) unit value of redundant at B applied to released structure.

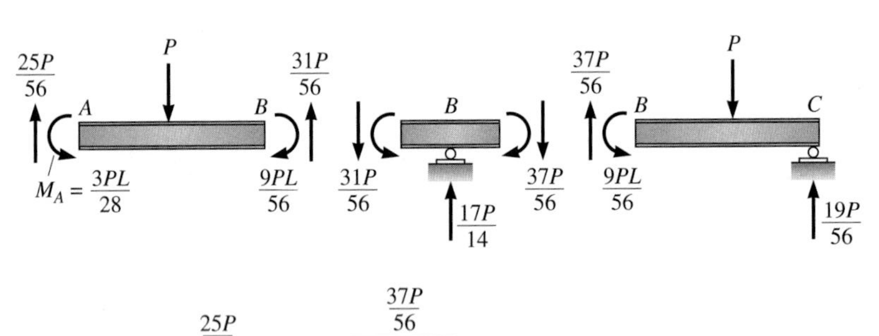

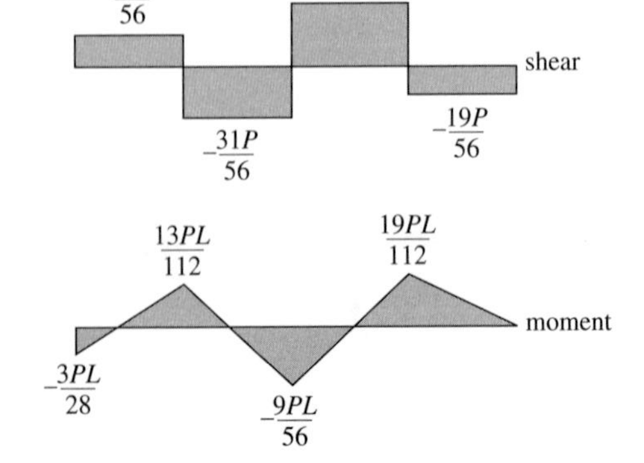

Figure 11.20: Free-body diagrams of beams used to evaluate shears as well as the shear and moment diagrams.

EXAMPLE 11.10

Determine the bar forces and reactions that develop in the indeterminate truss shown in Figure 11.21*a*.

Solution

Since $b + r = 10$ and $2n = 8$, the truss is indeterminate to the second degree. Select the force F_{AC} at section 1-1 and the horizontal reaction B_x as the redundants.

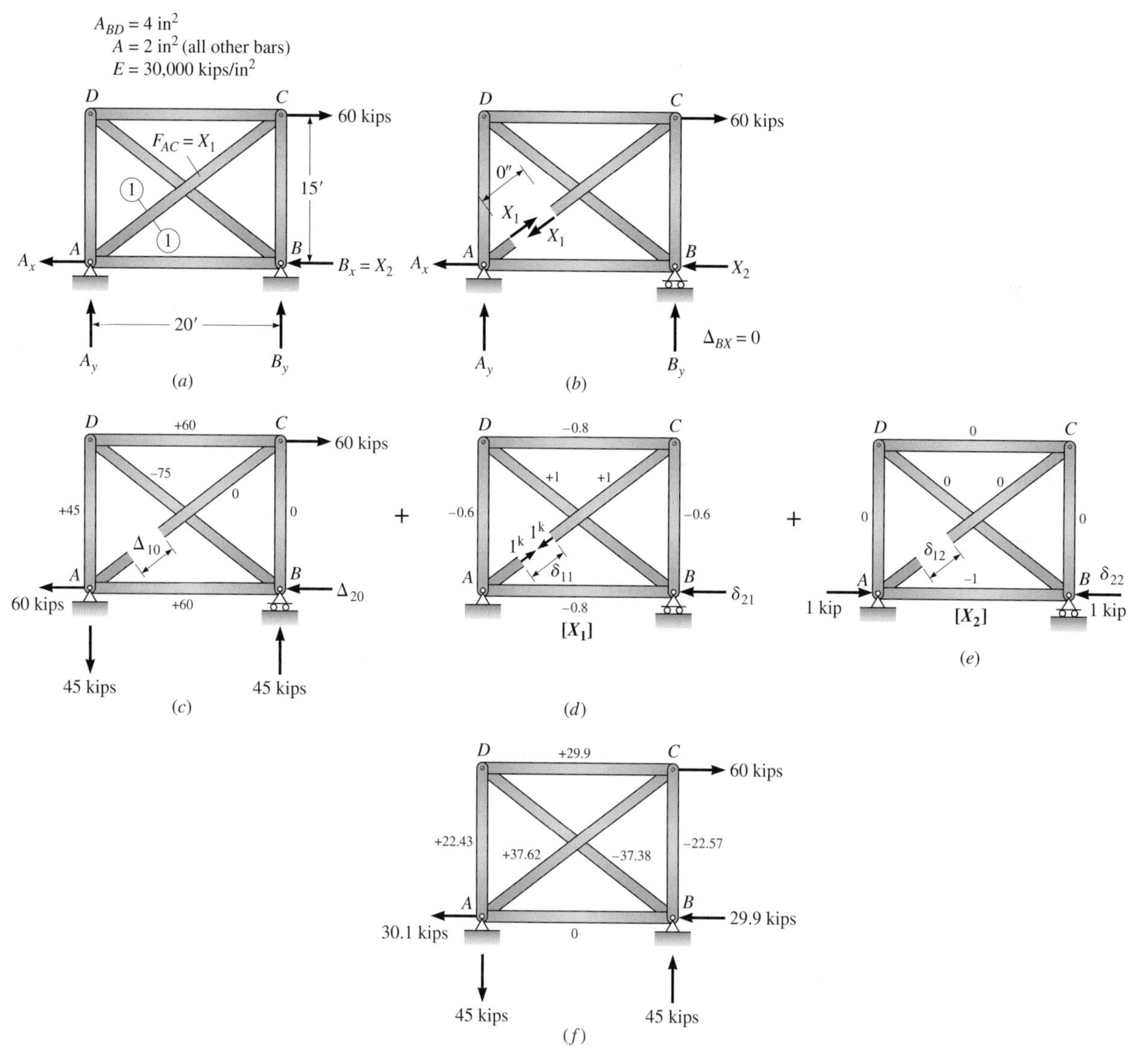

Figure 11.21: (*a*) Details of truss; (*b*) released structure loaded by redundants X_1 and X_2 and 60-kip load; (*c*) released structure with actual load; (*d*) released structure–forces and displacements due to a unit value of redundant X_1; (*e*) released structure–forces and displacements due to a unit value of redundant X_2; (*f*) final forces and reactions = (c) + X_1(*d*) + X_2(*e*).

[*continues on next page*]

Example 11.10 continues . . .

The released structure with the redundants applied as loads is shown in Figure 11.21*b*.

The compatibility equations are based on (1) no horizontal displacement at *B*

$$\Delta_{BX} = 0 \tag{1}$$

and (2) no relative displacement of the ends of bars at section 1-1

$$\Delta_{1,\text{Rel}} = 0 \tag{2}$$

Superimposing deflections at Section 1-1 and support *B* in the released structure (see Figure 11.21*c* to *e*), we can write the compatibility equations as

$$\Delta_{1,\text{Rel}} = 0: \qquad \Delta_{10} + X_1\delta_{11} + X_2\delta_{12} = 0 \tag{3}$$

$$\Delta_{BX} = 0: \qquad \Delta_{20} + X_1\delta_{21} + X_2\delta_{22} = 0 \tag{4}$$

To complete the solution, we must compute the six deflections Δ_{10}, Δ_{20}, δ_{11}, δ_{12}, δ_{21}, and δ_{22} in Equations 3 and 4 by virtual work.

Δ_{10}:

Use the force system in Figure 11.21*d* as the *Q* system.

$$\sum \delta_P Q = \sum F_Q \frac{F_P L}{AE} \tag{10.24}$$

$$1\text{ kip}(\Delta_{10}) = (-0.8)\frac{60(20 \times 12)}{2(30{,}000)}(2) + (-0.6)\frac{45(15 \times 12)}{2(30{,}000)}$$

$$+ (1)\frac{-75(25 \times 12)}{4(30{,}000)}$$

$$\Delta_{10} = -0.6525 \text{ in} \qquad \text{(gap opens)}$$

Δ_{20}:

Use the force system in Figure 11.21*e* as the *Q* system for the *P* system shown in Figure 11.21*c*.

$$1\text{ kip}(\Delta_{20}) = (-1)\frac{60(20 \times 12)}{2(30{,}000)}$$

$$\Delta_{20} = -0.24 \text{ in} \rightarrow$$

δ_{11}:

The force system in Figure 11.21*d* serves as both the *P* and *Q* systems. Since $F_Q = F_P, U_Q = F_Q^2 L/(AE)$,

$$1 \text{ kip}(\delta_{11}) = \frac{(-0.8)^2(20 \times 12)}{2(30{,}000)}(2) + \frac{(-0.6)^2(15 \times 12)}{2(30{,}000)}(2) + \frac{1^2(25 \times 12)}{2(30{,}000)} + \frac{1^2(25 \times 12)}{4(30{,}000)}$$

$$\delta_{11} = +\,0.0148 \text{ in} \qquad \text{(gap closes)}$$

δ_{12}:

Use the force system in Figure 11.21*d* as the *Q* system for the *P* system in Figure 11.21*e*.

$$1 \text{ kip}(\delta_{12}) = (-0.8)\frac{-1(20 \times 12)}{2(30{,}000)}$$

$$\delta_{12} = 0.0032 \text{ in}$$

δ_{21}:

Use the force system in Figure 11.21*e* as the *Q* system for the *P* system in Figure 11.21*d*.

$$1 \text{ kip}(\delta_{21}) = (-1)\frac{-0.8(20 \times 12)}{2(30{,}000)}$$

$$\delta_{21} = 0.0032 \text{ in}$$

(Alternately, use the Maxwell-Betti law, which gives $\delta_{21} = \delta_{12} = 0.0032$ in.)

δ_{22}:

The force system in Figure 11.21*e* serves as both the *P* and *Q* systems.

$$1 \text{ kip}(\delta_{22}) = (-1)\frac{(-1)(20 \times 12)}{2(30{,}000)}$$

$$\delta_{22} = 0.004 \text{ in}$$

Substituting the displacements above into Equations 3 and 4 and solving for X_1 and X_2 give

$$X_1 = 37.62 \text{ kips} \qquad X_2 = 29.9 \text{ kips} \qquad \textbf{Ans.}$$

The final forces and reactions are shown in Figure 11.21*f*.

EXAMPLE 11.11

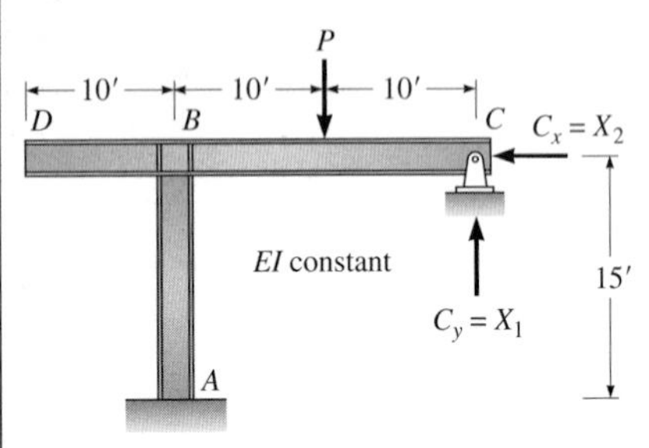

(*a*)

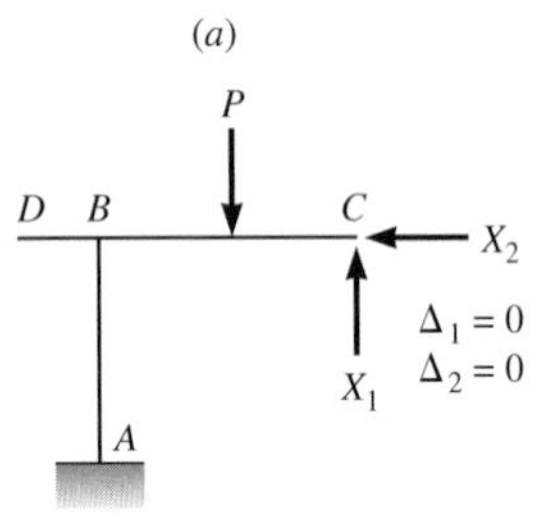

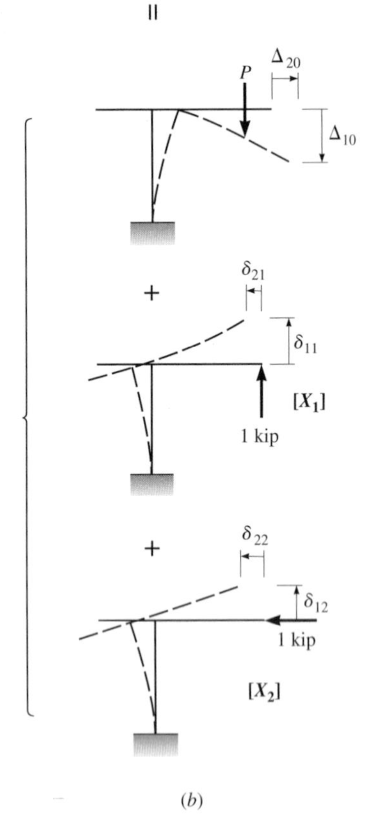

(*b*)

(*a*) Choose the horizontal and vertical reactions at *C* (Figure 11.22*a*) as redundants. Draw all the released structures, and clearly label all displacements needed to write the equations of compatibility. Write the equations of compatibility in terms of displacements, but *do not calculate* the values of displacement.

(*b*) Modify the equations in part (*a*) to account for the following support movements: 0.5-in vertically upward displacement of *C* and 0.002-rad clockwise rotation of *A*.

Solution

(*a*) As a sign convention, displacements in the direction of redundants in Figure 11.22(*a*) are positive. See Figure 11.22*b*; note that sign is contained within the symbol for displacements.

$$\Delta_1 = 0 = \Delta_{10} + \delta_{11}X_1 + \delta_{12}X_2$$
$$\Delta_2 = 0 = \Delta_{20} + \delta_{21}X_1 + \delta_{22}X_2 \qquad \textbf{Ans.}$$

where 1 denotes the vertical and 2 the horizontal direction at *C*.

(*b*) Modify compatibility equation for support movements. See Figure 11.22*c*.

$$\Delta_1 = 0.5 = \Delta_{10} + (-0.48) + \delta_{11}X_1 + \delta_{12}X_2$$
$$\Delta_2 = 0 = \Delta_{20} + (-0.36) + \delta_{21}X_1 + \delta_{22}X_2 \qquad \textbf{Ans.}$$

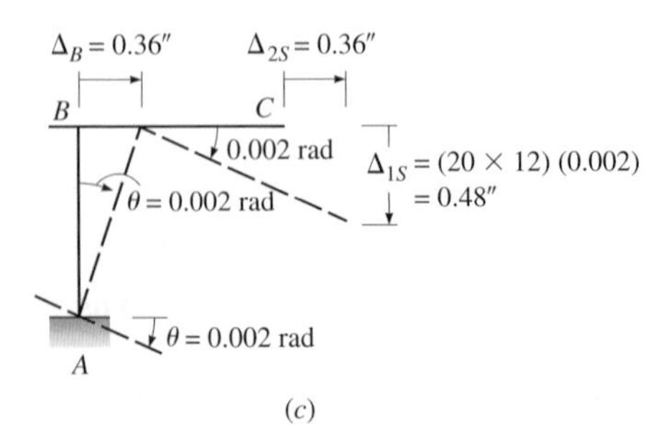

(*c*)

Figure 11.22: (*c*) Displacements produced by clockwise rotation of support *A*.

11.8 Beam on Elastic Supports

The supports of certain structures deform when they are loaded. For example, in Figure 11.23*a* the support for the right end of girder *AB* is beam *CD*, which deflects when it picks up the end reaction from beam *AB*. If beam *CD* behaves elastically, it can be idealized as a spring (see Figure 11.23*b*). For the spring the relationship between the applied load *P* and the deflection Δ is given as

$$P = K\Delta \tag{11.12}$$

where *K* is the stiffness of the spring in units of force per unit displacement. For example, if a 2-kip force produces a 0.5-in deflection of the spring, $K = P/\Delta = 2/0.5 = 4$ kips/in. Solving Equation 11.12 for Δ gives

$$\Delta = \frac{P}{K} \tag{11.13}$$

The procedure to analyze a beam on an elastic support is similar to that for a beam on an unyielding support, with one difference. If the force *X* in the spring is taken as the redundant, the *compatibility equation* must state that the deflection Δ of the beam at the location of the redundant equals

$$\Delta = -\frac{X}{K} \tag{11.14}$$

The minus sign accounts for the fact that the deformation of the spring is opposite in sense to the force it exerts on the member it supports. For example, if a spring is compressed, it exerts an upward force but displaces downward. If the spring stiffness is large, Equation 11.14 shows that the deflection Δ will be small. In the limit, as *K* approaches infinity, the right side of Equation 11.14 approaches zero and Equation 11.14 becomes identical to the compatibility equation for a beam on a simple support. We will illustrate the use of Equation 11.14 in Example 11.12.

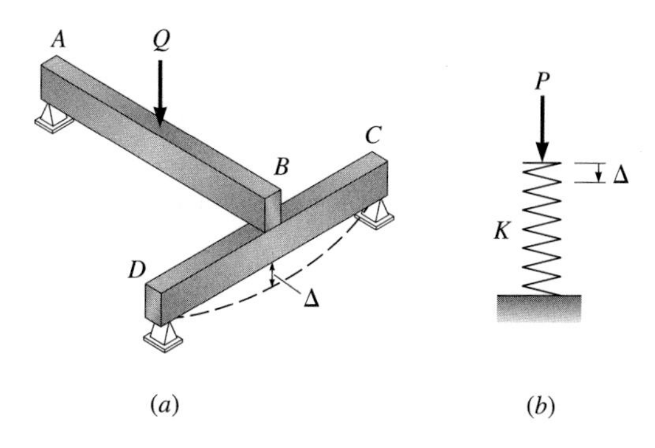

Figure 11.23: (*a*) Beam *AB* with an elastic support at *B*; (*b*) elastic support idealized as a linear elastic spring ($P = K\Delta$).

EXAMPLE 11.12

Set up the compatibility equation for the beam in Figure 11.24*a*. Determine the deflection of point *B*. The spring stiffness $K = 10$ kips/in, $w = 2$ kips/ft, $I = 288$ in^4, and $E = 30{,}000$ kips/in^2.

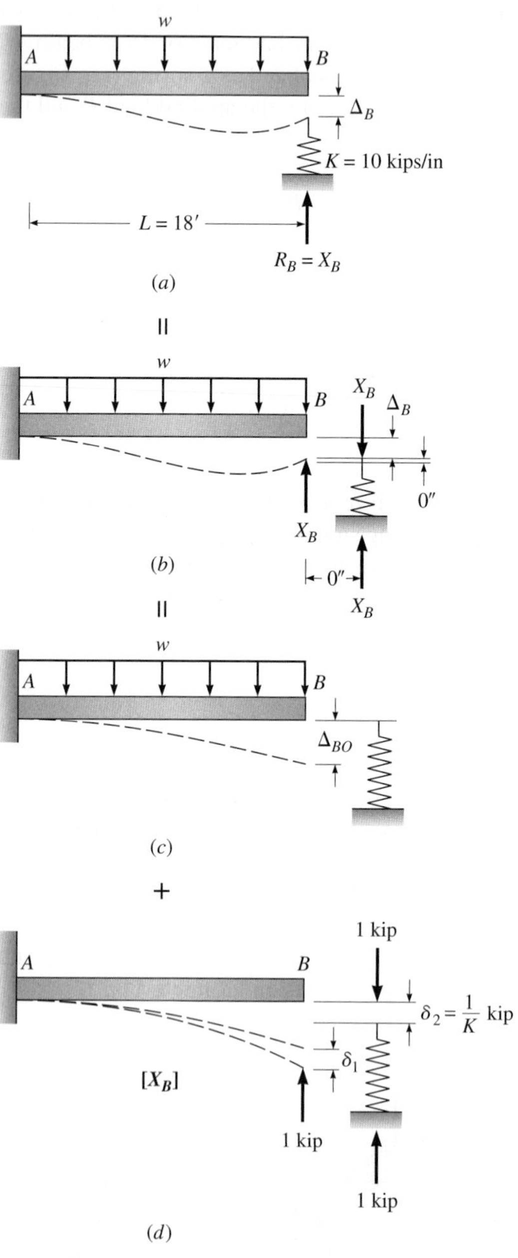

Figure 11.24: (*a*) Uniformly loaded beam on an elastic support, indeterminate to the first degree; (*b*) released structure with uniform load and redundant X_B applied as an external load to both the beam and the spring; (*c*) released structure with actual load; (*d*) released structure, forces and displacements by a unit value of the redundant X_B.

Solution

Figure 11.24*b* shows the released structure loaded with the applied load and the redundant. For clarity the spring is displaced laterally to the right, but the displacement is labeled zero to indicate that the spring is actually located directly under the tip of the beam. Following the previously established sign convention (i.e., the direction of the redundant establishes the positive direction for the displacements), displacements of the right end of the beam are positive when up and negative when down. Deflection of the spring is positive downward. Because the tip of the beam and the spring are connected, they both deflect the same amount Δ_B, that is,

$$\Delta_{B,\text{beam}} = \Delta_{B,\text{spring}} \tag{1}$$

Using Equation 11.13, we can write Δ_B of the spring as

$$\Delta_{B,\text{spring}} = \frac{X_B}{K} \tag{2}$$

and substituting Equation 2 into Equation 1 gives us

$$\Delta_{B,\text{beam}} = -\frac{X_B}{K} \tag{3}$$

The minus sign is added to the right side of Equation 3 because the end of the beam displaces downward.

If $\Delta_{B,\text{beam}}$ (the left side of Equation 3) is evaluated by superimposing the displacements of the *B* end of the beam in Figure 11.24*c* and *d*, we can write Equation 3 as

$$\Delta_{B0} + \delta_1 X_B = -\frac{X_B}{K} \tag{4}$$

Using Figure 11.3 to evaluate Δ_{B0} and δ_1 in Equation 4, we compute X_B:

$$-\frac{wL^4}{8EI} + \frac{L^3}{3EI}X_B = -\frac{X_B}{K}$$

Substituting the specified values of the variables into the equation above, we obtain

$$-\frac{2(18)^4(1728)}{8(30{,}000)(288)} + \frac{(18)^3(1728)}{3(30{,}000)(288)}X_B = -\frac{X_B}{10}$$

$$X_B = 10.71 \text{ kips}$$

If support *B* had been a roller and no settlement had occurred, the right side of Equation 4 would equal zero and X_B would increase to 13.46 kips

and $$\Delta_{B,\text{spring}} = -\frac{X_B}{K} = -\frac{10.71}{10} = 1.071 \text{ in} \qquad \textbf{Ans.}$$

Summary

- The flexibility method of analysis, also called the method of *consistent deformations*, is one of the oldest classical methods of analyzing indeterminate structures.
- Before the development of general-purpose computer programs for structural analysis, the flexibility method was the only method available for analyzing indeterminate trusses. The flexibility method is based on removing restraints until a stable determinate released structure is established. Since the engineer has alternate choices with respect to which restraints to remove, this aspect of the analysis does not lend itself to the development of a general-purpose computer program.
- The flexibility method is still used to analyze certain types of structures in which the general configuration and components of the structure are standardized but the dimensions vary. For this case the restraints to be removed are established, and the computer program is written for their specific value.

PROBLEMS

P11.1. Compute the reactions, draw the shear and moment curves, and locate the point of maximum deflection for the beam in Figure P11.1. *EI* is constant.

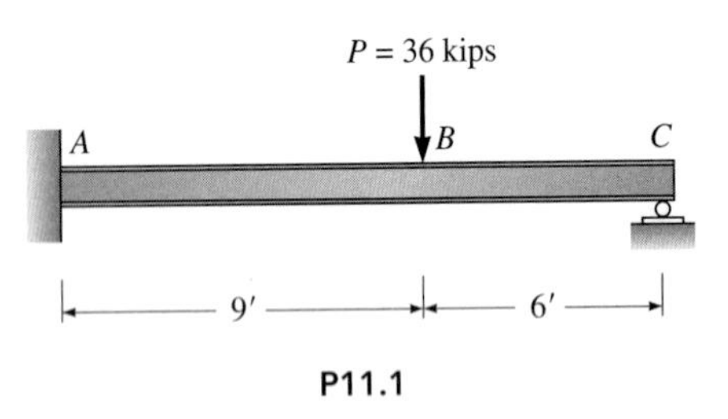

P11.1

P11.2. For the beam in Figure P11.2, compute the reactions, drawn the shear and moment curves, and compute the deflection of the hinge at *C*. $E = 29{,}000$ ksi and $I = 180$ in^4.

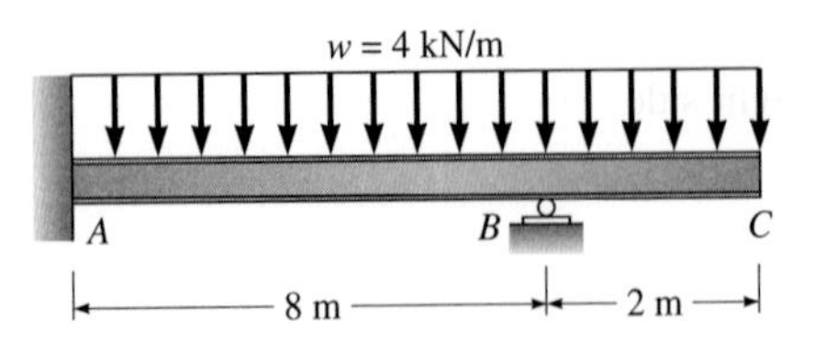

P11.2

P11.3. Compute the reactions and draw the shear and moment curves for the beam in Figure P11.3. *EI* is constant.

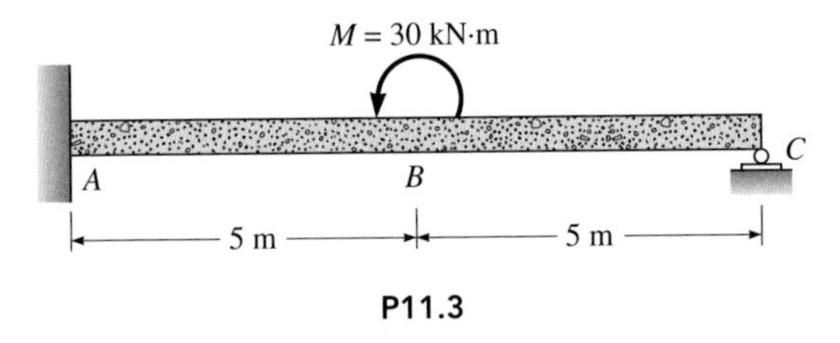

P11.3

P11.4. Compute the reactions and draw the shear and moment curves for the beam in Figure P11.4. *EI* is constant.

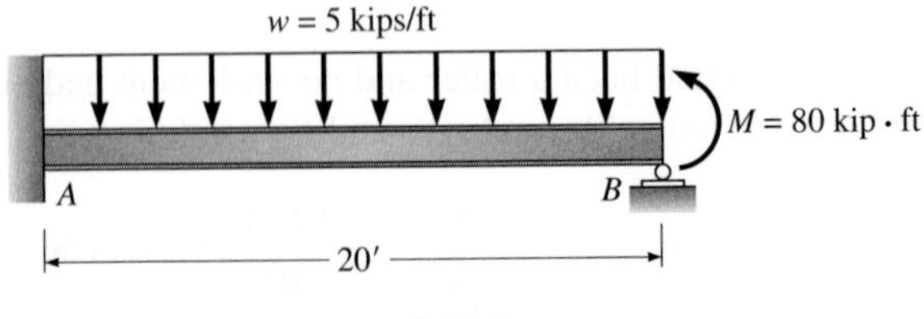

P11.4

P11.5. Compute the reactions, draw the shear and moment curves, and locate the point of maximum deflection for the beam in Figure P11.5. Repeat the computation if I is constant over the entire length. E is constant. Express answer in terms of E, I, and L.

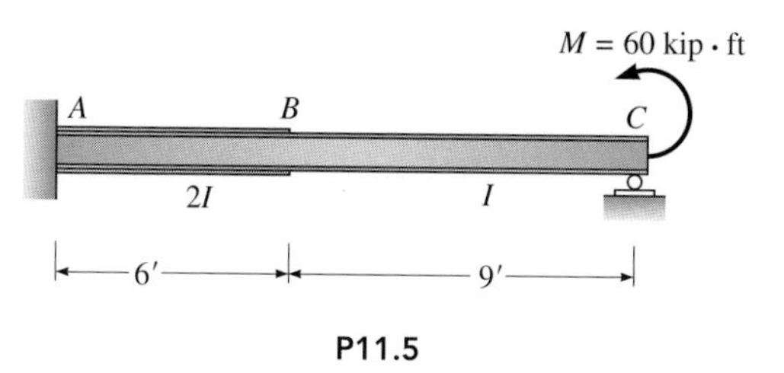

P11.5

P11.6. Compute the reactions and draw the shear and moment curves for the beam in Figure P11.6. EI is constant.

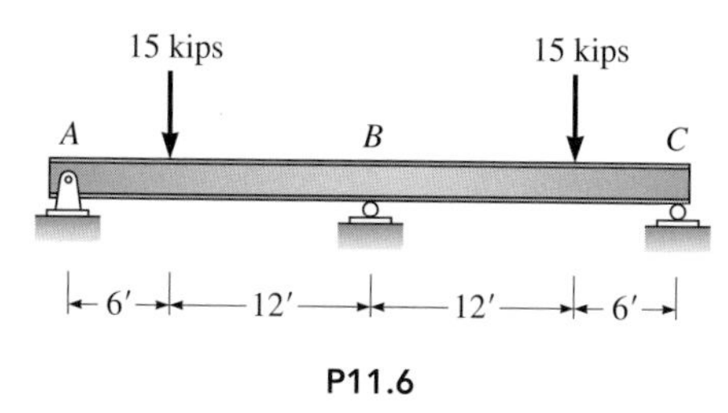

P11.6

P11.7. Determine the reactions for the beam in Figure P11.7. When the uniform load is applied, the fixed support rotates clockwise 0.003 rad and support B settles 0.3 in. Given: $E = 30{,}000$ kips/in^2 and $I = 240$ in^4.

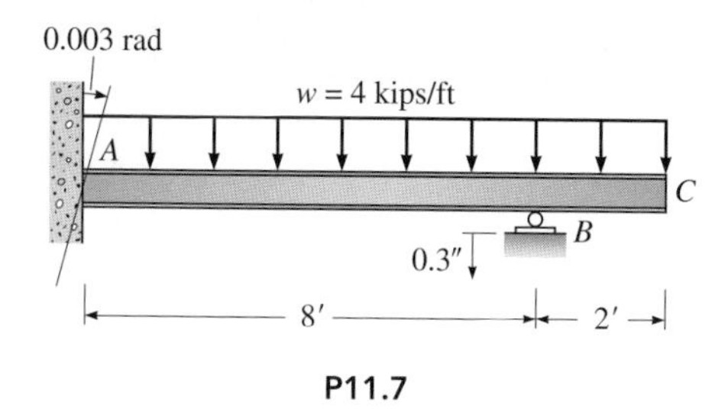

P11.7

P11.8. Solve Problem P11.1 for the loading shown if support C settles 0.25 in when the load is applied. $E = 30{,}000$ kips/in^2 and $I = 320$ in^4.

P11.9. Assuming that no load acts, compute the reactions and draw the shear and moment curves for the beam in Figure P11.1 if support A settles 0.5 in and support C settles 0.75 in. Given: $E = 29{,}000$ kips/in^2 and $I = 150$ in^4.

P11.10. Compute the reactions and draw the shear and moment curves for the beam in Figure P11.10. E is constant.

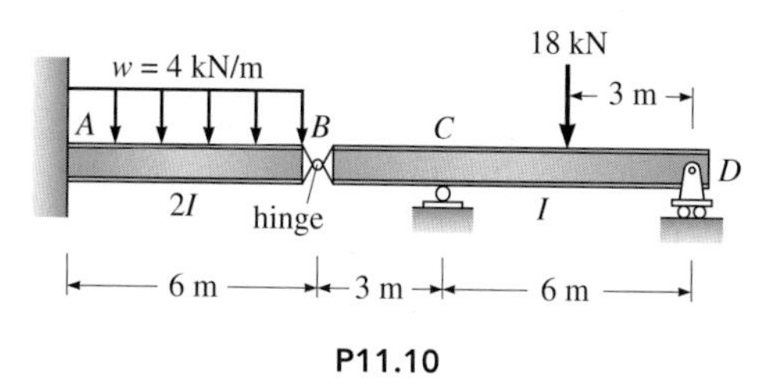

P11.10

P11.11. Compute the reactions and draw the shear and moment curves for the beam in Figure P11.11. EI is constant. The bolted web connection at B may be assumed to act as a hinge. Use the shear at hinge B as the redundant. Express answer in terms of E, I, L, and w.

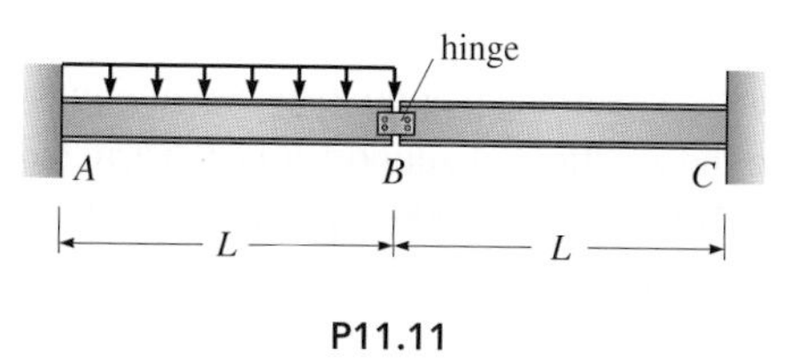

P11.11

P11.12. (*a*) Determine the reactions and draw the shear and moment curves for the beam in Figure P11.12. Given: EI is constant, $E = 30{,}000$ kips/in^2, and $I = 288$ in^4. (*b*) Repeat the computations if, in addition to the applied loads, support B settles 0.5 in and support D settles 1 in.

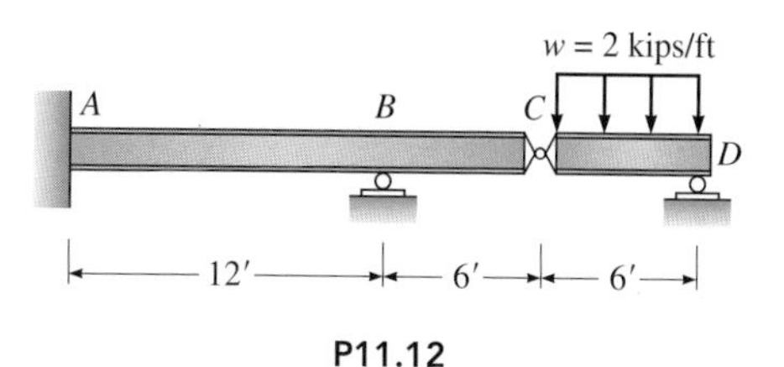

P11.12

P11.13. (*a*) Compute all reactions for the beam in Figure P11.13 assuming that the supports do not move; *EI* is constant. (*b*) Repeat computations given that support *C* moves upward a distance of $288/(EI)$ when the load is applied.

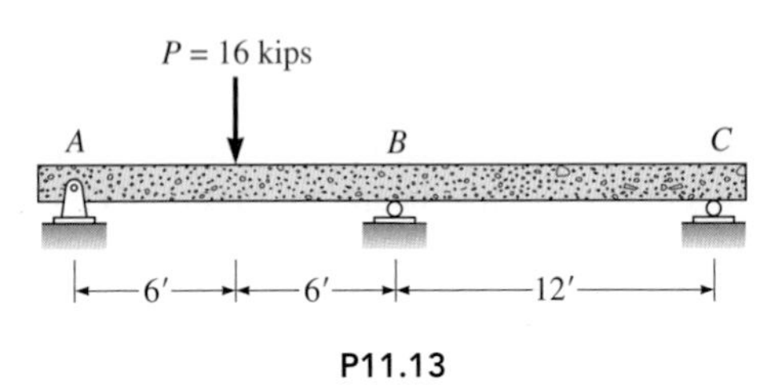

P11.13

P11.14. Determine all reactions and draw the shear and moment curves for the beam in Figure P11.14. *EI* is constant.

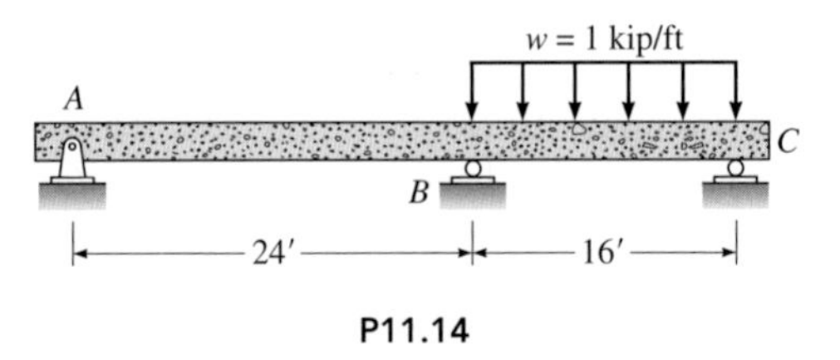

P11.14

P11.15. (*a*) Assuming that no loads act in Figure P11.14, compute the reactions if support *B* is constructed 0.48 in too low. Given: $E = 29{,}000$ kips/in^2, $I = 300$ in^4. (*b*) If support *B* settles $\frac{3}{2}$ in under the applied loads, compute the reactions.

P11.16. Compute the reactions and draw the shear and moment curves for the beam in Figure P11.16. Given: *EI* is constant. Take advantage of symmetry and use the end moment as the redundant.

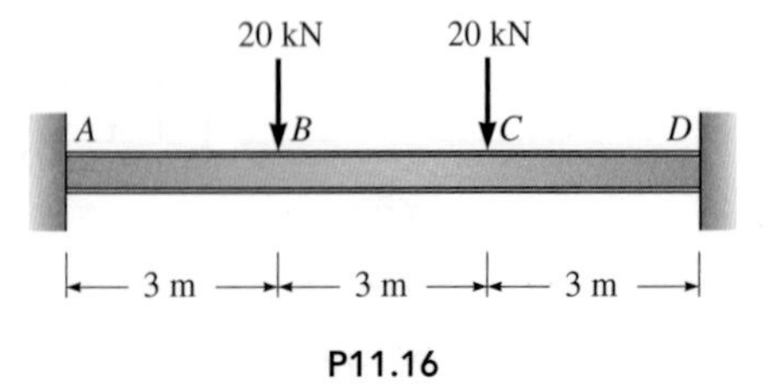

P11.16

P11.17. Compute the reactions and draw the shear and moment curves for the beam in Figure P11.17. Given: *EI* is constant. Use the reactions at *B* as the redundants.

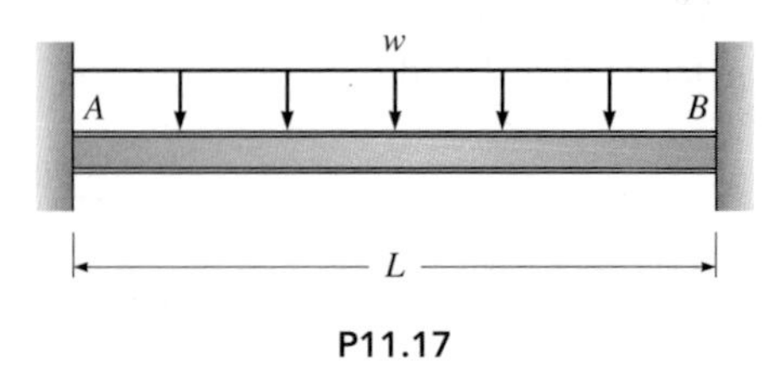

P11.17

P11.18. Compute the reactions and draw the shear and moment curves for the beam in Figure P11.18. Given: *EI* is constant for the beam. $E = 200$ GPa, $I = 40 \times 10^6$ mm^4.

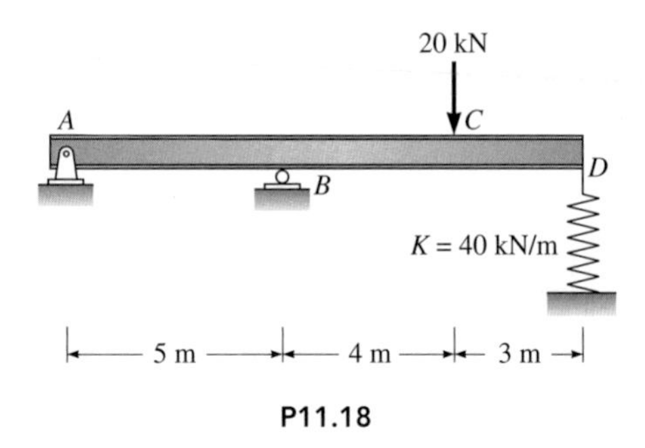

P11.18

P11.19. Compute the reactions and draw the shear and moment curves for the beam in Figure P11.19. In addition to the applied load, the support at *D* settles by 0.1 m. *EI* is constant for the beam. $E = 200$ GPa, $I = 60 \times 10^6$ mm^4.

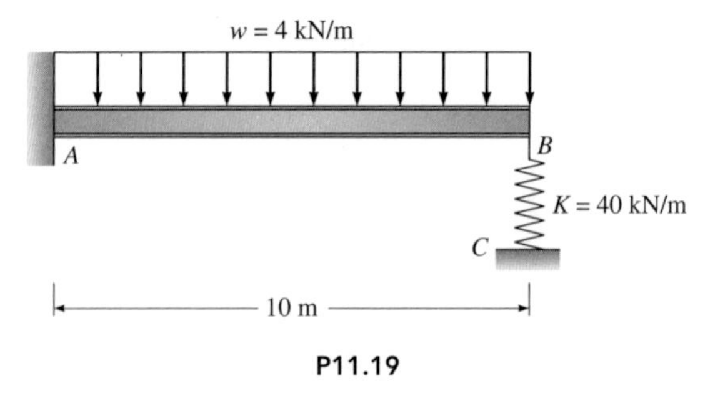

P11.19

P11.20. Consider the beam in Figure P11.19 without the applied load and support settlement. Compute the reactions and draw the shear and moment curves for the beam if support *A* rotates clockwise by 0.005 rad.

P11.21. Compute the reactions and bar forces in all members for the truss in Figure P11.21. The area of all bars is 5 in^2 and $E = 30{,}000$ kips/in^2.

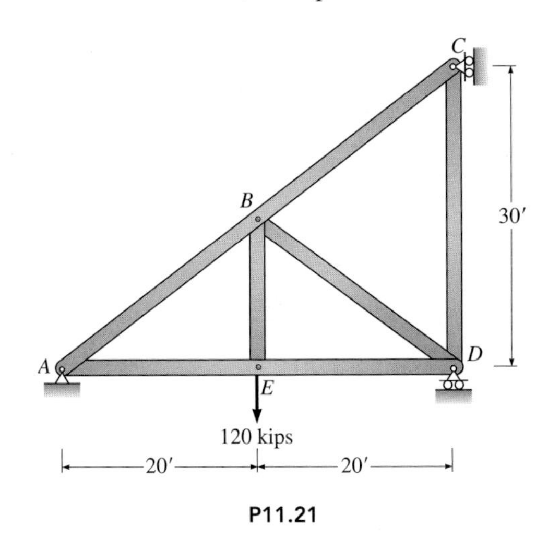

P11.21

P11.22. Assuming that the 120-kip load is removed from the truss in Figure P11.21, compute the reactions and bar forces if the temperature of bars *AB* and *BC* increases 60°F; the coefficient of temperature expansion $\alpha = 6 \times 10^{-6}$ (in/in)/°F.

P11.23 to **P11.25.** For the trusses in Figures P11.23 through P11.25, compute the reactions and bar forces produced by the applied loads. Given: AE = constant, A = 1000 mm^2, and E = 200 GPa.

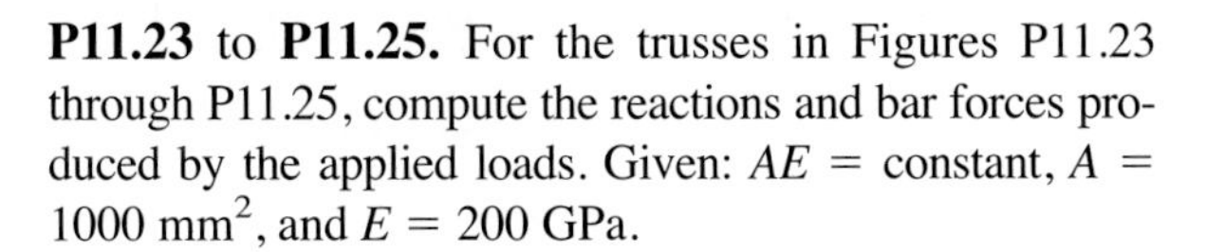

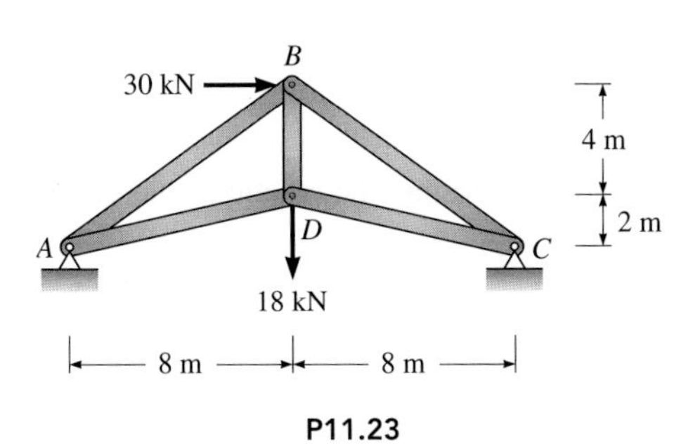

P11.23

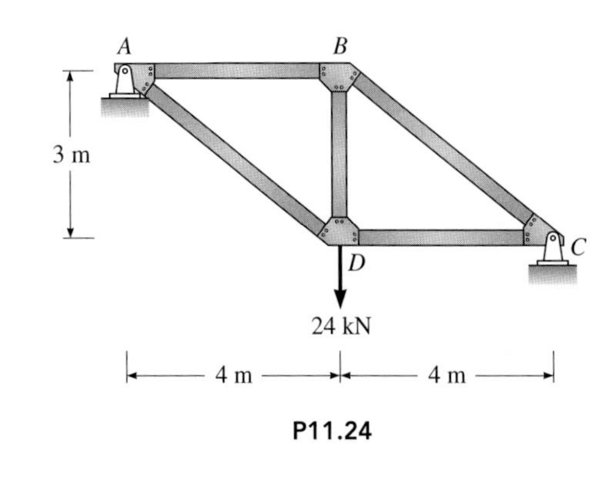

P11.24

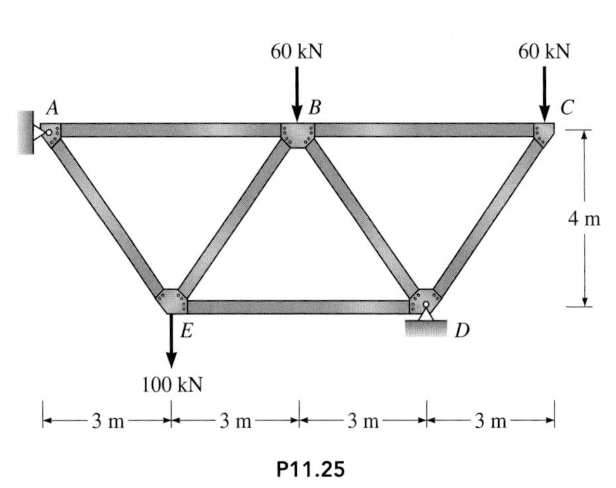

P11.25

P11.26. Determine the reactions and bar forces that are created in the truss in Figure P11.26 when the top chords (*ABCD*) are subjected to a 50°F temperature increase. Given: AE is constant for all bars, A = 10 in^2, E = 30,000 kips/in^2, $\alpha = 6.5 \times 10^{-6}$ (in/in)/°F.

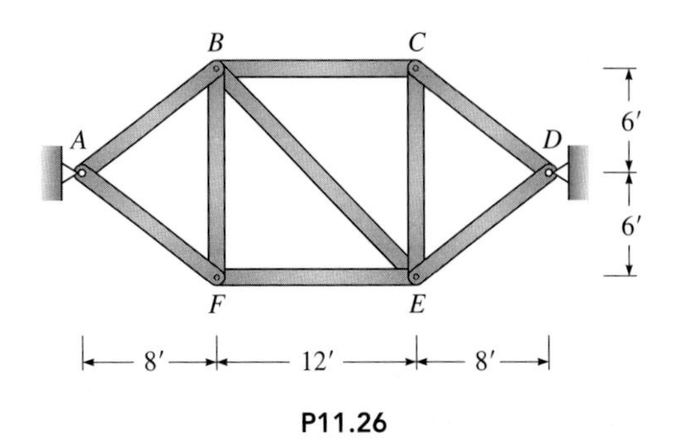

P11.26

P11.27. Determine the vertical and horizontal displacements at A of the pin-connected structure in Figure P11.27. Given: $E = 200$ GPa and $A = 500\ \text{mm}^2$ for all members.

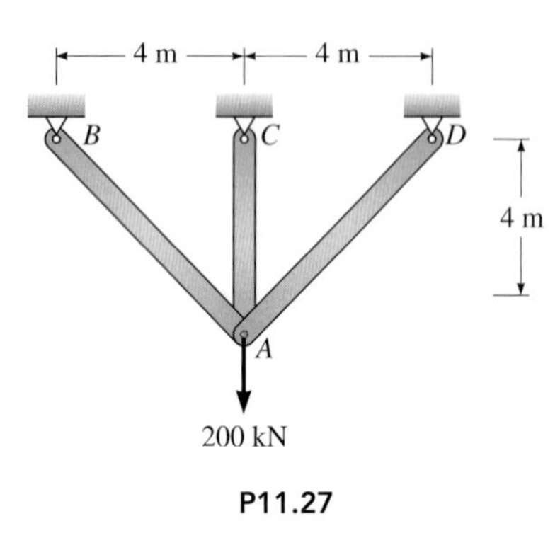

P11.27

P11.28. Determine the vertical and horizontal displacements at A of the pin-connected structure in Figure P11.27. Given: $E = 200$ GPa, $A_{AB} = 1000\ \text{mm}^2$, and $A_{AC} = A_{AD} = 500\ \text{mm}^2$.

P11.29. (*a*) Determine all reactions and bar forces produced by the applied load in Figure P11.29. (*b*) If support B settles 1 in and support C settles 0.5 in while the load acts, recompute the reactions and bar forces. For all bars the area $= 2\ \text{in}^2$ and $E = 30{,}000\ \text{kips/in}^2$.

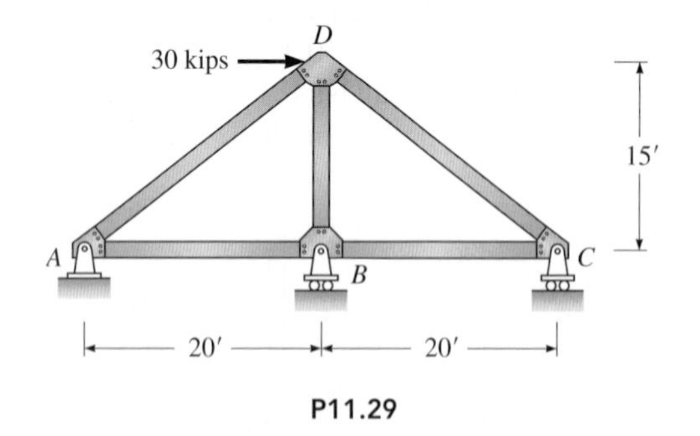

P11.29

P11.30. Determine the reactions and all bar forces for the truss in Figure P11.30. $E = 200$ GPa and $A = 1000\ \text{mm}^2$ for all bars.

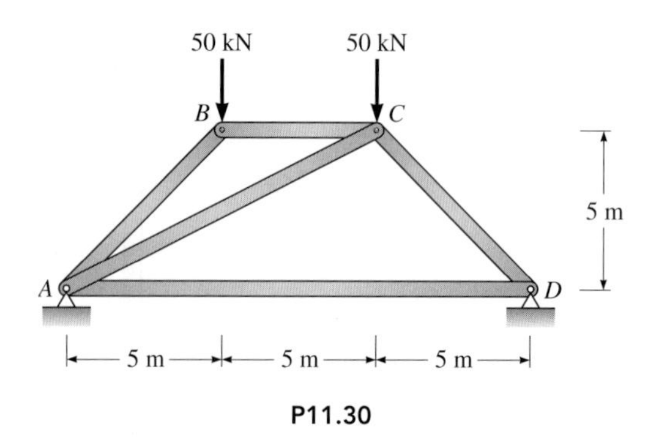

P11.30

P11.31. Consider the truss in Figure P11.30 without the applied loads. Determine the reactions and all bar forces for the truss if member AC is fabricated 10 mm too short.

P11.32. Determine the reactions and all bar forces for the truss in Figure P11.32. Given: $E = 200$ GPa and $A = 1000\ \text{mm}^2$ for all bars.

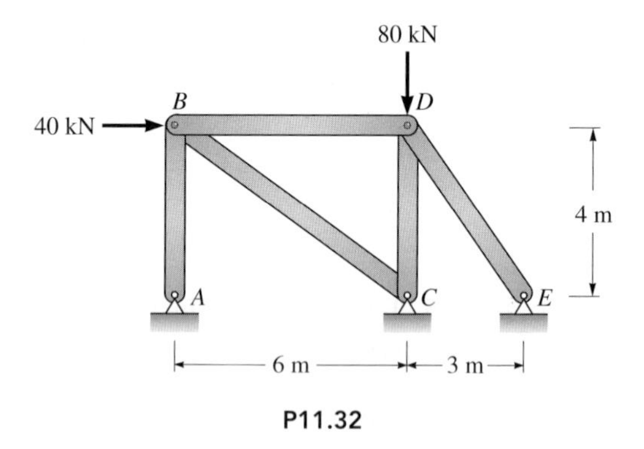

P11.32

P11.33. Consider the truss in Figure P11.32 without the applied loads. Determine the reactions and all bar forces for the truss if support A settles vertically 20 mm.

P11.34. Determine all bar forces and reactions for the truss in Figure P11.34. Given: area of bar $BD = 4\text{ in}^2$, all other bars $= 2\text{ in}^2$, and $E = 30{,}000\text{ kips/in}^2$.

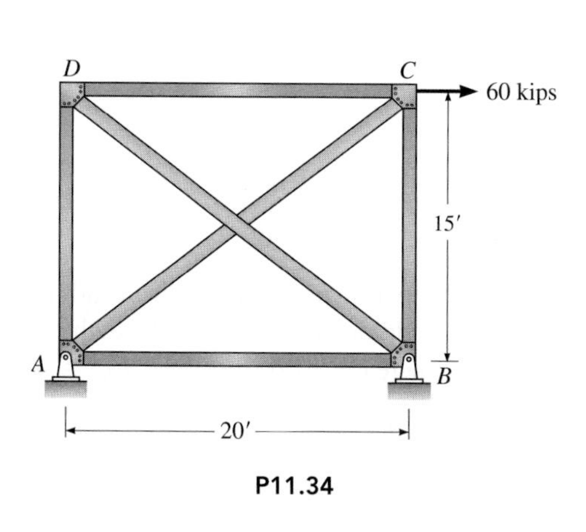

P11.34

P11.35. Determine the reactions at A and C in Figure P11.35. EI is constant for all members.

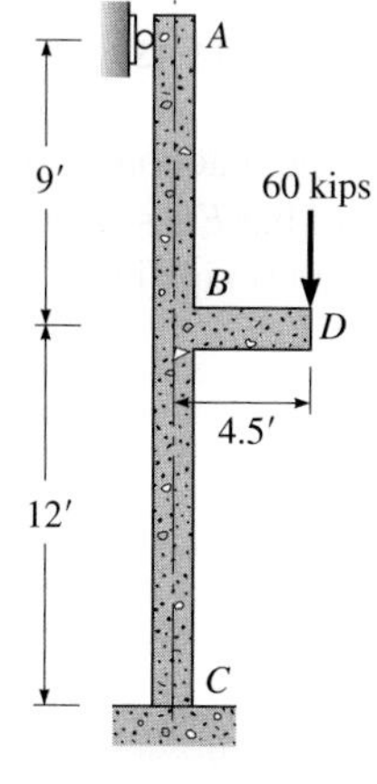

P11.35

P11.36. Determine all reactions for the frame in Figure P11.36 given $I_{AB} = 600\text{ in}^4$, $I_{BC} = 900\text{ in}^4$, and $E = 29{,}000\text{ kips/in}^2$. Neglect axial deformations.

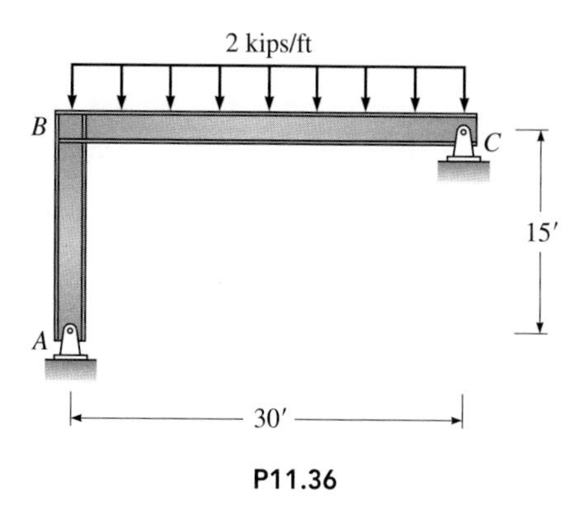

P11.36

P11.37. Assuming that the load is removed, compute all reactions for the frame in Figure P11.36 if member BC is fabricated 1.2 in too long.

P11.38. (*a*) Compute the reaction at C in Figure P11.38. EI is constant. (*b*) Compute the vertical deflection of joint B.

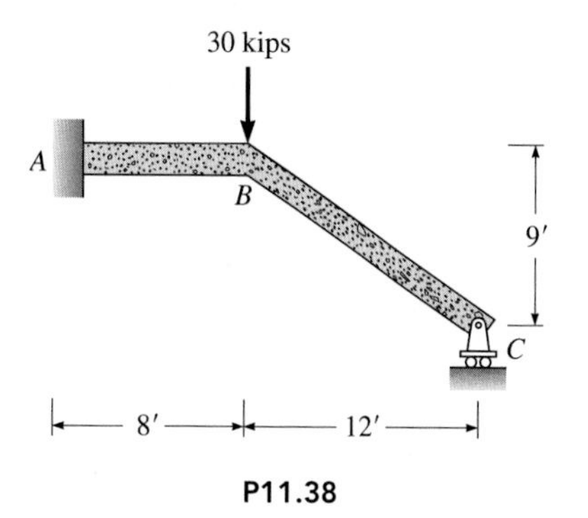

P11.38

P11.39. Determine all reactions and draw the shear and moment diagrams for beam *BC* in Figure P11.39. *EI* is constant.

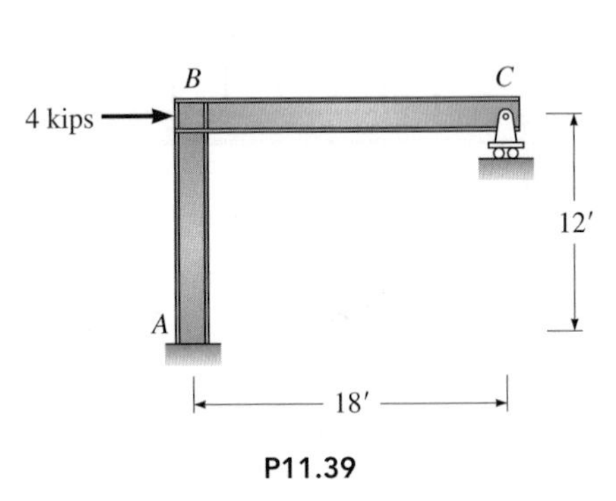

P11.39

P11.40. Recompute the reactions for the frame in Figure P11.39 if support *C* settles 0.36 in when the load acts and support *A* is constructed 0.24 in above its intended position. $E = 30{,}000$ kips/in^2, $I = 60$ in^4.

P11.41. (*a*) Determine the reactions and draw the shear and moment curves for all members of the frame in Figure P11.41. Given: EI = constant. (*b*) Compute the vertical deflection of the girder at point *C* produced by the 60-kip load.

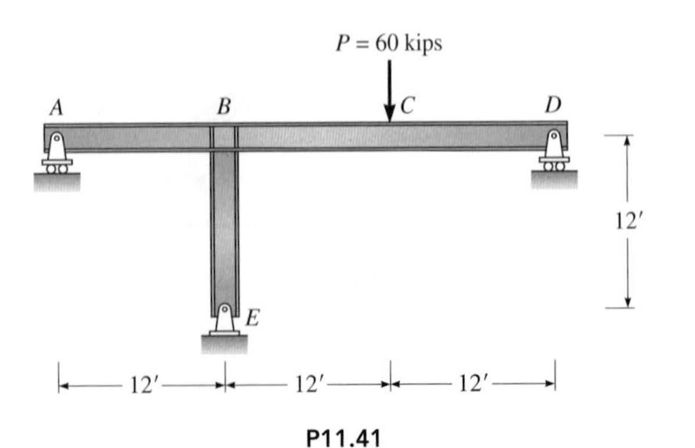

P11.41

P11.42. Determine the reactions at supports *A* and *E* in Figure P11.42; *EI* is constant for all members.

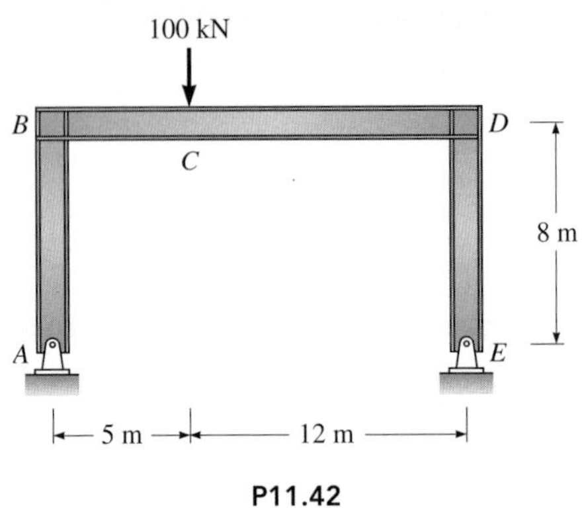

P11.42

P11.43. Determine the reactions in the rigid frame in Figure P11.43. In addition to the applied load, the temperature of beam *BC* increases by 60°F. Given: $I_{BC} = 3600$ in^4, $I_{AB} = I_{CD} = 1440$ in^4, $\alpha = 6.5 \times 10^{-6}$ (in/in)/°F, and $E = 30{,}000$ kips/in^2.

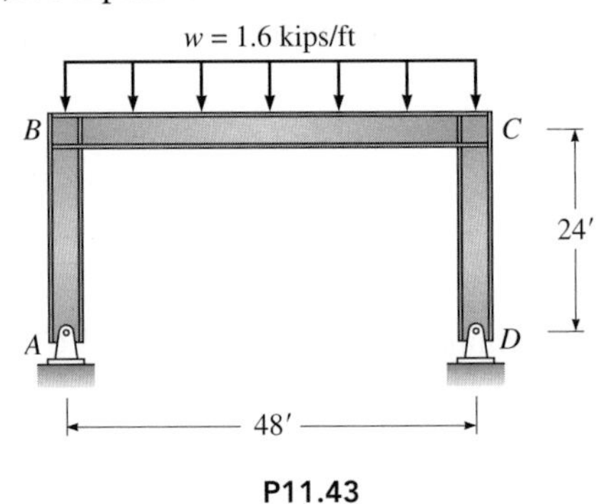

P11.43

P11.44. Determine the reactions at supports *A* and *E* in Figure P11.44. Area of bar $EC = 2$ in^2, $I_{AD} = 400$ in^4, and $A_{AD} = 8$ in^2; $E = 30{,}000$ kips/in^2.

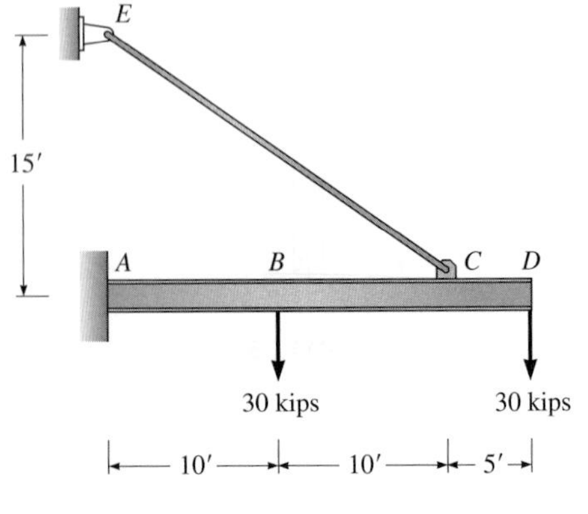

P11.44

P11.45. Practical Design Example

The tall building in Figure P11.45 is constructed of structural steel. The exterior columns, which are uninsulated, are exposed to the outside ambient temperature. To reduce the differential vertical displacements between the interior and exterior columns due to temperature differences between the interior and exterior of the building, a *bonnet truss* has been added at the top of the building. For example, if a bonnet truss was not used to restrain the outer columns from shortening in the winter due to a 60°F temperature difference between the interior and exterior columns, points D and F at the top of the exterior columns would move downward 1.68 inches relative to the top of the interior column at point E. Displacements of this magnitude in the upper stories would produce excessive slope of the floor and would damage the exterior facade.

If the temperature of interior column BE is 70°F at all times but the temperature of the exterior columns in winter drops to 10°F, determine (*a*) the forces created in the columns and the truss bars by the temperature differences and (*b*) the vertical displacements of the tops of the columns at points D and E. Slotted truss connections at D and F have been designed to act as rollers and transmit vertical force only, and the connection at E is designed to act as a pin. The shear connections between the beam webs and the columns may be assumed to act as hinges.

Given: $E = 29{,}000$ kips/in^2. The *average area* of the interior column is 42 in^2 and 30 in^2 for the exterior columns. The areas of all members of the truss are 20 in^2. The coefficient of temperature expansion $\alpha = 6.5 \times 10^{-6}$ (in/in)/°F. *Note*: The interior columns must be designed for both the floor loads and the compression force created by the temperature differential.

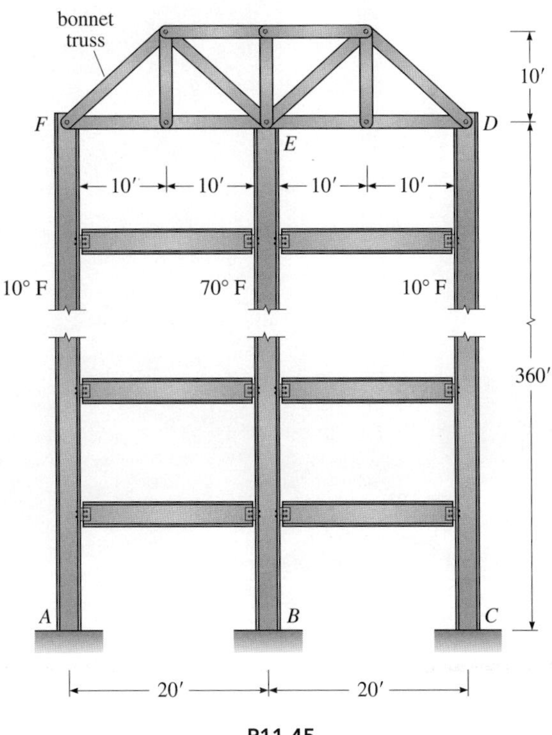

P11.45

Collapse of a reinforced concrete building during construction

The collapse was attributed primarily to lack of shoring and understrength concrete in the lower floors. Because the building was constructed in the winter without adequate heat, much of the freshly placed concrete in the forms froze and failed to gain its design strength.

CHAPTER 12

Analysis of Indeterminate Beams and Frames by the Slope-Deflection Method

Chapter Objectives

- Learn a stiffness method using the joint displacements–both rotations and translations–as the unknowns.
- Identify all the unknown joint displacements and the associated degree of kinematic indeterminacy.
- Establish the slope-deflection equations of individual members.
- Set up the equilibrium equations such that the unknown displacements can be solved.

12.1 Introduction

The *slope-deflection method* is a procedure for analyzing indeterminate beams and frames. It is known as a *displacement method* since *equilibrium equations*, which are used in the analysis, are expressed in terms of *unknown joint displacements*.

The slope-deflection method is important because it introduces the student to the *stiffness method* of analysis. This method is the basis of many general-purpose computer programs for analyzing all types of structures—beams, trusses, shells, and so forth. In addition, *moment distribution*—a commonly used hand method for analyzing beams and frames rapidly—is also based on the stiffness formulation.

In the slope-deflection method an expression, called the *slope-deflection equation*, is used to relate the moment at each end of a member both to the end displacements of the member and to the loads applied to the member between its ends. End displacements of a member can include both a rotation and a translation perpendicular to the member's longitudinal axis.

12.2 Illustration of the Slope-Deflection Method

Slope-Deflection Method

Reinforced concrete building and bridge structures became popular in the beginning of the 20th century. To analyze these indeterminate structures that were primarily governed by the effects of bending, W.M. Wilson and G.A. Maney in 1915 developed the slope-deflection method. This method was widely used until the moment distribution method was developed in the early 1930s (see Chapter 13). However, the slope-deflection method represents a turning point for the development of the matrix stiffness method (see Chapters 16 to 18), and is the basis of modern computer structural analysis software.

To introduce the main features of the slope-deflection method, we briefly outline the analysis of a two-span continuous beam. As shown in Figure 12.1*a*, the structure consists of a single member supported by rollers at points *A* and *B* and a pin at *C*. We imagine that the structure can be divided into beam segments *AB* and *BC* and joints *A*, *B*, and *C* by passing planes through the beam an infinitesimal distance before and after each support (see Figure 12.1*b*). Since the joints are essentially points in space, the length of each member is equal to the distance between joints. In this problem θ_A, θ_B, and θ_C, the rotational displacements of the joints (and also the rotational displacements of the ends of the members), are the unknowns. These displacements are shown to an exaggerated scale by the dashed line in Figure 12.1*a*. Since the supports do not move vertically, the lateral displacements of the joints are zero; thus there are no unknown joint translations in this example.

To begin the analysis of the beam by the slope-deflection method, we use the *slope-deflection equation* (which we will derive shortly) to express the moments at the ends of each member in terms of the unknown joint

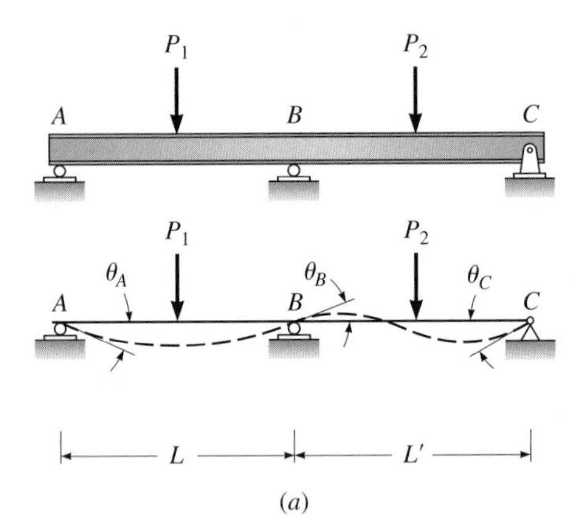

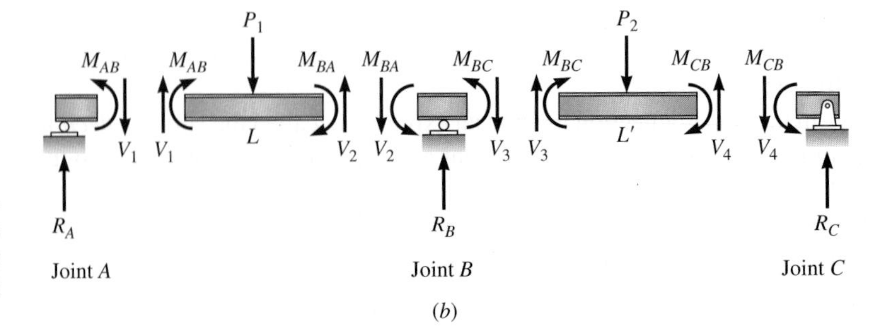

Figure 12.1: (*a*) Continuous beam with applied loads (deflected shape shown by dashed line); (*b*) free bodies of joints and beams (sign convention: clockwise moment on the end of a member is positive).

displacements and the applied loads. We can represent this step by the following set of equations:

$$\begin{aligned} M_{AB} &= f(\theta_A,\ \theta_B,\ P_1) \\ M_{BA} &= f(\theta_A,\ \theta_B,\ P_1) \\ M_{BC} &= f(\theta_B,\ \theta_C,\ P_2) \\ M_{CB} &= f(\theta_B,\ \theta_C,\ P_2) \end{aligned} \tag{12.1}$$

where the symbol $f(\ )$ stands for *a function of.*

We next write equilibrium equations that express the condition that the joints are in equilibrium with respect to the applied moments; that is, the sum of the moments applied to each joint by the ends of the beams framing into the joint equals zero. As a sign convention we assume that all *unknown moments* are *positive* and act *clockwise* on the *ends of members*. Since the moments applied to the ends of members represent the action of the joint on the member, equal and oppositely directed moments must act on the joints (see Figure 12.1*b*). The three joint equilibrium equations are

$$\begin{aligned} &\text{At joint } A\text{:} & M_{AB} &= 0 \\ &\text{At joint } B\text{:} & M_{BA} + M_{BC} &= 0 \\ &\text{At joint } C\text{:} & M_{CB} &= 0 \end{aligned} \tag{12.2}$$

By substituting Equations 12.1 into Equations 12.2, we produce three equations that are functions of the three unknown displacements (as well as the applied loads and properties of the members that are specified). These three equations can then be solved simultaneously for the values of the unknown joint rotations. After the joint rotations are computed, we can evaluate the member end moments by substituting the values of the joint rotations into Equations 12.1. Once the magnitude and direction of the end moments are established, we apply the equations of statics to free bodies of the beams to compute the end shears. As a final step, we compute the support reactions by considering the equilibrium of the joints (i.e., summing forces in the vertical direction).

In Section 12.3 we derive the slope-deflection equation for a typical flexural member of constant cross section using the moment-area method developed in Chapter 9.

12.3 Derivation of the Slope-Deflection Equation

To develop the slope-deflection equation, which relates the moments at the ends of members to the end displacements and the applied loads, we will analyze span *AB* of the continuous beam in Figure 12.2*a*. Since differential settlements of supports in continuous members also create end moments, we will include this effect in the derivation. The beam, which is initially straight, has

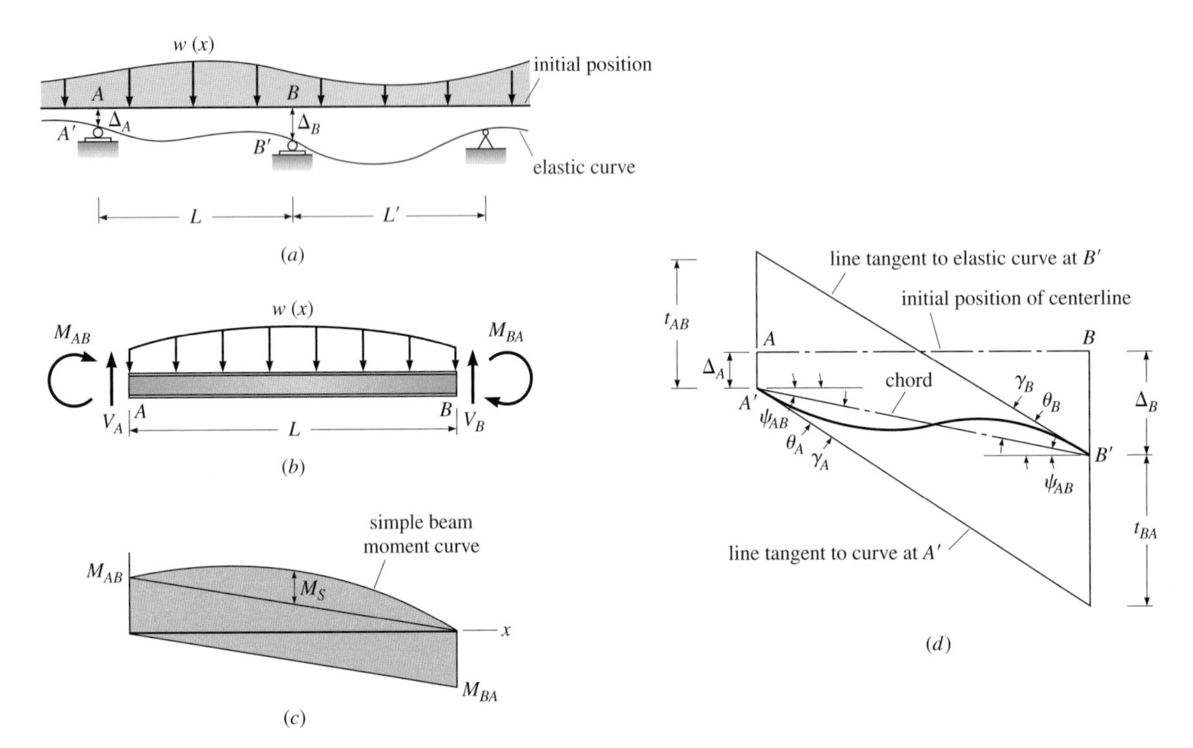

Figure 12.2: (*a*) Continuous beam whose supports settle under load; (*b*) free body of member *AB*; (*c*) moment curve plotted by parts, M_S equals the ordinate of the simple beam moment curve; (*d*) deformations of member *AB* plotted to an exaggerated vertical scale.

a constant cross section; that is, *EI* is constant along the longitudinal axis. When the distributed load $w(x)$, which can vary in any arbitrary manner along the beam's axis, is applied, supports *A* and *B* settle, respectively, by amounts Δ_A and Δ_B to points A' and B'. Figure 12.2*b* shows a free body of span *AB* with all applied loads. The moments M_{AB} and M_{BA} and the shears V_A and V_B represent the forces exerted by the joints on the ends of the beam. Although we assume that no axial load acts, the presence of small to moderate values of axial load (say, 10 to 15 percent of the member's buckling load) would not invalidate the derivation. On the other hand, a large compression force would reduce the member's flexural stiffness by creating additional deflection due to the secondary moments produced by the eccentricity of the axial load—the P-Δ effect. As a sign convention, we assume that moments acting at the ends of members in the *clockwise direction* are *positive*. Clockwise rotations of the ends of members will also be considered positive.

In Figure 12.2*c* the moment curves produced by both the distributed load $w(x)$ and the end moments M_{AB} and M_{BA} are drawn by parts. The moment curve associated with the distributed load is called the *simple beam moment curve*. In other words, in Figure 12.2*c*, we are superimposing the moments produced by three loads: (1) the end moment M_{AB}, (2) the end moment M_{BA}, and (3) the load $w(x)$ applied between ends of the beam. The moment curve for each force has been plotted on the side of the beam that is placed in compression by that particular force.

Figure 12.2*d* shows the deflected shape of span *AB* to an *exaggerated scale*. All angles and rotations are shown in the positive sense; that is, all have undergone clockwise rotations from the original horizontal position of the axis. The slope of the chord, which connects the ends of the member at points A' and B' in their deflected position, is denoted by ψ_{AB}. To establish if a chord angle is positive or negative, we can draw a horizontal line through either end of the beam. If the horizontal line must be rotated clockwise through an acute angle to make it coincide with the chord, the slope angle is positive. If a counterclockwise rotation is required, the slope is negative. Notice, in Figure 12.2*d*, that ψ_{AB} is positive regardless of the end of the beam at which it is evaluated. And θ_A and θ_B represent the end rotations of the member. At each end of span AB, tangent lines are drawn to the elastic curve; t_{AB} and t_{BA} are the tangential deviations (the vertical distance) from the tangent lines to the elastic curve.

To derive the slope-deflection equation, we will now use the second moment-area theorem to establish the relationship between the member end moments M_{AB} and M_{BA} and the rotational deformations of the elastic curve shown to an exaggerated scale in Figure 12.2*d*. Since the deformations are small, γ_A, the angle between the chord and the line tangent to the elastic curve at point A, can be expressed as

$$\gamma_A = \frac{t_{BA}}{L} \tag{12.3a}$$

Similarly, γ_B, the angle between the chord and the line tangent to the elastic curve at B, equals

$$\gamma_B = \frac{t_{AB}}{L} \tag{12.3b}$$

Since $\gamma_A = \theta_A - \psi_{AB}$ and $\gamma_B = \theta_B - \psi_{AB}$, we can express Equations 12.3*a* and 12.3*b* as

$$\theta_A - \psi_{AB} = \frac{t_{BA}}{L} \tag{12.4a}$$

$$\theta_B - \psi_{AB} = \frac{t_{AB}}{L} \tag{12.4b}$$

where

$$\psi_{AB} = \frac{\Delta_B - \Delta_A}{L} \tag{12.4c}$$

To express t_{AB} and t_{BA} in terms of the applied moments, we divide the ordinates of the moment curves in Figure 12.2*c* by EI to produce M/EI curves and, applying the second moment-area principle, sum the moments of the area under the M/EI curves about the A end of member AB to give t_{AB} and about the B end to give t_{BA}.

$$t_{AB} = \frac{M_{BA}}{EI}\frac{L}{2}\frac{2L}{3} - \frac{M_{AB}}{EI}\frac{L}{2}\frac{L}{3} - \frac{(A_M\bar{x})_A}{EI} \tag{12.5}$$

$$t_{BA} = \frac{M_{AB}}{EI}\frac{L}{2}\frac{2L}{3} - \frac{M_{BA}}{EI}\frac{L}{2}\frac{L}{3} + \frac{(A_M\bar{x})_B}{EI} \tag{12.6}$$

The first and second terms in Equations 12.5 and 12.6 represent the first moments of the triangular areas associated with the end moments M_{AB} and

M_{BA}. The last term—$(A_M\bar{x})_A$ in Equation 12.5 and $(A_M\bar{x})_B$ in Equation 12.6—represents the first moment of the area under the simple beam moment curve about the ends of the beam (the subscript indicates the end of the beam about which moments are taken). As a sign convention, we assume that the contribution of each moment curve to the tangential deviation is positive if it increases the tangential deviation and negative if it decreases the tangential deviation.

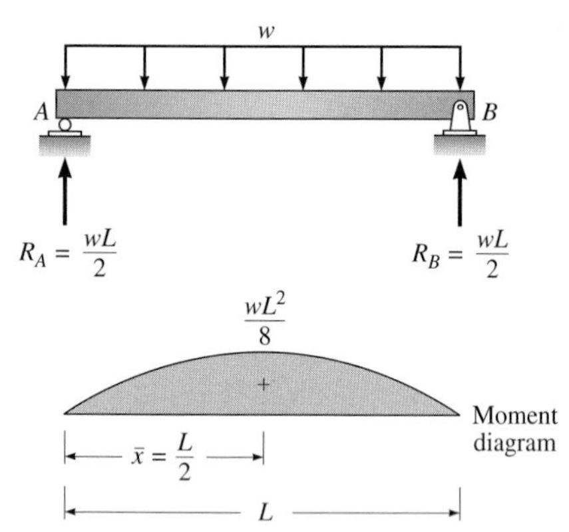

Figure 12.3: Simple beam moment curve produced by a uniform load.

To illustrate the computation of $(A_M\bar{x})_A$ for a beam carrying a uniformly distributed load w (see Figure 12.3), we draw the simple beam moment curve, a parabolic curve, and evaluate the product of the area under the curve and the distance $\bar{x}$ between point A and the centroid of the area:

$$(A_M\bar{x})_A = \text{area} \cdot \bar{x} = \frac{2L}{3}\frac{wL^2}{8}\left(\frac{L}{2}\right) = \frac{wL^4}{24} \tag{12.7}$$

Since the moment curve is symmetric, $(A_M\bar{x})_B$ equals $(A_M\bar{x})_A$.

If we next substitute the values of t_{AB} and t_{BA} given by Equations 12.5 and 12.6 into Equations 12.4*a* and 12.4*b*, we can write

$$\theta_A - \psi_{AB} = \frac{1}{L}\left[\frac{M_{BA}}{EI}\frac{L}{2}\frac{2L}{3} - \frac{M_{AB}}{EI}\frac{L}{2}\frac{L}{3} - \frac{(A_M\bar{x})_A}{EI}\right] \tag{12.8}$$

$$\theta_B - \psi_{AB} = \frac{1}{L}\left[\frac{M_{AB}}{EI}\frac{L}{2}\frac{2L}{3} - \frac{M_{BA}}{EI}\frac{L}{2}\frac{L}{3} - \frac{(A_M\bar{x})_B}{EI}\right] \tag{12.9}$$

To establish the slope-deflection equations, we solve Equations 12.8 and 12.9 simultaneously for M_{AB} and M_{BA} to give

$$M_{AB} = \frac{2EI}{L}(2\theta_A + \theta_B - 3\psi_{AB}) + \frac{2(A_M\bar{x})_A}{L^2} - \frac{4(A_M\bar{x})_B}{L^2} \tag{12.10}$$

$$M_{BA} = \frac{2EI}{L}(2\theta_B + \theta_A - 3\psi_{AB}) + \frac{4(A_M\bar{x})_A}{L^2} - \frac{2(A_M\bar{x})_B}{L^2} \tag{12.11}$$

In Equations 12.10 and 12.11, the last two terms that contain the quantities $(A_M\bar{x})_A$ and $(A_M\bar{x})_B$ are a function of the loads applied between ends of the member only. We can give these terms a physical meaning by using Equations 12.10 and 12.11 to evaluate the moments in a fixed-end beam that has the same dimensions (cross section and span length) and supports the same load as member AB in Figure 12.2*a* (see Figure 12.4). Since the ends of the beam in Figure 12.4 are fixed, the member end moments M_{AB} and M_{BA}, which are also termed *fixed-end moments*, may be designated FEM_{AB} and FEM_{BA}. Because the ends of the beam in Figure 12.4 are fixed against rotation and because no support settlements occur, it follows that

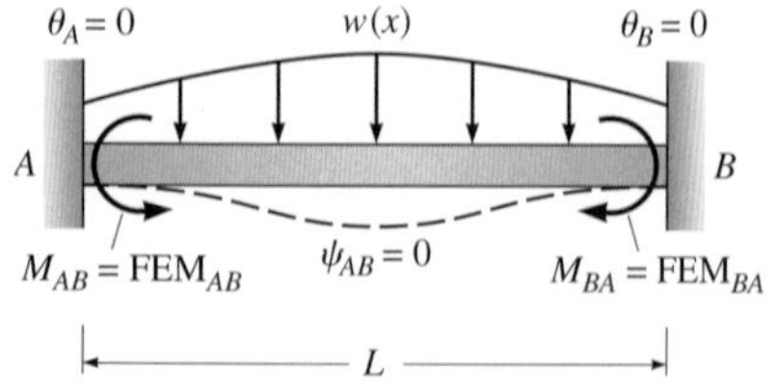

Figure 12.4

$$\theta_A = 0 \qquad \theta_B = 0 \qquad \psi_{AB} = 0$$

Substituting these values into Equations 12.10 and 12.11 to evaluate the member end moments (or fixed-end moments) in the beam of Figure 12.4, we can write

$$\text{FEM}_{AB} = M_{AB} = \frac{2(A_M\bar{x})_A}{L^2} - \frac{4(A_M\bar{x})_B}{L^2} \tag{12.12}$$

$$\text{FEM}_{BA} = M_{BA} = \frac{4(A_M\bar{x})_A}{L^2} - \frac{2(A_M\bar{x})_B}{L^2} \tag{12.13}$$

Using the results of Equations 12.12 and 12.13, we can write Equations 12.10 and 12.11 more simply by replacing the last two terms by FEM_{AB} and FEM_{BA} to produce

$$M_{AB} = \frac{2EI}{L}(2\theta_A + \theta_B - 3\psi_{AB}) + \text{FEM}_{AB} \tag{12.14}$$

$$M_{BA} = \frac{2EI}{L}(2\theta_B + \theta_A - 3\psi_{AB}) + \text{FEM}_{BA} \tag{12.15}$$

Since Equations 12.14 and 12.15 have the same form, we can replace them with a single equation in which we denote the end where the moment is being computed as the near end (N) and the opposite end as the far end (F). With this adjustment we can write the slope-deflection equation as

$$M_{NF} = \frac{2EI}{L}(2\theta_N + \theta_F - 3\psi_{NF}) + \text{FEM}_{NF} \tag{12.16}$$

In Equation 12.16 the proportions of the member appear in the ratio I/L. This ratio, which is called the *relative flexural stiffness* of member NF, is denoted by the symbol K.

$$\text{Relative flexural stiffness } K = \frac{I}{L} \tag{12.17}$$

Substituting Equation 12.17 into Equation 12.16, we can write the slope-deflection equation as

$$M_{NF} = 2EK(2\theta_N + \theta_F - 3\psi_{NF}) + \text{FEM}_{NF} \tag{12.16a}$$

The value of the fixed-end moment (FEM_{NF}) in Equation 12.16 or 12.16*a* can be computed for any type of loading by Equations 12.12 and 12.13. The use of these equations to determine the fixed-end moments produced by a single concentrated load at midspan of a fixed-ended beam is illustrated in Example 12.1. See Figure 12.5. Values of fixed-end moments for other types of loading as well as support displacements are also given on the back cover.

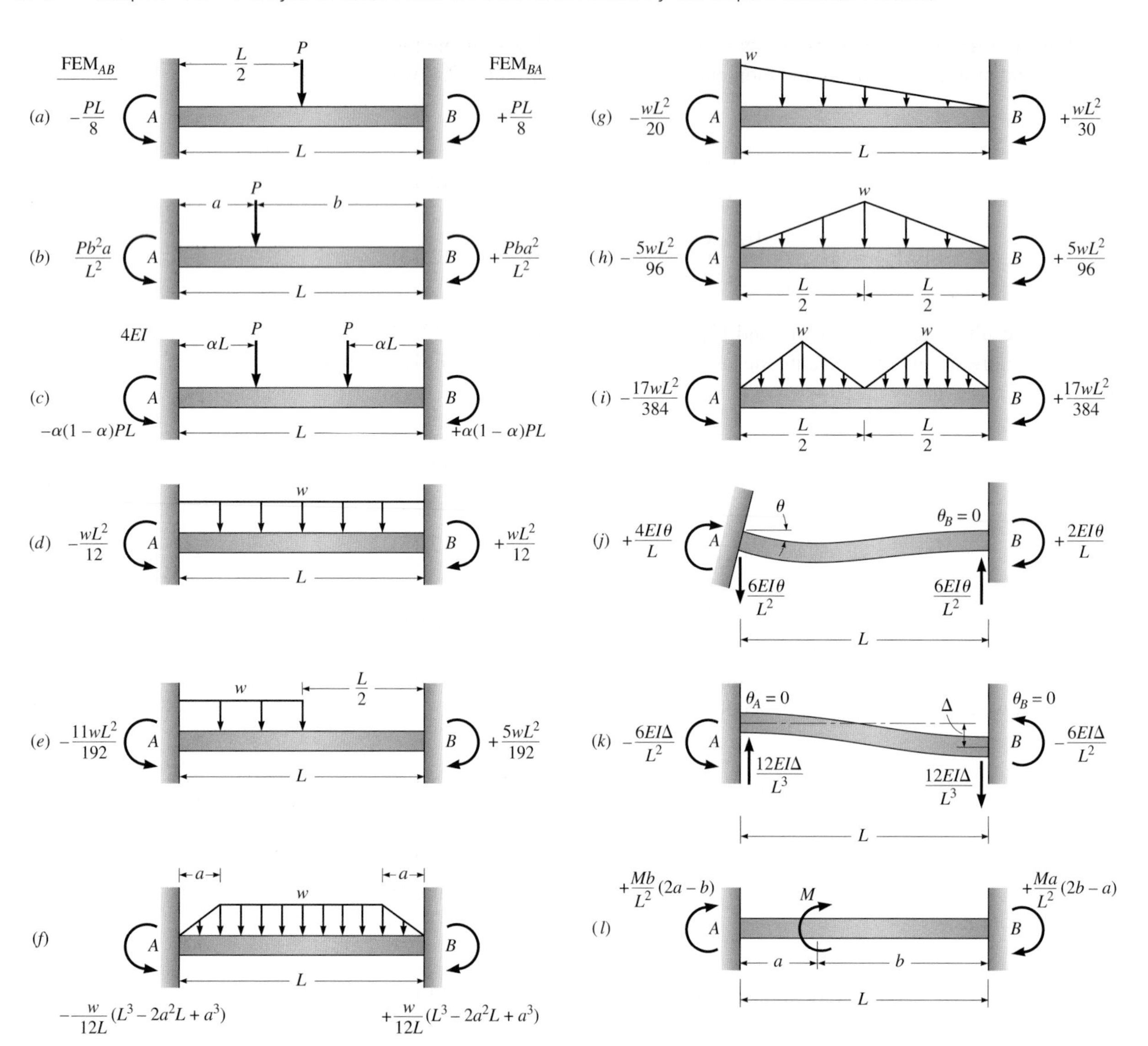

Figure 12.5: Fixed-end moments.

EXAMPLE 12.1

Using Equations 12.12 and 12.13, compute the fixed-end moments produced by a concentrated load P at midspan of the fixed-ended beam in Figure 12.6*a*. We know that EI is constant.

Solution

Equations 12.12 and 12.13 require that we compute, with respect to both ends of the beam in Figure 12.6*a*, the moment of the area under the simple beam moment curve produced by the applied load. To establish the simple beam moment curve, we imagine the beam AB in Figure 12.6*a* is removed from the fixed supports and placed on a set of simple supports, as shown in Figure 12.6*b*. The resulting simple beam moment curve produced by the concentrated load at midspan is shown in Figure 12.6*c*. Since the area under the moment curve is symmetric,

$$(A_M\bar{x})_A = (A_M\bar{x})_B = \frac{1}{2}L\frac{PL}{4}\left(\frac{L}{2}\right) = \frac{PL^3}{16}$$

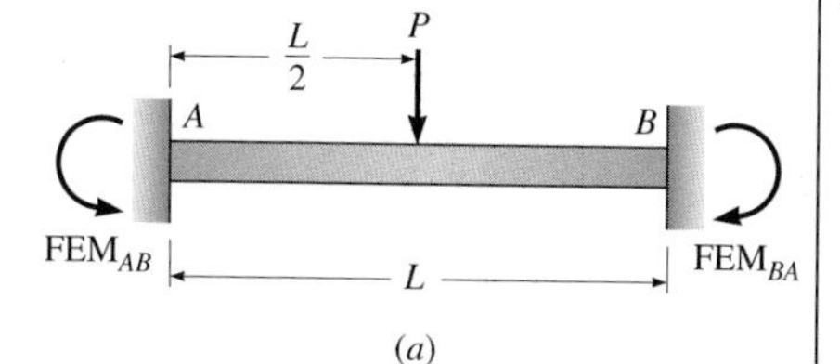

(*a*)

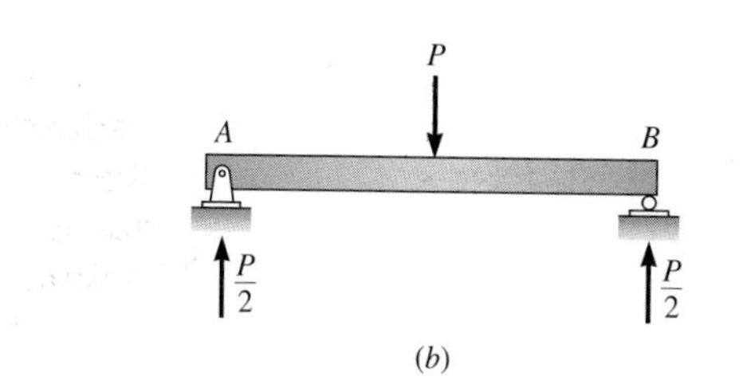

(*b*)

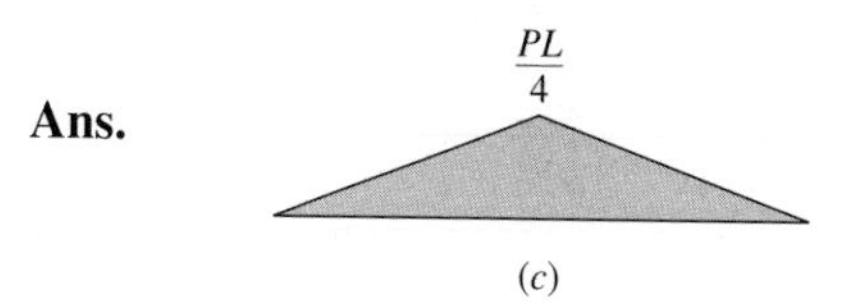

(*c*)

Figure 12.6

Using Equation 12.12 yields

$$\text{FEM}_{AB} = \frac{2(A_M\bar{x})_A}{L^2} - \frac{4(A_M\bar{x})_B}{L^2}$$

$$= \frac{2}{L^2}\left(\frac{PL^3}{16}\right) - \frac{4}{L^2}\left(\frac{PL^3}{16}\right)$$

$$= -\frac{PL}{8} \quad \text{(the minus sign indicates a counterclockwise moment)} \qquad \textbf{Ans.}$$

Using Equation 12.13 yields

$$\text{FEM}_{BA} = \frac{4(A_M\bar{x})_A}{L^2} - \frac{2(A_M\bar{x})_B}{L^2}$$

$$= \frac{4}{L^2}\left(\frac{PL^3}{16}\right) - \frac{2}{L^2}\left(\frac{PL^3}{16}\right) = +\frac{PL}{8} \quad \text{clockwise} \qquad \textbf{Ans.}$$

12.4 Analysis of Structures by the Slope-Deflection Method

Although the slope-deflection method can be used to analyze any type of indeterminate beam or frame, we will initially limit the method to indeterminate beams whose supports do not settle and to *braced* frames whose joints are free to rotate but are restrained against the displacement—restraint can be supplied by bracing members (Figure 3.23*g*) or by supports. For these types of structures, the chord rotation angle ψ_{NF} in Equation 12.16 equals *zero*. Examples of several structures whose joints do not displace laterally but are free to rotate are shown in Figure 12.7*a* and *b*. In Figure 12.7*a* joint *A* is restrained against displacement by the fixed support and joint *C* by the pin support. Neglecting second-order changes in the length of members produced by bending and axial deformations, we can assume that joint *B* is restrained against horizontal displacement by member *BC*, which is connected to an immovable support at *C* and against vertical displacement by member *AB*, which connects to the fixed support at *A*. The approximate deflected shape of the loaded structures in Figure 12.7 is shown by dashed lines.

Figure 12.7*b* shows a structure whose configuration and loading are symmetric with respect to the vertical axis passing through the center of member *BC*. Since a symmetric structure under a symmetric load must deform in a symmetric pattern, no lateral displacement of the top joints can occur in either direction.

Figure 12.7*c* and *d* shows examples of frames that contain joints that are free to displace laterally as well as to rotate under the applied loads. Under the lateral load *H*, joints *B* and *C* in Figure 12.7*c* displace to the right. This displacement produces chord rotations $\psi = \Delta/h$ in members *AB* and *CD*. Since no vertical displacements of joints *B* and *C* occur—neglecting second-order bending and axial deformations of the columns—the chord rotation of the girder ψ_{BC} equals zero. Although the frame in Figure 12.7*d* supports

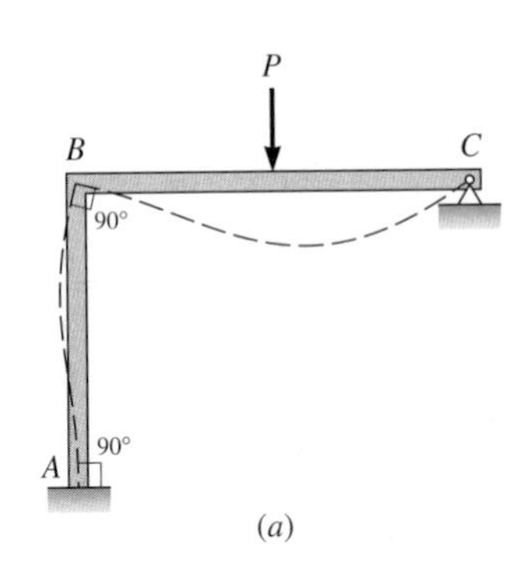

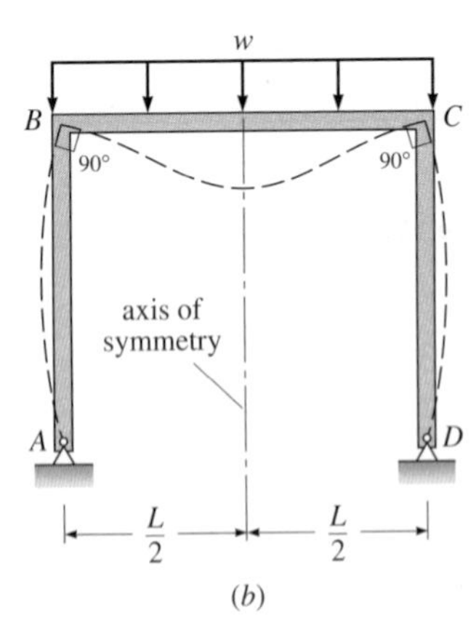

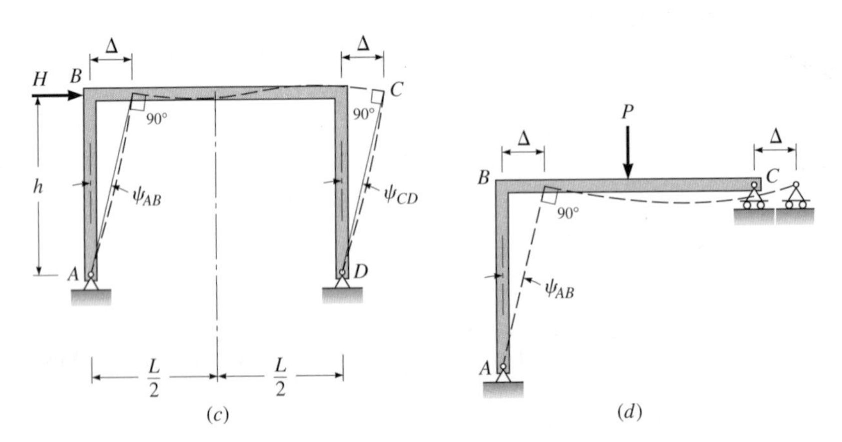

Figure 12.7: (*a*) All joints restrained against displacement; all chord rotations ψ equal zero; (*b*) due to symmetry of structure and loading, joints free to rotate but not translate; chord rotations equal zero; (*c*) and (*d*) unbraced frames with chord rotations.

a vertical load, joints B and C will displace laterally to the right a distance Δ because of the bending deformations of members AB and BC. We will consider the analysis of structures that contain one or more members with chord rotations in Section 12.5.

The basic steps of the slope-deflection method, which were discussed in Section 12.2, are summarized briefly below:

1. Identify all unknown joint displacements (rotations) to establish the number of unknowns.
2. Use the slope-deflection equation (Equation 12.16) to express all member end moments in terms of joint rotations and the applied loads.
3. At each joint, except fixed supports, write the moment equilibrium equation, which states that the sum of the moments (applied by the members framing into the joint) equals zero. An equilibrium equation at a fixed support, which reduces to the identity $0 = 0$, supplies no useful information. The number of equilibrium equations must equal the number of unknown displacements.

   As a sign convention, *clockwise moments on the ends of the members are assumed to be positive*. If a moment at the end of a member is unknown, it must be shown clockwise on the end of a member. The moment applied by a member to a joint is always equal and opposite in direction to the moment acting on the end of the member. If the magnitude and direction of the moment on the end of a member are known, they are shown in the actual direction.
4. Substitute the expressions for moments as a function of displacements (see step 2) into the equilibrium equations in step 3, and solve for the unknown displacements.
5. Substitute the values of displacement in step 4 into the expressions for member end moment in step 2 to establish the value of the member end moments. Once the member end moments are known, the balance of the analysis—drawing shear and moment curves or computing reactions, for example—is completed by statics.

Examples 12.2 and 12.3 illustrate the procedure outlined above.

EXAMPLE 12.2

Using the slope-deflection method, determine the member end moments in the indeterminate beam shown in Figure 12.8*a*. The beam, which behaves elastically, carries a concentrated load at midspan. After the end moments are determined, draw the shear and moment curves. If $I = 240\ \text{in}^4$ and $E = 30{,}000\ \text{kips/in}^2$, compute the magnitude of the slope at joint B.

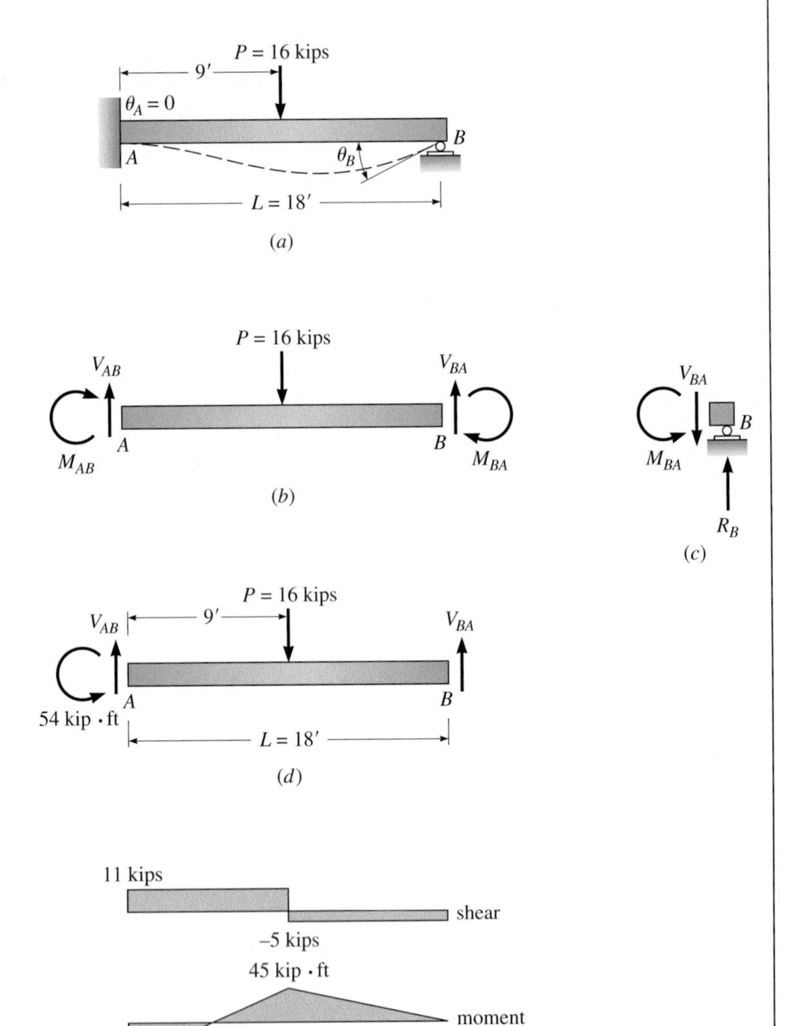

Figure 12.8: (*a*) Beam with one unknown displacement θ_B; (*b*) free body of beam AB; unknown member end moments M_{AB} and M_{BA} shown clockwise; (*c*) free body of joint B; (*d*) free body used to compute end shears; (*e*) shear and moment curves.

Solution

Since joint A is fixed against rotation, $\theta_A = 0$; therefore, the only unknown displacement is θ_B, the rotation of joint B (ψ_{AB} is, of course, zero since no support settlements occur). Using the slope-deflection equation

$$M_{NF} = \frac{2EI}{L}(2\theta_N + \theta_F - 3\psi_{NF}) + \text{FEM}_{NF} \tag{12.16}$$

and the values in Figure 12.5a for the fixed-end moments produced by a concentrated load at midspan, we can express the member end moments shown in Figure 12.8b as

$$M_{AB} = \frac{2EI}{L}(\theta_B) - \frac{PL}{8} \tag{1}$$

$$M_{BA} = \frac{2EI}{L}(2\theta_B) + \frac{PL}{8} \tag{2}$$

To determine θ_B, we next write the equation of moment equilibrium at joint B (see Figure 12.8c):

$$\circlearrowleft + \quad \Sigma M_B = 0$$

$$M_{BA} = 0 \tag{3}$$

Substituting the value of M_{BA} given by Equation 2 into Equation 3 and solving for θ_B give

$$\frac{4EI}{L}\theta_B + \frac{PL}{8} = 0$$

$$\theta_B = -\frac{PL^2}{32EI} \tag{4}$$

where the minus sign indicates both that the B end of member AB and joint B rotate in the counterclockwise direction. To determine the member end moments, the value of θ_B given by Equation 4 is substituted into Equations 1 and 2 to give

$$M_{AB} = \frac{2EI}{L}\left(\frac{-PL^2}{32EI}\right) - \frac{PL}{8} = -\frac{3PL}{16} = -54 \text{ kip} \cdot \text{ft} \qquad \textbf{Ans.}$$

$$M_{BA} = \frac{4EI}{L}\left(\frac{-PL^2}{32EI}\right) + \frac{PL}{8} = 0$$

[*continues on next page*]

Example 12.2 continues . . .

Although we know that M_{BA} is zero since the support at B is a pin, the computation of M_{BA} serves as a check.

To complete the analysis, we apply the equations of statics to a free body of member AB (see Figure 12.8*d*).

$$\circlearrowright^{+} \quad \Sigma M_A = 0$$

$$0 = (16 \text{ kips})(9 \text{ ft}) - V_{BA}(18 \text{ ft}) - 54 \text{ kip} \cdot \text{ft}$$

$$V_{BA} = 5 \text{ kips}$$

$$\overset{+}{\uparrow} \quad \Sigma F_y = 0$$

$$0 = V_{BA} + V_{AB} - 16$$

$$V_{AB} = 11 \text{ kips}$$

To evaluate θ_B, we express all variables in Equation 4 in units of inches and kips.

$$\theta_B = -\frac{PL^2}{32EI} = -\frac{16(18 \times 12)^2}{32(30{,}000)240} = -0.0032 \text{ rad}$$

Expressing θ_B in degrees, we obtain

$$\frac{2\pi \text{ rad}}{360^\circ} = \frac{-0.0032}{\theta_B}$$

$$\theta_B = -0.183^\circ \qquad \textbf{Ans.}$$

where the slope θ_B is very small and not discernible to the naked eye.

Note that when you analyze a structure by the slope-deflection method, you must follow a rigid format in formulating the equilibrium equations. There is no need to guess the direction of unknown member end moments since the solution of the equilibrium equations will automatically produce the correct direction for displacements and moments. For example, in Figure 12.8*b* we show the moments M_{AB} and M_{BA} clockwise on the ends of member AB even though intuitively we may recognize from a sketch of the deflected shape in Figure 12.8*a* that moment M_{AB} must act in the counterclockwise direction because the beam is bent concave downward at the left end by the load. When the solution indicates M_{AB} is -54 kip · ft, we know from the negative sign that M_{AB} actually acts on the end of the member in the counterclockwise direction.

EXAMPLE 12.3

Using the slope-deflection method, determine the member end moments in the braced frame shown in Figure 12.9*a*. Also compute the reactions at support *D*, and draw the shear and moment curves for members *AB* and *BD*.

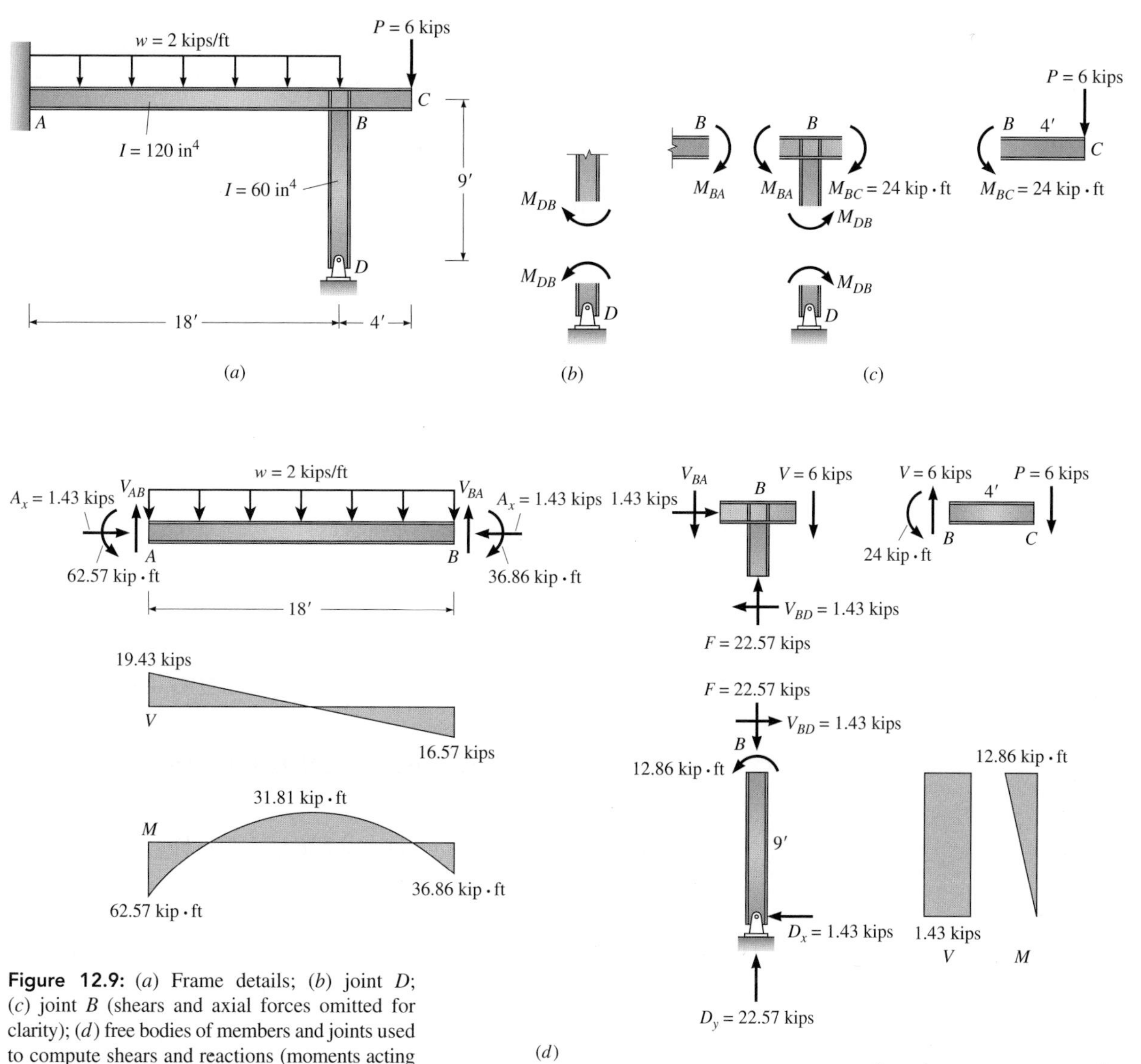

Figure 12.9: (*a*) Frame details; (*b*) joint *D*; (*c*) joint *B* (shears and axial forces omitted for clarity); (*d*) free bodies of members and joints used to compute shears and reactions (moments acting on joint *B* omitted for clarity).

[*continues on next page*]

Example 12.3 continues . . .

Solution

Since θ_A equals zero because of the fixed support at A, θ_B and θ_D are the only unknown joint displacements we must consider. Although the moment applied to joint B by the cantilever BC must be included in the joint equilibrium equation, there is no need to include the cantilever in the slope-deflection analysis of the indeterminate portions of the frame because the cantilever is determinate; that is, the shear and the moment at any section of member BC can be determined by the equations of statics. In the slope-deflection solution, we can treat the cantilever as a device that applies a vertical force of 6 kips and a clockwise moment of 24 kip · ft to joint B.

Using the slope-deflection equation

$$M_{NF} = \frac{2EI}{L}(2\theta_N + \theta_F - 3\psi_{NF}) + \text{FEM}_{NF} \tag{12.16}$$

where all variables are expressed in units of kip·inches and the fixed-end moments produced by the uniform load on member AB (see Figure 12.5d) equal

$$\text{FEM}_{AB} = -\frac{wL^2}{12}$$

$$\text{FEM}_{BA} = +\frac{wL^2}{12}$$

we can express the member end moments as

$$M_{AB} = \frac{2E(120)}{18(12)}(\theta_B) - \frac{2(18)^2(12)}{12} = 1.11E\theta_B - 648 \tag{1}$$

$$M_{BA} = \frac{2E(120)}{18(12)}(2\theta_B) + \frac{2(18)^2(12)}{12} = 2.22E\theta_B + 648 \tag{2}$$

$$M_{BD} = \frac{2E(60)}{9(12)}(2\theta_B + \theta_D) = 2.22E\theta_B + 1.11E\theta_D \tag{3}$$

$$M_{DB} = \frac{2E(60)}{9(12)}(2\theta_D + \theta_B) = 2.22E\theta_D + 1.11E\theta_B \tag{4}$$

To solve for the unknown joint displacements θ_B and θ_D, we write equilibrium equations at joints D and B.

At joint D (see Fig. 12.9b): $\circlearrowleft^{+} \quad \Sigma M_D = 0$

$$M_{DB} = 0 \tag{5}$$

At joint B (see Fig. 12.9c): $\circlearrowleft^{+} \quad \Sigma M_B = 0$

$$M_{BA} + M_{BD} - 24(12) = 0 \tag{6}$$

Since the magnitude and direction of the moment M_{BC} at the B end of the cantilever can be evaluated by statics (summing moments about point B), it is applied in the correct sense (counterclockwise) on the end of member BC, as shown in Figure 12.9*c*. On the other hand, since the magnitude and direction of the end moments M_{BA} and M_{BD} are unknown, they are assumed to act in the positive sense—clockwise on the ends of the members and counterclockwise on the joint.

Using Equations 2 to 4 to express the moments in Equations 5 and 6 in terms of displacements, we can write the equilibrium equations as

At joint D: $$2.22E\theta_D + 1.11E\theta_B = 0 \qquad (7)$$

At joint B: $$(2.22E\theta_B + 648) + (2.22E\theta_B + 1.11E\theta_D) - 288 = 0 \qquad (8)$$

Solving Equations 7 and 8 simultaneously gives

$$\theta_D = \frac{46.33}{E}$$

$$\theta_B = -\frac{92.66}{E}$$

To establish the values of the member end moments, the values of θ_B and θ_D above are substituted into Equations 1, 2, and 3, giving

$$M_{AB} = 1.11E\left(-\frac{92.66}{E}\right) - 648$$

$$= -750.85 \text{ kip} \cdot \text{in} = -62.57 \text{ kip} \cdot \text{ft} \qquad \textbf{Ans.}$$

$$M_{BA} = 2.22E\left(-\frac{92.66}{E}\right) + 648$$

$$= 442.29 \text{ kip} \cdot \text{in} = +36.86 \text{ kip} \cdot \text{ft} \qquad \textbf{Ans.}$$

$$M_{BD} = 2.22E\left(-\frac{92.66}{E}\right) + 1.11E\left(\frac{46.33}{E}\right)$$

$$= -154.28 \text{ kip} \cdot \text{in} = -12.86 \text{ kip} \cdot \text{ft} \qquad \textbf{Ans.}$$

Now that the member end moments are known, we complete the analysis by using the equations of statics to determine the shears at the ends of all members. Figure 12.9*d* shows free-body diagrams of both members and joints: Except for the cantilever, all members carry axial forces as well as shear and moment. After the shears are computed, axial forces and reactions can be evaluated by considering the equilibrium of the joints. For example, vertical equilibrium of the forces applied to joint B requires that the vertical force F in column BD equal the sum of the shears applied to joint B by the B ends of members AB and BC.

EXAMPLE 12.4

Use of Symmetry to Simplify the Analysis of a Symmetric Structure with a Symmetric Load

Determine the reactions and draw the shear and moment curves for the columns and girder of the rigid frame shown in Figure 12.10*a*. Given: $I_{AB} = I_{CD} = 120$ in^4, $I_{BC} = 360$ in^4, and E is constant for all members.

Solution

Although joints B and C rotate, they do not displace laterally because both the structure and its load are symmetric with respect to a vertical axis of symmetry passing through the center of the girder. Moreover, θ_B and θ_C are equal in magnitude; however, θ_B, a clockwise rotation, is positive, and θ_C, a counterclockwise rotation, is negative. Since the problem contains only one unknown

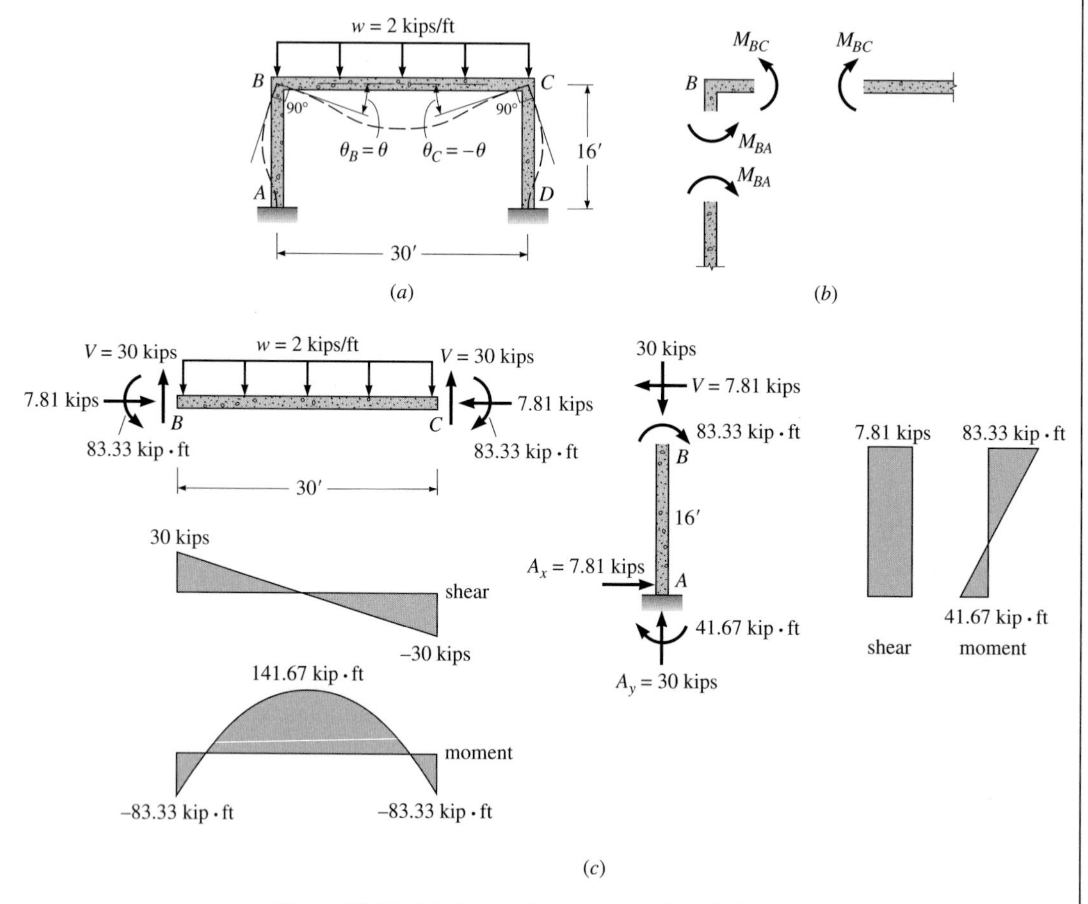

Figure 12.10: (*a*) Symmetric structure and load; (*b*) moments acting on joint B (axial forces and shears omitted); (*c*) free bodies of girder BC and column AB used to compute shears; final shear and moment curves also shown.

joint rotation, we can determine its magnitude by writing the equilibrium equation for either joint B or joint C. We will arbitrarily choose joint B.

Expressing member end moments with Equation 12.16, reading the value of fixed-end moment for member BC from Figure 12.5d, expressing units in kips·inch, and substituting $\theta_B = \theta$ and $\theta_C = -\theta$, we can write

$$M_{AB} = \frac{2E(120)}{16(12)}(\theta_B) = 1.25E\theta_B \quad (1)$$

$$M_{BA} = \frac{2E(120)}{16(12)}(2\theta_B) = 2.50E\theta_B \quad (2)$$

$$M_{BC} = \frac{2E(360)}{30(12)}(2\theta_B + \theta_C) - \frac{wL^2}{12}$$

$$= 2E[2\theta + (-\theta)] - \frac{2(30)^2(12)}{12} = 2E\theta - 1800 \quad (3)$$

Writing the equilibrium equation at joint B (see Figure 12.10b) yields

$$M_{BA} + M_{BC} = 0 \quad (4)$$

Substituting Equations 2 and 3 into Equation 4 and solving for θ produce

$$2.5E\theta + 2.0E\theta - 1800 = 0$$

$$\theta = \frac{400}{E} \quad (5)$$

Substituting the value of θ given by Equation 5 into Equations 1, 2, and 3 gives

$$M_{AB} = 1.25E\left(\frac{400}{E}\right)$$

$$= 500 \text{ kip} \cdot \text{in} = 41.67 \text{ kip} \cdot \text{ft}$$ **Ans.**

$$M_{BA} = 2.5E\left(\frac{400}{E}\right)$$

$$= 1000 \text{ kip} \cdot \text{in} = 83.33 \text{ kip} \cdot \text{ft}$$ **Ans.**

$$M_{BC} = 2E\left(\frac{400}{E}\right) - 1800$$

$$= -1000 \text{ kip} \cdot \text{in} = -83.33 \text{ kip} \cdot \text{ft}$$ **Ans.**

The final results of the analysis are shown in Figure 12.10c.

EXAMPLE 12.5

Using symmetry to simplify the slope-deflection analysis of the frame in Figure 12.11*a*, determine the reactions at supports *A* and *D*. *EI* is constant for all members.

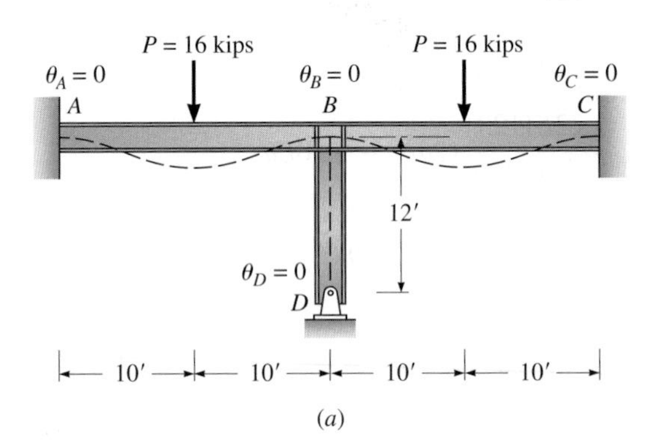

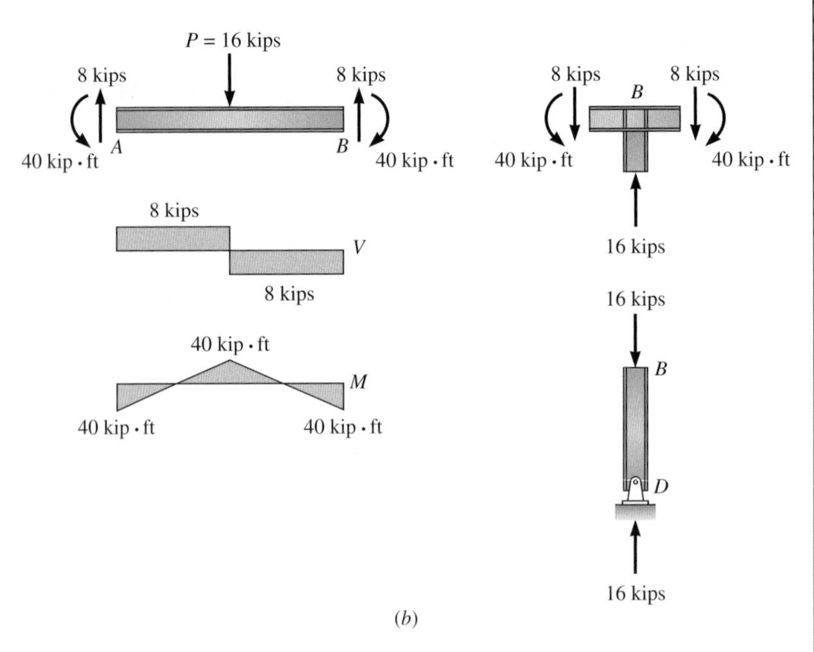

Figure 12.11: (*a*) Symmetric frame with symmetric load (deflected shape shown by dashed line); (*b*) free body of beam *AB*, joint *B*, and column *BD*. Final shear and moment diagrams for beam *AB*.

Solution

Examination of the frame shows that all joint rotations are zero. Both θ_A and θ_C are zero because of the fixed supports at *A* and *C*. Since column *BD* lies on the vertical axis of symmetry, we can infer that it must remain straight since the deflected shape of the structure with respect to the axis of symmetry must be symmetric. If the column were to bend in either direction, the requirement that the pattern of deformations be symmetric would be violated.

Since the column remains straight, neither the top nor bottom joints at *B* and *D* rotate; therefore, both θ_B and θ_D equal zero. Because no support settlements occur, chord rotations for all members are zero. Since all joint and chord rotations are zero, we can see from the slope-deflection equation (Equation 12.16) that the member end moments at each end of beams *AB* and *BC* are equal to the fixed-end moments *PL*/8 given by Figure 12.5*a*:

$$\text{FEM} = \pm\frac{PL}{8} = \frac{16(20)}{8} = \pm 40 \text{ kip} \cdot \text{ft} \qquad \textbf{Ans.}$$

Free bodies of beam *AB*, joint *B*, and column *BD* are shown in Figure 12.11(*b*).

NOTE. The analysis of the frame in Figure 12.11 shows that column *BD* carries only axial load because the moments applied by the beams to each side of the joint are the same. A similar condition often exists at the interior columns of multistory buildings whose structure consists of either a continuous reinforced concrete or a welded rigid-jointed steel frame. Although a rigid joint has the capacity to transfer moments from the beams to the column, it is the *difference* between the moments applied by the girders on either side of a joint that determines the moment to be transferred. When the span lengths of the beams and the loads they support are approximately the same (a condition that exists in most buildings), the difference in moment is small. As a result, in the preliminary design stage of rigid frames for gravity loads, most columns can be sized reasonably by considering only the magnitude of the axial load produced by the gravity load from the tributary area supported by the column.

EXAMPLE 12.6

Determine the reactions and draw the shear and moment curves for the beam in Figure 12.12. The support at A has been accidentally constructed with a slope that makes an angle of 0.009 rad with the vertical y-axis through support A, and B has been constructed 1.2 in below its intended position. Given: EI is constant, $I = 360 \text{ in}^4$, and $E = 29{,}000 \text{ kips/in}^2$.

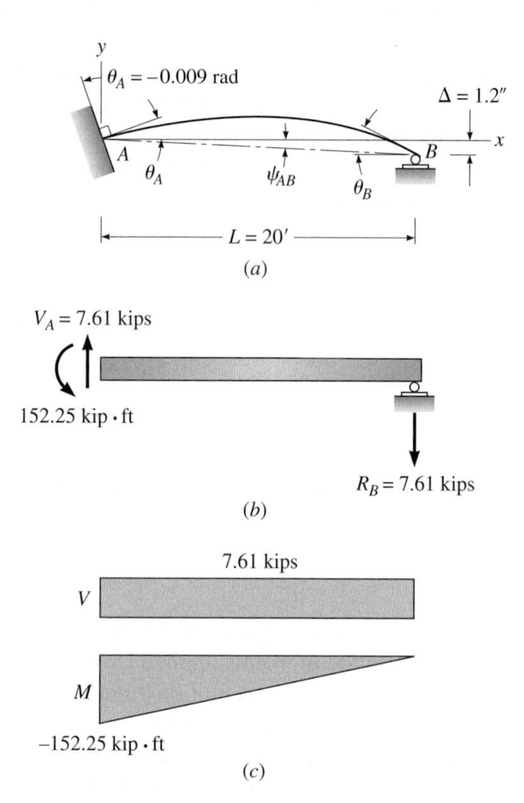

Figure 12.12: (*a*) Deformed shape; (*b*) free body used to compute V_A and R_B; (*c*) shear and moment curves.

Solution

The slope at A and the chord rotation ψ_{AB} can be determined from the information supplied about the support displacements. Since the end of the beam is rigidly connected to the fixed support at A, it rotates counterclockwise with the support; and $\theta_A = -0.009$ rad. The settlement of support B relative to support A produces a clockwise chord rotation

$$\psi_{AB} = \frac{\Delta}{L} = \frac{1.2}{20(12)} = 0.005 \text{ radians}$$

Angle θ_B is the only unknown displacement, and the fixed-end moments are zero because no loads act on beam. Expressing member end moments with the slope-deflection equation (Equation 12.16), we have

$$M_{AB} = \frac{2EI_{AB}}{L_{AB}}(2\theta_A + \theta_B - 3\psi_{AB}) + \text{FEM}_{AB}$$

$$M_{AB} = \frac{2E(360)}{20(12)}[2(-0.009) + \theta_B - 3(0.005)] \quad (1)$$

$$M_{BA} = \frac{2E(360)}{20(12)}[2\theta_B + (-0.009) - 3(0.005)] \quad (2)$$

Writing the equilibrium equation at joint B yields

$$\circlearrowleft^{+} \quad \Sigma M_B = 0$$

$$M_{BA} = 0 \quad (3)$$

Substituting Equation 2 into Equation 3 and solving for θ_B yield

$$3E(2\theta_B - 0.009 - 0.015) = 0$$

$$\theta_B = 0.012 \text{ radians}$$

To evaluate M_{AB}, substitute θ_B into Equation 1:

$$M_{AB} = 3(29{,}000)[2(-0.009) + 0.012 - 3(0.005)]$$

$$= -1827 \text{ kip}\cdot\text{in} = -152.25 \text{ kip}\cdot\text{ft}$$

Complete the analysis by using the equations of statics to compute the reaction at B and the shear at A (see Figure 12.12*b*).

$$\circlearrowright^{+} \quad \Sigma M_A = 0$$

$$0 = R_B(20) - 152.25$$

$$R_B = 7.61 \text{ kips}$$ **Ans.**

$$\overset{+}{\uparrow} \quad \Sigma F_y = 0$$

$$V_A = 7.61 \text{ kips}$$ **Ans.**

EXAMPLE 12.7

Although the supports are constructed in their correct position, girder *AB* of the frame shown in Figure 12.13 is fabricated 1.2 in too long. Determine the reactions created when the frame is connected into the supports. Given: *EI* is a constant for all members, $I = 240$ in^4, and $E = 29{,}000$ kips/in^2.

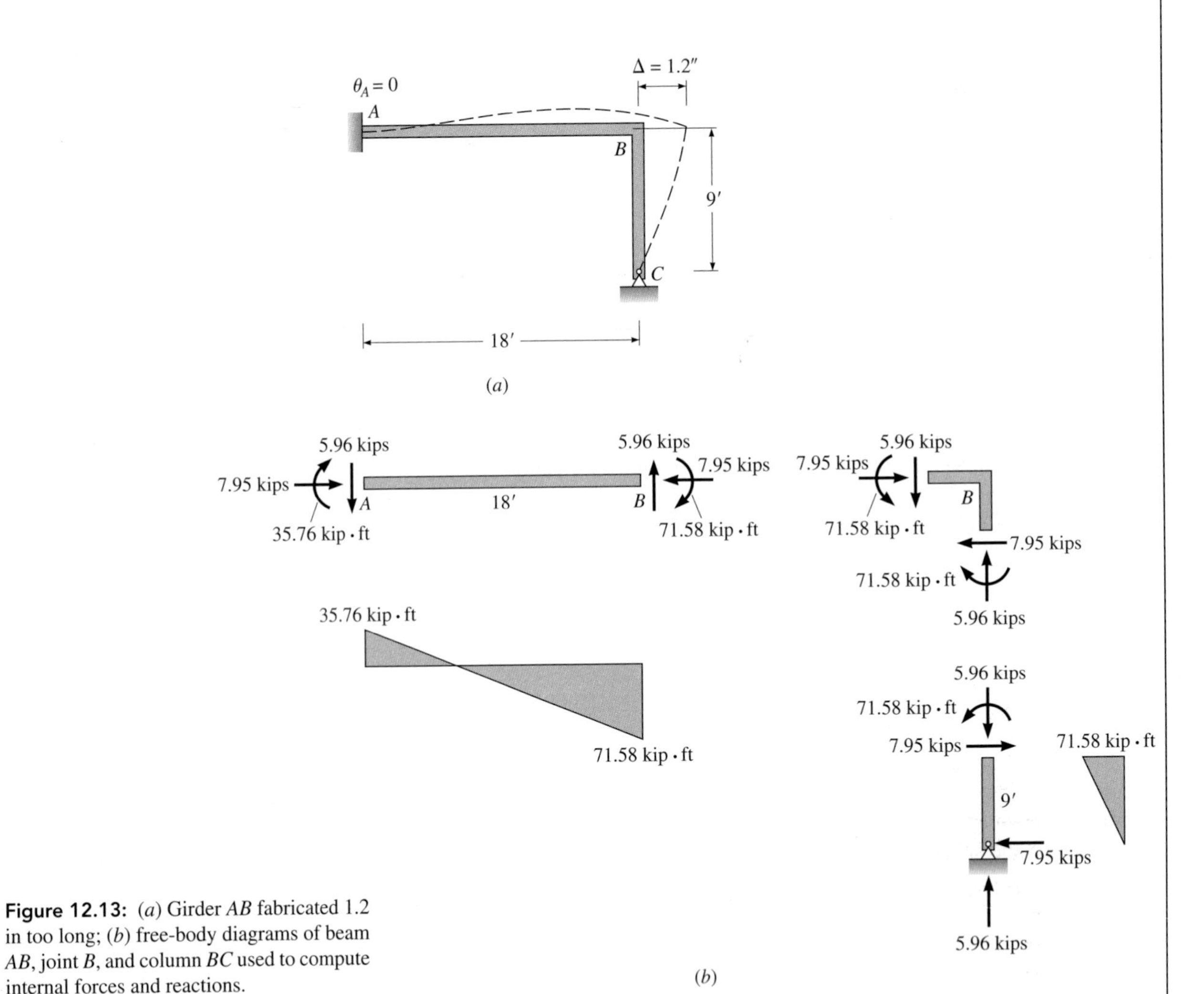

Figure 12.13: (*a*) Girder *AB* fabricated 1.2 in too long; (*b*) free-body diagrams of beam *AB*, joint *B*, and column *BC* used to compute internal forces and reactions.

Solution

The deflected shape of the frame is shown by the dashed line in Figure 12.13*a*. Although internal forces (axial, shear, and moment) are created when the frame is forced into the supports, the deformations produced by these forces are neglected since they are small compared to the 1.2-in fabrication error; therefore, the chord rotation ψ_{BC} of column BC equals

$$\psi_{BC} = \frac{\Delta}{L} = \frac{1.2}{9(12)} = \frac{1}{90} \text{ rad}$$

Since the ends of girder AB are at the same level, $\psi_{AB} = 0$. The unknown displacements are θ_B and θ_C.

Using the slope-deflection equation (Equation 12.16), we express member end moments in terms of the unknown displacements. Because no loads are applied to the members, all fixed-end moments equal zero.

$$M_{AB} = \frac{2E(240)}{18(12)}(\theta_B) = 2.222E\theta_B \tag{1}$$

$$M_{BA} = \frac{2E(240)}{18(12)}(2\theta_B) = 4.444E\theta_B \tag{2}$$

$$M_{BC} = \frac{2E(240)}{9(12)}\left[2\theta_B + \theta_C - 3\left(\frac{1}{90}\right)\right]$$

$$= 8.889E\theta_B + 4.444E\theta_C - 0.1481E \tag{3}$$

$$M_{CB} = \frac{2E(240)}{9(12)}\left[2\theta_C + \theta_B - 3\left(\frac{1}{90}\right)\right]$$

$$= 8.889E\theta_C + 4.444E\theta_B - 0.1481E \tag{4}$$

Writing equilibrium equations gives

Joint C: $$M_{CB} = 0 \tag{5}$$

Joint B: $$M_{BA} + M_{BC} = 0 \tag{6}$$

Substituting Equations 2 to 4 into Equations 5 and 6 solving for θ_B and θ_C yield

$$8.889E\theta_C + 4.444E\theta_B - 0.1481E = 0$$

$$4.444E\theta_B + 8.889E\theta_B + 4.444E\theta_C - 0.1481E = 0$$

$$\theta_B = 0.00666 \text{ rad} \tag{7}$$

$$\theta_C = 0.01332 \text{ rad} \tag{8}$$

Substituting θ_C and θ_B into Equations 1 to 3 produces

$$M_{AB} = 35.76 \text{ kip}\cdot\text{ft} \qquad M_{BA} = 71.58 \text{ kip}\cdot\text{ft}$$

$$M_{BC} = -71.58 \text{ kip}\cdot\text{ft} \qquad M_{CB} = 0 \qquad \textbf{Ans.}$$

The free-body diagrams used to compute internal forces and reactions are shown in Figure 12.13*b*, which also shows moment diagrams.

12.5 Analysis of Structures That Are Free to Sidesway

Thus far we have used the slope-deflection method to analyze indeterminate beams and frames with joints that are free to rotate but which are restrained against displacement. We now extend the method to frames whose joints are also free to *sidesway*, that is, to displace laterally. For example, in Figure 12.14*a* the horizontal load results in girder *BC* displacing laterally a distance Δ. Recognizing that the axial deformation of the girder is insignificant, we assume that the horizontal displacement of the top of both columns equals Δ. This displacement creates a clockwise chord rotation ψ in both legs of the frame equal to

$$\psi = \frac{\Delta}{h}$$

where h is the length of column.

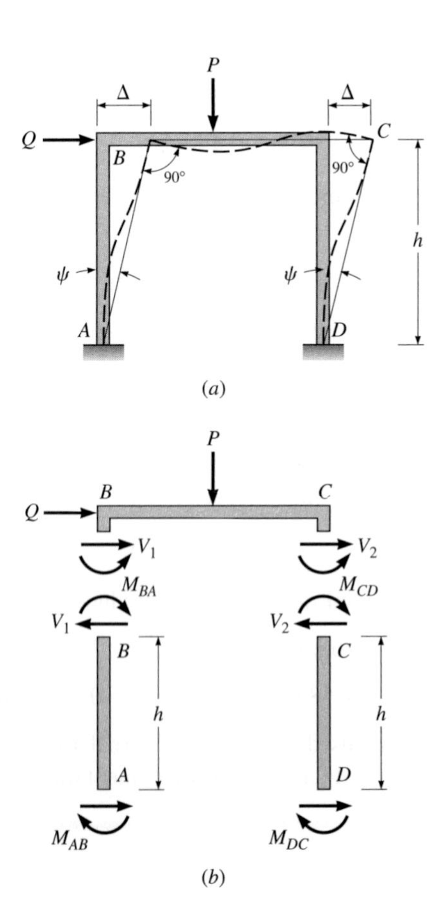

Figure 12.14: (*a*) Unbraced frame, deflected shape shown to an exaggerated scale by dashed lines, column chords rotate through a clockwise angle ψ; (*b*) free-body diagrams of columns and girders; unknown moments shown in the positive sense, that is, clockwise on ends of members (axial loads in columns and shears in girder omitted for clarity).

Since three independent displacements develop in the frame [i.e., the rotation of joints B and C (θ_B and θ_C) and the chord rotation ψ], we require three equilibrium equations for their solution. Two equilibrium equations are supplied by considering the equilibrium of the moments acting on joints B and C. Since we have written equations of this type in the solution of previous problems, we will only discuss the second type of equilibrium equation—the *shear equation*. The shear equation is established by summing in the horizontal direction the forces acting on a free body of the girder. For example, for the girder in Figure 12.14b we can write

$$\xrightarrow{+} \quad \Sigma F_x = 0$$

$$V_1 + V_2 + Q = 0 \tag{12.18}$$

In Equation 12.18, V_1, the shear in column AB, and V_2, the shear in column CD, are evaluated by summing moments about the bottom of each column of the forces acting on a free body of the column. As we established previously, the unknown moments on the ends of the column must always be shown in the positive sense, that is, acting clockwise on the end of the member. Summing moments about point A of column AB, we compute V_1:

$$\circlearrowright^{+} \quad \Sigma M_A = 0$$

$$M_{AB} + M_{BA} - V_1 h = 0$$

$$V_1 = \frac{M_{AB} + M_{BA}}{h} \tag{12.19}$$

Similarly, the shear in column CD is evaluated by summing moments about point D.

$$\circlearrowright^{+} \quad \Sigma M_D = 0$$

$$M_{CD} + M_{DC} - V_2 h = 0$$

$$V_2 = \frac{M_{CD} + M_{DC}}{h} \tag{12.20}$$

Substituting the values of V_1 and V_2 from Equations 12.19 and 12.20 into Equation 12.18, we can write the third equilibrium equation as

$$\frac{M_{AB} + M_{BA}}{h} + \frac{M_{CD} + M_{DC}}{h} + Q = 0 \tag{12.21}$$

Examples 12.8 and 12.9 illustrate the use of the slope-deflection method to analyze frames that carry lateral loads and are free to sidesway. Frames that carry only vertical load will also undergo small amounts of sidesway unless both the structure and the loading pattern are symmetric. Example 12.10 on page 499 illustrates this case.

EXAMPLE 12.8

Analyze the frame in Figure 12.15a by the slope-deflection method. E is constant for all members; $I_{AB} = 240 \text{ in}^4$, $I_{BC} = 600 \text{ in}^4$, and $I_{CD} = 360 \text{ in}^4$.

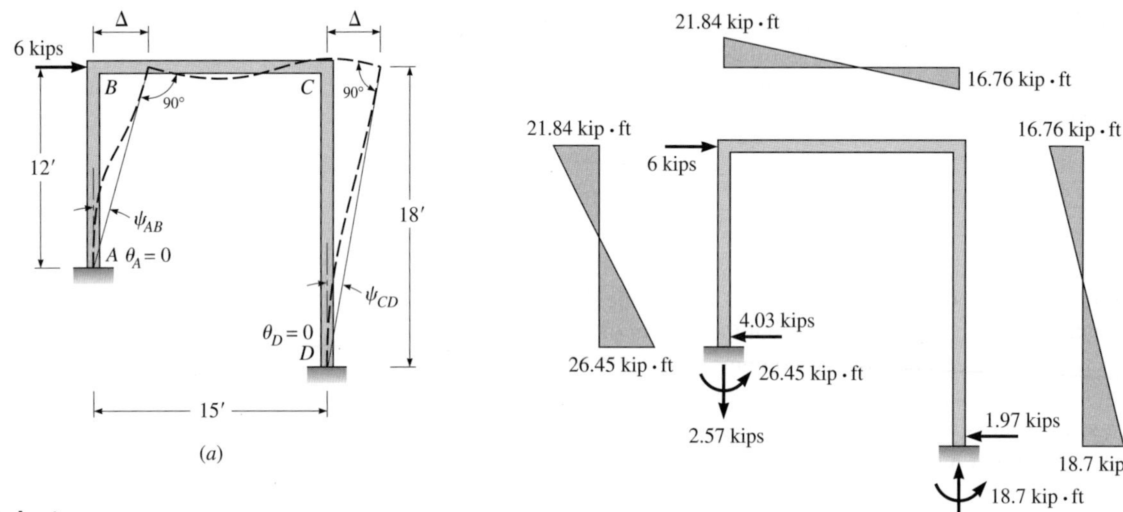

Figure 12.15: (a) Details of frame; (b) reactions and moment diagrams.

Solution

Identify the unknown displacements θ_B, θ_C, and Δ.
Express the chord rotations ψ_{AB} and ψ_{CD} in terms of Δ:

$$\psi_{AB} = \frac{\Delta}{12} \quad \text{and} \quad \psi_{CD} = \frac{\Delta}{18} \quad \text{so} \quad \psi_{AB} = 1.5\psi_{CD} \tag{1}$$

Compute the relative bending stiffness of all members.

$$K_{AB} = \frac{EI}{L} = \frac{240E}{12} = 20E$$

$$K_{BC} = \frac{EI}{L} = \frac{600E}{15} = 40E$$

$$K_{CD} = \frac{EI}{L} = \frac{360E}{18} = 20E$$

If we set $20E = K$, then

$$K_{AB} = K \qquad K_{BC} = 2K \qquad K_{CD} = K \tag{2}$$

Express member end moments in terms of displacements with slope-deflection Equation 12.16: $M_{NF} = (2EI/L)(2\theta_N + \theta_F - 3\psi_{NF}) + \text{FEM}_{NF}$. Since no loads are applied to members between joints, all $\text{FEM}_{NF} = 0$.

$$\begin{aligned} M_{AB} &= 2K_{AB}(\theta_B - 3\psi_{AB}) \\ M_{BA} &= 2K_{AB}(2\theta_B - 3\psi_{AB}) \\ M_{BC} &= 2K_{BC}(2\theta_B + \theta_C) \\ M_{CB} &= 2K_{BC}(2\theta_C + \theta_B) \\ M_{CD} &= 2K_{CD}(2\theta_C - 3\psi_{CD}) \\ M_{DC} &= 2K_{CD}(\theta_C - 3\psi_{CD}) \end{aligned} \tag{3}$$

In the equations above, use Equations 1 to express ψ_{AB} in terms of ψ_{CD}, and use Equations 2 to express all stiffness in terms of the parameter K.

$$\begin{aligned} M_{AB} &= 2K(\theta_B - 4.5\psi_{CD}) \\ M_{BA} &= 2K(2\theta_B - 4.5\psi_{CD}) \\ M_{BC} &= 4K(2\theta_B + \theta_C) \\ M_{CB} &= 4K(2\theta_C + \theta_B) \\ M_{CD} &= 2K(2\theta_C - 3\psi_{CD}) \\ M_{DC} &= 2K(\theta_C - 3\psi_{CD}) \end{aligned} \qquad (4)$$

The equilibrium equations are:

Joint B: $$M_{BA} + M_{BC} = 0 \qquad (5)$$

Joint C: $$M_{CB} + M_{CD} = 0 \qquad (6)$$

Shear equation (see Eq. 12.21): $$\frac{M_{BA} + M_{AB}}{12} + \frac{M_{CD} + M_{DC}}{18} + 6 = 0 \qquad (7)$$

Substitute Equations 4 into Equations 5, 6, and 7 and combine terms.

$$12\theta_B + 4\theta_C - 9\psi_{CD} = 0 \qquad (5a)$$

$$4\theta_B + 12\theta_C - 6\psi_{CD} = 0 \qquad (6a)$$

$$9\theta_B + 6\theta_C - 39\psi_{CD} = -\frac{108}{K} \qquad (7a)$$

Solving the equations above simultaneously gives

$$\theta_B = \frac{2.257}{K} \qquad \theta_C = \frac{0.97}{K} \qquad \psi_{CD} = \frac{3.44}{K}$$

Also, $$\psi_{AB} = 1.5\psi_{CD} = \frac{5.16}{K}$$

Since all angles are positive, all joint rotations and the sidesway angles are clockwise.

Substituting the values of displacement above into Equations 4, we establish the member end moments.

$$\begin{aligned} M_{AB} &= -26.45 \text{ kip}\cdot\text{ft} & M_{BA} &= -21.84 \text{ kip}\cdot\text{ft} \\ M_{BC} &= 21.84 \text{ kip}\cdot\text{ft} & M_{CB} &= 16.78 \text{ kip}\cdot\text{ft} \\ M_{CD} &= -16.76 \text{ kip}\cdot\text{ft} & M_{DC} &= -18.7 \text{ kip}\cdot\text{ft} \end{aligned}$$ **Ans.**

The final results are summarized in Figure 12.15*b*.

EXAMPLE 12.9

Analyze the frame in Figure 12.16*a* by the slope-deflection method. Given: EI is constant for all members.

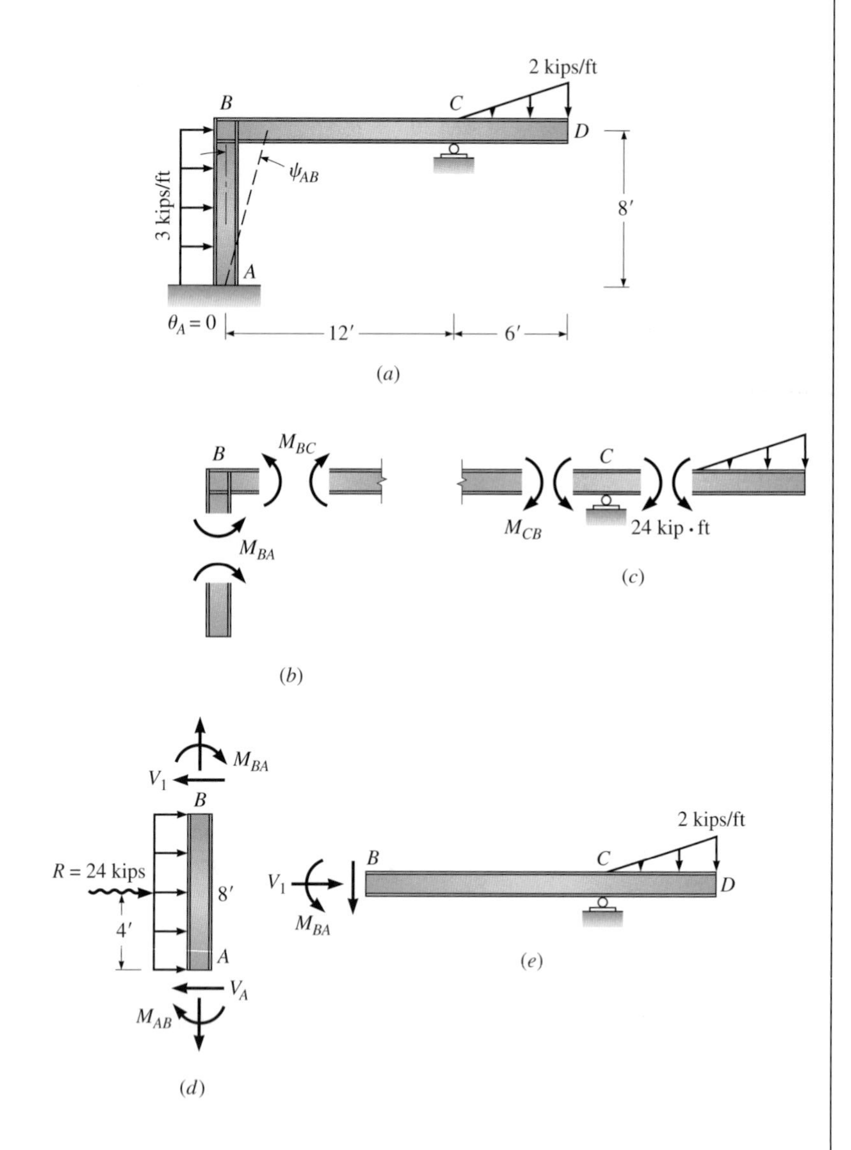

Figure 12.16: (*a*) Details of frame: rotation of chord ψ_{AB} shown by dashed line; (*b*) moments acting on joint B (shear and axial forces omitted for clarity); (*c*) moments acting on joint C (shear forces and reaction omitted for clarity); (*d*) free body of column AB; (*e*) free body of girder used to establish third equilibrium equation (*continues on page* 498).

Solution

Identify the unknown displacements; θ_B, θ_C, and ψ_{AB}. Since the cantilever is a determinate component of the structure, its analysis does not have to be included in the slope-deflection formulation. Instead, we consider the cantilever a device to apply a vertical load of 6 kips and a clockwise moment of 24 kip · ft to joint C.

Express member end moments in terms of displacements with Equation 12.16 (all units in kip·feet).

$$
\begin{aligned}
M_{AB} &= \frac{2EI}{8}(\theta_B - 3\psi_{AB}) - \frac{3(8)^2}{12} \\
M_{BA} &= \frac{2EI}{8}(2\theta_B - 3\psi_{AB}) + \frac{3(8)^2}{12} \\
M_{BC} &= \frac{2EI}{12}(2\theta_B + \theta_C) \\
M_{CB} &= \frac{2EI}{12}(2\theta_C + \theta_B)
\end{aligned}
\qquad (1)
$$

Write the joint equilibrium equations at B and C.
Joint B (see Figure 12.16*b*):

$$\stackrel{+}{\circlearrowleft} \quad \Sigma M_B = 0: \qquad M_{BA} + M_{BC} = 0 \qquad (2)$$

Joint C (see Figure 12.16*c*):

$$\stackrel{+}{\circlearrowleft} \quad \Sigma M_C = 0: \qquad M_{CB} - 24 = 0 \qquad (3)$$

Shear equation (see Figure 12.16*d*):

$$\circlearrowright^{+} \quad \Sigma M_A = 0 \qquad M_{BA} + M_{AB} + 24(4) - V_1(8) = 0$$

solving for V_1 gives $\qquad V_1 = \dfrac{M_{BA} + M_{AB} + 96}{8} \qquad (4a)$

Isolate the girder (See Figure 12.16*e*) and consider equilibrium in the horizontal direction.

$$\rightarrow + \quad \Sigma F_x = 0: \quad \text{therefore} \quad V_1 = 0 \qquad (4b)$$

Substitute Equation 4*a* into Equation 4*b*:

$$M_{BA} + M_{AB} + 96 = 0 \qquad (4)$$

Express equilibrium equations in terms of displacements by substituting Equations 1 into Equations 2, 3, and 4. Collecting terms and simplifying, we find

$$
\begin{aligned}
10\theta_B - 2\theta_C - 9\psi_{AB} &= -\frac{192}{EI} \\
\theta_B - 2\theta_C &= \frac{144}{EI} \\
3\theta_B - 6\psi_{AB} &= -\frac{384}{EI}
\end{aligned}
$$

Solution of the equations above gives

$$\theta_B = \frac{53.33}{EI} \qquad \theta_C = \frac{45.33}{EI} \qquad \psi_{AB} = \frac{90.66}{EI}$$

[*continues on next page*]

Example 12.9 continues . . .

Establish the values of member end moments by substituting the values of θ_B, θ_C, and ψ_{AB} into Equations 1.

$$M_{AB} = \frac{2EI}{8}\left[\frac{53.33}{EI} - \frac{(3)(90.66)}{EI}\right] - 16 = -70.67 \text{ kip} \cdot \text{ft}$$

$$M_{BA} = \frac{2EI}{8}\left[\frac{(2)(53.33)}{EI} - \frac{(3)(90.66)}{EI}\right] + 16 = -25.33 \text{ kip} \cdot \text{ft}$$

$$M_{BC} = \frac{2EI}{12}\left[\frac{(2)(53.33)}{EI} + \frac{45.33}{EI}\right] = 25.33 \text{ kip} \cdot \text{ft}$$

$$M_{CB} = \frac{2EI}{12}\left[\frac{(2)(45.33)}{EI} + \frac{53.33}{EI}\right] = 24 \text{ kip} \cdot \text{ft} \qquad \textbf{Ans.}$$

After the end moments are established, we compute the shears in all members by applying the equations of equilibrium to free bodies of each member. Final results are shown in Figure 12.16*f*.

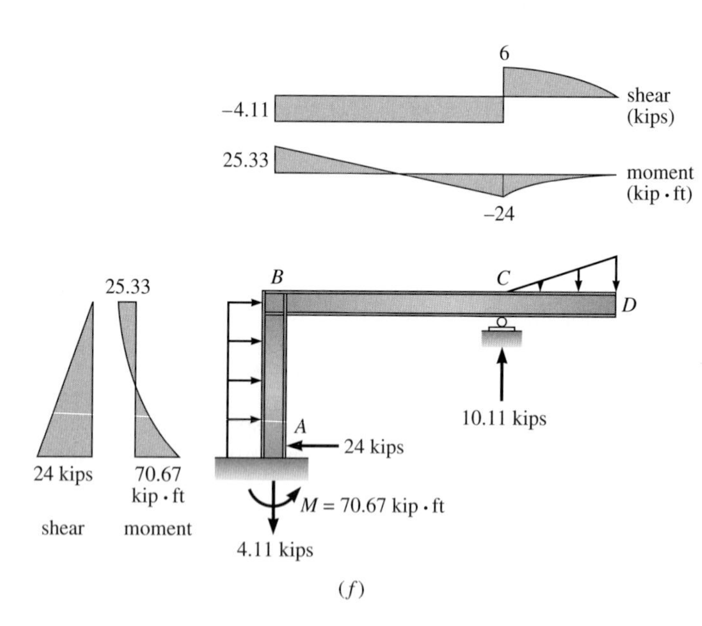

Figure 12.16: (*Continued*) (*f*) Reactions and shear and moment curves.

EXAMPLE 12.10

Analyze the frame in Figure 12.17*a* by the slope-deflection method. Determine the reactions, draw the moment curves for the members, and sketch the deflected shape. If $I = 240 \text{ in}^4$ and $E = 30{,}000 \text{ kips/in}^2$, determine the horizontal displacement of joint *B*.

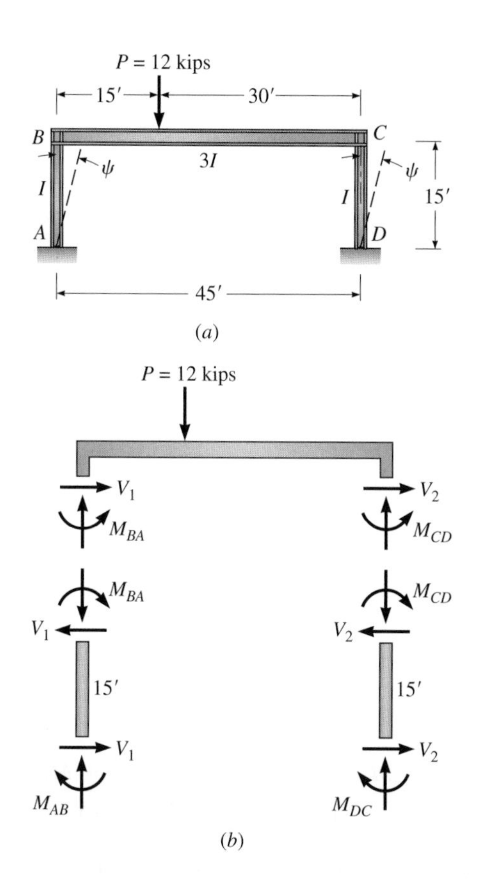

Figure 12.17: (*a*) Unbraced frame positive chord rotations assumed for columns (see the dashed lines), deflected shape shown in (*d*); (*b*) free bodies of columns and girder used to establish the shear equation (*continues on page* 501).

Solution

Unknown displacements are θ_B, θ_C, and ψ. Since supports at *A* are fixed, θ_A and θ_D equal zero. There is no chord rotation of girder *BC*.

Express member end moments in terms of displacements with the slope-deflection equation. Use Figure 12.5 to evaluate FEM_{NF}.

$$M_{NF} = \frac{2EI}{L}(2\theta_N + \theta_F - 3\psi_{NF}) + \text{FEM}_{NF} \qquad (12.16)$$

$$\text{FEM}_{BC} = -\frac{Pb^2a}{L^2} = \frac{12(30)^2(15)}{(45)^2} = -80 \text{ kip}\cdot\text{ft} \qquad \text{FEM}_{CD} = \frac{Pa^2b}{L^2} = \frac{12(15)^2(30)}{(45)^2} = 40 \text{ kip}\cdot\text{ft}$$

[*continues on next page*]

Example 12.10 continues . . .

To simplify slope-deflection expressions, set $EI/15 = K$.

$$
\begin{aligned}
M_{AB} &= \frac{2EI}{15}(\theta_B - 3\psi) &&= 2K(\theta_B - 3\psi) \\
M_{BA} &= \frac{2EI}{15}(2\theta_B - 3\psi) &&= 2K(2\theta_B - 3\psi) \\
M_{BC} &= \frac{2EI}{45}(2\theta_B + \theta_C) - 80 &&= \frac{2}{3}K(2\theta_B + \theta_C) - 80 \\
M_{CB} &= \frac{2EI}{45}(2\theta_C + \theta_B) + 40 &&= \frac{2}{3}K(2\theta_C + \theta_B) + 40 \\
M_{CD} &= \frac{2EI}{15}(2\theta_C - 3\psi) &&= 2K(\theta_C - 3\psi) \\
M_{DC} &= \frac{2EI}{15}(\theta_C - 3\psi) &&= 2K(\theta_C - 3\psi)
\end{aligned} \tag{1}
$$

The equilibrium equations are:

Joint B: $$M_{BA} + M_{BC} = 0 \tag{2}$$

Joint C: $$M_{CB} + M_{CD} = 0 \tag{3}$$

Shear equation (see the girder in Figure 12.17b):

$$\rightarrow + \quad \Sigma F_x = 0 \qquad V_1 + V_2 = 0 \tag{4a}$$

where $$V_1 = \frac{M_{BA} + M_{AB}}{15} \qquad V_2 = \frac{M_{CD} + M_{DC}}{15} \tag{4b}$$

Substituting V_1 and V_2 given by Equations 4b into 4a gives

$$M_{BA} + M_{AB} + M_{CD} + M_{DC} = 0 \tag{4}$$

Alternatively, we can set $Q = 0$ in Equation 12.21 to produce Equation 4.

Express equilibrium equations in terms of displacements by substituting Equations 1 into Equations 2, 3, and 4. Combining terms and simplifying give

$$
\begin{aligned}
8K\theta_B + K\theta_C - 9K\psi &= 120 \\
2K\theta_B + 16K\theta_C - 3K\psi &= -120 \\
K\theta_B + K\theta_C - 4K\psi &= 0
\end{aligned}
$$

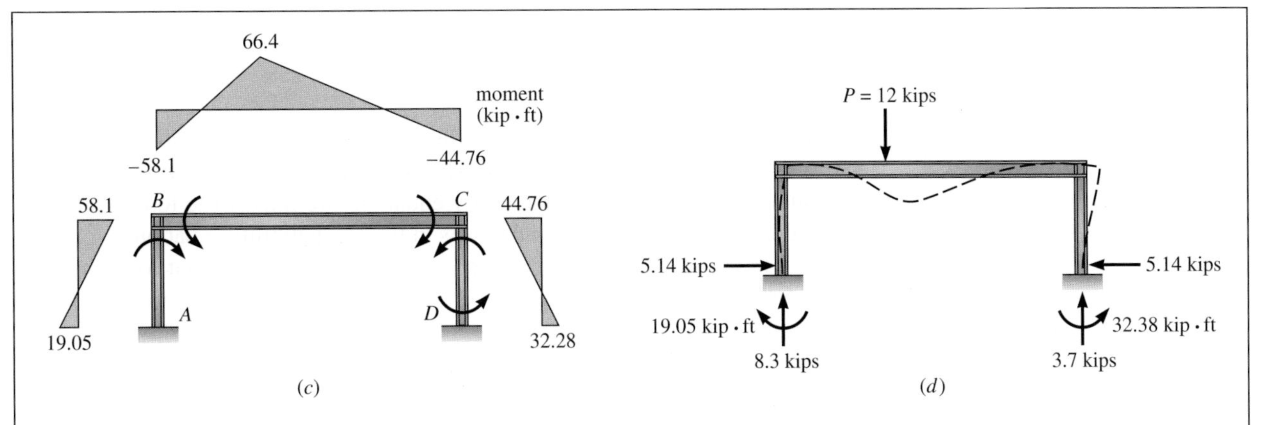

Figure 12.17: (*Continued*) (*c*) Member end moments and moment curves (in kip·ft); (*d*) reactions and deflected shape.

Solving the equations above simultaneously, we compute

$$\theta_B = \frac{410}{21K} \qquad \theta_C = -\frac{130}{21K} \qquad \psi = \frac{10}{3K} \tag{5}$$

Substituting the values of the θ_B, θ_C, and ψ into Equations 1, we compute the member end moments below.

$$M_{AB} = 19.05 \text{ kip}\cdot\text{ft} \qquad M_{BA} = 58.1 \text{ kip}\cdot\text{ft}$$

$$M_{CD} = -44.76 \text{ kip}\cdot\text{ft} \qquad M_{DC} = -32.38 \text{ kip}\cdot\text{ft} \tag{6}$$

$$M_{BC} = -58.1 \text{ kip}\cdot\text{ft} \qquad M_{CB} = 44.76 \text{ kip}\cdot\text{ft} \qquad \textbf{Ans.}$$

Member end moments and moment curves are shown on the sketch in Figure 12.17*c*; the deflected shape is shown in Figure 12.17*d*.

Compute the horizontal displacement of joint *B*. Use Equation 1 for M_{AB}. Express all variables in units of inches and kips.

$$M_{AB} = \frac{2EI}{15(12)}(\theta_B - 3\psi) \tag{7}$$

From the values in Equation 5 (p. 485), $\theta_B = 5.86\psi$; substituting into Equation 7, we compute

$$19.05(12) = \frac{2(30{,}000)(240)}{15(12)}(5.86\psi - 3\psi)$$

$$\psi = 0.000999 \text{ rad}$$

$$\psi = \frac{\Delta}{L} \qquad \Delta = \psi L = 0.000999(15 \times 12) = 0.18 \text{ in} \qquad \textbf{Ans.}$$

12.6 Kinematic Indeterminacy

To analyze a structure by the flexibility method, we first established the degree of indeterminacy of the structure. The degree of *statical indeterminacy* determines the number of compatibility equations we must write to evaluate the redundants, which are the unknowns in the compatibility equations.

In the slope-deflection method, displacements—both joint rotations and translations—are the unknowns. As a basic step in this method, we must write equilibrium equations equal in number to the independent joint displacements. The number of independent joint displacements is termed the degree of *kinematic indeterminacy*. To determine the kinematic indeterminacy, we simply count the number of independent joint displacements that are free to occur. For example, if we neglect axial deformations, the beam in Figure 12.18*a* is kinematically indeterminate to the first degree. If we were to analyze this beam by slope-deflection, only the rotation of joint *B* would be treated as an unknown.

If we also wished to consider axial stiffness in a more general stiffness analysis, the axial displacement at *B* would be considered an additional unknown, and the structure would be classified as kinematically indeterminate to the second degree. Unless otherwise noted, we will neglect axial deformations in this discussion.

In Figure 12.18*b* the frame would be classified as kinematically indeterminate to the fourth degree because joints *A*, *B*, and *C* are free to rotate and the girder can translate laterally. Although the number of joint rotations is simple to identify, in certain types of problems the number of independent joint displacements may be more difficult to establish. One method to determine the number of independent joint displacements is to introduce imaginary rollers as joint restraints. The number of rollers required to restrain the joints of the structure from translating equals the number of independent joint displacements. For example, in Figure 12.18*c* the structure would be classified as kinematically indeterminate to the eighth degree, because six joint rotations and two joint displacements are possible. Each imaginary roller (noted by the numbers 1 and 2) introduced at a floor prevents all joints in that floor from displacing laterally. In Figure 12.18*d* the Vierendeel truss would be classified as kinematically indeterminate to the eleventh degree (i.e., eight joint rotations and three independent joint translations). Imaginary rollers (labeled 1, 2, and 3) added at joints *B*, *C*, and *H* prevent all joints from translating.

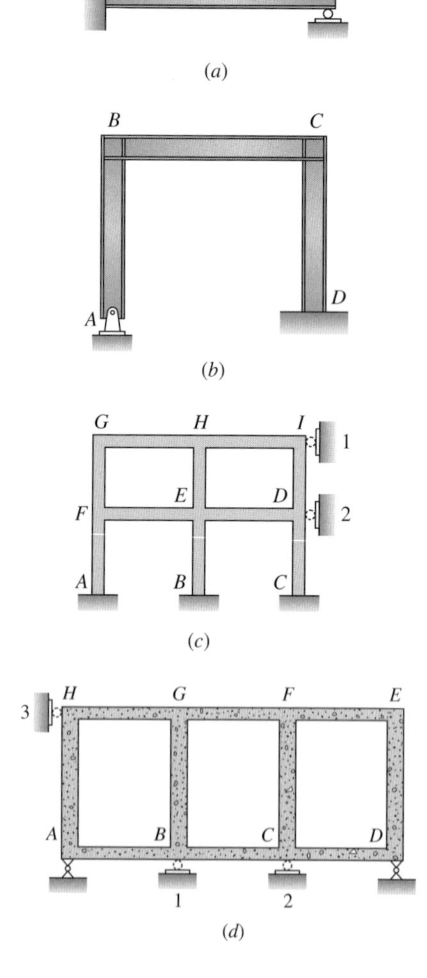

Figure 12.18: Evaluating degree of kinematic indeterminacy: (*a*) indeterminate first degree, neglecting axial deformations; (*b*) indeterminate fourth degree; (*c*) indeterminate eighth degree, imaginary rollers added at points 1 and 2; (*d*) indeterminate eleventh degree, imaginary rollers added at points 1, 2, and 3.

Summary

- The *slope-deflection* procedure is a classical method for analyzing indeterminate beams and rigid frames. In this method joint displacements are the unknowns.
- A step-by-step procedure to analyze an indeterminate beam or frame based on the slope-deflection method is summarized in Section 12.4.
- For highly indeterminate structures with a large number of joints, the slope-deflection solution requires that the engineer solve a series of simultaneous equations equal in number to the unknown displacements—a time-consuming operation. While the use of the slope-deflection method to analyze structures is impractical given the availability of computer programs, familiarity with the method provides students with valuable insight into the behavior of structures.
- As an alternate to the slope-deflection method, *moment distribution* was developed in the 1930s to analyze indeterminate beams and frames by distributing unbalanced moments at joints in an artificially restrained structure. While this method eliminates the solution of simultaneous equations, it is still relatively long, especially if a large number of loading conditions must be considered. Nevertheless, moment distribution is a useful tool as an approximate method of analysis both for checking the results of a computer analysis and in making preliminary studies. We will use the slope-deflection equation to develop the moment distribution method in Chapter 13.
- A variation of the slope-deflection procedure, the *general stiffness* method, used to prepare general-purpose computer programs, is presented in Chapter 16. This method utilizes stiffness coefficients—forces produced by unit displacements of joints.

PROBLEMS

P12.1 and **P12.2.** Using Equations 12.12 and 12.13, compute the fixed-end moments for the fixed-ended beams. See Figures P12.1 and P12.2.

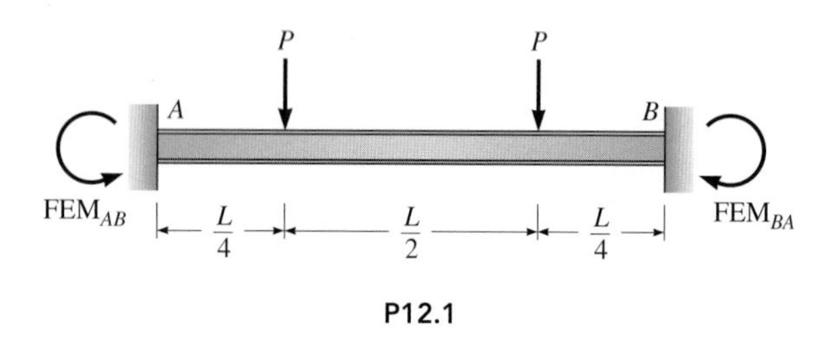

P12.1

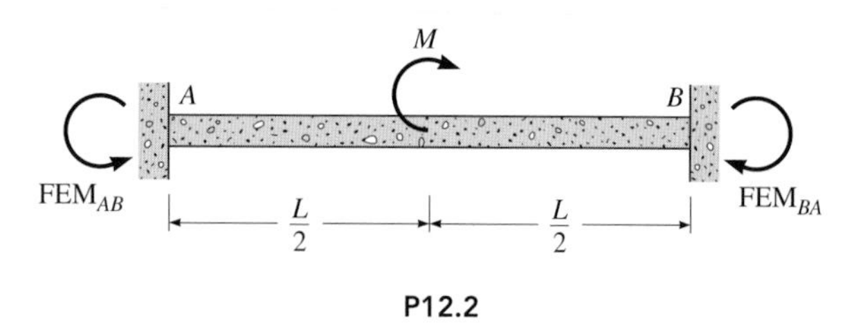

P12.2

P12.3. Analyze by slope-deflection and draw the shear and moment curves for the beam in Figure P12.3. Given: EI = constant.

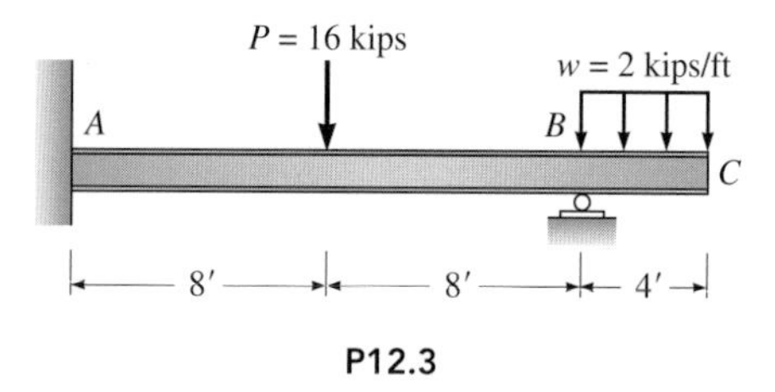

P12.3

P12.4. Analyze the beam in Figure P12.4 by slope-deflection and draw the shear and moment diagrams for the beam. EI is constant.

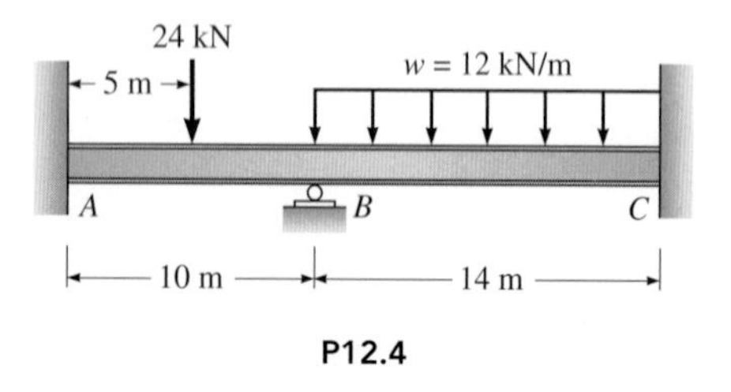

P12.4

P12.5. Compute the reactions at A and C in Figure P12.5. Draw the shear and moment diagram for member BC. Given: $I = 2{,}000$ in^4 and $E = 3{,}000$ kips/in^2.

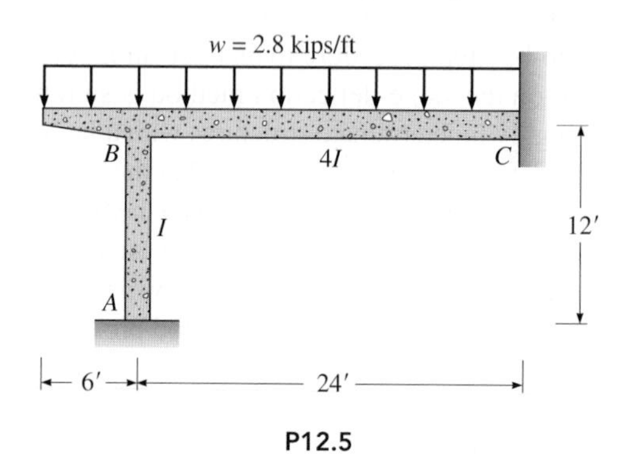

P12.5

P12.6. Draw the shear and moment curves for the frame in Figure P12.6. Given: EI is constant.

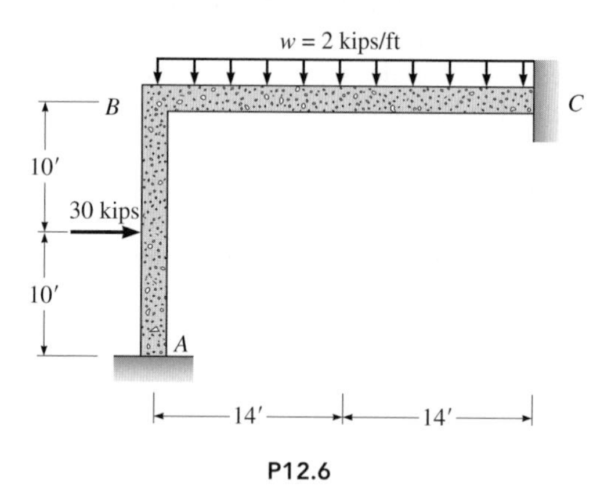

P12.6

P12.7. Analyze the beam in Figure P12.7. Draw the shear and moment curves. Given: $E = 29{,}000$ ksi and $I = 100$ in^4.

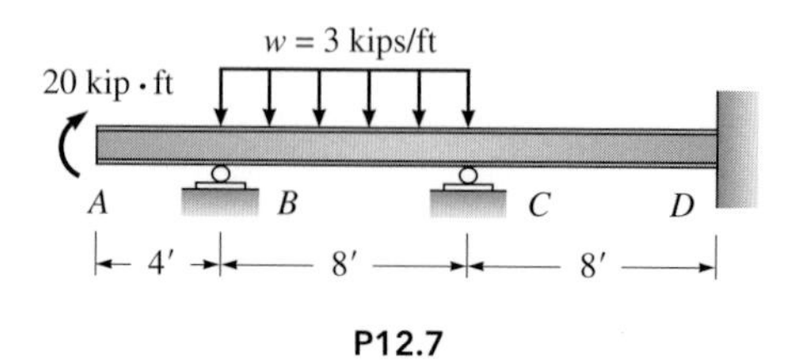

P12.7

P12.8. If no vertical deflection is permitted at end A for the beam in Figure P12.8, compute the required weight W that needs to be placed at midspan of CD. Given: $E = 29{,}000$ ksi and $I = 100$ in^4.

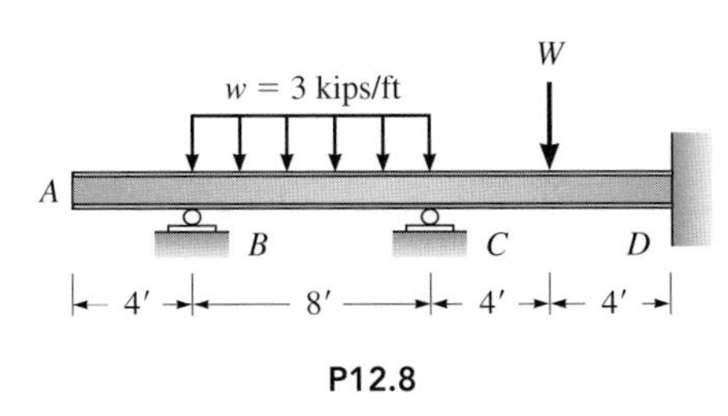

P12.8

P12.9. (*a*) Under the applied loads support B in Figure P12.9 settles 0.5 in. Determine all reactions. Given: $E = 30{,}000$ kips/in^2, $I = 240$ in^4. (*b*) Compute the deflection of point C.

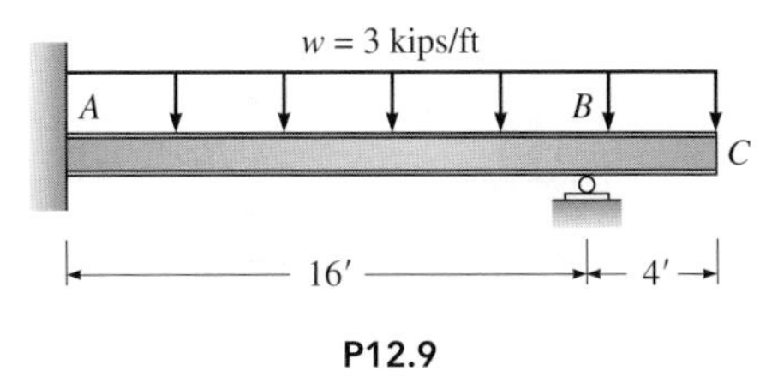

P12.9

P12.10. In Figure P12.10, support A rotates 0.002 rad and support C settles 0.6 in. Draw the shear and moment curves. Given: $I = 144$ in^4 and $E = 29{,}000$ kips/in^2.

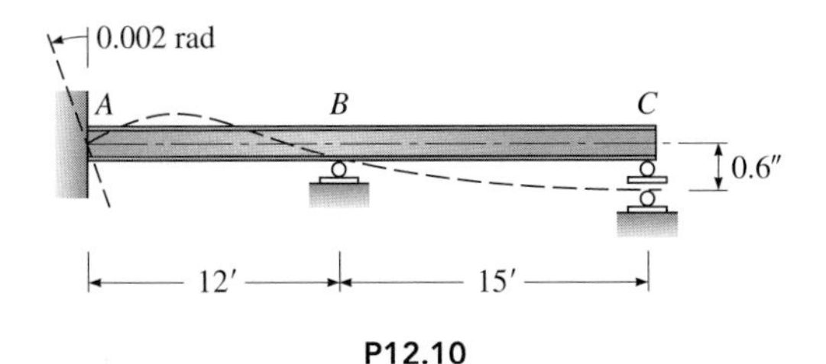

P12.10

In Problems P12.11 to P12.14, take advantage of symmetry to simplify the analysis by slope deflection.

P12.11. (*a*) Compute all reactions and draw the shear and moment curves for the beam in Figure P12.11. Given: EI is constant. (*b*) Compute the deflection under the load.

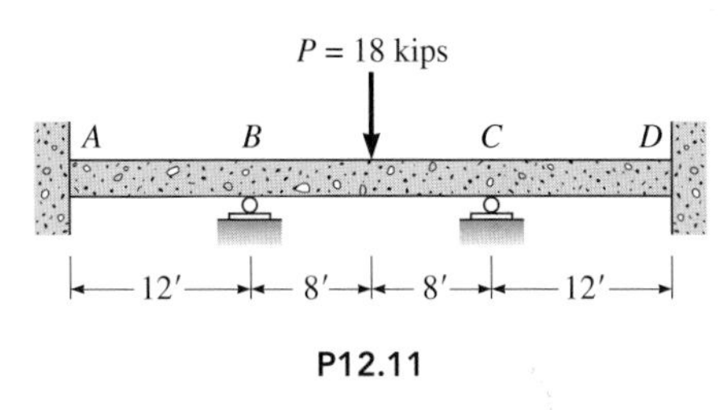

P12.11

P12.12. (*a*) Determine the member end moments for the rectangular ring in Figure P12.12, and draw the shear and moment curves for members AB and AD. The cross section of the rectangular ring is 12 in × 8 in and $E = 3000$ kips/in^2. (*b*) What is the axial force in member AD and in member AB?

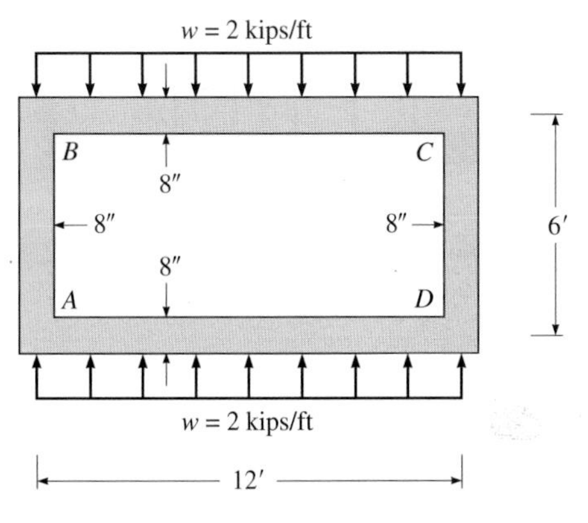

P12.12

P12.13. Figure P12.13 shows the forces exerted by the soil pressure on a typical 1-ft length of a concrete tunnel as well as the design load acting on the top slab. Assume that a fixed-end condition at the bottom of the walls at A and D is produced by the connection to the foundation mat. Determine the member end moments and draw the shear and moment curves. Also draw the deflected shape. EI is constant.

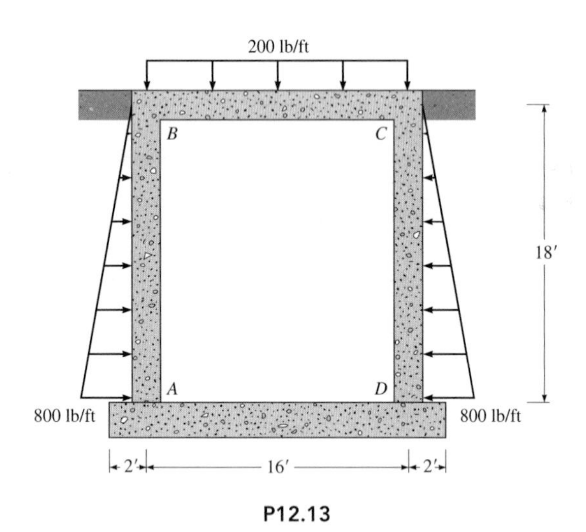

P12.13

P12.14. Compute the reactions and draw the shear and moment curves for the beam in Figure P12.14. Given: $E =$ 200 GPa and $I = 120 \times 10^6$ mm^4.

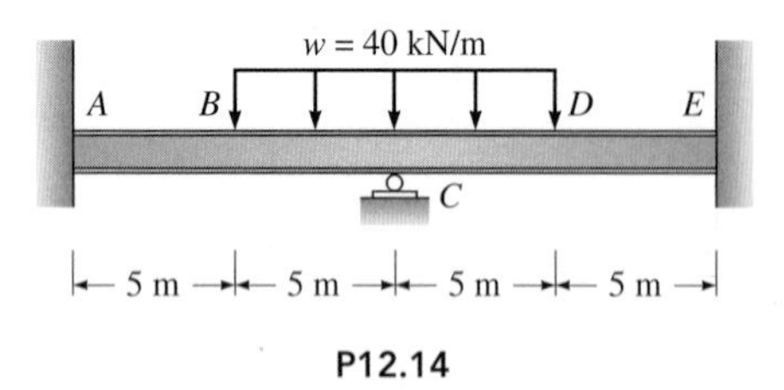

P12.14

P12.15. Consider the beam in Figure P12.14 without the applied load. Compute the reactions and draw the shear and moment curves for the beam if support C settles 24 mm and support A rotates counterclockwise 0.005 rad.

P12.16. Analyze the frame in Figure P12.16. In addition to the applied loads, supports A and D settle by 2.16 in. $EI = 36{,}000$ kip·ft^2 for beams and $EI = 72{,}000$ kip·ft^2 for columns. Use symmetry to simplify the analysis.

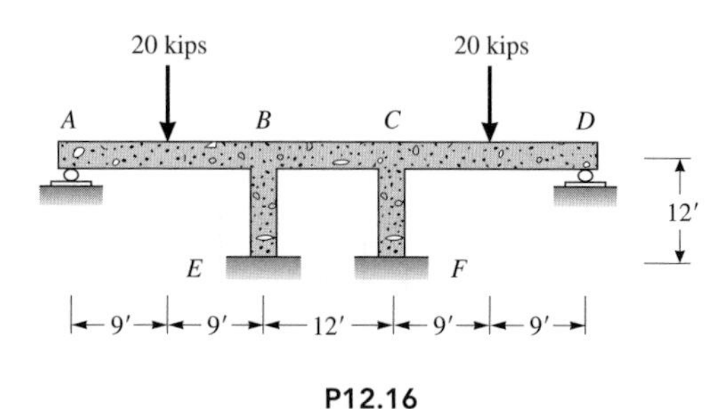

P12.16

P12.17. Analyze the structure in Figure P12.17. In addition to the applied load, support A rotates clockwise by 0.005 rad. Also $E = 200$ GPa and $I = 25 \times 10^6$ mm^4 for all members.

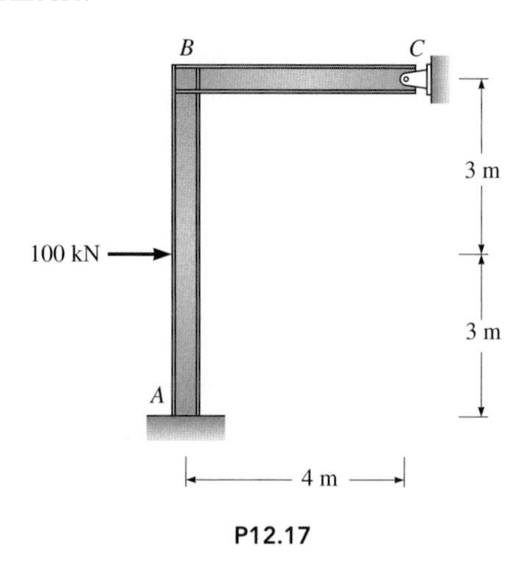

P12.17

P12.18. Analyze the frame in Figure P12.18. Compute all reactions. Given: EI is constant.

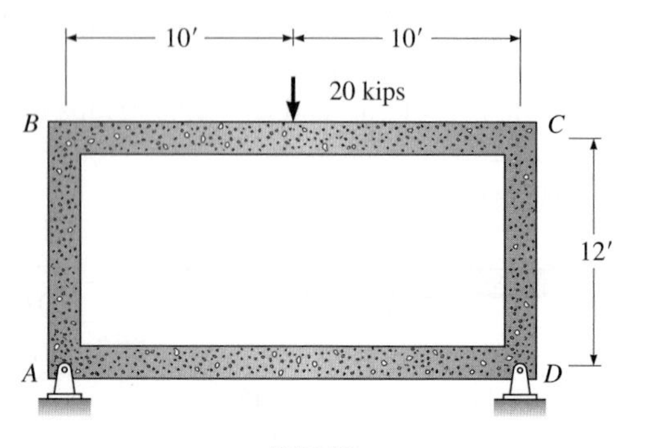

P12.18

P12.19. Analyze the frame in Figure P12.19. Given: *EI* is constant.

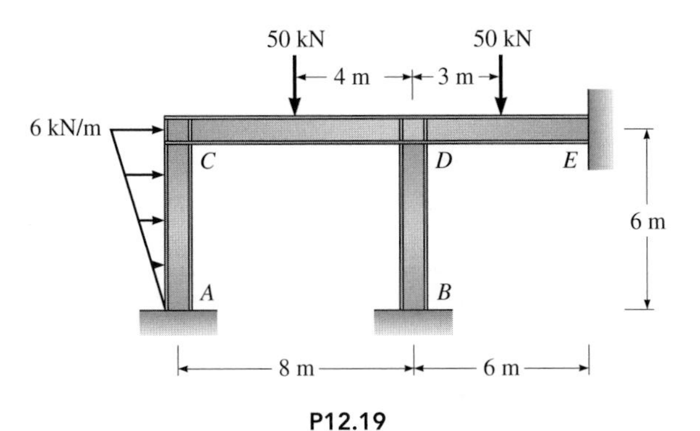

P12.19

P12.20. Analyze the frame in Figure P12.20. Compute all reactions. Given: *EI* is constant.

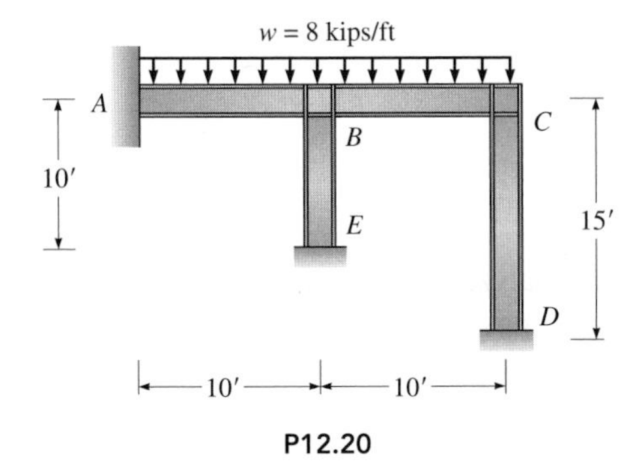

P12.20

P12.21. Analyze the frame in Figure P12.20. Ignore the applied load. But support *E* settles by 1 in. Use $E = 29{,}000$ ksi and $I = 100\ \text{in}^4$.

P12.22. Compute the reactions and draw the shear and moment diagrams for beam *BD* in Figure P12.22. *EI* is constant.

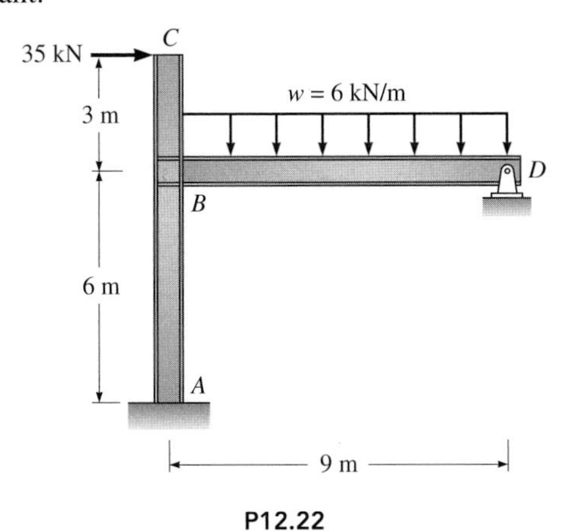

P12.22

P12.23. Analyze the frame in Figure P12.23. Compute the reactions and draw the shear and moment diagrams for members *AB* and *BD*. Given: *EI* is constant.

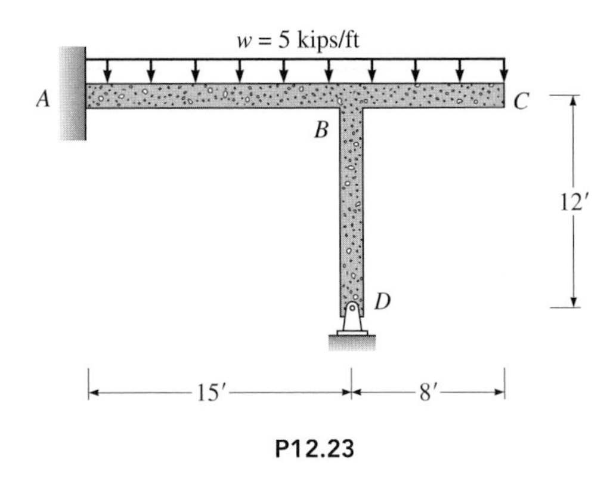

P12.23

P12.24. If support *A* in Figure P12.24 is constructed 0.48 in too low and the support at *C* is accidentally constructed at a slope of 0.016 rad clockwise from a vertical axis through *C*, determine the moment and reactions created when the structure is connected to its supports. Given: $E = 29{,}000\ \text{kips/in}^2$.

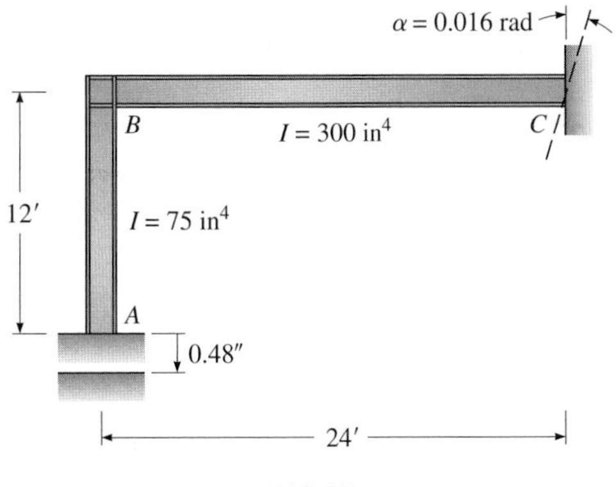

P12.24

P12.25. If member *AB* in Figure P12.25 is fabricated $\frac{3}{4}$ in too long, determine the moments and reactions created in the frame when it is erected. Sketch the deflected shape. Given: $E = 29{,}000$ kips/in^2.

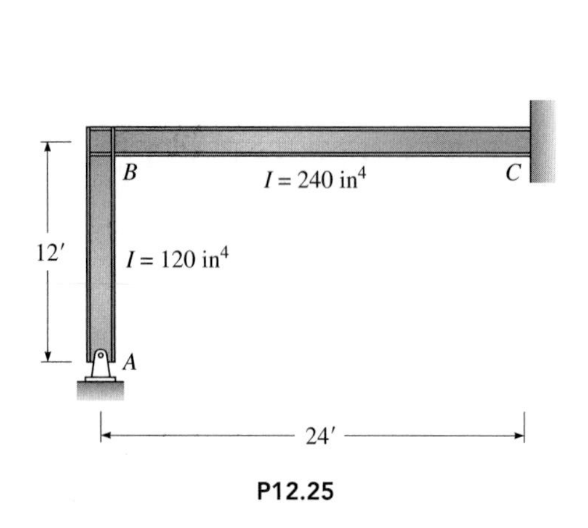

P12.25

P12.26. Analyze the frame in Figure P12.26. Given: *EI* is constant.

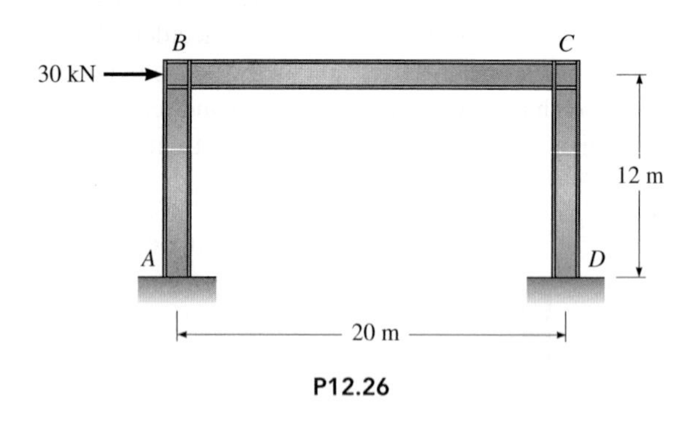

P12.26

P12.27. Analyze the frame in Figure P12.27. Note that support *D* can translate in the horizontal direction only. Compute all reactions and draw the shear and moment curves. Given: $E = 29{,}000$ ksi and $I = 100$ in^4.

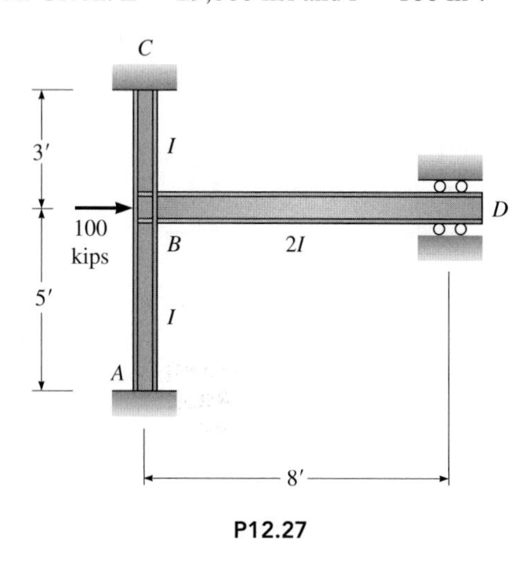

P12.27

P12.28. Analyze the frame in Figure P12.28. Notice that sidesway is possible because the load is unsymmetric. Compute the horizontal displacement of joint *B*. Given: $E = 29{,}000$ kips/in^2 and $I = 240$ in^4 for all members.

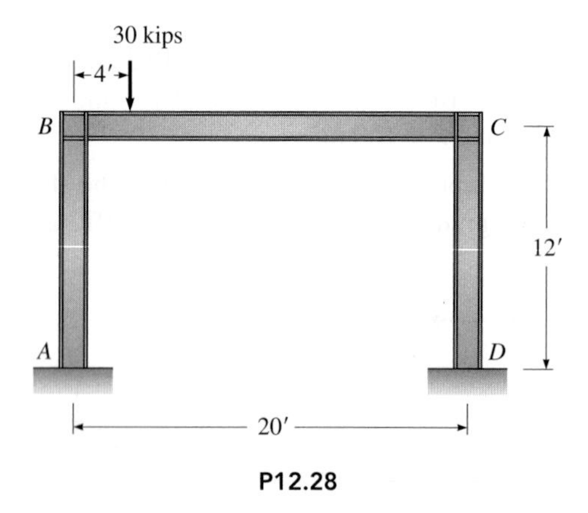

P12.28

P12.29. Analyze the frame in Figure P12.29. Compute all reactions. Also $I_{BC} = 200$ in^4 and $I_{AB} = I_{CD} = 150$ in^4. E is constant.

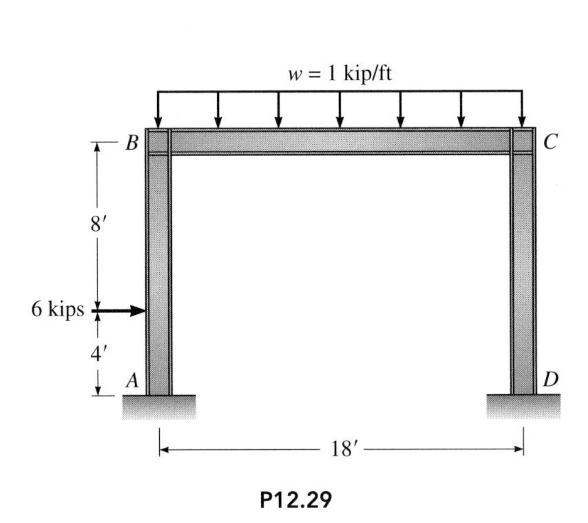

P12.29

P12.30. Determine all reactions at points A and D in Figure P12.30. EI is constant.

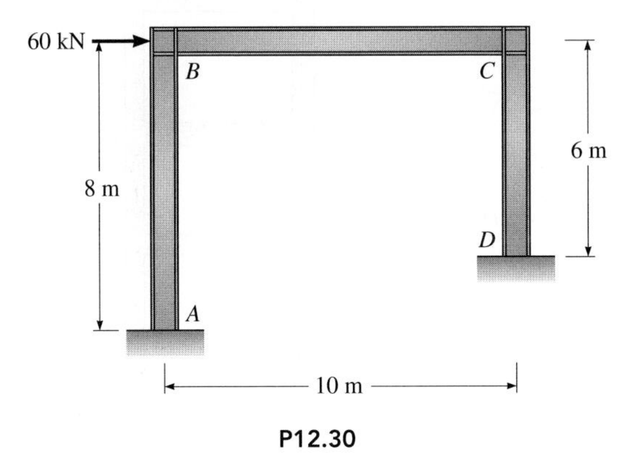

P12.30

P12.31. Analyze the frame in Figure P12.31. EI is constant.

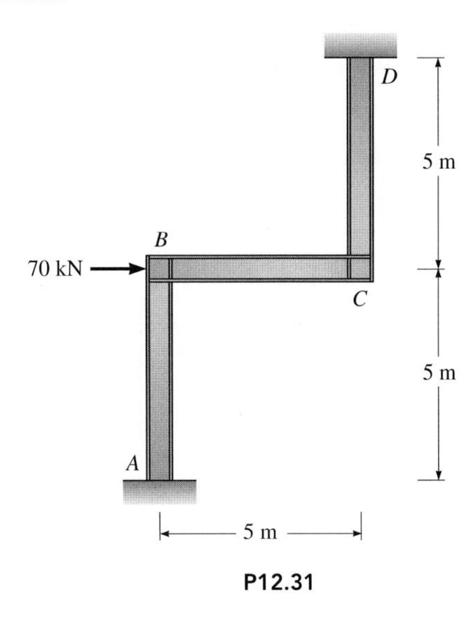

P12.31

P12.32. Set up the equilibrium equations required to analyze the frame in Figure P12.32 by the slope deflection method. Express the equilibrium equations in terms of the appropriate displacements; EI is constant for all members.

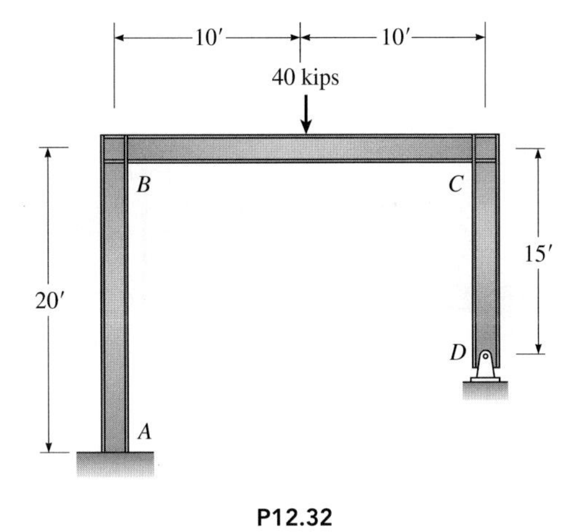

P12.32

P12.33. Set up the equilibrium equations required to analyze the frame in Figure P12.33 by the slope-deflection method. Express the equilibrium equations in terms of the appropriate displacements; *EI* is constant for all members.

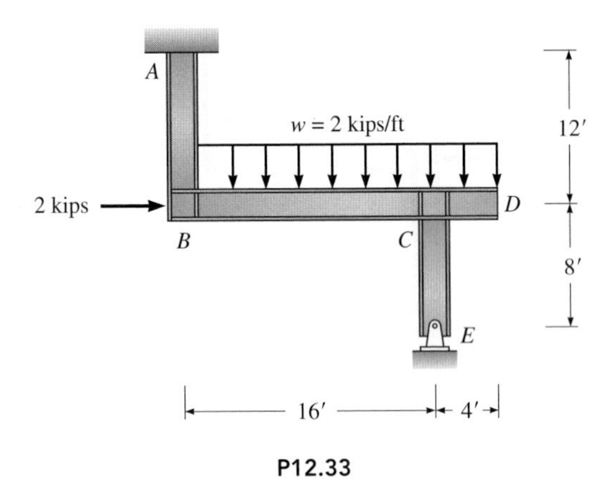

P12.33

P12.34. Set up the equilibrium equations required to analyze the frame in Figure 12.34 by the slope-deflection method. Express the equilibrium equations in terms of the appropriate displacements; *EI* is constant for all members.

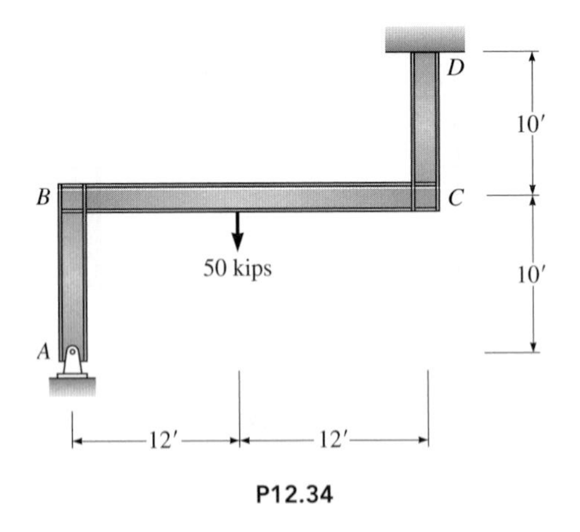

P12.34

P12.35. Determine the degree of kinematic indeterminacy for each structure in Figure P12.35. Neglect axial deformations.

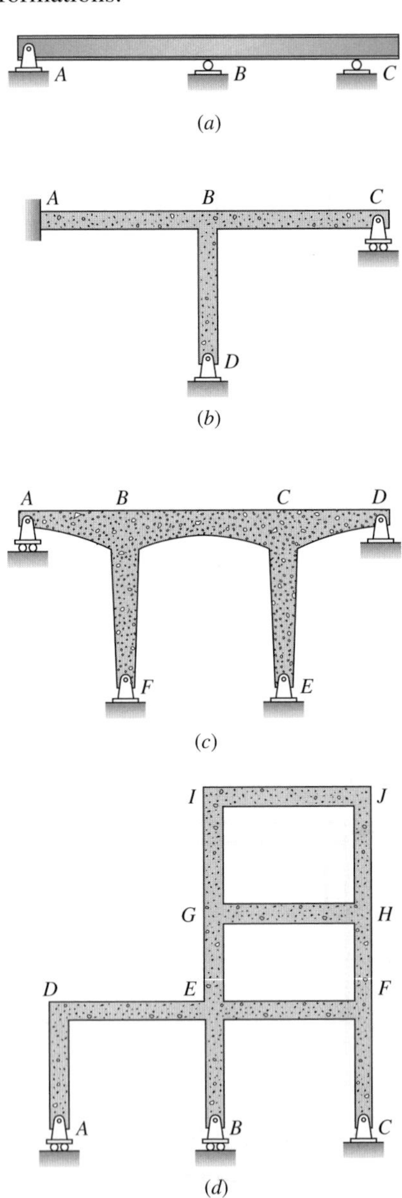

P12.35

Federal Reserve Office Building, Boston, MA

The Federal Reserve Bank Building in Boston, nicknamed "the washboard", is a 33-story office tower. Spanning between two end towers, a 36-ft deep steel truss rising 140 ft above the street level supports the tower office floors.

C H A P T E R 13

Moment Distribution

Chapter Objectives

- Learn the moment distribution method, which is an approximate procedure for analyzing indeterminate beams and frames and eliminates the need to write and solve the simultaneous equations.
- Understand how joint equilibrium is achieved by unlocking and locking joints in succession and distributing moments to both ends of all members framing to the joint until all joints achieve equilibrium.
- Learn the procedure to analyze beams and frames with sway inhibited, and then extend the procedure to frames with sway uninhibited.
- Extend the use of moment distribution method to beams and frames with nonprismatic members.

13.1 Introduction

Moment distribution, developed by Hardy Cross in the early 1930s, is a procedure for establishing the end moments in members of indeterminate beams and frames with a series of simple computations. The method is based on the idea that the sum of the moments applied by the members framing into a joint must equal zero because the joint is in equilibrium. In many cases moment distribution eliminates the need to solve large numbers of simultaneous equations such as those produced in the analysis of highly indeterminate structures by either the flexibility or slope-deflection method. While continuous rigid-jointed structures—welded steel or reinforced concrete frames and continuous beams—are routinely and rapidly analyzed for multiple loading conditions by computer, moment distribution remains a valuable tool for (1) checking the results of a computer analysis or (2) carrying out an approximate analysis in the preliminary design phase when members are initially sized.

In the moment distribution method, we imagine that temporary restraints are applied to all joints of a structure that are free to rotate or to displace. We apply hypothetical clamps to prevent rotation of joints and introduce imaginary rollers to prevent lateral displacements of joints (the rollers are required

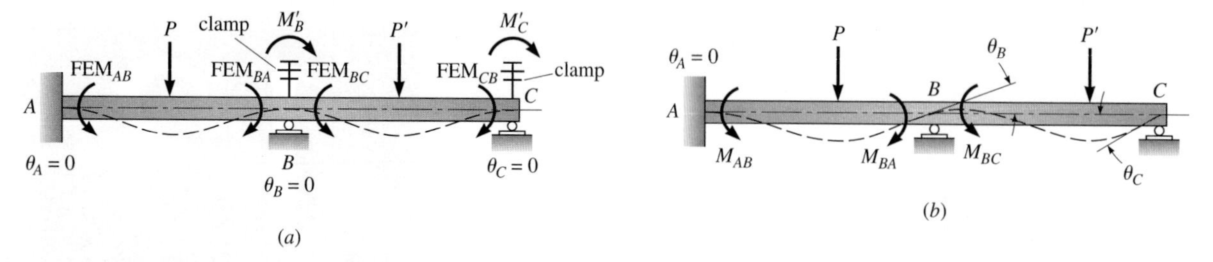

Figure 13.1: Continuous beam analyzed by moment distribution: (*a*) temporary clamps added at joints *B* and *C* to produce a restrained structure consisting of two fixed-end beams; (*b*) clamps removed and beam deflected into its equilibrium position.

Hardy Cross

Before the advent of computer, analyzing a multistory building frame was a formidable task because the structure is highly indeterminate. Hardy Cross (1885–1959) overcame this obstacle and achieved an international reputation by developing the moment distribution method. Bypassing the solution of simultaneous equations, his iterative method combined reasonable precision with speed. This stiffness method is rooted in a numerical analysis procedure called the Jacobi method. Cross also extended his method to solving complex water pipe network problems.

only for structures that sidesway). The initial effect of introducing restraints is to produce a structure composed entirely of fixed-end members. When we apply the design loads to the restrained structure, moments are created in the members and clamps.

For a structure restrained against sidesway (the most common case), the analysis is completed by removing clamps—one by one—from successive joints and distributing moments to the members framing into the joint. Moments are distributed to the ends of members in proportion to their flexural stiffness. When the moments in all clamps have been absorbed by the members, the indeterminate analysis is complete. The balance of the analysis—constructing shear and moment curves, computing axial forces in members, or evaluating reactions—is completed with the equations of statics.

For example, as the first step in the analysis of the continuous beam in Figure 13.1*a* by moment distribution, we apply imaginary clamps to joints *B* and *C*. Joint *A*, which is fixed, does not require a clamp. When loads are applied to the individual spans, fixed-end moments develop in the members and restraining moments (M'_B and M'_C) develop in the clamps. As the moment distribution solution progresses, the clamps at supports *B* and *C* are alternately removed and replaced in a series of iterative steps until the beam deflects into its equilibrium position, as shown by the dashed line in Figure 13.1*b*. After you learn a few simple rules for distributing moments among the members framing into a joint, you will be able to analyze many types of indeterminate beams and frames rapidly.

Initially we consider structures composed of straight, prismatic members only, that is, members whose cross sections are constant throughout their entire length. Later we will extend the procedure to structures that contain members whose cross section varies along the axis of the member.

13.2 Development of the Moment Distribution Method

To develop the moment distribution method, we will use the slope-deflection equation to evaluate the member end moments in each span of the continuous beam in Figure 13.2*a* after an imaginary clamp that prevents rotation of joint *B* is removed and the structure deflects into its final equilibrium position.

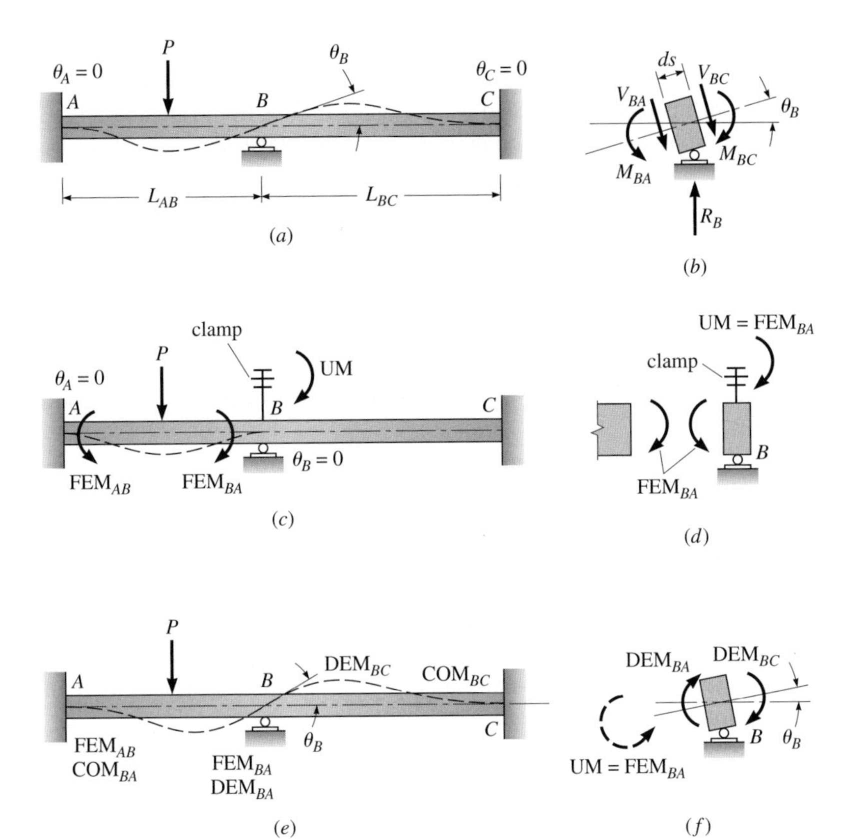

Figure 13.2: Various stages in the analysis of a beam by moment distribution; (*a*) loaded beam in deflected position; (*b*) free-body diagram of joint *B* in deflected position; (*c*) fixed-end moments in restrained beam (joint *B* clamped); (*d*) free-body diagram of joint *B* before clamp removed; (*e*) moments in beam after clamp removed; (*f*) distributed end moments (DEMs) produced by joint rotation θ_B to balance the unbalanced moment (UM).

Although we introduce moment distribution by analyzing a simple structure that has only one joint that is free to rotate, this case will permit us to develop the most important features of the method.

When the concentrated load *P* is applied to span *AB*, the initially straight beam deflects into the shape shown by the dashed line. At support *B* a line tangent to the elastic curve of the deformed beam makes an angle of θ_B with the horizontal axis. The angle θ_B is shown greatly exaggerated and would typically be less than 1°. At supports *A* and *C*, the slope of the elastic curve is zero because the fixed ends are not free to rotate. In Figure 13.2*b* we show a detail of the joint at support *B* after the loaded beam has deflected into its equilibrium position. The joint, which consists of a differential length *ds* of beam segment, is loaded by shears and moments from beam *AB* and *BC*, and by the support reaction R_B. If we sum moments about the centerline of support *B*, equilibrium of the joint with respect to moment requires that $M_{BA} = M_{BC}$, where M_{BA} and M_{BC} are the moments applied to joint *B* by members *AB* and *BC*, respectively. Since the distance between the faces of the element and the centerline of the support is infinitesimally small, the moment produced by the shear forces is a second-order quantity and does not have to be included in the moment equilibrium equation.

We now consider in detail the various steps of the moment distribution procedure that permits us to calculate the values of member end moments in spans *AB* and *BC* of the beam in Figure 13.2. In the first step (see Figure 13.2*c*) we imagine that joint *B* is locked against rotation by a large clamp. The application of the clamp produces two fixed-end beams. When P is applied to the midspan of member AB, fixed-end moments (FEMs) develop at each end of the member. These moments can be evaluated using Figure 12.5 or from Equations 12.12 and 12.13. No moments develop in beam BC at this stage because no loads act on the span.

Figure 13.2*d* shows the moments acting between the end of beam AB and joint B. The beam applies a counterclockwise moment FEM_{BA} to the joint. To prevent the joint from rotating, the clamps must apply a moment to the joint that is equal and opposite to FEM_{BA}. The moment that develops in the clamp is called the *unbalanced moment* (UM). If span BC were also loaded, the unbalanced moment in the clamp would equal the *difference* between the fixed-end moments applied by the two members framing into the joint.

If we now remove the clamp, joint B will rotate through an angle θ_B in the counterclockwise direction into its equilibrium position (see Figure 13.2*e*). As joint B rotates, additional moments, labeled DEM_{BC}, COM_{BC}, DEM_{BA}, and COM_{BA}, develop at the ends of members AB and BC. At joint B these moments, called the *distributed end moments* (DEMs), are opposite in sense to the unbalanced moment (see Figure 13.2*f*). In other words, when the joint reaches equilibrium, the sum of the distributed end moments equals the unbalanced moment, which was formerly equilibrated by the clamp. We can state this condition of joint equilibrium as

$$\circlearrowright^{+} \quad \Sigma M_B = 0$$
$$\text{DEM}_{BA} + \text{DEM}_{BC} - \text{UM} = 0 \tag{13.1}$$

where DEM_{BA} = moment at B end of member AB produced by rotation of joint B

DEM_{BC} = moment at B end of member BC produced by rotation of joint B

UM = unbalanced moment applied to joint

In all moment distribution computations, the sign convention will be the same as that used in the slope-deflection method: *Rotations of the ends of members and moments applied to the ends of members are positive in the clockwise direction and negative in the counterclockwise direction*. In Equation 13.1 and in the sketches of Figure 13.2, the plus or minus sign is not shown but is contained in the abbreviations used to designate the various moments.

The moments produced at the A end of member AB and at the C end of member BC by the rotation of joint B are called *carryover moments* (COMs). As we will show next:

1. The final moment at the end of each member equals the algebraic sum of the distributed end moment (or the carryover moment) and fixed-end moment (if the span is loaded).
2. For members of constant cross section, the carryover moment in each span has the same sign as the distribution end moment, but is one-half as large.

To verify the magnitude of the final moments at each end of the members AB and BC in Figure 13.2*e*, we will use the slope-deflection equation (Equation 12.16) to express the member end moments in terms of the properties of the members, the applied load, and the rotation of joint B: For $\theta_A = \theta_C = \psi = 0$, Equation 12.16 yields

Member AB:

$$M_{BA} = \frac{2EI_{AB}}{L_{AB}}(2\theta_B) + \text{FEM}_{BA} = \underbrace{\frac{4EI_{AB}}{L_{AB}}\theta_B}_{(\text{DEM}_{BC})} + \text{FEM}_{BA} \tag{13.2}$$

$$M_{AB} = \underbrace{\frac{2EI_{AB}}{L_{AB}}\theta_B}_{(\text{COM}_{BA})} + \text{FEM}_{AB} \tag{13.3}$$

Member BC:

$$M_{BC} = \frac{2EI_{BC}}{L_{BC}}(2\theta_B) = \underbrace{\frac{4EI_{BC}}{L_{BC}}\theta_B}_{(\text{DEM}_{BC})} \tag{13.4}$$

$$M_{CB} = \underbrace{\frac{2EI_{BC}}{L_{BC}}\theta_B}_{(\text{COM}_{BC})} \tag{13.5}$$

Equation 13.2 shows that the total moment M_{BA} at the B end of member AB (Figure 13.2*e*) equals the sum of (1) the fixed-end moment FEM_{BA} and (2) the distributed end moment DEM_{BA}. DEM_{BA} is given by the first term on the right side of Equation 13.2 as

$$\text{DEM}_{BA} = \frac{4EI_{AB}}{L_{AB}}\theta_B \tag{13.6}$$

In Equation 13.6 the term $4EI_{AB}/L_{AB}$ is termed the *absolute flexural stiffness* of the B end of member AB. It represents the moment required to produce a rotation of 1 rad at B when the far end at A is fixed against rotation. If the beam is not prismatic, that is, if the cross section varies along the axis of the member, the numerical constant in the absolute flexural stiffness will not equal 4 (see Section 13.9).

Equation 13.3 shows that the total moment at the A end of member AB equals the sum of the fixed-end moment FEM_{AB} and the carryover moment COM_{BA}. COM_{BA} is given by the first term of Equation 13.3 as

$$\text{COM}_{BA} = \frac{2EI_{AB}}{L_{AB}}\theta_B \tag{13.7}$$

If we compare the values of DEM_{BA} and COM_{BA} given by Equations 13.6 and 13.7, we see that they are identical except for the numerical constants 2 and 4. Therefore, we conclude that

$$\text{COM}_{BA} = \frac{1}{2}(\text{DEM}_{BA}) \tag{13.8}$$

Since both the carryover moment and the distributed end moment given by Equations 13.6 and 13.7 are functions of θ_B—the only variable that has a plus or minus sign—*both moments have the same sense*, that is, positive if θ_B is clockwise and negative if θ_B is counterclockwise.

Equation 13.4 shows that the moment at the B end of member BC is due only to rotation θ_B of joint B since no loads act on span BC. Similarly, Equation 13.5 indicates the carryover moment at the C end of member BC is due only to rotation θ_B of joint B. If we compare the value of M_{BC}, the distributed end moment at the B end of member BC, with M_{CB}, the carryover moment at the C end of member BC, we reach the same conclusion given by Equation 13.8; that is, the *carryover moment equals one-half the distributed end moment*.

We can establish the magnitude of the distributed end moments at joint B (see Figure 13.2*f*) as a percentage of the unbalanced moment in the clamp at joint B by substituting their values, given by the first term of Equation 13.2 and by Equation 13.4, into Equation 13.1:

$$\text{DEM}_{BA} + \text{DEM}_{BC} - \text{UM} = 0 \tag{13.1}$$

$$\frac{4EI_{BC}}{L_{BC}}\theta_B + \frac{4EI_{AB}}{L_{AB}}\theta_B = \text{UM} \tag{13.9}$$

Solving Equation 13.9 for θ_B yields

$$\theta_B = \frac{\text{UM}}{4EI_{AB}/L_{AB} + 4EI_{BC}/L_{BC}} \tag{13.10}$$

If we let

$$K_{AB} = \frac{I_{AB}}{L_{AB}} \quad \text{and} \quad K_{BC} = \frac{I_{BC}}{L_{BC}} \tag{13.11}$$

where the ratio I/L is termed the *relative flexural stiffness*, we may write Equation 13.10 as

$$\theta_B = \frac{\text{UM}}{4EK_{AB} + 4EK_{BC}} = \frac{\text{UM}}{4E(K_{AB} + K_{BC})} \tag{13.12}$$

If $K_{AB} = I_{AB}/L_{AB}$ (see Equation 13.11) and θ_B given by Equation 13.12 are substituted into Equation 13.6, we may express the distributed end moment DEM_{BA} as

$$\text{DEM}_{BA} = 4EK_{AB}\frac{\text{UM}}{4E(K_{AB} + K_{BC})} \tag{13.13}$$

If the modulus of elasticity E of all members is the same, Equation 13.13 can be simplified (by canceling the constants $4E$) to

$$\text{DEM}_{BA} = \frac{K_{AB}}{K_{AB} + K_{BC}}\text{UM} \tag{13.14}$$

the term $K_{AB}/(K_{AB} + K_{BC})$, which gives the ratio of the relative flexural stiffness of member AB to the sum of the relative flexural stiffnesses of the members (AB and BC) framing into joint B, is called the *distribution factor* (DF_{BA}) for member AB.

$$DF_{BA} = \frac{K_{AB}}{K_{AB} + K_{BC}} = \frac{K_{AB}}{\Sigma K} \tag{13.15}$$

where $\Sigma K = K_{AB} + K_{BC}$ represents the sum of the relative flexural stiffnesses of the members framing into joint B. Using Equation 13.15, we can express Equation 13.14 as

$$DEM_{BA} = DF_{AB}(UM) \tag{13.16}$$

Similarly, the distributed end moment to member BC may be expressed as

$$DEM_{BC} = DF_{BC}(UM) \tag{13.16a}$$

where

$$DF_{BC} = \frac{K_{BC}}{K_{AB} + K_{BC}} = \frac{K_{BC}}{\Sigma K}$$

13.3 Summary of the Moment Distribution Method with No Joint Translation

We have now discussed in detail the basic moment distribution principles for analyzing a continuous structure in which joints are free to rotate but not to translate. Before we apply the procedure to specific examples, we summarize the method below.

1. Draw a line diagram of the structure to be analyzed.
2. At each joint that is free to rotate, compute the distribution factor for each member and record in a box on the line diagram adjacent to the joint. *The sum of the distribution factors at each joint must equal 1.*
3. Write down the fixed-end moments at the ends of each loaded member. As the sign convention we take clockwise moments on the ends of members as positive and counterclockwise moments as negative.
4. Compute the unbalanced moment at the first joint to be unlocked. The unbalanced moment at the first joint is the algebraic sum of the fixed-end moments at the ends of all members framing into the joint. After the first joint is unlocked, the unbalanced moments at the adjacent joints will equal the algebraic sum of fixed-end moments and any carryover moments.
5. Unlock the joint and distribute the unbalanced moment to the ends of each member framing into the joint. The distributed end moments are computed by multiplying the unbalanced moment by the distribution factor of each member. The sign of the distributed end moments is *opposite* to the sign of the unbalanced moment.

6. Write the carryover moments at the other end of the member. The carryover moment has the same sign as the distributed end moment but is one-half as large.
7. Replace the clamp and proceed to the next joint to distribute moments there. The analysis is finished when the unbalanced moments in all clamps are either zero or close to zero.

13.4 Analysis of Beams by Moment Distribution

To illustrate the moment distribution procedure, we will analyze the two-span continuous beam in Figure 13.3 of Example 13.1. Since only the joint at support B is free to rotate, a complete analysis requires only a single distribution of moments at joint B. In succeeding problems we consider structures that contain multiple joints that are free to rotate.

To begin the solution in Example 13.1, we compute member stiffness, the distribution factors at joint B, and the fixed-end moments in span AB. This information is recorded on Figure 13.4, where the moment distribution computations are carried out. *The 15-kip load on span AB and the clamp on joint B are not shown, to keep the sketch simple*. No distribution factors are computed for joints A and C because these joints are never unlocked. The unbalanced moment in the clamp at B is equal to the algebraic sum of the fixed-end moments at joint B. Since only span AB is loaded, the unbalanced moment—not shown on the sketch—equals +30 kip·ft. We now assume that the clamp at joint B is removed. The joint now rotates and distributed end moments of −10 and −20 kip·ft develop at the ends of member AB and BC. These moments are recorded directly below support B on the line below the fixed-end moments. Carryover moments of −5 kip·ft at joint A and −10 kip·ft at joint C are recorded on the third line. Since joints A and C are fixed supports, they never rotate, and the analysis is complete. The final moments at the ends of each member are computed by summing moments in each column. Note that at joint B the moments on each side of the support are equal but opposite in sign because the joint is in equilibrium. Once the end moments are established, the shears in each beam can be evaluated by cutting free bodies of each member and using the equations of statics. After the shears are calculated, the shear and moment curves are constructed. The final results are shown in Figure 13.5.

EXAMPLE 13.1

Determine the member end moments in the continuous beam shown in Figure 13.3 by moment distribution. Note that EI of all members is constant.

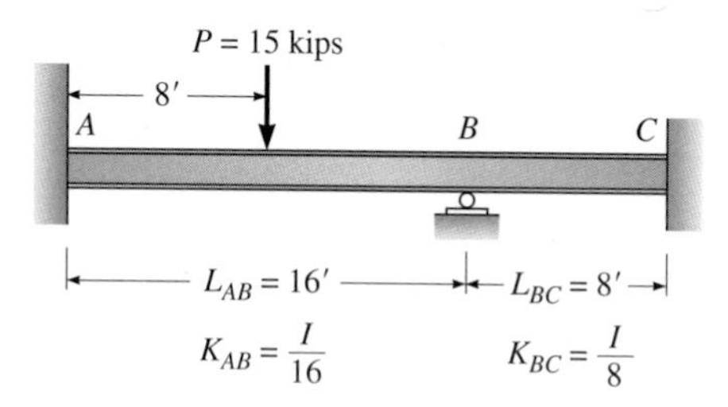

Figure 13.3

Solution

Compute the stiffness K of each member connected to joint B.

$$K_{AB} = \frac{I}{L_{AB}} = \frac{I}{16} \qquad K_{BC} = \frac{I}{L_{BC}} = \frac{I}{8}$$

$$\Sigma K = K_{AB} + K_{BC} = \frac{I}{16} + \frac{I}{8} = \frac{3I}{16}$$

Evaluate the distribution factors at joint B and record on Figure 13.4.

$$\text{DF}_{BA} = \frac{K_{AB}}{\Sigma K} = \frac{I/16}{3I/16} = \frac{1}{3}$$

$$\text{DF}_{BC} = \frac{K_{BC}}{\Sigma K} = \frac{I/8}{3I/16} = \frac{2}{3}$$

Compute the fixed-end moments at each end of member AB (see Figure 12.5) and record on Figure 13.4.

[*continues on next page*]

Example 13.1 continues . . .

$$\text{FEM}_{AB} = \frac{-PL}{8} = \frac{-15(16)}{8} = -30 \text{ kip} \cdot \text{ft}$$

$$\text{FEM}_{BA} = \frac{+PL}{8} = \frac{15(16)}{8} = +30 \text{ kip} \cdot \text{ft}$$

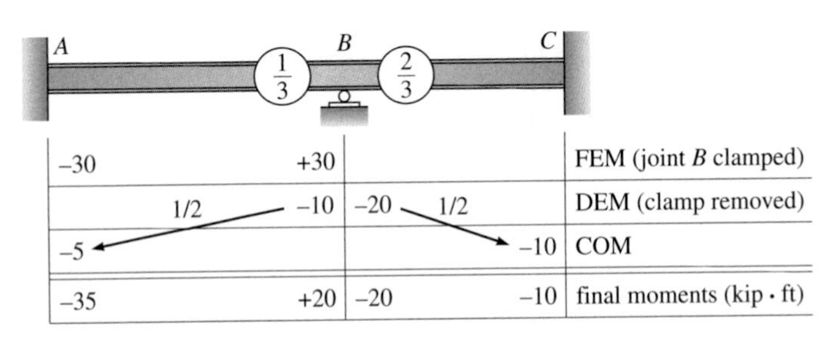

A		B	C	
–30	+30			FEM (joint B clamped)
1/2	–10	–20 1/2		DEM (clamp removed)
–5			–10	COM
–35	+20	–20	–10	final moments (kip · ft)

Figure 13.4: Moment distribution computations.

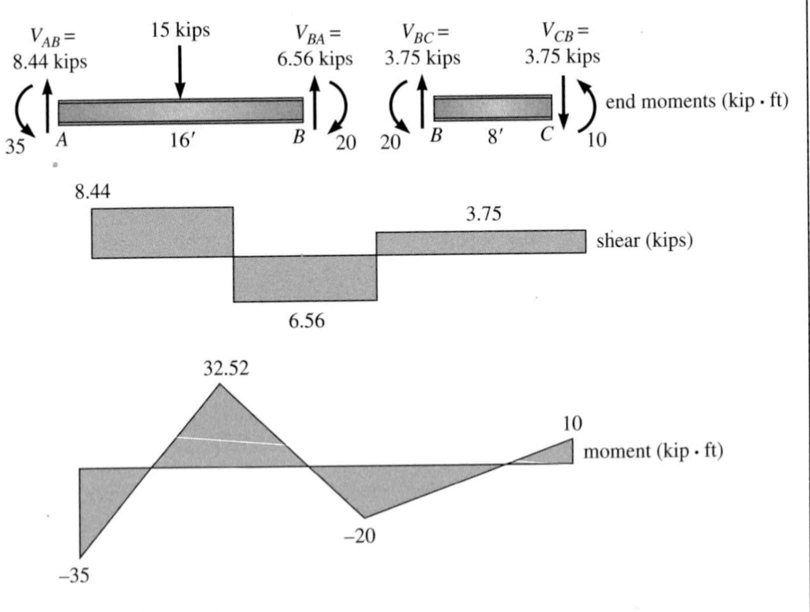

Figure 13.5: Shear and moment curves.

In Example 13.2 we extend the moment distribution method to the analysis of a beam that contains two joints—B and C—that are free to rotate (see Figure 13.6). As you can observe in Figure 13.7 where the moments distributed at each stage of the analysis are tabulated, the clamps on joints B and C must be locked and unlocked several times because each time one of these joints is unlocked, the moment changes in the clamp of the other joint because of the carryover moment. We begin the analysis by clamping joints B and C. Distribution factors and fixed-end moments are computed and recorded on the diagram of the structure in Figure 13.7. To help you follow the various steps in the analysis, a description of each operation is noted to the right of each line in Figure 13.7. As you become more familiar with moment distribution, this aid will be discontinued.

Although we are free to begin the distribution of moments by unlocking either joint B or joint C, we will assume that the imaginary clamp at joint B is removed first. The unbalanced moment at joint B—the algebraic sum of the fixed-end moments on either side of the joint—equals UM $= -96 + 48 = -48$ kip · ft. To compute the distributed end moments in each member, we reverse the sign of the unbalanced moment and multiply it by the member's distribution factor (each $\frac{1}{2}$ at joint B). Distributed end moments of +24 kip·ft are entered on the second line, and carryover moments of +12 kip · ft at supports A and C are recorded on the third line of Figure 13.7. To show that moments have been distributed and joint B is in equilibrium, we draw a short line under the distributed end moments at that joint. The imaginary clamp at joint B is now reapplied. Because joint B is now in equilibrium, the moment in the clamp is zero. Next we move to joint C, where the clamp equilibrates an unbalanced moment of +108 kip · ft. The unbalanced moment at C is the sum of the fixed-end moment of +96 kip · ft and the carryover moment of +12 kip · ft from joint B. We next remove the clamp at joint C. As the joint rotates, distributed end moments of −36 kip · ft and −72 kip · ft develop in the ends of the members to the left and right of the joint, and carryover moments of −36 kip · ft and −18 kip · ft develop at joints D and B, respectively. Since all joints that are free to rotate have been unlocked once, we have completed *one cycle* of moment distribution. At this point the clamp is replaced at joint C. Although no moment exists in the clamp at C, a moment of −18 kip · ft has been created in the clamp at B by the carryover moment from joint C; therefore, we must continue the moment distribution process. We now remove the clamp at B for the second time and distribute +9 kip · ft to each side of the joint and carryover moments of +4.5 kip · ft to joints A and C. We continue the distribution procedure until the moment in the clamps is inconsequential. Normally, the designer terminates the distribution when the distributed end moments have reduced to approximately 0.5 percent of the final value of the member end moment. In this problem we end the analysis after three cycles of moment distribution. The final member end moments, computed by summing algebraically the moments in each column, are listed on the last line in Figure 13.7.

EXAMPLE 13.2

Analyze the continuous beam in Figure 13.6 by moment distribution. The *EI* of all members is constant.

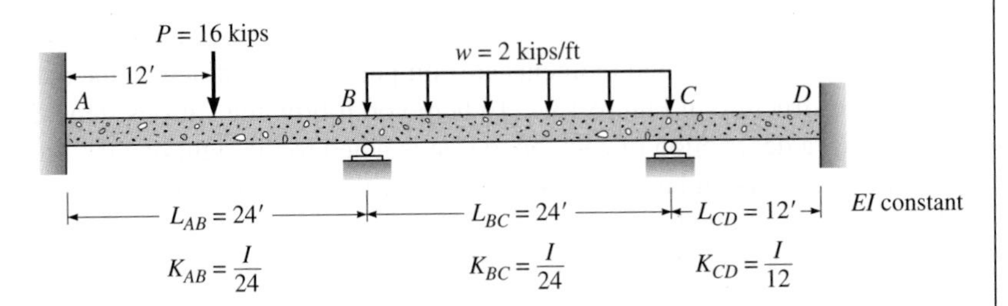

Figure 13.6

Solution

Compute distribution factors at joints *B* and *C* and record on Figure 13.7. At joint *B*:

$$K_{AB} = \frac{I}{24} \qquad K_{BC} = \frac{I}{24} \qquad \Sigma K = K_{AB} + K_{BC} = \frac{2I}{24}$$

$$\mathrm{DF}_{BA} = \frac{K_{AB}}{\Sigma K} = \frac{I/24}{2I/24} = 0.5$$

$$\mathrm{DF}_{BC} = \frac{K_{BC}}{\Sigma K} = \frac{I/24}{2I/24} = 0.5$$

At joint *C*:

$$K_{BC} = \frac{I}{24} \qquad K_{CD} = \frac{I}{12} \qquad \Sigma K = K_{BC} + K_{CD} = \frac{3I}{24}$$

$$\mathrm{DF}_{BC} = \frac{K_{BC}}{\Sigma K} = \frac{I/24}{3I/24} = \frac{1}{3} \qquad \mathrm{DF}_{CD} = \frac{K_{CD}}{\Sigma K} = \frac{I/12}{3I/24} = \frac{2}{3}$$

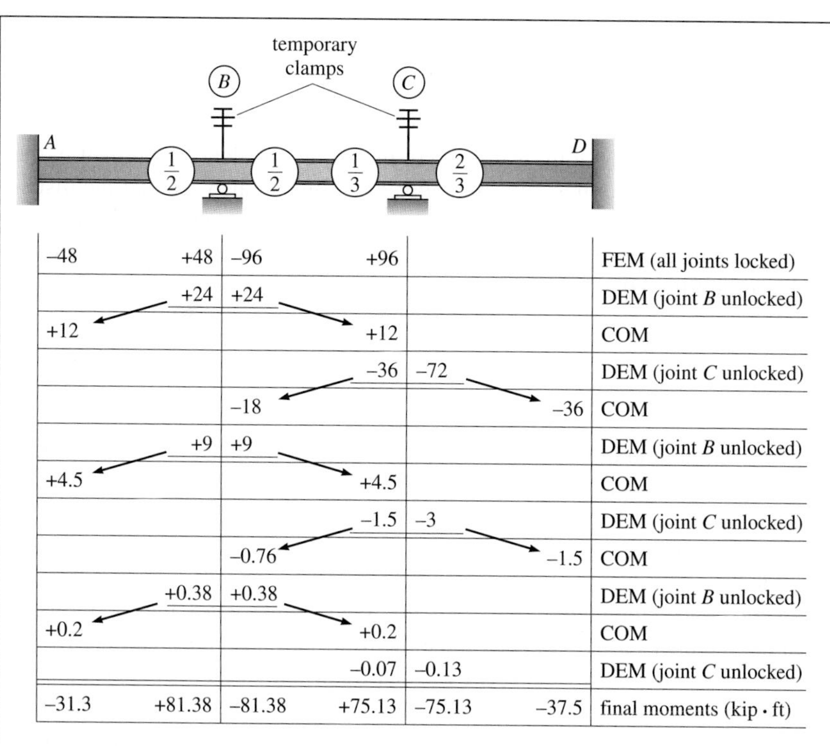

A		B		C		D	
−48	+48	−96	+96			FEM (all joints locked)
	+24	+24				DEM (joint *B* unlocked)
+12			+12			COM
			−36	−72		DEM (joint *C* unlocked)
		−18			−36	COM
	+9	+9				DEM (joint *B* unlocked)
+4.5			+4.5			COM
			−1.5	−3		DEM (joint *C* unlocked)
		−0.76			−1.5	COM
	+0.38	+0.38				DEM (joint *B* unlocked)
+0.2			+0.2			COM
			−0.07	−0.13		DEM (joint *C* unlocked)
−31.3	+81.38	−81.38	+75.13	−75.13	−37.5	final moments (kip · ft)

Figure 13.7: Details of moment distribution (all moments in kip · ft).

Fixed-end moments (see Figure 12.5):

$$\text{FEM}_{AB} = \frac{-PL}{8} = \frac{-16(24)}{8} = -48 \text{ kip} \cdot \text{ft}$$

$$\text{FEM}_{BA} = \frac{+PL}{8} = +48 \text{ kip} \cdot \text{ft}$$

$$\text{FEM}_{BC} = \frac{-wL^2}{12} = \frac{-2(24)^2}{12} = -96 \text{ kip} \cdot \text{ft}$$

$$\text{FEM}_{CB} = \frac{+wL^2}{12} = +96 \text{ kip} \cdot \text{ft}$$

Since span CD is not loaded, $\text{FEM}_{CD} = \text{FEM}_{DC} = 0$.

Example 13.3 covers the analysis of a continuous beam supported by a roller at C, an exterior support (see Figure 13.8). To begin the analysis (Figure 13.9), joints B and C are clamped and the fixed-end moments computed in each span. At joint C the distribution factor DF_{CB} is set equal to 1 because when this joint is unlocked, the entire unbalanced moment in the clamp is applied to the end of member BC. You can also see that the distribution factor at joint C must equal 1, recognizing that $\Sigma K = K_{BC}$ because only one member extends into joint C. If you follow the standard procedure for computing DF_{CB},

$$\mathrm{DF}_{CB} = \frac{K_{BC}}{\Sigma K} = \frac{K_{BC}}{K_{BC}} = 1$$

The computation of the distribution factor at joint B follows the same procedure as before because joints A and C will always be clamped when joint B is unclamped.

Although we have the option of starting the analysis by unlocking either joint B or joint C, we begin at joint C by removing the clamp which carries an unbalanced moment of $+16.2$ kN·m. As the joint rotates, the end moment in the member reduces to zero since the roller provides no rotational resistance to the end of the beam. The angular deformation that occurs is equivalent to that produced when a counterclockwise distributed end moment of -16.2 kN·m acts at joint C. The rotation of joint C also produces a carryover moment of -8.1 kN·m at joint B. The balance of the analysis follows the same steps as previously described. Shear and moment curves are shown in Figure 13.10.

EXAMPLE 13.3

Analyze the beam in Figure 13.8 by moment distribution, and draw the shear and moment curves.

Solution

$$K_{AB} = \frac{1.5I}{6} \qquad K_{BC} = \frac{I}{6} \qquad \text{then} \qquad \Sigma K = K_{AB} + K_{BC} = \frac{2.5I}{6}$$

Compute distribution factors at joint B:

$$\text{DF}_{AB} = \frac{K_{AB}}{\Sigma K} = \frac{1.5I/6}{2.5I/6} = 0.6 \qquad \text{DF}_{BC} = \frac{K_{BC}}{\Sigma K} = \frac{I/6}{2.5I/6} = 0.4$$

$$\text{FEM}_{AB} = -\frac{wL^2}{12} = -\frac{3(6)^2}{12} = -9\ \text{kN}\cdot\text{m}$$

$$\text{FEM}_{BA} = -\text{FEM}_{AB} = +9\ \text{kN}\cdot\text{m}$$

$$\text{FEM}_{BC} = -\frac{wL^2}{12} = -\frac{5.4(6)^2}{12} = -16.2\ \text{kN}\cdot\text{m}$$

$$\text{FEM}_{CB} = -\text{FEM}_{BC} = +16.2\ \text{kN}\cdot\text{m}$$

Analysis. See Figure 13.9.

Shear and Moment Curves. See Figure 13.10.

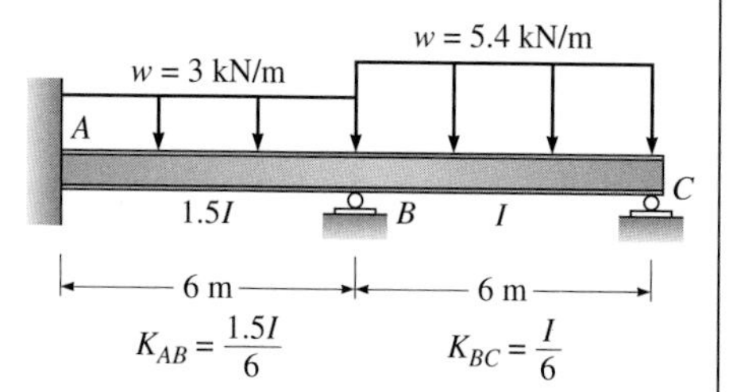

Figure 13.8

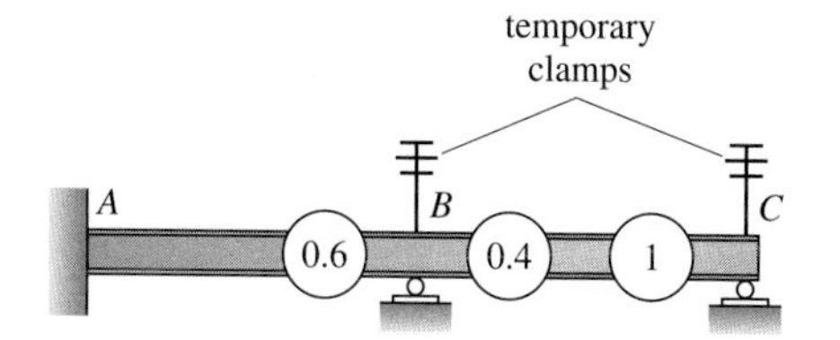

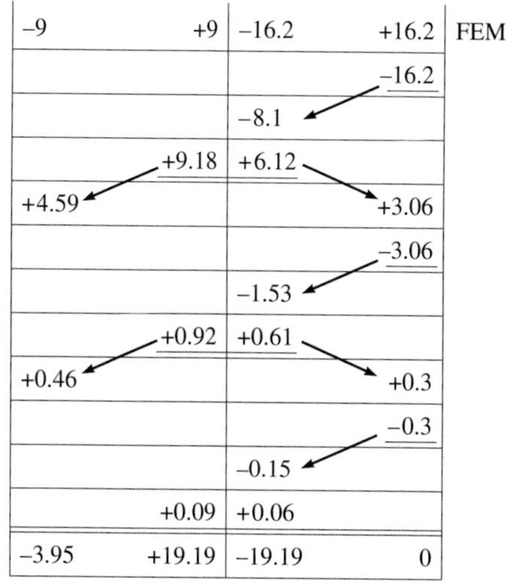

Figure 13.9: Details of moment distribution (all moments in kN · m).

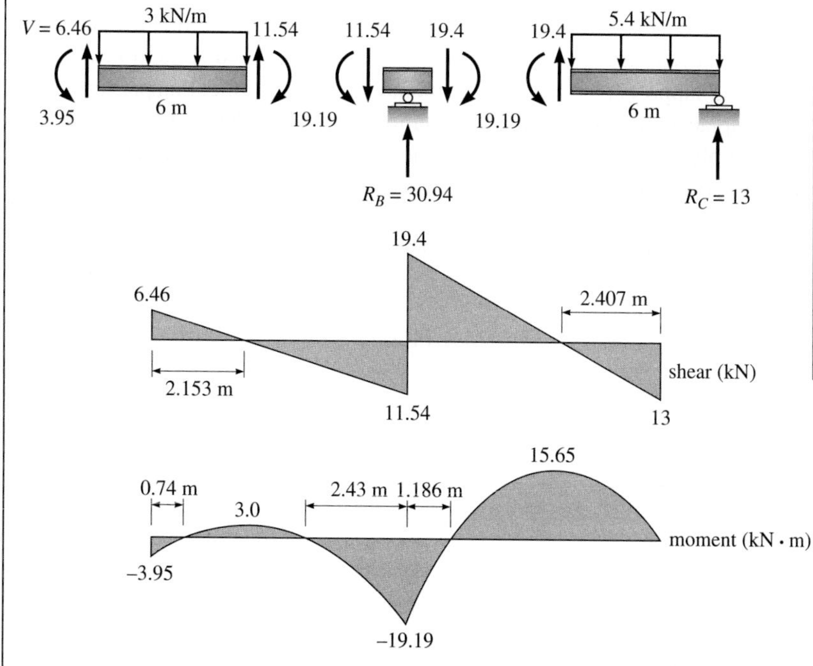

Figure 13.10: Shear and moment curves.

13.5 Modification of Member Stiffness

We can often reduce the number of cycles of moment distribution required to analyze a continuous structure by adjusting the flexural stiffness of certain members. In this section we consider members whose ends terminate at an *exterior* support consisting of either a pin or roller (e.g., see members *AB*, *BF*, and *DE* in Figure 13.11). We will also establish the influence of a variety of end conditions on the flexural stiffness of a beam.

To measure the influence of end conditions on the flexural stiffness of a beam, we can compare the moment required to produce a *unit rotation* (1 radian) of the end of a member for various end conditions. For example, if the far end of a beam is fixed against rotation as shown in Figure 13.12*a*, where $\theta_A = 1$ radian and $\theta_B = 0$, we can use the slope-deflection equation to express the applied moment in terms of the beams properties. Since no support settlements occur and no loads are applied between ends, $\psi_{AB} = 0$ and $\text{FEM}_{AB} = \text{FEM}_{BA} = 0$.

Substituting the above terms into Equation 12.16, we compute

$$M_{AB} = \frac{2EI}{L}(2\theta_A + \theta_B - 3\psi_{AB}) + \text{FEM}$$

$$= \frac{2EI}{L}[2(1) + 0 - 0] + 0$$

$$M_{AB} = \frac{4EI}{L} \tag{13.17}$$

Previously we have seen that $4EI/L$ represents the absolute flexural stiffness of a beam acted upon by a moment whose far end is fixed (Equation 13.6).

If the support at the *B* end of the member is a pin or roller that prevents vertical displacement, but provides no rotational restraint (Figure 13.12*b*), we can again apply the slope-deflection equation to evaluate the member's flexural stiffness. For this case:

$\theta_A = 1$ radian $\quad \theta_B = -\frac{1}{2}$ radian $\quad$ (see Figure 11.3*e* for the relationship between θ_A and θ_B)

$\psi_{AB} = 0 \quad$ and $\quad \text{FEM}_{AB} = \text{FEM}_{BA} = 0$

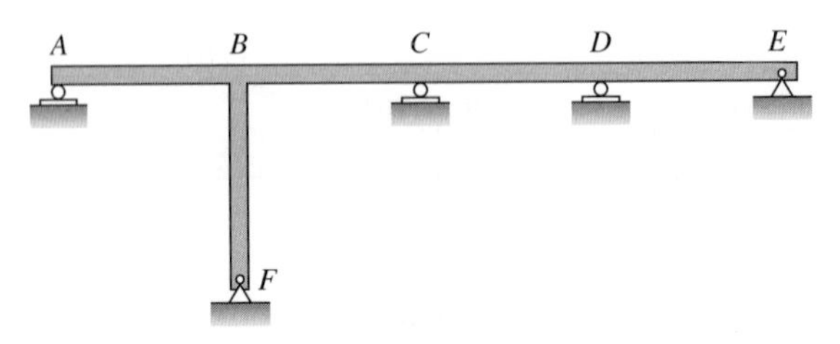

Figure 13.11

Substituting into Equation 12.16 gives

$$M_{AB} = \frac{2EI}{L}[2(1) - \tfrac{1}{2} + 0] + 0$$

$$M_{AB} = \frac{3EI}{L} \tag{13.18}$$

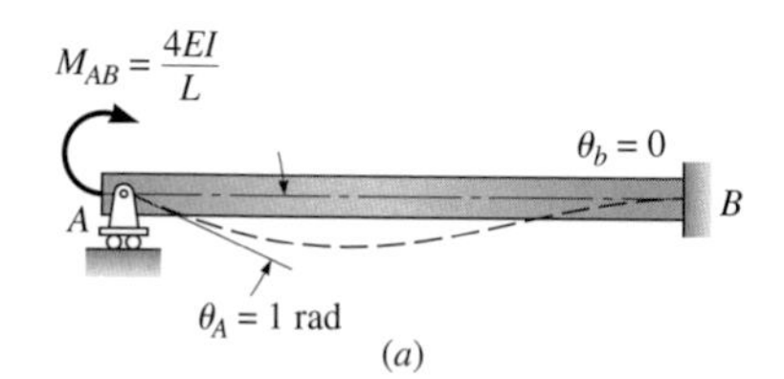

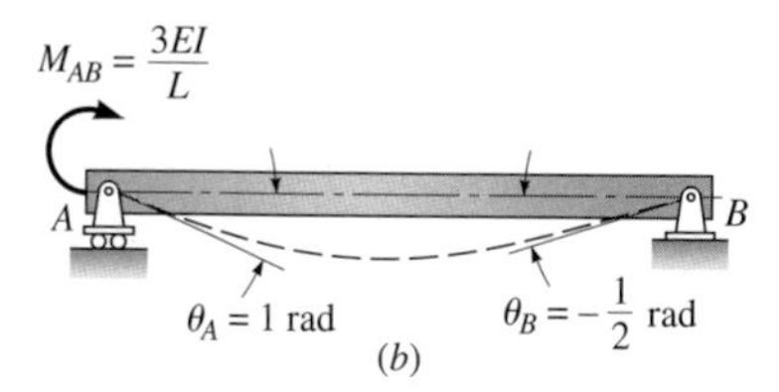

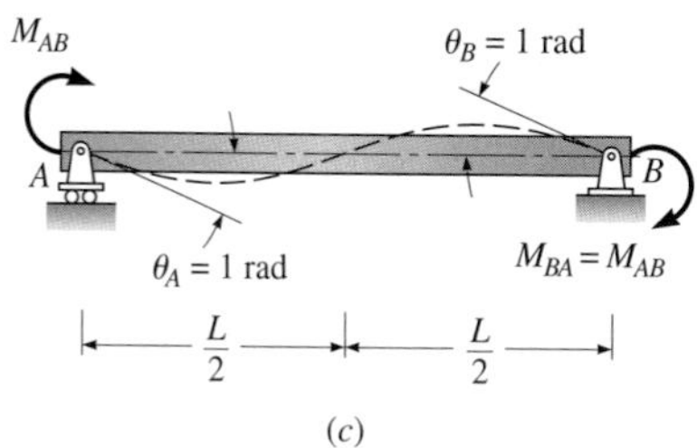

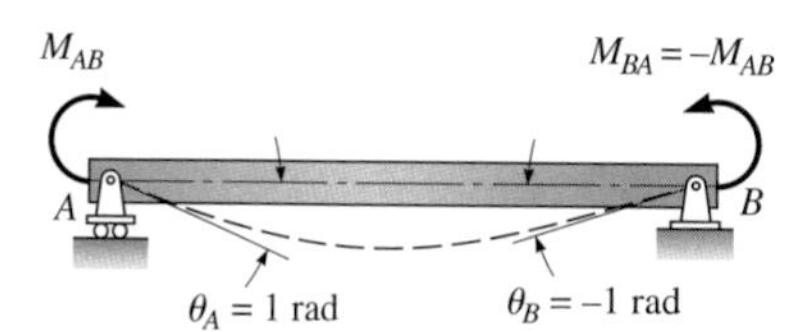

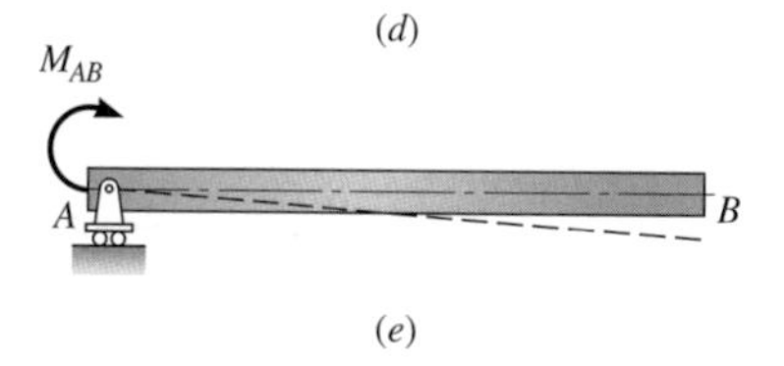

Figure 13.12: (*a*) beam with far end fixed; (*b*) beam with far end unrestrained against rotation; (*c*) equal values of clockwise moment at each end; (*d*) single curvature bending by equal values of end moments; (*e*) cantilever loaded at supported end.

Comparing Equations 13.17 and 13.18, we see that a beam loaded by a moment at one end whose far end is pinned is *three-fourths as stiff with respect to resistance to joint rotation as a beam of the same dimensions whose far end is fixed*.

If a member is bent into double curvature by equal end moments (Figure 13.12*c*), the resistance to rotation increases because the moment at *B*, the far end, rotates the near end *A* in a direction opposite in sense to the moment at *A*. We can relate the magnitude of M_{AB} to the rotation at *A* by using the slope-deflection equation with $\theta_A = \theta_B = 1$ rad, $\psi_{AB} = 0$, and $\text{FEM}_{AB} = 0$. Substituting the above values into the slope-deflection equation gives

$$M_{AB} = \frac{2EI}{L}(2\theta_A + \theta_B - 3\psi_{AB}) \pm \text{FEM}_{AB} \tag{12.16}$$

$$M_{AB} = \frac{2EI}{L}[2(1) + 1] = \frac{6EI}{L}$$

where the absolute stiffness is

$$K_{AB} = \frac{6EI}{L} \tag{13.19}$$

Comparing Equation 13.19 with Equation 13.17, we find that the absolute stiffness for a member bent in *double curvature by equal end moments* is 50 percent greater than the stiffness of a beam whose far end is fixed against rotation.

If a flexural member is acted on by equal values of end moments (Figure 13.12*d*), producing single-curvature bending, the effective bending stiffness with respect to the *A* end is reduced because the moment at the far end (the *B* end) contributes to the rotation at the *A* end.

Using the slope-deflection equation with $\theta_A = 1$ radian, $\theta_B = -1$ radian, $\psi_{AB} = 0$, and $\text{FEM}_{AB} = 0$, we get

$$\begin{aligned} M_{AB} &= \frac{2EI}{L}(2\theta_A + \theta_B - 3\psi_{AB}) \pm \text{FEM}_{AB} \\ &= \frac{2EI}{L}[2 \times 1 + (-1) - 0] \pm 0 \\ &= \frac{2EI}{L} \end{aligned}$$

where the absolute stiffness

$$K_{AB} = \frac{2EI}{L} \tag{13.20}$$

Comparing Equation 13.20 to Equation 13.17, we find that the absolute stiffness K_{AB} of a member bent into single curvature by equal values of end moments has an effective stiffness K_{AB} that is 50 percent smaller than that of a beam whose far end is fixed against rotation.

Members, when acted upon by equal values of end moment that produce single-curvature bending, are located at the *axis of symmetry* of *symmetric structures* that are *loaded symmetrically* (see members *BC* in Figure 13.13*a* and *b*). In the symmetrically loaded box beam in Figure 13.13*c*, the end moments act to produce single-curvature bending on all four sides. Of course if transverse loads also act, there can be regions of both positive and negative moments. As we will demonstrate in Example 13.6, taking advantage of this modification in a moment distribution analysis of a symmetric structure simplifies the analysis significantly.

Stiffness of a Cantilever

In Figures 13.12*a* to *d*, the fixed and the pin supports at *B* provide vertical restraint that prevents the beam from rotating clockwise as a rigid body about support *A*. Since each of these beams is supported in a stable manner, they are able to resist the moment applied at joint *A*. On the other hand, if a moment is applied to the *A* end of the cantilever beam in

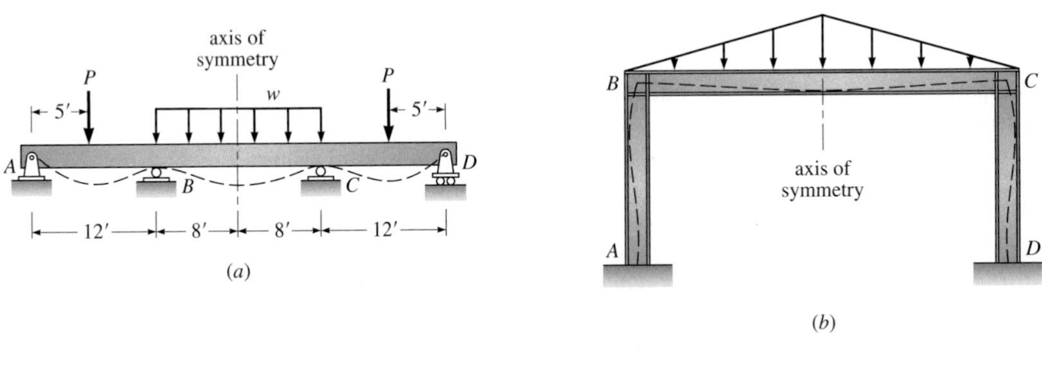

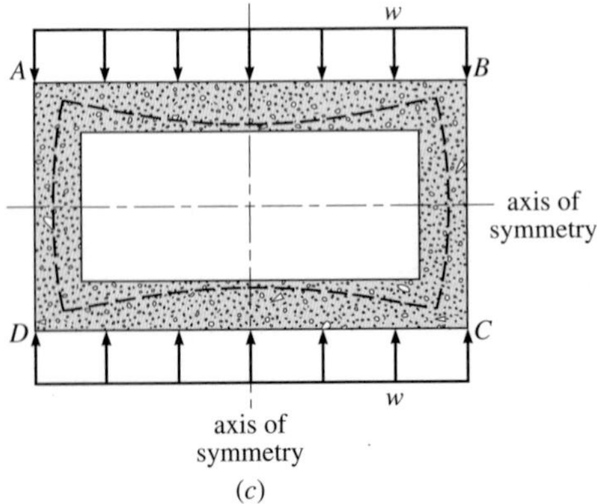

Figure 13.13: Examples of symmetric structures, symmetrically loaded, that contain members whose end moments are equal in magnitude and produce single-curvature bending. (*a*) Beam *BC* of the continuous beam; (*b*) beam *BC* of the rigid frame; (*c*) all four members of the box beam.

Figure 13.12*e*, the cantilever is not able to develop any flexural resistance to the moment because no support exists at the right to prevent the beam from rotating clockwise about support *A*. Therefore, you can see that a cantilever has zero resistance to moment. When you compute the distribution factors at a joint that contains a cantilever, the distribution factor for the cantilever is zero, and *no unbalanced moment is ever distributed to the cantilever*.

Of course, if a cantilever is loaded, it can transmit both a shear and a moment to the joint where it is supported; however, this is a separate function and has nothing to do with its ability to absorb unbalanced moment.

In Example 13.4 we illustrate the use of the factor $\frac{3}{4}$ to modify the stiffness of pin-ended members of the continuous beam in Figure 13.14*a*. In the analysis of the beam in Figure 13.14, the flexural stiffness I/L of members *AB* and *CD* can both be reduced by $\frac{3}{4}$ since both members terminate at pin or roller supports. You may have some concern that the factor $\frac{3}{4}$ is applicable to span *CD* because of the cantilever extension *DE* to the right of the support. However, as we just discussed, the cantilever has zero stiffness as far as absorbing any unbalanced moment that is carried by a clamp on joint *D*; therefore, after the clamp is removed from joint *D*, the cantilever has no influence on the rotational restraint of member *CD*.

We begin the analysis in Figure 13.15*a* with all joints locked against rotation. The loads are next applied, producing the fixed-end moments tabulated on the first line. From the free-body diagram of cantilever *DE* in Figure 13.14*b*, you can see that equilibrium of the member requires that the moment at the *D* end of member *DE* act counterclockwise and equal -60 kip·ft. Since the flexural stiffnesses of members *AB* and *CD* have been reduced by $\frac{3}{4}$, the clampsat joints *A* and *D must be removed first*. When the clamp is removed at *A*, a distributed end moment of $+33$ kip·ft and a carryover moment of $+16.7$ kip·ft develop in span *AB*. The total moment at joint *A* is now zero. In the balance of the analysis, joint *A* will remain unclamped. Since joint *A* is now free to rotate, no carryover moment will develop there whenever joint *B* is unclamped.

We next move to joint *D* and remove the clamp, which initially carries an unbalanced moment equal to the difference in fixed-end moments at the joint

$$\text{UM} = +97.2 - 60 = +37.2 \text{ kip} \cdot \text{ft}$$

As joint *D* rotates, a distributed end moment of -37.2 kip·ft develops at *D* and a carryover moment of -18.6 kip·ft at *C* develops in member *CD*. *Note*: Joint *D* is now in balance, and the -60 kip·ft applied by the cantilever is balanced by the $+60$ kip·ft at the *D* end of member *CD*. For the balance of the analysis, joint *D* will remain unclamped and no carryover moment will develop there when joint *C* is unclamped. The analysis is completed by distributing moments between joints *B* and *C* until the magnitude of carryover moment is negligible. By using freebodies of beam elements between supports, reactions are computed by statics and shown in Figure 13.15*b*.

EXAMPLE 13.4

Analyze the beam in Figure 13.14*a* by moment distribution, using modified flexural stiffnesses for members *AB* and *CD*. Given: *EI* is constant.

Solution

$$K_{AB} = \frac{3}{4}\left(\frac{360}{15}\right) = 18 \qquad K_{BC} = \frac{480}{20} = 24$$

$$K_{CD} = \frac{3}{4}\left(\frac{480}{18}\right) = 20 \qquad K_{DE} = 0$$

Compute the distribution factors.

Joint *B*:

$$\Sigma K = K_{AB} + K_{BC} = 18 + 24 = 42$$

$$DF_{BA} = \frac{K_{AB}}{\Sigma K} = \frac{18}{42} = 0.43 \qquad DF_{BC} = \frac{K_{BC}}{\Sigma K} = \frac{24}{42} = 0.57$$

Joint *C*:

$$\Sigma K = K_{BC} + K_{CD} = 24 + 20 = 44$$

$$\text{DF}_{BC} = \frac{K_{BC}}{\Sigma K} = \frac{24}{44} = 0.55 \qquad \text{DF}_{CD} = \frac{K_{CD}}{\Sigma K} = \frac{20}{44} = 0.45$$

Compute the fixed-end moments (see Figure 12.5).

$$\text{FEM}_{AB} = -\frac{Pab^2}{L^2} = -\frac{30(10)(5^2)}{15^2} = -33.3\ \text{kip}\cdot\text{ft} \qquad \text{FEM}_{BA} = \frac{Pba^2}{L^2} = \frac{30(5)(10^2)}{15^2} = +66.7\ \text{kip}\cdot\text{ft}$$

$$\text{FEM}_{BC} = -\frac{wL^2}{12} = -120\ \text{kip}\cdot\text{ft} \qquad \text{FEM}_{CB} = -\text{FEM}_{BC} = 120\ \text{kip}\cdot\text{ft}$$

$$\text{FEM}_{CD} = -\frac{wL^2}{12} = -97.2\ \text{kip}\cdot\text{ft} \qquad \text{FEM}_{DC} = -\text{FEM}_{CD} = 97.2\ \text{kip}\cdot\text{ft}$$

$$\text{FEM}_{DE} = -60\ \text{kip}\cdot\text{ft} \quad \text{(see Figure 13.14}b\text{)}$$

The minus sign is required because the moment acts counterclockwise on the end of the member.

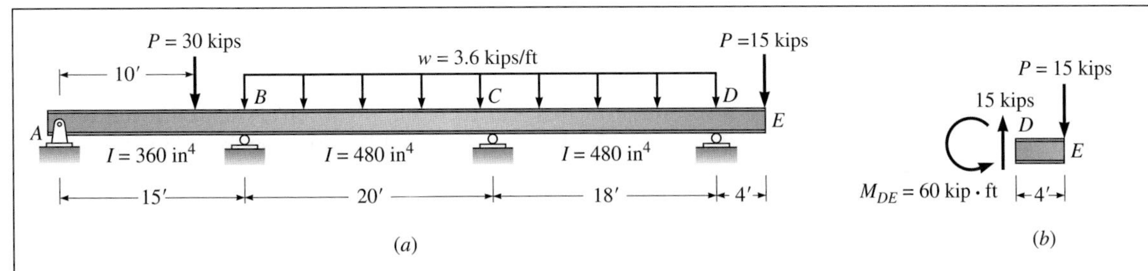

Figure 13.14: (*a*) Continuous beam; (*b*) free body of cantilever *DE*.

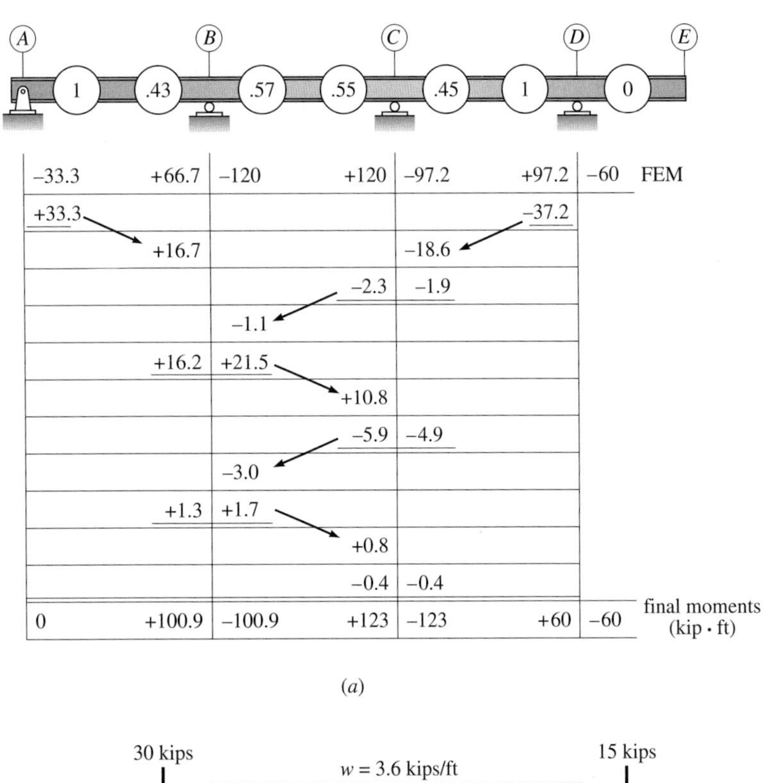

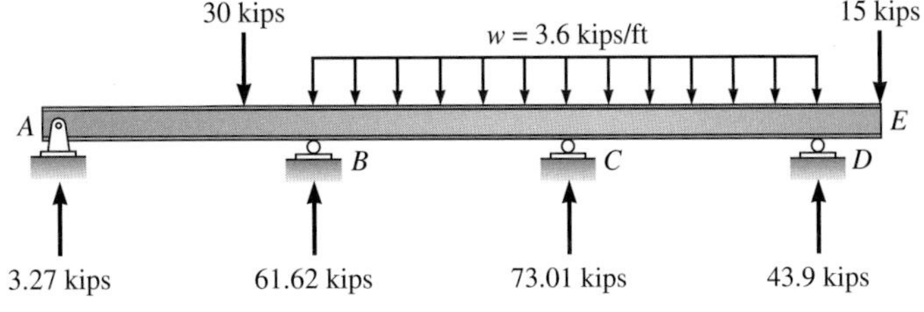

Figure 13.15: (*a*) Moment distribution details; (*b*) reactions.

The use of moment distribution to analyze a frame, whose joints are restrained against displacement but free to rotate, is illustrated in Example 13.5 by the analysis of the structure shown in Figure 13.16. We begin by computing the distribution factors and recording them on the line drawing of the frame in Figure 13.17*a*. Joints *A*, *B*, *C*, and *D*, which are free to rotate, are initially clamped. Loads are then applied and produce fixed-end moments of ± 120 kip·ft in span *AB* and ± 80 kip·ft in span *BC*. These moments are recorded on Figure 13.17*a* above the girders. To begin the analysis, joints *A* and *D* must be unlocked first because the stiffnesses of members *AB* and *CD* have been modified by factor $\frac{3}{4}$. As joint *A* rotates, a distributed end moment of $+120$ kip·ft at joint *A* and a carryover moment of $+60$ kip·ft at joint *B* develop in span *AB*. Since no transverse loads act on member *CD*, there are no fixed-end moments in member *CD*; therefore, no moments develop in member *CD* when the clamp is removed from joint *D*. Since joints *A* and *D* remain unclamped for the balance of the analysis, no carryover moments are made to these joints.

At joint *B* the unbalanced moment equals 100 kip·ft —the algebraic sum of the fixed-end moments of $+120$ and -80 kip·ft and the carryover moment of $+60$ kip·ft from joint *A*. The sign of the unbalanced moment is reversed, and distributed end moments of -33, -22, and -45 kip·ft, respectively, are made to the *B* end of members *BA*, *BC*, and *BF*. In addition, carryover moments of -11 kip·ft to the *C* end of member *BC* and -22.5 kip·ft to the base of column *BF* are made. Next, joint *C* is unlocked, and the unbalanced moment in the clamp of $+69$ kip·ft —the algebraic sum of the fixed-end moment of $+80$ kip·ft and the carryover moment of -11 kip·ft —is distributed. Unlocking of joint *C* also produces carryover moments of -7.2 kip·ft to joint *B* and -14.85 kip·ft to the base of column *CE*. After a second cycle of moment distribution is completed, the carryover moments are insignificant and the analysis can be terminated. A double line is drawn, and the moments in each member are summed to establish the final values of member end moment. Reactions, computed from free bodies of individual members, are shown in Figure 13.17*b*.

EXAMPLE 13.5

Analyze the frame in Figure 13.16 by moment distribution.

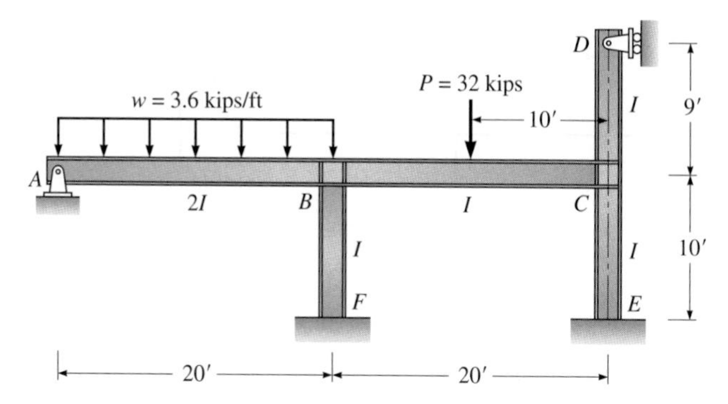

Figure 13.16: Details of rigid frame.

Solution

Compute the distribution factors at joint B.

$$K_{AB} = \frac{3}{4}\left(\frac{2I}{20}\right) = \frac{3I}{40} \qquad K_{BC} = \frac{I}{20} \qquad K_{BF} = \frac{I}{10} \qquad \Sigma K = \frac{9I}{40}$$

$$\text{DF}_{BA} = \frac{K_{AB}}{\Sigma K} = 0.33 \qquad \text{DF}_{BC} = \frac{K_{BC}}{\Sigma K} = 0.22 \qquad \text{DF}_{BF} = \frac{K_{BF}}{\Sigma K} = 0.45$$

Compute the distribution factors at joint C.

$$K_{CB} = \frac{I}{20} \qquad K_{CD} = \frac{3}{4}\left(\frac{I}{9}\right) \qquad K_{CE} = \frac{I}{10} \qquad \Sigma K = \frac{14I}{60}$$

$$\text{DF}_{CB} = 0.21 \qquad \text{DF}_{CD} = 0.36 \qquad \text{DF}_{CE} = 0.43$$

Compute the fixed-end moments in spans AB and BC (see Figure12.5).

$$\text{FEM}_{AB} = \frac{wL^2}{12} = \frac{-3.6(20)^2}{12} = -120 \text{ kip} \cdot \text{ft}$$

$$\text{FEM}_{BA} = -\text{FEM}_{AB} = +120 \text{ kip} \cdot \text{ft}$$

$$\text{FEM}_{BC} = \frac{-PL}{8} = \frac{-32(20)}{8} = -80 \text{ kip} \cdot \text{ft}$$

$$\text{FEM}_{CB} = -\text{FEM}_{BC} = +80 \text{ kip} \cdot \text{ft}$$

[*continues on next page*]

Example 13.5 continues . . .

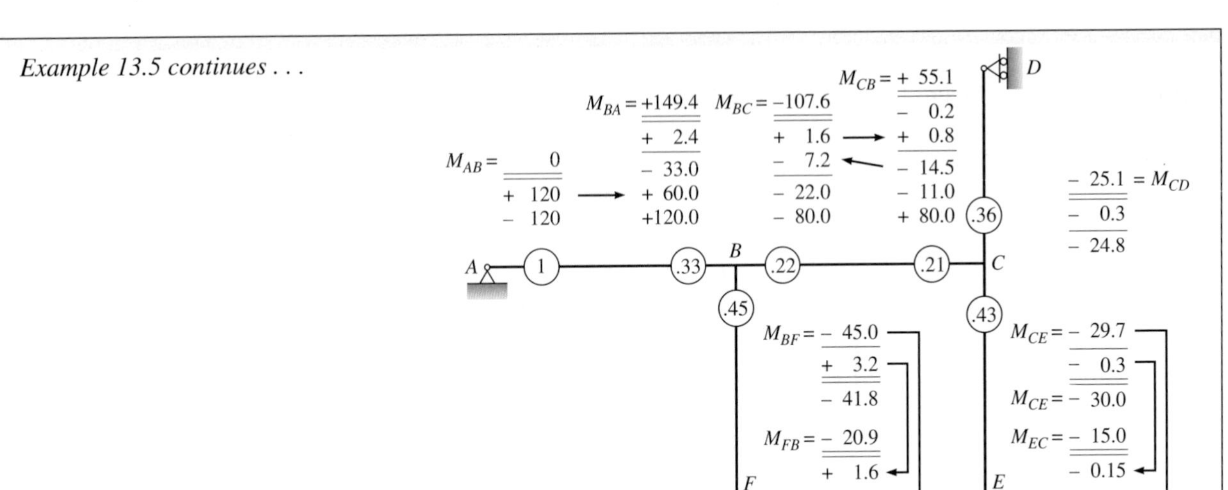

(*a*)

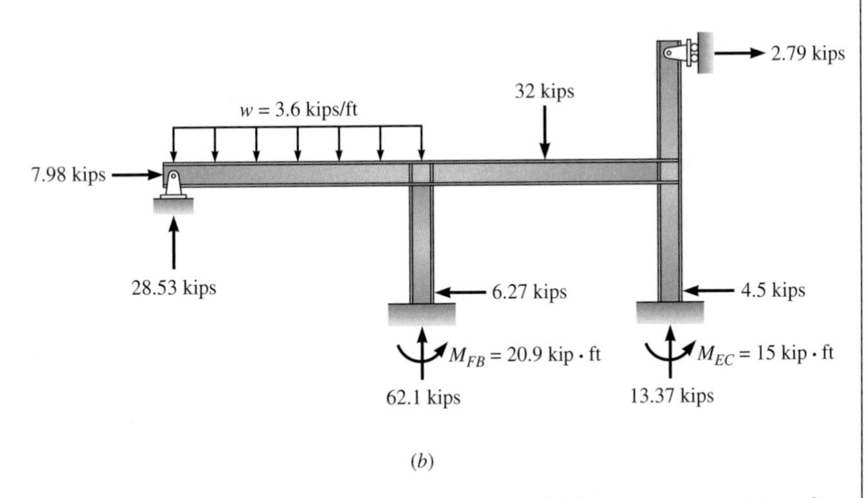

(*b*)

Figure 13.17: (*a*) Analysis by moment distribution; (*b*) Reactions computed from free bodies of members.

EXAMPLE 13.6

Analyze the frame in Figure 13.18*a* by moment distribution, modifying the stiffness of the columns and girder by the factors discussed in Section 13.5 for a symmetric structure, symmetrically loaded.

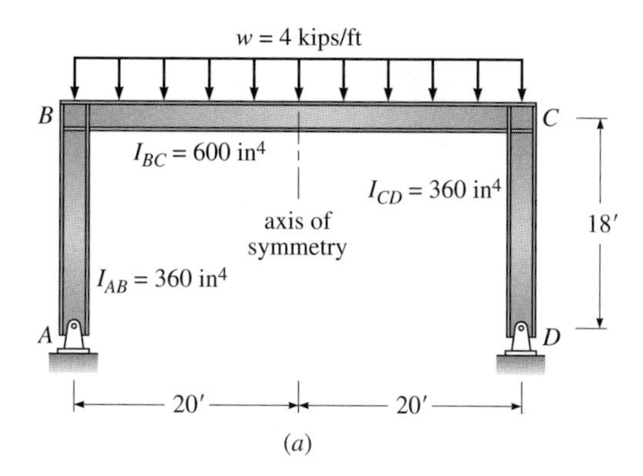

Figure 13.18: (*continues*)

Solution

STEP 1 Modify the stiffness of the columns by $\frac{3}{4}$ for a pin support at points *A* and *D*.

$$K_{AB} = K_{CD} = \frac{3}{4}\frac{I}{L} = \frac{3}{4}\,\frac{360}{18} = 15$$

Modify the stiffness of girder *BC* by $\frac{1}{2}$ (joints *B* and *C* will be unclamped simultaneously and no carryover moments are distributed).

$$K_{BC} = \frac{1}{2}\frac{I}{L} = \frac{1}{2}\,\frac{600}{40} = 7.5$$

STEP 2 Compute the distribution factors at joints *B* and *C*.

$$\text{DF}_{BA} = \text{DF}_{CD} = \frac{K_{AB}}{\Sigma K_s'} = \frac{15}{15 + 7.5} = \frac{2}{3}$$

$$\text{DF}_{BC} = \text{DF}_{CB} = \frac{K_{BC}}{\Sigma K_s'} = \frac{7.5}{15 + 7.5} = \frac{1}{3}$$

$$\text{FEM}_{BC} = \text{FEM}_{CB} = \frac{WL^2}{12} = \frac{4(40)^2}{12} = \pm 533.33 \text{ kip} \cdot \text{ft}$$

[*continues on next page*]

Example 13.6 continues . . .

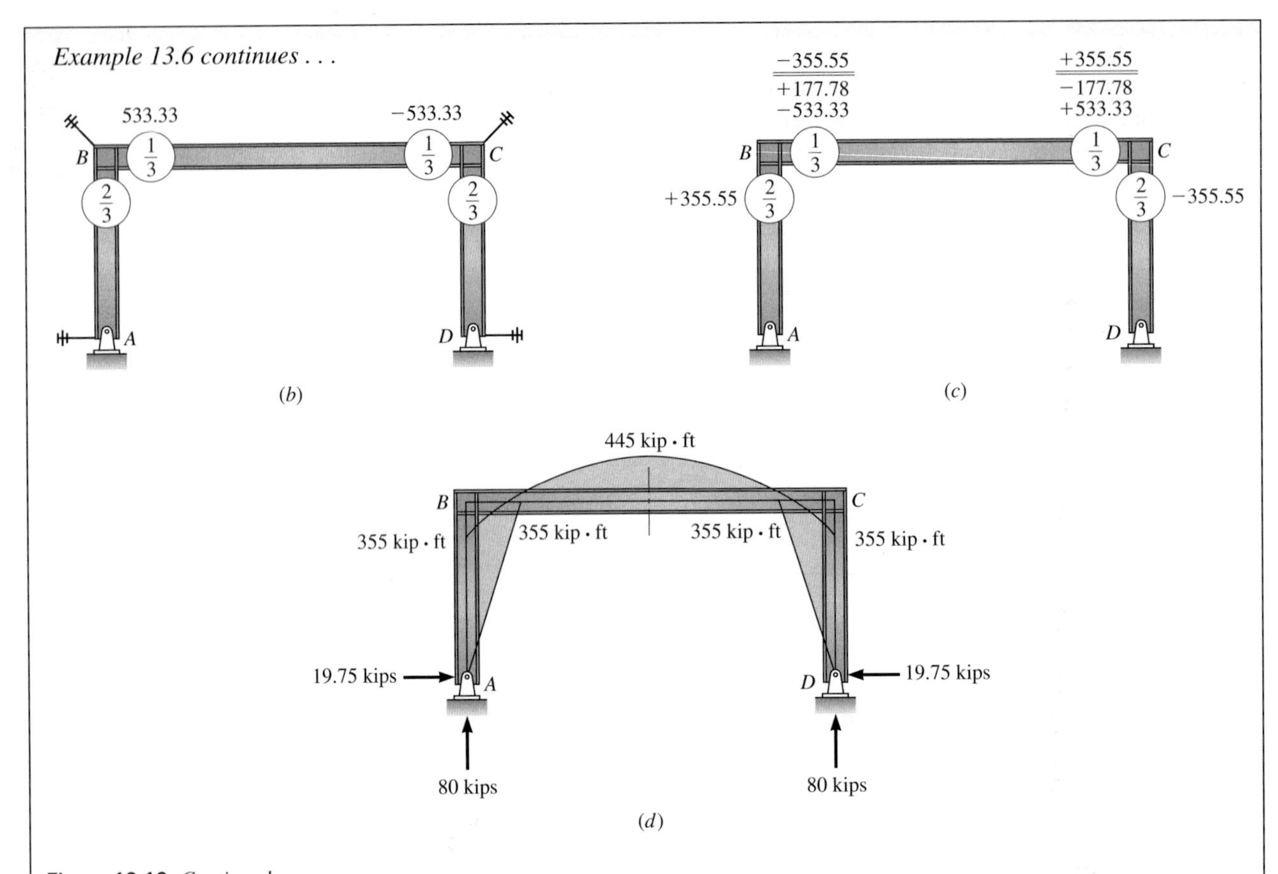

Figure 13.18: *Continued*

STEP 3 (a) Clamp all joints and apply the uniform load to girder *BC* (see Figure 13.18*b*).
(b) Remove clamps at supports *A* and *D*. Since no loads act on the columns, there are no moments to distribute. The joint at the supports will remain unclamped. Since the base of each column is free to rotate if the far end is unclamped, the stiffness of each column may be reduced by a factor of $\frac{3}{4}$.

STEP 4 Clamps at joints *B* and *C* are next removed simultaneously. Joints *B* and *C* rotate equally (the condition required for the $\frac{1}{2}$ factor applied to the girder stiffness), and equal values of end moment develop at each end of girder *BC* (see Figure 13.18*c*). Final results of the analysis are shown in Figure 13.18*d*.

Support Settlements, Fabrication Errors, and Temperature Change

Moment distribution and the slope-deflection equation provide an effective combination for determining the moments created in indeterminate beams and frames by fabrication errors, support settlements, and temperature change. In this application the appropriate displacements are introduced into the structure while simultaneously all joints that are free to rotate are locked by clamps against rotation in their initial orientation. Locking the joints against rotation ensures that the changes in slope at the ends of all members are *zero* and permits the end moments produced by specified values of displacement to be evaluated by the slope-deflection equation. To complete the analysis, the clamps are removed and the structure is allowed to deflect into its final equilibrium position.

In Example 13.7 we use this procedure to determine the moments in a structure whose supports are not located in their specified position—a common situation that frequently occurs during the construction. In Example 13.8 on page 542 the method is used to establish the moments created in an indeterminate frame by a fabrication error.

EXAMPLE 13.7

Determine the reactions and draw the shear and moment curves for the continuous beam in Figure 13.19*a*. The fixed support at *A* is accidentally constructed incorrectly at a slope of 0.002 radian counterclockwise from a vertical axis through *A*, and the support at *C* is accidentally constructed 1.5 in below its intended position. Given: $E = 29{,}000$ kips/in^2 and $I = 300$ in^4.

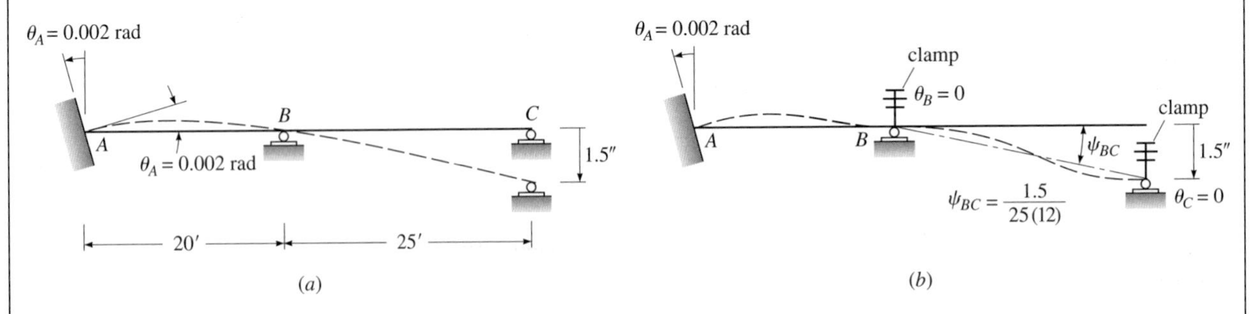

Figure 13.19: (*a*) Beam with supports constructed out of position, deflected shape shown by dashed line; (*b*) restrained beam locked in position by temporary clamps at joints *B* and *C*.

Solution

With the supports located in their as-built position (see Figure 13.19*b*), the beam is connected to the supports. Since the unloaded beam is straight but the supports are no longer in a straight line and correctly aligned, external forces must be applied to the beam to bring it into contact with its supports. After the beam is connected to its supports, reactions must develop to hold the beam in its bent configuration. Also at both joints *B* and *C* we imagine that clamps are applied at these joints to hold the ends of the beam in a horizontal position; that is, θ_B and θ_C are zero. We now use the slope-deflection equation to compute the moments at each end of the restrained beams in Figure 13.19*b*.

$$M_{NF} = \frac{2EI}{L}(2\theta_N + \theta_F - 3\psi) + \text{FEM}_{NF} \qquad (12.16)$$

Compute moments in span *AB*: $\theta_A = -0.002$ rad, $\theta_B = 0$, and $\psi_{AB} = 0$. Since no transverse loads are applied to span *AB*, $\text{FEM}_{AB} = \text{FEM}_{BA} = 0$.

$$M_{AB} = \frac{2(29{,}000)(300)}{20(12)}[2(-0.002)] = -290 \text{ kip}\cdot\text{in} = -24.2 \text{ kip}\cdot\text{ft}$$

$$M_{BA} = \frac{2(29{,}000)(300)}{20(12)}(-0.002) = -145 \text{ kip}\cdot\text{in} = -12.1 \text{ kip}\cdot\text{ft}$$

Compute moments in span BC: $\theta_B = 0, \theta_C = 0, \psi = 1.5 \text{ in}/[25(12)] = 0.005$.

$\text{FEM}_{BC} = \text{FEM}_{CB} = 0$ since no transverse loads applied to span BC.

$$M_{BC} = M_{CB} = \frac{2(29{,}000)(300)}{12(25)}[2(0) + 0 - 3(0.005)]$$

$$= -870 \text{ kip}\cdot\text{in} = -72.5 \text{ kip}\cdot\text{ft}$$

Compute the distribution factors at joint B.

$$K_{AB} = \frac{300}{20} = 15 \qquad K_{BC} = \frac{3}{4}\left(\frac{300}{25}\right) = 9 \qquad \Sigma K = 24$$

$$\text{DF}_{BA} = \frac{K_{AB}}{\Sigma K} = \frac{15}{24} = 0.625 \qquad \text{DF}_{BC} = \frac{K_{BC}}{\Sigma K} = \frac{9}{24} = 0.375$$

The moment distribution is carried out in Figure 13.20*a*, shears and reactions are computed in Figure 13.20*b*, and the moment curve is shown in Figure 13.20*c*.

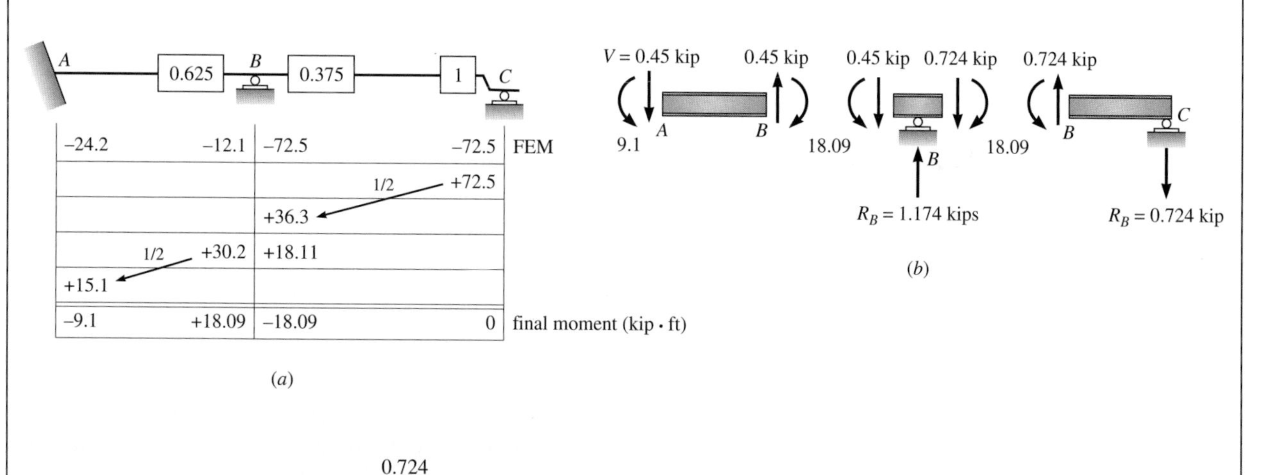

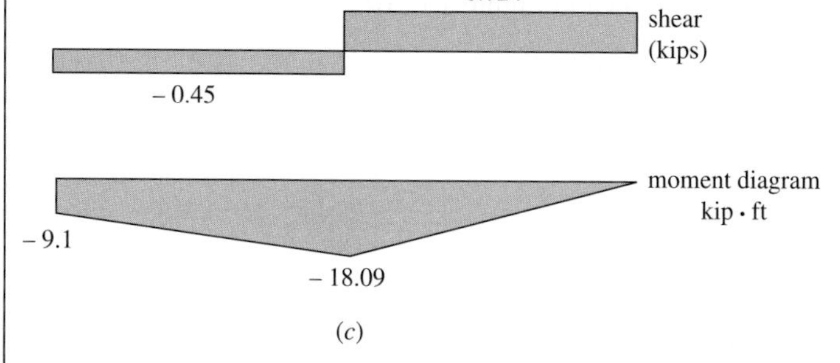

Figure 13.20: (*a*) Moment distribution; (*b*) free bodies used to evaluate shears and reactions; (*c*) moment curve produced by support movements.

EXAMPLE 13.8

If girder *AB* of the rigid frame in Figure 13.21*a* is fabricated 1.92 in too long, what moments are created in the frame when it is erected? Given: $E = 29{,}000$ kips/in^2.

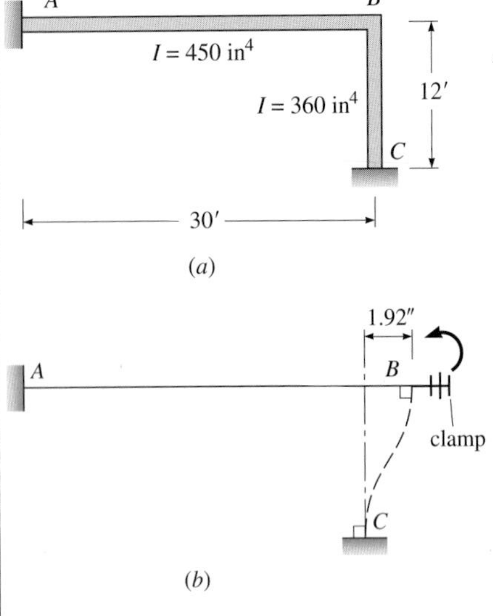

Figure 13.21: (*a*) Frame; (*b*) deformation introduced and joint *B* clamped against rotation ($\theta_B = 0$); (*c*) analysis by moment distribution (moments in kip·ft); (*d*) reactions and deflected shape; (*e*) moment diagrams.

Solution

Add 1.92 in to the end of girder *AB*, and erect the frame with a clamp at joint *B* to prevent rotation (see Figure 13.21*b*). Compute the fixed-end moments in the clamped structure using the slope-deflection equation.

$$\text{Column } BC\text{:}\quad \theta_B = 0 \qquad \theta_C = 0 \qquad \psi_{BC} = \frac{1.92}{12(12)} = +0.0133 \text{ rad}$$

And $\text{FEM}_{BC} = \text{FEM}_{CB} = 0$ since no loads are applied between joints.

$$M_{BC} = M_{CB} = \frac{2EI}{L}(-3\psi_{BC})$$

$$= \frac{2(29{,}000)(360)}{12(12)}[-3(0.0133)]$$

$$= -5785.5 \text{ kip} \cdot \text{in} = -482.13 \text{ kip} \cdot \text{ft}$$

No moments develop in member *AB* because $\psi_{AB} = \theta_A = \theta_B = 0$.

Compute the distribution factors.

$$K_{AB} = \frac{I}{L} = \frac{450}{30} = 15 \qquad K_{BC} = \frac{360}{12} = 30 \qquad \Sigma K = 15 + 30 = 45$$

$$\text{DF}_{BA} = \frac{K_{AB}}{\Sigma K} = \frac{15}{45} = \frac{1}{3} \qquad \text{DF}_{BC} = \frac{K_{BC}}{\Sigma K} = \frac{30}{45} = \frac{2}{3}$$

Analysis by moment distribution is carried out on Figure 13.21*c*. Member end moments and reactions are computed by cutting out free bodies of each member and using equations of statics to solve for the shears. Reactions and the deflected shape are shown in Figure 13.21*d*.

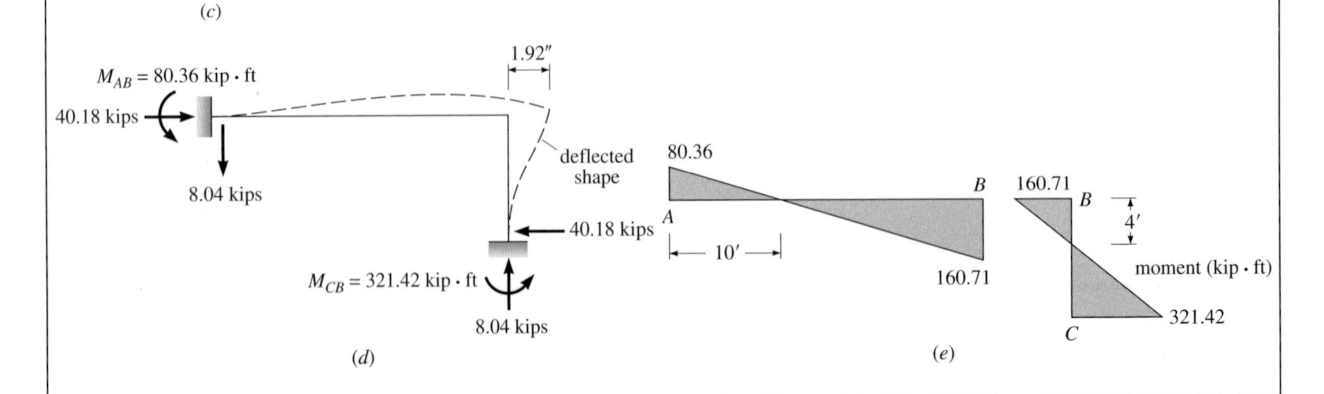

13.6 Analysis of Frames That Are Free to Sidesway

All structures that we have analyzed thus far contained joints that were free to rotate but not translate. Frames of this type are called *braced* frames. In these structures we were always able to compute the initial moments to be distributed because the final position of the joints was known (or specified in the case of a support movement).

When certain joints of an *unbraced* frame are free to translate, the designer must include the moments created by chord rotations. Since the final positions of the unrestrained joints are unknown, the sidesway angles cannot be computed initially, and the member end moments to be distributed cannot be determined. To introduce the analysis of unbraced frames, we will first consider the analysis of a frame with a lateral load applied at a joint that is free to sidesway (see Figure 13.22*a*). In Section 13.7, we will extend the method of analysis to an unbraced frame whose members are loaded between joints or whose supports settle.

Under the action of a lateral load P at joint B, girder BC translates horizontally to the right a distance Δ. Since the magnitude of Δ and the joint rotations are unknown, we cannot compute the end moments to be distributed in a moment distribution analysis directly. However, an indirect solution is possible if the structure behaves in a linear elastic manner, that is, if all deflections and internal forces vary linearly with the magnitude of the lateral load P at joint B. For example, if the frame behaves elastically, doubling the value of P will double the value of all forces and displacements (see Figure 13.22*b*). Engineers typically assume that the majority of structures behave elastically. This assumption is reasonable as long as deflec-tions are small and stresses do not exceed the proportional limit of the material.

If a linear relationship exists between forces and displacements, the following procedures can be used to analyze the frame:

1. The girder of the frame is displaced an arbitrary distance to the right while the joints are prevented from rotating. Typically, a *unit displacement* is introduced. To hold the structure in the deflected position, temporary restraints are introduced (see Figure 13.22*c*). These restraints consist of a roller at B to maintain the 1-in displacement and clamps at A, B, and C to prevent joint rotation.
   Since all displacements are known, we can compute the member end moments in the columns of the restrained frame with the slope-deflection equation. Because all joint rotations equal zero ($\theta_N = 0$ and $\theta_F = 0$) and no fixed-end moments are produced by loads applied to members between joints ($\text{FEM}_{NF} = 0$), with $\psi_{NF} = \Delta/L$, the slope-deflection equation (Equation 12.16) reduces to

$$M_{NF} = \frac{2EI}{L}(-3\psi_{NF}) = -\frac{6EI}{L}\frac{\Delta}{L} \tag{13.20}$$

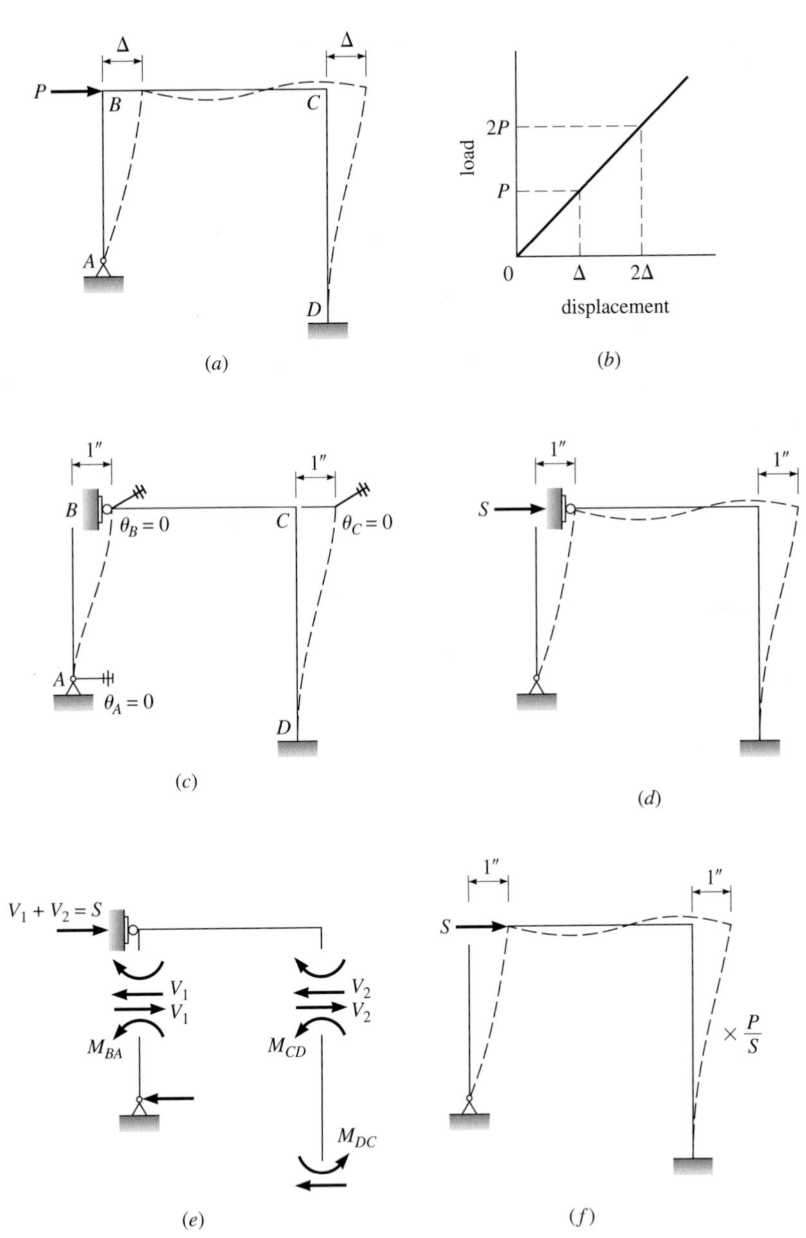

Figure 13.22: (*a*) Displacement of loaded frame; (*b*) linear elastic load displacement curve; (*c*) unit displacement of frame, temporary roller, and clamps introduced to restrain frame; (*d*) displaced frame with clamps removed, joints rotated into equilibrium position; all member end moments are known; (*e*) computation of reaction (*S*) at roller after column shears computed; axial forces in columns omitted for clarity; (*f*) frame displaced 1 in by a horizontal force *S*, multiply all forces by P/S to establish forces and deflections produced in (*a*) by force *P*.

For $\Delta = 1$, we can write Equation 13.20 as

$$M_{NF} = -\frac{6EI}{L^2} \tag{13.21}$$

At this stage with the joints clamped and prevented from rotating, the moments in the girder are zero because no loads act on this member.

2. Clamps are now removed and moments distributed until the structure relaxes into its equilibrium position (see Figure 13.22*d*). In the equilibrium position, the temporary roller at B applies a lateral force S to the frame. The force required to produce a unit displacement of the frame, denoted by S, is termed a *stiffness coefficient*.
3. The force S can be computed from a free-body diagram of the girder by summing forces in the horizontal direction (see Figure 13.22*e*). Axial forces in columns and the moments acting on the girder are omitted from Figure 13.22*e* for clarity. The column shears V_1 and V_2 applied to the girder are computed from free-body diagrams of the columns.
4. In Figure 13.22*f* we redraw the frame shown in Figure 13.22*d* in its deflected position. We imagine that the roller has been removed, but show the force S applied by the roller as an external load. At this stage we have analyzed the frame for a horizontal force S rather than P. However, since the frame behaves linearly, the forces produced by P can be evaluated by multiplying all forces and displacements in Figure 13.22*f* by the ratio P/S. For example, if P is equal to 10 kips and S is equal to 2.5 kips, the forces and displacements in Figure 13.22*f* must be multiplied by a factor of 4 to produce the forces induced by the 10-kip load. Example 13.9 illustrates the analysis of a simple frame of the type discussed in this section.

EXAMPLE 13.9

Determine the reactions and the member end moments produced in the frame shown in Figure 13.23*a* by a load of 5 kips at joint *B*. Also determine the horizontal displacement of girder *BC*. Given: $E = 30{,}000$ kips/in^2. Units of I are in in^4.

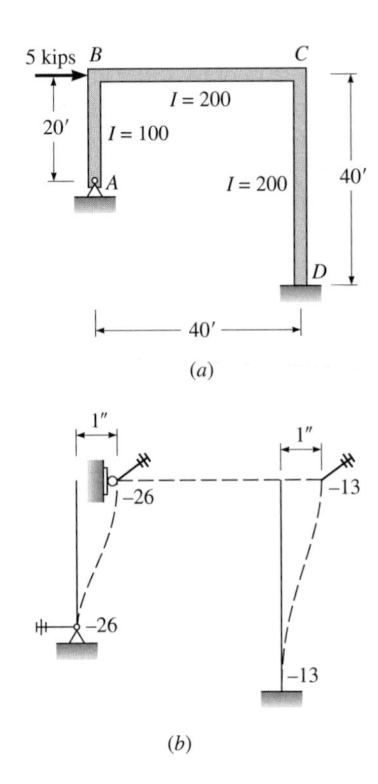

Figure 13.23: (*a*) Frame details; (*b*) moments in units of kip · ft induced in restrained frame (joints clamped to prevent rotation) by a unit displacement (*continues on page* 548).

Solution

We first displace the frame 1 in to the right with all joints clamped against rotation (see Figure 13.23*b*) and introduce a temporary roller at *B* to provide horizontal restraint. The column moments in the restrained structure are computed using Equation 13.21.

$$M_{AB} = M_{BA} = -\frac{6EI}{L^2} = -\frac{6(30{,}000)(100)}{(20 \times 12)^2} = -312 \text{ kip} \cdot \text{in}$$

$$= -26 \text{ kip} \cdot \text{ft}$$

$$M_{CD} = M_{DC} = -\frac{6EI}{L^2} = -\frac{6(30{,}000)(200)}{(40 \times 12)^2} = -166 \text{ kip} \cdot \text{in}$$

$$= -13 \text{ kip} \cdot \text{ft}$$

The clamps are now removed (but the roller remains) and the column moments distributed until all joints are in equilibrium. Details of the analysis are shown in Figure 13.23c. The distribution factors at joints B and C are computed below.

Joint B:

Distribution factors

$$K_{AB} = \frac{3}{4}\left(\frac{I}{L}\right) = \frac{3}{4}\left(\frac{100}{20}\right) = \frac{15}{4} \qquad \frac{K_{AB}}{\Sigma K} = \frac{3}{7}$$

$$K_{BC} = \frac{I}{L} = \frac{200}{40} = \frac{20}{4} \qquad \frac{K_{BC}}{\Sigma K} = \frac{4}{7}$$

$$\Sigma K = \frac{35}{4}$$

Joint C:

Distribution factors

$$K_{CB} = \frac{I}{L} = \frac{200}{40} = 5 \qquad \frac{5}{10} = \frac{1}{2}$$

$$K_{CD} = \frac{I}{L} = \frac{200}{40} = 5 \qquad \frac{5}{10} = \frac{1}{2}$$

$$\Sigma K = 10$$

We next compute the column shears by summing moments about an axis through the base of each column (see Figure 13.23d).

Compute V_1.

$$\circlearrowright^{+} \quad \Sigma M_A = 0 \qquad 20V_1 - 8.5 = 0 \qquad V_1 = 0.43 \text{ kip}$$

Compute V_2.

$$\circlearrowright^{+} \quad \Sigma M_D = 0 \qquad 40V_2 - 8.03 - 10.51 = 0 \qquad V_2 = 0.46 \text{ kip}$$

Considering horizontal equilibrium of the free body of the girder (in Figure 13.23d), compute the roller reaction at B.

$$\rightarrow + \quad \Sigma F_x = 0 \qquad S - V_1 - V_2 = 0$$

$$S = 0.46 + 0.43 = 0.89 \text{ kip}$$

At this stage we have produced a solution for the forces and reactions produced in the frame by a lateral load of 0.89 kip at joint B. (The results of the analysis in Figure 13.23c and d are summarized in Figure 13.23e.)

To compute the forces and displacements produced by a 5-kip load, we scale up all forces and displacements by the ratio of $P/S = 5/0.89 = 5.62$. Final results are shown in Figure 13.23f. The displacement of the girder = (P/S) (1 in) = 5.62 in.

[*continues on next page*]

Example 13.9 continues . . .

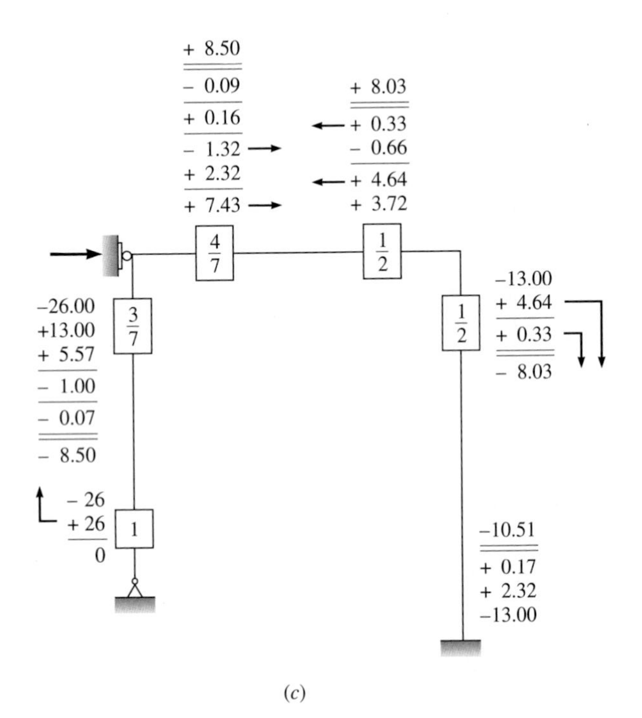

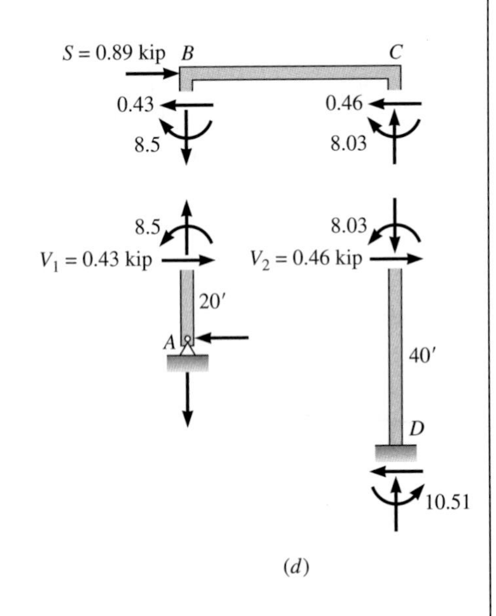

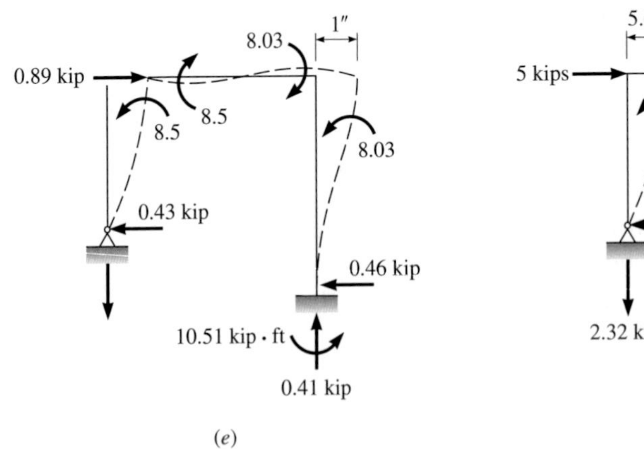

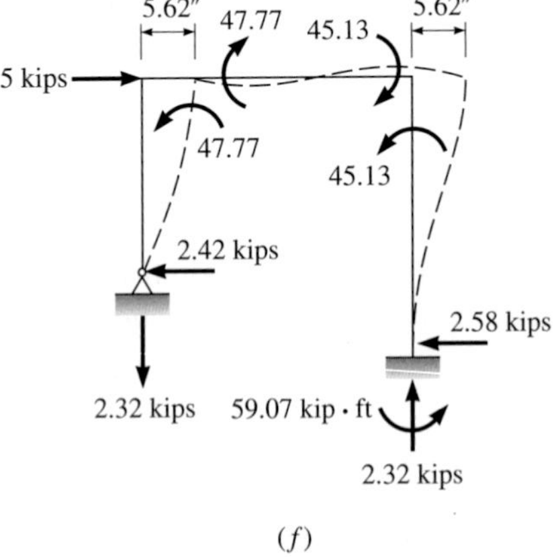

Figure 13.23: *Continued* (*c*) Moment distribution computations; (*d*) computation of roller force; (*e*) forces created in the frame by a unit displacement after clamps in (*b*) removed (moments in kip · ft and forces in kips); (*f*) reactions and member end moments produced by 5-kip load.

13.7 Analysis of an Unbraced Frame for General Loading

If a structure that is loaded between joints undergoes sidesway (Figure 13.24*a*), we must divide its analysis into several cases. We begin the analysis by introducing temporary restraints (holding forces) to prevent joints from translating. The number of restraints introduced must equal the number of independent joint displacements or degrees of sidesway (see Section 12.16). The restrained structure is then analyzed by moment distribution for the loads applied between joints. After the shears in all members are computed from free bodies of individual members, the holding forces are evaluated using the equations of statics by considering the equilibrium of members and/or joints. For example, to analyze the frame in Figure 13.24*a*, we introduce a temporary roller at C (or B) to prevent sidesway of the upper joints (see Figure 13.24*b*). We then analyze the structure by moment distribution in the standard manner for the applied loads (P and P_1) and determine the reaction R supplied by the roller. This step constitutes the Case A analysis.

Since no roller exists in the real structure at joint C, we must remove the roller and allow the structure to absorb the force R supplied by the roller. To eliminate R, we carry out a second analysis—the Case B analysis shown in Figure 13.24*c*. In this analysis we apply a force to joint C equal to R but acting in the opposite direction (to the right). Superposition of the Case A and Case B analyses produces results equivalent to the original case in Figure 13.24*a*.

Example 13.10 illustrates the foregoing procedure for a simple one-bay frame. Since this frame was previously analyzed for a lateral load at the top joint in Example 13.9, we will make use of these results for the Case B analysis (sidesway correction).

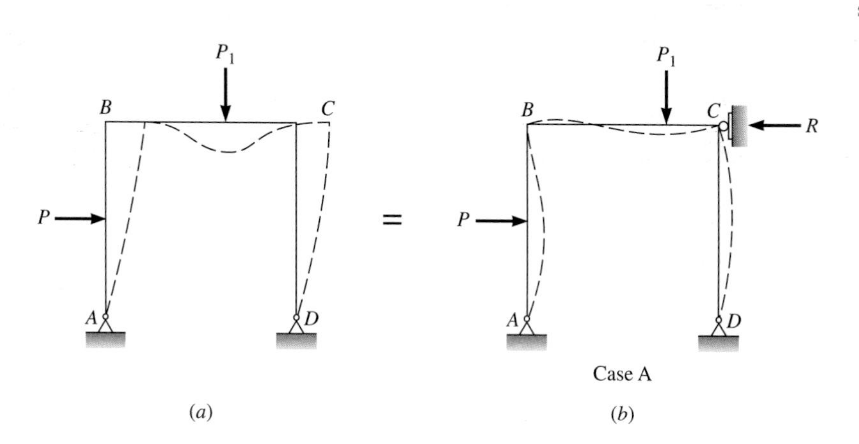

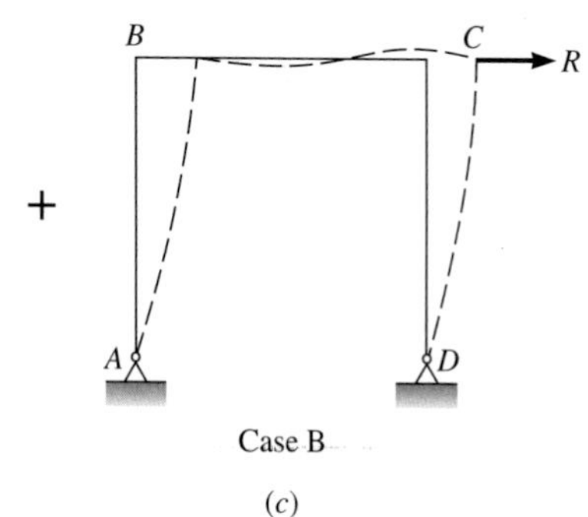

Figure 13.24: (*a*) Deformations of an unbraced frame; (*b*) sidesway prevented by adding a temporary roller that provides a holding force R at C; (*c*) sidesway correction, holding force reversed and applied to structure at joint C.

EXAMPLE 13.10

Determine the reactions and member end moments produced in the frame shown in Figure 13.25*a* by the 8-kip load. Also determine the horizontal displacement of joint *B*. Values of moment of inertia of each member in units of in^4 are shown on Figure 13.23*a*. $E = 30{,}000$ kips/in^2.

Solution

Since the frame in Figure 13.25 is the same as that in Example 13.9, we will refer to that example for the forces produced by the lateral load (Case B) analysis. Because the frame is free to sidesway, the analysis is broken into two cases. In the case A analysis, an imaginary roller is introduced at support *B* to prevent sidesway (see Figure 13.25*b*). The analysis of the restrained frame for the 8-kip load is carried out in Figure 13.25*d*. The fixed-end moments produced by the 8-kip load are equal to

$$\text{FEM} = \pm\frac{PL}{8} = \pm\frac{8(20)}{8} = \pm 20 \text{ kip}\cdot\text{ft}$$

The distribution factors were previously computed in Example 13.9. After the moment distribution is completed, the column shears, the axial forces, and the reaction *R* at the temporary support at *B* are computed from the free-body diagrams in Figure 13.25*e*. Since the roller force at *B* equals 4.97 kips, we must add the Case B sidesway correction shown in Figure 13.25*c*.

We have previously determined in Figure 13.23*e* the forces created in the frame by a horizontal force of $S = 0.89$ kip applied at *B*. This force produces a 1-in horizontal displacement of the girder. Since the frame is assumed to be elastic, we can establish the forces and displacement produced by a horizontal force of 4.97 kips by direct proportion; that is, all forces and displacements in Figure 13.23*e* are multiplied by a scale factor $4.97/0.89 = 5.58$. The results of this computation are shown in Figure 13.25*f*.

The final forces in the frame produced by summing the Case A and Case B solutions are shown in Figure 13.25*g*. The displacement of the girder is 5.58 in to the right.

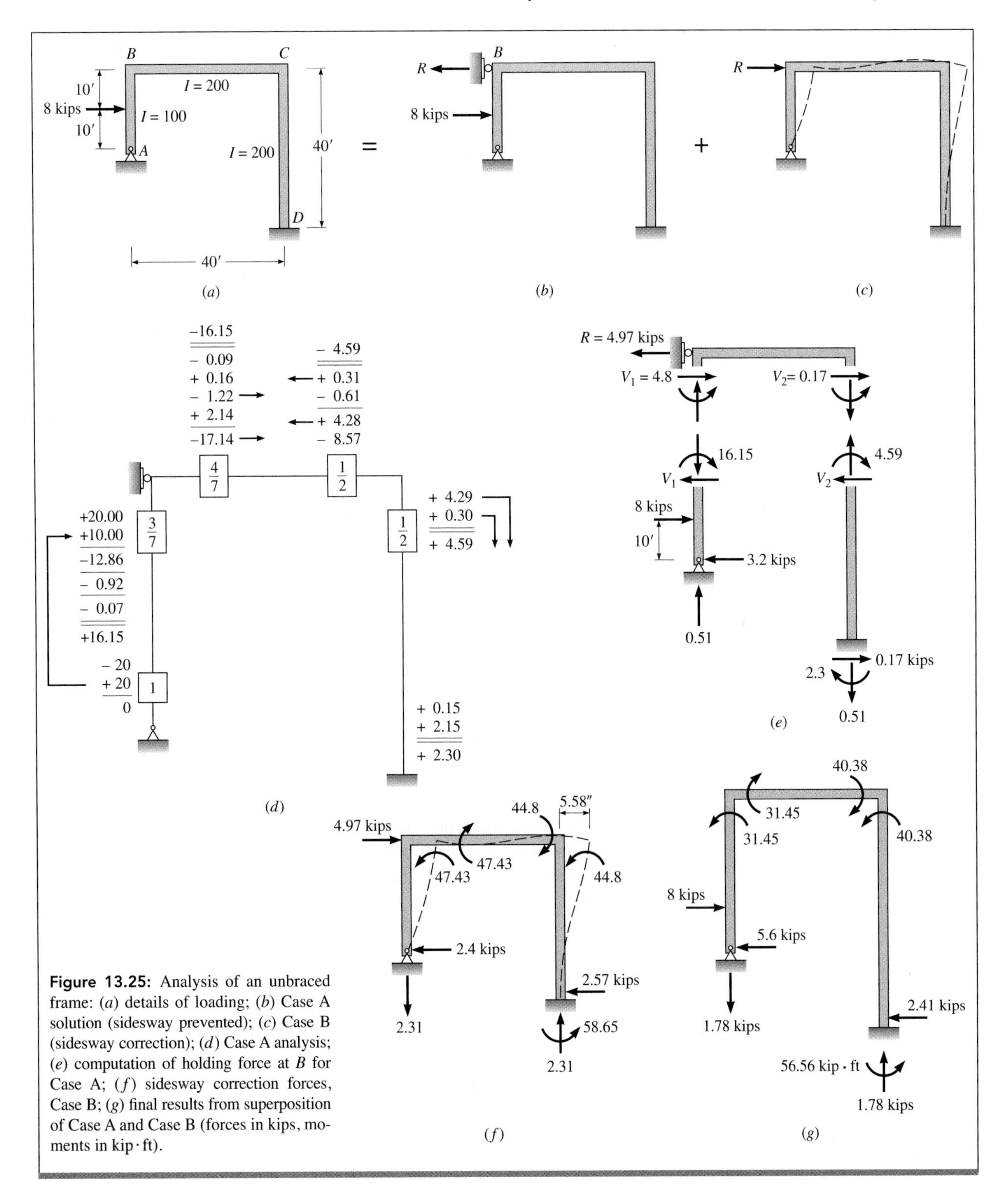

Figure 13.25: Analysis of an unbraced frame: (*a*) details of loading; (*b*) Case A solution (sidesway prevented); (*c*) Case B (sidesway correction); (*d*) Case A analysis; (*e*) computation of holding force at *B* for Case A; (*f*) sidesway correction forces, Case B; (*g*) final results from superposition of Case A and Case B (forces in kips, moments in kip · ft).

EXAMPLE 13.11

If a member BC of the frame in Example 13.9 is fabricated 2 in too long, determine the moments and reactions that are created when the frame is connected to its supports. Properties, dimensions of the frame, distribution factors, and so forth are specified or computed in Example 13.9.

Solution

If the frame is connected to the fixed support at D (see Figure 13.26*a*), the bottom of column AB will be located 2 in to the left of support A because of the fabrication error. Therefore, we must force the bottom of column AB to the right in order to connect it to the support at A. Before we bend the frame to connect the bottom of column AB to the pin support at A, we will fix the position of joints B and C by adding a roller at B and clamps at B and C. We then translate the bottom of column AB laterally 2" without allowing joint A to rotate ($\theta_A = 0$) and connect it to the pin support. A clamp is then added at A to prevent the bottom of the column from rotating. We now compute the end moments in column AB due to the chord rotation, using the modified form of the slope-deflection equation given by Equation 13.20. Since the chord rotation is counterclockwise, ψ_{AB} is negative and equal to

$$\psi_{AB} = -\frac{2}{20(12)} = -\frac{1}{20} \text{ rad}$$

$$M_{AB} = M_{BA} = -\frac{6EI}{L}\psi_{AB} = -\frac{6(30{,}000)(100)}{20 \times 12}\left(-\frac{1}{120}\right)$$

$$= 625 \text{ kip} \cdot \text{in} = 52.1 \text{ kip} \cdot \text{ft}$$

To analyze for the effect of removing the clamps in the restrained structure (Figure 13.26*a*), we carry out a moment distribution until the frame has absorbed the clamp moments—the roller at B remains in position during this phase of the analysis. Details of the distribution are shown in Figure 13.26*b*. The reaction at the roller is next computed from the free-body diagrams of the columns and girder (in Figure 13.26*c*). Since the rollers exerts a reaction on the frame of 0.85 kip to the left (Figure 13.26*d*), we must add the sidesway correction shown in Figure 13.26*e*. The forces associated with the correction are determined by proportion from the basic case in Figure 13.23*e*. Final reactions, shown in Figure 13.26*f*, are determined by superimposing the forces in Figure 13.26*d* and *e*.

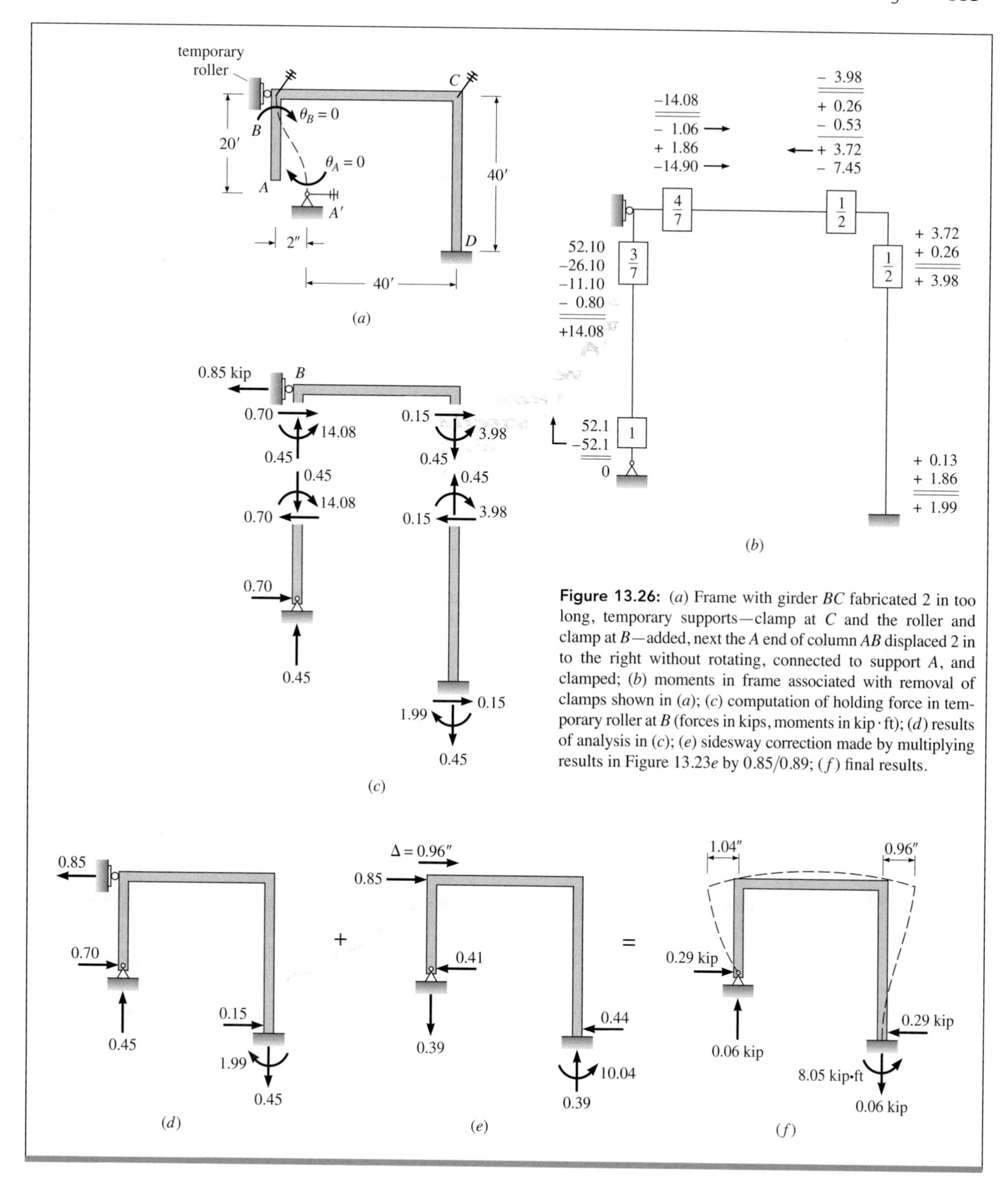

Figure 13.26: (*a*) Frame with girder *BC* fabricated 2 in too long, temporary supports—clamp at *C* and the roller and clamp at *B*—added, next the *A* end of column *AB* displaced 2 in to the right without rotating, connected to support *A*, and clamped; (*b*) moments in frame associated with removal of clamps shown in (*a*); (*c*) computation of holding force in temporary roller at *B* (forces in kips, moments in kip · ft); (*d*) results of analysis in (*c*); (*e*) sidesway correction made by multiplying results in Figure 13.23*e* by 0.85/0.89; (*f*) final results.

13.8 Analysis of Multistory Frames

To extend moment distribution to the analysis of multistory frames, we must add one sidesway correction for each independent degree of sidesway. Since the repeated analysis of the frame for the various cases becomes time-consuming, we will only outline the method of analysis, so the student is aware of the complexity of the solution. In practice, engineers today use computer programs to analyze frames of all types.

Figure 13.27*a* shows a two-story frame with two independent sidesway angles ψ_1 and ψ_2. To begin the analysis, we introduce rollers as temporary restraints at joint *D* and *E* to prevent sidesway (see Figure 13.27*b*). We then use moment distribution to analyze the restrained structure for the loads applied between joints (Case A solution). After the column shears are computed, we compute reactions R_1 and R_2 at the rollers using free bodies of the girders. Since the real structure is not restrained by forces at joints *D* and *E*, we must eliminate the roller forces. For this purpose we require two independent solutions (sidesway corrections) of the frame for lateral loads at joints *D* and *E*. One of the most convenient sets of sidesway corrections is produced by introducing a unit displacement that corresponds to one of the roller reactions while preventing all other joints from displacing laterally. These two cases are shown in Figure 13.27*c* and *d*. In Figure 13.27*c* we restrain joint *E* and introduce a 1-in displacement at joint *D*. We then analyze the frame and compute the holding forces S_{11} and S_{21} at joints *D* and *E*. In Figure 13.27*d* we introduce a unit displacement at joint *E* while restraining joint *D* and compute the holding forces S_{12} and S_{22}.

The final step in the analysis is to superimpose the forces at the rollers in the restrained structure (see Figure13.27*b*) with a certain fraction *X* of Case I (Figure 13.27*c*) and certain fraction *Y* of Case II (Figure 13.27*d*). The amount of each case to be added must eliminate the holding forces at joints *D* and *E*.

Figure 13.27: (*a*) Building frame with two degrees of sidesway; (*b*) restraining forces introduced at joints *D* and *E*; (*c*) Case I correction unit displacement introduced at joint *D*; (*d*) Case II correction, unit displacement introduced at joint *E*.

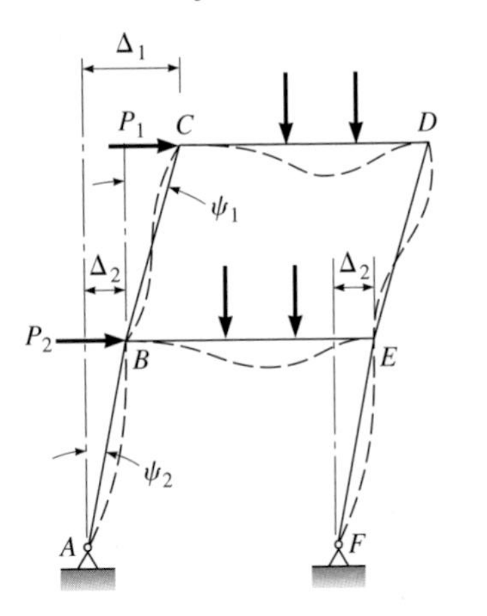

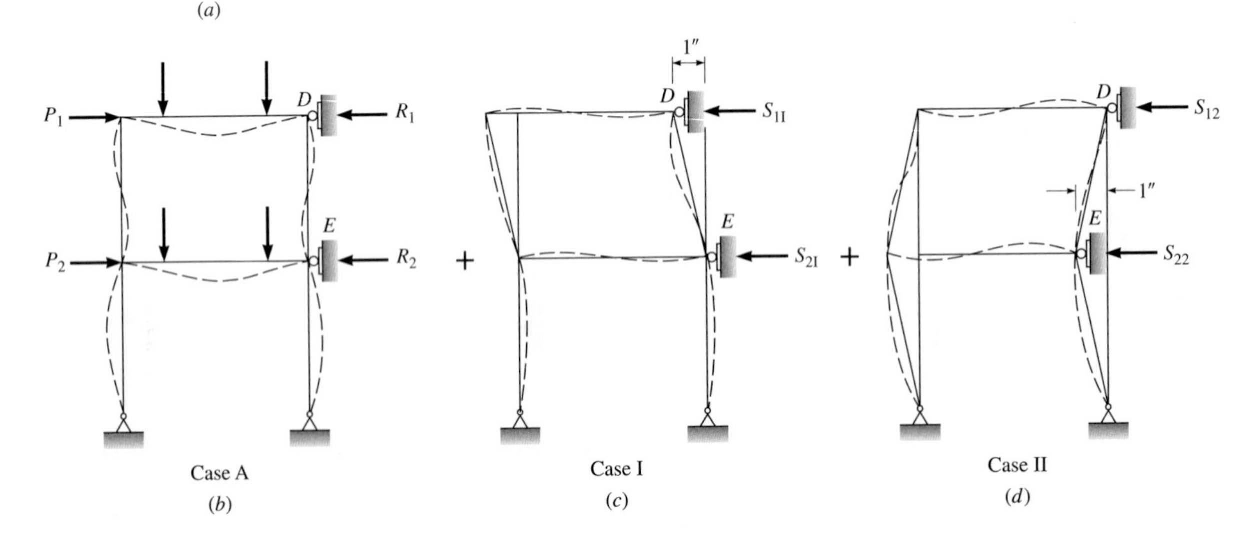

To determine the values of X and Y, two equations are written expressing the requirement that the sum of the lateral forces at joints D and E equal zero when the basic case and the two corrections are superimposed. For the frame in Figure 13.27*a*, these equations state

$$\text{At } D: \quad \Sigma F_x = 0 \tag{13.22}$$

$$\text{At } E: \quad \Sigma F_x = 0 \tag{13.23}$$

Expressing Equations 13.22 and 13.23 in terms of the forces shown in Figure 13.27*b* to *d* gives

$$R_1 + XS_{11} + YS_{12} = 0 \tag{13.24}$$

$$R_2 + XS_{21} + YS_{22} = 0 \tag{13.25}$$

By solving Equations 13.24 and 13.25 simultaneously, we can determine the values of X and Y. Examination of Figure 13.27 shows that X and Y represent the magnitude of the deflections at joints D and E, respectively. For example, if we consider the magnitude of the deflection Δ_1 at joint D, it is evident that all the displacement must be supplied by the Case I correction in Figure 13.27*c* since joint D is restrained in the Case A and Case II solutions.

13.9 Nonprismatic Members

Many continuous structures contain members whose cross sections vary along the length of the members. Some members are tapered to conform to the moment curve; other members, although the depth remains constant for a certain distance, are thickened where the moments are largest (see Figure 13.28). Although moment distribution can be used to analyze these structures, the fixed-end moments, carryover moments, and member stiffness are different from those we have used to analyze structures composed of prismatic members. In this section we discuss procedures for evaluating the various terms required to analyze structures with nonprismatic members. Since these terms and factors require considerable effort to evaluate, design tables (for example, see Tables 13.1 and 13.2) have been prepared to facilitate these computations.

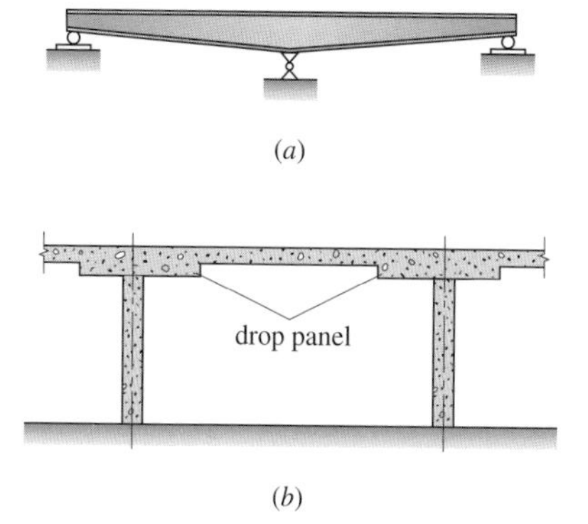

Figure 13.28: (*a*) Tapered beam; (*b*) floor slab with drop panels that is designed as a continuous beam with a variable depth.

Computation of the Carryover Factor

When a clamp is removed from a joint during a moment distribution, a portion of the unbalanced moment is distributed to each member framing into the joint. Figure 13.29*a* shows the forces applied to a typical member (i.e., the end at which the moment is applied is free to rotate but not to translate, and the far end is fixed). The moment M_A represents the distributed end moment, and the moment M_B equals the carryover moment. As we have seen in Section 13.2, the carryover moment is related to the distributed end moment; for example, for a prismatic member COM $= \frac{1}{2}$(DEM). We can express the carryover moment M_B as

$$M_B = \text{COM}_{AB} = C_{AB}(M_A) \tag{13.26}$$

where C_{AB} is the carryover factor from A to B. To evaluate C_{AB}, we will apply the M/EI curves associated with the loading in Figure 13.29*a* by "parts" to the conjugate beam in Figure 13.29*b*. If the computation is simplified further by setting $M_A = 1$ kip·ft in Equation 13.22, we find

$$M_B = C_{AB}$$

If we assume (to simplify the computations) that the member is prismatic (that is, *EI* is constant), we can compute C_{AB} by summing moments of the areas under the M/EI curve about support A of the conjugate beam.

$$\circlearrowright^{+} \quad \Sigma M_A = 0$$

$$\left(\frac{1}{2}L\right)\left(\frac{1}{EI}\right)\left(\frac{L}{3}\right) - \left(\frac{1}{2}L\right)\left(\frac{C_{AB}}{EI}\right)\left(\frac{2L}{3}\right) = 0$$

$$C_{AB} = \frac{1}{2}$$

The value above, of course, confirms the results of Section 13.2. In Example 13.2 we use this procedure to compute the carryover factor for a beam with a variable moment of inertia. Since the beam is not symmetric, the carryover factors are different for each end.

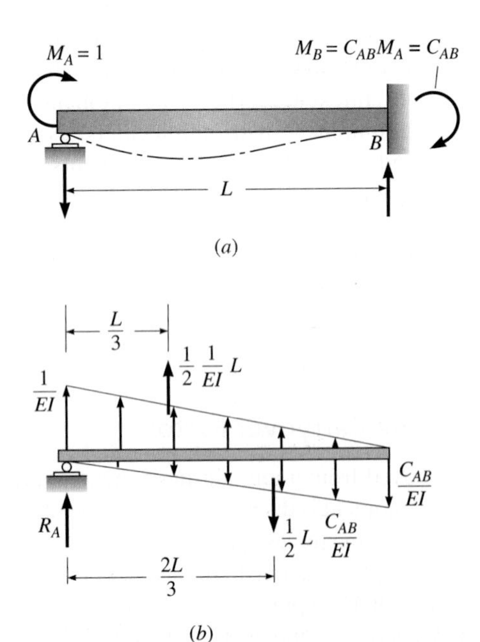

Figure 13.29: (*a*) Beam loaded by a unit moment at *A*; (*b*) conjugate structure loaded with *M*/*EI* curve by parts.

Computation of Absolute Flexural Stiffness

To compute the distribution factors at a joint where nonprismatic members intersect, we must use absolute flexural stiffness K_{ABS} of the members. The absolute flexural stiffness of a member is measured by the magnitude of the moment required to produce a specified value of rotation—typically 1 rad. Moreover, to compare one member with another, the boundary conditions of the members must also be standardized. Since one end of a member is free to rotate and the other end is fixed in the moment distribution method, these boundary conditions are used.

To illustrate the method used to compute the absolute flexural stiffness of a beam, we consider the beam of constant cross section in Figure 13.30. To the A end of the beam, we apply a moment K_{ABS} that produces a rotation of 1 rad at support A. If we assume that C_{AB} has been previously computed, the moment at the fixed end equals $C_{AB}K_{\text{ABS}}$. Using the slope-deflection equation, we can express the moment K_{ABS} in terms of the properties of the member as

$$K_{\text{ABS}} = \frac{2EI}{L}(2\theta_A) = \frac{4EI\theta_A}{L}$$

Substituting $\theta_A = 1$ rad gives

$$K_{\text{ABS}} = \frac{4EI}{L} \tag{13.27}$$

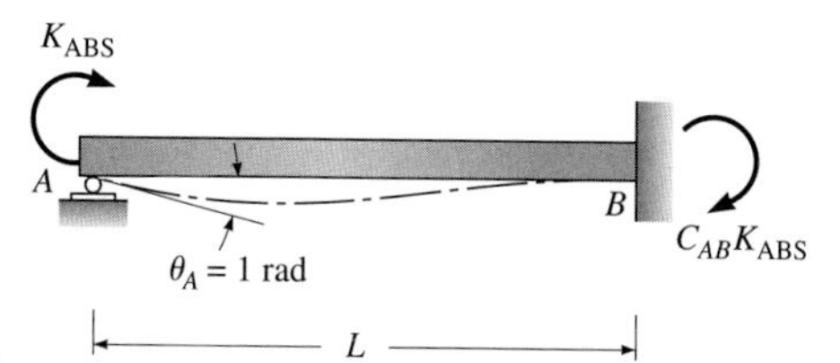

Figure 13.30: Support conditions used to establish the flexural stiffness of the A end of beam AB. The flexural stiffness is measured by the moment K_{ABS} required to produce a unit rotation at end A.

Since the slope-deflection equation applies only to prismatic members, we must use a different procedure to express the absolute flexural stiffness K_{ABS} of a nonprismatic member in terms of the properties of the member. Although a variety of methods can be used, we will use the moment-area method. Since the slope at B is zero and the slope at A is 1 rad, the area under the M/EI curve between the two points must equal 1. To produce an M/EI curve when the moment of inertia varies, we will express the moment of inertia at all sections as a multiple of the smallest moment of inertia. The procedure is illustrated in Example 13.12.

Reduced Absolute Flexural Stiffness

Once the carryover factors and the absolute flexural stiffness are established for a nonprismatic member, they can be used to evaluate the reduced absolute flexural stiffness K^R_{ABS}, for a beam with its far end pinned. To establish the expression for K^R_{ABS}, we consider the simply supported beam in Figure 13.31a. If a temporary clamp is applied to joint B, a moment applied at A equal to K_{ABS} will produce a rotation of 1 rad at A and carryover moment of $C_{AB}K_{\text{ABS}}$ at joint B. If we now clamp joint A and unclamp joint B (see Figure 13.31b), the moment at B reduces to zero and the moment at A, which now represents K^R_{ABS}, equals

$$K^R_{\text{ABS}} = K_{\text{ABS}} - C_{BA}C_{AB}K_{\text{ABS}}$$

$$= K_{\text{ABS}}(1 - C_{BA}C_{AB}) \tag{13.28}$$

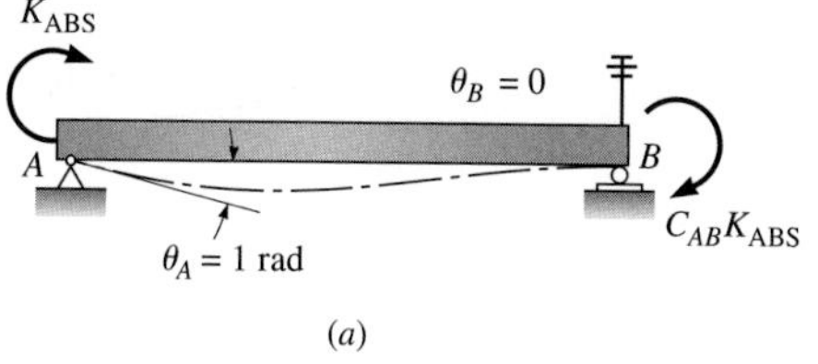

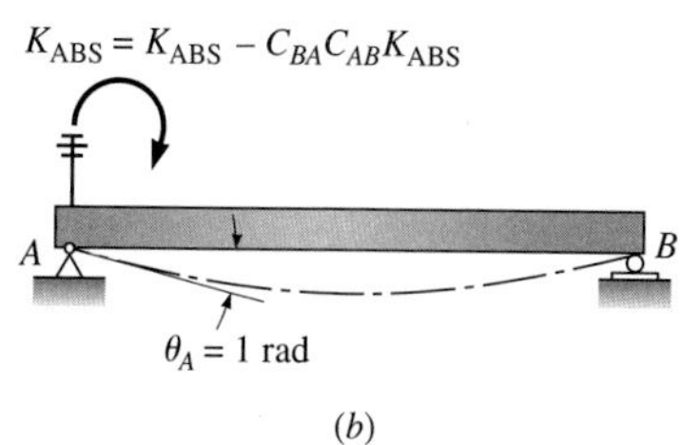

Figure 13.31

Computation of Fixed-End Moments

To compute the fixed-end moments that develop in a nonprismatic beam, we load the conjugate beam with the M/EI curves. When a real beam has fixed ends, the supports in the conjugate beam are free ends. To facilitate the computations, the moment curves should be drawn by "parts" to produce simple geometric shapes. At this stage the values of the fixed-end moments are *unknown*. To solve for the fixed-end moments, we must write two equilibrium equations. For the conjugate beam to be in equilibrium, the algebraic sum of the areas under the M/EI diagrams (loads) must equal zero. Alternatively, the moments of the areas under the M/EI curves about each end of the conjugate beam must also equal zero. To establish the fixed-end moments, we solve simultaneously any two of the three equations above.

To illustrate the basic principles of the method, we will compute the fixed-end moments produced in a prismatic beam (EI is constant) by a concentrated load at midspan. This same procedure (with the M/EI diagrams modified to account for the variations in moment of inertia) will be used in Example 13.12 to evaluate the fixed-end moments at the ends of the nonprismatic beam.

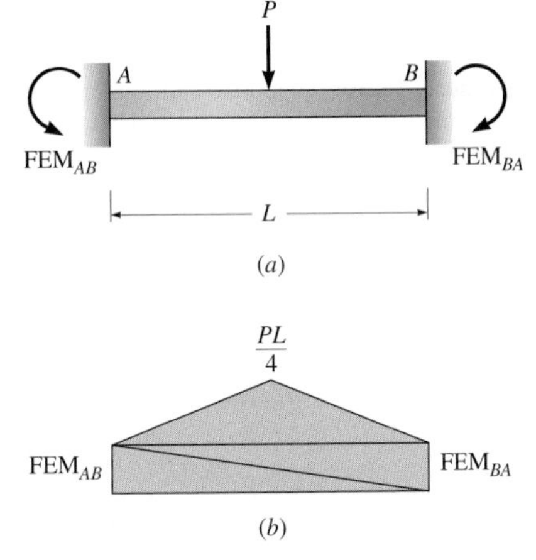

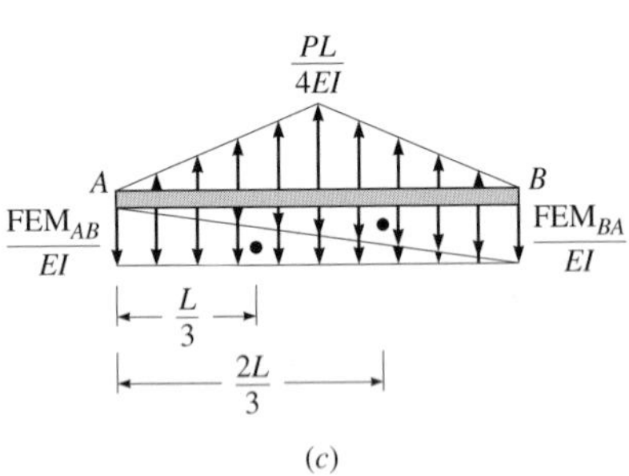

Figure 13.32: (*a*) Fixed-end beam with EI constant; (*b*) moment curves by parts; (*c*) conjugate beam loaded with the M/EI diagrams.

Computation of Fixed-End Moments for the Beam in Figure 13.32*a*

Load the conjugate beam with the M/EI curves (see Figure 13.32*c*), and sum moments about A, giving

$$\circlearrowright^{+} \quad \Sigma M_A = 0$$

$$-\frac{1}{2}\frac{PL}{4EI}L\frac{L}{2} + \frac{1}{2}\text{FEM}_{AB}\,L\frac{L}{3} + \frac{1}{2}\text{FEM}_{BA}\,L\frac{2L}{3} = 0 \tag{1}$$

Recognizing that the structure and load are symmetric, we set $\text{FEM}_{AB} = \text{FEM}_{BA}$ in Equation 1 and solve for FEM_{BA}:

$$\text{FEM}_{BA} = \frac{PL}{8}$$

EXAMPLE 13.12

The beam in Figure 13.33*a* has a variable moment of inertia. Determine (*a*) the carryover factor from *A* to *B*, (*b*) the absolute flexural stiffness of the left end, and (*c*) the fixed-end moment produced by a concentrated load *P* at midspan. Over the length of the beam *E* is constant.

Solution

(a) Computation of the Carryover Factor. We apply a unit moment of 1 kip · ft to the end of the beam at *A* (Figure 13.33*b*), producing the carryover moment C_{AB} at *B*. The moment curves are drawn by parts, producing two triangular moment diagrams. The ordinates of the moment curve are then divided by *EI* on the left half and by 2*EI* on the right half to produce the *M*/*EI* diagrams, which are applied as loads to the conjugate beam (see Figure 13.33*c*). Since the moment of inertia of the right half of the beam is twice as large as that on the left side, a discontinuity in the *M*/*EI* curve is created at midspan. Positive moment is applied as an upward load and negative moment as a downward load. To express C_{AB} in terms of the properties of the member, we divide the areas under the *M*/*EI* diagram into rectangles and triangles and sum moments of these areas about the support at *A* to be equal to zero. In the moment-area method, this step is equivalent to the condition that the tangential deviation of point *A* from the tangent drawn at *B* is zero.

$$\circlearrowleft^{+} \quad \Sigma M_A = 0$$

$$\frac{1}{2EI}\frac{L}{2}\frac{L}{4} + \frac{1}{2}\frac{1}{2EI}\frac{L}{2}\frac{L}{6} + \frac{1}{2}\frac{1}{4EI}\frac{L}{2}\left(\frac{L}{2} + \frac{L}{6}\right)$$

$$-\frac{1}{2}\frac{L}{2}\frac{C_{AB}}{2EI}\left(\frac{2}{3}\frac{L}{2}\right) - \frac{C_{AB}}{4EI}\frac{L}{2}\left(\frac{L}{2} + \frac{L}{4}\right) - \frac{1}{2}\frac{L}{2}\frac{C_{AB}}{4EI}\left(\frac{L}{2} + \frac{2}{3}\frac{L}{2}\right) = 0$$

Simplifying and solving for C_{AB} give

$$C_{AB} = \frac{2}{3}$$

If the supports are switched (the fixed support moved to *A* and the roller to *B*) and a unit moment applied at *B*, we find the carryover factor $C_{BA} = 0.4$ from *B* to *A*.

(b) Computation of Absolute Flexural Stiffness K_{ABS}. The absolute flexural stiffness of the left end of the beam is defined as the moment K_{ABS} required to produce a unit rotation ($\theta_A = 1$ rad) at *A* with the right end fixed and the left end restrained against vertical displacement by a roller (see Figure 13.33*d*). Figure 13.33*e* shows the *M*/*EI* curves for the loading in Figure 13.33*d*. Because the slope at *B* is zero, the change in slope between ends of the beam (equal to the area under the *M*/*EI* curve by the first

[*continues on next page*]

Example 13.12 continues . . .

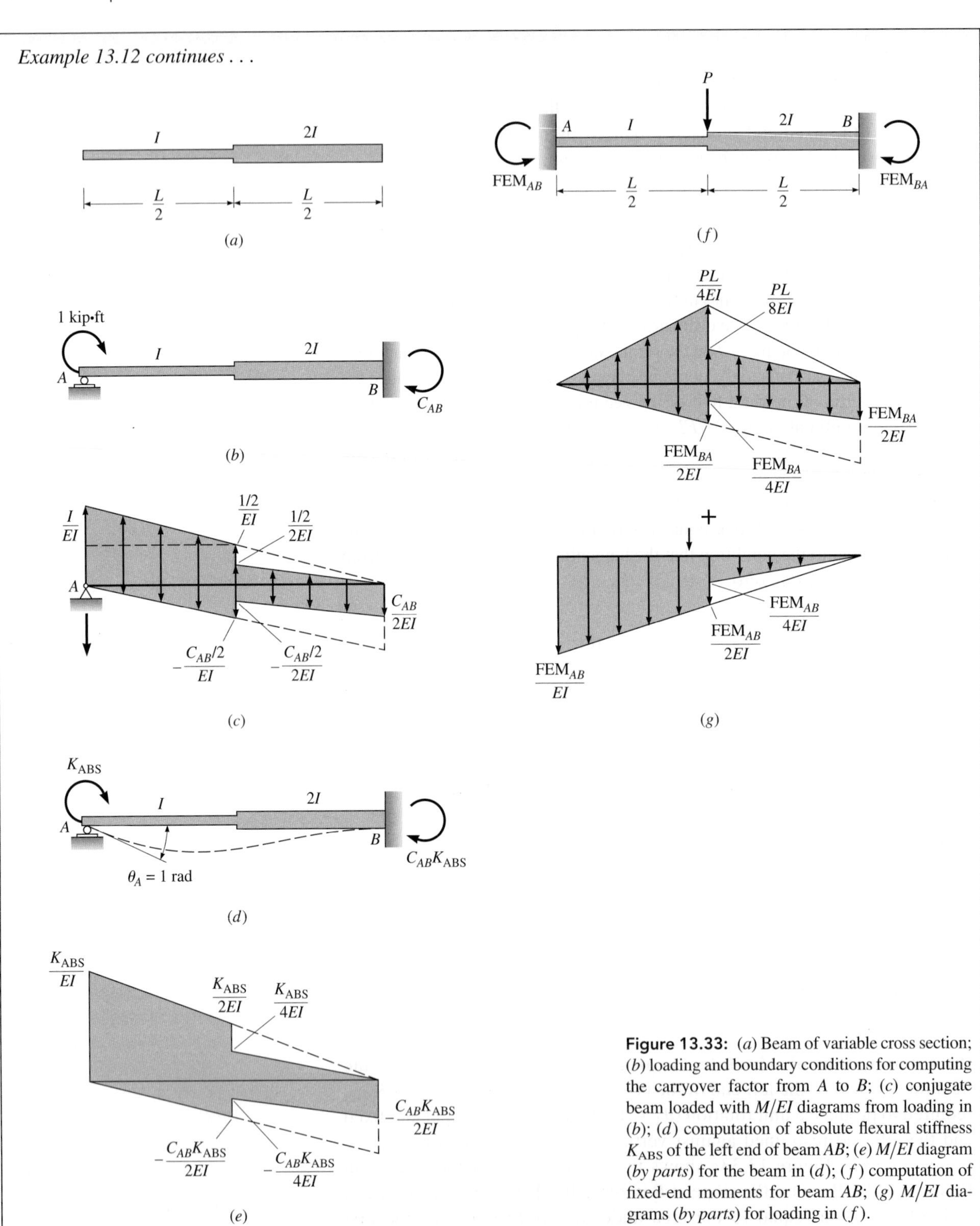

Figure 13.33: (*a*) Beam of variable cross section; (*b*) loading and boundary conditions for computing the carryover factor from A to B; (*c*) conjugate beam loaded with M/EI diagrams from loading in (*b*); (*d*) computation of absolute flexural stiffness K_{ABS} of the left end of beam AB; (*e*) M/EI diagram (*by parts*) for the beam in (*d*); (*f*) computation of fixed-end moments for beam AB; (*g*) M/EI diagrams (*by parts*) for loading in (*f*).

moment-area principle) equals 1. To evaluate the area under the M/EI curves, we divide it into triangles and a rectangle

$$\Sigma \text{ areas} = 1$$

$$\frac{1}{2}\frac{L}{2}\frac{K_{\text{ABS}}}{EI} + \frac{1}{2}\frac{L}{2}\frac{K_{\text{ABS}}}{2EI} + \frac{1}{2}\frac{L}{2}\frac{K_{\text{ABS}}}{4EI}$$

$$-\frac{1}{2}\frac{L}{2}\frac{C_{AB}K_{\text{ABS}}}{2EI} - \frac{C_{AB}K_{\text{ABS}}}{4EI}\frac{L}{2} - \frac{1}{2}\frac{C_{AB}K_{\text{ABS}}}{4EI}\frac{L}{2} = 1$$

Substituting $C_{AB} = \frac{2}{3}$ from (a) and solving for K_{ABS} give

$$K_{\text{ABS}} = 4.36\frac{EI}{L}$$

(c) Computation of Fixed-End Moments Produced by a Concentrated Load at Midspan. To compute the fixed-end moments, we apply the concentrated load to the beam with its ends clamped (see Figure 13.33f). Moment curves are drawn by parts and converted to M/EI curves that are applied as loads to the conjugate beam, as shown in Figure 13.33g. (The M/EI curve, produced by the fixed-end moment FEM_{AB} at the left end, is drawn below the conjugate beam for clarity.) Since both fixed-end moments are unknown, we write two equations for their solution:

$$\Sigma F_y = 0 \tag{1}$$

$$\Sigma M_A = 0 \tag{2}$$

Expressing Equation 1 in terms of the areas of the M/EI diagrams gives

$$\frac{1}{2}\frac{L}{2}\frac{PL}{4EI} + \frac{1}{2}\frac{L}{2}\frac{PL}{8EI} - \frac{1}{2}\frac{\text{FEM}_{BA}}{2EI}\frac{L}{2} - \frac{\text{FEM}_{BA}}{4EI}\frac{L}{2} - \frac{1}{2}\frac{L}{2}\frac{\text{FEM}_{BA}}{4EI}$$

$$-\left(\frac{1}{2}\frac{\text{FEM}_{AB}}{EI}L - \frac{1}{2}\frac{\text{FEM}_{AB}}{4EI}\frac{L}{2}\right) = 0$$

Simplifying and collecting terms yield

$$\frac{5}{16}\text{FEM}_{BA} + \frac{7}{16}\text{FEM}_{AB} = \frac{3PL}{32} \tag{1a}$$

Expressing Equation 2 in terms of the moments of areas by multiplying each of the areas above by the distance between point A and the respective centroids gives

$$\frac{9}{48}\text{FEM}_{BA} + \frac{1}{8}\text{FEM}_{AB} = \frac{PL}{24} \tag{2a}$$

Solving Equations 1a and 2a simultaneously gives

$$\text{FEM}_{AB} = 0.106PL \qquad \text{FEM}_{BA} = 0.152PL$$ **Ans.**

As expected, the fixed-end moment on the right is larger than that on the left because of the greater stiffness of the right side of the beam.

EXAMPLE 13.13

Analyze the rigid frame in Figure 13.34 by moment distribution. All members 12 in thick are measured perpendicular to the plane of the structure.

Solution

Since the girder has a variable moment of inertia, we will use Table 13.2 to establish the carryover factor, the stiffness coefficient, and the fixed-end moments. The parameters to enter in Table 13.2 are

$$aL = 10 \text{ ft} \qquad \text{since } L = 50 \text{ ft}, a = \frac{10}{50} = 0.2$$

$$rh_c = 6 \text{ in} \qquad \text{since } h_c = 10 \text{ in}, r = 0.6$$

Read in Table 13.2:

$$C_{CB} = C_{BC} = 0.674$$

$$k_{BC} = 8.8$$

$$\begin{aligned} \text{FEM}_{CB} = -\text{FEM}_{BC} &= 0.1007wL^2 \\ &= 0.1007(2)(50)^2 \\ &= 503.5 \text{ kip} \cdot \text{ft} \end{aligned}$$

$$I_{\text{min girder}} = \frac{bh^3}{12} = \frac{12(10)^3}{12} = 1000 \text{ in}^4$$

$$I_{\text{column}} = \frac{bh^3}{12} = \frac{12(16)^3}{12} = 4096 \text{ in}^4$$

Compute distribution factors at joint B or C:

$$K_{\text{girder}} = \frac{8.8EI}{L} = \frac{8.8E(1000)}{50} = 176E$$

$$K_{\text{column}} = \frac{4EI}{L} = \frac{4E(4096)}{16} = 1024E$$

$$\Sigma K = 1200E$$

$$\text{DF}_{\text{column}} = \frac{1024E}{1200E} = 0.85$$

$$\text{DF}_{\text{girder}} = \frac{176E}{1200E} = 0.15$$

See Figure 13.34b for distribution. Reactions are shown in Figure 13.34c.

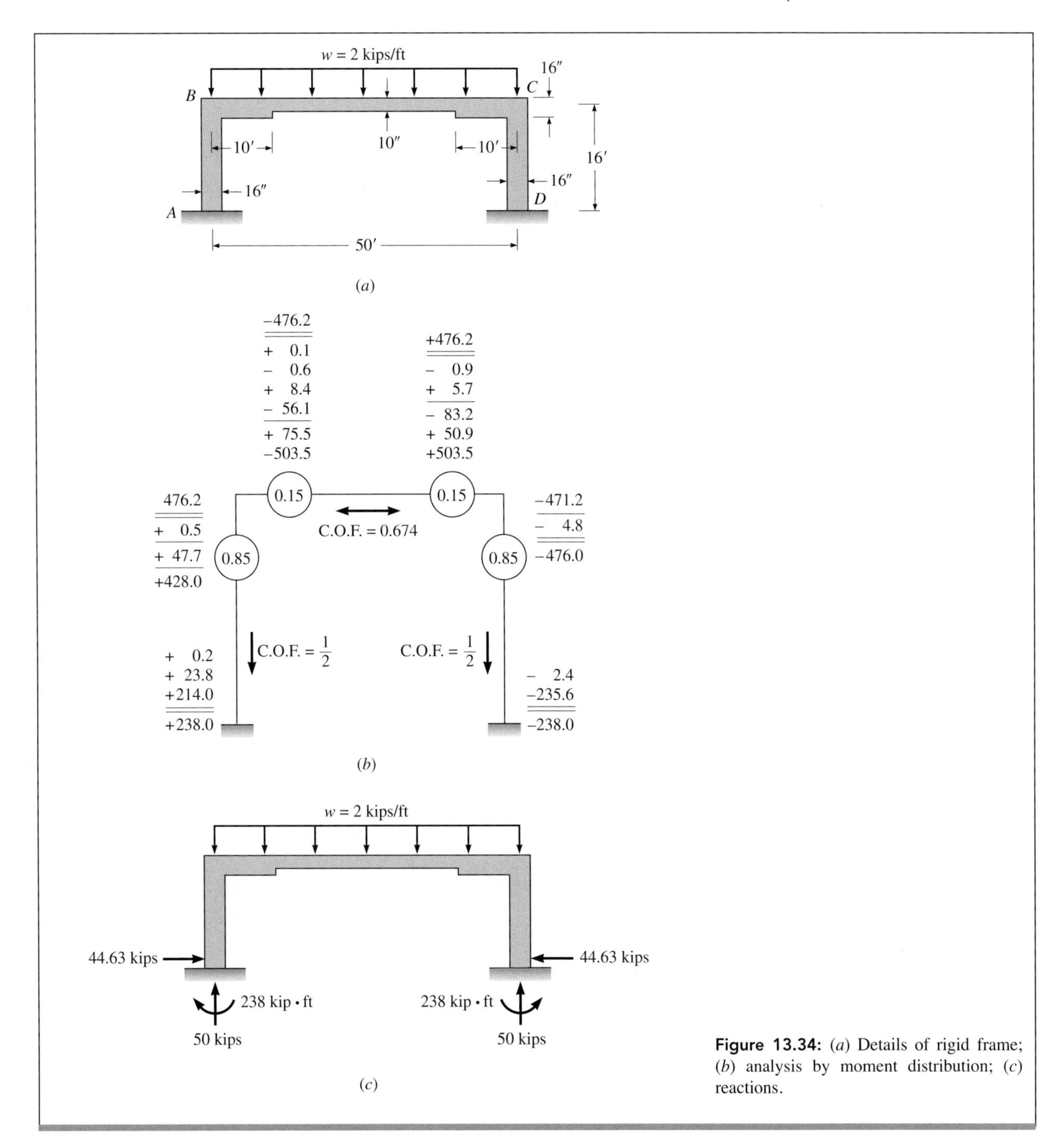

Figure 13.34: (*a*) Details of rigid frame; (*b*) analysis by moment distribution; (*c*) reactions.

TABLE 13.1

Prismatic Haunch at One End (from *Handbook of Frame Constants* by the Portland Cement Association)

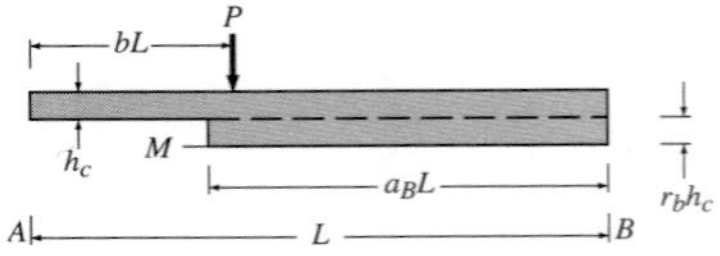

Note: All carryover factors are negative and all stiffness factors are positive. All fixed-end moment coefficients are negative except where plus sign is shown.

Right haunch		Carryover factors		Stiffness factors		Unif. load FEM coef. × wL^2		Concentrated load FEM—coef. × PL, b = 0.1		0.3		0.5		0.7		0.9		$1-a_B$		Moment M at $b = 1-a_B$ FEM coef. × M		Haunch load FEM coef. × wL^2	
a_B	r_B	C_{AB}	C_{BA}	k_{AB}	k_{BA}	M_{AB}	M_{BA}	M_{AB}	M_{BA}	M_{AB}	M_{BA}	M_{AB}	M_{BA}	M_{AB}	M_{BA}	M_{AB}	M_{BA}	M_{AB}	M_{BA}	M_{AB}	M_{BA}	M_{AB}	M_{BA}
				$r_A = 0$											$a_A = 0$								
	0.4	0.593	0.491	4.24	5.12	0.0749	0.1016	0.0799	0.0113	0.1397	0.0788	0.1110	0.1553	0.0478	0.1798	0.0042	0.0911	0.0042	0.0911	0.0793	0.8275	0.0001	0.0047
	0.6	0.615	0.490	4.30	5.40	0.0727	0.1062	0.0797	0.0119	0.1378	0.0828	0.1074	0.1630	0.0439	0.1881	0.0029	0.0937	0.0029	0.0937	0.0561	0.8780	0.0001	0.0048
0.1	1.0	0.639	0.488	4.37	5.72	0.0703	0.1114	0.0794	0.0125	0.1358	0.0873	0.1035	0.1716	0.0396	0.1974	0.0016	0.0966	0.0016	0.0966	0.0304	0.9339	0.0001	0.0049
	1.5	0.652	0.487	4.40	5.89	0.0690	0.1143	0.0792	0.0129	0.1346	0.0898	0.1012	0.1764	0.0373	0.2026	0.0008	0.0982	0.0008	0.0982	0.0161	0.9651	0.0000	0.0049
	2.0	0.658	0.487	4.42	5.97	0.0684	0.1156	0.0791	0.0131	0.1341	0.0910	0.1002	0.1786	0.0361	0.2050	0.0005	0.0990	0.0005	0.0990	0.0094	0.9795	0.0000	0.0050
	0.4	0.677	0.469	4.42	6.37	0.0706	0.1126	0.0791	0.0134	0.1345	0.0925	0.1020	0.1788	0.0409	0.1975	0.0050	0.0890	0.0182	0.1581	0.1640	0.6037	0.0013	0.0171
	0.6	0.730	0.463	4.56	7.18	0.0664	0.1225	0.0785	0.0149	0.1302	0.1025	0.0942	0.1972	0.0335	0.2148	0.0037	0.0917	0.0137	0.1684	0.1241	0.7005	0.0010	0.0178
0.2	1.0	0.793	0.458	4.74	8.22	0.0510	0.1353	0.0777	0.0168	0.1248	0.1154	0.0843	0.2207	0.0242	0.2368	0.0022	0.0951	0.0080	0.1815	0.0728	0.8245	0.0006	0.0187
	1.5	0.831	0.455	4.86	8.88	0.0576	0.1434	0.0772	0.0180	0.1214	0.1235	0.0781	0.2355	0.0182	0.2507	0.0012	0.0973	0.0044	0.1897	0.0403	0.9029	0.0003	0.0193
	2.0	0.849	0.453	4.91	9.20	0.0559	0.1473	0.0769	0.0186	0.1197	0.1276	0.0750	0.2429	0.0153	0.2576	0.0007	0.0984	0.0026	0.1939	0.0242	0.9418	0.0002	0.0196
	0.4	0.741	0.439	4.52	7.63	0.0698	0.1155	0.0787	0.0149	0.1319	0.1013	0.0987	0.1899	0.0420	0.1929	0.0056	0.0868	0.0420	0.1929	0.2371	0.3457	0.0045	0.0338
	0.6	0.831	0.427	4.75	9.24	0.0542	0.1296	0.0777	0.0175	0.1255	0.1182	0.0877	0.2185	0.0338	0.2130	0.0045	0.0893	0.0338	0.2130	0.1935	0.4682	0.0036	0.0359
0.3	1.0	0.954	0.415	5.09	11.69	0.0559	0.1511	0.0762	0.0215	0.1158	0.1440	0.0711	0.2621	0.0217	0.2436	0.0028	0.0930	0.0217	0.2436	0.1261	0.6548	0.0023	0.0391
	1.5	1.036	0.409	5.34	13.53	0.0497	0.1673	0.0751	0.0245	0.1085	0.1633	0.0587	0.2948	0.0128	0.2665	0.0017	0.0959	0.0128	0.2665	0.0750	0.7952	0.0014	0.0415
	2.0	1.078	0.407	5.48	14.54	0.0464	0.1762	0.0745	0.0262	0.1045	0.1740	0.0520	0.3129	0.0080	0.2792	0.0010	0.0974	0.0080	0.2792	0.0467	0.8725	0.0008	0.0448
	0.4	0.774	0.405	4.55	8.70	0.0703	0.1117	0.0786	0.0156	0.1315	0.1035	0.0992	0.1855	0.0445	0.1773	0.0059	0.0849	0.0713	0.1938	0.2780	0.0876	0.0106	0.0509
	0.6	0.901	0.386	4.83	11.28	0.0646	0.1269	0.0774	0.0192	0.1240	0.1254	0.0875	0.2182	0.0377	0.1932	0.0049	0.0869	0.0611	0.2204	0.2456	0.2035	0.0089	0.0547
0.4	1.0	1.102	0.367	5.33	16.03	0.0549	0.1548	0.0752	0.0257	0.1105	0.1658	0.0671	0.2780	0.0267	0.2222	0.0034	0.0904	0.0438	0.2689	0.1817	0.4177	0.0063	0.0616
	1.5	1.260	0.357	5.79	20.46	0.0462	0.1807	0.0732	0.0319	0.0982	0.2035	0.0485	0.3339	0.0173	0.2491	0.0022	0.0938	0.0284	0.3142	0.1198	0.6183	0.0037	0.0579
	2.0	1.349	0.352	6.09	23.32	0.0407	0.1975	0.0719	0.0358	0.0903	0.2278	0.0367	0.3699	0.0113	0.2664	0.0014	0.0959	0.0187	0.3434	0.0793	0.7479	0.0027	0.0720
	0.4	0.768	0.371	4.56	9.45	0.0700	0.1048	0.0786	0.0154	0.1312	0.0993	0.0983	0.1679	0.0442	0.1663	0.0059	0.0836	0.0983	0.1679	0.2710	+0.1319	0.0189	0.0556
	0.6	0.919	0.343	4.84	12.94	0.0651	0.1176	0.0774	0.0193	0.1240	0.1218	0.0884	0.1935	0.0386	0.1769	0.0051	0.0849	0.0884	0.1935	0.2593	+0.0493	0.0167	0.0702
0.5	1.0	1.200	0.316	5.42	20.61	0.0561	0.1451	0.0749	0.0280	0.1096	0.1709	0.0706	0.2486	0.0299	0.1993	0.0038	0.0877	0.0705	0.2486	0.2203	0.1356	0.0131	0.0802
	1.5	1.470	0.301	6.10	29.74	0.0466	0.1777	0.0720	0.0384	0.0934	0.2290	0.0516	0.3137	0.0215	0.2255	0.0027	0.0909	0.0516	0.3137	0.1663	0.3579	0.0094	0.0918
	2.0	1.647	0.295	6.63	37.04	0.0393	0.2036	0.0698	0.0466	0.0807	0.2755	0.0370	0.3655	0.0153	0.2463	0.0019	0.0934	0.0370	0.3655	0.1209	0.5361	0.0067	0.1011
	0.4	0.726	0.341	4.62	9.84	0.0675	0.0986	0.0782	0.0146	0.1280	0.0916	0.0923	0.1519	0.0419	0.1603	0.0056	0.0829	0.1154	0.1276	0.2103	+0.2862	0.0283	0.0769
	0.6	0.872	0.305	4.88	13.97	0.0630	0.1072	0.0771	0.0183	0.1214	0.1096	0.0835	0.1664	0.0368	0.1666	0.0048	0.0837	0.1068	0.1463	0.2221	+0.2453	0.0254	0.0813
0.6	1.0	1.196	0.267	5.43	24.35	0.0560	0.1277	0.0748	0.0274	0.1092	0.1537	0.0705	0.1999	0.0299	0.1804	0.0038	0.0854	0.0926	0.1910	0.2190	+0.1321	0.0212	0.0913
	1.5	1.588	0.247	6.18	39.79	0.0482	0.1572	0.0718	0.0408	0.0939	0.2183	0.0572	0.2478	0.0237	0.1997	0.0030	0.0878	0.0762	0.2559	0.1926	0.0433	0.0171	0.1055
	2.0	1.905	0.237	6.92	55.51	0.0412	0.1870	0.0688	0.0544	0.0792	0.2839	0.0455	0.2960	0.0186	0.2189	0.0023	0.0901	0.0611	0.3215	0.1589	0.2243	0.0136	0.1197
	0.4	0.657	0.321	4.86	9.96	0.0631	0.0954	0.0770	0.0138	0.1175	0.0846	0.0844	0.1461	0.0392	0.1582	0.0053	0.0827	0.1175	0.0846	0.0959	+0.3666	0.0372	0.0854
	0.6	0.770	0.275	5.14	14.39	0.0580	0.1006	0.0758	0.0167	0.1097	0.0955	0.0745	0.1543	0.0335	0.1621	0.0045	0.0832	0.1097	0.0955	0.1322	+0.3615	0.0330	0.0890
0.7	1.0	1.056	0.224	5.62	26.45	0.0516	0.1122	0.0738	0.0243	0.0992	0.1203	0.0626	0.1710	0.0269	0.1694	0.0035	0.0841	0.0992	0.1213	0.1655	+0.3228	0.0280	0.0965
	1.5	1.491	0.196	6.24	47.48	0.0463	0.1304	0.0714	0.0371	0.0890	0.1633	0.0537	0.1959	0.0223	0.1796	0.0028	0.0854	0.0890	0.1633	0.1731	+0.2367	0.0241	0.1076
	2.0	1.944	0.183	6.95	73.85	0.0417	0.1523	0.0687	0.0530	0.0793	0.2149	0.0468	0.2255	0.0191	0.1915	0.0024	0.0869	0.0793	0.2149	0.1646	+0.1219	0.0210	0.1210
	0.4	0.583	0.319	5.46	9.97	0.0585	0.0951	0.0741	0.0137	0.1040	0.0837	0.0793	0.1456	0.0380	0.1580	0.0053	0.0826	0.1023	0.0461	0.0804	+0.3734	0.0452	0.0917
	0.6	0.645	0.263	5.89	14.44	0.0516	0.0990	0.0721	0.0160	0.0921	0.0907	0.0667	0.1520	0.0311	0.1614	0.0043	0.0831	0.0950	0.0517	0.0150	+0.3956	0.0388	0.0951
0.8	1.0	0.818	0.196	6.47	27.06	0.0435	0.1053	0.0696	0.0211	0.0781	0.1025	0.0521	0.1615	0.0232	0.1660	0.0031	0.0838	0.0863	0.0628	0.0588	+0.4118	0.0314	0.1004
	1.5	1.128	0.155	6.98	50.85	0.0385	0.1130	0.0676	0.0296	0.0692	0.1175	0.0432	0.1715	0.0184	0.1705	0.0024	0.0844	0.0802	0.0793	0.0990	+0.4009	0.0268	0.1064
	2.0	1.533	0.135	7.47	84.60	0.0355	0.1222	0.0658	0.0412	0.0638	0.1357	0.0384	0.1824	0.0159	0.1750	0.0020	0.0849	0.0759	0.1009	0.1150	+0.3684	0.0242	0.1133
	0.4	0.524	0.356	6.87	10.10	0.0604	0.0948	0.0674	0.0157	0.1031	0.0835	0.0844	0.1439	0.0418	0.1568	0.0059	0.0824	0.0674	0.0157	0.3652	+0.2913	0.0550	0.0942
	0.6	0.542	0.295	7.95	14.58	0.0497	0.0991	0.0623	0.0184	0.0866	0.0913	0.0691	0.1510	0.0339	0.1605	0.0048	0.0830	0.0523	0.0184	0.2658	+0.3364	0.0460	0.0985
0.9	1.0	0.594	0.206	9.44	27.16	0.0372	0.1052	0.0553	0.0226	0.0642	0.1023	0.0484	0.1603	0.0231	0.1656	0.0032	0.0837	0.0553	0.0226	0.1311	+0.3969	0.0337	0.1044
	1.5	0.695	0.142	10.48	51.25	0.0289	0.1098	0.0506	0.0266	0.0492	0.1105	0.0346	0.1680	0.0159	0.1692	0.0021	0.0842	0.0505	0.0266	0.0410	+0.4351	0.0255	0.1089
	2.0	0.842	0.107	11.07	86.80	0.0245	0.1147	0.0481	0.0305	0.0414	0.1159	0.0274	0.1723	0.0121	0.1714	0.0016	0.0845	0.0481	0.0306	0.0049	+0.4515	0.0213	0.1117

TABLE 13.2

Prismatic Haunch at Both Ends (from *Handbook of Frame Constants* by the Portland Cement Association)

Note: All carryover factors and fixed-end moment coefficients are negative and all stiffness factors are positive.

		Carryover factors	Stiffness factors	Unif. load	Concentrated load FEM—coef. × PL										Haunch load, both haunches
					b										
				FEM	0.1		0.3		0.5		0.7		0.9		FEM coef. × wL^2
a	r	C_{AB} = C_{BA}	k_{AB} = k_{BA}	coef. × wL^2	M_{AB}	M_{BA}	M_{AB}	M_{BA}	M_{AB}	M_{BA}	M_{AB}	M_{BA}	M_{AB}	M_{BA}	M_{AB} = M_{BA}
	0.4	0.583	5.49	0.0921	0.0905	0.0053	0.1727	0.0606	0.1396	0.1396	0.0606	0.1727	0.0053	0.0905	0.0049
	0.6	0.603	5.93	0.0940	0.0932	0.0040	0.1796	0.0589	0.1428	0.1428	0.0589	0.1796	0.0040	0.0932	0.0049
0.1	1.0	0.624	6.45	0.0961	0.0962	0.0023	0.1873	0.0566	0.1462	0.1462	0.0566	0.1873	0.0023	0.0962	0.0050
	1.5	0.636	6.75	0.0972	0.0980	0.0013	0.1918	0.0551	0.1480	0.1480	0.0551	0.1918	0.0013	0.0980	0.0050
	2.0	0.641	6.90	0.0976	0.0988	0.0008	0.1939	0.0543	0.1489	0.1489	0.0543	0.1939	0.0008	0.0988	0.0050
	0.4	0.634	7.32	0.0970	0.0874	0.0079	0.1852	0.0623	0.1506	0.1506	0.0623	0.1852	0.0079	0.0874	0.0187
	0.6	0.674	8.80	0.1007	0.0899	0.0066	0.1993	0.0584	0.1575	0.1575	0.0584	0.1993	0.0066	0.0899	0.0191
0.2	1.0	0.723	11.09	0.1049	0.0935	0.0046	0.2193	0.0499	0.1654	0.1654	0.0499	0.2193	0.0046	0.0935	0.0195
	1.5	0.752	12.87	0.1073	0.0961	0.0029	0.2338	0.0420	0.1699	0.1699	0.0420	0.2338	0.0029	0.0961	0.0197
	2.0	0.765	13.87	0.1084	0.0976	0.0018	0.2410	0.0372	0.1720	0.1720	0.0372	0.2410	0.0018	0.0976	0.0198
	0.4	0.642	9.02	0.0977	0.0845	0.0097	0.1763	0.0707	0.1558	0.1558	0.0707	0.1763	0.0097	0.0845	0.0397
	0.6	0.697	12.09	0.1027	0.0861	0.0095	0.1898	0.0700	0.1665	0.1665	0.0700	0.1898	0.0095	0.0861	0.0410
0.3	1.0	0.775	18.68	0.1091	0.0890	0.0094	0.2136	0.0627	0.1803	0.1803	0.0627	0.2136	0.0084	0.0890	0.0426
	1.5	0.828	26.49	0.1132	0.0920	0.0065	0.2376	0.0492	0.1891	0.1891	0.0492	0.2376	0.0065	0.0920	0.0437
	2.0	0.855	32.77	0.1153	0.0943	0.0048	0.2555	0.0366	0.1934	0.1934	0.0366	0.2555	0.0048	0.0943	0.0442
	0.4	0.599	10.15	0.0937	0.0825	0.0101	0.1601	0.0732	0.1509	0.1509	0.0732	0.1601	0.0101	0.0825	0.0642
	0.6	0.652	14.52	0.0986	0.0833	0.0106	0.1668	0.0776	0.1632	0.1632	0.0776	0.1668	0.0106	0.0833	0.0668
0.4	1.0	0.744	26.06	0.1067	0.0847	0.0112	0.1790	0.0835	0.1833	0.1833	0.0835	0.1790	0.0112	0.0847	0.0711
	1.5	0.827	45.95	0.1131	0.0862	0.0113	0.1919	0.0852	0.1995	0.1995	0.0852	0.1919	0.0113	0.0862	0.0746
	2.0	0.878	71.41	0.1169	0.0876	0.0108	0.2033	0.0822	0.2089	0.2089	0.0822	0.2033	0.0108	0.0876	0.0766
0.5	0.0	0.500	4.00	0.0833	0.0810	0.0090	0.1470	0.0630	0.1250	0.1250	0.0630	0.1470	0.0090	0.0810	0.0833

Summary

- Moment distribution is an approximate procedure for analyzing indeterminate beams and frames that eliminates the need to write and solve the simultaneous equations required in the slope-deflection method.
- The analyst begins by assuming that all joints free to rotate are restrained by clamps, producing fixed-end conditions. When loads are applied, fixed-end moments are induced. The solution is completed by unlocking and relocking joints in succession and distributing moments to both ends of all the members framing into the joint until all joints are in equilibrium. The time required to complete the analysis increases significantly if frames are free to sidesway. The method can be extended to nonprismatic members if standard tables of fixed-end moments are available (see Table 13.1).
- Once end moments are established, free bodies of members are analyzed to determine shear forces. After shears are established, axial forces in members are computed using free bodies of joints.
- Although moment distribution provides students with an insight into the behavior of continuous structures, its use is limited in practice because a computer analysis is much faster and more accurate.
- However, moment distribution does provide a simple procedure to verify the results of the computer analysis of large multistory, multibay, continuous frames under vertical load. In this procedure (illustrated in Section 15.7), a free-body diagram of an individual floor (including the attached columns above and below the floor) is isolated, and the ends of the columns are assumed to be fixed or the column stiffness is adjusted for boundary conditions. Because the influence of forces on floors above and below has only a small effect on the floor being analyzed, the method provides a good approximation of forces in the floor system in question.

PROBLEMS

P13.1 to **P13.7.** Analyze each structure by moment distribution. Determine all reactions and draw the shear and moment diagrams, locating points of inflection and labeling values of maximum shear and moment in each span. Unless otherwise noted, EI is constant.

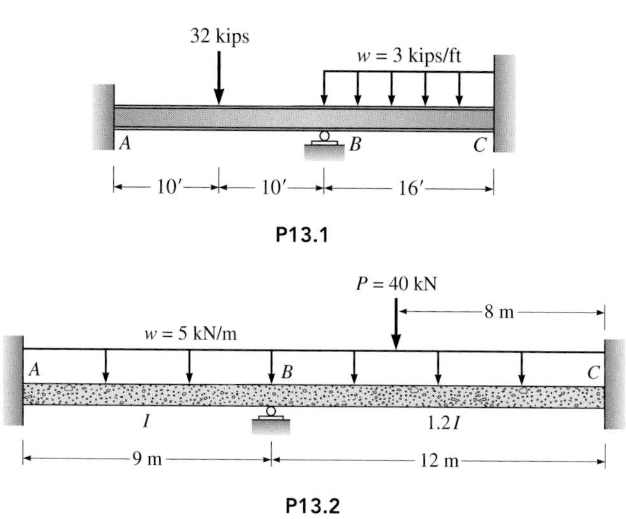

P13.1

P13.2

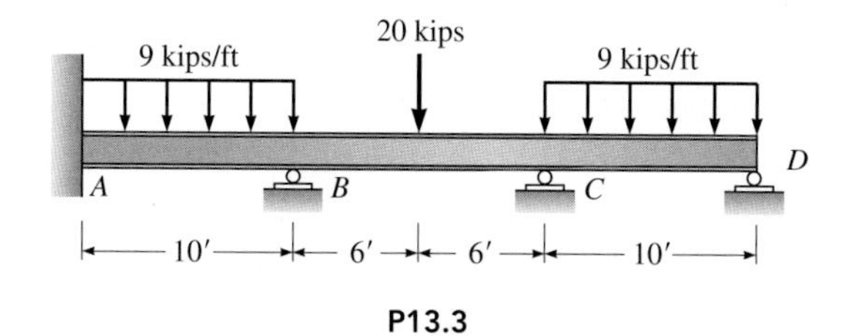

P13.3

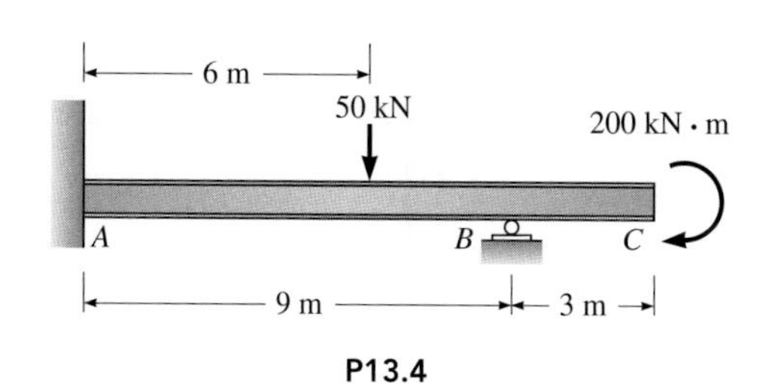

P13.4

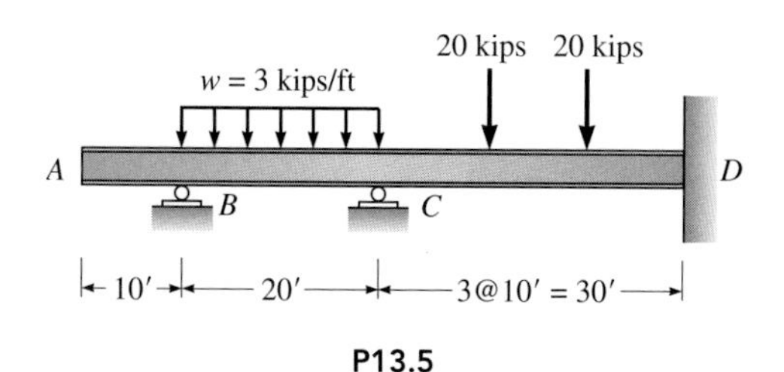

P13.5

P13.6

P13.7

P13.8 to **P13.10.** Analyze by moment distribution. Modify stiffness as discussed in Section 13.5. EI is constant. Draw the shear and moment diagrams.

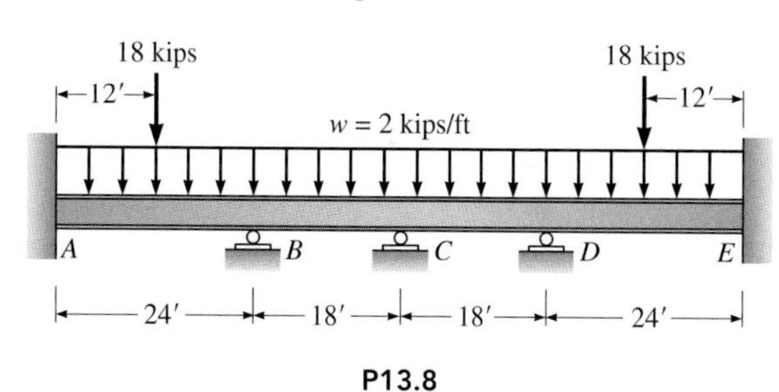

P13.8

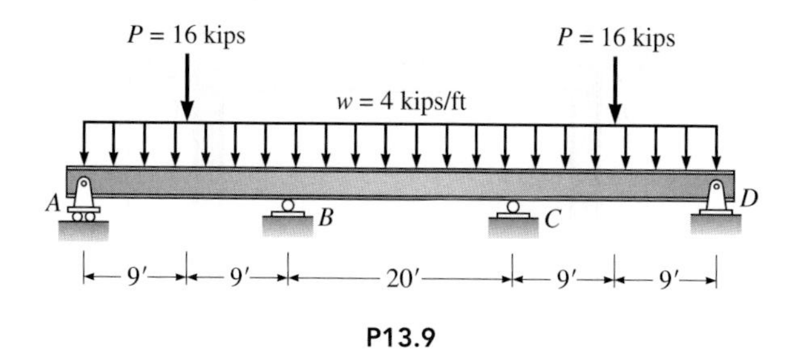

P13.9

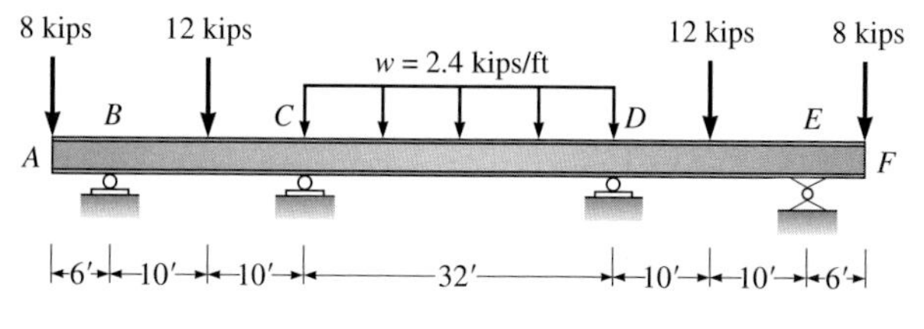

P13.10

P13.11. Analyze the frame in Figure P13.11 by moment distribution. Determine all reactions and draw the shear and moment diagrams, locating points of inflection and labeling values of maximum shear and moment in each span. Given: *EI* is constant.

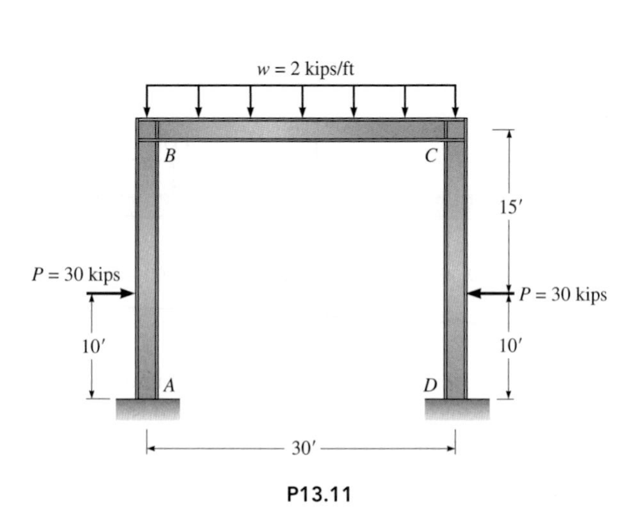

P13.11

P13.12. Analyze the frame in Figure P13.12 by moment distribution. Determine all reactions and draw the shear and moment diagrams. Given: *EI* is constant.

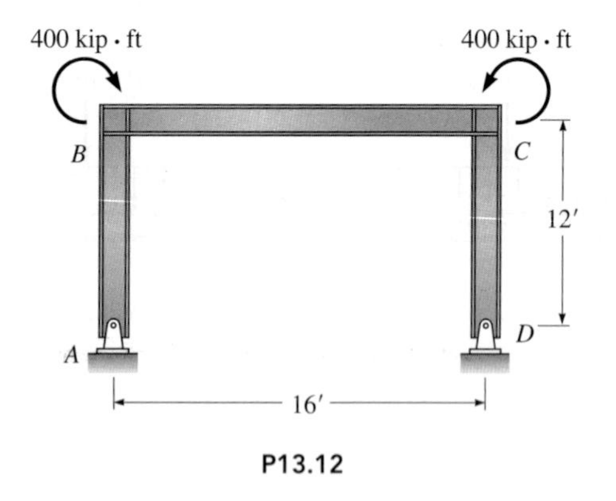

P13.12

P13.13. Analyze the reinforced concrete box in Figure P13.13 by moment distribution. Modify stiffnesses as discussed in Section 13.5. Draw the shear and moment diagrams for the top slab *AB*. Given: *EI* is constant.

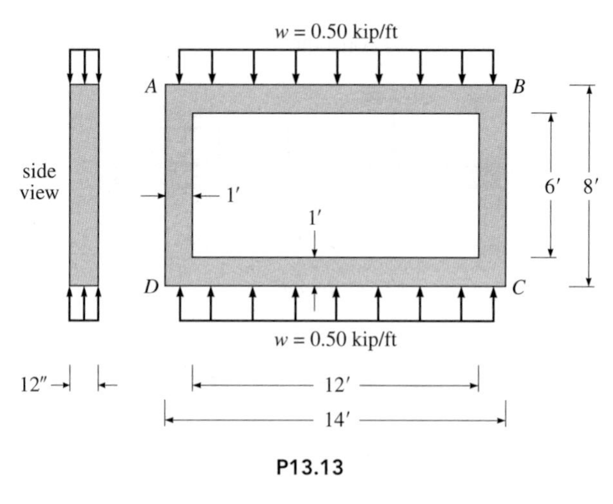

P13.13

P13.14. Analyze the frame in Figure P13.14 by the moment distribution method. Determine all reactions and draw the moment and shear curves. Given: *E* is constant. Fixed supports at *A* and *D*.

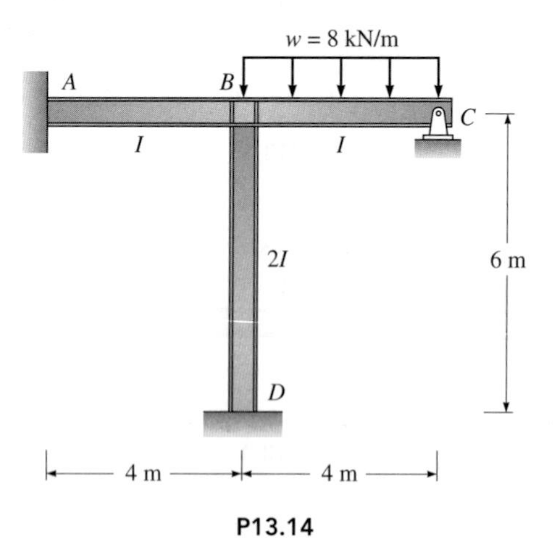

P13.14

P13.15. The cross section of the rectangular ring in Figure P13.15 is 12 in × 8 in. Draw the moment and shear curves for the ring; $E = 3{,}000$ kips/in^2.

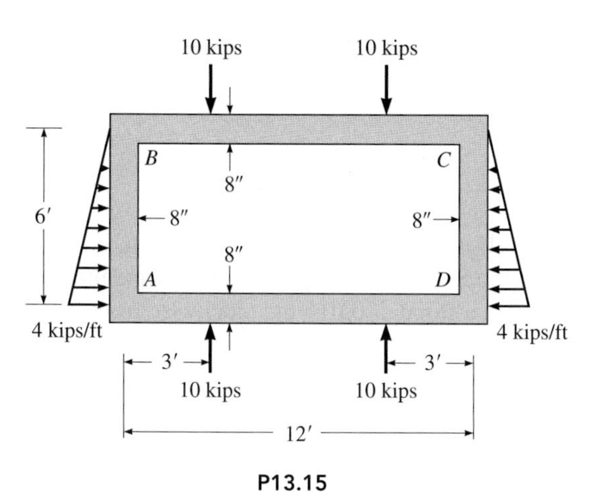

P13.15

P13.16. Analyze the frame in Figure P13.16 by moment distribution. Determine all reactions and draw the shear and moment diagrams, locating points of inflection and labeling values of maximum shear and moment in each span. E is constant, but I varies as indicated below.

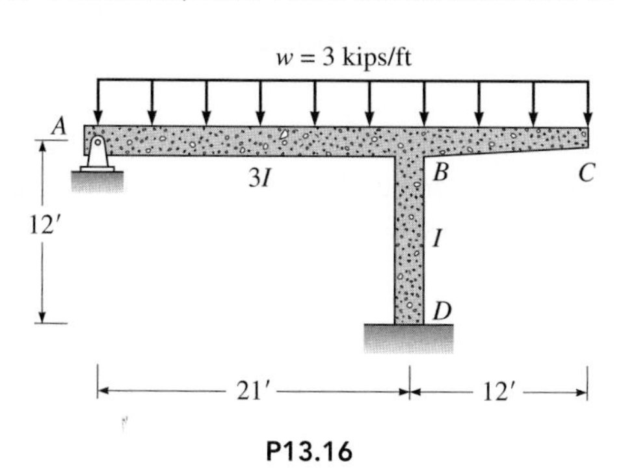

P13.16

P13.17. Analyze the frame in Figure P13.17 by moment distribution. Determine all reactions and draw the shear and moment diagrams, locating points of inflection and labeling values of maximum shear and moment in each span. Given: EI is constant.

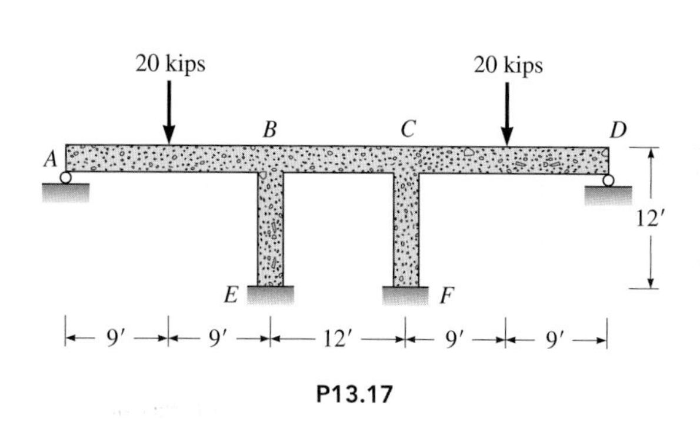

P13.17

P13.18. Analyze the frame in Figure P13.18 by moment distribution. Determine all reactions and draw the shear and moment diagrams, locating points of inflection and labeling values of maximum shear and moment in each span. E is constant, but I varies as noted.

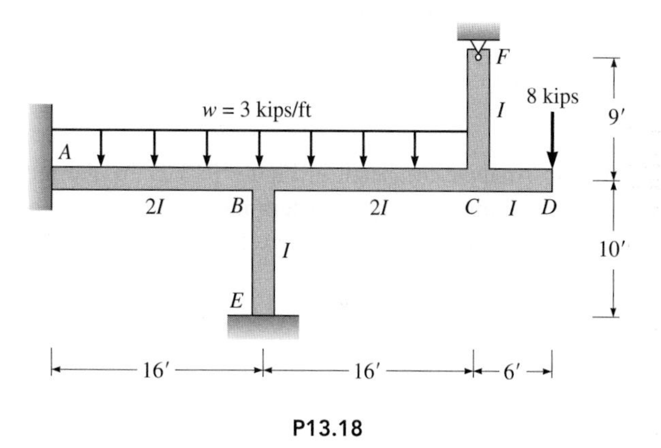

P13.18

P13.19. Analyze the frame in Figure P13.19 by moment distribution. Determine all reactions and draw the shear and moment curves. Given: EI is constant.

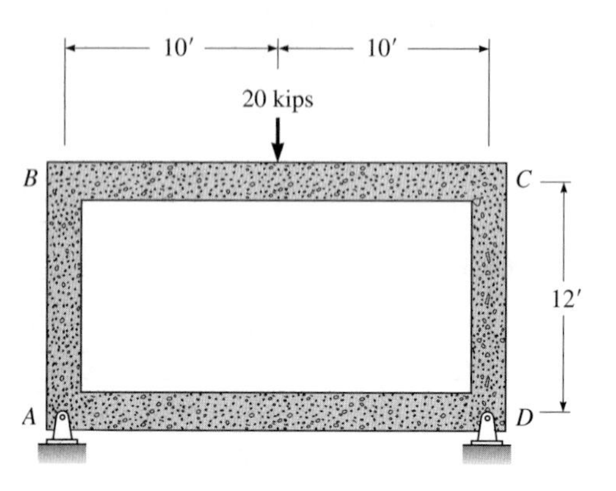

P13.19

P13.20. Analyze the frame in Figure P13.20 by the moment distribution method. Determine all reactions and draw the shear and moment curves. E is constant, but I varies as noted.

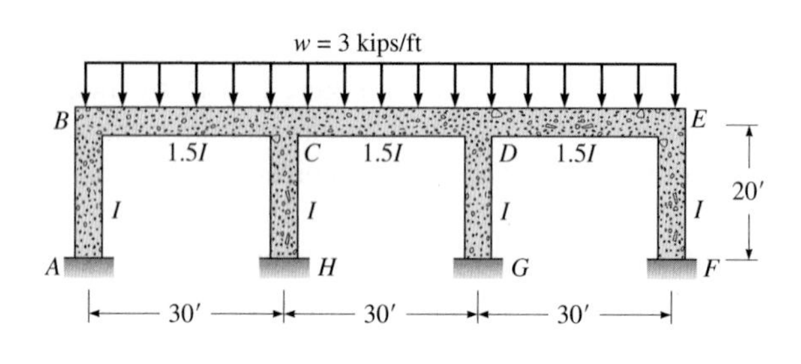

P13.20

P13.21. Analyze the beam in Figure P13.21 by the moment distribution method. Determine all reactions and draw the moment and shear curves for beam $ABCDE$; EI is constant.

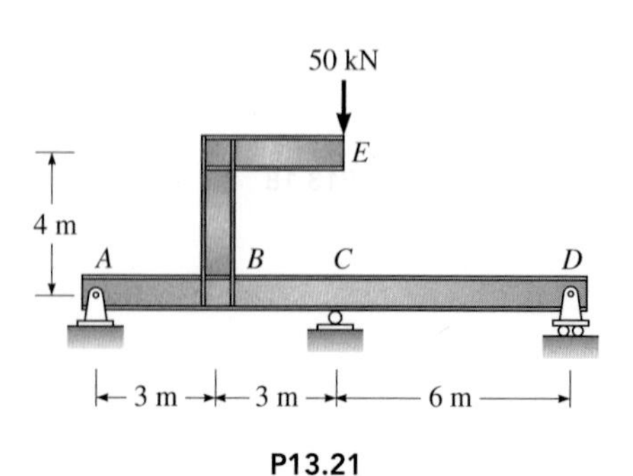

P13.21

P13.22. If support B in Figure P13.22 settles $\frac{1}{2}$ in under the 16-kip load, determine the reactions and draw the shear and moment curves for the beam. Given: $E = 30{,}000$ kips/in^2, $I = 600$ in^4.

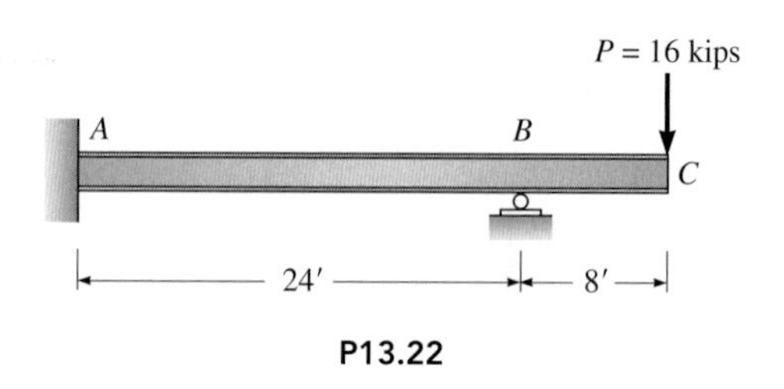

P13.22

P13.23. If support A in Figure P13.23 is constructed 0.48 in too low and the support at C is accidentally constructed at a slope of 0.016 rad clockwise from a vertical axis through C, determine the moment and reactions created when the structure is connected to its supports. Given: $E = 29{,}000$ kips/in^2.

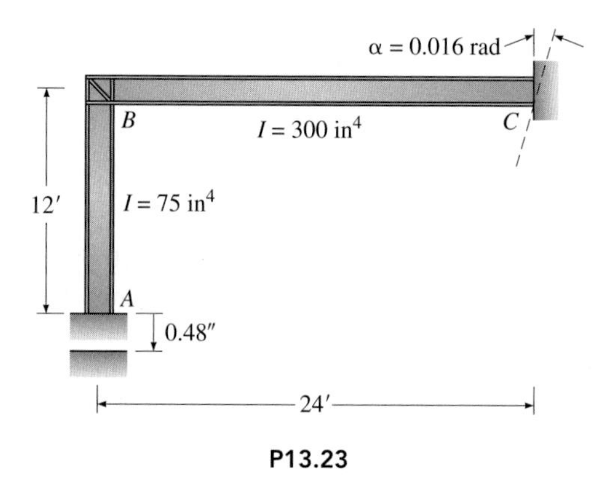

P13.23

P13.24. Analyze the frame in Figure P13.24 by moment distribution. Determine all reactions and draw the shear and moment curves. Given: EI is constant.

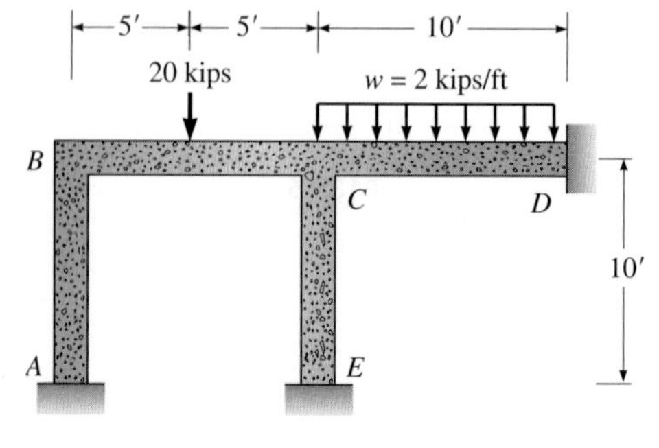

P13.24

P13.25. Due to a construction error, the support at D has been constructed 0.6 in to the left of column BD. Using moment distribution, determine the reactions that are created when the frame is connected to the support and the uniform load is applied to member BC. Draw the shear and moment diagrams and sketch the deflected shape. $E = 29{,}000$ kips/in^2, $I = 240$ in^4 for all members.

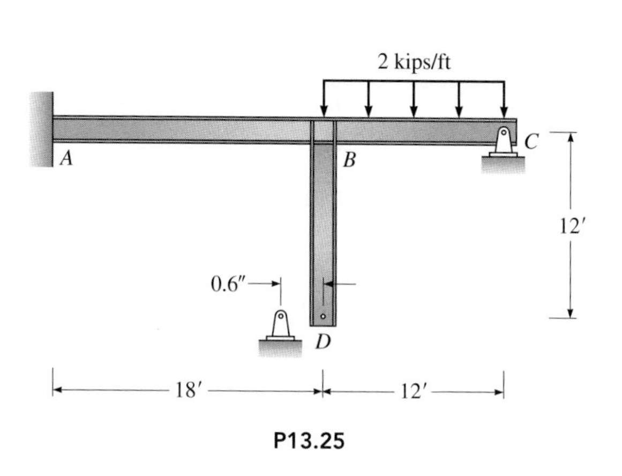

P13.25

P13.26. What moments are created in the frame in Figure P13.26 by a temperature change of +80°F in girder ABC? The coefficient of temperature expansion $\alpha_t = 6.6 \times 10^{-6}$ (in/in)/°F and $E = 29{,}000$ kips/in^2.

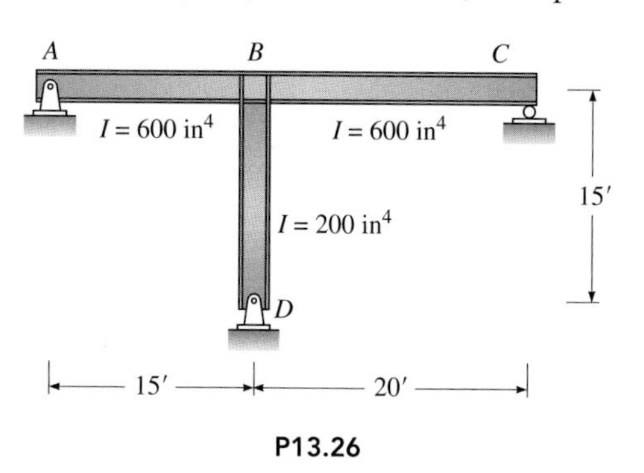

P13.26

P13.27. Determine the reactions and the moments induced in the members of the frame in Figure P13.27. Also determine the horizontal displacement of joint B. Given: EI is constant for all members, $I = 1{,}500$ in^4, and $E = 3{,}000$ kips/in^2.

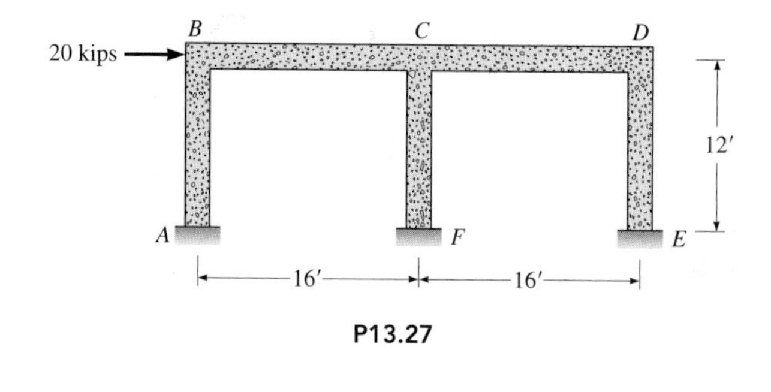

P13.27

P13.28. Analyze the structure in Figure P13.28 by moment distribution. Draw the shear and moment diagrams. Sketch the deflected shape. Also compute the horizontal displacement of joint B. E is constant and equals 30,000 kips/in^2.

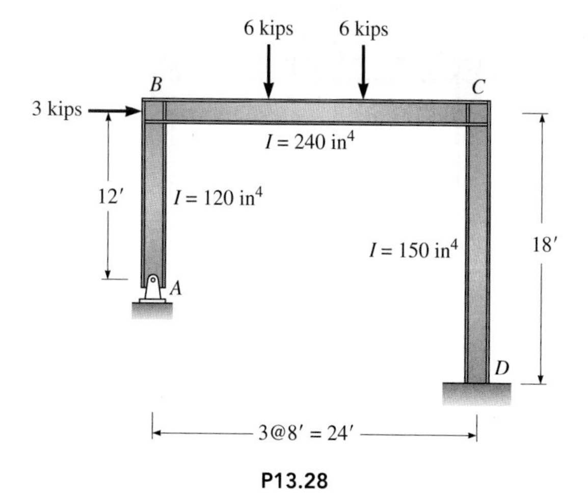

P13.28

P13.29. Analyze the frame in Figure P13.29 by moment distribution. Draw the shear and moment curves. Sketch the deflected shape. E is constant and equals 30,000 kips/in^2. $I = 300$ in^4 for all members.

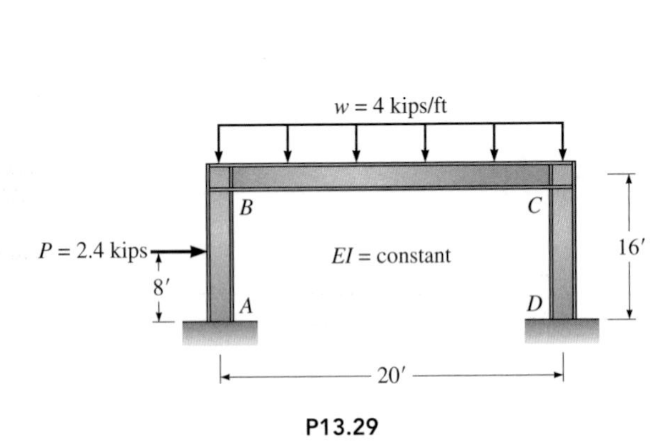

P13.29

P13.30. Analyze the frame in Figure P13.30 by moment distribution. Determine all reactions and draw the shear and moment diagrams. Sketch the deflected shape. E is constant and equals 30,000 kips/in^2.

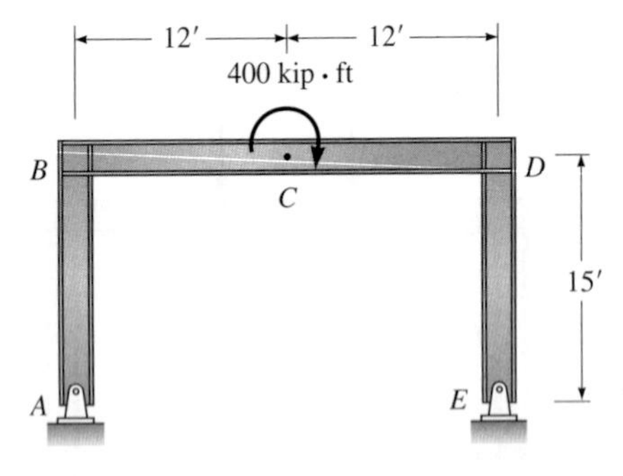

P13.30

P13.31. Analyze the Vierendeel truss in Figure P13.31 by moment distribution. Draw the shear and moment diagrams for members AB and AF. Sketch the deflected shape, and determine the deflection at midspan. Given: EI is constant, $E = 200$ GPa, and $I = 250 \times 10^6$ mm^4.

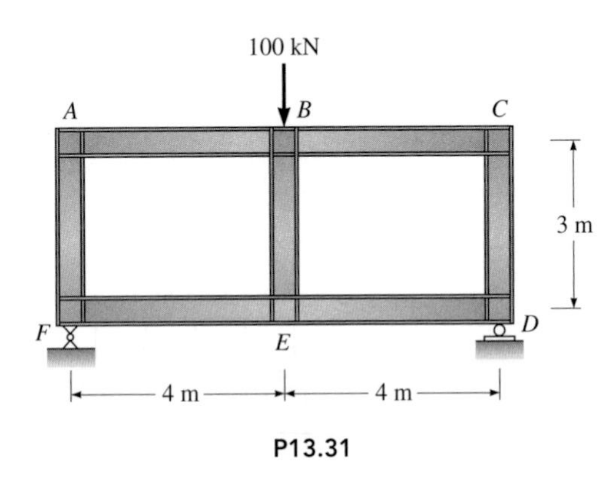

P13.31

P13.32. Analyze the frame in Figure P13.32 by moment distribution. Draw the shear and moment curves. Compute the horizontal deflection of joint B. Sketch the deflected shape. E is constant and equals 30,000 kips/in^2.

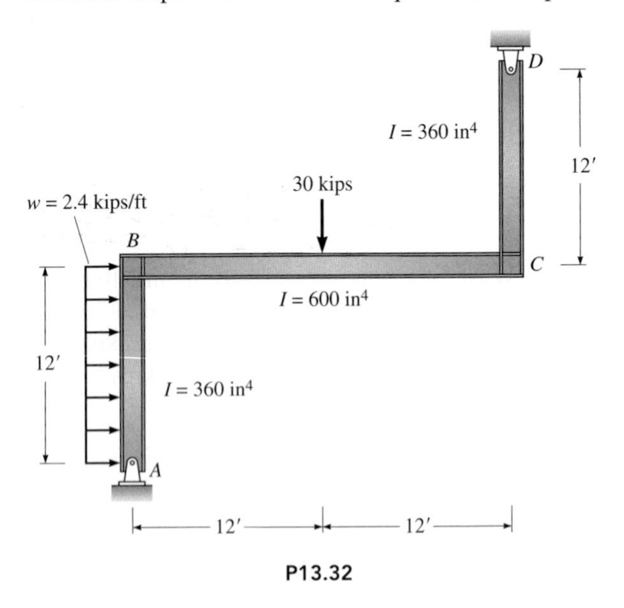

P13.32

George Washington Bridge spanning the Hudson River between Manhattan and Fort Lee, New Jersey

The center span is 3,500 ft, the towers rise 604 ft above the water, and the overall distance between anchorages is 4,760 ft. Built at a cost of $59 million, the original structure, shown here, was opened to traffic in 1931. A six-lane lower level was added in 1962.

C H A P T E R 14

Indeterminate Structures: Influence Lines

Chapter Objectives

- Extend the Müller–Breslau principle learned in Chapter 8 to construct qualitative influence lines for indeterminate structures.
- Apply the indeterminate structural analysis techniques (e.g., moment distribution, slope-deflection, or flexibility method) to establish ordinates of the influence lines.
- Learn to position live loads to maximize the internal forces and reactions based on the influence lines.

14.1 Introduction

To establish how a particular internal force at a designated point varies as live load passes over a structure, we construct influence lines. The construction of influence lines for indeterminate structures follows the same procedure as that in Chapter 8 for determinate structures; that is, a unit load is moved across the structure, and values of a particular reaction or internal force are plotted below successive positions of the load. Since computer programs for analyzing structures are generally available to practicing engineers, even highly indeterminate structures can be analyzed for many positions of the unit load rapidly and inexpensively. Therefore, certain of the traditional time-consuming *hand methods*, formerly used to construct influence lines, are of limited value to contemporary engineers. Our main goals in this chapter are

1. To become familiar with the shape of influence lines for the support reactions and forces in continuous beams and frames
2. To develop an ability to sketch the approximate shape of influence lines for indeterminate beams and frames rapidly
3. To establish how to position distributed loads on continuous structures to maximize shear and moment at critical sections of beams and columns

We begin this chapter by constructing influence lines for the reactions, shears, and moments in several simple indeterminate beams. Although the influence lines for determinate structures consist of straight segments, the influence lines for indeterminate beams and frames are curved. Therefore, to define clearly the shape of the influence lines of an indeterminate beam, we must often evaluate the ordinates at more points than is necessary for a determinate beam. In the case of an indeterminate truss or girder loaded at panel points by a floor beam and stringer system composed of simply supported members, the influence lines will consist of straight segments between panel points.

We will also discuss the use of the Müller–Breslau principle to sketch qualitative influence lines for both internal forces and reactions for a variety of indeterminate beams and frames. Based on these influence lines, we will establish guidelines for positioning live loads to produce maximum values of shear and moments at critical sections (adjacent to supports or at midspan) of these structures.

14.2 Construction of Influence Lines Using Moment Distribution

Moment distribution provides a convenient technique for constructing influence lines for continuous beams and frames of constant cross section. Moreover, with appropriate design charts, the method can easily be extended to structures that contain members of variable depth (for example, see Table 13.1).

For each position of the unit load, the moment distribution analysis supplies all member end moments. After the end moments are determined, reactions and internal forces at critical sections can be established by cutting free bodies and using the equations of statics to compute internal forces. Example 14.1 illustrates the use of moment distribution for constructing the influence lines for the reactions of a beam indeterminate to the first degree. To simplify the computations in this example, the ordinates of the influence lines (see Figure 14.1*c* to *e*) are evaluated at intervals of one-fifth the span length. In an actual design situation (for example, a bridge girder) a smaller increment—one-twelfth to one-fifteenth of the span length—would be more appropriate.

EXAMPLE 14.1

(*a*) Using a moment distribution, construct the influence lines for the reactions at supports *A* and *B* of the beam in Figure 14.1*a*.
(*b*) Given $L = 25$ ft, determine the moment created at support *B* by the 16- and 24-kip set of wheel loads shown in Figure 14.1*a* when they are positioned at points 3 and 4. *EI* is constant.

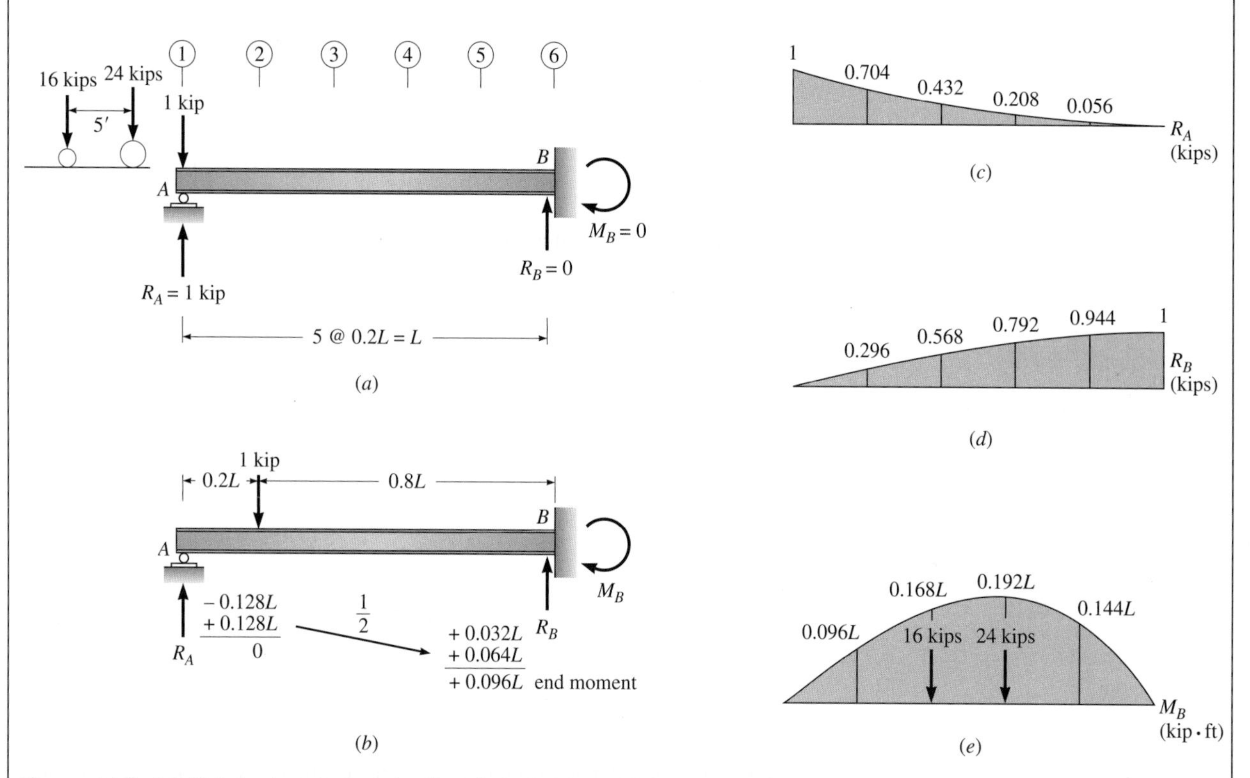

Figure 14.1: (*a*) Unit load at support *A*; (*b*) unit load 0.2*L* to right of support *A*; (*c*) influence line for reaction at *A*; (*d*) influence line for vertical reaction at *B*; (*e*) influence line for moment at support *B*.

Solution

(*a*) Influence lines will be constructed by placing the unit load at six points—a distance 0.2*L* apart—along the axis of the beam. The points are indicated by the circled numbers in Figure 14.1*a*. We will discuss the computations for points 1, 2, and 6 to illustrate the procedure.

To establish the influence line ordinate at the left end (point 1), the unit load is placed on the beam directly over support *A* (see Figure 14.1*a*). Since the entire load passes directly into the support, the beam is unstressed;

[*continues on next page*]

Example 14.1 continues . . .

therefore, $R_A = 1$ kip, $R_B = 0$, and $M_B = 0$. Similarly, if the unit load is moved to point 6 (applied directly to the fixed support), $R_B = 1$ kip, $R_A = 0$, and $M_B = 0$. The above reactions, which represent the ordinates of the influence line at points 1 and 6, are plotted in Figure 14.1*c*, *d*, and *e*.

We next move the unit load a distance $0.2L$ to the right of support A and determine the moment at B by moment distribution (see Figure 14.1*b*). Compute fixed-end moments (see Figure 12.5):

$$\text{FEM}_{AB} = -\frac{Pab^2}{L^2} = -\frac{1(0.2L)(0.8L)^2}{L^2} = -0.128L$$

$$\text{FEM}_{BA} = \frac{Pba^2}{L^2} = \frac{1(0.8L)(0.2L)^2}{L^2} = +0.032L$$

The moment distribution is carried out on the sketch in Figure 14.1*b*. After the end moment of $0.096L$ is established at support B, we compute the vertical reaction at A by summing moments about B of the forces on a free body of the beam:

$$\circlearrowright^{+} \quad \Sigma M_B = 0$$

$$R_A L - 1(0.8L) + 0.096L = 0$$

$$R_A = 0.704 \text{ kip}$$

Compute R_B:

$$\overset{+}{\uparrow} \quad \Sigma F_y = 0$$

$$R_A + R_B - 1 = 0$$

$$R_B = 0.296 \text{ kip}$$

To compute the balance of the influence line ordinates, we move the unit load to points 3, 4, and 5 and reanalyze the beam for each position of the load. The computations, which are not shown, establish the remaining influence line ordinates. Figure 14.1*c* to *e* shows the final influence lines.

(*b*) Moment at B due to wheel loads (see Figure 14.1*e*) is

$$\begin{aligned} M_B &= \Sigma \text{ influence line ordinate} \times (\text{load}) \\ &= 0.168L(16 \text{ kips}) + 0.192L(24 \text{ kips}) \\ &= 7.296L = 7.296(25) = 182.4 \text{ kip}\cdot\text{ft} \end{aligned}$$ **Ans.**

EXAMPLE 14.2

Construct the influence lines for shear and moment at section 4 of the beam in Figure 14.1*a*, using the influence line in Figure 14.1*c* to evaluate the reaction at *A* for various positions of the unit load.

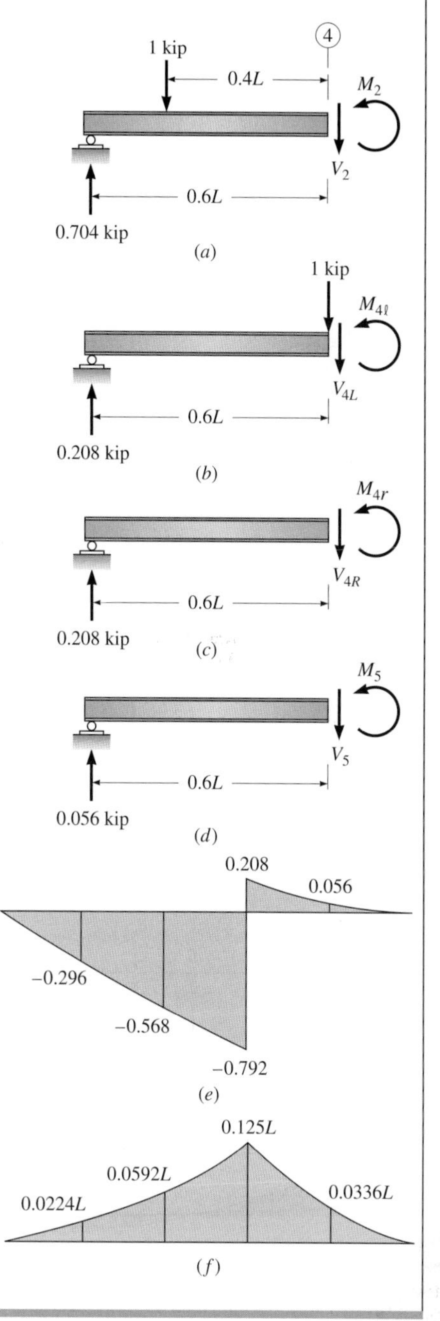

Solution

With the unit load at either support *A* or *B* (points 1 and 6 in Figure 14.1*a*), the beam is unstressed; therefore, the shear and moment at point 4 are zero, and the ordinates of the influence lines in Figure 14.2*e* and *f* begin and end at zero.

To establish the ordinates of the influence lines for other positions of the unit load, we will use the equations of statics to evaluate the internal forces on a free body of the beam to the left of a section through point 4. The free body in Figure 14.2*a* shows the unit load at point 2. The reaction at *A* of 0.704 kip is read from Figure 14.1*c*.

$$\overset{+}{\uparrow} \quad \Sigma F_y = 0$$

$$0.704 - 1 - V_2 = 0$$

$$V_2 = -0.296 \text{ kip}$$

$$\circlearrowright^{+} \quad \Sigma M_4 = 0$$

$$(0.704 \text{ kip})(0.6L) - (1 \text{ kip})(0.4L) - M_2 = 0$$

$$M_2 = 0.0224L \text{ kip} \cdot \text{ft}$$

Figure 14.2*b* shows the unit load just to the left of point 4. For this position of the unit load, the equations of equilibrium give $V_{4L} = -0.792$ kip and $M_{4L} = 0.125L$ kip·ft. If the unit load is moved a distance *dx* across the cut to the free body on the right of section 4, the reaction at *A* does not change, but the unit load is no longer on the free body (see Figure 14.2*c*). Writing the equations of equilibrium, we compute $V_{4R} = 0.208$ kip and $M_{4r} = 0.125L$ kip·ft. Figure 14.2*d* shows the forces on the free body when the unit load is at point 5 (off the free body). Computations give $V_5 = 0.056$ kip and $M_5 = 0.0336L$ kip·ft. Using the computed values of shear and moment at section 4 for the various positions of the unit load, we plot the influence lines for shear in Figure 14.2*e* and for moment in Figure 14.2*f*.

Figure 14.2: Influence lines for shear and moment at section 4; (*a*) unit load at section 2; (*b*) unit load to left of section 4; (*c*) unit load to right of section 4; (*d*) unit load at section 5; (*e*) influence line for shear; (*f*) influence line for moment.

14.3 Müller–Breslau Principle

The Müller–Breslau principle (previously introduced and applied to determinate structures in Sec. 8.4) states:

> **The influence line for any reaction or internal force (shear, moment) corresponds to the deflected shape of the structure produced by removing the capacity of the structure to carry that force and then introducing into the modified (or released) structure a unit deformation that corresponds to the restraint removed.**

We begin this section by using Betti's law to demonstrate the validity of the Müller–Breslau principle. We will then use the Müller–Breslau principle to construct qualitative and quantitative influence lines for several common types of indeterminate beams and frames.

To demonstrate the validity of the Müller–Breslau principle, we will consider two procedures to construct an influence line for the reaction at support A of the continuous beam in Figure 14.3*a*. In the conventional procedure, we apply a unit load to the beam at various points along the span, evaluate the corresponding value of R_A, and plot it below the position of the unit load. For example, Figure 14.3*a* shows a unit load, used to construct an influence line, at an arbitrary point x on the beam; R_A is assumed positive in the direction shown (vertically upward).

If the Müller–Breslau principle is valid, we can also produce the correct shape of the influence line for the reaction at A simply by removing the support at A (to produce the released structure) and introducing into the structure at that point a vertical displacement which corresponds to reaction R_A supplied by the roller (see Figure 14.3*b*). We introduce the displacement that corresponds to R_A by arbitrarily applying a 1-kip load vertically at A.

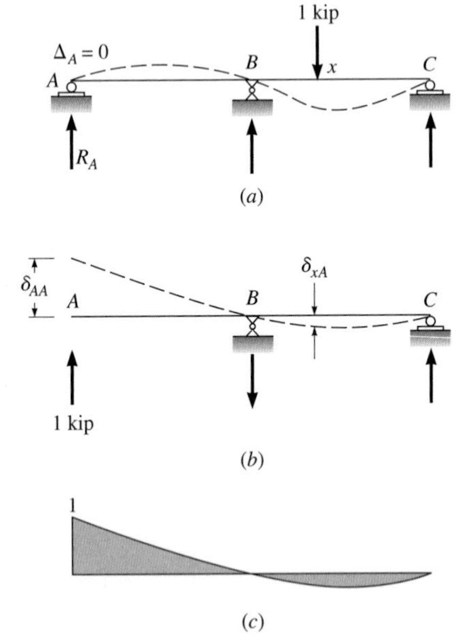

Figure 14.3: (*a*) Unit load used to construct influence line for R_A; (*b*) unit load used to introduce a displacement into the released structure; (*c*) influence line for R_A.

Denoting the loaded beam in Figure 14.3*a* as system 1 and the loaded beam in Figure 14.3*b* as system 2, we now apply Betti's law, given by Equation 10.41, to the two systems

$$\Sigma F_1 \Delta_2 = \Sigma F_2 \Delta_1 \tag{10.41}$$

where Δ_2 is the displacement in system 2 that corresponds to the loading F_1 in system 1 and Δ_1 is the displacement in system 1 that corresponds to the loading F_2 in system 2. If a force in one of the systems is a moment, the corresponding displacement is a rotation. Substituting into Equation 10.41, we find

$$R_A \delta_{AA} + (1 \text{ kip})(\delta_{xA}) = 1(0) \tag{14.1}$$

Since the reactions at supports B and C in both systems do no virtual work because the supports in the other system do not displace, these terms are omitted from both sides of Equation 14.1. Solving Equation 14.1 for R_A, we compute

$$R_A = -\frac{\delta_{xA}}{\delta_{AA}} \tag{14.2}$$

Since δ_{AA} has a constant value but the value of δ_{xA} varies along the axis of the beam, Equation 14.2 shows that R_A is proportional to the ordinates of the deflected shape in Figure 14.3*b*. Therefore, the shape of the influence line for R_A is the same as that of the deflected shape of the released structure produced by introducing the displacement δ_{AA} at point *A*, and we verify the Müller–Breslau principle. The final influence line for R_A is shown in Figure 14.3*c*. The ordinate at *A* equals 1 because the unit load on the real structure at that point produces a 1-kip reaction at *A*.

A qualitative influence line, of the type shown in Figure 14.3*c*, is often adequate for many types of analysis; however, if a quantitative influence line is required, Equation 14.2 shows that it can be constructed by dividing the ordinates of the deflected shape by the magnitude of the displacement δ_{AA} introduced at point *A*.

Significance of the Minus Sign in Equation 14.2. As a first step in the construction of an influence line, we must assume a positive direction for the function. For example, in Figure 14.3*a*, we assume that the positive direction for R_A is vertically upward. The first virtual work term in Equation 14.1 is always positive because both the displacement δ_{AA} and R_A are in the same direction. The vertical work represented by the second term [(1 kip)(δ_{xA})] is also positive because the 1-kip force and the displacement δ_{xA} are both directed downward. When we transfer the second term to the right side of Equation 14.1, a minus sign is introduced. The minus sign indicates that R_A is actually directed downward. If the 1-kip load had been located on span *AB*—a region where the influence line ordinates are positive—the virtual work terms containing δ_{xA} would have been negative, and when the term was transferred to the right side of Equation 14.1, the expression for R_A would be positive, indicating that R_A was directed upward.

In summary, we conclude that where an influence line is positive, downward load will always produce a value of the function directed in the positive direction. On the other hand, in regions where the influence line is negative, downward load will always produce a value of the function directed in the negative direction.

14.4 Qualitative Influence Lines for Beams

In this section we illustrate the use of the Müller–Breslau method to construct *qualitative* influence lines for a variety of forces in continuous beams and frames. As described in Section 14.3 in the Müller–Breslau method, we first remove the capacity of the structure to carry the function represented by the influence line. At the location of the release, we introduce a displacement that corresponds to the restraint released. The resulting deflected shape is the influence line to some scale. If you are uncertain about the type of displacement to introduce, imagine a force that corresponds to the function is applied at the location of the release and creates the displacement.

As an example, we will draw the influence line for positive moment at point C of the two-span continuous beam in Figure 14.4*a*. Point C is located at the midpoint of span BD. To remove the flexural capacity of the beam, we insert a hinge at point C. Since the original structure was indeterminate to the first degree, the released structure shown in Figure 14.4*b* is stable and determinate. We next introduce a displacement at C that corresponds to a positive moment, as indicated by the two curved arrows on either side of the hinge. The effect of the positive moments at C is to rotate the ends of each member in the direction of the moment and to displace the hinge upward. Figure 14.4*c* shows the deflected shape of the beam, which is also the shape of the influence line.

Although it is evident that a positive moment at C rotates the ends of the members, the vertical displacement that also occurs may not be obvious. To clarify the displacements produced by the moments on each side of the hinge, we will examine the free bodies of the beam on each side of the hinge (see Figure 14.4*d*). We first compute the reaction at D by summing moments, about the hinge at C, of the forces on member CD.

$$\circlearrowleft^{+} \quad \Sigma M_C = 0$$

$$M - R_D \frac{L}{2} = 0$$

$$R_D = \frac{2M}{L}$$

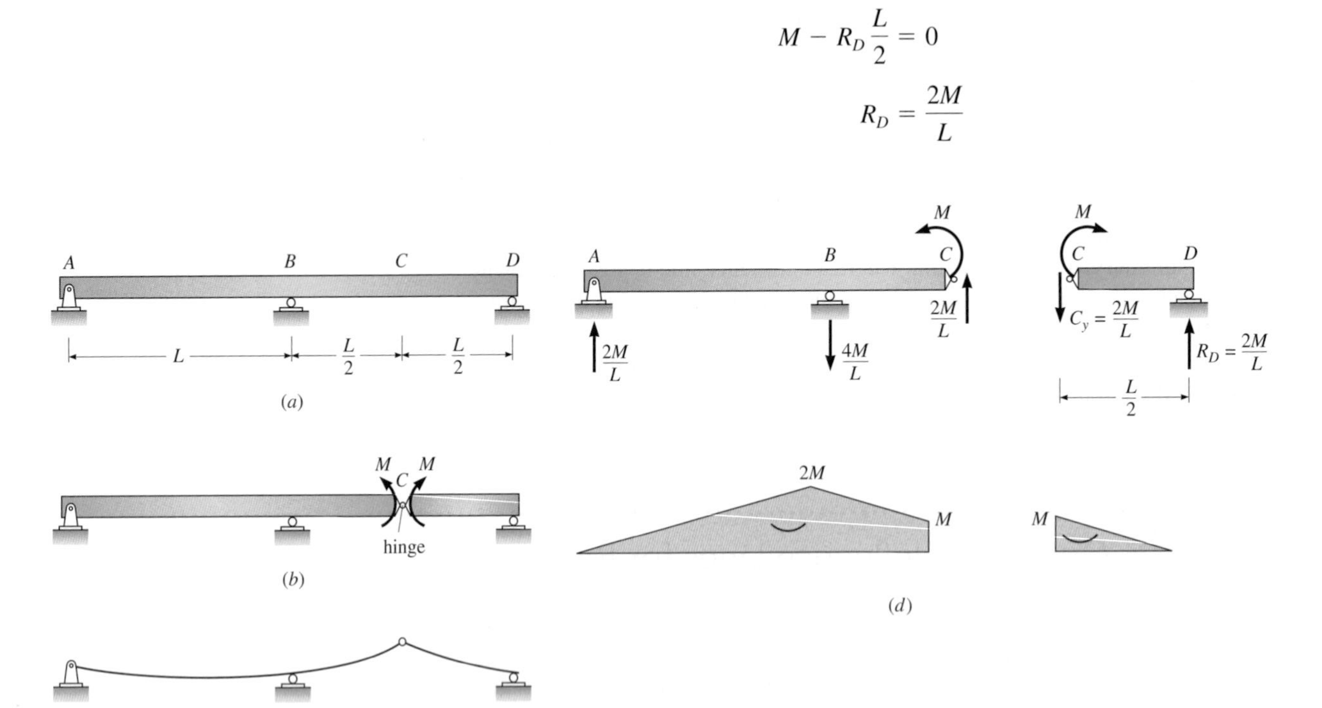

Figure 14.4: Construction of the influence line for moment at C by the Müller–Breslau method: (*a*) two-span beam; (*b*) released structure; (*c*) deflected shape produced by a displacement to the restraint removed at C; (*d*) moment curves to establish deflected shape of released structure; (*e*) influence line for moment C.

For equilibrium to exist in the y-direction for member CD, the vertical force at the hinge C_y must be equal in magnitude and opposite in sense to R_D. Since C_y represents the action of the free body on the left, an equal and opposite force—acting upward—must act at joint C of member ABC.

We next compute the reactions at supports A and B of member ABC, and we draw the moment curves for each member. Since the moment is positive along the entire length of both members, they bend concave upward, as indicated by the curved lines under the moment diagrams. When member ABC is placed on supports A and B (see Figure 14.4c), point C must move vertically upward to be consistent with the restraints supplied by the supports and the curvature created by the moment. The final shape of the influence line is shown in Figure 14.5e. Although the magnitude of the positive and negative ordinates is unknown, we can reason that the ordinates are greatest in the span that contains the hinge and the applied loads. As a general rule, the influence of a force in one span drops off rapidly with distance from the loaded span. Moreover, a span that contains a hinge is much more flexible than a span that is continuous.

Additional Influence Lines for Continuous Beams

In Figure 14.5 we use the Müller–Breslau principle to sketch qualitative influence lines for a variety of forces and reactions in a three-span continuous beam. In each case the restraint corresponding to the function represented by the influence lines is removed, and a displacement corresponding to the restraint is introduced into the structure. Figure 14.5b shows the influence line for the reaction at C. The roller and plate device that removes the shear capacity of the cross section in Figure 14.5c is able to transmit both axial load and moment. Since the plates must remain parallel as the shear deformation occurs, the slopes of the members attached to each side of the plate must be the same, as shown by the detail to the right of the beam. In Figure 14.5d the influence line for negative moment is constructed by introducing a hinge into the beam at C. Since the beam is attached to the support at that point, the ends of the members, under the action of the moments, on each side of the hinge are free to rotate but not to move vertically. The influence line for the reaction at F is generated by removing the vertical support at F and introducing a vertical displacement (see Figure 14.5f).

In Example 14.3 we illustrate the use of a qualitative influence line to establish where to load a continuous beam to produce the maximum value of shear at a section.

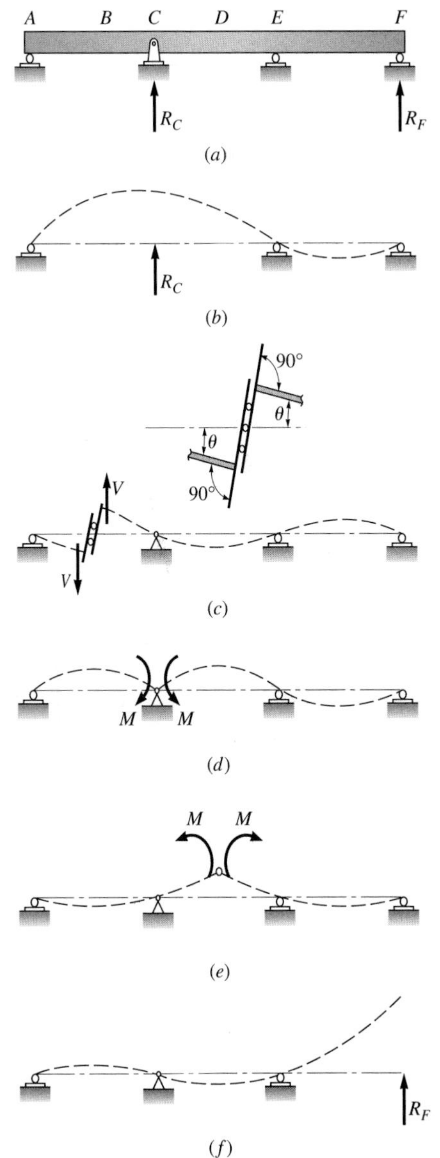

Figure 14.5: Construction of influence lines by the Müller–Breslau method for the three-span continuous beam in (a); (b) influence line for R_C; (c) influence line for shear at B; (d) influence line for negative moment at C; (e) influence line for positive moment at D; (f) influence line for reaction R_F.

EXAMPLE 14.3

The continuous beam in Figure 14.6a carries a uniformly distributed live load of 4 kips/ft. The load can be located over all or a portion of each span. Compute the maximum value of shear at midspan (point B) of member AC. Given: EI is constant.

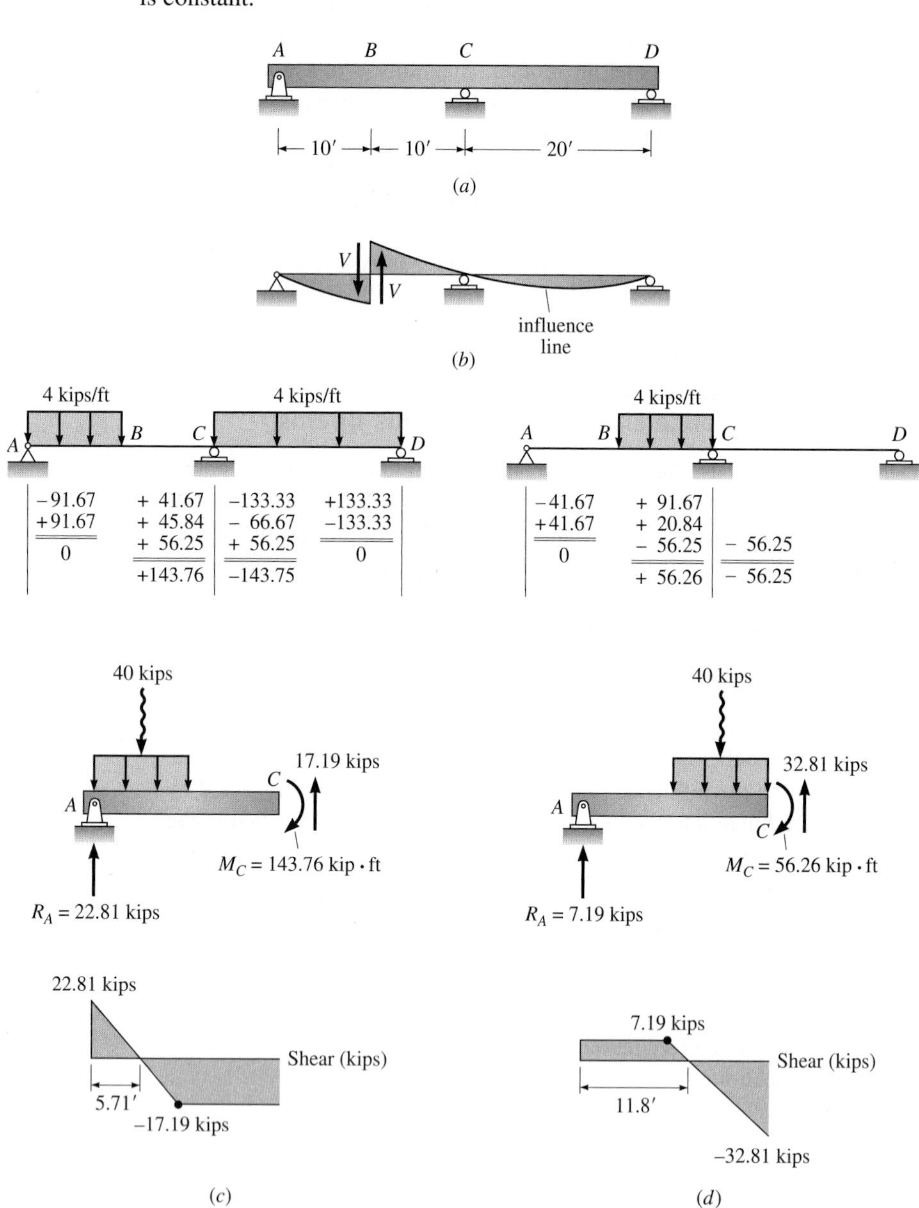

Figure 14.6: Computation of maximum shear at section B: (a) continuous beam; (b) influence line for shear at B; (c) analysis of beam with distributed load placed to produce maximum negative shear of 17.19 kips at B; (d) analysis of beam with distributed load positioned to produce maximum positive shear of 7.19 kips at B.

Solution

To establish the position of the live load to maximize the shear, we first construct a qualitative influence line for shear at point B. Using the Müller–Breslau principle, we introduce displacements corresponding to positive shear forces into the beam at section B to produce the influence line shown in Figure 14.6*b*. Since the influence line contains both positive and negative regions, we must investigate two loading conditions. In the first case (see Figure 14.6*c*) we distribute the uniform load over all sections where the ordinates of the influence line are negative. In the second case (see Figure 14.6*d*) we load the continuous beam between points B and C where the influence line ordinates are positive. Using moment distribution, we next determine the moment in the beam at support C. Since the beam is symmetric about the center support, both members have the same stiffness, and the distribution factors at joint C are identical and equal to $\frac{1}{2}$. Using Figure 12.5, we computed fixed-end moments for members AC and CD in Figure 14.6*c*.

$$\text{FEM}_{AC} = -\frac{11wL^2}{192} = -\frac{11(4)(20^2)}{192} = -91.67 \text{ kip}\cdot\text{ft}$$

$$\text{FEM}_{CA} = \frac{5wL^2}{192} = \frac{5(4)(20^2)}{192} = 41.67 \text{ kip}\cdot\text{ft}$$

$$\text{FEM}_{CD} = -\text{FEM}_{DC} = \frac{wL^2}{12} = \frac{4(20)^2}{12} = \pm 133.33 \text{ kip}\cdot\text{ft}$$

The moment distribution, which is carried out under the sketch of the beam in Figure 14.6*c*, produces a value of moment in the beam at C equal to 143.76 kip·ft. Because of roundoff error in the analysis, a small difference exists in the values of the moments on each side of joint C. We next compute the reaction at A by summing moments about C of the forces acting on a free body of beam AC. After the reaction at A is computed, the shear diagram (see the bottom sketch in Figure 14.6*c*) is drawn. The analysis shows that $V_B = -17.19$ kips. A similar analysis for the loading in Figure 14.6*d* gives $V_B = +7.19$ kips. Since the magnitude of the shear rather than its sign determines the greatest value of the shear stresses at B, the section must be sized to carry a shear force of 17.19 kips.

EXAMPLE 14.4

The continuous beam in Figure 14.7*a* carries a uniformly distributed live load of 3 kips/ft. Assuming that the load can be located over all or a portion of any span, compute the maximum values of positive and negative moment that can develop at midspan of member *BD*. Given: *EI* is constant.

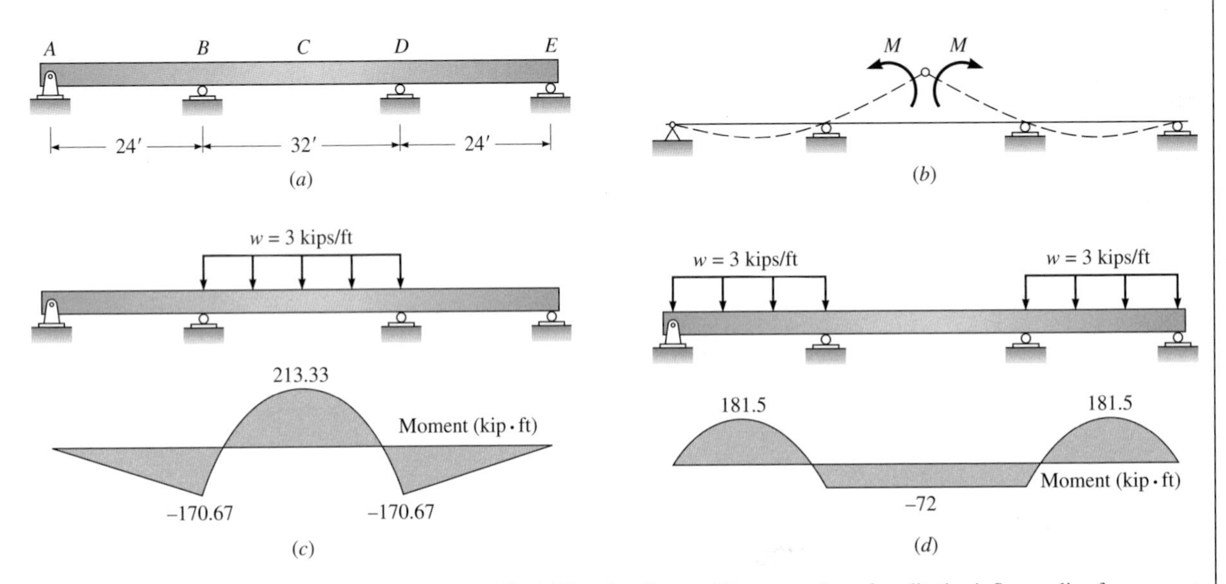

Figure 14.7: (*a*) Details of beam; (*b*) construction of qualitative influence line for moment at *C*; (*c*) load positioned to maximize positive moment at *C*; (*d*) load positioned to maximize negative moment at *C*.

Solution

The qualitative influence line for moment at point *C*, located at midspan of *BD*, is constructed using the Müller–Breslau principle. A hinge is inserted at *C*, and a deformation associated with positive moment is introduced at that point (see Figure 14.7*b*). Figure 14.7*c* shows the load positioned over the section of the beam in which the influence line ordinates are positive. Using moment distribution (the computations are not shown), we compute the member end moments and construct the moment curve. The maximum positive moment equals 213.33 kip · ft.

To establish the maximum value of negative moment at point *C*, the load is positioned on the beam in those sections in which the influence line ordinates are negative (see Figure 14.7*d*). The moment curve for this loading is shown below the beam. The maximum value of negative moment is −72 kip · ft.

NOTE. To establish the *total* moment at section *C*, we must also combine each of the live load moments with the positive moment at *C* produced by dead load.

EXAMPLE 14.5

The frame in Figure 14.8*a* is loaded only through girder *ABC*. If the frame carries a uniformly distributed load of 3 kips/ft that can act over part or all of spans *AB* and *BC*, determine the maximum value of horizontal thrust D_x that develops in each direction at support *D*. For all members *EI* is a constant.

Solution

The positive sense of the thrust D_x is shown in Figure 14.8*a*. To construct the influence line for the horizontal reaction at support *D* by the Müller–Breslau principle, we remove the horizontal restraint by introducing a roller at *D* (see Figure 14.8*b*). A displacement corresponding to D_x is introduced by applying a horizontal force *F* at *D*. The deflected shape, shown by the dashed line, is the influence line.

In Figure 14.8*c* we apply the uniform load to span *BC*, where the ordinates of the influence line are positive. Analyzing the frame by moment distribution, we compute a clockwise moment of 41.13 kip·ft at the top of the column. Applying statics to a free body of column *BD*, we compute a horizontal reaction of 3.43 kips.

To compute the maximum thrust in the negative direction, we load the frame in the region where the ordinates of the influence line are negative (see Figure 14.8*d*). Analysis of the frame produces a thrust of 2.17 kips to the left.

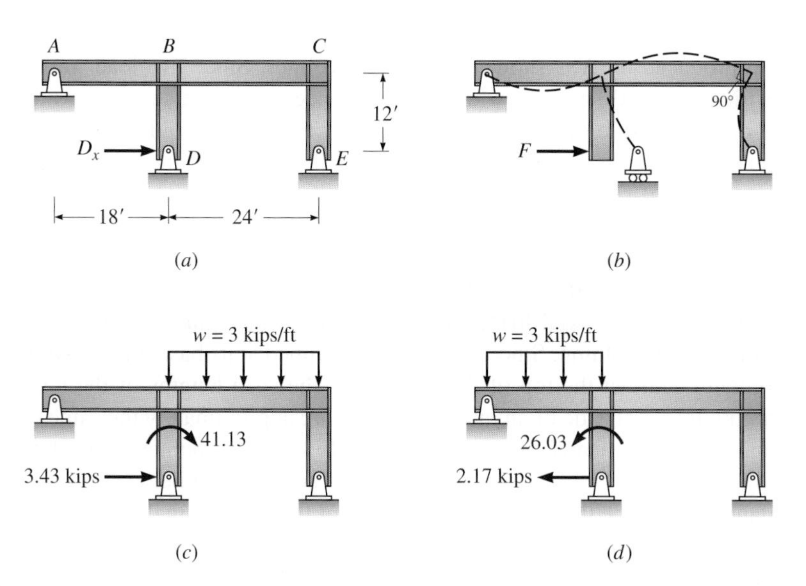

Figure 14.8: (*a*) Dimensions of frame; (*b*) establishing the shape of the influence line, horizontal restraint removed by replacing pin with a roller, dashed lines show the influence line; (*c*) position of load to establish maximum lateral thrust in positive sense (to the right); (*d*) position of load to produce maximum thrust in negative sense.

14.5 Live Load Patterns to Maximize Forces in Multistory Buildings

Building codes specify that members of multistory buildings be designed to support a uniformly distributed live load as well as the dead load of the structure and the nonstructural elements. Nonstructural elements include walls, ceilings, ducts, pipe, light fixtures, and so forth. Normally, we analyze for dead and live loads separately. While the dead load is fixed in position, the position of the live load must be varied to maximize a particular force at a certain section. In most cases, the greatest live load force at a section is produced by pattern loading; that is, live load is placed on certain spans or portions of spans but not on other spans. By using the Müller–Breslau principle to construct qualitative influence lines, we can establish the spans or portions of a span that should be loaded to maximize the force or forces at critical design sections of individual members.

For example, to establish the loading pattern to maximize the axial force in a column, we imagine the capacity of the column to carry axial load is removed and an axial displacement is introduced into the structure. If we wished to determine the spans on which live load should be placed to maximize the axial force in column AB of the structure in Figure 14.9*a*, we would disconnect the column from its support at A and introduce a vertical displacement Δ at that point. The deflected shape, which is the influence line, produced by Δ is shown by the dashed lines. Since live load must be positioned on all spans in which the influence line ordinates are positive, we must place the distributed live load over the entire length of all beams connected directly to the column on all floors above the column (see Figure 14.9*b*). Since all floors displace by the same amount, a given value of live load on the third or fourth floor (the roof) produces the same increment of axial load in column AB as that load positioned on the second floor (i.e., directly above the column).

In addition to axial load, the loading shown in Figure 14.9*b* produces moment in the column. Since the column is pinned at its base, the maximum moment occurs at the top of the column. If the span lengths of the beams framing into each side of an interior column are approximately the same (the usual case), the nearly equal but oppositely directed moment each beam applies to the joint directly at the top of the column will balance or nearly balance out. Since the unbalanced moment at the joint is small, the moment in the column will also be small. *Therefore, in a preliminary design of an interior column, the engineer can size the column accurately by considering only the axial load.*

Although the forces produced by the loading pattern in Figure 14.9*b* control the dimensions of most interior columns, under certain conditions—for example, a large difference in adjacent span lengths, or a high live-to-dead load ratio—we may wish to verify that the capacity of the column is also adequate for the loading pattern that maximizes the moment (rather than

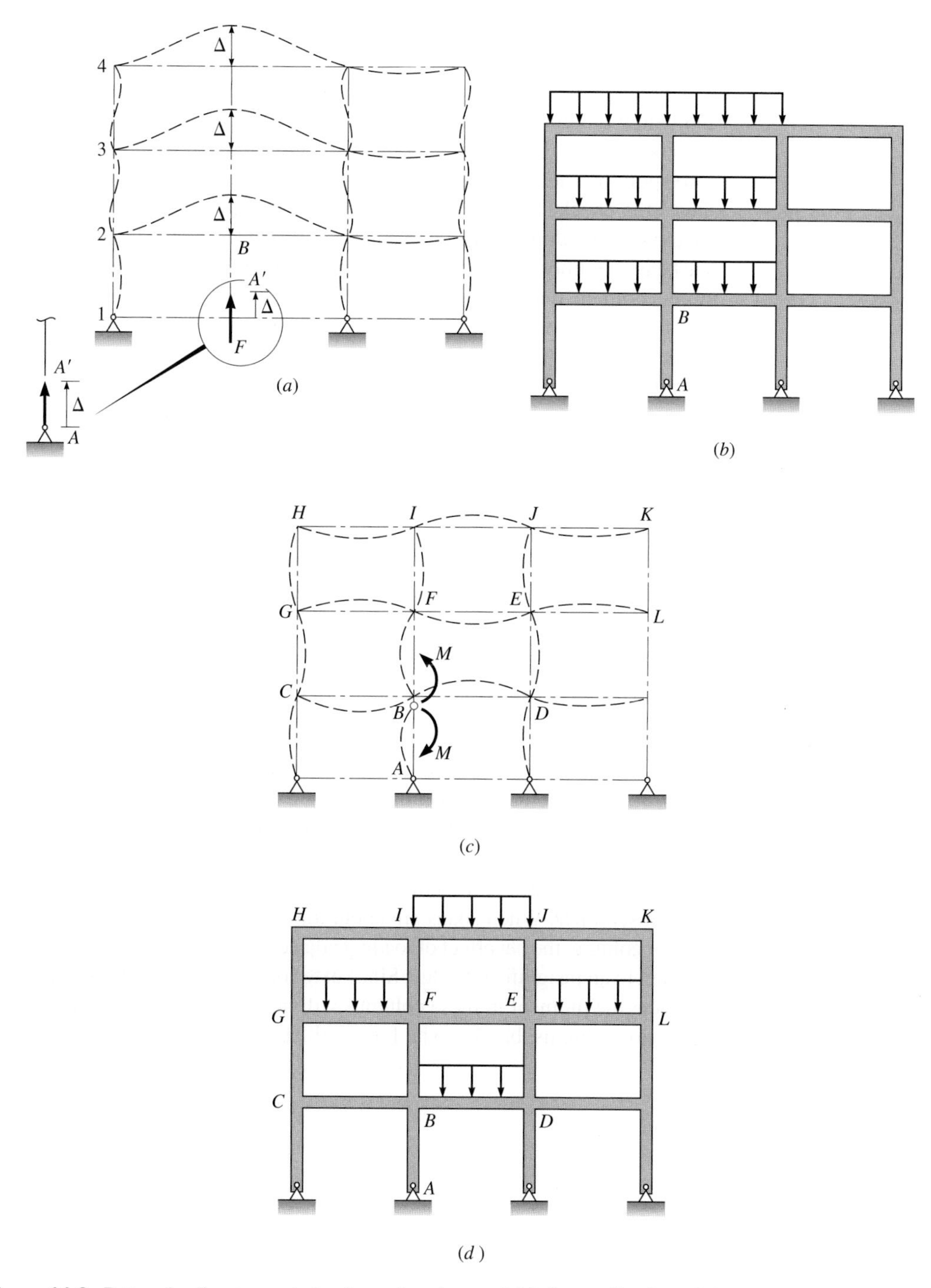

Figure 14.9: Pattern loading to maximize forces in columns: (*a*) influence line for axial load in column *AB*; (*b*) live load pattern to maximize axial force in column *AB*; (*c*) influence line for moment in column *AB*; (*d*) position of live load to maximize moment in column *AB*, and the axial force associated with maximum moment is approximately one-half that shown in (*b*) since a checkerboard pattern of loading is required.

the axial load). To construct the qualitative influence line for moment in the column, we insert a hinge into the column just below the floor beams at point *B* and then apply a rotational displacement to the ends of the structure above and below the hinge (see Figure 14.9*c*). We can imagine that this displacement is produced by applying moments of magnitude *M* to the structure. The corresponding deflected shape is shown by the dashed line. Figure 14.9*d* shows the checkerboard pattern of live load that maximizes the moment at the top of the column. Since this pattern is produced by load applied to only one beam per floor above the column, the axial load associated with the maximum moment will be approximately one-half as large as that associated with the loading in Figure 14.9*b* that maximizes axial load. Because the magnitudes of the influence line ordinates produced by the moments at *B* reduce rapidly with distance from the hinge, the greatest portion (on the order of 90 percent) of the column moment at *B* is produced by loading only span *BD*. Therefore, we can usually neglect the contribution to the moment at *B* (but not the axial load) produced by the load on all spans except *BD*. For example, Section 8.8.1 of the American Concrete Institute Building Code, which controls the design of reinforced concrete buildings in the United States, specifies that: "Columns shall be designed to resist . . . the maximum moment from factored loads on a *single adjacent span* of the floor or roof under consideration."

Moments Produced by Dead Load

In addition to live load, we must consider the forces produced in a column by dead load, which is present on every span. If we consider spans *BC* and *BD* in Figure14.9*c*, we can see that the influence line is negative in span *BC* and positive in span *BD*. Vertical load on span *BD* produces moments in the direction shown on the sketch. On the other hand, load on span *BC* produces moment in the opposite direction and reduces the moment produced by the load on span *BD*. When spans are about the same length on either side of an interior column, the net effect of loading adjacent spans is to reduce the column moment to an insignificant value. Since exterior columns are loaded from only one side, the moment in these columns will be much larger than the moment in interior columns, but the axial force will be much smaller.

EXAMPLE 14.6

Using the Müller–Breslau principle, construct the influence lines for positive moment at the center of span BC in Figure 14.10a and for negative moment in the girder adjacent to joint B. The frames have rigid joints. Indicate the spans on which a uniformly distributed live load should be positioned to maximize these forces.

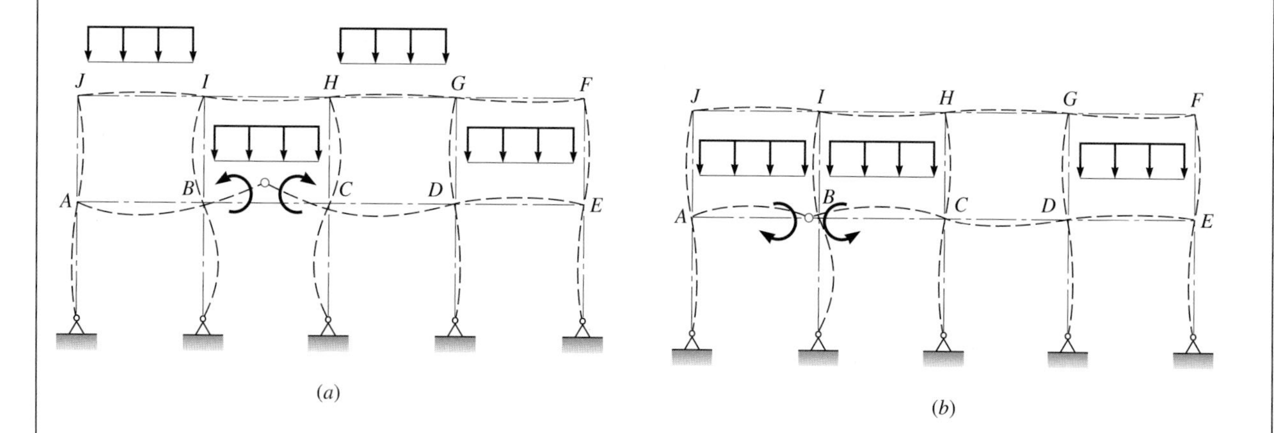

Solution

The influence line for positive moment is constructed in Figure 14.10a by inserting a hinge at the midspan of member BC and introducing a displacement associated with a positive moment. The deflected shape, shown by the dashed lines, is the influence line. As indicated on the sketch, the ordinates of the influence line reduce rapidly on either side of span BC, and the bending of the girders in the top floor is small. The influence line indicates that in a multistory building vertical load (also termed gravity load) applied to one floor has very little effect on the *moments* created in adjacent floors. Moreover, as we have noted previously, the moments created in the girders of a particular floor by loading one span reduce rapidly with distance from the span. Therefore, the contribution to the positive moment in span BC by load on span DE is small—on the order of 5 or 6 percent of that produced by the load on girder BC. To maximize the positive moment in span BC, we position live load on all spans where the influence line is positive.

Figure 14.10b shows the influence line for negative moment in the girder and the spans to be loaded. Figure 14.10c shows a detail of the joint B to clarify the deformation introduced in Figure 14.10b. As discussed previously, the major contribution to the negative moment in the girder at B is produced by load on spans AB and BC. The contribution to the negative moment from load on span DE is small. Recognizing that the negative moment produced at B by load on other floors is small, we position the distributed load on spans AB, BC, and DE to compute the maximum negative moment at B.

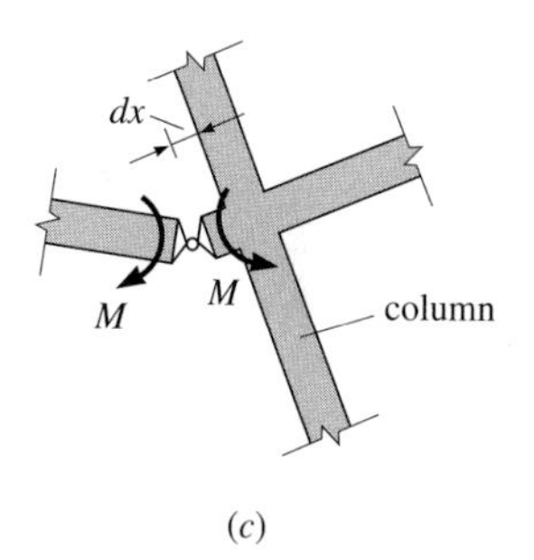

Figure 14.10: Positioning uniformly distributed loads to maximize positive and negative moments in continuous frames; (a) influence line for positive moment at midspan of beam BC; (b) influence line for negative moment in beam adjacent to a column; (c) detail of position of hinge for frame in (b).

EXAMPLE 14.7

(*a*) Using the Müller–Breslau principle stated by Equation 14.2, construct the influence line for moment at support *C* for the beam in Figure 14.11*a*. (*b*) Show the computations for the ordinate of the influence line at point *B*. Given: *EI* is constant.

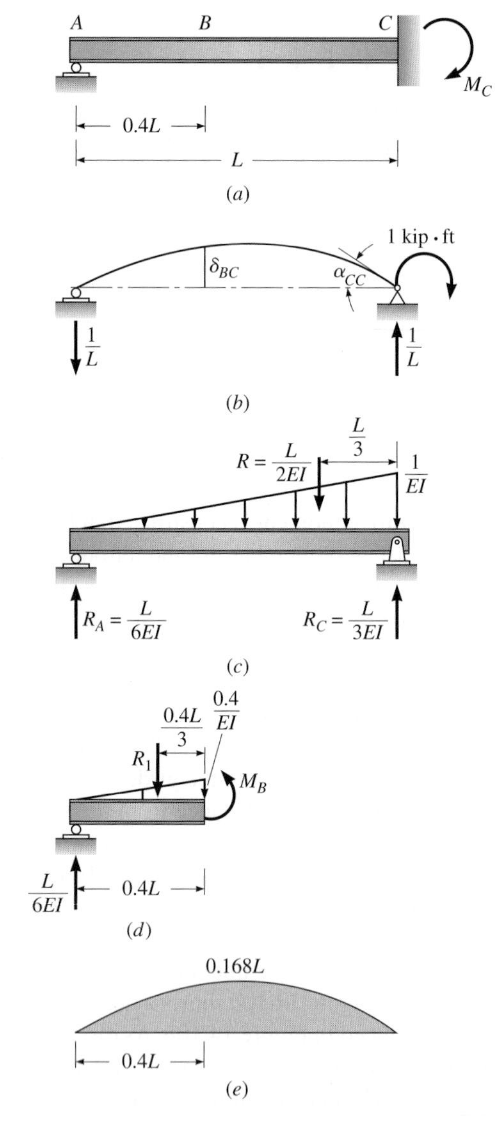

Figure 14.11: Influence line for M_C: (*a*) beam showing positive sense of M_C; (*b*) displacement α_{CC} introduced into released structure; (*c*) conjugate beam loaded with M/EI diagram; (*d*) moment in conjugate beam equals deflection at *B* in real structure; (*e*) influence line for M_C.

Solution

(*a*) Assume that the positive sense of M_C is clockwise, as shown in Figure 14.11*a*. Produce the *released* structure by introducing a pin support at *C*. Introduce a rotational displacement at *C* by applying a unit moment to the right end of the beam, as shown in Figure 14.11*b*. The deflected shape is the influence line for M_C.

(*b*) Compute the ordinate of the influence line at *B* by using the conjugate beam method to evaluate the deflections in Equation 14.2. Figure 14.11*c* shows the conjugate beam loaded by the M/EI curve associated with the unit value of M_C in Figure 14.11*b*. To determine the reactions of the conjugate beam, we compute the resultant *R* of the triangular loading diagram:

$$R = \frac{1}{2} L \frac{1}{EI} = \frac{L}{2EI}$$

Since the slope at *C* in the released structure equals the reactions at *C* in the conjugate beam, we compute R_C by summing moments about the roller at *A* to give

$$\alpha_{CC} = R_C = \frac{L}{3EI}$$

To compute the deflection at *B*, we evaluate the moment in the conjugate beam at *B*, using the free body shown in Figure 14.11*d*:

$$\delta_{BC} = M_B = \frac{L}{6EI}(0.4L) - R_1 \frac{0.4L}{3}$$

where $\quad R_1 = \text{area under } M/EI \text{ curve} = \dfrac{1}{2}(0.4L)\dfrac{0.4}{EI} = \dfrac{0.08L}{EI}$

$$\delta_{BC} = \frac{0.4L^2}{6EI} - \frac{0.08L}{EI}\frac{0.4L}{3} = \frac{0.336L^2}{6EI}$$

Evaluate the influence line ordinate at point *B*, using Equation 14.2:

$$M_C = \frac{\delta_{BC}}{\alpha_{CC}} = \frac{0.336L^2/(6EI)}{L/(3EI)} = 0.168L$$

The influence line, which was constructed in Example 14.1 (see Figure 14.1*e*), is shown in Figure 14.11*e*.

EXAMPLE 14.8

Using the Müller–Breslau principle stated by Equation 14.2, construct the influence line for the reaction at B for the beam in Figure 14.12a. Evaluate the ordinates at midspan of AB, at B and at C. Given: EI is constant.

Solution

The positive sense of R_B is taken as upward, as shown in Figure 14.12a. Figure 14.12b shows the released structure with a unit value of R_B applied to introduce the displacement that produces the influence line. The influence line is shown by the dashed line. In Figure 14.12c the conjugate beam for the released structure is loaded by the M/EI curve associated with the released structure in Figure 14.12b. The slope in the released structure, given by the shear in the conjugate beam, is shown in Figure 14.12d. This curve indicates that the maximum deflection in the conjugate beam, which occurs where the shear is zero, is located a small distance to the right of support B. The deflection of the released structure, represented by moment in the conjugate beam, is shown in Figure 14.12e. To compute the ordinates of the influence line, we use Equation 14.2.

$$R_B = \frac{\delta_{XB}}{\delta_{BB}}$$

where both $\delta_{BB} = 204/EI$ and δ_{XB} are shown in Figure 14.12e.

The influence line is shown in Figure 14.12f.

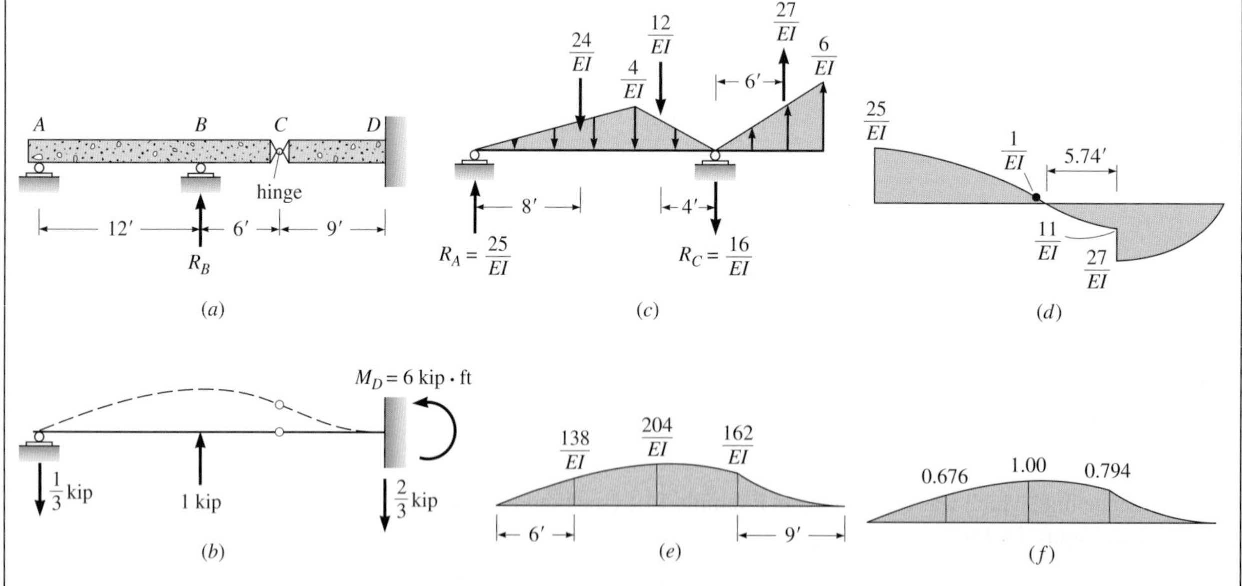

Figure 14.12: Influence line for R_B using the Müller–Breslau principle: (a) dimensions of beam; (b) released structure displaced by unit value of R_B; (c) conjugate beam loaded by M/EI curve for loading in (b); (d) shear in conjugate beam (slope of released structure); (e) moment in conjugate beam (deflection of released structure); (f) influence line for R_B.

EXAMPLE 14.9

For the indeterminate truss shown in Figure 14.13, construct the influence lines for the reactions at I and L and for the force in upper chord member DE. The truss is loaded through the lower chord panel points, and AE is constant for all members.

Solution

The truss will be analyzed for a 1-kip load at successive panel points. Since the truss is indeterminate to the first degree, we use the method of consistent deformations for the analysis. Because of symmetry, we only have to consider the unit load at panel points N and M. Only the computations for the unit load at panel point N are shown.

We begin by establishing the influence lines for reaction R_L at the center support. After this force is established for each position of the unit load, all other reactions and bar forces can be computed by statics.

Select R_L as the redundant. Figure 14.13*b* shows the bar forces produced in the released structure by the unit load at panel point N. The deflection at support L is denoted by Δ_{LN}. Figure 14.13*c* shows the bar forces and vertical deflection δ_{LL} at point L produced by a unit value of the redundant. Since the roller support at L does not deflect, the compatibility equation is

$$\Delta_{LN} + \delta_{LL}R_L = 0 \tag{1}$$

where the positive direction for displacements is upward.

Using the method of virtual work, we compute Δ_{LN}:

$$1 \cdot \Delta_{LN} = \sum \frac{F_P F_Q L}{AE} \tag{2}$$

Since AE is a constant, we can factor it out of the summation:

$$\Delta_{LN} = \frac{1}{AE}\sum F_P F_Q L = -\frac{64.18}{AE} \tag{3}$$

where the quantity $\Sigma F_P F_Q L$ is evaluated in Table 14.1 (see column 5).

Compute δ_{LL} by virtual work:

$$(1 \text{ kip})(\delta_{LL}) = \frac{1}{AE}\sum F_Q^2 L = \frac{178.72}{AE} \tag{4}$$

The quantity $\Sigma F_Q^2 L$ is evaluated in column 6 of Table 14.1.

Substituting the values of Δ_{LN} and δ_{LL} above into Equation 1, we compute R_L:

$$-\frac{64.18}{AE} + R_L\frac{178.72}{AE} = 0$$

$$R_L = 0.36 \text{ kip}$$

[*continues on next page*]

Example 14.9 continues . . .

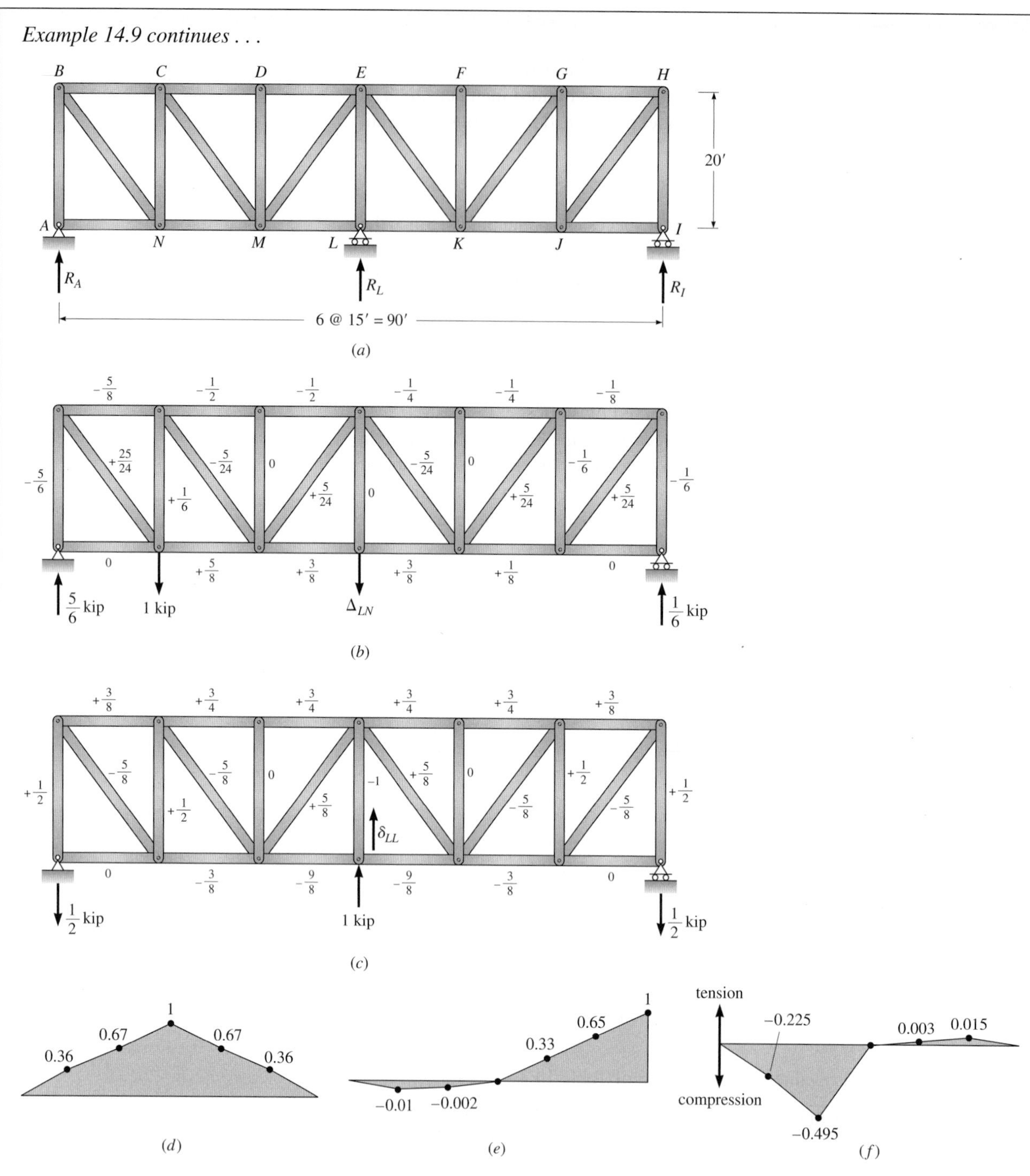

Figure 14.13: (*a*) Details of truss; (*b*) unit load on released structure produces F_P forces; (*c*) unit value of redundant R_L produces F_Q forces; (*d*) influence line for R_L; (*e*) influence line R_I; (*f*) influence line for force in upper chord F_{DE}.

TABLE 14.1

Bar (1)	F_P (2)	F_Q (3)	L (4)	F_QF_PL (5)	F_Q^2L (6)
AB	$-\frac{5}{6}$	$\frac{1}{2}$	20	−8.33	5.00
BC	$-\frac{5}{8}$	$\frac{3}{8}$	15	−3.52	2.11
CD	$-\frac{1}{2}$	$\frac{3}{4}$	15	−5.63	8.44
DE	$-\frac{1}{2}$	$\frac{3}{4}$	15	−5.63	8.44
EF	$-\frac{1}{4}$	$\frac{3}{4}$	15	−2.81	8.44
FG	$-\frac{1}{4}$	$\frac{3}{4}$	15	−2.81	8.44
GH	$-\frac{1}{8}$	$\frac{3}{8}$	15	−0.70	2.11
HI	$-\frac{1}{6}$	$-\frac{1}{2}$	20	−1.67	5.00
IJ	0	0	15	0	0
JK	$\frac{1}{8}$	$-\frac{3}{8}$	15	−0.70	2.11
KL	$\frac{3}{8}$	$-\frac{9}{8}$	15	−6.33	18.98
LM	$\frac{3}{8}$	$-\frac{9}{8}$	15	−6.33	18.98
MN	$\frac{5}{8}$	$-\frac{3}{8}$	15	−3.52	2.11
NA	0	0	15	0	0
BN	$\frac{25}{24}$	$-\frac{5}{8}$	25	−16.28	9.76
CN	$\frac{1}{6}$	$\frac{1}{2}$	20	1.67	5.00
CM	$-\frac{5}{24}$	$-\frac{5}{8}$	25	3.26	9.76
DM	0	0	20	0	0
EM	$\frac{5}{24}$	$\frac{5}{8}$	25	3.26	9.76
EL	0	−1	20	0	20.00
EK	$-\frac{5}{24}$	$\frac{5}{8}$	25	−3.26	9.76
FK	0	0	20	0	0
KG	$\frac{5}{24}$	$-\frac{5}{8}$	25	−3.26	9.76
GJ	$-\frac{1}{6}$	$\frac{1}{2}$	20	−1.67	5.00
JH	$\frac{5}{24}$	$-\frac{5}{8}$	25	−3.26	9.76
				$\Sigma F_QF_PL = -64.18$	$\Sigma F_Q^2L = 178.72$

If the unit load is next moved to panel point M and the computations repeated using the method of consistent deformations, we find

$$R_L = 0.67 \text{ kip}$$

The influence line for R_L, which is symmetric about the centerline of the structure, is drawn in Figure 14.13*d*. When the unit load is at support L, it is carried into support L; thus $R_L = 1$. The remaining influence lines can now be constructed by the equations of statics for each position of the unit load. Figure 14.13*e* shows the influence line for R_I. Because of symmetry, the influence line for R_A is the mirror image of that for R_I.

[*continues on next page*]

Example 14.9 continues . . .

As you can see, the influence lines for bar forces and reactions of the truss are nearly linear. Moreover, because the number of panel points between supports is small, the trusses, which are relatively short and deep, are very stiff. Therefore, the forces in members, produced by applied loads, are largely limited to the span in which the load acts. For example, the axial force in bar *DE* in the left span is nearly zero when the unit load moves to span *LI* (see Figure 14.13*f*). If additional panels were added to each span, increasing the flexibility of the structure, the bar forces produced in an adjacent span, by a load in the other span, would be larger.

Summary

- Qualitative influence lines for indeterminate structures can be constructed by using the Müller–Breslau principle previously introduced in Chapter 8.
- Quantitative influence lines can be most easily generated by a computer analysis in which a unit load is positioned at intervals of one-fifteenth to one-twentieth of the span of individual members. As an alternate to constructing influence lines, the designer can position the live load at successive positions along the span and use a computer analysis to establish the forces at critical sections. Influence lines for indeterminate structures are composed of curved lines.
- Influence lines for multistory buildings with continuous frames (Section 14.5) clarify standard building code provisions that specify how uniformly distributed live loads are to be positioned on floors to maximum moments at critical sections.

PROBLEMS

Unless otherwise noted, EI is constant for all problems.

P14.1. Construct the influence lines for the vertical reaction at support A and the moment at support C. Evaluate the ordinates at 6-ft intervals of the influence line. EI is constant.

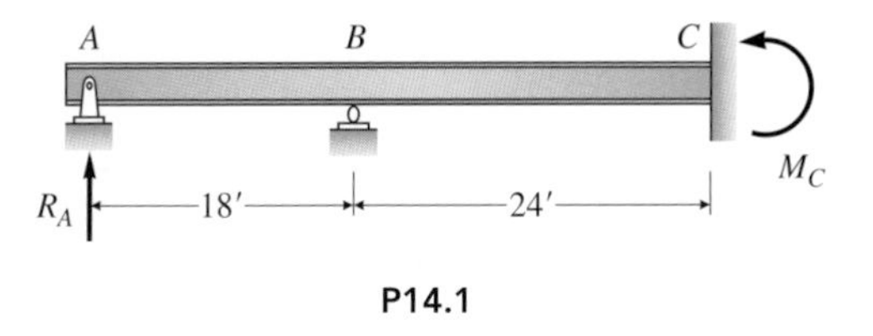

P14.1

P14.2. (*a*) Construct the influence lines for the moment and the vertical reaction R_A at support A for the beam in Figure P14.2. Evaluate the influence line ordinates at the quarter points of the span. (*b*) Using the influence lines for reactions, construct the influence line for moment at point B. Compute the maximum value of R_A produced by the set of wheel loads.

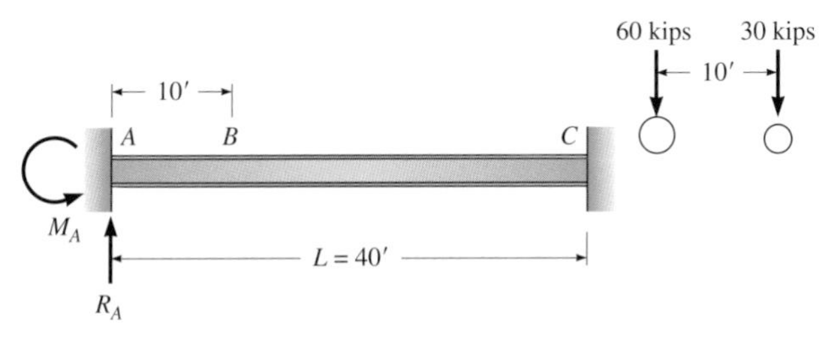

P14.2

P14.3. Using moment distribution, construct the influence lines for the reaction at A and the shear and moment at section B (Figure P14.3). Evaluate influence line ordinates at 8-ft intervals in span AC and CD and at E.

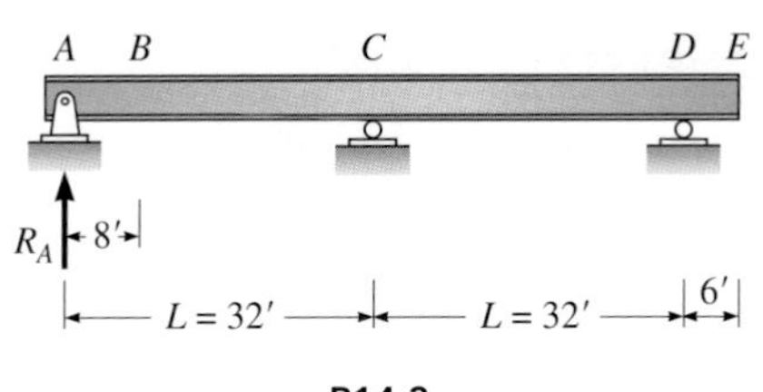

P14.3

P14.4. (*a*) Draw the influence line for the moment at B (Figure P14.4). (*b*) If the beam carries a uniformly distributed live load of 2 kips/ft that can act on all or part of each span as well as a concentrated live load of 20 kips that can act anywhere, compute the maximum moment at B. (*c*) Determine the maximum value of live load moment at B produced by the set of wheel loads.

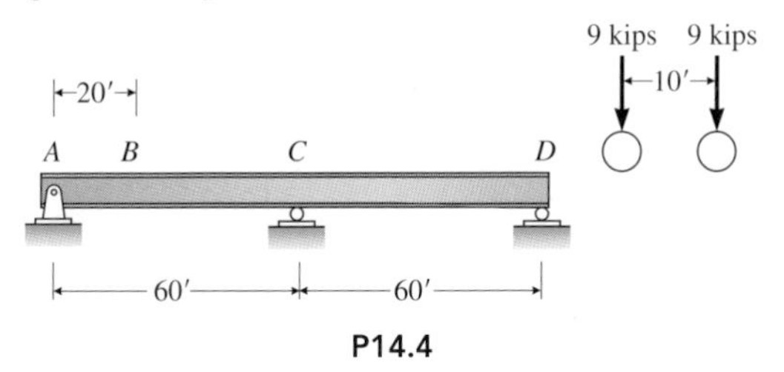

P14.4

P14.5. Draw the qualitative influence lines for R_A, R_B, M_C, V_C, and shear to the left of support D for the beam in Figure P14.5.

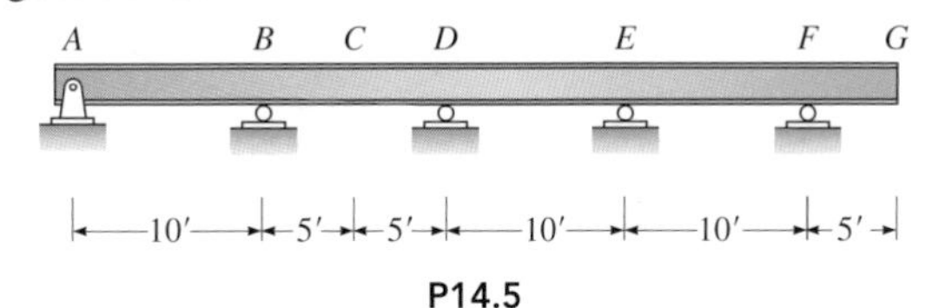

P14.5

P14.6. (*a*) Draw the qualitative influence lines for the moment at a section located at the top of the first-floor column BG and the vertical reaction at support C. Columns are equally spaced. (*b*) Indicate the spans on which a uniformly distributed load should be placed to maximize the moment on a section at the top of column BG. (*c*) Draw a qualitative influence line for negative moment on a vertical section through the floor beam at E.

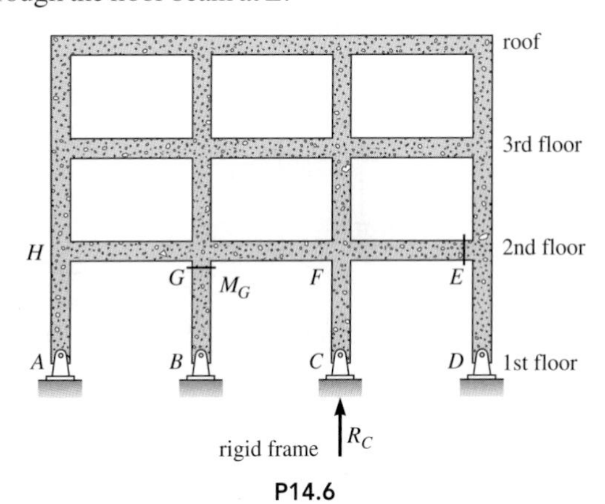

P14.6

P14.7. The ordinates of the moment influence line at midspan B of a 2-span continuous beam are provided at every one-tenth of the span in Figure P14.7. (*a*) Position the HS 20–44 truck load in Figure 8.25(a) to produce the maximum positive moment. (*b*) Position the HS 20–44 lane loads shown in Figure 8.25(b) to produce the maximum positive moment. Which loading is more critical?

A B C D

35′ 35′ $L = 70'$

+ −

0 0.038L 0.076L 0.116L 0.158L 0.203L 0.152L 0.105L 0.064L 0.028L 0 0.021L 0.036L 0.045L 0.048L 0.047L 0.042L 0.034L 0.024L 0.012L 0

P14.7

P14.8. (*a*) Draw a qualitative influence line for the reaction at support A for the beam in Figure P14.8. Using *moment distribution*, calculate the ordinate of the influence line at section 4. (*b*) Draw the qualitative influence line for the moment at B. Using the *conjugate beam or moment distribution method*, calculate the ordinate of the influence line at Section 8. EI is constant.

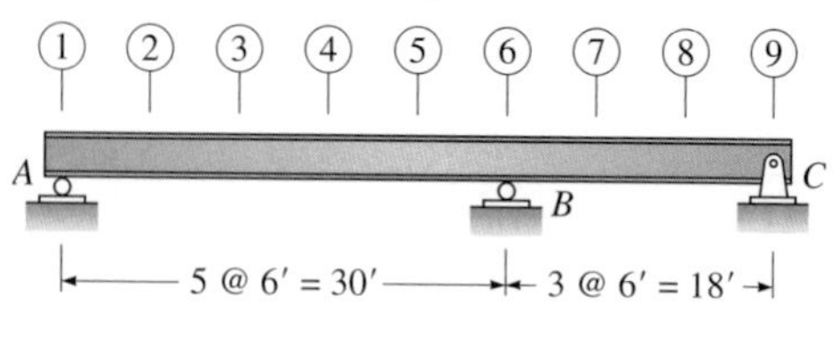

P14.8

P14.9. Construct the influence lines for R_A and M_C in Figure P14.9, using the Müller–Breslau method. Evaluate the ordinates at points A, B, C, and D.

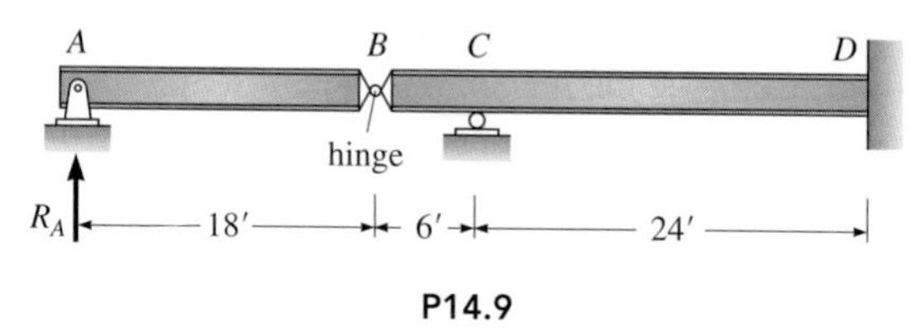

P14.9

P14.10. *Computer analysis of beam of varying depth.* The reinforced concrete bridge girder, attached to the massive end wall as shown in Figure P14.10, may be treated as a fixed-ended beam of varying depth. (*a*) Construct the influence lines for the reactions R_A and M_A at support A. Evaluate the ordinates at 15-ft intervals. (*b*) Evaluate the moment M_A and the vertical reaction R_A at end A produced by the loaded ore-carrier when its 30-kip rear wheel is positioned at point B. $E = 3000$ kips/in².

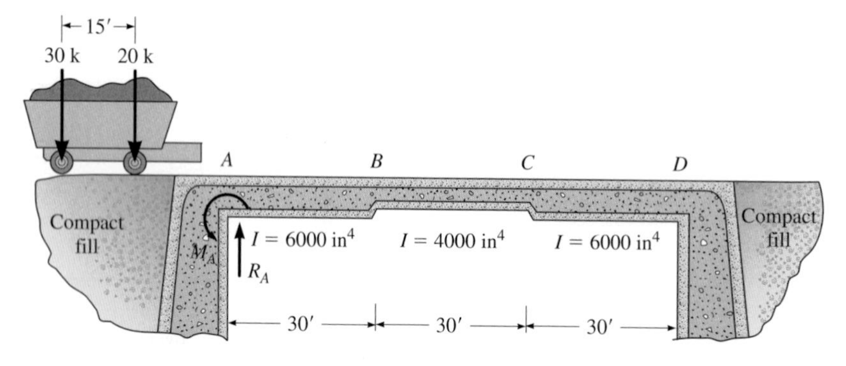

P14.10

Bayonne Bridge that links Staten Island and Bayonne

The Bayonne Bridge that links Staten Island in New York and Bayonne in New Jersey features a 1,675-ft long steel arch. It was the longest arch bridge in the world when it was opened in 1931. The bridge was erected using falsework instead of the cantilever method over the strait. Heavy truss bracing in the plane of the top chord is used to stiffen the side arches and to transmit the lateral component of the wind forces into the end supports of the arches.

CHAPTER

15

Approximate Analysis of Indeterminate Structures

Chapter Objectives

- Learn the importance of approximate analysis for estimating member forces with reasonable accuracy, checking hand calculations, and to avoid an overlook in computer-generated structural analysis results.
- Approximate analysis is also a very useful tool to estimate design forces that can be used for the selection of preliminary member sizes of a structure.
- Approximate analysis methods include estimating locations of points of inflections for beams and frames, and utilizing beam action as an analogy for approximate analysis of trusses.
- The *portal method* and *cantilever method* are very useful approximate analysis methods for multi-story frames with lateral loads, such as wind and earthquake forces.

15.1 Introduction

Thus far we have used exact methods to analyze indeterminate structures. These methods produce a structural solution that satisfies the equilibrium of forces and compatibility of deformations at all joints and supports. If a structure is highly indeterminate, an exact analysis (for example, consistent deformations or slope deflection) can be time-consuming. Even when a structure is analyzed by computer, the solution may take a great deal of time and effort to complete if the structure contains many joints or if its geometry is complex.

If designers understand the behavior of a particular structure, they can often use an approximate analysis to *estimate* closely, with a few simple computations, the approximate magnitude of the forces at various points in the structure. In an approximate analysis, we make simplifying assumptions about structural action or about the distribution of forces to various members.

In the mid-1800s, rigid frames of steel and reinforced concrete were introduced for resisting lateral loads. As structures became more highly indeterminate with the development of multi-bay mill buildings and high-rise structures, a need for approximate analysis methods grew. In 1913, Robin Fleming, a bridge engineer for the American Bridge Company in New York, fulfilled that need with his development of the cantilever and portal methods. When Hardy Cross' method of moment distribution was developed in 1930, it became the choice of analysis for indeterminate frames. However, cantilever and portal methods continue to be useful tools to estimate forces for preliminary sizing of framing members and checking computer analysis results.

These assumptions often permit us to evaluate forces by using only the equations of statics without considering compatibility requirements.

Although the results of an approximate solution may sometimes deviate as much as 10 or 20 percent from those of an exact solution, they are useful at certain design stages. Designers use the results of an approximate analysis for the following purposes:

1. To size the main members of a structure during the preliminary design phase—the stage when the initial configuration and proportions of the structure are established. Since the distribution of forces in an indeterminate structure is influenced by the stiffness of individual members, the designer must estimate the size of members closely before the structure can be analyzed accurately.
2. To verify the accuracy of an exact analysis. As you have discovered from solving homework problems, computational errors are difficult to eliminate in the analysis of a structure. Therefore, it is essential that a designer always use an *approximate* analysis to verify the results of an *exact* analysis. If a gross error in computations is made and the structure is sized for forces that are too small, it may fail. The penalty for a structural failure is incalculable—loss of life, loss of investment, loss of reputation, lawsuits, inconvenience to the public, and so forth. On the other hand, if a structure is sized for values of force that are too large, it will be excessively costly.

If radical assumptions are required to model a complex structure, the results of an exact analysis of the simplified model are often no better than those of an approximate analysis. In this situation the designer can base the design on the approximate analysis with an appropriate factor of safety.

Designers use a large variety of techniques to carry out an approximate analysis. These include the following:

1. Guessing the location of points of inflection in continuous beams and frames.
2. Using the solution of one type of structure to establish the forces in another type of structure whose structural action is similar. For example, the forces in certain members of a continuous truss may be estimated by assuming that the truss acts as a continuous beam.
3. Analyzing a portion of a structure instead of the entire structure.

In this chapter we discuss methods to make an approximate analysis of the following structures:

1. Continuous beams and trusses for vertical loads.
2. Simple rigid frames and multistory building frames for both vertical and lateral loads.

15.2 Approximate Analysis of a Continuous Beam for Gravity Load

The approximate analysis of a continuous beam is normally made by one of the following two methods:

1. Guessing the location of points of inflection (points of zero moment).
2. Estimating the values of the member end moments.

Method 1. Guessing the Location of Inflection Points

Since the moment is zero at a point of inflection (the point where the curvature reverses), we can treat a point of inflection as if it were a hinge for the purposes of analysis. At each point of inflection we can write a condition equation (that is, $\Sigma M = 0$). Therefore, each hinge we introduce at a point of inflection reduces the degree of indeterminacy of the structure by 1. By adding hinges equal in number to the degree of indeterminacy, we can convert an indeterminate beam to a determinate structure that can be analyzed by statics.

To serve as a guide for locating the approximate position of points of inflection in a continuous beam, we observe the position of the points of inflection for the idealized cases shown in Figure 15.1. We can then use our judgment to modify these results to account for deviations of the actual end conditions from those of the idealized cases.

For the case of a uniformly loaded beam whose ends are completely fixed against rotation (see Figure 15.1*a*), the points of inflection are located $0.21L$ from each end. If a fixed-end beam carries a concentrated load at midspan (see Figure 15.1*b*), the points of inflection are located $0.25L$ from each end. If a beam is supported on either a roller or a pin, the end restraint is zero (see Figure 15.1*c*). For this case the points of inflection shift outward to the ends of the member. Support conditions in Figures 15.1*a* (full restraint) and 15.5*c* (no restraint) establish the range of positions in which a point of inflection may be located. For the case of a uniformly loaded beam fixed at one end and simply supported at the other, the point of inflection is located a distance $0.25L$ from the fixed support (see Figure 15.1*d*).

As a preliminary step in the approximate analysis of a continuous beam, you may find it helpful to draw a sketch of the deflected shape to locate the approximate position of the points of inflection. Examples 15.1 and 15.2 illustrate the use of the cases in Figure 15.1 to analyze continuous beams by guessing the location of the points of inflection.

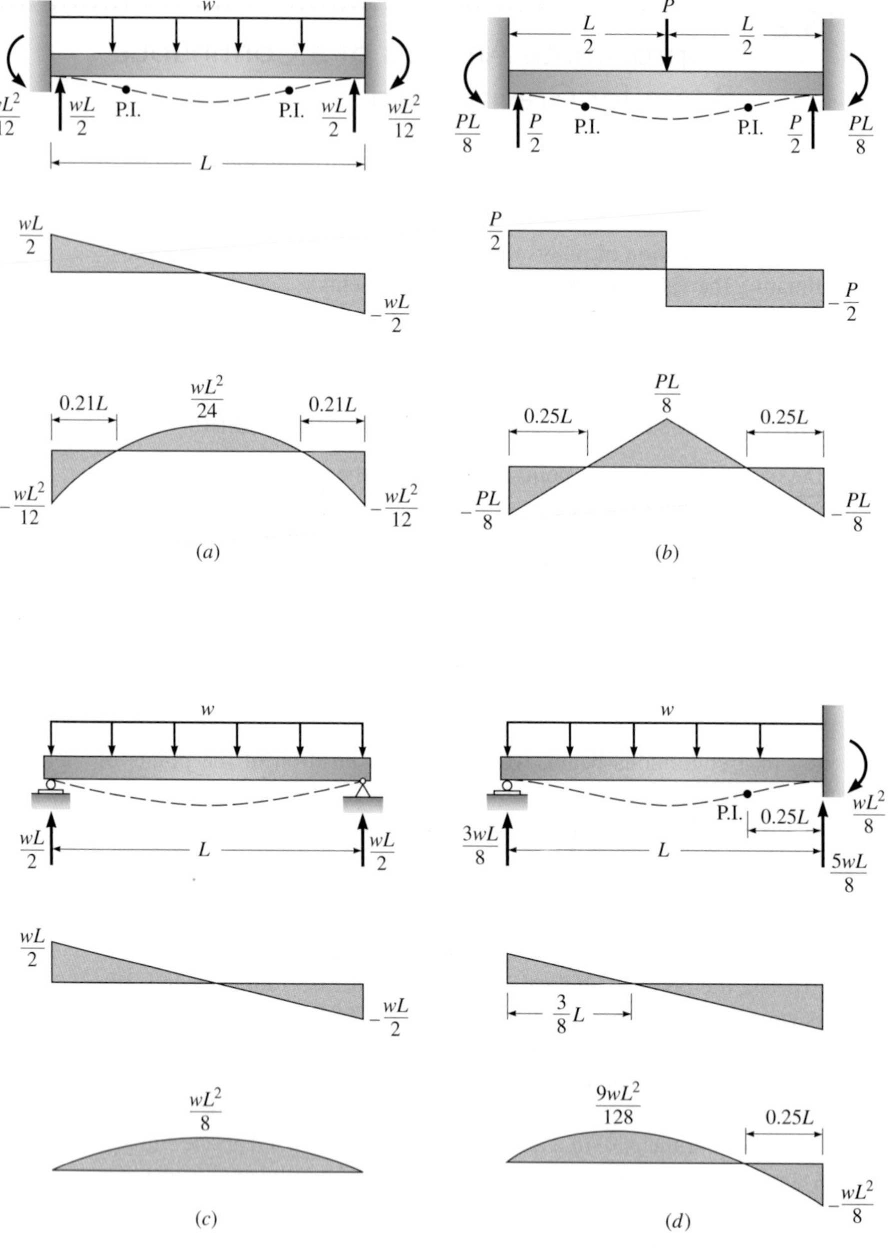

Figure 15.1: Location of points of inflection and shear and moment curves for beams with various idealized end conditions.

EXAMPLE 15.1

Carry out an approximate analysis of the continuous beam in Figure 15.2a by assuming the location of a point of inflection.

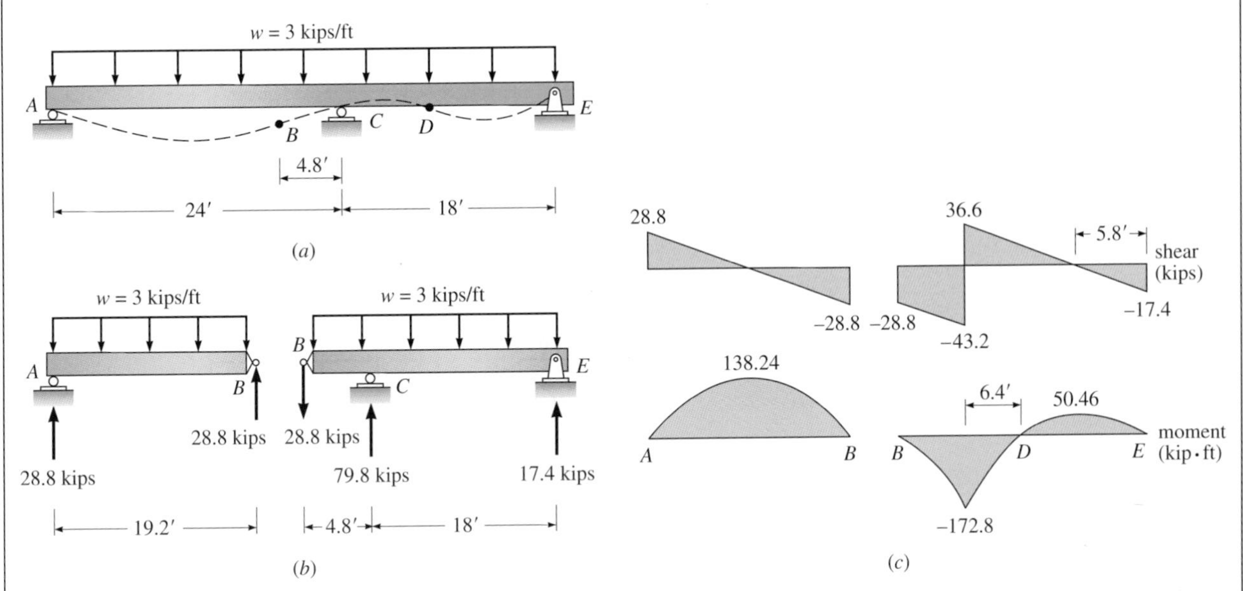

Figure 15.2: (a) Continuous beam, points of inflection indicated with a black dot; (b) free bodies of beam on either side of the point of inflection; (c) shear and moment curves based on the approximate analysis. *Note*: An exact analysis gives $M_C = -175.5$ kip·ft.

Solution

The approximate location of each point of inflection is indicated by a small black dot on the sketch of the deflection shape shown by the dashed line in Figure 15.2a. Although the continuous beam has a point of inflection in each span, we only have to guess the location of one point because the beam is indeterminate to the first degree. Since the shape of the longer-span AC is probably more accurately drawn than the shorter span, we will guess the position of the point of inflection in that span.

If joint C did not rotate, the deflected shape of member AC would be identical to that of the beam in Figure 15.1d, and the point of inflection would be located $0.25L$ to the left of support C. Because span AC is longer than span CE, it applies a greater fixed-end moment to joint C than span CE does. Therefore, joint C rotates counterclockwise. The rotation of joint C causes the point of inflection at B to shift a short distance to the right toward support C. We will arbitrarily guess that the point of inflection is located $0.2L_{AC} = 4.8$ ft to the left of support C.

We now imagine that a hinge is inserted into the beam at the location of the point of inflection, and we compute the reactions using the equations of statics. Figure 15.2b represents the results of this analysis. The shear and moment curves in Figure 15.2c show the results of the approximate analysis.

EXAMPLE 15.2

Estimate the values of moment at midspan of member *BC* as well as at support *B* of the beam in Figure 15.3*a*.

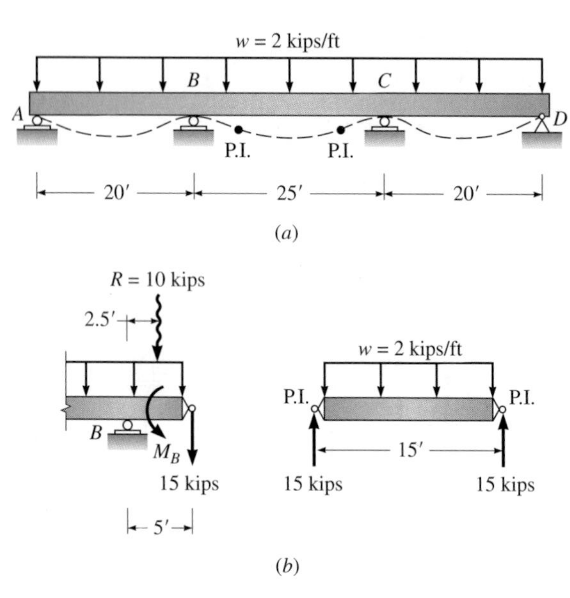

Figure 15.3: (*a*) Uniformly loaded continuous beam showing assumed location of points of inflection; (*b*) free bodies of the center span.

Solution

Since the beam in Figure 15.3*a* is indeterminate to the second degree, we must assume the location of two points of inflection to analyze by the equations of statics. Because all spans are about the same length and carry the same load, the slope of the beam at supports *B* and *C* will be zero or nearly zero. Therefore, the deflected shape, as shown by the dashed line, will be similar to that of the fixed-end beam in Figure 15.1*a*. Consequently, we can assume that points of inflection develop at a distance of $0.2L = 5$ ft from each support. If we imagine that hinges are inserted at both points of inflection, the 15-ft segment between the two points of inflection can be analyzed as a simply supported beam. Accordingly, the moment at midspan equals

$$M \approx \frac{wL^2}{8} = \frac{2(15)^2}{8} = 56.25 \text{ kip} \cdot \text{ft}$$

Treating the 5-ft segment of beam between the hinge and the support at *B* as a cantilever, we compute the moment at *B* as

$$M_B \approx 15(5) + (2)5(2.5) = 100 \text{ kip} \cdot \text{ft}$$

Method 2. Estimating Values of End Moments

As we have seen from our study of indeterminate beams in Chapters 12 and 13, the shear and moment curves for the individual spans of a continuous beam can be constructed after the member end moments are established. The magnitude of the end moments is a function of the rotational restraint supplied by either the end support or the adjacent members. Depending on the magnitude of the rotational restraint at the ends of a member, the end moments produced by a *uniform* load can vary from zero (simple supports) at one extreme to $wL^2/8$ (one end fixed and the other pinned) at the other.

To establish the influence of end restraint on the magnitude of the positive and negative moments that can develop in a span of continuous beam, we can again consider the various cases shown in Figure 15.1. From examining Figure 15.1*a* and *c*, we observe that the shear curves are identical for *uniformly loaded* beams with *symmetric* boundary conditions. Since the area under the shear curve between the support and midspan equals the simple beam moment $wL^2/8$, we can write

$$M_s + M_c = \frac{wL^2}{8} \tag{15.1}$$

where M_s is the absolute value of the negative moment at each end, and M_c is the positive moment at midspan.

In a continuous beam the rotational restraint supplied by adjacent members depends on how they are loaded as well as on their flexural stiffness. For example, in Figure 15.4*a* the spans of the exterior beams have been selected so that the rotations of joints B and C are zero when uniform load acts on all spans. Under this condition the moments in member BC are equal to those in a fixed-end beam of the same span (see Figure 15.4*b*). On the other hand, if the exterior spans are unloaded when the center span is loaded (see Figure 15.4*c*), the joints at B and C rotate, and the end moments are reduced by 35 percent. Because rotation at the ends increases the curvature at midspan, the positive moment increases 70 percent. The change in moment at midspan—associated with the end rotation—is twice as large as that at the supports because the initial moments (assuming we start with the ends fixed and allow the end joints to rotate) at the ends are twice as large as the moment at midspan. We also observe that rotation of the ends of the members results in the points of inflection moving outward toward the supports (from $0.21L_2$ to $0.125L_2$).

We will now use the results of Figures 15.1 and 15.4 to carry out an approximate analysis of the uniformly loaded beam of equal spans in Figure 15.5. Because all spans are about the same length and carry uniform load, all beams will be concave up in the center—indicating positive moment at or near midspan—and concave down—indicating negative moment—over the supports.

We begin by considering interior span CD. Since the end moments applied to each side of an interior joint are about the same, the joint undergoes no significant rotation, and the slope of the beam at supports C and D will be nearly horizontal—a condition similar to that of the fixed-ended beam in Figure 15.1*a*; therefore, we can assume that the negative moments at supports C and D are approximately equal to $wL^2/12$. In addition, Figure 15.1*a* shows

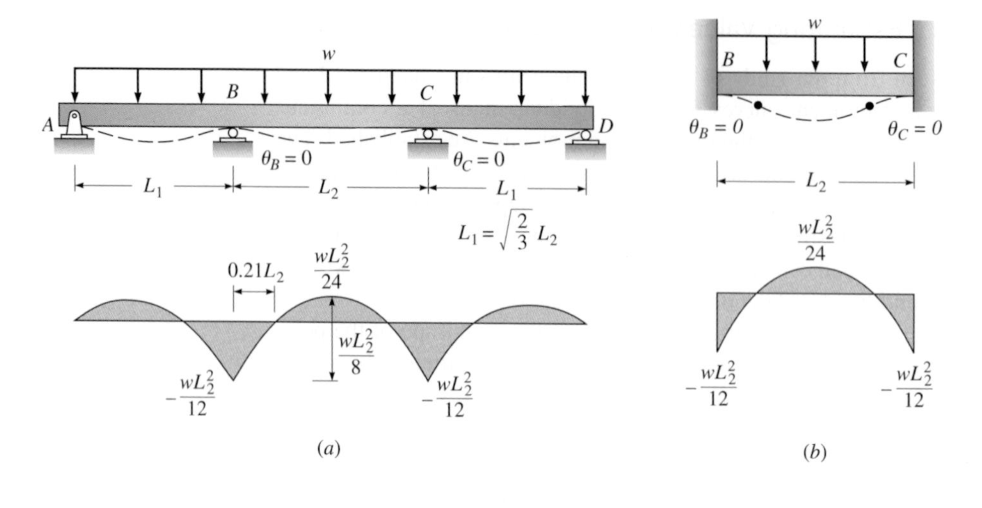

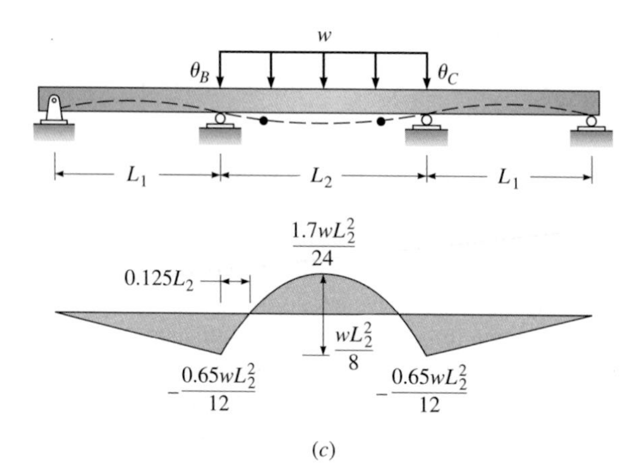

Figure 15.4

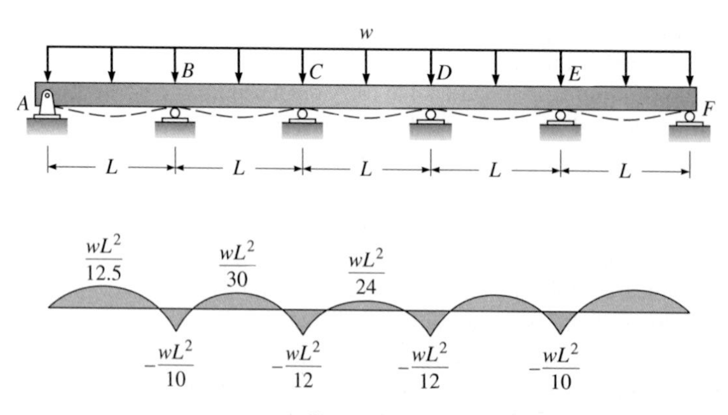

Figure 15.5

that the positive moment at midspan of span *CD* will be approximately $wL^2/24$.

To estimate the moments in span *AB*, we will use the moment curve for the beam in Figure 15.1*d* as a guide. If the support at *B* were completely fixed, the negative moment at *B* would equal $wL^2/8$. Since some *counterclockwise* rotation of joint *B* occurs, the negative moment will reduce moderately. Assuming a 20 percent reduction in the negative moment occurs, we estimate the value of a negative moment at *B* equals $wL^2/10$. After the negative moment is estimated, analysis of a free body of the exterior span gives a positive value of moment near midspan equal to $wL^2/12.5$. In a similar manner, computations show the positive moment in span *BC* is approximately equal to $wL^2/30$.

The value of the shear at the ends of a continuous beam is influenced by the difference in the magnitudes of the end moments as well as the size and position of the loading. If the end moments are equal and the beam is loaded symmetrically, the end reactions are equal. The greatest difference in the magnitude of the reactions in Figure 15.1 occurs when one end is fixed and the other end pinned, that is, when $(3/8)wL$ goes to the pin support and $(5/8)wL$ to the fixed support (see Figure 15.1*d*).

15.3 Approximate Analysis of a Rigid Frame for Vertical Load

The design of the columns and girder of a rigid frame used to support the roof of a field house or a warehouse is controlled by moment. Since the axial force in both the legs and the girder of a rigid frame is typically small, it can be neglected, and in an approximate analysis the members are sized for moment.

The magnitude of the negative moment at the ends of the girder in a rigid frame will depend on the relative stiffness between the columns (the legs) and the girder. Typically, girders are 4 or 5 times longer than the columns. On the other hand, the moment of inertia of the girder is often much larger than that of the columns. Since the relative stiffness between the legs and the girder of a rigid frame can vary over a wide range, the end moment in the girder can range from 20 to 75 percent of the fixed-end moment. As a result, the values of the moment predicted by an approximate analysis may deviate considerably from the values of an exact analysis.

If the members of a uniformly loaded rigid frame are constructed of the same-size members, the flexural stiffness of the shorter legs will be relatively large compared to the stiffness of the girder. For this condition we can assume that the rotational restraint supplied by the legs produces an end moment in a uniformly loaded girder that is on the order of 70 to 85 percent of the moment that occurs in a fixed-end beam of the same span (see Figure 15.1*a*). On the other hand, if for architectural reasons the frame is constructed with shallow columns and a deep girder, the rotational restraint supplied by the flexible legs will be small. For this condition the end moments that develop in the girder may

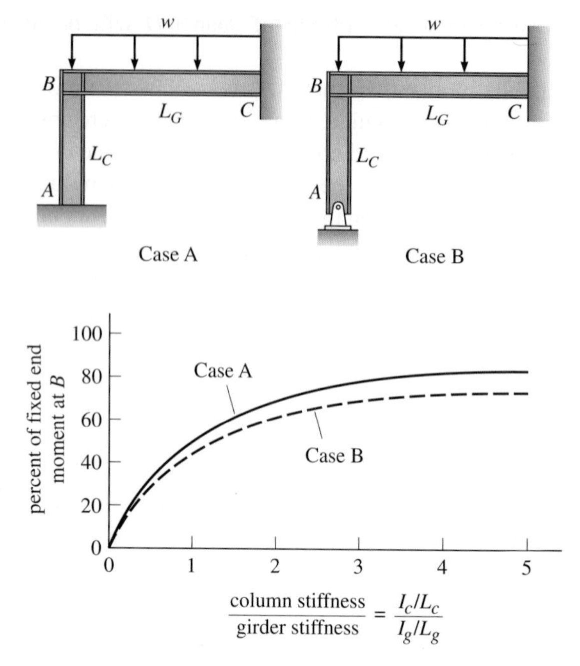

Figure 15.6: Influence of column stiffness on the end moment at joint *B* in a girder whose far end is fixed. Case A: base of column fixed; case B: base of column pinned.

be on the order of 15 to 25 percent of those that develop in a fixed-end beam. Figure 15.6 shows the variation of negative moment at the end of a girder (fixed at *C*) as a function of the ratio between the flexural stiffness of the column and the girder.

A second procedure for estimating the moments in a frame is to guess the location of the points of inflection (the points of zero moments) in the girder. Once these points are established, the balance of the forces in the frame can be determined by statics. If the columns are stiff and supply a large rotational restraint to the girder, the points of inflection will be located at about the same position as those in a fixed-ended beam (i.e., about $0.2L$ from each end). On the other hand, if the columns are flexible relative to the girder, the points of inflection will move toward the ends of the girder. For this case the designer might assume that the point of inflection is located between $0.1L$ and $0.15L$ from the ends of the girder. Use of this method to estimate the forces in a rigid frame is illustrated in Example 15.4 on page 614.

As a third method of determining the moments in a rigid frame, the designer can estimate the ratio between the positive and negative moments in the girder. Typically, the negative moments are 1.2 to 1.6 times greater than the positive moment. Since the sum of the positive and negative moments in a girder that carries a uniformly distributed load must equal $wL^2/8$, once the ratio of moments is assumed, the values of positive and negative moments are established.

EXAMPLE 15.3

Analyze the symmetric frame in Figure 15.7*a* by estimating the values of negative moments at joints *B* and *C*. Columns and girders are constructed from the same-size members—that is, *EI* is constant.

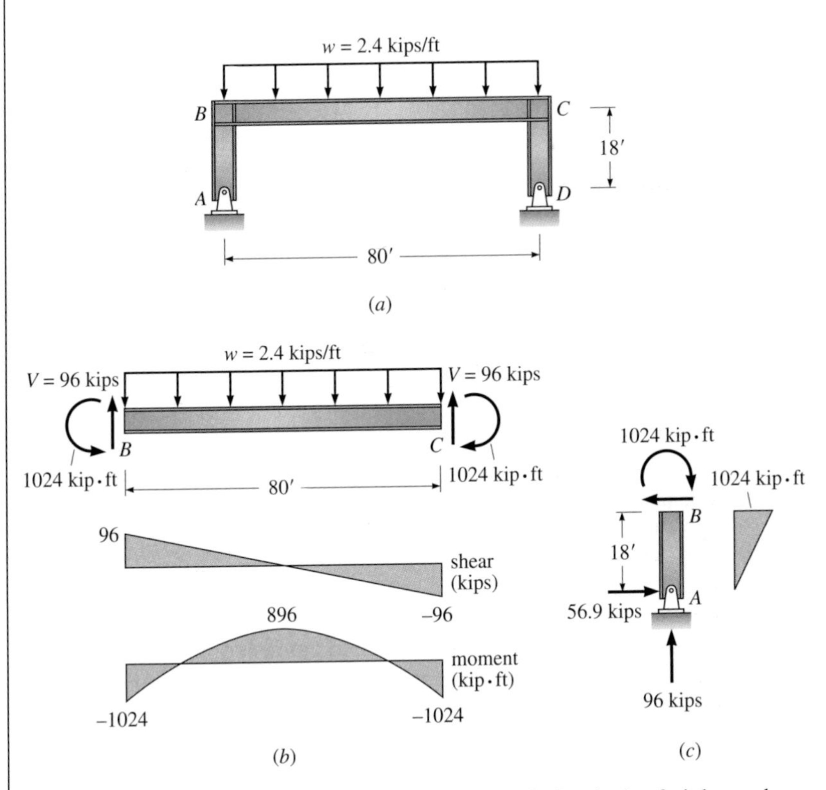

Figure 15.7: (*a*) Symmetric frame with uniform load; (*b*) free body of girder; and approximate shear and moment diagrams; (*c*) free body of column with estimated value of the end moment.

Solution

Since the shorter columns are much stiffer than the longer girders (flexural stiffness varies inversely with length), we will assume the negative moments at joints *B* and *C* are equal to 80 percent of the end moments in a fixed-end beam of the same span.

$$M_B = M_C = -0.8\frac{wL^2}{12} = -\frac{0.8(2.4)80^2}{12} = -1024 \text{ kip}\cdot\text{ft}$$

We next isolate the girder (Figure 15.7*b*) and the column (Figure 15.7*c*), compute the end shears using the equations of statics, and draw the shear and moment curves.

An exact analysis of the structure indicates that the end moment in the girder is 1113.6 kip·ft and the moment at midspan is 806 kip·ft.

EXAMPLE 15.4

Estimate the moments in the frame shown in Figure 15.8*a* by guessing the location of the points of inflection in the girder.

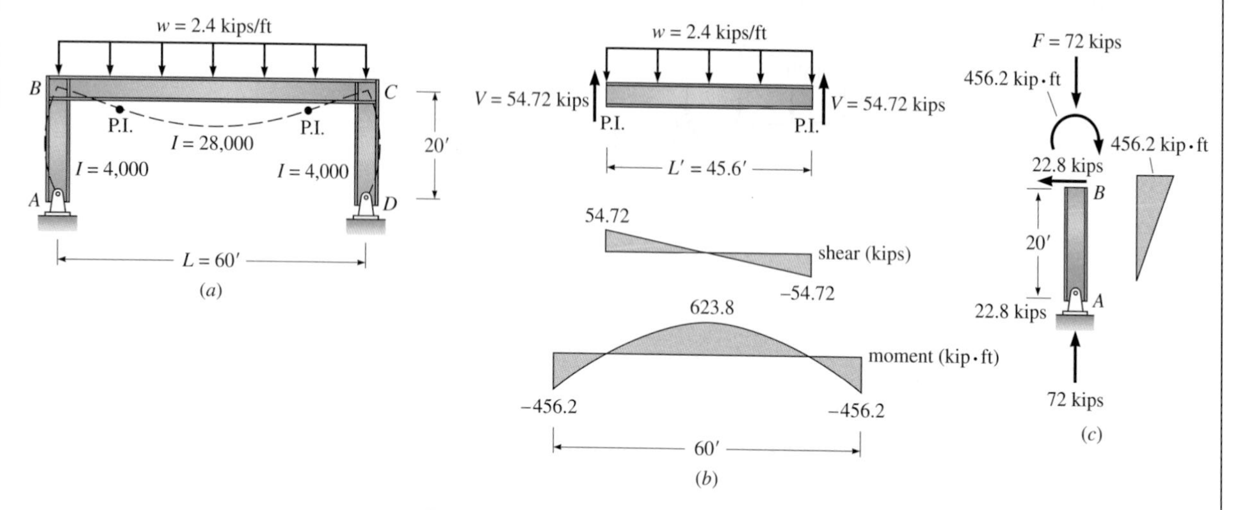

Figure 15.8: (*a*) Details of frame. (*b*) Free body of girder between points of inflection. *Note*: Moment curve in units of kip · feet is for the entire girder, the shear curve in units of kips runs between points of inflection. (*c*) Free body of column *AB*.

Solution

If we consider the influence of both length and moment of inertia on the flexural stiffness of the columns and the girder, we observe that the columns, because of a smaller I, are more flexible than the girder. Therefore, we will assume arbitrarily that the points of inflection in the girder are located $0.12L$ from the ends of the girder.

Compute the distance L' between points of inflection in the girder.

$$L' = L - (0.12L)(2) = 0.76L = 45.6 \text{ ft}$$

Since the segment of girder between points of inflection acts as a simply supported beam (i.e., the moments are zero each end), the moment at midspan equals

$$M_c = \frac{wL'^2}{8} = \frac{2.4(45.6)^2}{8} = 623.8 \text{ kip} \cdot \text{ft} \qquad \textbf{Ans.}$$

Using Equation 15.1, we compute the girder end moments M_s:

$$M_s + M_c = \frac{wL^2}{8} = \frac{2.4(60)^2}{8} = 1080 \text{ kip} \cdot \text{ft}$$

$$M_s = 1080 - 623.8 = 456.2 \text{ kip} \cdot \text{ft} \qquad \textbf{Ans.}$$

The moment curves for the girder and column are shown in Figure 15.8*b* and *c*. The exact value of moment at the ends of the girder is 404.64 kip · ft.

15.4 Approximate Analysis of a Continuous Truss

As we discussed in Section 4.1, the structural action of a truss is similar to that of a beam (see Figure 15.9). The chords of the truss, which act as the flanges of a beam, carry the bending moment, and the diagonals of the truss, which perform the same function as the web of a beam, carry the shear. Since the behavior of a truss and a beam is similar, we can evaluate the forces in a truss by treating it as a beam instead of using the method of joints or sections. In other words, we apply the panel loads acting on the truss to an imaginary beam whose span is equal to that of the truss, and we construct conventional shear and moment curves. By equating the internal couple M_I produced by the forces in the chords to the internal moment M at the section produced by the external loads (and given by the moment curve), we can compute the approximate value of axial force in the chord. For example, in Figure 15.9*b* we can express the internal moment on Section 1 of the truss by summing moments of the horizontal forces acting on the section about point o at the level of the bottom chord to give

$$M_I = Ch$$

Setting $M_I = M$ and solving the expression above for C give

$$C = \frac{M}{h} \tag{15.2}$$

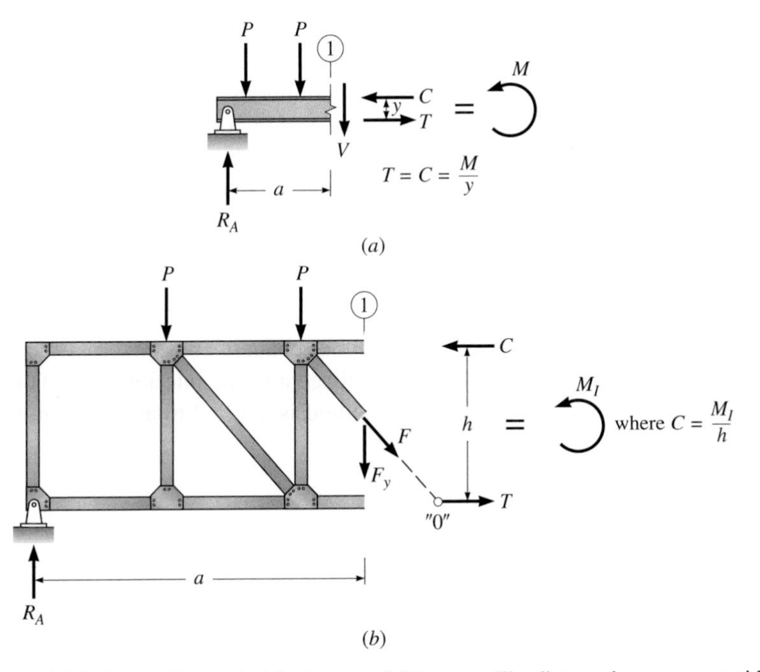

Figure 15.9: Internal forces in (*a*) a beam and (*b*) a truss. The distance between centroids of the flanges is y, and h is the distance between centroids of the chords.

where h equals the distance between centroids of the top and bottom chords and M equals the moment in the beam at Section 1 in Figure 15.9*a*.

When the panel loads acting on a truss are equal in magnitude, we can simplify the beam analysis by replacing the concentrated loads by an equivalent uniform load w. To make this computation, we divide the sum of the panel loads ΣP_n by the span length L:

$$w = \frac{\Sigma P_n}{L} \tag{15.3}$$

If the truss is long compared to its depth (say, the span-to-depth ratio exceeds 10 or more), this substitution should have little influence on the results of the analysis. We will use this substitution when analyzing a continuous truss as a beam, because the computation of fixed-end moments for a uniform load acting over the entire span is simpler than computing the fixed-end moments produced by a series of concentrated loads.

Continuing the analogy, we can compute the force in the diagonal of a truss by assuming that the vertical component of force F_y in the diagonal equals the shear V at the corresponding section of the beam (see Figure 15.9).

To illustrate the details of the beam analogy and to check its accuracy, we will use the method to compute the forces in several members of the determinate truss in Example 15.5. We will then use the method to analyze the indeterminate truss in Example 15.6 on page 619.

Example 15.5 shows that the bar forces in a determinate truss computed by the beam analogy are exact. This result occurs because the distribution of forces in a determinate structure does not depend on the stiffness of the individual members. In other words, the forces in a determinate beam or truss are computed by applying the equations of statics to free bodies of the truss. On the other hand, the forces in a continuous truss will be influenced by the dimensions of the chord members, which correspond to the flanges of a beam. Since the forces in the chords are much larger adjacent to an interior support, the cross section of the members in that location will be larger than those between the center of each span and the exterior supports. Therefore, the truss will act as a beam with a variable moment of inertia. To adjust for the variable stiffness of the equivalent beam in an approximate analysis, the designer can arbitrarily increase by 15 or 20 percent the forces (produced by analyzing the truss as a continuous beam of constant cross section) in the chords. Forces in the diagonals adjacent to the interior supports may be increased about 10 percent. The method is applied to an indeterminate truss in Example 15.6.

EXAMPLE 15.5

By analyzing the truss in Figure 15.10*a* as a beam, compute the axial forces in the top chord (member *CD*) and bottom chord (member *JK*) at midspan and in diagonal *BK*. Compare the values of force to those computed by the method of joints or sections.

Solution

Apply the loads acting at the bottom panel points of the truss to a beam of the same span, and construct the shear and moment curves (see Figure 15.10*b*).

Compute the axial force in member *CD* of the truss, using Equation 15.2 and the beam moment at *D*. (see Figure 15.10*c*).

$$\Sigma M_J = 0$$

$$F_{CD} = C = \frac{M_D}{h} = \frac{810}{12} = 67.5 \text{ kips} \qquad \textbf{Ans.}$$

Similarly, compute the axial force in member *JK* of the truss, using Equation 15.2 and the beam moment at *C* (see Figure 15.10*d*).

$$F_{JK} = C = \frac{M_C}{h} = \frac{720}{12} = 60 \text{ kips} \qquad \textbf{Ans.}$$

Compute the force in diagonal *BK*. Equate the shear of 30 kips between *BC* of the beam to the vertical component F_y of the axial force in bar *BK* (see Figure 15.10*e*).

$$F_y = V$$

$$= 30 \text{ kips}$$

$$F_{BK} = \frac{5}{4} F_y = 37.5 \text{ kips} \qquad \textbf{Ans.}$$

Values of force are identical to those produced by an exact analysis of the truss.

[*continues on next page*]

Example 15.5 continues . . .

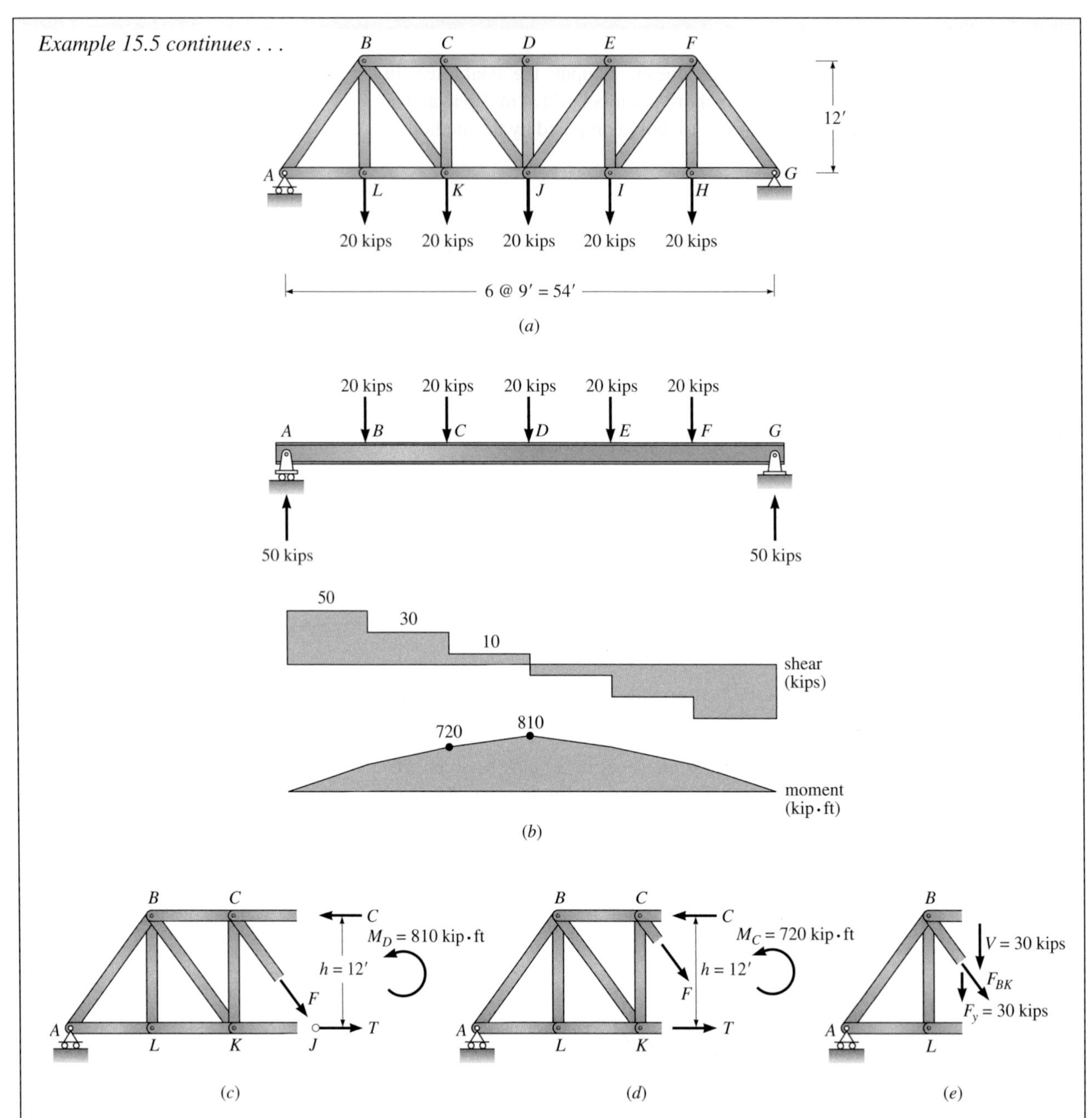

Figure 15.10: Analysis of a truss by beam analogy: (*a*) details of truss; (*b*) loads from truss applied to beam of same span; (*c*) free body of truss cut by a vertical section an infinitesimal distance to the left of midspan; (*d*) free body of truss cut by a vertical section at infinitesimal distance to the right of joint *K*; (*e*) free body of truss cut by a vertical section through panel *BC*.

EXAMPLE 15.6

Estimate the forces in bars *a*, *b*, *c*, and *d* of the continuous truss in Figure 15.11.

Solution

The truss will be analyzed as a continuous beam of constant cross section (see Figure 15.11*b*). Using Equation 15.3, we convert the panel loads to a statically equivalent uniform load.

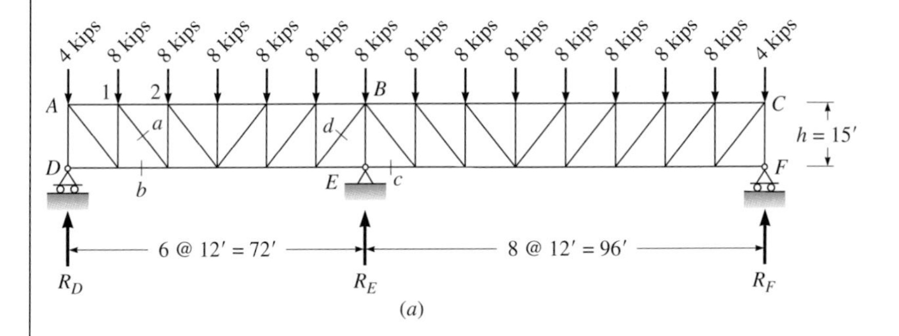

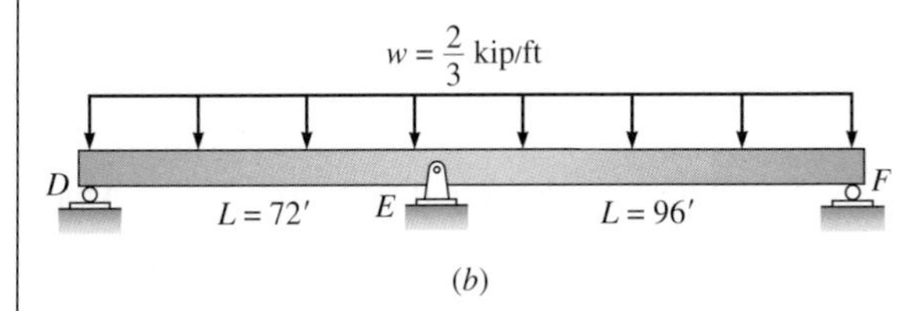

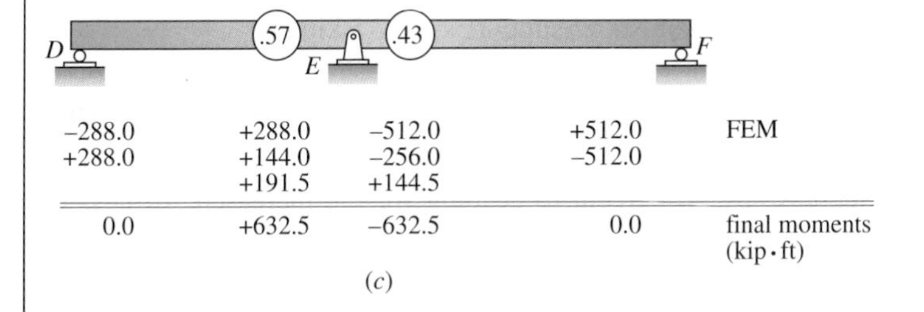

–288.0	+288.0	–512.0	+512.0	FEM
+288.0	+144.0	–256.0	–512.0	
	+191.5	+144.5		
0.0	+632.5	–632.5	0.0	final moments (kip · ft)

(*c*)

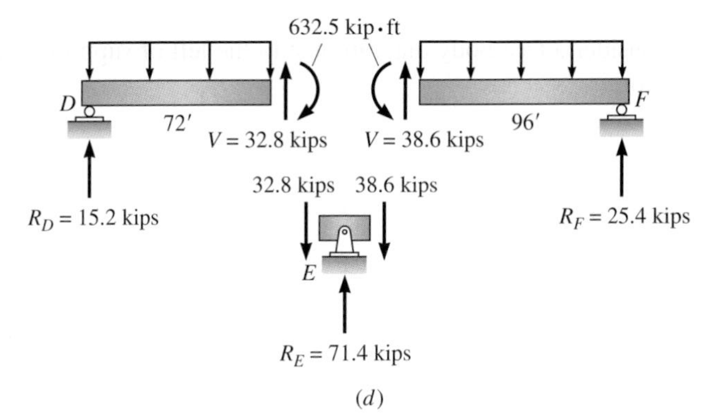

Figure 15.11: (*a*) Details of truss and loads; (*b*) beam loaded by an equivalent uniform load; (*c*) analysis of beam in (b) by moment distribution (moments in kip · ft); (*d*) computation of reactions using free-body diagrams of beams and support at *E* (*continues*).

[*continues on next page*]

Example 15.6 continues . . .

$$w = \frac{\Sigma P}{L} = \frac{(8\text{ kips})(13) + (4\text{ kips})(2)}{72 + 96} = \frac{2}{3}\text{ kip/ft}$$

Analyze the beam by moment distribution (see Figure 15.11*c* for details). Compute reactions using the free bodies shown in Figure 15.11*d*.

To compute bar forces, we will pass vertical sections through the beam; alternatively, after the reactions are established, we can analyze the truss directly.

For bar *a* (see free-body in Figure 15.11*e*),

$$\overset{+}{\uparrow} \quad \Sigma F_y = 0$$

$$15.2 - 4 - 8 - F_{ay} = 0$$

$$F_{ay} = 3.2\text{ kips}$$

$$F_a = \frac{5}{4}F_{ay} = \frac{5}{4}(3.2) = 4\text{ kips} \qquad \textbf{Ans.}$$

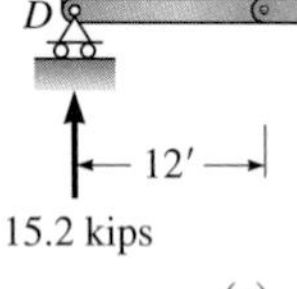

For bar *b*, sum moments about point 1, 12 ft to the right of support *D* (Figure 15.11*f*):

$$\circlearrowright^{+} \quad \Sigma M_1 = 0$$

$$(15.2)12 - 4(12) - 15F_b = 0$$

$$F_b = \frac{134.4}{15} = 8.96\text{ kips tension, round to 9 kips} \qquad \textbf{Ans.}$$

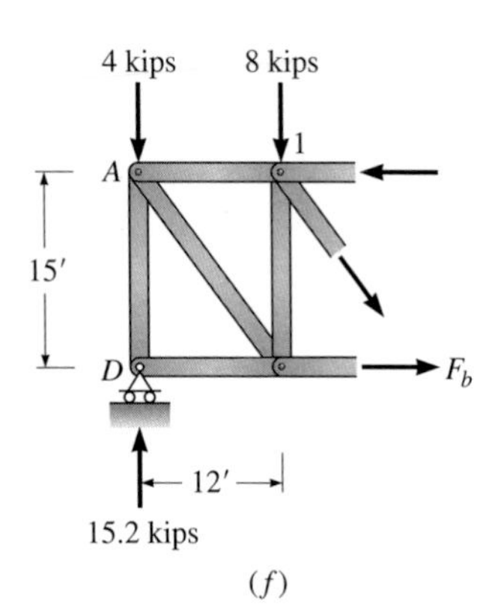

Figure 15.11: (*Continued*) (*e*) Computation of force in diagonal bar; (*f*) computation of force F_b.

For bar *c*,

$$\text{Moment at center support} = 632.5\text{ kip}\cdot\text{ft}$$

$$F_c = \frac{M}{h} = \frac{623.5}{15} = 42.2\text{ kips} \qquad \textbf{Ans.}$$

Arbitrarily increase by 10 percent to account for the increased stiffness of heavier chords adjacent to the center support in the real truss.

$$F_c = 1.1(42.2) = 46.4\text{ kips compression}$$

For bar *d*, consider a free-body diagram just to the left of support *E* cut by a vertical section.

$$\overset{+}{\uparrow} \quad \Sigma F_y = 0$$

$$15.2\text{ kips} - 4\text{ kips} - 5(8\text{ kips}) + F_{dy} = 0$$

$$F_{dy} = 28.8\text{ kips (tension)}$$

$$F_d = \frac{5}{4}F_{dy} = \frac{5}{4}(28.8) = 36\text{ kips}$$

Increase by 10 percent: $\quad F_d = 39.6\text{ kips}$ **Ans.**

15.5 Estimating Deflections of Trusses

Virtual work, which requires that we sum the strain energy in all bars of a truss, is the only method available for computing exact values of truss deflections. To verify that deflections computed by this method are of the *correct order of magnitude*, we can carry out an approximate analysis of the truss by treating it as a beam and by using standard beam deflection equations such as those given in Figure 11.3.

Deflection equations for beams are derived on the assumption that all deformations are produced by moment. These equations all contain the moment of inertia I in the denominator. Since shear deformations in shallow beams are normally small, they are neglected.

Unlike a beam the deformations of the vertical and diagonal members of a truss contribute nearly as much to the total deflection as do the deformations of the top and bottom chords. Therefore, if we use a beam equation to predict the deflection of a truss, the value will be approximately 50 percent too small. Accordingly, to account for the contribution of the web members to the deflection of the truss, the designer should double the value of the deflection given by a beam equation.

Example 15.7 illustrates the use of a beam equation to estimate the deflection of a truss. The value of moment of inertia I in the beam equation is based on the area of the chords at midspan. If the chord areas are smaller at the ends of a truss (where the magnitude of the forces is smaller), use of the midspan properties overestimates the stiffness of the truss and produces values of deflection that are smaller than the true values.

EXAMPLE 15.7

Estimate the midspan deflection of the truss in Figure 15.12 by treating it as a beam of constant cross section. The truss is symmetric about a vertical axis at midspan. The area of the top and bottom chords in the four center panels is 6 in^2. The area of all other chords equals 3 in^2. The area of all diagonals equals 2 in^2; the area of all verticals equals 1.5 in^2. Also E = 30,000 kips/in^2.

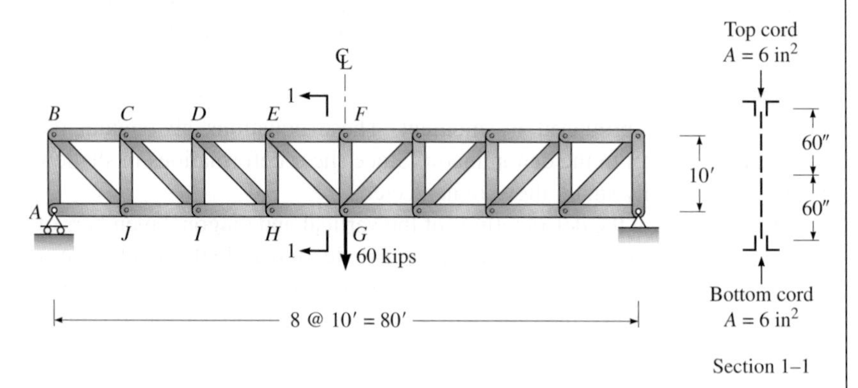

Figure 15.12

Solution

Compute the moment of inertia I of the cross section at midspan. Base your computation on the area of the top and bottom chords. Neglecting the moment of inertia of the chord area about its own centroid (I_{na}), we evaluate I with the standard equation (see Section 1-1)

$$I = \Sigma(I_{na} + Ad^2)$$

$$= 2[6(60)^2] = 43{,}200 \text{ in}^4$$

Compute the deflection at midspan (see Figure 11.3d for the equation).

$$\Delta = \frac{PL^3}{48EI}$$

$$= \frac{60(80 \times 12)^3}{48(30{,}000)(43{,}200)}$$

$$= 0.85 \text{ in}$$

Double Δ to account for contribution of web members:

$$\text{Estimated } \Delta_{\text{truss}} = 2\Delta = 2(0.85) = 1.7 \text{ in} \qquad \textbf{Ans.}$$

Solution by virtual work, which accounts for the reduced area of chords at each end and the actual contribution of the diagonals and verticals to the deflection, gives Δ_{truss} = 2.07 in.

15.6 Trusses with Double Diagonals

Trusses with double diagonals are a common structural system. Double diagonals are typically incorporated into the roofs and walls of buildings and into the floor systems of bridges to stabilize the structure or to transmit wind or other lateral loads (for example, sway of trains) into the end supports. Each panel containing a double diagonal adds 1 degree of indeterminacy to the truss; therefore, the designer must make one assumption per panel to carry out an approximate analysis.

If the diagonals are fabricated from heavy structural shapes and have sufficient flexural stiffness to resist buckling, the *shear in a panel may be assumed to divide equally between diagonals.* Resistance to buckling is a function of the member's slenderness ratio—the length divided by the radius of gyration of the cross section as well as the restraint supplied by the boundaries. Example 15.8 illustrates the analysis of a truss in which both diagonals are effective.

If the diagonals are slender—constructed from small-diameter steel rods of light structural shapes—the designer can assume that the diagonals only carry tension and buckle under compression. Because the slope of a diagonal determines if it acts in tension or compression, the designer must establish the diagonal in each panel that is effective, and must assume that the force in the other diagonal is zero. Since wind or other lateral forces can act in either transverse direction, both sets of diagonals are essential. Example 15.9 on page 625 illustrates the analysis of a truss with tension diagonals.

EXAMPLE 15.8

Analyze the indeterminate truss in Figure 15.13. Diagonals in each panel are identical and have sufficient strength and stiffness to carry loads in either tension or compression.

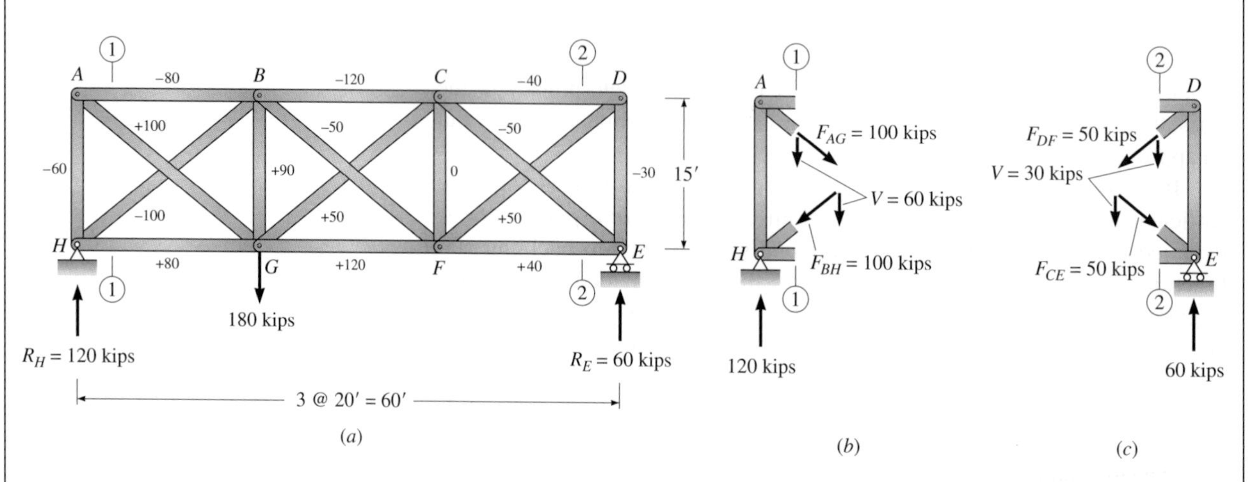

Figure 15.13: (*a*) Truss with two effective diagonals; (*b*) free body of truss cut by Section 1-1; (*c*) free body of truss cut by Section 2-2. All bar forces in units of kips.

Solution

Pass a vertical Section 1-1 through the first panel of the truss cutting the free body shown in Figure 15.13*b*. Assume each diagonal carries one-half the shear in the panel (120 kips produced by the reaction at support *H*). Since the reaction is up, the vertical component of force in each diagonal must act downward and equal 60 kips. To be consistent with this requirement, member *AG* must be in tension and member *BH* in compression. Since the resultant bar force is $\frac{5}{3}$ of the vertical component, the force in each bar equals 100 kips.

We next pass Section 2-2 through the end panel on the right. From a summation of forces in the vertical direction, we observe that a shear of 60 kips acting downward is required in the panel to balance the reaction on the right; therefore, the vertical component of force in each diagonal equals 30 kips acting downward. Considering the slope of the bars, we compute a tension force of 50 kips in member *DF* and a compression force of 50 kips in member *CE*. If we consider a free body of the truss to the right of a vertical section through the center panel, we observe that the unbalance shear is 60 kips and the forces in the diagonals act in the same direction as those shown in Figure 15.13*c*. After the forces in all diagonals are evaluated, the forces in the chords and verticals are computed by the method of joints. The final results are summarized in Figure 15.13*a*.

EXAMPLE 15.9

Small-diameter rods form the diagonal members of the truss in Figure 15.14*a*. The diagonals can transmit tension but buckle if compressed. Analyze the truss for the loading shown.

Solution

Since the truss is externally determinate, we first compute the reactions. We next pass vertical sections through each panel and establish the direction of the internal force in the diagonal bars required for vertical equilibrium of the shear in each panel. The tension and compression diagonals are next identified as discussed in Example 15.8 (the compression diagonals are indicated by the dashed lines in Figure 15.14*b*). Since the compression diagonals buckle, the entire shear in a panel is assigned to the tension diagonal, and the force in the compression diagonals is set equal to zero. Once the compression diagonals are identified, the truss may be analyzed by the methods of joints or sections. The results of the analysis are shown in Figure 15.14*b*.

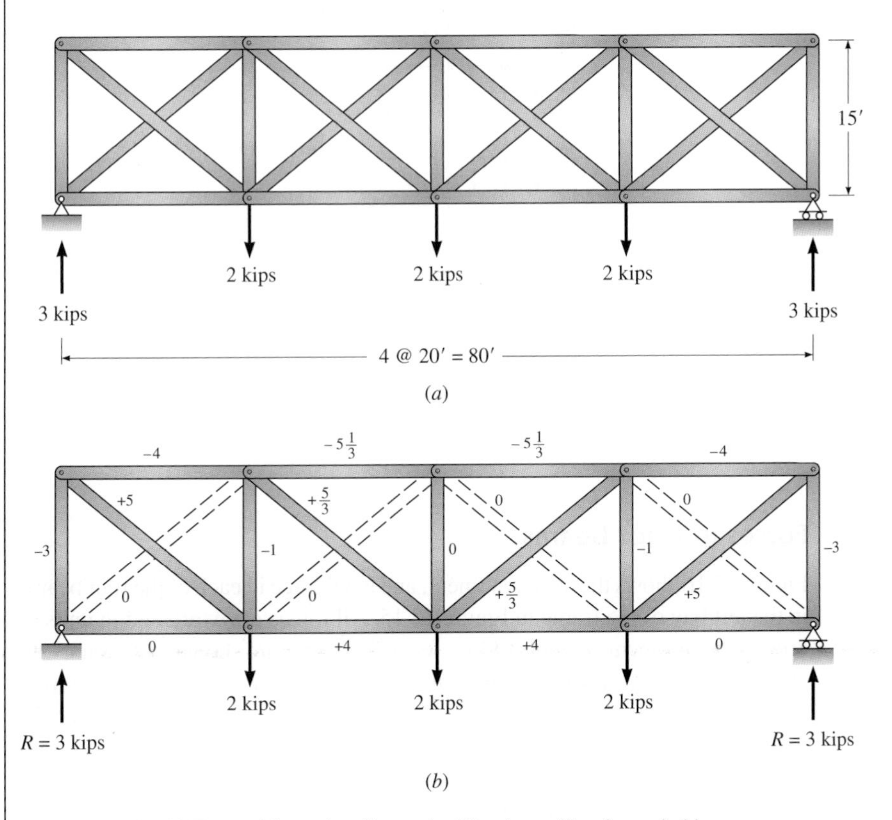

Figure 15.14: (*a*) Truss with tension diagonals; (*b*) values of bar forces in kips, compression diagonals indicated by dashed lines.

15.7 Approximate Analysis of a Multistory Rigid Frame for Gravity Load

To establish a set of guidelines for estimating the force in members of highly indeterminate multistory frames with rigid joints, we will examine the results of a computer analysis of the symmetric reinforced concrete building frame in Figure 15.15. The computer analysis considers both the axial and flexural stiffness of all members. The dimensions and properties of the members in the frame are representative of those typically found in small office or apartment buildings. In this study all beams in the frame carry a uniform load of $w = 4.3$ kips/ft to simplify the discussion. In practice, building codes permit the engineer to reduce values of live load on lower floors because of the low probability that the maximum values of live load will act simultaneously on all floors at any given time.

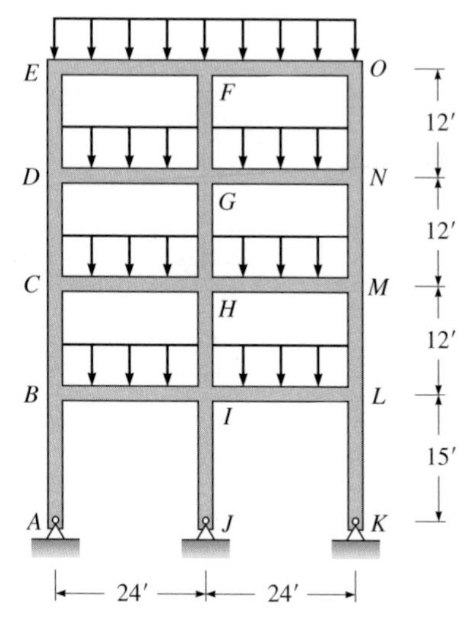

Properties of Members

Member	A in²	I in⁴
Exterior columns	100	1000
Interior columns	144	1728
Girders	300	6000

Figure 15.15: Dimensions and member properties of a vertically loaded multistory building frame.

Forces in Floor Beams

Figure 15.16 shows the shear, moment, and axial force in each of the four beams in the left bay of the frame in Figure 15.15. All forces are expressed in units of kips and all moments in units of kip · feet. The beams are shown in the same relative position they occupy in the frame (i.e., the top beam is located at the roof, the next at the fourth floor, and so on). We observe in each beam that the moment is greater at the right end—where the beams connect to the interior column—than at the left end, where the beams connect to the exterior column. Larger moments develop on the right because the interior joint, which does not rotate, acts as a fixed support. The interior joint does not rotate because the moments, applied by the beams on each side of the joint, are equal in magnitude and opposite in direction (see the curved arrows in Figure 15.18*b* on page 631). On the other hand, at the exterior joints where beams frame into one side of the

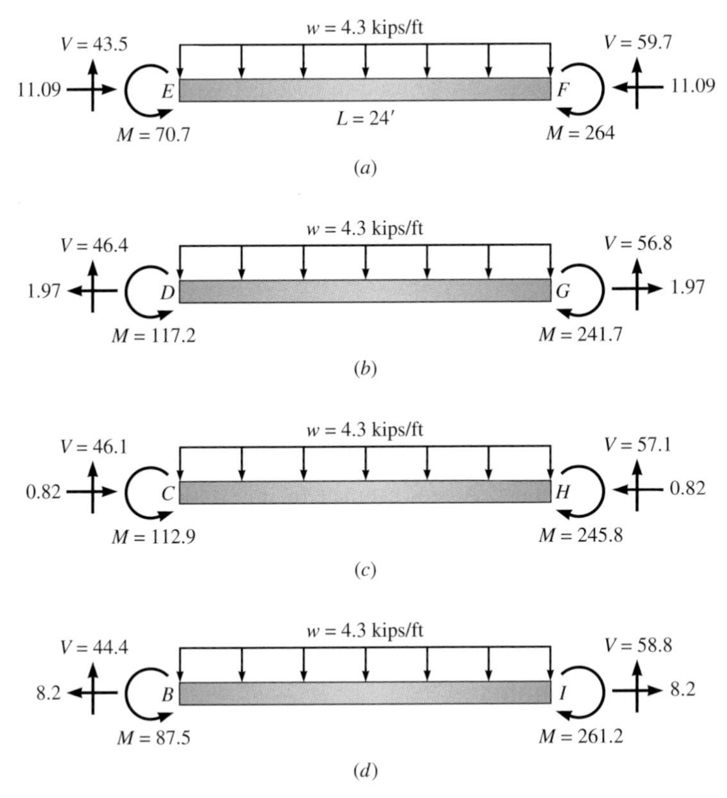

Figure 15.16: Free bodies of floor beams showing forces from an exact analysis: (*a*) roof; (*b*) fourth floor; (*c*) third floor; (*d*) second floor (load in kips per foot, forces in kips, and moments in kip · ft).

column only, the exterior joint—subjected to an unbalanced moment—will rotate in the clockwise direction. As the joint rotates, the moment in the left end of the beam reduces and the moment at the right end builds up due to the carryover moment. Therefore, the negative moment at the first interior support will always be larger than the fixed-end moment. For uniformly loaded beams the negative moment at the first interior support will usually range between $wL^2/9$ and $wL^2/10$. As the flexibility of the exterior column increases, the moments in the beam approach those shown in Figure 15.1*d*.

The moment of 70.7 kip·ft at the exterior end of the roof beam in Figure 15.16*a* is smaller than the exterior moment in the floor beams below because the roof beam is restrained by a single column at joint *E*, whereas the floor beams are restrained by two columns (i.e., one below and one above the floor). Two columns apply twice the rotational restraint of one column, assuming that they have the same dimensions and end conditions. The moment at joint *B* of the second floor beam in Figure 15.16*d* is smaller than that in the upper floor beams because the bottom column, which is pinned at its base and 15 ft long, is more flexible than the shorter columns in the upper floors that are bent in double curvature.

We also observe that the reactions and consequently the shear and moment curves of the beams on the third and fourth floors are approximately the same because they have identical spans and loadings and are supported by the same size columns. Therefore, if we design the beams for a typical floor, the same members can be used in all other typical floors. Since the dimensions of columns supporting the lower floors of tall buildings have larger cross sections than those in the upper floors where the column loads are smaller, their flexural stiffness is larger than that of the smaller columns. As a result, the exterior moment in the floor beams will increase as the stiffness of the columns increases. This effect, which is often moderate, is generally neglected in practice.

Estimating Values of End Shear in Beams

Because the end moments on the beams (in Figure 15.16) are greater on the right than on the left end, the end shears are not equal. The *difference* in end moments reduces the shear produced by the uniform load at the left end and increases it at the right end. A good estimate for all exterior beams (beams that connect to an exterior column) is to assume 45 percent of the total uniform load wL is carried to the exterior column and 55 percent to the interior column. If a beam spans between two interior columns, the shears are approximately equal at both ends (that is, $V = wL/2$).

Axial Loads in Beams

Although axial forces develop in all beams because of shear in the columns, the stresses produced by these forces are small and may be neglected. For example, the axial stress, which is greatest in the roof beams, produced by 11.09 kips (see Figure 15.16*a*) is about 37 psi.

Computation of Approximate Values of Shear and Moment in Floor Beams

The shears and moments that develop from gravity loads applied to the beams of a typical floor are due almost entirely to the loads acting directly on that floor. Therefore, we can estimate the moments in the floor beams closely by analyzing an individual floor instead of the entire building. To determine the shear and moment in a floor of the frame in Figure 15.15, we will analyze a frame composed of the floor beams and the attached columns. The frame used to analyze the roof beams is shown in Figure 15.17*a*. Figure 15.17*b* shows the frame used to analyze the beams of the third floor.

We normally assume that the ends of the columns are fixed at the point where they attach to the floors above or below the floor being analyzed (for example, this is the assumption specified in Section 8.9 of the American Concrete Institute Building Code). Since the rotation of the interior joints is small, this assumption appears reasonable. On the other hand, since the exterior joints at each floor level rotate in the same direction, the exterior columns are bent into double curvature (see Figure 15.18*c*). As we established in Figure 13.12*c*, the flexural stiffness of a member bent into double curvature is 50 percent greater than that of a member fixed at one end. As a result, the values of moment in the exterior columns from an approximate analysis of the frames in

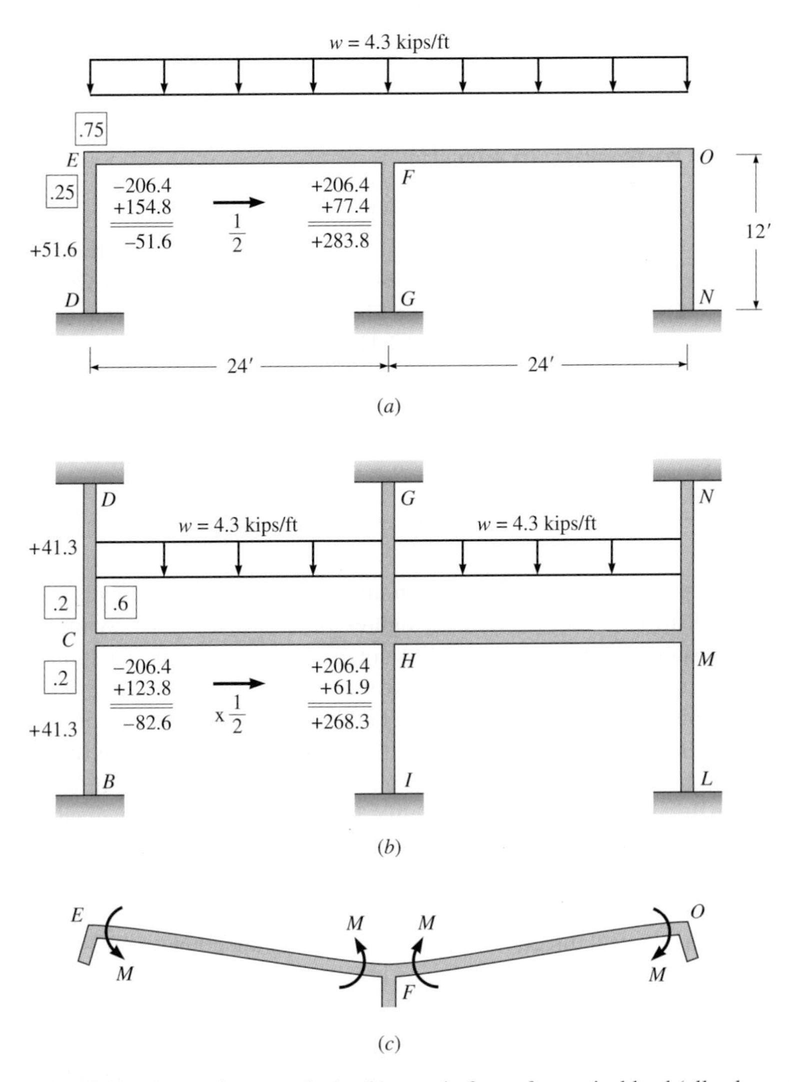

Figure 15.17: Approximate analysis of beams in frame for vertical load (all values of moment in kip · ft): (*a*) rigid frame composed of roof beams and attached columns; (*b*) rigid frame composed of floor beams and attached columns; (*c*) moments created by differential displacement of interior and exterior joints (these moments are not included in the approximate analysis).

Figure 15.17*a* and *b* will be much smaller than those produced by an analysis that considers the entire building frame, unless the engineer arbitrarily increases the stiffness of the exterior columns by a factor of 1.5.

Since building owners often want the exterior columns as small as possible for architectural reasons (small columns are easier to conceal in the exterior walls and simplify the wall details), the fixed-end assumption for columns is retained as the standard in the design of reinforced concrete buildings.

The analysis of the frames in Figure 15.17 is carried out by moment distribution. Since sidesway produced by gravity loads is either zero (if the structure and loading are symmetric) or very small in other cases, we neglect the

TABLE 15.1

Comparison Between Exact and Approximate Values of Girder End Moment (all moments in kip·ft)

Moment	Exact Analysis (Fig. 15.16)	Approximate Analysis	
		Ends of Columns Assumed Fixed (Fig. 15.17)	Double Curvature Bending, Exterior Column Stiffness Increased 50 Percent
M_{EF}	70.7	51.6	68.8
M_{FE}	264.0	283.6	275.2
M_{CH}	112.9	82.6	103.2
M_{HC}	245.8	268.3	258.0

moments produced by sidesway in an approximate analysis. Details of the moment distribution are shown in the figures. Since the structure is symmetric, we can assume that the center joint does not rotate and treat it as a fixed support. Therefore, only one-half of the frame has to be analyzed. The moments produced by analyzing the frames (see Table 15.1) compare closely to the more exact values of the computer analysis. If the stiffness of the exterior columns (excluding column *AB*, which is pin-ended) is increased by 50 percent, the difference between the exact and approximate values is on the order of 5 or 6 percent (see the last column in Table 15.1).

In the roof beams, most of the difference between the approximate and the exact values of moments is due to the differential displacement of the end joints in the vertical direction. The interior column undergoes a greater axial deformation than the exterior columns because it carries more than twice as much load but has an area that is only 44 percent greater. Figure 15.17*c* shows the deformation and the direction for the member end moments produced in the roof beams by the differential displacement of the ends of the beams. The effect—a function of the length of the column—is greatest in the top floor and diminishes toward the bottom of the column.

In the computer analysis the properties of the members (area and moment of inertia) are based on the gross area of the members' cross section (a standard assumption). If the influence of the reinforcing steel area on axial stiffness is considered by transforming the stiffer steel into equivalent concrete, the difference in axial deformations of the various columns would be largely eliminated. Since the moments induced in the beams by the differential axial deformations of the columns are typically small, they are neglected in an *approximate* analysis.

Axial Forces in Columns

Loads, applied to columns at each floor, are produced by the shears and moments at the ends of the beams. In Figure 15.18*a* the arrows at the end of

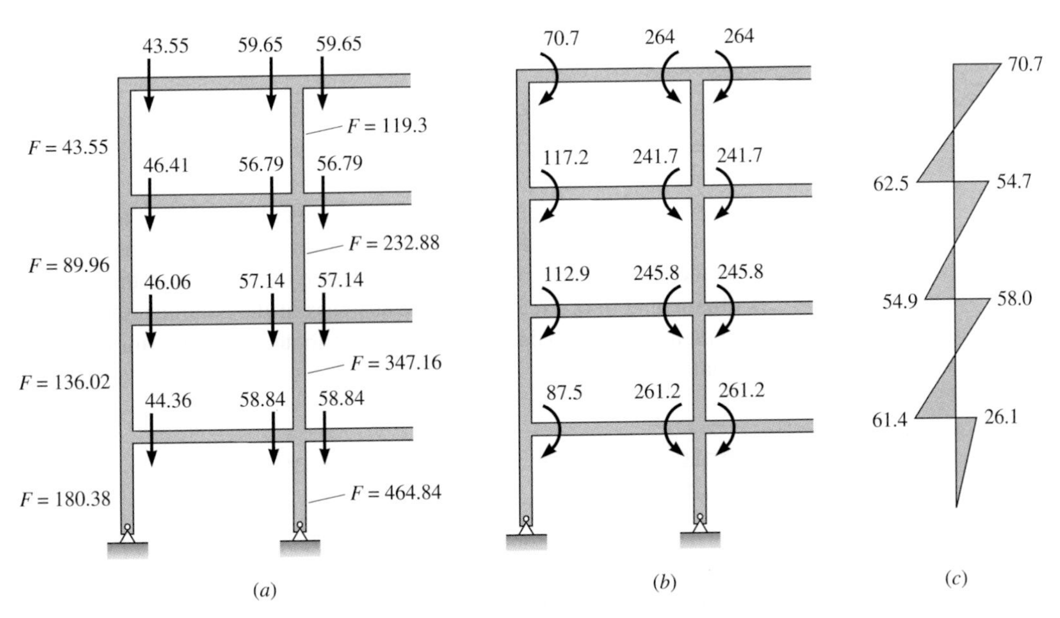

Figure 15.18: Results of computer analysis of frame in Figure 15.15: (*a*) axial force (kips) in columns created by reactions of beams supporting a uniformly distributed load of 4.3 kips/ft. (*b*) Moments (kip · ft) applied to columns by beams; this moment divides between top and bottom columns (*c*) Moment curve for exterior column (kip · ft). *Note:* Moments are not cumulative as the axial load is.

each beam indicate the force (end shears in the beams) applied to the column by the ends of the beam (the uniformly distributed load of 4.3 kips/ft acting on all beams is not shown on the figure for clarity). The axial force F in the column at any level is equal to the sum of the beam shears above that level. Since the axial force in columns varies with the number of floors supported, the column loads increase nearly linearly with the number of floors supported. Engineers often increase the size of the column's cross section or use higher-strength materials to carry the larger loads in the lower sections of multistory columns. Axial forces in interior columns, which carry the load from beams on each side, are typically more than twice as large as those in exterior columns—unless the weight of the exterior wall is large (see Figure 15.18*a*).

The moments applied by the ends of the beams to the columns in the building frame are shown in Figure 15.18*b*. Since the beams framing into the interior column are the same length and carry the same value of uniform load, they apply equal values of end moments to the column at an interior joint. Because the moments on each side of the column act in opposite directions, the joint does not rotate. As a result, no bending moments are created in the interior column. Therefore, when we make an approximate analysis of an interior column, we consider only the axial load. If we considered pattern loading of the live load (i.e., total load placed on the longer span and dead load on the shorter span framing into the sides of a column), moment would develop in the column, but the axial load would reduce. Even if the beams are not the same length or carry different values of load, the moments induced in an interior column will be small and typically can be neglected in an approximate analysis. Moments are small for the following reasons:

1. The unbalanced moment applied to the column equals the difference between the beam moments. Although the moments may be large, the difference in moments is usually small.
2. The unbalanced moment is distributed to the columns above and below the joint as well as to the beams on each side of the joint in proportion to the flexural stiffness of each member. Since the stiffness of the beams is often equal to or greater than the stiffness of the columns, the increment of the unbalanced moment distributed to an interior column is small.

Moments in Exterior Columns Produced by Gravity Loads

Figure 15.18*b* shows the moments applied by the girders at each floor to the interior and exterior columns. In the exterior columns these moments—resisted by the columns above and below each floor (except at the roof where only one column exists)—bend the column into double curvature, producing the moment curve shown in Figure 15.18*c*. From an examination of the moment curve, we can reach the following conclusions:

1. Moments do not build up in the lower floors.
2. All exterior columns (except the bottom column, which is pinned at the base) are bent into double curvature, and a point of contraflexure develops near *midheight* of the column.
3. The greatest moment develops at the top of the column supporting the roof beam because the entire moment at the end of the beam is applied to a single column. In the lower floors the moment applied by the beam to the joint is resisted by two columns.
4. The most highly stressed section in a column segment (between floors) occurs at either the top or the bottom; that is, the axial load is constant throughout the length of the column, but the maximum moment occurs at one of the ends.

EXAMPLE 15.10

Using an approximate analysis, estimate the axial forces and moments in columns *BG* and *HI* of the frame in Figure 15.19*a*. Also draw the shear and moment curves for beam *HG*. Assume that *I* of all exterior columns equals 833 in^4, *I* of interior columns equals 1728 in^4, and *I* of all girders equals 5000 in^4. Circled numbers represent column lines.

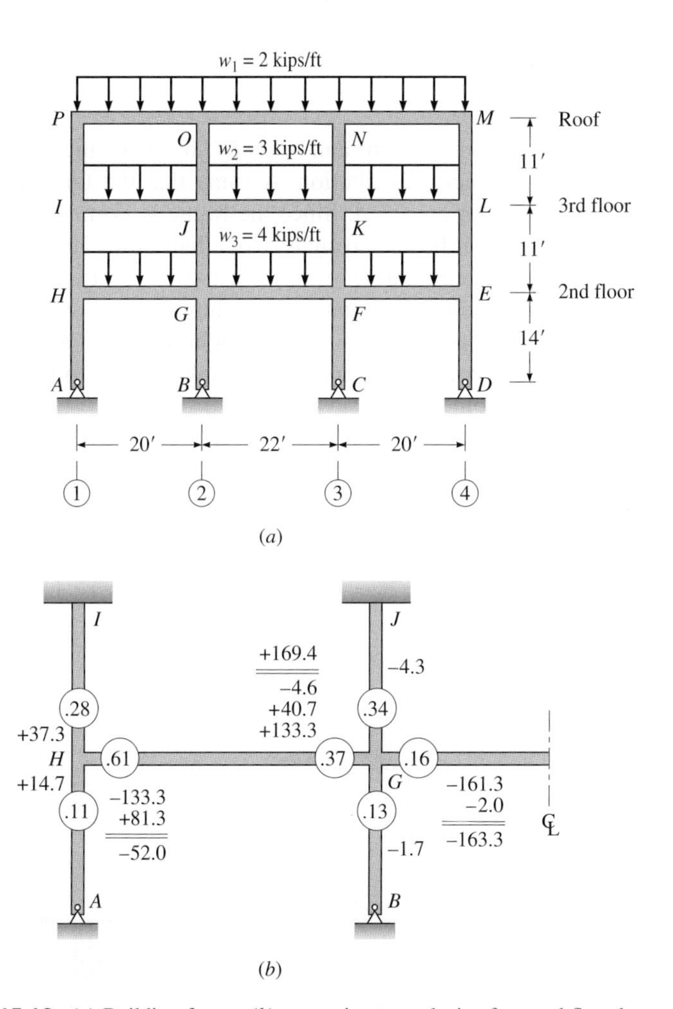

Figure 15.19: (*a*) Building frame; (*b*) approximate analysis of second floor by moment distribution to establish moments in beams and columns; only one cycle used because carryover moments small (moments in kip·ft).

Solution

Axial Load in Column HI Assume that 45 percent of the uniform load on beams *PO* and *IJ* is carried to the exterior column.

$$F_{HI} = 0.45(w_1L + w_2L) = 0.45[2(20) + 3(20)] = 45 \text{ kips} \qquad \textbf{Ans.}$$

[*continues on next page*]

Example 15.10 continues . . .

Axial Load in Column BG Assume that 55 percent of the load from exterior beams on the left side of the column and 50 percent of the load from the interior beams on the right side of the column are carried into the column.

$$F_{BG} = 0.55[2(20) + 3(20) + 4(20)] + 0.5[2(22) + 3(22) + 4(22)]$$

$$= 198 \text{ kips} \qquad \textbf{Ans.}$$

Compute the moments in columns and beam *HG* by analyzing the frame in Figure 15.19*b* by moment distribution. Assume that the far ends of the columns above the floor are fixed. Since the frame is symmetric, modify the stiffness of the center beam and analyze one-half of the structure. Also, increase the stiffness of column *HI* by 50 percent to account for double curvature bending. The results of the analysis are shown in Figure 15.20. Since the end moments are approximately the same at both ends of a column, the moment at the top of column *HI* may also be taken equal to the value of 37.3 kip · ft at the bottom.

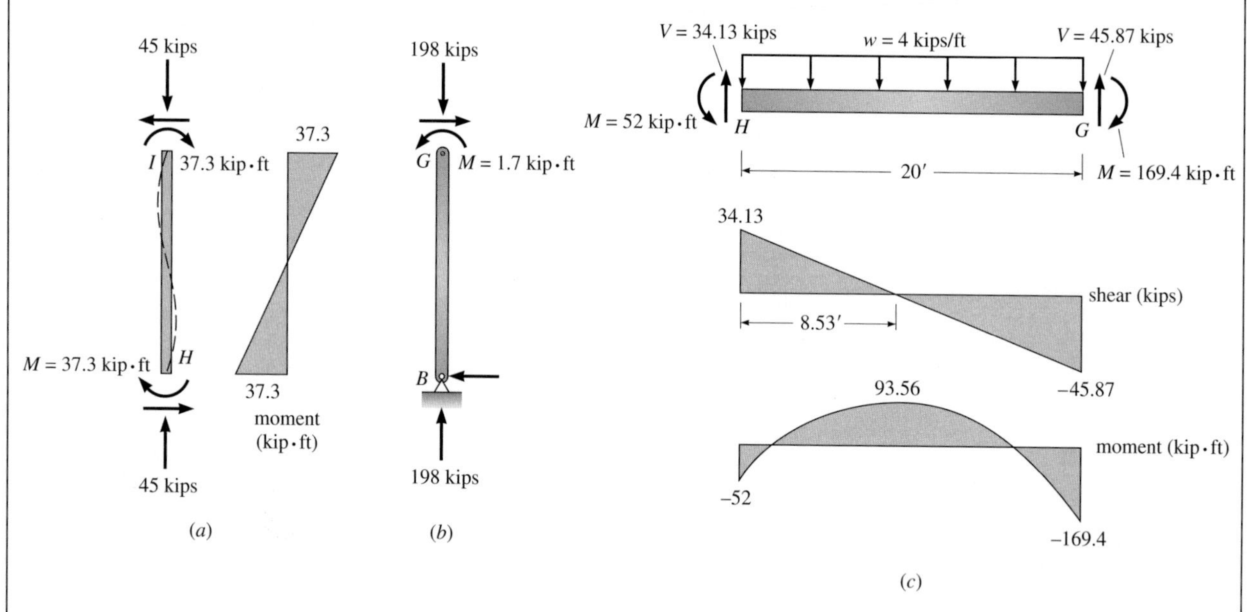

Figure 15.20: Results of approximate analysis of frame: (*a*) column *HI*; (*b*) column *BG*; (*c*) shear and moment curves for beam *HG*.

15.8 Analysis of Unbraced Frames for Lateral Load

Although we are primarily interested in approximate methods to analyze *multistory* unbraced frames with rigid joints, we begin our discussion with the analysis of a simple one-story rectangular unbraced frame. The analysis of this simple structure will (1) provide an understanding of how lateral forces stress and deform a rigid frame and (2) introduce the basic assumptions required for the approximate analysis of more complex multistory frames. Lateral loads on buildings are typically produced by either wind or inertia forces created by ground movements during an earthquake.

When gravity loads are much larger than lateral loads, designers initially size a building frame for gravity loads. The resulting frame is then checked for various combinations of gravity and lateral loads as specified by the governing building code.

As we have seen in Section 15.7, except for exterior columns, gravity loads produce mostly axial force in columns. Since columns carry axial load efficiently in direct stress, relatively small cross sections are able to support large values of axial load; moreover, designers tend to use compact column sections for architectural reasons. A compact section is easier to conceal in a building than a deep section. Since a compact section has a smaller bending stiffness than a deep section, the flexural stiffness of a column is often relatively small compared to its axial stiffness. As a result, small to moderate values of lateral load, which are resisted primarily by bending of the columns, produce significant lateral displacements of unbraced multistory frames. Therefore, as a general rule, knowledgeable engineers make every effort to avoid designing unbraced building frames that must resist lateral loads. Instead, they incorporate shear walls or diagonal bracing into the structural system to transmit lateral loads efficiently.

In Section 15.9 we describe procedures for evaluating the force produced by lateral loads in unbraced multistory building frames. These procedures include the *portal* and the *cantilever* methods. The portal method is considered best for low buildings (say five or six stories) in which shear is resisted by double curvature bending of the columns. For taller buildings the cantilever method, which considers that the building frame behaves as a vertical cantilever beam, produces the best results. Although both methods produce reasonable estimates of the forces in members of a building frame, neither method provides an estimate of the lateral deflections. Since lateral deflections can be large in tall buildings, a deflection computation should also be made as part of a complete design.

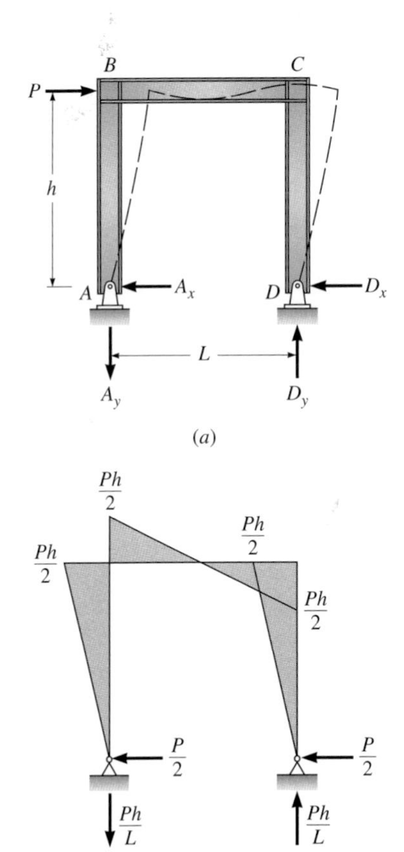

Figure 15.21: (*a*) Laterally loaded frame; (*b*) reactions and moment curves; point of inflection occurs at midspan of girder.

Approximate Analysis of a Simple Pin-Supported Frame

The rigid frame in Figure 15.21*a*, supported by pins at *A* and *D*, is indeterminate to the first degree. To analyze this structure, we must make one assumption about the distribution of forces. If the legs of the frame are identical, the flexural stiffness of both members is identical (both members also have the same end restraint). Since the lateral load divides in proportion to the flexural stiffness of the columns, we can assume that the lateral load divides equally between the columns, producing equal horizontal reactions of $P/2$ at the base. Once this assumption is made, the vertical reactions and the internal forces can be computed by statics. To compute the vertical reaction at D, we sum moments about A (Figure 15.21*a*):

$$\circlearrowright^{+} \quad \Sigma M_A = 0$$

$$Ph - D_y L = 0$$

$$D_y = \frac{Ph}{L} \uparrow$$

Compute A_y.

$$\overset{+}{\uparrow} \quad \Sigma F_y = 0$$

$$-A_y + D_y = 0 \quad \text{and} \quad A_y = D_y = \frac{Ph}{L} \downarrow$$

The moment curves for the members are shown in Figure 15.21*b*. Since the moment at midspan of the girder is zero, a point of inflection occurs there and the girder bends into double curvature. (The deflected shape is shown by the dashed line in Figure 15.21*a*.)

Approximate Analysis of a Frame Whose Columns Are Fixed at the Base

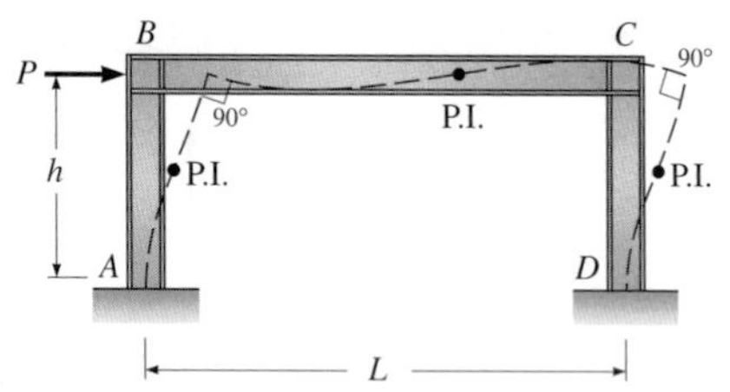

Figure 15.22: A laterally loaded rigid frame with fixed-end columns.

If the base of the columns in a rigid frame is fixed against rotation, the legs will bend in double curvature (see Figure 15.22). In the columns the position of the point of inflection depends on the ratio of the flexural stiffness of the girder to that of the column. The point of inflection will never be located below mid-height of the column, and even then this lower limit is theoretically possible only when the girder is infinitely stiff. As the girder stiffness reduces relative to the column stiffness, the point of inflection moves upward. For a typical frame the designer can assume the point of inflection is located a distance of approximately 60 percent of the column height above the base. In practice, a fixed support is difficult to construct because most foundations are not completely rigid. If the fixed support rotates, the point of inflection will rise.

Because the frame in Figure 15.22 is indeterminate to the third degree, we must make *three* assumptions about the distribution of the forces and the location of the points of inflection. Once these assumptions are made, the approximate magnitude of the reactions and the forces in the members can be computed by statics. If the columns are identical in size, we can assume the lateral load divides equally between the columns, producing horizontal reactions at the base (and shears in each column) equal to $P/2$. As we discussed previously, points of inflection in the columns may be assumed to develop at 0.6 of the column height above the base. Finally, although not actually required for a solution (if the first three assumptions are used), we can assume a point of inflection develops at midspan of the girder. These assumptions are used to analyze the frame in Example 15.11.

EXAMPLE 15.11

Estimate the reactions at the base of the frame in Figure 15.23*a* produced by the horizontal load of 4 kips at joint *B*. The column legs are identical.

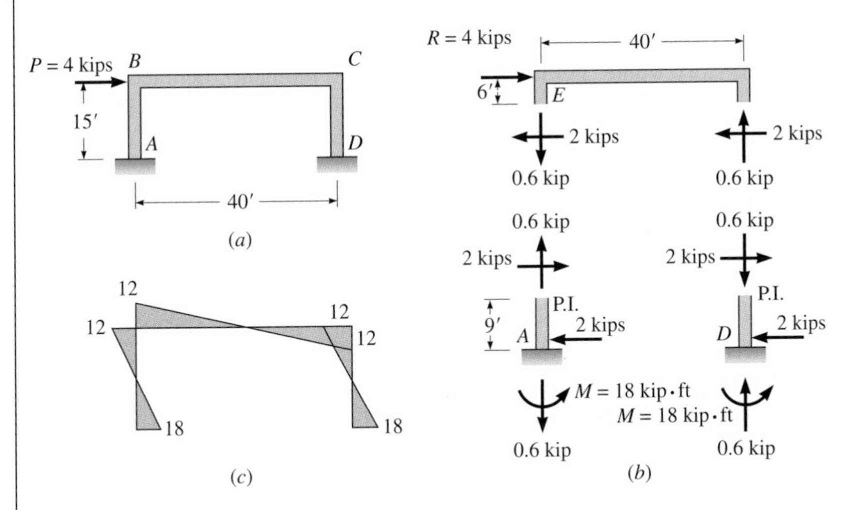

Figure 15.23: (*a*) Dimensions of frame; (*b*) free bodies above and below the points of inflections in the columns (forces in kips and moments in kip·ft); (*c*) moment diagram (kip·ft).

Solution

Assume that the 4-kip load divides equally between the two columns, producing shears of 2 kips in each column and horizontal reactions of 2 kips at *A* and *D*. Assume that the points of inflection (P.I.) in each column are located 0.6 of the column height, or 9 ft, above the base. Free bodies of the frame above and below the points of inflection are shown in Figure 15.23*b*. Considering the upper free body, we sum moments about the point of inflection in the left column (point *E*) to compute an axial force $F = 0.6$ kip in the column on the right. We next reverse the forces at the points of inflection on the upper free body and apply them to the lower column segments. We then use the equations of statics to compute the moments at the base.

$$M_A = M_D = (2 \text{ kips})(9 \text{ ft}) = 18 \text{ kip} \cdot \text{ft}$$

15.9 Portal Method

Under lateral load, the floors of multistory frames with rigid joints deflect horizontally as the beams and columns bend in double curvature. If we neglect the small axial deformations of the girders, we can assume all joints in a given floor deflect laterally the same distance. Figure 15.24 shows the deformations of a two-story frame. Points of inflection (zero moment), denoted by small dark circles, are located at or near the midpoints of all members. The figure also shows typical moment curves for both columns and girders (moments plotted on the compression side).

The portal method, a procedure for estimating forces in members of laterally loaded multistory frames, is based on the following three assumptions:

1. The shears in interior columns are twice as large as the shears in exterior columns.
2. A point of inflection occurs at midheight of each column.
3. A point of inflection occurs at midspan of each girder.

The first assumption recognizes that interior columns are usually larger than exterior columns because they support greater load. Interior columns typically support about twice as much floor area as exterior columns do. However, exterior columns also carry the load of exterior walls in addition to floor loads. If window areas are large, the weight of exterior walls is minimal. On the other hand, if exterior walls are constructed of heavy masonry and window areas are small, loads supported by the exterior columns may be similar in magnitude to those carried by the interior columns. Under these conditions, the designer may wish to modify the distribution of shear specified in assumption 1. The shear distributed to columns supporting a particular floor will be approximately proportional to their flexural stiffness (EI/L).

Since all columns supporting a given floor are the same length and presumably constructed of the same material, their flexural stiffness will be proportional to the moment of inertia of the cross section. Therefore, if the cross

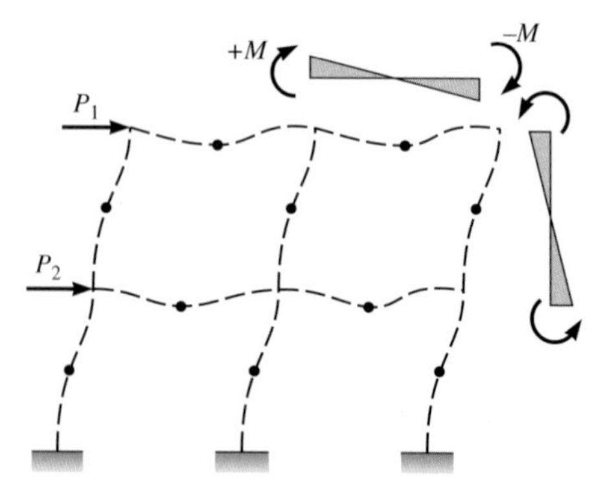

Figure 15.24: Deflected shape of rigid frame; points of inflection shown at center of all members by black dots.

sections of the columns can be estimated, the designer may wish to distribute the shears in proportion to the moments of inertia of the columns.

The second assumption recognizes that columns in lateral loaded frames bend in double curvature. Since the floors above and below a column are usually similar in size, they apply about the same restraint to the top and bottom ends of each column. Therefore, inflection points develop at or near midheight of columns.

If the columns in the bottom floor are connected to pins, the column bends in single curvature. For this case the point of inflection (zero moment) is at the base.

The final assumption recognizes that points of inflection occur at or near midspan of girders in laterally loaded frames. Since the shear is constant through the length, the girder bends in double curvature, and the moments at each end are of the same magnitude and act in the same sense. We have previously observed this behavior in the girders of Figures 15.21 and 15.22. The steps in the analysis of multistory rigid frame by the portal method are outlined below:

1. Pass an imaginary section between any two floors through the columns at their midheight. Since the section passes through the points of inflection of all columns, only shear and axial load act on the cut. The total shear distributed to all columns equals the sum of all lateral loads above the cut. Assume that the shear on interior columns is twice as large as the shear on exterior columns unless properties of the columns indicate that some other distribution of forces is more appropriate.
2. Compute the moments at the ends of the columns. The column end moments equal the product of the column shear and the half-story height.
3. Compute the moment at the end of the girders by considering equilibrium of the joints. Start with an exterior joint and proceed systematically across the floor, considering free bodies of the girders and joints. Since all girders are assumed to have a point of inflection at midspan, the moments at each end of a girder are equal and act in the same sense (clockwise or counterclockwise). At each joint the moments in the girders balance those in the columns.
4. Compute the shear in each girder by dividing the sum of the girder end moments by the span length.
5. Apply the girder shears to the adjacent joints and compute the axial force in the columns.
6. To analyze an entire frame, start at the top and work down. The procedure is illustrated in Example 15.12.

EXAMPLE 15.12

Analyze the frame in Figure 15.25*a*, using the portal method. Assume the reinforced baseplates at supports *A*, *B*, and *C* produce fixed ends.

Solution

Pass horizontal Section 1 (see number in circle) through the middle of the row of columns supporting the roof, and consider the upper free body shown in Figure 15.25*b*. Establish the shear in each column by equating the lateral load above the cut (3 kips at joint *L*) to the sum of the column shears. Let V_1 represent the shear in the exterior columns and $2V_1$ equal the shear in the interior column.

$$\xrightarrow{+} \quad \Sigma F_x = 0$$

$$3 - (V_1 + 2V_1 + V_1) = 0 \qquad \text{and} \qquad V_1 = 0.75 \text{ kip}$$

Compute moments at the tops of the columns by multiplying the shear forces at the points of inflection by 6 ft, the half-story height. Moments applied by the column to the upper joints are shown by curved arrows. The reaction of the joint on the column is equal and opposite.

Isolate joint *L* (see Figure 15.25*c*). Compute $F_{LK} = 2.25$ kips by summing forces in the *x* direction. Since the girder moment must be equal and opposite to the moment in the column for equilibrium, $M_{LK} = 4.5$ kip·ft. Both V_L and F_{LG} are calculated after the shear in girder *LK* is computed (see Figure 15.25*d*). Apply equal and oppositely directed values of F_{LK} and M_{LK} to the free body of the beam in Figure 15.25*d*. Since the shear is constant along the entire length and a point of inflection is assumed to be located at midspan, the moment M_{KL} at the right end of the girder equals 4.5 kip·ft and acts clockwise on the end of the girder. We observe that all end moments on all girders at all levels act in the same direction (clockwise). Compute the shear in the girder by summing moments about *K*.

$$V_L = \frac{\Sigma M}{L} = \frac{4.5 + 4.5}{24} = 0.375 \text{ kip}$$

Return to joint *L* (Figure 15.25*c*). Since the axial load in the column equals the shear in the girder, $F_{LG} = 0.375$ kip tension. Proceed to joint *K* (see Figure 15.25*e*) and use the equilibrium equations to evaluate all unknown forces acting on the joint. Isolate the next row of girders and columns between Sections 1 and 2 (see Figure 15.25*f*). Evaluate shears at points of inflection in the columns along section 2.

$$\xrightarrow{+} \quad \Sigma F_x = 0$$

$$3 + 5 - 4V_2 = 0$$

$$V_2 = 2 \text{ kips}$$

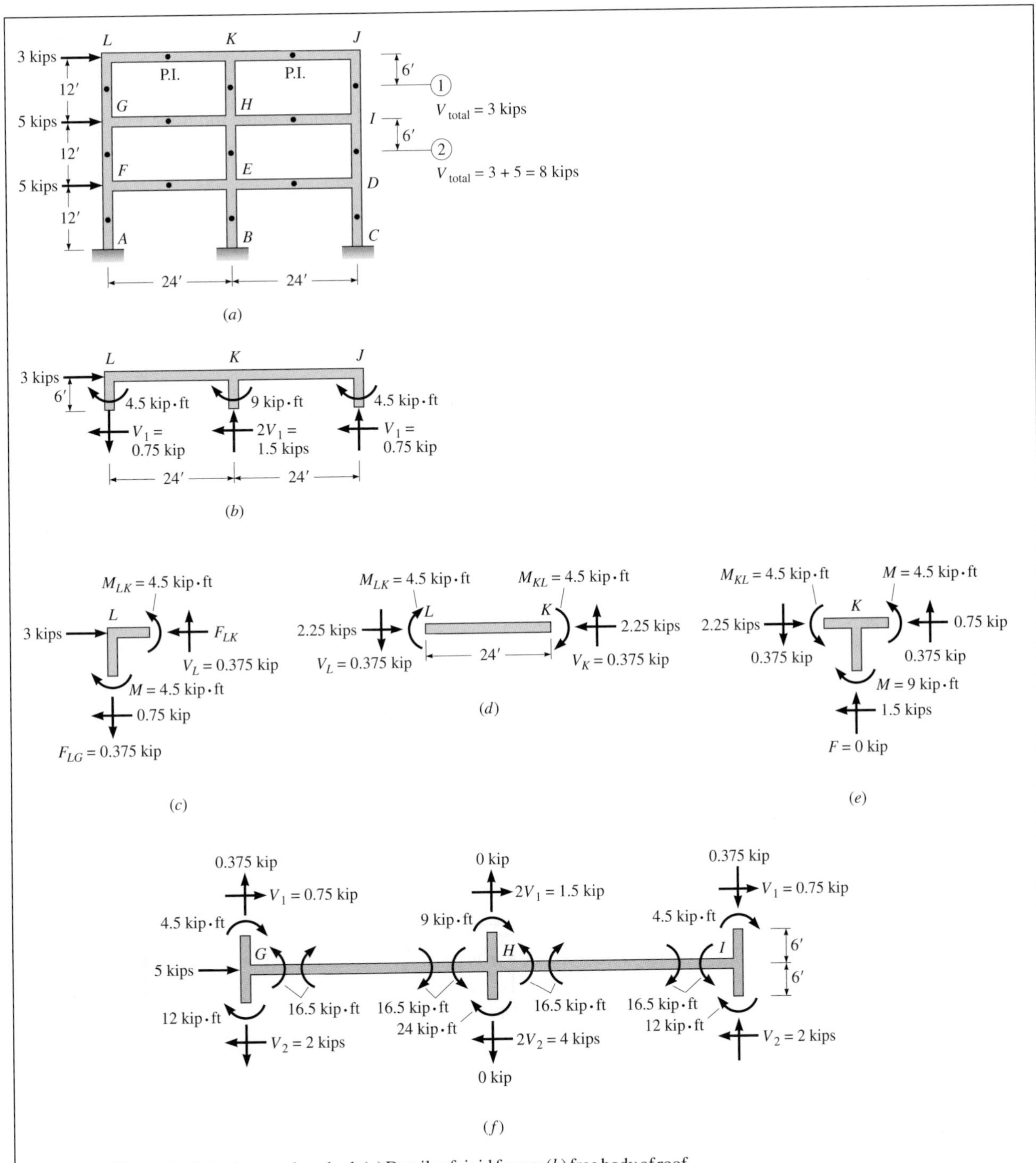

Figure 15.25: Analysis by the portal method. (*a*) Details of rigid frame; (*b*) free body of roof and columns cut by Section 1, which passes through points of inflection of columns; (*c*) free body of joint *L* (forces in kips and moments in kip·ft); (*d*) free body of girder *LK* used to compute shears in girders; (*e*) free body of joint *K*; (*f*) free body of floor and columns located between Sections 1 and 2 in (*a*) (moments in kip·ft).

[*continues on next page*]

Example 15.12 continues . . .

Evaluate moments applied to joints G, H, and I by multiplying the shear by the half-column length (see curved arrows). Starting with an exterior joint (G, for example), compute the forces in girders and axial loads in columns following the procedure previously used to analyze the top floor. Final values of shear, axial load, and moment are shown on the sketch of the building in Figure 15.26.

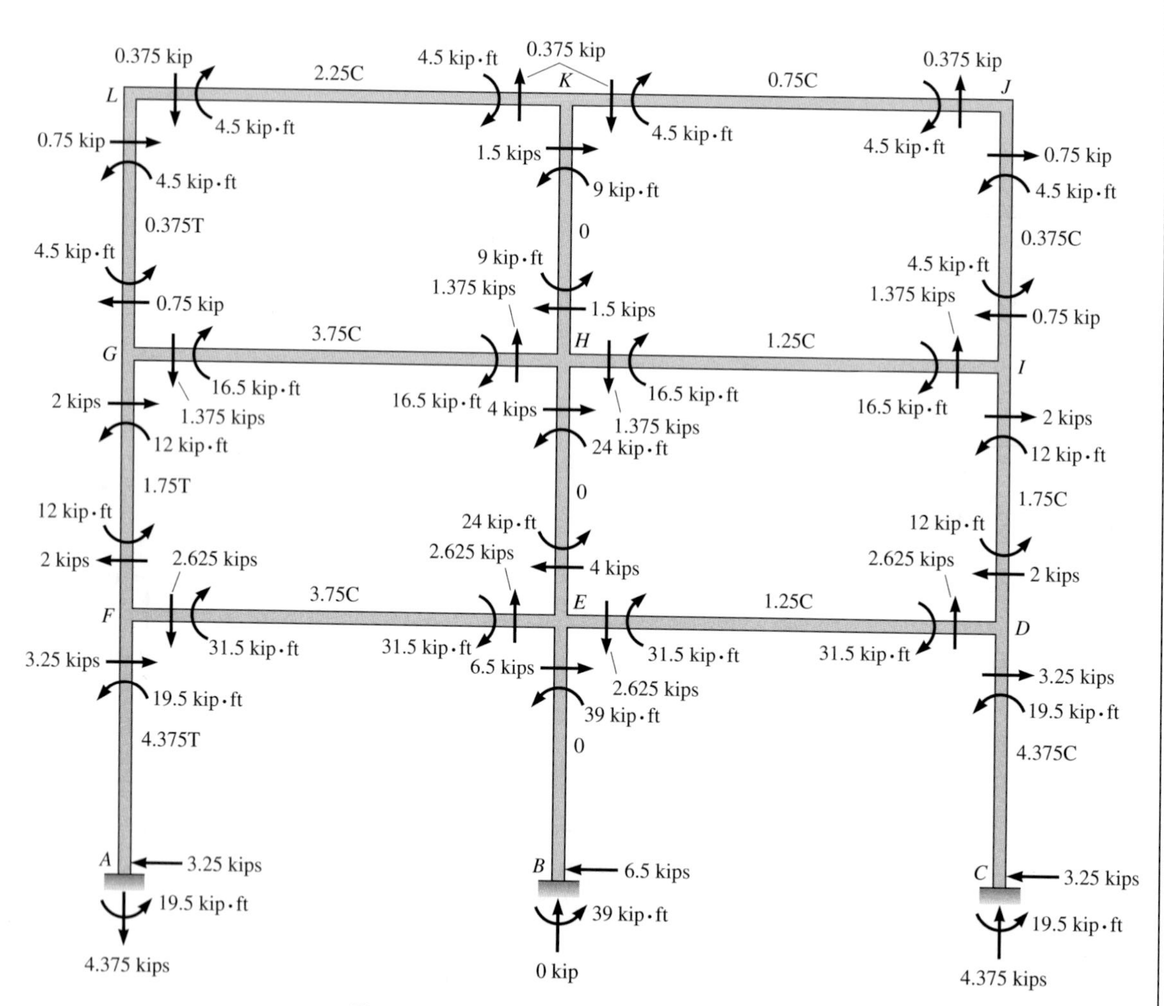

Figure 15.26: Summary of portal analysis. Arrows indicate the direction of the forces applied to the members by the joints. Reverse forces to show the action of members on joints. Axial forces are labeled with a C for compression and a T for tension. All forces in kips; all moments in kip · ft.

Analysis of a Vierendeel Truss

The portal method can also be used for an approximate analysis of a Vierendeel truss (see Figure 15.27*a*). In a Vierendeel truss the diagonals are omitted to provide a clear, open rectangular area between chords and verticals. When the diagonals are removed, a significant portion of truss action is lost (i.e., forces are no longer transmitted exclusively by creating axial forces in members). The shear force, which must be transmitted through the top and bottom chords, creates bending moments in these members. Since the main function of the vertical members is to supply a resisting moment at the joints to balance the sum of the moments applied by the chords, they are most heavily stressed.

For the analysis of the Vierendeel truss we assume that (1) the top and bottom chords are the same size, and therefore, shear divides equally between the chords; and (2) all members bend in double curvature, and a point of inflection develops at midspan. In the case of the symmetrically loaded, four-panel truss in Figure 15.27, no bending moments develop in the vertical member at midspan because it lies on the axis of symmetry. The deflected shape is shown in Figure 15.27*d*.

To analyze a Vierendeel truss by the portal method, we pass vertical sections through the center of each panel (through the points of inflection where $M = 0$). We then establish the shear and axial forces at the points of inflection. Once the forces at the points of inflection are known, all other forces can be computed by statics. Details of the analysis are illustrated in Example 15.13.

EXAMPLE 15.13

Carry out an approximate analysis of the Vierendeel truss in Figure 15.27, using the assumptions of the portal method.

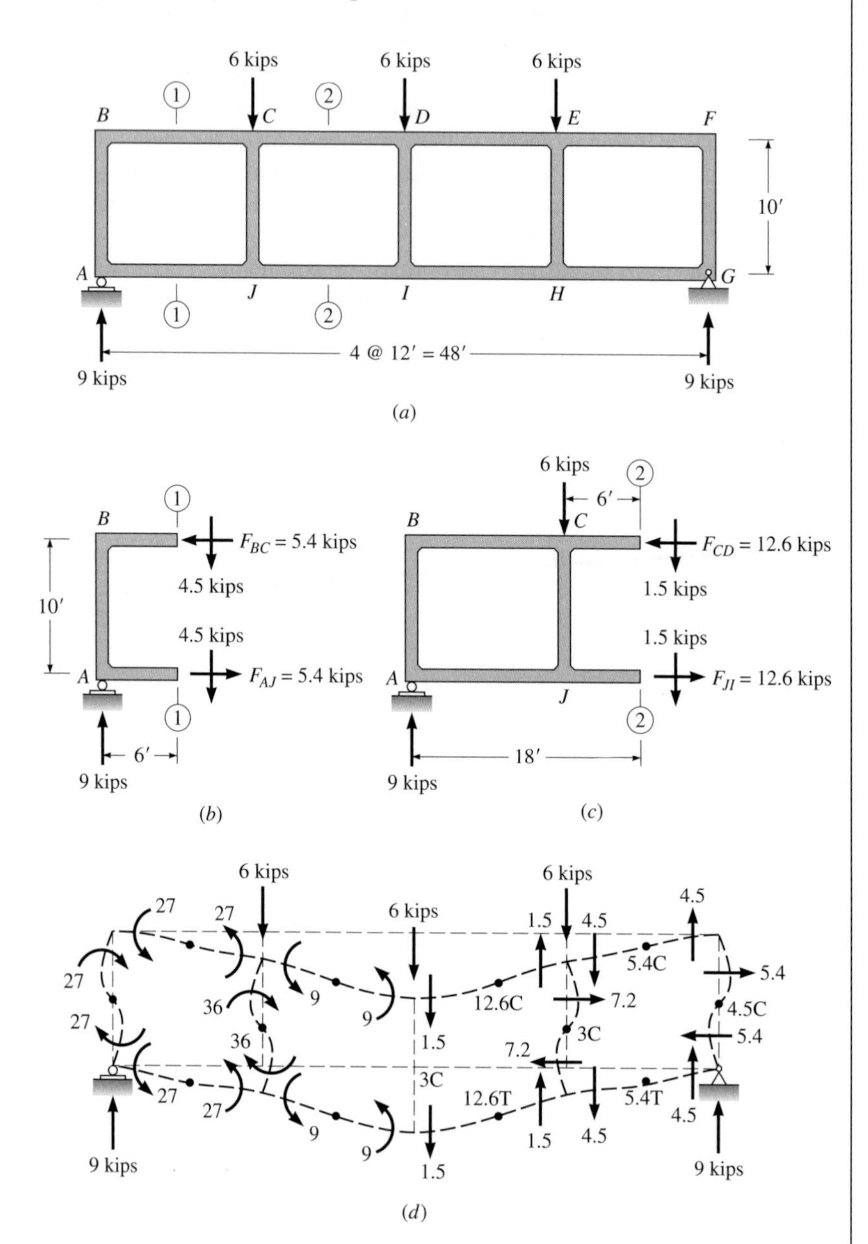

Figure 15.27: (*a*) Details of Vierendeel truss; (*b*) free body used to establish the forces at the points of inflection in the first panel; (*c*) free body to compute forces at points of inflection in second panel; (*d*) deflected shape: points of inflection denoted by black dots, moments acting on the ends of member indicated by curved arrows, shears and axial forces in kips, moments in kip · ft. Structure symmetric about centerline.

Solution

Since the structure is externally determinate, the reactions are computed by statics. Next, Section 1-1 is passed through the center of the first panel, producing the free body shown in Figure 15.27*b*. Because the section passes through the points of inflection in the chords, no moments act on the ends of the members at the cut. Assuming the shear is equal in each chord, equilibrium in the vertical direction requires that shear forces of 4.5 kips develop to balance the 9-kip reaction at support *A*. We next sum moments about an axis through the bottom point of inflection (at the intersection of Section 1-1 and the longitudinal axis of the bottom chord) to compute an axial force of 5.4 kips compression in the top chord:

$$\circlearrowleft^{+} \ \Sigma M = 0$$

$$9(6) - F_{BC}(10) = 0$$

$$F_{BC} = 5.4 \text{ kips}$$

Equilibrium in the x direction establishes that a tension force of 5.4 kips acts in the bottom chord.

To evaluate the internal forces at the points of inflection in the second panel, we cut the free body shown in Figure 15.27*c* by passing Section 2-2 through the midpoint of the second panel. As before, we divide the unbalanced shear of 3 kips between the two chords and compute the axial forces in the chords by summing moments about the bottom point of inflection:

$$\circlearrowleft^{+} \ \Sigma M = 0$$

$$9(18) - 6(6) - F_{CD}(10) = 0$$

$$F_{CD} = 12.6 \text{ kips}$$

The results of the analysis are shown on the sketch of the deflected shape in Figure 15.27*d*. The moments applied by the joints to the members are shown on the left half of the figure. The shears and axial forces are shown on the right half. Because of symmetry, forces are identical in corresponding members on either side of the centerline.

A study of the forces in the Vierendeel truss in Figure 15.27*d* indicates that the structure acts partially as a truss and partially as a beam. Since the moments in the chords are produced by the shear, they are greatest in the end panels where the shear has its maximum value, and the smallest in the panels at midspan where the minimum shear exists. On the other hand, because part of the moment produced by the applied loads is resisted by the axial forces in the chords, the axial force is maximum in the center panels where the moment produced by the panel loads is greatest.

15.10 Cantilever Method

The cantilever method, a second procedure for estimating forces in laterally loaded frames, is based on the *assumption that a building frame behaves as a cantilever beam.* In this method we assume that the cross section of the imaginary beam is composed of the cross-sectional areas of the columns. For example, in Figure 15.28*b* the cross section of the imaginary beam (cut by Section *A-A*) consists of the four areas A_1, A_2, A_3, and A_4. On any horizontal section through the frame, we assume that the longitudinal stresses in the columns—like those in a beam—vary linearly from the centroid of the cross section. The forces in the columns created by these stresses make up the internal couple that balances the overturning moment produced by the lateral loads. The cantilever method, like the portal method, assumes that points of inflection develop at the middle of all beams and columns.

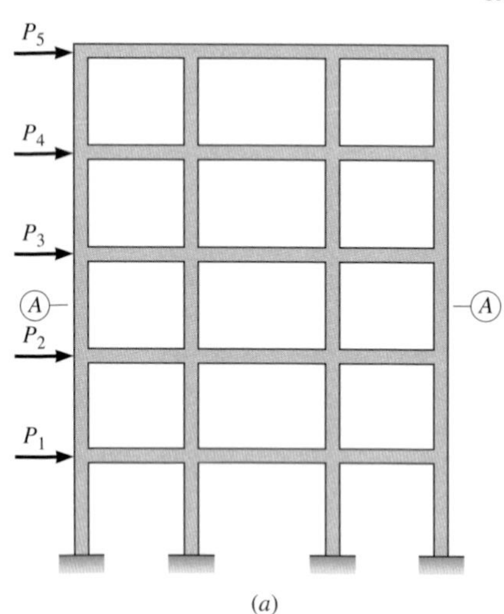

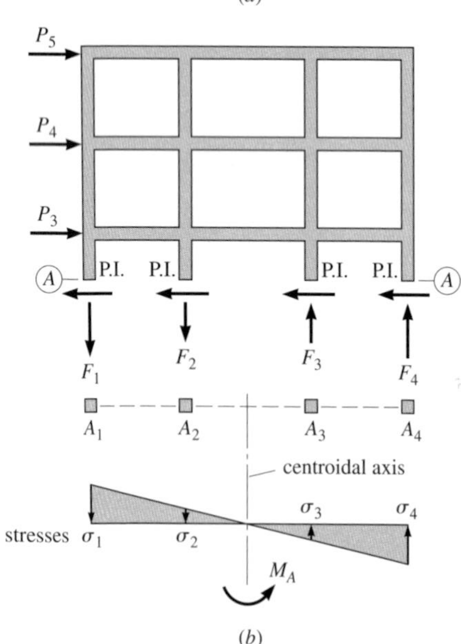

Figure 15.28: (*a*) Laterally loaded frame; (*b*) free body of frame cut by Section *A-A*, axial stresses in columns (σ_1 through σ_4) assumed to vary linearly from centroid of the four column areas.

To analyze a frame by the cantilever method, we carry out the following steps:

1. Cut free bodies of each story together with the upper and lower halves of the attached columns. The free bodies are cut by passing sections through the middle of the columns (midway between floors). Since the sections pass through the points of inflection, only axial and shear forces act on each column at that point.
2. Evaluate the axial force in each column at the points of inflection in a given story by equating the internal moments produced by the column forces to the moment produced by all lateral loads above the section.
3. Evaluate the shears in the girders by considering vertical equilibrium of the joints. The shear in the girders equals the difference in axial forces in the columns. Start at an exterior joint and proceed laterally across the frame.
4. Compute the moments in the girders. Since the shear is constant, the girder moment equals

$$M_G = V\left(\frac{L}{2}\right)$$

5. Evaluate the column moments by considering equilibrium of joints. Start with the exterior joints of the top floor and proceed downward.
6. Establish the shears in the columns by dividing the sum of the column moments by the length of the column.
7. Apply the column shears to the joints and compute the axial forces in the girders by considering equilibrium of forces in the *x* direction.

The details of the method are illustrated in Example 15.14.

EXAMPLE 15.14

Use the cantilever method to estimate the forces in the laterally loaded frame shown in Figure 15.29*a*. Assume that the area of the interior columns is twice as large as the area of the exterior columns.

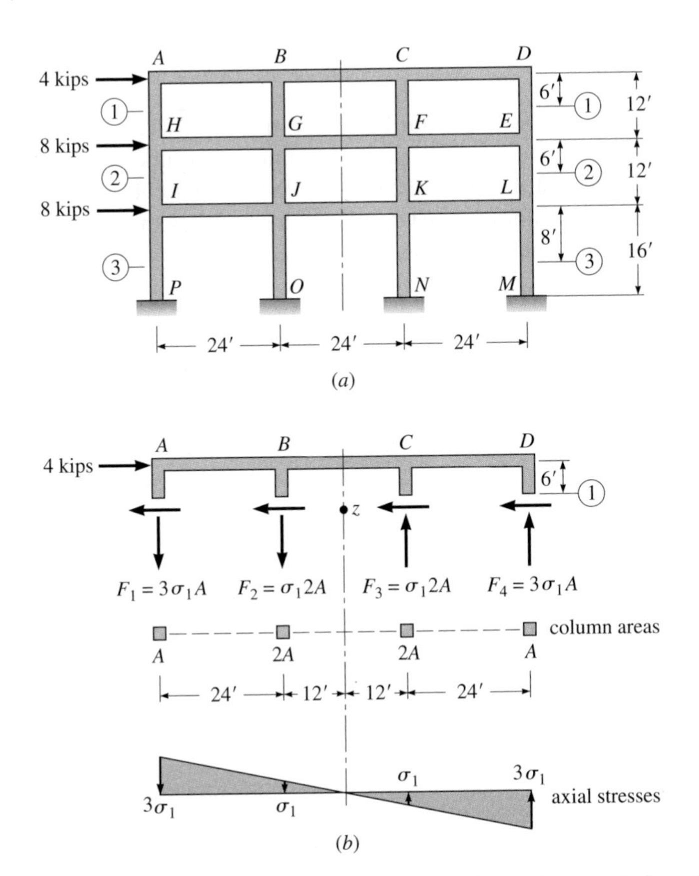

Figure 15.29: Analysis by the cantilever method: (*a*) continuous frame under lateral load; (*b*) free body of roof and attached columns cut by Section 1-1, axial stress in columns assumed to vary linearly with distance from centroid of column areas.

Solution

Establish the axial forces in the columns. Pass Section 1-1 through the frame at midheight of the upper floor columns. The free body above Section 1-1 is shown in Figure 15.29*b*. Since the cut passes through the points of inflection, only shear and axial force act on the ends of each column. Compute the moment on Section 1-1 produced by the external force of 4 kips at *A*. Sum moments about point *z* located at the intersection of the axis of symmetry and Section 1-1:

$$\text{External moment } M_{\text{ext}} = (4 \text{ kips})(6 \text{ ft}) = 24 \text{ kip} \cdot \text{ft} \qquad (1)$$

Compute the internal moment on Section 1-1 produced by axial forces in columns. The assumed variation of axial stress on the columns is shown in

[*continues on next page*]

Example 15.14 continues . . .

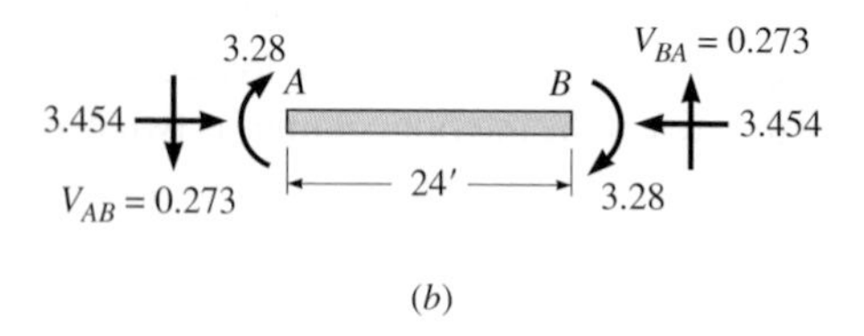

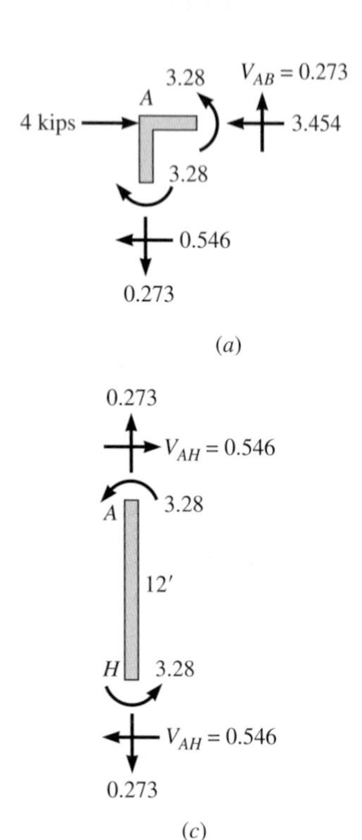

Figure 15.30: (*a*) Free body of joint *A* initially used to establish $V_{AB} = 0.273$ kip; (*b*) free body of beam *AB*, used to establish end moments in beam; (*c*) free body of column used to compute shear. All moments expressed in kip·ft and all forces in kips.

Figure 15.29*b*. We will arbitrarily denote the axial stress in the interior columns as σ_1. Since the stress in the columns is assumed to vary linearly from the centroid of the areas, the stress in the exterior columns equals $3\sigma_1$. To establish the axial force in each column, we multiply the area of each column by the indicated axial stress. Next, we compute the internal moment by summing moments of the axial forces in the columns about an axis passing through point *z*.

$$M_{\text{int}} = 36F_1 + 12F_2 + 12F_3 + 36F_4 \tag{2}$$

Expressing the forces in Equation 2 in terms of the stress σ_1 and the column areas, we can write

$$\begin{aligned} M_{\text{int}} &= 3\sigma_1 A(36) + 2\sigma_1 A(12) + 2\sigma_1 A(12) + 3\sigma_1 A(36) \\ &= 264\sigma_1 A \end{aligned} \tag{3}$$

Equating the external moment given by Equation 1 to the internal moment given by Equation 3, we find

$$24 = 264\sigma_1 A; \quad \sigma_1 A = \frac{1}{11}$$

Substituting the value of $\sigma_1 A$ into the expressions for column force gives

$$F_1 = F_4 = 3\sigma_1 A = \frac{3}{11} = 0.273 \text{ kip}$$

$$F_2 = F_3 = 2\sigma_1 A = \frac{2}{11} = 0.182 \text{ kip}$$

Compute the axial force in the second-floor columns. Pass Section 2-2 through the points of inflection of the second-floor columns, and consider the free body of the entire structure above the section. Compute the moment on Section 2-2 produced by the external loads.

$$M_{\text{ext}} = (4 \text{ kips})(12 + 6) + (8 \text{ kips})(6) = 120 \text{ kip}\cdot\text{ft} \tag{4}$$

Compute the internal moment on Section 2-2 produced by the axial forces in the columns. Since the variation of stress in the columns cut by Section 2-2 is the same as that along Section 1-1 (see Figure 15.29*b*), the internal moment at any section can be expressed by Equation 3. To indicate the stresses act on

Section 2-2, we will change the subscript on the stress to a 2. Equating the internal and external moments, we find

$$120 \text{ kip} \cdot \text{ft} = 264\sigma_2 A; \quad \sigma_2 A = \frac{5}{11}$$

Axial forces in columns are

$$F_1 = F_4 = 3\sigma_2 A = \frac{15}{11} = 1.364 \text{ kips}$$

$$F_2 = F_3 = 2\sigma_2 A = \frac{10}{11} = 0.91 \text{ kip}$$

To find the axial forces in the first-floor columns, pass Section 3-3 through the points of inflection, and consider the entire building above the section as a free body. Compute the moment on Section 3 produced by all external loads acting above the section.

$$M_{\text{ext}} = (4 \text{ kips})(32) + (8 \text{ kips})(20) + (8 \text{ kips})(8) = 352 \text{ kip} \cdot \text{ft}$$

Equate the external moment of 352 kip·ft to the internal moment given by Equation 3. To indicate the stresses act on Section 3-3, the symbol for stress in Equation 3 is subscripted with a 3.

$$264\sigma_3 A = 352; \quad \sigma_3 A = \frac{3}{4}$$

Compute the forces in the columns.

$$F_1 = F_4 = 3\sigma_3 A = 3\left(\frac{4}{3}\right) = 4 \text{ kips}$$

$$F_2 = F_3 = 2\sigma_3 A = 2\left(\frac{4}{3}\right) = 2.67 \text{ kips}$$

With the axial forces established in all columns, the balance of the forces in the members of the frame can be computed by applying the equations of static equilibrium to free bodies of joints, columns, and girders in sequence. To illustrate the procedure, we will describe the steps required to compute the forces in girder *AB* and column *AH*.

Compute the shear in girder *AB* by considering equilibrium of vertical forces applied to joint *A* (see Figure 15.30*a*).

$$\overset{+}{\uparrow} \quad \Sigma F_y = 0 \qquad 0 = -0.273 + V_{AB} \qquad V_{AB} = 0.273 \text{ kip}$$

Compute the end moments in girder *AB*. Since a point of inflection is assumed to exist at midspan, the end moments are equal in magnitude and act in the same sense.

$$M = V_{AB}\frac{L}{12} = 0.273(12) = 3.28 \text{ kip} \cdot \text{ft}$$

Apply the girder end moment to joint *A*, and sum moments to establish that the moment at the top of the column equals 3.28 kip·ft (the moment at the bottom of the column has the same value).

[*continues on next page*]

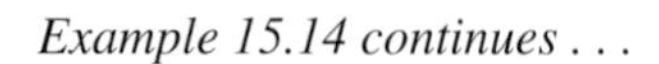

Example 15.14 continues . . .

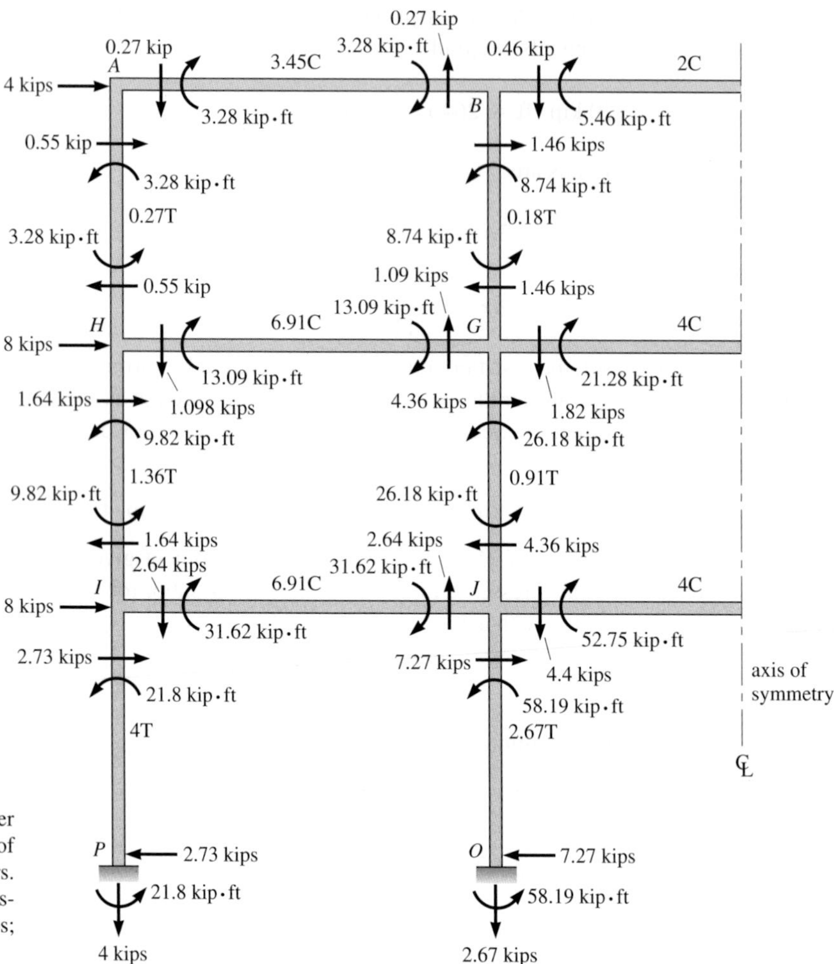

Figure 15.31: Summary of cantilever analysis. Arrows indicate the direction of the forces acting on the ends of members. Axial forces labeled with a C for compression and a T for tension. All forces in kips; all moment in kip·ft.

Compute the shear in column *AH*. Since a point of inflection is assumed to occur at the center of the column, the shear in the column equals

$$V_{AH} = \frac{M}{L/2} = \frac{3.28}{6} = 0.547 \text{ kip}$$

To compute the axial force in the girder *AB*, we apply the value of column shear from above to joint *A*. The equilibrium of forces in the *x* direction establishes that the axial force in the girder equals the difference between 4 kips and the shear in column *AH*.

The final values of force—applied by the joints to the members—are summarized in Figure 15.31. Because of symmetry of structure and antisymmetry of load, shears and moments at corresponding points on either side of the vertical axis of symmetry must be equal. The small differences that occur in the value of forces—that should be equal—are due to roundoff error.

Summary

- Since it is difficult to avoid mistakes when analyzing highly indeterminate structures with many joints and members, designers typically check the results of a computer analysis (or occasionally the result of an analysis by one of the classical methods previously discussed) by making an approximate analysis. In addition, during the initial design phase when the member proportions are established, designers use an approximate analysis to estimate the design forces to enable them to select the initial proportions of members.
- This chapter covers several of the most common methods used to make an approximate analysis. As designers acquire a greater understanding of structural behavior, they will be able to estimate forces within 10 to 15 percent of the exact values in most structures by using a few simple computations.
- A simple procedure to analyze a continuous structure is to estimate the location of the points of inflection (where the moment is zero) in a particular span. This permits the designer to cut out a free-body diagram that is statically determinate. To help locate points of inflection (where the curvature changes from concave up to concave down), the designer can sketch the deflected shape.
- Force in the chords and the diagonal and vertical members of continuous trusses can be estimated by treating the truss as a continuous beam. After the shear and moment diagrams are constructed, the chord forces can be estimated by dividing the moment at a given section by the depth of the truss. Vertical components of forces in diagonal members are assumed to be equal to the shear in the beam at the same section.
- The classical methods for approximate analysis of multistory frames for lateral wind loads or earthquake forces by the portal and cantilever methods are presented in Sections 15.9 and 15.10.

PROBLEMS

P15.1. Use an approximate analysis (assume the location of a point of inflection) to estimate the moment in the beam at support *B* (Figure P15.1). Draw the shear and moment curves for the beam. Check results by moment distribution. *EI* is constant.

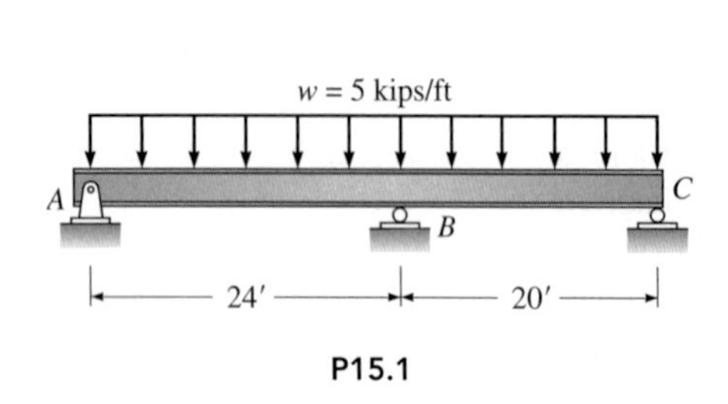

P15.1

P15.2. Guess the location of the points of inflection in each span in Figure P15.2. Compute the values of moment at supports *B* and *C*, and draw the shear and moment curves. *EI* is constant.

Case 1: $L_1 = 3$ m
Case 2: $L_1 = 12$ m

Check your results by using moment distribution.

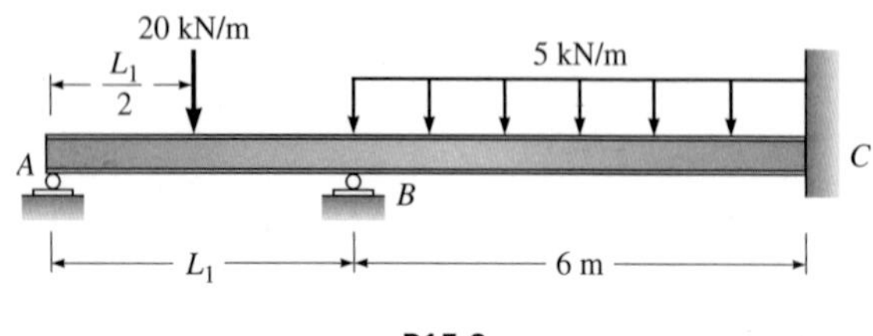

P15.2

P15.3. Assume values for member end moments and compute all reactions in Figure P15.3 based on your assumption. Given: *EI* is constant. If $I_{BC} = 8I_{AB}$, how would you adjust your assumptions of member end moments?

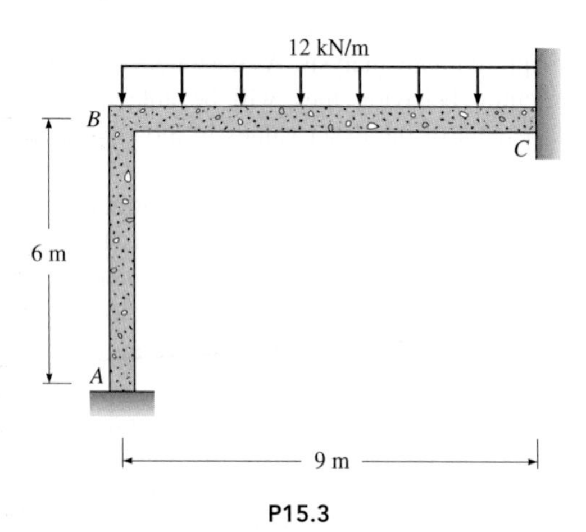

P15.3

P15.4. Assuming the location of the point of inflection in the girder in Figure P15.4, estimate the moment at *B*. Then compute the reactions at *A* and *C*. Given: *EI* is constant.

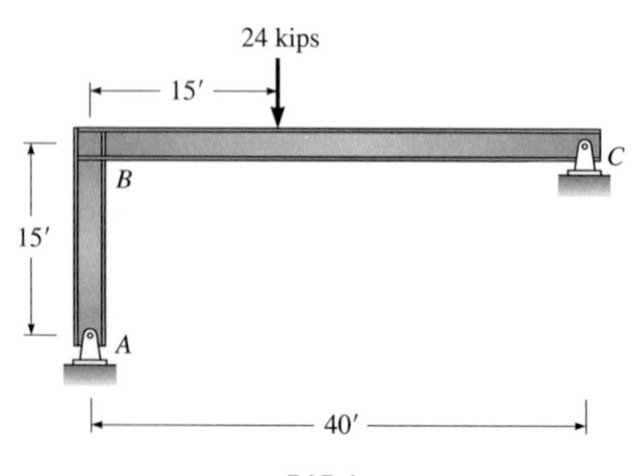

P15.4

P15.5. Estimate the moment in the beam in Figure P15.5 at support *C* and the maximum positive moment in span *CD* by guessing the location of one of the points of inflection in span *CD*. Check the results by moment distribution. *EI* is constant.

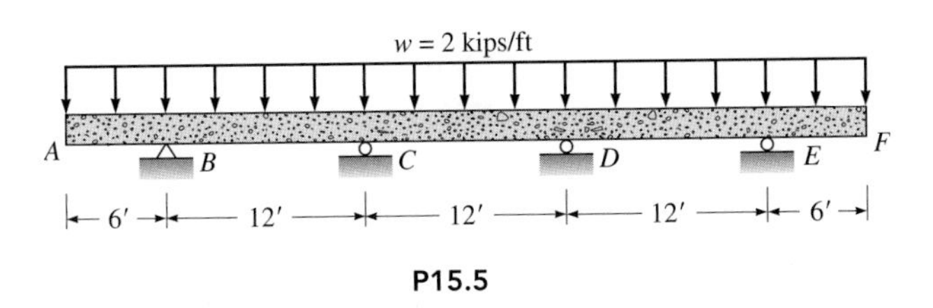

P15.5

P15.6. Estimate the moment at support *C* in Figure P15.6. Based on your estimate, compute the reactions at *B* and *C*.

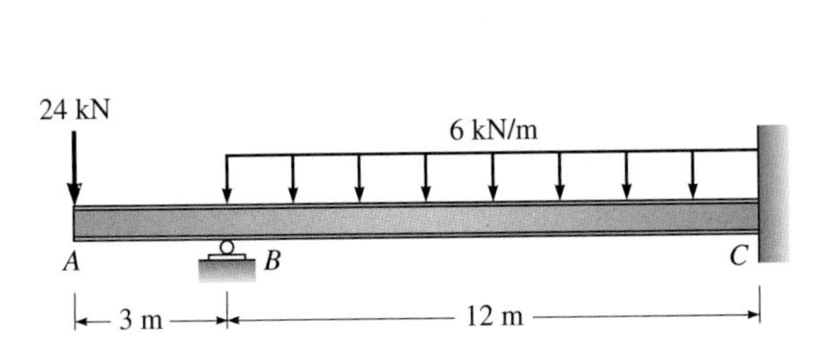

P15.6

P15.7. The beam is indeterminate to the second degree. Assume the location of the minimum number of points of inflection required to analyze the beam. Compute all reactions and draw the shear and moment diagrams. Check the results using moment distribution.

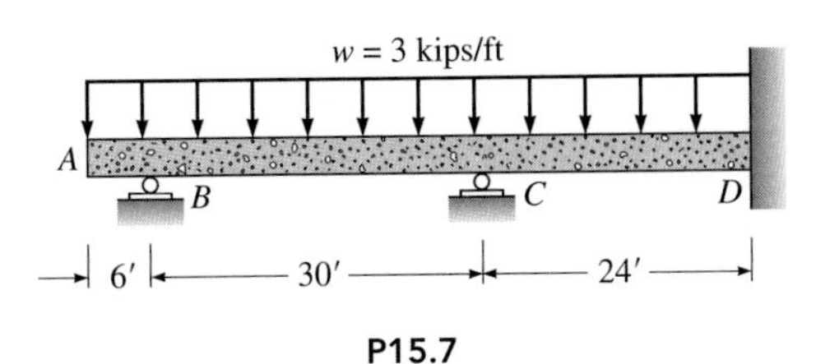

P15.7

P15.8. The frame in Figure P15.8 is to be constructed with a deep girder to limit deflections. However, to satisfy architectural requirements, the depth of the columns will be as small as possible. Assuming that the moments at the ends of the girder are 25 percent of the fixed-ended moments, compute the reactions and draw the moment curve for the girder.

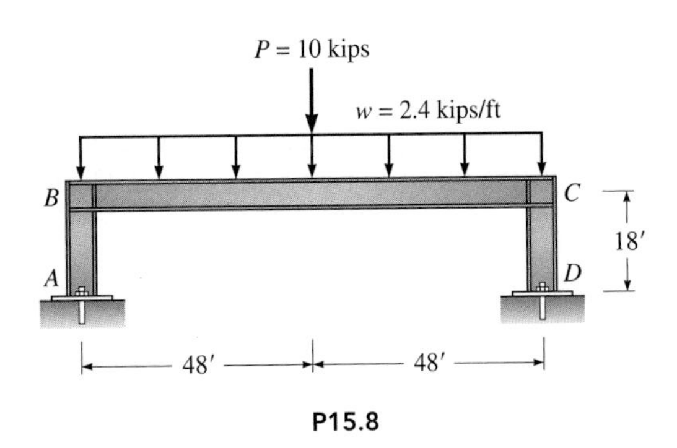

P15.8

P15.9. The cross sections of the columns and girder of the frame in Figure P15.9 are identical. Carry out an approximate analysis of the frame by estimating the location of the points of inflection in the girder. The analysis is to include evaluating the support reactions and drawing the moment curves for column *AB* and girder *BC*. Check the results by moment distribution. *EI* is constant.

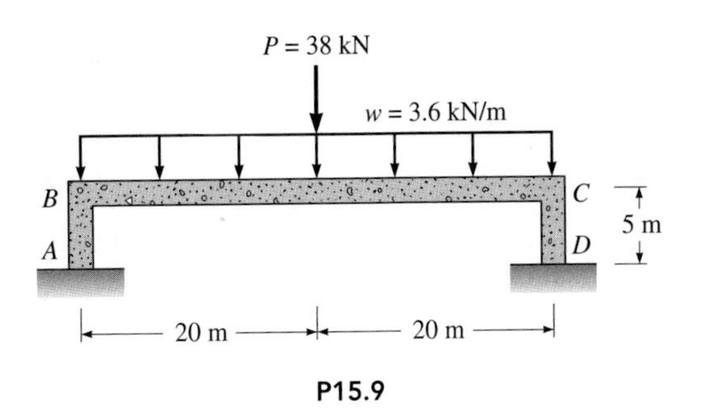

P15.9

P15.10. Carry out an approximate analysis of the truss in Figure P15.10 by treating it as a continuous beam of constant cross section. As part of the analysis, evaluate the forces in members DE and EF and compute the reactions at A and K.

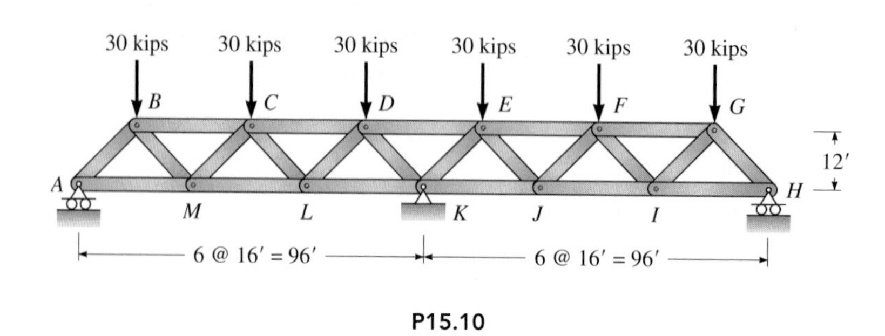

P15.10

P15.11. Use an approximate analysis of the continuous truss in Figure P15.11 to determine the reactions at A and B. Also evaluate the forces in bars a, b, c, and d. Given: $P = 9$ kN.

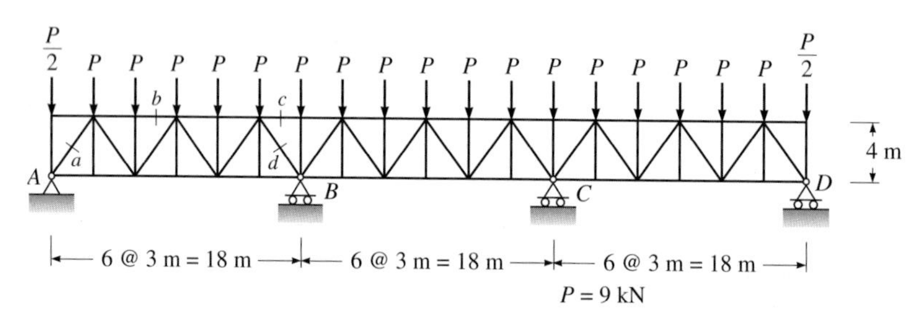

P15.11

P15.12. Estimate the deflection at midspan of the truss in Figure P15.12, treating it as a beam of constant cross section. The area of both the top and bottom chords is 10 in^2. $E = 29{,}000$ kips/in^2. The distance between the centroids of the top and bottom chords equals 9 ft.

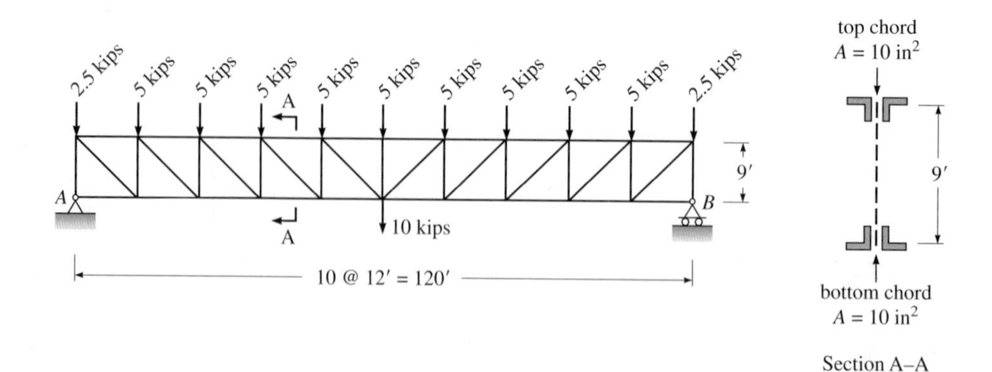

P15.12

P15.13. Determine the approximate values of force in each member of the truss in Figure P15.13. Assume that the diagonals can carry either tension or compression.

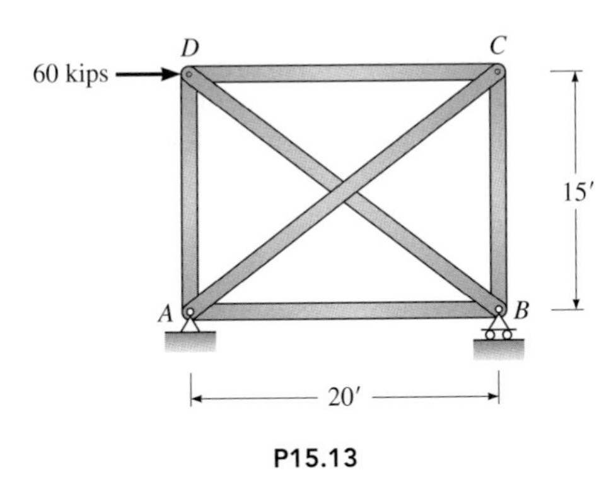

P15.13

P15.14. Determine the approximate values of bar force in the members of the truss in Figure P15.14 for the following two cases.

(*a*) Diagonal bars are slender and can carry only tension.

(*b*) Diagonal bars do not buckle and may carry either tension or compression.

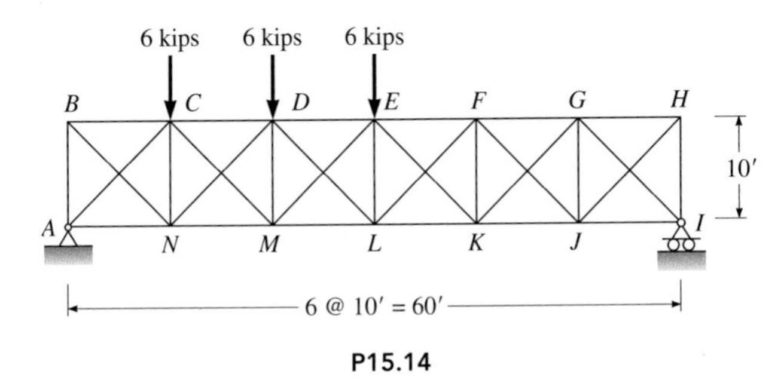

P15.14

P15.15. (*a*) All beams of the frame in Figure P15.15 have the same cross section and carry a uniformly distributed gravity load of 3.6 kips/ft. Estimate the approximate value of axial load and the moment at the top of columns *AH* and *BG*. Also estimate the shear and moment at each end of beams *IJ* and *JK*. (*b*) Assuming that all columns are 12 in square ($I = 1728\text{ in}^4$) and the moment of inertia of all girders equals $12{,}000\text{ in}^4$, carry out an approximate analysis of the second floor by analyzing the second-floor beams and the attached columns (above and below) as a rigid frame.

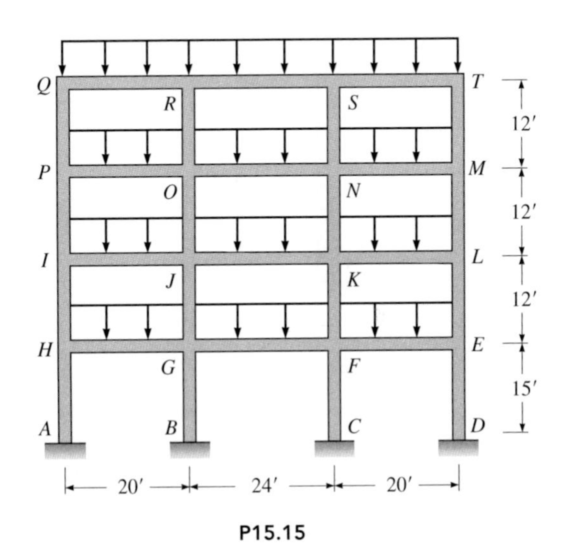

P15.15

P15.16. Using an approximate analysis of the Vierendeel truss in Figure P15.16, determine the moments and axial forces acting on free bodies of members *AB*, *BC*, *IB*, and *HC*.

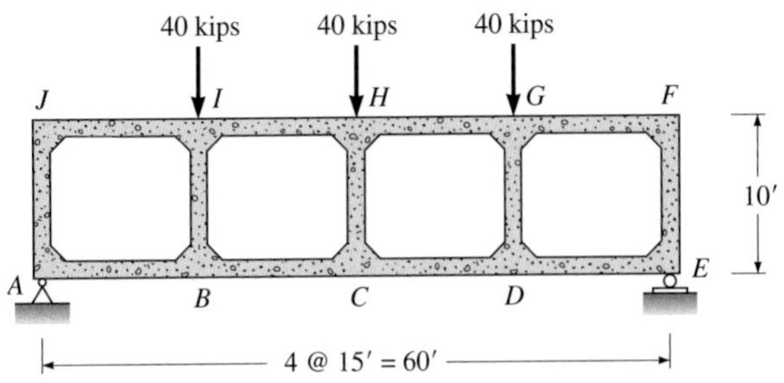

P15.16

P15.17. *Computer Study – comparison of cantilever and portal methods with an exact analysis.* (*a*) Determine the moments, shear, and axial forces in the members of the frame in Figure P15.17 using the portal method. (*b*) Repeat the analysis by using the cantilever method. (*c*) Perform an exact analysis using a computer software. (*d*) Prepare a summary table and compare the results of all three methods. Use $E = 6{,}240$ kips/in^2 and $I = 18{,}000$ in^4 for all members.

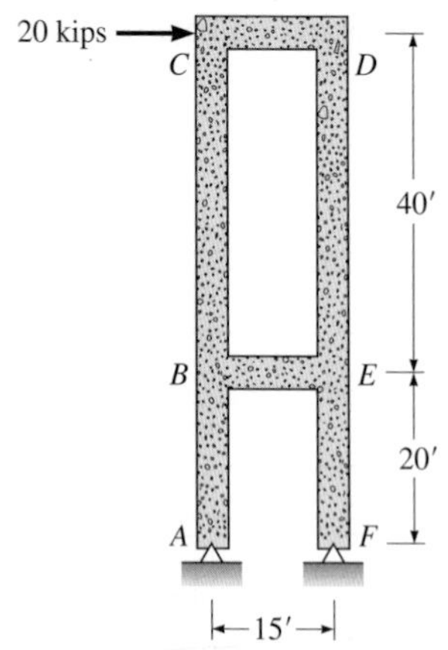

P15.17

P15.18. *Computer Study – comparison of cantilever and portal methods with an exact analysis.* (*a*) Determine the moments, shear, and axial forces in the members of the frame in Figure P15.18, using the portal method. (*b*) Repeat the analysis using the cantilever method. Assume the area of the interior columns is twice the area of the exterior columns. (*c*) Compare the results with an exact analysis using a computer software. Use $E = 29{,}000$ kips/in^2 for all members; $A = 12$ in^2 and $I = 600$ in^4 for beams and interior columns; $A = 8$ in^2 and $I = 400$ in^4 for exterior columns.

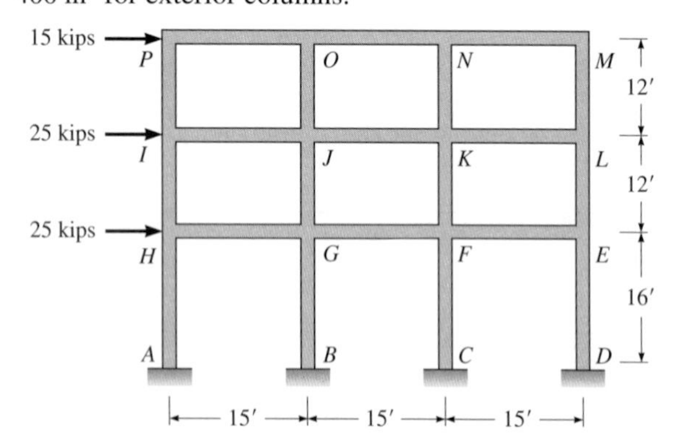

P15.18

P15.19. *Computer Study – comparison of cantilever and portal methods with an exact analysis.* (*a*) Analyze the two-story frame in Figure P15.19 by the portal method. (*b*) Repeat the analysis using the cantilever method. Assume the area of the interior columns is twice the area of the exterior columns. Assume the baseplates connecting all columns to the foundations can be treated as a pin support. (*c*) Compare the results with an exact analysis using a computer software. Use $E = 200$ GPa for all members; for beams and interior columns use $A = 10{,}000$ mm^2 and $I = 50 \times 10^6$ mm^4; for exterior columns use $A = 5{,}000$ mm^2 and $I = 25 \times 10^6$ mm^4.

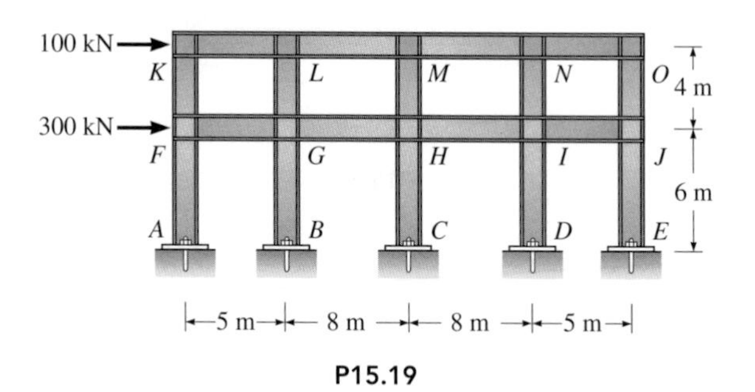

P15.19

P15.20. *Computer Study – comparison of approximate analysis with an exact analysis.* (*a*) Use approximate analysis to compute the reactions and draw the moment diagrams for column *AB* and girder *BC* of the frame in Figure P15.20, and draw the deflected shape. Consider column bases are fixed. (*b*) Repeat the computations with pinned column bases at *A* and *D*. (*c*) Compare the results with the exact analysis using a computer software. Consider columns $A = 13.1$ in^2, $I = 348$ in^4; Girder $A = 16.2$ in^2, $I = 1350$ in^4. Use $E = 29{,}000$ kips/in^2 for all members.

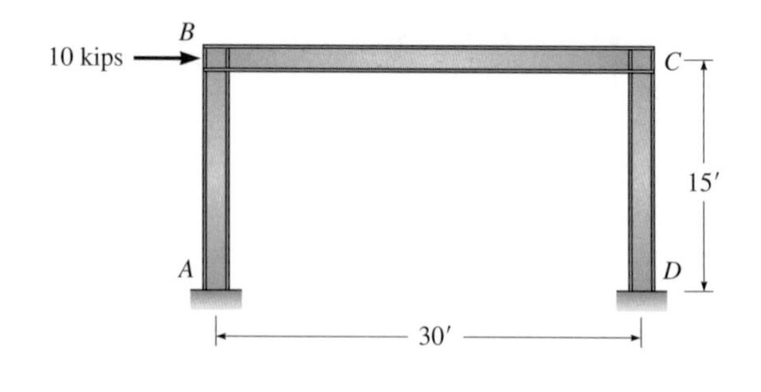

P15.20

P15.21. and **P15.22.** *Computer Study – comparison of approximate analysis with exact analysis.* Consider the structures in Figures P15.21 and P15.22, respectively. (*a*) Use approximate analysis to compute the reactions and draw moment diagram for the column *AB* and draw the approximate deflected shape of the frame. (*b*) Determine truss bar forces. All truss joints are pinned. (*c*) Compare the results with the exact analysis using a computer software. Truss member properties are $A = 4$ in^2, columns $A = 13.1$ in^2, $I = 348$ in^4 and $E = 29{,}000$ kips/in^2 for all members.

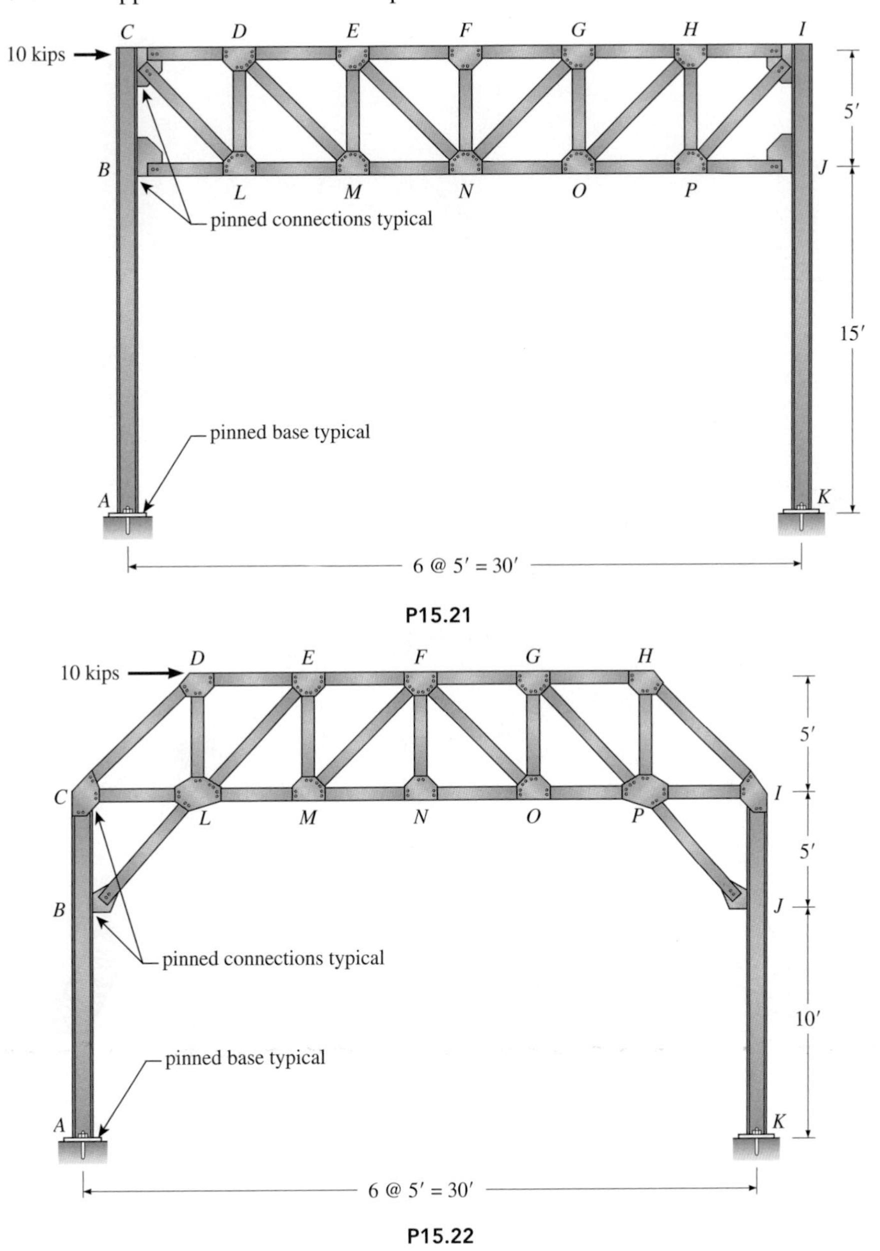

P15.21

P15.22

P15.23. Compare the results of Problem P15.20(*a*) and (*c*) with Problem P15.21(*a*) and (*c*).

P15.24. Compare the results of Problem 15.21(*a*) and (*c*) with Problem 15.22(*a*) and (*c*).

Space truss support for a radar antenna

A three-dimensional space truss used to support a 150-ft diameter radar antenna was under erection. A computer program using a matrix formulation was used to analyze this complex structure for a variety of static and dynamic loading conditions.

CHAPTER

16

Introduction to the General Stiffness Method

Chapter Objectives

- This chapter provides a transition from the classical to matrix methods of structural analysis.
- A comparison is first made between the classical flexibility method and stiffness (slope-deflection) method. Then the latter method is extended to a general stiffness method for analyzing an indeterminate structure with only one degree of kinematic indeterminacy.

16.1 Introduction

This chapter provides a transition from classical methods of hand analysis, such as the flexibility method (Chapter 11) or slope-deflection method (Chapter 12), to analysis by computer, which follows a set of programmed instructions. Before computers first became available in the 1950s, teams of engineers could require several months to produce an approximate analysis of a highly indeterminate three-dimensional space frame. Today, however, once the engineer specifies joint coordinates, type of joint (such as pinned or rigid), member properties, and the distribution of applied loads, the computer program can produce an exact analysis within minutes. The computer output specifies the forces in all members, reactions, and the displacement components of joints.

Although sophisticated computer programs are now available to analyze the most complex structures composed of shells, plates, and space frames, in this introductory chapter we will limit the discussion to planar structures (trusses, beams, and frames) composed of linear elastic members. To minimize computations and clarify concepts, we will only consider structures that are kinematically indeterminate to the first degree. Later in Chapters 17 and 18, using matrix notation, we extend the stiffness method to more complex structures with multiple degrees of kinematic indeterminacy.

To set up the analytical procedures used in a computer analysis, we will use a modified form of the slope-deflection method—a stiffness method in which equilibrium equations at joints are written in terms of unknown joint displacements. The stiffness method eliminates the need to select redundants and a released structure, as discussed in Chapter 11.

We begin the study of the stiffness method in Section 16.2 by comparing the basic steps required to analyze a simple indeterminate, pin-connected, two-bar system by both the flexibility and stiffness methods. Next, we extend the stiffness method to the analysis of indeterminate beams, frames, and trusses. A brief review of matrix operations, which provide a convenient format for programming the computations required to analyze indeterminate structures by computers, is available from the following web site: http://www.mhhe.com/leet.

16.2 Comparison between Flexibility and Stiffness Methods

The flexibility and stiffness methods represent two basic procedures that are used to analyze indeterminate structures. We discussed the flexibility method in Chapter 11. The slope-deflection method, covered in Chapter 12, is a stiffness formulation.

In the *flexibility method*, we write *compatibility equations* in terms of unknown *redundant forces*. In the *stiffness method*, we write *equilibrium equations* in terms of unknown *joint displacements*. We will illustrate the main characteristic of each method by analyzing the two-bar structure in Figure 16.1*a*. In this system, which is statically indeterminate to the first degree, the axially loaded bars connect to a center support that is free to displace horizontally but not vertically. In this structure, joints are designated by a number in a square, and members are identified by a number in a circle.

Flexibility Method

To analyze the structure in Figure 16.1*a*, we select the horizontal reaction F_1 at joint 1 as the *redundant*. We produce a stable determinate *released structure* by imagining that the pin at joint 1 is replaced by a roller. To analyze the structure, we load the released structure separately with (1) the applied load (Figure 16.1*b*) and (2) the redundant F_1 (Figure 16.1*c*). We then superimpose the displacements at joint 1 and solve for the redundant.

Since support 3 in the released structure is the only support able to resist a horizontal force, the entire 30-kip load in Figure 16.1*b* is transmitted through member 2. As member 2 compresses, joints 1 and 2 displace to the right a distance Δ_{10}. This displacement is computed by Equation 10.8. See Figure 16.1*a* for member properties.

$$\Delta_{10} = \frac{F_{20}L_2}{A_2E_2} = \frac{-30(150)}{0.6(20{,}000)} = -0.375 \text{ in} \qquad (16.1)$$

where the minus sign indicates that Δ_{10} is opposite in direction to the redundant.

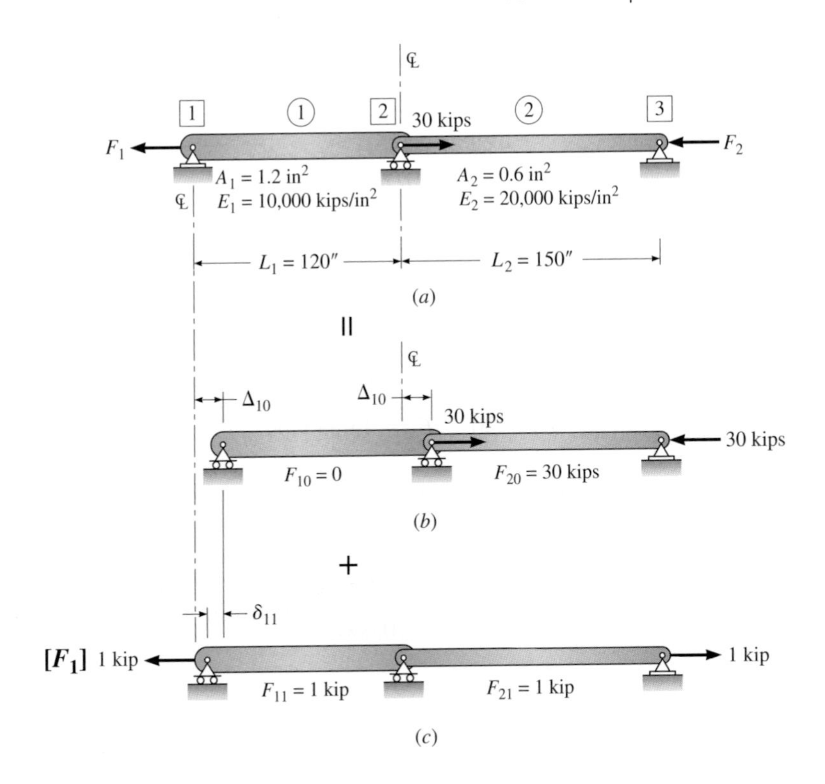

Figure 16.1: Analysis by the flexibility method: (*a*) details of the structure; (*b*) load applied to the released structure; (*c*) redundant F_1 applied to the released structure at joint 1; (*d*) forces acting on support 2.

We now apply a unit value of the redundant to the released structure (Figure 16.1*c*) and use Equation 16.1 to compute the horizontal displacement δ_{11} due to the elongation of bars 1 and 2.

$$\delta_{11} = \frac{F_{11}L_1}{A_1E_1} + \frac{F_{21}L_2}{A_2E_2} \tag{16.2}$$

$$= \frac{1(120)}{1.2(10{,}000)} + \frac{1(150)}{0.6(20{,}000)} = 0.0225 \text{ in}$$

To determine the reaction F_1, we write an *equation of compatibility* based on the geometric requirement that the horizontal displacement at support 1 must be zero:

$$\Delta_1 = 0 \tag{16.3}$$

Expressing Equation 16.3 in terms of the displacements yields

$$\Delta_{10} + \delta_{11}F_1 = 0 \tag{16.4}$$

Substituting the numerical values of Δ_{10} and δ_{11} into Equation 16.4 and solving for F_1, we compute

$$F_1 = \frac{-\Delta_{10}}{\delta_{11}} = \frac{0.375}{0.0225} = 16.67 \text{ kips}$$

To compute F_2, we consider equilibrium in the horizontal direction of the center support (Figure 16.1*d*).

$$\rightarrow + \quad \Sigma F_x = 0$$

$$30 - F_1 - F_2 = 0$$

$$F_2 = 30 - F_1 = 13.33 \text{ kips}$$

The actual displacement of joint 2 can be found by computing either the elongation of bar 1 or the shortening of bar 2.

$$\Delta L_1 = \frac{F_1 L_1}{A_1 E_1} = \frac{16.67(120)}{1.2(10{,}000)} = 0.167 \text{ in}$$

$$\Delta L_2 = \frac{F_2 L_2}{A_2 E_2} = \frac{13.33(150)}{0.6(20{,}000)} = 0.167 \text{ in}$$

Stiffness Method

The structure in Figure 16.1*a* (repeated in Figure 16.2*a*) will now be reanalyzed by the stiffness method. Since only joint 2 is free to displace, the structure is *kinematically indeterminate* to the first degree. Under the action of the 30-kip load in Figure 16.2*b*, joint 2 moves a distance Δ_2 to the right. Since *compatibility of deformations* requires that the elongation of bar 1 equal the shortening of bar 2, we can write

$$\Delta L_1 = \Delta L_2 = \Delta_2 \tag{16.5}$$

Using Equations 16.1 and 16.5, we express the forces in each bar in terms of the displacement of joint 2 and the properties of the members.

$$\begin{aligned} F_1 &= \frac{A_1 E_1}{L_1} \Delta L_1 = \frac{1.2(10{,}000)}{120} \Delta_2 = 100\,\Delta_2 \\ F_2 &= \frac{A_2 E_2}{L_2} \Delta L_2 = \frac{0.6(20{,}000)}{150} \Delta_2 = 80\,\Delta_2 \end{aligned} \tag{16.6}$$

Horizontal equilibrium of joint 2 (see Figure 16.2*c*) gives

$$\Sigma F_x = 0 \tag{16.7}$$

$$30 - F_1 - F_2 = 0$$

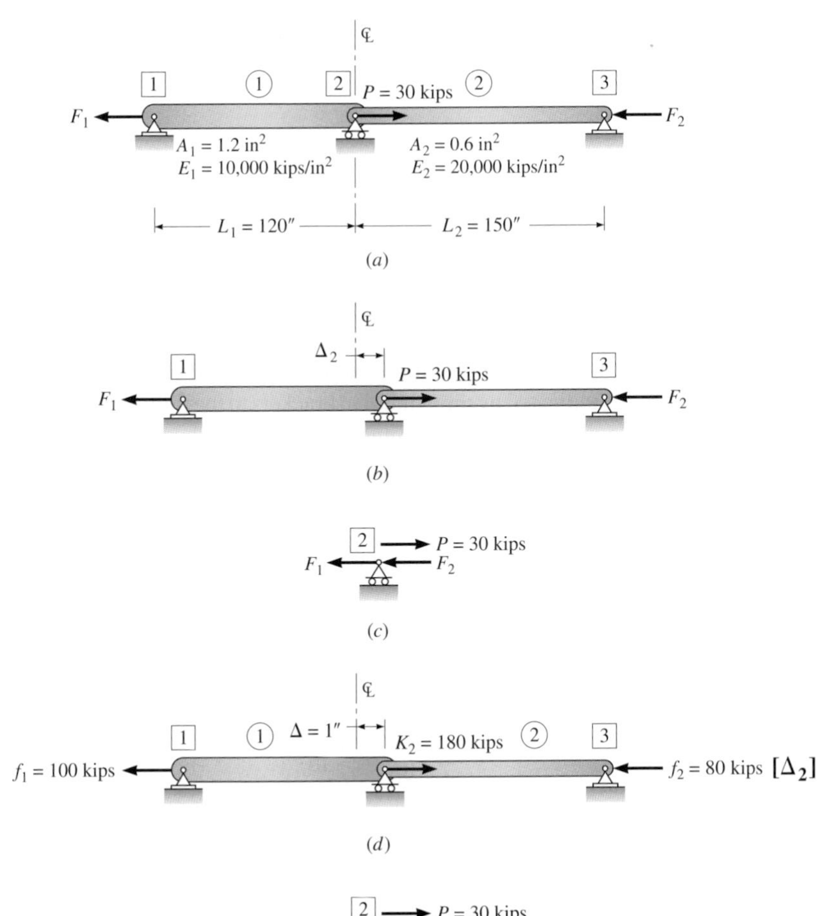

Figure 16.2: (*a*) Structure kinematically indeterminate to first degree; (*b*) deflected position of loaded structure; (*c*) free body of joint 2; (*d*) forces produced by a unit displacement of joint 2; (*e*) free body of center support.

Expressing the forces in Equation 16.7 in terms of the displacement Δ_2 given by Equation 16.6 and solving for Δ_2 give

$$30 - 100\,\Delta_2 - 80\,\Delta_2 = 0 \tag{16.8}$$

$$\Delta_2 = \frac{1}{6}\text{ in}$$

To establish the bar forces, we substitute the value of Δ_2 above into Equation 16.6.

$$F_1 = 100\,\Delta_2 = 100\left(\frac{1}{6}\right) = 16.67\text{ kips}$$

$$F_2 = 80\,\Delta_2 = 80\left(\frac{1}{6}\right) = 13.33\text{ kips} \tag{16.9}$$

Equation 16.8 can also be set up in a slightly different way. Let us introduce a unit displacement of 1 in at joint 2, as shown in Figure 16.2*d*. Using

Equation 16.1, the force K_2 required to hold the joint in this position can be computed by summing the forces needed to elongate bar 1 and compress bar 2 by 1 in.

$$K_2 = \frac{A_1E_1}{L_1}(1 \text{ in}) + \frac{A_2E_2}{L_2}(1 \text{ in}) \tag{16.10}$$

$$= 180 \text{ kips/in}$$

Note that K_2 represents the force required to produce a unit displacement at joint 2. So the unit of K_2 is kips/in. Since the actual displacement of joint 2 is not 1 in but Δ_2, we must multiply all forces and deflections (Figure 16.2) by the magnitude of Δ_2, as indicated by the symbol in brackets to the right of joint 3. For the block to be in equilibrium, the magnitude of Δ_2, the displacement of joint 2, must be large enough to develop only 30 kips of resistance. Since the restraining force exerted by the bars is a linear function of the displacement of joint 2, the actual joint displacement Δ_2 can be determined by writing the *equilibrium equation* for forces in the horizontal direction at joint 2 (Figure 16.2*e*).

$$\rightarrow + \quad \Sigma F_x = 0$$

$$30 - K_2\,\Delta_2 = 0$$

and

$$\Delta_2 = \frac{30}{180} = \frac{1}{6} \text{ in}$$

The quantity K_2 is called a *stiffness coefficient*. If the two bars are treated as a large spring, the stiffness coefficient measures the resistance (or stiffness) of the system to deformation.

Most computer programs are based on the stiffness method. This method eliminates the need for the designer to select a released structure and permits the analysis to be automated. Once the designer identifies the joints that are free to displace and specifies the joint coordinates, the computer is programmed to introduce unit displacements and to generate the required stiffness coefficients, set up and solve the joint equilibrium equations, and compute all reactions, joint displacements, and member forces.

16.3 Analysis of an Indeterminate Structure by the General Stiffness Method

In the example in Figure 16.3 we extend the *general stiffness* method to the analysis of an indeterminate beam—a structural element whose deformations are produced by bending moments. This example will also provide the background for the analysis of indeterminate frames (with the matrix formulation, covered in Chapter 18). As you will observe, the method utilizes procedures and equations previously developed in Chapters 12 and 13, which introduced the slope-deflection and moment distribution methods.

Figure 16.3*a* shows a continuous beam of constant cross section. Since the only unknown displacement of the continuous beam is the rotation θ_2 that occurs at joint 2, the structure is *kinematically indeterminate to the first degree* (Section 12.6).

As the first step in the analysis, before loads are applied, we clamp joint 2 to prevent rotation, thereby producing two fixed-end beams (Figure 16.3*b*). Next we apply the 15-kip load, which produces fixed-end moments FEM_{12} and FEM_{21}. Using Figure 12.5*a* to evaluate these moments gives

$$\text{FEM}_{12} = -\frac{PL}{8} = -\frac{15(16)}{8} = -30 \text{ kip}\cdot\text{ft}$$

$$\text{FEM}_{21} = \frac{PL}{8} = \frac{15(16)}{8} = 30 \text{ kip}\cdot\text{ft}$$

We will adopt the previous sign convention used in Chapters 12 and 13; that is, *clockwise moments and rotations at the ends of members are positive, and counterclockwise moments and rotations are negative*.

Figure 16.3*c* shows the forces on a free body of joint 2. Since no loads act on the 8-ft span at this stage, it remains unstressed and applies no forces to the right side of joint 2.

To account for the rotation θ_2 that occurs in the actual beam (Figure 16.3*d*), we next, in a separate step, induce a unit clockwise rotation of +1 rad at joint 2 and clamp the beam in its deflected position. This rotation produces member end moments that can be evaluated using the first two terms of the slope-deflection equation (Equation 12.16). We will denote these moments with the superscript JD, which stands for a joint displacement, in this case, a joint rotation.

In span 1-2

$$M_{12}^{\text{JD}} = \frac{2EI}{L}[2(0) + 1] = \frac{2EI}{16}[0 + 1] = \frac{EI}{8} \tag{16.11}$$

$$M_{21}^{\text{JD}} = \frac{2EI}{L}[2(1) + 0] = \frac{2EI}{16}[2(1) + 0] = \frac{EI}{4} \tag{16.12}$$

In span 2-3

$$M_{23}^{\text{JD}} = \frac{2EI}{L}[2(1) + 0] = \frac{2EI}{8}(2) = \frac{EI}{2} \tag{16.13}$$

$$M_{32}^{\text{JD}} = \frac{2EI}{L}[2(0) + (1)] = \frac{2EI}{8}(1) = \frac{EI}{4} \tag{16.14}$$

From the free-body diagram of joint 2 shown in Figure 16.3*e*, we observe that the moment K_2 (the stiffness coefficient) applied by the clamp

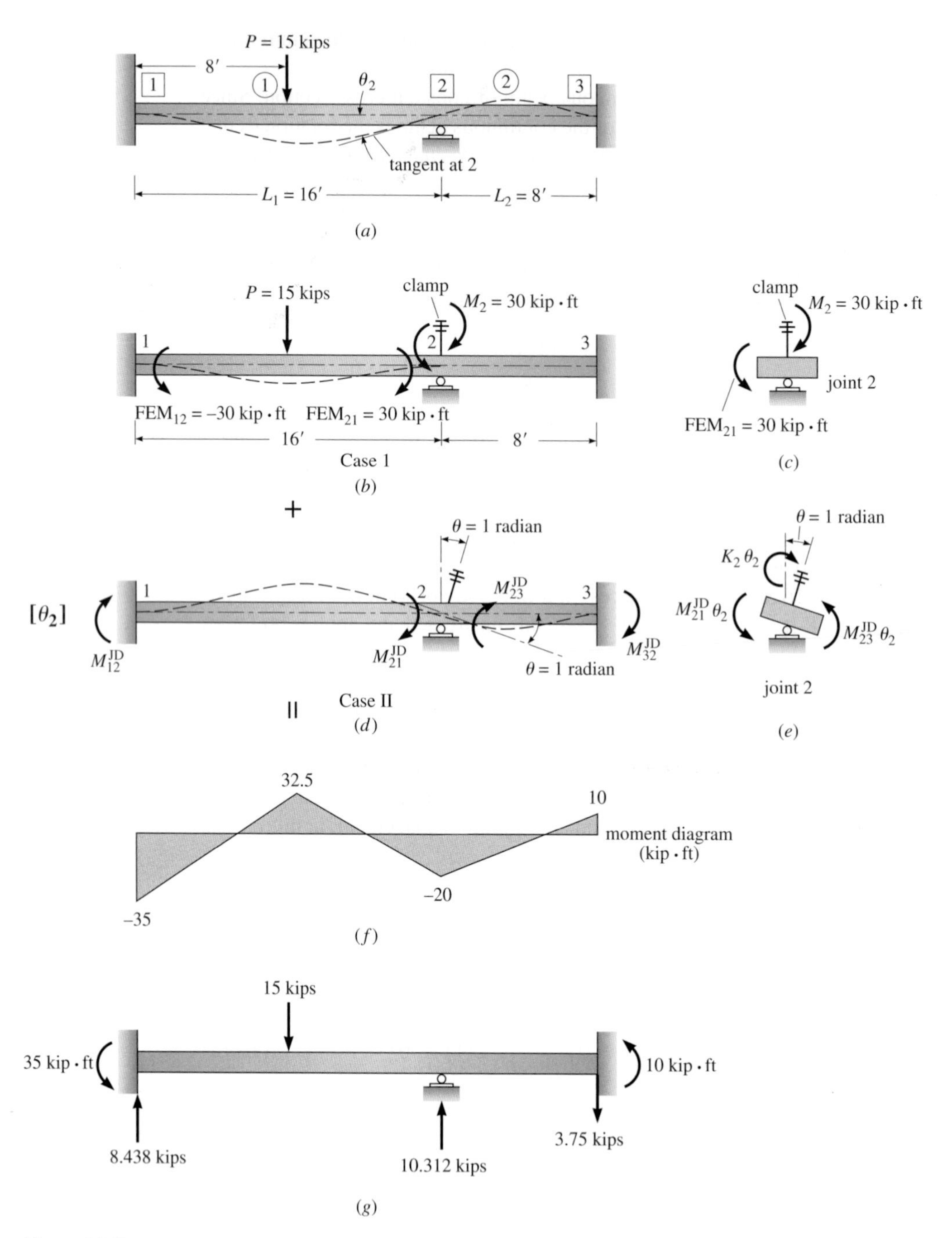
P = 15 kips
8′
1
1
θ2
2
2
3
tangent at 2
L1 = 16′
L2 = 8′
(a)
P = 15 kips
clamp
M2 = 30 kip · ft
1
2
3
FEM12 = −30 kip · ft
FEM21 = 30 kip · ft
16′
8′
Case 1
(b)
clamp
M2 = 30 kip · ft
joint 2
FEM21 = 30 kip · ft
(c)
+
θ = 1 radian
[θ2]
1
2
3
M23JD
M12JD
M21JD
θ = 1 radian
M32JD
Case II
=
(d)
θ = 1 radian
K2 θ2
M21JD θ2
M23JD θ2
joint 2
(e)
32.5
10
moment diagram
(kip · ft)
−20
−35
(f)
15 kips
35 kip · ft
10 kip · ft
8.438 kips
10.312 kips
3.75 kips
(g)

Figure 16.3

to maintain the unit rotation equals the sum of $M_{21}^{JD} + M_{23}^{JD}$ (given by Equations 16.12 and 16.13); that is,

$$K_2 = M_{21}^{JD} + M_{23}^{JD} = \frac{EI}{4} + \frac{EI}{2} = \frac{3EI}{4} \tag{16.15}$$

Since the behavior is linearly elastic, to establish both the actual deformation and the member end moments, we must multiply the unit rotation and the moments it produces (Figure 16.3*d*) by the actual rotation θ_2. We denote this operation by showing θ_2 in brackets to the left of the fixed support at joint 1.

Since no external moments or a clamp exist at joint 2 in the real beam, it must follow that M_2 in Figure 16.3*c* equals $\theta_2 K_2$ in Figure 16.3*e*; that is, for the joint to be in equilibrium

$$\circlearrowright^{+} \quad \Sigma M_2 = 0$$
$$30 + K_2\theta_2 = 0 \tag{16.16}$$

Substituting the value of K_2 given by Equation 16.15 into Equation 16.16 gives

$$30 + \frac{3EI\theta_2}{4} = 0$$

Solving for θ_2 gives

$$\theta_2 = -\frac{40}{EI} \quad \text{radian} \tag{16.17}$$

Since the value of θ_2 is negative, the rotation at joint 2 is opposite to that assumed in Figure 16.3(*d*) to define the stiffness coefficient, i.e., the rotation at joint 2 is counterclockwise. Once θ_2 is determined, the member end moments can be evaluated by superposition of the cases shown in Figures 16.3*b* and *d*. For example, to evaluate the moment in the beam just to the left of joint 2, we write the following *superposition equation*, substituting into Equation 16.18 the value of M_{21}^{JD} given by Equation 16.12 and θ_2 given by Equation 16.17; we find

$$M_{21} = \text{FEM}_{21} + M_{21}^{JD}\theta_2 \tag{16.18}$$

$$M_{21} = 30 + \frac{EI}{4}\left(-\frac{40}{EI}\right) = 20 \text{ kip}\cdot\text{ft} \quad \text{clockwise}$$

At fixed support (joint 3),

$$M_{32} = 0 + M_{32}^{JD}\theta_2 = 0 + \frac{EI}{4}\left(-\frac{40}{EI}\right) = -10 \text{ kip}\cdot\text{ft}$$

where the minus sign indicates that the direction for M_{32} is counterclockwise.

After the member end moments are computed, shear forces and reactions can be calculated by using free-body diagrams of each beam. The complete moment diagram is shown in Figure 16.3*f*. The final reactions are shown in Figure 16.3*g*.

Summary of the General Stiffness Method

The analysis of the continuous beam in Figure 16.3*a* is based on the superposition of two cases. In case 1, we clamp all joints that are free to rotate and apply the load. The load creates fixed-end moments in the beam and an equal moment in the clamp. Had there been loads on both spans, the moment in the clamp would have been equal to the difference of the fixed-end moment acting on joint 2. At this point the structure is in equilibrium with the load; however, the joint has been restrained by a clamp and not allowed to rotate.

To eliminate the clamp, we must remove it and allow the joint to rotate. This rotation will produce additional moments in the members. We are primarily interested at this stage in the magnitude of the moments at the ends of each member. Since we do not know the magnitude of the rotation, in a separate case 2, we arbitrarily introduce a unit rotation of 1 radian and clamp the beam in the deflected position. The case 2 clamp now applies a moment, termed a *stiffness coefficient*, which holds the beam in the rotated position. Since we have induced a specific value of rotation (that is, 1 rad), we are able to compute the moments at the ends of each member by using the slope-deflection equation. The moment in the clamp is computed from a free body of the joint. If we now multiply the forces and displacements in case 2 by the *actual magnitude of the joint rotation* θ_2, all forces and displacements (including the moment in the clamp and the rotation at joint 2) will be scaled proportionally to the correct value. Since no clamp exists in the actual beam, it follows that the sum of the moments in the clamp from the two cases must equal zero. Accordingly the value of θ_2 can now be determined by writing an equilibrium equation that states the sum of the moments in the clamp, from case 1 and case 2, must equal zero. Once θ_2 is known, all forces in case 2 can be evaluated and added directly to those of case 1.

EXAMPLE 16.1

Analyze the rigid frame in Figure 16.4*a* by the general stiffness method. *EI* is constant.

Solution

Since the only unknown displacement is the rotation θ_2 at joint 2, the frame is kinematically indeterminate to the first degree; therefore, a solution requires one joint equilibrium equation, written at joint 2. In the first step, we imagine a clamp is applied to joint 2 that prevents rotation and produces two fixed-end members (Figure 16.4*b*). When the loads are applied, fixed-end moments develop in the beam but not in the column because the clamp prevents rotation of the top of the column. Using the equation given in Figure 12.5*c*, these fixed-end moments in the beam are

$$\text{FEM} = \pm\frac{2PL}{9} = \pm\frac{2(24)(18)}{9} = \pm 96 \text{ kN}\cdot\text{m} \tag{1}$$

Figure 16.4*c* shows a detail of the fixed-end moment acting on a free body of joint 2.

We next introduce a clockwise unit rotation of 1 rad at joint 2 and clamp the joint in the deflected position. The moments produced by the unit rotation are superscripted with a JD (for joint displacement). Since we want the effect of the actual rotation θ_2 produced by the 24-kN loads, we must multiply this case by θ_2, as indicated by the symbol θ_2 in brackets at the left of Figure 16.4*d*. We express the moments induced by the unit rotation at joint 2 in terms of the member properties, using the *slope-deflection equation* (Equation 12.16). Since joint 2 cannot translate, the terms ψ_{NF} and FEM_{NF} in Equation 12.16 equal zero, and the slope-deflection equation reduces to

$$M_{NF} = \frac{2EI}{L}(2\theta_N + \theta_F) \tag{2}$$

Using Equation 2, we next evaluate the member end moments produced by the unit joint rotation.

$$M_{12}^{\text{JD}} = \frac{2EI}{6}(0 + 1) = \frac{EI}{3} \tag{3}$$

$$M_{21}^{\text{JD}} = \frac{2EI}{6}[2(1) + 0] = \frac{2EI}{3} \tag{4}$$

$$M_{23}^{\text{JD}} = \frac{2EI}{18}[2(1) + 0] = \frac{2EI}{9} \tag{5}$$

$$M_{32}^{\text{JD}} = \frac{2EI}{18}[2(0) + 1] = \frac{EI}{9} \tag{6}$$

[*continues on next page*]

Example 16.1 continues . . .

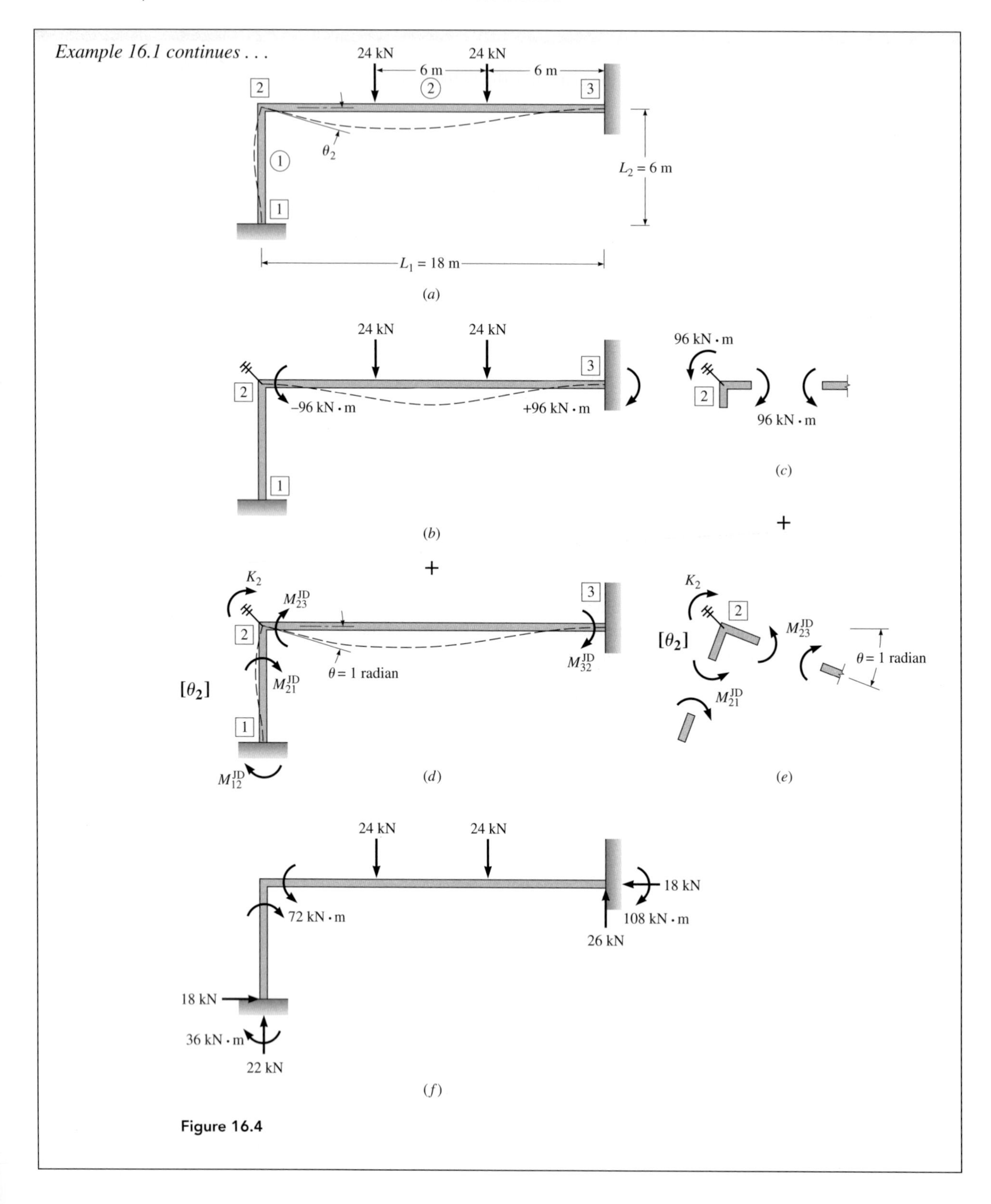

Figure 16.4

The total moment K_2 applied by the clamp equals the sum of the moments applied to the ends of the beams framing into joint 2 (Figure 16.4*e*).

$$K_2 = M_{21}^{\text{JD}} + M_{23}^{\text{JD}}$$

$$K_2 = \frac{2EI}{3} + \frac{2EI}{9} = \frac{8EI}{9} \tag{7}$$

For the clamp to be removed, equilibrium requires that the sum of the moments acting on the clamp at joint 2 (Figure 16.4*c* and *e*) equal zero.

$$\circlearrowright^{+} \quad \Sigma M_2 = 0$$

$$K_2\theta_2 - 96 = 0 \tag{8}$$

Substituting the value of K_2 given by Equation 7 into Equation 8 and solving for θ_2 give

$$\left(\frac{8EI}{9}\right)\theta_2 - 96 = 0$$

$$\theta_2 = \frac{108}{EI} \tag{9}$$

To establish the magnitude of the moment at the end of each member, we superimpose the forces at each joint shown in Figure 16.4*b* and *d*; that is, we multiply the values of moment due to the unit rotation (Equations 3, 4, 5, and 6) by the actual rotation θ_2 and add any fixed-end moments.

$$M_{12} = \theta_2 M_{12}^{\text{JD}} = \frac{108}{EI}\left(\frac{EI}{3}\right) = 36 \text{ kN}\cdot\text{m} \quad \text{clockwise}$$

$$M_{21} = \theta_2 M_{21}^{\text{JD}} = \frac{108}{EI}\left(\frac{2EI}{3}\right) = 72 \text{ kN}\cdot\text{m} \quad \text{clockwise}$$

$$M_{23} = \theta_2 M_{23}^{\text{JD}} + \text{FEM}_{23} = \frac{108}{EI}\left(\frac{2EI}{9}\right) - 96 = -72 \text{ kN}\cdot\text{m} \quad \text{counterclockwise}$$

$$M_{32} = \theta_2 M_{32}^{\text{JD}} + \text{FEM}_{32} = \frac{108}{EI}\left(\frac{EI}{9}\right) + 96 = 108 \text{ kN}\cdot\text{m} \quad \text{clockwise}$$

The remainder of the analysis is carried out using free-body diagrams of each member to establish shears and reactions. The final results are summarized in Figure 16.4*f*.

EXAMPLE 16.2

The pin-connected bars in Figure 16.5*a* are connected at joint 1 to a roller support. Determine the force in each bar and the magnitude of the horizontal displacement Δ_x of joint 1 produced by the 60-kip force. Area of bar 1 = 3 in^2, area of bar 2 = 2 in^2, and E = 30,000 kips/in^2.

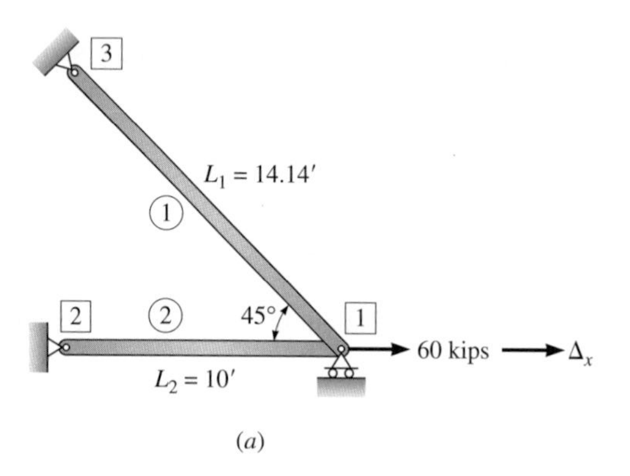

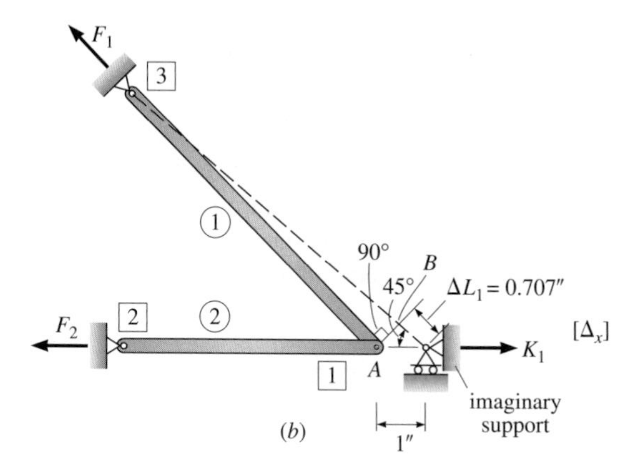

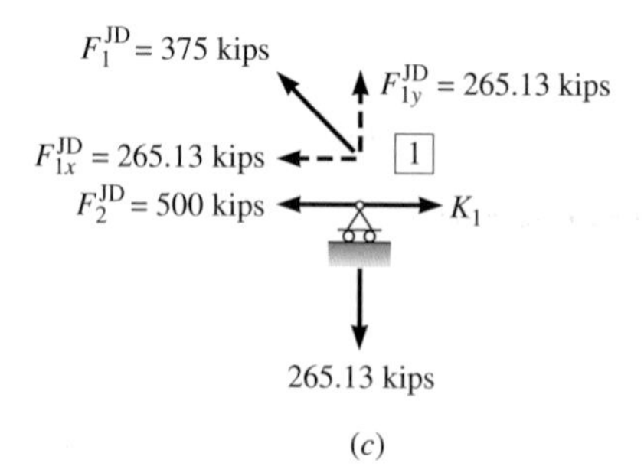

Figure 16.5: (*a*) Details of structure; (*b*) joint 1 displaced 1 in to the right and attached to imaginary support; (*c*) forces at joint 1 produced by a 1-in horizontal displacement.

Solution

We first displace the roller 1 in to the right and connect it to an imaginary pin support (Figure 16.5*b*) that develops a reaction of K_1 kips to hold the joint in its new position. Because the horizontal displacement of joint 1, shown to an exaggerated scale in Figure 16.5*b*, is very small compared to the length of the bars, we assume its slope remains 45° in the deflected position. To establish the elongation of bar 1, we mark its original unstressed length on the displaced bar by rotating the original length about the pin at joint 3. Since the end of the unstressed bar moves on the arc of a circle, from point A to B, the initial displacement of its end is perpendicular to the original position of the bar's axis. Since we require the bar forces due to the actual displacement, which is a fraction of an inch, we multiply the forces and displacements shown in Figure 16.5*b* by Δ_x.

From the geometry of the displacement triangle at joint 1 (Figure 16.5*b*), we compute ΔL_1:

$$\Delta L_1 = (1 \text{ in})(\cos 45°) = 0.707 \text{ in}$$

With the elongation of each bar established, we can use Equation 16.1 to compute the force in each bar.

$$F_1^{\text{JD}} = \frac{A_1 E \Delta L_1}{L_1} = \frac{3(30{,}000)(0.707)}{14.14 \times 12} = 375 \text{ kips}$$

$$F_2^{\text{JD}} = \frac{A_2 E \Delta L_2}{L_2} = \frac{2(30{,}000)(1)}{10 \times 12} = 500 \text{ kips}$$

We then compute the horizontal and vertical components of F_1.

$$F_{1x}^{\text{JD}} = F_1^{\text{JD}}(\cos 45°) = 375(0.707) = 265.13 \text{ kips}$$

$$F_{1y}^{\text{JD}} = F_1^{\text{JD}}(\sin 45°) = 375(0.707) = 265.13 \text{ kips}$$

To evaluate K_1, we sum forces applied to the pin (Figure 16.5*c*) in the horizontal direction.

$$\Sigma F_x = 0$$

$$K_1 - F_{1x}^{\text{JD}} - F_2^{\text{JD}} = 0$$

$$K_1 = F_{1x}^{\text{JD}} + F_2^{\text{JD}} = 265.13 + 500 = 765.13 \text{ kips}$$

To compute the actual displacement, we multiply the force K_1 in Figure 16.5*c* by Δ_x, the actual displacement and consider the horizontal force equilibrium at joint 2.

$$K_1\,\Delta_x - 60 = 0$$

$$765.13\,\Delta_x - 60 = 0$$

$$\Delta_x = 0.0784 \text{ in}$$

Compute the force in each bar.

$$F_1 = F_1^{\text{JD}}(\Delta_x) = 375(0.0784) = 29.4 \text{ kips}$$

$$F_2 = F_2^{\text{JD}}(\Delta_x) = 500(0.0784) = 39.2 \text{ kips}$$

EXAMPLE 16.3

Analyze the rigid frame in Figure 16.6*a* by the general stiffness method.

Solution

The rigid frame in Figure 16.6*a* is kinematically indeterminate to the third degree because joints 2 and 3 can rotate and the beam can displace laterally. However, because the structure and load are symmetric with respect to a vertical axis through the center of the frame, the deflections form a symmetric pattern. Therefore, rotations θ_2 and θ_3 of joints 2 and 3 are equal in magnitude, and no lateral displacement of the frame occurs. These conditions permit a solution based on a single equilibrium equation, arbitrarily written at joint 2.

We begin the analysis by clamping joints 2 and 3 to prevent rotation (Figure 16.6*b*), and we apply the load, producing fixed-end moments in the beam where

$$\text{FEM} = \pm\frac{PL}{8} = \pm\frac{20(36)}{8} = \pm 90 \text{ kip}\cdot\text{ft} \tag{1}$$

Figure 16.6*c* shows the moments acting on joint 2 from the beam and column as well as the clamp (shear forces are omitted for clarity).

We next introduce simultaneously rotations of 1 rad clockwise at joint 2 and -1 rad counterclockwise at joint 3, and we clamp the joints in the deflected position (Figure 16.6*d*). The moments in the beam and columns at joints 2 and 3 produced by the rotations are identical in magnitude but act in opposite directions. Using the first two terms of the slope-deflection equation at joint 2, we compute the moments at the left end of the beam and the moments at the top and bottom of the left column.

$$M_{23}^{\text{JD}} = \frac{2EI}{36}[2(1) + (-1)] = \frac{EI}{18} \tag{2}$$

$$M_{21}^{\text{JD}} = \frac{2EI}{12}[2(1) + 0] = \frac{EI}{3} \tag{3}$$

$$M_{12}^{\text{JD}} = \frac{2EI}{12}[2(0) + 1] = \frac{EI}{6} \tag{4}$$

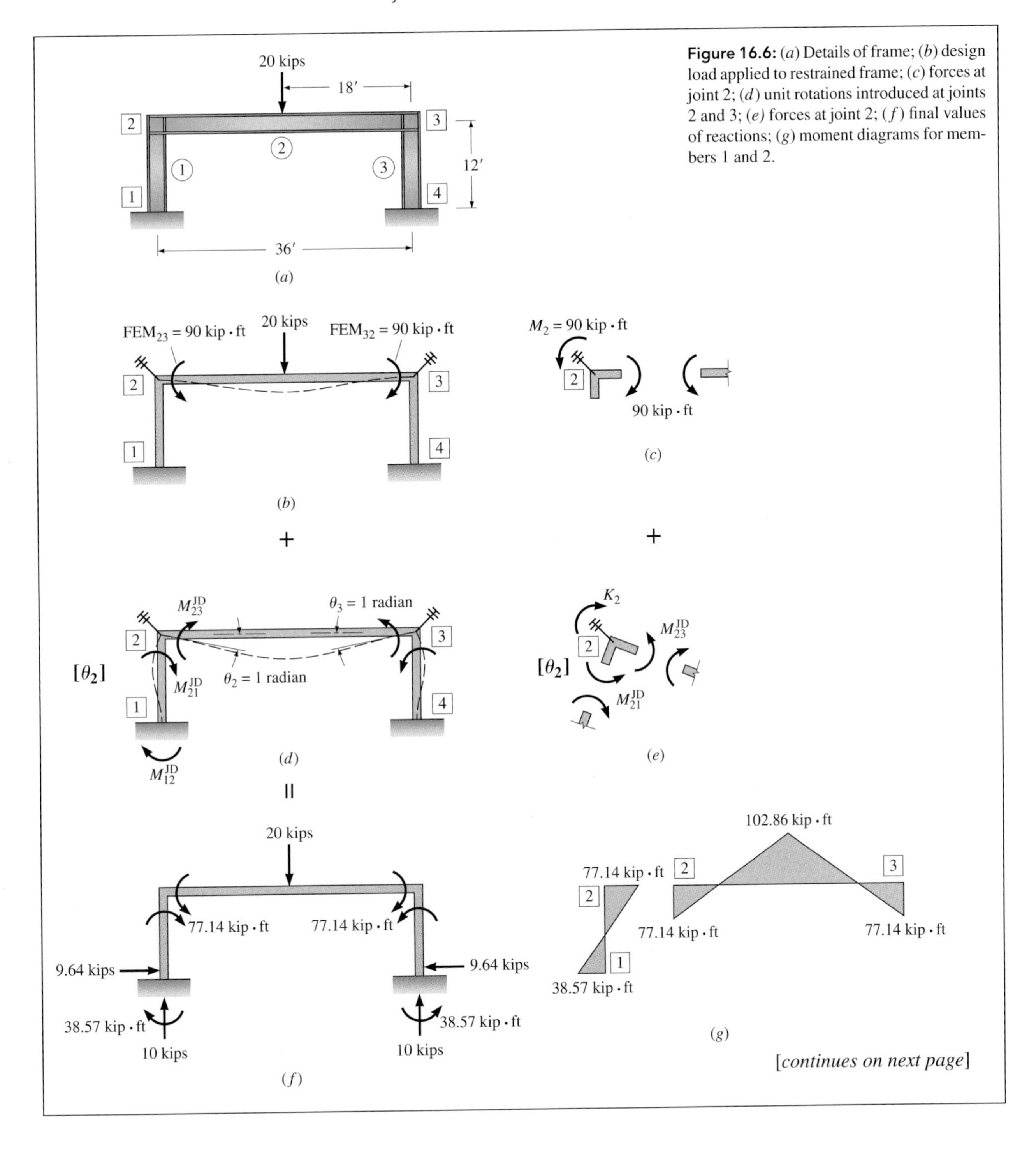

Figure 16.6: (*a*) Details of frame; (*b*) design load applied to restrained frame; (*c*) forces at joint 2; (*d*) unit rotations introduced at joints 2 and 3; (*e*) forces at joint 2; (*f*) final values of reactions; (*g*) moment diagrams for members 1 and 2.

[*continues on next page*]

Example 16.3 continues . . .

The moment K_2 exerted by the clamp at joint 2 (Figure 16.6*e*) equals the sum of the applied moments at joint 2.

$$\circlearrowright^{+} \quad \Sigma M_2 = 0 \tag{5}$$

$$K_2 = M_{21}^{\text{JD}} + M_{23}^{\text{JD}} \tag{6}$$

Substituting Equations 2 and 3 into Equation 6 gives

$$K_2 = \frac{EI}{3} + \frac{EI}{18} = \frac{7EI}{18} \tag{7}$$

To establish the moment produced by the actual rotation, we multiply all forces and displacements in Figure 16.6*d* by θ_2.

Since the sum of the moments acting on the clamp at joint 2 in Figures 16.6*c* and *e* must equal zero, we write the equilibrium equation

$$\circlearrowright^{+} \quad \Sigma M_2 = 0$$

$$\theta_2 K_2 - 90 = 0 \tag{8}$$

Substituting the value of K_2 given by Equation 7 into Equation 8 gives

$$\theta_2\left(\frac{7EI}{18}\right) = 90$$

$$\theta_2 = \frac{231.42}{EI} \tag{9}$$

The final moment at any section is computed by combining moments at corresponding sections in Figures 16.6*b* and *d*.

At joint 2 in the beam,

$$M_{23} = \text{FEM}_{23} + \theta_2 M_{23}^{\text{JD}}$$

$$= -90 + \frac{231.42}{EI}\left(\frac{EI}{18}\right) = -77.14 \text{ kip}\cdot\text{ft} \quad \text{counterclockwise}$$

From symmetry,

$$M_{32} = -M_{23} = 77.14 \text{ kip}\cdot\text{ft} \quad \text{clockwise}$$

$$M_{21} = \theta_2 M_{21}^{\text{JD}} = \frac{231.42}{EI}\left(\frac{EI}{3}\right) = 77.14 \text{ kip}\cdot\text{ft} \quad \text{clockwise}$$

$$M_{12} = \theta_2 M_{12}^{\text{JD}} = \frac{231.42}{EI}\left(\frac{EI}{6}\right) = 38.57 \text{ kip}\cdot\text{ft} \quad \text{clockwise}$$

Final results are shown in Figures 16.6*f* and *g*.

Summary

- The *general stiffness method* introduced in this chapter is the basis of the majority of computer programs used to analyze all types of determinate and indeterminate structures including planar structures and three-dimensional trusses, frames, and shells. The stiffness method eliminates the need to select redundants and a released structure, as required by the flexibility method.
- In the general stiffness method, joint displacements are the unknowns. With all joints initially artificially restrained, unit displacements are introduced at each joint and the forces associated with the unit displacements (known as *stiffness coefficients*) computed. In this introductory discussion, we consider beams, frames, and trusses with a single unknown linear or rotational displacement. In structures with multiple joints that are free to displace, the number of unknown displacements will be equal to the degree of *kinematic indeterminacy*. If programs are written for three-dimensional structures with rigid joints, six unknown displacements (three linear and three rotational) are possible at each unrestrained joint. For these situations the torsional stiffness as well as the axial and bending stiffness of members must be considered when evaluating stiffness coefficients.
- In a typical computer program, the designer must select a coordinate system to establish the location of joints, specify member properties (such as area, moment of inertia, and modulus of elasticity), and specify the type of loading. An approximate analysis (see Chapter 15) can be carried out to size members initially.

PROBLEMS

P16.1. The structure in Figure P16.1 is composed of three pin-connected bars. The bar areas are shown in the figure. Given: $E = 30{,}000 \text{ kips/in}^2$.

(*a*) Compute the stiffness coefficient K associated with a 1-in vertical displacement of joint A. (*b*) Determine the vertical displacement at A produced by a vertical load of 24 kips directed downward. (*c*) Determine the axial forces in all bars.

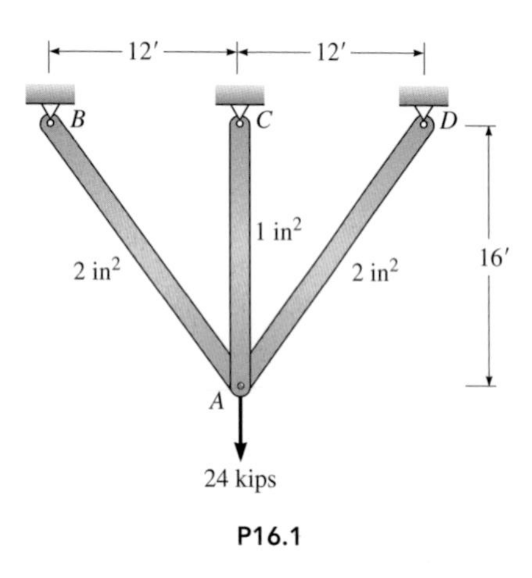

P16.1

P16.2. The cantilever beam in Figure P16.2 is supported on a spring at joint B. The spring stiffness is 10 kips/in. Given: $E = 30{,}000$ kips/in

(*a*) Compute the stiffness coefficient associated with a 1-in vertical displacement at joint B. (*b*) Compute the vertical deflection of the spring produced by a vertical load of 15 kips acting downward at B. (*c*) Determine all support reactions produced by the 15-kip load.

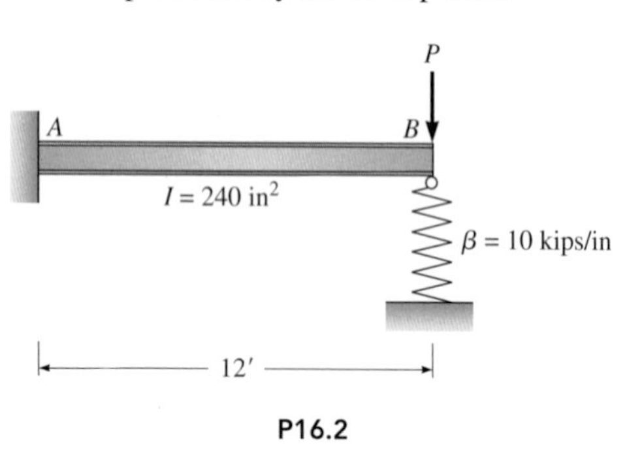

P16.2

P16.3. The pin-connected bar system in Figure P16.3 is stretched 1 in horizontally and connected to the pin support 4. Determine the horizontal and vertical components of force that the support must apply to the bars. Area of bar 1 = 2 in², area of bar 2 = 3 in², and $E = 30{,}000 \text{ kips/in}^2$. K_{2x} and K_{2y} are stiffness coefficients.

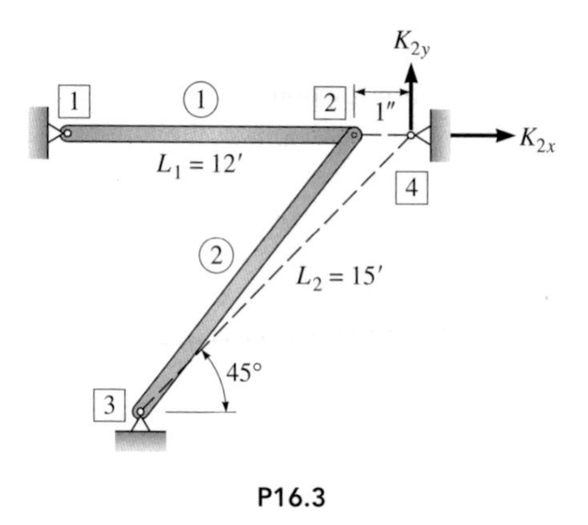

P16.3

P16.4. Analyze the beam in Figure P16.4. After member end moments are determined, compute all reactions and draw the moment diagrams. EI is constant.

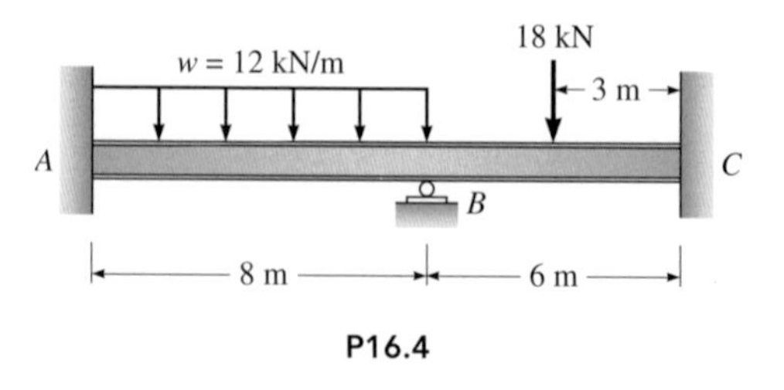

P16.4

P16.5. Analyze the steel rigid frame in Figure P16.5. After member end moments are evaluated, compute all reactions and the moment diagram for beam BC. Supports at A and C are detailed to produce fixed ends.

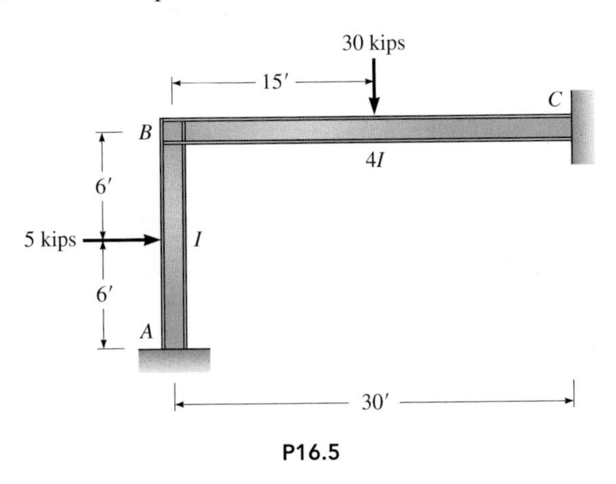

P16.5

P16.6. Analyze the beam in Figure P16.6. Compute all reactions and draw the shear and moment diagrams. Given: EI is constant.

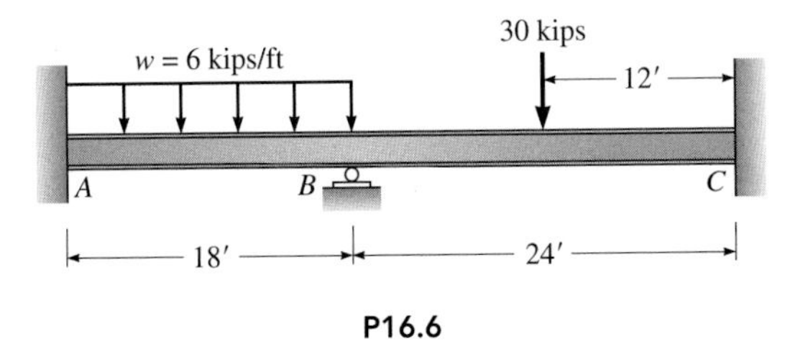

P16.6

P16.7. Analyze the reinforced concrete frame in Figure P16.7. Determine all reactions. E is constant.

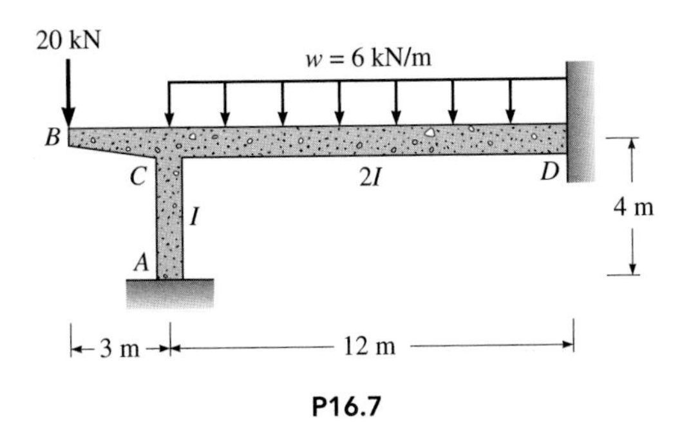

P16.7

P16.8. The structure in Figure P16.8 is composed of a beam supported by two struts at the cantilever end. Compute all reactions and the strut member forces. Use $E = 30{,}000$ kips/in^2.

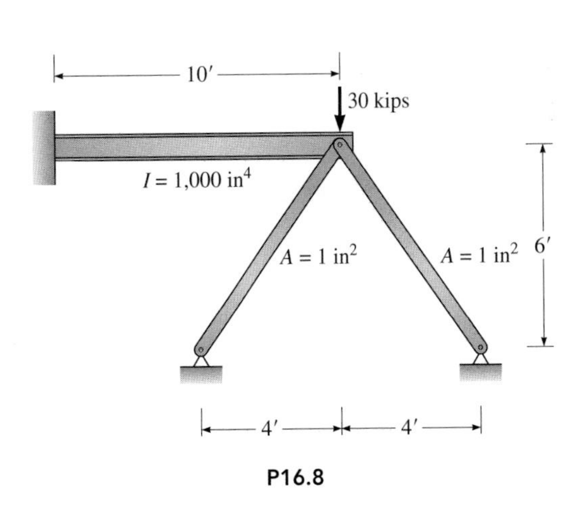

P16.8

P16.9. The cantilever beam in Figure P16.9 is connected to a bar at joint 2 by a pin. Compute all reactions. Given: $E = 30{,}000$ kips/in^2. Ignore axial deformation of the beam.

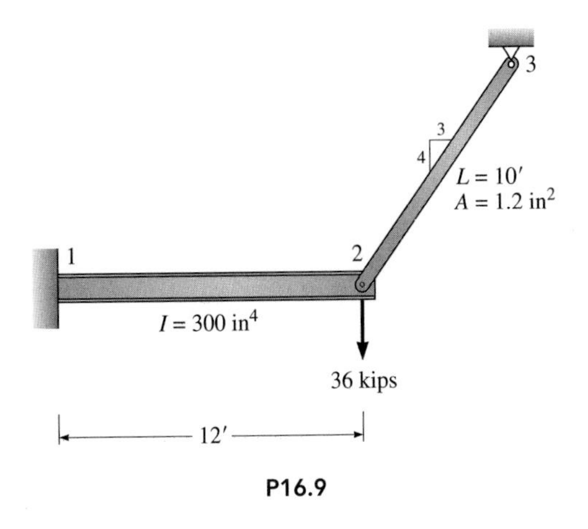

P16.9

P16.10 and **P16.11.** Analyze the rigid frames in Figures P16.10 and P16.11, using symmetry to simplify the analysis. Compute all reactions and draw the moment diagrams for all members. Also E is constant.

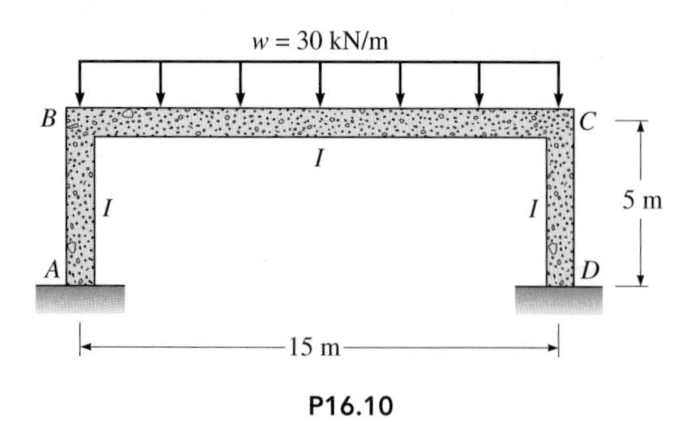

P16.10

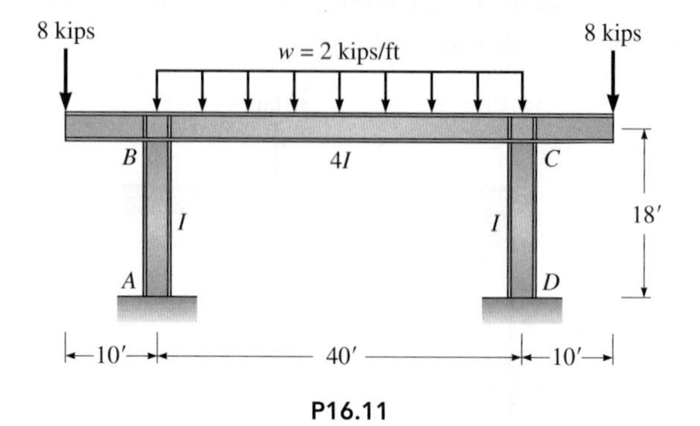

P16.11

U.S. Pavilion at Expo '67 in Montreal, Canada

The United States Pavilion was enclosed by a 250-ft diameter geodesic dome. The dome is a beautifully proportioned, three-quarter sphere which is enclosed by a space frame of steel pipes with 1,900 molded acrylic panels. The designer distributed the structure's weight over the whole surface of the dome with minimum use of materials.

C H A P T E R

17

Matrix Analysis of Trusses by the Direct Stiffness Method

Chapter Objectives

- Learn how to establish the matrix form of equilibrium equations for a determinate or an indeterminate truss and to partition the matrices such that both the unknown joint displacements and unknown reactions can be solved by matrix operation.
- Learn how to establish the structure stiffness matrix, which can be constructed by either basic mechanics or, more conveniently, individual member stiffness matrices. The latter method, which is suitable for computer implementation, is called the direct stiffness method.
- Construct member stiffness matrix by using either the local or global coordinate system. Learn how to convert a member stiffness matrix from local to global coordinate system by the concept of coordinate transformation.

17.1 Introduction

In this chapter we introduce the *direct stiffness method*, a procedure that provides the basis for most computer programs used to analyze structures. The method can be applied to almost any type of structure, for example, trusses, continuous beams, indeterminate frames, plates, and shells. When the method is applied to plates and shells (or other types of problems that can be subdivided into two- and three-dimensional elements), it is called the *finite element method*.

As in the flexibility method of Chapter 11, the direct stiffness method requires that we divide the analysis of a structure into a number of basic cases that, when superimposed, are equivalent to the original structure. However, instead of writing compatibility equations in terms of unknown redundant forces and flexibility coefficients, we write joint equilibrium equations in terms of unknown joint displacements and *stiffness coefficients* (forces produced by unit displacements). Once the joint displacements are known, the

Matrix Structural Analysis

John Hadji Argyris (1913–2004), a professor in aeronautical engineering, pioneered the matrix structural analysis and finite element analysis. In his own words (1957), "We have known for some years that none of the conventional statical methods are really suitable for determining the stress distribution and flexibility matrices of highly statically indeterminate systems of modern aircraft designs. ... We have overcome these difficulties by way of the matrix formulation of statics in conjunction with automatic electronic digital computers."

forces in the members of the structure can be calculated from force-displacement relationships.

To illustrate the method, we analyze the two-bar truss in Figure 17.1*a*. We identify truss joints or *nodes* by numbers in circles and bars by numbers in squares. Under the action of the 10-kip vertical load at joint 2, the bars deform, and joint 2 displaces a distance Δ_x horizontally and Δ_y vertically. These displacements are the *unknowns* in the stiffness method. To establish the positive and negative sense of forces and displacements in the horizontal and vertical directions, we introduce a global *xy* coordinate system at joint 2. The *x direction* is denoted by the number 1 and the *y direction* by the number 2. The positive directions are indicated by the arrowheads.

In the stiffness method, we carry out the truss analysis by superimposing the following two loading cases:

Case I. The structure is loaded at joint 2 by a set of forces that displace joint 2 a unit distance to the right but permit no vertical displacement. The forces and displacements associated with unit displacements are then multiplied by the magnitude of Δ_x to produce the forces and displacements associated with the actual displacement Δ_x. This multiplication is indicated by Δ_x in brackets to the right of the sketch in Figure 17.1*b*.

Case II. The structure is loaded at joint 2 by a set of forces that displace joint 2 a unit distance vertically but permit no horizontal displacement. The forces and displacements are then multiplied by the magnitude of Δ_y, to produce the forces and displacements associated with the actual displacement Δ_y (see Figure 17.1*c*).

If the structure responds to load in a linear, elastic manner, superposition of the two cases above is equivalent to the actual case. Case I supplies the required horizontal displacement, and Case II supplies the required vertical displacement.

In Figure 17.1*b*, forces K_{11} and K_{21} represent the forces required to displace joint 2 by 1 in to the right. In Figure 17.1*c*, forces K_{22} and K_{12} denote the forces required to displace joint 2 by 1 in upward. Subscripts are used

Figure 17.1: (*a*) Horizontal and vertical displacements Δ_x and Δ_y produced by the 10-kip load at joint 2; initially bar 1 is horizontal: bar 2 slopes upward at 45°; (*b*) forces (stiffness coefficients) K_{21} and K_{11} required to produce a unit horizontal displacement of joint 2; (*c*) forces K_{22} and K_{12} required to produce a unit vertical displacement of joint 2.

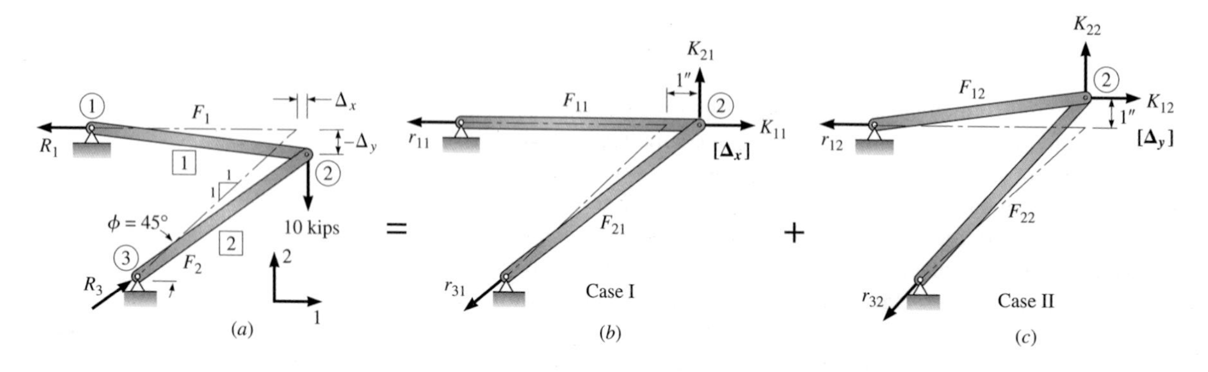

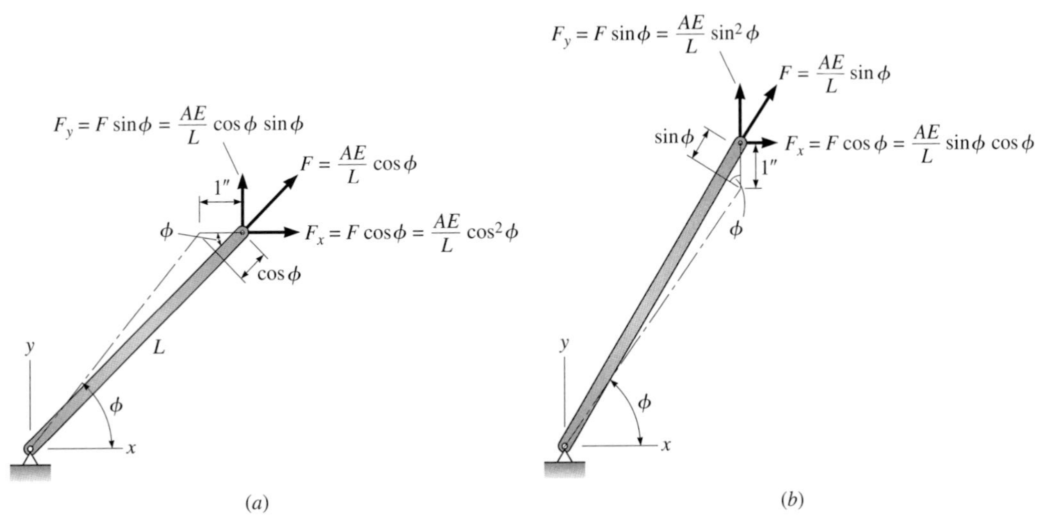

Figure 17.2: Stiffness coefficients for an axially loaded bar with area A, length L, and modulus of elasticity E. (*a*) Forces created by a unit horizontal displacement; (*b*) forces created by a unit vertical displacement.

to denote the direction of both the forces and the unit displacement with reference to the local x-y coordinate system at joint 2. The first subscript specifies the direction of the force. The second subscript denotes the direction of the unit displacement. The forces associated with a *unit displacement* are termed *stiffness coefficients*. These coefficients can be evaluated by referring to the member oriented with respect to the horizontal axis by an angle ϕ in Figure 17.2. In Figure 17.2*a*, the initial position of the unstressed member is shown by a dashed line. A unit horizontal displacement is induced at one end of the member, while vertical displacement is prevented. This displacement causes the member to elongate by an amount $\cos\phi$, which results in an axial force F equal to $(AE/L)\cos\phi$. The horizontal component F_x and vertical component F_y of the axial force represent the contribution of this member to K_{11} and K_{12}, respectively, in Figure 17.1*b*. Similarly, to evaluate the contribution of the member to K_{12} and K_{22}, a unit vertical displacement is induced, which produces an axial deformation $\sin\phi$. See Figure 17.2*b* for the corresponding force components. These expressions relate longitudinal force in an axially loaded bar (restrained at one end) to unit displacements in the horizontal and the vertical direction at the opposite end.

There is no need to guess the direction of the actual joint displacements. We arbitrarily specify the positive sense of the unit displacements. (In this book we assume that positive displacements are in the same direction as the positive sense of the local coordinate axes.) If the solution of the equilibrium equations (a step in the analysis that we discuss shortly) produces a positive value of displacement, the displacement is in the same direction as the unit displacement. Conversely, a negative value of displacement indicates that the actual displacement is opposite in direction to the unit displacement.

To establish the values of Δ_x and Δ_y for the truss in Figure 17.1*a*, we solve two equilibrium equations. These equations are established by superimposing the forces at joint 2 in Figure 17.1*b* and *c* and then equating their sum to the values of the actual joint forces in the original structure (see Figure 17.1*a*).

$$\xrightarrow{+} \quad \Sigma F_x = 0 \qquad K_{11}\Delta_x + K_{12}\Delta_y = 0 \tag{17.1}$$

$$\overset{+}{\uparrow} \quad \Sigma F_y = 0 \qquad K_{21}\Delta_x + K_{22}\Delta_y = -10 \tag{17.2}$$

Equations 17.1 and 17.2 can be written in matrix form as

$$\mathbf{K\Delta} = \mathbf{F} \tag{17.3}$$

where

$$\mathbf{K} = \begin{bmatrix} K_{11} & K_{12} \\ K_{21} & K_{22} \end{bmatrix} \qquad \mathbf{\Delta} = \begin{bmatrix} \Delta_x \\ \Delta_x \end{bmatrix} \qquad \mathbf{F} = \begin{bmatrix} F_1 \\ F_2 \end{bmatrix} = \begin{bmatrix} 0 \\ -10 \end{bmatrix} \tag{17.4}$$

where $\mathbf{K}$ = structure stiffness matrix (i.e., its elements are stiffness coefficients)
$\mathbf{\Delta}$ = column matrix of unknown joint displacements
$\mathbf{F}$ = column matrix of applied joint forces

To determine the values of Δ_x and Δ_y (the elements in the $\mathbf{\Delta}$ matrix), we premultiply both sides of Equation 17.3 by $\mathbf{K}^{-1}$, the inverse of $\mathbf{K}$.

$$\mathbf{K}^{-1}\mathbf{K\Delta} = \mathbf{K}^{-1}\mathbf{F}$$

Since $\mathbf{K}^{-1}\mathbf{K} = 1$,

$$\mathbf{\Delta} = \mathbf{K}^{-1}\mathbf{F} \tag{17.5}$$

After Δ_x and Δ_y are computed, reactions and bar forces can be calculated by superposition of corresponding forces acting at the supports and in the members shown in Cases I and II; that is, we multiply the forces in Case I by Δ_x and add the product to the corresponding forces in Case II multiplied by Δ_y. For example,

Reaction at support 1: $$R_1 = r_{11}\Delta_x + r_{12}\Delta_y \tag{17.6a}$$

Force in bar 1: $$F_1 = F_{11}\Delta_x + F_{12}\Delta_y \tag{17.6b}$$

To illustrate the details of the stiffness method, we will analyze the truss in Figure 17.1*a*, assuming the following member properties:

Bar areas: $A_1 = A_2 = A$

Modulus of elasticity: $E_1 = E_2 = E$

Length of bar: $L_1 = L_2 = L$

We evaluate the stiffness coefficients in Figure 17.1*b* with the aid of Figure 17.2*a*, where $\phi = 0°$ for bar 1 and $\phi = 45°$ for bar 2. For these angles, the respective values of sin ϕ and cos ϕ are

Bar 1: $\cos 0° = 1 \qquad \sin 0° = 0$

Bar 2: $\cos 45° = \dfrac{\sqrt{2}}{2} \qquad \sin 45° = \dfrac{\sqrt{2}}{2}$

Although the properties (A, E, and L) of both bars are identical, we will initially identify the terms that apply to each bar by using subscripted variables. Using Figure 17.2*a* to evaluate the stiffness coefficients in Figure 17.1*b* yields

$$K_{11} = \sum \frac{AE}{L} \cos^2 \phi = \frac{A_1E_1}{L_1}(1)^2 + \frac{A_2E_2}{L_2}\left(\frac{\sqrt{2}}{2}\right)^2 \tag{17.7}$$

$$K_{21} = \sum \frac{AE}{L} \cos \phi \sin \phi = \frac{A_1E_1}{L_1}(1)(0) + \frac{A_2E_2}{L_2}\left(\frac{\sqrt{2}}{2}\right)^2 \tag{17.8}$$

We evaluate the stiffness coefficients in Figure 17.1*c* with Figure 17.2*b*.

$$K_{22} = \sum \frac{AE}{L} \sin^2 \phi = \frac{A_1E_1}{L_1}(0)^2 + \frac{A_2E_2}{L_2}\left(\frac{\sqrt{2}}{2}\right)^2 \tag{17.9}$$

$$K_{12} = \sum \frac{AE}{L} \sin \phi \cos \phi = \frac{A_1E_1}{L_1}(0)(1) + \frac{A_2E_2}{L_2}\left(\frac{\sqrt{2}}{2}\right)^2 \tag{17.10}$$

Writing the stiffness coefficients in Equations 17.7 to 17.10 in terms of A, E, and L; combining terms; and substituting them into Equation 17.4, we can write the structure stiffness matrix $\mathbf{K}$ as

$$\mathbf{K} = \begin{bmatrix} \dfrac{3AE}{2L} & \dfrac{AE}{2L} \\ \dfrac{AE}{2L} & \dfrac{AE}{2L} \end{bmatrix} = \frac{AE}{2L}\begin{bmatrix} 3 & 1 \\ 1 & 1 \end{bmatrix} \tag{17.11}$$

The inverse of the $\mathbf{K}$ matrix is

$$\mathbf{K}^{-1} = \frac{L}{AE}\begin{bmatrix} 1 & -1 \\ -1 & 3 \end{bmatrix} \tag{17.12}$$

Substituting $\mathbf{K}^{-1}$ given by Equation 17.12 and $\mathbf{F}$ given by Equation 17.4 into Equation 17.5 and multiplying give

$$\begin{bmatrix} \Delta_x \\ \Delta_y \end{bmatrix} = \frac{L}{AE}\begin{bmatrix} 1 & -1 \\ -1 & 3 \end{bmatrix}\begin{bmatrix} 0 \\ -10 \end{bmatrix} = \frac{L}{AE}\begin{bmatrix} 10 \\ -30 \end{bmatrix}$$

that is,

$$\Delta_x = \frac{10L}{AE} \qquad \Delta_y = -\frac{30L}{AE} \tag{17.13}$$

Bar forces are now computed by superimposing Cases I and II. To evaluate the axial forces produced by unit displacements, we use Figure 17.2. For bar 1 ($\phi = 0°$),

$$F_1 = F_{11}\Delta_x + F_{12}\Delta_y \tag{17.6b}$$

where $F_{11} = F = (AE/L)\cos\phi$ (Figure 17.2*a*) and $F_{12} = F = (AE/L)\sin\phi$ (Figure 17.2*b*).

$$F_1 = \frac{AE}{L}(1)\left(\frac{10L}{AE}\right) + \frac{AE}{L}(0)\left(-\frac{30L}{AE}\right) = 10 \text{ kips}$$

For bar 2 ($\phi = 45°$),

$$F_2 = \Delta_x F_{21} + \Delta_y F_{22}$$

where $F_{21} = F = (AE/L)\cos\phi$ in Figure 17.2*a*, and $F_{22} = F = (AE/L)$ $\sin\phi$ in Figure 17.2*b*.

$$F_2 = \frac{AE}{L}\left(\frac{\sqrt{2}}{2}\right)\left(\frac{10L}{AE}\right) + \left(\frac{AE}{L}\right)\left(\frac{\sqrt{2}}{2}\right)\left(-\frac{30L}{AE}\right) = -10\sqrt{2} \text{ kips}$$

17.2 Member and Structure Stiffness Matrices

To permit the stiffness method (introduced in Section 17.1) to be programmed automatically from the input data (i.e., joint coordinates, member properties, joint loads, and so forth), we now introduce a slightly different procedure for generating the *structure stiffness matrix* **K**. In this modified procedure we generate the *member stiffness matrix* **k** of individual truss members and then combine these matrices to form the structure stiffness matrix **K**.

The member stiffness matrix for an axially loaded bar relates the axial forces at the ends of the member to the axial displacements at each end. The elements of the member stiffness matrix are initially expressed in terms of a *local* or *member coordinate system* whose x' axis is collinear with that of the member's longitudinal axis. Since the inclination of the longitudinal axes of individual bars usually varies, before we can combine the member stiffness matrices, we must transform their properties from the individual member coordinate systems to that of a single *global coordinate system* for the structure. Although the orientation of the global coordinate system is arbitrary, typically we locate its origin at an exterior joint on the base of the structure. For a planar structure we position its x and y axes in the horizontal and vertical directions.

In Section 17.3, we introduce a procedure to construct the member stiffness matrix $\mathbf{k}'$ in terms of a local coordinate system. When the local coordinate system of all truss bars coincides with the global coordinate system, Section 17.4 presents a procedure to assemble the structure stiffness matrix from the member stiffness matrices. After the structure stiffness matrix is established, Section 17.5 describes a procedure to determine the unknown nodal displacements, reactions, member deformations, and forces. Section 17.6 discusses the more general case of truss bars that are inclined with respect to the global coordinate system; for this case a procedure to establish the member stiffness matrix **k** in terms of the global coordinate system is presented. Section 17.7 describes an alternate approach to construct **k** from **k**′using a transformation matrix.

17.3 Construction of a Member Stiffness Matrix for an Individual Truss Bar

To generate the member stiffness matrix of an axially loaded bar, we will consider member n with length L, area A, and modulus of elasticity E in Figure 17.3*a*. The nodes (or joints) of the member are denoted by the numbers 1 and 2. We also show a local coordinate system with origin at 1 and x' and y' axes superimposed on the bar. We assume that the positive direction for horizontal forces and displacements is in the positive direction of the x' axis (i.e., directed to the right). As shown in Figure 17.3*b*, we first introduce a displacement Δ_1 at joint 1, while joint 2 is assumed to be restrained by a temporary pin support. Expressing the end forces in terms of Δ_1 using Equation 16.6 yields

$$Q_{11} = \frac{AE}{L}\Delta_1 \quad \text{and} \quad Q_{21} = -\frac{AE}{L}\Delta_1 \tag{17.14}$$

The end forces produced by the displacement Δ_1 are identified by two subscripts. The first subscript denotes the location of the joint at which the force acts, and the second subscript indicates the location of the displacement. The minus sign for Q_{21} is required because it acts in the negative x' direction. As we have seen in Section 17.1, the end forces Q_{11} and Q_{21} could also have been generated by introducing a unit displacement at joint 1 and multiplying the stiffness coefficients $K_{11} = AE/L$ and $K_{21} = -AE/L$ by the actual displacement Δ_1.

Similarly, if joint 1 is restrained while joint 2 is displaced in the positive direction a distance Δ_2, the end forces are

$$Q_{12} = -\frac{AE}{L}\Delta_2 \quad \text{and} \quad Q_{22} = \frac{AE}{L}\Delta_2 \tag{17.15}$$

To evaluate the resultant forces Q_1 and Q_2 at each end of the member in terms of the end displacements Δ_1 and Δ_2 (see Figure 17.3*d*), we add corresponding terms of Equations 17.14 and 17.16, producing

$$\begin{aligned} Q_1 &= Q_{11} + Q_{12} = \frac{AE}{L}(\Delta_1 - \Delta_2) \\ Q_2 &= Q_{21} + Q_{22} = \frac{AE}{L}(-\Delta_1 + \Delta_2) \end{aligned} \tag{17.16}$$

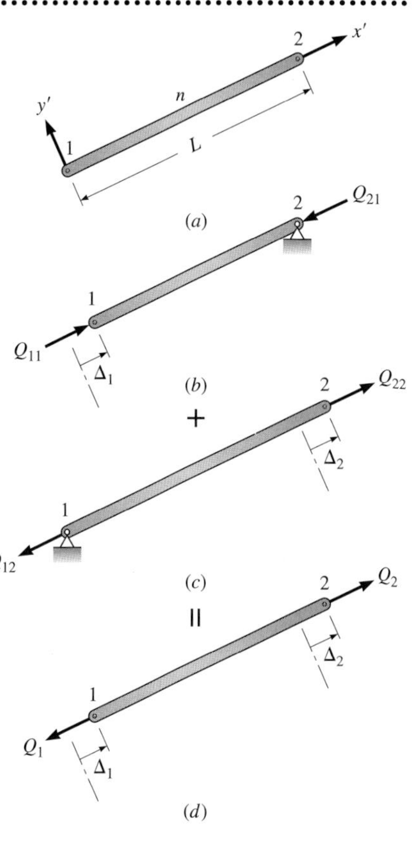

Figure 17.3: Stiffness coefficients for an axially loaded bar; (*a*) bar showing local coordinate system with origin at node 1; (*b*) displacement introduced at node 1 with node 2 restrained; (*c*) displacement introduced at node 2 with node 1 restrained; (*d*) end forces and displacements of the actual bar produced by superposition of (*b*) and (*c*).

Equation 17.16 can be expressed in matrix notation as

$$\begin{bmatrix} Q_1 \\ Q_2 \end{bmatrix} = \begin{bmatrix} \dfrac{AE}{L} & -\dfrac{AE}{L} \\ -\dfrac{AE}{L} & \dfrac{AE}{L} \end{bmatrix} \begin{bmatrix} \Delta_1 \\ \Delta_2 \end{bmatrix} \tag{17.17}$$

or

$$\mathbf{Q} = \mathbf{k}'\mathbf{\Delta} \tag{17.18}$$

where the member stiffness matrix in the local coordinate system is

$$\mathbf{k}' = \begin{bmatrix} \dfrac{AE}{L} & -\dfrac{AE}{L} \\ -\dfrac{AE}{L} & \dfrac{AE}{L} \end{bmatrix} = \frac{AE}{L}\begin{bmatrix} 1 & -1 \\ -1 & 1 \end{bmatrix} \tag{17.19}$$

and $\mathbf{\Delta}$ is the displacement vector. The prime is added to $\mathbf{k}'$ to indicate that the formulation is in terms of the member's local coordinates x' and y'. Since all elements AE/L in matrix $\mathbf{k}'$ can be interpreted as the force associated with a unit axial displacement of one end of the member when the opposite end is restrained, they are stiffness coefficients and may be denoted as

$$k = \frac{AE}{L} \tag{17.20}$$

We also observe that the sum of the elements in each column of $\mathbf{k}'$ equals zero. This condition follows because the coefficients in each column represent the forces produced by a unit displacement of one joint while the other joint is restrained. Since the bar is in equilibrium in the x' direction, the forces must sum to zero. In addition, all coefficients along the main diagonal must be positive because these terms are associated with the force acting at that joint at which a positive displacement is introduced into the structure, and correspondingly the force is in the same (positive) direction as the displacement.

Note that the displacement matrix $\mathbf{\Delta}$ in Equation 17.17 contains only displacements Δ_1 and Δ_2 along the axis of the member. End displacements in the y' direction do not have to be included in the formulation because these transverse movements do not produce internal force in truss members based on small-deformation theory.

17.4 Assembly of the Structure Stiffness Matrix

If a structure consists of several bars and the local coordinate system of these bars coincides with the global coordinate system, then the stiffness matrix $\mathbf{K}$ of the structure can be generated by either of the two following methods:

1. Introducing displacements at each joint with all other joints restrained
2. Combining the stiffness matrices of the individual bars

We will illustrate the use of both methods by generating the structure stiffness matrix of the two-bar system shown in Figure 17.4*a*.

Method 1. Superimposing Forces Produced by Nodal Displacements

As shown in Figure 17.4*b* to *d*, we introduce displacements at each joint while all other joints are restrained and compute the joint forces using Equation 16.6 [that is, $Q = (AE/L)\Delta = k\Delta$]. Displacements and forces are positive when directed to the right. Define $k_1 = A_1E_1/L_1$ and $k_2 = A_2E_2/L_2$.

Case 1. Joint 1 displaces Δ_1; joints 2 and 3 are restrained (see Figure 17.4*b*). Since bar 2 does not deform, no reaction develops at joint 3.

$$Q_{11} = k_1\Delta_1 \qquad Q_{21} = -k_1\Delta_1 \qquad Q_{31} = 0 \tag{17.21}$$

Case 2. Joint 2 displaces Δ_2; joints 1 and 3 are restrained (see Figure 17.4*c*).

$$Q_{12} = -k_1\Delta_2 \qquad Q_{22} = (k_1 + k_2)\Delta_2 \qquad Q_{32} = -k_2\Delta_2 \tag{17.22}$$

Case 3. Joint 3 displaces Δ_3; joints 2 and 3 are restrained (see Figure 17.4*d*).

$$Q_{13} = 0 \qquad Q_{23} = -k_2\Delta_3 \qquad Q_{33} = k_2\Delta_3 \tag{17.23}$$

To express the resultant joint forces Q_1, Q_2, and Q_3 in terms of nodal displacements, we sum the Q forces at each joint given by Equations 17.21, 17.22, and 17.23.

$$\begin{aligned} Q_1 &= Q_{11} + Q_{12} + Q_{13} = k_1\Delta_1 \quad -k_1\Delta_2 \\ Q_2 &= Q_{21} + Q_{22} + Q_{23} = -k_1\Delta_1 + (k_1 + k_2)\Delta_2 - k_2\Delta_3 \\ Q_3 &= Q_{31} + Q_{32} + Q_{33} = \quad -k_2\Delta_2 \quad k_2\Delta_3 \end{aligned} \tag{17.24}$$

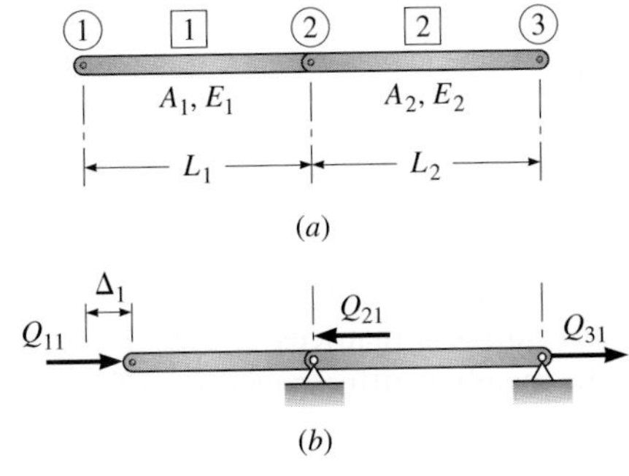

(*b*)

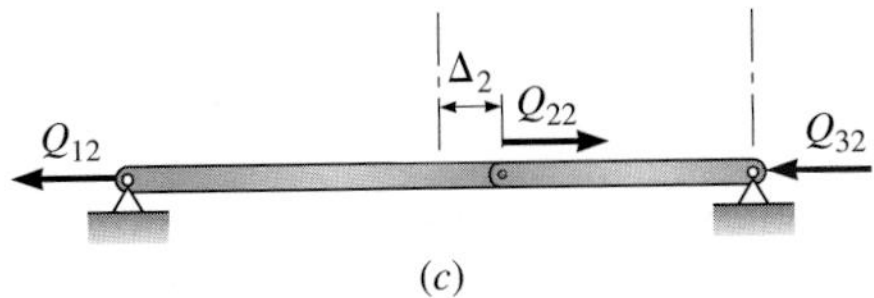

(*c*)

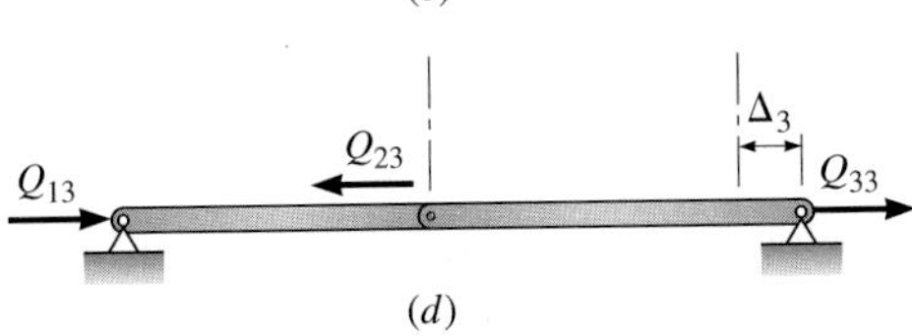

(*d*)

Figure 17.4: Loading conditions used to generate the structure stiffness matrix: (*a*) properties of two-bar system; (*b*) node forces produced by a positive displacement Δ_1 of joint 1 with nodes 2 and 3 restrained; (*c*) node forces produced by a positive displacement of node 2 with nodes 1 and 3 restrained; (*d*) node forces produced by a positive displacement of node 3 with nodes 1 and 2 restrained.

Expressing the three equations above in matrix notation yields

$$\begin{bmatrix} Q_1 \\ Q_2 \\ Q_3 \end{bmatrix} = \begin{bmatrix} k_1 & -k_1 & 0 \\ -k_1 & k_1 + k_2 & -k_2 \\ 0 & -k_2 & k_2 \end{bmatrix} \begin{bmatrix} \Delta_1 \\ \Delta_2 \\ \Delta_3 \end{bmatrix} \tag{17.25}$$

or

$$\mathbf{Q} = \mathbf{K\Delta} \tag{17.26}$$

where $\mathbf{Q}$ = column matrix of nodal forces
$\mathbf{\Delta}$ = column matrix of nodal displacements
$\mathbf{K}$ = structure stiffness matrix

As we discussed previously, the coefficients in each column of the stiffness matrix of Equation 17.25 sum to zero because they constitute a set of forces in equilibrium. Since the matrix is symmetric (the Maxwell-Betti principle), the sum of the coefficients in each row must also equal zero.

If the nodal forces in vector **Q** of Equation 17.26 are specified, it appears initially that we can determine the joint displacements by premultiplying both sides of Equation 17.26 by the inverse of the structure stiffness matrix **K**. However, the three equations represented by Equation 17.25 are not independent since row 2 is a linear combination of rows 1 and 3. To prove this, we can produce row 2 by adding rows 1 and 3 after they are multiplied by -1. Since only two independent equations are available to solve for three unknowns, the **K** matrix is singular and cannot be inverted. The fact that we are not able to solve the three equilibrium equations indicates that the structure is unstable (i.e., not in equilibrium). The instability occurs because no supports were specified for the structure (see Figure 17.4*a*). As we will discuss shortly, if sufficient supports are provided to produce a stable structure, we can partition the matrix into submatrices that can be solved for the unknown nodal displacements.

Method 2. Construction of the Structure Stiffness Matrix by Combining Member Stiffness Matrices

The stiffness matrix of the structure in Figure 17.4 can also be generated by combining the member stiffness matrices of bars 1 and 2. Using Equation 17.19, we can write the member stiffness matrices of the two bars as

$$\mathbf{k}_1' = \begin{bmatrix} \overset{1}{k_1} & \overset{2}{-k_1} \\ -k_1 & k_1 \end{bmatrix}\begin{matrix} 1 \\ 2 \end{matrix} \qquad \mathbf{k}_2' = \begin{bmatrix} \overset{2}{k_2} & \overset{3}{-k_2} \\ -k_2 & k_2 \end{bmatrix}\begin{matrix} 2 \\ 3 \end{matrix} \tag{17.27}$$

Subscripts are added to the stiffness coefficients to identify the bar whose properties they represent. We also label the top of each column with a number that identifies the particular joint displacement associated with the elements in the column, and we number the rows to the right of each bracket to identify the nodal force associated with the elements in the row.

We construct a global xy coordinate system at joint 1 such that this system coincides with the local $x'y'$ coordinate system of individual bars. Because the x' axis of each bar coincides with the x axis in the global coordinate system, so $\mathbf{k}_1 = \mathbf{k}_1'$ and $\mathbf{k}_2 = \mathbf{k}_2'$. Since the elements in the first and second columns of each matrix in Equation 17.27 refer to different joints, adding these two matrices directly has no physical significance. To permit addition of the matrices, we expand them to the same order as that of the structure stiffness matrix (3 in this case for horizontal displacements at three joints) by adding an extra row and an extra column.

$$\mathbf{k}_1 = \begin{bmatrix} k_1 & -k_1 & 0 \\ -k_1 & k_1 & 0 \\ 0 & 0 & 0 \end{bmatrix} \qquad \mathbf{k}_2 = \begin{bmatrix} 0 & 0 & 0 \\ 0 & k_2 & -k_2 \\ 0 & -k_2 & k_2 \end{bmatrix} \tag{17.28}$$

For example, the coefficients in matrix $\mathbf{k}_1$ (Equation 17.27) relate the forces at joints 1 and 2 to the displacement of the same joints. To eliminate in the expanded matrix (Equation 17.28) the effect of displacements at joint 3 on the forces at joints 1, 2, and 3, the elements in the third column of the expanded matrix must be set equal to zero because these terms will be multiplied by the displacement of joint 3. Similarly, since the original 2×2 $\mathbf{k}_1$ matrix does not influence the force at joint 3, the elements in the bottom row of the matrix must all be set equal to zero. Similar reasoning requires that we expand matrix $\mathbf{k}_2$ to a 3×3 matrix by adding zeros in the first row and column. Since the expanded matrices given by Equation 17.28 are of the same order, we can add their elements directly to produce the structure stiffness matrix $\mathbf{K}$.

$$\mathbf{K} = \mathbf{k}_1 + \mathbf{k}_2 = \begin{bmatrix} k_1 & -k_1 & \\ -k_1 & k_1 & \\ 0 & & \end{bmatrix} + \begin{bmatrix} & & 0 \\ & k_2 & -k_2 \\ & -k_2 & k_2 \end{bmatrix} = \begin{bmatrix} k_1 & -k_1 & 0 \\ -k_1 & k_1 + k_2 & -k_2 \\ 0 & -k_2 & k_2 \end{bmatrix} \tag{17.29}$$

The stiffness matrix given by Equation 17.29 is identical to that produced by method 1 (see Equation 17.25).

It is not necessary in actual application to expand the individual member stiffness matrices to construct the structure stiffness matrix. More simply, we insert the stiffness coefficients of the member stiffness matrix into the appropriate rows and columns of the structure stiffness matrix. In Equation 17.29 the individual member stiffness matrix is enclosed in dashed lines to show its position in the structure stiffness matrix.

17.5 Solution of the Direct Stiffness Method

Once the structure stiffness matrix $\mathbf{K}$ is assembled and the force-displacement relationship (Equation 17.26) established, we describe in this section how to evaluate the unknown joint displacement vector $\mathbf{\Delta}$ and support reactions of a structure. As we discussed in Section 17.1, the first step in the stiffness analysis is to compute the unknown nodal displacements. This step consists of solving a set of equilibrium equations (for example, see Equations 17.1 and 17.2) in which the nodal displacements are the unknowns. The terms that make up these equilibrium equations are submatrices of the three matrices **Q**, **K**, and **Δ** of Equation 17.26. These submatrices can be established by partitioning the matrices in Equation 17.26 so that terms associated with the nodes that are free to displace are separated from terms that are associated with nodes restrained by the supports. This step requires that all rows associated with the degrees of freedom be shifted to the top of the matrix. (When a row is shifted upward, the corresponding column also needs to be shifted forward to the left in a similar manner.) If the matrix analysis is done by hand, we can accomplish this step by numbering the unrestrained joints before the restrained joints. The result of this reorganization and partitioning will permit us to express Equation 17.26 in terms of the following submatrices:

$$\begin{bmatrix} \mathbf{Q}_f \\ \hline \mathbf{Q}_s \end{bmatrix} = \left[\begin{array}{c|c} \mathbf{K}_{11} & \mathbf{K}_{12} \\ \hline \mathbf{K}_{21} & \mathbf{K}_{22} \end{array}\right] \begin{bmatrix} \mathbf{\Delta}_f \\ \hline \mathbf{\Delta}_s \end{bmatrix} \tag{17.30}$$

where $\mathbf{Q}_f$ = matrix containing values of load at joints free to displace
$\mathbf{Q}_s$ = matrix containing unknown support reactions
$\mathbf{\Delta}_f$ = matrix containing unknown joint displacements
$\mathbf{\Delta}_s$ = matrix containing support displacements

Multiplying the matrices in Equation 17.30 gives

$$\mathbf{Q}_f = \mathbf{K}_{11}\mathbf{\Delta}_f + \mathbf{K}_{12}\mathbf{\Delta}_s \tag{17.31}$$

$$\mathbf{Q}_s = \mathbf{K}_{21}\mathbf{\Delta}_f + \mathbf{K}_{22}\mathbf{\Delta}_s \tag{17.32}$$

If the supports do not move (i.e., $\mathbf{\Delta}_s$ is a null matrix), the equations above reduce to

$$\mathbf{Q}_f = \mathbf{K}_{11}\mathbf{\Delta}_f \tag{17.33}$$

$$\mathbf{Q}_s = \mathbf{K}_{21}\mathbf{\Delta}_f \tag{17.34}$$

Since the elements in $\mathbf{Q}_f$ and $\mathbf{K}_{11}$ are known, Equation 17.33 can be solved for $\mathbf{\Delta}_f$ by premultiplying both sides of the equation by $\mathbf{K}_{11}^{-1}$, to give

$$\mathbf{\Delta}_f = \mathbf{K}_{11}^{-1}\mathbf{Q}_f \tag{17.35}$$

Substituting the value of $\mathbf{\Delta}_f$ into Equation 17.34 gives the support reactions

$$\mathbf{Q}_s = \mathbf{K}_{21}\mathbf{K}_{11}^{-1}\mathbf{Q}_f \tag{17.36}$$

In Example 17.1 we apply the stiffness method to the analysis of a simple truss. The method does not depend on the degree of indeterminacy of the structure and is applied in the same way to both determinate and indeterminate structures.

EXAMPLE 17.1

Determine the joint displacements and reactions for the structure in Figure 17.5 by partitioning the structure stiffness matrix.

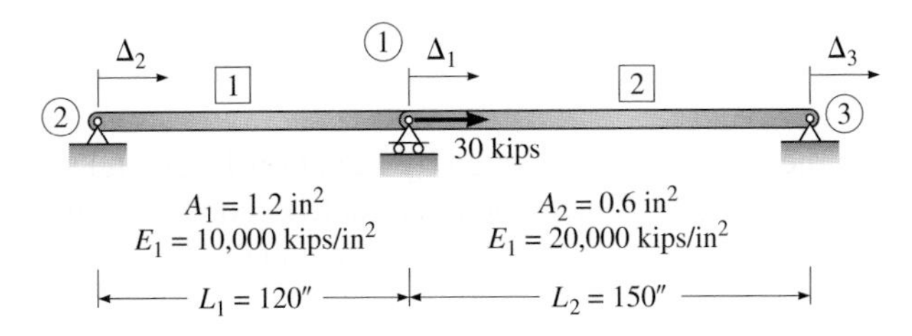

Figure 17.5

Solution

Number the joints, starting with those that are free to displace. The positive sense of displacements and forces at each joint are indicated by arrows. Since the bars carry only axial force, we only consider displacements in the horizontal direction.

Compute the stiffness $k = AE/L$ for each member.

$$k_1 = \frac{1.2(10{,}000)}{120} = 100 \text{ kips/in}$$

$$k_2 = \frac{0.6(20{,}000)}{150} = 80 \text{ kips/in}$$

Evaluate member stiffness matrices, using Equation 17.19. Because the local coordinate system of each bar coincides with the global coordinate system, $\mathbf{k}' = \mathbf{k}$.

$$\mathbf{k}_1 = k_1\begin{bmatrix} 1 & -1 \\ -1 & 1 \end{bmatrix} = \begin{bmatrix} \overset{1}{100} & \overset{2}{-100} \\ -100 & 100 \end{bmatrix}\begin{matrix} 1 \\ 2 \end{matrix}$$

$$\mathbf{k}_2 = k_2\begin{bmatrix} 1 & -1 \\ -1 & 1 \end{bmatrix} = \begin{bmatrix} \overset{1}{80} & \overset{3}{-80} \\ -80 & 80 \end{bmatrix}\begin{matrix} 1 \\ 3 \end{matrix}$$

[continues on next page]

Example 17.1 continues . . .

Set up the structure stiffness matrix **K** by combining terms of the member stiffness matrices $\mathbf{k}_1$ and $\mathbf{k}_2$. Establish Equation 17.30 as follows:

$$\begin{bmatrix} Q_1 = 30 \\ \hline Q_2 \\ Q_3 \end{bmatrix} = \left[\begin{array}{c|cc} \overset{1}{100+80} & \overset{2}{-100} & \overset{3}{-80} \\ \hline -100 & 100 & 0 \\ -80 & 0 & 80 \end{array}\right] \begin{bmatrix} \Delta_1 \\ \hline \Delta_2 = 0 \\ \Delta_3 = 0 \end{bmatrix}$$

Partition the matrices as indicated by Equation 17.30 and solve for Δ_1 using Equation 17.35. Since each submatrix contains one element, Equation 17.35 reduces to a simple algebraic equation.

$$\mathbf{\Delta}_f = \mathbf{K}_{11}^{-1}\mathbf{Q}_f$$

$$\Delta_1 = \frac{1}{180}(30) = \frac{1}{6} \text{ in}$$

Solve for the reactions, using Equation 17.36.

$$\mathbf{Q}_s = \mathbf{K}_{21}\mathbf{K}_{11}^{-1}\mathbf{Q}_f$$

$$\begin{bmatrix} Q_2 \\ Q_3 \end{bmatrix} = \begin{bmatrix} -100 \\ -80 \end{bmatrix} \begin{bmatrix} \dfrac{1}{180} \end{bmatrix} [30] = \begin{bmatrix} -16.67 \\ -13.33 \end{bmatrix}$$

where

$$Q_2 = \frac{1}{180}(-100)30 = -16.67 \text{ kips}$$

$$Q_3 = \frac{1}{180}(-80)30 = -13.33 \text{ kips}$$

Therefore, the reactions at joints 2 and 3 are -16.67 and -13.33 kips, respectively. The minus signs indicate that the forces act to the left.

17.6 Member Stiffness Matrix of an Inclined Truss Bar

To illustrate the construction of the structure stiffness matrix in Section 17.4, we analyzed a simple truss with horizontal bars. Since the orientation of both the member and the global coordinate systems for these bars is identical, $\mathbf{k}'$ equals $\mathbf{k}$ and we are able to insert the 2×2 member stiffness matrices directly into the structure stiffness matrix. This method, however, cannot be applied to a truss with inclined bars. In this section, we develop the member stiffness matrix $\mathbf{k}$ for an inclined bar in terms of global coordinates so that the direct stiffness method can be extended to trusses with diagonal members.

In Figure 17.6*a* we show an inclined member ij. Joint i is denoted the *near* end and joint j the *far* end. The initial position of the unstressed member is shown by a dashed line. The member's local axis, x', makes an angle ϕ with the x axis of the global coordinate system whose origin is located at joint i. We assign a *positive direction to the bar by placing an arrow directed from joint i to joint j along the axis of the bar*. By assigning a positive direction to each bar, we will be able to account for the sign (plus or minus) of the sine and cosine functions that appear in the elements of the member stiffness matrix.

To generate the force-displacement relationships for an inclined bar in the global coordinate system, we introduce, in sequence, displacements in the x and y directions at each end of the member. These displacements are labeled with two subscripts. The first identifies the location of the joint where the displacement occurs; the second denotes the direction of the displacement with respect to the global axes.

The components of force at the ends of the bar and the magnitude of the joint displacement along the axis of the bar created by the respective displacements in Figure 17.6 are evaluated using Figure 17.2. Since the forces and deformations in Figure 17.2 are produced by unit displacements, they must be multiplied by the actual magnitude of the displacements in Figure 17.6. Displacements in Figure 17.6 are shown to an exaggerated scale to show the geometric relationships clearly. Since the displacements are actually small, we can assume that the slope of the bar is not changed by the end displacements. Treating x_i, y_i, x_j, and y_j as the coordinates of joints i and j, respectively, $\sin\phi$ and $\cos\phi$ can be expressed in terms of the coordinates of nodes i and j as

$$\sin\phi = \frac{y_j - y_i}{L} \qquad \cos\phi = \frac{x_j - x_i}{L} \tag{17.37}$$

where

$$L = \sqrt{(x_j - x_i)^2 + (y_j - y_i)^2} \tag{17.38}$$

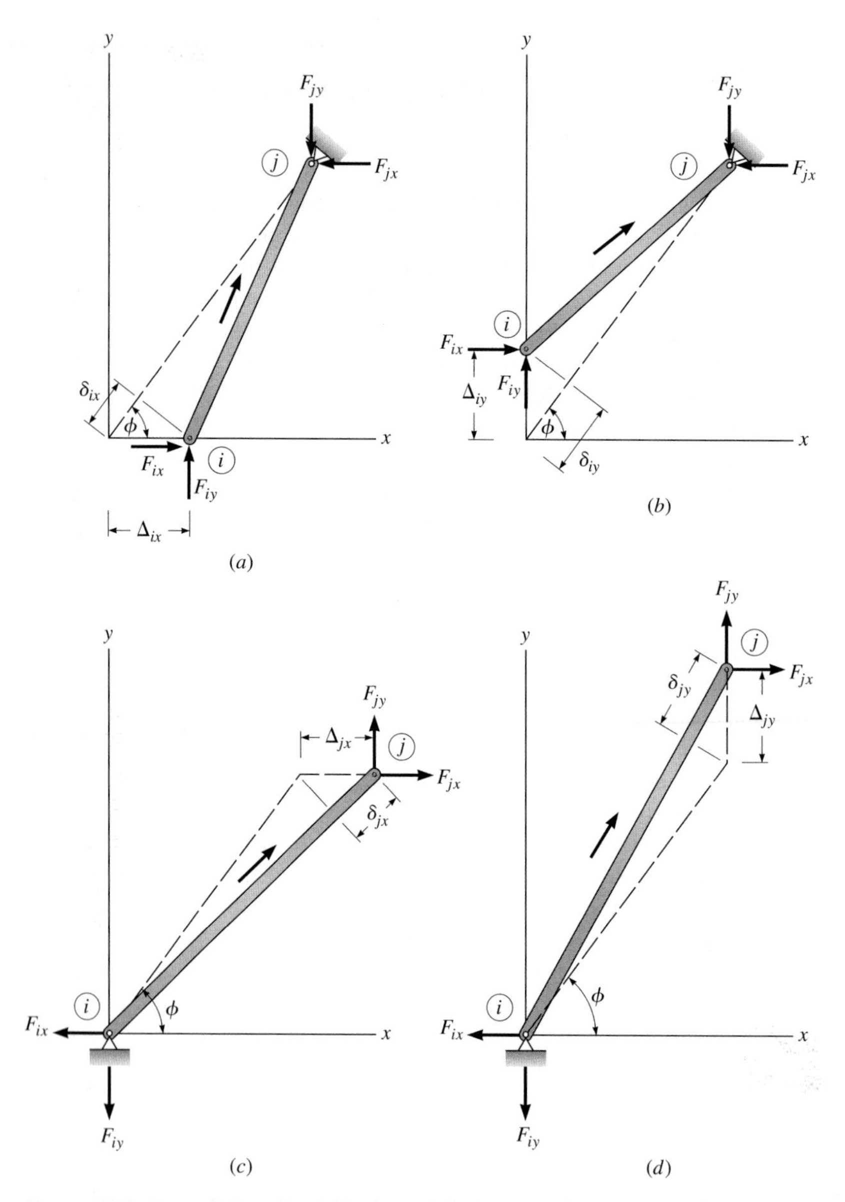

Figure 17.6: Forces induced by (*a*) horizontal displacement Δ_{ix}; (*b*) vertical displacement Δ_{iy}; (*c*) horizontal displacement Δ_{jx}; (*d*) vertical displacement Δ_{jy}.

Case 1. Introduce a horizontal displacement Δ_{ix} at node i with the j end of the bar restrained, producing an axial force F_i in the bar (see Figure 17.6*a*).

$$F_i = \frac{AE}{L}\delta_{ix} \qquad \text{where } \delta_{ix} = (\cos\phi)\Delta_{ix} \tag{17.39}$$

$$
\begin{aligned}
F_{ix} &= F_i \cos\phi = \frac{AE}{L}(\cos^2\phi)\Delta_{ix} \\
F_{iy} &= F_i \sin\phi = \frac{AE}{L}(\cos\phi)(\sin\phi)\Delta_{ix} \\
F_{jx} &= -F_{ix} = -\frac{AE}{L}(\cos^2\phi)\Delta_{ix} \\
F_{jy} &= -F_{iy} = -\frac{AE}{L}(\cos\phi)(\sin\phi)\Delta_{ix}
\end{aligned}
\tag{17.40}
$$

Case 2. Introduce a vertical displacement Δ_{iy} at node i with the j end of the bar restrained (see Figure 17.6*b*).

$$F_i = \frac{AE}{L}\delta_{iy} \qquad \text{where } \delta_{iy} = (\sin\phi)\Delta_{iy} \tag{17.41}$$

$$
\begin{aligned}
F_{ix} &= \frac{AE}{L}(\sin\phi)(\cos\phi)\Delta_{iy} \\
F_{iy} &= \frac{AE}{L}(\sin^2\phi)\Delta_{iy} \\
F_{jx} &= -F_{ix} = -\frac{AE}{L}(\sin\phi)(\cos\phi)\Delta_{iy} \\
F_{jy} &= -F_{iy} = -\frac{AE}{L}(\sin^2\phi)\Delta_{iy}
\end{aligned}
\tag{17.42}
$$

Case 3. Introduce a horizontal displacement Δ_{jx} at node j with the i end of the bar restrained (Figure 17.6*c*).

$$\delta_{jx} = (\cos\phi)\Delta_{jx} \tag{17.43}$$

Values of joint force are identical to those given by Equations 17.40 but with Δ_{jx} substituted for Δ_{ix} and the signs reversed; that is, the forces at joint j act upward and to the right, and the reactions at joint i act downward and to the left.

$$
\begin{aligned}
F_{ix} &= -\frac{AE}{L}(\cos^2\phi)\Delta_{jx} \\
F_{iy} &= -\frac{AE}{L}(\sin\phi)(\cos\phi)\Delta_{jx} \\
F_{jx} &= \frac{AE}{L}(\cos^2\phi)\Delta_{jx} \\
F_{jy} &= \frac{AE}{L}(\sin\phi)(\cos\phi)\Delta_{jx}
\end{aligned}
\tag{17.44}
$$

Case 4. Introduce a vertical displacement Δ_{jy} at node j with the i end of the bar restrained (Figure 17.6d).

$$\delta_{jy} = (\sin \phi)\Delta_{jy} \tag{17.45}$$

Values of joint forces are identical to those given by Equations 17.42 but with Δ_{jy} substituted for Δ_{iy} and the signs reversed.

$$\begin{aligned}
F_{ix} &= -\frac{AE}{L}(\sin \phi)(\cos \phi)\Delta_{jy} \\
F_{iy} &= -\frac{AE}{L}(\sin^2 \phi)\Delta_{jy} \\
F_{jx} &= \frac{AE}{L}(\sin \phi)(\cos \phi)\Delta_{jy} \\
F_{jy} &= \frac{AE}{L}(\sin^2 \phi)\Delta_{jy}
\end{aligned} \tag{17.46}$$

If horizontal and vertical displacements occur at both joints i and j, the components of member force Q at each end can be evaluated by summing the forces given by Equations 17.40, 17.42, 17.44, and 17.46; that is,

$$\begin{aligned}
Q_{ix} = \Sigma F_{ix} &= \frac{AE}{L}[(\cos^2 \phi)\Delta_{ix} + (\sin \phi)(\cos \phi)\Delta_{iy} - (\cos^2 \phi)\Delta_{jx} - (\sin \phi)(\cos \phi)\Delta_{jy}] \\
Q_{iy} = \Sigma F_{iy} &= \frac{AE}{L}[(\sin \phi)(\cos \phi)\Delta_{ix} + (\sin^2 \phi)\Delta_{iy} - (\sin \phi)(\cos \phi)\Delta_{jx} - (\sin^2 \phi)\Delta_{jy}] \\
Q_{jx} = \Sigma F_{jx} &= \frac{AE}{L}[-(\cos^2 \phi)\Delta_{ix} - (\sin \phi)(\cos \phi)\Delta_{iy} + (\cos^2 \phi)\Delta_{jx} + (\sin \phi)(\cos \phi)\Delta_{jy}] \\
Q_{jy} = \Sigma F_{jy} &= \frac{AE}{L}[-(\sin \phi)(\cos \phi)\Delta_{ix} - (\sin^2 \phi)\Delta_{iy} + (\sin \phi)(\cos \phi)\Delta_{jx} + (\sin^2 \phi)\Delta_{jy}]
\end{aligned} \tag{17.47}$$

Letting $\cos \phi = c$ and $\sin \phi = s$, we can write the foregoing set of equations in matrix notation as

$$\begin{bmatrix} Q_{ix} \\ Q_{iy} \\ Q_{jx} \\ Q_{jy} \end{bmatrix} = \frac{AE}{L} \begin{bmatrix} c^2 & sc & -c^2 & -sc \\ sc & s^2 & -sc & -s^2 \\ -c^2 & -sc & c^2 & sc \\ -sc & -s^2 & sc & s^2 \end{bmatrix} \begin{bmatrix} \Delta_{ix} \\ \Delta_{iy} \\ \Delta_{jx} \\ \Delta_{jy} \end{bmatrix} \tag{17.48}$$

or

$$\mathbf{Q} = \mathbf{k}\boldsymbol{\Delta} \tag{17.49}$$

where $\mathbf{Q}$ = vector of member end forces referenced to the global coordinate system

$\mathbf{k}$ = member stiffness matrix in terms of global coordinates

$\boldsymbol{\Delta}$ = matrix of joints displacements referenced to the global coordinate system

The axial displacement δ_i of joint i in the direction of the member's longitudinal axis can be expressed in terms of the horizontal and vertical components of displacement at joint i by summing Equations 17.39 and 17.41. Similarly, Equations 17.43 and 17.45 can be summed to establish the axial displacement at joint j.

$$\begin{aligned} \delta_i &= \delta_{ix} + \delta_{iy} = (\cos\phi)\Delta_{ix} + (\sin\phi)\Delta_{iy} \\ \delta_j &= \delta_{jx} + \delta_{jy} = (\cos\phi)\Delta_{jx} + (\sin\phi)\Delta_{jy} \end{aligned} \tag{17.50}$$

The expressions above also can be represented by the matrix equation

$$\begin{bmatrix} \delta_i \\ \delta_j \end{bmatrix} = \begin{bmatrix} c & s & 0 & 0 \\ 0 & 0 & c & s \end{bmatrix} \begin{bmatrix} \Delta_{ix} \\ \Delta_{iy} \\ \Delta_{jx} \\ \Delta_{jy} \end{bmatrix} \tag{17.51}$$

or

$$\boldsymbol{\delta} = \mathbf{T}\boldsymbol{\Delta} \tag{17.52}$$

where $\mathbf{T}$ is a transformation matrix that converts the components of member end displacements in global coordinates to the axial displacements in the direction of the member's axis.

The axial force F_{ij} in bar ij depends on the net axial deformation of the member, that is, the difference in the end displacements $\delta_j - \delta_i$. This force can be expressed in terms of the member's stiffness AE/L as

$$F_{ij} = \frac{AE}{L}(\delta_j - \delta_i) \tag{17.53}$$

EXAMPLE 17.2

Determine the joint displacements and bar forces of the truss in Figure 17.7 by the direct stiffness method. Member properties: $A_1 = 2\text{ in}^2$, $A_2 = 2.5\text{ in}^2$, and $E = 30{,}000\text{ kips/in}^2$.

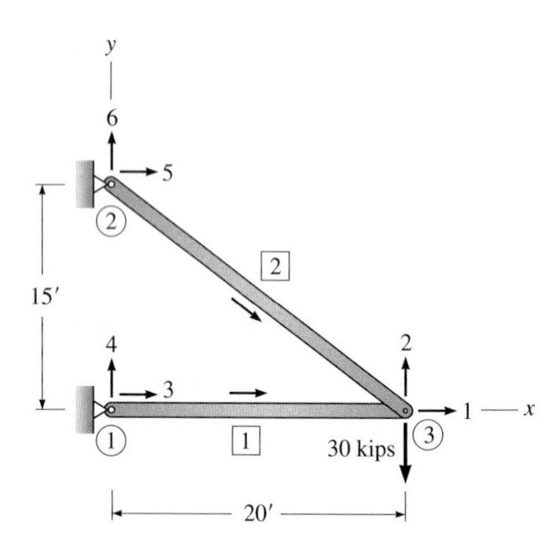

Figure 17.7

Solution

Members and joints of the truss are identified by numbers in squares and circles, respectively. We arbitrarily select the origin of the global coordinate system at joint 1. Arrows are shown along the axis of each member to indicate the direction from the near to far joints. At each joint we establish the positive direction for the components of global displacements and forces with a pair of numbered arrows. The coordinate in the x direction is assigned the lower number because the rows of the member stiffness matrix in Equation 17.48 are generated by introducing displacements in the x direction before those in the y direction. As we discussed in Section 17.4, we number the directions in sequence, starting with the joints that are free to displace. For example, in Figure 17.7, we begin at joint 3 with direction components 1 and 2. After we number the displacement components at the unrestrained joints, we number the coordinates at the restrained joints. This sequence of numbering produces a structure stiffness matrix that can be partitioned according to Equation 17.30 without shifting the rows and columns.

Construct member stiffness matrices (see Equation 17.48). For member 1, joint 1 is the near joint and joint 3 is the far joint. Compute the sine and cosine of the slope angle with Equation 17.37.

$$\cos\phi = \frac{x_j - x_i}{L} = \frac{20 - 0}{20} = 1 \quad \text{and} \quad \sin\phi = \frac{y_i - y_i}{L} = \frac{0 - 0}{20} = 0$$

$$\frac{AE}{L} = \frac{2(30{,}000)}{20(12)} = 250 \text{ kips/in}$$

$$\mathbf{k}_1 = 250 \begin{bmatrix} \overset{1}{1} & \overset{2}{0} & \overset{3}{-1} & \overset{4}{0} \\ 0 & 0 & 0 & 0 \\ -1 & 0 & 1 & 0 \\ 0 & 0 & 0 & 0 \end{bmatrix}$$

For member 2, joint 2 is the near joint and joint 3 is the far joint:

$$\cos\phi = \frac{20 - 0}{25} = 0.8 \qquad \sin\phi = \frac{0 - 15}{25} = -0.6$$

$$\frac{AE}{L} = \frac{2.5(30{,}000)}{25(12)} = 250 \text{ kips/in}$$

$$\mathbf{k}_2 = 250 \begin{bmatrix} \overset{1}{0.64} & \overset{2}{-0.48} & \overset{5}{-0.64} & \overset{6}{0.48} \\ -0.48 & 0.36 & 0.48 & -0.36 \\ -0.64 & 0.48 & 0.64 & -0.48 \\ 0.48 & -0.36 & -0.48 & 0.36 \end{bmatrix}$$

Set up the matrices for the force-displacement relationship of Equation 17.30 (that is, $\mathbf{Q} = \mathbf{K}\boldsymbol{\Delta}$). The structure stiffness matrix is assembled by inserting the elements of the member stiffness matrices $\mathbf{k}_1$ and $\mathbf{k}_2$ into the appropriate rows and columns.

$$\begin{bmatrix} Q_1 = 0 \\ Q_2 = -30 \\ \hline Q_3 \\ Q_4 \\ Q_5 \\ Q_6 \end{bmatrix} = 250 \left[\begin{array}{cc|cccc} \overset{1}{1.64} & \overset{2}{-0.48} & \overset{3}{-1} & \overset{4}{0} & \overset{5}{0.64} & \overset{6}{0.48} \\ -0.48 & 0.36 & 0 & 0 & 0.48 & -0.36 \\ \hline -1 & 0 & 1 & 0 & 0 & 0 \\ 0 & 0 & 0 & 0 & 0 & 0 \\ -0.64 & 0.48 & 0 & 0 & 0.64 & -0.48 \\ 0.48 & -0.36 & 0 & 0 & -0.48 & 0.36 \end{array}\right] \begin{bmatrix} \Delta_1 \\ \Delta_2 \\ \hline \Delta_3 = 0 \\ \Delta_4 = 0 \\ \Delta_5 = 0 \\ \Delta_6 = 0 \end{bmatrix}$$

[continues on next page]

Example 17.2 continues . . .

Partition the matrices above as indicated in Equation 17.30, and solve for the unknown displacements Δ_1 and Δ_2 by using Equation 17.33.

$$\mathbf{Q}_f = \mathbf{K}_{11}\mathbf{\Delta}_f$$

$$\begin{bmatrix} 0 \\ -30 \end{bmatrix} = 250\begin{bmatrix} 1.64 & -0.48 \\ -0.48 & 0.36 \end{bmatrix}\begin{bmatrix} \Delta_1 \\ \Delta_2 \end{bmatrix}$$

Solving for the displacements gives

$$\mathbf{\Delta}_f = \begin{bmatrix} \Delta_1 \\ \Delta_2 \end{bmatrix} = \begin{bmatrix} -0.16 \\ -0.547 \end{bmatrix}$$

Substitute the values of Δ_1 and Δ_2 into Equation 17.34 and solve for the support reactions $\mathbf{Q}_s$.

$$\mathbf{Q}_s = \mathbf{K}_{21}\mathbf{\Delta}_f$$

$$\begin{bmatrix} Q_3 \\ Q_4 \\ Q_5 \\ Q_6 \end{bmatrix} = 250\begin{bmatrix} -1 & 0 \\ 0 & 0 \\ -0.64 & 0.48 \\ 0.48 & -0.36 \end{bmatrix}\begin{bmatrix} -0.16 \\ -0.547 \end{bmatrix} = \begin{bmatrix} 40 \\ 0 \\ -40 \\ 30 \end{bmatrix}$$

A minus sign indicates a force or displacement is opposite in sense to the direction indicated by the direction arrows at the joints.

Compute member end displacements $\boldsymbol{\delta}$ in terms of member coordinates with Equation 17.51. For bar 1, i = joint 1 and j = joint 3, $\cos\phi = 1$, and $\sin\phi = 0$.

$$\begin{bmatrix} \delta_1 \\ \delta_3 \end{bmatrix} = \begin{bmatrix} 1 & 0 & 0 & 0 \\ 0 & 0 & 1 & 0 \end{bmatrix}\begin{bmatrix} \Delta_3 = 0 \\ \Delta_4 = 0 \\ \Delta_1 = -0.16 \\ \Delta_2 = -0.547 \end{bmatrix} = \begin{bmatrix} 0 \\ -0.16 \end{bmatrix}$$ **Ans.**

Substituting these values of δ into Equation 17.53, we compute the bar force in member 1 as

$$F_{13} = 250[0\ -0.16]\begin{bmatrix} -1 \\ 1 \end{bmatrix} = -40 \text{ kips (compression)}$$ **Ans.**

For bar 2, i = joint 2 and j = joint 3, $\cos\phi = 0.8$, and $\sin\phi = 0.6$.

$$\begin{bmatrix} \delta_2 \\ \delta_3 \end{bmatrix} = \begin{bmatrix} 0.8 & -0.6 & 0 & 0 \\ 0 & 0 & 0.8 & -0.6 \end{bmatrix}\begin{bmatrix} \Delta_5 = 0 \\ \Delta_6 = 0 \\ \Delta_1 = -0.16 \\ \Delta_2 = -0.547 \end{bmatrix} = \begin{bmatrix} 0 \\ 0.20 \end{bmatrix}$$

Substituting into Equation 17.53 yields

$$F_{23} = 250[0\ 0.20]\begin{bmatrix} -1 \\ 1 \end{bmatrix} = 50 \text{ kips (tension)}$$ **Ans.**

EXAMPLE 17.3

Analyze the truss in Figure 17.8 by the direct stiffness method. Construct the structure stiffness matrix without considering if joints are restrained or unrestrained against displacement. Then, rearrange the terms and partition the matrix so that the unknown joint displacements $\mathbf{\Delta}_f$ can be determined by Equation 17.30. Use $k_1 = k_2 = AE/L = 250$ kips/in and $k_3 = 2AE/L = 500$ kips/in.

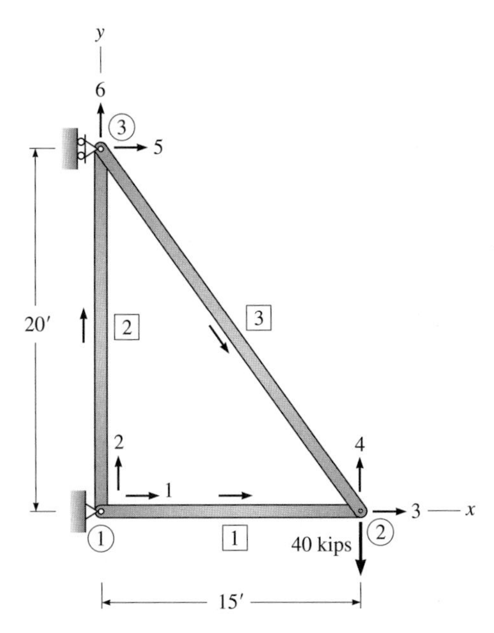

Figure 17.8: Truss with origin of global coordinate system at joint 1.

Solution

Number the joints arbitrarily as shown in Figure 17.8. Arrows are shown along the axis of each truss bar to indicate the direction from the near end to the far end of the member. We then establish, for each joint sequentially, the positive direction of the components of global displacements and forces with a pair of numbered arrows without considering if the joint is restrained from movement. Superimpose on the truss a global coordinate system with origin at joint 1. Form the member stiffness matrices using Equation 17.48. For bar 1, i = joint 1 and j = joint 2. Using Equation 17.37,

$$\cos\phi = \frac{x_j - x_i}{L} = \frac{15 - 0}{15} = 1$$

[continues on next page]

Example 17.3 continues . . .

$$\sin\phi = \frac{y_j - y_i}{L} = \frac{0-0}{15} = 0$$

$$\mathbf{k}_1 = 250\begin{bmatrix} 1 & 0 & -1 & 0 \\ 0 & 0 & 0 & 0 \\ -1 & 0 & 1 & 0 \\ 0 & 0 & 0 & 0 \end{bmatrix}\begin{matrix}1\\2\\3\\4\end{matrix}$$

(columns 1, 2, 3, 4)

For bar 2, i = joint 1 and j = joint 3.

$$\cos\phi = \frac{0-0}{20} = 0 \qquad \sin\phi = \frac{20-0}{20} = 1$$

$$\mathbf{k}_2 = 250\begin{bmatrix} 0 & 0 & 0 & 0 \\ 0 & 1 & 0 & -1 \\ 0 & 0 & 0 & 0 \\ 0 & -1 & 0 & 1 \end{bmatrix}\begin{matrix}1\\2\\5\\6\end{matrix}$$

(columns 1, 2, 5, 6)

For bar 3, i = joint 3 and j = joint 2.

$$\cos\phi = \frac{15-0}{25} = 0.6 \qquad \sin\phi = \frac{0-20}{25} = -0.8$$

$$\mathbf{k}_3 = 500\begin{bmatrix} 0.36 & -0.48 & -0.36 & 0.48 \\ -0.48 & 0.64 & 0.48 & -0.64 \\ -0.36 & 0.48 & 0.36 & -0.48 \\ 0.48 & -0.64 & -0.48 & 0.64 \end{bmatrix}\begin{matrix}5\\6\\3\\4\end{matrix}$$

(columns 5, 6, 3, 4)

Add $\mathbf{k}_1$, $\mathbf{k}_2$, and $\mathbf{k}_3$ by inserting the elements of the member stiffness matrices into the structure stiffness matrix at the appropriate locations. Multiply the elements of $\mathbf{k}_3$ by 2 so that all matrices are multiplied by the same scalar AE/L, i.e. 250.

$$\mathbf{K} = 250\begin{bmatrix} 1 & 0 & -1 & 0 & 0 & 0 \\ 0 & 1 & 0 & 0 & 0 & -1 \\ -1 & 0 & 1.72 & -0.96 & -0.72 & 0.96 \\ 0 & 0 & -0.96 & 1.28 & 0.96 & -1.28 \\ 0 & 0 & -0.72 & 0.96 & 0.72 & -0.96 \\ 0 & -1 & 0.96 & -1.28 & -0.96 & 2.28 \end{bmatrix}\begin{matrix}1\\2\\3\\4\\5\\6\end{matrix}$$

(columns 1, 2, 3, 4, 5, 6)

Establish the force-displacement matrices of Equation 17.30 by shifting the rows and columns of the structure stiffness matrix so that elements associated with the joints that displace (i.e., direction components 3, 4, and 6) are located in the upper left corner. This can be achieved by first shifting the third row to the top and then shifting the third column to the first column. The procedure is then repeated for the direction components 4 and 6.

$$\begin{bmatrix} Q_3=0 \\ Q_4=-40 \\ Q_6=0 \\ \hline Q_1 \\ Q_2 \\ Q_5 \end{bmatrix} = 250 \left[\begin{array}{ccc|ccc} \overset{3}{1.72} & \overset{4}{-0.96} & \overset{6}{0.96} & \overset{1}{-1} & \overset{2}{0} & \overset{5}{-0.72} \\ -0.96 & 1.28 & -1.28 & 0 & 0 & 0.96 \\ 0.96 & -1.28 & 2.28 & 0 & -1 & -0.96 \\ \hline -1 & 0 & 0 & 1 & 0 & 0 \\ 0 & 0 & -1 & 0 & 1 & 0 \\ -0.72 & 0.96 & -0.96 & 0 & 0 & 0.72 \end{array}\right] \begin{bmatrix} \Delta_3 \\ \Delta_4 \\ \Delta_6 \\ \hline \Delta_1=0 \\ \Delta_2=0 \\ \Delta_5=0 \end{bmatrix} \begin{matrix} 3 \\ 4 \\ 6 \\ 1 \\ 2 \\ 5 \end{matrix}$$

Partition the matrix and solve for the unknown joint displacements, using Equation 17.33.

$$\mathbf{Q}_f = \mathbf{K}_{11}\boldsymbol{\Delta}_f$$

$$\begin{bmatrix} 0 \\ -40 \\ 0 \end{bmatrix} = 250 \begin{bmatrix} 1.72 & -0.96 & 0.96 \\ -0.96 & 1.28 & -1.28 \\ 0.96 & -1.28 & 2.28 \end{bmatrix} \begin{bmatrix} \Delta_3 \\ \Delta_4 \\ \Delta_6 \end{bmatrix}$$

Solving the set of equations above gives

$$\begin{bmatrix} \Delta_3 \\ \Delta_4 \\ \Delta_6 \end{bmatrix} = \begin{bmatrix} -0.12 \\ -0.375 \\ -0.16 \end{bmatrix}$$

Ans.

Solve for the support reactions, using Equation 17.34.

$$\mathbf{Q}_s = \mathbf{K}_{21}\boldsymbol{\Delta}_f \tag{17.34}$$

$$\begin{bmatrix} Q_1 \\ Q_2 \\ Q_5 \end{bmatrix} = 250 \begin{bmatrix} -1 & 0 & 0 \\ 0 & 0 & -1 \\ -0.72 & 0.96 & -0.96 \end{bmatrix} \begin{bmatrix} -0.12 \\ -0.375 \\ -0.16 \end{bmatrix} = \begin{bmatrix} 30 \\ 40 \\ -30 \end{bmatrix}$$

Ans.

EXAMPLE 17.4

If the horizontal displacement of joint 2 of the truss in Example 17.3 is restrained by the addition of a roller (see Figure 17.9), determine the reactions.

Solution

The structure stiffness matrix of the truss was established in Example 17.3. Although the addition of an extra support creates an indeterminate structure, the solution is carried out in the same manner. The rows and columns associated with the degrees of freedom that are free to displace are shifted to the upper left corner of the structure stiffness matrix. This operation produces the following force-displacement matrices:

$$\begin{bmatrix} Q_4=-40 \\ Q_6=0 \\ \hline Q_1 \\ Q_2 \\ Q_3 \\ Q_5 \end{bmatrix} = 250 \left[\begin{array}{cc|cccc} \overset{4}{1.28} & \overset{6}{-1.28} & \overset{1}{0} & \overset{2}{0} & \overset{3}{-0.96} & \overset{5}{0.96} \\ -1.28 & 2.28 & 0 & 0 & 0.96 & -0.96 \\ \hline 0 & 0 & 1 & 0 & -1 & 0 \\ 0 & -1 & 0 & 1 & 0 & 0 \\ -0.96 & 0.96 & -1 & 0 & 1.72 & -0.72 \\ 0.96 & -0.96 & 0 & 0 & -0.72 & 0.72 \end{array}\right] \begin{bmatrix} \Delta_4 \\ \Delta_6 \\ \hline \Delta_1=0 \\ \Delta_2=0 \\ \Delta_3=0 \\ \Delta_5=0 \end{bmatrix} \begin{matrix} 4 \\ 6 \\ 1 \\ 2 \\ 3 \\ 5 \end{matrix}$$

Partition the matrix above, and solve for the unknown joint displacements, using Equation 17.33.

$$\mathbf{Q}_f = \mathbf{K}_{11}\mathbf{\Delta}_f$$

$$\begin{bmatrix} -40 \\ 0 \end{bmatrix} = 250 \begin{bmatrix} 1.28 & -1.28 \\ -1.28 & 2.28 \end{bmatrix} \begin{bmatrix} \Delta_4 \\ \Delta_6 \end{bmatrix}$$

Solution of the set of equations above gives

$$\begin{bmatrix} \Delta_4 \\ \Delta_6 \end{bmatrix} = \begin{bmatrix} -0.285 \\ -0.160 \end{bmatrix}$$

Solve for the reactions using Equation 17.34.

$$\mathbf{Q}_s = \mathbf{K}_{21}\mathbf{\Delta}_f$$

$$\begin{bmatrix} Q_1 \\ Q_2 \\ Q_3 \\ Q_5 \end{bmatrix} = 250 \begin{bmatrix} 0 & 0 \\ 0 & -1 \\ -0.96 & 0.96 \\ 0.96 & -0.96 \end{bmatrix} \begin{bmatrix} -0.285 \\ -0.160 \end{bmatrix} = \begin{bmatrix} 0 \\ 40 \\ 30 \\ -30 \end{bmatrix} \qquad \textbf{Ans.}$$

Results are shown in Figure 17.9*b*. Bar forces can be computed using Equations 17.52 and 17.53.

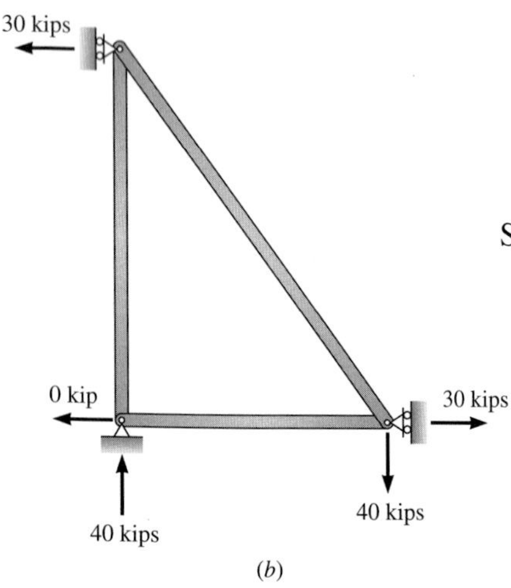

Figure 17.9: (*a*) Details of truss; (*b*) results of analysis.

17.7 Coordinate Transformation of a Member Stiffness Matrix

In Section 17.3 we derived the 2×2 stiffness member matrix $\mathbf{k}'$ of a truss bar with respect to a local coordinate system (see Equation 17.19). In the analysis of a truss composed of members inclined at various angles of inclination, it was shown in Section 17.6 that the assembly of the structure stiffness matrix $\mathbf{K}$ requires that we express all member stiffness matrices in terms of a common global coordinate system. For an individual truss bar whose axis forms an angle ϕ with the global x axis (see Figure 17.10), the 4×4 member stiffness matrix $\mathbf{k}$ in global coordinates is given by the middle matrix in Equation 17.48. Although we derived this matrix from basic principles in Section 17.6, it is more commonly generated from the member stiffness matrix $\mathbf{k}'$ formulated in local coordinates, using a transformation matrix $\mathbf{T}$ constructed from the geometric relationship between the local and global coordinate systems. The equation used to perform the coordinate transformation is

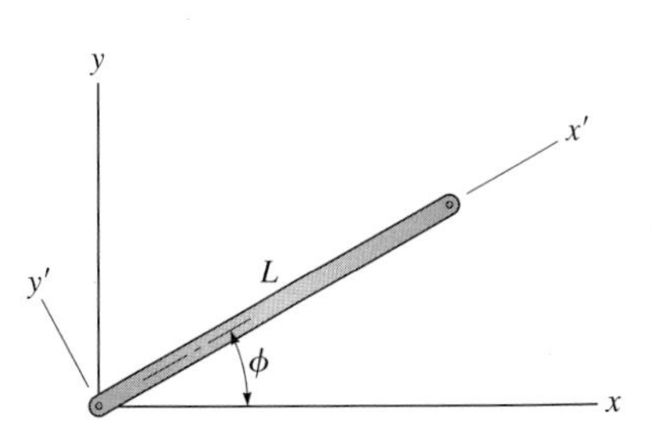

Figure 17.10: Global coordinates shown by xy system; member or local coordinates shown by $x'y'$ system.

$$\mathbf{k} = \mathbf{T}^{\mathrm{T}}\mathbf{k}'\,\mathbf{T} \tag{17.54}$$

where $\mathbf{k}$ = 4×4 member stiffness matrix referenced to global coordinates
$\mathbf{k}'$ = 2×2 member stiffness matrix referenced to local coordinate system
$\mathbf{T}$ = transformation matrix, that is, matrix that converts 4×1 displacement vector $\boldsymbol{\Delta}$ in global coordinates to the 2×1 axial displacement vector $\boldsymbol{\delta}$ in the direction of bar's longitudinal axis

The matrix $\mathbf{T}$ was previously derived in Section 17.6 and appears in Equation 17.51.

EXAMPLE 17.1

Show that the member stiffness matrix $\mathbf{k}$ in global coordinates that appears in Equation 17.48 can be generated from the member stiffness matrix $\mathbf{k}'$ in local coordinates (see Equation 17.19) by using Equation 17.54.

Solution

$$\mathbf{k} = \mathbf{T}^{\mathrm{T}}\mathbf{k}'\,\mathbf{T}$$

$$= \begin{bmatrix} c & 0 \\ s & 0 \\ 0 & c \\ 0 & s \end{bmatrix} \frac{AE}{L} \begin{bmatrix} 1 & -1 \\ -1 & 1 \end{bmatrix} \begin{bmatrix} c & s & 0 & 0 \\ 0 & 0 & c & s \end{bmatrix} = \begin{bmatrix} c^2 & sc & -c^2 & -sc \\ sc & s^2 & -sc & -s^2 \\ -c^2 & -sc & c^2 & sc \\ -sc & -s^2 & sc & s^2 \end{bmatrix} \qquad \textbf{Ans.}$$

As we observe, the product of this operation produces the member stiffness matrix that appears initially in Equation 17.48.

Summary

- Structural analysis computer software is generally programmed using the direct stiffness matrix. In matrix form, the equilibrium equation is

$$\mathbf{K}\boldsymbol{\Delta} = \mathbf{F}$$

where $\mathbf{K}$ is the structure stiffness matrix, $\mathbf{F}$ is a column vector of forces acting on the joints of a truss, and $\boldsymbol{\Delta}$ is a column vector of unknown joint displacements.
- The element $\mathbf{k}_{ij}$, which is located in the ith row and jth column of the $\mathbf{K}$ matrix, is called a stiffness coefficient. Coefficient $\mathbf{k}_{ij}$ represents the joint force in the direction (or degree of freedom) of i due to a unit displacement in the direction of j. With this definition, the $\mathbf{K}$ matrix can be constructed by basic mechanics. For computer applications, however, it is more convenient to assemble the structure stiffness matrix from the member stiffness matrices.
- A local x'-y' coordinate system can be constructed for each truss member (see Figure 17.3). With one axial deformation at each joint in the longitudinal (x') direction, a 2×2 member stiffness matrix $\mathbf{k}'$ in local coordinates is presented in Equation 17.19. If the structure does not have inclined members and if the local coordinates of the members coincide with the global (x-y) coordinates of the truss, Section 17.4 illustrates a procedure to construct the structure stiffness matrix by combining member stiffness matrices (see Equation 17.29).
- The equilibrium equation needs to be partitioned to separate the degrees of freedom that are allowed to move from those that cannot move (i.e., those restrained by supports); the joint forces corresponding to the degrees of freedom that cannot move are the support reactions. Once the equilibrium equation is partitioned as in Equation 17.30, two equations result. The first one, Equation 17.33, is used to calculate the unknown joint displacements, $\boldsymbol{\Delta}_f$. Once $\boldsymbol{\Delta}_f$ is determined, the support reactions, $\mathbf{Q}_s$, can be determined using Equation 17.34.
- When inclined members exist in a truss, it is more useful to express the member stiffness matrix using a global coordinate system. The general form of such a 4×4 member stiffness matrix, $\mathbf{k}$, is presented in Equation 17.48. The matrix $\mathbf{k}$ can be constructed from the basic mechanics described in Section 17.6. Alternatively, $\mathbf{k}$ can be obtained from $\mathbf{k}'$ using the coordinate transformation matrix described in Section 17.7.
- Once the unknown joint displacements are computed from the equilibrium equation, axial deformations at both ends of a member can be determined from Equation 17.52. With this information, the axial force of the member is computed using Equation 17.53.

PROBLEMS

P17.1. Using the stiffness method, write and solve the equations of equilibrium required to determine the horizontal and vertical components of deflection at joint 1 in Figure P17.1. For all bars E = 200 GPa and A = 800 mm^2.

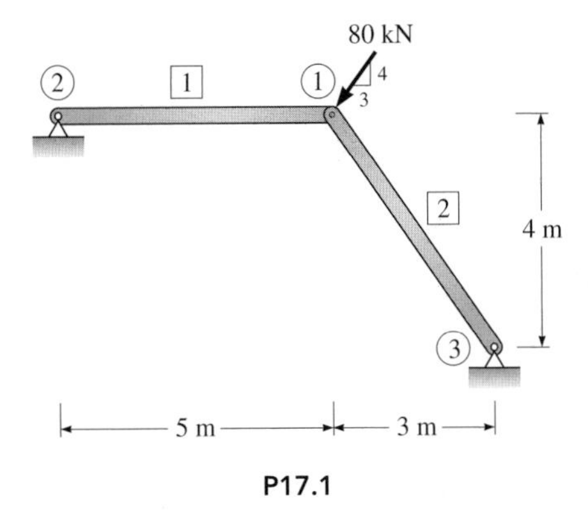

P17.1

P17.3. Form the structure stiffness matrix for the truss in Figure P17.3. Partition the matrix as indicated by Equation 17.30. Compute all joint displacements and reactions using Equations 17.34 and 17.35. For all bars, A = 2 in^2 and E = 30,000 kips/in^2.

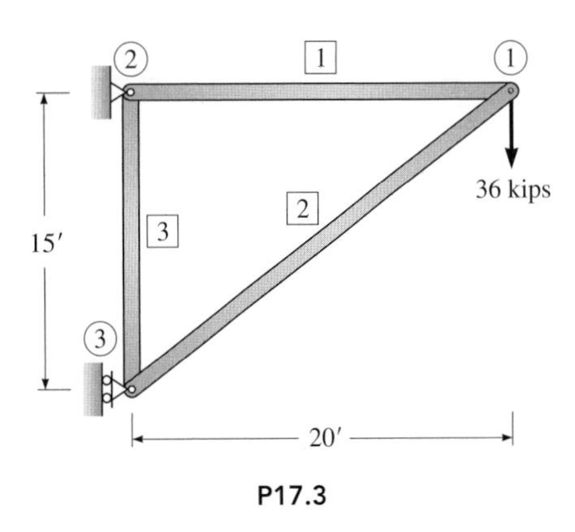

P17.3

P17.2. Using the stiffness method, determine the horizontal and vertical components of displacement of joint 1 in Figure P17.2. Also compute all bar forces. For all bars, L = 20 ft, E = 30,000 kips/in^2, and A = 3 in^2.

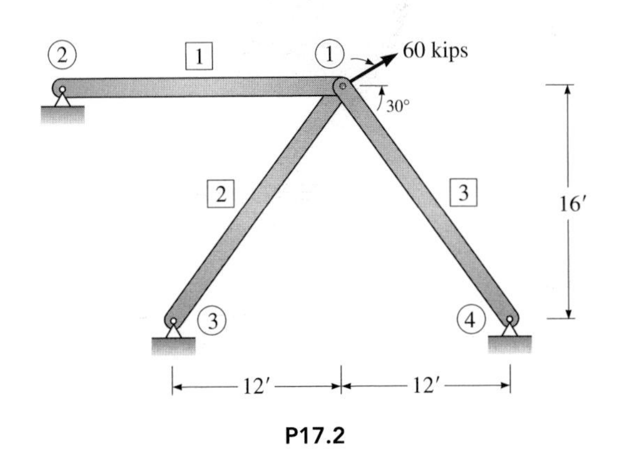

P17.2

P17.4. Form the structure stiffness matrix for the truss in Figure P17.4. Use the partitioned matrix to compute the displacement of all joints and reactions. Also compute the bar forces. Area of bars 1 and 2 = 2.4 in^2, area of bar 3 = 2 in^2, and E = 30,000 kips/in^2.

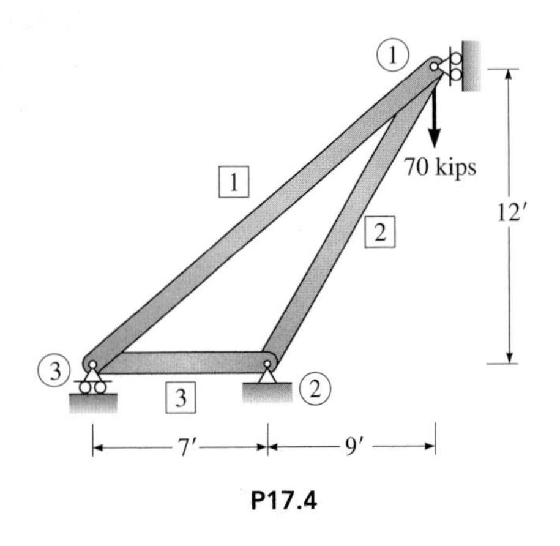

P17.4

P17.5. Determine all joint displacements, reactions, and bar forces for the truss in Figure P17.5. AE is constant for all bars. $A = 2$ in^2, $E = 30{,}000$ kips/in^2.

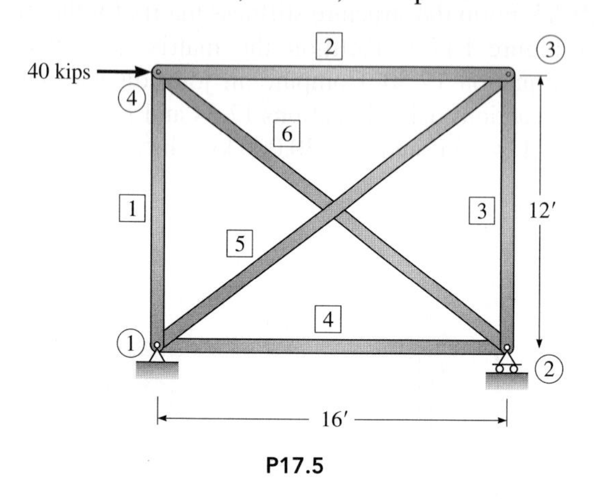

P17.5

P17.6. Determine all joint displacements, reactions, and bar forces for the truss in Figure P17.6. For all bars, $A = 1500$ mm^2 and $E = 200$ GPa.

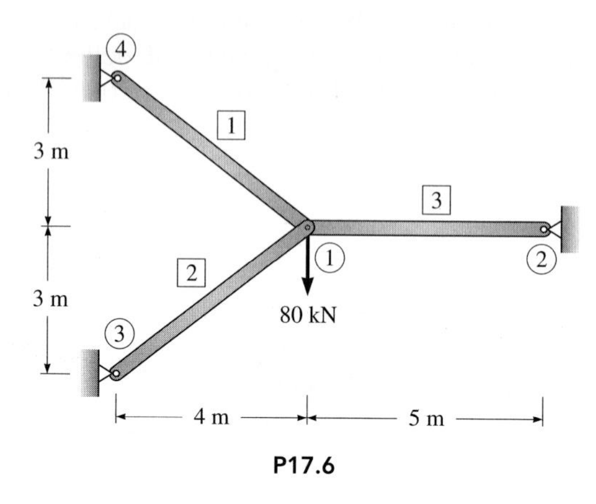

P17.6

Collapse of Hartford Civic Center Arena roof truss in Connecticut

Failure of the roof supported by the space truss shown in the photo at the beginning of Chapter 3, reminds us that the results of a computer analysis are no better than the information supplied by the engineer (see Section 1.7 for details). Although engineers nowadays have access to powerful computer programs that can analyze very complex structure, they must still exercise great care in modeling the structure and selecting the proper loads.

CHAPTER

18

Matrix Analysis of Beams and Frames by the Direct Stiffness Method

Chapter Objectives

- Extend the direct stiffness method learned in Chapter 17 for trusses to indeterminate beams and frames.
- Learn how to establish the structure stiffness matrix, which can be constructed by either basic mechanics or, more conveniently, individual member stiffness matrices.
- Construct member stiffness matrix as a 2×2, 4×4, or 6×6 matrix, depending on whether joint translation is allowed and whether axial deformation of the member is considered.
- Learn how to convert a member stiffness matrix from local to global coordinate system by using the concept of coordinate transformation.

18.1 Introduction

In Chapter 17 we discussed the analysis of trusses using the direct stiffness method. In this chapter we extend the method to structures in which loads may be applied to joints as well as to members between joints, and induce both axial forces and shears and moments. Whereas in the case of trusses we had to consider only joint displacements as unknowns in setting up the equilibrium equations, for frames we must add joint rotations. Consequently, a total of three equations of equilibrium, two for forces and one for moment, can be written for each joint in a plane frame.

Even though the analysis of a plane frame using the direct stiffness method involves three displacement components per joint (θ, Δ_x, Δ_y), we can often reduce the number of equations to be solved by neglecting the change in length of the members. In typical beams or frames, this simplification introduces little error in the results.

In the analysis of any structure using the stiffness method, the value of any quantity (for example, shear, moment, or displacement) is obtained from the sum of two parts. The first part is obtained from the analysis of a *restrained structure* in which all the joints are restrained against movement. The moments induced at the ends of each member are fixed-end moments. This procedure is similar to that used in the moment distribution method in Chapter 13. After the net restraining forces are computed and the signs reversed at each joint, these restraining forces are applied to the original structure in the second part of the analysis to determine the effect induced by joint displacements.

The superposition of forces and displacements from two parts can be explained using as an example the frame in Figure 18.1*a*. This frame is composed of two members connected by a rigid joint at *B*. Under the loading shown, the structure will deform and develop shears, moments, and axial loads

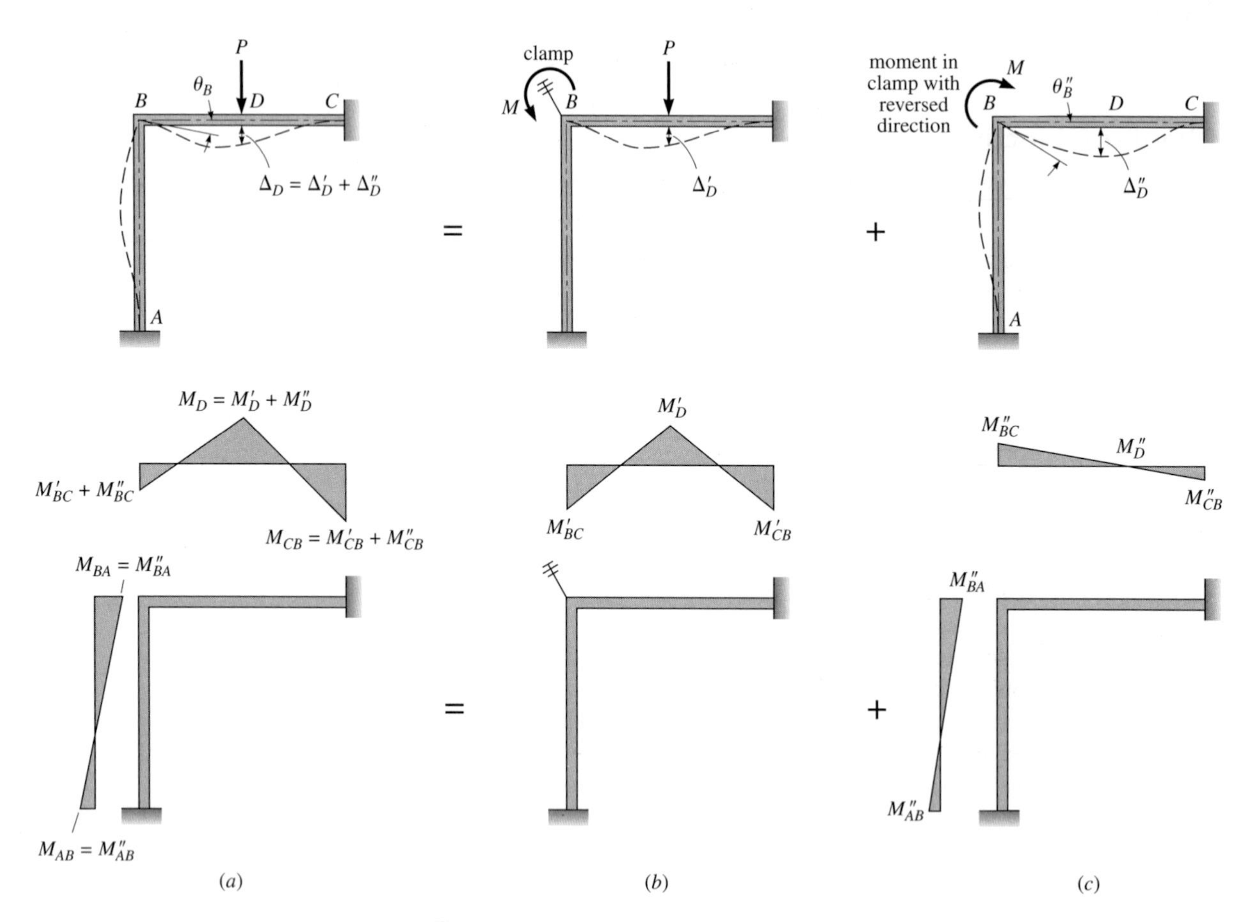

Figure 18.1: Analysis by the stiffness method. (*a*) Deflected shape and moment diagrams (bottom of figure) produced by the vertical load at *D*; (*b*) loads applied to the restrained structure; imaginary clamp at *B* prevents rotation, producing two fixed-end beams; (*c*) deflected shape and moment diagrams produced by a moment opposite to that applied by the clamp at *B*.

in both members. Because of the changes in length induced by the axial forces, joint B will experience, in addition to a rotation θ_B, small displacements in the x and y directions. Since these displacements are small and do not appreciably affect the member forces, we neglect them. With this simplification we can analyze the frame as having only *one degree of kinematic indeterminacy* (i.e., the rotation of joint B).

In the first part of the analysis, which we designate as the *restrained condition*, we introduce a rotational restraint (an imaginary clamp) at joint B (see Figure 18.1*b*). The addition of the clamp transforms the structure into two fixed-end beams. The analysis of these beams can be readily carried out using established equations (e.g., see Figure 12.5). The deflected shape and the corresponding moment diagrams (directly under the sketch of the frame) are shown in Figure 18.1 *b*. Forces and displacements associated with this case are superscripted with a prime.

Since the counterclockwise moment M applied by the clamp at B does not exist in the original structure, we must eliminate its effect. We do this in the second part of the analysis by solving for the rotation θ_B of joint B produced by an applied moment that is equal in magnitude but *opposite* in sense to the moment applied by the clamp. The moments and displacements in the members for the second part of analysis are superscripted with a double prime, as shown in Figure 18.1*c*. The final results, shown in Figure 18.1*a*, follow by direct superposition of the cases in Figure 18.1 *b* and *c*.

We note that not only are the final moments obtained by adding the values in the restrained case to those produced by the joint rotation θ_B, but also any other force or displacement can be obtained in the same manner. For example, the deflection directly under the load Δ_D equals the sum of the corresponding deflections at D in Figure 18.1*b* and *c*, that is,

$$\Delta_D = \Delta'_D + \Delta''_D$$

18.2 Structure Stiffness Matrix

In the analysis of a structure using the direct stiffness method, we start by introducing sufficient restraints (i.e., clamps) to prevent movement of all unrestrained joints. We then calculate the forces in the restraints as the sum of fixed-end forces for the members meeting at a joint. The internal forces at other locations of interest along the elements are also determined for the restrained condition.

In the next step of the analysis we determine values of joint displacements for which the restraining forces vanish. This is done by first applying the joint restraining forces, but with the sign reversed, and then solving a set of equilibrium equations that relate forces and displacements at the joints. In matrix form we have

$$\mathbf{K}\boldsymbol{\Delta} = \mathbf{F} \tag{18.1}$$

where $\mathbf{F}$ is the column matrix or vector of forces (including moments) in the fictitious restraints but with the sign reversed, $\boldsymbol{\Delta}$ is the column vector of joint displacements selected as degrees of freedom, and $\mathbf{K}$ is the structure stiffness matrix.

The term *degree of freedom* (*DOF*) refers to the independent joint displacement components that are used in the solution of a particular problem by the direct stiffness method. The number of degrees of freedom may equal the number of all possible joint displacement components (for example, three times the number of free joints in planar frames) or may be smaller if simplifying assumptions (such as neglecting axial deformations of members) are introduced. In all cases, the number of degrees of freedom and the degree of kinematic indeterminacy are identical.

Once the joint displacements Δ are calculated, the member actions (i.e., the moments, shears, and axial loads produced by these displacements) can be readily calculated. The final solution follows by adding these results to those from the restrained case.

The individual elements of the structure stiffness matrix **K** can be computed by introducing successively unit displacements that correspond to one of the degrees of freedom while all other degrees of freedom are restrained. The external forces at the location of the degrees of freedom required to satisfy equilibrium of the deformed configuration are the elements of the matrix **K**. More explicitly, a typical element k_{ij} of the structure stiffness matrix **K** is defined as follows: k_{ij} = force at degree of freedom i due to a unit displacement of degree of freedom j; when degree of freedom j is given a unit displacement, all others are restrained.

18.3 The 2 × 2 Rotational Stiffness Matrix for a Flexural Member

In this section we derive the member stiffness matrix for an individual flexural element using only joint rotations as degrees of freedom. The 2 × 2 matrix that relates moments and rotations at the ends of the member is important because it can be used directly in the solution of many practical problems, such as continuous beams and braced frames where joint translations are prevented. Furthermore, it is a basic item in the derivation of the more general 4 × 4 member stiffness matrix to be presented in Section 18.4.

Figure 18.2 shows a beam of length L with end moments M_i and M_j. As a sign convention the end rotations θ_i and θ_j are positive when clockwise and

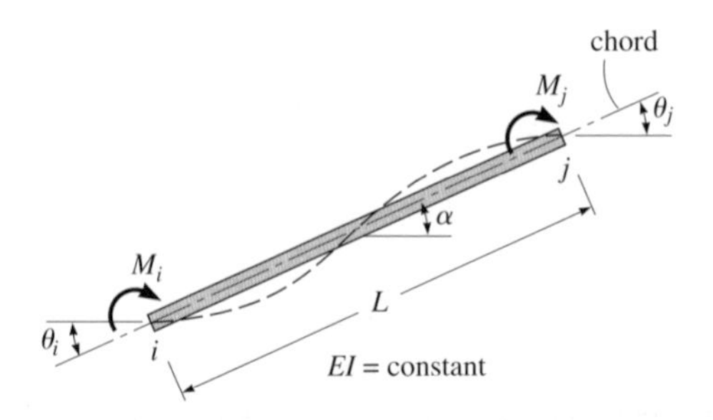

Figure 18.2: End rotations produced by member end moments.

negative when counterclockwise. Similarly, clockwise end moments are also positive, and counterclockwise moments are negative. To highlight the fact that the derivation to follow is independent of the member orientation, the axis of the element is drawn with an arbitrary inclination α.

In matrix notation, the relationship between the end moments and the resulting end rotations can be written as

$$\begin{bmatrix} M_i \\ M_j \end{bmatrix} = \bar{\mathbf{k}} \begin{bmatrix} \theta_i \\ \theta_j \end{bmatrix} \tag{18.2}$$

where $\bar{\mathbf{k}}$ is the 2×2 member rotational stiffness matrix.

To determine the elements of this matrix, we use the slope-deflection equation to relate end moments and rotations (see Equations 12.14 and 12.15). The sign convention and the notation in this formulation are identical to those used in the original derivation of the slope-deflection equation in Chapter 12. Since no loads are applied along the member's axis and no chord rotation ψ occurs (both ψ and the FEM equal zero), the end moments can be expressed as

$$M_i = \frac{2EI}{L}(2\theta_i + \theta_j) \tag{18.3}$$

and

$$M_j = \frac{2EI}{L}(\theta_i + 2\theta_j) \tag{18.4}$$

Equations 18.3 and 18.4 can be written in matrix notation as

$$\begin{bmatrix} M_i \\ M_j \end{bmatrix} = \frac{2EI}{L} \begin{bmatrix} 2 & 1 \\ 1 & 2 \end{bmatrix} \begin{bmatrix} \theta_i \\ \theta_j \end{bmatrix} \tag{18.5}$$

By comparing Equations 18.2 and 18.5 it follows that the member rotational stiffness matrix $\bar{\mathbf{k}}$ is

$$\bar{\mathbf{k}} = \frac{2EI}{L} \begin{bmatrix} 2 & 1 \\ 1 & 2 \end{bmatrix} \tag{18.6}$$

We will now illustrate the use of the preceding equations by solving a number of examples. To analyze a structure, it is necessary to identify the degree of freedom first. After the degree of freedom has been identified, the solution process can be conveniently broken down into the following five steps:

1. Analyze the restrained structure and calculate the clamping forces at the joints.
2. Assemble the structure stiffness matrix.
3. Apply the joint clamping forces but with the sign reversed to the original structure, and then calculate the unknown joint displacements using Equation 18.1.
4. Evaluate the effects of joint displacements (for example, deflections, moments, shears).
5. Sum the results of steps 1 and 4 to obtain the final solution.

EXAMPLE 18.1

Using the direct stiffness method, analyze the frame shown in Figure 18.3*a*. The change in length of the members may be neglected. The frame consists of two members of constant flexural rigidity EI connected by a rigid joint at B. Member BC supports a concentrated load P acting downward at midspan. Member AB carries a uniform load w acting to the right. The magnitude of w (in units of load per unit length) is equal to $3P/L$.

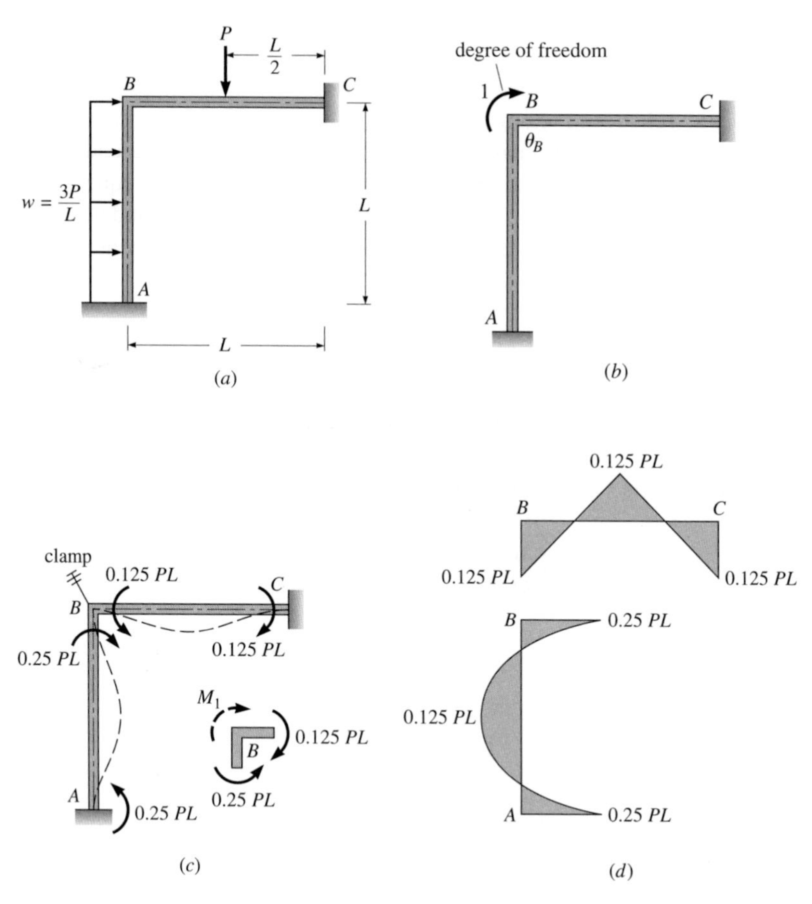

Figure 18.3: (*a*) Details of frame; (*b*) Curved arrow indicates positive sense of joint rotation at B; (*c*) fixed-end moments in restrained structure produced by applied loads (loads omitted from sketch for clarity); the clamp at B applies moment M_1 to the structure (see detail in lower right corner of figure); (*d*) moment diagrams for restrained structure (*continues on page* 722).

Solution

With axial deformations neglected, the degree of kinematic indeterminacy equals 1 (this structure is discussed in Section 18.1). Figure 18.3*b* illustrates the positive direction (clockwise) selected for the rotational degree of freedom at joint B.

Step 1: Analysis of the Restrained Structure With the rotation at joint B restrained by a temporary clamp, the structure is transformed into two fixed-end beams (Figure 18.3*c*). The fixed-end moments (see Figure 12.5*d*) for member AB are

$$M'_{AB} = -\frac{wL^2}{12} = -\frac{3P}{L}\left(\frac{L^2}{12}\right) = -\frac{PL}{4} \tag{18.7}$$

$$M'_{BA} = -M'_{AB} = \frac{PL}{4} \tag{18.8}$$

and for member BC (see Figure 12.5*a*),

$$M'_{BC} = -\frac{PL}{8} \tag{18.9}$$

$$M'_{CB} = -M'_{BC} = \frac{PL}{8} \tag{18.10}$$

Figure 18.3*c* shows the fixed-end moments and the deflected shape of the *restrained* frame. To illustrate the calculation of the restraining moment M_1, a free-body diagram of joint B is also shown in the lower right corner of Figure 18.3*c*. For clarity, shears acting on the joint are omitted. From the requirement of rotational equilibrium of the joint ($\Sigma M_B = 0$) we obtain

$$-\frac{PL}{4} + \frac{PL}{8} + M_1 = 0$$

from which we compute

$$M_1 = \frac{PL}{8} \tag{18.11}$$

In this 1-degree of freedom problem, the value of M_1 with its sign *reversed* is the only element in the restraining force vector **F** (see Equation 18.1). Figure 18.3*d* shows the moment diagrams for the members in the restrained structure.

Step 2: Assembly of the Structure Stiffness Matrix To assemble the stiffness matrix, we introduce a unit rotation at joint B and calculate the moment required to maintain the deformed configuration. The deflected shape of the frame produced by a unit rotation at joint B is shown in Figure 18.3*e*.

[*continues on next page*]

Example 18.1 continues . . .

Substituting $\theta_A = \theta_C = 0$ and $\theta_B = 1$ rad into Equation 18.5, we compute the moments at the ends of members *AB* and *BC* as

$$\begin{bmatrix} M_{AB} \\ M_{BA} \end{bmatrix} = \frac{2EI}{L}\begin{bmatrix} 2 & 1 \\ 1 & 2 \end{bmatrix}\begin{bmatrix} 0 \\ 1 \end{bmatrix} = \begin{bmatrix} \frac{2EI}{L} \\ \frac{4EI}{L} \end{bmatrix}$$

and

$$\begin{bmatrix} M_{BC} \\ M_{CB} \end{bmatrix} = \frac{2EI}{L}\begin{bmatrix} 2 & 1 \\ 1 & 2 \end{bmatrix}\begin{bmatrix} 1 \\ 0 \end{bmatrix} = \begin{bmatrix} \frac{4EI}{L} \\ \frac{2EI}{L} \end{bmatrix}$$

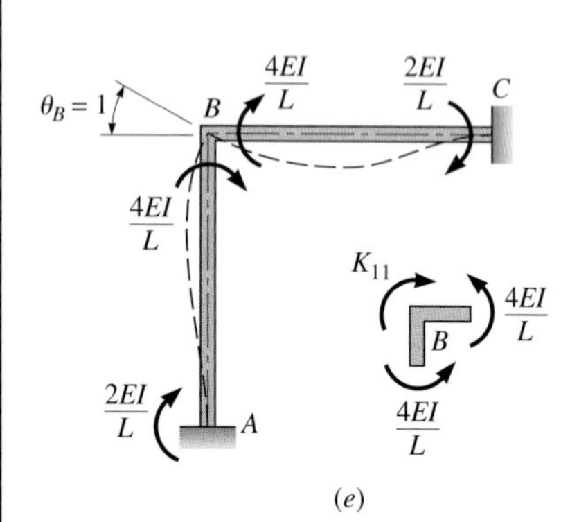

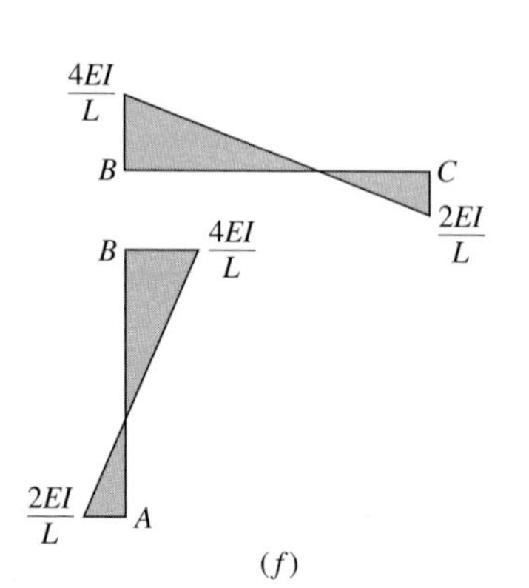

Figure 18.3: *Continued* (*e*) Moments produced by a unit rotation of joint *B*; the stiffness coefficient K_{11} represents the moment required to produce the unit rotation; (*f*) moment diagrams produced by the unit rotation of joint *B* (*continues*).

These moments are shown on the sketch of the deformed structure in Figure 18.3*e*. The moment required at joint *B* to satisfy equilibrium can be easily determined from the free-body diagram shown in the lower right corner of Figure 18.3*e*. Summing moments at joint *B*, we compute the stiffness coefficient K_{11} as

$$K_{11} = \frac{4EI}{L} + \frac{4EI}{L} = \frac{8EI}{L} \tag{18.12}$$

In this problem the value given by Equation 18.12 is the only element of the stiffness matrix **K**. The moment diagrams for the members corresponding to the condition $\theta_B = 1$ rad are shown in Figure 18.3*f*.

Step 3: Solution of Equation 18.1 Because this problem has only one degree of freedom, Equation 18.1 is a simple algebraic equation. Substituting the previously calculated values of **F** and **K** given by Equations 18.11 and 18.12, respectively, yields

$$\mathbf{K}\Delta = \mathbf{F} \tag{18.1}$$

$$\frac{8EL}{L}\theta_B = -\frac{PL}{8} \tag{18.13}$$

Solving for θ_B yields

$$\theta_B = -\frac{PL^2}{64EI} \tag{18.14}$$

The minus sign indicates that the rotation of joint *B* is counterclockwise, that is, opposite in sense to the direction defined as positive in Figure 18.3*b*.

Step 4: Evaluation of the Effects of Joint Displacements Since the moments produced by a unit rotation of joint B are known from step 2 (see Figure 18.3f), the moments produced by the actual joint rotation are readily obtained by multiplying the forces in Figure 18.3f by θ_B given by Equation 18.14; proceeding, we find

$$M''_{AB} = \frac{2EI}{L}\theta_B = -\frac{PL}{32} \tag{18.15}$$

$$M''_{BA} = \frac{4EI}{L}\theta_B = -\frac{PL}{16} \tag{18.16}$$

$$M''_{BC} = \frac{4EI}{L}\theta_B = -\frac{PL}{16} \tag{18.17}$$

$$M''_{CB} = \frac{2EI}{L}\theta_B = -\frac{PL}{32} \tag{18.18}$$

The double prime indicates that these moments are associated with the joint displacement condition.

Step 5: Calculation of Final Results The final results are obtained by adding the values from the restrained condition (step 1) with those produced by the joint displacements (step 4).

$$M_{AB} = M'_{AB} + M''_{AB} = -\frac{PL}{4} + \left(-\frac{PL}{32}\right) = -\frac{9PL}{32}$$

$$M_{BA} = M'_{BA} + M''_{BA} = \frac{PL}{4} + \left(-\frac{PL}{16}\right) = \frac{3PL}{16}$$

$$M_{BC} = M'_{BC} + M''_{BC} = -\frac{PL}{8} + \left(-\frac{PL}{16}\right) = -\frac{3PL}{16}$$

$$M_{CB} = M'_{CB} + M''_{CB} = \frac{PL}{8} + \left(-\frac{PL}{32}\right) = \frac{3PL}{32}$$

The member moment diagrams can also be evaluated by combining the diagrams from the restrained case with those corresponding to the joint displacements. Once the end moments are known, however, it is much easier to construct the individual moment diagrams using basic principles of statics. The final results are shown in Figure 18.3g.

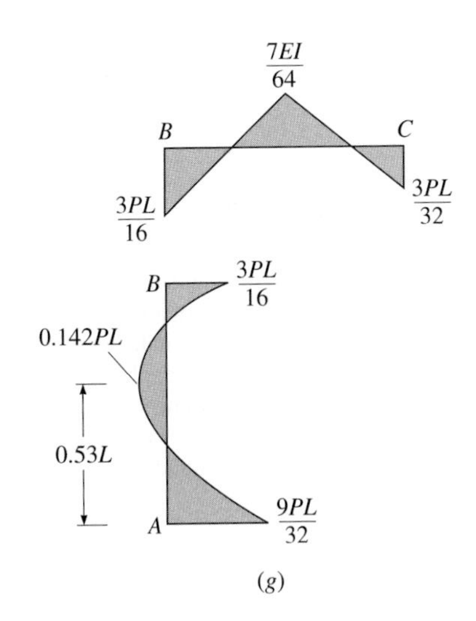

Figure 18.3: *Continued* (g) Final moment diagrams produced by superimposing moments in (d) with those in (f) multiplied by θ_B.

EXAMPLE 18.2

Construct the bending moment diagram for the three-span continuous beam shown in Figure 18.4*a*. The beam, which has a constant flexural rigidity *EI*, supports a 20-kip concentrated load acting at the center of span *BC*. In addition, a uniformly distributed load of 4.5 kips/ft acts over the length of span *CD*.

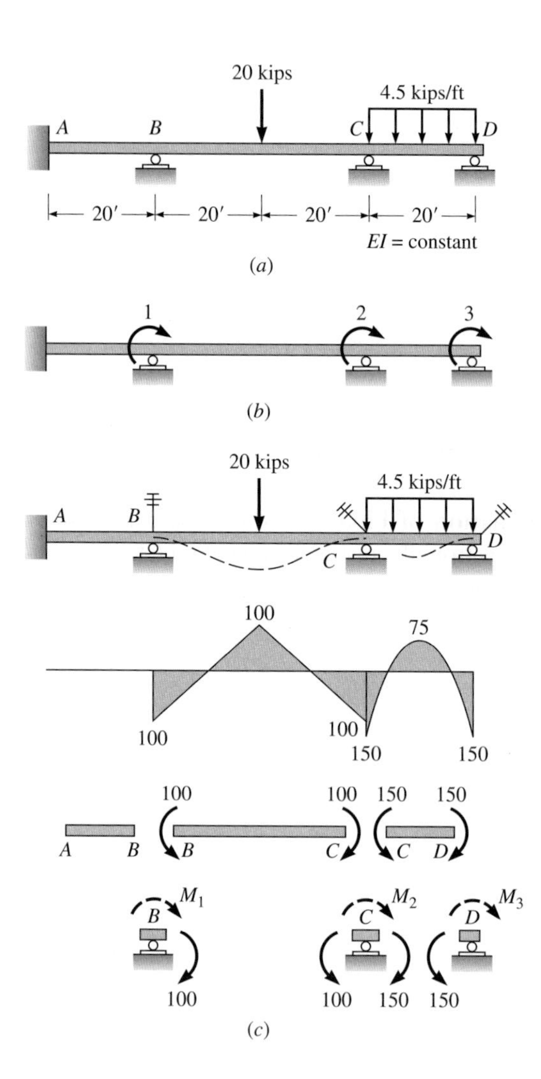

Figure 18.4: (*a*) Details of continuous beam; (*b*) curved arrows indicate the positive direction of the unknown joint rotations at *B*, *C*, and *D*; (*c*) moments induced in the restrained structure by the applied loads; bottom figures show the moments acting on free-body diagrams of the clamped joints (shears and reactions omitted for clarity) (*continues*).

Solution

An inspection of the structure indicates that the degree of kinematic indeterminacy is three. The positive directions selected for the three degrees of freedom (rotations at joints *B*, *C*, and *D*) are shown with curved arrows in Figure 18.4*b*.

Step 1: Analysis of the Restrained Structure The fixed-end moments induced in the restrained structure by the applied loads are calculated using the formulas in Figure 12.5. Figure 18.4*c* shows the moment diagram for the restrained condition and the free-body diagrams of the joints that are used to calculate the forces in the restraints. Considering moment equilibrium, we compute the restraining moments as follows:

Joint B: $\quad M_1 + 100 = 0 \qquad M_1 = -100 \text{ kip} \cdot \text{ft}$

Joint C: $\quad -100 + M_2 + 150 = 0 \qquad M_2 = -50 \text{ kip} \cdot \text{ft}$

Joint D: $\quad -150 + M_3 = 0 \qquad M_3 = 150 \text{ kip} \cdot \text{ft}$

Reversing the sign of these restraining moments, we construct the force vector **F**:

$$\mathbf{F} = \begin{bmatrix} 100 \\ 50 \\ -150 \end{bmatrix} \text{kip} \cdot \text{ft} \tag{18.19}$$

Step 2: Assembly of the Structure Stiffness Matrix The forces at the ends of the members resulting from the introduction of unit displacements at each one of the degrees of freedom are shown in Figure 18.4*d* to *f*. The elements of the structure stiffness matrix are readily calculated from the free-body diagrams of the joints. Summing moments, we calculate from Figure 18.4*d*:

$$-0.2EI - 0.1EI + K_{11} = 0 \quad \text{and} \quad K_{11} = 0.3EI$$

$$-0.05EI + K_{21} = 0 \quad \text{and} \quad K_{21} = 0.05EI$$

$$K_{31} = 0 \quad \text{and} \quad K_{31} = 0$$

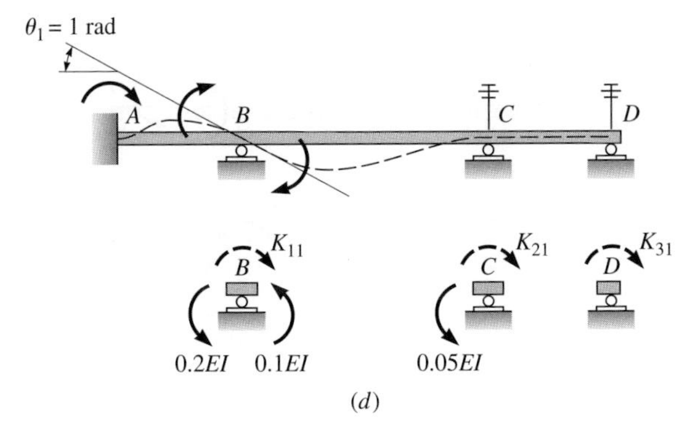

Figure 18.4: *Continued* (*d*) Stiffness coefficients produced by a unit rotation of joint B with joints C and D restrained (*continues*).

[*continues on next page*]

Example 18.2 continues . . .

From Figure 18.4*e*,

$$-0.05EI + K_{12} = 0 \quad \text{and} \quad K_{12} = 0.05EI$$
$$-0.1EI - 0.2EI + K_{22} = 0 \quad \text{and} \quad K_{22} = 0.3EI$$
$$-0.1EI + K_{32} = 0 \quad \text{and} \quad K_{32} = 0.1EI$$

From Figure 18.4*f*,

$$K_{13} = 0 \quad \text{and} \quad K_{13} = 0$$
$$-0.1EI + K_{23} = 0 \quad \text{and} \quad K_{23} = 0.1EI$$
$$-0.2EI + K_{33} = 0 \quad \text{and} \quad K_{33} = 0.2EI$$

Arranging these stiffness coefficients in matrix form, we produce the following structure stiffness matrix **K**:

$$\mathbf{K} = EI\begin{bmatrix} 0.3 & 0.05 & 0 \\ 0.05 & 0.3 & 0.1 \\ 0 & 0.1 & 0.2 \end{bmatrix} \tag{18.20}$$

As we would anticipate from Betti's law, the structure stiffness matrix **K** is symmetric.

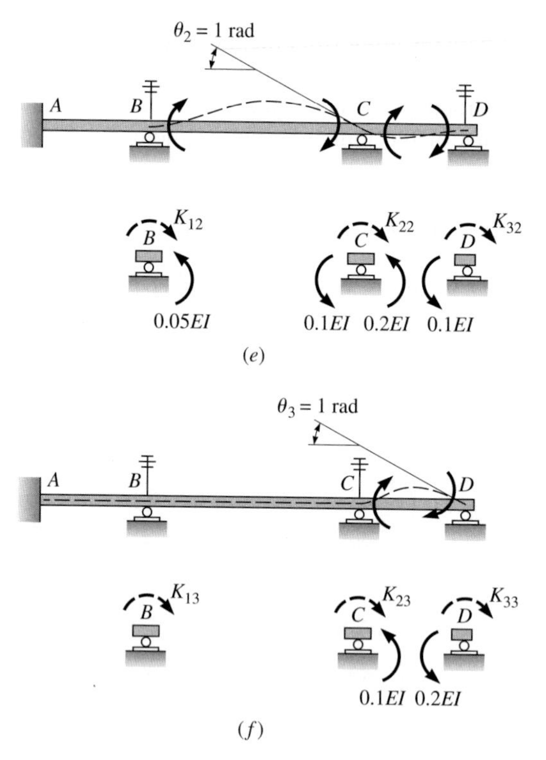

Figure 18.4: *Continued* (*e*) Stiffness coefficients produced by a unit rotation of joint *C* with joints *B* and *D* restrained; (*f*) Stiffness coefficients produced by a unit rotation of joint *D* with joints *B* and *C* restrained (*continues on page* 728).

Step 3: Solution of Equation 18.1 Substituting the previously calculated values of **F** and **K** (given by Equations 18.19 and 18.20) into Equation 18.1 gives

$$EI\begin{bmatrix} 0.3 & 0.05 & 0 \\ 0.05 & 0.3 & 0.1 \\ 0 & 0.1 & 0.2 \end{bmatrix}\begin{bmatrix} \theta_1 \\ \theta_2 \\ \theta_3 \end{bmatrix} = \begin{bmatrix} 100 \\ 50 \\ -150 \end{bmatrix} \tag{18.21}$$

Solving Equation 18.21, we compute

$$\begin{bmatrix} \theta_1 \\ \theta_2 \\ \theta_3 \end{bmatrix} = \frac{1}{EI}\begin{bmatrix} 258.6 \\ 448.3 \\ -974.1 \end{bmatrix} \tag{18.22}$$

Step 4: Evaluation of the Effect of Joint Displacements The moments produced by the actual joint rotations are determined by multiplying the moments produced by the unit displacements (see Figure 18.4d to f) by the actual displacements and superimposing the results. For example, the end moments in span BC are

$$M''_{BC} = \theta_1(0.1EI) + \theta_2(0.05EI) + \theta_3(0) = 48.3 \text{ kip}\cdot\text{ft} \tag{18.23}$$

$$M''_{CB} = \theta_1(0.05EI) + \theta_2(0.1EI) + \theta_3(0) = 57.8 \text{ kip}\cdot\text{ft} \tag{18.24}$$

The evaluation of the member end moments produced by joint displacements using superposition requires that for an n degree of freedom structure we add n appropriately scaled unit cases. This approach becomes increasingly cumbersome as the value of n increases. Fortunately, we can evaluate these moments in one step by using the individual member rotational stiffness matrices. For example, consider span BC, for which the end moments due to joint displacements were calculated previously by using superposition. If we substitute the end rotations θ_1 and θ_2 (given by Equation 18.22) into Equation 18.5 with $L = 40$ ft, we obtain

$$\begin{bmatrix} M''_{BC} \\ M''_{CB} \end{bmatrix} = \frac{2EI}{40}\begin{bmatrix} 2 & 1 \\ 1 & 2 \end{bmatrix}\frac{1}{EI}\begin{bmatrix} 258.6 \\ 448.3 \end{bmatrix} = \begin{bmatrix} 48.3 \\ 57.8 \end{bmatrix} \tag{18.25}$$

These results are, of course, identical to those obtained by superposition in Equations 18.23 and 18.24.

[*continues on next page*]

Example 18.2 continues . . .

Proceeding in a similar manner for spans *AB* and *CD*, we find that

$$\begin{bmatrix} M''_{AB} \\ M''_{BA} \end{bmatrix} = \frac{2EI}{20}\begin{bmatrix} 2 & 1 \\ 1 & 2 \end{bmatrix}\frac{1}{EI}\begin{bmatrix} 0 \\ 258.6 \end{bmatrix} = \begin{bmatrix} 25.9 \\ 51.7 \end{bmatrix} \tag{18.26}$$

$$\begin{bmatrix} M''_{CD} \\ M''_{DC} \end{bmatrix} = \frac{2EI}{20}\begin{bmatrix} 2 & 1 \\ 1 & 2 \end{bmatrix}\frac{1}{EI}\begin{bmatrix} 448.3 \\ -974.1 \end{bmatrix} = \begin{bmatrix} -7.8 \\ -150.0 \end{bmatrix} \tag{18.27}$$

The results are plotted in Figure 18.4*g*.

Step 5: Calculation of Final Results The complete solution is obtained by adding the results from the restrained case in Figure 18.4*c* to those produced by the joint displacements in Figure 18.4*g*. The resulting moment diagrams are plotted in Figure 18.4*h*.

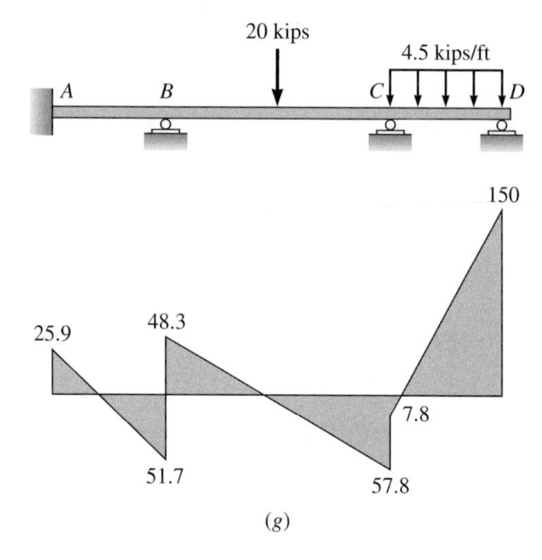

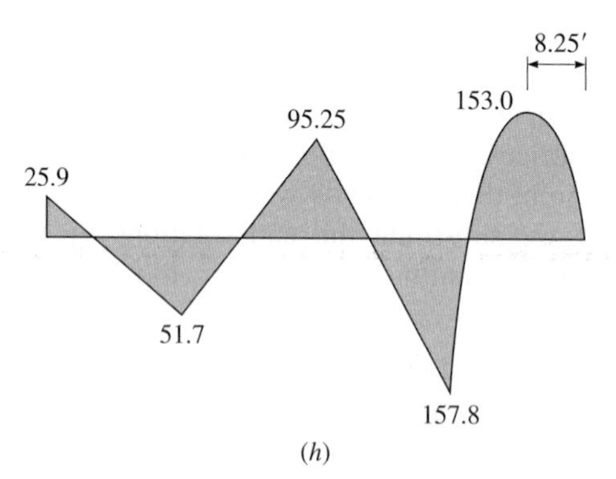

Figure 18.4: *Continued* (*g*) Moments produced by actual joint rotations; (*h*) Final moment diagrams (in units of kip · ft).

18.4 The 4 × 4 Member Stiffness Matrix in Local Coordinates

In Section 18.3 we derived a 2 × 2 member rotational stiffness matrix for the analysis of a structure in which joints can only rotate, but not translate. We now derive the member stiffness matrix for a flexural element considering both joint rotations and transverse joint displacements as degrees of freedom; the axial deformation of the member is still ignored. With the resulting 4 × 4 matrix we can extend the application of the direct stiffness method to the solution of structures with joints that both translate and rotate as a result of applied loading.

For educational purposes, the 4 × 4 member stiffness matrix in local coordinates will be derived in three different ways.

Derivation 1: Using the Slope-Deflection Equation

Figure 18.5*a* shows a flexural element of length L with end moments and shears; Figure 18.5*b* illustrates the corresponding joint displacements. The sign convention is as follows: Clockwise moments and rotations are positive. Shears and transverse joint displacements are positive when in the direction of the positive y axis.

The positive directions for local coordinates are as follows: The local x' axis runs along the member from the near joint i to the far joint j. The positive z' axis is always directed into the paper, and y' is such that the three axes form a right-handed coordinate system.

Setting the fixed-end moment (FEM) equal to zero in Equations 12.14 and 12.15 (assuming no load between joints) yields

$$M_i = \frac{2EI}{L}(2\theta_i + \theta_j - 3\psi) \tag{18.28}$$

and

$$M_j = \frac{2EI}{L}(2\theta_j + \theta_i - 3\psi) \tag{18.29}$$

where the chord rotation ψ from Equation 12.4*c* is

$$\psi = \frac{\Delta_j - \Delta_i}{L} \tag{18.30}$$

Equilibrium ($\Sigma M_j = 0$) requires that the end shears and moments in Figure 18.5*a* be related as follows:

$$V_i = -V_j = \frac{M_i + M_j}{L} \tag{18.31}$$

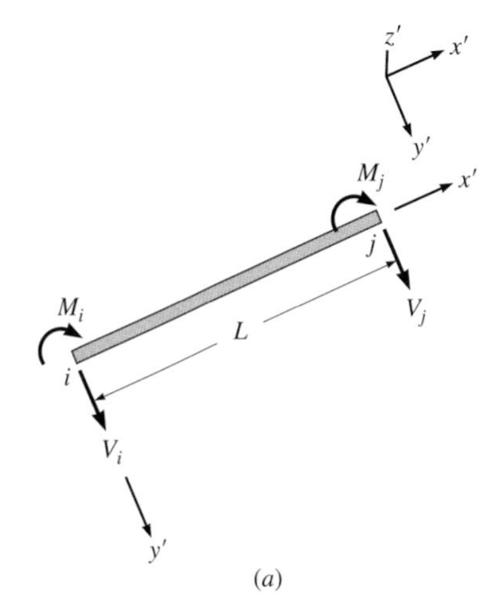

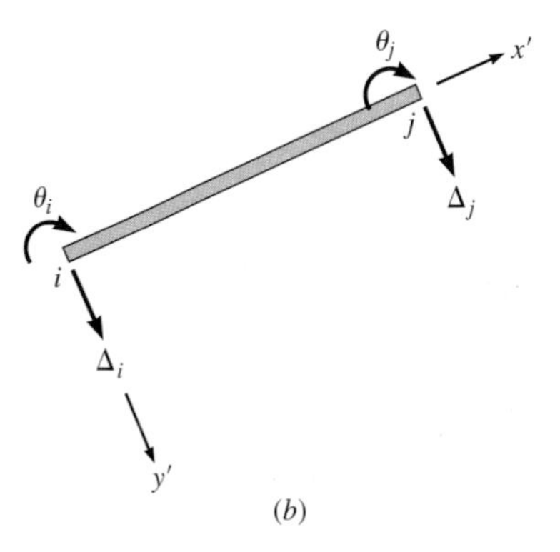

Figure 18.5: (*a*) Convention for positive end shears and moments; (*b*) convention for positive joint rotations and end displacements.

Substituting Equation 18.30 into Equations 18.28 and 18.29 and then substituting these equations into Equations 18.31, we produce the following four equations:

$$M_i = \frac{2EI}{L}\left(2\theta_i + \theta_j + \frac{3}{L}\Delta_i - \frac{3}{L}\Delta_j\right) \tag{18.32}$$

$$M_j = \frac{2EI}{L}\left(\theta_i + 2\theta_j + \frac{3}{L}\Delta_i - \frac{3}{L}\Delta_j\right) \tag{18.33}$$

$$V_i = \frac{2EI}{L}\left(\frac{3}{L}\theta_i + \frac{3}{L}\theta_j + \frac{6}{L^2}\Delta_i - \frac{6}{L^2}\Delta_j\right) \tag{18.34}$$

$$V_j = -\frac{2EI}{L}\left(\frac{3}{L}\theta_i + \frac{3}{L}\theta_j + \frac{6}{L^2}\Delta_i - \frac{6}{L^2}\Delta_j\right) \tag{18.35}$$

We can write these equations in matrix notation as

$$\begin{bmatrix} M_i \\ M_j \\ V_i \\ V_j \end{bmatrix} = \frac{2EI}{L}\begin{bmatrix} 2 & 1 & \frac{3}{L} & -\frac{3}{L} \\ 1 & 2 & \frac{3}{L} & -\frac{3}{L} \\ \frac{3}{L} & \frac{3}{L} & \frac{6}{L^2} & -\frac{6}{L^2} \\ -\frac{3}{L} & -\frac{3}{L} & -\frac{6}{L^2} & \frac{6}{L^2} \end{bmatrix}\begin{bmatrix} \theta_i \\ \theta_j \\ \Delta_i \\ \Delta_j \end{bmatrix} \tag{18.36}$$

where the 4 × 4 matrix together with the multiplier $2EI/L$ is the 4 × 4 member stiffness matrix $\mathbf{k}'$.

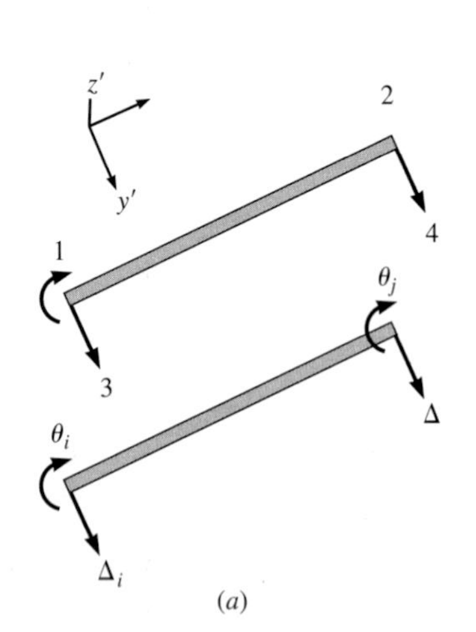

Figure 18.6: (*a*) Positive sense of unknown joint displacements indicated by numbered arrows; (*b*) stiffness coefficients produced by a unit clockwise rotation of the left end of the beam with all other joint displacements prevented; (*c*) stiffness coefficients produced by a unit clockwise rotation of the right end of the beam with all other joint displacements prevented (*continues*).

Derivation 2: Using the Basic Definition of Stiffness Coefficient

The 4 × 4 member stiffness matrix can also be derived using the basic approach of introducing unit displacements at each one of the degrees of freedom. The external forces, at the DOF, required to satisfy equilibrium in each deformed configuration are the elements of the member stiffness matrix in the column corresponding to that DOF. Refer to Figure 18.6 for the following derivations.

Unit Displacement at DOF 1 ($\theta_i = 1$ rad)

The corresponding sketch is shown in Figure 18.6*b*; the end moments computed with Equation 18.5 are the usual $4EI/L$ and $2EI/L$. The shears at the ends are readily calculated from statics. (The positive sense of displacements is indicated by the numbered arrows in Figure 18.6*a*.) From these computations we get

$$k'_{11} = \frac{4EI}{L} \qquad k'_{21} = \frac{2EI}{L} \qquad k'_{31} = \frac{6EI}{L^2} \qquad k'_{41} = -\frac{6EI}{L^2} \qquad (18.37)$$

These four elements constitute the first column of matrix $\mathbf{k}'$.

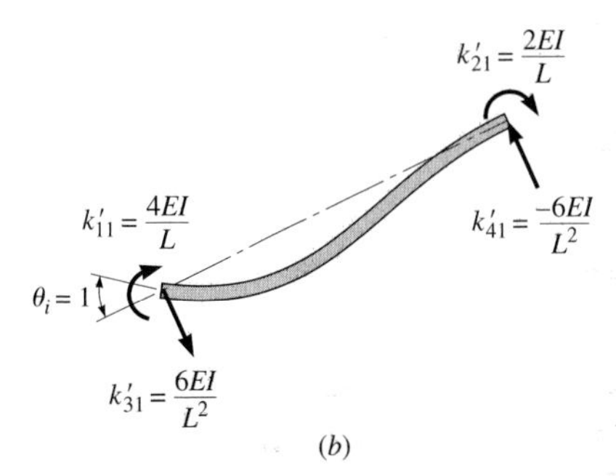

Unit Displacement at DOF 2 (θ_j = 1 rad)

The sketch for this condition is illustrated in Figure 18.6*c*; proceeding as before, we obtain

$$k'_{12} = \frac{2EI}{L} \qquad k'_{22} = \frac{4EI}{L} \qquad k'_{32} = \frac{6EI}{L^2} \qquad k'_{42} = -\frac{6EI}{L^2} \qquad (18.38)$$

The four elements constitute the second column of matrix $\mathbf{k}'$.

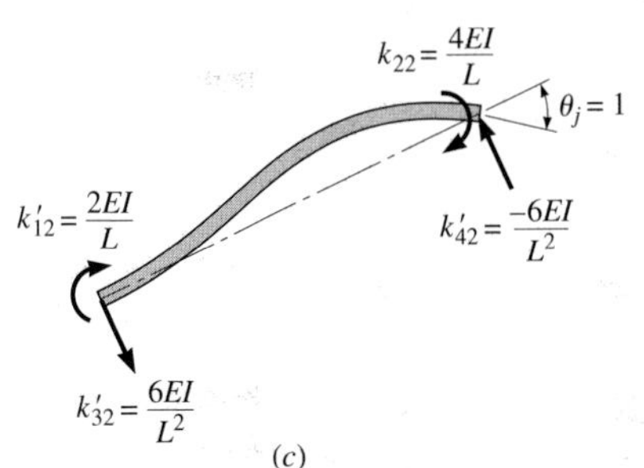

Unit Displacement at DOF 3 (Δ_i = 1)

From the sketch in Figure 18.6*d* we can see that this displacement pattern, as far as member distortions go, is equivalent to a positive rotation of $1/L$ measured from the beam chord to the deformed configuration of the beam. (Note that rigid-body motions do not introduce moments or shears in the beam element.) Substituting these rotations in Equation 18.5, we obtain the following end moments:

$$\begin{bmatrix} M_i \\ M_j \end{bmatrix} = \frac{2EI}{L}\begin{bmatrix} 2 & 1 \\ 1 & 2 \end{bmatrix}\frac{1}{L}\begin{bmatrix} 1 \\ 1 \end{bmatrix} = \frac{6EI}{L^2}\begin{bmatrix} 1 \\ 1 \end{bmatrix} \qquad (18.39)$$

The end moments and corresponding shears (calculated from statics) are depicted in Figure 18.6*d*; again we have

$$k'_{13} = \frac{6EI}{L^2} \qquad k'_{23} = \frac{6EI}{L^2} \qquad k'_{33} = \frac{12EI}{L^3} \qquad k'_{43} = -\frac{12EI}{L^3} \qquad (18.40)$$

These four elements constitute the third column of matrix $\mathbf{k}'$.

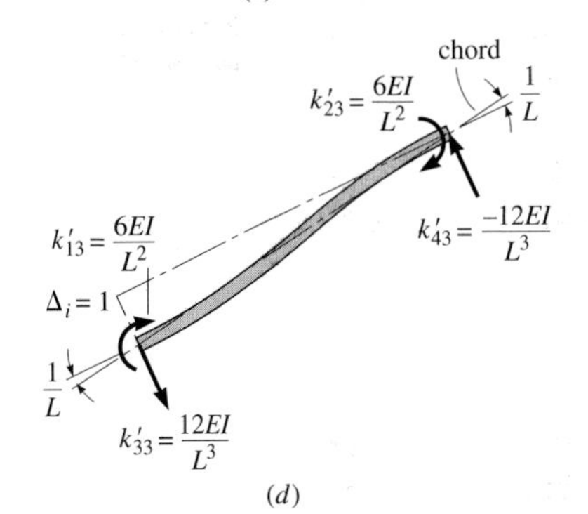

Unit Displacement at DOF 4 (Δ_j = 1)

In this case the rotation from the beam chord to the final configuration of the member, as shown in Figure 18.6*e*, is counterclockwise and, therefore, negative. Proceeding in exactly the same manner as before, the result is

$$k'_{14} = -\frac{6EI}{L^2} \qquad k'_{24} = -\frac{6EI}{L^2} \qquad k'_{34} = -\frac{12EI}{L^3} \qquad k'_{44} = \frac{12EI}{L^3} \qquad (18.41)$$

These four elements constitute the fourth column of matrix $\mathbf{k}'$.

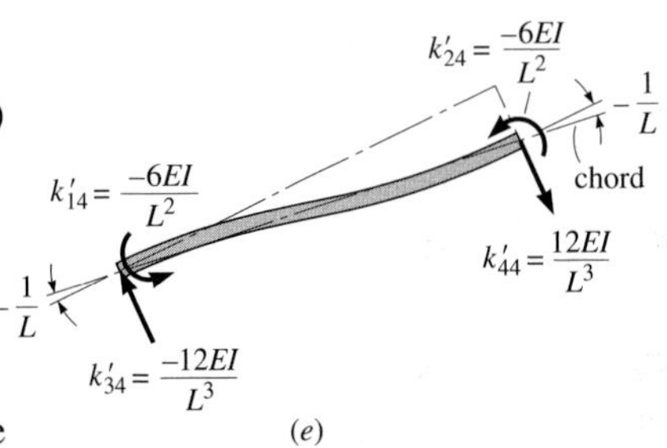

Figure 18.6: *Continued* (*d*) Stiffness coefficients produced by a unit vertical displacement of the left end with all other joint displacements prevented; (*e*) stiffness coefficients produced by a unit vertical displacement of the right end with all other joint displacements prevented.

Organizing these coefficients in a matrix format for the member stiffness matrix yields

$$\mathbf{k}' = \frac{2EI}{L}\begin{bmatrix} 2 & 1 & \frac{3}{L} & -\frac{3}{L} \\ 1 & 2 & \frac{3}{L} & -\frac{3}{L} \\ \frac{3}{L} & \frac{3}{L} & \frac{6}{L^2} & -\frac{6}{L^2} \\ -\frac{3}{L} & -\frac{3}{L} & -\frac{6}{L^2} & \frac{6}{L^2} \end{bmatrix} \tag{18.42}$$

Equation 18.42 is identical to the matrix derived previously using the slope-deflection equation (see Equation 18.36).

Derivation 3: Using the 2 × 2 Rotational Stiffness Matrix with a Coordinate Transformation

As we saw in the preceding derivation, as far as distortions go, the transverse displacements of the flexural member are equivalent to end rotations with respect to the chord. Since the rotations with respect to the chord are a function of both the rotations with respect to the local axis x' and the transverse displacements, we can write

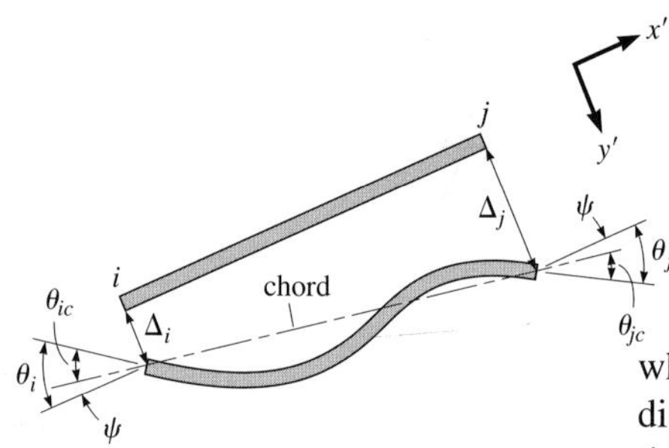

Figure 18.7: Deflected shape of a beam element whose joints rotate and displace laterally.

$$\begin{bmatrix} \theta_{ic} \\ \theta_{jc} \end{bmatrix} = \mathbf{T}\begin{bmatrix} \theta_i \\ \theta_j \\ \Delta_i \\ \Delta_j \end{bmatrix} \tag{18.43}$$

where $\mathbf{T}$ is the transformation matrix and the subscript c has been added to distinguish between rotations measured with respect to the chord and rotations with respect to the local axis x'.

The elements of the transformation matrix $\mathbf{T}$ can be obtained with the aid of Figure 18.7. From there we have

$$\theta_{ic} = \theta_i - \psi \tag{18.44}$$

$$\theta_{jc} = \theta_j - \psi \tag{18.45}$$

where the chord rotation ψ is given by

$$\psi = \frac{\Delta_j - \Delta_i}{L} \tag{18.30}$$

Substituting Equation 18.30 into Equations 18.44 and 18.45, we obtain

$$\theta_{ic} = \theta_i + \frac{\Delta_i}{L} - \frac{\Delta_j}{L} \tag{18.46}$$

$$\theta_{jc} = \theta_j + \frac{\Delta_i}{L} - \frac{\Delta_j}{L} \tag{18.47}$$

Writing Equations 18.46 and 18.47 in matrix notation produces

$$\begin{bmatrix} \theta_{ic} \\ \theta_{jc} \end{bmatrix} = \begin{bmatrix} 1 & 0 & \frac{1}{L} & -\frac{1}{L} \\ 0 & 1 & \frac{1}{L} & -\frac{1}{L} \end{bmatrix} \begin{bmatrix} \theta_i \\ \theta_j \\ \Delta_i \\ \Delta_j \end{bmatrix} \tag{18.48}$$

The 2 × 4 matrix in Equation 18.48 is, by comparison with Equation 18.43, the transformation matrix **T**.

From Section 17.7 we know that if two sets of coordinates are geometrically related, then if the stiffness matrix is known in one set of coordinates, it can be transformed to the other by the following operation:

$$\mathbf{k}' = \mathbf{T}^{\mathrm{T}}\,\bar{\mathbf{k}}\,\mathbf{T} \tag{18.49}$$

where $\bar{\mathbf{k}}$ is the 2 × 2 rotational stiffness matrix (Equation 18.6) and $\mathbf{k}'$ is the 4 × 4 member stiffness matrix in local coordinates. Substituting the **T** matrix in Equation 18.48 and the rotational stiffness matrix of Equation 18.6 for $\bar{\mathbf{k}}$, we get

$$\mathbf{k}' = \begin{bmatrix} 1 & 0 \\ 0 & 1 \\ \frac{1}{L} & \frac{1}{L} \\ -\frac{1}{L} & -\frac{1}{L} \end{bmatrix} \frac{2EI}{L} \begin{bmatrix} 2 & 1 \\ 1 & 2 \end{bmatrix} \begin{bmatrix} 1 & 0 & \frac{1}{L} & -\frac{1}{L} \\ 0 & 1 & \frac{1}{L} & -\frac{1}{L} \end{bmatrix}$$

The multiplication of the matrices shown above yields the same beam element stiffness matrix as derived previously and presented as Equation 18.42; the verification is left as an exercise for the reader.

EXAMPLE 18.3

Analyze the plane frame shown in Figure 18.8*a*. The frame is made up of two columns of moment of inertia I, rigidly connected to a horizontal beam whose moment of inertia is $3I$. The structure supports a concentrated load of 80 kips acting horizontally to the right at the midheight of column AB. Neglect the deformations due to axial forces.

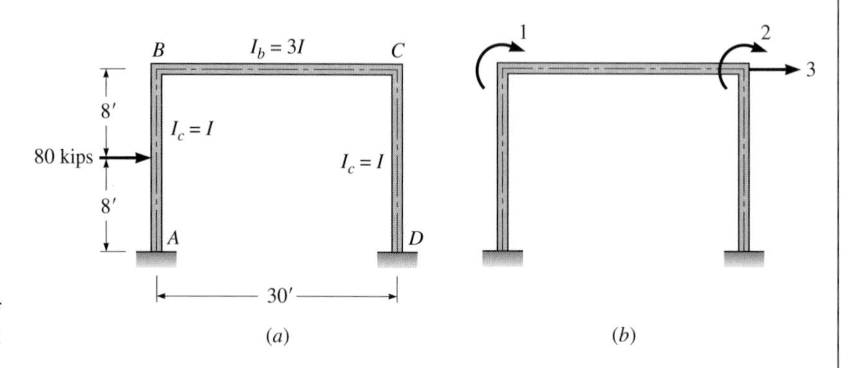

Figure 18.8: Analysis of an unbraced frame. (*a*) Details of frame; (*b*) positive sense of unknown joint displacements defined (*continues*).

Solution

Because axial deformations are neglected, joints B and C do not move vertically but have the same horizontal displacement. In Figure 18.8*b* we use arrows to show the positive sense of the three independent joint displacement components. We now apply the five-step solution procedure utilized in the preceding examples.

Step 1: Analysis of the Restrained Structure With the degrees of freedom restrained by a clamp at B as well as a clamp and horizontal support at C, the frame is transformed to three independent fixed-end beams. The moments in the restrained structure are shown in Figure 18.8*c*. The restraining forces are calculated using the free-body diagrams shown at the bottom of Figure 18.8*c*.

We note that the horizontal restraint at joint C that prevents sway of the frame (DOF 3) can be placed at either joint B or C without affecting the results. The selection of joint C in the sketch of Figure 18.8*c* is thus arbitrary. We also note that the simplification introduced by neglecting axial deformations does not imply that there are no axial forces. It only means that axial loads are assumed to be carried without producing shortening or elongation of the members.

From the free-body diagrams in Figure 18.8*c* we compute the restraining forces as

$$-160.0 + M_1 = 0 \qquad M_1 = 160.0$$

$$M_2 = 0$$

$$40.0 + F_3 = 0 \qquad F_3 = -40.0$$

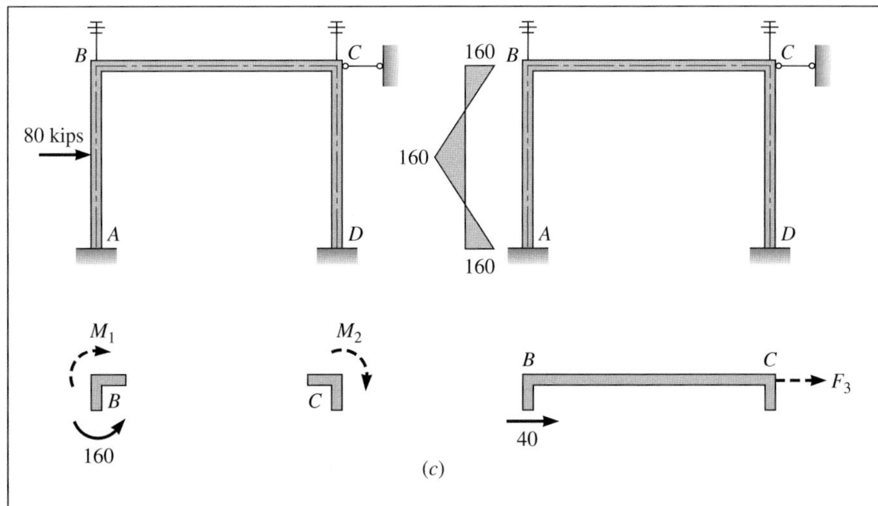

Figure 18.8: *Continued* (*c*) Computation of restraining forces corresponding to three unknown joint displacements; moments in kip·ft (*continues*).

Reversing the sign of restraining forces to construct the force vector **F** gives

$$\mathbf{F} = \begin{bmatrix} -160.0 \\ 0 \\ 40.0 \end{bmatrix} \qquad (18.50)$$

where forces are in kips and moments are in kip·ft.

Step: 2 Assembly of the Structure Stiffness Matrix The deformed configurations, corresponding to unit displacements at each degree of freedom, are shown in Figure 18.8*d*. The moments at the end of the members, in the sketches corresponding to unit rotations of joints *B* and *C* (i.e., DOF 1 and 2, respectively), are most easily calculated from the 2 × 2 member rotational stiffness matrix of Equation 18.5. Using the appropriate free-body diagrams, we compute

$$-0.25EI - 0.4EI + K_{11} = 0 \quad \text{or} \quad K_{11} = 0.65EI$$

$$-0.2EI + K_{21} = 0 \quad \text{or} \quad K_{21} = 0.20EI$$

$$0.0234EI + K_{31} = 0 \quad \text{or} \quad K_{31} = -0.0234EI$$

$$-0.4EI - 0.25EI + K_{22} = 0 \quad \text{or} \quad K_{22} = 0.65EI$$

$$0.0234EI + K_{32} = 0 \quad \text{or} \quad K_{32} = -0.0234EI$$

The elements of the third row of the structure stiffness matrix are evaluated by introducing a unit horizontal displacement at the top of the frame (DOF 3). The forces in the members are calculated as follows. From Figure 18.8*d* we see that for this condition member *BC* remains undeformed, thus having no

[*continues on next page*]

Example 18.3 continues . . .

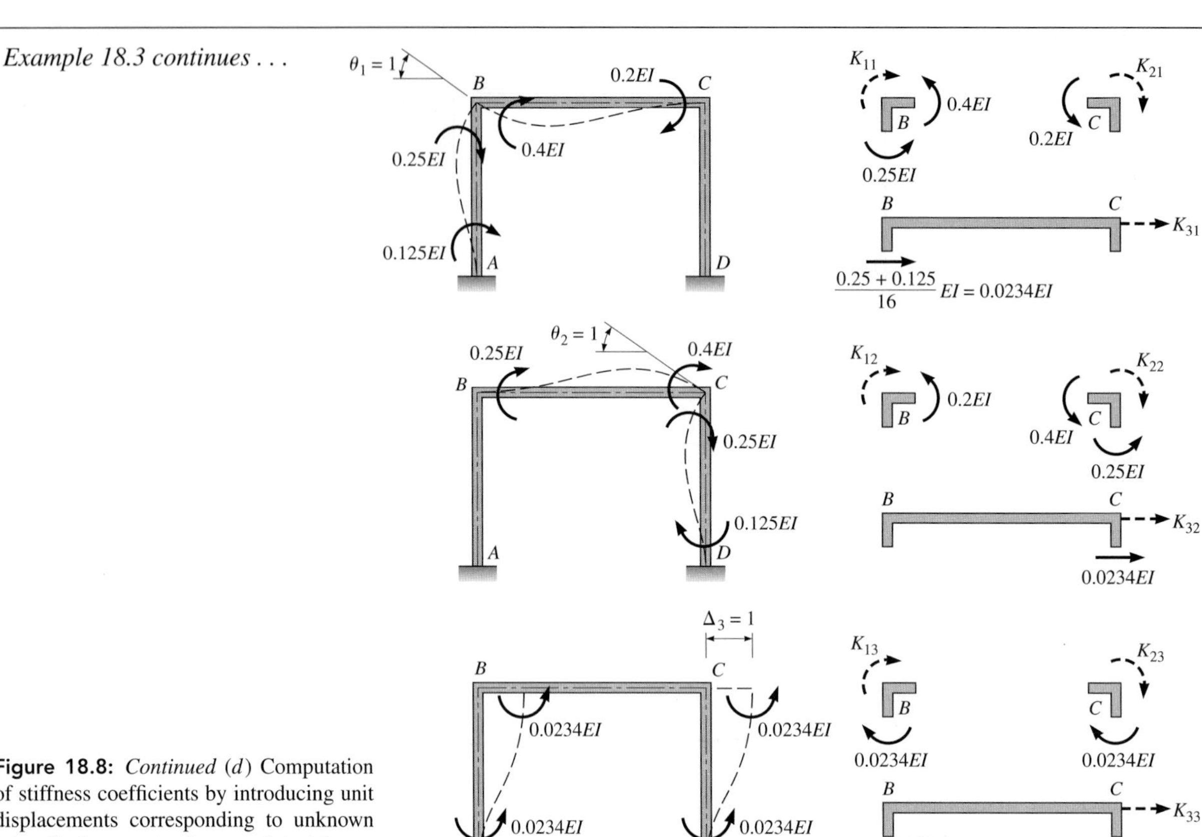

Figure 18.8: *Continued* (*d*) Computation of stiffness coefficients by introducing unit displacements corresponding to unknown joint displacements; the restraints (clamps and the lateral support at joint *C*) are omitted to simplify the sketches (*continues on page* 738).

moments or shears. The columns, members *AB* and *DC*, are subjected to the deformation pattern given by

$$\begin{bmatrix} \theta_i \\ \theta_j \\ \Delta_i \\ \Delta_j \end{bmatrix} = \begin{bmatrix} 0 \\ 0 \\ 0 \\ 1 \end{bmatrix}$$

where the subscripts i and j are used to designate the near and the far joints, respectively. Notice that by defining the columns as going from A to B and from D to C, both local y axes are in accordance with the previously established sign convention, directed to the right, thus making the displacement $\Delta = 1$ positive.

The moments and shears in each column are obtained by substituting the displacements shown above into Equation 18.36, that is,

$$\begin{bmatrix} M_i \\ M_j \\ V_i \\ V_j \end{bmatrix} = \frac{2EI}{L} \begin{bmatrix} 2 & 1 & \frac{3}{L} & -\frac{3}{L} \\ 1 & 2 & \frac{3}{L} & -\frac{3}{L} \\ \frac{3}{L} & \frac{3}{L} & \frac{6}{L^2} & -\frac{6}{L^2} \\ -\frac{3}{L} & -\frac{3}{L} & -\frac{6}{L^2} & \frac{6}{L^2} \end{bmatrix} \begin{bmatrix} 0 \\ 0 \\ 0 \\ 1 \end{bmatrix}$$

Substituting $L = 16$ ft gives

$$\begin{bmatrix} M_i \\ M_j \\ V_i \\ V_j \end{bmatrix} = EI \begin{bmatrix} -0.0234 \\ -0.0234 \\ -0.0029 \\ 0.0029 \end{bmatrix}$$

These results are shown in Figure 18.8*d*. From equilibrium of forces in the horizontal direction on the beam, we compute

$$-0.0029EI - 0.0029EI + K_{33} = 0 \qquad \text{or} \qquad K_{33} = 0.0058EI$$

Equilibrium of moments at joints *B* and *C* requires that $K_{13} = K_{23} = -0.0234EI$.

Arranging these coefficients in matrix form, we produce the structure stiffness matrix

$$\mathbf{K} = EI \begin{bmatrix} 0.65 & 0.20 & -0.0234 \\ 0.20 & 0.65 & -0.0234 \\ -0.0234 & -0.0234 & 0.0058 \end{bmatrix}$$

As a check of the computations, we observe the structure stiffness matrix **K** is symmetric (Betti's law).

Step 3: Solution of Equation 18.1 Substituting **F** and **K** into Equation 18.1, we generate the following set of simultaneous equations:

$$EI \begin{bmatrix} 0.65 & 0.20 & -0.0234 \\ 0.20 & 0.65 & -0.0234 \\ -0.0234 & -0.0234 & 0.0058 \end{bmatrix} \begin{bmatrix} \theta_1 \\ \theta_2 \\ \Delta_3 \end{bmatrix} = \begin{bmatrix} -160.0 \\ 0.0 \\ 40.0 \end{bmatrix}$$

Solving yields

$$\begin{bmatrix} \theta_1 \\ \theta_2 \\ \Delta_3 \end{bmatrix} = \frac{1}{EI} \begin{bmatrix} -57.0 \\ 298.6 \\ 7793.2 \end{bmatrix}$$

The units are radians and feet.

[*continues on next page*]

Example 18.3 continues . . .

Step 4: Evaluation of the Effect of Joint Displacements As explained in Example 18.2, the effects of the joint displacements are most easily calculated using the individual element stiffness matrices. These computations produce the following values of displacement at the ends of each member. For member *AB*,

$$\begin{bmatrix} \theta_A \\ \theta_B \\ \Delta_A \\ \Delta_B \end{bmatrix} = \frac{1}{EI} \begin{bmatrix} 0 \\ -57.0 \\ 0.0 \\ 7793.2 \end{bmatrix}$$

for member *BC*,

$$\begin{bmatrix} \theta_B \\ \theta_C \\ \Delta_B \\ \Delta_C \end{bmatrix} = \frac{1}{EI} \begin{bmatrix} -57.0 \\ 298.6 \\ 0 \\ 0 \end{bmatrix}$$

and for member *DC*,

$$\begin{bmatrix} \theta_D \\ \theta_C \\ \Delta_D \\ \Delta_C \end{bmatrix} = \frac{1}{EI} \begin{bmatrix} 0 \\ 298.6 \\ 0 \\ 7793.2 \end{bmatrix}$$

The results obtained by substituting these displacements into Equation 18.36 (with the appropriate values of L and flexural stiffness EI) are shown graphically in Figure 18.8*e*.

Step 5: Calculation of Final Results The complete solution is obtained by superimposing the results of the restrained case (Figure 18.8*c*) and the effects of the joint displacements (Figure 18.8*e*). The final moment diagrams for the members of the frame are plotted in Figure 18.8*f*.

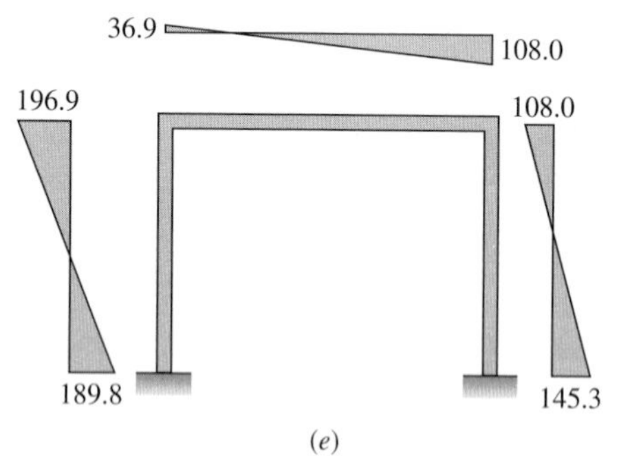

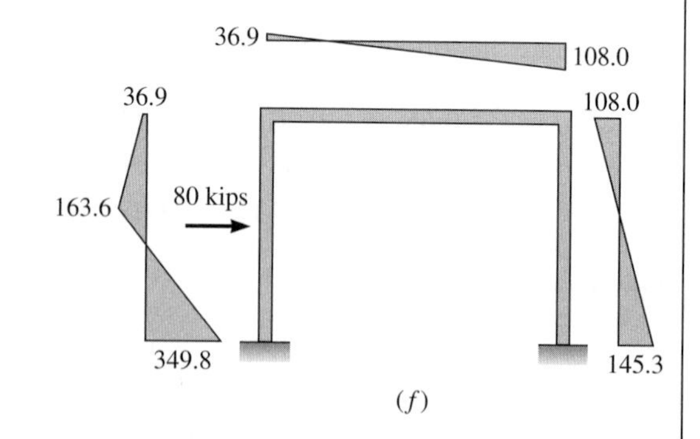

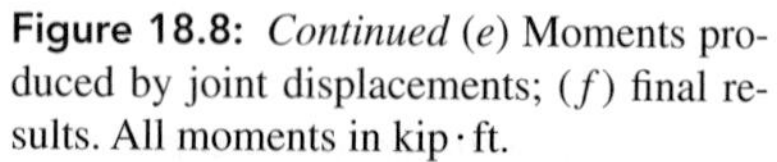

Figure 18.8: *Continued* (*e*) Moments produced by joint displacements; (*f*) final results. All moments in kip · ft.

18.5 The 6 × 6 Member Stiffness Matrix in Local Coordinates

While virtually all members in real structures are subject to both axial and flexural deformations, it is often possible to obtain accurate solutions using analytical models in which only one deformation mode (flexural or axial) is considered. For example, as we showed in Chapter 17, the analysis of trusses can be carried out using a member stiffness matrix that relates axial loads and deformations; bending effects, although present (since real joints do not behave as frictionless pins, and the dead weight of a member produces moment), are negligible. In other structures, such as beams and frames treated in the previous sections of this chapter, often the axial deformations have a negligible effect, and the analysis can be carried out considering bending deformations only. When it is necessary to include both deformation components, in this section we derive a member stiffness matrix in local coordinates that will allow us to consider both axial and bending effects simultaneously.

When bending and axial deformations are considered, each joint has three degrees of freedom; thus the order of the member stiffness matrix is 6. Figure 18.9 shows the positive direction of the degrees of freedom (joint displacements) in local coordinates; notice that the sign convention for end rotations and transverse displacements (degrees of freedom 1 through 4) is identical to that previously used in the derivation of the member stiffness matrix given by Equation 18.36. The displacements in the axial direction (degrees of freedom 5 and 6) are positive in the direction of the positive x' axis, which, as stated previously, runs from the near to the far joint.

The coefficients in the 6 × 6 member stiffness matrix can readily be obtained from information derived previously for the beam and truss elements.

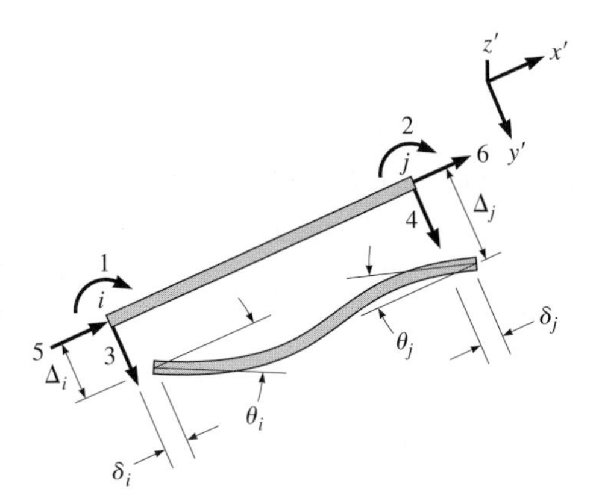

Figure 18.9: Positive sense of joint displacement for a flexural member.

Unit Displacements at DOF 1 through 4

These displacement patterns were shown in Figure 18.6; the results were calculated in Section 18.4 and are contained in Equations 18.37, 18.38, 18.40, and 18.41. We also notice that since these displacements do not introduce any axial elongations,

$$k'_{51} = k'_{52} = k'_{53} = k'_{54} = k'_{61} = k'_{62} = k'_{63} = k'_{64} = 0 \tag{18.51}$$

Unit Displacements at DOF 5 and 6

These conditions were considered in the derivation of the 2×2 member stiffness matrix for a truss bar in Chapter 17. From Equation 17.15 we compute

$$k'_{55} = k'_{66} = -k'_{56} = -k'_{65} = \frac{AE}{L} \tag{18.52}$$

Since no moments or shears are induced by these axial deformations, it follows that

$$k'_{15} = k'_{25} = k'_{35} = k'_{45} = k'_{16} = k'_{26} = k'_{36} = k'_{46} = 0 \tag{18.53}$$

Notice that the coefficients in Equations 18.51 and 18.53 satisfy symmetry (Betti's law).

Organizing all the stiffness coefficients in a matrix, we obtain the 6×6 member stiffness matrix in local coordinates as

$$\begin{array}{c} \text{DOF:} \\ \\ \mathbf{k}' = \end{array} \begin{array}{c} \begin{array}{cccccc} 1 & 2 & 3 & 4 & 5 & 6 \end{array} \\ \begin{bmatrix} \frac{4EI}{L} & \frac{2EI}{L} & \frac{6EI}{L^2} & -\frac{6EI}{L^2} & 0 & 0 \\ \frac{2EI}{L} & \frac{4EI}{L} & \frac{6EI}{L^2} & -\frac{6EI}{L^2} & 0 & 0 \\ \frac{6EI}{L^2} & \frac{6EI}{L^2} & \frac{12EI}{L^3} & \frac{-12EI}{L^3} & 0 & 0 \\ -\frac{6EI}{L^2} & -\frac{6EI}{L^2} & -\frac{12EI}{L^3} & \frac{12EI}{L^3} & 0 & 0 \\ 0 & 0 & 0 & 0 & \frac{AE}{L} & -\frac{AE}{L} \\ 0 & 0 & 0 & 0 & -\frac{AE}{L} & \frac{AE}{L} \end{bmatrix} \end{array} \begin{array}{c} 1 \\ 2 \\ 3 \\ 4 \\ 5 \\ 6 \end{array} \tag{18.54}$$

We illustrate the use of Equation 18.54 in Example 18.4.

EXAMPLE 18.4

Analyze the frame in Figure 18.10*a*, considering both axial and flexural deformations. The flexural and axial stiffnesses *EI* and *AE* are the same for both members and equal 24×10^6 kip·in^2 and 0.72×10^6 kips, respectively. The structure supports a concentrated load of 40 kips that acts vertically down at the center of span *BC*.

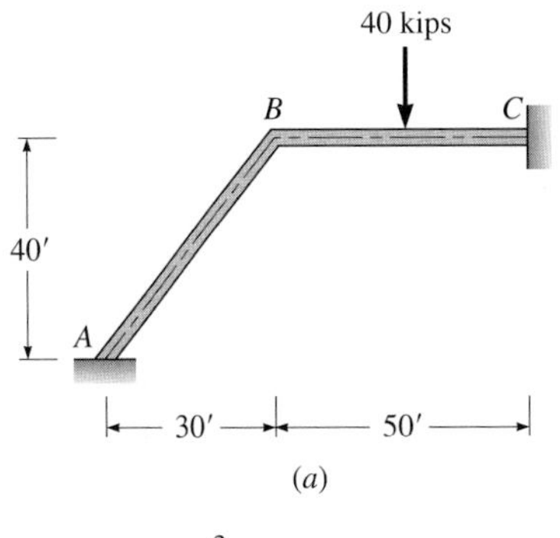

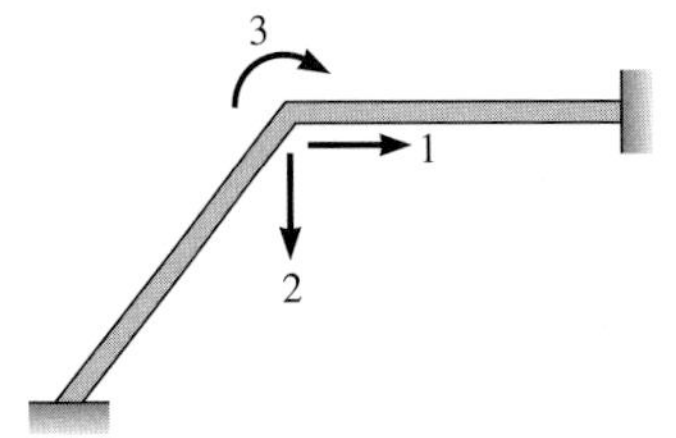

Figure 18.10: (*a*) Details of frame; (*b*) positive sense of unknown joint displacements.

Solution

With axial elongations considered, the structure has three degrees of kinematic indeterminacy, as shown in Figure 18.10*b*. The five-step solution procedure follows:

Step 1: Analysis of the Restrained Structure With the three degrees of freedom restrained at joint *B*, the frame is transformed to two fixed-end beams. The moments for this case are shown in Figure 18.10*c*. From equilibrium of the free-body diagram of joint *B*,

$$X_1 = 0 \qquad \text{or} \qquad X_1 = 0$$

$$Y_2 + 20.0 = 0 \qquad \text{or} \qquad Y_2 = -20.0$$

$$M_3 + 250.0 = 0 \qquad \text{or} \qquad M_3 = -250.0 \text{ kip} \cdot \text{ft} = -3000 \text{ kip} \cdot \text{in}$$

Reversing the sign of these restraining forces to construct the force vector **F** gives

$$\mathbf{F} = \begin{bmatrix} 0 \\ 20.0 \\ 3000.0 \end{bmatrix} \tag{18.55}$$

The units are kips and inches.

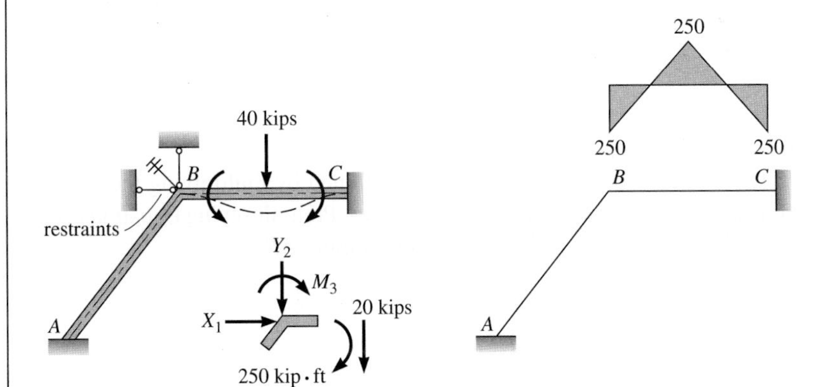

Figure 18.10: (*c*) Forces in the restrained structure produced by the 40-kip load; only member *BC* is stressed. Moments in kip·ft (*continues*).

[*continues on next page*]

Example 18.4 continues . . .

Step 2: Assembly of the Structure Stiffness Matrix The stiffness matrices in local coordinates for members AB and BC are identical because their properties are the same. Substituting the numerical values for EI, AE, and the length L, which is 600 in., into Equation 18.54 gives

$$\mathbf{k}' = 10^2\begin{bmatrix} 1600 & 800 & 4 & -4 & 0 & 0 \\ 800 & 1600 & 4 & -4 & 0 & 0 \\ 4 & 4 & 0.0133 & -0.0133 & 0 & 0 \\ -4 & -4 & -0.0133 & 0.0133 & 0 & 0 \\ 0 & 0 & 0 & 0 & 12 & -12 \\ 0 & 0 & 0 & 0 & -12 & 12 \end{bmatrix} \tag{18.56}$$

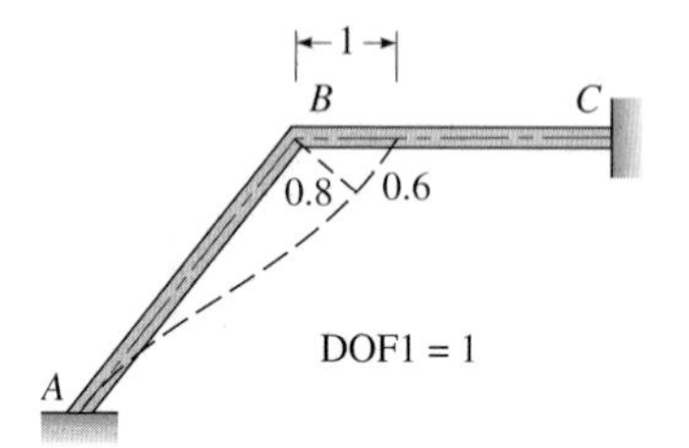

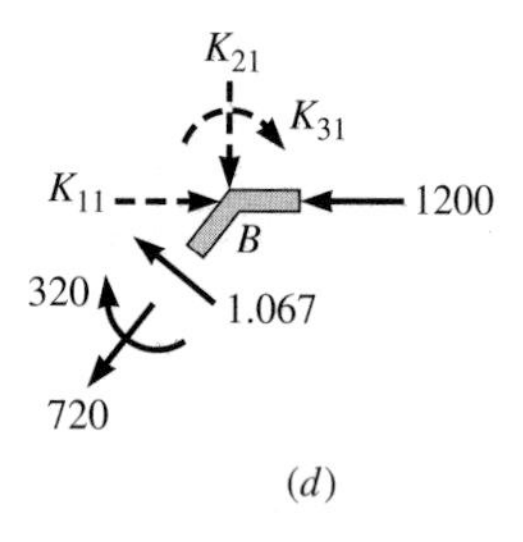

Figure 18.10: *Continued* (*d*) Stiffness coefficients associated with a unit horizontal displacement of joint B (*continues*).

The deformed configuration corresponding to a 1-in displacement of degree of freedom 1 is shown in Figure 18.10*d*. The deformations expressed in local coordinates for member AB are

$$\begin{bmatrix} \theta_A \\ \theta_B \\ \Delta_A \\ \Delta_B \\ \delta_A \\ \delta_B \end{bmatrix} = \begin{bmatrix} 0 \\ 0 \\ 0 \\ 0.8 \\ 0 \\ 0.6 \end{bmatrix} \tag{18.57}$$

and for member BC are

$$\begin{bmatrix} \theta_B \\ \theta_C \\ \Delta_B \\ \Delta_C \\ \delta_B \\ \delta_C \end{bmatrix} = \begin{bmatrix} 0 \\ 0 \\ 0 \\ 0 \\ 1 \\ 0 \end{bmatrix} \tag{18.58}$$

The units are radians and inches.

The forces in the members are then obtained by multiplying the member deformations by the element stiffness matrices. Premultiplying Equations 18.57 and 18.58 by Equation 18.56, we get for member AB,

$$\begin{bmatrix} M_i \\ M_j \\ V_i \\ V_j \\ F_i \\ F_j \end{bmatrix} = \begin{bmatrix} -320.0 \\ -320.0 \\ -1.064 \\ 1.064 \\ -720.0 \\ 720.0 \end{bmatrix} \tag{18.59}$$

and for member *BC*,

$$\begin{bmatrix} M_i \\ M_j \\ V_i \\ V_j \\ F_i \\ F_j \end{bmatrix} = \begin{bmatrix} 0 \\ 0 \\ 0 \\ 0 \\ 1200.0 \\ -1200.0 \end{bmatrix} \qquad (18.60)$$

In Equations 18.59 and 18.60 subscripts i and j are used to designate the near and far joints, respectively. These member end forces, with the sign reversed, can be used to construct the free-body diagram of joint *B* in Figure 18.10*d*. We compute from this diagram the forces required for equilibrium of this deformed configuration.

$$K_{11} - 1200 - (720 \times 0.6) - (1.067 \times 0.8) = 0 \quad \text{or} \quad K_{11} = 1632.85$$

$$K_{21} + (720 \times 0.8) - (1.067 \times 0.6) = 0 \quad \text{or} \quad K_{21} = -575.36$$

$$K_{31} + 320.0 = 0 \quad \text{or} \quad K_{31} = -320.0$$

In Figure 18.10*e* we show the deformed configuration for a unit displacement at degree of freedom 2. Proceeding as before, we find the member deformations. For member *AB*,

$$\begin{bmatrix} \theta_A \\ \theta_B \\ \Delta_A \\ \Delta_B \\ \delta_A \\ \delta_B \end{bmatrix} = \begin{bmatrix} 0 \\ 0 \\ 0 \\ 0.6 \\ 0 \\ -0.8 \end{bmatrix} \qquad (18.61)$$

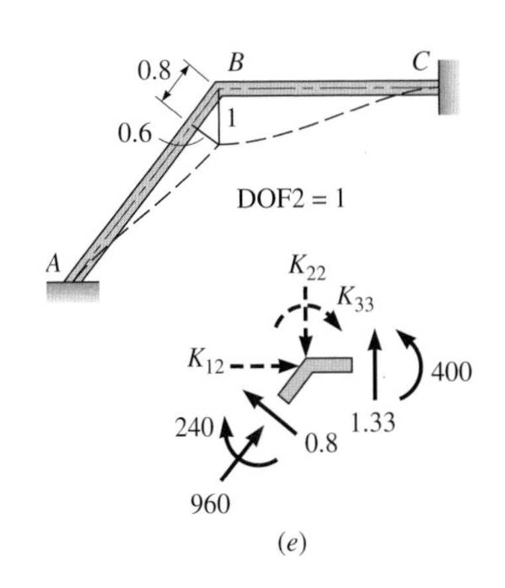

Figure 18.10: *Continued* (*e*) Stiffness coefficients produced by a unit vertical displacement of joint *B* (*continues on page* 745).

and for member *BC*,

$$\begin{bmatrix} \theta_B \\ \theta_C \\ \Delta_B \\ \Delta_C \\ \delta_B \\ \delta_C \end{bmatrix} = \begin{bmatrix} 0 \\ 0 \\ 1 \\ 0 \\ 0 \\ 0 \end{bmatrix} \qquad (18.62)$$

Multiplying the deformations in Equations 18.61 and 18.62 by the element stiffness matrices, we obtain the following member forces. For member *AB*,

[*continues on next page*]

Example 18.4 continues . . .

$$\begin{bmatrix} M_i \\ M_j \\ V_i \\ V_j \\ F_i \\ F_j \end{bmatrix} = \begin{bmatrix} -240.0 \\ -240.0 \\ -0.8 \\ 0.8 \\ 960.0 \\ -960.0 \end{bmatrix} \tag{18.63}$$

and for member BC,

$$\begin{bmatrix} M_i \\ M_j \\ V_i \\ V_j \\ F_i \\ F_j \end{bmatrix} = \begin{bmatrix} 400.0 \\ 400.0 \\ 1.333 \\ -1.333 \\ 0 \\ 0 \end{bmatrix} \tag{18.64}$$

Given the internal member forces, the external forces required for equilibrium at the degrees of freedom are readily found; referring to the free-body diagram of joint B in Figure 18.10e, we calculate the following stiffness coefficients:

$$K_{12} + (960 \times 0.6) - (0.8 \times 0.8) = 0 \quad \text{or} \quad K_{12} = -575.36$$

$$K_{22} - (960 \times 0.8) - (0.8 \times 0.6) - 1.33 = 0 \quad \text{or} \quad K_{22} = 769.81$$

$$K_{32} + 240 - 400 = 0 \quad \text{or} \quad K_{32} = -160.0$$

Finally, introducing a unit displacement at degree of freedom 3, we obtain the following results (see Figure 18.10f). The deformations for member AB are

$$\begin{bmatrix} \theta_A \\ \theta_B \\ \Delta_A \\ \Delta_B \\ \delta_A \\ \delta_B \end{bmatrix} = \begin{bmatrix} 0 \\ 1 \\ 0 \\ 0 \\ 0 \\ 0 \end{bmatrix} \tag{18.65}$$

and for member BC,

$$\begin{bmatrix} \theta_B \\ \theta_C \\ \Delta_B \\ \Delta_C \\ \delta_B \\ \delta_C \end{bmatrix} = \begin{bmatrix} 1 \\ 0 \\ 0 \\ 0 \\ 0 \\ 0 \end{bmatrix} \tag{18.66}$$

The member forces for member *AB* are

$$\begin{bmatrix} M_i \\ M_j \\ V_i \\ V_j \\ F_i \\ F_j \end{bmatrix} = \begin{bmatrix} 8000 \\ 160{,}000 \\ 400 \\ -400 \\ 0 \\ 0 \end{bmatrix} \tag{18.67}$$

and for member *BC*,

$$\begin{bmatrix} M_i \\ M_j \\ V_i \\ V_j \\ F_i \\ F_j \end{bmatrix} = \begin{bmatrix} 160{,}000 \\ 80{,}000 \\ 400 \\ -400 \\ 0 \\ 0 \end{bmatrix} \tag{18.68}$$

From the free-body diagram of joint *B* in Figure 18.10*f* we get the following stiffness coefficients:

$$K_{13} + 400 \times 0.8 = 0 \quad \text{and} \quad K_{13} = -320$$

$$K_{23} + 400 \times 0.6 - 400 = 0 \quad \text{and} \quad K_{23} = 160$$

$$K_{33} - 160{,}000 - 160{,}000 = 0 \quad \text{and} \quad K_{33} = 320{,}000$$

Organizing the stiffness coefficients in matrix notation, we obtain the following structure stiffness matrix:

$$\mathbf{K} = \begin{bmatrix} 1632.85 & -575.36 & -320.0 \\ -575.36 & 769.81 & 160.0 \\ -320.0 & 160.0 & 320{,}000.0 \end{bmatrix} \tag{18.69}$$

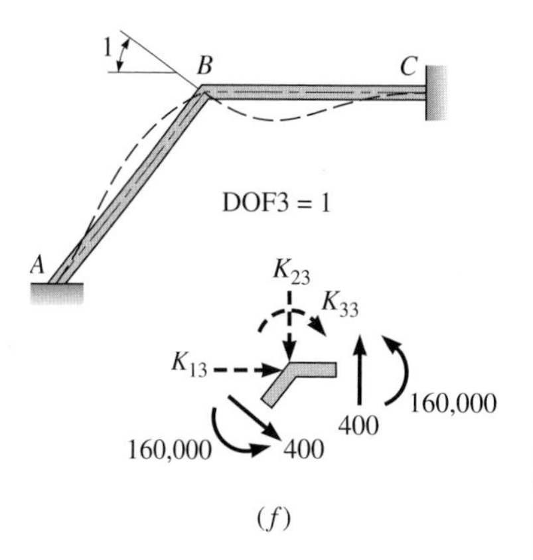

Figure 18.10: *Continued* (*f*) Stiffness coefficients produced by a unit rotation of joint *B* (*continues on page* 747).

Step 3: Solution of Equation 18.1 Substituting **F** and **K** into Equation 18.1, we produce the following system of simultaneous equations:

$$\begin{bmatrix} 1632.85 & -575.36 & -320.0 \\ -575.36 & 769.81 & 160.0 \\ -320.0 & 160.0 & 320{,}000.0 \end{bmatrix} \begin{bmatrix} \Delta_1 \\ \Delta_2 \\ \theta_3 \end{bmatrix} = \begin{bmatrix} 0 \\ 20.0 \\ 3000.0 \end{bmatrix} \tag{18.70}$$

Solving Equation 18.70 gives

$$\begin{bmatrix} \Delta_1 \\ \Delta_2 \\ \theta_3 \end{bmatrix} = \begin{bmatrix} 0.014 \\ 0.0345 \\ 0.00937 \end{bmatrix} \tag{18.71}$$

The units are radians and inches.

[*continues on next page*]

Example 18.4 continues . . .

Step 4: Evaluation of the Effect of Joint Displacements The effects of joint displacements are calculated by multiplying the individual member stiffness matrices by the corresponding member deformations in local coordinates, which are defined in Figure 18.9. Member deformations can be computed from global displacements (Equation 18.71) using the geometric relationships established in Figures 18.10*d*, *e*, and *f*. Consider the axial deformation of member *AB* for example. The axial deformation δ_A at joint *A* is zero because it is a fixed end. The axial deformations δ_B produced by a unit displacement in the horizontal, vertical, and rotational directions of joint *B* are 0.6, −0.8, and 0.0, respectively. Therefore, joint displacements calculated in Equation 18.71 produce the following axial deformation at joint *B*:

$$\delta_B = (0.014 \times 0.6) + (0.0345 \times -0.8) + (0.00937 \times (0.0) = -0.0192$$

Following this procedure, the six components of the local deformations for member *AB* are

$$\theta_A = 0$$

$$\theta_B = 0.00937$$

$$\Delta_A = 0$$

$$\Delta_B = (0.014 \times 0.8) + (0.0345 \times 0.6) = -0.0319$$

$$\delta_A = 0$$

$$\delta_B = (0.014 \times 0.6) + (0.0345 \times -0.8) = -0.0192$$

Similarly, for member *BC*,

$$\theta_B = 0.00937$$

$$\theta_C = 0$$

$$\Delta_B = 0.0345$$

$$\Delta_C = 0$$

$$\delta_B = 0.014$$

$$\delta_C = 0$$

Multiplying these deformations by the member stiffness matrix (Equation 18.54), we get the member forces from joint displacements. For member *AB*,

$$\begin{bmatrix} M''_{AB} \\ M''_{BA} \\ V''_{AB} \\ V''_{BA} \\ F''_{AB} \\ F''_{BA} \end{bmatrix} = \begin{bmatrix} 736.98 \\ 1486.71 \\ 3.706 \\ -3.706 \\ 23.04 \\ -23.04 \end{bmatrix} \tag{18.72}$$

and for bar *BC*,

$$\begin{bmatrix} M''_{BC} \\ M''_{CB} \\ V''_{BC} \\ V''_{CB} \\ F''_{BC} \\ F''_{CB} \end{bmatrix} = \begin{bmatrix} 1513.29 \\ 763.54 \\ 3.79 \\ -3.79 \\ 16.80 \\ -16.80 \end{bmatrix} \tag{18.73}$$

The results given by Equations 18.72 and 18.73 are plotted in Figure 18.10*g*. Note that the units of moment in the figure are kip · feet.

Step 5: Calculation of Final Results The complete solution is obtained as usual by adding the restrained case (Figure 18.10*c*) to the effects of joint displacements (Figure 18.10*g*). The results are plotted in Figure 18.10*h*.

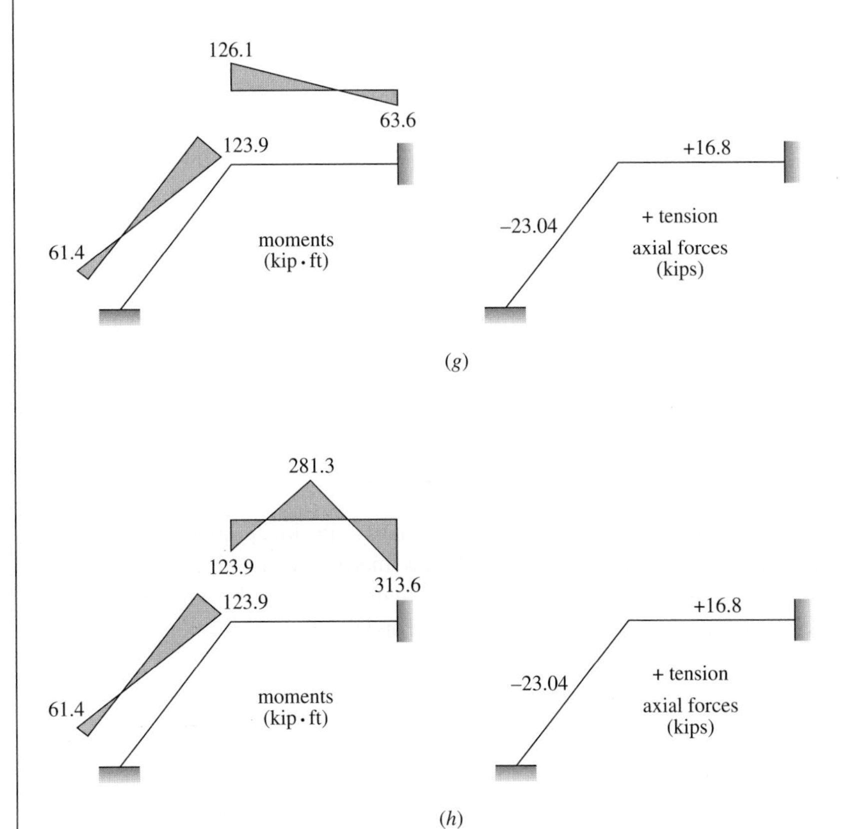

Figure 18.10: *Continued* (*g*) Moment diagrams and axial forces produced by the actual displacements of joint *B*; (*h*) final results.

18.6 The 6 × 6 Member Stiffness Matrix in Global Coordinates

The stiffness matrix of a structure can be assembled by introducing unit displacement at the selected degrees of freedom (with all other joints restrained) and then calculating the corresponding joint forces required for equilibrium. This approach, although most efficient when using hand calculators, is not well suited to computer applications.

The technique actually utilized to assemble the structure stiffness matrix in computer applications is based on the addition of the individual member stiffness matrices in a global coordinate system. In this approach, initially discussed in Section 17.2 for the case of trusses, the individual member stiffness matrices are expressed in terms of a common coordinate system, usually referred to as the global coordinate system. Once expressed in this form, the individual member stiffness matrices are expanded to the size of the structure stiffness matrix (by adding columns and rows of zeros where necessary) and then added directly.

In this section we derive the general beam-column member stiffness matrix in global coordinates. In Section 18.7, the direct summation process by which these matrices are combined to give the total stiffness matrix for the structure is illustrated with an example.

The 6 × 6 member stiffness matrix for a beam-column element is derived in local coordinates in Section 18.5 and is presented as Equation 18.54. A derivation in global coordinates can be carried out in much the same manner by using the basic approach of introducing unit displacement at each node and calculating the required joint forces. The process is, however, rather cumbersome because of the geometric relationships involved. A simpler, more concise derivation can be made using the member stiffness matrix in local coordinates and the coordinate transformation expression presented in Section 17.7. For convenience in this development, the equation for the transformation of coordinates, originally denoted as Equation 17.54, is repeated below as Equation 18.74.

$$\mathbf{k} = \mathbf{T}^{\mathrm{T}}\mathbf{k}'\mathbf{T} \tag{18.74}$$

where $\mathbf{k}'$ is the member stiffness matrix in local coordinates (Equation 18.54), $\mathbf{k}$ is the member stiffness matrix in global coordinates, and $\mathbf{T}$ is the transformation matrix. The $\mathbf{T}$ matrix is formed from the geometric relationships that exist between the local and the global coordinates. In matrix form

$$\boldsymbol{\delta} = \mathbf{T}\boldsymbol{\Delta} \tag{18.75}$$

where $\boldsymbol{\delta}$ and $\boldsymbol{\Delta}$ are the vectors of local and global joint displacements, respectively.

Refer to Figure 18.11*a* and *b* for the member *ij* expressed in the local and global coordinate systems, respectively. Note that the components of translation are different at each end, but the rotation is identical. The relationship between the local displacement vector $\boldsymbol{\delta}$ and the global displacement vector $\boldsymbol{\Delta}$ is established as follows. Figure 18.11*c* and *d* shows the displacement components in the local coordinate system produced by global displacements Δ_{ix} and Δ_{iy}, at joint *i*, respectively. From the figure,

$$\delta_i = (\cos\phi)(\Delta_{ix}) - (\sin\phi)(\Delta_{iy}) \tag{18.76}$$

$$\Delta_i = (\sin\phi)(\Delta_{ix}) + (\cos\phi)(\Delta_{iy}) \tag{18.77}$$

Similarly, by introducing Δ_{jx} and Δ_{jy} respectively, to joint *j* (see Figure 18.11*e* and *f*), the following expressions can be established:

$$\delta_j = (\cos\phi)(\Delta_{jx}) - (\sin\phi)(\Delta_{jy}) \tag{18.78}$$

$$\Delta_j = (\sin\phi)(\Delta_{jx}) + (\cos\phi)(\Delta_{jy}) \tag{18.79}$$

Together with two identity equations for joint rotations ($\theta_i = \theta_i$ and $\theta_j = \theta_j$), the relationship between $\boldsymbol{\delta}$ and $\boldsymbol{\Delta}$ is

$$\begin{bmatrix} \theta_i \\ \theta_j \\ \Delta_i \\ \Delta_j \\ \delta_i \\ \delta_j \end{bmatrix} = \begin{bmatrix} 0 & 0 & 1 & 0 & 0 & 0 \\ 0 & 0 & 0 & 0 & 0 & 1 \\ s & c & 0 & 0 & 0 & 0 \\ 0 & 0 & 0 & s & c & 0 \\ c & -s & 0 & 0 & 0 & 0 \\ 0 & 0 & 0 & c & -s & 0 \end{bmatrix} \begin{bmatrix} \Delta_{ix} \\ \Delta_{iy} \\ \theta_i \\ \Delta_{jx} \\ \Delta_{jy} \\ \theta_j \end{bmatrix} \tag{18.80}$$

where $s = \sin\phi$, $c = \cos\phi$, and the 6 × 6 matrix is the transformation matrix $\mathbf{T}$.

From Equation 18.74, the member stiffness matrix in global coordinates is

$$\mathbf{k} = \mathbf{T}^{\mathrm{T}}\mathbf{k}'\mathbf{T}$$

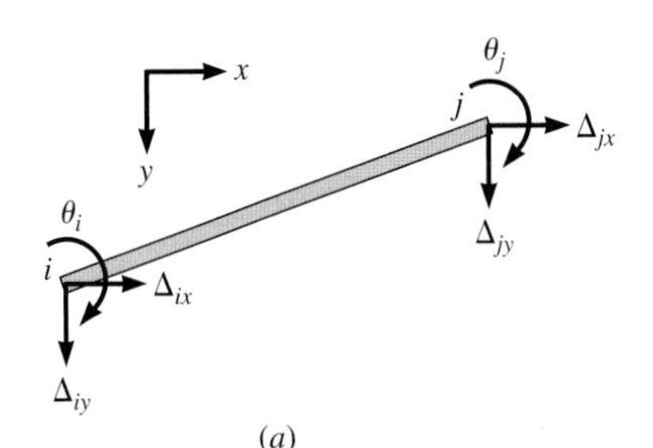

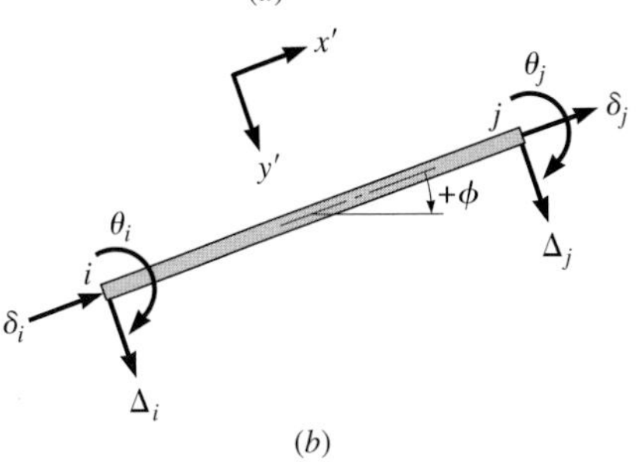

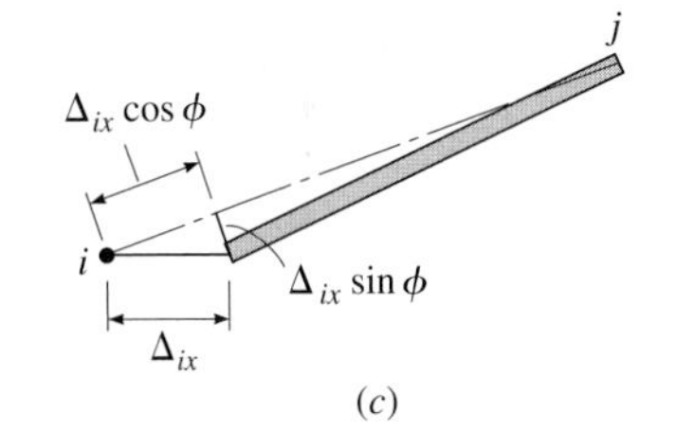

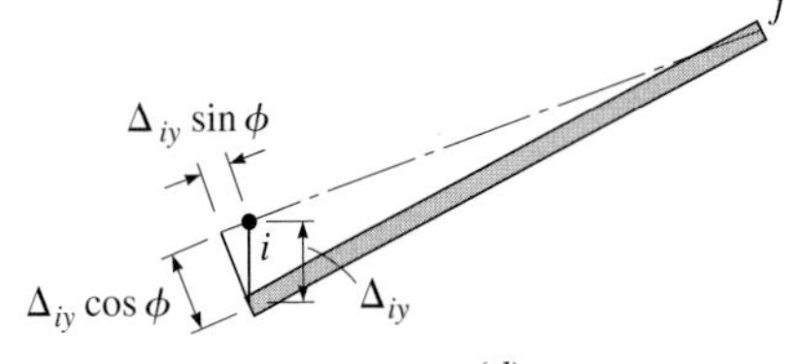

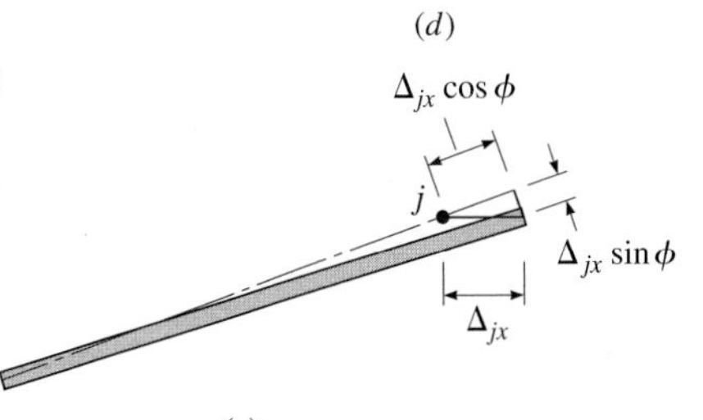

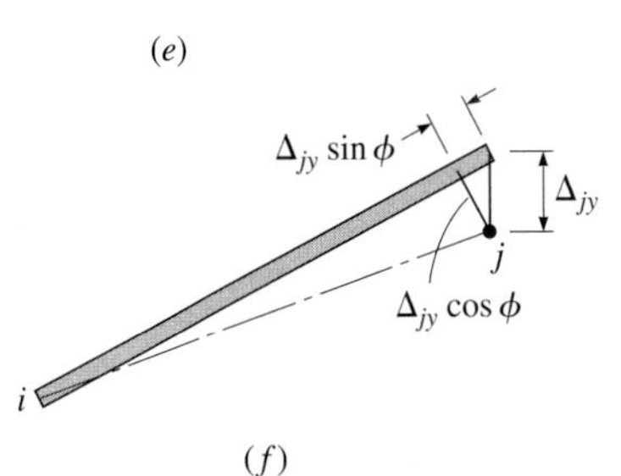

Figure 18.11: (*a*) Member displacement components in global coordinates; (*b*) member displacement components in local coordinates; (*c*) local displacement components produced by a global displacement Δ_{ix}; (*d*) local displacement components produced by a global displacement Δ_{iy}; (*e*) Local displacement components produced by a global displacement Δ_{jx}; and (*f*) local displacement components produced by a global displacement Δ_{jy}.

$$
= \begin{bmatrix} 0 & 0 & s & 0 & c & 0 \\ 0 & 0 & c & 0 & -s & 0 \\ 1 & 0 & 0 & 0 & 0 & 0 \\ 0 & 0 & 0 & s & 0 & c \\ 0 & 0 & 0 & c & 0 & -s \\ 0 & 1 & 0 & 0 & 0 & 0 \end{bmatrix}
\begin{bmatrix} \frac{4EI}{L} & \frac{2EI}{L} & \frac{6EI}{L^2} & -\frac{6EI}{L^2} & 0 & 0 \\ \frac{2EI}{L} & \frac{4EI}{L} & \frac{6EI}{L^2} & -\frac{6EI}{L^2} & 0 & 0 \\ \frac{6EI}{L^2} & \frac{6EI}{L^2} & \frac{12EI}{L^3} & \frac{-12EI}{L^3} & 0 & 0 \\ -\frac{6EI}{L^2} & -\frac{6EI}{L^2} & -\frac{12EI}{L^3} & \frac{12EI}{L^3} & 0 & 0 \\ 0 & 0 & 0 & 0 & \frac{AE}{L} & -\frac{AE}{L} \\ 0 & 0 & 0 & 0 & -\frac{AE}{L} & \frac{AE}{L} \end{bmatrix}
\begin{bmatrix} 0 & 0 & 1 & 0 & 0 & 0 \\ 0 & 0 & 0 & 0 & 0 & 1 \\ s & c & 0 & 0 & 0 & 0 \\ 0 & 0 & 0 & s & c & 0 \\ c & -s & 0 & 0 & 0 & 0 \\ 0 & 0 & 0 & c & -s & 0 \end{bmatrix}
$$

$$
\mathbf{k} = \frac{EI}{L} \begin{bmatrix} Nc^2 + Ps^2 & sc(-N+P) & Qs & -(Nc^2+Ps^2) & -sc(-N+P) & Qs \\ & Ns^2 + Pc^2 & Qc & sc(N-P) & -(Ns^2+Pc^2) & Qc \\ & & 4 & -Qs & -Qc & 2 \\ & \text{Symmetric about main diagonal} & & Nc^2+Ps^2 & sc(-N+P) & -Qs \\ & & & & Ns^2+Pc^2 & -Qc \\ & & & & & 4 \end{bmatrix} \tag{18.81}
$$

where $\mathbf{k}'$ is from Equation 18.54, $N = A/I$, $P = 12/L^2$, and $Q = 6/L$.

18.7 Assembly of a Structure Stiffness Matrix—Direct Stiffness Method

Once the individual member stiffness matrices are expressed in global coordinates, they can be summed directly using the procedure described in Chapter 17. The combination of individual member stiffness matrices to form the structure stiffness matrix can be simplified by the introduction of the following notation in Equation 18.81. Partitioning after the third column (and row), we can write Equation 18.81 in compact form as

$$
\mathbf{k} = \begin{bmatrix} \mathbf{k}_N^m & \mathbf{k}_{NF}^m \\ \mathbf{k}_{FN}^m & \mathbf{k}_F^m \end{bmatrix} \tag{18.82}
$$

where the subscripts N and F refer to near and far joints for the member, respectively, and the superscript m is the number assigned to the member in question in the structural sketch. The terms in each of the submatrices of Equation 18.82 are readily obtained from Equation 18.81 and are not repeated here.

To illustrate the assembly of the structure stiffness matrix by direct summation, let's consider once again the frame shown in Figure 18.10. The stiffness matrix for this structure is derived in Example 18.4 and is labeled Equation 18.69.

EXAMPLE 18.5

Using the direct stiffness method, assemble the structure stiffness matrix for the frame in Figure 18.10*a*.

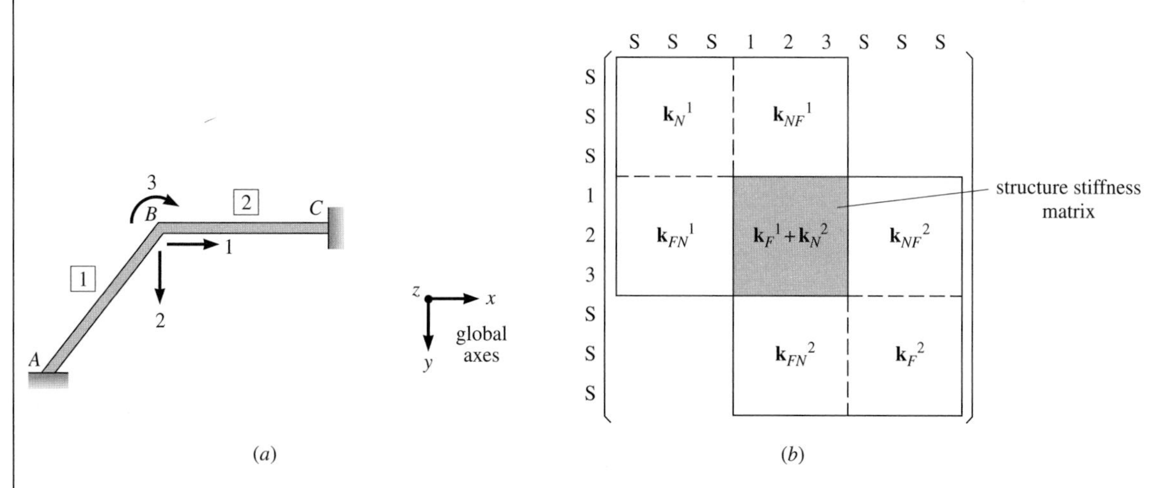

Figure 18.12: (*a*) Frame with three degrees of freedom; (*b*) assembly of structure stiffness matrix from member stiffness matrices.

Solution

Figure 18.12*a* illustrates the structure and identifies the degrees of freedom. Note that the degrees of freedom are numbered in the order x, y, z and are shown in the sense of the positive direction of the global axes; this order is necessary to take advantage of the special form of Equation 18.82.

Since the frame considered has three joints, the total number of independent joint displacement components, before any supports are introduced, is 9. Figure 18.12*b* shows the stiffness matrices for the two members (expressed in the format of Equation 18.82), properly located within the 9×9 matrix space. Because of the specific support conditions, the columns and rows labeled S (support) can be deleted, thus leaving only a 3×3 structure stiffness matrix.

As can be seen in Figure 18.12*b*, the structure stiffness matrix, in terms of the individual members, is given by

$$\mathbf{K} = \mathbf{k}_F^1 + \mathbf{k}_N^2 \qquad (18.83)$$

where $\mathbf{k}_F^1$ refers to the submatrix of member 1 at far end, and $\mathbf{k}_N^2$ refers to the submatrix of member 2 at near end. The matrices in Equation 18.83 are evaluated from Equation 18.81 as follows. For member 1, $\alpha = 53.13°$ (positive

[*continues on next page*]

Example 18.5 continues . . .

since clockwise from local to global x axes); so $s = 0.8$ and $c = 0.6$. From the data in Example 18.4,

$$N = \frac{A}{I} = \frac{0.72}{24.0} = 0.03 \text{ in}^{-2}$$

$$P = \frac{12}{L^2} = \frac{12}{600^2} = 33.33 \times 10^{-6} \text{ in}^{-2}$$

$$Q = \frac{6}{L} = \frac{6}{600} = 0.01 \text{ in}^{-1}$$

$$\frac{EI}{L} = \frac{24.0 \times 10^6}{600} = 40{,}000 \text{ kip} \cdot \text{in}$$

For member 2, $\alpha = 0°$, $s = 0$, $c = 1$, and the values of N, P, Q, and EI are the same as in member 1. Substituting these numerical results into Equation 18.81, we compute

$$\mathbf{k}_F^1 = \begin{bmatrix} 432.85 & -575.36 & -320 \\ -575.36 & 768.48 & -240 \\ -320 & -240 & 160{,}000 \end{bmatrix} \tag{18.84}$$

(columns and rows numbered 1, 2, 3)

and

$$\mathbf{k}_N^2 = \begin{bmatrix} 1200 & 0 & 0 \\ 0 & 1.33 & 400 \\ 0 & 400 & 160{,}000 \end{bmatrix} \tag{18.85}$$

(columns and rows numbered 1, 2, 3)

Finally, substituting Equations 18.84 and 18.85 into Equation 18.83, we obtain the structure stiffness matrix by direct summation.

$$\mathbf{K} = \begin{bmatrix} 1632.85 & -575.36 & -320 \\ -575.36 & 769.81 & 160 \\ -320 & 160 & 320{,}000 \end{bmatrix} \quad \textbf{Ans.} \tag{18.86}$$

The $\mathbf{K}$ matrix in the above equation is identical to Equation 18.69, which was derived in Example 18.4 using the unit displacement approach.

Summary

- For the analysis of an indeterminate beam or frame structure by the matrix stiffness method, a five-step procedure is presented in this chapter. The procedure requires that the structure be analyzed first as a restrained system. After the joint restraining forces are determined, the second part of the analysis requires the solution of the following equilibrium equation for the unrestrained (or original) structure:

$$\mathbf{K\Delta} = \mathbf{F}$$

  where **K** is the structure stiffness matrix, **F** is the column vector of joint restraining forces but with the signs reversed, and **Δ** is the column vector of unknown joint displacements.
- The structure stiffness matrix **K** can be assembled from the member stiffness matrices by the direct stiffness method. When only rotations at two end joints are considered, the 2×2 member stiffness matrix $\bar{\mathbf{k}}$ is expressed by Equation 18.6, and the five-step solution process presented in Section 18.3 can be used to analyze an indeterminate beam or a braced frame when joint translations are prevented.
- When joint translations are present, but axial deformation of the member can be ignored, the 4×4 member stiffness matrix based on the local coordinate system in Figure 18.5 given by Equation 18.42 is used.
- When both bending and axial deformations are considered, each joint has three degrees of freedom. The 6×6 member stiffness matrix $\mathbf{k}'$ based on the local coordinate system in Figure 18.9 is presented in Equation 18.54.
- For computerized applications, however, it is desirable to express the member stiffness matrix in a common (or global) coordinate system, such that a direct summation process can be used to establish the structure stiffness matrix **K**. The member stiffness matrix **k**, presented in Equation 18.81, in the global coordinate system can be constructed from $\mathbf{k}'$ using the concept of coordinate transformation. Once the stiffness matrix **k** is established for each member, the structure stiffness matrix **K** is formed by summing the member stiffness matrices (see Section 18.7).

PROBLEMS

P18.1. Using the stiffness method, analyze the two-span continuous beam shown in Figure P18.1 and draw the shear and moment diagrams. *EI* is constant.

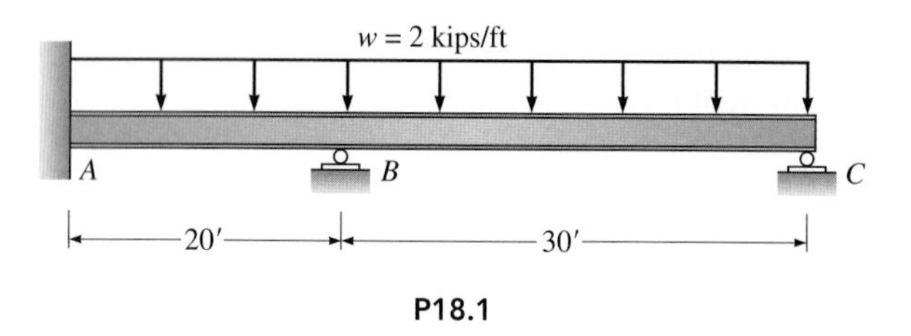

P18.1

P18.2. Write the stiffness matrix corresponding to the degrees of freedom 1, 2, and 3 of the continuous beam shown in Figure P18.2.

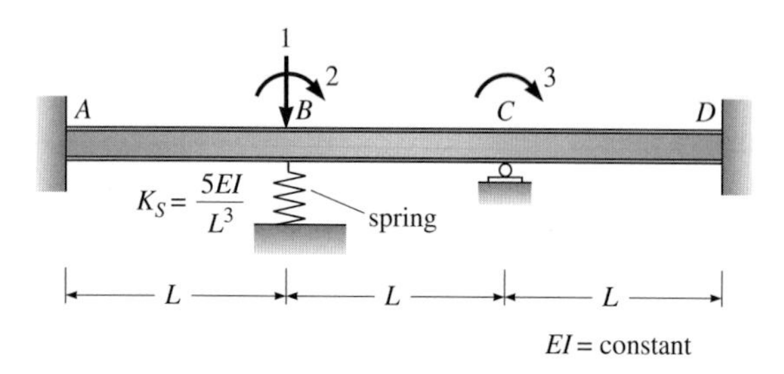

P18.2

P18.3. In Problem P18.2, find the force in the spring located at *B* if beam *ABCD* supports a downward uniform load *w* along its entire length.

P18.4. Neglecting axial deformations, find the end moments in the frame shown in Figure P18.4. The load acts at midspan on member *BC*.

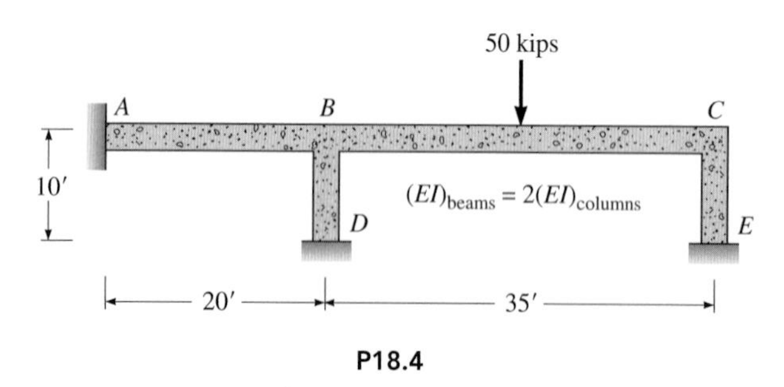

P18.4

P18.5. Analyze the frame in Figure P18.5 and draw the shear and moment diagrams for the members. Neglect axial deformation. *EI* is constant.

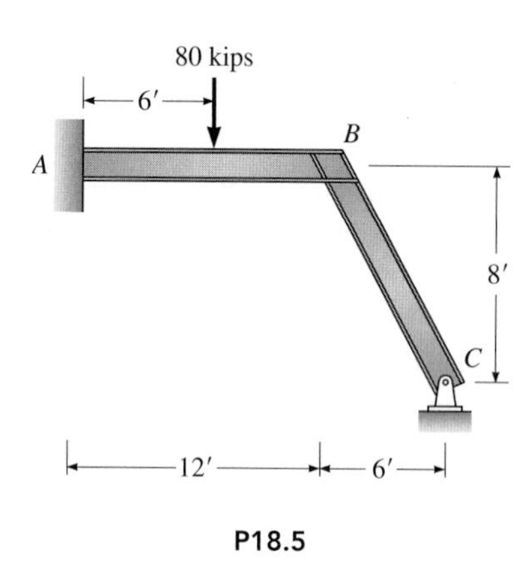

P18.5

P18.6. Using the stiffness method, analyze the frame in Figure P18.6 and draw the shear and moment diagrams for the members. Neglect axial deformations. *EI* is constant.

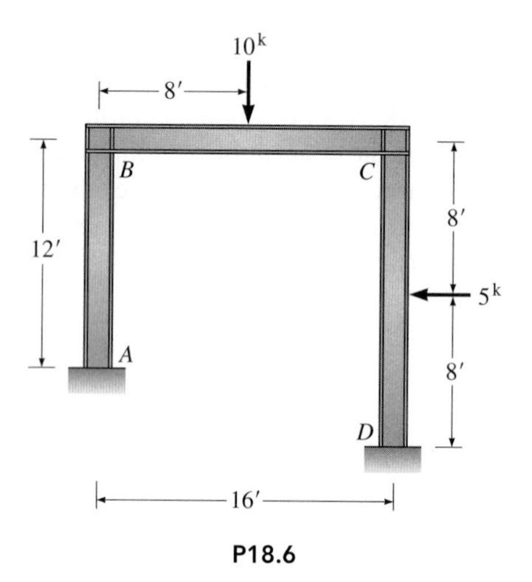

P18.6

P18.7. For the frame shown in Figure P18.7, write the stiffness matrix in terms of the three degrees of freedom indicated. Use both the method of introducing unit displacements and the member stiffness matrix of Equation 18.36. Given: $E = 30{,}000$ kips/in^2; $I = 500$ in^4, and $A = 15$ in^2.

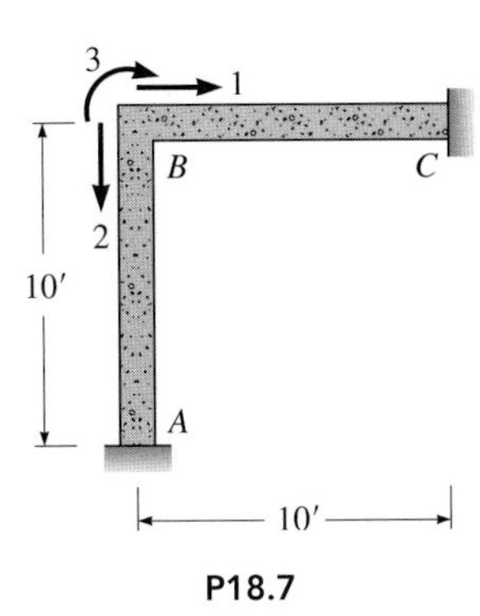

P18.7

P18.8. Solve Problem P18.7 using the direct summation of global element stiffness matrices.

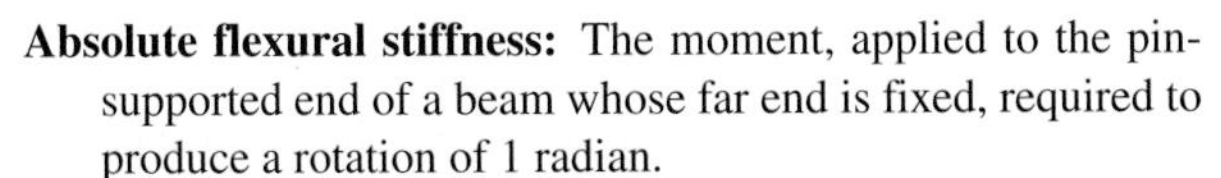

GLOSSARY

Absolute flexural stiffness: The moment, applied to the pin-supported end of a beam whose far end is fixed, required to produce a rotation of 1 radian.

Abutment: An end wall or element that transfers load from the end of a structural member into the foundation.

Base shear: The total lateral inertia or wind forces acting on all floors of a building that are transmitted to the foundations.

Beam column: Column that carries both axial force and moment. When the axial load is large, it reduces the flexural stiffness of the column.

Bearing wall: A structural wall, usually constructed of reinforced masonry or concrete, that supports floor and roof loads.

Bernoulli's principle: A reduction in air pressure is produced by an increase in wind velocity as it flows around obstructions in its path. Building codes consider this effect when they establish design wind forces on building walls and roofs.

Box beam: A hollow, rectangular beam. By eliminating material at the center of the member, weight is reduced but the bending stiffness is not significantly affected.

Braced frame: A structural frame whose joints are free to rotate, but not to displace laterally. Its resistance to lateral displacement is supplied by cross-bracing or by connection to shear walls or fixed supports.

Buckling: A failure mode of columns, plates, and shells when loaded in compression. When the buckling load is reached, the initial shape is no longer stable, and a bent configuration develops.

Building code: A set of provisions that controls design and construction in a given region. Its provisions establish minimum architectural, structural, mechanical, and electrical design requirements for buildings and other structures.

Cable sag: The vertical distance between the cable and its chord.

Cooper E 80 loading: The loading contained in the AREMA Manual for railroad engineering. The loading consists of the wheel loads of two locomotives followed by a uniform load representing the weight of freight cars.

Cross-bracing: Light diagonal members in the shape of an X that run from the top of a column to the bottom of an adjacent column. The bracing acts together with floor beams and columns as a truss to carry lateral loads into the foundations and reduce lateral displacements.

Dead load: Also called *Gravity load*. The load associated with the weight of a structure and its components such as walls, floors, utility pipes, air ducts, and so forth.

Diagonal bracing: See *Cross-bracing*.

Diaphragm action: The ability of shallow floor and roof slabs to transfer in-plane loads into supporting members.

Ductility: The ability of materials or structures to undergo large deformation without rupture. Ductility is the opposite of brittle behavior.

Dynamic analysis: An analysis that considers the inertia forces created by the motion of a structure. A dynamic analysis requires that a structure be modeled to account for its stiffness, mass, and the effect of damping.

Factored load: Load that is established by multiplying design load by a load factor—typically greater than 1 (part of the factor of safety).

First-order analysis: An analysis based on the original geometry of the structure in which deflections are assumed to be insignificant.

Flexibility coefficient: The deflection produced by a unit value of load or moment.

Floor beam: A member of a floor system positioned transversely to the direction of the span. Floor beams typically pick up the load from stringers and transfer it to the panel points of the main structural members, such as trusses, girders, or arches.

Free-body diagram: A sketch of a structure or a part of a structure showing all forces and dimensions required for an analysis.

Geometrically unstable: Refers to a support configuration that is not able to restrain rigid body displacements in all directions.

Girder: A large beam that often supports one or more secondary beams.

Gravity load: See *Dead load.*

Gusset plate: Connection plates that are used to form the joints of a truss. The forces between the members running into the joint are transferred through the gusset plate.

Hurricane region: Coastal regions where winds of large velocity (90 mph and larger) occur.

Idealized structure: A simplified sketch of a structure—usually a line drawing—that shows the loads and dimensions and assumed support conditions.

Impact: The force applied by moving bodies as kinetic energy is converted to additional force. The magnitude of the kinetic energy is a function of the body's mass and the velocity squared.

Indeterminate structure: A structure whose reactions and internal forces cannot be determined by the equations of statics.

Inertia forces: Forces produced on a moving structure by its own mass.

Kinetic energy: Energy possessed by a moving body. Its magnitude varies with the square of the velocity and its mass.

Leeward side: The side of a building opposite to the side impacted by the wind.

Link: See *Two-force member.*

Live load: Load that can be moved on or off a structure, such as furniture, vehicles, people, and supplies.

Load factor: Part of the factor of safety applied to members that are sized in strength design where design is based on the failure strength of members.

Membrane stress: In-plane stress that develops in shells and plates from applied loads.

Modulus of elasticity: A measure of a material's stiffness, that is, defined as the ratio of stress divided by strain and represented by the variable E.

Moment curves by parts: Moment diagrams are plotted for individual forces to produce simple geometric shapes whose areas and centroids are known (see back endsheet).

Moment of inertia: A property of a cross-sectional area that is a measure of a section's bending capacity.

Monolithic construction: Structure in which all parts act as a continuous unit.

Natural period: The time that it takes a structure to move through a full cycle of motion.

Nonprismatic: Refers to a member whose cross-sectional area varies along the length of its longitudinal axis.

***P*-delta effect:** Additional moments created by axial force due to lateral displacements of a member's longitudinal axis.

Panel point: Points where floor beams frame into girders or trusses. Also the joints of trusses.

Pattern loading: Positioning of live load in those locations that maximize the internal forces at a particular section of a structure. Influence lines are used for this purpose.

Pier: A wall of reinforced concrete or masonry that is loaded by the supports of a structure and transfers the loads into the foundations.

Planar structure: A structure all of whose members are located in the same plane.

Point of inflection: The point along a beam's axis where curvature changes from positive to negative.

Prestressing: Inducing beneficial stresses into a member by tensioned bars or cables anchored to the member.

Principle of superposition: The stresses and deflections produced by a set of forces are identical to those produced by the addition of the effects of the individual forces.

Rigid frame: A structure composed of flexural members connected by rigid joints.

Second-order analysis: An analysis that accounts for the effect of joint displacements on the forces in a structure that undergoes significant displacements.

Section modulus: A property of the cross-sectional area that measures a member's capacity to carry moment.

Seismic loads: Loads produced by the ground motion associated with earthquakes.

Service loads: Design loads specified by building codes.

Serviceability: The ability of a structure to function safely under all loading conditions.

Shear connection: A connection that can transfer shear but no significant moment. Typically it refers to load transferred by clip angles connected to the webs of the beams being joined to columns or other beams.

Shear wall: A deep stiff structural wall that carries lateral loads from all floors into the foundations.

Sidesway: Freedom of the joints of a structure to displace laterally when loaded.

Slenderness ratio: Parameters l/r (in which l is the length of member and r is the radius of gyration) that measure the slenderness of a member. The compressive strength of columns reduces as the slenderness ratio increases.

Static wind pressure: A value of uniformly distributed load listed in a building code that represents the pressure exerted on walls or roofs by the wind. The pressure is a function of wind speed in a given region, elevation above grade, and ground surface roughness.

Strain: The ratio of a change in length divided by the original length.

Stress: Force per unit area.

Stringer: A beam running in the longitudinal direction of a bridge that supports a floor slab on its upper flanges and transfers the load to the transverse floor beams.

Tributary area: The area of a slab or wall that is supported by a particular beam or column. Typically for columns the surrounding area is bounded by panel centerlines.

Two-force member: A member that carries axial load only. No loads are applied between ends of the member.

Unbraced frame: A frame whose lateral stiffness depends on the bending stiffness of its members.

Vierendeel truss: A truss with rigid joints that contains no diagonal members. For this structure, shear is carried by the chord members and creates large bending stresses.

Virtual displacement: A displacement by an outside force, used in the method of virtual work.

Virtual work: A technique based on work-energy for computing a single component of displacement.

Vortex shedding: A phenomenon caused by wind that is restrained by friction from the surface of the member it is passing over. Small masses of air particles that are initially restrained speed up as they leave the member, creating cycles of variation in air pressure that cause members to vibrate.

Web connection: See *Shear connection*.

Wind bracing: A bracing system whose purpose is to transfer lateral wind loads into the ground and to reduce lateral displacements produced by wind forces.

Windward side: The side of a building that faces the wind. The wind produces a direct load on the windward wall.

Work-energy: A law that states the following: The energy stored in a deformable structure equals the work done by the forces acting on the structure.

Zero bar: A bar of a truss that remains unstressed under a particular loading condition.

ANSWERS TO ODD-NUMBERED PROBLEMS

CHAPTER 2

P2.1 900 lb/ft
P2.3 1.66 kips/ft
P2.5 (*a*) Method 1 $A_{B1} = 600\,\text{ft}^2$, method 2 $A_{B1} = 500\,\text{ft}^2$, (*b*) $A_{C2} = 2{,}200\,\text{ft}^2$
P2.7 43.03 kips
P2.9 (*a*) I = 20%, (*b*) 2,300 lb, (*c*) 1,000 lb
P2.11 Windward wall pressure for 0–15′ is 8.43 psf and 15′–16′ is 9.17 psf. Windward roof pressure is $p = 3.34$ psf ↘
P2.13 (*a*) Windward wall pressure $p_{0'} = 34.4$ psf, $p_{35'} = 39.7$ psf, $p_{70'} = 44.7$ psf, $p_{105'} = 48$ psf, $p_{140'} = 50.7$ psf.
P2.15 $V = 810$ kips
2.17 313.6 lb/ft

CHAPTER 3

P3.1 $R_{AX} = 6$ kips →, $R_{AY} = 8.62$ kips ↑, $R_{BY} = 19.38$ kips ↑
P3.3 $R_{AX} = 4.2$ kN →, $R_{AY} = 34.4$ kN ↑
P3.5 $R_{AX} = 0$, $R_{AY} = 0.83$ kip ↓, $R_{CY} = 0.83$ kip ↑
P3.7 $M_A = 12$ kN · m ↻, $R_{CY} = 7$ kN ↑, $R_{DY} = 3$ kN ↓
P3.9 $R_{AX} = 1.33$ kips ←, $R_{AY} = 5$ kips ↑, $R_{EX} = 4.67$ kips ←, $R_{EY} = 11$ kips ↑
P3.11 $R_{AX} = 15$ kips ←, $R_{AY} = 7.5$ kips ↑, $R_{CY} = 81.5$ kips ↑, $R_{DY} = 56$ kips ↑, $R_{BY} = 13$ kips ↑, $R_{BX} = 0$
P3.13 $R_{AX} = 9$ kN →, $R_{AY} = 12$ kN ↑, $R_{GX} = 9$ kN →, $R_{GY} = 0$
P3.15 $R_{AY} = 4$ kN ↑, $R_{CY} = 80$ kN ↑, $R_{EY} = 4$ kN ↑, $M_E = 16$ kN · m ↻
P3.17 $R_{AX} = 75$ kN ←, $R_{BY} = 152.25$ kN ↑, $R_{AY} = 39.75$ kN ↑
P3.19 $R_{AX} = 450$ kips →, $R_{AY} = 675$ kips ↑
P3.21 $R_{AX} = 5$ kips ←, $R_{AY} = 10.44$ kips ↓, $R_{DX} = 6.6$ kips ←, $R_{DY} = 2.44$ kips ↑
P3.23 $R_{AX} = 21.6$ kips →, $R_{AY} = 5.13$ kips ↓, $R_{CY} = 0.27$ kips ↓
P3.25 $R_{AX} = 6.25$ kips →, $R_{AY} = 20$ kips ↑, $M_A = 0$, $R_{FX} = 6.25$ kips ←, $R_{FY} = 20$ kips ↑
P3.27 $R_{AX} = 8$ kips ←, $R_{AY} = 65.83$ kips ↑, $R_{DY} = 121.37$ kips ↑
P3.29 $R_{AX} = 10$ kN ←, $R_{AY} = 90$ kN ↑, $R_{BY} = 70$ kN ↓, $E_X = 30$ kN, $E_Y = 105$ kN
P3.31 $R_{AX} = 5.6$ kips →, $R_{AY} = 5.6$ kips ↑, $R_{CX} = 25.6$ kips ←, $R_{CY} = 38.4$ kips ↑, $R_{EX} = 20$ kips →, $R_{EY} = 40$ kips ↑
P3.33 (*a*) Indeterminate 1°; (*b*) indeterminate 3°; (*c*) unstable; (*d*) indeterminate 2°; (*e*) indeterminate 3°; (*f*) indeterminate 4°
P3.35 $R_{AX} = 20$ kips ←, $R_{AY} = 75$ kips ↑, $M_A = 760$ kip · ft ↻ $F_{BF} = 29.73$ kips (C), $F_{CG} = 11$ kips (C) $F_{DE} = 64$ kips (T)
P3.37 $R_{AX} = 0$, $R_{AY} = 100.8$ kips ↑, $R_{By} = 259.2$ kips ↑, $R_{EY} = 257.3$ kips ↑, $R_{FY} = 132.7$ kips ↑

CHAPTER 4

P4.1 (*a*) Stable, indeterminate to second degree; (*b*) indeterminate second degree; (*c*) unstable; (*d*) determinate; (*e*) geometrically unstable; (*f*) stable determinate; (*g*) stable, determinate
P4.3 $F_{AB} = 20$ kN, $F_{AG} = 15$ kN, $F_{DF} = 0$, $F_{EF} = 25$ kN
P4.5 $F_{AJ} = 17.5$ kips, $F_{CD} = -15$ kips, $F_{DG} = -45.96$ kips
P4.7 $F_{BC} = 125$ kips, $F_{BD} = -125$ kips
P4.9 $F_{AB} = 38.67$ kips, $F_{AC} = 4.81$ kips
P4.11 $F_{AB} = -14.12$ kips, $F_{CE} = 30$ kips
P4.13 $F_{BH} = -26.5$ kips, $F_{CG} = 6.5$ kips, $F_{EF} = 4.7$ kips
P4.15 $F_{AB} = 104$ kips, $F_{CG} = 42.67$ kips, $F_{CF} = -20.87$ kips
P4.17 $F_{AB} = 0$, $F_{GF} = 17.5$ kips, $F_{IC} = -3.54$ kips
P4.19 $F_{AB} = -42$ kN, $F_{AD} = 0$, $F_{BF} = 59.4$ kN
P4.21 $F_{AB} = -34.67$ kN, $F_{BG} = -2$ kN, $F_{ED} = 46.67$ kN
P4.23 $F_{AB} = 123.8$ kN, $F_{AF} = -39.6$ kN
P4.25 $F_{AB} = 124.9$ kips, $F_{CQ} = 0$, $F_{PO} = -49.5$ kips
P4.27 Unstable
P4.29 $F_{AB} = 24$ kips, $F_{FE} = 0$
P4.31 $F_{CB} = -7.2$ kips, $F_{DB} = 5.625$ kips
P4.33 $F_{AB} = 14.85$ kips, $F_{CF} = -17.57$ kips

P4.35 $F_{AB} = -18$ kips, $F_{BD} = 18$ kips, $F_{AD} = -30$ kips

P4.37 $F_{BF} = 40$ kips, $F_{BI} = -135$ kips, $F_{CD} = 145$ kips

P4.39 $F_{IJ} = -13.33$ kN, $F_{MC} = 6.67$ kN

P4.41 $F_{AB} = 40$ kips, $F_{IH} = -50$ kips, $F_{GF} = -40$ kips

P4.43 $F_{AB} = 5$ kN, $F_{IE} = -48.47$ kN, $F_{CG} = 12$ kN

P4.45 $F_{AB} = -25$ kN, $F_{BC} = -20$ kN, $F_{CF} = -5$ kN

P4.47 $F_{AB} = -30$ kips, $F_{DE} = -40$ kips, $F_{CI} = 8$ kips

P4.49 $F_{AB} = -60$ kN, $F_{CE} = -150$ kN

P4.51 $F_{AJ} = 30$ kN, $F_{JI} = 108.66$ kN, $F_{EH} = 40.75$ kN

P4.53 Case 1, joint 1: $\delta_x = 0.0$ in,
joint 2: $\delta_x = 0.492$ in, $\delta_y = 0.11$ in
Case 2: for $A = 6$ in^2, $\delta_X = 0.217$ inches

P4.55 (*a*) $F_1 = 64.8$ kips, $F_2 = 71.9$ kips, $F_{8,9} = 54$ kips, $F_{10} = -24$ kips, $F_{11} = 21.5$ kips, $F_{12} = 0$, $\Delta_{MIDSPAN} = 0.892$ in.; (*b*) $F_{5,6} = 57$ kips, $M_{@jt.6} = 7.22$ ft-kips, $\sigma_{MAX.} = 63.2$ ksi

CHAPTER 5

P5.1 $V_{B\text{-}C} = -53.75$ kips, $M_B = -53.5$ ft·kips, $M_C = -187.5$ ft-kips

P5.3 $V = 1 - \dfrac{x^2}{4}$; $M = 12 + x - \dfrac{x^3}{12}$

P5.5 Origin at B, SEGMENT BC;
$V = -4 - 3x$; $M = -16 - 4x - \dfrac{3}{2}x^2$

P5.7 SEGMENT BC; $0 \le x \le 3$; Origin at B;
$V = 17.83 - 5x$; $M = -40 + 37.83x - \dfrac{5}{2}(4 + x)^2$

P5.9 MEMBER BC; $0 \le x \le 16$; Origin at B;
$M = -60 + 48x - 3x^2$

P5.11 $V_{BC} = \dfrac{2}{9}x^2 - 8.67$, $M_{BC} = 8.67x - \dfrac{2x^3}{27}$

P5.13 $M_{max} = 218.4$ kip · ft

P5.15 $M_{max} = -650$ kip · ft at D

P5.17 V_{max} at $D = 87.7$ kips, $M_{max} = 481.3$ kip · ft at 11.87 ft from D

P5.19 V_{max} left of support $C = 92$ kips, $M_{max} = -462$ kip · ft at support C

P5.21 $R_{DY} = 32$ kips ↑, $R_{EX} = 6$ kips ←, $R_{EY} = 22$ kips ↑, $M_{max} = 170.67$ kip · ft at 10.67 ft from D

P5.23 $M_A = 120$ kN · m ↺, $R_{AY} = 15$ kN ↑, $R_{DY} = 15$ kN ↑

P5.25 $M_A = 140$ kip · ft ↻, $R_{AX} = 4$ kips →, $R_{AY} = 42$ kips ↑

P5.27 $R_{AX} = 2$ kips →, $R_{AY} = 8$ kips ↑, $R_{CX} = 2$ kips ←

P5.29 $M_A = 33.36$ kN · m ↻, $R_{AY} = 13.33$ kN ↑, $B_Y = 11.67$ kN ↓ $R_{CY} = 76.67$ kN ↑, $R_{EY} = 0$

P5.31 $R_{BY} = 15.19$ kips ↑, $R_{CY} = 10.5$ kips ↑, $M_{max} = 13.76$ kip · ft at 2.62 ft from B on segment BC

P5.33 $R_{AY} = 18.85$ kips ↑, $R_{BY} = 85.49$ kips ↑, $R_{CY} = 27.66$ kips ↑

P5.35 max $+ M = 52.12$ kip · ft, max $- M = 47.96$ kip · ft

P5.37 $R_{AY} = 10.4$ kips ↓, $R_{BY} = 23.4$ kips ↑, $R_{EY} = 18.2$ kips ↑, $R_{FY} = 5.2$ kips ↓, $M_{max} = -104$ kip · ft

P5.39 $R_{AY} = R_{HY} = 6$ kips ↑, $M_{max} = \pm 18$ kip · ft

P5.41 $R_{AY} = 21$ kips ↑, $R_{DX} = 24$ kips ←, $R_{DY} = 3$ kips ↑

P5.43 $R_{BY} = R_{CY} = 10$ kips ↑, $M_{max} = -42.7$ kip · ft

P5.45 $M_{CB} = 120$ kip · ft, $M_{CE} = 200$ kip · ft, $M_{CD} = -80$ kip · ft, $M_A = 120$ kip · ft ↺, $R_{AY} = 20$ kips ↓, $R_{EY} = 40$ kips ↑

P5.47 Member BE: $M_{max} = 34.03$ kip · ft, $M_{BA} = 18$ kip · ft, $M_{BC} = 0$, $M_{BE} = -18$ kip · ft

P5.49 $M_{max} = 908.3$ kip · ft, $V_{max} = 244.8$ kips.

P5.51 (*a*) indeterminate 2°, (*b*) unstable, (*c*) indeterminate 3°, (*d*) indeterminate 6° (*e*) determinate, (*f*) indeterminate 9°

P5.53 Beam 1: $R_{AY} = R_{BY} = 10.5$ kips ↑
Beam 2: $R_{AY} = R_{CY} = 7.5$ kips ↑

P5.55 (*a*) $R_{1X} = 25.395$ kips, $R_{1Y} = 77.75$ kips, $M_1 = 0$
$R_{4X} = -31.395$ kips, $R_{4Y} = 82.25$ kips, $M_4 = 0$
(*b*) Vertical deflection at midspan of girder = 1.179 in. Round the camber to $1\frac{1}{4}$ in for practical applications

CHAPTER 6

P6.1 $A_Y = 60$ kips, $A_X = 75$ kips, $T_{AB} = 96$ kips, $T_{BC} = 80.78$ kips, cable length = 114.3 ft.

P6.3 $A_Y = 400$ kips, $A_X = 447.4$ kips, $h_B = 44.7$ ft

P6.5 $A_X = B_X = 2160$ kips, $A_Y = 0$, $B_Y = 1440$ kips, $T_{max} = 2531.4$ kips

P6.7 $T = 28.02$ kips

P6.9 Cable force = 38.2 kips, post force = 15 kips

P6.11 $A_Y = 37.67$ kN, $T_{max} = 100.65$ kN, $H = 93.33$ kN, $B_Y = 14.33$ kN

P6.13 $A_X = 78.75$ kN, $A_Y = 18$ kN, $T_{max} = 80.78$ kips

P6.15 Required weight of tension ring = 11.78 kips, T_{max} = 25.28 kips, $A_{CABLE\ REQUIRED} = 0.23$ in^2

CHAPTER 7

P7.1 $T = 969.33$ kips for $h = 12$ ft;
$T = 576.28$ kips for $h = 24$ ft;
$T = 486.6$ kips for $h = 36$ ft;
$T = 424.53$ kips for $h = 48$ ft;
$T = 402.5$ kips for $h = 60$ ft

P7.3 $A_X = 48.3$ kN $\rightarrow$, $A_Y = 48.3$ kN $\uparrow$,
$C_X = 48.3$ kN $\leftarrow$, $C_Y = 88.26$ kN $\uparrow$

P7.5 $A_Y = 27.29$ kips, $C_Y = 12.71$ kips, $T = 16.95$ kips

P7.7 $A_X = 20$ kips $\rightarrow$, $A_Y = 30$ kips $\uparrow$,
$E_X = 20$ kips $\leftarrow$, $E_Y = 30$ kips $\uparrow$,
$F_B = 25$ kips $\swarrow$, $V_B = 0$, $M_B = 0$,
$F_D = 34$ kips $\searrow$, $V_D = 12$ kips $\swarrow$, $M_D = 75$ kip · ft $\circlearrowright$

P7.9 $A_X = 30.5$ kN $\rightarrow$, $A_Y = 38.75$ kN $\uparrow$, $C_X = 12.5$ kN $\leftarrow$,
$C_Y = 21.25$ kN $\uparrow$

P7.11 Case A: $A_X = 67.5$ kN $\rightarrow$, $A_Y = G_Y = 45$ kN $\uparrow$,
$G_X = 67.5$ kN $\leftarrow$, $F_{AM} = 22.5$ kN, $F_{BL} = 15$ kN,
$F_{ML} = 22.5$ kN, $F_{LK} = F_{KD} = 67.5$ kN
Case B: $A_X = 37.5$ kN $\rightarrow$, $A_Y = 25$ kN $\uparrow$,
$G_X = 97.5$ kN $\leftarrow$, $G_Y = 65$ kN $\uparrow$, $F_{DE} = 205.55$ kN,
$F_{EF} = 156.21$ kN, $F_{FG} = 137.88$ kN

P7.13 $P = 46.67$ kN, $y_1 = 8$ m

P7.15 $y_B = 7.73$ m, $y_C = 11.7$ m, $y_E = 5.4$ m

P7.17 max. $\Delta_X = 4.23$ in $\rightarrow$ at joint 4,
max. $\Delta_Y = 5.88$ in $\downarrow$ at joint 18

CHAPTER 8

P8.1 R_A, ordinates: 1 at A, 0 at D; M_C: 0 at A, 5 kip · ft at midspan

P8.3 R_A: 1 at A, $-\frac{2}{7}$ at D; M_B: 0 at A, $\frac{24}{7}$ at B; V_C: $-\frac{4}{7}$ at B,
$-\frac{2}{7}$ at D

P8.5 V_E: 0.5 at C, $-\frac{1}{2}$ at G

P8.7 R_A, ordinates: $\frac{3}{2}$ at B, 1 at C, 0 at D, $-\frac{1}{2}$ at E;
R_D, ordinates: $-\frac{1}{2}$ at B, 0 at C, 1 at D, $\frac{3}{2}$ at E;
M_D: -5 at E; M_C: -5 at B; V_C: $\frac{1}{2}$ at B, $-\frac{1}{2}$ at E

P8.9 F_{CE}: 0 at A, -2.29 at D;
R_{AY}: 1 at A, $\frac{1}{2}$ at B, 0 at C, -0.375 at D;
M_B: 0 at A, 2 at B, 0 at C, -1.5 at D

P8.11 M_A: 0 at A, -12 kip · ft at B, 6 kip · ft at D;
R_A: 1 at A, 1 at B, $-\frac{1}{2}$ at D;

P8.13 R_C: 0 at A, $\frac{7}{5}$ at B, $\frac{1}{2}$ at D;
M_D: 0 at A, -8 kip · ft at B, 5 kip · ft at D

P8.15 V_{AB}: $\frac{4}{5}$ at B, $\frac{3}{5}$ at C;
M_E: $\frac{48}{5}$ at C, $\frac{96}{5}$ at E

P8.17 V_{BC}: -2 at A, 0.625 at hinge, 0.25 at D;
M_C: -8 at A, 10 at hinge

P8.19 R_I: 1 at B, $\frac{2}{3}$ at C; V(to the right of I): $\frac{2}{3}$ at C;
V_{CE}: $-\frac{1}{2}$ at D, $-\frac{1}{3}$ at C, $\frac{1}{3}$ at E

P8.21 R_A: 0.8 at B, 0.4 at D; M_D: 2 at B, 6 at D;
V_A: 0.8 at B, 0.4 at D

P8.23 A_Y: 1.0 at A, 0.342 at B, 0 at C;
A_X: 0 at A, 0.658 at B, 0 at C

P8.25 R_A: 1 at A, -1 at B, 0 at C; R_F: 0 at A, 2 at B, 0 at C;
V_1: -0.75 and 0.25 at Section 1, -1 at B;
M_1: 0 at A, 0.375 at Section 1, -15 at B;
$R_A = 200$ kN $\downarrow$, $R_F = 800$ kN $\uparrow$

P8.27 At B: post axial force $= -1$ kip,
cable force $= 1.346$ kips

P8.29 Ordinates for A_X: 0 at B, 0.28 at Section 1, 0.667 at C,
0 at D;
Ordinates for A_Y: 1 at B, 0.979 at Section 1, 0.5 at C, 0 at D;
Ordinates for $M_{\text{Section 1}}$: 0 at B, 0.479 at Section 1, –11.5
at C, 0 at D

P8.31 Ordinates for F_{DE}: 0, $-\frac{1}{4}$, $-\frac{1}{2}$, $-\frac{3}{4}$, -1, $-\frac{1}{2}$, 0;
Ordinates for F_{DI}: 0, -0.208, -0.417, $-\frac{5}{8}$,
0.417, 0.208, 0;
Ordinates for F_{EI}: 0, 0.083, 0.167, 0.25, 0.33, 0.167, 0;
Ordinates for F_{IJ}: 0, $\frac{3}{8}$, $\frac{3}{4}$, 1.12, $\frac{3}{4}$, $\frac{3}{8}$, 0

P8.33 $F_{AD} = -\frac{5}{11}$ at B, $F_{EF} = -0.566$ at B, $F_{EM} = 0.884$ at M,
$F_{NM} = -\frac{3}{4}$ at B

P8.35 $F_{CD} = -\frac{2}{3}$ at L and $+\frac{2}{3}$ at J; $F_{BL} = -\sqrt{2}/3$ at M and J

P8.37 Load at C: $F_{BC} = 0$, $F_{CA} = -0.938$ kip,
$F_{CD} = 0.375$ kip, $F_{CG} = 0.375$ kip

P8.39 Load at C: $F_{AL} = 0$, $F_{KJ} = 0.75$ kips

P8.41 $M_{\max} = 208.75$ kip · ft, $V_{\max} = 33.33$ kips

P8.43 (*a*) $V_{\max} = 49.67$ kN, $M_{\max} = 280.59$ kN · m;
(*b*) at midspan $M_{\max} = 276$ kN · m

P8.45 $M_{\max} = 323.26$ kip · ft, $V_{\max} = 40.2$ kips

P8.47 at B, $V = 60$ kN; at C, $V = 39$ kN; at D, $V = 24$ kN

P8.49 (*a*) $\Delta_{\max} = 107{,}400{,}000/EI \downarrow$ at 2.4 ft right of C;
(*b*) $\Delta_{\max} = 108{,}749{,}000/EI \downarrow$ at midspan

CHAPTER 9

P9.1 $\theta_B = -PL^2/2EI$, $\Delta_B = PL^3/3EI$

P9.3 $\Delta_{\max}$ at $x = 0.4725L$;
$\Delta_{\max} = -0.094ML^2/EI$

P9.5 $\theta_A = \frac{ML}{24}$, $\theta_C = \frac{ML}{24}$

P9.7 $\theta_B = \theta_C = -960/EI$, $v_B = 3840/EI \uparrow$, $v_C = 7680/EI \uparrow$

P9.9 $\theta_A = -40/EI$, $\theta_C = 40/EI$, $v_B = 106.67/EI \downarrow$

P9.11 $\theta_A = 360/EI$, $\Delta_A = 1800/EI \downarrow$,
$\Delta_E = 540/EI \uparrow$ at midspan

P9.13 $\theta_A = 114PL^2/768EI$, $v_B = 50PL^3/1536EI \downarrow$

P9.15 $\theta_C = -282/EI$, $\Delta_C = 1071/EI \downarrow$

P9.17 $\theta_B = 0$, $\Delta_B = 0.269$ in ↓, $\theta_D = 0$, $\Delta_D = 0.269$ in ↓
P9.19 $P = 5.184$ kips
P9.21 $\Delta_{DH} = 216/EI$ →, $\Delta_{BV} = 0$
P9.23 $\theta_C = 0.00732$ rad, $\Delta_{CV} = -0.903$ in ↓, $\Delta_{DH} = 0.309$ in →
P9.25 $\Delta_C = 1728/EI$ ↑
P9.27 $\theta_A = 450/EI$, $\Delta_{DH} = 2376/EI$ →, $\Delta_{DV} = -1944/EI$ ↓
P9.29 $\theta_B = 0.0075$ rad, $v_D = 0.07$ m ↑
P9.31 $\theta_B = -607.5/EI$, $\Delta_B = 3645/EI$ ↓
P9.33 $\theta_C = -67.5/EI$, $\Delta_C = 175.5/EI$ ↓, $\Delta_{max} = 54/EI$ ↑
P9.35 $K = 3.20\ EI/L$ for M@A, $K = 5.33\ EI/L$ for M@B
P9.37 $\Delta_{max} = -444.8/EI$, $\theta_{BL} = -72/EI$, $\theta_{BR} = -48/EI$
P9.39 $\theta_{BL} = 90/EI$, $\theta_{BR} = 95/EI$, $\Delta_B = 720/EI$ ↓, $\Delta_{max} = 1272/EI$ ↓
P9.41 $\theta_B = -87.2/EI$, $\Delta_C = 1628/EI$ ↓
P9.43 0.312 in ↑
P9.45 $I_{REQ'D} = 2038.6\ in^4$

CHAPTER 10

P10.1 $\delta_{BH} = 0.70$ in →, $\delta_{BV} = 0.28$ in ↑
P10.3 $\delta_{CH} = 0.02$ m →, $\delta_{CV} = 0$
P10.5 $\delta_{BH} = 0.298$ in →, $\delta_{BV} = 0.198$ in ↓
P10.7 $P = 1.488$ kips
P10.9 $\delta_{BV} = 1.483$ in ↓
P10.11 $\delta_{CV} = 0.41$ in ↓, $\delta_{CH} = 0$
P10.13 (*a*) $\delta_{DV} = 0.895$ in; (*b*) $\delta_{BH} = 8/3$ in
P10.15 (*a*) $\delta_{EH} = 0.18$ in →, $\delta_{EV} = 0.135$ in ↑; (*b*) $\delta_{EV} = 0.81$ in ↑
P10.17 $\delta_{CV} = 8.6$ in ↑, $\delta_{CH} = 15.4$ in →
P10.19 $\delta = 0.86$ in ↓ at midspan, $\theta_A = 0.00745$ rad.
P10.21 $\delta_B = 24{,}468.7/EI$ ↓, $\theta_C = 2568.75/EI$
P10.23 $\delta_C = 1.034$ in ↓
P10.25 $\delta_{AH} = 2$ in →
P10.27 (*a*) $\delta_{BH} = 1$ in →, $\delta_{BV} = 3/4$ in ↓; (*b*) $\Delta\theta_{BC} = 0.004167$ rad
P10.29 $\delta_{CH} = 25.4$ mm →, $\delta_{CV} = 30.3$ mm ↓
P10.31 $\theta_B = 0.00031$ rad, $\delta_{CH} = 44.1$ mm →
P10.33 $\delta_{BH} = 1.175$ in →, $\delta_{BV} = 0.883$ in ↓
P10.35 $\delta_{CV} = 76.3$ mm ↓
P10.37 $\delta_{BV} = 1.13$ in ↓, $\delta_{CH} = 0.096$ in ←
P10.39 (*a*) $\delta_{BV} = 0.59$ in ↓; (*b*) $\Delta L_{DE} = 4$ in (shorten)
P10.41 $\delta_{BH} = 92.5$ mm →
P10.43 $\delta_C = 206.2$ mm ↓

CHAPTER 11

P11.1 $M_A = 90.72$ kip · ft ↺, $R_{AY} = 20.45$ kips ↑, $R_{CY} = 15.55$ kips ↑
P11.3 $M_A = 3.75$ kN · m ↺, $R_{AY} = 3.375$ kN ↑, $R_{CY} = 3.375$ kN
P11.5 $R_{AY} = 6.71$ kips ↑, $M_A = 40.65$ kip · ft ↺, $R_{CY} = 6.71$ kips ↓ If I is constant, $M_A = 30$ kip · ft ↺
P11.7 $R_{AY} = 18.9$ kips ↑, $M_A = 30.8$ kip · ft ↺, $R_{BY} = 21.15$ kips ↑
P11.9 $R_{AY} = 0.559$ kips ↑, $M_A = 8.39$ kip · ft ↺
P11.11 $M_A = 5wL^2/16$ ↺, $R_{AY} = 13wL/16$ ↑, $R_{CY} = 3wL/16$ ↑, $M_C = 3wL^2/16$ ↻
P11.13 (*a*) $R_{AY} = 6.5$ kips ↑, $R_{BY} = 11$ kips ↑, $R_{CY} = 1.5$ kips ↓; (b) $R_{AY} = 6.75$ kips ↑, $R_{BY} = 10.5$ kips ↑, $R_{CY} = 1.25$ kips ↓
P11.15 (*a*) $R_{AY} = 0.787$ kips ↑, $R_{BY} = 1.967$ kips ↓, $R_{CY} = 1.18$ kips ↑ (*b*) $R_{AY} = 1.925$ kips ↑, $R_{BY} = 3.187$ kips ↑, $R_{CY} = 10.89$ kips ↑
P11.17 $R_{AY} = R_{BY} = wL/2$ ↑, $M_A = wL^2/12$ ↺, $M_B = wL^2/12$ ↻
P11.19 $M_A = 140$ kN · m ↺, $R_{AY} = 34$ kN ↑, $R_{BY} = 6$ kN
P11.21 $R_{AX} = 32.9$ kips ←, $R_{AY} = 35.33$ kips ↑, $R_{DY} = 84.67$ kips ↑, $R_{CX} = 32.9$ kips →, $F_{AB} = -58.9$ kips, $F_{BC} = 41.1$ kips, $F_{AE} = F_{ED} = 80$ kips, $F_{BE} = 120$ kips, $F_{BD} = -100$ kips, $F_{CD} = -24.67$ kips
P11.23 $R_{AX} = 1.89$ kN →, $R_{AY} = 2.25$ kN ↓, $R_{CX} = 31.89$ kN ←, $R_{CY} = 20.25$ kN ↑, $F_{AB} = 6.8$ kN, $F_{BC} = -30.7$ *kN*, $F_{BD} = 14.34$ kN, $F_{AD} = F_{CD} = -7.54$ kN
P11.25 $R_{AX} = 30.95$ kN ←, $R_{AY} = 80.42$ kN ↑, $R_{CX} = 30.95$ kN →, $R_{CY} = 139.58$ kN ↑, $F_{AB} = -29.4$ kN, $F_{BC} = 45$ kN, $F_{CD} = -75$ kN, $F_{AE} = 100.53$ kN, $F_{BE} = 24.47$ kN, $F_{BD} = -99.47$ kN, $F_{DE} = 45.63$ kN
P11.27 $\Delta_{AH} = 0$, $\Delta_{AV} = 4.69$ mm ↓
P11.29 (*a*) $R_{AX} = 30$ kips ←, $R_{AY} = 14.2$ kips ↓, $R_{BY} = 5.9$ kips ↑, $R_{CY} = 8.3$ kips ↑ $F_{AB} = F_{BC} = 11.07$ kips, $F_{AD} = 23.7$ kips, $F_{CD} = -13.83$ kips, $F_{BD} = -5.9$ kips; (*b*) $R_{AX} = 30$ kips ←, $R_{AY} = 13.57$ kips ↑, $R_{BY} = 49.64$ kips ↓, $R_{CY} = 36.07$ kips ↑ $F_{AB} = F_{BC} = 48.1$ kips, $F_{AD} = -22.6$ kips, $F_{CD} = -60.1$ kips, $F_{BD} = 49.64$ kips

P11.31 A change in length will produce movement of joints but no stresses. No bars are stressed

P11.33 $R_{AY} = 45.4$ kN $\downarrow$, $R_{CY} = 136.1$ kN $\uparrow$, $R_{CX} = 68$ kN $\leftarrow$, $R_{EY} = 90.7$ kN $\downarrow$, $R_{EX} = 68$ kN $\rightarrow$, $F_{AB} = 45.4$ kN, $F_{BC} = -81.78$ kN, $F_{BD} = 68$ kN, $F_{CD} = -90.7$ kN, $F_{DE} = 113.3$ kN

P11.35 $R_{AX} = 15.74$ kips $\leftarrow$, $R_{CX} = 15.74$ kips $\rightarrow$, $R_{CY} = 60$ kips $\uparrow$, $M_C = 60.54$ kip·ft ↻

P11.37 $R_{AX} = 4.6$ kips $\rightarrow$, $R_{AY} = 2.3$ kips $\uparrow$, $R_{CX} = 4.6$ kips $\leftarrow$, $R_{CY} = 2.3$ kips $\downarrow$

P11.39 $R_{AX} = 4$ kips $\leftarrow$, $M_A = 31.98$ kips·ft ↻, $R_{AY} = 0.89$ kips $\downarrow$, $R_{CY} = 0.89$ kips $\uparrow$

P11.41 (*a*) $R_{AY} = 15$ kips $\downarrow$, $R_{EY} = 52.5$ kips $\uparrow$, $R_{DY} = 22.5$ kips $\uparrow$, (b) $\Delta_C = 10800/EI \downarrow$

P11.43 $R_{AY} = 38.4$ kips $\uparrow$, $R_{AX} = 7.26$ kips $\rightarrow$, $R_{DX} = 7.26$ kips $\leftarrow$, $R_{DY} = 38.4$ kips $\uparrow$

P11.45 $R_{EY} = 232.18$ kips $\uparrow$, $R_{DY} = R_{FY} = 116.09$ kips $\downarrow$

CHAPTER 12

P12.1 $\text{FEM}_{AB} = -3PL/16$, $\text{FEM}_{BA} = 3PL/16$

P12.3 $M_A = 40$ kip·ft ↺, $R_{AY} = 9.5$ kips $\uparrow$, $R_{BY} = 14.5$ kips $\uparrow$

P12.5 $R_{AX} = 3.5$ kips $\rightarrow$, $M_A = 14$ kip·ft ↻, $R_{AY} = 46.9$ kips $\uparrow$, $R_{CX} = 3.5$ kips $\leftarrow$, $R_{CY} = 37.1$ kips $\uparrow$, $M_C = 162.4$ kip·ft ↻

P12.7 $R_{BY} = 7.07$ kips $\uparrow$, $R_{CY} = 20.57$ kips $\uparrow$, $R_{DY} = 3.64$ kips $\downarrow$, $M_D = 9.71$ kip·ft ↺

P12.9 $R_{AY} = 29.27$ kips $\uparrow$, $M_A = 108.4$ kip·ft, $R_{BY} = 30.73$ kips $\uparrow$, $\Delta_C = 0.557$ in $\downarrow$

P12.11 $M_{AB} = 13.09$ kip·ft, $M_{BA} = -26.18$ kip·ft, $R_{AY} = 3.27$ kips $\downarrow$, $R_{BY} = 12.27$ kips $\uparrow$, $\Delta = 698.1/EI \downarrow$

P12.13 $M_A = 14.36$ kip·ft ↺, $R_{AX} = 5.27$ kips $\leftarrow$, $R_{AY} = 1.6$ kips $\uparrow$, $M_B = 5.84$ kip·ft ↻

P12.15 $M_A = 76.56$ kN·m ↺, $R_{AY} = 12.312$ kN $\uparrow$, $R_{CY} = 21.024$ kN $\downarrow$

P12.17 $M_A = 77.94$ kN·m ↺, $R_{AX} = 55.636$ kN $\leftarrow$, $R_{AY} = 11.031$ kN $\uparrow$, $R_{CX} = 44.364$ kN $\leftarrow$, $R_{CY} = 11.031$ kN $\downarrow$

P12.19 $R_{AX} = 0.62$ kN $\rightarrow$, $R_{AY} = 22.715$ kN $\uparrow$, $M_A = 4.84$ kN·m ↻, $R_{BX} = 1.96$ kN $\leftarrow$, $R_{BY} = 54.245$ kN $\uparrow$, $M_B = 3.92$ kN·m ↺

P12.21 $R_{AX} = 2.53$ kips $\rightarrow$, $R_{AY} = 18.29$ kips $\uparrow$, $M_A = 94.12$ kip·ft ↺, $R_{EX} = 1.62$ kips $\rightarrow$, $R_{EY} = 30.25$ kips $\downarrow$, $M_E = 5.4$ kip·ft ↻ $R_{DX} = 4.15$ kips $\leftarrow$, $R_{DY} = 11.96$ kips $\uparrow$, $M_D = 20.7$ kip·ft ↺

P12.23 $R_{AX} = 2.67$ kips $\leftarrow$, $R_{AY} = 34.08$ kips $\uparrow$, $M_A = 76.66$ kip·ft ↺, $R_{DX} = 2.67$ kips $\rightarrow$, $R_{DY} = 40.92$ kips $\uparrow$

P12.25 $R_{AX} = 1.12$ kips $\rightarrow$, $R_{AY} = 1.495$ kips $\uparrow$, $M_{BA} = 13.45$ kip·ft

P12.27 $M_A = 61.2$ kip·ft ↺, $R_{AX} = 26.7$ kips $\leftarrow$, $M_C = 119.9$ kip·ft ↻, $R_{CX} = 73.4$ kips $\leftarrow$, $M_D = 14.2$ kip·ft ↺, $R_{DY} = 5.3$ kips $\downarrow$

P12.29 $R_{AX} = 3.1$ kips $\leftarrow$, $R_{AY} = 8.8$ kips $\uparrow$, $M_A = 7.23$ kip·ft ↺, $R_{DX} = 2.9$ kips $\leftarrow$, $R_{DY} = 9.2$ kips $\uparrow$, $M_D = 13.57$ kip·ft ↺

P12.31 $M_{AB} = -116.66$ kN·m, $M_{BA} = -58.33$ kN·m, $M_{DC} = 116.66$ kN·m

P12.33 $M_{BA} + M_{BC} = 0$, $M_{CB} + M_{CE} - 16 = 0$, $M_{EC} = 0$,
$$2 - \frac{M_{AB} + M_{BA}}{12} + \frac{M_{CE}}{8} = 0$$

P12.35 (*a*) Indeterminate 3°: $\theta_A, \theta_B, \theta_C$; (*b*) Indeterminate 3°: $\theta_B, \theta_C, \theta_D$; (*c*) Indeterminate 6°: $\theta_A, \theta_B, \theta_C, \theta_D, \theta_E, \theta_F$; (*d*) Indeterminate 13°: 10 joint rotations and 3 degrees of sidesway

CHAPTER 13

P13.1 $R_{AY} = 16.53$ kips $\uparrow$, $M_A = 83.56$ kip·ft ↺, $M_B = -72.89$ kip·ft, $M_C = 59.56$ kip·ft ↻, $R_{CY} = 23.17$ kips $\uparrow$, $R_{BY} = 40.3$ kips $\uparrow$

P13.3 $R_{AY} = 50.81$ kips $\uparrow$, $M_A = 94.4$ kip·ft ↺, $R_{BY} = 46.74$ kips $\uparrow$, $R_{CY} = 64.04$ kips $\uparrow$, $R_{DY} = 38.42$ kips $\uparrow$

P13.5 $R_{BY} = 22.94$ kips $\uparrow$, $R_{CY} = 57.45$ kips $\uparrow$, $R_{DY} = 19.61$ kips $\uparrow$, $M_D = 12.94$ kip·ft ↻

P13.7 $R_{AY} = 4.64$ kips $\downarrow$, $M_A = 13.9$ kip·ft ↻, $R_{BY} = 17.97$ kips $\uparrow$, $M_B = -27.86$ kip·ft, $R_{CY} = 40$ kips $\uparrow$, $M_C = -47.96$ kip·ft, $R_{DY} = 12.67$ kips $\uparrow$

P13.9 $R_{AY} = 34.87$ kips $\uparrow$, $R_{BY} = R_{CY} = 93.13$ kips $\uparrow$, $R_{DY} = 34.87$ kips $\uparrow$, $M_B = M_C = -164.33$ kip·ft

P13.11 $M_A = 80.47$ kip·ft ↺, $M_D = 80.47$ kip·ft ↻, $R_{AX} = 16.14$ kips $\leftarrow$, $R_{AY} = R_{DY} = 30$ kips $\uparrow$

P13.13 $V_A = V_B = 3.25$ kips, $M_A = M_B = -4.58$ kip·ft

P13.15 $M_A = M_D = -17.4$ kip·ft, $M_B = M_C = -16.8$ kip·ft

P13.17 $R_{AY} = 7.2$ kips $\uparrow$, $R_{EY} = 12.8$ kips $\uparrow$, $R_{EX} = 4.2$ kips $\leftarrow$, $M_E = 16.88$ kip·ft ↺

P13.19 $R_{AX} = 3.5$ kips $\rightarrow$, $R_{AY} = 10$ kips $\uparrow$, $R_{DX} = 3.5$ kips $\leftarrow$, $R_{DY} = 10$ kips $\uparrow$, $M_B = M_C = -36.4$ kip · ft

P13.21 $R_{AY} = 6.25$ kN $\downarrow$, $R_{CY} = 62.5$ kN $\uparrow$, $R_{DY} = 6.25$ kN $\leftarrow$

P13.23 $M_A = 17.62$ kip · ft $\circlearrowleft$, $M_B = 35.24$ kip · ft, $M_C = 151$ kip · ft $\circlearrowright$, $R_{AX} = 4.4$ kips $\leftarrow$, $R_{AY} = 7.76$ kips $\downarrow$

P13.25 $R_{AY} = 2.21$ kips $\downarrow$, $R_{AX} = 0.69$ kip $\rightarrow$, $M_A = 13.25$ kip · ft $\circlearrowright$, $R_{DX} = 1.71$ kips $\leftarrow$, $R_{DY} = 14.71$ kips $\uparrow$, $R_{CX} = 1.03$ kips $\rightarrow$, $R_{CY} = 11.5$ kips $\uparrow$

P13.27 $R_{AX} = 5.99$ kips $\leftarrow$, $R_{AY} = 3.17$ kips $\downarrow$, $M_A = 43.6$ kip · ft $\circlearrowleft$, $R_{FX} = 8.02$ kips $\leftarrow$, $R_{FY} = 0$, $M_F = 51.69$ kip · ft $\circlearrowleft$, $M_{CB} = 22.25$ kip · ft, $M_{CF} = 44.49$ kip · ft, $\Delta_{BH} = 0.543$ in $\rightarrow$

P13.29 $R_{AX} = 7$ kips $\rightarrow$, $R_{AY} = 39.8$ kips $\uparrow$, $M_A = 36.96$ kip · ft $\circlearrowright$, $R_{DX} = 9.4$ kips $\leftarrow$, $R_{DY} = 40.2$ kips $\uparrow$, $M_D = 52.14$ kip · ft $\circlearrowleft$

P13.31 $R_{DY} = R_{FY} = 50$ kN $\uparrow$, $M_A = -44.44$ kN · m, $M_B = 55.56$ kN · m, $\Delta = 3.56$ mm

CHAPTER 14

P14.1 Ordinates for R_A: 1, 0.593, 0.241, 0, −0.083; Ordinates for M_C: 0, −0.667, −0.833, 0, 3.75

P14.3 Ordinates for R_A: 1, 0.691, 0.406, 0.168, 0, −0.082, −0.094, −0.059, 0, 0.047
Ordinates for M_B: 0, 5.53, 3.25, 1.344, 0, −0.66, −0.75, −0.472, 0, 0.376
Ordinates for V_B: 0, −0.309, 0.691, 0.406, 0.168, 0, −0.082, −0.0938, −0.059, 0, 0.047

P14.5 Ordinates for R_A: 1, 0, −0.074, 0, 0, 0, 0.009
Ordinates for R_B: 0, 1, 0.567, 0, 0, 0, −0.054
Ordinates for M_C: 0, 0, 0.497, 0, 0, 0, −0.022

P14.7 (*a*) $M_{max} = 773.36$ kip · ft (*b*) $M_{max} = 550.56$ kip · ft

P14.9 Ordinates for R_A: $A = 1$ kip, $B = 0$, $C = 0$, $D = 0$
Ordinates for M_C: $A = 0$, $B = -6$ kip · ft, $C = 0$, $D = 0$

CHAPTER 15

Note: Since the approximate analysis for Problems P15.1 through P15.9 requires an assumption, individual answers will vary.

P15.1 For assumption P.I. in span $AB = 0.25L = 6$ ft, $M_B = -360$ kip · ft. By moment distribution: $M_B = -310$ kip · ft

P15.3 For assumption P.I. $= 0.2L = 8$ ft to right of joint B: $A_X = 8.48$ kips, $A_Y = 18.18$ kips, $M_B = 127.2$ kip · ft, and $C_Y = 5.82$ kips. By moment distribution: $C_X = 8.85$ kips, $C_Y = 5.68$ kips, $M_B = 132.95$ kip · ft

P15.5 For assumption P.I. $= 0.2L = 2.4$ ft to supports C and D in span CD: max + moment = 13.0 kip·ft, $M_C = 23.0$ kip · ft. By moment distribution, max. + moment = 14.4 kip · ft, $M_C = 21.6$ kip·ft

P15.7 For assumption P.I. $= 0.25L$ left side of center support and P.I. $= 0.2L$ out from wall; $R_B = 54.15$ kips, $R_C = 99.17$ kips, and $M_D = 95.9$ kip · ft. By moment distribution: $R_B = 56.53$ kips, $R_C = 93.79$ kips, and $M_D = 91.97$ kip · ft

P15.9 For assumption P.I. $= 0.2L$ in grider: $M_A = 306.4$ kip · ft, $A_X = 183.84$ kips, $A_Y = 91$ kips. By moment distribution: $M_A = 315.29$ kip · ft, $A_X = 189.18$ kips, $A_Y = 91$ kips

P15.11 Analyze truss as a continuous beam: $R_B = 59.4$ kips, $F_B = 18.9$ kips compr, $F_D = 34.88$ kips

P15.13 BD: $F = 37.5$ kips compr; CB: $F = 22.5$ kips compr; CD: $F = 30$ kips compr.

P15.15 For assumption P.I. $= 0.2L = 2.4$ ft to supports C and D in span CD: max. + moment = 13.0 kip · ft, $M_c = 23.0$ kip · ft. By moment distribution, max. + moment = 14.4 kip · ft, $M_c = 21.6$ kip · ft.

P15.17 (*a*) $M_{BE} = 400$ kip · ft, (*b*) $M_{BE} = 400$ kip · ft, (*c*) $M_{BE} = 390$ kip · ft,

P15.19 Top end of column AF (*a*) $M = 300$ kN · m, shear = 50 kN, $P = -140$ kN, (*b*) $M = 131.3$ kN · m, $V = 21.9$ kN, $P = -61.3$ kN, (*c*) $M = 312.3$ kN · m, $V = 52.1$ kN, $P = -161.9$ kN

P15.21 (*a*) $A_x = 5$ kips, $A_y = 6.67$ kips, Column moment at $B = 75$ kip · ft (*b*) $F_{BL} = +20$ kips, $F_{CD} = -18.33$ kips (*c*) $A_x = 4.9$ kips, $A_y = 6.67$ kips, Column moment at $B = 73.8$ kip · ft, $F_{BL} = +19.7$ kips, $F_{CD} = -18.10$ kips

CHAPTER 16

P16.1 (*a*) $K = 476.25$ kips/in, (*b*) $\Delta = 0.050$ in, (*c*) $F_{AB} = F_{AD} = 10.08$ kips, $F_{AC} = 7.87$ kips

P16.3 $K_{2x} = 666.6$ kips, $K_{2y} = 249.93$ kips

P16.5 $M_A = 12.69$ kip · ft $\circlearrowright$, $M_C = 144.81$ kip · ft $\circlearrowright$, $R_{AX} = 2.55$ kips $\rightarrow$, $R_{AY} = 11.77$ kips$\uparrow$, $R_{CX} = 7.55$ kips $\leftarrow$, $R_{CY} = 18.23$ kips$\uparrow$

P16.7 $K_2 = -\frac{5}{3}EI$, $M_{CD} = -67.2$ kN · m, $A_X = 2.7$ kN, $M_{DC} = 74.4$ kN · m

P16.9 Joint 3: $F = 42.96$ kips; joint 1: $R_x = 25.78$ kips, $R_Y = 1.62$ kips; $M = 19.42$ kip · ft

P16.11 $R_{AX} = 8.187$ kips $\rightarrow$, $R_{AY} = R_{Dy} = 48$ kips $\uparrow$, $R_{DX} = 8.187$ kips $\leftarrow$, $M_A = 49.12$ kip · ft ↻, $M_D = 49.12$ kip · ft ↺

CHAPTER 17

P17.1 $\Delta_X = -96L/AE$; $\Delta_Y = -172L/AE$

P17.3 Joint 1: $\Delta_X = 0.192$ in $\rightarrow$, $\Delta_Y = 0.865$ in down

P17.5 Joint 3: $\Delta_X = 0.152$ in $\rightarrow$, $\Delta_Y = 0.036$ in $\downarrow$; Joint 4: $\Delta_X = 0.216$ in $\rightarrow$, $\Delta_Y = 0.036$ in $\uparrow$

CHAPTER 18

P18.1 $M_A = 13.89$ kip · ft, $A_Y = 12.08$ kips, $B_Y = 63.66$ kips, $C_Y = 24.26$ kips

P18.3 Force in the Spring $= 0.208\ wL$

P18.5 $M_A = 151.579$ kip · ft ↺, $R_{AY} = 47.895$ kips $\uparrow$ $R_{AX} = 31.184$ kips $\rightarrow$, $V_{BC} = 5.684$ kips

P18.7
$$[K] = \begin{bmatrix} 3854.2 & 0 & -6250 \\ 0 & 3854.2 & 6250 \\ -6250 & 6250 & 1{,}000{,}000 \end{bmatrix}$$

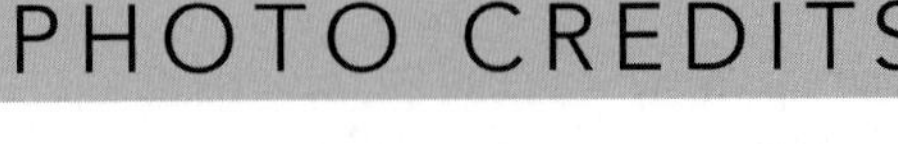

PHOTO CREDITS

CHAPTER 1

Opener: Library of Congress; **1.1:** © Kenneth Leet; **1.2:** Courtesy of the Godden Collection, NISEE, University of California, Berkeley; **1.3:** © Michael MaslanHistoric Photographs/Corbis; **1.4a–b:** Courtesy of the Godden Collection, NISEE, University of California, Berkeley

CHAPTER 2

Opener: Courtesy of Frieder Seible, University of California, San Diego; **2.1:** © AP/Wide World Photos; **2.2:** Courtesy of California Department of Transportation; **2.3:** © Chia-Ming Uang; **2.4a:** © Chia-Ming Uang; **2.4b:** Courtesy of Robert Reitherman.

CHAPTER 3

Opener: © Howard Epstein, University of Connecticut; **3.1:** © Chia-Ming Uang; **3.2:** © Chia-Ming Uang; **3.3:** Courtesy of the Godden Collection, NISEE, University of California, Berkeley.

CHAPTER 4

Opener: Courtesy of Port Authority of New York and New Jersey; **4.1:** Courtesy of Ewing Cole Cherry Brott Architects and Engineers, Philadelphia, PA; **4.2:** Courtesy of the Godden Collection, NISEE, University of California, Berkeley.

CHAPTER 5

Opener: Photo by Banks Photo Service, Courtesy of Simpson Gumpetz and Heger, Inc.; **5.1–5.2:** © Kenneth Leet; **5.3:** Courtesy of Bergmann Associates.

CHAPTER 6

Opener: © Vince Streano/TY Lin International; **6.1–6.2:** Courtesy of Portland Cement Association.

CHAPTER 7

Opener: Courtesy of the Godden Collection, NISEE, University of California, Berkeley; **7.1:** Courtesy of the Godden Collection, NISEE, University of California, Berkeley.

CHAPTER 8

Opener: GEFYRA S.A. (2. Rizariou street—152 33 Halandri/Greece). Nikos Daniilidis (107. Zoodohou Pigis street—114 73 Athens/Greece)

CHAPTER 9

Opener: Courtesy of Kevin Chang, Evergreen Consulting Engineering, Inc.; **Opener Inset:** Courtesy of Rowan Williams Davies & Irwin Inc. (RWDI).

CHAPTER 10

Opener: Courtesy of Mark McCrindle, Prospect Steel Company.

CHAPTER 11

Opener: Courtesy of Arvid Grant and Associates.

CHAPTER 12

Opener: Courtesy of Simpson Gumpetz and Heger, Inc.

CHAPTER 13

Opener: Courtesy of Federal Reserve Bank of Boston.

CHAPTER 14

Opener: Courtesy of Port Authority of New York and New Jersey.

CHAPTER 15

Opener: Courtesy of Port Authority of New York and New Jersey.

CHAPTER 16

Opener: Courtesy of Simpson Gumpetz and Heger, Inc.

CHAPTER 17

Opener: Courtesy of Simpson Gumpetz and Heger, Inc.

CHAPTER 18

Opener: © The Hartford Courant, Arman Hatsian.

INDEX

A

B

C

D

E

F

G

H

I

J

K

L

M

N

O

P

Q

R

S

T

U

V

W

Z

Table 3: Properties of Areas

Shape	Figure	Area	Centroidal Distance $\bar{x}$
(a) Triangle	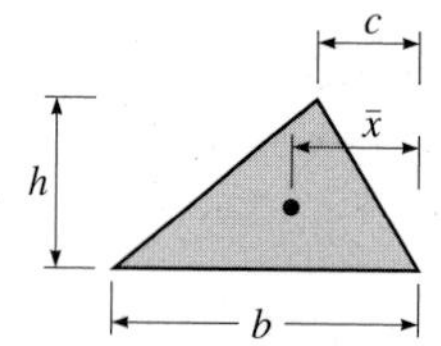	$\frac{bh}{2}$	$\frac{b+c}{3}$
(b) Right triangle	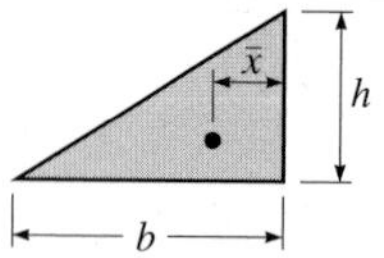	$\frac{bh}{2}$	$\frac{b}{3}$
(c) Parabola	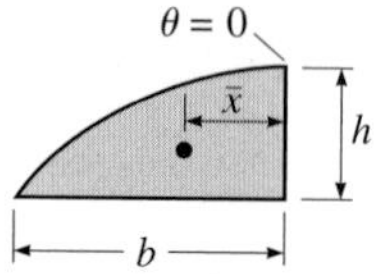	$\frac{2bh}{3}$	$\frac{3b}{8}$
(d) Parabola	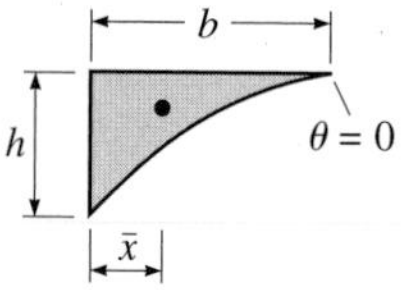	$\frac{bh}{3}$	$\frac{b}{4}$
(e) Third-degree parabola	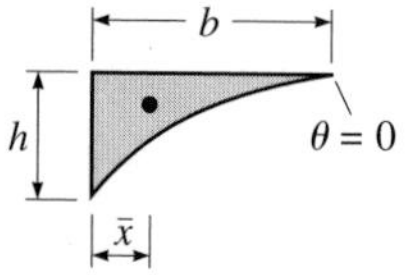	$\frac{bh}{4}$	$0.2b$
(f) Rectangle	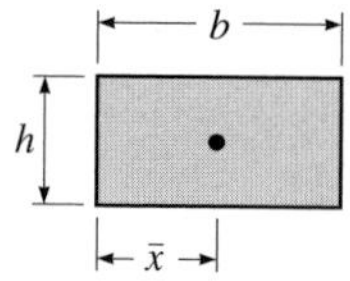	bh	$\frac{b}{2}$
(g) Trapezoid	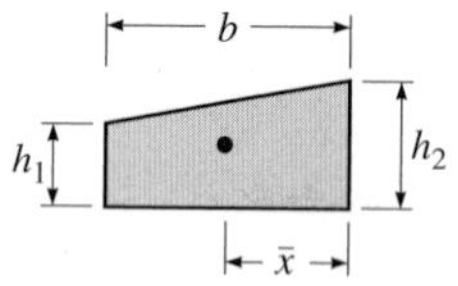	$\frac{b}{2}(h_1+h_2)$	$\frac{b(2h_1+h_2)}{3(h_1+h_2)}$